The Reticuloendothelial System

A COMPREHENSIVE TREATISE

Volume 1

Morphology

The Reticuloendothelial System
A COMPREHENSIVE TREATISE

Series Editors:
Herman Friedman, *University of South Florida, Tampa, Florida*
Mario Escobar, *Medical College of Virginia, Richmond, Virginia*
and
Sherwood M. Reichard, *Medical College of Georgia, Augusta, Georgia*

Volume 1 MORPHOLOGY
Edited by Ian Carr and W. T. Daems

Volume 2 BIOCHEMISTRY AND METABOLISM
Edited by Anthony J. Sbarra and Robert R. Strauss

Volume 3 PHYLOGENY AND ONTOGENY
Edited by Nicholas Cohen and M. Michael Sigel

Volume 4 PHYSIOLOGY
Edited by Sherwood M. Reichard and James P. Filkins

Volume 5 PHARMACOLOGY
Edited by Andor Szentivanyi and Jack R. Battisto

Volume 6 IMMUNOLOGY
Edited by Joseph A. Bellanti and Herbert B. Herscowitz

Volume 7 IMMUNOPATHOLOGY
Edited by Noel R. Rose and Benjamin V. Siegel

Volume 8 CANCER
Edited by Herman Friedman and Ronald B. Herberman

Volume 9 RES AND ALLERGIC DISEASES
Edited by Peter Abramoff and S. Michael Phillips

Volume 10 INFECTION
Edited by John P. Utz and Mario R. Escobar

The Reticuloendothelial System

A COMPREHENSIVE TREATISE

Volume 1
Morphology

Edited by
IAN CARR
University of Saskatchewan
Saskatoon, Saskatchewan, Canada

and

W. T. DAEMS
University of Leiden
Leiden, The Netherlands

With the Editorial Assistance of
AVA LOBO

SPRINGER SCIENCE+BUSINESS MEDIA, LLC

Library of Congress Cataloging in Publication Data

Main entry under title:

The Reticuloendothelial system.

Includes index.
CONTENTS: v. 1. Carr, I., Daems, W. T., and Lobo, A. Morphology.
1. Reticulo-endothelial system. 2. Macrophages. I. Friedman, Herman, 1931-
II. Escobar, Mario R. III. Reichard, Sherwood M. [DNLM: 1. Reticuloendothelial system. WH650 R437]
QP115.R47 591.2'95 79-25933

ISBN 978-1-4899-5033-8 ISBN 978-1-4899-5031-4 (eBook)
DOI 10.1007/978-1-4899-5031-4

Originally published by Plenum Press, New York in 1980
Softcover reprint of the hardcover 1st edition 1980

SAMUEL K. ACKERMAN • Department of Medicine, University of Minnesota Medical School, Minneapolis, Minnesota

IAN Y. R. ADAMSON • Department of Pathology, University of Manitoba, Winnipeg, Canada

FRANK BALDWIN • Department of Veterinary Anatomy, Western College of Veterinary Medicine, University of Saskatchewan, Saskatoon, Saskatchewan, Canada

BRIGID M. BALFOUR • National Institute for Medical Research, London, England

ROBERT H. J. BEELEN • Department of Electron Microscopy, Medical Faculty, Free University, Amsterdam, The Netherlands

DRUMMOND H. BOWDEN • Department of Pathology, University of Manitoba, Winnipeg, Canada

P. BREDEROO • Laboratory for Electron Microscopy, University of Leiden, Leiden, The Netherlands

IAN CARR • Department of Pathology, University of Saskatchewan, Saskatoon, Saskatchewan, Canada

K. E. CARR • Department of Anatomy, University of Glasgow, Glasgow, Scotland, United Kingdom

W. T. DAEMS • Laboratory for Electron Microscopy, University of Leiden, Leiden, The Netherlands

PETER P. H. DE BRUYN • Department of Anatomy, University of Chicago, Chicago, Illinois

K. DONALD • Department of Pathology, The University of Queensland Medical School, Herston, Queensland, Australia

STEVEN D. DOUGLAS • Departments of Medicine and Microbiology, University of Minnesota Medical School, Minneapolis, Minnesota

J. J. EMEIS • Gaubius Institute, Health Research Organization TNO, Leiden, The Netherlands

ANDREW G. FARR • Department of Pathology, Harvard Medical School, Boston, Massachusetts. Present affiliation: National Jewish Hospital and Research Center, Denver, Colorado

HANS R. HENDRIKS • Department of Electron Microscopy, Medical Faculty, Free University, Amsterdam, The Netherlands

ELISABETH C. M. HOEFSMIT • Department of Electron Microscopy, Medical Faculty, Free University, Amsterdam, The Netherlands

G. HUDSON • University Department of Hematology, Royal Hallamshire Hospital, Sheffield, England

EDWIN KAISERLING • Department of Pathology, Christian-Albrecht University, Kiel, West Germany

ED W. A. KAMPERDIJK • Department of Electron Microscopy, Medical Faculty, Free University, Amsterdam, The Netherlands

GILLA KAPLAN • Institute for Medical Biology, University of Tromsø, Tromsø, Norway

ULF LUCHT • Department of Cell Biology, Institute of Anatomy, University of Aarhus, Aarhus, Denmark

PAUL E. MCKEEVER • Laboratory of Microbial Immunity, National Institute of Allergy and Infectious Diseases, Bethesda, Maryland; and Surgical Neurology Branch, National Institute of Neurological and Communicative Disorders and Stroke, Bethesda, Maryland

AMY C. MUNTHE-KAAS • Laboratory for Immunology, Department of Pathology, Norsk Hydro's Institute for Cancer Research, The Norwegian Radium Hospital, Oslo, Norway

J. R. SHORTLAND • Department of Pathology, University of Sheffield, Sheffield, England

INGA SILBERBERG-SINAKIN • Department of Dermatology, New York University School of Medicine, New York, New York

Gerard T. Simon • Department of Pathology, University of Toronto, Toronto, Canada. Present affiliation: Electron Microscopy Facility and Department of Pathology, McMaster University, Hamilton, Ontario, Canada

Samuel S. Spicer • Department of Pathology, Medical University of South Carolina, Charleston, South Carolina

G. Jeanette Thorbecke • Department of Pathology, New York University School of Medicine, New York, New York

Jos W. M. van der Meer • Department of Infectious Diseases, University Hospital, Leiden, The Netherlands

Jan E. Veldman • Laboratory of Histophysiology and Experimental Pathology, Department of Otorhinolaryngology, University of Utrecht, Utrecht, The Netherlands

E. Wisse • Laboratory for Cell Biology and Histology, Free University of Brussels, Brussels-Jette, Belgium

Foreword

This comprehensive treatise on the reticuloendothelial system is a project jointly shared by individual members of the Reticuloendothelial (RE) Society and biomedical scientists in general who are interested in the intricate system of cells and molecular moieties derived from these cells which constitute the RES. It may now be more fashionable in some quarters to consider these cells as part of what is called the mononuclear phagocytic system or the lymphoreticular system. Nevertheless, because of historical developments and current interest in the subject by investigators from many diverse areas, it seems advantageous to present in one comprehensive treatise current information and knowledge concerning basic aspects of the RES, such as morphology, biochemistry, phylogeny and ontogeny, physiology, and pharmacology as well as clinical areas including immunopathology, cancer, infectious diseases, allergy, and hypersensitivity. It is anticipated that by presenting information concerning these apparently heterogeneous topics under the unifying umbrella of the RES attention will be focused on the similarities as well as interactions among the cell types constituting the RES from the viewpoint of various disciplines. The treatise editors and their editorial board, consisting predominantly of the editors of individual volumes, are extremely grateful for the enthusiastic cooperation and enormous task undertaken by members of the biomedical community in general and especially by members of the American as well as European and Japanese Reticuloendothelial Societies. The assistance, cooperation, and great support from the editorial staff of Plenum Press are also valued greatly. It is hoped that this unique treatise, the first to offer a fully comprehensive treatment of our knowledge concerning the RES, will provide a unified framework for evaluating what is known and what still has to be investigated in this actively growing field. The various volumes of this treatise provide extensive in-depth and integrated information on classical as well as experimental aspects of the RES. It is expected that these volumes will serve as a major reference for day-to-day examination of various subjects dealing with the RES from many different viewpoints.

Herman Friedman
Mario R. Escobar
Sherwood M. Reichard

Introduction

What is the reticuloendothelial system? As indicated in this first book of a multivolume treatise on the RES, it is quite apparent that many investigators believe the central cell type involved in the reticuloendothelial system is the macrophage, or the "wandering" cell which Metchnikoff so elegantly described when he observed mobile cells surrounding a splinter introduced into the body of a starfish larva. Thus it seems appropriate that the first volume in this treatise be co-edited by Ian Carr, who so "irreverently" stated in a talk presented at the Seventh International Congress of the Reticuloendothelial Society that "the term RES is an interesting and historical term, a charming and picturesque relic of a bygone age, redolent of Aschoff, Frieburg, chintz and 'lavender water.'" Although Dr. Carr believes the major use of the term RES is in naming societies and books—and there is a biblical injunction against removing ancient boundary stones—it is certainly timely to attempt to focus attention on this very important cell system which has so many important functions.

The RES obviously consists of a working complex of macrophages, dendritic reticular cells, lymphocytes, and plasma cells as they are arrayed in lymph nodes, spleen, thymus, bone marrow, gut-associated lymphoid tissue, inflammatory lesions, and elsewhere. It is widely accepted that these cells and this system are involved not only in inflammation and clearance of particulate agents, but also contain the cellular elements involved in immune responses, control of neoplasia, iron metabolism, lipid metabolism, and secretion of a wide variety of substances. Among the many triumphs of the biological revolution over the last few decades is the rapidly developing understanding of how various cell systems interact in the individual. It is now recognized widely that the RES functions as a primary defense system against invading agents and also provides a wide repertoire of cells and secretions involved not only in host defense mechanisms but also in metabolism—catabolic and anabolic.

This system of cells is now being interpreted in molecular, physiological, and endocrinological terms. With this surging new information about the system comes a newer understanding of how these cells interact in a wide variety of functions and how these interactions influence other cell functions and activities. It is the purpose of this comprehensive treatise on the RES to bring together in one place and in an encyclopedic form current and relevant knowledge about this system. This first volume deals with the morphology of the RES and reviews in detail various aspects of structure as related to function with regard to macrophages, histiocytes, and other cells. Although it is likely that the "purist" would believe that only macrophages constitute the mononuclear phagocytic system, a general chapter on lymphocytes and plasmacytes is also presented since these important cells, which are a part of the lymphoid cell system, interact closely with macrophages. The following volume deals with the biochemistry of the RES. A third volume deals with the ontogeny and phylogeny of the RES, since it is widely recognized that there are not only differences but also much similarity in RE cell functions among various species of animals and that ontological development of the RES is an important aspect of elemental biology. Following the first three basic volumes are volumes related to functional activities, namely one on physiology, emphasizing pathophysiological aspects, and the other on pharmacology, dealing with immunopharmacological reactions of the RES. The next five volumes in this treatise deal with more "practical" or applied subjects, including immunology in general, immunopathology, cancer, hypersensitivity, and infectious diseases.

Our understanding of the molecular and cellular basis of development, as well as our knowledge on interactions and function of the RE cells is progressing at a logarithmic rate. The role of cell membranes, receptors, nucleotides, hormones, soluble mediators, nucleic acids, and other macromolecules is being further elucidated almost on a daily basis. It is certainly unique that bioscientists from many disciplines, including anatomy, physiology, biochemistry, pharmacology, microbiology, immunology, and infectious diseases, all have a common interest in various aspects of the RES. It is toward this goal that this treatise has been designed, viz., to bring together in one place new and relevant information concerning this all important biological system. Most bioscientists should look forward with great excitement to further development in understanding of the RES. It is anticipated that publication of this treatise will provide not only a source of current information concerning the RES but will also be a stimulus for further investigative work.

Herman Friedman
Mario R. Escobar
Sherwood M. Reichard

Preface

This, the first volume of a multivolume treatise, sets the morphological stage on which the drama of the function, both normal and abnormal, of that curious complex, the reticuloendothelial system, will be set. It is a controversial drama; argument enters already with the use of the phrase "reticuloendothelial system." This prefatory essay states the scope of the volume, defines or describes its terms, and gives a discursive overview of the subject so that a novice does not wander into the experts' maze without a clue.

The volume describes morphology in relation to function. Morphology tends to be a dirty word—conjuring up myopic professors peering at minutiae and drowning any little significances they have in a prolixity of paper. The word has not been cleaned up by the modern morphologist who, substituting the ultrastructural appearance of a few nanometers of osmicated shadow for the lesser superficial petrosal nerve, is as far from reality as his predecessor. The morphological structure is like the biochemical formula, an epiphenomenon of life; and real only when it is remembered to be so. And much more real when related to function, normal or abnormal.

The scope of the volume is to describe the structure, in the main as seen by electron microscopy, of macrophages in the various sites around the body where they are to be found. No account of the subject could omit the name of Elie Metchnikoff, the Russian zoologist who first described the macrophage as the larger of two groups of avidly phagocytic cells, or of Ludwig Aschoff, the German pathologist who first recognized the aggregation of these cells as a functional system, the reticuloendothelial system. An excellent historical account of the subject is given by Stuart (1970), and some useful general references on the subject are given in the Bibliography following the References.

An account of these cells cannot, of course, be given without reference to the kindred framework cells of the lymphoid organs, the reticulum or reticular cells. The lymphocytes are in themselves a major field and will be described only in their relationships to macrophages. The accounts that follow are essentially accounts of cells as seen in the framework of a particular organ and not accounts of the structure of that organ.

Some of the terms which will be used again and again must be defined or at least described. Of these a few are controversial; the writers of the essays that will follow define their own terms, and no uniformity has been imposed. There is no implication that the present glossary is more correct; it is merely a systematized delusion. As in all such volumes the authors of individual essays express views for which the editors are not responsible. While an attempt has been made to cover the morphological appearances and function of macrophages in most of the major organs, it is a set of essays by authorities in different fields and not a systematic text. There are therefore inevitable lacunae. The bias is toward the macrophage as seen in the living organism. No attempt has been made to cover the wide area of tissue culture studies—relevant as such studies may sometimes be to reality in the living animal.

Macrophage: An avidly phagocytic cell of ultimate monocytic origin, possessed of the apparatus for further synthesis of lysosomes and with Fc and Ig receptors on the surface.

Reticular cell: A framework cell of lymphoreticular tissue.

Dendritic reticular cell: A reticular cell with long processes capable of binding antigen.

Reticulum cell: An unidentified cell in lymphoreticular tissue; the term is best reserved for pathological description and will disappear as cell classification improves.

Reticuloendothelial system: The complex of leaky sinusoids and endovascular and paravascular macrophages which results in clearance of colloids from the blood stream. This term is neither defended nor attacked here. It is still widely used, despite its manifest deficiencies. The abbreviation RES will be used throughout the book.

Mononuclear phagocyte system: The phagocytic cells of monocyte origin scattered throughout the body. The abbreviation MPS will be used throughout the book.

Lymphoreticular system: The total complex of macrophages, reticular and reticulum cells and lymphocytes, and their derivatives, as they are aggregated into lymph nodes, spleen, bone marrow, gut- and bronchus-associated lymphoid tissue, liver, and elsewhere.

Ian Carr
W. T. Daems

Contents

5. Surface Receptors of Mononuclear Phagocytes

PAUL E. MCKEEVER and SAMUEL S. SPICER

6. Scanning Electron Microscopy of Macrophages

K. E. Carr

7. Monocytes

Samuel K. Ackerman and Steven D. Douglas

8. Bone Marrow Macrophages

G. Hudson and J. R. Shortland

9. On the Fine Structure and Function of Rat Liver Kupffer Cells

E. Wisse

10. Interdigitating Cells

JAN E. VELDMAN and EDWIN KAISERLING

11. Lymph Node Macrophages

ELISABETH C. M. HOEFSMIT, ED W. A. KAMPERDIJK, HANS R. HENDRIKS, ROBERT H. J. BEELEN, and BRIGID M. BALFOUR

18. Macrophages in Connective Tissues: The Granuloma Macrophage

Ian Carr

19. Osteoclasts: Ultrastructure and Functions

Ulf Lucht

1

The Macrophage
A Bird's-Eye View

IAN CARR and W. T. DAEMS

1. A TYPICAL MACROPHAGE

This introduction describes briefly, and for the complete newcomer to the field, some of the current ideas on the form and function of the macrophage. A typical macrophage, such as a peritoneal macrophage (see Chapter 3) is about 12 μm in diameter and has small peripheral cytoplasmic processes (Carr, 1973a; Daems and Brederoo, 1973). The cell center may be eosinophilic. Cytochemical staining shows a wide variety of hydrolytic enzymes—acid phosphatase, naphthol AS acetate esterase (almost completely inhibited by sodium fluoride), glucuronidase, and others. At the electron microscopic (EM) level, the cell surface is rough due to the presence of numerous processes, both fingerlike and flaplike, and there are numerous deep invaginations, often labyrinthine in nature. The cytoplasm contains good evidence of secretory activity in the form of granular endoplasmic reticulum and a well developed Golgi zone, often with related centriole. There are numerous vacuoles in the cell ranging from large vacuoles resembling pinocytic vacuoles (about 0.1 μm or more in diameter) to small micropinocytic vacuoles less than 100 nm in diameter, some of them "coated" vesicles. Membrane-bound inclusions fall into two groups, large heterogeneous structures containing residual debris of various kinds and smaller homogeneous dense bodies which probably represent primary lysosomes. Some of these cells contain peroxidase in dense bodies while others contain peroxidase in the channels of the endoplasmic reticulum, Golgi zone, and the perinuclear space. These latter probably represent cells resident in the peritoneal cavity or other regions (Daems *et al.*, 1975). Within the cytoplasmic sap are varying numbers of free RNP particles, sometimes arranged in evident polysomes, and some glycogen. Microtubules are particularly evident near the Golgi zone. Scattered

IAN CARR • Department of Pathology, University of Saskatchewan, Saskatoon, Saskatchewan, S7N OWO, Canada. W. T. DAEMS • Laboratory for Electron Microscopy, University of Leiden, 2333 AA Leiden, The Netherlands.

throughout the cytoplasm are microfilaments (or microfibrils) falling into two classes, those 10–12 nm in diameter, of uncertain function and those 6–8 nm in diameter; the latter form arrowhead complexes with heavy meromyosin and represent actin, the major contractile protein of the macrophage (Allison *et al.*, 1971). Microfilaments are closely related and probably attached to vacuoles and other organelles and are present in high concentration, though not always clearly evident in the significant part of the cytoplasm where they are related to cytoplasmic movement. The cell coat of the macrophage (see Chapter 4) is particularly prominent and is composed of a superficial protein layer loosely attached and electron-lucent and an adherent mucosubstance layer, stained, for example, by ruthenium red (Carr *et al.*, 1970; Emeis and Wisse, 1975). On the cell surface are receptors for the Fc end of the immunoglobulin molecule and for complement, to be more precise, the third component of complement, and for the attachment of lymphocytes; these are demonstrable immunocytochemically as distinct, localized sites probably within the cell coat. The immunoglobulin binding site has been localized to a decapeptide component of the carbohydrate domain. Immune complex receptors have been demonstrated by staining with peroxidase–antiperoxidase complex. They occur at intervals of 20–120 nm on the surface of peritoneal macrophages to a total number of about 200,000 per mature stimulated cell. They are present in rather larger numbers than immunoglobulin receptors, but it is not currently clear whether they represent a separate class of receptor. Immune complex receptors are ingested during pinocytosis and resynthesized within 30 min (McKeever *et al.*, 1976). The nucleus of the macrophage may show a fibrous peripheral lamina and may show spheridia; the chromatin is usually marginated and a nucleolus of conventional structure may be evident.

The monocyte, from which, according to the mononuclear phagocyte hypothesis, all macrophages derive, is smaller than the peritoneal macrophage and with a smooth surface, few processes or invaginations, scanty granular endoplasmic reticulum, and few lysosomes (Nichols and Bainton, 1975, reviewed in Kay and Douglas, 1977). It has, however, more cytoplasmic membrane than a lymphocyte (see Chapter 5).

Monocytes are produced in the bone marrow; a stem cell pool gives rise to monoblasts. These are rapidly dividing small, round cells with macrophage markers (cytoplasmic esterase, lysozyme, and surface receptors for complement and immunoglobulin). They mature into promonocytes and then monocytes. The precursor cells are, unlike mature macrophages, radiosensitive. It is likely that the production of monocytes is controlled by a humoral factor; a specific thermolabile factor has been identified in mice. It is likely that maturation and cell division in the marrow last 24–48 hr, that the newly divided cells remain in the marrow for up to 24 hr, and that the peripheral monocyte after circulating for one and a half to four days, then randomly migrates into the tissues (van Furth, 1975; Whitelaw and Batho, 1975).

Macrophages derive their energy from glycolysis, the hexose monophosphate shunt, and/or by aerobic metabolism, the latter especially in the case of alveolar macrophages. The activated or stimulated cell phagocytizes, digests, and kills microorganisms more rapidly and uses more glucose. Phagocytosis is accompanied by increased respiratory activity and hexose monophosphate shunt activity; this perturbation of metabolism is very evident in the activated macrophage; alveolar macrophages do not, however, have the usual increase in oxygen consumption following a phagocytic event (Karnovsky *et al.*, 1975; Karnovsky *et al.*, 1976; Rossi *et al.*, 1976). Only monoblasts and promonocytes show extensive DNA synthesis; fewer mature macrophages synthesize DNA in the resting state, though they will do so *in vitro* or *in vivo* under stimulation (Wynne *et al.*, 1975; Carr *et al.*, 1976). Macrophages synthesize RNA and a wide variety of proteins. Macrophages synthesize and esterify fatty acids and synthesize phospholipids, but probably not cholesterol (Day, 1964). New membrane components are readily resynthesized. New surface receptors are resynthesized 6 hr after removal of surface receptors. After phagocytosis, in which 50% of the plasma membrane may be interiorized, it is 4–8 hr before new membrane is formed; this involves RNA and protein synthesis (Karnovsky, 1962; Werb, 1975). Macromolecular membrane components are not directly reused. The metabolism of bacterial killing by macrophages is still unclear; factors incriminated include acidity, lysozyme, the peroxide and superoxide radicals, and the chlorinium ion (MPO-H_2O_2-Cl^-) (Sbarra *et al.*, 1976).

2. THE SITES AND ORIGIN OF MACROPHAGES

Macrophages are found in the body in a wide variety of sites—liver, spleen, lymph nodes, as well as thymus-, bronchus-, and gut-associated lymphoreticular tissues, the marrow, the brain (microglia), and the connective tissues, as well as in sites where physiological repair processes are occurring, like the uterus and inflammatory lesions. There are also a few in such sites as the arterial intima where they play a part of some significance in the atheromatous process.

An important part of the function of many of these static macrophages is to monitor streams of fluid which pass them or pass near them, whether these be streams of blood or of lymph. The precise site of the macrophages in relation to the sinusoid is important. The liver macrophages dangle into the stream of the hepatic sinusoid and are apparently not attached by any morphologically visible means to the adjacent sinusoidal endothelium (Wisse, 1974) (see Chapter 9). Splenic macrophages are entirely extravascular so material must first leak through patent gaps in the endothelium of the splenic sinusoids before they can be phagocytized (see Chapter 12). Marrow macrophages are similarly extrasinusoidal (see Chapter 8). Lymph node macrophages line the sinusoids, subcapsular, radial, and medullary, with the exception of the capsular side of the subcapsular sinus which is lined by endothelial cells. There are also extrasinusoidal macrophages which are not primarily responsible for clearance of the lymph (see Chapters 10 and 11).

A series of studies using cells marked either by a chromosomal marker or with tritiated thymidine makes it clear that in inflammatory lesions macrophages derive recently from blood monocytes and thence from a rapidly dividing marrow precursor (Volkman and Gowans, 1965a,b; Spector and Lykke, 1966).

There is also in many situations local proliferation of macrophages. In many chronic inflammatory lesions fusion of recently arrived monocytes with preexisting macrophages lead to the formation of macrophage-derived giant cells or macrophage polykaryons (Spector and Mariano, 1975). It is also clear that in many sites, for example, in the Kupffer cells, the tissue macrophages proper to the tissue can divide readily (North, 1969; Wisse, 1977), but the quantitative significance of this is in dispute. Under conditions of extreme stress on population replacement mechanisms, there may be large-scale immigration of marrow-derived monocytes. It is not yet universally agreed as to how peripheral macrophage populations are replaced in steady-state situations (Volkman, 1976).

The differences between macrophages in different sites are less than their similarities. When removed from the body and reinjected they show a relative, though not absolute, tendency to home to their original sites; for instance, Kupffer cells home to the liver (Roser, 1968). The reason for this homing is not clear but may relate to organ-specific surface-antigenic differences. Other differences are relative and may relate to different local endocytic stimuli. For instance splenic macrophages contain much red cell debris and ferritin, while the latter is hard to find in peritoneal macrophages. Deep, tubular invaginations of the surface of the cell ("micropinocytosis vermiformis"), or wormlike bodies (Wisse, 1974), are more common in Kupffer cells than elsewhere; possibly this is related to active protein ingestion by Kupffer cells. Elongated cytoplasmic inclusions containing an electron-dense core with a periodic striation 10 nm apart are found in the Langerhans cells within the epidermis (Breathnach, 1965). These inclusions may represent a similar process to micropinocytosis vermiformis and are found occasionally in similar cells in lymph nodes, and in macrophages in histiocytosis X (Nezelof *et al.*, 1977) (see Chapter 15). The microglial cell is a small macrophage which enlarges and matures in various reactive lesions and is joined by recently marrow-derived monocytes (see Chapter 17). The pulmonary macrophage contains much ingested surfactant, and has large mitochondria, being more dependent on oxidative metabolism.

Large populations of macrophages are found in chronic granulomas (see Chapter 18). In a granuloma recently arrived monocytes mature into large macrophages with prominent granular endoplasmic reticulum and numerous lysosomes; there is also local proliferation of macrophages. Macrophage polykaryons or giant cells form by fusion of recently emigrated monocytes with adherent and then fused macrophages (Spector and Mariano, 1975). In both mononuclear macrophages and giant cells phagocytosis may decline and the cell come to contain numerous inclusions probably of secretory nature. These are so called epithelioid cells and are thought to be secretory (Spector and Mariano, 1975). Macrophages are immobilized in chronic inflammatory lesions by a migration inhibition factor, released in the presence of antigen by sensitized lymphocytes. This fact is used as an *in vitro* test for cellular immunity.

Two important varieties of macrophage or macrophage-related cell are the interdigitating cell, which probably is a macrophage, and the dendritic cell, which may not be (see Chapters 10 and 11).

The interdigitating cell occurs in the thymus-dependent areas of lymph node and spleen. It is a fairly typical large macrophage, into whose cytoplasm either the processes of a lymphocyte or a whole lymphocyte are deeply invaginated. It has been shown that T lymphocytes home to these cells, and insert themselves into them; this is apparently necessary for the full maturation of the T lymphocyte. This "nurse cell" function has been likened to that of the Sertoli cell in relation to spermatids (Veerman, 1974; Veldman, 1970; van Ewijk *et al.*, 1974).

The dendritic reticular cell is a cell with long processes present in germinal centers; onto its processes antigen adheres (Nossal *et al.*, 1968; Carr, 1973b). It may provide a framework for the interaction of antigen with lymphocytes, but whether or not it is really a macrophage or indeed a truly separate cell type is currently not agreed upon. Cells with long, slender processes variously described as dendritic cells or reticular cells have been grown in cultures from the peritoneal cavity and from lymphoreticular organs (Stuart and Davidson, 1971; Steinman and Cohn, 1975). It is not clear what these cells do and what is their relation to antigen-retaining dendritic reticular cells.

A macrophage can be recognized in several ways—by its content of acid phosphatase and esterase, by its ultrastructure, by phagocytosis, notably of red blood cells (RBC), and by rosette tests. Typically when macrophages are incubated *in vitro* under appropriate conditions with sheep erythrocytes, the erythrocytes form a rosette around the macrophage (E rosettes); similar rosettes form when the indicator cells are coated with IgG-Fc, or C3 (Fc and C3 rosettes). This pattern distinguishes macrophages from lymphocytes (Green *et al.*, 1975).

3. THE FUNCTION OF MACROPHAGES

3.1. PHAGOCYTOSIS (SEE CHAPTER 2)

There are two sets of phagocytic circumstances; either the free macrophage moves toward and engulfs the particle, or the fixed macrophage removes particles from a passing stream of fluid.

The free macrophage is attracted toward many particles by chemotaxis (Wilkinson, 1976). By definition this process occurs *in vitro;* it probably also happens *in vivo.* Among the things which attract macrophages are bacteria and antigen–antibody complexes. They do this by releasing substances which interact with serum components to form active chemotactic factors. Such factors are present in diminishing amounts with distance forming a chemical gradient up which the cell moves. A purified phospholipid derived from corynebacteria is a specific chemotactic factor for macrophages active at concentrations of less than 1 μg/ml. Presumably the messenger substance interacts with the locomotor system of the cell at one pole. The regulation of chemotaxis is complex. Chemotactic mediators are generated *in vivo* from inactive circulating substances in serum; they include complement components (C3 and C4), kallikrein, and lymphokines released by lymphocytes. Serum contains both inhibitors which interact directly with the monocyte or macrophage, and inactivators which react with the chemotactic mediators (Wilkinson, 1976). When contact is established, the particle adheres to the cell coat, often by interaction between the Fc receptor and an immunoglobulin coat around the particle. The particle attaches sequentially to surface receptors causing a purely local movement of the cell surface, and pseudopodia protrude around and enclose the particle in a membranous vacuole. This process involves contraction of actin filaments, is energy dependent, and sets off a burst of respiration. The vacuole is pulled within the cell by contraction of actin filaments (Griffin and Silverstein, 1974; Griffin *et al.*, 1975; Stossel and Hartwig, 1976). The phagocytic vacuole then fuses with primary and or secondary lysosomes, allowing the extensive variety of hydrolytic enzymes in the macrophage to break down the ingested material. If the latter is a bacterium it must first of all be killed. How this happens is not clear but may involve peroxidase and/or active hydroxyl or superoxide radicals generated within the cell. Some bacteria notably mycobacteria appear to survive, possibly because of the waxes present in the bacterial cell wall; this intracellular parasitism allows transport of the bacterium through the tissues of the host. The phagocytosis of such toxic substances as silica may kill the macrophage, but more commonly phagocytosis far from killing the macrophage stimulates it to the synthesis of more lysosomal hydrolytic enzymes.

Phagocytosis is the characteristic way in which macrophages ingest foreign material. There are other kinds of endocytosis—pinocytosis, micropinocytosis in either a smooth or a coated small vesicle, and micropinocytosis vermiformis. Pinocytosis, or cell drinking, is the energy-dependent ingestion of microscopically visible droplets of fluid by cells in tissue culture; macrophages pinocytize very freely in tissue culture (Cohn and Benson, 1965; Cohn, 1970), but it is not certain how important pinocytosis is in the life of the intact mammalian organism.

Micropinocytosis refers to the ingestion of materials in much smaller (< 100 nm) vesicles, without dependence on metabolic energy. Considerable quantities of material may be ingested in these small vesicles, which later fuse. Some micropinocytic vesicles may be surrounded by a corona of fine microfilaments forming so-called "coated" vesicles; these have been specially associated with protein ingestion. Lastly, some macrophages ingest materials by means of tubular channels—so-called micropinocytosis vermiformis. This is especially obvious in the case of Kupffer cells.

The fixed macrophages of the liver and lymph node clear the streams of fluid which pass them. The major organ responsible for clearance of the systemic blood is the liver. This has been entitled "reticuloendothelial clearance" (Stuart, 1970). A colloidal suspension is introduced into the blood stream and its disappearance from the blood stream plotted. This clearance can be artifically stimulated or depressed by a wide variety of experimental manipulations. Reticuloendothelial function is stimulated by such nonspecific stimuli as BCG, bacterial infection, and triglyceride suspensions, and depressed by treatment with such simple lipids as ethyl oleate after radiotherapy, in advanced neoplastic disease, and in terminal illness. Such measurements of clearance are not simple measurements of macrophage function, since the clearance of colloids is affected by such factors as plasma opsonin levels. Moreover, particles can be cleared from plasma by leakage between the endothelial cells of the splenic sinusoid or by platelet phagocytosis (Tennent and Donald, 1976) (see Chapter 14). However, an increase in the number and maturation of Kupffer cells does play a part in increasing clearance of colloids. Individual Kupffer cells have been shown, at least *in vitro,* to phagocytize in at least two ways. Phagocytosis may depend on the presence of receptors in the Kupffer cell membrane for, respectively, the Fc end of the immunoglobulin molecule and the third component of complement. In the former case, macrophage cytoplasm is protruded up towards the particle, while in the latter situation, the particle sinks into a membrane-bound vacuole in the cytoplasm. Micropinocytosis almost certainly plays some part in clearance of colloids but the relative importance of this is not known (Munthe-Kaas, 1976).

It has been known for some time that reticuloendothelial function affects susceptibility to shock. Stimulation leads to increased resistance; this is probably due to ingestion of such substances as antigen/antibody complexes and bacterial endotoxin. The ingestion of antigen/antibody complexes in anaphylactic shock may lead to disintegration of Kupffer cells and release of lysosomal enzymes (Treadwell and Santos-Buch, 1967; Santos-Buch and Treadwell, 1967).

3.2. SECRETION

Renaut (1907) early suggested that macrophages might be secretory cells, or "rhagiocrine cells." The introduction of histochemical techniques led to the early observation that macrophages contained acid phosphatase and other esterases. The isolation of macrophage and polymorph lysosomes was followed by the demonstration that they contained a very wide range of hydrolytic enzymes—including acid phosphatase, β-glucuronidase, and various esterases and, indeed, enzymes which break down a wide variety of biological macromolecules.

More recently, macrophages have been shown to release a wide variety of substances *in vitro.* All the experiments essentially involved culturing macrophages *in vitro* and assaying supernatants either biochemically or biologically for the substance concerned. The substances which have been alleged to be produced in this way include: lysosomal hydrolases, neutral proteinases (plasminogen activator, collagenase, elastase), colony stimulating factor, transferrin, erythropoietin, lysozyme, hyaluronidase, pyrogen, prostaglandins, cytolytic factor(s), T-cell stimulant, B-cell stimulant, factors modulating fibroblast proliferation, inhibitor of lymphocyte transformation, stimulator of polymorph migration, and interferon (Allison and Davies, 1975; Kay and Douglas, 1977). In such experiments it is not certain, of course, whether macrophages, as opposed to contaminant cells, are releasing these substances or whether they are releasing them by active secretion or by disintegration. And the demonstration of such a secretion *in vitro* leaves unclear its role in the living animal.

It seems highly likely that the macrophage is a major source of pyrogen in febrile disease in which there is no polymorphonuclear leukocytosis and that products released by macrophages modulate the activity of lymphocytes in chronic inflammatory lesions and perpetuate the chronic inflammatory reaction. Probably, products are released into the lymph and blood that drain chronic granulomatous lesions. It is reasonable to suppose that macrophages in such lesions release a wide variety of substances, some with a purely local action, perhaps locally broken down, and others with a systemic effect. It is likely that populations of macrophages in different organs have different secretory abilities—for instance, alveolar macrophages are a richer source of lysozyme than macrophages in other sites. Moreover, even within one organ, the secretory ability of macrophages differs from one site to another; for instance, the macrophages within lymph node germinal centers show much less lysozyme histochemically than those in lymph node sinuses. Resident macrophages in the peritoneal cavity have a distribution of peroxidase different from that of recently immigrated mononuclear cells. These findings raise the question as to whether there are several distinct populations of macrophages, perhaps of different derivation and with different functions, or whether, in a given population, some cells are switched on to make one product and some another.

3.3. MACROPHAGES IN THE IMMUNE RESPONSE

The part played by macrophages in the immune response is of great importance (Nelson, 1976); this is an area in which ideas change frequently and the current ideas will not necessarily persist. The macrophage has an important role in both the afferent and efferent limbs of the immune response. The afferent functions are dependent on the close anatomical proximity which occurs between macrophage (or dendritic cell) and lymphocyte in lymphoreticular tissue (see Chapter 13), and, notably, in germinal centers. Possible mechanisms of functional cellular interaction as summarized by Sell (1975) include interaction between macrophage and B cell, transfer to T-cell receptors on to macrophages allowing subsequent B-cell stimulation, and interaction between T cell, B cell, and macrophage in various ways. These processes may involve binding of antigen on or near the surface of the macrophage or ingestion of the antigen; thereafter most of the antigen may be degraded and a small amount released intact. The antigen may be modified to become more immunogenic, complexed to low molecular-weight RNA, or it may induce formation of a specific messenger RNA. The term "antigen-focusing" has become popular to describe the role of the macrophage in relation to the lymphocyte. More recent work on macrophages emphasizes that different subgroups may have considerably different functions (Walker, 1976). In addition, their function may vary greatly with the nature and state of the antigen. They present antigens to T and B cells in a manner that allows a suitable response, possibly by altering concentration, presenting a multivalent form, or merely by providing a rigid substrate. They also improve the viability of lymphocytes. Under certain circumstances they may eliminate tolerogenic antigen, thus inhibiting the development of immunological tolerance (Pierce and Kapp, 1976; Mosier, 1976). This kind of function occurs in cell-mediated as well as in humoral immunity.

The role of the macrophage in some immune responses is currently thought to be dependent on the immune response (*Ir*) genes, part of the major histocompatibility complex (Benacerraf, 1977). These code for self markers on cell surfaces, such as the Ia antigens present on the surface of macrophages among other cells (Neiderhuber and Shreffler, 1977; Moraes *et al.*, 1977). These receptors may be of "broad specificity that focus or orient antigen for T cells to recognize it" (Feldman, 1977). An initial antigen-independent association of lymphocyte and macrophage is followed by specific interaction of lymphocyte and macrophage if the two cell types share surface determinants coded by H-linked loci. The presence of antigen may then stabilize the interaction and lead to a signal to the lymphocyte to proliferate; this signal may be electrical or chemical (Rosenthal *et al.*, 1976). There may be a second signal to the lymphocyte which is distinct from antigenic contact (Rosenstreich and Oppenheim, 1976). It has become clear that among the many secretory substances produced by macrophages are factors which stimulate lymphocytes to proliferate and mature and increase lymphokine production (Unanue *et al.*, 1976; Epstein, 1976). Other factors may inhibit lymphocyte proliferation.

In addition to these functions on the afferent side of the immune response, macrophages ingest and destroy large amounts of immunogenic substances; when antigens are introduced intravenously, they are cleared mainly by Kupffer cells. Normally the Kupffer cells largely clear the gut-derived antigens from the portal circulation. In hepatic cirrhosis there may be an increase in circulating gamma-globulin levels conceivably due to shunting of blood to the spleen and excessive immunological stimulation (Bradfield, 1974). On the efferent side of the immune response, macrophages accumulate at the site of cell-mediated immune responses as a result of interaction between antigen and sensitized T cell. This specific interaction leads to the release of lymphokines which attract monocytes, immobilize them at the site of inflammation, and activate them, causing maturation. The mature, activated cell is nonspecifically more active in phagocytosis and the killing of bacteria (and also tumor cells) (Dumonde *et al.*, 1975; Adams, 1976). This is of major importance in chronic inflammatory lesions. Quite distinct from this, macrophages have a strange ability to kill other cells *in vitro* by extracellular action. This has been most clearly demonstrated in the case of red blood cells; macrophages from several species, including the mouse, lyse mouse red blood cells *in vitro* by an ATP-dependent process which involves osmotic shock of the red cell. The mediator is a labile factor secreted by the macrophage, with a molecular weight of less than 1000, and, possibly, an oligopeptide acting on the cell membrane by lipid peroxidation (Seljelid, 1975; Melsom *et al.*, 1975). Similar mechanisms may occur in the case of other nonneoplastic cells (Gallily, 1975).

Macrophages play an important part in the immune response to neoplasia. They are present in considerable numbers in many tumors and the growth and metastasis of tumors may be inversely related to their macrophage content. It is likely that in several experimental systems macrophages can kill tumor cells. How this happens is not certain, but at least two different sets of conditions operate. Mature stimulated or activated macrophages can kill tumor cells nonspecifically, and macrophages which have been "armed" with specific cytocidal factors produced by T lymphocytes can kill tumor cells specifically. This process may involve release of a soluble mediator, possibly of lysosomal origin. In addition, macrophages can phagocytize tumor cells, but it is not clear whether this happens before or after the cell is killed. (Alexander, 1976; Alexander *et al.*, 1976). Human neoplasia differs widely from experimental animal neoplasia in many ways and the role of the macrophage in human neoplasia is much less clear (Carr, 1977). There is no doubt that most human neoplasms have a lymphoreticular infiltrate, sometimes scanty, sometimes dense, and composed of lymphocytes, plasma cells, reticulum cells, and macrophages. The latter are often in close contact with both tumor cells and lymphocytes, but usually there is no sign of any cytotoxic interaction (Underwood and Carr, 1972; Carr and Underwood, 1974). The exceptions to this statement are malignant melanoma and lymphoma. In certain melanomas it has been suggested that circulating cells participate in antibody-mediated cytotoxic interactions against the tumor cells, and that similar cells are present in the tumor, in contact with degenerating tumor cells. Such cytotoxic cells are, however, more like lymphocytes than macrophages (Roubin *et al.*, 1975). Again in Hodgkin's disease, macrophages are present in the lesion but the cytotoxic cells are probably lymphocytes. It is likely that in the case of many tumors an extensive lymphoreticular infiltrate implies a better prognosis, but this relationship pertains to the total (and mixed) lymphoreticular reaction. A recent study of breast cancer, however, convincingly related high-macrophage and plasmacyte content to good prognosis and absence of metastasis (Lauder *et al.*, 1977). There is a relation between changes in the lymph nodes draining a tumor and prognosis, but such changes are essentially in the lymphocyte population.

Neoplasia affects reticuloendothelial clearance of colloids, probably by nonspecific depletion of nonimmunoglobulin opsonins or recognition factors. It has been suggested that this may imply a defect in initial surveillance mechanisms, allowing the emergence of a neoplastic clone. This is a rather indirect inference (Di Luzio, 1975).

Several abnormalities in monocyte function occur in human neoplasia. The most interesting of these is an apparently specific impairment of monocyte chemotaxis. In bladder cancer, chemotaxis recovers after surgical resection or BCG stimulation, while in melanoma, impaired chemotaxis is correlated with lymph node metastasis and a poorer prognosis. Such impaired chemotaxis may be due to release of a specific inhibitor by the tumor cells (Snyderman *et al.*, 1973; Snyderman *et al.*, 1977; Boetcher and Leonard, 1974; Hausman *et al.*, 1975; Hausman and Brosman, 1976; Rubin *et al.*, 1976; Snyderman *et al.*, 1976).

3.4. THE METABOLIC ROLE OF MACROPHAGES

Macrophages are involved in iron metabolism and in lipid and other intermediary metabolism. The macrophages which play a major role in iron metabolism are those of the spleen (Edwards and Simon, 1970) (see Chapter 12). Red pulp macrophages ingest either whole effete RBC or fragments thereof. It is likely that elderly or otherwise effete RBC are preselected for phagocytosis because they respond differently to the deformation stress of passing through the narrow interendothelial cell gap. The red cell is rapidly broken down into fragments and digested by lysosomal enzymes. Iron exists in the macrophage in two compartments—a slow turnover storage pool and a rapidly turning over labile pool. Iron which goes into the stable pool is attached to preexisting apoferritin molecules to form ferritin. These are found both free in the cytoplasm and within membrane-bound lysosomal inclusions. Some lysosome inclusions contain iron but no evident ferritin. The term hemosiderin applies to masses of cytoplasmic material demonstrable at the light-microscopic level as containing iron. There is an exchange between the iron in the stable iron pool and the labile iron pool. Labile iron is poorly bound to transferrin which is synthesized in the macrophage. Iron is excreted from the macrophage partly bound to transferrin. The saturated transferrin molecule carries two atoms of iron but probably each atom of iron as it leaves the macrophage does not have its own transferrin molecule. Unsaturated plasma transferrin molecules can probably attach to a receptor site on the macrophage surface and extract iron. Iron passes from plasma transferrin either directly to the reticulocyte which has a transferrin receptor, or by uptake by marrow macrophages which resynthesize ferritin. This is then taken up by the erythroblast by micropinocytosis (Haurani and Meyer, 1976). Macrophages therefore play an important part in the recycling of used iron. After splenectomy or during excess hemolysis, Kupffer cells may also break down red cells. It is not clear whether splenic and marrow macrophages differ in a way which can be related to their very different situations in iron metabolism, other than in their accumulation of iron.

Macrophages possess numerous esterases which are involved in the breakdown and synthesis of a variety of lipids. This is of at least some significance in the arterial wall where they are mobilized to cope with the lipid accumulation in atheroma; while they are best regarded as being involved in the clearance of lipids in this site, they may, when they break up, release factors which promote fibrosis in the lesion. The Kupffer cells have a significant role in lipid metabolism, since they are the first macrophages to monitor the chylomicron-rich blood from the small intestine and to ingest dietary cholesterol. They do not, however, apparently, clear endogenous cholesterol (Bessis and Breton-Gorius, 1959). Kupffer cells also metabolize steroids—for instance ring reduction of Δ^4 ketocorticosteroids (Favarger, 1962), acetylate sulfonamide (Nabors *et al.*, 1967), synthesize urea, and bind and degrade insulin (Govier, 1965). It is not clear to what extent macrophages at other sites carry out these functions. Circulating monocytes bind insulin; this binding is less in the diabetic and in the obese.

The functions of macrophages in disease are extensive and are reviewed elsewhere (Territo and Cline, 1976), and in a later volume in this series.

This outline has highlighted salient features of the biology of the macrophage—the common structural pattern reflecting mobility, avid ingestion, secretory ability, the putative common origin, the similarities and differences between macrophages at different sites, and the way in which the fundamental phagocytic and secretory properties of the cell fit into the inflammatory and

TABLE 1. TERMINOLOGY OF MONOCYTES AND MACROPHAGES[a]

A macrophage
Activated macrophage
Adherent cell
Adventitial cell (Marchand)[b]
Alveolar macrophage
Alveolar phagocyte
AM (= alveolar macrophage)
AM (= activated macrophage)
Angry macrophage
Anode cell
Anitschkow cell
Armed macrophage

B macrophage
Blood histiocyte (Aschoff)[c]

Clasmatocyte (Ranvier)[b,d]
Chromatophore

Dendritic cell
Dictyocyte
Dust cell

Elicited macrophage
Endothelial cell
Endothelial leukocyte (Mallory)[c,d]
Endotheliocyte
Endothelioid cell
Epithelioid cell
Exudate macrophage

Fixed macrophage
Fixed tissue macrophage
Free macrophage
Free macrophage

Germinal center macrophage
Germinoblast
Giant cell
Gitter cell

Heart failure cell
Hemendotheliocyte
Hemocytoblast
Hemohistioblast
Histioblast
Histiocyte (Kiyono[b,d], Aschoff[e])
Histogenic monocyte
Hofbauer cell
Histogenic wandering cell

Immune macrophage
Induced macrophage
Inflammatory cell
Interstitial cell

Kupffer cell

Large mononuclear cell (Ehrlich)[c]
Lepra cell
Leukocytoid cell
Lipophage
Littoral cell
Littoral cell of the spleen
Lymphendotheliocyte
Lymphocytoform histiocyte

Macrophage (Metchnikoff)[e]
Melanophage
Metalocyte
Microglia
Monocyte (Ehrlich, Naegeli)[b]
Monocyte-derived macrophage
Mononuclear phagocyte
Multinucleate giant cell

Noninduced macrophage
Normal macrophage

Oligodendroglia

Pericyte
Perithelial cell
Polyblast (Maximov)[d,e]
Primed macrophage
Professional phagocyte
Promonocyte
Pyrral cell
Pyrrhol blue cell
Pyrrhol cell (Goldmann)[b]

Resident macrophage
Resting wandering cell
Reticular cell
Reticuloendothelial cell (Aschoff)[e]
Reticuloendotheliocyte
Reticulum cell
Rhagiocrine cell (Renaut)[b]
Rhagiocyte (Renaut)[b]

Satellite cell
Schwann cell
Splenocyte
Stimulated macrophage

Tingible-body macrophage
Tissue macrophage
Transitional cell

Virchow cell

Wandering cell (Maximov)[b]

[a] From various sources and Craddock (1972) and Weiss (1972).
[b] Albegger (1976).
[c] Carrell and Ebeling (1926).
[d] Clark and Clark (1930).
[e] Sabin *et al.* (1925).

immune response and the general metabolism of the body. It is merely a pointer to the more detailed information in the chapters which follow. Following the Reference section to this chapter there is a brief chronological bibliography of accessible literature in the field. Table 1 is a list of monocyte and macrophage terminology.

ACKNOWLEDGMENT. We are grateful to Mrs. M. Hopewell for much secretarial help in the preparation of this volume.

REFERENCES

Adams, D. O., 1976, The granulomatous inflammatory response: A review, *Am. J. Pathol.* **84**:161.

Albegger, K. W., 1976, Zur Morphologie und Bedeutung des Mononukleären-Phagozyten-Systems (MPS) bei der chronischen Rhinosinusitis. Eine licht- und elektronenmikroskopische Untersuchung (Engl. Abstr.), *Arch. Oto-Rhino-Laryngol.* **214:**27.

Alexander, P., 1976, The function of the macrophage in malignant disease, *Annu. Rev. Med.* **27**:207.

Alexander, P., Eccles, S. A., and Gauci, C. L. L., 1976, The significance of macrophages in human and experimental tumors, *Ann. N.Y. Acad. Sci.* **276**:124.

Allison, A. C., and Davies, P. E., 1975, Increased biochemical and biological activities of mononuclear phagocytes exposed to various stimuli, with special reference to secretion of lysosomal enzymes, in: *Mononuclear Phagocytes in Immunity, Infection and Pathology* (R. van Furth, ed.), pp. 487–504, Blackwell, London.

Allison, A. C., Davies, P., and de Petris, S., 1971, Role of contractile microfilaments in macrophage movement and endocytosis, *Nature New Biol.* **232**:153.

Benacerraf, B., 1977, Role of major histocompatibility complex in genetic regulation of immunologic responsiveness, *Transplant. Proc.* **9**:825.

Bessis, M., and Breton-Gorius, J., 1959, Aspects différents du fer dans l'organisme. I. Ferritin et micelles ferrugineuse, *J. Biophys. Biochem. Cytol.* **6**:231.

Boetcher, D. A., and Leonard, E. J., 1974, Abnormal monocyte chemotaxis in cancer patients, *J. Natl. Cancer Inst.* **52**:109.

Bradfield, J. W. B., 1974, Control of spillover: The importance of Kupffer cell function in clinical medicine, *Lancet* **2**:883.

Breathnach, A. S., 1965, The cell of Langerhans, *Int. Rev. Cytol.* **18**:1.

Carr, I., 1973a, *The Macrophage: A Review of Ultrastructure and Function*, Academic Press, London.

Carr, I., 1973b, The reticulum cell and the reticular cell in the mouse popliteal lymph-node: An electronmicroscopic autoradiographic study, *Virchows Arch. B* **15**:1.

Carr, I., 1977, Macrophages in human cancer, in: *The Macrophage and Cancer* (K. James, B. McBride, and A. Stuart, eds.), pp. 364–374, Blackwell, Oxford.

Carr, I., and Underwood, J. C. E., 1974, The ultrastructure of the local cellular reaction to neoplasic, *Int. Rev. Cytol.* **37**:329.

Carr, I., Everson, G., Rankin, A., and Rutherford, J., 1970, The fine structure of the cell coat of the peritoneal macrophage and its role in the recognition of foreign material, *Z. Zellforsch.* **105**:339.

Carr, I., Price, P., and Westby, S., 1976, The effect of tumor extracts on macrophage proliferation in lumph nodes, *J. Pathol.* **120**:251.

Carrel, A., and Ebeling, A. H., 1926, The fundamental properties of the fibroblast and the macrophage. II. The macrophage, *J. Exp. Med.* **44**:285.

Clark, E. R., and Clark, E. L., 1930, Relation of monocytes of the blood to the tissue macrophages, *Am. J. Anat.* **46**:149.

Cohn, Z. A., 1970, Endocytosis and intracellular digestion, in: *Mononuclear Phagocytes* (R. van Furth, ed.), pp. 121–129, Blackwell, London.

Cohn, Z. A., and Benson, B., 1965, The differentiation of mononuclear phagocytes: Morphology, cytochemistry and biochemistry, *J. Exp. Med.* **121**:153.

Craddock, C. G., 1972, Cellular kinetics of monocytes and macrophages, in: *Hematology* (P. K. Schneider and S. D. Boyston, eds.), pp. 748–751, McGraw-Hill, New York.

Daems, W. Th., and Brederoo, P., 1973, Electron microscopical studies on the structure, phagocytic properties and peroxidatic activity of resident and exudate peritoneal macrophages in the guinea pig, *Z. Zellforsch. Mikroskop. Anat.* **144**:247.

Daems, W. Th., Wisse, E., Brederoo, P., and Emeis, J. J., 1975, Peroxidatic activity in monocytes and macrophages, in: *Immunity, Infection and Pathology* (R. van Furth, ed.), pp. 57–77, Blackwell, London.

Day, A. J., 1964, The macrophage system, lipid metabolism and atherosclerosis, *J. Atheroscler. Res.* **4**:117.

DiLuzio, N. R., 1975, Macrophages, recognition factors, and neoplasia, in: *The Reticuloendothelial System* (J. W. Rebuck, C. W. Berard, M. and R. Abell, eds.), pp. 49–64, William and Wilkins, Baltimore.

Dumonde, D. C., Kelly, R. H., Preston, P. M., and Wolstencroft, R. A., 1975, Lymphokines and macrophage function in the immunological response, in: *Mononuclear Phagocytes in Immunity, Infection and Pathology* (R. van Furth, ed.), pp. 675–699, Blackwell, London.

Edwards, V. D., and Simon, S. T., 1970, Ultrastructural aspects of red cell destruction in the normal rat spleen, *J. Ultrastruct. Res.* **33**:187.

Emeis, J. J., and Wisse, E., 1975, On the cell coat of rat liver Kupffer cells, in: *Mononuclear Phagocytes in Immunity, Infection and Pathology* (R. van Furth, ed.), pp. 315–325, Blackwell, London.

Epstein, L. B., 1976, The ability of macrophages to augment *in vitro* mitogen- and antigen-stimulated production of interferon and other mediators of cellular immunity by lymphocytes, in: *Immunobiology of the Macrophage* (D. S. Nelson, ed.), pp. 202–234, Academic Press, New York.

Favarger, P., 1962, The liver and lipid metabolism, in: *The Liver* (C. H. Rouiller, ed.), pp. 549–604, Academic Press, New York.

Feldman, M., 1977, Are the *Ir* genes expressed by macrophages? *Nature* **267**:105.

Gallily, R., 1975, The killing capacity of immune macrophages, in: *Mononuclear Phagocytes in Immunity, Infection and Pathology* (R. van Furth, ed.), pp. 895–909, Blackwells, London.

Govier, W. C., 1965, Reticuloendothelial cell as the site of sulfanilamide acetylation in the rat, *J. Pharmacol. Exp. Ther.* **150**:305.

Green, I., Jaffe, E. S., Shevach, E. M., Edelson, R. L., Frank, M. M., and Berard, C. W., 1975, Determination of the origin of malignant reticular cells by the use of surface membrane markers, in: *The Reticuloendothelial System* (J. W. Rebuck, C. W. Berard, and M. R. Abell, eds.), pp. 282–300, William and Wilkins, Baltimore.

Griffin, F. M., and Silverstein, S. C., 1974, Segmental response of the macrophage plasma membrane to a phagocytic stimulus, *J. Exp. Med.* **139**:323.

Griffin, F. M., Griffin, J. A., Leider, J. E., and Silverstein, F. M., 1975, Studies on the mechanism of phagocytosis-I. Requirements for circumferential attachment of particle bound ligands to specific receptors on the macrophage plasma membrane, *J. Exp. Med.* **142**:1263.

Haurani, F. I., and Meyer, A., 1976, Iron and the reticuloendothelial system, in: *The Reticuloendothelial System in Health and Disease, Functions and Characteristics* (S. M. Reichard, M. R. Escobar, and H. Friedman, eds.), pp. 171–187, Plenum Press, New York.

Hausman, M. S., and Brosman, S. A., 1976, Abnormal monocyte function in bladder cancer patients, *J. Urol.* **115**:537.

Hausman, M. S., Brosman, S., Snyderman, R., Mickey, M. R., and Fahey, J., 1975, Defective monocyte function in patients with genitourinary carcinoma, *J. Natl. Cancer. Inst.* **55**:1047.

Karnovsky, M. L., 1962, Metabolic basis of phagocytic activity, *Physiol. Rev.* **42**:143.

Karnovsky, M. L., Lazdins, J., and Simmons, S. R., 1975, Metabolism of activated mononuclear phagocytes at rest and during phagocytosis, in: *Immunity, Infection and Pathology* (R. van Furth, ed.), pp. 423–438, Blackwell, London.

Karnovsky, M. L., Drath, D., and Lazdins, J., 1976, Biochemical aspects of the function of the reticuloendothelial system, in: *The Reticuloendothelial System in Health and Disease: Functions and Characteristics* (S. M. Reichard, M. R. Escobar, and H. Friedman, eds.), pp. 121–130, Plenum Press, New York.

Kay, N. E., and Douglas, D. S., 1977, Mononuclear phagocyte: Development, structure, function and involvement in immune response, *N.Y. State J. Med.* **77**:327.
Lauder, I., Aherne, W., Stewart, J., and Sainsbury, R., 1977, Macrophage infiltration of breast tumours: A prospective study, *J. Clin Pathol.* **30**:563.
McKeever, P. E., Garvin, A. J., Hardin, D. H., and Spicer, S. S., 1976, Immune complex receptors on cell surfaces - II. Cytochemical evaluation of their abundance on different immune cells: Distribution, uptake and regeneration, *Am. J. Pathol.* **84**:437.
Melsom, H., Sanner, T., and Seljelid, R., 1975, Macrophage cytolytic factor: Some observations on its physicochemical properties and site of action, *Exp. Cell. Res.* **94**:221.
Moraes, M. E., Moraes, J. R., and Stastny, P., 1977, Separate Ia-like determinants in human lymphocytes and macrophages, *Transplant. Proc.* **9**:1211.
Mosier, D. E., 1976, The role of macrophages in the specific determination of immunogenicity and tolerogenicity, in: *Immunobiology of the Macrophage* (D. S. Nelson, ed.), pp. 35–44, Academic Press, New York.
Munthe-Kaas, A. C., 1976, Phagocytosis in rat Kupffer cells *in vitro, Exp. Cell Res.* **99**:319.
Nabors, C. J., Berliner, D. L., and Dougherty, T. F., 1967, The functions of hepatic reticuloendothelial cells in steroid hormone transformation, *J. Reticuloendothel Soc.* **4**:237.
Neiderhuber, J. E., and Shreffler, D. C., 1977, Anti Ia serum blocking of macrophage function in the *in vitro* humoral response, *Transplant. Proc.* **9**:875.
Nelson, D. S. (ed.), 1976, *Immunobiology of the Macrophage,* Academic Press, London.
Nezelof, C., Diebold, N., Rousseau-Merck, M. F., 1977, Ig surface receptors and erythrophagocytic activity of histiocytosis X cells *in vitro, J. Pathol.* **122**:105.
Nichols, B. A., and Bainton, D. F., 1975, Ultrastructure and cytochemistry of mononuclear phagocytes, in: *Mononuclear Phagocytes in Immunity, Infection and Pathology* (R. van Furth, ed.), pp. 17–55, Blackwell, London.
North, R. J., 1969, The mitotic potential of fixed phagocytes in the liver as revealed during the development of cellular immunity, *J. Exp. Med.* **130**:315.
Nossal, G. J. V., Abbot, A., Mitchell, J., and Lummus, Z., 1968, Antigens in immunity. XV. Ultrastructural features of antigen capture in primary and secondary lymphoid follicles, *J. Exp. Med.* **127**:227.
Pierce, C. W., and Kapp, J. A., 1976, The role of macrophage in antibody responses *in vitro,* in: *Immunobiology of the Macrophage* (D. S. Nelson, ed.), pp. 2–33, Academic Press, New York.
Renaut, J., 1907, Les cellules connectives rhagiocrines, *Arch. Anat. Microsc.* **9**:495.
Rosenstreich, D. L., and Oppenheim, J. J., 1976, The role of macrophages in the activation of T and B lymphocytes *in vitro,* in: *Immunobiology of the Macrophage* (D. S. Nelson, ed.), pp. 162–199, Academic Press, New York.
Rosenthal, A. S., Blake, J. T., Ellner, J. J., Greineder, D. K., and Lipsky, P. E., 1976, Macrophage function in antigen recognition by T lymphocytes, in: *Immunobiology of the Macrophage* (D. S. Nelson, ed.), pp. 131–160, Academic Press, New York.
Roser, B., 1968, The distribution of intravenously injected Kupffer cells in the mouse, *J. Reticuloendothel. Soc.* **5**:455.
Rossi, F., Patriarca, R., Romeo, D., and Zabucchi, G., 1976, The mechanism of control of phagocytic metabolism, in: *The Reticuloendothelial System in Health and Disease: Function and Characteristics* (S. M. Reichard, M. R. Escobar, and H. Friedman, eds.), pp. 205–223, Plenum Press, New York.
Roubin, R., Cesarini, J-P., Fridman, W. H., Pavie-Fischer, J., and Peter, H. H., 1975, Characterization of the mononuclear cell infiltrate in human malignant melanoma, *Int. J. Cancer.* **16**:61.
Rubin, R. H., Cosimi, A. B., and Goetzl, E. J., 1976, Defective human mononuclear leucocyte chemotaxis as an index of host resistance in malignant melanoma, *Clin. Immunol. Immunopathol.* **6**:376.
Sabin, F. R., Doan, C. A., and Cunningham, R. S., 1925, Discrimination of two types of phagocytic cells in the connective tissues by the supravital technique, *Contrib. Embryol.* **82**:125.
Santos-Buch, C. A., and Treadwell, P. E., 1967, Distribution of Kupffer cells during systemic anaphylaxis in the mouse. II. Changes in tissue and plasma heat- and formaldehyde-stable *p*-nitrophenylphosphatase, *Am. J. Pathol.* **51**:505.

Sbarra, A. J., Selvaraj, R. J., Paul, B. B., Zgliczynski, J. M., Poskitt, P. K. F., Mitchell, G. W. Jr., and Louis, F., 1976, Chlorination, decarboxylation and bactericidal activity mediated by the MPO-H_2O_2-Cl^- system, in: *The Reticuloendothelial System in Health and Disease: Function and Characteristics* (S. M. Reichard, M. R. Escobar, and H. Friedman, eds.), pp. 191–203, Plenum Press, New York.

Seljelid, R., 1975, Cytotoxic effect of macrophages on mouse red cells, in: *Mononuclear Phagocytes in Immunity, Infection and Pathology* (R. van Furth, ed.), pp. 911–925, Blackwell, London.

Sell, S., 1975, *Immunology, Immunopathology and Immunity*, 2nd ed., Harper and Row, Hagerstown.

Snyderman, R., and Pike, M. C., 1976, An inhibitor of macrophage chemotaxis produced by neoplasms, *Science* **192**:370.

Snyderman, R., Allman, L. C., Frankel, A., and Blaese, R. N., 1973, Defective mononuclear leucocyte chemotaxis: A previously unrecognized immune dysfunction, *Ann. Intern. Med.* **78**:509.

Snyderman, R., Pike, M. C., Blaylock, B. L., and Weinstein, P., 1976, Effects of neoplasms on inflammation: Depression of macrophage accumulation after tumour implantation, *J. Immunol.* **116**:585.

Snyderman, R., Seigler, H. F., and Meadows, L., 1977, Abnormalities of monocyte chemotaxis in patients with melanoma: Effects of immunotherapy and tumor removal, *J. Natl. Cancer. Inst.* **58**:37.

Spector, W. G., and Lykke, A. W. J., 1966, The cellular evolution of inflammatory granulomata, *J. Pathol. Bact.* **92**:163.

Spector, W. G., and Mariano, M., 1975, Macrophage behaviour in experimental granulomas, in: *Mononuclear Phagocytes in Immunity, Infection and Pathology* (R. van Furth, ed.), pp. 927–938, Blackwell, London.

Steinman, R. M., and Cohn, Z. A., 1975, Dendritic cells, reticular cells and macrophages, in: *Mononuclear Phagocytes in Immunity, Infection and Pathology* (R. van Furth, ed.), pp. 95–107, Blackwell, London.

Stossel, T. P., and Hartwig, J. H., 1976, Interactions of actin, myosin and a new actin-binding protein of rabbit pulmonary macrophages. II. Role in cytoplasmic movement and phagocytosis, *J. Cell Biol.* **68**:602.

Stuart, A. E., 1970, *The Reticulo-endothelial System*, E. and S. Livingstone, Edinburgh.

Stuart, A. E., and Davidson, A. E., 1971, The human reticular cell: Morphology and cytochemistry, *J. Pathol.* **103**:41.

Tennent, R. J., and Donald, K. J., 1976, The ultrastructure of platelets and macrophages in particle clearance stimulated by zymosan, *J. Reticuloendothel. Soc.* **19**:269.

Territo, M., and Cline, M. J., 1976, Macrophages and their disorders in man, in: *Immunobiology of the Macrophage* (D. S. Nelson, ed.), pp. 593–616, Academic Press, London.

Treadwell, P. E., and Santos-Buch, C. A., 1967, Disruption of Kupffer cells during systemic anaphylaxis in the mouse. I. Properties of distribution of heat and formaldehyde-stable liver acid *p*-nitrophenylphosphatase, *Am. J. Pathol.* **51**:483.

Unanue, E. R., Beller, D. I., Calderon, J., Kiely, J. M., and Stadecker, M. J., 1976, Regulation of immunity and inflammation by mediators from macrophages, *Am. J. Pathol.* **85**:456.

Underwood, J. C. E., and Carr, I., 1972, The ultrastructure of the lymphoreticular cells in non-lymphoid human neoplasm, *Virchows Arch. B* **12**:39.

Van Ewijk, W., Verizuden, J. H. M., van der Kwast, Th. H., and Luijex-Meijer, S. W. M., 1974, Reconstruction of the thymus dependent area in the spleen of lethally irradiated mice, *Cell Tissue Res.* **149**:43.

van Furth, R., 1975, Modulation of monocyte production in: *Mononuclear Phagocytes in Immunity, Infection and Pathology* (R. van Furth, ed.), pp. 161–172, Blackwell, London.

Veerman, A. J. P., 1974, On the interdigitating cells in the thymus-dependent area of the rat spleen: A relation between the mononuclear phagocyte system and T lymphocytes, *Cell Tissue Res.* **148**:247.

Veldman, J. E., 1970, Histophysiology and electron microscopy of the immune response, Thesis, University of Groningen.

Volkman, A., 1976, Disparity in origin of mononuclear phagocyte populations, *J. Reticuloendothel. Soc.* **19**:249.

Volkman, A., and Gowans, J. L., 1965a, The production of macrophages in the rat, *Br. J. Exp. Pathol.* **46**:50.
Volkman, A., and Gowans, J. L., 1965b, The origin of macrophages from bone marrow in the rat, *Br. J. Exp. Pathol.* **46**:62.
Walker, W. S., 1976, Functional heterogeneity of macrophages, in: *Immunobiology of the Macrophage* (D. S. Nelson, ed.), pp. 91–110, Academic Press, New York.
Weiss, L., 1972, *The Cells and Tissues of the Immune System. Structure, Functions, Interactions,* Prentice Hall, New Jersey.
Werb, Z., 1975, Macrophage membrane synthesis, in: *Mononuclear Phagocytes in Immunity, Infection and Pathology* (R. van Furth, ed.), pp. 331–345, Blackwell, London.
Whitelaw, D. M., and Batho, H. F., 1975, Kinetics of monocytes, in: *Mononuclear Phagocytes in Immunity, Infection and Pathology* (R. van Furth, ed.), pp. 175–187, Blackwell, London.
Wilkinson, P. C., 1976, Cellular and molecular aspects of chemotaxis of macrophages and monocytes, in: *Immunobiology of the Macrophage* (D. S. Nelson, ed.), pp. 350–365, Academic Press, New York.
Wisse, E., 1974, Observations on the fine structure and peroxidase cytochemistry of normal rat liver Kupffer cells, *J. Ultrastruct. Res.* **46**:393.
Wisse, E., 1977, Ultrastructure and function of Kupffer cells and other sinusoidal cells, in the liver in: *The Kupffer Cells* (D. L. Knook and E. Wisse, eds.), pp. 33–60, Elsevier, Amsterdam.
Wynne, K. M., Spector, W. G., and Willoughby, D. A. 1975, Macrophage proliferation *in vitro* induced by exudate, *Nature* **253**:636.

BIBLIOGRAPHY

Metchnikoff, E., 1893, *Lectures on the Comparative Pathology of Inflammation,* Kegan Paul, London, 1893, reprinted Dover, New York, 1968.
Policard, A., 1963, *Physiologie et Pathologie du Système Lymphoide,* Masson, Paris.
Pearsall, N. N., and Weiser, R. S., 1970, *The Macrophage,* Lea and Febiger, Philadelphia.
Stuart, A. E., 1970, *The Reticulo-endothelial System,* E and S Livingstone, Edinburgh.
van Furth, R. (ed.), 1970, *Mononuclear Phagocytes,* Blackwell, London.
Vernon-Roberts, B., 1972, *The Macrophage,* Cambridge University Press, Cambridge.
Carr, I., 1973, *The Macrophage: A Review of Ultrastructure and Function,* Academic Press, London.
van Furth, R. (ed.), 1975, *Mononuclear Phagocytes in Immunity, Infection and Pathology,* Blackwell, London.
Fink, M. A. (ed.), 1976, *The Macrophage in Neoplasia,* Academic Press, New York.
Reichard, S. M., Escobar, M. R., and Friedman, H. (eds.), 1976, *The Reticuloendothelial System in Health and Disease: Functions and Characteristics,* Vol. 73A, *Advances in Experimental Medicine and Biology,* Plenum Press, New York.
Friedman, H., Escobar, M. R., and Reichard, S. M. (eds.), 1976, *The Reticuloendothelial System in Health and Disease: Immunologic and Pathogenic Aspects,* Vol. 73B, *Advances in Experimental Medicine and Biology,* Plenum Press, New York.
Nelson, D. S. (ed.), 1976, *Immunobiology of the Macrophage,* Academic Press, London.
Carr, I., Hancock, B. W., Henry, L., and Ward, A. W., 1977, *Lymphoreticular Disease,* Blackwell, Oxford.
James, K., McBride, B., and Stuart, A. (eds.), 1977, *The Macrophage and Cancer,* Proceedings of the European Reticuloendothelial Society, Edinburgh.
Sanders, C. L., Schneider, R. P., Dagle, G. E., and Raga, H. A. (eds.), 1977, *Pulmonary Macrophage and Epithelial Cells,* Technical Information Center, Springfield.
Weiss, L., 1977, *The Blood Cells and Hematopoietic Tissues,* McGraw-Hill, New York.
Wisse, E., and Knook, D. L. (eds.), 1977, *Kupffer Cells and Other Liver Sinusoidal Cells,* Elsevier/North Holland, Amsterdam.
Möller, G. (ed.), 1978, *Role of Macrophages in the Immune Response,* Immunological Reviews 40, Munksgaard, Copenhagen.

2

Endocytosis by Macrophages

AMY C. MUNTHE-KAAS
and GILLA KAPLAN

1. INTRODUCTION

1.1. DEFINITIONS

Endocytosis is a cellular function which regulates the intake of extracellular materials by enclosure into plasma-membrane-derived invaginations. As a result of endocytosis, exogenous substances are found in intracellular membrane-limited vesicles or vacuoles. Although many eukaryotic cells demonstrate this function, it is particularly prominent in leukocytes, macrophages, oocytes, capillary endothelial cells, and thyroid epithelial cells (Silverstein *et al.*, 1977). Traditionally it has been convenient to distinguish between two general categories of endocytosis: phagocytosis, the uptake of particulate material with diameters over 1 μm, and pinocytosis, the uptake of soluble substances. The consensus of opinion, as mentioned by Jacques in his review (1969), is that the differences between these two categories are superficial, and that the basic cellular mechanisms are similar. For example, viruses, which are small particles visible only in the electron microscope, enter cells by means of an endocytic mechanism (Dales, 1973), but it is difficult to say whether this is phagocytosis or pinocytosis. However, studies in recent years have shown that different mechanisms may be responsible for different types of endocytosis. It has now become necessary to divide pinocytosis into separate categories with distinct characteristics. One way of subdivision has been based on the size of the pinocytic vesicle formed, as well as the requirement of metabolic energy (Allison and Davies, 1974). Micropinocytosis is the term used to designate the uptake of substances into membranous vesicles of diameter less than 0.1 μm and visible only in the electron microsope. Macropinocytosis is the uptake of substances visible by light microscopy into vesicles of diameter greater than 0.2 μm.

AMY C. MUNTHE-KAAS • Laboratory for Immunology, Department of Pathology, Norsk Hydro's Institute for Cancer Research, The Norwegian Radium Hospital, Oslo 3, Norway. GILLA KAPLAN • Institute for Medical Biology, University of Tromsø, Tromsø, Norway.

Another way of subgrouping pinocytosis is to distinguish between adsorptive pinocytosis and nonadsorptive or fluid phase pinocytosis. Adsorptive pinocytosis refers to cases where the substance first interacts with the cell membrane, attaches, and is then folded into the cell as the plasma membrane invaginates. The type of interaction between the substances and the plasma membrane can vary from sheer physical electrovalent bonds (Nagura *et al.*, 1977) to specific chemical interactions. Nonadsorptive pinocytosis is more or less passive, and its extent depends on the concentration of the substances in the immediate extracellular milieu and on the level of activity of the plasma membrane. Using this criterion of attachment prior to internalization we can perhaps more logically divide endocytosis into nonreceptor-mediated endocytosis (mainly nonadsorptive pinocytosis) and receptor-mediated endocytosis (phagocytosis and adsorptive pinocytosis).

1.2. SCOPE

The study of endocytosis has until recent years mainly been concerned with nutrition in unicellular organisms or with the interaction between microbes and polymorphonuclear leukocytes. It is now evident that endocytosis is of basic importance in many aspects of mammalian physiology and metabolism, and can provide cell biologists with excellent systems to study the structures and functions of subcellular components—such as membrane chemistry and behavior, the contractile elements in the cytoplasm, the secretory apparatus of the cell, and different compartments of cellular metabolism. Excellent reviews on endocytosis have been written with emphasis on the identification of different stages in the process (Rabinovitch, 1967; Stossel, 1974a,b,c), on the mechanisms of particle uptake (Allison and Davies, 1974; Silverstein *et al.*, 1977), on the kinetics of the process (Jacques, 1969), and the physiology and biochemistry of the phagocyte (Cohn, 1977). In this present chapter we will concentrate on *in vitro* studies of receptor-mediated endocytosis in macrophages, with particular attention to the morphological aspects of the phenomenon. Phase contrast microscopy enables direct observation of live cells, but it is limited in resolution; electron microscopy, on the other hand, allows good resolution, but the extensive processing of the samples introduces many problems in the interpretation of the images attained. With careful interpretation, however, much important information can be obtained with the electron microscope.

2. SYSTEMS FOR STUDYING ENDOCYTOSIS

2.1. STUDIES *IN VIVO*

Studies *in vivo* on macrophage endocytosis have concentrated on the ability of liver and spleen macrophages to clear substances introduced into the circulation. A multitude of substances have been used, varying in nature, size, surface charge, and quantity. They range from colloidal carbon (Donald and Tennent, 1975), colloidal gold and mercury (Parks and Chiquoine, 1957; Scheeberger-Keeley and Burger, 1970), Thorotrast (Tessmer and Chang, 1967; Carr, 1968), and polystyrene latex particles (Matter *et al.*, 1970; Saba and Scovill, 1972) to macromolecules such as bacterial proteases (Debanne *et al.*, 1973), viruses (Inchley, 1969; Jahrling and Gorelkin, 1975), denatured serum albumin (Bouveng *et al.*, 1975), and others.

Morphological studies *in vivo* include the demonstration of particle ingestion intraperitoneally by resident macrophages (Orenstein and Shelton, 1977) and ultrastructure of *in situ* phagocytosis by liver macrophages (Muto and Fujita, 1977; Satodate *et al.*, 1977). Critical evaluation of clearance data is complicated by the uncertain effects of variables such as dosage (Donald and Tennent, 1975), the presence of serum immunoglobulins or other opsonic factors (Saba, 1970), and hormones (Cordingley, 1968).

Studies *in vivo* remain a useful tool for clinical investigations and for studying the patterns of endocytic functions of the organism as a whole. However, if the complex aspects of endocytosis are to be analyzed separately and in detail, *in vitro* studies under well-controlled conditions are obviously necessary.

2.2. STUDIES *IN VITRO*

Until a few years ago, almost all information concerning macrophage endocytosis had been obtained from studies on the mouse peritoneal macrophage (reviewed by Gordon and Cohn, 1973) or monocytes from blood (Huber *et al.*, 1968; Huber and Fudenberg, 1970; Abramson *et al.*, 1971; Schmidt and Douglas, 1972; Lobuglio, 1973; Ødegaard *et al.*, 1974) which mature into macrophages in culture. Earlier studies on the phagocytic and bactericidal functions of alveolar macrophages (Pavillard and Rowley, 1962; Pavillard, 1963) have now been supplemented with more detailed studies on the ultrastructure of their phagocytic mechanism (Nichols, 1976a), isolation of their phagocytic vesicles (Stossel *et al.*, 1972), and cytochemistry of their postphagocytic lysosomal vesicles (Nichols, 1976b). The concentration on these macrophage types is due to the relative ease with which they can be obtained and cultured. Macrophages from the rat peritoneum have been found difficult to keep in culture, but technical obstacles here have also been overcome and quantitative studies on their endocytic and degradative functions have been reported recently (Pratten *et al.*, 1977). Improved techniques in cell dispersion (Berry and Friend, 1969), selective enzymic digestion (Mills and Zucker-Franklin, 1969; Berg and Boman, 1973), and differential centrifugation (Knook and Sleyster, 1976) have enabled mass culture of pure liver macrophages from both rats (Munthe-Kaas *et al.*, 1975) and mice (Knook and Sleyster, 1977) and studies *in vitro* on their endocytic capacity (Munthe-Kaas, 1976, 1977) and lysosomal apparatus (Berkel *et al.*, 1975; Munthe-Kaas *et al.*, 1976a; Berg and Munthe-Kaas, 1977).

The use of cells *in vitro* also has a very significant advantage for morphological studies. Monolayers of cells are easily visible and relatively easy to process for electron microscopy. The cell surface is exposed and particle binding and ingestion can be synchronized within any population of cells.

In addition to these mammalian sources of macrophages, studies on invertebrate phagocytes such as those recently reported on insect (Anderson, 1977) and echinoid phagocytes (Kaplan and Bertheussen, 1977; Bertheussen and Seljelid, 1978) are interesting not only in that they give additional evidence to a general pattern of macrophage behavior, but also provide a perspective into the phylogenetic development of macrophages, and stimulate research which may lead to a better understanding into the multifaceted behavior of these cells.

2.3. THE USE OF CELL LINES

Since some nonmacrophages are also capable of endocytosis, the purification of macrophages to obtain cultures uncontaminated by other cell types is of great concern. In this respect, established macrophage-like cell lines can be a useful tool. One example is the reticulum cell sarcoma J774 (Ralph *et al.*, 1975), adapted to *in vitro* culture by L. Frank and M. Scharff, which has shown macrophage characteristics insofar as the cells have phagocytic receptors, can phagocytize particles by means of both foreign surface and immune receptors, kill ingested bacteria, and release lysosomal enzymes into the culture medium (Kaplan and Mørland, 1978; Mørland and Kaplan, 1978). Muschel *et al.* (1977a,b) have isolated variants defective in phagocytosis from this cell line and have begun experimental analysis into the genetics and biochemistry of phagocytosis. The cell line P388D1, which originated from a murine lymphoid tumor, and was described as having most of the characteristics of macrophages (Koren *et al.*, 1975), has been used for membrane receptor studies (Unkeless, 1977) (see Section 3.3.1). Similarly, Walker (1976) has used a macrophage cell line IC-21 for studying these membrane receptors (see Section 3.3.1). Another cell line which has been employed is M1/436–7, originally established *in vitro* from a spontaneous myeloid leukemia of SL strain mice by Ichikawa (1969). This cell line can be induced to differentiate into macrophages *in vitro*. Nagata *et al.* (1976) have characterized phagocytic activities of these cells, followed their maturation process, and analyzed certain factors necessary for differentiation of phagocytic receptors (Nagata *et al.*, 1977).

Studies carried out in the above mentioned systems, and others, have contributed extensively to the understanding of endocytosis in general and phagocytosis by macrophages in particular.

3. PHAGOCYTOSIS

Two initial requirements for macrophage phagocytosis are the proximity of the target substance, and its palatability. Most macrophages *in vivo* are attached to the surface of other cells, basement membranes, or various matrix materials. Efficient surveillance of areas adjacent to the immediate vicinity of the phagocytes require the ability for directional movement. Macrophages achieve this by being able to respond to concentration gradients of certain chemicals generated in the cellular environment when homeostasis is disturbed. This is called chemotaxis. The palatability of the target substance, on the other hand, is determined not only by the nature and surface properties of the substance, but also by various opsonic factors in the serum. These opsonic factors can be of immunoglobulin or nonimmunoglobulin nature.

3.1. CHEMOTAXIS

Chemotaxis is of particular interest in the analysis of the mechanisms behind the congregation of phagocytes at the site of an inflammatory response. It has been known for some years that the interaction between microorganisms and host would generate substances which are either directly chemotactic for phagocytes or are able to activate complement factors giving rise to protein fragments which are chemotactic for phagocytes (Sorkin *et al.*, 1970). The early contact phase of the intrinsic system of blood coagulation may also lead to production of chemotactic factors (Ward, 1968; Bianco *et al.*, 1976). This may be related to the generation of complement fragments since plasmin has been reported to cleave C3 and generate C3a (Ward, 1967).

Ward (1968) reported two factors unrelated to complement fragments which are selectively chemotactic for macrophages. One of these is generated when serum is treated with immune complexes, and the other is released from neutrophils and is akin to cationic peptides from lysosomes. A lipid factor from some bacteria was also reported to be selectively chemotactic for macrophages (Russel *et al.*, 1976).

Macrophage chemotactic factors can also be lymphocyte products. One such lymphocyte-derived factor has been characterized after gel filtration and ion-exchange chromatography to have a molecular weight of 40,000 on Sephadex G-100 (Meltzer *et al.*, 1977). This same group of workers also described a separate factor from murine sarcomas, chemotactic for syngeneic peritoneal macrophages, with a molecular weight of 15,000. However, although much is now known about the sources, number, and characteristics of chemotactic factors, it has not yet been determined how they exert their effect on macrophages.

Many studies concerned with chemotaxis have used morphological methods as the main way of monitoring phagocyte movement. In these studies the phagocytes can be seen to move from one area of the culture surface to another, or from one side of a filter to another, in response to a chemotactic stimulus. Other aspects of macrophage chemotaxis that have also been studied are the electrophysiology of the macrophage membrane in response to a chemotactic stimulus (Gallin and Gallin, 1977), the role of cytoplasmic contractile elements in the response (Crispe, 1976), and the metabolic requirements involved in the directional movements (Dohlman and Goetzl, 1978).

The significance of macrophage chemotaxis in the endocytic process and the homeostatic balance of the organism as a whole is obvious. This is reflected in the instances where that ability is increased to meet particular needs or decreased in various diseases (Pike *et al.*, 1977). When guinea pig macrophages are stimulated by administration *in vivo* of BCG (*Mycobacterium bovis*, strain BCG), their chemotactic response to complement factors is enhanced (Poplack *et al.*, 1976). In rats, a release of chemotactic factors in the peritoneal exudate as a result of peritoneal inflammation resulting from *Listeria monocytogenes* induction has also been reported, stimulating the accumulation of macrophages (Jungi and McGregor, 1977). It was also mentioned earlier that several murine sarcomas were shown to generate specific macrophage chemotactic factors (Meltzer *et al.*, 1977). The interpretation of this last example is difficult, as the authors stated, since the accumulation of macrophages at tumor sites may be both cytopathic for tumor cells and trophic for other cells providing a prerequisite for tumor cell proliferation. The bulk of evidence from studies of neoplastic diseases shows that the effect of the tumor is to inhibit macrophage chemotaxis (Stevenson and Meltzer, 1976; Snyderman and Pike, 1977; Normann and Sorkin, 1977; Pasternak *et al.*, 1978; Meltzer and Stevenson, 1978). It remains an enigma, however, that although chemotactic mobility of macrophages is inhibited by such tumor cell products, their phagocytic ability is often stimulated by these same substances (Snyderman and Pike, 1977; Meltzer and Stevenson, 1978). Further investigation into the interaction between chemotactic factors and macrophages would certainly add to a better understanding of macrophage endocytosis as a whole.

3.2. SERUM OPSONINS

Not all substances are readily phagocytized by macrophages. Slight surface modifications, or opsonization, of test particles can transform them from being totally ignored by macrophages to being avidly ingested (Rabinovitch, 1970). Mammalian serum contains factors, or opsonins, which can bring about these surface modifications. Three main categories of opsonization are known. The first of these is the fixation of the third component of complement, C3, onto the surface of the particle. This is achieved either by way of the particle interacting with a natural antibody, which then initiates the classical pathway of complement activation, or by a direct action of the particle on the properdins, resulting in triggering of the alternate pathway of complement activation. In either case C3 becomes fixed on the particle surface (Nelson, 1963; Huber *et al.*, 1968; Bianco and Nussenzweig, 1977).

The second type of particle opsonization is dependent on previous exposure of the animal to the particles, and involves the interaction of the particle with an antibody of immunoglobulin G (IgG) nature. The structure in the antigen–antibody complex which is recognized by the macrophages is the Fc region of the IgG. Papain and trypsin digestion of the IgG, which splits away the Fc part of the molecule while leaving its antigen-binding capacity intact, will remove its opsonic effect on macrophages (Berken and Benacerraf, 1966; Steinman and Cohn, 1972).

The ability of the C3 and the Fc portion of IgG to opsonize particles is due to the existence of specific macrophage surface receptors for C3 (Mantovani *et al.*, 1972; Reynolds *et al.*, 1975) and Fc (Huber *et al.*, 1968). These will be discussed later in Section 3.3.

The third type of opsonization is by serum proteins other than immunoglobulins and complement. One such protein which stimulates macrophage phagocytosis has been purified from rat serum and characterized as an α-2-globulin with a molecular weight at 4°C of approximately 800,000 (Blumenstock *et al.*, 1976). It is present also in human serum at about 300–350 μg/ml, and at 37°C is composed of two subunits, each about 220,000 daltons (Saba, 1978). This α-2-globulin is a glycoprotein, and has been shown to be antigenically related to, and possibly identical with, the cold-insoluble globulin, also referred to as fibronectin and LETS protein. The mechanism whereby this protein enhances macrophage phagocytosis is not known, nor have specific macrophage membrane receptors for this type of protein been identified.

Another protein, referred to as C-reactive protein (CRP), which becomes greatly elevated in human serum concentration during inflammatory processes (Claus *et al.*, 1976), also has opsonic effects for macrophages. The effect of CRP may be exerted through its interaction with the macrophage Fc receptor (Mortensen and Duszkiewicz, 1977).

3.3. RECOGNITION BY MACROPHAGE RECEPTORS

Macrophage membranes contain recognition structures which enable them to distinguish opsonized particles and bind them for ingestion. Particles opsonized with IgG are recognized by the Fc receptors, and particles opsonized by C3 are recognized by the C3 receptors. Membrane structures that mediate phagocytosis of other substances, such as latex, silica, denatured protein aggregates, or glutaraldehyde-fixed red cells, are less specifically defined. It is usual to refer to these as foreign surface receptors.

Phagocytic receptors have been visualized only through the presence of appropriately attached ligands. Thus, any study of the location and function of receptors must involve the use of markers which bind to these receptors. For morphological purposes, the particles attached to the receptors must be easily visible as well as quantifiable.

3.3.1. The Fc Receptors

Although Fc receptors are also found on lymphocytes (Basten *et al.*, 1971), mast cells (Tigelaar *et al.*, 1971), herpes-virus-infected cells (Watkins, 1964), and others, only those on macrophages and polymorphonuclear leukocytes (PMN) mediate phagocytosis. Distinct Fc receptors for different subclasses of IgG have been reported for the mouse macrophage (Walker, 1976; Heusser *et al.*, 1977; Unkeless, 1977; Andersen and Grey, 1978). Walker (1976) used the murine macrophage cell line IC-21 and found one Fc receptor which binds both aggregated and monomeric IgG2a, and one which binds aggregated IgG2b, neither of which is sensitive to trypsin digestion. Heusser *et al.*, (1977) and Unkeless (1977), using P388D1 and peritoneal exudate macrophages, found that the receptor for monomeric IgG2a is trypsin-sensitive and the receptor for aggregated IgG2b is trypsin-resistant. The discrepancy in the trypsin sensitivities may be due to different cell lines used. The difference in function between these two types of Fc receptors is not known. Walker (1977) suggested that these two types of Fc receptors mediate distinct functions: That for IgG2a mediates phagocytosis, and that for IgG2b mediates cytolytic activities.

The distribution of Fc receptors on macrophage membranes has been studied with the scanning electron microscope on mouse peritoneal macrophages (Tizard *et al.*, 1974; Kaplan *et al.*, 1975), rat liver macrophages (Munthe-Kaas *et al*, 1976b), and rabbit alveolar macrophages (McKeever *et al.*, 1977). They appear to be located over the entire surface and are particularly abundant in perinuclear regions and the cytoplasmic veils and pseudopodia. The sensitivity of this identification is limited by the size of the test particles, and, therefore, the use of soluble immune complexes (McKeever *et al.*, 1977) is obviously more sensitive than IgG-opsonized red cells (Munthe-Kaas *et al.*, 1976b). Transmission electron microscopy, although more laborious, is probably superior in both resolution and sensitivity for studying receptor localization (McKeever *et al.*, 1976a,b).

The number of Fc receptors on mouse macrophages and rabbit alveolar macrophages has also been computed (Phillips-Quagliata *et al.*, 1971; Arend and Mannik, 1973; Unkeless and Eisen, 1975). There is a tenfold difference between the number of IgG2a receptors on mouse macrophages (1.5×10^5/cell) (Unkeless and Eisen, 1975) and that on rabbit alveolar macrophages (2×10^6/cell) (Phillips-Quagliata *et al.*, 1971). Whether this is due to species variation or technical differences will not be resolved until more systematic studies are carried out.

Recent studies have shed some light on the mechanism of Fc receptor function. Atkinson and Parker (1977) found that the macrophage subplasmalemmal microfilaments are probably involved in the interaction of the rabbit alveolar macrophage Fc receptor and IgG-opsonized sheep red cells since this interaction can be modulated by cytochalasins. These authors support the hypothesis that Fc receptors are connected to a subplasma membrane microfilament network which provides the proper spatial pattern for receptor activity on the membrane surface. Alexander *et al.* (1978) attempted to identify the IgG domain which is actually bound by the guinea pig macrophage Fc receptor. By testing the binding of proteolytic fragments of IgG, they showed that macrophage Fc receptors recognize the CH2 domain of the guinea pig IgG. Recently reported techniques to purify plasma membrane from macrophages (Remold-O'Donell, 1977) and to isolate plasma membrane vesicles bearing the Fc receptors (Scott and Rosenthal, 1977) are also promising in the study of the biochemical characteristics of macrophage membranes.

3.3.2. The C3 Receptors

Trypsin-sensitive membrane receptors for the third component of complement, C3, are found on lymphocytes (Bianco *et al.*, 1970; Eden *et al.*, 1971) and

PMN (Lay and Nussenzweig, 1968), as well as monocytes and macrophages (Huber and Fudenberg, 1970; Mantovani *et al.*, 1972; Reynolds *et al.*, 1975). The macrophage C3 receptors have affinity for both C3b and C3d fragments, whereas those on PMN bind only C3b (Ehlenberger and Nussenzweig, 1977). It is not yet resolved whether the C3b and C3d receptors are identical receptors with dual affinity or whether they are distinct from each other.

Unlike the Fc receptors, the C3 receptor of macrophages normally mediates only binding but not ingestion of opsonized particles (Lay and Nussenzweig, 1969; Ehlenberger and Nussenzweig, 1977). However, macrophages in inflammatory exudates can also ingest C3b-coated particles (Bianco *et al.*, 1975). This suggests that the function of this receptor can change with the physiological state of the cell.

Particle binding by the C3 receptor is temperature dependent (Lay and Nussenzweig, 1968; Munthe-Kaas and Berg, 1977; Kaplan and Mørland, 1978), requires the presence of Ca^{2+} (Lay and Nussenzweig, 1969), and can be modulated by corticosteroids (Schreiber *et al.*, 1975) and substances which disrupt microfilaments (Atkinson *et al.*, 1977). When mediating ingestion, C3 receptors function independently of the Fc receptors (Griffin *et al.*, 1975a). The two types of receptors, however, have a synergistic effect on binding and ingestion: particles opsonized with suboptimal doses of either IgG or C3 will not be ingested, whereas particles coated with the same suboptimal amounts of both IgG and C3 will be avidly ingested (Ehlenberger and Nussenzweig, 1977).

The number of C3 receptors on macrophages has not been studied. Their distribution on the cell membrane appears to differ slightly from that of the Fc receptors when opsonized red cells are used as markers (Munthe-Kaas *et al.*, 1976b; Mørland and Kaplan, 1977): receptor activity is absent from the extreme periphery of macrophages spread on culture surfaces. Studies on macrophages seeded on immobilized ligands (culture plates coated with antigen–antibody complexes or complements) yield data which may support the hypothesis that, whereas the Fc receptors are motile in the transverse plane of the membrane, the C3 receptor may be anchored and less motile (Kaplan *et al.*, 1978). The mechanism whereby C3 receptors mediate ingestion has also been found to differ from that of the Fc receptors; this will be discussed in Section 3.4.1.

3.3.3. The Foreign Surface Receptors

The presence of the Fc and C3 receptors on vertebrate macrophages represents a highly specialized function which probably succeeded the evolution of immunoglobulins and complement factors. A more primitive form of cellular recognition system (Weir and Øgmundsdottir, 1977) for phagocytosis may have existed farther back in phylogeny, controlling physiological processes such as removal of dead cells during the metamorphosis of insects. The ability of vertebrate macrophages to bind and ingest a variety of particles—such as zymosan, latex, or glutaraldehyde-treated red cells—without previous opsonization by IgG or C3 is also a well-known phenomenon (Rabinovitch and deStefano, 1973; Stossel, 1975). This ability is often referred to as phagocytosis through foreign surface receptors or nonspecific receptors (Silverstein *et al.*, 1977). The role of divalent cations in the functioning of these receptors has been established in some cases (Stossel, 1973), and trypsin apparently removes certain of these nonspecific receptors while leaving others intact (Rabinovitch, 1968).

An opsonin-independent recognition mechanism in human monocyte cultures for activators of the alternate complement pathway has been reported recently (Czop *et al.*, 1978). In the case of mouse peritoneal macrophage recognition and attachment of certain gram-negative bacteria, the macrophage membrane component responsible was reported to be D-mannose residues (Bar-Shavit *et al.*, 1977). In addition, there is a group of receptors which are responsible for the receptor-mediated pinocytosis which will be discussed in Section 4.

While, in some cases, foreign surface receptor-mediated phagocytosis has been shown to function quite distinctly from that mediated by the Fc and C3 receptors (Michl *et al.*, 1976a,b), in others—such as the phagocytosis of senescent red cells—it has been proposed that the Fc receptor is eventually involved (Kay, 1975).

3.4. INGESTION

The binding of particles on the macrophage membrane is not necessarily followed by the ingestion of the particle. For phagocytosis to proceed, the macrophage plasma membrane must enclose the attached particle in a vesicle for subsequent intracellular disposal. How does the interaction between an external ligand and a membrane receptor lead to the proper cellular response? In this section we will first discuss some models based on ultrastructural studies concerning the responses of plasma membrane and the cytoplasmic contractile cytoskeletal elements to ligand binding. Then we will briefly discuss the metabolic requirements for the ingestion process.

3.4.1. Mechanism of Ingestion

3.4.1a. Membrane Phenomena. Ultrastructural studies on particle ingestion by phagocytes date back many years (Mudd *et al.*, 1934; Essner, 1960; Törö *et al.*, 1962; Horn *et al.*, 1969). Attempts to analyze the exact mechanism, however, have only recently been successful. A very significant observation, made by Griffin and Silverstein (1974), was that the plasma membrane engulfing response in the ingestion process is a very localized phenomenon. When mouse macrophages phagocytize latex particles or opsonized pneumococci, only localized segments of plasma membrane binding these particles are involved. Mouse red cells bound to the same macrophage via anti-mouse macrophage $F(ab')_2$ fragments (which are normally not ingested), remain on the surface of the macrophage. Thus, the ingestion of one particle does not trigger a generalized response to internalize all particles attached to the plasma membrane. This discriminatory segmental response of the plasma membrane suggests that the signal for ingestion originates from the attached particle, and the quality and magnitude of the membrane response are decided by the nature of the particle.

When an opsonized particle displays the proper ligands recognizable by macrophage receptors, is it the distribution of the ligands on the particle or the number of ligands that triggers ingestion by the macrophage membrane? Griffin and his co-workers showed that a step-by-step interaction of membrane receptors with ligands bound to the entire surface of the particle is necessary for ingestion (Griffin *et al.*, 1975b). Mouse macrophages were first allowed to bind IgG or C3 opsonized sheep red cells at low temperature. If then, either blocking or removal of the ligands or the receptors not involved in the initial binding took place, the red cells failed to be ingested when the temperature was raised to 37°C. Furthermore, by manipulating the opsonization of bone marrow-derived lymphocytes with anti-Ig molecules, Griffin *et al.* (1976) showed that only lymphocytes diffusedly coated with anti-Ig were ingested; the same number of anti-Ig molecules on the lymphocyte surface, redistributed to one pole of the cell by the capping process, will not mediate ingestion. The authors consider these data unambiguous proof for a zipper mechanism of ingestion, whereby macrophage receptors stepwise attach ligands distributed over the entire surface of the opsonized particle, creeping over or raising cytoplasmic arms around the particle, and finally enclosing it in a phagocytic vesicle.

The zipper mechanism, however, does not seem to apply to ingestion of all types of particles. When mouse macrophages phagocytize formalin-treated rabbit red cells and latex particles, ingestion is achieved through formation of large, hemispherical craters on the macrophage surface into which the particles sink (Polliack and Gordon, 1975). When guinea pig alveolar macrophages ingest *Mycoplasma pneumoniae,* thin cytoplasmic veils extending contiguously to the surface of the microorganism were rarely observed (Powell and Muse, 1977). An interesting proposal was put forward by Jones *et al.* (1977) when they studied ingestion of *Mycoplasma pulmonis* by mouse macrophages. They observed that the antibody-mediated ingestion of *M. pulmonis* was achieved by creeping-up of the macrophage membrane, in accordance with the zipper mechanism, whereas ingestion mediated by trypsinization of attached *M. pulmonis* was through a sinking into craters formed on the macrophage surface. These authors therefore proposed the term "immunologic mechanism" for the former type of ingestion, and "nonimmunologic mechanism" for the latter, and suggested that the difference between them is due to the different degrees to which contractile elements subjacent to the macrophage plasma membrane are involved.

Two distinct types of ingestion mechanisms have also been reported by other workers. Sheep red cells opsonized with IgG were ingested by rat liver macrophages and mouse peritoneal macrophages by the zipper mechanism (Munthe-Kaas *et al.*, 1976b; Kaplan, 1977). Those opsonized with C3, on the other hand, were ingested by the crater-formation mechanism by rat liver macrophages (Munthe-Kaas *et al.*, 1976b), stimulated mouse macrophages (Kaplan, 1977), and echinoid phagocytes (Kaplan and Bertheussen, 1977) (Figure 1). The two types of ingestion mechanisms differ in their sensitivities to cytochalasin B (see Section 3.4.1b) as well as in morphology. This suggests that the contractile elements could be involved to different extents and that there are probably several independent pathways whereby macrophage membrane receptors are linked to the ingestion machinery of the cell (Michl and Silverstein, 1978).

In addition, recent findings from studies on foreign surface phagocytosis

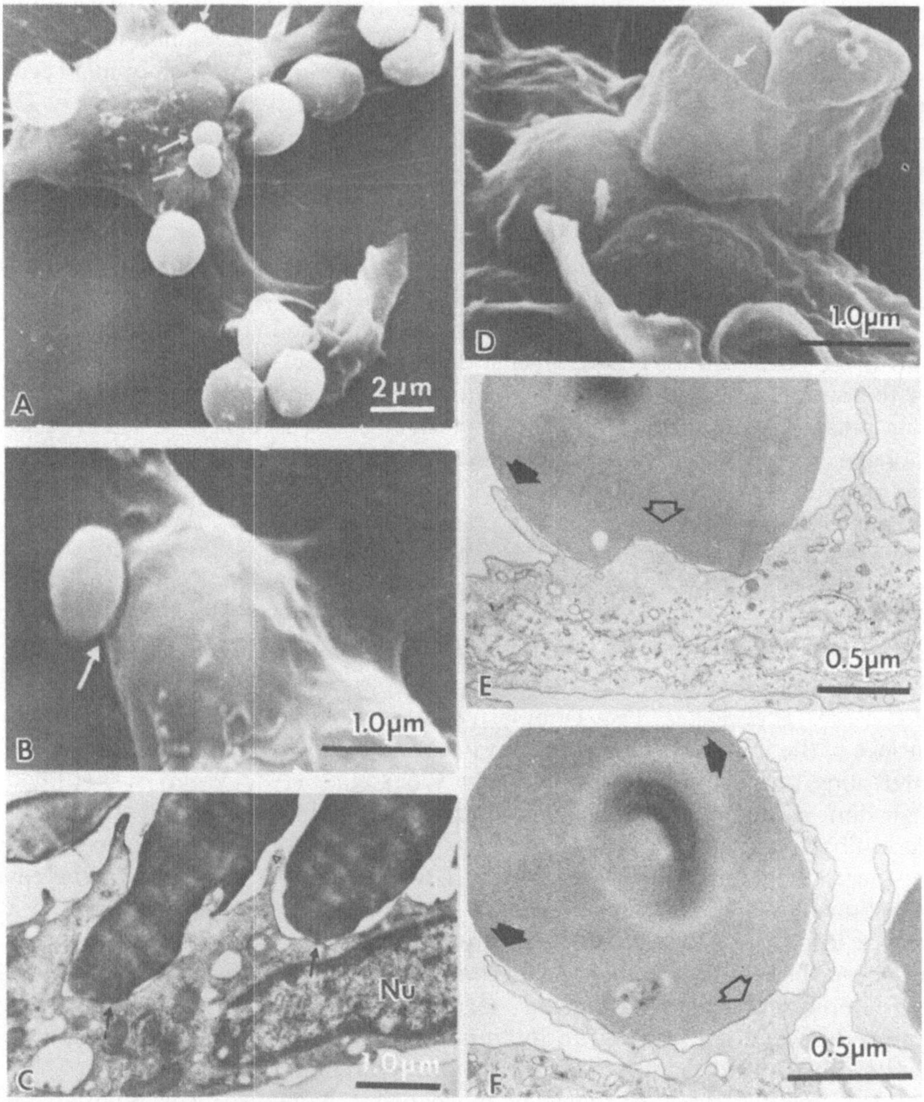

FIGURE 1. Phagocytosis of opsonized sheep red blood cells (SRBC) by mouse macrophages *in vitro*. (A) Scanning electron microscopy (SEM) micrograph of a macrophage ingesting C3b-opsonized SRBC (EIgMC) via the C3 receptors. Opsonized SRBC were incubated with macrophages activated *in vitro*. Many EIgMC can be seen attached to the macrophage. Three erythrocytes are partially ingested and can be seen in the process of sinking into the phagocyte cytoplasm. Only part of the red cells is seen protruding over the macrophage surface (arrows). (B) A SEM micrograph close-up of an EIgMC being ingested. The outline of the ingested portion of the particle can be seen, with only a bit of the red cell remaining above the macrophage surface (arrow). (C) Transmission electron microscopy (TEM) micrograph of phagocytosis by the C3 receptors. EIgMC were incubated with macrophages activated *in vitro*. The particles can be seen sinking into a crater on the macrophage surface (arrows). Membrane extensions are not found in close contact with the particles being ingested (▽). Nu, macrophage nucleus. (D) SEM of a macrophage ingesting an IgG opsonized SRBC (EIgG). Two particles can be seen partially enclosed by macrophage membrane extensions creeping up over them

also fall into two distinct types (Smedsrød and Seljelid, personal communication (1978); Seljelid, 1980). Red blood cells denatured by glutaraldehyde fixation are ingested by mouse macrophages and rat Kupffer cells by the C3 or crater-formation mechanism. Red cells treated by *n*-ethylmaleimide, on the other hand, are ingested by the Fc or zipper mechanism.

3.4.1b. Cytoplasmic Contractile Elements. The cytoskeleton is composed of two types of contractile elements, the microtubules and the microfilaments. Microtubules are composed of the protein tubulin, are about 250 Å in diameter, and traverse the cytoplasm from the centrosphere region towards the plasma membrane, or run parallel to and underneath the plasma membrane (Weber *et al.*, 1975). Microfilaments were demonstrated in macrophages to have a diameter of about 50 Å (DePetris *et al.*, 1962; Allison *et al.*, 1971). These filaments are composed of actin (Ishikawa *et al.*, 1969; Pollard *et al.*, 1970), are present in abundance underneath the plasma membrane (Allison *et al.*, 1971; Reaven and Axline, 1973; Singh, 1974), and can become gelated upon interaction with myosin and another high-molecular-weight actin-binding protein (Stossel and Hartwig, 1975, 1976). The participation of these contractile elements in various cell functions has been mostly studied through the use of substances which disrupt these elements: the cytochalasins for interfering with microfilament functions (Spooner and Wessels, 1970) and colchicine and related substances for interfering with microtubular functions (Wilson *et al.*, 1974). Since particle ingestion by macrophage involves movement of membrane components and translocation of the phagocytic vesicle into the cytoplasm, it is natural to believe that the cytoskeleton would be involved (Figure 2). This is supported by data showing altered rates of phagocytosis (Allison *et al.*, 1971; Malawista *et al.*, 1971; Dumont, 1972; Munthe-Kaas *et al.*, 1976b; Kaplan, 1977) and other cellular movements (Bhisey and Freed, 1971a,b, 1975; Sundharadas and Cheung, 1977) in the presence of colchicine or cytochalasins (Figure 3).

In an attempt to analyze the extent to which cytoplasmic contractile elements are involved in plasma membrane activities during phagocytosis, Reaven and Axline (1973) carried out ultrastructural studies on cultured macrophages at rest and during phagocytosis of polystyrene particles. They observed two different patterns of contractile element distribution in resting macrophages. Underneath the plasma surface attached to the culture dish, a layer of microfilament network (400–600 Å) was seen, and immediately subjacent to this layer, oriented bundles of microfilaments and a zone of microtubules were found; microtubules were also seen to extend radially from the plasma membrane to the interior of the cell. In contrast, under the surface, which was free and not

in a tight cuplike formation (arrow). [A–D reproduced with permission from Kaplan (1977), *Scand. J. Immunol.* **6**:787).] (E) TEM micrograph of macrophage ingesting an IgG-opsonized SRBC (early stage). The particle is tightly bound to the macrophage surface (open arrow). A macrophage membrane extension can be seen starting to creep up over the particle surface (black arrow). (Courtesy of Dr. S. C. Silverstein). (F) TEM micrograph of macrophage ingesting an IgG opsonized SRBC. Macrophage membrane extensions can be seen creeping up over the particle enclosing it in a cuplike formation (black arrows). Open arrow, probable site of initial contact between the phagocyte and the particle. (Courtesy of Dr. S. C. Silverstein.)

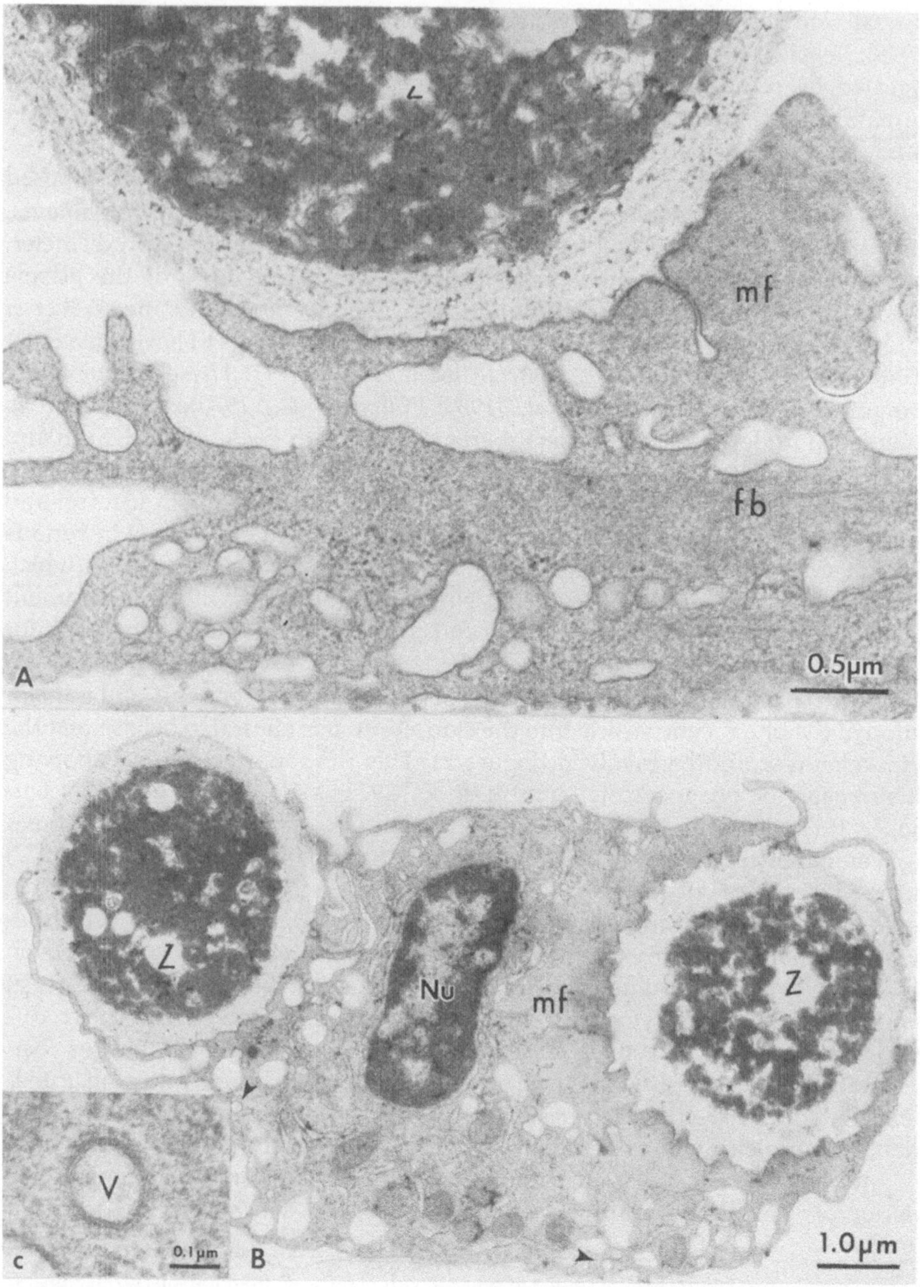

FIGURE 2. Contractile elements in phagocytosis of complement-(C3b)-opsonized zymosan particles by mouse macrophages *in vitro* as seen by TEM. (A) Zymosan particle (Z) at early stages of phagocytosis. In the area of contact between the particle and the macrophage a large amount of microfilaments (mf) and filament bundles (fb) can be seen excluding cell organelles from the area. (B) Macrophage in the process of ingesting two zymosan (Z) particles. In the area adjacent to the site of ingestion, microfilament (mf) gelation can be seen to exclude other cell organelles. Nu, macrophage nucleus. Some coated vesicles can be seen in the cell (arrows). (C) A close-up of a macrophage coated vesicle (V). The inner and the outer surfaces of the vesicle show bristle-like structures.

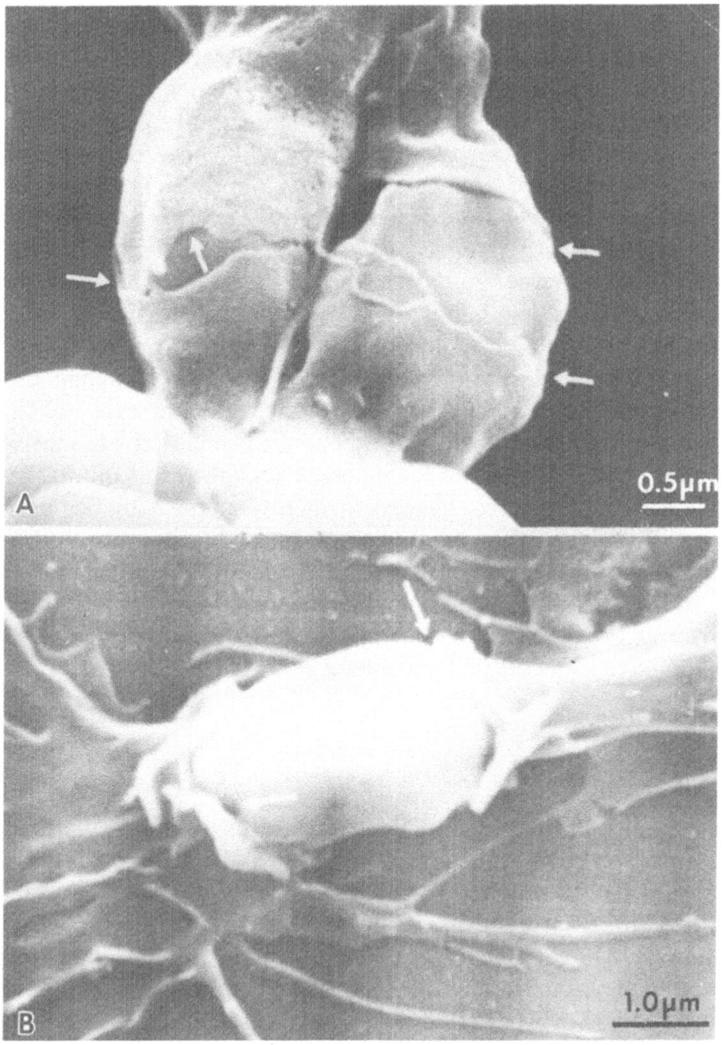

FIGURE 3. Phagocytosis of IgG-opsonized SRBC (via Fc receptors) in the presence of colchicine and cytochalasin B as seen by SEM. (A) Particle ingestion in the presence of 10^{-5} M colchicine. Two opsonized SRBC can be seen partially covered by membrane extensions creeping over them from the two macrophages they are attached to (arrows). The presence of colchicine does not affect the phagocytosis (compare to Figure 1). (B) Particle ingestion in the presence of 1 μg/ml of cytochalasin B. The macrophage membrane cannot creep up over the particle. Contact between the macrophages and the particle is maintained at the attachment sites only (arrows). Cytochalasin B inhibits phagocytosis. [Reproduced with permission from Kaplan (1977), *Scand. J. Immunol.* **6**:797.]

attached to the culture dish, only the layer of microfilament network was seen, without the subjacent oriented bundles and microtubules. When the macrophages were allowed to phagocytize, microtubules and oriented microfilaments, similar to those subjacent to the attached surface, appeared under the surface membrane surrounding the ingested polystyrene particles. Since attachment to culture surface is believed to be similar to an attempt by the macrophage to phagocytize (North, 1970), the authors believed that the organization of oriented bundles of microfilaments and the microtubules are probably associated with the ingestion process. Additional evidence for the redistribution and local condensation of contractile elements at pseudopods during phagocytosis is provided by Hartwig and others who showed that the contractile proteins were concentrated in the cell cortex, and their translocation into pseudopods followed by gelation caused the thrust of pseudopods during engulfment of particles (Hartwig *et al.*, 1977; Davies and Stossel, 1977).

The exact mechanism whereby contractile elements control the membrane activities is still obscure. The behavior of the lymphocyte surface receptors for concanavalin A (Con A) suggests that these membrane components are attached to colchicine-sensitive cytoplasmic structures which restrain their motility (in capping) in the transverse plane of the membrane (Edelman *et al.*, 1973; Berlin *et al.*, 1974). The microfilaments, on the other hand, enabled the motility of the Con A receptors, and cytochalasin B treatment abolishes this motility (DePetris, 1974). In alveolar macrophages, however, microfilament activities do not seem to oppose microtubular activity where receptor mobility is concerned, since disruption of either one led to capping of Con A receptors (Williams *et al.*, 1977). These authors suggested that alveolar macrophage Con A receptors may be controlled by a set of cytochalasin-B-insensitive microfilaments. Such a subset of microfilaments have indeed previously been reported to exist in these cells (Axline and Reaven, 1974).

3.4.2. Metabolic Requirements

Endocytosis is a temperature-dependent process. Particle binding by macrophage receptors can occur at low temperature (4°C) whereas the ingestion process seems to have a critical threshold below which it cannot occur (Kaplan and Mørland, 1978). Disturbances in the cellular concentration of certain cyclic nucleotides, which have been recognized to have intricate effects on cellular metabolism, have resulted in modulation of the ingestion process in macrophages and PMN (Seyberth *et al.*, 1973; Cox and Karnovsky, 1973 ; Ignarro *et al.*, 1974; Goodall *et al.*, 1978). Marked changes in oxygen consumption, glycolysis, hexose-monophosphate shunt, H_2O_2 production, etc., were observed when phagocytes ingested particles (reviewed Karnovsky, 1962; Karnovsky *et al.*, 1970; Axline, 1970). All these observations have been taken to mean that the ingestion of particles requires expenditure of cellular energy derived through the metabolic machinery.

By applying different metabolic inhibitors to macrophage cultures during phagocytosis studies, Karnovsky and his colleagues noted that anaerobic glycolysis was the main path of energy production for this process whereas oxidative phosphorylation was only of secondary significance except in alveolar macrophages (Karnovsky *et al.*, 1970). Studies by Michl and his colleagues showed that the relation between ingestion and cellular energy production measured as intracellular ATP content is not a direct or simple one (Michl *et al.*,

1976a,b). These investigators observed that 2-deoxyglucose, an analog of mannose and glucose, inhibited ingestion via the Fc and C3 receptors, but had no effect on the ingestion of latex particles, carbon, or boiled yeast cell walls (Michl *et al.*, 1976a). Since ingestion of latex, carbon, or boiled yeast would presumably require the same energy expenditure and contractile-element-associated locomotive forces to advance pseudopods as in the case of Fc and C3 ingestion, the inhibitory effect of 2-deoxyglucose must therefore be due to other prerequisites of ingestion. Furthermore, direct measurements revealed that 2-deoxyglucose reduced intracellular ATP content to about 20% of normal values without affecting latex ingestion, and addition of mannose and glucose restored the ability to ingest via the Fc and C3 receptors without raising intracellular ATP concentration back to normal levels (Michl *et al.*, 1976b). The authors postulated that the inhibitory effect of 2-deoxyglucose is probably due to its incorporation into membrane components instead of mannose or glucose, thus preventing further glycosylation of these precursors into glycolipids and/or glycoproteins.

4. PINOCYTOSIS

Pinocytosis, the cellular uptake of nonparticulate substances into membrane-limited vesicles, is a widespread phenomenon. Pinocytic vesicles arise by initial membrane invaginations of spherical, tubular, or cuplike structures. Two main types of vesicles have been recognized by electron microscopy: smooth or uncoated vesicles, and vesicles coated on both plasma membrane and cytoplasmic surfaces (Figure 2c).

4.1. NON-RECEPTOR-MEDIATED OR FLUID-PHASE PINOCYTOSIS

Available data suggest that smooth vesicles are associated with the nonselective transport of soluble substances into the cell in proportion to their concentration in the medium (Steinman *et al.*, 1976). As mentioned in the Introduction, there are apparently two distinct subtypes in this category: "micropinocytosis," designating the process whereby substances are taken into vesicles of diameter less than 0.1 μm, visible only in the electron microscope, and which is resistant to metabolic inhibitors and independent of microfilament activities, and "macropinocytosis," designating the subtype where vesicle size is over 0.2 μm, visible in the light microscope, and which is sensitive to metabolic inhibitors as well as cytochalasin B (Allison and Davies, 1974). The potential for and simultaneous occurrence of both these types of pinocytosis make their separate analyses very difficult (Munthe-Kaas, 1977).

Non-receptor-mediated pinocytosis is characterized by its nonselectivity, and is probably an activity partaken of by most other cell types as well as macrophages. Steinman and his co-workers have used horseradish peroxidase as a solute marker and systematically studied its non-receptor-mediated uptake by macrophages and other cells (Steinman and Cohn, 1972; Steinman *et al.*, 1974, 1976). A macrophage actively pinocytizing shows ruffling movements of its plasma membrane in the region where pinocytic vesicles seem to originate. The rate of uptake is directly proportional to the incubation temperature (2–38°C) and in this case is dependent on metabolic energy derived from either glycolysis or oxidative phosphorylation.

4.2. RECEPTOR-MEDIATED OR ADSORPTIVE PINOCYTOSIS

As the name implies, adsorptive pinocytosis involves binding of molecules to the plasma membrane prior to their internalization. The pinocytic vesicles formed have been described as "coated vesicles" in macrophages (Cotutiu and Ericsson, 1970) and other cells (Catt and Dufau, 1977; Anderson *et al.*, 1977), as opposed to the smooth vesicles formed during non-receptor-mediated pinocytosis. This type of pinocytosis is distinguished by its selectivity since it requires the presence of specific receptor sites on the cell membrane, and may be responsible for the uptake of various growth factors and regulatory proteins which depend on entry into cells for the exertion of their physiological functions (Ryser, 1968; Catt and Dufau, 1977). The most extensive studies of this uptake process are those on low-density lipoproteins (LDL), the major cholesterol-carrying lipoprotein of human plasma (Goldstein and Brown, 1977). Uptake of LDL by cells is facilitated by the binding of this molecule to receptors localized to discrete regions of the plasma membrane called "coated pits." The "coated pits" invaginate into the cell to form coated vesicles, which fuse with cellular lysosomes leading to degradation of vesicle contents. The degradation of LDL releases cholesterol from the lipoprotein so that it can be used by the cell for structural and regulatory purposes (Brown and Goldstein, 1976).

Other systems studied are those involving the uptake of the transport protein for vitamin B_{12} by human placenta cells (Friedman *et al.*, 1977) and fibroblasts (Youngdahl-Turner *et al.*, 1978), and the uptake of epidermal growth factors by human fibroblasts (Carpenter and Cohen, 1976).

Reports on this uptake process in macrophages are beginning to appear. Studies on liver macrophages have suggested that they have receptors for *N*-acetylglucosamine and mannose terminals on glycoproteins (Steer, 1978), and similar findings are reported on rat alveolar macrophages for uptake of lysosomal glycosidases (Stahl *et al.*, 1978). The full impact and physiological significance of this uptake mechanism in macrophages is still not clear.

5. POSTENDOCYTIC EVENTS

Although the engulfment of particles may be the most dramatic stage of phagocytosis from a morphologist's viewpoint, the act itself is only the prelude to a whole sequence of morphological and biochemical phenomena.

5.1. FATE OF THE PHAGOCYTIC VESICLE

Phagocytic vesicles, or phagosomes, once formed, detach from the plasma membrane, move towards the perinuclear region of the macrophage, and eventually fuse with the lysosomes of the cell which unload their hydrolytic enzyme contents into the fused phagolysosome (Cohn and Fedorko, 1969; Cohn, 1970; Pesanti and Axline, 1975b) (Figure 4). These vesicles congregate in the vicinity of the Golgi apparatus (Cohn, 1975). The factors which control this fusion between lysosome and phagosome are not clear. Disturbances of the membrane of phagosomes by binding of membrane-active agents such as Con A (Edelson and Cohn, 1974a,b) or the polybasic anion suramin (Hart and Young, 1975) can inhibit their fusion with lysosomes. Microtubules, which have been suggested to be involved in phagolysosome formation in PMN (Weissmann *et al.*, 1971), apparently are not responsible for this process in macrophages (Pesanti and Axline, 1975a,b). The fusion of phagosome and lysosome has been interpreted to signal the beginning of intracellular degradation of ingested substances (Cohn, 1963a); and indeed, once this has begun, metabolic inhibitors have been shown to have no effect on the breakdown of ingested substances (Cohn, 1963b). However, exceptions to this general rule do exist. When cultured mouse peritoneal macrophages were allowed to phagocytize the facultative intracellular parasite *Mycobacterium tuberculosis*, phagosomes containing intact bacteria often did not fuse with lysosomes, whereas those containing damaged bacteria did (Armstrong and Hart, 1971). Phagosomes containing viable bacilli pretreated with antiserum fused with lysosomes, but did not trigger bactericidal activities (Armstrong and Hart, 1975). Thus, it appears that neither phagolysosome formation nor subsequent degradation of its contents is a matter of course.

In a recent report on PMN, Rikihisa and Mizuna (1978) showed, by using specific markers for the two sides of the plasma membrane, that the phagosome membrane components were arranged quite differently depending on the type of substance ingested, and these differences were reflected also in their degree of fusion with lysosomes. The phagosome content has also been shown to affect fusion with lysosomes. Some intracellular parasites, although ingested by macrophages, survive by specifically inhibiting phagosome–lysosome fusion (see Section 6).

5.2. DEGRADATION OF INGESTED SUBSTANCES

When ingested substances are exposed to the battery of lysosomal hydrolytic enzymes after phagolysosome formation, enzymatic degradation of most biological macromolecules usually occurs. There have been several different approaches to studying digestion by macrophages. The release of macrophage-bound acid-soluble radioactivity into the medium after ingestion of radiolabeled proteins, for instance, has given some information on the digestive capacity of this cell. Macrophages from different sources have been found to degrade proteins at different rates (Inchley, 1969); and for a given type macrophage—such as

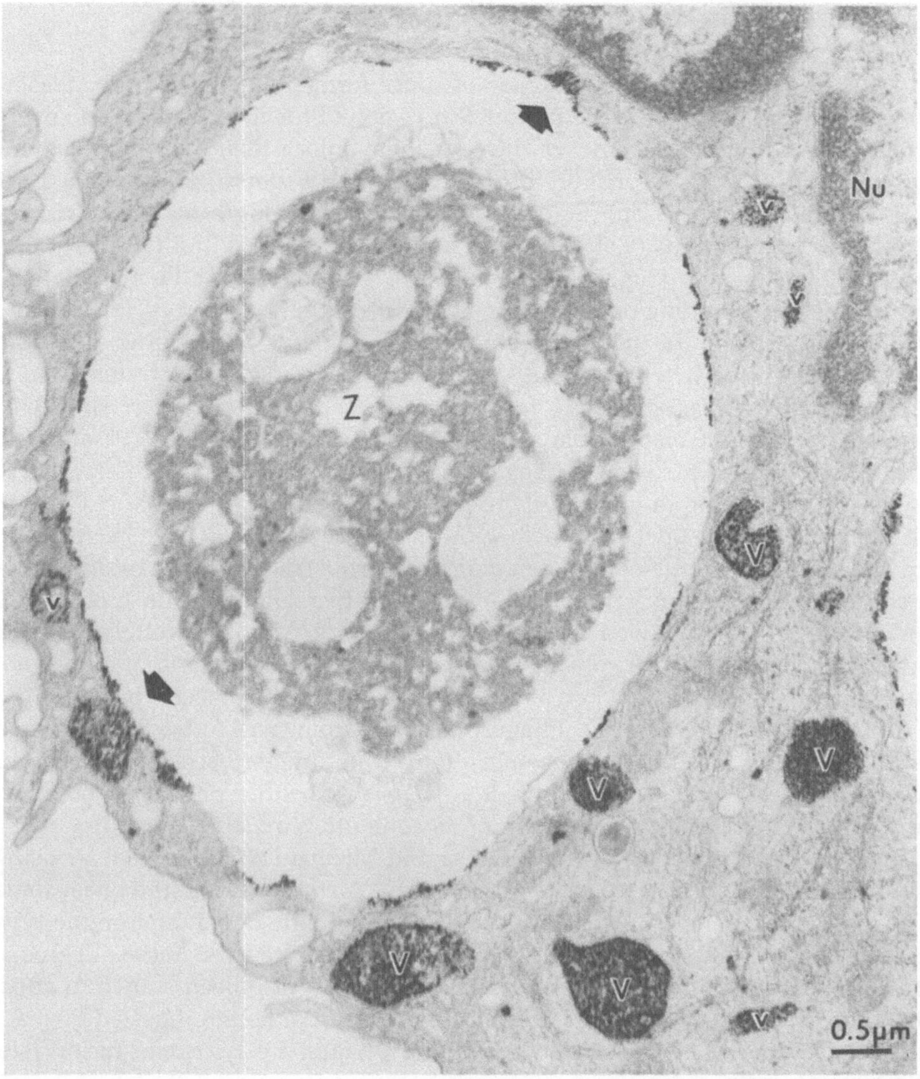

FIGURE 4. TEM micrograph showing fusion of Thorotrast-labeled lysosome with phagosome. Macrophages were cultured for 13 hr with medium containing Thorotrast which was taken up into the cell by pinocytosis. Further incubation of the cells (4 days) induced fusion of the pinocytic vesicles containing Thorotrast with lysosomes. The macrophages were then allowed to ingest zymosan particles (Z). Heavily Thorotrast-labeled lysosomes (V) are seen, two of which have just fused with the zymosan-containing phagosome and are releasing their content into the phagosome (arrows). Nu, macrophage nucleus. (Courtesy of Dr. M. C. Kielian.)

the mouse peritoneal macrophage—different substances are degraded at unequal rates. When human serum albumin was used as the test compound, 50% of the ingested isotope was released after 5 hr (Ehrenreich and Cohn, 1967); in the same period of time, over 80% of the hemoglobin in ingested sheep red cells is believed digested (Sohnle and Sussdorf, 1972), and, as for thyroglobulins, only 2% cell-associated radioactivity was recovered as digestive products after 18 hr (Hoddevik and Seljelid, 1975).

Another approach is to observe the morphological expression of intracellular degradation. Liver macrophages were examined with the electron microscope at different periods after ingestion of various cell organelles, and ultrastructural changes of such ingested organelles in the phagolysosomes were taken to represent corresponding rates of breakdown (Ericsson *et al.*, 1972; Glaumann *et al.*, 1975). Again different rates of structural changes were observed. Whereas ingested mitochondria were no longer recognizable 20 min after uptake, ribosomes were still identifiable after 24 hr.

A third approach is to study killing of bacteria after ingestion by macrophages by viability counts of bacteria after lysis of the phagocyte (Mackaness, 1970; Whaley and Singh, 1973; Cole and Brostoff, 1975; LaForce, 1976). In such systems it is important to keep in mind that some bacteria may remain alive through avoidance of phagosome–lysosome fusion (Section 6.3), and others may be killed by antibiotics internalized concurrently from the incubation medium rather than by the phagocyte's hydrolytic enzymes (Cole and Brostoff, 1975; Biroum-Noerjasin, 1977).

5.3. METABOLIC CONSEQUENCES OF INGESTION

Phagocytosis by macrophages is accompanied by a tremendous stimulation of the cell's metabolic machinery. Uptake of digestible substances was reported to induce lysosomal enzyme synthesis (Axline and Cohn, 1970) and secretion (Schnyder and Baggiolini, 1977). While, in some cases, increased intracellular lysosomal enzyme activities are not detected after phagocytosis (Berg and Munthe-Kaas, 1977), in other cases, lysosomal enzymes are released also when the substance ingested is nondigestible (Weissman *et al.*, 1971), or when mere perturbation of phagocyte membrane by complement fixation occurs unaccompanied by phagocytosis (Schorlemmer *et al.*, 1977a,b; Sorber, 1978). Thus, the causal link between digestion and lysosomal enzyme secretion is not yet certain.

A number of neutral proteases are also secreted by macrophages when triggered by phagocytosis. These include collagenase (Werb and Gordon, 1975a; Horwitz and Crystal, 1976), elastase (Werb and Gordon, 1975b; White *et al.*, 1977), and plasminogen activator (Schnyder and Baggiolini, 1977). The ingestion of zymosan and opsonized red cells have also been reported to lead to release of prostaglandin E from macrophages (Humes *et al.*, 1977; Gemsa *et al.*, 1978; Brune *et al.*, 1978). The secretion of these various substances during and following ingestion all indicate that phagocytosis plays a significant role in the organism during inflammatory processes (Davies, 1976; Davies and Allison, 1976).

Another aspect of metabolic stimulation accompanying phagocytosis, which has been drawing attention in the past few years, is the processes leading to the reported increases in H_2O_2 production and glucose oxidation through the hexosemonophosphate shunt. Recent reports (Johnston *et al.*, 1976, 1978) on the generation of superoxide anion and singlet oxygen (intermediate metabolites when O_2 is converted to H_2O_2) and the release of H_2O_2 (Nathan and Root, 1977) are interesting for the understanding of bactericidal mechanisms in macrophages parallel to those already known in PMN.

5.4. RECOVERY OF PLASMA MEMBRANE AND RECEPTORS

With each act of particle ingestion, the macrophage also removes a portion of its plasma membrane together with its receptors from its surface. Stereological studies have indeed shown that an actively endocytic macrophage internalizes an area of plasma membrane equivalent to its entire surface in about 30 min (Steinman *et al.*, 1976). What happens to the vast amount of membrane and how the macrophages replace their plasma membrane are topics of major interest. Morphological studies have shown that phagolysosomes shrink in size as the contents are digested (Steinman and Cohn, 1972) or the fluid is absorbed out of the vesicles (Cohn *et al.*, 1966).

The bulk of evidence so far available suggests that plasma membrane is recycled back to the cell surface. This hypothesis of membrane recycling is also supported by the report on amoeba (Goodall *et al.*, 1972), where data ruled out a selective *de nòvo* synthesis of plasma membrane during phagocytosis, and that on fibroblasts (Schneider *et al.*, 1977), where membrane markers traced into lysosomes during phagocytosis actually reappeared on the cell surface. In addition, the macrophage is able to synthesize new membrane in response to endocytic events in which large segments of membrane are internalized and trapped around nondigestible particles such as latex. Thus, by means of recycling and synthesis, the macrophage maintains the integrity of its plasma membrane during endocytosis.

As for the internalization and replacement of specific receptors, little systematic study has been carried out. The Fc receptors on human monocytes, internalized during latex ingestion, reappear fairly slowly: 6 hr elapsed before 78% of the cells displayed the receptors again (Schmidt and Douglas, 1972). In this case the authors also concluded that protein synthesis may be necessary since cycloheximide delayed the recovery of receptors. With rat liver macrophages, full Fc ingestion after an initial phagocytic stimulation required some 40 hr (Munthe-Kaas, 1976). More study is necessary before the problem of receptor replacement can be resolved.

6. ENDOCYTOSIS AND OBLIGATE INTRACELLULAR PARASITES

6.1. HOST CELL SURFACE RECEPTORS FOR INTRACELLULAR PARASITES

An interesting and yet little understood aspect of endocytosis is the use obligate intracellular parasites make of a cell's endocytic pathway to gain entry into the cell. The successful entry of the parasite, be it a bacterium, virus, or protozoan, is dependent on certain of their surface structures which can interact with appropriate receptor components on host cells, resulting in binding and subsequent endocytic uptake. For instance, human cells have genetically defined membrane receptors for polio virus, mouse cells do not and, therefore, are not infected by this virus (Silverstein, 1977). In a comprehensive review, Dales (1973) concludes that in most cases animal viruses attach to membrane receptors of host cells, and enter these cells by internalization into endocytic vesicles. The malarial parasite *Plasmodium knowlesi* apparently also enters erythrocytes through initial attachment to erythrocyte surface receptors (Duffy blood group), and Duffy-negative human erythrocytes, lacking these receptors, are resistant to infection (Miller *et al.*, 1975). Similarly, infection of mouse macrophages by *Trypanosoma cruzi* is mediated by a protease-sensitive membrane component of the macrophage, and trypsinization of the phagocyte abolishes the attachment and ingestion of both epimastigotes and trypomastigotes without impairment of the ability to ingest IgG-coated red cells (Nogueira and Cohn, 1976).

6.2. MORPHOLOGY OF PARASITE PENETRATION AND INTRACELLULAR LOCATION

The malarial parasite *Plasmodium knowlesi* has been shown to make initial contact with erythrocytes at the apical end of the merozoite. A depression on the erythrocyte membrane at the contact site is subsequently created, and the parasite enters into the invaginated vesicle by a typical endocytic mechanism (Lada *et al.*, 1969; Aikawa *et al.*, 1978). In a morphological study on the entry of *Toxoplasma gondii* into macrophages and fibroblasts, the parasites were seen to enter by a phagocytic mechanism (Jones *et al.*, 1972). The orientation of the toxoplasmas during this ingestion was found to be random. *Trypanosoma cruzi* trypomastigotes have been shown to attach at the posterior end to mouse macrophages and become, consequently, ingested into phagocytic vesicles (Nogueira *et al.*, 1980). These organisms are initially enclosed within a phagosome (Figure 5), but later escape into the cytosol where they replicate (Nogueira and Cohn, 1976).

Chang and Dwyer (1978) studied the endocytic uptake of *Leishmania donovani* by hamster macrophages, and were able to observe the fusion of parasite-containing phagosomes with lysosomes. The endocytosis of parasitic mycobacteria has also been documented, and fusion of bacteria-containing phagosomes and ferritin-prelabeled lysosomes have been shown to occur (Hart *et al.*, 1972).

6.3. FATE OF INTRACELLULAR PARASITES

Under the normal sequence of events in endocytosis, the parasite taken up by macrophages in the phagosome should eventually be exposed to the entire battery of hydrolytic enzymes in lysosomes. The survival and subsequent replication of the parasite are obviously dependent on its capacity to avoid digestion by lysosomal enzymes. Three main escape mechanisms have been described.

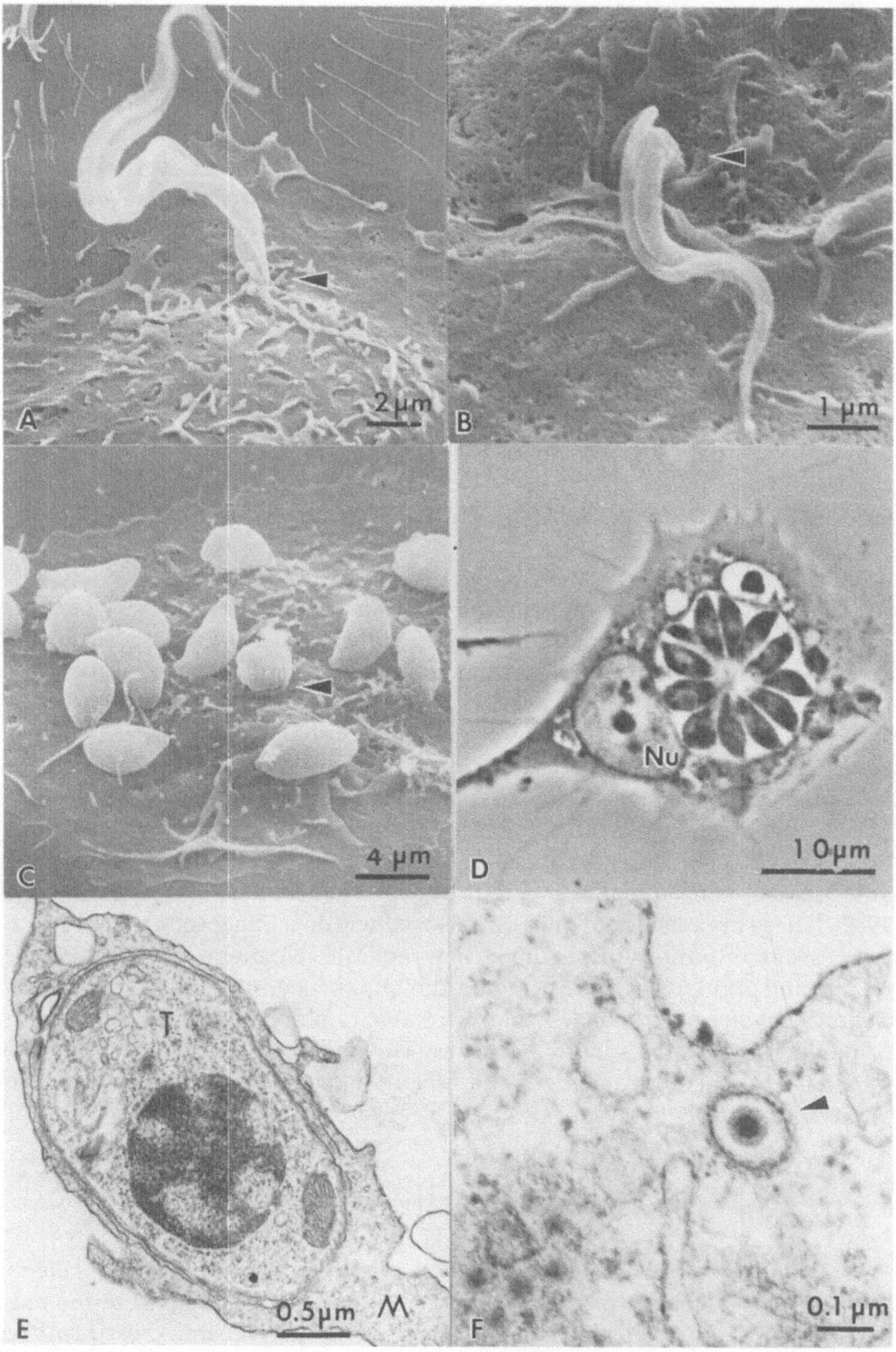

FIGURE 5. The penetration of intracellular parasites into mouse macrophages *in vitro*. (A) SEM micrograph of attachment of *Trypanosoma cruzi* strain trypomastigote to mouse macrophage. The parasite is attached to the macrophage surface at the posterior end (arrow). (B) SEM micrograph of

One escape mechanism is that the endocytic vesicle containing the parasite is somehow prevented from fusing with lysosomes, thereby avoiding exposure to lysosomal enzymes. Examples of this can be found when macrophages ingest *Mycobacterium tuberculosis* (Armstrong and Hart, 1971; Goren *et al.*, 1976), *Mycoplasma microti* (Hart *et al.*, 1972), *Toxoplasma gondii* (Jones and Hirsch, 1972; Goren *et al.*, 1976), *Encephalitozoon cuniculi* (Weidner, 1975), and some Chlamydiae (Goren *et al.*, 1976). How the prevention of phagosome–lysosome fusion is effected is unknown. Goren *et al.* (1976) proposed that the parasites may act from within the phagosome, rendering the vesicle membrane incapable of fusion with lysosomes. They showed that sulfatides (a type of glycolipids) from these parasites, when introduced directly into phagosomes by attachment to target yeast prior to their ingestion, can prevent fusion of these phagosomes with lysosomes. Another possibility is that the entry of these parasites is regulated by specific signals arising at the host cell surface upon receptor triggering which designates this type of endocytic vesicle into a cellular compartment not destined to fuse with lysosomes (Silverstein, 1977). Silverstein raises the possibility that if the parasite is coated with antibodies, and endocytic uptake is mediated through the Fc receptor instead, perhaps the phagosome would then be redirected into the lysosomal compartment. Examples of altering the fate of phagosomes by pretreatment of parasites with antibodies or immune serum have indeed been documented (Armstrong and Hart, 1975; Anderson *et al.*, 1976). Further studies would undoubtedly be of great interest and value.

A second escape mechanism is that some parasites seem to be able to withstand the digestive enzymes of lysosomes, and can indeed survive and replicate within phagolysosomes. Examples can be found with *Mycobacterium lepraemurium* (Hart *et al.*, 1972), *Leishmania* (Alexander and Vickerman, 1975; Chang and Dwyer, 1978), and some antibody-treated virulent *Mycobacterium tuberculosis* (Armstrong and Hart, 1971). An outstanding case is that of reovirus particles which are concentrated within host cell secondary lysosomes where they not only resist killing and digestion, but are uncoated and thereby converted to the biosynthetically active infectious forms (Silverstein *et al.*, 1976).

The third mechanism is that the parasite can avoid lysosomal digestion by somehow escaping from the phagosomes into the cytosol where it replicates. This is examplified by *Trypanosoma cruzi* trypomastigotes, but how this occurs is not yet known (Nogueira and Cohn, 1976).

Despite the multiple escape mechanisms parasites can display to achieve

---

trypomastigote during entry into the mouse macrophage. Most of the parasite body is within the macrophage; only the flagellum can be seen (arrow). Trypomastigotes are always seen to be endocytized via the posterior end. Though binding to the macrophage is observed, uptake of viable parasites is never initiated at the flagellar region. (C) SEM micrograph of *Toxoplasma gondii* bound to a mouse macrophage. In many cases internalization of the particles can be seen to be initiated at the apical end (arrow). (D) Phase contrast light microscopy micrograph of 12 *Toxoplasma gondii* which have replicated inside a macrophage vacuole. Nu, Macrophage nucleus. (Courtesy of Dr. H. Murray.) (E) TEM micrograph of *Trypanosoma cruzi* trypomastigote (T) inside a vacuole in a mouse macrophage (M). (Courtesy of Dr. N. Nogueira.) (F) TEM micrograph of reovirus particle within a vesicle (arrow) of the host cell. (Courtesy of Dr. S. C. Silverstein.)

intracellular survival and proliferation, it seems that with proper manipulation, the macrophage can still be rendered microbicidal, or at least microbistatic, toward many of these parasites. Thus, while normal macrophages cannot prevent intracellular proliferation of *Listeria monocytogenes, Toxoplasma gondii* (Figure 5D), *Trypanosoma cruzi, Leishmania enriettii,* and others, macrophages activated by immunization, lymphocyte factors, or other means, have enhanced toxic effects against these bacteria (Mackaness, 1962; North and Mackaness, 1963; Anderson *et al.*, 1976; Williams *et al.*, 1976; Williams and Remington, 1977; Nogueria *et al.*, 1977; Biroum-Noerjasin, 1977; Mauel *et al.*, 1978; Nogueira *et al.*, 1980). Macrophage activation, however, is a complex phenomenon [see Cohn (1978) and Allison (1978) for reviews].

7. SUMMARY AND PERSPECTIVES

Endocytosis has been defined as a mechanism whereby extracellular substances are transported into the cytoplasm within membrane-limited vesicles. These endocytic vesicles fuse with lysosomes as a rule, and their contents are thus subjected to the activities of acid hydrolases in their lysosomes. Endocytosis is an efficient and economical way to bring external environmental factors under cellular control.

Different categories of endocytic mechanisms have been described. Nonreceptor-mediated endocytosis or fluid-phase pinocytosis lacks specificity, and its significance, aside from the nutritive aspects, is difficult to assess. Receptor-mediated endocytosis implies a selectivity on the part of the endocytic cell, and as such, is better defined and easier to evaluate by the experimental approach. We have concentrated on this latter type of endocytosis in our discussion.

Endocytosis begins with the encounter of the phagocyte and the target substance. Selectivity operates already at this stage in that the first contact is promoted by both environmental factors (chemotactic substances and opsonins) and cellular recognition mechanisms (membrane receptors). The chemical moieties responsible for chemotaxis and opsonization still require detailed biochemical analyses; the recognition mechanisms on the macrophage surface, on the other hand, lend themselves to morphological study when appropriate markers are used. Thus in the case of the macrophage receptor for IgG, much is already known about their number per cell, their subtype specificities, their motility in the transverse plane of the membrane, and their mode of action.

Once contact is established between the macrophage receptor and the target substance, ingestion usually follows. The mechanisms of ingestion can most readily be studied morphologically. Ingestion seems to involve both plasma membrane activities and the contractile cytoskeleton of the macrophage. Two separate mechanisms have been described. One mechanism involves a stepwise interaction between membrane receptors and ligands on the target surface, effecting close apposition of the two surfaces and eventually resulting in total enclosure of the target substance into a membranous vesicle. The other mechanism is effected through the sinking of the attached substance into craters formed on the macrophage surface. That two fundamentally different mechanisms operate is further evidenced by the fact that the so-called foreign surface phagocytosis also falls into two such distinct categories.

While the exact metabolic requirements for endocytosis seem to require further analysis, much has been described of the cellular metabolic consequences following endocytosis. The metabolic rate and pattern of the macrophage become altered, and numerous enzymes are released into the cellular environment.

Many of the postendocytic phenomena can be studied morphologically. With proper labeling of target substance and membrane components, the fate of endocytic vesicles—both the membrane and the contents—can be followed with microscopy.

Finally, we have discussed some aspects of the interaction between macrophages and intracellular parasites. Most of these parasites enter their host cells by triggering the endocytic mechanism of these cells. Thus, this interaction is also characterized by host cell membrane receptors and formation of membrane-limited endocytic vesicles, and can be subjected to microscopic analyses. How the parasites survive the next step—lysosomal digestion—raises very interesting questions regarding adaptive mechanisms which have not yet been resolved.

Macrophage endocytosis has relevance to many other physiological phenomena which we are unable to discuss here in detail. The significance of enzyme secretion after endocytosis in the pathology of inflammation has been mentioned as well as the experimental manipulation of macrophage endocytosis in resistance to intracellular parasites and other infectious agents. One aspect which has not been discussed is the endocytic uptake and intracellular processing of antigenic substances by macrophages which obviously has far-reaching consequences in the immune response. Another interesting probelm is the macrophage recognition of and interaction with neoplastic cells in the organism. These subjects will be dealt with in other parts of this volume (or series). Thus, our understanding of the macrophage has progressed a long way since the first recognition of its role as a nonspecific scavenger cell; it is most intriguing that many of its multifaceted physiological functions involve one or more aspects of its capacity as an avidly endocytic cell.

ACKNOWLEDGMENTS. We are grateful to Drs. Victoria Freedman and Jay Unkeless for helpful discussion in the preparation of the manuscript. We also thank Judy Adams for help with the micrographs and Liv Årnes for typing the manuscript.

REFERENCES

Abramson, N., Lobuglio, A. F., Gaude, J. H., and Cotran, R. S., 1971, The interaction between human monocytes and red cells, *J. Exp. Med.* **132**:1191.

Aikawa, M., Miller, L. H., Johnson, J., and Rabbega, J., 1978, Erythrocyte entry by malarial parasites. A moving junction between erythrocytes and parasites, *J. Cell Biol.* **77**:72.

Alexander, J., and Vickerman, K., 1975, Fusion of host cell secondary lysosomes with the parasitophorous vacuoles of *Leishmania mexicana*-infected macrophages, *J. Protozool.* **22**(4):502.

Alexander, M. D., Andrews, J. A., Leslie, R. G. Q., and Wood, N. J., 1978, The binding of human and guinea pig IgG subclasses to homologous macrophage and monocyte Fc receptors, *Immunology* **35**:115.

Allison, A. C., 1978, Mechanisms by which activated macrophages inhibit lymphocyte responses, *Immunol. Rev.* **40**:3.

Allison, A. C., and Davies, P., 1974, Mechanisms of endocytosis and exocytosis, *Symp. Soc. Exp. Biol.* **28**:419.

Allison, A. C. P., Davies, P., and DePetris, S., 1971, Role of contractile microfilaments in macrophage movement and endocytosis, *Nature New Biol.* **232**:153.

Anderson, C. L., and Grey, M. H., 1978, Physicochemical separation of 2 distinct Fc receptors on murine macrophage-like cell lines, *J. Immunol.* **121**:648.

Anderson, G. R. W., Brown, M. S., and Goldstein, J. L., 1977, Role of coated endocytic vesicles in the uptake of receptor-bound low density lipoprotein in human fibroblasts, *Cell* **10**:351.

Anderson, R. S., 1977, Rosette formation by insect macrophages. Inhibition by cytochalasin B, *Cell. Immunol.* **29**:331.

Anderson, S. E., Bautista, S. C., and Remington, J. S., 1976, Specific antibody-dependent killing of *Toxoplasma gondii* by normal macrophages, *Clin. Exp. Immunol.* **26**:375.

Arend, W. P., and Mannik, M., 1973, The macrophage receptor for IgG: Number and affinity of binding sites, *J. Immunol.* **110**:1455.

Armstrong, J. A., and Hart, P. D., 1971, Response of cultured macrophages to *Mycobacterium tuberculosis,* with observations on fusion of lysosomes with phagosomes, *J. Exp. Med.* **134**:713.

Armstrong, J. A., and Hart, P. D., 1975, Phagosome-lysosome interactions in cultured macrophages infected with virulent tubercle bacilli, *J. Exp. Med.* **142**:1.

Atkinson, J. P., and Parker, C. W., 1977, Modulation of macrophage Fc receptor function by cytochalasin-sensitive structures, *Cell. Immunol.* **33**:353.

Atkinson, J. P., Michael, J. M., Chaplin, H., and Parker, C. W., 1977, Modulation of macrophage C_{3b} receptor function by cytochalasin-sensitive structures, *J. Immunol.* **118**:1292.

Axline, S. G., 1970, Functional biochemistry of the macrophage, *Semin. Hematol.* **7**:142.

Axline, S. G., and Cohn, Z. A., 1970, *In vitro* induction of lysosomal enzymes by phagocytosis, *J. Exp. Med.* **131**:1239.

Axline, S. G., and Reaven, E. P., 1974, Inhibition of phagocytosis and plasma membrane mobility of the cultivated macrophage by cytochalasin B. Role of subplasmalemmal microfilaments, *J. Cell Biol.* **61**:647.

Bar-Shavit, Z., Ofek, I., Goldman, R., Mirelman, D., and Sharon, N., 1977, Mannose residues on phagocytes as receptors for the attachment of *E. coli* and *S. typhi, Biochem. Biophys. Res. Comm.* **78**:455.

Basten, A., Miller, J. F. A. P., Sprent, G., and Pye, G., 1971, A receptor for antibodies on B lymphocytes, *J. Exp. Med.* **135**:610.

Berg, T., and Boman, D., 1973, Distribution of lysosomal enzymes between parenchymal and Kupffer cells of rat liver, *Biochim. Biophys. Acta* **321**:585.

Berg, T., and Munthe-Kaas, A. C., 1977, Lysosomal enzymes in cultured rat Kupffer cells, *Exp. Cell Res.* **109**:119.

Berkel, T. J. C. van, Kruijt, J. K., and Koster, J. F., 1975, Identity and activities of lysosomal enzymes in parenchymal and non-parenchymal cells of rat liver, *Eur. J. Biochem.* **58**:145.

Berken, A., and Benacerraf, B., 1966, Properties of antibodies cytophilic for macrophages, *J. Exp. Med.* **123**:119.

Berlin, R. D., Oliver, J. M., Ukena, T. E., and Yin, H. H., 1974, Control of cell surface topography, *Nature* **247**:45.

Berry, M. D., and Friend, D. S., 1969, High yield preparation of isolated rat liver parenchymal cells. A biochemical and fine structural study, *J. Cell Biol.* **43**:506.

Bertheussen, K., and Seljelid, R., 1978, Echinoid phagocytes *in vitro, Exp. Cell Res.* **111**:401.

Bhisey, A. N., and Freed, J. J., 1971a, Amoeboid movement induced in cultured macrophages by colchicine or vinblastine, *Exp. Cell Res.* **64**:419.

Bhisey, A. N., and Freed, J. J., 1971b, Altered movements of endosomes in colchicine-treated cultured macrophages, *Exp. Cell Res.* **64**:430.

Bhisey, A. N., and Freed, J. J., 1975, Remnant motility of macrophages treated with cytochalasin B in the presence of colchicine, *Exp. Cell Res.* **95**:376.

Bianco, C., and Nussenzweig, V., 1977, Complement receptors, in: *Contemporary Topics in Molecular Immunology* (R. R. Porter, ed.), pp. 145–176, Plenum Press, New York.

Bianco, C., Patrick, R., and Nussenzweig, V., 1970, A population of lymphocytes bearing a membrane receptor for antigen-antibody-complement complexes. Separation and characterization, *J. Exp. Med.* **132**:702.

Bianco, C., Griffin, F. M., and Silverstein, S. C., 1975, Studies of the macrophage complement receptor. Alteration of receptor function upon macrophage activation, *J. Exp. Med.* **141**:1278.

Bianco, C., Eden, A., and Cohn, Z. A., 1976, The induction of macrophage spreading: Role of coagulation factors and the complement system, *J. Exp. Med.* **144**:1531.

Biroum-Noerjasin, 1977, Listericidal activity of nonstimulated and stimulated human macrophages *in vitro, Clin. Exp. Immunol.* **28**:138.

Blumenstock, F., Saba, T. M., Weber, P., and Cho, E., 1976, Purification and biochemical characterization of a macrophage stimulating alpha-2-globulin opsonic protein, *J. Reticuloendothel. Soc.* **19**:157.

Bouveng, R., Schildt, R., and Sjøquist, J., 1975, Estimation of RES phagocytosis and catabolism in man by the use of ^{125}I-labeled microaggregates of human serum albumin, *J. Reticuloendothel. Soc.* **18**:151.

Brown, M. S., and Goldstein, J. L., 1976, Receptor mediated control of cholesterol metabolism, *Science* **191**:150.

Brune, K., Glatt, M., Kålin, H., and Peskar, B. A., 1978, Pharmacological control of prostaglandin and thromboxane release from macrophages, *Nature* **274**:261.

Carpenter, G., and Cohen, S., 1976, ^{125}I-labelled human epidermal growth factor, *J. Cell Biol.* **71**:159.

Carr, I., 1968, Some aspects of the fine structure of the reticuloendothelial system: The cells which clear colloids from the blood stream, *Z. Zellforsch.* **89**:355.

Catt, K. J., and Dufau, M. L., 1977, Peptide hormone receptors, *Annu. Rev. Physiol.* **39**:529.

Chang, K. P., and Dwyers, D. M., 1978, *Leishmania donovani.* Hamster macrophage interactions *in vitro:* Cell entry, intracellular survival, and multiplication of amastigotes, *J. Exp. Med.* **147**:515.

Claus, D., Osmand, A. P., and Gewurz, H., 1976, Radioimmunoassay of human C-reactive protein and levels in normal sera, *J. Lab. Clin. Med.* **87**:120.

Cohn, Z. A., 1963a, The fate of bacteria within phagocytic cells. I. The degradation of isotopically labelled bacteria by polymorphonuclear leucocytes and macrophages, *J. Exp. Med.* **117**:27.

Cohn, Z. A., 1963b, The fate of bacteria within phagocytic cells. II. The modification of intracellular degradation, *J. Exp. Med.* **117**:43.

Cohn, Z. A., 1970, Endocytosis and intracellular digestion, in: *Mononuclear Phagocytes* (R. van Furth, ed.), pp. 121–128, Blackwell, Oxford.

Cohn, Z. A., 1975, Macrophage physiology, *Fed. Proc.* **34**:1725.

Cohn, Z. A., 1977, Physiology and biochemistry of phagocytic cells, in: *Frontiers of Medicine* (G. M. Meade, ed.), p. 95, Plenum Press, New York.

Cohn, Z. A., 1978, The activation of mononuclear phagocytes: Fact, fancy and future, *J. Immunol.* **121**:813.

Cohn, Z. A., and Fedorko, M. E., 1969, The formation and fate of lysosomes, in: *Lysosomes in Biology and Pathology* (J. J. Dingle and H. B. Fell, eds.), pp. 43–63, North-Holland, Amsterdam.

Cohn, Z. A., Fedorko, M. E., and Hirsch, J. G., 1966, The *in vitro* differentiation of mononuclear phagocytes. V. The formation of macrophage lysosomes, *J. Exp. Med.* **123**:757.

Cole, P., and Brostoff, J., 1975, Intracellular killing of *Listeria monocytogenes* by activated macrophages (Mackaness system) is due to antibiotic, *Nature* **256**:515.

Cordingley, J. L., 1968, Kupffer cell response to oestrogen stimulation, *J. Reticuloendothel. Soc.* **5**:591

Cotutiu, C. C., and Ericsson, J. L. E., 1970, Two types of "microendocytosis" in Kupffer cells studied with electron opaque tracers, *Septième Congrès International de Microscopie Electronique, Grenoble,* p. 53, Société Française de Microscopie Electronique, Paris.

Cox, J. P., and Karnovsky, M. L., 1973, The depression of phagocytosis by exogenous cyclic nucleotides, prostaglandins, and theophylline, *J. Cell Biol.* **59**:480.

Crispe, I. N., 1976, The effect of vinblastine, colchicine and hexylene glycol on migration of human monocytes, *Exp. Cell Res.* **100**:443.

Czop, J. K., Fearon, D. T., and Austen, K. F., 1978, Opsonin-independent phagocytosis of activators of the alternative complement pathway by human monocytes, *J. Immunol.* **120**:1132.

Dales, S., 1973, Early events in cell-animal virus interactions, *Bact. Rev.* **37**:103.

Davies, P., 1976, Essential role of macrophages in chronic inflammatory processes, *Schweiz. Med. Wochenschr.* **106**:1351.

Davies, P., and Allison, A. C., 1976, Secretion of macrophage enzymes in relation to the pathogenesis of chronic inflammation, in: *Immunobiology of the Macrophage* (D. S. Nelson, ed.), pp. 427–461, Academic Press, New York.

Davies, W. A., and Stossel, 1977, Peripheral hyaline blebs (podosomes) of macrophages, *J. Cell Biol.* **75**:941.

Debanne, M. T., Regolozi, E., and Dolovich, J. C., 1973, Serum protease inhibition in the blood clearance of subtilisin A, *Br. J. Exp. Pathol.* **54**:571.

DePetris, S., 1974, Inhibition and reversal of capping by cytochalasin B, vinblastin and colchicine, *Nature* **250**:54.

DePetris, S., Karlbad, G., and Pernis, B., 1962, Filamentous structures in the cytoplasma of normal mononuclear phagocytes, *J. Ultrastruct. Res.* **7**:39.

Dohlman, J. G., and Goetzl, E. J., 1978, Unique determinants of alveolar macrophage spontaneous and chemokinetically stimulated migration, *Cell. Immunol.* **39**:36.

Donald, K. J., and Tennent, R. J., 1975, The relative roles of platelets and macrophages in clearing particles from the blood: The value of carbon clearance as a measure of reticulo-endothelial phagocytosis, *J. Pathol.* **117**:235.

Dumont, A., 1972, Ultrastructural aspects of phagocytosis of facultative intracellular parasite by hamster peritoneal macrophages, *J. Reticuloendothel. Soc.* **11**:469.

Edelman, G. M., Yahara, I, and Wang, J. L., 1973, Receptor mobility and receptor-cytoplasmic interactions in lymphocytes, *Proc. Natl. Acad. Sci. USA* **70**:1442.

Edelson, P. J., and Cohn, Z. A., 1974a, Effects of Con A on mouse peritoneal macrophages. I. Stimulation of endocytic activity and inhibition of phagolysosome formation, *J. Exp. Med.* **140**:1364.

Edelson, P. J., and Cohn, Z. A., 1974b, Effects of Con A on mouse peritoneal macrophages. II. Metabolism of endocytized proteins and reversibility of the effects by mannose, *J. Exp. Med.* **140**:1387.

Eden, A., Bianco, C., and Nussenzweig, V., 1971, A population of lymphocytes bearing a membrane receptor for antigen–antibody–complement complexes. II. Specific isolation, *Cell. Immunol.* **2**:658.

Ehlenberger, A. G., and Nussenzweig, V., 1977, The role of membrane receptors for C_{3b} and C_{3d} in phagocytosis, *J. Exp. Med.* **145**:357.

Ehrenreich, B. A., and Cohn, Z. A., 1967, The uptake and digestion of iodinated human serum albumin by macrophages *in vitro, J. Exp. Med.* **126**:941.

Ericsson, J. L. E., Arbough, B., and Glaumann, H., 1972, Sequential conformational changes in mitochondria, endoplasmic reticulum, plasma membrane, lysosomes, and glycogen particles during digestion in lysosomes, *Proceedings of the 4th International Congress of Histochemistry and Cytochemistry, Tokyo*, p. 27.

Essner, E., 1960, An electron microscopic study of erythrophagocytosis, *J. Biophys. Biochem. Cytol.* **7**:329.

Friedman, P. A., Shia, M. A., and Wallace, J. K., 1977, A saturable high affinity binding site for transcobalamin 11-vitamin B_{12} complexes in human placental membrane preparation, *J. Clin. Invest.* **59**:51.

Gallin, E. K., and Gallin, J. I., 1977, Interaction of chemotactic factors with human macrophages, *J. Cell Biol.* **75**:277.

Gemsa, D., Seitz, M., Kramer, W., Till, G., and Resch, K., 1978, The effects of phagocytosis, dextran sulfate, and cell damage on PGE_1 sensitivity and PGE_1 production of macrophages, *J. Immunol.* **120**:1187.

Glaumann, H., Berezesky, E. K., Ericsson, J. L. E., and Trump, B. R., 1975, Lysosomal degradation of cell organelles. II. Ultrastructural analysis of uptake and digestion of intravenously injected microsomes and ribosomes by Kupffer Cells, *Lab. Invest.* **33**:252.

Goldstein, J. L., and Brown, M. S., 1977, The low-density lipoprotein pathway and its relation to arteriosclerosis, *Annu. Rev. Biochem.* **46**:897.

Goodall, R. J., Lai, Y. F., and Thompson, J. E., 1972, Turnover of plasma membrane during phagocytosis, *J. Cell Sci.* **11**:569.

Goodell, E. M., Bilgin, S., and Carchman, R. A., 1978, Biochemical characteristics of phagocytosis in the P388 D_1 cell, *Exp. Cell Res.* **114**:57.

Gordon, S., and Cohn, Z. A., 1973, The macrophage, *Int. Rev. Cytol.* **36**:171.

Goren, M. B., Hart, P. D., Young, M. R., and Armstrong, J. A., 1976, Prevention of phagosome-lysosome fusion in cultured macrophages by sulfatides of *Mycobacterium tuberculosis, Proc. Natl. Acad. Sci. USA* **73**:2510.

Griffin, F. M., and Silverstein, S. C., 1974, Segmental response of the macrophage plasma membrane to a phagocytic stimulus, *J. Exp. Med.* **139**:323.

Griffin, F. M., Bianco, C., and Silverstein, S. C., 1975a, Characterization of the macrophage receptor for complement and demonstration of its functional independence from the receptor for the Fc portion of immunoglobulin G, *J. Exp. Med.* **141**:1269.

Griffin, F. M., Griffin, J. A., Leider, J. E., and Silverstein, S. C., 1975b, Studies on the mechanism of phagocytosis. I. Requirement for circumferential attachment of particle-bound ligands to specific receptors on the macrophage plasma membrane, *J. Exp. Med.* **142**:1263.

Griffin, F. M., Griffin, J. A., and Silverstein, S. C., 1976, Studies on the mechanism of phagocytosis. II. The interaction of macrophages with anti-immunoglobulin IgG-coated bone marrow-derived lymphocytes, *J. Exp. Med.* **144**:788.

Hart, P. D., and Young, M. R., 1975, Interference with normal phagosome-lysosome fusion in macrophages, using ingested yeast cells and suramin, *Nature* **256**:47.

Hart, P. D., Armstrong, J. A., Brown, C. A., and Draper, P., 1972, Ultrastructural study of the behaviour of macrophages toward parasitic Mycrobacteria, *Infec. Immunity* **5**:803.

Hartwig, J. H., Davies, W. A., and Stossel, T. P., 1977, Evidence for contractile protein translocation in macrophage spreading, phagocytosis and phagolysosome formation, *J. Cell Biol.* **75**:956.

Heusser, C. H., Anderson, C. L., and Grey, H. M., 1977, Receptors for IgG: Subclass specificity of receptors on different mouse cell types and the definition of 2 distinct receptors on a macrophage cell line, *J. Exp. Med.* **145**:1316.

Hoddevik, G., and Seljelid, R., 1975, The uptake and degradation of sheep thyroglobulin by macrophages *in vitro, Exp. Cell Res.* **93**:152.

Horn, R. G., Koenig, M. G., Goodman, J. S., and Collins, R. D., 1969, Phagocytosis of *Staphylococcus aureus* by hepatic reticuloendothelial cells: An ultrastructural study, *Lab. Invest.* **21**:406.

Horwitz, A. L., and Crystal, R. G., 1976, Collagenase from rabbit pulmonary alveolar macrophages, *Biochem. Biophys. Res. Commun.* **69**:296.

Huber, H., and Fudenberg, H. H., 1970, The interaction of monocytes and macrophages with immunoglobulins and complement, *Ser. Haemat.* **111**(2):160.

Huber,.H., Polley, M. L., Lenscoti, W. D., Fudenberg, H. H., Müller-Eberhard, H. J., 1968, Human monocytes: Distinct receptor sites for the third component of complement and for IgG, *Science* **162**:1281.

Humes, J. L., Bonney, R. J., Pelus, L., Dahlgren, M. E., Sadowsky, S. J., Kuehl, F. A., and Davies, P., 1977, Macrophages synthesize and release prostaglandins in response to inflammatory stimuli, *Nature* **269**:149.

Ichikawa, Y., 1969, Differentiation of a cell line of myeloid leukemia, *J. Cell Physiol.* **74**:223.

Ignarro, L. J., Lint, T. F., Geroge, W. J., 1974, Hormonal control of lysosomal enzyme release from human neutrophils, *J. Exp. Med.* **139**:1395.

Inchley, C. J., 1969, The activity of mouse Kupffer cells following intravenous injection of T4 bacteriophage, *Clin. Exp. Immunol.* **5**:173.

Ishikawa, H. Bischoff, R., and Holtzer, H., 1969, Formation of arrowhead complexes with heavy meromyosin in a variety of cell types, *J. Cell Biol.* **38**:312.

Jacques, P. J., 1969, Endocytosis, in: *Lysosomes in Biology and Pathology* (J. J. Dingle and H. B. Fell, ed.), pp. 395–420, North-Holland, Amsterdam.

Jahrling, P. B., and Gorelkin, L., 1975, Selective clearance of a benign clone of Venezuelan equine encephalitis virus from hamster plasma by hepatic reticuloendothelial cells, *J. Infect. Dis.* **132**:667.

Johnston, R. B., Lehmeyer, J. E., and Guthrie, L. A., 1976, Generation of superoxide anion and chemiluminescence by human monocytes during phagocytosis and on contact with surface-bound immunoglobulin G, *J. Exp. Med.* **143**:1551.

Johnston, R. B., Godzik, C. A., and Cohn, Z. A., 1978, Increased superoxide anion production by immunologically activated and chemically elicited macrophages, *J. Exp. Med*, **148**:115.

Jones, T. C., and Hirsch, J. G., 1972, The interaction between *Toxoplasma gondii* and mammalian cells. II. The absence of lysosomal fusion with phagocytic vacuoles containing living parasites, *J. Exp. Med.* **136**:1173.

Jones, T. C., Minick, R., and Yang, L., 1977, Attachment and ingestion of *Mycoplasma* by mouse macrophages, *Am. J. Pathol.* **87**:347.

Jungi, T. W., and McGregor, D. D., 1977, Generation of macrophage chemotactic activity *in situ* in *Listeria*-immune rats, *Cell. Immunol.* **33**:322.

Kaplan, G., 1977, Differences in the mode of phagocytosis with Fc and C_3 receptors in macrophages, *Scand. J. Immunol.* **6**:797.

Kaplan, G., and Bertheussen, K., 1977, The morphology of echinoid phagocytes and mouse peritoneal macrophages during phagocytosis *in vitro, Scand. J. Immunol.* **6**:1289.

Kaplan, G., and Mørland, B., 1978, Properties of a murine monocytic tumor cell line J-774 *in vitro.* I. Morphology and endocytosis, *Exp. Cell Res.* **115**:53.

Kaplan, G., Gaudernack, G., and Seljelid, R., 1975, Localization of receptors and early events of phagocytosis in the macrophage, *Exp. Cell Res.* **95**:365.

Kaplan, G., Eskeland, T., and Seljelid, R., 1978, Difference in the effect of immobilized ligands on the Fc and C_3 receptors of mouse peritoneal macrophages *in vitro, Scand. J. Immunol.* **7**:19.

Karnovsky, M. L., 1962, Metabolic basis of phagocytic activity, *Physiol. Rev.* **42**:143.

Karnovsky, M. L., Simmons, S., Glass, E. A., Shafer, A. W., and D'Arcy Hart, P., 1970, Metabolism of macrophages, in: *Mononuclear Phagocytes* (R. van Furth, ed.), pp. 103–117, Blackwell, Oxford.

Kay, M. M. B., 1975, Mechanism of removal of senescent cells by human macrophages *in situ, Proc. Natl. Acad. Sci. USA* **72**:3521.

Knook, D. L., and Sleyster, E. C., 1976, Separation of Kupffer and endothelial cells of the rat liver by centrifugal elutriation, *Exp. Cell Res.* **99**:444.

Knook, D. L., and Sleyster, E. C., 1977, Preparation and characterization of Kupffer cells from rat and mouse livers, in: *Kupffer cells and other sinusoidal cells* (E. Wisse and D. L. Knook, eds.), pp. 273–288, Elsevier/North Holland, Amsterdam.

Koren, H. S., Handwerger, B. S., and Wunderlich, J. R., 1975, Identification of macrophage-like characteristics in a cultured murine tumor line, *J. Immunol.* **114**:894.

Lada, R., Aikawa, M., and Sprinz, H., 1969, Penetration of erythrocytes by merozoites of mammalian and avian malarial parasites, *J. Parasitol.* **55**:633.

LaForce, F. M., 1976, Effect of alveolar lining material on phagocytic and bactericidal activity of lung macrophages against *Staphylococcus aureus, J. Lab. Clin. Med.* **88**:691.

Lay, W. H., and Nussenzweig, V., 1968, Receptors for complement on leucocytes, *J. Exp. Med.* **128**:991.

Lay, W. H., and Nussenzweig, V., 1969, Ca^{++}-dependent binding of antigen-19s antibody complexes to macrophages, *J. Immun.* **102**:1172.

Lobuglio, A. F., 1973, The monocyte: new concepts of function, *New Engl. J. Med.* **288**:212.

Mackaness, G., 1962, Cellular resistance to infection, *J. Exp. Med.* **116**:381.

Mackaness, G. B., 1970, The mechanism of macrophage activation, in: *Infectious Agents and Host Reactions* (S. Mudd, ed.), pp. 61–75, W. B. Saunders, Philadelphia.

Malawista, S. E., Gee, J. B. L., and Bensch, K. G., 1971, Cytochalasin B reversibly inhibits phagocytosis: functional, metabolic, and ultrastructural effects in human blood leucocytes and rabbit alveolar macrophages, *Yale J. Biol. Med.* **44**:286.

Mantovani, B., Rabinovitch, M., and Nussenzweig, V., 1972, Phagocytosis of immune complexes by macrophages, *J. Exp. Med.* **135**:780.

Matter, A., Orci, L., Karnovsky, M. J., Rouiller, C., 1970, Digestive events in the Kupffer cells of rat liver, *Septième Congrès International de Microscopie Electronique, Grenoble* **111**:55.
Mauel, J., Buchmuller, Y., and Behin, R., 1978, Studies on the mechanisms of macrophage activation. I. Destruction of intracellular *Leishmania enriettii* in macrophages activated by cocultivation with stimulated lymphocytes, *J. Exp. Med.* **148**:393.
McKeever, P. E., Garvin, A. J., and Spicer, S. S., 1976a, Immune complex receptors on cell surfaces. I. Ultrastructural demonstration on macrophages, *J. Histochem. Cytochem.* **24**:948.
McKeever, P. E., Garvin, A. J., Hardin, D. H., and Spicer, S. S., 1976b, Immune complex receptors on cell surfaces. II. Cytochemical evaluation of their abundance on different cells, distribution, uptake and regeneration, *Am. J. Pathol.* **84**:437.
McKeever, P. E., Spicer, S. S., Brissie, N. T., and Garvin, A. J., 1977, Immune complex receptors on cell surfaces. III. Topography of macrophage receptors demonstrated by new scanning electron microscopic peroxidase marker, *J. Histochem. Cytochem.* **25**:1063.
Meltzer, M. S., and Stevenson, M. M., 1978, Macrophage function in tumor-bearing mice. Dissociation of phagocytic and chemotactic responsiveness, *Cell. Immunol.* **35**:99.
Meltzer, M. S., Stevenson, M. M., and Leonard, E. J., 1977, Characterization of macrophage chemotaxis in tumor cell cultures and comparison with lymphocyte-derived chemotactic factors, *Cancer Res.* **37**:721.
Michl, J., and Silverstein, S. C., 1978, Role of macrophage receptors in the ingestion phase of phagocytosis, *Birth Defects: Original Article Series* **14**:(2):99.
Michl, J., Ohlbaum, D. J., and Silverstein, S. C., 1976a, 2-deoxyglucose selectively inhibits Fc and complement receptor-mediated phagocytosis in mouse peritoneal macrophages. I. Description of the inhibitory effect, *J. Exp. Med.* **144**:1465.
Michl, J., Ohlbaum, D. J., and Silverstein, S. C., 1976b, 2-deoxyglucose selectively inhibits Fc and complement receptor-mediated phagocytosis in mouse peritoneal macrophages. II. Dissociation of the inhibitory effects of 2-deoxyglucose on phagocytosis and ATP generation, *J. Exp. Med.* **144**:1484.
Miller, L. H., Mason, S. J., Dvorak, J. A., McGuinniss, M. H., and Rothman, I. K., 1975, Erythrocyte receptors for (*Plasmodium knowlesi*) malaria. Duffy blood group determinants, *Science* **189**:561.
Mills, D. M., and Zucker-Franklin, D., 1969, Electron microscopic studies of isolated Kupffer cells, *Am. J. Pathol.* **54**:147.
Mørland, B., and Kaplan, G., 1977, Macrophage activation *in vivo* and *in vitro*, *Exp. Cell Res.* **108**:279.
Mørland, B., and Kaplan, G., 1978, Properties of a murine monocytic tumor cell line J-774 *in vitro*. II. Enzyme activities, *Exp. Cell Res.* **115**:63.
Mortensen, R. F., and Duszkiewicz, G. A., 1977, Mediation of CRP-dependent phagocytosis through mouse macrophage Fc-receptors, *J. Immunol.* **119**:1611.
Mudd, S., McCutcheon, M., and Lucke, B., 1934, Phagocytosis, *Physiol. Rev.* **4**:210.
Munthe-Kaas, A. C., 1976, Phagocytosis in rat Kupffer cells *in vitro*, *Exp. Cell Res.* **99**:319.
Munthe-Kaas, A. C., 1977, Uptake of macromolecules by rat Kupffer cells *in vitro*, *Exp. Cell Res.* **107**:55.
Munthe-Kaas, A. C., and Berg, T., 1977, *In vitro* studies on phagocytosis and lysosomal enzyme patterns in rat Kupffer cells, in: *Movement, metabolism and Bactericidal Mechanisms of Phagocytes* (F. Rossi, P. L. Patriarca, and D. Romeo, eds.), pp. 65–73, Piccin, Padua.
Munthe-Kaas, A. C., Berg, T., Seglen, P. O., and Seljelid, R., 1975, Mass isolation and culture of rat Kupffer cells, *J. Exp. Med.* **141**:1.
Munthe-Kaas, A. C., Berg, T., and Seljelid, R., 1976a, Distribution of lysosomal enzymes in different types of rat liver cells, *Exp. Cell Res.* **99**:146.
Munthe-Kaas, A. C., Kaplan, G., and Seljelid, R., 1976b, On the mechanism of internationalization of opsonized particles by rat Kupffer cells *in vitro*, *Exp. Cell Res.* **103**:201.
Muschel, R. J., Rosen, N., Rosen, O. M., and Bloom B. R., 1977a, Modulation of Fc-mediated phagocytosis by cyclic AMP and insulin in a macrophage-like cell line, *J. Immunol.* **119**:1813.
Muschel, R. J., Rosen, N., and Bloom, B. R., 1977b, Isolation of variants in phagocytosis of a macrophage-like continuous cell line, *J. Exp. Med.* **145**:175.
Muto, M., and Fujita, T., 1977, Phagocytic activities of the Kupffer cell: A scanning electron microscope study, in: *Kupffer Cells and Other Liver Sinusoidal Cells* (E. Wisse and D. L. Knook, eds.), pp. 109–121, Elsevier/North Holland, Amsterdam.

Nagata, K., Takahashi, E., Saito, M., Ono, J., Kuboyama, M., and Ogasa, K., 1976, Differentiation of a cell line of mouse myeloid leukemia. I. Simultaneous induction of lysosomal enzyme activities and phagocytosis, *Exp. Cell Res.* **100**:322.

Nagata, K., Ooguro, K., Saito, M., Kuboyama, M., and Ogasa, K., 1977, A factor inducing differentiation of mouse myeloid leukemia cells in human amniotic fluid, *Gann* **68**:757.

Nagura, H., Asai, J., and Kojima, K., 1977, Studies on the mechanism of phagocytosis. I. Effect of electric surface charge on phagocytic activity of macrophages for fixed red cells, *Cell Struc. Func.* **2**:21.

Nathan, C. F., and Root, R. K., 1977, Hydrogen peroxide release from mouse peritoneal macrophages, *J. Exp. Med.* **146**:1648.

Nelson, D. S., 1963, Immune adherence, *Adv. Immunol.* **3**:131.

Nichols, B. A., 1976a, Normal rabbit alveolar macrophages. I. The phagocytosis of tubular myelin, *J. Exp. Med.* **144**:906.

Nichols, B. A., 1976b, Normal rabbit alveolar macrophages. II. Their primary and secondary lysosomes as revealed by electron microscopy and cytochemistry, *J. Exp. Med.* **144**:920.

Nogueira, N., and Cohn, Z. A., 1976, *Trypanosoma cruzi:* Mechanism of entry and intracellular fate in mammalian cells. *J. Exp. Med.* **143**:1402.

Nogueira, N., Gordon, S., and Cohn, Z. A., 1977, *Trypanosoma cruzi:* Modification of macrophage function during infection, *J. Exp. Med.* **146**:157.

Nogueira, N., Kaplan, G., and Cohn, Z. A., 1980, Induction of microbicidal activity, in: *Proceedings of the Third Symposium on the Mononuclear Phagocytic System, Leiden,* Blackwell, Oxford (in press).

Normann, S. J., and Sorkin, E., 1977, Inhibition of macrophage chemotaxis by neoplastic and other rapidly proliferating cells *in vitro, Cancer Res.* **37**:705.

North, R. J., 1970, Endocytosis, *Semin. Hematol.* **7**:161.

North, R. J., and Mackaness, G., 1963, Electron microscopic observations on the peritoneal macrophages of normal mice and mice immunized with *Listeria monocytogenes, Br. J. Exp. Pathol.* **44**:601.

Ødegaard, A., Viken, K. E., and Lamvik, J., 1974, Structural and functional properties of blood monocytes culture *in vitro, Acta Pathol. Microbiol. Scand. Sect. B* **82:**223.

Orenstein, J. M., and Shelton, E., 1977, Membrane phenomena accompanying erythrophagocytosis. A scanning electron microscopy study, *Lab. Invest.* **36**:363.

Parks, H. F., and Chiquoine, A. D., 1957, Observations on early stages of phagocytosis of colloidal particles by hepatic phagocytes of the mouse, in: *Electron Microscopy. Proceedings of the Stockholm Conference, September 1956,* p. 154, Almquist and Wiksell, Stockholm.

Pasternak, G. R., Snyderman, R., Pike, M. C., Johnson, R. J., and Shin, H. S., 1978, Resistance of neoplasms to immunological destruction: Role of a macrophage chemotaxis inhibitor, *J. Exp. Med.* **148**:93.

Pavillard, E. R. S., 1963, *In vitro* phagocytosis and bactericidal ability of alveolar and peritoneal macrophages of normal rats, *Aust. J. Exp. Biol.* **41**:265.

Pavillard, E. R. S., and Rowley, D., 1962, A comparison of the phagocytosis and bacterial ability of guinea pig alveolar and mouse peritoneal macrophages, *Aust. J. Exp. Biol.* **40**:207.

Pesanti, E. L., and Axline, S. G., 1975a, Colchicine effects on lysosomal enzyme induction and intracellular degradation in the cultivated macrophage, *J. Exp. Med.* **141**:1030.

Pesanti, E. L., and Axline, S. G., 1975b, Phagolysosome formation in normal and colchicine-treatea macrophages, *J. Exp. Med.* **142**:903.

Phillips-Quagliata, J. M., Levine, B. B., Quagliata, F., and Uhr, J. W., 1971, Mechanisms underlying binding of immune complexes to macrophages, *J. Exp. Med.* **113**:589.

Pike, M. C., Daniels, C. A., and Snyderman, R., 1977, Influenza-induced depression of monocyte chemotaxis: Reversal by levamisole, *Cell. Immunol.* **32**:234.

Pollard, T. D., Shelton, E., Weihing, R. R., and Korn, E. D., 1970, Ultrastructural characterization of F-actin isolated from *Acanthamoeba castellanii* and identification of cytoplasmic filaments as F-actin by reaction with rabbit heavy meromyosin, *J. Molec. Biol.* **50**:91.

Polliack, A., and Gordon, S., 1975, Scanning electron microscopy of murine macrophages. Surface characteristics during maturation, activation and phagocytosis, *Lab. Invest.* **33**:469.

Poplack, D. G., Sher, N. A., Chaparas, S. D., and Blaese, R. M., 1976, The effect of *Mycobacterium bovis* (BCG) on macrophage random migration, chemotaxis and pinocytosis, *Cancer Res.* **36**:1233.

Powell, D. A., and Muse, K. A., 1977, Scanning electron microscopy of guinea pig alveolar macrophages. *In vitro* phagocytosis of *Mycoplasma pneumoniae, Lab. Invest.* **37**:535.

Pratten, M. K., Williams, K. E., and Lloyd, J. B., 1977, A quantitative study of pinocytosis and intracellular proteolysis in rat peritoneal macrophages, *Biochem. J.* **168**:365.

Rabinovitch, M., 1967, The dissociation of the attachment and ingestion phases of phagocytosis by macrophages, *Exp. Cell Res.* **46**:19.

Rabinovitch, M., 1968, Effect of antiserum on the attachment of modified erythrocytes to normal or to trypsinized macrophages, *Proc. Soc. Exp. Biol. Med.* **127**:351.

Rabinovitch, M., 1970, Phagocytic recognition, in: *Mononuclear Phagocytes* (R. van Furth, ed.), pp. 299–315, Blackwell, Oxford.

Rabinovitch, M., and deStefano, M. J., 1973, Particle recognition by cultivated macrophages, *J. Immun.* **110**:695.

Ralph, P., Prichard, J., and Cohn, M., 1975, Reticulum cell sarcoma: an effector cell in antibody-dependent cell-mediated immunity, *J. Immunol.* **114**:898.

Reaven, E. P., and Axline, S. G., 1973, Subplasmalemmal microfilaments and microtubules in resting and phagocytizing cultivated macrophages, *J. Cell Biol.* **59**:12.

Remold-O'Donnell, E., 1977, Purification of plasma membrane of guinea pig peritoneal macrophages, *Prep. Biochem* **7**(6):441.

Reynolds, H. Y., Atkinson, J. P., Newball, H. H., and Frank, M. M., 1975, Receptors for immunoglobulin and complement on human alveolar macrophages, *J. Immunol.* **114:**1813.

Rikihisa, Y., and Mizuna, D., 1978, Different arrangements of phagolysosomal membranes which depend upon the particles phagocytized. Observation with markers of the two sides of plasma membranes, *Exp. Cell Res.* **111**:437.

Russel, R. J., McInroy, R. J., Wilkinson, P. C., and White, R. G., 1976, A lipid chemotactic factor from anaerobic coryneform bacteria including *Corynebacterium parvum* with activity for macrophages and monocytes, *Immunology* **30**:935.

Ryser, H. J.-P., 1968, Uptake of protein by mammalian cells: An underdeveloped area, *Science* **159**:390.

Saba, T. M., 1970, Physiology and physiopathology of the reticuloendothelial system, *Arch. Intern. Med.* **126**:1031.

Saba, T. M., 1978, Humoral control of Kupffer cell function after injury, *Bull. Kupffer Cell Found.* **1**(2):12.

Saba, T. M., and Scovill, W. A., 1972, Leucocyte level as a potential factor determining hepatic phagocytosis following surgery, *Proc. Soc. Exp. Biol. Med.* **140**:1210.

Satodate, R., Sason, S., Oikawa, K., Hatakeyama, N., and Katsura, S., 1977, Scanning electron microscopical studies on Kupffer cell in phagocytic activity, in: *Kupffer Cells and Other Liver Sinusoidal Cells* (D. L. Knook and E. Wisse, eds.), pp. 121–131, Elsevier/North Holland, Amsterdam.

Scheeberger-Keeley, E. F., and Burger, E. J., 1970, Intravascular macrophages in cat lungs after open chest ventilation. An electron microscopic study, *Lab. Invest.* **22**:361.

Schmidt, M. E., and Douglas, S. D., 1972, Disappearance and recovery of human monocyte IgG receptor activity after phagocytosis, *J. Immunol.* **109**:914.

Schneider, Y.-J., Tulkens, P., and Trouet, A., 1977, Recycling of fibroblast plasma membrane antigens internalized during endocytosis, *Biochem. Soc. Transac.* **5**:1164.

Schnyder, J., and Baggiolini, M., 1977, *In vitro* activation of macrophages by phagocytosis, *Experientia* **33**:829.

Schorlemmer, H. U., Bitter-Suermann, D., and Allison, A. C., 1977a, Complement activation by the alternative pathway and macrophage enzyme secretion in the pathogenesis of chronic inflammation, *Immunology* **32**:929.

Schorlemmer, H. U., Burger, R., Hylton, W., and Allison, A. C., 1977b, Induction of lysosomal enzyme release from cultured macrophages by dextran sulfate, *Clin. Immunol. Immunopathol.* **7**:88.

Schreiber, A. D., Parsons, J., McDermott, P., and Cooper, R. A., 1975, Effect of corticosteroids on the human monocyte IgG and complement receptors, *J. Clin. Invest.* **56**:1189.

Scott, R. E., and Rosenthal, A. S., 1977, Isolation of receptor-bearing plasma membrane vesicles from guinea pig macrophages, *J. Immunol.* **119**:143.

Seljelid, R., 1980, Properties of Kupffer cells, in: *Proceedings of the Third Symposium on the Mononuclear Phagocytic System, Leiden,* Blackwell, Oxford (in press).

Seyberth, H. W., Schmidt-Gayk, H., Jakobs, K. H., and Hackenthal, E., 1973, Cyclic adenosine monophosphate in phagocytizing granulocytes and alveolar macrophages, *J. Cell Biol.* **57**:567.

Silverstein, S. C., 1977, Endocytic uptake of particles by mononuclear phagocytes and the penetration of obligate intracellular parasites, *Am. J. Trop. Med. Hyg.* **26**:161.

Silverstein, S. C., Christman, J. K., and Acs, G., 1976, The reovirus replicate cycle, *Annu. Rev. Biochem.* **45**:375.

Silverstein, S. C., Steinman, R. M., and Cohn, Z. A., 1977, Endocytosis, *Annu. Rev. Biochem.* **46**:669.

Singh, A., 1974, The subplasmalemmal microfilaments in Kupffer cells, *J. Ultrastruct. Res.* **48**:67.

Snyderman, R., and Pike, M. C., 1977, Macrophage migratory dysfunction in cancer, *Am. J. Pathol.* **88**:727.

Sohnle, P. G., and Sussdorf, D. H., 1972, Processing of normal antibody-coated sheep erythrocytes by mouse peritoneal macrophages, *Immunology* **23**:361.

Sorber, W. A., 1978, Effect of complement fixation on the release of lysosomal enzymes from rabbit alveolar macrophages, *Infect. Immun.* **19**:799.

Sorkin, F., Stecher, V. J., and Borel, J. F., 1970, Chemotaxis of leucocytes and inflammation, *Ser. Haemat.* **3**:131.

Spooner, B. S., and Wessells, N. K., 1970, Effects of cytochalasin B upon microfilaments involved in morphogenesis of salivary epithelium, *Proc. Natl. Acad. Sci. USA* **66**:360.

Stahl, P. D., Rodman, J. S., Miller, M. J., and Schlesinger, P. H., 1978, Evidence for receptor-mediated binding of glycoproteins, glycoconjugates, and lysosomal glycosidases by alveolar macrophages, *Proc. Natl. Acad. Sci. USA* **75**:1399.

Steer, C., 1978, Kupffer cells and glycoproteins: Does a recognition phenomenon exist? *Bull. Kupffer Cell Found* **1**(3):26.

Steinman, R., and Cohn, Z. A., 1972, The interaction of particulate horseradish peroxidase (HRP)-anti HRP immune complexes with mouse peritoneal macrophages *in vitro, J. Cell Biol.* **55**:616.

Steinman, R. M., Silver, J. M., and Cohn, Z. A., 1974, Pinocytosis in fibroblasts. Quantitative studies *in vitro, J. Cell Biol.* **63**:949.

Steinman, R. M., Brodie, S. E., and Cohn, Z. A., 1976, Membrane flow during pinocytosis. A stereologic analysis, *J. Cell Biol.* **68**:665.

Stevenson, M. M., and Meltzer, M. S., 1976, Depressed chemotactic responses *in vitro* of peritoneal macrophages from tumor-bearing mice, *J. Natl. Cancer Inst.* **57**:847.

Stossel, T. P., 1973, Quantitative studies of phagocytosis: kinetic effects of cations and heat labile opsonins, *J. Cell Biol.* **58**:346.

Stossel, T. P., 1974a, Phagocytosis. I, *New Engl. J. Med.* **290**:717.

Stossel, T. P., 1974b, Phagocytosis. II, *New Engl. J. Med.* **290**:774.

Stossel, T. P., 1974c, Phagocytosis. III, *New Engl. J. Med.* **290**:833.

Stossel, T. P., 1975, Phagocytosis: Recognition and ingestion, *Semin. Haematol.* **12**:83.

Stossel, T. P., and Hartwig, J. H., 1975, Interaction between actin, myosin, and an actin-binding protein from rabbit alveolar macrophages, *J. Biol. Chem.* **250**:5706.

Stossel, T. P., and Hartwig, J. H., 1976, Interaction of actin, myosin, and a new actin-binding protein of rabbit pulmonary macrophages. II. Role in cytoplasmic movement and phagocytosis, *J. Cell Biol.* **68**:602.

Stossel, T. P., Mason, R. J., Pollard, T. D., and Vaughan, M., 1972, Isolation and properties of phagocytic vesicles. II. Alveolar macrophages, *J. Clin. Invest.* **51**:604.

Sundharadas, G., and Cheung, H. T., 1977, Enhancement of movement of mouse peritoneal macrophages caused by colchicine treatment, *Cell Biol. Int. Rep.* **1**:439.

Tessmer, C. F., and Chang, J. P., 1967, Thorotrast localization by light and electron microscopy, *Ann. N.Y. Acad. Sci.* **145**:545.

Tigelaar, R. E., Vaz, N. M., and Ovary, Z., 1971, Immunoglobulin receptors on mouse mast cells, *J. Immunol.* **106**:661.

Tizard, I. R., Holmes, W. L., and Parappally, N. P., 1974, Phagocytosis of sheep erythrocytes by

macrophages: A study of the attachment phase by scanning electron microscopy, *J. Reticuloendothel. Soc.* **15**:225.
Törö, L, Ruzsa, P., and Röhlich, P., 1962, Ultrastructure of early phagocytic stages in sinus endothelial and Kupffer cells of the liver, *Exp. Cell Res.* **26**:601.
Unkeless, J. C., 1977, The presence of two Fc receptors on mouse macrophages: evidence from a variant cell line and differential trypsin sensitivity, *J. Exp. Med.* **145**:931.
Unkeless, J. C., and Eisen, H. N., 1975, Binding of monomeric immunoglobulins to Fc receptors of mouse macrophages, *J. Exp. Med.* **142**:1520.
Walker, W. S., 1976, Separate Fc-receptors for immunoglobulin IgG 2a and IgG 2b on an established cell line of mouse macrophages, *J. Immunol.* **116**:911.
Walker, W. S., 1977, Mediation of macrophage cytolytic and phagocytic activities by antibodies of different classes and class-specific Fc-receptors, *J. Immunol.* **119**:367.
Ward, P. A., 1967, A plasmin split fragment of C_3 as a new chemotactic factor, *J. Exp. Med.* **126**:189.
Ward, P. A., 1968, Chemotaxis of mononuclear cells, *J. Exp. Med.* **128**:1201.
Watkins, J. F., 1964, Adsorption of sensitized sheep erythrocytes to HeLa cells infected with Herpes simplex virus, *Nature (London)* **202**:1364.
Weber, K., Pollack, R, and Bibring, T., 1975, Antibody against tubulin: the specific visualization of cytoplasmic microtubules in tissue culture cells, *Proc. Natl. Acad. Sci. USA* **72**:459.
Weidner, E., 1975, Interactions between *Encephalitozoon cuniculi* and macrophages, *J. Parasitol.* **47**:1.
Weir, D. M., and Øgmundsdottir, H. M., 1977, Non-specific recognition mechanisms by mononuclear phagocytes, *Clin. Exp. Immunol.* **30**:323.
Weissman, G., Dukor, P., and Zurier, R. B., 1971, Effect of cyclic AMP on release of lysosomal enzymes from phagocytes, *Nature New Biol.* **231**:131.
Werb, Z., and Gordon, S., 1975a, Secretion of a specific collagenase by stimulated macrophages, *J. Exp. Med.* **142**:346.
Werb, Z., and Gordon, S., 1975b, Elastase secretion by stimulated macrophages, *J. Exp. Med.* **142**:361.
Whaley, K., and Singh, H., 1973, *In vitro* studies on the phagocytosis of *Staphylococcus aureus* by peritoneal macrophages of New Zealand mice, *Immunology* **24**:25.
White, R., Lin, H. S., and Kuhn, C., 1977, Elastase secretion by peritoneal exudative and alveolar macrophages, *J. Exp. Med.* **146**:802.
Williams, D. M., and Remington, J. S., 1977, Effect of human monocytes and macrophages on *Trypanosoma cruzi, Immunology* **32**:19.
Williams, D. M., Sawyer, S., and Remington, J. S., 1976, Role of activated macrophages in resistance of mice to infection with *Trypanosoma cruzi, J. Infect. Dis.* **134**:610.
Williams, D. M., Boxer, L. A., Oliver, J. M., and Baehner, R. L., 1977, Cytoskeletal regulation of concanavalin A capping in pulmonary alveolar macrophages, *Nature* **267**:255.
Wilson, L., Bamburg, J. R., Mizel, S. B., Grishema, L. M., and Creswell, K. M., 1974, Interaction of drugs with microtubule proteins, *Fed. Proc.* **33**:158.
Youngdahl-Turner, P., Rosenberg, L. E., and Allen, R. H., 1978, Binding and uptake of transcobalamin II by human fibroblasts, *J. Clin. Invest.* **61**:133.

3

Peritoneal Macrophages

W. T. DAEMS

1. INTRODUCTION

The peritoneal cavity is generally considered to offer a convenient approach for the study of phagocytes under normal and pathological conditions (e.g., Cappell, 1930), because, as clearly outlined by Padawer (1973):

1. It contains a finite cell population characterizable in terms of cell types.
2. The total cell population is relatively manageable.
3. The cells are evenly dispersed in the fluid and aliquots are characteristically representative of the total population.
4. The cell numbers are easily quantified by routine hematologic methods, and shifts in cell populations are readily ascertained.

Another important advantage of the use of peritoneal macrophages is that they can be obtained easily. In contrast, other types of resident macrophages, for instance those in the liver, can only be isolated by enzymatic digestion, the effect of which on the function of the cell is not yet fully established. For these reasons the peritoneal cavity, especially that of the mouse, is frequently used as a source of macrophages for studies on the morphology, differentiation, phagocytosis, cytotoxicity, immune response, and other aspects of these cells.

Lavage of the unstimulated peritoneal cavity provides a cell suspension containing a large number of macrophages, together with lymphocytes, eosinophils, and mast cells (Table 1). Monocytes occur in the unstimulated peritoneal cavity of some species. Mesothelial cells are rarely present. With respect to the proportions of all types of cell in the suspensions, species differences are considerable (Table 1) (Davis and McGowan, 1968).

The number of cells obtained from the peritoneal cavity can be increased by the intraperitoneal administration of compounds which induce an inflammatory process locally. It must be kept in mind, however, that when the normal steady state of an experimental animal is disturbed, a rapid influx of cells and other

W. T. DAEMS • Laboratory for Electron Microscopy, University of Leiden, 2333 AA Leiden, The Netherlands.

TABLE 1. COMPOSITION OF CELL POPULATIONS IN THE UNSTIMULATED PERITONEAL CAVITY (IN PER CENT OF TOTAL CELL YIELD)

Type of cell	Mouse	Guinea pig	Rat[c]
Resident macrophages[a]	54 ± 9	74 ± 8	66 ± 3
Monocytes[a,b]	9 ± 3	<1	6 ± 2
Lymphocytes	25 ± 8	1 ± 2	2 ± 1
Neutrophils	<1	<1	<1
Eosinophils	4 ± 3	19 ± 10	20 ± 3
Mast cells	6 ± 5	<1	5 ± 2

[a] As defined in the text.
[b] Including PO-negative monocytes and macrophages.
[c] Data from R.H.J. Beelen [personal communication (1979)].

components into the peritoneal cavity can take place, especially in experiments lasting longer than a few hours. Furthermore, in the collection of cells it cannot be excluded that under certain conditions part of the population in the peritoneal cavity is left behind, despite the usual massage of the peritoneal wall (Smith and Goldman, 1972a). Indeed, it is considered that free-floating peritoneal macrophages become adherent when foreign particles are introduced into the peritoneal cavity (for a discussion of this phenomenon, see Padawer, 1973). In this respect it is important that the injection of foreign substances into the peritoneal cavity not only increases the number of peritoneal macrophages, but also evokes a peritoneal macrophage population that differs from the normal resident population both morphologically and functionally. In addition, the peritoneal macrophages can be influenced qualitatively and quantitatively by subcutaneous or intravenous injection of agents activating the immune system and that have a remote effect on the peritoneal cells (Mackaness, 1970; Stubbs *et al.*, 1973; Chin and Hudson, 1974; Güttner *et al.*, 1975; Karnovsky *et al.*, 1975; Reikvam *et al.*, 1975; Rhodes *et al.* 1975).

In this chapter a description will be given of the cell populations obtained from the unstimulated and the stimulated peritoneal cavities, restricted mainly to the resident macrophages, the monocytes, and the monocyte-derived macrophages, epithelioid cells, and multinucleated giant cells. Special attention will be given to the morphological and cytochemical aspects of these cells. The heterogeneity of peritoneal macrophage populations will be discussed, as well as the origin of resident peritoneal macrophages.

2. TERMINOLOGY

Many names have been proposed for the group of highly phagocytic cells found in organisms, for example the reticuloendothelial system (Aschoff, 1913, 1924), the reticulohistiocytic system (Thomas, 1949), the mononuclear system (Langevoort *et al.*, 1970), the lymphoreticular system (as discussed by Carr, 1970), the monocyte–macrophage system (Meuret, 1976), and the macrophage system (as discussed by Vernon Roberts, 1972). For a discussion of the various terms, reference can be made to Carr (1976) and Begemann and Kaboth (1976). The nomenclature used in the field of study of monocytes and macrophages to describe the individual cells has also been prolific. Attempts have been made to clarify the terminology, and distinction is made (Karnovsky *et al.*, 1975) between "activated" macrophages, e.g., resulting from the intravenous (i.v.) administration of *Listeria monocytogenes*, and "elicited" macrophages resulting from the intraperitoneal (i.p.) administration of, e.g., sodium caseinate or peptone. The term "stimulated macrophages" is also used, i.e., for macrophages obtained after the i.p. injection of endotoxin or thioglycollate broth. Silverstein (1977) rightly emphasizes that the term *mononuclear phagocytes* encompasses a group of cells with widely differing properties. Until information about the effect of various stimuli on macrophage populations becomes available, it seems wise to drop such adjectives as activated, stimulated, elicited, armed, primed, and angry, and to confine ourselves to purely descriptive terms (for a further discussion, see Silverstein, 1977, and also Allison and Davies, 1974). A similar recommendation was made by a World Health Organization (WHO) scientific group in 1973: "The terms "stimulated" or "activated" must not be used loosely but should always be defined in terms of the agent utilized to produce the effect and the method by which the effect is assayed" (WHO Scientific Group, 1973; see also Allison and Davies, 1975). Sheagren and Hahn (1974) pointed out that "... the parameters by which activation was evaluated vary so much as to make it mandatory that each worker precisely defines his parameters" (see also Chapter 5 of this volume). In this context the following quotation is still up-to-date: "We prefer the original term macrophage ..., since it emphasizes the mode of behavior of these cells, in regard to which there is practically universal agreement, instead of their origin, in regard to which there is still no unanimity of opinion" (Clark and Clark, 1930).

In the present paper use will be made of the following terms: monocyte, macrophage, epithelioid cell, and multinucleated giant cell. A distinction will be made between tissue (= resident) macrophages and exudate (= monocyte-derived) macrophages (Daems *et al.*, 1976) in the relevant sections, and if required, the term reticuloendothelial system will be used.

3. UNSTIMULATED PERITONEAL CAVITY

3.1. COMPOSITION OF THE CELL POPULATION

The cell suspension obtained from an unstimulated peritoneal cavity contains chiefly three types of cell: resident macrophages, eosinophilic granulocytes, and lymphoid cells (Table 1). Monocytes occur occasionally, depending on still unknown factors, in the unstimulated peritoneal cavity of some animal species. In contrast to the guinea pig, where monocytes are generally absent (Daems and Brederoo, 1973), the peritoneal cavity of mice and rats contains a relatively high number of monocytes (up to 10% in the case of the mouse; Daems and Koerten, 1978). Neutrophilic granulocytes are rarely observed in unstimulated peritoneal cavities, but occasionally a variable but always low number of neutrophils can be found, often in association with an increase in the number of monocytes, which might reflect an unhealthy state of the animal (Forbes, 1966). Eosinophilic granulocytes occur regularly in the unstimulated peritoneal cavity of the rat and the guinea pig. In the peritoneal cavity of mice and rats, mast cells are relatively numerous, whereas in guinea pigs they are rarely or never observed (Padawer and Gordon, 1956; Davis and McGowan, 1968; Bültman *et al.*, 1971; Daems and Brederoo, 1973). Observation of mesothelial cells is exceptional in all cases.

The number of cells in the peritoneal cavity and their relative proportions vary according to species (Davis and McGowan, 1968), the age of the animal (Padawer, 1973; Yang and Skinsness, 1973; Lin *et al.*, 1978), the personal vagaries of the observer (Carr, 1973), and the sex, at least of mice (Forbes, 1966; Davis and McGowan, 1968). In addition, a strong influence is exerted by the method of harvesting (Carr, 1973).

In general, gentle abdominal massage is used before aspirating the washing fluid. It can be assumed that most of the cells thus obtained are free floating or only slightly adherent to the walls of the peritoneal cavity (Mims, 1963; Chin and Hudson, 1974). Since the number of cells that can be isolated from the unstimulated peritoneal cavity is generally low, it is often necessary to pool the cells of several animals.

The values given for the total number of cells in the unstimulated peritoneal cavity of the mouse by various authors (Table 2) show considerable divergence, the mean being 3.55 with a standard deviation of 1.94. Although the criteria used to identify a cell as a macrophage vary widely (McLaughlin *et al.*, 1972; Olivotto and Bomford, 1974; Boetcher and Meltzer, 1975; Stewart *et al.*, 1975; Nelson, 1976; Morahan and Kaplan, 1978), the absolute number of macrophages found in the unstimulated peritoneal cavities as given by the various authors lies within remarkably narrow limits, the mean being 1.52 ± 0.49.

It has been postulated that resident peritoneal macrophages form a self-renewing population (see Section 6). The small percentage of cells with the morphological and cytochemical characteristics of monocytes in the unstimulated peritoneal cavity should be considered (Daems and Brederoo, 1973; Volkman, 1976) to be monocytes that are passing through the peritoneal cavity (Figure 43a), and should not be looked upon as precursors of the resident peritoneal macrophages. It became clear from studies on monocyte kinetics in the rat (Whitelaw, 1966; Whitelaw and Batho, 1972), and can be deduced from similar studies in mice (van Furth and Cohn, 1968), that with respect to monocyte distribution the circulating blood and the extravascular pool are in equilibrium, random interchange taking place between them. This view is supported by the finding that in mice a depletion of blood monocytes caused by hydrocortisone acetate (Thompson and van Furth, 1973) also led to the complete disappearance of monocytes from the unstimulated peritoneal cavity, whereas the number of resident macrophages remained almost unchanged [Daems and Koerten, unpublished observations (1977)]. In this respect the findings of Harper and Wolf (1972) are relevant. These authors found that after sublethal irradiation of mice, which initially

TABLE 2. PUBLISHED DATA ON THE NUMBER OF MACROPHAGES IN THE UNSTIMULATED PERITONEAL CAVITY OF MICE

References	Total number of peritoneal cells	Total number of macrophages
Albrecht *et al.* (1978)	5	1.0–1.5
Argyris (1967)	2.0	1.7
Bar-Eli and Gallily (1975)	5.7	2.0
Boetcher and Meltzer (1975)	4–5	1.5–1.6
Burgaleta and Golde (1977)	2.4	1.8
Burgaleta *et al.* (1978)	2.4	1.8
Casciato *et al.* (1976)	1.4	1.3
Cowing *et al.* (1978)	4	2
Daems and Koerten (1978)	2.6	1.8
Dimitriu *et al.* (1975)	2–4	0.9–1.2
Fishel *et al.* (1976)	1.8	1.2
Gallily and Eliahu (1974)	2.4	0.8–1.0
Gordon *et al.* (1974)	5–8	2.0–2.4
Harper and Wolf (1972)	5	1.7
Kondo and Kanai (1977)	1.5	1.4
Kornfeld and Greenman (1966)	7.8	2.4
Lin and Stewart (1974)	2–3	1–1.5
McLaughlin *et al.* (1972)	8.2	2.0
Meltzer *et al.* (1975)	3–5	1.2–1.5
Morahan *et al.* (1977)	3–5	1.2–2
Nelson and Nelson (1978)	2.9	1.2
Ratzan *et al.* (1972)	3	1.9–2.1
Raz *et al.* (1977)	1.6	0.8
Schnyder and Baggiolini (1978)		2.3
Stewart *et al.* (1975)	1.6	0.24
Unkeless *et al.* (1974)	5–8	2.0–2.4
Yang and Skinsness (1973)	6.9	2.0

led to a decrease in the total number of peritoneal macrophages, a numerically stable macrophage population remained. The absolute number of macrophages comprising this population was roughly the same as that of the population of resident peritoneal macrophages. Similar findings were reported by Kornfeld and Greenman (1966). This phenomenon can be considered to reflect the disappearance of monocytes from the circulation and hence from the peritoneal cavity as the result of the destruction of their precursors in the bone marrow. Indeed, when X-irradiation preceded stimulation of the peritoneal cavity with thioglycollate (Harper and Wolf, 1972), the normally occurring migration of great numbers of monocytes into the peritoneal cavity did not occur. It therefore seems quite possible that the effect of irradiation on the cell population of the unstimulated peritoneal cavity can be explained by the inhibition of monocyte renewal and thus the disappearance of these cells from the peritoneal cavity, leaving the resident peritoneal macrophages unaffected, at least quantitatively. In this respect it is of interest that an increase in the number of blood monocytes, induced for example by i.v. administration of *Listeria monocytogenes*, leads to an increase in the number of monocytes in the unstimulated peritoneal cavity, whereas the number of resident macrophages is unchanged [Daems and Koerten, unpublished observations (1977)]. Furthermore, systemic infection with parasites or bacteria often results in a peritoneal macrophage population which differs morphologically and functionally from the normal resident population (see Section 5).

These data and others obtained in studies on the stimulated mouse peritoneal cavity (Daems and Koerten, 1978), indicate that the resident peritoneal macrophages form, in a quantitative sense, a relatively stable population (see Section 4 and Figure 43).

3.2. MORPHOLOGY AND CYTOCHEMISTRY OF THE RESIDENT PERITONEAL MACROPHAGES

The macrophages in the unstimulated peritoneal cavity (resident peritoneal macrophages) are relatively large cells whose morphology is to a certain extent species dependent (Daems *et al.*, 1976). In addition, the appearance of the cells is strongly influenced by the preparative procedure (compare Figures 15 and 16).

In contrast to the macrophages in the stimulated peritoneal cavity, which show wide variations in fine structure (see Section 4.2), the resident macrophages in the unstimulated peritoneal cavity form a homogeneous population morphologically, especially with respect to size (Williams and Mayhew, 1973), but probably also functionally. They have, if fixed immediately after harvesting, a round shape (Figure 1). In the scanning electron microscope they also appear—before or shortly after attachment—as round cells (Figure 2), in contrast to exudate macrophages, which are reported to spread out rapidly and become flat (Polliack and Gordon, 1975; Robertson *et al.*, 1977; Burgaleta *et al.*, 1978; see also Section 5). The cytoplasm of resident peritoneal macrophages contains the usual cell organelles, as will be described below in detail. These structures show a certain polarity in that the Golgi apparatuses—up to six—and the lysosomes are located at one side of the nucleus and at the other, surrounding the nucleus, the elements of the rough endoplasmic reticulum (RER) (Figure 1).

3.2.1. The Cell Surface

The surface of resident peritoneal macrophages is seen in the scanning electron microscope (Figure 2) to be covered by numerous pleomorphic ridges, veils (lamellipodia), and ruffles, as well as by fingerlike extrusions varying in length (Güttner *et al.*, 1975; Polliack and Gordon, 1975; Orenstein and Shelton, 1976a,b; Robertson *et al.*, 1977; Burgaleta *et al.*, 1978). The latter structures (Figure 6) are reported to have a constant thickness of about 100 nm (Takayama *et al.*, 1975) and in ultrathin sections have the appearance of microvilli. The surface structure of macrophages is dependent on the nature of the substratum to which the cells are attached (Allen and Dexter, 1976; Lajtha, 1976; Orenstein and Shel-

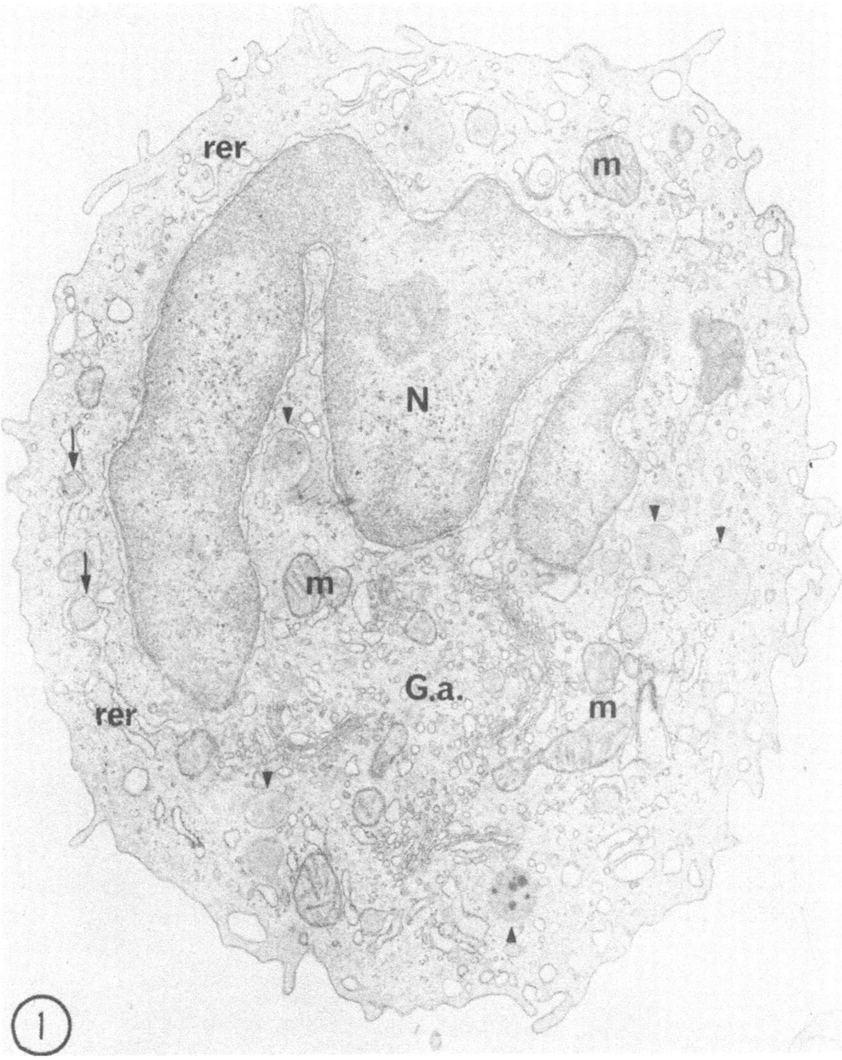

FIGURE 1. Macrophages from the unstimulated peritoneal cavity of a guinea pig. The nucleus is multilobular, the cytoplasm contains the usual cell organelles, i.e., mitochondria (m), rough ER (rer), microbodies (arrows), lysosomes (arrowheads), and an extensive Golgi apparatus (G.a.). The cell surface is covered with pleomorphic microvillous extensions. Immediately under the cell surface, numerous lacunae are present, most of which are probably, and some of which are distinctly, continuous with the extracellular space. ×12,000.

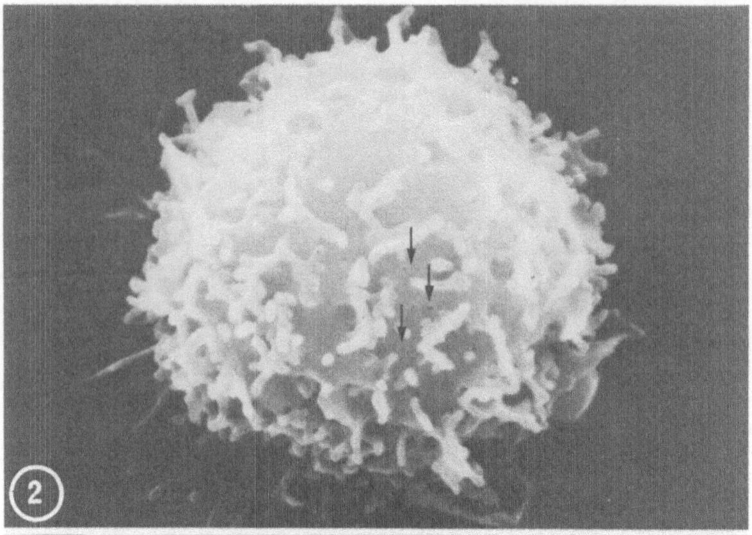

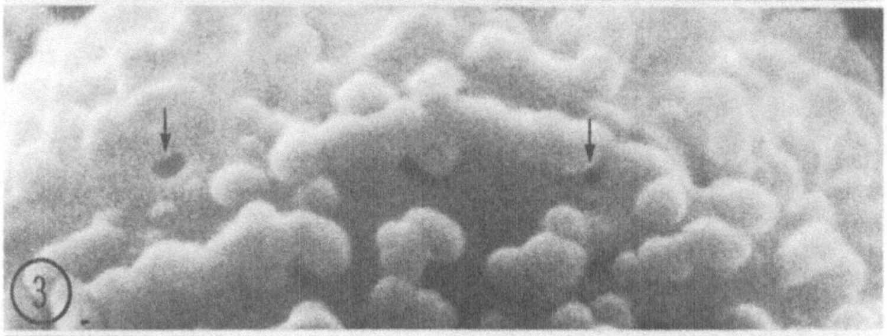

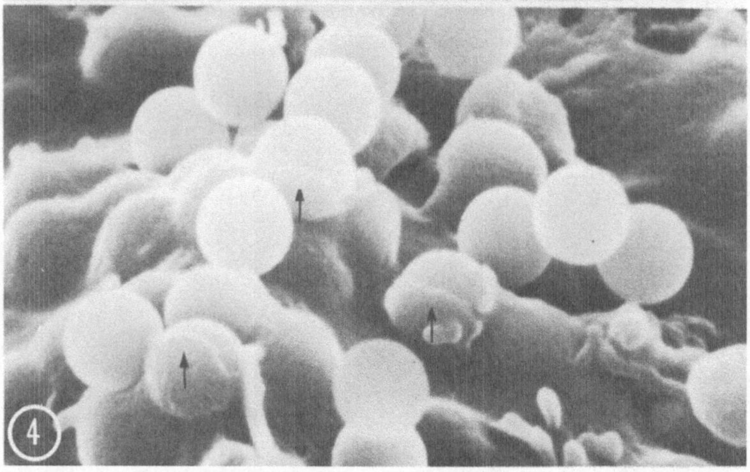

FIGURE 2. Scanning electron micrograph of a macrophage from an unstimulated mouse peritoneal cavity. The cell surface is covered with ridges and microvilli. A number of indentations on the cell surface represent endocytotic pits (arrows). ×10,000. Reduced 10% for reproduction.

FIGURE 3. Scanning electron micrograph of a macrophage from an unstimulated mouse peritoneal cavity showing, at high magnification, a number of endocytic invaginations of the cell surface (arrows). ×45,000. Reduced 10% for reproduction.

FIGURE 4. Scanning electron micrograph of a macrophage from an unstimulated mouse peritoneal cavity after 24 hr in culture. One hour before fixation, latex beads (diameter: 0.8 μm) were added to the culture medium. Note how the cell surface has been creeping over the latex particles (arrows) in the first stage of endocytosis. ×20,000. Reduced 10% for reproduction.

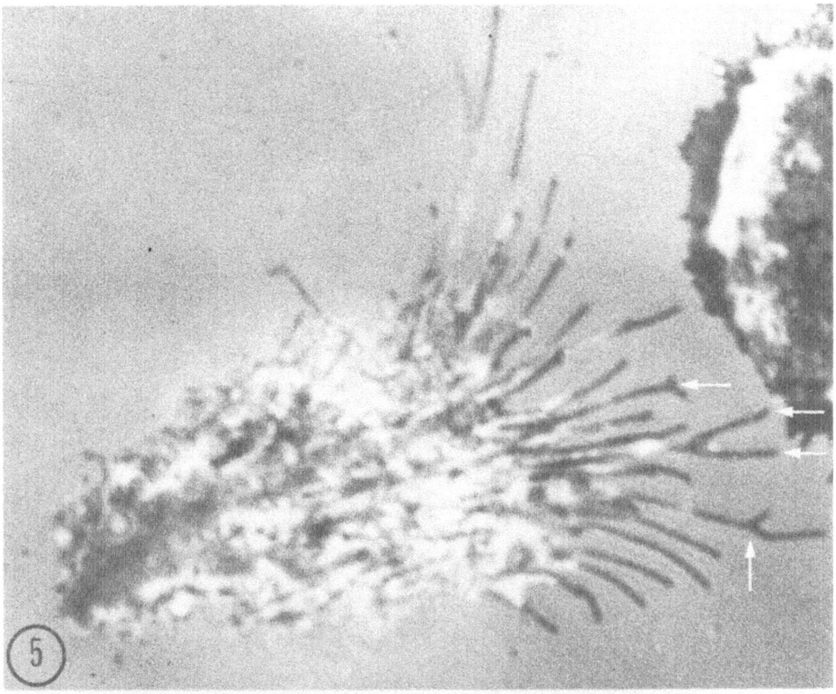

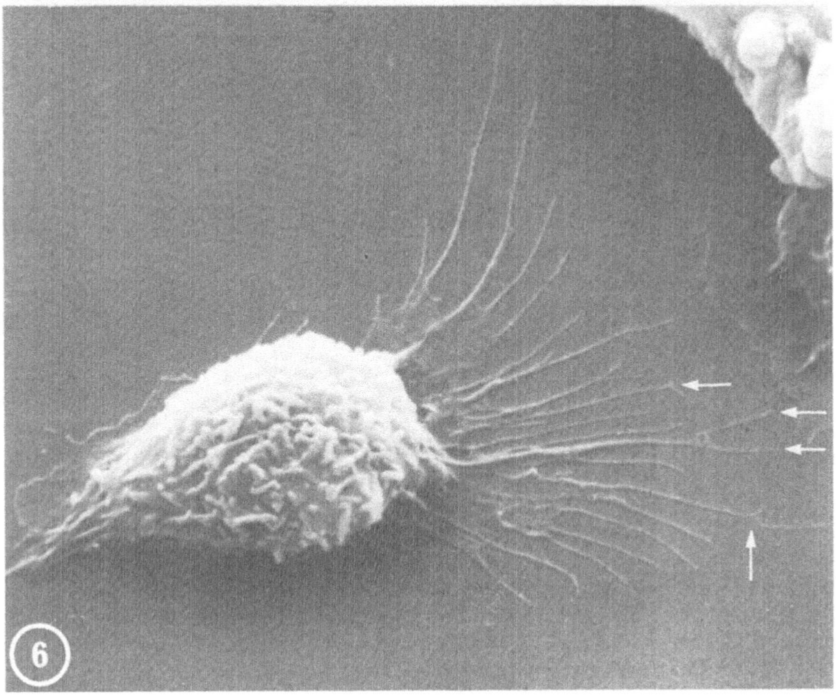

FIGURE 5. Reflection contrast micrograph of a macrophage from an unstimulated mouse peritoneal cavity after 24 hr in culture. Compare the shape of the filopodia in this mode of microscopy with that shown by the scanning electron microscope in Figure 6. ×5000.
FIGURE 6. Scanning electron micrograph of the same cell as in Figure 5. Note that fixation and dehydration in the critical-point procedure did not change the fine structure of the cell surface and particularly not that of the filopodia. ×5000.

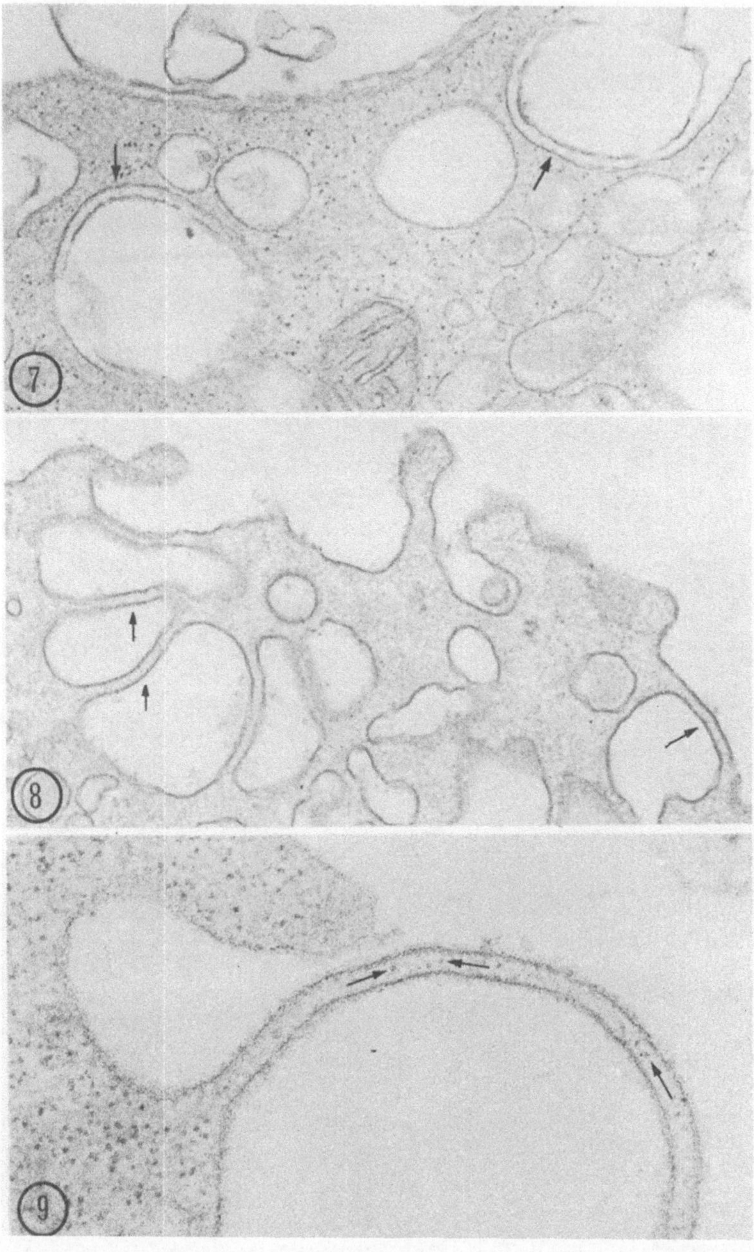

FIGURE 7. Part of a resident peritoneal macrophage from a mouse. Due to membrane-like sheets attached to the cell surface, an electron-lucid area is visible on the plasma membrane as well as within phagocytic vacuoles; this area probably represents the thin variety of cell coat (see also Chapter 4). ×77,000. Reduced 10% for reproduction.

FIGURE 8. Part of a macrophage from an unstimulated mouse peritoneal cavity. The trilaminar plasma membrane shows "loop"-like configurations (arrows) in which the plasma membrane is folded back onto itself. These loops have a constant width of about 40 nm. ×47,000. Reduced 10% for reproduction.

FIGURE 9. A loop on the cell surface of a mouse peritoneal macrophage isolated 16 days after i.p. administration of iron-dextran, shown at higher magnification than in Figure 8. Note that ferritin particles, which are abundantly present in the cell's cytoplasm, are also present within the loop (arrows) and are situated equidistant from the two plasma membranes. ×118,000. Reduced 10% for reproduction.

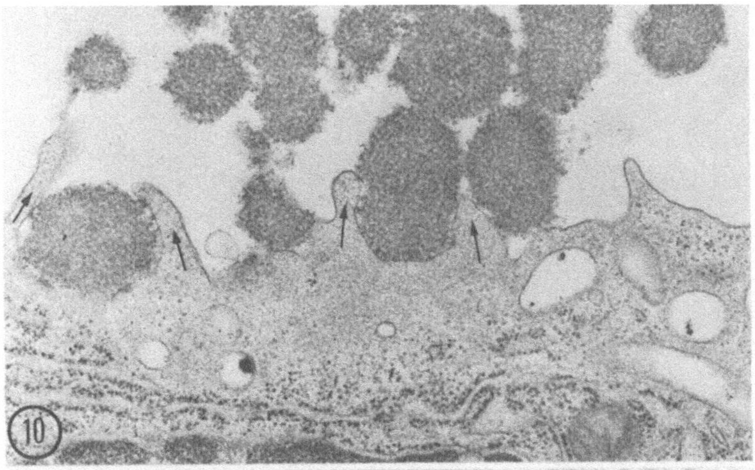

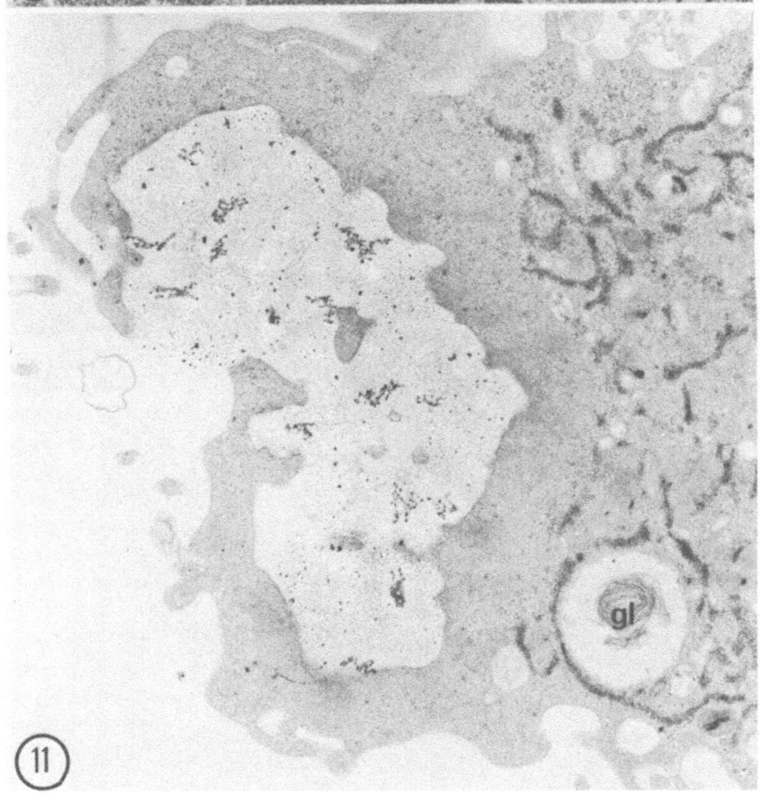

FIGURE 10. Part of an exudate macrophage from a mouse peritoneal cavity, 2 hr after the i.p. administration of iron-dextran. The cell was in the process of endocytosis of mast cell granules when fixed. Note the extension of the cell surface (arrows) surrounding the material to be ingested. The zone of filamentous material underlying the plasma membrane is considerably larger than that of a resting macrophage (compare with Figures 15 and 16). ×25,000. Reduced 10% for reproduction.

FIGURE 11. Part of a resident peritoneal macrophage from a mouse, 28 days after the i.p. administration of glucan (gl). When fixed, the cell was in the process of endocytizing a relatively large amount of filamentous material. Note that the extensions of the cell surface and the greatly enlarged zone of the cytoplasm immediately adjacent to the material to be endocytized both contain filamentous material and are devoid of cell organelles. ×15,000. Reduced 10% for reproduction.

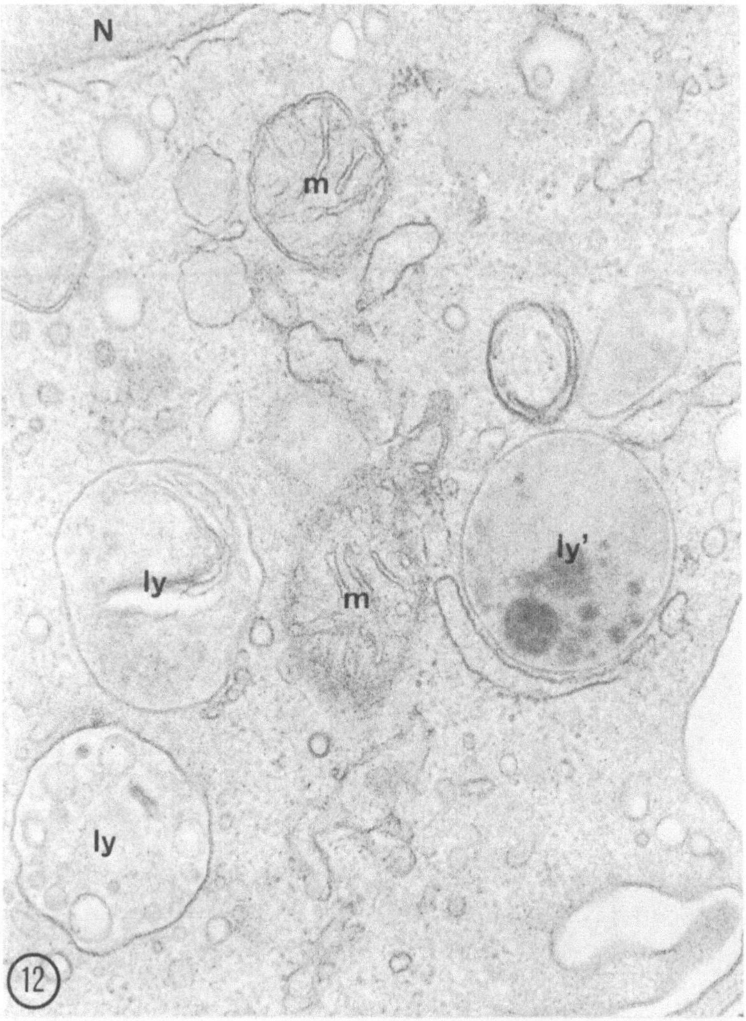

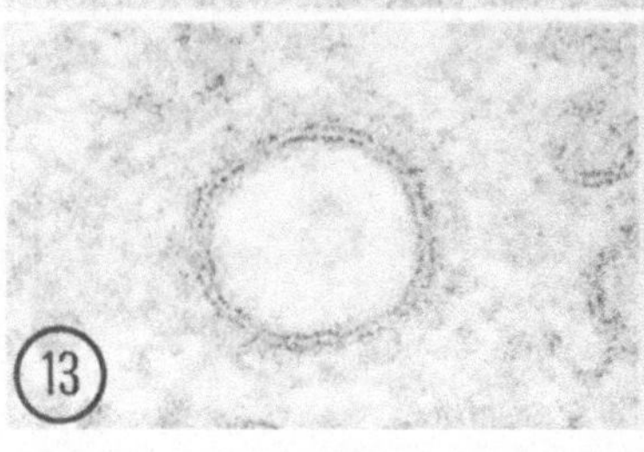

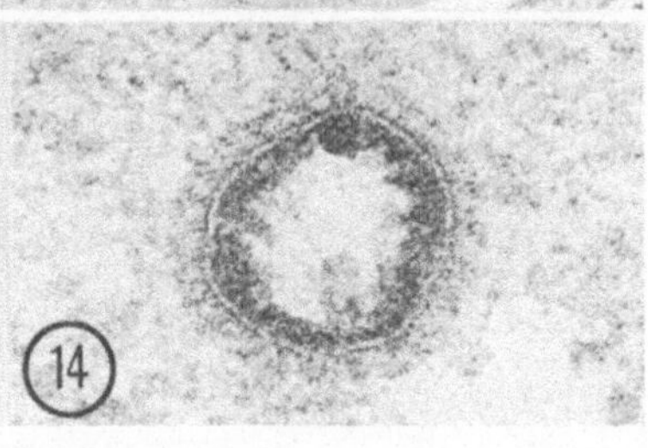

FIGURE 12. Part of a macrophage from the unstimulated guinea pig peritoneal cavity, postfixed in uranyl acetate (see also Figure 14). In addition to part of the nucleus and some of the mitochondria, the variety in fine structure of the lysosomes can be seen. Between the limiting membrane of the lysosomes (which has the same thickness as the plasma membrane) and the matrix, a clear halo is present. Note the close relationship between one of the lysosomes (ly) and a cisterna of the endoplasmic reticulum. ×55,000. Reduced 10% for reproduction.

FIGURE 13. Coated micropinocytic vesicle at high magnification, showing the trilaminar structures of the plasma membrane. Uranyl postfixation. ×150,000. Reduced 10% for reproduction.

FIGURE 14. Coated micropinocytotic vesicle after staining with ruthenium red. The presence of ruthenium-red-stainable material indicates that the vesicle is, outside of the plane of the section, connected with the cell surface. ×150,000. Reduced 10% for reproduction.

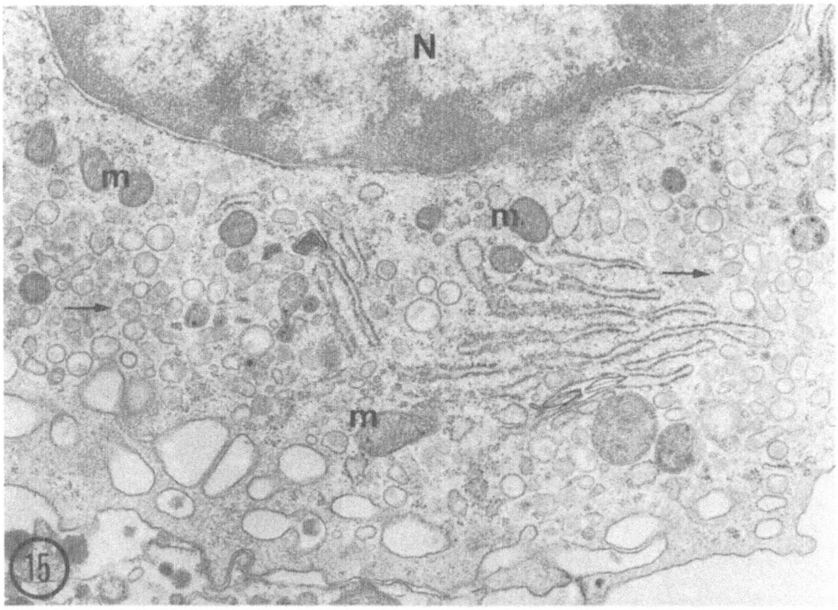

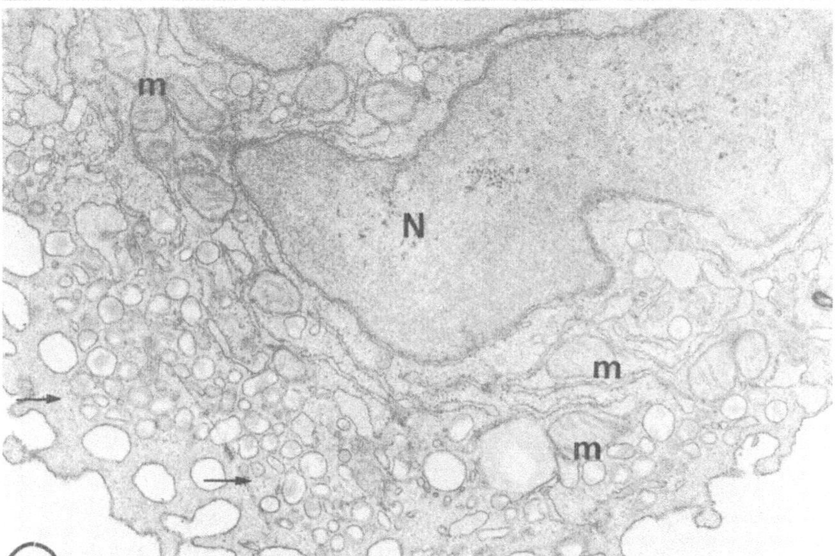

FIGURE 15. Part of a macrophage from an unstimulated mouse peritoneal cavity, fixed in cacodylate-buffered glutaraldehyde followed by fixation in 1% phosphate-buffered osmium tetroxide. Numerous vesicles are present in the cytoplasm (arrows), many of them containing amorphous material of moderate density. The nucleus (N) shows densely stained chromatin. The mitochondria (m) also have a marked density. Compare with Figure 16. ×16,500.

FIGURE 16. Part of a macrophage obtained from the same unstimulated mouse peritoneal cavity as the cell shown in Figure 15, but fixed in a mixture of glutaraldehyde and osmium tetroxide (Hirsch and Fedorko, 1968) and postfixed in uranyl acetate. Note the differences in the appearance of the two cells, which mainly concern the lower density of the various cell organelles in the uranyl-acetate-postfixed cell as well as the high electron density of the membranes in the latter (see also Figure 15; for a discussion of this phenomenon, see Daems and Brederoo, 1973). ×16,500.

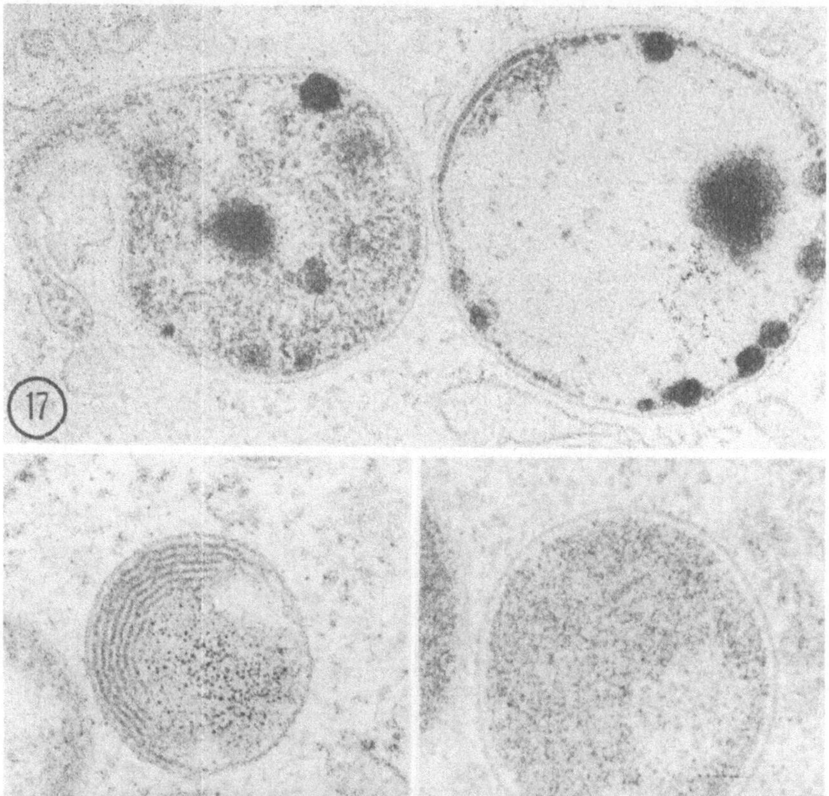

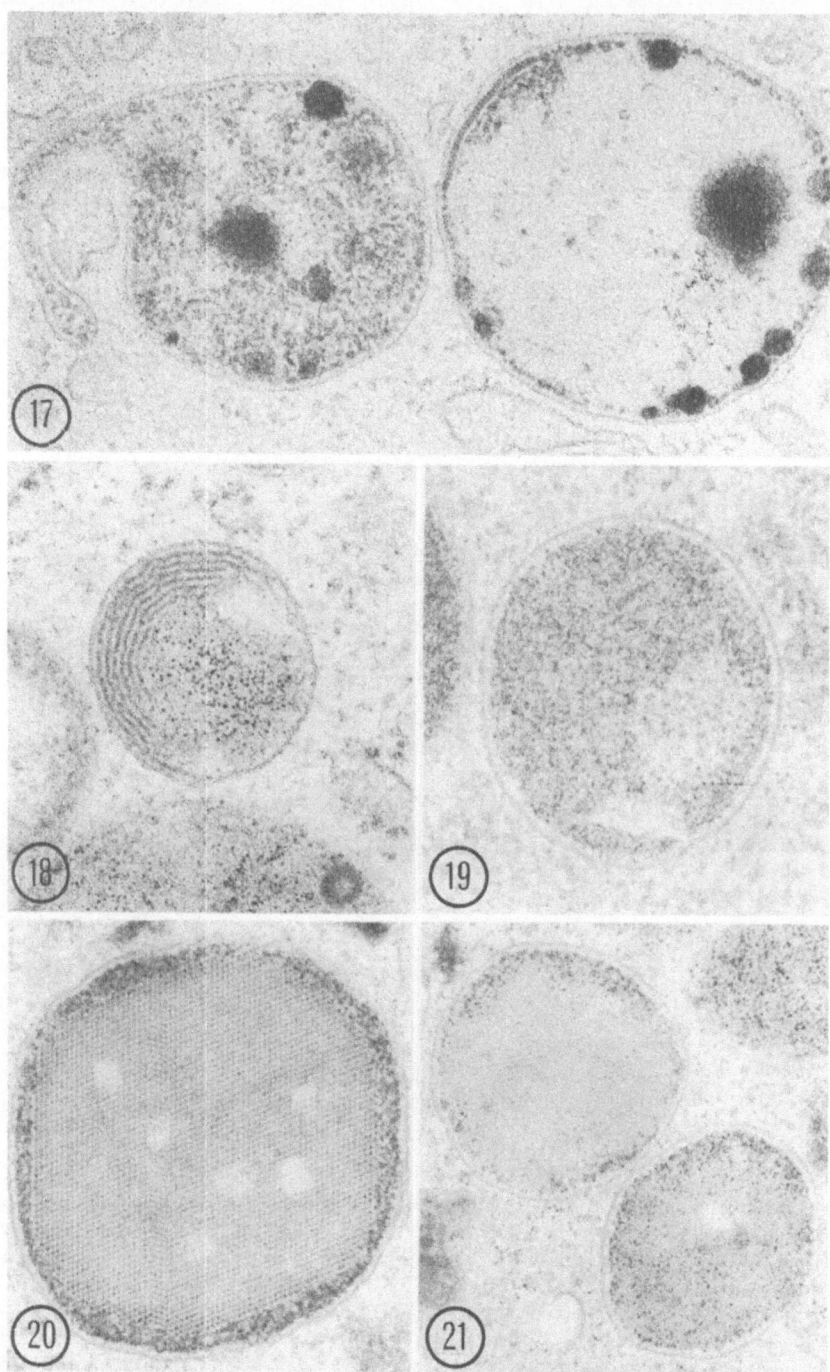

FIGURES 17–21. Lysosomes from mouse resident peritoneal macrophages, showing the variety of fine structure. Those shown in Figures 18–21 are from macrophages isolated at different times after the i.p. administration of iron-dextran: Figure 18 after 30 min, Figure 19 after 5 days, Figure 21 after 2

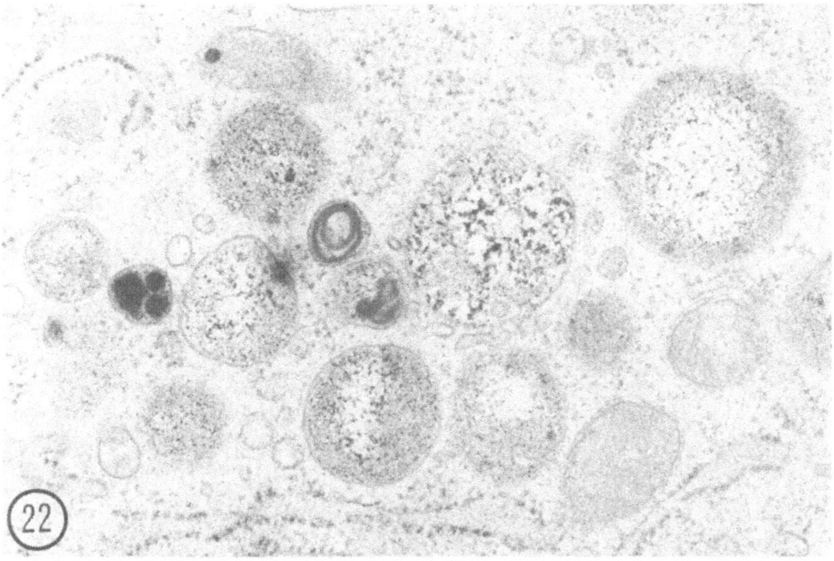

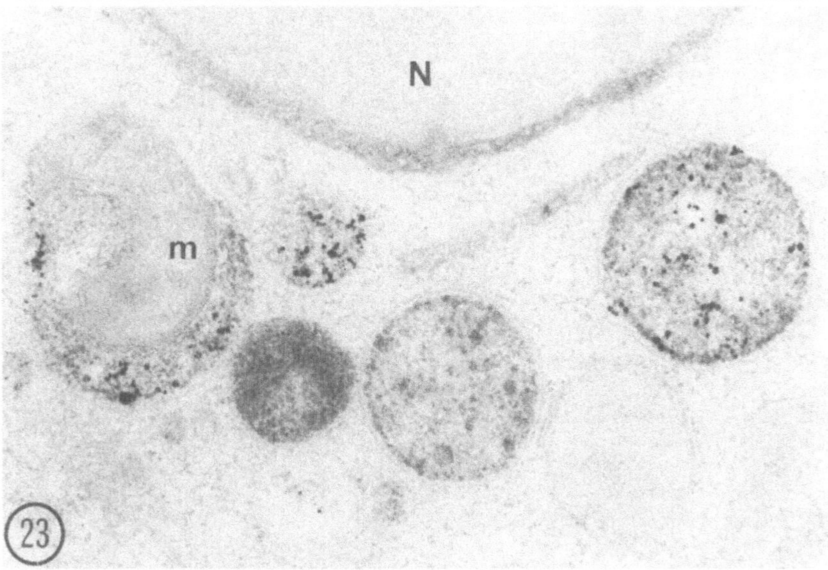

FIGURE 22. Part of a mouse peritoneal macrophage 30 min after the i.p. administration of iron-dextran. Note the presence of iron-dextran in the preexistent lysosomes. ×25,000.
FIGURE 23. Part of a mouse peritoneal resident macrophage fixed 3 hr after the i.p. administration of colloidal gold and 127 days after the i.p. administration of iron-dextran. Some of the lysosomes, including the autophagic vacuole with a mitochondrion (m), contain gold particles. ×25,000.

---

months. Note the crystalline structure of the apoferritin in the lysosome shown in Figure 20; belonging to a macrophage fixed 127 days after the administration of iron-dextran. ×73,000.

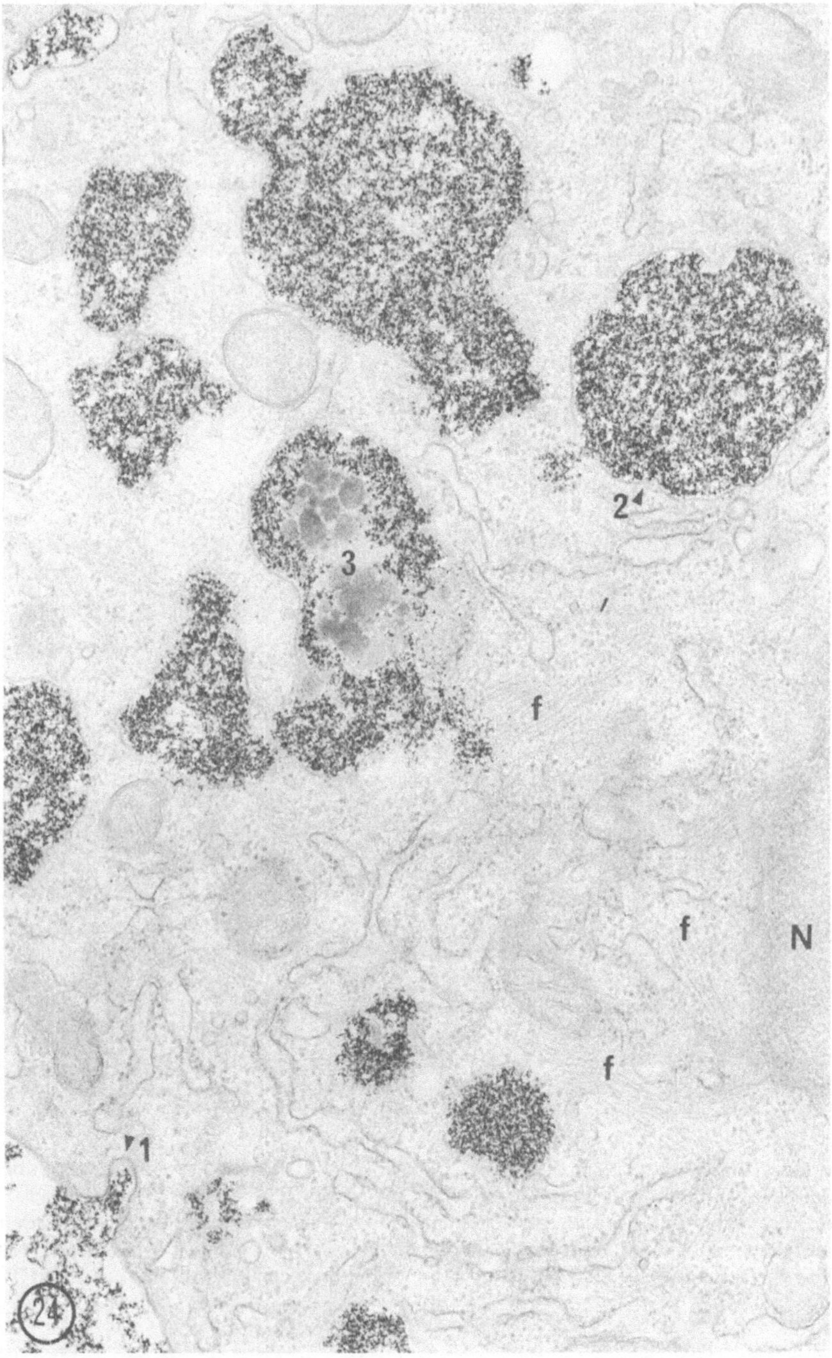

FIGURE 24. Part of the cytoplasm of a guinea pig peritoneal macrophage fixed 3 hr after the i.p. administration of Thorotrast. The various stages of endocytosis and subsequent processes can be seen, i.e., the endocytic invagination of the cell surface (1), the presence of Thorotrast in large

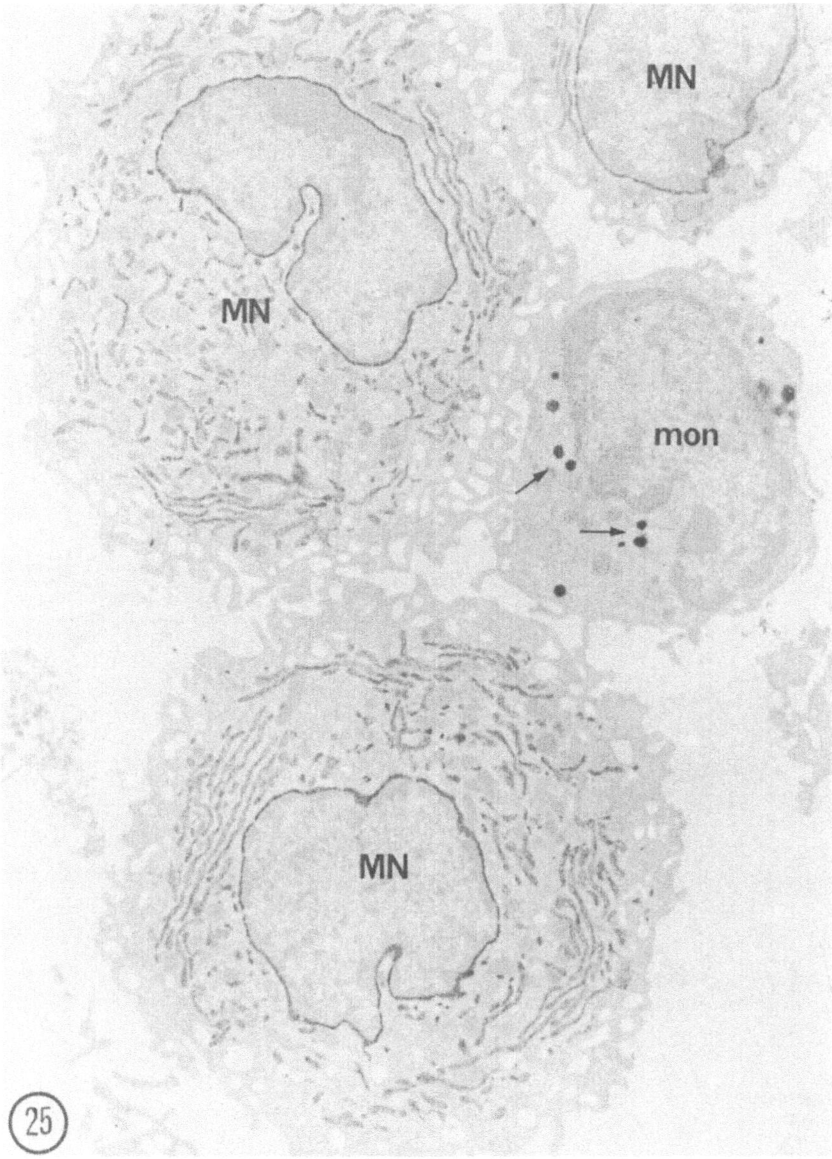

FIGURE 25. Three resident peritoneal macrophages (MN) and one monocyte (mon) from an unstimulated mouse peritoneal cavity after incubation for the demonstration of peroxidase activity. Reaction product is present in the nuclear envelope and the RER of the resident macrophages and in cytoplasmic granules of the monocyte (arrows). ×10,000.

---

membrane-limited vacuoles (2), and the presence of Thorotrast in a lysosome (3) comparable to that shown in Figure 17. Note the bundles of 10-nm-thick filaments, which indicate that this cell is a resident macrophage. ×31,000.

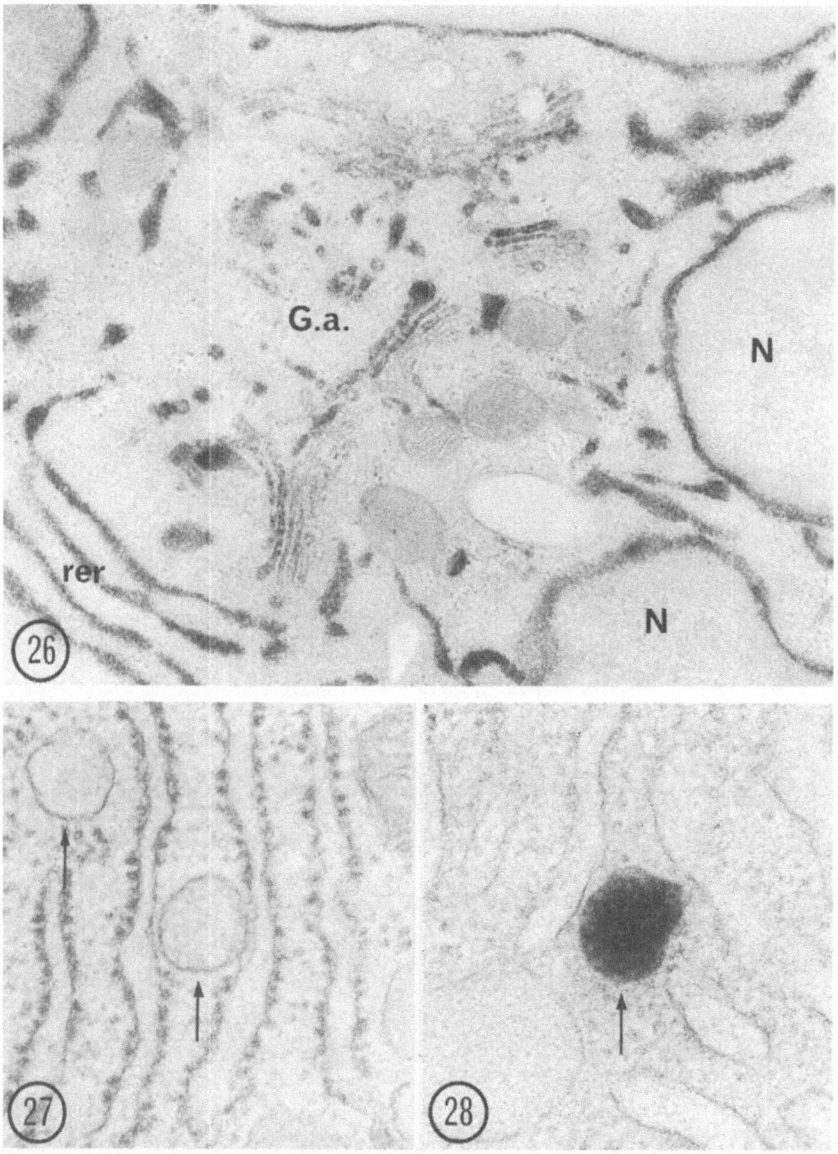

FIGURE 26. Part of a resident peritoneal macrophage of a guinea pig after incubation for the demonstration of peroxidase activity. In addition to the nuclear envelope and the RER, reaction product is present in elements of the Golgi apparatus (G.a.). ×29,000.

FIGURE 27. Part of a guinea pig resident peritoneal macrophage showing two microperoxisomes (arrows) in close proximity to the RER. ×67,000.

FIGURE 28. Part of a guinea pig resident peritoneal macrophage incubated for the demonstration of catalase activity. Reaction product is present in a microperoxisome (arrow). ×65,000.

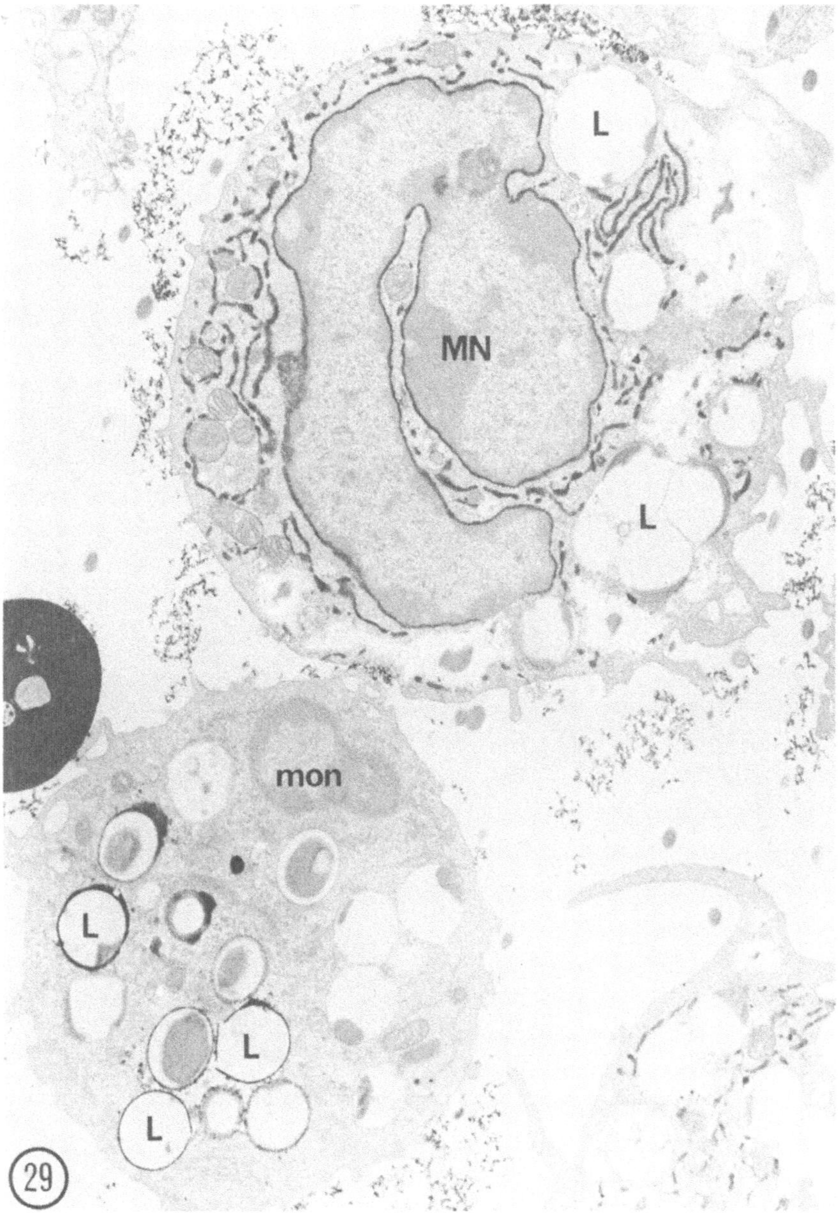

FIGURE 29. Resident peritoneal macrophage (MN) and monocyte-derived exudate macrophage (mon) obtained from a guinea pig peritoneal cavity 16 hr after the i.p. administration of latex beads and incubation for the demonstration of peroxidase activity. Both types of cell have endocytized latex particles (L). Note the differences in the distribution of peroxidase activity: in the resident macrophage reaction product is present in the nuclear envelope and in the RER, whereas the vacuoles containing latex particles are devoid of PO activity. In the exudate macrophage reaction product is only present in the vacuoles containing latex particles; the nuclear envelope and the RER are PO negative. ×9,000.

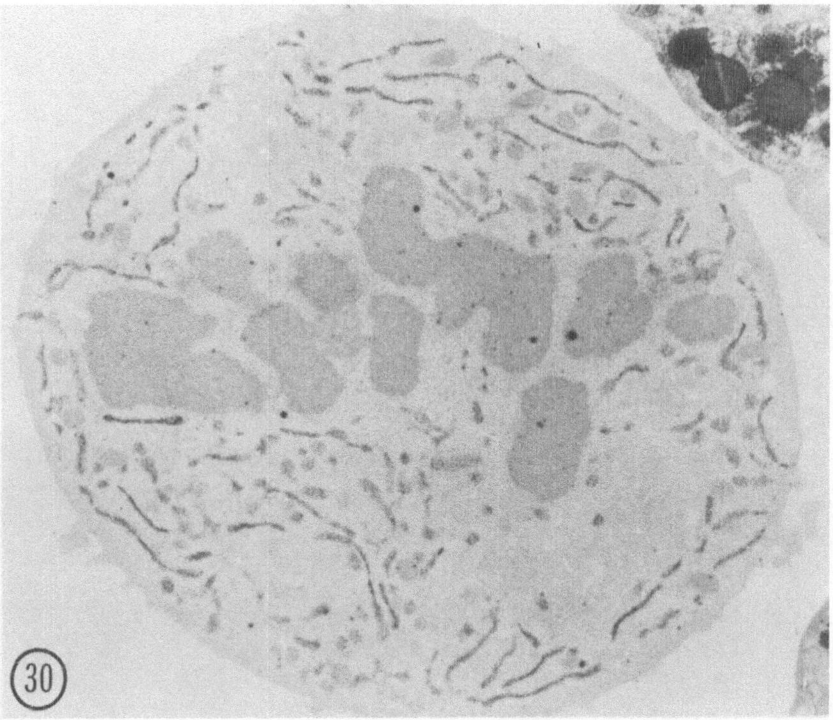
(30)

FIGURE 30. Resident peritoneal macrophages in mitosis isolated 28 days after i.p. administration of glucan. ×16,000.

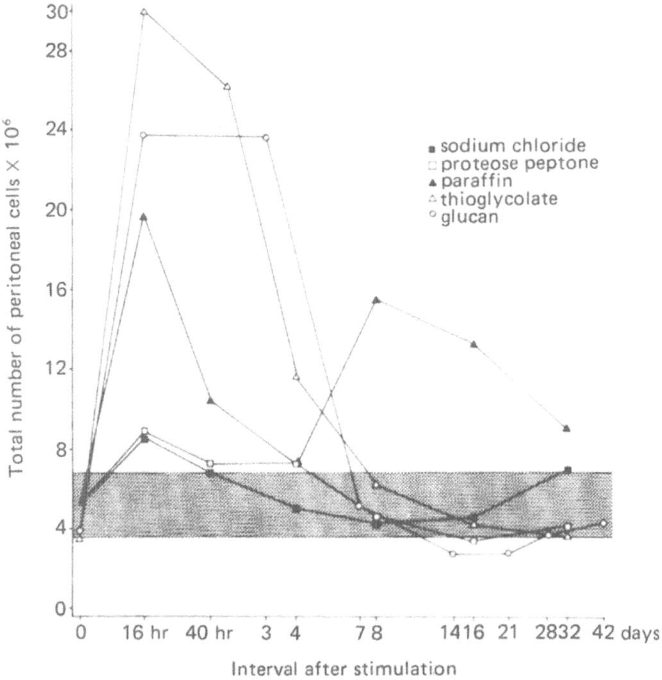

FIGURE 31. Total exudate cell yield from mouse peritoneal cavity as a function of time after the i.p. injection of various stimulants. The gray areas indicate control values (unstimulated peritoneal cavity) with standard deviation.

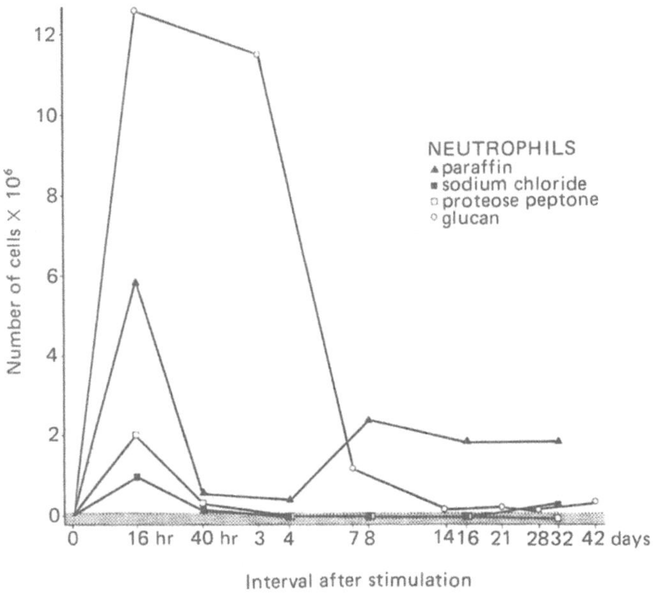

FIGURE 32. Yield of neutrophils from mouse peritoneal cavity as a function of time after the i.p. injection of various stimulants. The gray areas indicate the control values with standard deviation.

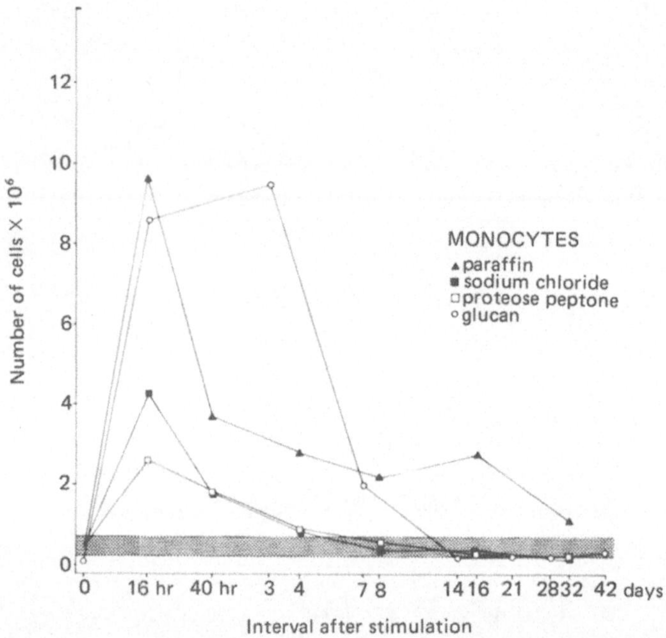

FIGURE 33. Yield of monocytes from mouse peritoneal cavity as a function of time after the i.p. injection of various stimulants. The gray areas indicate the control values with standard deviation.

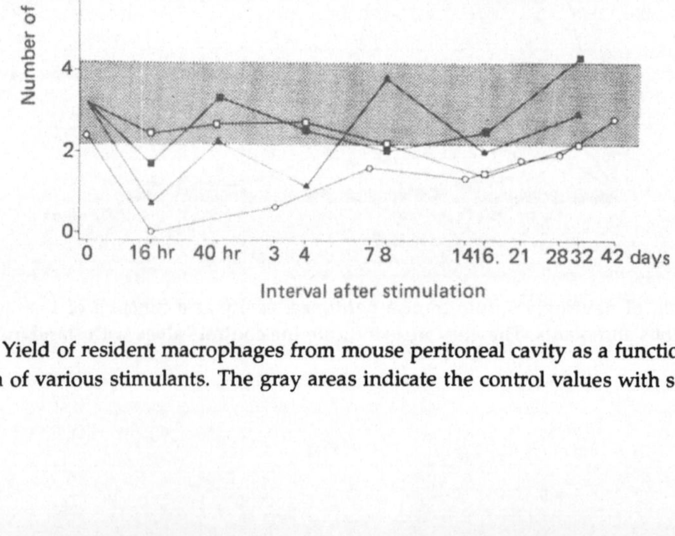

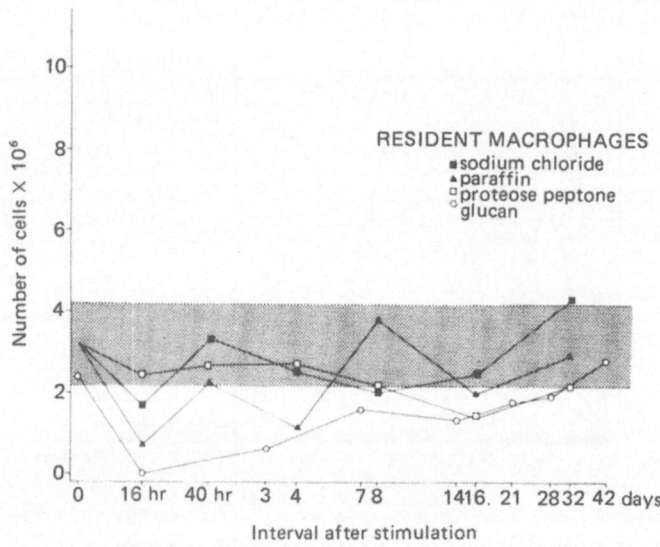

FIGURE 34. Yield of resident macrophages from mouse peritoneal cavity as a function of time after i.p. injection of various stimulants. The gray areas indicate the control values with standard deviation.

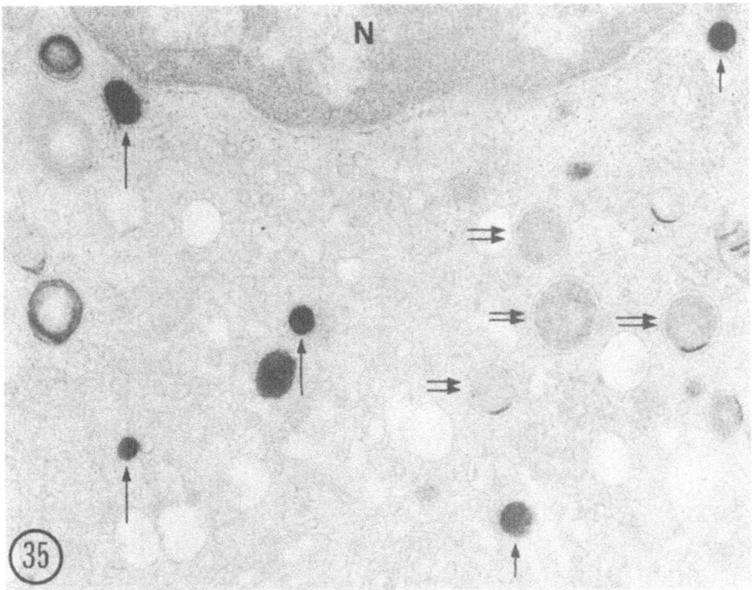

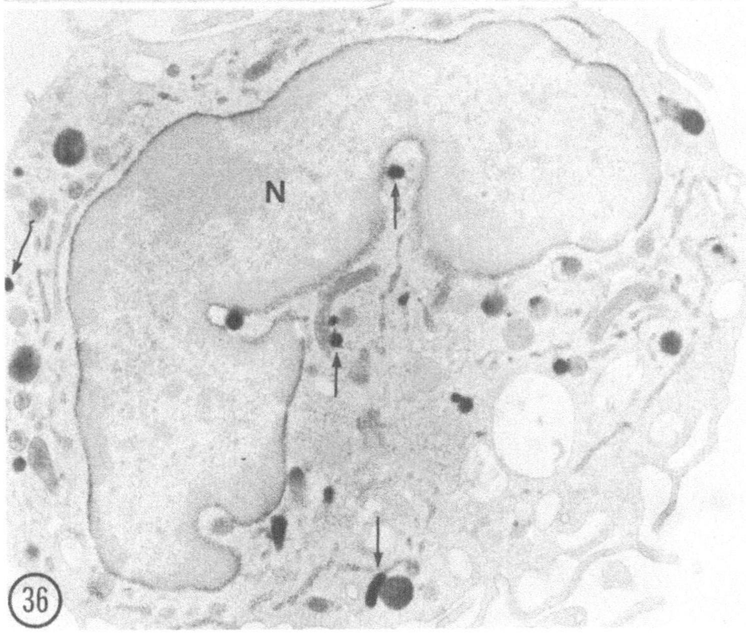

FIGURE 35. Part of a monocyte-derived exudate macrophage obtained from a mouse peritoneal cavity 32 days after the i.p. adminstration of iron-dextran and 24 hr after the i.p. administration of saline, incubated for the demonstration of peroxidase activity. Reaction product is present in primary granules (arrows). The macrophage granules (double arrows), which are characterized by a clear halo underneath the limiting membrane and/or electron dense membrane-like material, are devoid of reaction product. The cytoplasm and the lysosomes are devoid of ferritin particles. ×30,000. Reduced 10% for reproduction.

FIGURE 36. Rat peritoneal exudate macrophage adhering to Melinex implanted intraperitoneally 3 days previously; incubated for the demonstration of peroxidase activity. Reaction product is present in primary granules (arrows) and occurs as a transient phenomenon in the RER and the nuclear envelope. ×35,000. Reduced 10% for reproduction.

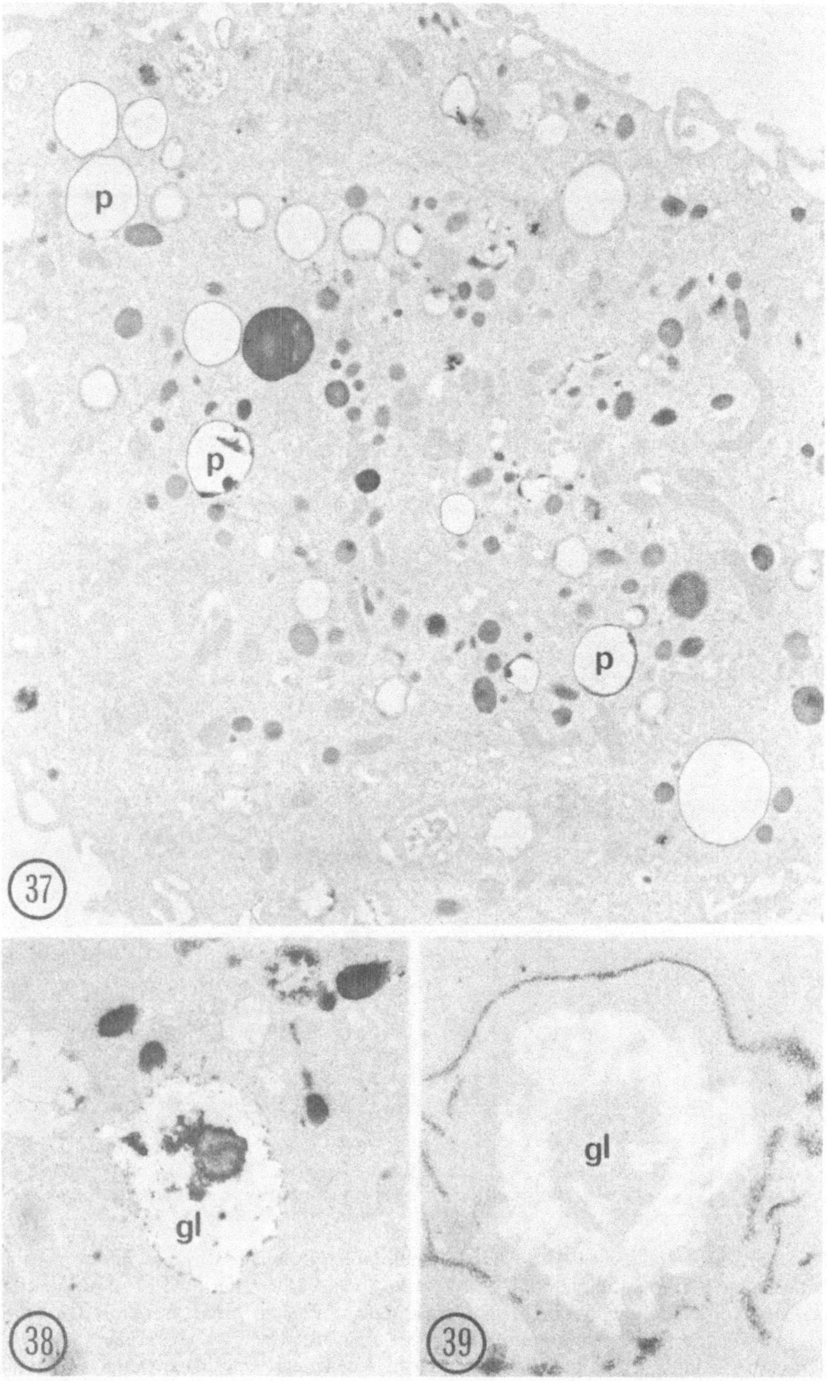

FIGURE 37. Part of an exudate macrophage from a mouse peritoneal cavity obtained 32 days after the i.p. administration of paraffin oil and incubated for the demonstration of peroxidase activity. The cytoplasm contains droplets of paraffin oil with a PO-positive rim. ×11,000.

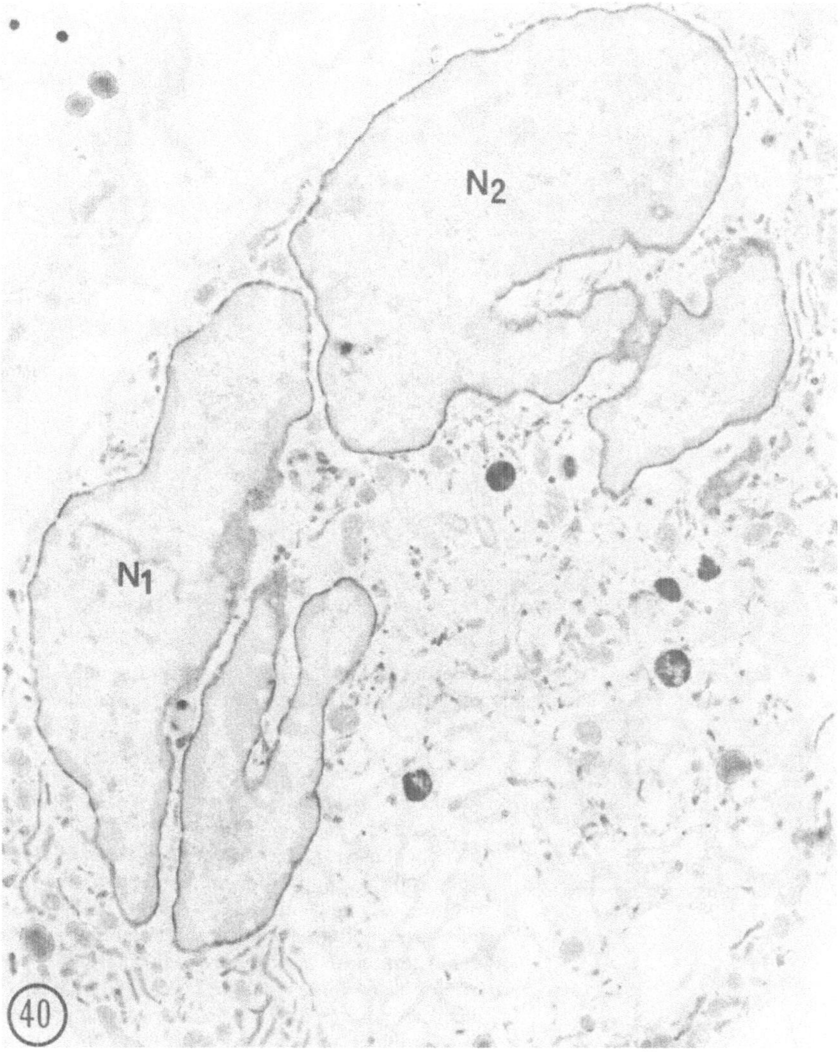

FIGURE 40. Rat binucleated giant cell adhering to Melinex implanted intraperitoneally 3 days previously; incubated for the demonstration of peroxidase activity. Reaction product is present in the nuclear envelope and the RER. ×9,200.

FIGURE 38. Part of a mouse exudate macrophage isolated 28 days after the i.p. administration of glucan and incubated for the demonstration of PO activity. Reaction product is only present in primary granules and around the vacuole containing glucan. ×20,000.

FIGURE 39. Part of a mouse resident peritoneal macrophage isolated 28 days after the i.p. administration of glucan (gl) and incubated for the demonstration of PO activity. Reaction product is restricted to the RER. ×20,000.

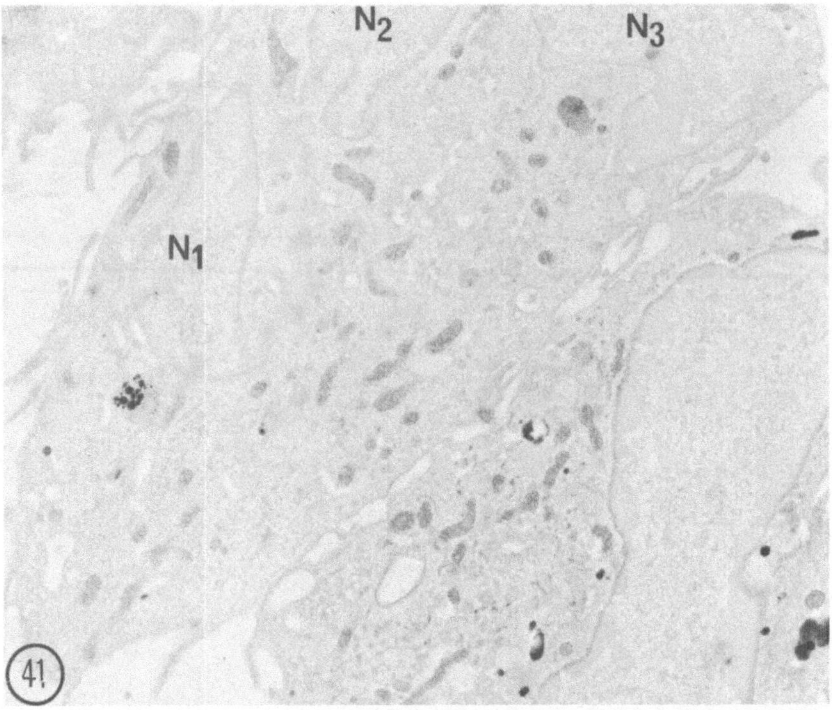

FIGURE 41. Rat multinucleated giant cell adhering to Melinex implanted intraperitoneally 3 days previously; incubated for the demonstration of peroxidase activity. Except for two primary granules the cell is PO-negative. ×8,000.

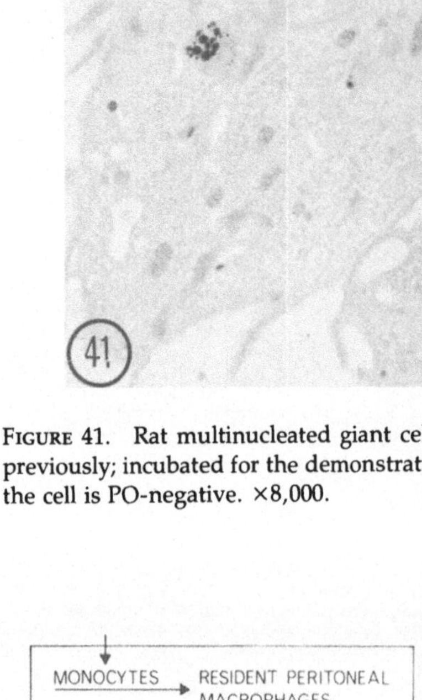

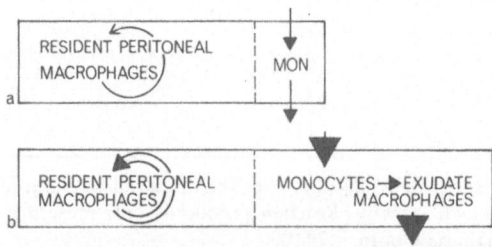

FIGURE 42. Schematic representation of the two prevailing views on the relation between monocytes and resident peritoneal macrophages under normal steady state conditions. (a) The population of peritoneal macrophages is continuously replenished by blood monocytes which enter the peritoneal cavity and differentiate there to resident peritoneal macrophages. (b) Resident peritoneal macrophages form a self-sustaining population of cells; blood monocytes form a transient (travelling) population of cells. Blood monocytes do not transform into resident peritoneal macrophages.

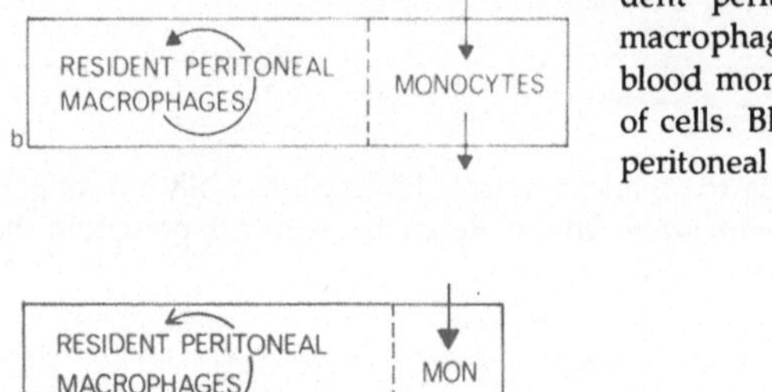

FIGURE 43. Schematic representation of the composition of peritoneal exudates with respect to resident peritoneal macrophages and monocytes. Induction of a peritoneal exudate leads, among other things, to an increased influx of monocytes from the peripheral blood into the peritoneal cavity. These cells differentiate locally to become exudate macrophages and leave the peritoneal cavity as such. They do not transform into resident peritoneal macrophages. The population of resident peritoneal macrophages remains numerically stable under these conditions but is replenished, if necessary, by means of an increased rate of mitosis.

ton, 1976b) and has furthermore been shown to be dependent upon the degree of macrophage stimulation. Cytochalasin B causes diminution of the number of microvillous extensions of the cell surface, as can be determined from scanning electron micrographs (see Takayama *et al.*, 1975). In addition, the number of microvilli is assumed to be dependent on the type of fixation used (Daems and Brederoo, 1973), although the absence and presence of such extensions have been assigned a functional qualification (Carr, 1968, 1970).

The number of microvilli is also considered to be related to the endocytic activity of the cell. Rat peritoneal macrophages that had ingested asbestos particles had a more intricate surface structure than macrophages from a glycogen-stimulated peritoneal cavity and spread more rapidly on glass (Miller and Kagan, 1976). Walters *et al.* (1976), who performed a scanning electron microscope study, also describe the development of filopodia and lamellipodia accompanying the process of endocytosis. The involvement of filopodia in the process of endocytosis has recently been described on the basis of reflection contrast and scanning electron microscopy (Koerten *et al.*, 1979). Güttner *et al.* (1975) found that some subcutaneously administered chemotherapeutic agents lead to pronounced cell-surface variations in mouse peritoneal macrophages. Comparable studies on possible differences in surface morphology between resident peritoneal macrophages and macrophages derived from blood monocytes are still hampered by the lack of reliable means to identify such cells in the scanning electron microscope (for a discussion, see Albrecht *et al.*, 1978).

After staining with ruthenium red, the plasma membrane of macrophages is covered with an 8–16 nm thick layer of dense material (Carr *et al.*, 1970; Daems and Brederoo, 1973) extending into surface-connected invaginations. Similarly, ferrocyanide-reduced osmium stains material related to the cell surface (Dvorak *et al.*, 1972). After appropriate precautions (Brederoo and Daems, 1972), peritoneal macrophages are seen to be covered with a cell coat roughly 60 nm thick. A layer approximately 20 nm thick, thought to contain membrane glycoproteins, was also observed by Hardy *et al.* (1976) in mouse peritoneal macrophages and can be observed on the outside of the plasma membrane as well as in phagocytic vacuoles (Figure 7). The thick cell coat continues into the so-called wormlike structures (Brederoo and Daems, 1972; Emeis and Wisse, 1975; Emeis, 1976; Emeis and Planqué, 1976), which originate from invaginations of the plasma membrane and are also a characteristic feature of resident and exudate peritoneal macrophages (see Chapter 4 in this volume). These wormlike structures were shown to contain i.v.-injected metal colloids, thus suggesting that they play a role in endocytosis (Emeis, 1976).

The cell surface of resident peritoneal cells shows a certain "stickiness" with respect to Thorotrast particles, similar to that known for Kupffer cells (Emeis and Wisse, 1971). The absence of Thorotrast on granulocytes or lymphoid cells points to a more than accidental relationship between Thorotrast and the peritoneal macrophages.

Cytochemical studies have demonstrated adenosine triphosphatase activity at the level of the plasma membrane of peritoneal macrophages (North, 1966; Puvion *et al.*, 1976), although other reports discuss the specific nature of the cytochemical reaction product (Poelmann and Daems, 1973).

The plasma membrane of the resident peritoneal macrophages has a trilaminar fine structure (Figure 12) with a center-to-center spacing of about 8 nm. The fine structure of the plasma membrane is particularly distinct after uranyl acetate postfixation (Daems and Brederoo, 1973). A regular finding is "loop"-like membrane-to-membrane interrelations on the surface of guinea pig resident peritoneal macrophages (Figures 8 and 9) (Daems and Brederoo, 1973). A characteristic feature is the constant distance of about 40 nm between the two limiting membranes, these membranes being separated by a line equidistant from both, which is especially evident when particulate material, such as ferritin, is present within the "loop" (Figure 9). In agreement with this phenomenon is the presence, immediately under the plasma membrane of the resident peritoneal macrophages, of a so-called intracellular coat (Daems and Brederoo, 1973), a narrow zone with a width of about 20 nm. This coat is especially evident in ferritin-laden macrophages, where the ferritin particles are separated from the plasma membrane by a distance corresponding to the width of this coat.

Underlying the plasma membrane of the macrophages there is a zone of uniform width which is devoid of cell organelles (Figures 1, 15, and 16). This zone contains a network of filamentous material; the individual filaments have a diameter of 5–6 nm and represent actin-like material (Allison, 1973). During endocytosis, this zone increases considerably in size (Figures 10 and 11) (see Michl and Silverstein, 1978) and lacks cell organelles such as the RER, mitochondria, and lysosomes. The pseudopods involved in the ingestion of particles have numerous microfilaments (Figure 11), and the presence of actin can be shown by immunofluorescence techniques (Berlin and Oliver, 1978). A cooperative interaction between such filaments and microtubules in the control of macrophage movement has been suggested (Cheung et al., 1978). Cytochalasin B treatment leads to the disappearance of this zone (Carter and Chir, 1972) and inhibits receptor-mediated as well as nonspecific particle ingestion, possibly by interfering with contractile protein function (see Michl and Silverstein, 1978).

Endocytosis is an important function resident peritoneal macrophages share with other types of phagocytes. Depending on the nature of the material to be taken up, these cells have at their disposal of a variety of mechanisms, ranging from micropinocytic vesicles with a constant diameter and fine structure (Figures 13 and 14) to phagocytic vacuoles whose size is highly dependent on the size of the material to be ingested (Figure 11) (see Wisse *et al.*, 1974). The process of phagocytosis can be studied in detail with the scanning electron microscope (Takayama *et al.*, 1975; Walters *et al.*, 1976; Kaplan and Bertheussen, 1977; Orenstein and Shelton, 1977). Such studies have shown that, with respect to phagocytosis, at least two modes of ingestion can be distinguished, depending on the nature of the particles and of the receptor involved: on the one hand, the "sinking-in" of the material to be ingested into the cytoplasm of the macrophage in the case of C3 receptors, and on the other, the covering of the particle to be ingested by the membrane of the macrophage (Figure 4) in the case of Fc-receptor-mediated attachment (Kaplan and Bertheussen, 1977; Kaplan and Morland, 1978).

In ultrathin sections, deep indentations or lacunae extending from the cell surface into the cytoplasm are often encountered (Figures 1 and 24). Under the plasma membrane a varying number of vesicles and vacuoles are present, some of which are continuous with the cell surface, as can be clearly seen after staining with ruthenium red (Carr *et al.*, 1970; Daems and Brederoo, 1973). With the scanning electron microscope, the openings of endocytic vesicles to the exterior can be observed (Figures 2 and 3) (Polliack and Gordon, 1975).

The small vesicles, which have a diameter of about 100 nm, are of the bristle-coated type (Figures 13 and 14). Some of these vesicles appear to be connected to the plasma membrane by a neck or channel, sometimes of considerable length, as is especially clear after ruthenium red staining (Daems and Brederoo, 1973). Such flask-shaped caveolae might represent the mechanism by which the macrophage secretes its coat substance (Carr *et al.*, 1970). possibly in a way similar to that occurring in intestinal epithelial cells (Ginsel *et al.*, 1975). In cells harvested after the intraperitoneal injection of various compounds, such as colloidal gold or silver and Thorotrast, the micropinocytic vesicles contained the injected material (Figure 24).

In the mouse the cytoplasm on one side of most of the macrophages characteristically shows numerous vesicles containing moderately dense material (Fig. ures 15 and 16) (Hard, 1969; Hirsch and Fedorko, 1970; Daems *et al.*, 1976). There is no evidence that these vesicles are involved in endocytic processes. In all probability they represent elements of the smooth endoplasmic reticulum (SER) (Dvorak *et al.*, 1966; Carr, 1967), although they have also been considered to be primary lysosomes (Carr, 1973).

Guinea pig and mouse peritoneal macrophages possess what has been called a labyrinth, a spongelike area located directly under the cell surface and continuous with the plasma membrane (Brederoo and Daems, 1972).

Especially on scanning electron micrographs, close spatial relationships are found between peritoneal macrophages and lymphoid cells (see Albrecht *et al.*, 1978).

3.2.2. Nucleus

Resident peritoneal macrophages have, in specimens prepared for transmission electron microscopy, a nucleus with a characteristically irregular shape (Robbins *et al.*, 1971; Watanabe and Enomoto, 1972; Daems and Brederoo, 1973; Mayhew and Williams, 1974), and sometimes so deeply indented that it seems slender and folded (Figure 1). It should be mentioned that in specimens prepared for electron microscopy, the shape of the nuclei is more irregular than the light-microscopical appearance. The nucleus has dispersed chromatin and a large number of nuclear pores, which are observed most clearly in freeze-etched preparations (Daems and Brederoo, 1970). One or two nucleoli are often conspicuous. In the hamster, resident peritoneal macrophages have—like occasional Kupffer cells (Wisse, 1972)—one or two nuclear bodies (Dumont, 1972).

3.2.3. Lysosomes

Compared with other types of resident macrophages, such as those in liver, spleen, and bone marrow, the resident peritoneal macrophages from the unstimulated peritoneal cavity generally have only a limited number of lysosomes which vary in size and fine structure (Figures 12 and 17–23). They all have a single limited membrane which, after uranyl acetate postfixation (Daems and Brederoo, 1973), clearly shows a three-layered structure and which has dimensions similar to those of the plasma membrane (Figure 12). They often show an electron-lucid halo between the limiting membrane and the matrix (Figures 12, 17, and 19–23) (Carr, 1967; Daems and Brederoo, 1973), and in this respect strongly resemble the so-called macrophage granules (Figure 35) (van der Rhee *et al.*, 1979a,c). The fine structure of the lysosomes in resident peritoneal macrophages is to a certain extent dependent on the species or strain. However, there is a commonly found type of lysosome which is characterized by a varying number of osmiophilic globules dispersed in a moderately dense matrix (Figures 12, 17, and 24).

After i.p. administration, various compounds are found in the heterogeneous inclusion bodies of resident peritoneal macrophages (Figures 18–24), thus showing—in addition to the cytochemical findings—their lysosomal nature (Daems, 1976; Schellens *et al.*, 1977). In the case of endocytized iron compounds the ingested material was found in the lysosomes for periods of up to several months (Figures 20 and 21) (see also Section 3.2.9 and De Bakker *et al.*, 1980). Multivesicular bodies are almost always present. Autophagic vacuoles containing mitochondria (Figure 23), rough endoplasmic reticulum (RER), or other material derived from the cells' own cytoplasm are rare in resident peritoneal macrophages of the unstimulated peritoneal cavity (Komiyama *et al.*, 1975), but may become more numerous after intraperitoneal or subcutaneous administration of various compounds (Komiyama *et al.*, 1975; Rhodes *et al.*, 1975; Raz and Goldman, 1976) or in pathological conditions (Gordon and Cohn, 1973). Numerous cytochemical studies have shown acid phosphatase activity in the lysosomes or resident peritoneal macrophages. Other lysosomal enzymes are demonstrated by the electron microscope to be present in these cell organelles, among them an aminopeptidase (DAP II) (Sannes *et al.*, 1977). This enzyme was also found by the authors in foci in the nuclear envelope and the RER, thus suggesting the synthesis at these loci. Acid phosphatase activity was also found to be associated cytochemically with the RER (de Jong *et al.*, 1979). Lysosomes in resident peritoneal macrophages have no endogenous peroxidatic (PO) activity (Figures 25, 29, and 39) (Daems and Brederoo, 1973; for a review, see Daems *et al.*, 1979), in contrast to monocytes and—monocyte-derived—exudate macrophages (Daems *et al.*, 1975) (Figures 25, 29, 35, and 38). In a few cases resident macrophages containing ingested granulocytes have been observed in the unstimulated peritoneal cavity. Phagocytosis of other macrophages by resident macrophages has never been described.

3.2.4. Golgi and RER

The Golgi area of the resident peritoneal macrophages contains up to six crescent-shaped Golgi apparatuses lying in a radial pattern (Figure 1) around the centrioles (Daems and Brederoo, 1973). Golgi cisternae or vesicles with peroxidatic activity are regularly seen (Figure 26) (Daems *et al.*, 1979). The Golgi apparatus also has reaction product after incubation for the demonstration of acid phosphatase activity (de Jong *et al.*, 1979).

In general, the endoplasmic reticulum of resident peritoneal macrophages, which is of the rough (i.e., ribosome-covered) type, is well developed. It is located in the cytoplasmic area surrounding the nucleus and the Golgi apparatus (Figures 1, 24, and 25). Free ribosomes are dispersed through the cytoplasm as well. Cylindrical inclusions consisting of membranes of the RER surrounding pieces of cytoplasm are regularly observed in resident peritoneal macrophages (Daems and Brederoo, 1973; Chin and Hudson, 1974; Komiyama *et al.*, 1975).

Characteristically, the rough endoplasmic reticulum and the nuclear envelope of resident peritoneal macrophages have PO activity (Figures 26, 29, and 30), a feature these cells have in common with other types of resident macrophages (see Daems *et al.*, 1979). The elements of the SER, such as the vesicles found in mouse resident macrophages (see Section 3.2.3), are PO negative. Resident peritoneal macrophages retain their PO activity in the nuclear envelope and the RER when they adhere *in vivo* (van der Rhee *et al.*, 1979c) or *in vitro* (Lepper and D'Arcy Hart, 1976; Bodel *et al.*, 1978) to glass, Melinex, or other materials. Resident peritoneal macrophages with PO-positive granules identical to those occurring in monocytes are not found in the unstimulated peritoneal cavity (Daems *et al.*, 1975, 1979). However, after stimulation of the peritoneal cavity with newborn calf serum (Beelen *et al.*, 1978) or on intraperitoneally implanted Melinex (van der Rhee *et al.*, 1979c) for a limited period, exudate macrophages with PO-positive granules are found with transient PO activity in the RER and nuclear envelope (Figure 36) (see also Section 4.2.1e). Other stimulants do not lead to the appearance of such cells in the peritoneal cavity (Daems *et al.*, 1975; Abe, 1977; Daems and Koerten, 1978; Daems and van der Rhee, 1980).

The requirements for optimal cytochemical demonstrability of PO activity in resident peritoneal macrophages differ from species to species and these differences are considerable (see also Watanabe *et al.*, 1971; Robbins *et al.*, 1971; Nichols and Bainton, 1975; Daems *et al.*, 1975). In fact, the optimal conditions for the cytochemical demonstration of PO activity must be determined for each species individually (see Daems *et al.*, 1979). The PO activity in guinea pig macrophages is easily demonstrable with the usual methods for both light and electron microscopy. The PO activity of mouse resident peritoneal macrophages has, however, been denied (van Furth *et al.*, 1970, Watanabe *et al.*, 1971; Simmons and Karnovsky, 1973; Goud *et al.*, 1975; Karnovsky *et al.*, 1975; Klebanoff and Hamon, 1975; van Furth and Fedorko, 1976), although it can be demonstrated under appropriate conditions (Daems *et al.*, 1976, 1979; Lepper and D'Arcy Hart, 1976).

With respect to the role of this enzyme in the RER and the Golgi apparatus, it is of interest to note that endogenous peroxidase participates in the synthesis of prostaglandins (Cavallo, 1974, 1976; Gerrard *et al.*, 1976; Litwin, 1977) and is located in endoplasmic-reticulum-derived microsomes (Ohki *et al.*, 1978). Macrophages are known to release prostaglandin-like substances during their participation in inflammatory reactions (see Davies and Allison, 1976; Humes *et al.*, 1977; Kurland *et al.*, 1977; Brune *et al.*, 1978; Gemsa *et al.*, 1978; Kurland and Bockman, 1978; Page *et al.*, 1978). For a further discussion of the functional significance of PO activity in resident macrophages, the reader is referred to a recent review (Daems *et al.*, 1979).

3.2.5. Microperoxisomes

In a number of cases small round to oval structures limited by a single membrane and containing a moderately electron-dense material are found in close association with the cisternae of the RER (Figures 1 and 27). When incubated under appropriate conditions (Daems *et al.*, 1976, 1979), these organelles can be shown to contain catalase (Figure 28) and can be regarded as (micro)peroxisomes (Novikoff and Novikoff, 1973).

Microperoxisomes seem to increase in size and number during the maturation of macrophages (van der Rhee *et al.*, 1977). Furthermore, they are also found in Kupffer cells (Gray *et al.*, 1973; Novikoff *et al.*, 1973; Fahimi *et al.*, 1976) and alveolar macrophages (Petrik, 1971; Biggar and Sturgess, 1976). The presence of microperoxisomes in macrophages is in agreement with the biochemical finding that catalase is present in such cells (Miller, 1971; Simmons and Karnovsky, 1973; van Berkel, 1974).

The role of catalase in macrophages can potentially be of importance for the control of the intracellular level of hydrogen peroxide (Fahimi *et al.*, 1976). Phagocytosis is known to be associated with increased levels of hydrogen peroxide production and lipid peroxidation. It is conceivable that catalase protects the macrophages from the lethal effects of the latter.

3.2.6. Mitochondria

Resident peritoneal macrophages contain a relatively large number of mitochondria dispersed through the cytoplasm and often found in close relation to the RER (Figures 1, 15, and 16).

3.2.7. Glycogen

Especially in the guinea pig, resident peritoneal macrophages have been reported by some authors to show relatively large quantities of glycogen randomly distributed throughout the cytoplasm (Dvorak *et al.*, 1972; Daems and Brederoo, 1973), which is in contrast to the biochemical findings of other authors (Cline, 1970; Roos, 1970). Culture of the macrophages and inhibition by sensitized lymphocytes both influence the distribution of the glycogen particles (Dvorak *et al.*, 1972), which leads to the formation of large glycogen aggregates.

3.2.8. Filaments and Microtubules

In addition to the 5 nm-filament system adjacent to the plasma membrane, resident peritoneal macrophages have filaments with a diameter of 10 nm which are combined into perinuclear bundles (Figures 1 and 24) (see Daems and Brederoo, 1973). It has been suggested that these filaments are involved, as contractile elements, in the process of endocytosis (Simson and Spicer, 1973). However, such a role does not seem probable. Although, compared with actin filaments and microtubules, these filaments are less well understood, it has recently been shown that the main component of similar filaments in smooth muscle cells is a protein called desmin (for a review, see Lazarides, 1978).

Microtubules found in the cytoplasm of macrophages appear to play a role in determining the direction of the movement of phagocytic vesicles in the cell. However, they are not likely to play a critical role in the fusion of phagosomes with lysosomes (Pesanti and Axline, 1975). They have a function in maintaining the polarity and the structural organization of the macrophage (Albrecht and Hong, 1976; see also Porter, 1973, and Ginsel *et al.*, 1975). Indeed, when colchicine is added to macrophages they round off (Takayama *et al.*, 1975) and acquire bizarre shapes (Pesanti and Axline, 1975). In addition, the characteristic localization of the various cell organelles is lost and the cytoplasmic organelles in particular are withdrawn from the cell periphery, a phenomenon also found in other types of cell under these conditions (Ginsel *et al.*, 1975). Furthermore, under the influence of colchicine the ingestion of particles takes place on only one side of the cell, whereas under normal conditions the endocytic process occurs over the entire cell surface.

3.2.9. Mitoses

Macrophages in mitosis are found in the unstimulated peritoneal cavity (for references, see Daems and Brederoo, 1973; Padawer, 1973; Shands *et al.*, 1974). When incubated for the demonstration of PO activity, such cells have the cytochemical characteristics of resident peritoneal macrophages (Daems and Brederoo, 1973). Local cell division is considered an important factor in the normal turnover of peritoneal macrophages (Forbes and Mackaness, 1963; Imai *et al.*, 1973; Padawer, 1973; Shands *et al.*, 1974; Ohta and Shimizu, 1975; Volkman, 1976; Shands and Axelrod, 1977; see Daems *et al.*, 1975, and Spector, 1977). In agreement with this, the macrophages in the unstimulated peritoneal cavity proved to be resident cells (De Bakker *et al.*, 1979). In these experiments the resident peritoneal macrophages were labeled by intraperitoneal injection with iron-dextran (Figure 22). Up to four months later, the macrophages in the peritoneal cavity, which had the morphological and cytochemical characteristics of resident peritoneal macrophages, still contained iron, either as cytoplasmic ferritin or in cytoplasmic inclusion bodies (Figure 20). The fact that in these bodies the ferritin was often packed in a crystalline array may be interpreted as indicating that the ferritin had been present in these cells for a considerable time. In contrast, the blood or peritoneal monocytes—which, as mentioned above, can occur in low numbers in the unstimulated peritoneal cavity—did not contain iron in either form (Figure 35). Felix and Dalton (1956) too reported that after the i.p. administration of carbon the majority of the peritoneal macrophages continued to contain carbon throughout the experimental period of 250 days. In agreement with these findings, Padawer (1973) reported that peritoneal macrophages can and do remain *in situ* for long periods, possibly throughout the animal's life.

Higher mitotic rates are found after various stimuli [Felix and Dalton, 1956;

Mackaness, 1962; Forbes, 1966; Burgaleta *et al.*, 1978; Daems and de Water, unpublished results (1978)], and, e.g., after i.p. administration of glucan, the cells in mitosis have the cytochemical characteristics of resident peritoneal macrophages (Figure 30). Many dividing cells occur in the milky spots (*tâches laiteuses*) which are regarded as an important source of peritoneal macrophages in mice (Cappell, 1930; Mims, 1963; Watanabe *et al.*, 1971; Imai *et al.*, 1973). In this respect it should be mentioned that several authors (Lin and Stewart, 1973; Lin, 1974; Lin and Devaraj, 1974; Lin *et al.*, 1975; Stewart *et al.*, 1975; Chu and Lin, 1976; Stanley *et al.*, 1976; Burgaleta and Golde, 1977) have described the presence in the stimulated peritoneal cavity of colony-forming cells that gave rise to adherent mononuclear cells with a high phagocytic activity. These authors emphasize that committed precursors of macrophages migrate to the inflammatory site and proliferate locally.

4. STIMULATED PERITONEAL CAVITY

As already mentioned, peritoneal exudates are commonly used for the study of various properties of macrophages. To elicit these exudates, use can be made of the i.p. administration of a variety of inflammatory agents or irritants such as glycogen, proteose peptone, thioglycollate, and mineral oils (see Daems and Brederoo, 1973; Conrad *et al.*, 1977; Daems and Koerten, 1978), all of which lead to an increase in the number of peritoneal cells (Figure 31). Polysaccharides such as levan (Leibovici *et al.*, 1975; Robertson *et al.*, 1977), pyran (Morahan and Kaplan, 1978), and glucan (DiLuzio, 1976; Burgaleta and Golde, 1977; Burgaleta *et al.*, 1978), are also used frequently at present to modify the function and numbers of peritoneal macrophages.

Intraperitoneal injection of any of these substances leads, among other things, to the migration and intraperitoneal accumulation of blood monocytes which differentiate *in loco* to mature macrophages. Since inflammatory exudates also contain a large number of granulocytes, particularly in the earlier stages after induction (see Figure 32), peritoneal exudates too are frequently used for studies on granulocytes. The peritoneal cavity itself is considered an excellent site to study changes in inflammatory cells and cell populations, and many methods have been used to initiate a mild inflammatory reaction in the peritoneal cavity, ranging from i.p. administration of saline to partial hepatectomy. The obvious reason to use peritoneal exudates instead of cell populations retrieved from the unstimulated peritoneal cavity is the considerably higher number of cells in exudates. However, as will be discussed below in some detail, serious doubts have been raised as to whether cells from peritoneal exudates—macrophages or granulocytes—may be considered to represent resting cells of the intact, normal organism.

It must also be kept in mind that most of the methods used for the collection of peritoneal cells require killing of the animal. Hirsch (1956) and Casciato *et al.* (1976) described techniques permitting repeated collection from individual animals, but it remains possible that the composition of the peritoneal cell population is influenced by the previous collection of cells, which might act as a stimulus.

4.1. COMPOSITION OF THE CELL POPULATION IN THE STIMULATED PERITONEAL CAVITY

Relatively few comparative studies are available in which the cell population in the peritoneal cavity was analyzed quantitatively at various intervals after the i.p. administration of a number of compounds (Felix and Dalton, 1956; Whaley *et al.*, 1972; Lin, 1974; Stewart *et al.*, 1975; Casciato *et al.*, 1976; Fishel *et al.*, 1976; Conrad *et al.*, 1977; Beelen *et al.*, 1978; Daems and Koerten, 1978; Lin *et al.*, 1980). Most of the reports dealing with the composition of the cells in the stimulated peritoneal cavity are based on the use of only one irritant and cells isolated rather soon after the induction of the exudate. Often, detailed information about the composition of the population—either in a quantitative sense or with respect to the various types of cell present—are lacking. Moreover, comparison of results is hampered by the lack of uniformity in the terminology used by the various authors and because an exact classification of cell types based on the use of well-described criteria is not attempted. Although it must be assumed that such terms as *histiocyte, macrophage, monocyte, monocytoid cell,* and *adherent cell* all refer to the same type of cell, there can be no certainty that this is the case. Furthermore, a distinction is rarely made between resident macrophages and monocytes or monocyte-derived exudate macrophages, in combination with a quantitative analysis. For these reasons, the following does not pretend to give anything more than an approximation of the data in the literature.

In connection with the subject of this section, it seems worth underlining the fact that a change or the absence of a change in the percentage of a particular type of cell in the peritoneal cavity as the result of a given treatment, gives no indication as to changes—or their absence—in the absolute number of that type of cell, because even a change in the number of only one type of cell will cause the percentages of all the other cell types to change. Fruhmann (1973), for instance, noted that although on a percentage basis male mice had more peritoneal macrophages than did female mice, whereas on the basis of the absolute numbers the difference could not be considered significant. Similarly, Davies and Thompson (1975) and Briggs *et al.* (1966) found that although cortisol reduced the number of cells in an exudate, the proportions of monocytes and neutrophils were not altered. In our studies on the effect of various stimuli on the composition of peritoneal exudates (Daems and Koerten, 1978) too, the use of irritants changed the percentage of resident peritoneal macrophages considerably but their absolute number only changed slightly (Figure 34). It therefore seems preferable to describe changes in cell populations recovered from the peritoneal cavity in quantitative terms only on the basis of the absolute number of cells.

In general, it becomes clear from the data on the composition of peritoneal exudates that the same types of cell as those present in the unstimulated peritoneal cavity are found there after stimulation with any of the great variety of agents used to increase the number of peritoneal cells. The main difference between the unstimulated and the stimulated states is in the total number of cells, which changes mainly due to the appearance of increased numbers of neutrophilic granulocytes and monocytes in response to stimulation. With respect to the origin of these additional cells, it has been demonstrated convincingly that they are bone marrow derived (Volkman and Gowans, 1965a,b; van Furth, 1976). Van Waarde *et al.* (1976) found that after the i.p. injection of newborn calf serum, murine serum contains a factor that not only causes monocytosis but also leads to an increase in the number of promonocytes and monocytes in the bone marrow. The increase in the number of monocytes in the circulation and the peritoneal cavity is the result of augmented production of these cells in the bone marrow, caused by an increase in the number of promonocytes (van Furth *et al.*, 1973; Volkman and Collins, 1974; Meuret *et al.*, 1975).

The patterns of response to the i.p. administration of the many materials used to produce a peritoneal exudate (for a survey, see Daems and Brederoo, 1973), are in general the same. An initial and immediate invasion by neutrophils, starting after 1–2 hr, is followed by a slow but continuous rise in the number of monocytes. The latter contribute, after differentiation, to the population of peritoneal macrophages. The number of neutrophils begin to decline in 24 hr, followed by a decrease in the number of monocytes. Both slowly return to control levels. In the case of proteose peptone and saline, by the 4th day after the i.p. administration of the stimulant the total number of cells has returned to a normal level. In animals injected with paraffin oil the total number rises again after 4 days and only starts to decrease on the 8th day (Figure 31) (Daems and Koerten, 1978). The time at which the return to normal values occurs varies with the substance used. For instance, with thioglycollate, a type of stimulant that gives an extremely high number of cells after i.p. administration (Figure 31) (see Raz *et al.*, 1977), the increase in the total number of cells is prolonged as with latex (Joos *et al.*, 1969) or glucan (Figure 31) [Daems and de Water, unpublished results (1978)]. After i.p. administration of melanin granules, 50 days elapsed before the population of peritoneal cells returned to normal numbers (Felix and Dalton, 1956).

The total number of cells harvested from the stimulated peritoneal cavity varies from species to species in strong dependence on the nature and concentration of the stimulus (see Lin, 1974) and the duration of the interval after stimulation (Daems and Koerten, 1978). The reaction to a given stimulus also differs in various strains of one species, with respect to both the number of cells present in the peritoneal exudate and time required for the restoration of normal values (Wachsmuth and Stoye, 1977b; Lin *et al.*, 1978). After the administration of most stimulants the number of resident macrophages remains essentially unaltered throughout the entire experimental period (Figure 34). However, the number of (resident) macrophages is reported to decrease shortly after the i.p. administration of certain compounds such as melanin (Felix and Dalton, 1956), Concanavalin A (Smith and Goldman, 1972a,b), newborn calf serum (Beelen *et al.*, 1978), carbon (Whaley *et al.*, 1972), and glucan (Figure 34) [Daems and de Water, unpublished results (1978)] a behavior which resembles the macrophage disappearance reaction (MDR) found by Nelson and Boyden (1963) to occur after the injection of antigen into guinea pigs previously immunized with the same antigen. In experiments with glucan the number of resident macrophages reached control values only after 42 days (Figure 34) [Daems and de Water, unpublished results (1978)]. An increase in macrophage adhesiveness, both to each other and to the mesothelium lining of the peritoneal cavity, is considered to be the most likely explanation for the disappearance of macrophages from peritoneal exudates (Joos *et al.*, 1969; Tomazic *et al.*, 1977), and, indeed, Felix and Dalton (1956) describe an increased clumping of cells after stimulation of the peritoneal cavity. The MDR is dependent on the choice of the irritant and might be induced by a mucopeptide–carbohydrate complex (Bültman *et al.*, 1971).

For each of the stimuli used, the number of monocytes increased significantly 16 hr after the onset of stimulation (Figure 33). The number of monocytes remained elevated until the 8th day after stimulation, except for the animals injected with paraffin oil, where the number of monocytes remained higher than normal throughout the experimental period, or with glucan, where the number of monocytes only returned to normal values at 14 days (Figure 33) [Daems and de Water, unpublished results (1978)]. The increase in the total number of macrophages after i.p. administration of glucan reported by Burgaleta and Golde (1977) is the result of the highly elevated numbers of monocytes (Figure 33).

Neutrophilic granulocytes are usually absent in the unstimulated peritoneal cavity, but after i.p. administration of all stimulants there is a strong and stimulant-dependent increase in the number of these cells (Figure 32) which was the strongest and most lasting after glucan administration. In general, the number of neutrophils returns to normal 4 days after stimulation (Daems and Koerten, 1978), but after paraffin oil a second peak occurs on the 8th day in mice (Figure 32) (Daems and Koerten, 1978) and on the 4th day in rats (Parwaresch *et al.*, 1971). Ten days after the i.p. administration of latex to mice, the number of neutrophils was still above normal (Joos *et al.*, 1969).

The number of lymphocytes in the peritoneal cavity is on average not influenced by i.p. administration of most of the stimulants. However, a selective immigration of small and medium-sized blood lymphocytes is reported to occur after i.p. administration of latex in the rat (Joos *et al.*, 1969; Slonecker, 1971).

A significant increase in the number of eosinophilic granulocytes is found after a number of types of stimulation (Daems and Koerten, 1978). The number of eosinophilic granulocytes returns to normal on the 16th day, except in the animals injected with paraffin oil. The level reached by the numerical increase of the eosinophils is also dependent on the nature of the stimulant, paraffin oil giving the strongest effect.

The number of mast cells is in general not significantly influenced by stimulation of the peritoneal cavity, except for paraffin oil, which leads to a significant decrease (Daems and Koerten, 1978).

In conclusion it may be said that the increase in the total number of cells in the peritoneal cavity after i.p. administration of various compounds is due to an increase in the number of monocytes and neutrophilic, and in some cases eosinophilic granulocytes. The degree of this response is dependent on the nature of the stimulant used and on the time elapsed after stimulation. The number of resident peritoneal macrophages, mast cells, and lymphocytes remained essentially unaltered.

The foregoing makes it evident that any increase or decrease in the total number of peritoneal macrophages resulting from experimental procedures can be assumed to reflect changes in the number of monocytes or monocyte-derived macrophages. In other words, the routine harvesting of peritoneal exudates 3–4 days after induction means the use of a mixed population of resident macrophages on the one hand and monocytes or monocyte-derived macrophages on the other, the latter in general predominating strongly. In view of the functional differences between resident macrophages and monocytes or monocyte-derived macrophages referred to in Section 5, it is, in our opinion, hazardous to use such exudates for studies on the function of macrophages under steady-state conditions. The use of macrophages from peritoneal exudates at best provides data on monocytes or monocyte-derived macrophages. This holds all the more because recent studies with glucan have provided evidence that shortly after the induction of a sterile inflammation in the peritoneal cavity the number of resident peritoneal macrophages sharply decreases and returns to normal values only after a considerable period [Daems and de Water, unpublished results (1978)].

Although the data presented above point to the fact that the cell populations in peritoneal exudates induced in different ways are not directly comparable, there are other grounds for doubt concerning the use of peritoneal exudates as a source of macrophages. The nature and behavior of the macrophages in peritoneal exudates appear to be dependent on the nature of the stimulus used (Forbes, 1966; McLaughlin *et al.*, 1972; Smith and Goldman, 1972a,b; Bar-Eli and Gallily, 1975; Fishel *et al.*, 1976; Conrad *et al.*, 1977) in that the use of irritants yields macrophage populations whose metabolism is altered as well as the phagocytic and bactericidal potentials (Jenkin and Rowley, 1963), and which therefore may no longer represent normal cells (Rowley, 1966; Williams and Mayhew, 1973; Lin, 1974; Boetcher and Meltzer, 1975; Dimitriu *et al.*, 1975; Ohta and Shimizu, 1975). These alterations occur either directly—by the uptake of the irritant, especially when it is particulate in nature (Figures 29 and 36–37) (Fedorko and Hirsch, 1970; Bültman *et al.*, 1971; Dvorak *et al.*, 1972; Edelson and Cohn, 1974; Raz *et al.*, 1977; Daems and Koerten, 1978), because in such cells the lysosomal apparatus is stimulated (see Allison, 1978)—or indirectly due to changes such as "activation." For instance, after Con A and thioglycollate stimulation the macrophages show large electron-dense lysosomes and an intense generation of lipid droplets (Raz *et al.*, 1977). Mineral oil too is reported to be ingested by the macrophages in large amounts (Figure 37) (Bültman *et al.*, 1971; Rhodes, 1975; Conrad *et al.*, 1977; Daems and Koerten, 1978) and is therefore not recommendable as a stimulant even though it causes a steady increase in the number of cells (Figure 31). The same holds for the peritoneal cell population isolated after the i.p. administration of glucan: all types of phagocytic cells, i.e., macrophages and granulocytes, contained glucan material (Figures 11 and 38–39) [Daems and de Water, unpublished results (1978)]. Macrophages isolated from thioglycollate-induced peritoneal exudates have large vacuoles containing undegradable material (Allison, 1978), and for this and other reasons (see also Section 5) the use of macrophages recovered by lavage of the unstimulated peritoneal cavity should be preferred (Allison, 1978; Canonico *et al.*, 1978).

But even when the cell population obtained by intraperitoneal administration of an irritant returns to normal quantitative values and normal composition, the constituent cells may still differ in important respects from those originally present. For instance, 16 days after the i.p. administration of iron-dextran, resident peritoneal macrophages showed ferritin dispersed in the cytoplasm (Figure 9), and four months afterward inclusion bodies were laden with ferritin in a crystalline form (Figure 20). Paraffin droplets (Figure 37) and glucan (Figures 38 and 39) were present in peritoneal macrophages almost a month after i.p. administration. Thus, it is quite possible that such cells are not fully comparable in all respects with the resident peritoneal macrophages of control animals.

The formation of autophagic vacuoles in peritoneal macrophages after the i.p. administration of various compounds (Figure 23)—including chloroquine (Sannes *et al.*, 1977)—has also been described (Komiyama *et al.*, 1975). Furthermore, as the result of the ingestion of the material used for the induction of the exudate, monocytes lose their peroxidase-containing primary granules, either by fusion with phagosomes (Figures 29 and 38) (Daems *et al.*, 1973; see also Daems *et al.*, 1979) or by exocytosis (van der Rhee *et al.*, 1979c). Such cells cannot replenish their stores of the lost enzyme, which is assumed to play an important role in the cytotoxic activity against, for instance, tumor cells (Edelson and Cohn, 1973; Clark and Klebanoff, 1975) and erythrocytes (Klebanoff and Clark, 1975). As a consequence, irreversible changes occur in important and often-tested properties of these cells.

These phenomena make it hazardous to compare macrophages from peritoneal exudates with relatively inactive cells, such as resident macrophages. It has been stated that this also holds for peritoneal lymphocytes, i.e., peritoneal exudate lymphocytes are a population functionally distinguishable from lymphocytes in the unstimulated peritoneal cavity (Catanzaro and Graham, 1974; Catanzaro *et al.*, 1974a,b).

Apart from the fact that the nature of the inducing agent itself can considerably influence the metabolic, phagocytic, and microbicidal properties of macrophages, the way in which the exudate is induced has a strong effect on the ratio of resident macrophages to monocytes [see McLaughlin *et al.*, 1972; Daems and Brederoo, 1973; Fishel *et al.*, 1976; Puvion *et al.*, 1976; Daems and Koerten, 1978; Daems and de Water, unpublished results (1978)]. In view of the considerable diversity of the induction methods applied and of the considerable differences in the interval between induction and harvesting reported in the literature, comparison of the results of studies on so-called "peritoneal macrophages" should be done with caution.

Finally, the question has to be answered whether the changes in the overall properties of the peritoneal cell population resulting from the induction of an inflammation, are caused by a change in the morphology, metabolism, or function of individual cells, or by the addition to the preexisting cells of a new population of cells with different properties. It seems quite possible that changes in the overall properties of peritoneal macrophages are not necessarily based on changes in the properties of all cells, but might equally well be due to the entry into the peritoneal cavity of a type of macrophage with properties and origin differing essentially from those of the preexisting resident population. In the latter case the macrophage population is a heterogeneous one, as can be demonstrated (see Section 5). Such considerations lead to the conclusion that changes in the properties of a population of peritoneal macrophages as a whole should be studied at the level of single cells (Cooper and Houston, 1964; McIntyre *et al.*, 1967; Nathan *et al.*, 1976) in order to determine whether subpopulations of cells with different functions exist (see Walker, 1976a,b). To avoid the problem of the heterogeneity of macrophages, use is made of homogeneous cell lines of defined origin and function possessing macrophage properties such as phagocytic capacity, Fc and C3 receptors, lysozyme production, and the ability to mediate antibody-dependent cellular cytotoxicity (Rosenstreich and Mizel, 1978; see also Chapter 2 of this volume).

4.2. MORPHOLOGY AND CYTOCHEMISTRY OF MONOCYTES AND MONOCYTE-DERIVED MACROPHAGES, EPITHELIOID CELLS, AND MULTINUCLEATED GIANT CELLS

4.2.1. Monocytes and Monocyte-Derived Macrophages

In peritoneal exudates a differentiation of monocytes into macrophages takes place: they enter the peritoneal cavity as cells having all of the characteristics of blood monocytes and then, during the development of the inflammatory process, gradually acquire the characteristics of mature macrophages (compare Figures 25 and 35). As a result, the macrophage population in peritoneal exudates is much more pleomorphic than that in the unstimulated peritoneal cavity (see also Section 5). Together with an increase in size, differentiating monocytes show changes in cell surface and cell organelles that will be described in the following on the basis of a comparison between blood monocytes and mature monocyte-derived exudate macrophages.

4.2.1a. Cell Surface. In ultrathin sections the cell surface shows microvillous projections that reflect the membrane undulations seen in phase contrast microscopy. In the scanning electron microscope, monocytes show a surface covered by ridgelike profiles and ruffled membranes, with a few short microvilli (Michaelis *et al.*, 1971; Parakkal *et al.*, 1974; Golomb *et al.*, 1975). Monocytes that mature into macrophages develop numerous filopodia and large undulating membranes (lamellipodia) (Parakkal *et al.*, 1974). The latter are stated to play a major role in the phagocytosis of microorganisms, whereas the filopodia are assumed to function in the attachment of the cell to the substratum prior to endocytosis (Walters *et al.*, 1976; Chang, 1978). In contrast, the filopodia of resident peritoneal macrophages are involved in the endocytosis of latex beads (Koerten *et al.*, 1979). Oil-induced macrophages (Albrecht *et al.*, 1972) or macrophages from thioglycollate-stimulated peritoneal cavities (Polliack and Gordon, 1975) have extensive membranous and ridge structures over the entire surface. Carr *et al.* (1969) reported that macrophages obtained from peritoneal cavities stimulated with trioleate 5 days earlier are more irregular in shape than macrophages from unstimulated peritoneal cavities. In addition, their surfaces became rougher and showed more prominent ridgelike processes and large flangelike processes. Fingerlike processes were seldom seen. However, in the interpretation of scanning electron microscope images of monocytes and macrophages it should be kept in mind that the appearance of the surface of these cells depends on whether they are attached to a natural or an artificial substrate (Orenstein and Shelton, 1976b) and is also strongly dependent on prefixation and preparation procedures (Brederoo and Daems, 1972; Daems and Brederoo, 1973).

The cell surface of exudate macrophages is found to be covered by a cell coat 650 nm thick if certain precautions are taken in handling the cells (Brederoo and Daems, 1972). Under such conditions, wormlike structures invaginating the cell surface were observed, as well as occasional cytoplasmic vacuoles containing material similar to the cell coat (see also Chapter 4 of this volume). Blood monocytes with a similar coat have not been observed. Labyrinths like those found in resident peritoneal macrophages have not been seen in exudate macrophages (Brederoo and Daems, 1972). The number of endocytic invaginations shown by blood monocytes depends on the animal species studied, but is generally low. However, monocyte-derived macrophages occurring in inflammatory areas display a great variety of mechanisms for the ingestion of material. With the scanning electron microscope the sites of formation of endocytic pits are distinctly observable (Polliack and Gordon, 1975).

4.2.1b. Nucleus. The monocytes of the peripheral blood and those in the peritoneal exudates have an eccentrically placed nucleus, commonly of a reniform shape and having, in addition to a thin layer of condensed chromatin marginating the nuclear envelope, a finely dispersed chromatin and one or two small nucleoli. During the differentiation to a macrophage, the nucleus of the monocyte changes slightly and obtains a more irregular shape without, however, acquiring the highly irregular shape of the nucleus of a resident macrophage.

4.2.1c. Filaments and Microtubules. Immediately beneath the plasma membrane there is a network of fine filamentous material in which filaments with a diameter of 5–6 nm can occasionally be distinguished (Sutton, 1967). Microtubules are virtually absent in monocytes. Filaments with a diameter of 10 nm are present in exudate macrophages, especially close to the nucleus (de Petris *et al.*, 1962; Sutton, 1967; Hirsch and Fedorko, 1968; Daems and Brederoo, 1973; Breton-Gorius and Reyes, 1976; Bentfeld *et al.*, 1977; van der Rhee *et al.*, 1977), but are less numerous than in resident peritoneal macrophages, where they often occur in perinuclear bundles (Daems and Brederoo, 1973).

4.2.1d. Mitochondria. The mitochondria in blood monocytes are small and, particularly in comparison with lymphocytes, relatively numerous (in human blood, 14.1 ± 1.0 per cell per ultrathin section). Morphometric data show that in human blood monocytes the relative surface area of the mitochondria is 2.97 ± 0.20 (Roos *et al.*, 1980). During differentiation to macrophages the number of mitochondria increases.

4.2.1e. RER and Golgi. RER is present in relatively small amounts and has relatively small dimensions in blood monocytes. During the differentiation from monocyte to macrophage the amount of RER increases considerably. The Golgi complex is situated close to the large nuclear indentation and is, although not conspicuous in blood monocytes, well-developed in monocyte-derived macrophages.

Neither monocytes nor monocyte-derived macrophages show detectable peroxidase activity in the endoplasmic reticulum, the nuclear envelope, or the Golgi cisternae (Figures 25, 29, and 35) (Breton-Gorius and Guichard, 1969; Daems and Brederoo, 1973; Nichols and Bainton, 1973), as seen characteristically in tissue macrophages (Figures 25, 26, and 29). Only under very specific circumstances in peritoneal exudates are there appreciable numbers of macrophages having PO activity in both the RER and nuclear envelope as well as in cytoplasmic granules, i.e., after the i.p. administration of newborn calf serum (NBCS) (Beelen *et al.*, 1978) and after the i.p. implantation of Melinex (Figure 36) (van der Rhee *et al.*, 1979c). After i.p. injection of glucan [Daems and de Water, unpublished results (1978)] only a very limited number of such cells was found. They do not occur, or only rarely, in the unstimulated peritoneal cavity, nor are they found after the administration of any of the many other compounds used to produce peritoneal exudates (see Daems and van der Rhee, 1980). There is reason to assume that these cells represent exudate macrophages with PO activity in the RER and nuclear envelope similar to that shown in adherent monocytes *in vitro* (Bodel *et al.*, 1977, 1978). It could be demonstrated that in the case of i.p.-implanted Melinex the PO activity in the RER of exudate macrophages was a transient phenomenon occurring during a limited period of the inflammatory response (van der Rhee *et al.*, 1979c). It was further found that this type of PO activity, which is as yet of unknown nature, differs cytochemically from that in resident macrophages (van der Rhee *et al.*, 1978, 1979c), at least in the sense that its demonstrability is more strongly dependent on the cytochemical conditions applied (for a discussion see Daems *et al.*, 1979). In addition, adherent human monocytes do acquire PO activity in the RER *in vitro* (Bodel *et al.*, 1977), whereas human resident macrophages are never reported to have such PO activity (see Daems *et al.*, 1979). These data do not support the assumption that monocytes are the precursors of resident peritoneal macrophages, although superficial consideration of the cytochemical data could lead to such an interpretation (Bodel *et al.*, 1977, 1978; Beelen *et al.*, 1978; van Furth *et al.*, 1977).

Finally, it should be kept in mind here that in contrast to transient appearance of PO activity in monocyte-derived macrophages (Bodel *et al.*, 1977, 1978; van der Rhee *et al.*, 1977, 1979c), the PO activity in resident macrophages is a permanent feature, both *in vivo* (Ogawa *et al.*, 1978) and *in vitro* (Bodel *et al.*, 1978). It seems possible that the endogenous PO activity in macrophages is related to the synthesis of prostaglandins (for a discussion, see Daems *et al.*, 1979), because the course of production of prostaglandins by macrophages seems to correspond with the appearance and disappearance of endogenous PO activity (see Section 3.2.4).

4.2.1f. Lysosomes. Close to the Golgi apparatus there are cytoplasmic granules of homogeneous, moderate density and surrounded by a single limiting membrane (Figure 25). These granules can be divided into two groups on the basis of their size and shape (Sonoda and Kobayashi, 1970; Daems and Brederoo, 1973; Daems *et al.*, 1975; Nichols and Bainton, 1975; van der Rhee *et al.*, 1977) as well as their fine structure (van der Rhee *et al.*, 1977). In accordance with the terminology used for neutrophilic or heterophilic granulocytes (see Brederoo and Daems, 1978), van der Rhee *et al.* (1977) distinguish between primary and secondary granules in monocytes. The primary or azurophil granules are oval to elongated and are smaller in diameter than the round secondary granules. The latter are characterized by a clear halo under the limiting membrane. Both types of granule have PO activity but with different cytochemical characteristics: the primary granules have peroxidase activity (Figures 25 and 35), whereas the secondary granules most probably contain catalase (van der Rhee *et al.*, 1977). With respect to morphology, staining characteristics, and relation to the RER, however, the latter differ characteristically from microperoxisomes (compare Figures 28 and 35). Biochemical and cytochemical studies have revealed that the peroxidase in primary granules is a myeloperoxidase similar to that present in the azurophil granules of neutrophilic granulocytes (Bos *et al.*, 1978; for a review, see Daems *et al.*, 1979).

In biochemical and cytochemical respects, too, the PO activity in monocytes differs from that in resident macrophages (for a discussion, see Daems *et al.*, 1979). The peroxidase-positive primary granules, which also contain aryl sulfatase and acid phosphatase, appear early in monocyte development. The peroxidase-negative secondary granules appear later (Nichols and Bainton, 1973). Nichols and Bainton (1975) point to the fact that the production of secondary granules continues in blood monocytes, whereas in neutrophils the production of all types of granules is completed in the bone marrow.

Both the primary and secondary granules can be considered to be primary lysosomes (Daems *et al.*, 1975) and both fuse with phagosomes containing endocytized material (Daems *et al.*, 1973; Nichols and Bainton, 1975).

During the transformation of monocytes into exudate macrophages the primary and secondary granules are translocated from the Golgi area to the periphery of the cell (van der Rhee *et al.*, 1979a,b). Simultaneously, the number of primary and secondary granules is decreased (Nichols and Bainton, 1975; van der Rhee *et al.*, 1979a,b) by fusion with phagosomes (Figure 29) (Nichols *et al.*, 1971; Daems and Brederoo, 1973; Daems *et al.*, 1975; Nichols and Bainton, 1975) and possibly exocytosis (van der Rhee *et al.*, 1979a,b). The release of granule enzymes can readily be demonstrated by cytochemical tests showing reaction product of peroxidase (Daems and Brederoo, 1973; Daems *et al.*, 1975) or arylsulphatase (Nichols and Bainton, 1975; Bainton *et al.*, 1976) around endocytized material (Figures 29 and 37–38). Since the production of peroxidase-containing granules is restricted to the early stages of monocytopoiesis in the bone marrow, phagocytizing monocytes gradually lose their peroxidase activity. This might explain why monocyte-derived macrophages are peroxidase-negative (for references, see Daems and Koerten, 1978), a property which characteristically distinguishes these cells from the PO-positive tissue macrophages (see p. 106).

More or less simultaneously with the disappearance of the primary and secondary granules, the so-called macrophage granules (Carr, 1967, 1968; van der Rhee *et al.*, 1979a) appear; these granules are characterized by their larger size and round shape, as well as by the presence of a halo separating the limiting membrane from the homogeneously moderately dense matrix (Figure 35). The matrix of most macrophage granules shows dense membrane-like material close to the halo. In other, usually larger, macrophage granules the amount of such dense material is increased.

A distinction between macrophage granules and the secondary granules of monocytes is not always easy to make and will usually be based on the higher density of the matrix in the macrophage granules and the smaller size of the secondary granules, as well as the absence of the dense membrane-like material in the latter. There is little doubt, however, that both types of granule accumulate material endocytized by the cell and can therefore (Daems, 1976; Schellens *et al.*, 1977) be considered lysosomes.

Multivesicular bodies are occasionally seen in monocytes. Globule-containing lysosomes, such as those found in resident peritoneal macrophages (see Section 3.2.3), never occur in monocytes or exudate macrophages.

When soluble stimulants are used, the lysosomes of the exudate macrophages are found to participate only in the digestion of endocytosed neutrophils (for a detailed description, see Daems and Brederoo, 1973), but when a particulate substance or material in suspension is injected i.p., for instance mineral oil or glucan, the exudate macrophages, like the resident macrophages, endocytize, accumulate, and, if possible, digest these compounds in the lysosomes (Figures 29 and 37–38). Depending on its digestibility, the ingested material can be found in the macrophages for varying periods. Exudate macrophages often show PO activity deriving from the primary granules and associated with the ingested material (Figures 29 and 37–38) in contrast to the situation in resident macrophages where the secondary lysosomes are, without exception, devoid of endogenous PO activity (Figures 29 and 39).

4.2.1g. Microperoxisomes. Unlike blood monocytes, in which microperoxisomes could not be demonstrated (van der Rhee *et al.*, 1977), exudate macrophages have a limited number of microperoxisomes (Novikoff and Novikoff, 1973; Daems *et al.*, 1975, 1976), i.e., membrane-bound cytoplasmic particles, usually with a diameter of 0.15–0.25 μm, which are in close spatial relationship to and sometimes continuous with the membranes of the RER. Microperoxisomes stain positively in a catalase variant of the DAB method. They differ from the peroxisomes in, for example, liver parenchymal cells by their smaller dimensions and the absence of a dense core or crystalline nucleoid. The number of microperoxisomes increases in the developmental series comprising monocytes, macrophages, epithelioid cells, and multinucleated giant cells (van der Rhee *et al.*, 1977).

4.2.2. Epithelioid Cells and Multinucleated Giant Cells

The formation of epithelioid cells and/or multinucleated giant cells after the i.p. administration of soluble compounds has never been described. After the i.p. administration of sheetlike materials (Papadimitriou 1973; van der Rhee *et al.*, 1979c) or combined i.p. administration of talc and particulate prednisolone (Dreher *et al.*, 1978) or methylcellulose (Machado and Lair 1978), such cells were found in the peritoneal cavity. Three days after the i.p. implantation of Melinex, cells with the characteristics of epithelioid cells (van der Rhee *et al.*, 1979c) were found attached to the plastic, their number increasing up to 14 days. These cells are relatively rare compared with the number of giant cells found on subcutaneous Melinex implants (van der Rhee *et al.*, 1979a) and very rarely have more than five nuclei.

The multinucleated giant cells adhering to intraperitoneally implanted Melinex can be divided into two groups, one with (Figure 40) and the other without (Figure 41) PO acitivity in the RER and the nuclear envelopes (van der Rhee *et al.*, 1979b,c; Daems and van der Rhee, 1980). It was concluded from these studies that all giant cells are monocyte-derived and that the appearance of PO activity in the giant cells is a transient phenomenon, similar to that observed under specific conditions in exudate macrophages. This conclusion is based on the fact that PO activity occurred only in the young giant cells, the giant cells with a larger number of nuclei being PO-negative. In addition, the PO activity in resident macrophages is known to be permanent (Daems *et al.*, 1979) and not to disappear after adherence (Bodel *et al.*, 1978). It may therefore be concluded that resident macrophages cannot fuse and become multinucleated giant cells. Furthermore, evidence was obtained that upon adherence to Melinex, resident macrophages too acquire a PO activity differing cytochemically from that originally present (van der Rhee *et al.*, 1979c).

5. HETEROGENEITY OF PERITONEAL MACROPHAGES

As described above, resident peritoneal macrophages and monocyte-derived exudate macrophages have much in common. However, distinct differences between these cells are also found. Other evidence, too, raises the question whether resident peritoneal macrophages differ in function and origin from exudate macrophages and whether the former are a self-duplicating population of cells. As early as 1925, Sabin *et al.* pointed out that clasmatocytes [analogous with resident (= tissue) macrophages] are not identical to monocytes, but that from the point of view of origin and function there are two types of cell. In their description of peritoneal exudates (obtained by i.p. administration of blood) they, therefore, distinguish between clasmatocytes and monocytes, whose ratios are indeed roughly similar to those obtained later with more sophisticated methods. Remarkably, the authors underline the finding that the number of clasmatocytes remains unaltered during the development of the exudates, an observation which is in full agreement with much more recent data (Daems and Koerten, 1978; see also pp. 60–62). It is extremely interesting to read the reports of the very careful and elaborate studies done by these authors and to see with how much imagination they interpreted the results obtained with what seem to us relatively simple means, and how they describe the analogies between the Kupffer cells in the liver, the clasmatocytes in the spleen, and the resident peritoneal macrophages. They already postulate that the monocytes found in the liver of tuberculosis-infected animals derive *in situ* from primitive cells which in the bone marrow are the forerunners of white blood cells. This view was expressed about fifty years before Lin and Stewart (1973) and Lin (1974) expressed a similar opinion on the basis of extensive experimental studies (see Section 3.2.9). In view of the processes taking place in acquired immunity and, more generally, in inflammatory processes, Sabin *et al.* (1925) point out that an important difference between clasmatocytes and monocytes is that clasmatocytes derive *in loco* from the division of a mature cell and can function immediately after division, but that monocytes have to undergo maturation before they become typical adult cells. Confirming the views of Sabin *et al.* (1925), Seemann (1930) proved that "secondary histiocytes," which were derived from blood monocytes in inflammation, were distinguishable from "primary histiocytes," which are of local tissue origin.

The heterogeneity of macrophages is more or less explicitly recognized in many more recent papers (e.g., Jenkin and Rowley, 1963; McIntyre *et al.*, 1967; Joos *et al.*, 1969; Miller, 1971; Harper and Wolf, 1972; Boumsell and Meltzer, 1975; Karnovsky, 1975; Ohta and Shimizu, 1975; Rhodes, 1975; Defendi, 1976; Werb and Dingle, 1976; Kay and Douglas, 1977; Kondo and Kanai, 1977; Lee and Berry, 1977; Raz *et al.*, 1977; Shands and Axelrod, 1977; Stuart, 1977; Bianco and Edelson, 1978; Cowing *et al.*, 1978; Davies *et al.*, 1977; Lazdins *et al.*, 1978; Morahan and Kaplan, 1978; Thomas *et al.*, 1978; for reviews, see Walker, 1976a,b). There is increasing evidence that the macrophages from the unstimulated peritoneal cavity, the great majority (up to 98%) of which are resident macrophages, differ from the macrophages from the stimulated peritoneal cavity, the majority of which are monocyte-derived (Daems and Koerten, 1978). These studies indicate that macrophages are a heterogeneous population of cells which, although showing common morphological and functional properties, are by no means uniform in either of these respects. Important areas of divergence concern the cell surface, including its receptors, its surface charge, and the composition of the plasma membrane, the secretion of enzymes, and the digestive and killing capacity of the macrophages.

The data presented in the following provide evidence suggesting that fundamental functional differences exist between resident peritoneal macrophages and monocyte-derived exudate macrophages. To obtain definitive proof of this heterogeneity, further studies will have to be performed in a standardized way and either in pure populations of the respective types of cell, which might be realized by the use of a cell sorter (Kwan *et al.*, 1976) or at the single cell level (see also pp. 95–96).

The most important question with respect to the functional heterogeneity described below is whether it reflects differences in the physiological state of one type of macrophage or whether the divergence is based on the properties of different lineages of macrophages: resident macrophages on the one hand and monocyte-derived exudate macrophages on the other. In a number of cases the differences between resident and exudate macrophages do express differences in the functional state of the cell, which can be modified by the choice of appropriate experimental conditions, for example, for lysosomal enzyme activity. However, in other cases the differences remain even though the two types of cell are treated alike in all respects.

A number of studies showed differences between resident and exudate macrophages with respect to Fc receptors in both a quantitative sense (see Chapter 5 of this volume) and a functional sense: macrophages from the stimulated peritoneal cavity, like bone-marrow-derived blood monocytes, ingest red cells by virtue of IgG receptors, whereas resident peritoneal macrophages take up red cells by a more primitive, nonimmunological mechanism not involving IgG (Ohta and Shimizu, 1975). Lutton and Kiremidjian (1973) found that macrophages from a thioglycollate-stimulated peritoneal cavity differ from resident macrophages in their ability to bind rabbit gamma globulin. Rhodes (1975) found a sixfold increase in the proportion of cells with a high cellular affinity for IgG on the 4th day after the induction of a peritoneal exudate with mineral oil, but also demonstrated an increase in Fc receptor activation *in vivo* which, in her opinion, does not support the idea of functional subclasses. Differences were also found with respect to complement receptors: macrophages from thioglycollate-stimulated peritoneal cavities have complement receptors that mediate both attachment and ingestion, whereas in resident macrophages the complement receptors mediate only attachment (Bianco *et al.*, 1975; see also Michl and Silverstein, 1978). With respect to the secretion of factor B of the alternative complement pathway, differences have been found between macrophages from unstimulated peritoneal cavities and macrophages isolated 4 days after i.p. administration of thioglycollate. Culture supernatant from resident macrophages invariably contained more factor B than supernatants from exudate macrophages (Bentley *et al.*, 1977). On the other hand, when the macrophages from the stimulated peritoneal cavity were cultured on a surface covered with glutaraldehyde-like bovine serum albumin, the culture medium contained more factor B than did that of comparably cultured macrophages from the unstimulated peritoneal cavity. Whether any differences in complement receptors are related to the presence on exudate macrophages of cell-coat material, which is almost absent in resident macrophages (see Chapter 4), is an open question. It has been suggested that C receptors are present in the outer layer of the thick cell coat (Emeis, 1976) which most resident peritoneal macrophages lack.

In considering the monocytic origin of exudate macrophages it should be kept in mind that with respect to the Fc and C3 receptor functions, cells of a murine monocytic cell line have more resemblance to macrophages from an endotoxin-stimulated peritoneal exudate than to resident peritoneal macrophages (Kaplan and Morland, 1978). For a detailed discussion of the differences in receptors between resident and exudate macrophages, the reader is referred to Chapter 5.

In contrast to exudate macrophages, resident macrophages respond poorly to chemotactic stimuli (see Snyderman and Meyenhagen, 1976; Jungi and McGregor, 1978). Boumsell and Meltzer (1975) also described differences in both dose response and time course between blood monocytes and macrophages from the unstimulated peritoneal cavity of mice with respect to the chemotactic response to lymphokines and activated complement components. However, Meltzer *et al.* (1975) reported that increased chemotactic responsiveness was a functional property of macrophages activated by BCG, endotoxin, polynucleotides, or Con A, whereas macrophages from peritoneal exudates induced by oil or thioglycollate did not show this increase. The authors found a positive correlation between chemotactic responsiveness and tumoricidal activity. It should be noted, however, that the period after administration of the irritants they used was such that doubts arise as to whether the cell population in question was indeed a peritoneal exudate. In comparable experiments, the macrophage population had returned almost completely to the normal situation, both quantitatively and qualitatively, 7 days after the i.p. injection of the irritant (Daems and Koerten, 1978).

According to Lutton and Kiremidjian (1973), macrophages from the stimulated peritoneal cavity show a behavior in cell electrophoresis significantly different from that of resident peritoneal macrophages, which indicates differences in surface charge between the two types of cell. Whitley and Leu (1976) found that resident peritoneal macrophages are less responsive to migration inhibition factors (MIF) than oil-activated peritoneal macrophages are. Resident macrophages differ from pyran-activated macrophages in their mobility during incubation with Lewis lung carcinoma cells, the latter having a significantly greater incidence of lateral movements (Snodgrass *et al.*, 1977, 1978).

Differences were found between resident peritoneal macrophages and macrophages from peritoneal exudates in the composition of the plasma membrane of such cells. Kondo and Kanai (1977), who analyzed the phospholipid composition of the plasma membrane of resident macrophages and of macrophages from a casein-stimulated peritoneal cavity, found that the phospholipid content of the macrophages from the peritoneal exudate was about half that of the resident macrophages, mainly due to decreased amounts of phosphatidylcholine and phosphatidylethanolamine. In addition, there was an increased amount of cholesterol esters in the macrophages from the stimulated peritoneal cavity, the molar ratio to phospholipids being 1:3.3 compared with 1:10.5 in resident peritoneal macrophages. Lin *et al.* (1978) found specific quantitative and qualitative differences in plasma-membrane polypeptides between resident peritoneal macrophages and macrophages from peritoneal exudates induced with thioglycollate or endotoxin lipopolysaccharide.

With respect to attachment and spreading, too, resident peritoneal macrophages differ from macrophages from peritoneal exudates. McLaughlin *et al.* (1972) found that a substantial proportion of exudate cells showed stickiness, whereas the cells from the unstimulated peritoneal cavity did not show this property. Burgaleta *et al.* (1978) found that the ability of macrophages from glucan-induced peritoneal exudates to adhere to glass was 23% greater than that of resident peritoneal macrophages, and Raz *et al.* (1977) and Bianco and Edelson (1978) described a higher degree of spreading of macrophages from a Con-A-induced peritoneal exudate compared with the spreading of resident macrophages. BCG-activated macrophages require higher doses of cytochalasin B and higher-speed centrifugation to detach from the surface they adhere to than do macrophages from unstimulated peritoneal cavities (Helentjaris *et al.*, 1974). In agreement with these findings, Chang (1978) and Polliack and Gordon (1975) reported that macrophages from peritoneal exudates, in the latter study induced with thioglycollate, show more rapid and extensive spreading. Lazdins *et al.* (1978) found that macrophage from a casein-induced peritoneal exudate are more sensitive to the effects of mediators on cell adherence than resident peritoneal macrophages.

With respect to the activity, production, and release of various enzymes and other compounds, resident peritoneal macrophages differ characteristically from cells obtained from a peritoneal exudate. Specific 5′-nucleotidase activity is readily detected in association with the plasma membrane of resident peritoneal macrophages (Edelson and Cohn 1976a,b; Canonico *et al.*, 1978), whereas in cells from a thioglycollate-stimulated peritoneal cavity this activity was significantly lower (Edelson and Cohn, 1976a,b; Bianco and Edelson, 1978; Lazdins *et al.*, 1978). A similar difference was reported by Raz *et al.* (1977) for macrophages from a Con-A-stimulated peritoneal cavity, which had 10- to 40-fold less 5′-nucleotidase activity than resident peritoneal macrophages. Lazdins *et al.* (1978) too reported a 10-fold difference in macrophages from a casein-induced peritoneal exudate if compared to resident peritoneal macrophages and found that the exudate macrophages did not acquire the high 5′-nucleotidase after culturing. Furthermore, the resident macrophages did not lose enzyme activity when cultured.

Resident peritoneal macrophages release much more prostaglandin PGE_2 and 6-keto-PGF_{1a} when exposed to zymosan than do macrophages from exudates induced by thioglycollate (Humes *et al.*, 1977).

The reverse phenomenon was found for plasminogen activator. Elaborate studies by Unkeless *et al.* (1974) and Gordon *et al.* (1975) provided evidence that monocytes and macrophages from thioglycollate-induced peritoneal exudates synthesize, accumulate, and secrete considerable amounts of plasminogen activator. The release of plasminogen activator is also stimulated by phagocytosis. In contrast, macrophages from the unstimulated peritoneal cavity show only a very moderate increase in the release of plasminogen activators, even after a phagocytic stimulus, which gives, however, less than 10% of the activity of thioglycollate-stimulated cells. Unkeless *et al.* (1974) described this phenomenon as occurring in two steps: a primary reaction, which must be followed by a phagocytic trigger to obtain the plasminogen activator release. Indigestible material, when phagocytized, leads to prolonged secretion of plasminogen activator. Only in macrophages obtained from stimulated peritoneal cavities ("primed macrophages"; Werb and Dingle, 1976), does phagocytosis trigger the release of plasminogen activator. In view of data obtained from studies on the composition of peritoneal exudates (see Section 4.1) it seems possible that the plasminogen-activator-secreting cells are monocytes, for example, monocyte-derived macrophages, and that the resident (= tissue) macrophages characteristically differ from these cells in that they are not able to produce significant amounts of plasminogen activator.

Resident peritoneal macrophages secrete some collagenase after endocytosis, but considerably smaller amounts than are secreted by exudate macrophages obtained after i.p. administration of thioglycollate (see Page *et al.*, 1978). Resident peritoneal macrophages secrete barely detectable levels of elastase, but the amount increases after phagocytosis. However, cells from a stimulated peritoneal cavity secrete almost three times more elastase than phagocytizing resident macrophages, and after phagocytosis even show a further increase by a factor of 3.5 (White *et al.*, 1977). Monocyte-derived exudate macrophages also have a greater ability to synthesize aminopeptidase than do macrophages from the unstimulated peritoneal cavity (Wachsmuth and Stoye, 1977a,b).

Characteristic differences are found between resident macrophages and monocytes or monocyte-derived macrophages with respect to the intracellular distribution and nature of peroxidase activity. Monocytes and monocyte-derived macrophages, contain a peroxidase which is closely related to, and probably identical with, the myeloperoxidase present in neutrophils. In both types of cell the peroxidase is located in granules, neutrophils having about three times more of these granules than monocytes do. Moreover, peroxidase-containing granules are known to fuse with phagosomes, as a result of which peroxidase is released into these vesicles (see Cotran and Litt, 1969; Brederoo and Daems, 1970; Daems *et al.*, 1973; Lepper and D'Arcy Hart, 1976). Due to fusion of granules with incompletely closed phagosomes, peroxidase activity is also to some extent spilled into the medium outside the cells (Root *et al.*, 1972).

In contrast with these observations, resident macrophages contain a peroxidase which differs cytochemically from the enzyme found in neutrophils and monocytes (see Daems *et al.*, 1979) and which is located not in granules but in the RER and in the nuclear envelope (Figure 25). Since the latter cell compartments do not fuse with phagocytic vacuoles, peroxidase activity is not to be expected in these vesicles. Indeed, in resident peritoneal macrophages (Cotran and Litt, 1969, 1970; Daems *et al.*, 1973, 1979) PO activity has not been observed in association with endocytized material at any time during or after phagocytosis (Figure 29).

Thus, on the basis of the lack of peroxidase activity in the vicinity of the endocytized material, the peroxidase-mediated microbicidal system cannot play a role of any importance in these cells (Lepper and D'Arcy Hart, 1976; Daems *et al.*, 1979), although the other components of the microbicidal Klebanoff system such as chloride (or other halides) and hydrogen peroxide are available to (at least some) resident macrophages. The peroxidase-mediated antimicrobial system present in neutrophils and monocytes of many animals does not occur in resident macrophages. If the peroxidase of the RER and the nuclear envelope of resident macrophages has any antimicrobial activity, it must differ from the system catalyzed by the granular enzyme in neutrophils and monocytes. For instance, Paul *et al.* (1973) have found that the killing of *Escherichia coli* by the macrophage peroxidase takes place only with iodide as electron donor and not, as for the neutrophil enzyme, with either iodide or chloride. However, it cannot be excluded that the macrophage enzyme under study was actually catalase. Moreover, it is hard to imagine how killing can take place in the phagosome with the help of an enzyme located in the RER or nuclear envelope. On the other hand, a direct comparison has been made between the peroxidase content, the iodinating capacity, and the bactericidal activity of different types of phagocytic cell. For neutrophils and exudate macrophages, these activities show close correlation, especially at higher ratios of bacteria to phagocytes (Cohen and Cline, 1971; Simmons and Karnovksy, 1973). For resident macrophages, this correlation is not found: although these cells have very little or no peroxidase activity and show a very limited iodinating capacity, they kill various strains of bacteria almost as avidly as neutrophils do (Cohen and Cline, 1971; Simmons and Karnovsky, 1973; Biggar *et al.*, 1976). Again, we must conclude that resident macrophages must contain alternative bactericidal system(s).

With respect to the above, it should be noted that macrophages from thioglycollate- or endotoxin-induced peritoneal exudates have a significantly greater capacity to generate O_2^- compared with resident peritoneal macrophages, particularly when expressed as a function of cell number rather than as the amount of protein (Johnston *et al.*, 1978).

With respect to endocytosis, killing, and lysosomal activity resident peritoneal macrophages and macrophages from peritoneal exudates show a different behavior. The pinocytic rate of exudate macrophages is high compared with that of resident macrophages (Bianco and Edelson, 1978). Smith and Goldman (1972b) found that the macrophages from a Con-A-stimulated peritoneal cavity showed enhanced phagocytosis of rabbit and mouse red blood cells. Cooper and Houston (1964) found an enhancement of the phagocytic activity toward *E. coli* in macrophages from stimulated peritoneal cavities. It is of great importance that these authors emphasize the fact that this phenomenon is related to an increase in the number of functionally active cells rather than to an increase in the ingestive capacity of the cells already present. Rightly, the authors point out that the changes in the cell populations from unstimulated and stimulated peritoneal cavities must be studied at the single cell level. Support for their point of view is provided by Morahan *et al.* (1977), who point out how little is known about the exact cell type involved and the cellular activities in the peritoneal cavity after exposure to various stimuli. On the other hand, Smialowicz and Schwab (1977) found no significant difference in the indices of phagocytosis of streptococcal cell walls between macrophages collected from the unstimulated and thioglycollate-stimulated peritoneal cavities.

After the ingestion process, resident and exudate macrophages show differences. Macrophages from the unstimulated peritoneal cavity, when studied at the level of individual cells, show, on average, significantly less reduction of the viable count of *Listeria monocytogenes* than peritoneal exudate cells (Bast *et al.*,

1974a,b; Cole and Brostoff, 1975). The majority of *L. monocytogenes* ingested by monocyte-derived macrophages is degraded by these cells, whereas in resident peritoneal macrophages most of the microorganisms are morphologically unchanged [Ogawa, personal communication (1978)]. Also the rate of intracellular inactivation of *Salmonella typhimurium* is slower and less complete in normal cells than in activated macrophages from *Listeria-*, *Salmonella-*, or BCG-infected mice (Mackaness, 1962). Armstrong and D'Arcy Hart (1975) found that macrophages from the unstimulated peritoneal cavity of mice did not kill *Mycobacterium tuberculosis* even if phagosome–lysosome fusion was initiated by exposing viable bacilli to specific rabbit antiserum before their ingestion. Macrophages isolated from the peritoneal cavity of mice stimulated by the intraperitoneal injection of substances such as thioglycollate, Bayol-F, or sodium caseinate, degraded *Shigella* much faster than did macrophages from the unstimulated peritoneal cavity (Bar-Eli and Gallily, 1975). After the endocytosis of levan by resident macrophages, only a few lysosomes are found around the levan-filled phagosomes and there is no evident structural change in the ingested levan, whereas the levan-containing phagosomes of exudate macropahges are surrounded by numerous lysosomes and digestion of the levan occurs (Robertson *et al.*, 1977). Smith and Goldman (1972b) reported that neutral-red-dyed *Candida albicans* only turns red in the macrophages of the Con-A-stimulated peritoneal cavity, thus indicating a lowering of the pH resulting from fusion with lysosomes. Rhodes *et al.* (1977) found that macrophages from a proteose-peptone-stimulated peritoneal cavity showed greater capacity than resident macrophages to digest HSA–antibody complexes. On the other hand, Raz *et al.* (1977) found that the intracellular fusion of phagosomes containing interiorized Con A was prominent only in resident peritoneal macrophages, although the density of Con-A-binding sites was similar in resident and exudate macrophages. The killing capacity of peritoneal macrophages is not related to the state of activation of the lysosomal apparatus: despite similar morphology and increased lysosomal enzyme activity, *in vivo*-activated macrophages are much more effective in the restriction of intracellular growth of microorganisms than are *in vitro*-activated cells (Reikvam and Hoiby, 1975; Reikvam *et al.*, 1975). Kaiserling *et al.* (1972) found that, despite a normal lysosomal population and a normal ultrastructure, the bactericidal activity of the macrophages of a patient with impaired cellular immunity was really diminished. Puvion *et al.* (1976) reported that *Corynebacterium parvum*-activated macrophages have a reduced number of lysosomes, but nevertheless play an important role in tumor cell killing. Conversely, no direct correlation could be demonstrated between high lysosomal enzyme levels and the destructive capacity of macrophages (Bar-Eli and Gallily, 1975). On the other hand, evidence is available that supports the notion that lysosomes are important in nonspecific cytotoxicity (see Lejeune, 1975; Hibbs, 1976). This assumption is also supported by the phenomenon that cytotoxic activity is abolished by gold salts (Ghaffar and Cullen, 1976; Ghaffar *et al.*, 1976), trypan blue (Hibbs, 1974; Keller, 1974; Hibbs, 1975a,b; Morahan and Kaplan, 1976), silica (Chassoux and Salomon, 1975; Keller, 1976), cortisone (Hibbs, 1974; Gallily and Eliahu, 1976), and lysosomal overloading (Lejeune *et al.*, 1973), all of which interfere with lysosomal activation.

Bennedsen *et al.* (1977) found that macrophages from a proteose-peptone-stimulated peritoneal cavity showed some bactericidal mechanisms which were not present to the same degree in macrophages from the unstimulated peritoneal cavity. Intraperitoneal injection of sodium caseinate resulted in a marked increase in the number of peritoneal macrophages with antibacterial properties, as compared with the macrophages of the unstimulated peritoneal cavity (Janssen and Dangerfield, 1977).

The relationship of resident macrophages and macrophages from the stimulated peritoneal cavity with eukaryotic cells also differs. For example, contacts between macrophages and YC8 target cells were only observed for cells from the peritoneal cavity of mice stimulated with *C. parvum*. Such contacts were not found when target cells were incubated with adherent cells of the unstimulated peritoneal cavity (Puvion *et al.*, 1974). Resident peritoneal macrophages do not kill tumor cells even in the presence of high amounts of endotoxin, whereas macrophages from a proteose-peptone-induced exudate are made tumoricidal by relatively low amounts (Weinberg *et al.*, 1978). Bennett *et al.* (1964) reported that exudate peritoneal macrophages phagocytized tumor cells in greater numbers than did peritoneal macrophages from normal animals. Resident macrophages are not as cytotoxic as macrophages from peritoneal exudates (McLaughlin *et al.*, 1972). Compared with resident peritoneal macrophages, macrophages from peritoneal exudates induced by i.p. administration of glucan show a markedly enhanced cytostasis of malignant cells *in vitro*, as indicated by reduced [^{3}H]thymidine incorporation (Cook *et al.*, 1978). Such divergence cannot be explained by differences in lysosomal enzyme content (Rhodes *et al.*, 1975).

Morahan *et al.* (1977) found differences between BCG- or pyran-activated peritoneal macrophages and normal macrophages from the unstimulated peritoneal cavity, in that the former had antitumor and antiviral activities not shown by the latter. Macrophages from thioglycollate- or glycogen-stimulated peritoneal cavities showed an intermediate behavior, since they have antiviral but not antitumor activity. These authors too stress the necessity to determine the nature of the cell types responsible for these activities.

6. ORIGIN OF RESIDENT PERITONEAL MACROPHAGES

With respect to the origin of resident peritoneal macrophages, two views are currently held (Figure 42). In the first place, it is repeatedly argued that resident peritoneal macrophages as well as other types of resident (= tissue) macrophages derive from blood monocytes (Figure 42a) (see, e.g., van Furth, 1976). However, this assumption has been challenged in a number of papers (e.g., Daems and Brederoo, 1973; Imai *et al.*, 1973; Takahashi *et al.*, 1973; Wisse, 1974; Daems *et al.*, 1975, 1976, 1979; Kojima, 1976; Nelson, 1976; Volkman, 1976, 1977; Ohta and Shimizu 1975; Shands and Axelrod, 1977; van der Rhee *et al.*, 1978, 1979a,b,c) and the hypothesis has been put forward that resident macrophages form a self-renewing population (Figure 42b).

An important piece of evidence indicating that resident peritoneal macrophages are derived from peripheral blood monocytes is based on studies on the kinetics of peritoneal macrophages (see van Furth, 1976). Mice labeled by injection of [^{3}H]thymidine showed a peritoneal macrophage labeling index amounting maximally to 17% after 60 hr. These and more extended observations (for a review, see van Furth, 1976) led the authors to conclude that under normal conditions migration of monocytes into the peritoneal cavity occurs. In addition, the authors confirm earlier observations (Volkman and Gowans, 1965a,b) that intraperitoneal injection of a variety of stimuli inducing a sterile inflammation, leads to a rapid increase in the peritoneal cavity of labeled macrophages recruited from peripheral blood monocytes. On this basis the authors conclude that in animals under normal conditions as well as during an acute inflammation the peritoneal macrophages are derived from peripheral blood monocytes. These data can, however, also be explained by the observation that the proportion of labeled cells in the peritoneal cavity of mice can be relatively high if it is taken into consideration that, in addition to the labeled monocytes, the resident peritoneal macrophages incorporate [^{3}H]thymidine directly (Volkman, 1976). The identicality of the shape of the curves for labeled peritoneal macrophages and blood monocytes is also consistent with the assumption that the labeling of peritoneal macrophages actually reflects the "traffic pattern" of monocytes traversing the peritoneal cavity (Daems and Brederoo, 1973; Volkman, 1976). In this respect the observations of Thompson and van Furth (1973) are of interest, i.e., that one injection of hydrocortisone reduces the number of peritoneal macrophages by only about 20%, after which a plateau is reached, whereas the number of blood monocytes, which initially approaches zero, rises very slowly. If peritoneal macrophages indeed derived from blood monocytes, the effect of deprivation of monocytes in the blood should have had more prolonged and intense effects than were observed in these studies. The observations of Thompson and van Furth (1973) can, however, easily be explained by the assumption that, in fact, the only effect of the hydrocortisone acetate is to deprive the peritoneal cavity of traversing monocytes, leaving the number of resident peritoneal macrophages unaltered. Indeed, the absolute number of cells remaining in the peritoneal cavity in Thompson's study amounts to about 1.3×10^6, which is in good agreement with the number of resident macrophages in the unstimulated murine peritoneal cavity given in other reports (see Section 3.1 and Daems and Koerten, 1978).

In addition, studies done by Volkman (1976) provide strong experimental evidence that resident populations of macrophages in the peritoneal cavity are self-sustaining and linearly distinct from monocytes. Volkman's data fail to substantiate the notion that monocytes share the normal renewal of macrophage populations residing in the peritoneal cavity. Support for Volkman's results can also be found in experiments done by Shands and Axelrod (1977), Slonecker (1971), and Takahashi *et al.* (1973), which showed the proliferating character of peritoneal macrophages. The latter authors point out that the notion that peritoneal macrophages form a self-renewing population is not a new one and that Cappell (1930), Felix and Dalton (1956), Mims (1963), and Weiner (1967) have published data on dividing mature peritoneal macrophages. Mitotic figures are reported to occur in resident peritoneal macrophages (see also Section 3.2.9), and after a single intravenous injection of [^{3}H]thymidine, labeled peritoneal macrophages are found when very few labeled monocytes are present in the circulation (see Ohta and Shimizu, 1975). In addition, the mean grain count of peritoneal macrophages was more than double those of blood monocytes at every experimental point (Takahashi *et al.*, 1973). Imai *et al.* (1973) report that during fetal development, resident peritoneal macrophages are present in the peritoneal cavity before the appearance of the bone marrow. It should be kept in mind here that studies on other types of resident macrophages such as Kupffer cells have also provided strong evidence that these tissue macrophages are able to proliferate (Widmann *et al.*, 1972; Wisse, 1974; Sljivic and Warr, 1975; Widmann and Fahimi, 1976; Deimann and Fahimi, 1977, 1978; Naito and Wisse, 1977) and form a self-sustaining population of cells that do not derive from monocytes. It could be demonstrated unequivocally that after partial hepatectomy, regeneration of the Kupffer cell population was accomplished by the preexistent Kupffer cells (Wisse, 1974; Widmann and Fahimi, 1976). Transitional forms between monocytes and Kupffer cells were not observed; in fact, the latter showed a sharply increased number of mitoses. Further, the finding that macrophages in the fetal liver of the rat usually appear prior to the development of the bone marrow (Naito, 1975; Deimann and Fahimi, 1977, 1978; Naito and Wisse, 1977) also contradicts a monocyte origin of Kupffer cells. In agreement with studies in adult liver, results obtained in fetal liver showed the absence of transition forms between monocytes and Kupffer cells, whereas Kupffer cells in mitosis were regularly encountered.

In agreement with these latter data, it has been shown that the resident peritoneal macrophages are cells that reside for very long periods (up to 4 months) in the peritoneal cavity (Felix and Dalton, 1956; Whaley *et al.*, 1972; De Bakker *et al.*, 1980; see also Section 3.2.9). The other types of resident macrophages reside for considerable periods in the tissues (for a review, see Roos, 1970), as was demonstrated by, for instance, Roser (1970) who used radioactive gold and found that the size of macrophage populations of the liver and spleen remained constant for a period of two months. Whaley *et al.* (1972) found that after an i.p. injection of carbon, Kupffer cells were heavily laden and remained so up to 28 days.

Data obtained by investigation of the small intestine in patients with Whipple's disease led to the conclusion that in the lamina propria of the diseased intestine macrophages remained present for at least a year (Daems, unpublished observations). Similar findings have been reported by other authors (Martin *et al.*, 1972; Morningstar, 1975), who found in patients with Whipple's disease that although extracellular bacteria disappeared 6–8 weeks after the onset of tetracycline therapy, macrophages containing digestion products of ingested bacteria persisted for up to six years. Other types of tissue macrophages can also be assumed to survive for considerable periods. The presence of pigment-laden macrophages in tissues long after a local trauma suggests that individual cells can persist for periods of many years.

These findings are in agreement with those of Cappell (1930), who found after i.v. injection of India ink, that although a considerable proportion of the blood monocytes took up the particles, carbon did not occur in the peritoneal macrophages. This led the authors to conclude that peritoneal macrophages "do not arise to any considerable extent from circulating monocytes."

Another set of observations pointing strongly to the unlikelihood of the supposed origin of resident peritoneal macrophages from monocytes were made in studies on the development of peritoneal exudates [Daems and Koerten, unpublished observations (1977)]. It was found that in all cases the i.p. injection of stimulants led to a significant rise in the number of monocytes, reaching a peak 16 hr after injection. This often considerable increase in the number of monocytes was never followed by an increase in the number of resident peritoneal macrophages. In contrast, after a slight decrease, most probably caused by adherence of the cells to the peritoneal wall (Smith and Goldman, 1972a), the number of resident peritoneal macrophages showed a remarkable constancy.

It should be mentioned in this respect that peritoneal macrophages have been reported to originate from the so-called milky spots on the omentum (Hoefsmit, 1975; Takahashi *et al.*, 1973; Imai *et al.*, 1973; Cappell, 1930; Aronson and Elberg, 1962; Mims, 1963; Ohta and Shimizu, 1975, Hoefsmit and Smelt, 1978). Watanabe *et al.* (1971) raised the question of whether the milky spots are the sites of transformation of monocytes to peritoneal macrophages. However, it was found that macrophages are formed in considerable numbers in the isolated omentum *in vitro* (Aronson and Shahar, 1965).

Finally, in an excellent critical review of the various views on the origin of resident macrophages, Volkman (1977) reaches the conclusion that "... there does not appear to be any conclusive evidence to support the notion that... resident macrophage populations depend on the regular input of circulating monocytes for their renewal."

7. CONCLUSIONS

Present knowledge indicates that under normal steady-state conditions, peritoneal macrophages form a self-replicating population. The unstimulated peritoneal cavity may contain a number of monocytes which can be considered to form a transient (travelling) population of cells differing from that of the nontravelling and self-sustaining pool of resident peritoneal macrophages (Figure 43a). When an inflammation is induced in the peritoneal cavity there is an increased influx of bone-marrow-derived monocytes from the peripheral blood into the peritoneal cavity, these monocytes differentiating to mature (exudate) macrophages at the site of the inflammation (Figure 43b). Since the monocyte-derived macrophages differ in a number of respects (e.g., functionally) from the resident peritoneal macrophages, macrophages from a peritoneal exudate form a heterogeneous population of cells. Furthermore, the way in which the peritoneal exudate is induced and the nature of the inducing agent itself may change the properties of the macrophage population. For these reasons, doubt concerning the use of peritoneal exudates as a source of macrophages is unavoidable.

ACKNOWLEDGMENTS. The author is indebted to Drs. J. M. de Bakker, P. Brederoo, and H. J. van der Rhee, and to Mr. H. K. Koerten for their discussion of the manuscript. He also thanks Mrs. G. C. A. M. Spigt-van den Bercken, Mr. J. J. Beentjes, and Mr. L. D. C. Verschragen for their excellent technical assistance. The investigations were supported in part by the Foundation for Medical Research (FUNGO), which is subsidized by the Netherlands Organization for the Advancement of Pure Research (ZWO).

REFERENCES

Abe, 1977, cited by Kojima (1977).

Albrecht, R. M.. and Hong, R., 1976, Medical Progress. Basic and clinical considerations of the monocyte-macrophage system in man, *J. Pediatr.* **88**:751.

Albrecht, R. M., Hinsdill, R. D., Sandok, P. L., MacKenzie, A. P., and Sachs, I. D., 1972, A comparative study of the surface morphology of stimulated and unstimulated macrophages prepared without chemical fixation for scanning EM, *Exp. Cell Res.* **70**:230.

Albrecht, R. M., Hinsdill, R. D., Sandok, P. L., and Horowitz, S. D., 1978, Murine macrophage-lymphocyte interactions: Scanning electron microscopic study, *Infect. Immun.* **21**:254.

Allen, T. D., and Dexter, T. M., 1976, Surface morphology and ultrastructure of murine granulocytes and monocytes in long-term liquid culture, *Blood Cells* **2**:591.

Allison, A. C., 1973, The role of microfilaments and microtubules in cell movement, endocytosis and exocytosis, in: *Locomotion of Tissue Cells,* Ciba Foundation Symposium 14, pp. 109–148, Elsevier Excerpta Medica-North Holland, Amsterdam.

Allison, A. C., 1978, Mechanisms by which activated macrophages inhibit lymphocyte responses, *Immunol. Rev.* **40**:3.

Allison, A. C., and Davies, P., 1974, Mononuclear phagocyte activation in some pathological processes, in: *Activation of Macrophages* (W. J. Wagner and H. Hahn, eds.), pp. 141–156, Excerpta Medica, Amsterdam.

Allison, A. C., and Davies, P., 1975, Increased biochemical and biological activities of mononuclear phagocytes exposed to various stimuli, with special reference to secretion of lysosomal enzymes, in: *Mononuclear Phagocytes in Immunity, Infection and Pathology* (R. van Furth, ed.), pp. 487–506, Blackwell, Oxford.

Argyris, B. F., 1967, Role of macrophages in antibody production-immune response to sheep red blood cells, *J. Immunol.* **99**:744.

Armstrong, J. A., and D'Arcy Hart, P., 1975, Phagosome-lysosome interactions in cultured macrophages infected with virulent tubercle bacilli. Reversal of usual nonfusion pattern and observations on bacterial survival, *J. Exp. Med.* **142**:1.

Aronson, M., and Elberg, S., 1962, Proliferation of rabbit peritoneal histiocytes as revealed by autoradiography with tritiated thymidine, *Proc. Natl. Acad. Sci.* **48**:208.

Aronson, M., and Shahar, M., 1965, Formation of histiocytes by the omentum *in vitro, Exp. Cell Res.* **38**:133.

Aschoff, L., 1913, Ein Beitrag zur Lehre von den Makrophagen auf Grund von Untersuchungen des Herrn Dr. Kiyona, *Verh. Dtsch. Ges. Pathol.* **16**:107.

Aschoff, L., 1924, Das Reticuloendotheliale System, *Ergeb. Inn. Med. Kinderheilkd.* **26**:1.

Bainton, D. F., Nichols, B. A., and Farquhar, M. G., 1976, Primary lysosomes of blood leukocytes, in: *Lysosomes in Biology and Pathology,* Vol. 5 (J. T. Dingle and R. T. Dean, eds.), pp. 3–32, North-Holland, Amsterdam.

Bar-Eli, M., and Gallily, R., 1975, The effect of macrophage hydrolytic enzyme levels on the uptake and degradation of antigen and immune complexes, *J. Reticuloendothel. Soc.* **18**:317.

Bast, R. C., Cleveland, R. P., Littman, B. H., Zbar, B., and Rapp, H. J., 1974a, Acquired cellular immunity: Extracellular killing of *Listeria monocytogenes* by a product of immunologically activated macrophages, *Cell Immunol.* **10**:248.

Bast, R. C., Zbar, B., Borgos, T., and Rapp, H. J., 1974b, BCG and cancer, *N. Engl. J. Med.* **290**:1413.

Beelen, R. H. J., Broekhuis-Fluitsma, D. M., Korn, C., and Hoefsmit, E. C. M., 1978, Identification of exudate-resident macrophages on the basis of peroxidatic activity, *J. Reticuloendothel. Soc.* **23**:103.

Begemann, H., and Kaboth, W., 1976, Das Retikuloendotheliale System (RES) oder Retikulohistiozytäre System (RHS), in: *Handbuch der Inneren Medizin*, Vol. II/3, *Leukozytäres und retikuläres System I* (H. Begemann, ed.), pp. 439–470, Springer Verlag.

Bennedsen, J., Riisgaard, S., Rhodes, J. M., and Olesen Larsen, S., 1977, *In vitro* studies on normal, stimulated and immunologically activated mouse macrophages. III. Intracellular multiplication of *Listeria monocytogenes, Acta Pathol. Microbiol. Scand. Sect. C* **85**:246.

Bennett, B., Old, L. J., and Boyse, E. A., 1964, The phagocytosis of tumor cells *in vitro, Transplantation* **2**:183.

Bentfeld, M. E., Nichols, B. A., and Bainton, D. F., 1977, Ultrastructural localization of peroxidase in leukocytes of rat bone marrow and blood, *Anat. Rec.* **187**:219.

Bentley, C., Hadding, U., Bitter-Suermann, D., and Brade, V., 1977, Effect of *in vivo* stimulation of mice on the secretion of factor B of the alternative complement pathway by peritoneal macrophages, *Eur. J. Immunol.* **7**:188.

Berlin, R. D., and Oliver, J. M., 1978, Analogous ultrastructure and surface properties during capping and phagocytosis in leukocytes, *J. Cell Biol.* **77**:789.

Bianco, C., and Edelson, P. J., 1978, Plasma membrane expressions of macrophage differentiation, in: *The Molecular Basis of Cell-Cell Interaction* (R. A. Lerner and D. Bergsma, eds.), pp. 119–124, Alan R. Liss, New York.

Bianco, C., Griffin, F. M., and Silverstein, S. M., 1975, Studies on the macrophage complement receptor alteration of receptor function upon macrophage activation, *J. Exp. Med.* **141**:1278.

Biggar, W. D., and Sturgess, J. M., 1976, Peroxidase activity of alveolar macrophages, *Lab. Invest.* **34**:31.

Biggar, W. D., Buron, S., and Holmes, B., 1976, Bactericidal mechanisms in rabbit alveolar macrophages: Evidence against peroxidase and hydrogen peroxide bactericidal mechanisms, *Infect. Immun.* **14**:6.

Bodel, P. T., Nichols, B. A., and Bainton, D. F., 1977, Appearance of peroxidase reactivity within the rough endoplasmic reticulum of blood monocytes after surface adherence, *J. Exp. Med.* **145**:264.

Bodel, P. T. Nichols, B. A., and Bainton, D. F., 1978, Differences in peroxidase localization of rabbit peritoneal macrophages after surface adherence, *Am. J. Pathol.* **91**:107.

Boetcher, D. A., and Meltzer, M. S., 1975, Mouse mononuclear cell chemotaxis: Description of system, *J. Natl. Cancer Inst.* **54**:795.

Bos, A., Wever, R., and Roos, D., 1978, Characterization and quantification of the peroxidase in human monocytes, *Biochim. Biophys. Acta* **525**:37.

Boumsell, L., and Meltzer, M. S., 1975, Mouse mononuclear cell chemotaxis. I. Differential response of monocytes and macrophages, *J. Immunol.* **115**:1746.

Brederoo, P., and Daems, W. T., 1970, Submicroscopic cytology of guinea pig peritoneal exudates. I. The heterogeneity of the granules of neutrophilic granulocytes, in: *Microscopie Electronique 1970* (P. Favard, ed.), Résumés 7ième Congrès International, Grenoble 1970, Vol. 3, pp. 541–542, Societé Française de Microscopie Electronique, Paris.

Brederoo, P., and Daems, W. T., 1972, Cell coat, worm-like structures, and labyrinths in guinea pig resident and exudate peritoneal macrophages, as demonstrated by an abbreviated fixation procedure for electron microscopy, *Z. Zellforsch. Mikrosk. Anat.* **126**:135.

Brederoo, P., and Daems, W. T., 1978, The ultrastructure of guinea pig heterophil granulocytes and the heterogeneity of the granules. I. The development in the bone marrow, *Cell Tissue Res.* **194**:183.

Breton-Gorius, J., and Guichard, J., 1969, Etude au microscope électronique de la localization des peroxydases dans les cellules de la moelle osseux humaine, *Nouv. Rev. Fr. Hématol.* **9**:678.

Breton-Gorius, J., and Reyes, F., 1976, Ultrastructure of human bone marrow cell maturation, *Int. Rev. Cytol.* **46**:251.

Briggs, R. S., Perillie, P. E., and Finch, S. C., 1966, Lysozyme in bone marrow and peripheral blood cells, *J. Histochem. Cytochem.* **14**:167.

Brune, K., Glatt, M., Kälin, H., and Peskar, B. A., 1978, Pharmacological control of prostaglandin and thromboxane release from macrophages, *Nature* **274**:261.

Bültmann, B., Bigazzi, P. C., Heymer, B., and Haferkamp, O., 1971, Peritoneal macrophage disappearance reaction in the rat, *Z. Immunitaetsforsch.* **142**:267.

Burgaleta, C., and Golde, D. W., 1977, Effect of glucan on granulopoiesis and macrophage genesis in mice, *Cancer Res.* **37**:1739.
Burgaleta, C., Territo, M. C., Quan, S. G., and Golde, D. W., 1978, Glucan-activated macrophages: Functional characteristics and surface morphology, *J. Reticuloendothel. Soc.* **23**:195.
Canonico, P. G., Beaufay, H., and Nyssens-Jadin, M., 1978, Analytical fractionation of mouse peritoneal macrophages: Physical and biochemical properties of subcellular organelles from resident (unstimulated) and cultivated cells, *J. Reticuloendothel. Soc.* **24**:115.
Cappell, D. F., 1930, Intravitam and supravital staining. IV. The cellular reactions following mild irritation of the peritoneum in normal and vitally stained animals, with special reference to the origin and nature of the mononuclear cells, *J. Pathol. Bacteriol.* **33**:429.
Carr, I., 1967, The fine structure of the cells of the mouse peritoneum, *Z. Zellforsch. Mikrosk. Anat.* **80**:534.
Carr, I., 1968, Lysosome formation and surface changes in stimulated peritoneal cells, *Z. Zellforsch. Mikrosk. Anat.* **89**:328.
Carr, I., 1970, The fine structure of the mammalian lymphoreticular system, *Int. Rev. Cytol.* **27**:283.
Carr, I., 1973, *The Macrophage,* Academic Press, London.
Carr, I., 1976, The RES and the mononuclear phagocyte system, in: *The Reticuloendothelial System in Health and Disease* (S. M. Reichard, M. R. Escobar, and H. Friedman, eds.), pp. 3–9, *Adv. Exp. Biol. Med.* **73A.**
Carr, I., Clarke, J. A., and Salsbury, A. J., 1969, The surface structure of mouse peritoneal cells—a study with the scanning electron microscope, *J. Microsc.* **89**:105.
Carr, I., Everson, G., Rankin, A., and Rutherford, J., 1970, The fine structure of the cell coat of the peritoneal macrophage and its role in the recognition of foreign material, *Z. Zellforsch. Microsk. Anat.* **105**:339.
Carter, S. B., and Chir, B., 1972, The cytochalasins as research tools in cytology, *Endeavour* **31**:77.
Casciato, D. A., Goldberg, L. S., and Bluestone, R., 1976, Collection of peritoneal exudate cells from small laboratory animals, *Vox Sang.* **31**:25.
Catanzaro, P. J., and Graham, R. C., 1974, Normal peritoneal lymphocytes, *Am. J. Pathol.* **77**:23.
Catanzaro, P. J., Graham, R. C., and Burns, C. P., 1974a, Mouse peritoneal lymphocytes: General properties of normal peritoneal lymphocytes, *J. Reticuloendothel. Soc.* **16**:150.
Catanzaro, P. J., Graham, R. C., and Hogrefe, W. R., 1974b, Mouse peritoneal lymphocytes: A morphologic comparison of normal and exudate peritoneal lymphocytes, *J. Reticuloendothel. Soc.* **16**:161.
Cavallo, T., 1974, Fine structural localization of endogeneous peroxidase activity in inner medullary interstitial cells of the rat kidney, *Lab. Invest.* **31**:458.
Cavallo, T., 1976, Cytochemical localization of endogeneous peroxidase activity in renal medullary collecting tubules and papillary mucosa of the rat, *Lab. Invest.* **34**:223.
Chang, K. P., 1978, Hamster peritoneal macrophages *in vitro:* Substratum adhesion, spreading, phagocytosis and phagolysosome formation, *In Vitro* **14**:663.
Chassoux, D., and Salomon, J.-C., 1975, Therapeutic effect of intratumoral injection of BCG and other substances in rats and mice, *Int. J. Cancer* **16**:515.
Cheung, H. T., Cantarow, W. D., and Sundharadas, G., 1978, Colchicine and cytochalasine B(CB) effects on random movement, spreading and adhesion of mouse macrophages, *Exp. Cell Res.* **111**:95.
Chin, K. W., and Hudson, G., 1974, Ultrastructural changes in murine peritoneal cells following cyclophosphamide administration, *Br. J. Exp. Pathol.* **55**:554.
Chu, J. Y., and Lin, H. S., 1976, Induction of macrophage colony-forming cells in the pleural cavity, *J. Reticuloendothel. Soc.* **20**:299.
Clark, E. R., and Clark, E. L., 1930, Relation of monocytes of the blood to the tissue macrophages, *Am. J. Anat.* **46**:149.
Clark, R. A., and Klebanoff, S. J., 1975, Neutrophil-mediated tumor cell cytotoxicity: Role of the peroxidase system, *J. Exp. Med.* **141**:1442.
Cline, M. J., 1970, Leucocyte function in inflammation: The ingestion, killing, and digestion of microorganisms, in: *Aspects of Inflammation II* (K. G. Jensen and S. A. Kilmann, eds.), pp. 3–16, *Ser. Haematol.* **III/2.**

Cohen, A. B., and Cline, M. J., 1971, The human alveolar macrophage: Isolation, cultivation *in vitro*, and studies of morphologic and functional characters, *J. Clin. Invest.* **50**:1390.
Cole, P., and Brostoff, J., 1975, Intracellular killing of *Listeria monocytogenes* by activated macrophages (Mackaness system) is due to antibiotic, *Nature* **256**:515.
Conrad, R. E., Yang, L. C., and Herscowitz, H. B., 1977, Mononuclear phagocytic cells in peritoneal exudates in rabbits: A comparison of inducing agents, *J. Reticuloendothel. Soc.* **21**:103.
Cook, J. A., Taylor, D., Cohen, C., Rodrigue, J., Malshet, V., and Di Luzio, N. R., 1978, Comparative evaluation of the role of macrophages and lymphocytes in mediating antitumor action of glucan, in: *Progress in Cancer Research and Therapy*, Vol. 7, *Immune Modulation and Control of Neoplasia by Adjuvant Therapy* (M. A. Chirigos, ed.), pp. 183–194, Raven Press, New York.
Cooper, G. W., and Houston, B., 1964, Effects of simple lipids on the phagocytic properties of peritoneal macrophages. II. Studies on the phagocytic potential of cell populations, *Aus. J. Exp. Biol. Med. Sci.* **42**:429.
Cotran, R. S., and Litt, M., 1969, The entry of granule-associated peroxidase into the phagocytic vacuoles of eosinophils, *J. Exp. Med.* **129**:1291.
Cotran, R. S., and Litt, M., 1970, Ultrastructural localization of horse-radish peroxidase and endogeneous peroxidase activity in guinea pig peritoneal macrophages, *J. Immunol.* **105**:1536.
Cowing, C., Schwartz, B. D., and Dickler, H. B., 1978, Macrophage Ia antigens. I. Macrophage populations differ in their expression of Ia antigens, *J. Immunol.* **120**:378.
Daems, W. T., 1976, On the role of the electron microscope in the identification of lysosomes, *Verh. Dtsch. Ges. Pathol.* **60**:1.
Daems, W. T., and Brederoo, P., 1970, The fine structure of mononuclear phagocytes as revealed by freeze-etching, in: *Mononuclear Phagocytes* (R. van Furth, ed.), pp. 29–42, Blackwell, Oxford.
Daems, W. T., and Brederoo, P., 1973, Electron microscopical studies on the structure, phagocytic properties, and peroxidatic activity of resident and exudate peritoneal macrophages in the guinea pig, *Z. Zellforsch. Mikrosk. Anat.* **144**:247.
Daems, W. T., and Koerten, H. K., 1978, The effects of various stimuli on the cellular composition of peritoneal exudates in the mouse, *Cell Tissue Res.* **190**:47.
Daems, W. T., and van der Rhee, H. J., 1980, Peroxidase and catalase in monocytes, macrophages, epithelioid cells and giant cells of the rat, in: *Mononuclear Phagocytes—Functional Aspects* (R. van Furth, ed.), Proceedings of the Third Leiden Conference on Mononuclear Phagocytes, September 17–18, 1978, Noordwijk, The Netherlands, pp. 43–60, Martinus Nijhoff Publishing, The Hague.
Daems, W. T., Poelmann, R. E., and Brederoo, P., 1973, Peroxidatic activity in resident peritoneal macrophages and exudate monocytes of the guinea pig after ingestion of latex particles, *J. Histochem. Cytochem.* **21**:93.
Daems, W. T., Wisse, E., Brederoo, P., and Emeis, J. J., 1975, Peroxidatic activity in monocytes and macrophages, in: *Mononuclear Phagocytes in Immunity, Infection and Pathology* (R. van Furth, ed.), pp. 57–77, Blackwell, Oxford.
Daems, W. T., Koerten, H. K., and Soranzo, M. R., 1976, Differences between monocyte-derived and tissue macrophages, in: *The Reticuloendothelial System in Health and Disease: Functions and Characteristics* (S. M. Reichard, M. R. Escobar, and H. Friedman, eds.), pp. 27–40, Plenum Press, New York.
Daems, W. T., Roos, D., Berkel, T. J. C. van, and Rhee, H. J. van der, 1979, The subcellular distribution and biochemical properties of peroxidase in monocytes and macrophages, in: *Lysosomes in Applied Biology and Therapeutics,* Vol. 6 (J. T. Dingle, I. H. Shaw, and P. Jacques, eds.), pp. 463–514, Elsevier North-Holland, Amsterdam.
Davies, G. E., and Thompson, A., 1975, Effects of corticosteroid treatment and inflammation on the cellular content of blood and exudate in mice, *J. Pathol.* **115**:17.
Davies, P., and Allison, A. C., 1976, Secretion of macrophage enzymes in relation to the pathogenesis of chronic inflammation, in: *Immunobiology of the Macrophage* (D. S. Nelson, ed.), pp. 428–461, Academic Press, New York.
Davies, P., Bonney, R. J., Humes, J. L., and Kuehl, F. A., 1977, The activation of macrophages with special reference to biochemical changes: A review, in: *The Macrophage and Cancer* (K. James,

B. McBride, and A. Stuart, ed.), Proceedings of the European Reticuloendothelial Society Symposium, Edinburgh, Sept. 12–14, 1977, pp. 19–30, EURES, Edinburgh.

Davis, R. H., and McGowan, L., 1968, Comparative peritoneal cellular content as related to species and sex, *Anat. Rec.* **162**:357.

De Bakker, J. M., van't Noordende, J. M., and Daems, W. T., 1980, The existence of a resident mouse peritoneal macrophage population as demonstrated by Imferon, *Ultramicroscopy*, in press.

Defendi, V., 1976, Macrophage cell lines and their uses in immunobiology, in: *Immunobiology of the Macrophage* (D. S. Nelson, ed.), pp. 275–289, Academic Press, New York.

Deimann, W., and Fahimi, H. D., 1977, The ontogeny of mononuclear phagocytes in fetal rat liver using endogenous peroxidase as a marker, in: *Kupffer Cells and Other Liver Sinusoidal Cells* (E. Wisse and D. L. Knook, eds.), Proceedings of the International Kupffer Cell Symposium, Noordwijkerhout, The Netherlands, 1977, pp. 487–495, Elsevier North-Holland, Amsterdam.

Deimann, W., and Fahimi, D. H., 1978, Peroxidase cytochemistry and ultrastructure of resident macrophages in fetal rat liver. A developmental study, *Dev. Biol.* **66**:43.

de Jong, A. S. H., Hak, T. J., Duijn, P. van, and Daems, W. T., 1979, A new dynamic model system for the study of capture reaction for diffusable compounds in cytochemistry. II. Effect of the composition of the incubation medium on the trapping of phosphate ions in acid phosphatase cytochemistry, *Histochem. J.* **11**:145.

de Petris, S. Karlsbad, G., and Pernis, B., 1962, Filamentous structures in the cytoplasm of normal mononuclear phagocytes, *J. Ultrastruct. Res.* **7**:39.

Di Luzio, N. R., 1976, Pharmacology of the reticuloendothelial system: accent on glucan, in: *The Reticuloendothelial System in Health and Disease* (S. M. Reichard, M. R. Escobar, and H. Friedman, eds.), pp. 412–421, Plenum Press, New York.

Dimitriu, A., Dy, M., Thomson, N., and Hamburger, J., 1975, Macrophage cytotoxicity in the mouse immune response against a skin allograft, *J. Immunol.* **114**:195.

Dreher, R., Keller, H. U., Hess, M. W., Roos, B., and Cottier, H., 1978, Early appearance and mitotic activity of multinucleated giant cells in mice after combined injection of talc and prednisolone acetate. A model for studying rapid histiocytic polykaryon formation *in vivo*, *Lab. Invest.* **38**:149.

Dumont, A., 1972, Ultrastructural aspects of phagocytosis of facultative intracellular parasites by hamster peritoneal macrophages, *J. Reticuloendothel. Soc.* **11**:469.

Dvorak, M., Horky, D., and Lokaj, J., 1966, The submicroscopic structure of peritoneal cells of normal and immunized mice, *Folia Morphol. (Prague)* **14**:245.

Dvorak, A. M., Hammond, M. E., Dvorak, H. F., and Karnovsky, M. J., 1972, Loss of cell surface material from peritoneal exudate cells associated with lymphocyte-mediated inhibition of macrophage migration from capillary tubes, *Lab. Invest.* **27**:561.

Edelson, P. J., and Cohn, Z. A., 1973, Peroxidase-mediated mammalian cell cytotoxicity, *J. Exp. Med.* **138**:318.

Edelson, P. J., and Cohn, Z. A., 1974, Effects of concanavalin A on mouse peritoneal macrophages, I. Stimulation of endocytic activity and inhibition of phago-lysosome formation, *J. Exp. Med.* **140**:1364.

Edelson, P. J., and Cohn, Z. A., 1976a, 5′-Nucleotidase activity of mouse peritoneal macrophages. I. Synthesis and degradation in resident and inflammatory populations, *J. Exp. Med.* **144**:1581.

Edelson, P. J., and Cohn, Z. A., 1976b, 5′-Nucleotidase activity of mouse peritoneal macrophages. II. Cellular distribution and effects of endocytosis, *J. Exp. Med.* **144**:1596.

Emeis, J. J., 1976, Morphologic and cytochemical heterogeneity of the cell coat of rat liver Kupffer cells, *J. Reticuloendothel. Soc.* **20**:31.

Emeis, J. J., and Planqué, B., 1976, Heterogeneity of cells isolated from rat liver by pronase digestion: Ultrastructure, cytochemistry, and cell culture, *J. Reticuloendothel. Soc.* **20**:11.

Emeis, J. J., and Wisse, E., 1971, Electron microscopic cytochemistry of the cell coat of Kupffer cells in rat liver, in: *The RES and Immune Phenomena* (N. R. di Luzio and K. Flemming, eds.), *Adv. Exp. Med. Biol.* **15**:1.

Emeis, J. J., and Wisse, E., 1975, On the cell coat of the rat liver Kupffer cells, in: *Mononuclear Phagocytes in Immunity, Infection and Pathology* (R. van Furth, ed.), pp. 315–325, Blackwell, Oxford.

Fahimi, H. D., Gray, B. A., and Herzog, V. K., 1976, Cytochemical localization of catalase and peroxidase in sinusoidal cells of rat liver, *Lab. Invest.* **34**:192.

Fedorko, M. E., and Hirsch, J. G., 1970, Structure of monocytes and macrophages, *Semin. Hematol.* **7**:109.

Felix, M. D., and Dalton, A. J., 1956, A phase-contrast microscope study of free cells native to the peritoneal fluid of DBA/2 mice, *J. Natl. Cancer Inst.* **16**:415.

Fishel, C. W., Halkias, D. G., Klein, T. W., and Szentivanyi, A., 1976, Characteristics of cells present in peritoneal fluids of mice injected intraperitoneally with *Bordetella pertussis, Infect. Immun.* **13**:263.

Forbes, I. J., 1966, Mitosis in mouse peritoneal macrophages, *J. Immunol.* **96**:734.

Forbes, I. J., and Mackaness, G. B., 1963, Mitosis in macrophages, *Lancet* **2**:1203.

Fruhman, G. J., 1973, Peritoneal macrophages in male and female mice, *J. Reticuloendothel. Soc.* **14**:371.

Gallily, R., and Eliahu, H., 1974, Uptake and degradation of a polypeptide antigen by stimulated and unstimulated macrophages from responder and non-responder mice, *Immunology* **26**:603.

Gallily, R., and Eliahu, H., 1976, Mechanism and specificity of macrophage-mediated cytotoxity, *Cellular Immunol.* **25**:245.

Gemsa, D., Seitz, M., Kramer, W., Till, G., and Resch, K., 1978, The effects of phagocytosis, dextran sulphate, and cell damage on PGE sensitivity and PGE production of macrophages, *J. Immunol.* **120**:1187.

Gerrard, J. M., White, J. G., Rao, G. H. R., and Townsend, W., 1976, Localization of platelet prostaglandin production in the platelet dense tubular system, *Am. J. Pathol.* **83**:283.

Ghaffar, A., and Cullen, R. T., 1976, *In vitro* behaviour of *Corynebacterium parvum*-activated cytotoxic macrophages, *J. Reticuloendothel. Soc.* **20**:349.

Ghaffar, A., McBride, W. H., and Cullen, R. T., 1976, Interaction of tumor cells and activated macrophages *in vitro:* Modulation by *Corynebacterium parvum* and gold salt, *J. Reticuloendothel. Soc.* **20**:283.

Ginsel, L. A., Debets, W. F., and Daems, W. T., 1975, The effect of colchicine on the distribution of apical vesicles and tubules in absorptive cells of cultured human small intestine, in: *Microtubules and Microtubule Inhibitors* (M. Borgers and M. de Brabander, eds.), pp. 187–198, North-Holland Amsterdam.

Golomb, H. M., Braylan, R., and Polliack, A., 1975, "Hairy" cell leukaemia (leukaemic reticuloendotheliosis): A scanning electron microscopic study of eight cases, *Br. J. Haematol.* **29**:455.

Gordon, S., and Cohn, Z. A., 1973, The macrophage, *Int. Rev. Cytol.* **36**:171.

Gordon, S., Todd, J., and Cohn, Z. A., 1974, *In vitro* synthesis and secretion of lysozyme by mononuclear phagocytes, *J. Exp. Med.* **139**:1228.

Gordon, S., Unkeless, J. C., and Cohn, Z. A., 1975, The macrophage as secretory cell, in: *Immune Recognition* (A. S. Rosenthal, ed.), pp. 589–614, Academic Press, New York.

Goud, T. J. L. M., Scholte, C. G., and van Furth, R., 1975, Identification and characterization of the monoblast in mononuclear phagocyte colonies grown *in vitro, J. Exp. Med.* **142**:1180.

Gray, B., Herzog, V., and Fahimi, H. D., 1973, Localization of catalase in sinusoidal cells (SLC) of rat liver, *J. Cell Biol.* **59**:120a.

Güttner, J., Augusten, K., Bimberg, R., and Lange, P., 1975, Modification of the surface structure of murine peritoneal macrophages following chemotherapy, *Exp. Pathol.* **11**:209.

Hard, G. C., 1969, Electron microscopic study of the differentiation of mouse peritoneal macrophages stimulated by *Corynebacterium ovis* infection, *Lab. Invest.* **21**:309.

Hardy, B., Skutelsky, E., Globerson, A., and Danon, D., 1976, Ultrastructural differences between macrophages of newborn and adult mice, *J. Reticuloendothel. Soc.* **19**:291.

Harper, J. R., and Wolf, N. S., 1972, Effects of X-irradiation upon the response of mononuclear phagocytes to an induced sterile intraperitoneal inflammation, *J. Reticuloendothel. Soc.* **11**:368.

Helentjaris, T. G., Lombardi, P. S., and Glasgow, L. A., 1974, Cytochalasin B sensitivity of normal and activated macrophage adhesion to surfaces, *J. Reticuloendothel. Soc.* **16**:5a.

Hibbs, J. B., 1974, Heterocytolysis by macrophages activated by Bacillus Calmette-Guérin: Lysosome exocytosis into tumour cells, *Science* **184**:468.

Hibbs, J. B., 1975a, Activated macrophages as cytotoxic effector cells. I. Inhibition of specific and non-specific tumor resistance by trypan blue, *Transplantation* **19**:77.

Hibbs, J. B., 1975b, Activated macrophages as cytotoxic effector cells. II. Requirement for local persistence of inducing antigen, *Transplantation* 19:81.

Hibbs, J. B., 1976, Role of activated macrophages in nonspecific resistance to neoplasia, *J. Reticuloendothel. Soc.* **20**:223.

Hirsch, J. G., 1956, Phagocytin: a bactericidal substance from polymorphonuclear leucocytes, *J. Exp. Med.* **103**:589.

Hirsch, J. G., and Fedorko, M. E., 1968, Ultrastructure of human leukocytes after simultaneous fixation with glutaraldehyde and osmium tetroxide and "postfixation" in uranyl acetate, *J. Cell Biol.* **38**:615.

Hirsch, J. G., and Fedorko, M. E., 1970, Morphology of mouse mononuclear phagocytes, in: *Mononuclear Phagocytes* (R. van Furth, ed.), pp. 7–28, Blackwell, Oxford.

Hoefsmit, E. C. M., 1975, Mononuclear phagocytes, reticulum cells, and dendritic cells in lymphoid tissues, in: *Mononuclear Phagocytes in Immunity, Infection and Pathology* (R. van Furth, ed.), pp. 129–146, Blackwell, Oxford.

Hoefsmit, E. C. M., and Smelt, A. H. M., 1978, Milky spots and peritoneal exudate and resident macrophages in a state of cell mediated immunity, *Ultramicroscopy* **3**:135.

Humes, J. L., Bonney, R. J., Relus, L., Dahlgren, M. E., Sadowski, S. J., Kuehl, F. A., and Davies, P., 1977, Macrophages synthesize and release prostaglandins in response to inflammatory stimuli, *Nature(London)* **269**:149.

Imai, Y., Kasajima, T., and Matsuda, M., 1973, Electron microscopic study on the peritoneal macrophage and milky spot in omentum, *Rec. Adv. RES Res.* **11**:54.

Janssen, W. A., and Dangerfield, H. G., 1977, A rapid method for determining the percentage of antibacterial phagocytes in a sample population of leukocytes, *J. Reticuloendothel. Soc.* **21**:299.

Jenkin, C. R., and Rowley, D., 1963, Basis for immunity to typhoid in mice and the question of cellular immunity, *Bacterial. Rev.* **27**:391.

Joos, F., Roos, B., Bürki, H., Bürki, K., and Laissue, J., 1969, Umsatz, Proliferation und Phagozytosetätigkeit der freien Zellen im Peritonäalraum der Maus nach Injektion von Polystyren-Partikeln, *Z. Zellforsch. Mikrosk. Anat.* **95**:68.

Johnston, R. B., Godzik, C. A., and Cohn, Z. A., 1978, Increased superoxide anion production by immunologically activated and cytochemically elicited macrophages, *J. Exp. Med.* **148**:115.

Jungi, T. W., and McGregor, D. D., 1978, Impaired chemotactic responsiveness of macrophages from gnotobiotic rats, *Infect. Immun.* **19**:553.

Kaiserling, E., Lennert, K., Nitsch, K., and Drescher, J., 1972, Ultrastructure and pathogenesis of the BCG-histiocytosis (BCG-granulamatosis), *Virchows Arch. A* **355**:333.

Kaplan, G., and Bertheussen, K., 1977, The morphology of echinoid phagocytes and mouse peritoneal macrophages during phagocytosis *in vitro*, *Scand. J. Immunol.* **6**:1289.

Kaplan, G., and Morland, B., 1978, Properties of a murine monocytic tumor cell line J-774 *in vitro*. I. Morphology and endocytosis, *Exp. Cell Res.* **115**:53.

Karnovsky, M. L., 1975, Biochemical aspects of the function of polymorphonuclear and mononuclear leukocytes, in: *The Phagocytic Cell in Host Resistance* (J. A. Bellanti and D. H. Dayton, eds.), pp. 25–43, Raven Press, New York.

Karnovsky, M. L., Lazdins, J., Drath, D., and Harper, A., 1975, Biochemical characteristics of activated macrophages, *Ann. N.Y. Acad. Sci.* **256**:266.

Kay, N. E., and Douglas, S. D., 1977, Mononuclear phagocyte. Development, structure, function, and involvement in immune response, *N.Y. State J. Med.* **1977**:327.

Keller, R., 1974, Mechanisms by which activated normal macrophages destroy syngeneic rat tumour cells *in vitro*. Cytokinetics, non-involvement of T lymphocytes, and effect of metabolic inhibitors, *Immunology* **27**:285.

Keller, R., 1976, Promotion of tumor growth *in vivo* by anti-macrophage agents, *J. Natl. Cancer Inst.* **57**:1355.

Klebanoff, S. J., and Clark, R. A., 1975, Hemolysis and iodination of erythrocyte components by a myeloperoxidase-mediated system, *Blood* **45**:699.

Klebanoff, S. J., and Hamon, C. G., 1975, Antimicrobial systems of mononuclear phagocytes, in: *Mononuclear Phagocytes in Immunity, Infection and Pathology* (R. van Furth, ed.), pp. 507–531, Blackwell, Oxford.

Koerten, H. K., Ploem, J. S., and Daems, W. T., 1979, SEM and reflection contrast study of the filopodia of adherent mouse resident peritoneal macrophages, *Ultramicroscopy* **4**:150.

Kojima, M., 1976, Tumor growth of the reticuloendothelial system, *Acta Pathol. Jpn.* **26**:273.

Komiyama, A., Spicer, S. S., Bank, H., and Farrington, J., 1975, Induction of autophagic vacuoles in peritoneal cells, *J. Reticuloendothel. Soc.* **17**:146.

Kondo, E., and Kanai, K., 1977, Phospholipid distribution pattern in uninduced (resident) and casein-induced mouse peritoneal cells, *Jpn. J. Med. Sci. Biol.* **30**:269.

Kornfeld, L., and Greenman, V., 1966, Effects of total-body X-irradiation on peritoneal cells of mice, *Radiat. Res.* **29**:433.

Kurland, J. I., and Bockman, R., 1978, Prostaglandin E production by human blood monocytes and mouse peritoneal macrophages, *J. Exp. Med.* **147**:952.

Kurland, J., Broxmeijer, H., and Moore, M. A. A., 1977, Auto-regulation of committed granulocyte-macrophage progenitor cell proliferation by humoral mediators of the mononuclear phagocyte (abstr.), *Exp. Haematol* **5** (suppl. 2):5.

Kwan, D., Epstein, M. B., and Norman, A., 1976, Studies on human monocytes with a multiparameter cell sorter, *J. Histochem. Cytochem.* **24**:355.

Lajtha, L. G., 1976, Influence of surface on white cell movements *Blood Cells* **2**:411.

Langevoort, H. L., Cohn, Z. A., Hirsch, J. G., Humphrey, J. H., Spector, W. G., and van Furth, R., 1970, The nomenclature of mononuclear phagocytic cells. Proposal of a new classification, in: *Mononuclear Phagocytes* (R. van Furth, ed.), pp. 1–6, Blackwell, Oxford.

Lazarides, E., 1978, Comparison of the structure, distribution, and possible function of desmin (100 Å) filaments in various types of muscle and nonmuscle cells, in: *The Molecular Basis of Cell–Cell Interactions* (R. A. Lerner, and D. Bergsma, eds.), pp. 41–63, Alan R. Liss, New York.

Lazdins, J. K., Kühner, L., David, J. R., and Karnovsky, M. L., 1978, Alteration of some functional and metabolic characteristics of resident mouse peritoneal macrophages by lymphocyte mediators, *J. Exp. Med.* **148**:746.

Lee, K.-C., and Berry, D., 1977, Functional heterogeneity in macrophages activated by *Corynebacterium parvum:* Characterization of subpopulations with different activities in promoting immune responses and suppressing tumor cell growth, *J. Immunol.* **118**:1530.

Leibovici, J., Sinai, Y., Wolman, M., and Davidai, G., 1975, Effects of high-molecular levan on the growth and spread of lymphoma in AKR mice, *Cancer Res.* **35**:1921.

Lejeune, F. J., 1975, Role of macrophages in immunity, with special reference to tumour immunology. A review, *Biomed. J.* **22**:25.

Lejeune, F. J., Beaumont, E., and Garcia, J., 1973, Growth inhibitory effect of peritoneal macrophages on H.P.M. Its impairment by macrophage lysosome overloading, *Br. J. Cancer* **28**:80.

Lepper, A. W. D., and D'Arcy Hart, P., 1976, Peroxidase staining in elicited and non-elicited mononuclear peritoneal cells from BCG-sensitized and non-sensitized mice, *Infect. Immun.* **14**:522.

Lin, H., 1974, Peritoneal exudate cells, II. Kinetics of appearance of colony-forming cells, *J. Cell Physiol.* **84**:159.

Lin, H. S., and Devaraj, B., 1974, Effect of physical and chemical agents on the colony forming ability of peritoneal exudate cells, *J. Reticuloendothel. Soc.* **16**:30a.

Lin, H., and Stewart, C. C., 1973, Colony formation by mouse exudate peritoneal cells *in vitro, Nature New Biol.* **243**:176.

Lin, H., and Stewart, C. C., 1974, Peritoneal exudate cells. I. Growth requirement of cells capable of forming colonies in soft agar, *J. Cell Physiol.* **83**:369.

Lin, H., Freeman, P., and Devaraj, B., 1975, Characterization of mouse peritoneal colony-forming cells, *J. Reticuloendothel. Soc.* **18**:276.

Lin, H., Kuhn, C., and Stewart, C. C., 1978, Peritoneal exudate cells. V. Influence of age, sex, strain and species on the induction and the growth of macrophage colony forming cells, *J. Cell. Physiol.* **96**:133.

Lin, H. L., Bianco, C., and Cohn, Z. A., 1980, The iodination and turnover of macrophage plasma

membrane polypeptides, in: *Mononuclear Phagocytes—Functional Aspects* (R. van Furth, ed.), pp. 649–664, M. Nijhoff, The Hague.
Litwin, J. A., 1977, Does diaminobenzidine demonstrate prostaglandin synthetase?, *Histochemistry* **53**:301.
Lutton, J. D., and Kiremidjian, L., 1973, Electrokinetic studies on normal and thioglycollate stimulated peritoneal macrophages, *Exp. Cell Res.* **79**:492.
Machado, E. A., and Lair, S. V., 1978, Giant multinucleate macrophages in methyl-cellulose stimulated athymic nude mice, *J. Reticuloendothel. Soc.* **23**:383.
Mackaness, G. B., 1962, Cellular resistance to infection, *J. Exp. Med.* **116**:381.
Mackaness, G. B., 1970, The mechanisms of macrophage activation, in: *Infectious Agents and Host Reactions* (S. Mudd, ed.), pp. 61–75, Saunders, Philadelphia.
Martin, F. F., Vilseck, J., Dobbine, W. O., Buckley, C. E., and Tyor, M. P., 1972, Immunological alterations in patients with treated Whipple's disease, *Gastroenterology* **63**:6.
Mayhew, T. M., and Williams, M. A., 1974, A quantitative morphological analysis of macrophage stimulation. I. A study of subcellular compartments and of the cell surface, *Cell Tissue Res.* **147**:567.
McIntyre, J., Rowley, D., and Jenkin, C. R., 1967, The functional heterogeneity of macrophages at the single cell level, *Aust. J. Exp. Biol. Med.* **45**:675.
McLaughlin, J. F., Ruddle, N. H., and Waksman, B. H., 1972, Relationship between activation of peritoneal cells and their cytopathogenicity, *J. Reticuloendothel. Soc.* **12**:293.
Meltzer, M. S., Tucker, R. W., Sanford, K. K., and Leonard, E. J., 1975, Interaction of BCG-activated macrophages with neoplastic and nonneoplastic cell lines *in vitro:* Quantification of the cytotoxic reaction by release of tritiated thymidine from prelabeled target cells, *J. Natl. Cancer Inst.* **54**:1177.
Meuret, G., 1976, Das Monozyten-Makrophagen System, in: *Handbuch der Inneren Medizin, Leukozytäres und Retikuläres System I* (H. Begemann, ed.), Vol. II/3, pp. 361–437, Springer Verlag, Berlin.
Meuret, G., Detel, U., Kilz, H. P., Senn, H. J., and van Lessen, H., 1975, Human monocytopoiesis in acute and chronic inflammation, *Acta Haematol.* **54**:328.
Michaelis, T. W., Larrimer, N. R., Metz, E. N., and Balcerzak, S. P., 1971, Surface morphology of human leucocytes, *Blood* **37**:23.
Michl, J., and Silverstein, S. C., 1978, Role of macrophage receptors in ingestion phase of phagocytosis, in: *Molecular Basis of Cell–Cell Interaction. Birth Defects: Original Article Series* **14**:99.
Miller, T. E., 1971, Metabolic event involved in the bactericidal activity of normal mouse macrophages, *Infect. Immun.* **3**:390.
Miller, K., and Kagan, E., 1976, The *in vivo* effects of asbestos on macrophage membrane structure and population characteristics of macrophages: A scanning electron microscope study, *J. Reticuloendothel. Soc.* **20**:159.
Mims, C. A., 1963, The peritoneal macrophages of mice, *Br. J. Exp. Pathol.* **45**:37.
Morahan, P. S., and Kaplan, A. M., 1976, Macrophage activation and anti-tumor activity of biologic and synthetic agents, *Int. J. Cancer* **17**:82.
Morahan, P. S., and Kaplan, A. M., 1978, Antiviral and antitumor functions of activated macrophages, in: *Immune Modulation and Control of Neoplasia by Adjuvant Theory* (M. A. Chirigos, ed.), pp. 447–457, Raven Press, New York.
Morahan, P. S., Glasgow, L. A., Crane, J. L., and Kern, E. R., 1977, Comparison of antiviral and antitumor activity of activated macrophages, *Cell. Immunol.* **28**:404.
Morningstar, W. A., 1975, Whipple's disease: An example of the value of the electron microscope in diagnosis, follow-up, and correlation of a pathologic process, *Human Pathol.* **6**:443.
Naito, M., 1975, Electron microscopical study on the origin of Kupffer cells, *Rec. Adv. RES Res.* **15**:76.
Naito, M., and Wisse, E., 1977, Observations on the fine structure and cytochemistry of sinusoidal cells in fetal and neonatal rat liver, in: *Kupffer Cells and Other Liver Sinusoidal Cells* (E. Wisse and D. L. Knook, eds.), Proceedings of the International Kupffer Cell Symposium, Noordwijkerhout, The Netherlands, 1977, pp. 497–505, Elsevier North-Holland, Amsterdam.
Nathan, C. F., Hill, V. M., and Terry, W. D., 1976, Isolation of a subpopulation of adherent peritoneal cells with anti-tumor activity, *Nature* **260**:146.
Nelson, D. S. (ed.), 1976, *Immunobiology of the Macrophage*, Academic Press, New York.

Nelson, D. S., and Boyden, S. W., 1963, The loss of macrophages from peritoneal exudates following the injection of antigens into guinea pigs with delayed-type hypersensitivity, *Immunology* **6**:264.

Nelson, M., and Nelson, D. S., 1978, Macrophages and resistance to tumors. II. Influence of agents affecting macrophages and delayed-type hypersensitivity on resistance to a tumor inducing specific "sine comitant" immunity: Acquired resistance as an expression of delayed-type hypersensitivity, *Cancer Immunol. Immunother.* **4**:101.

Nichols, B. A., and Bainton, D. F., 1973, Differentiation of human monocytes in bone marrow and blood. Sequential formation of two granule populations, *Lab. Invest.* **29**:27.

Nichols, B. A., and Bainton, D. F., 1975, Ultrastructure and cytochemistry of mononuclear phagocytes, in: *Mononuclear Phagocytes in Immunity, Infection, and Pathology* (R. van Furth, ed.), pp. 17–55, Blackwell, Oxford.

Nichols, B. A., Bainton, D. F., and Farquhar, M. G., 1971, Differentiation of monocytes. Origin, nature and fate of their azurophil granules, *J. Cell Biol.* **50**:498.

North, R. J., 1966, The localization by electron microscopy of nucleoside phosphatase activity in guinea pig phagocytic cells, *J. Ultrastruct. Res.* **16**:83.

Novikoff, A. B., and Novikoff P. M., 1973, Microperoxisomes, *J. Histochem. Cytochem.* **21**:963.

Novikoff, A. B., Novikoff, P. M., Davis, D., and Quintana, N., 1973, Studies on microperoxisomes. V. Are microperoxisomes ubiquitors in mammalian cells?, *J. Histochem. Cytochem.* **21**:737.

Ogawa, T., Koerten, H. K., and Daems, W. T., 1978, Peroxidatic activity in monocytes and tissue macrophages of mice, *Cell Tissue Res.* **188**:361.

Ohki, S., Ogino, N., Yamamoto, S., and Hayashi, O., 1978, Peroxidase activity of prostaglandin endoperoxide synthetase from bovine vesicular gland microsomes, *Prostaglandins* **15**:718.

Ohta, H., and Shimizu, K., 1975, Maturation and interrelationship of mouse mononuclear phagocytes in bone marrow, peripheral blood and peritoneal cavity in terms of erythrophagocytic activity, *Tokohu J. Exp. Med.* **116**:111.

Olivotto, M., and Bomford, R., 1974, *In vitro* inhibition of tumour cell growth and DNA synthesis by peritoneal and lung macrophages from mice injected with *Corynebacterium parvum, Int. J. Cancer* **32**:478.

Orenstein, J. M., and Shelton, E., 1976a, Surface topography and interactions between mouse peritoneal cells allowed to settle on an artificial substrate: Observations by scanning electron microscopy, *Exp. Mol. Pathol.* **24**:201.

Orenstein, J. M., and Shelton, E., 1976b, Surface topography of leukocytes *in situ:* cells of mouse peritoneal milky spots, *Exp. Mol. Pathol.* **24**:415.

Orenstein, J. M., and Shelton, E., 1977, Membrane phenomena accompanying erythrophagocytosis. A scanning electron microscope study, *Lab. Invest.* **36**:363.

Padawer, J., 1973, The peritoneal cavity as a site for studying cell-cell and cell-virus interactions, *J. Reticuloendothel. Soc.* **14**:462.

Padawer, J., and Gordon, A. S., 1956, Cellular elements in the peritoneal fluid of some mammals, *Anat. Rec.* **124**:209.

Page, R. C., Davies, P., and Allison, A. C., 1978, The macrophage as a secretory cell, *Int. Rev. Cytol.* **52**:119.

Papadimitriou, J. M., 1973, Detection of macrophage receptors for heterologous IgG by scanning and transmission electron microscopy, *J. Pathol.* **110**:213.

Parakkal, P., Pinto, J., and Hanifin, J. M., 1974, Surface morphology of human mononuclear phagocytes during maturation and phagocytosis, *J. Ultrastr. Res.* **48**:216.

Parwaresch, M. R., Müller-Hermelink, H. K., Desaga, J. F., Zakari, V., and Lennert, K., 1971, Die Herkunft der Gewebsmastzellen bei der Ratte, zugleich ein Beitrag zur quantitativen Cytologie der sterilen Peritonitis, *Virch. Arch. B* **8**:20.

Paul, B. B., Strauss, R. R., Selvaraj, R. J., and Sbarra, A. J., 1973, Peroxidase mediated antimicrobial activities of alveolar macrophage granules, *Science* **181**:849.

Pesanti, E. L., and Axline, S. G., 1975, Phagolysosome formation in normal and colchicine-treated macrophages, *J. Exp. Med.* **142**:903.

Petrik, P., 1971, Fine structural identification of peroxisomes in mouse and rat bronchiolar and alveolar epithelium, *J. Histochem. Cytochem.* **19**:339.

Poelmann, R. E., and Daems, W. T., 1973, Problems associated with the demonstration by lead methods of adenosine triphosphatase activity in resident peritoneal macrophages and exudate monocytes of the guinea pig, *J. Histochem. Cytochem.* **21**:488.

Polliack, A., and Gordon, S., 1975, Scanning electron microscopy of murine macrophages. Surface characteristics during maturation, activation and pinocytosis, *Lab. Invest.* **33**:469.

Porter, K. R., 1973, Microtubules in intracellular locomotion, in: *Locomotion of Tissue Cells,* Ciba Foundation Symposium 14, pp. 149–169, Elsevier Excerpta Medica-North Holland, Amsterdam.

Puvion, F., Fray, A., and Halpern, B., 1974, A comparative, scanning electron microscope study of the interaction between stimulated or unstimulated mouse peritoneal macrophages and tumor cells, in: *Corynebacterium parvum* (B. Halpern, ed.), pp. 137–144, Plenum Press, New York.

Puvion, F., Fray, A., and B. Halpern, 1976, A cytochemical study of the *in vitro* interaction between normal and activated mouse peritoneal macrophages and tumor cells, *J. Ultrastruct. Res.* **54**:95.

Ratzan, K. R., Musher, D. M., Keusch, G. T., and Weinstein, L., 1972, Correlation of increased metabolic activity, resistance to infection, enhanced phagocytosis, and inhibition of bacterial growth by macrophages from *Listeria*- and BCG-infected mice, *Infect. Immun.* **5**:499.

Raz, A., and Goldman, R., 1976, Effect of hashish compounds on mouse peritoneal macrophages, *Lab. Invest.* **34**:69.

Raz, A., Shahar. A., and Goldman, R., 1977, Characterization of an *in vivo* induced peritoneal macrophage population following intraperitoneal injection of concanavalin A., *J. Reticuloendothel. Soc.* **22**:445.

Reikvam, A., and Hoiby, E. A., 1975, Phagocytosis and microbicidal capacity of mouse macrophages non-specifically activated *in vitro, Acta Pathol. Microbiol. Scand. Sect. C* **83**:121.

Reikvam, A., Grammeltdedt, R., and Hoiby, E. A., 1975, Activated mouse macrophages: Morphology, lysosomal biochemistry, and microbicidal properties of *in vivo* and *in vitro* activated cells, *Acta Pathol. Microbiol. Scand. Sect. C* **83**:129.

Rhodes, J., 1975, Macrophage heterogeneity in receptor activity: The activation of macrophage Fc receptor function *in vivo* and *in vitro, J. Immunol.* **114**:976.

Rhodes, J. M., Birch-Andersen, A., and Ravn, H., 1975, The effect of cytophosphamide, methotrexate and X-irradiation on the ultrastructure and endocytic capacity of murine peritoneal macrophages, *Acta Pathol. Microbiol. Scand. Sect. A* **83**:443.

Rhodes, J. M., Nielsen, G., Olesen Larsen, S., Bennedsen, J., and Riisgaard, S., 1977, *In vitro* studies on normal, stimulated and immunologically activated mouse macrophages. II. Degradation of radioactive antigen/antibody complexes, *Acta Pathol. Microbiol. Scand. Sect. C* **85**:239.

Robbins, D., Fahimi, H. D., and Cotran, R. S., 1971, Fine structural cytochemical localization of peroxidase activity in rat peritoneal cells, mononuclear cells, eosinophils, and mast cells, *J. Histochem. Cytochem.* **19**:571.

Robertson, T. A., Papadimitriou, J. M., Walters, M. N.-I., and Wolman, M., 1977, Effects of exposure of murine peritoneal exudate and resident macrophages to high molecular levan: A morphological study, *J. Pathol.* **123**:157.

Roos, B., 1970, Makrophagen: Herkunft, Entwicklung, und Funktion, in: *Handbuch Allgemeine Pathologie,* Vol. VII/3 (A. Studer and H. Cottier, eds.) pp. 1–128, Springer Verlag, Berlin.

Roos, D., Reiss, M., Balm, A. J. M., Palache, A. M., Cambier, P. H., and van der Stijl-Neijenhuis, J. S., 1980, A metabolic comparison between human blood monocytes and neutrophils, in: *Macrophages and Lymphocytes* (M. R. Escobar, ed.), Proceedings of the Eighth International RES Congress, Jerusalem, 1978, pp. 29–37, Plenum Press, New York.

Root, R. K., Rosenthal, A. S., and Balestra, 1972, Abnormal bactericidal, metabolic, and lysosomal functions of Chediak, Higashi syndrome leukocytes, *J. Clin. Invest.* **51**:649.

Rosenstreich, D. L., and Mizel, S. B., 1978, The participation of macrophages and macrophage cell lines in the activation of T lymphocytes by mitogens, *Immunol. Rev.* **40**:102.

Roser, B., 1970, The origin, kinetics, and fate of macrophage populations, *J. Reticuloendothel. Soc.* **8**:139.

Rowley, D., 1966, Phagocytosis and immunity, *Experientia* **22**(1):1.

Sabin, F. R., Doan, C. A., and Cunningham, B. S., 1925, Discrimination of two types of phagocytic cells in the connective tissues by the supravital technique, *Contrib. Embryol.* **82**:125.

Sannes, P. L., McDonald, J. K., and Spicer, S. S., 1977, Dipeptidyl aminopeptidase II in rat peritoneal wash cells. Cytochemical localization and biochemical characterization, *Lab. Invest.* **37**:243.

Schellens, J. P. M., Daems, W. T., Emeis, J. J., Brederoo, P., de Bruijn, W. C., and Wisse, E., 1977, Electron microscopical identification of lysosomes, in: *Lysosomes, A Laboratory Handbook* (J. T. Dingle, ed.), 2nd ed., pp. 147–208, North-Holland, Amsterdam.

Schnyder, J., and Baggiolini, M., 1978, Secretion of lysosomal hydrolases by stimulated and nonstimulated macrophages, *J. Exp. Med.* **148**:435.

Seeman, G., 1930, Über die Beziehungen zwischen Lymphocyten, Monocyten und Histiocyten, insbesondere bei Entzündung, *Beitr. Pathol. Anat.* **85**:304.

Shands, J. W., and Axelrod, B. J., 1977, Mouse peritoneal macrophages: Tritiated thymidine labeling and cell kinetics, *J. Reticuloendothel. Soc.* **21**:69.

Shands, J. W., Peavy, D. L., Gormus, B. J., and McGraw, J., 1974, *In vitro* and *in vivo* effects of endotoxin on mouse peritoneal cells, *Infect. Immun.* **9**:106.

Sheagren, J., and Hahn, H., 1974, Macrophage activation, in: *Progress in Immunology II*, Vol. 2 (L. Brant and J. Holborow, eds.), pp. 330–332, North-Holland, Amsterdam.

Silverstein, S. C., 1977, Endocytic uptake of particles by mononuclear phagocytes and the penetration of obligate intracellular parasites, *Am. J. Trop. Med. Hyg.* **26**:161.

Simmons, S. R., and Karnovsky, M. L., 1973, Iodinating ability of various leukocytes and their bactericidal activity, *J. Exp. Med.* **138**:44.

Simson, J. V., and Spicer, S. S., 1973, Activities of specific cell constituents in phagocytosis (endocytosis), *Int. Rev. Pathol.* **12**:79.

Sljivic, V. S., and Warr, G. W., 1975, Role of cellular proliferation in the stimulation of MPS phagocytic activity, *Br. J. Pathol.* **56**:314.

Slonecker, C. E., 1971, The cellular composition of an acute inflammatory exudate in rats, *J. Reticuloendothel. Soc.* **10**:269.

Smialowicz, R. J., and Schwab, J. H., 1977, Processing of streptococcal cell walls by rat macrophages and human monocytes *in vitro*, *Infect. Immun.* **17**:591.

Smith, C. W., and Goldman, A. S., 1972a, Effects of concanavalin A and pokeweed mitogen *in vivo* on mouse peritoneal macrophages, *Exp. Cell Res.* **73**:394.

Smith, C. W., and Goldman, A. S., 1972b, Selective effects of thermal injury on mouse peritoneal macrophages, *Infect. Immun.* **5**:938.

Snodgrass, M. J., Harris, T. M., Geeraets, R., and Kaplan, A. M., 1977, Motility and cytotoxicity of activated macrophages in the presence of carcinoma cells, *J. Reticuloendothel. Soc.* **22**:149.

Snodgrass, M. J., Harris, T. M., and Kaplan, A. M., 1978, Chemokinetic response of activated macrophages to soluble products of neoplastic cells, *Cancer Res.* **38**:2925.

Snyderman, R., and Meyenhagen, S. E., 1976, Chemotaxis of macrophages, in: *Immunobiology of the Macrophage* (D. S. Nelson, ed.), pp. 323–346, Academic Press, New York.

Sonoda, M., and Kobayashi, K., 1970, Monocytes of canine peripheral blood in electron microscopy, *Jpn. J. Vet. Res.* **18**:67.

Spector, W. G., 1977, Macrophage turnover and traffic, in: *The Macrophage and Cancer* (K. James, B. McBride, and A. Stuart, eds.), pp. 21–24, Proceedings of the European Reticuloendothelial Society Symposium, Edinburgh, Sept. 12–14, 1977, EURES, Edinburgh.

Stanley, E. R., Cifone, M., Heard, P. M., and Defendi, V., 1976, Factors regulating macrophage production and growth: Identity of colony-stimulating factor and macrophage growth factor, *J. Exp. Med.* **143**:631.

Stewart, C. C., Lin, A. S., and Adles, C., 1975, Proliferation and colony-forming ability of peritoneal exudate cells in liquid culture, *J. Exp. Med.* **141**:114.

Stuart, A. E., 1977, The heterogeneity of macrophages: a review, in: *The Macrophage and Cancer* (K. James, B. McBride, and A. Stuart, eds.), pp. 1–14, Proceedings of the European Reticuloendothelial Society Symposium, Edinburgh, Sept. 12–14, 1977, EURES, Edinburgh.

Stubbs, M., Kühner, A. V., Glass, E. A., David, J. R., and Karnovsky, M. L., 1973, Metabolic and functional studies on activated mouse macrophages, *J. Exp. Med.* **137**:537.

Sutton, J. S., 1967, Ultrastructural aspects of *in vitro* development of monocytes into macrophages, epithelioid cells, and multinucleated giant cells, *Natl. Cancer Inst. Monogr.* **26**:75.

Takahashi, H., Enzon, H., and Ohkita, T., 1973, Studies on cell kinetics of peritoneal macrophages, *Rec. Adv. RES Res.* **11**:85.

Takayama, H., Katsumoto, T., and Takagi, A., 1975, Ultrastructural studies on the phagocytosis of the cultured mouse macrophages and the effects of cytochalasin B., colchicine and pH conditions on their morphology, *J. Electr. Microsc.* **24**:145.

Thomas, J. A., 1949, Conception du système réticulo-histiocytaire: La régulation de l'état histiocytaire et la spécificité cellulaire, *Rev. Hémat.* **4**:639.

Thomas, M. A., Galbraith, I., and MacSween, R. N. M., 1978, Heterogeneity of rat peritoneal and alveolar macrophage populations: Spontaneous rosette formation using sheep and chicken red blood cells, *J. Reticuloendothel. Soc.* **23**:43.

Thompson, J., and van Furth, R., 1973, The effect of glucocorticosteroids on the proliferation and kinetics of promonocytes and monocytes of the bone marrow, *J. Exp. Med.* **137**:10.

Tomazic, V., Bigazzi, P. E., and Rose, N. R., 1977, The macrophage disappearance reaction (MDR) as an *in vivo* test of delayed hypersensitivity in mice, *Immunol. Comm.* **6**:49.

Unkeless, J. C., Gordon, S., and Reich, R., 1974, Secretion of plasminogen activator by stimulated macrophages, *J. Exp. Med.* **139**:834.

van Berkel, T. J. C., 1974, Difference spectra, catalase and peroxidase activities of isolated parenchymal and nonparenchymal cells from rat liver, *Biochem. Biophys. Res. Comm.* **61**:204.

van der Rhee, H. J., de Winter, C. P. M., and Daems, W. T., 1977, Fine structure and peroxidatic activity of rat blood monocytes, *Cell Tissue Res.* **185**:1.

van der Rhee, H. J., Brederoo, P., and Daems, W. T., 1978, *In vivo* adherence of monocytes does not lead to the appearance of peroxidatic activity in the rough endoplasmic reticulum, *Cell Biol. Int. Rep.* **2**:99.

van der Rhee, H. J., van der Burgh-de Winter, C. P. M., and Daems, W. T., 1979a, The differentiation of monocytes into macrophages, epithelioid cells, and multinucleated giant cells in subcutaneous granulomas. I. Fine structure, *Cell Tissue Res.* **197**:355.

van der Rhee, H. J., van der Burgh-de Winter, C. P. M., and Daems, W. T., 1979b, The differentiation of monocytes into macrophages, epithelioid cells, and multinucleated giant cells in subcutaneous granulomas. II. Peroxidatic activity, *Cell Tissue Res.* **197**:379.

van der Rhee, H. J., van der Burgh-de Winter, C. P. M., Tijssen, J. G. P., and Daems, W. T., 1979c, Comparative study on peroxidatic activity in inflammatory cells on cutaneous and peritoneal implants, *Cell Tiss. Res.* **197**:397.

van Furth, R., 1976, Origin and kinetics of mononuclear phagocytes, *Ann. N.Y. Acad. Sci.* **278**:161.

van Furth, R., and Cohn, Z. A., 1968, The origin and kinetics of mononuclear phagocytes, *J. Exp. Med.* **128**:415.

van Furth, R., and Fedorko, M. E., 1976, Ultrastructure of mouse mononuclear phagocytes in the bone marrow colonies grown *in vitro*, *Lab. Invest.* **34**:440.

van Furth, R., Hirsch, J. G., and Fedorko, M. E., 1970, Morphology and peroxidase cytochemistry of mouse promonocytes, monocytes, and macrophages, *J. Exp. Med.* **132**:794.

van Furth, R., Diesselhoff-den Dulk, M. M. C., and Mattie, H., 1973, Quantitative study on the production and kinetics of mononuclear phagocytes during an acute inflammatory reaction, *J. Exp. Med.* **138**:1314.

van Furth, R., Crofton, R. W., and Diesselhoff-den Dulk, M. M. C., 1977, The bone marrow origin of Kupffer cells, in: *Kupffer cells and Other Liver Sinusoidal Cells* (E. Wisse and D. L. Knook, eds.), Proceedings of the International Kupffer Cell Symposium, Noordwijkerhout, The Netherlands, 1977, pp. 471–480, Elsevier North-Holland, Amsterdam.

van Waarde, D., Hulsing-Hesselink, E., and van Furth, R., 1976, A serum factor inducing monocytosis during an acute inflammatory reaction, caused by newborn calf serum, *Cell Tissue Kinet.* **9**:51.

Vernon Roberts, B., 1972, *The Macrophage*, Cambridge University Press.

Volkman, A., 1976, Disparity in origin of mononuclear phagocyte populations, *J. Reticuloendothel. Soc.* **19**:249.

Volkman, A., 1977, The unsteady state of the Kupffer cell, in: *Kupffer Cells and Other Liver Sinusoidal Cells* (E. Wisse and D. L. Knook, eds.), Proceedings of the International Kupffer Cell Symposium, Noordwijkerhout, The Netherlands, 1977, pp. 459–470, Elsevier North-Holland, Amsterdam.

Volkman, A., and Collins, F. M., 1974, The cytokinetics of monocytosis in acute salmonella infection in the rat, *J. Exp. Med.* **139**:264.

Volkman, A., and Gowans, J. L., 1965a, The production of macrophages in the rat, *Br. J. Exp. Pathol.* **46**:50.

Volkman, A., and Gowans, J. L., 1965b, The origin of macrophages from bone marrow in the rat, *Brit. J. Exp. Pathol.* **46**:62.

Wachsmuth, E. D., and Stoye, J. P., 1977a, Aminopeptidase on the surface of differentiating macrophages: Induction and characterization of the enzyme, *J. Reticuloendothel. Soc.* **22**:469.

Wachsmuth, E. D., and Stoye, J. P., 1977b, Aminopeptidase on the surface of differentiating macrophages: Concentration changes on individual cells in culture, *J. Reticuloendothel. Soc.* **22**:485.

Walker, W. S., 1976a, Functional heterogeneity of macrophages in the induction and expression of acquired immunity, *J. Reticuloendothel. Soc.* **20**:57.

Walker, W. S., 1976b, Functional heterogeneity of macrophages, in: *Immunobiology of the Macrophage* (D. S. Nelson, ed.), pp. 91–108, Academic Press, New York.

Walters, M. N.-I., Papadimitriou, J. M., and Robertson, T. A., 1976, The surface morphology of the phagocytosis of micro-organisms by peritoneal macrophages, *J. Pathol.* **118**:221.

Watanabe, K., and Enomoto, Y., 1972, Ultrastructural localization of peroxidase in mononuclear phagocytes from the peritoneal cavity, blood, bone marrow and omentum: Origin of the peritoneal macrophages, in: *Histochemistry and Cytochemistry* (T. Takeuchi, K. Ogawa, and S. Fujita, eds.), pp. 433–434, Proceedings of the Fourth International Congress on Histochemistry and Cytochemistry, Kyoto.

Watanabe, K., Masubuchi, S., and Kageyama, K., 1971, Ultrastructural localization of peroxidase in mononuclear phagocytes from the peritoneal cavity, blood, bone marrow, and omentum, with a special reference to the origin of peritoneal macrophages, *Rec. Adv. RES Res.* **11**:120.

Weinberg, J. B., Chapman, H. A., and Hibbs, J. B., 1978, Characterization of the effects of endotoxin on macrophage tumor cell killing, *J. Immunol.* **121**:72.

Weiner, E., 1967, DNA-synthesis in peritoneal mononuclear phagocytes, *Exp. Cell Res.* **45**:450.

Werb, Z., and Dingle, J. T., 1976, Lysosomes as modulators of cellular functions. Influence on the synthesis and secretion of non-lysosomal materials, in: *Lysosomes in Biology and Pathology 5* (J. T. Dingle and R. T. Dean, eds.), pp. 127–156, North-Holland, Amsterdam.

Whaley, K., Singh, H., and Webb, J., 1972, Phagocytosis of colloidal carbon by the fixed tissue and peritoneal macrophages of New Zealand mice, *Scot. Med. J.* **17**:383.

White, R., Lin, H. S., and Kuhn, C., 1977, Elastase secretion by peritoneal exudative and alveolar macrophages, *J. Exp. Med.* **146**:802.

Whitelaw, D. M., 1966, The intravascular lifespan of monocytes, *Blood* **28**:455.

Whitelaw, D. M., and Batho, H. F., 1972, The distribution of monocytes in the rat, *Cell Tissue Kinet.* **5**:215.

Whitley, S. B., and Leu, R. W., 1976, Role of macrophage activation on the expression of MIF receptors by guinea pig peritoneal and alveolar macrophages, *J. Reticuloendothel. Soc.* **20**:9a.

WHO Scientific Group, 1973, Cell mediated immunity and resistance to infection, *World Health Organization Technical Report Series* No. 519, World Health Organization, Geneva.

Widmann, J. J., and Fahimi, H. D., 1976, Proliferation of endothelial cells in estrogen-stimulated rat liver: A light and electron microscopic cytochemical study, *Lab. Invest.* **34**:141.

Widmann, J. J., Cotran, R. S., and Fahimi, H. D., 1972, Mononuclear phagocytes (Kupffer cells) and endothelial cells. Identification of two functional cell types in rat liver sinusoids by endogenous peroxidase activity, *J. Cell Biol.* **52**:159.

Williams, M. A., and Mayhew, T. M., 1973, Quantitative microscopical studies of the mouse peritoneal macrophage following stimulation *in vivo*, *Z. Zellforsch. Mikrosk. Anat.* **140**:187.

Wisse, E., 1972, An ultrastructural characterization of the endothelial cell in the rat liver sinusoid under normal and various experimental conditions, as a contribution to the distinction between endothelial and Kupffer cells, *J. Ultrastruct. Res.* **38**:528.

Wisse, E., 1974, Observations on the fine structure and peroxidase cytochemistry of normal rat liver Kupffer cells, *J. Ultrastruct. Res.* **46**:393.

Wisse, E., Roels, F., de Prest, B., van der Meulen, J., Emeis, J. J., and Daems, W. T., 1974, Peroxidatic reaction of Kupffer and parenchymal cells compared in rat liver, in: *Electron Microscopy and Cytochemistry* (E. Wisse, W. T. Daems, I. Molenaar, and P. van Duijn, eds.), pp. 119–122, North-Holland, Amsterdam.

Yang, H. Y., and Skinsness, O. K., 1973, Peritoneal macrophage response in neonatal mice, *J. Reticuloendothel. Soc.* **14**:181.

4

The Cell Coat of Macrophages

J. J. EMEIS
and P. BREDEROO

1. INTRODUCTION

It is now well established, both morphologically and cytochemically, that the outer surface of animal cells is formed by a layer containing glycoproteins and called the cell coat, fuzzy coat, or, according to Bennett (1963), glycocalyx. Seen with the electron microscope, this cell coat has a variable appearance, differing from cell type to cell type. On the one hand, a fine-filamentous glycocalyx occurs on the plasma membrane of ameba (Hausmann, 1975) and on the microvilli of intestinal epithelial cells (Ito, 1974), and on the other, a glycocalyx, which is almost uniform in width and subdivided into layers parallel to the plasma membrane, is found under certain circumstances on peritoneal macrophages (Brederoo and Daems, 1972) and Kupffer cells (Emeis, 1976). Endocytic activity of macrophages may lead to invagination of the plasma membrane, which results in the formation of the so-called wormlike structures showing doubling of the cell coat.

The following review concerns the morphological and cytochemical aspects of the cell coat on macrophages. For information on the cell coat of animal cells in general, the reader is referred to Revel and Ito (1967), Bennett (1969), Martínez-Palomo (1970), Winzler (1970), Rambourg (1971), Parsons and Subjeck (1972), and Luft (1976).

J. J. EMEIS • Gaubius Institute, Health Research Organization TNO, 2313 AD Leiden, The Netherlands. P. BREDEROO • Laboratory for Electron Microscopy, University of Leiden, 2333 AA Leiden, The Netherlands.

2. MORPHOLOGY

2.1. CELL COAT AND WORMLIKE STRUCTURES

An outer layer on the plasma membrane of macrophages was first described by Törö *et al.* (1962). In their experiments on phagocytosis of India ink particles by rat liver sinusoidal cells, the authors observed the uptake of these particles via invagination of the plasma membrane followed by the formation of typical structures, which they called wormlike bodies or wormlike structures. The elongated, sometimes branching, invaginations of the plasma membrane were characterized by a constant width of 100–150 nm and a limiting membrane with the same thickness as the plasma membrane, as well as a filamentous cross-striation and a dense median line. These wormlike structures contained India ink particles but disappeared in later stages of the experiments, leaving large vacuoles in the cytoplasm in which the India ink particles were located. On the basis of their observations, the authors postulated the existence of an "absorptive layer" on the plasma membrane (cell coat) which becomes duplicated during the formation of the wormlike structures. Similar observations were made by the same authors in reticular cells of mesenteric lymph nodes of the rat (Röhlich and Törö, 1964). Here, wormlike structures were involved in the phagocytosis of chylomicron particles. The cell coat, in this case, had a uniform width of 50 nm and was composed of fine 2-nm-thick filaments lying perpendicular to the plasma membrane. Since Törö and co-workers made no distinction between types of liver sinusoidal cells, they considered these structures to occur in endothelial as well as Kupffer cells. However, Melis and Orci (1967) and Orci *et al.* (1967) reported that the presence of wormlike structures in liver sinusoidal cells was restricted to Kupffer cells, whereas Wisse and Daems (1970a) mention the wormlike structures as one of the criteria for the distinction between endothelial and Kupffer cells.

Because Törö *et al.* (1962) did not find round or oval profiles in the cytoplasm, they considered the wormlike structures to be flat lamellar and not tubular structures. However, on the basis of the study of serial sections and three-dimensional reconstruction, Matter *et al.* (1968) concluded that wormlike structures represent a continuous system of tubular as well as lamellar structures to which coated vesicles were attached. They also introduced the term "micropinocytosis vermiformis" for this type of endocytosis, which was called "surface endocytosis" by Pfeifer (1970).

2.2. OCCURRENCE OF WORMLIKE STRUCTURES

Wormlike structures have been found mainly in tissue macrophages of the liver and spleen. In a study on peritoneal macrophages present in a peritoneal lavage fluid (resident macrophages) as well as macrophages present in the peritoneal cavity after injection of an irritant (exudate macrophages), cell coat (Figures 1 and 2) and wormlike structures (Figures 3 and 5) were found in both

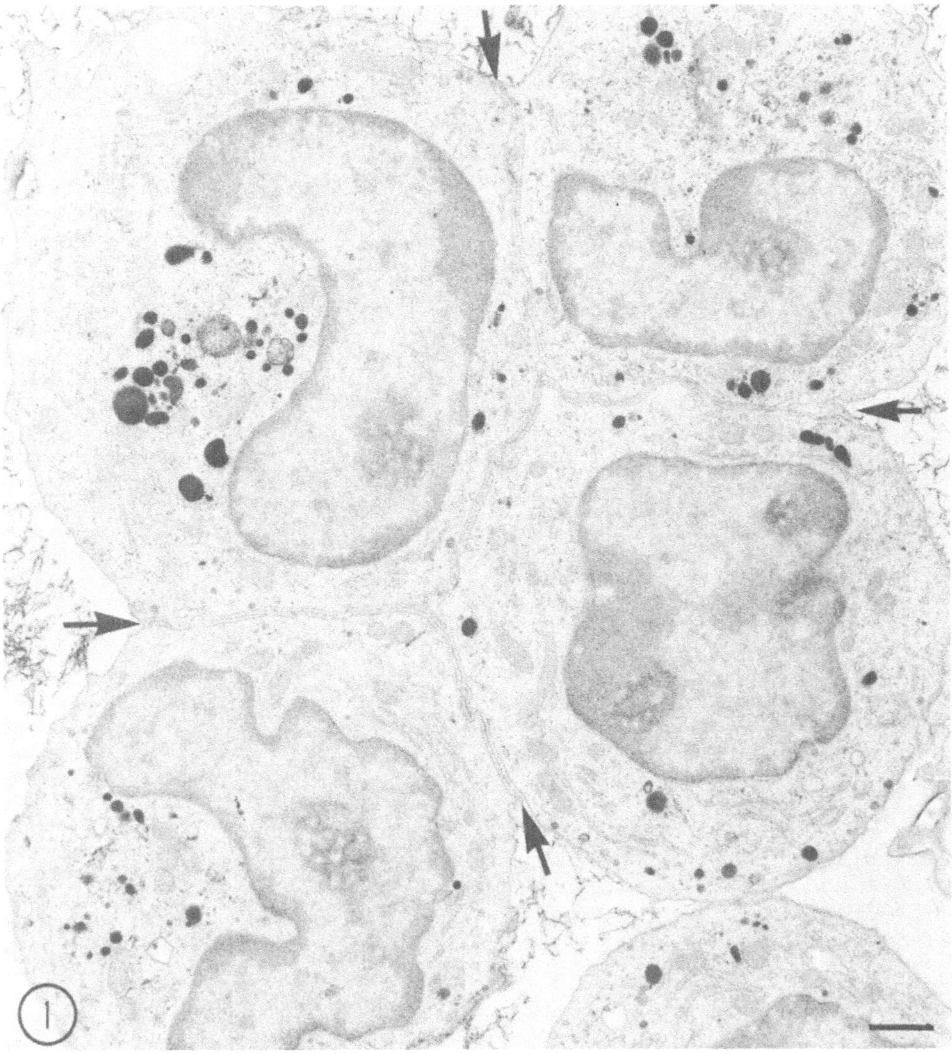

FIGURE 1. A cluster of four exudate peritoneal macrophages from a guinea pig. Adjacent cells show a continuous cell coat on their plasma membranes separated by an electron-dense median line (arrows). Free parts of the cell surface have no coat. Osmium fixation, silvermethenamine staining. Bar: 1 μm. (The authors wish to thank Springer Verlag for permission to reproduce this figure.)

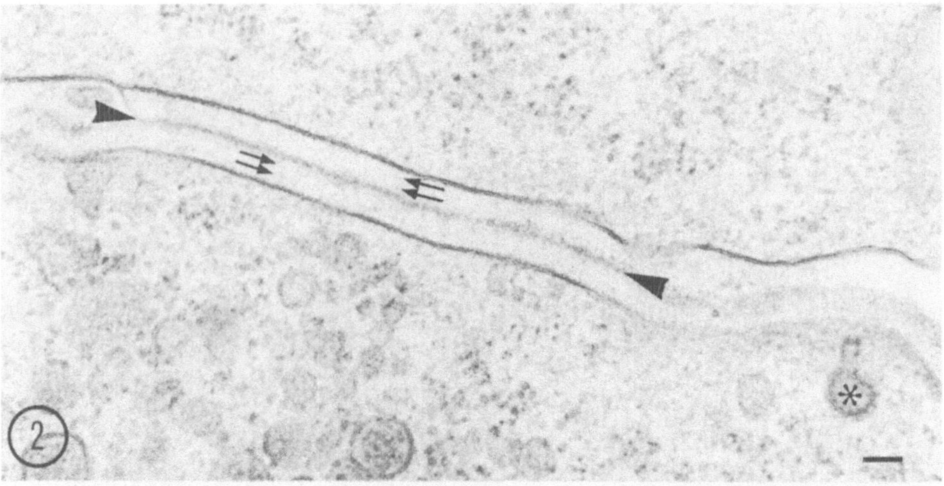

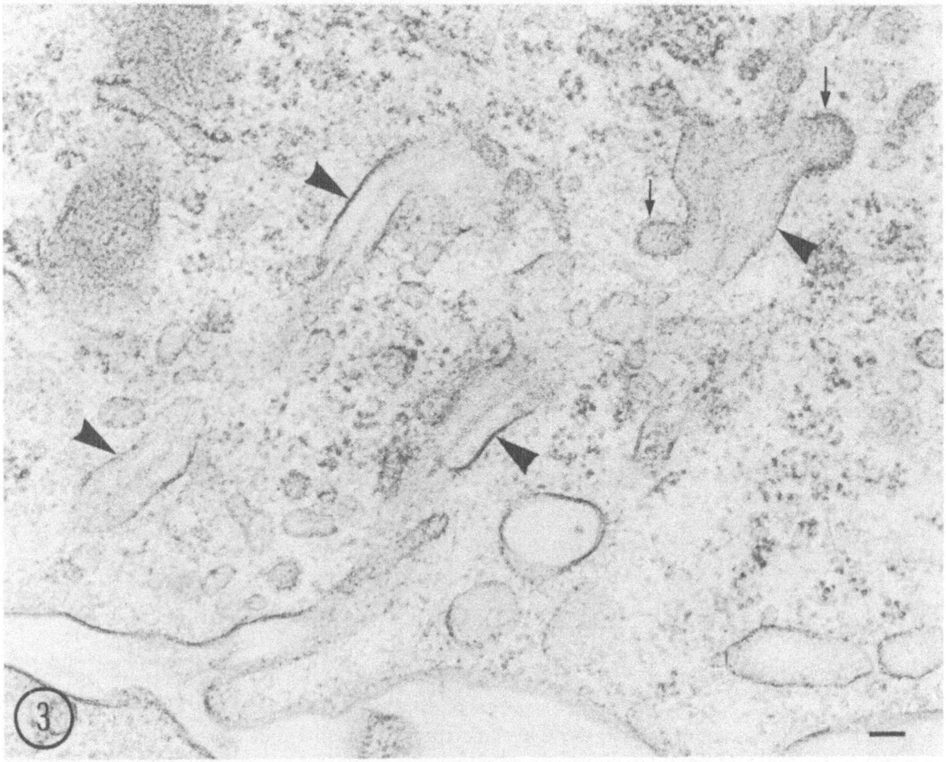

FIGURE 2. Detail from a cluster of exudate peritoneal macrophages showing cell coat on the plasma membranes of two adjacent cells. Arrowheads point to the median line. Arrows indicate dotted lines dividing each cell coat into three layers. Asterisk, bristle-coated vesicle. Osmium fixation. Bar: 0.1 μm. (The authors wish to thank Springer Verlag for permission to reproduce this figure.)

FIGURE 3. Detail of an exudate peritoneal macrophage of the guinea pig showing bristle-coated vesicles (arrows) connected to wormlike structures (arrowheads). Osmium fixation. Bar: 0.1 μm. (The authors wish to thank Springer Verlag for permission to reproduce this figure.)

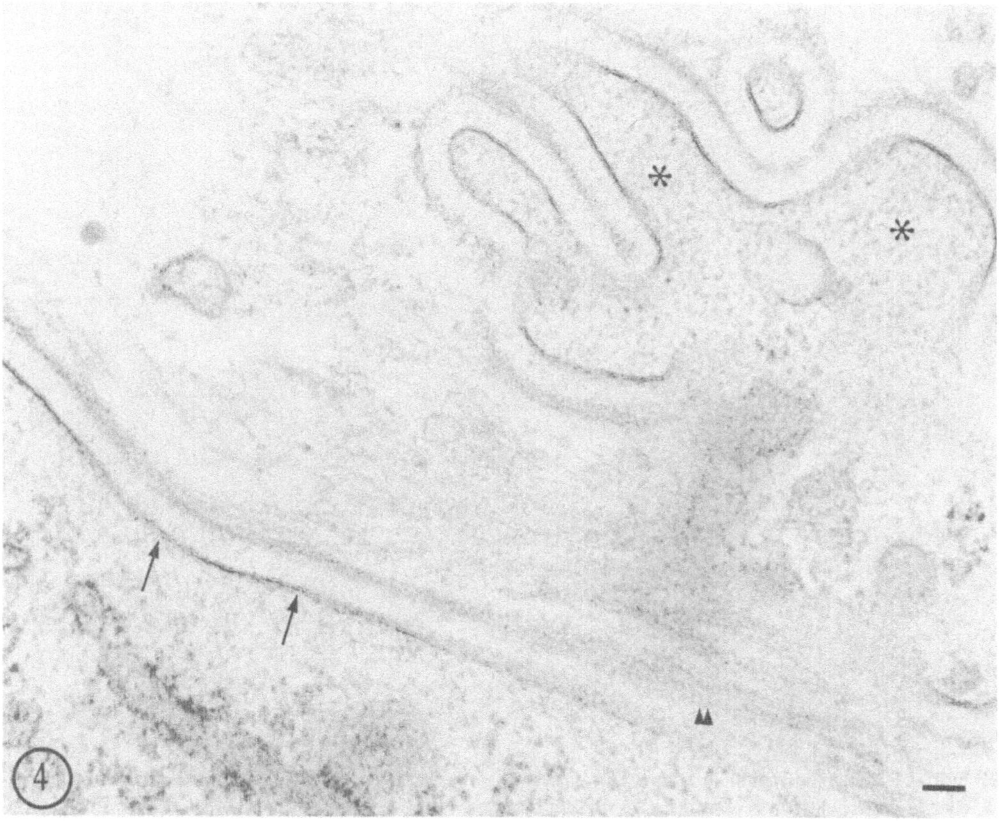

FIGURE 4. Detail of an exudate peritoneal macrophage showing cell coat on a straight part of the plasma membrane (arrows) and on cytoplasmic extrusions (asterisks). The cell coat is covered by fibrin with the characteristic cross-striation (arrowheads). Bar: 0.1 μm.

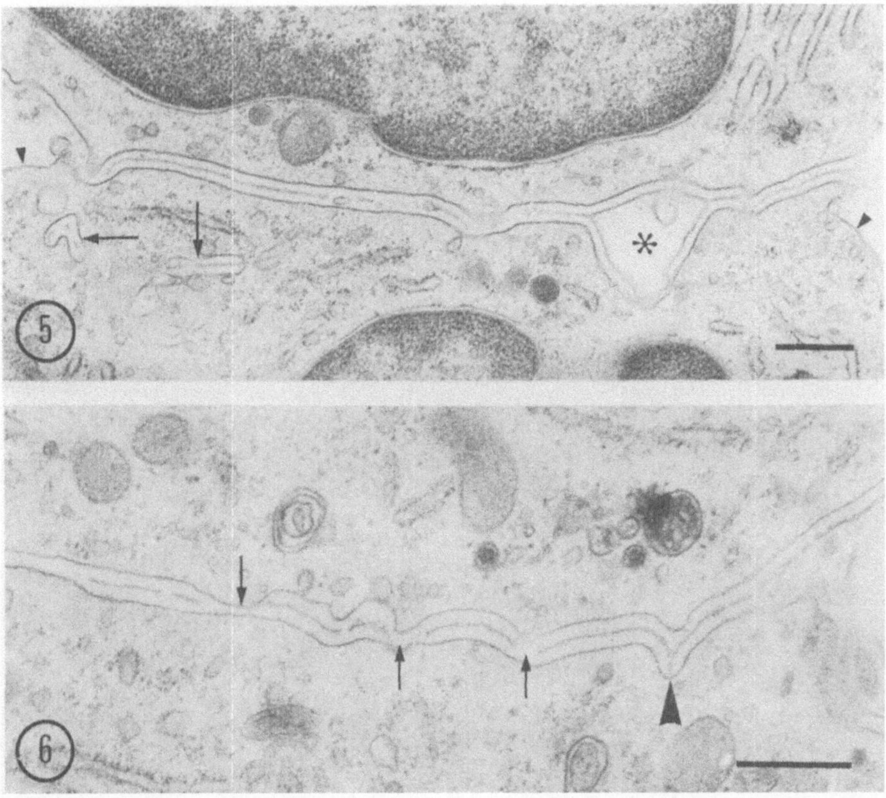

FIGURE 5. Two adjacent exudate macrophages, one of them with wormlike structures (arrows). Cell coat is absent (arrowheads) outside the contact zone between the two cells but is still present (asterisk) on each of the plasma membranes in a wider part of the contact zone. Osmium fixation. Bar: 0.5 μm.

FIGURE 6. Two adjacent cells in a cluster of exudate peritoneal macrophages. At sites of evagination of the plasma membrane, the cell coat is interrupted (arrows). Where a wormlike structure is being formed, the cell coat remains intact (arrowhead). Osmium fixation. Bar: 0.5 μm.

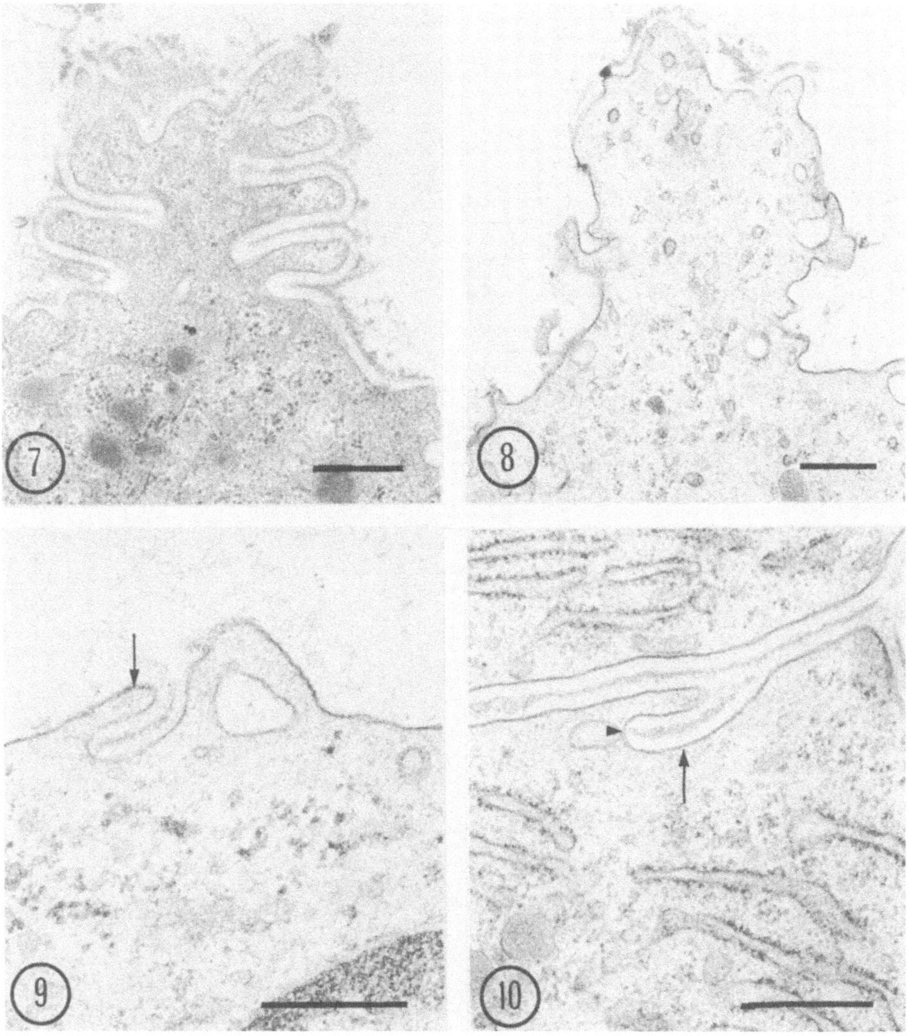

FIGURE 7. Pseudopodium of an exudate macrophage present in the peritoneal cavity after injection of a 3.6% solution of sodium chloride. Note the continuous cell coat over the entire pseudopod (compare with Figure 8). Bar: 0.5 μm.

FIGURE 8. Pseudopodium of an exudate macrophage evoked by an intraperitoneal injection of a 0.9% solution of sodium chloride. Note the discontinuous cell coat on part of the plasma membrane (compare with Figure 7). Bar: 0.5 μm. (The authors wish to thank Springer Verlag for permission to reproduce this figure.)

FIGURE 9. Free cell surface of an exudate peritoneal macrophage (0.9% solution of sodium chloride). Cell coat is only present underneath an overlying cell extrusion (arrow) (compare with Figure 10). Bar: 0.5 μm. (The authors wish to thank Springer Verlag for permission to reproduce this figure.)

FIGURE 10. Formation of a wormlike structure (arrow) in one of two adjacent exudate peritoneal macrophages (3.6% solution of sodium chloride). Cell coat is present between the two cells. Note the midline of the wormlike structures touching the plasma membrane (arrowhead). (Compare with Figure 9.) Bar: 0.5 μm. (The authors wish to thank Springer Verlag for permission to reproduce this figure.)

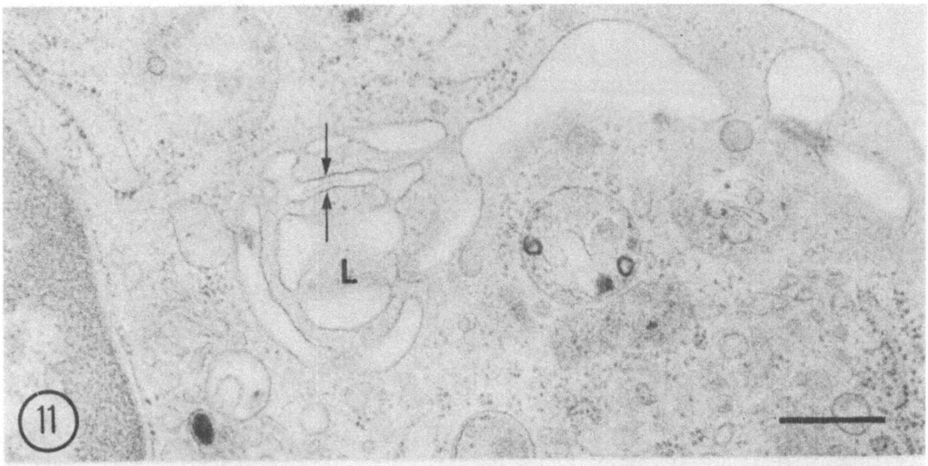

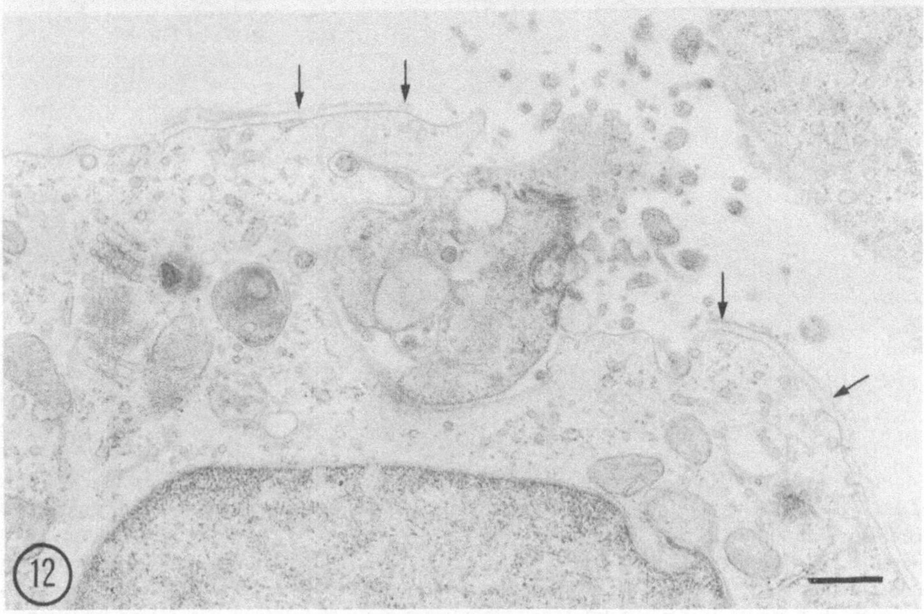

FIGURE 11. Thin cell coat present in part of a labyrinth (L) in a mouse resident peritoneal macrophage (arrow). Bar: 0.5 μm.

FIGURE 12. Exudate macrophage phagocytizing cell debris from the peritoneal cavity. Note that cell coat is present on the part of plasma membrane not involved in the process of phagocytosis (arrows). Bar: 0.5 μm. (The authors wish to thank Springer Verlag for permission to reproduce this figure.)

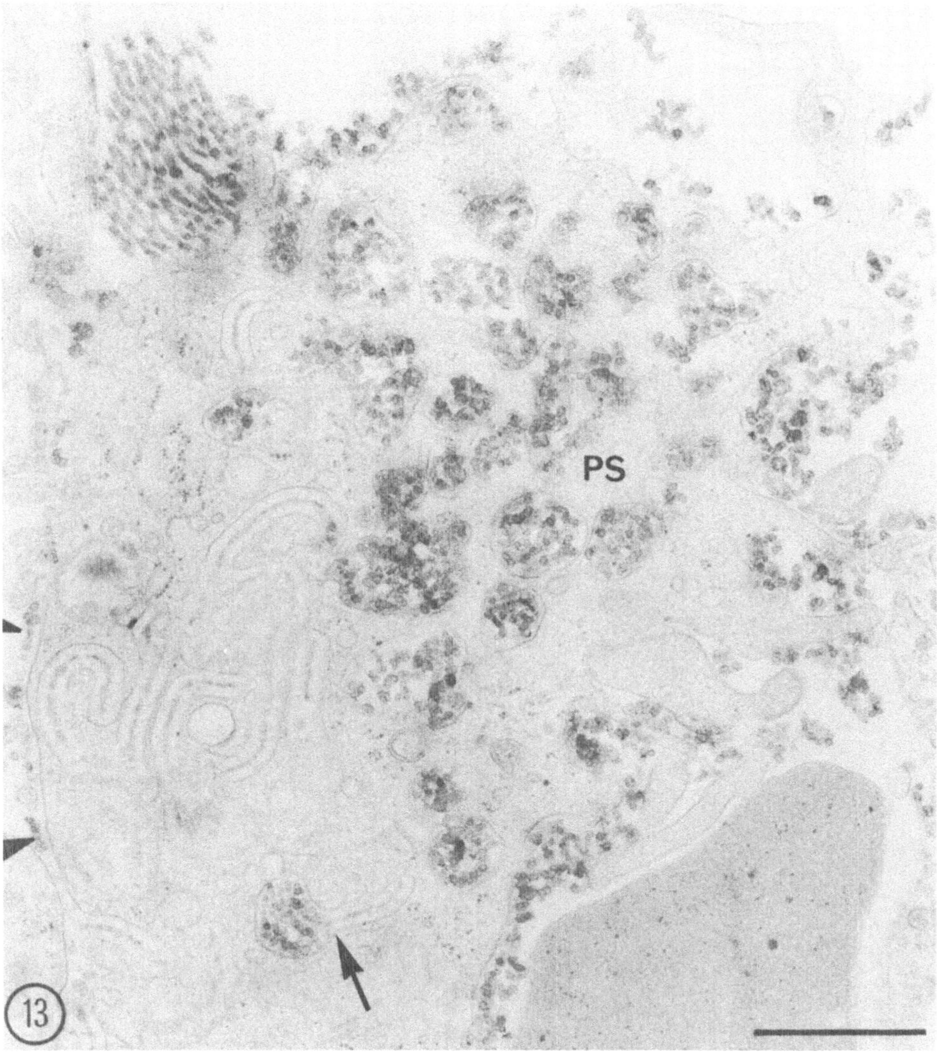

FIGURE 13. Part of a Kupffer cell after intravenous injection of colloidal carbon. A few carbon particles are attached to the thick coat (at arrowheads) near a wormlike structure. The most carbon occurs on or in an (organelle-poor) pseudopodium (ps) which seems to lack thick cell coat. Note continuity of a wormlike structure with a carbon-containing vacuole lacking thick coat (arrow). Bar: 1 μm.

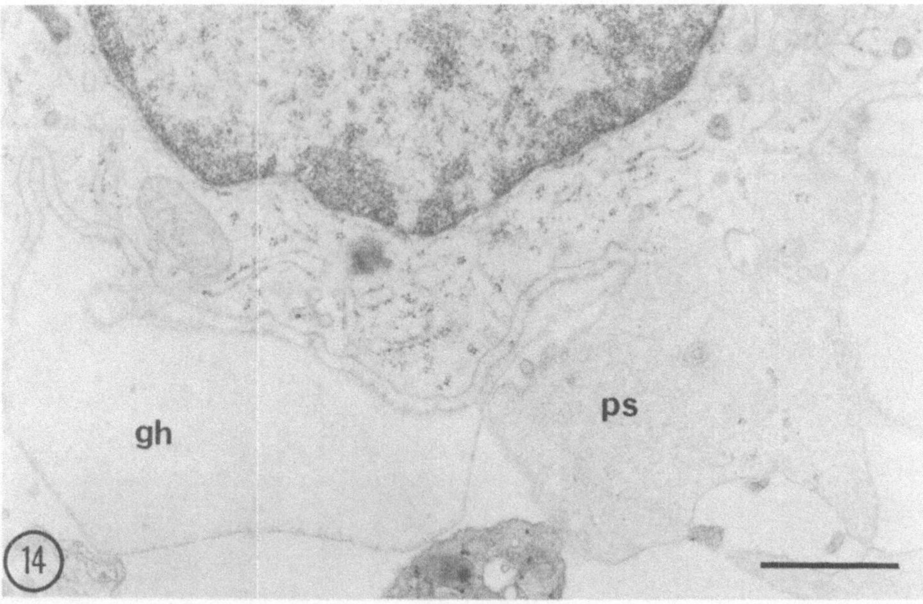

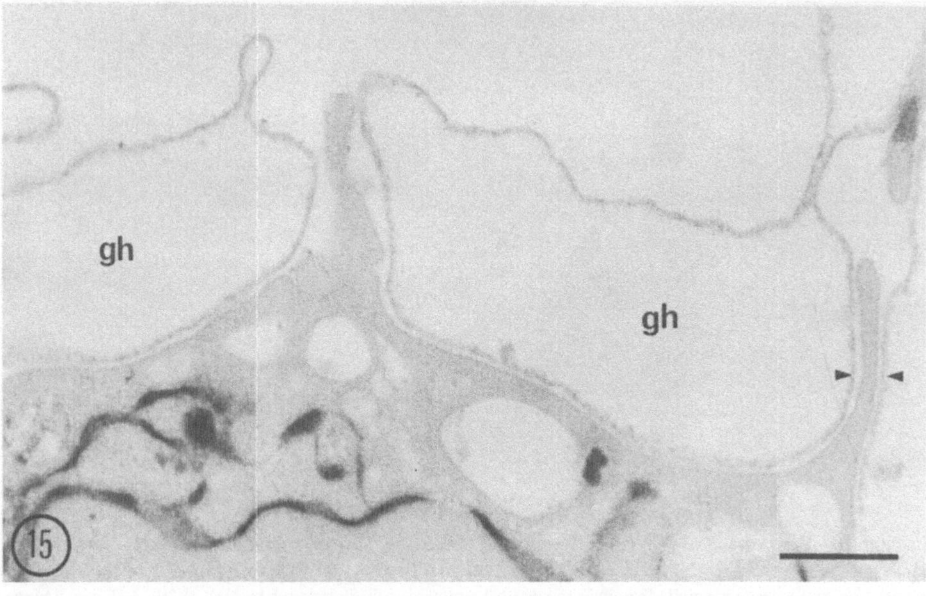

FIGURE 14. Erythrocyte ghosts (gh) attached to the thick coat of a Kupffer cell but not to a pseudopodium (ps) lacking thick coat. Bar: 1 μm.

FIGURE 15. Erythrocyte ghosts (gh) attached to the thin coat of mouse resident peritoneal macrophages, and also to one of the plasma membrane extrusions (arrowheads). Bar: 0.5 μm.

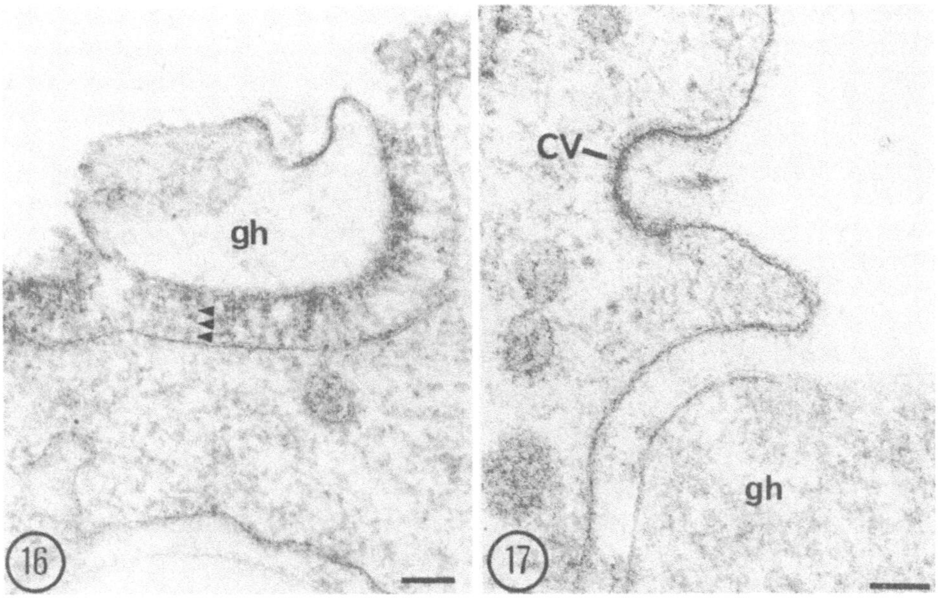

FIGURE 16. Detail of the thick coat of a Kupffer cell, to which an erythrocyte ghost (gh) is attached. The cell coat has three layers (arrowheads). Glutaraldehyde-tannic acid fixation. Bar: 0.1 μm.
FIGURE 17. First stage of formation of a wormlike structure on (or by?) a coated vesicle (cv). Bar: 0.1 μm.

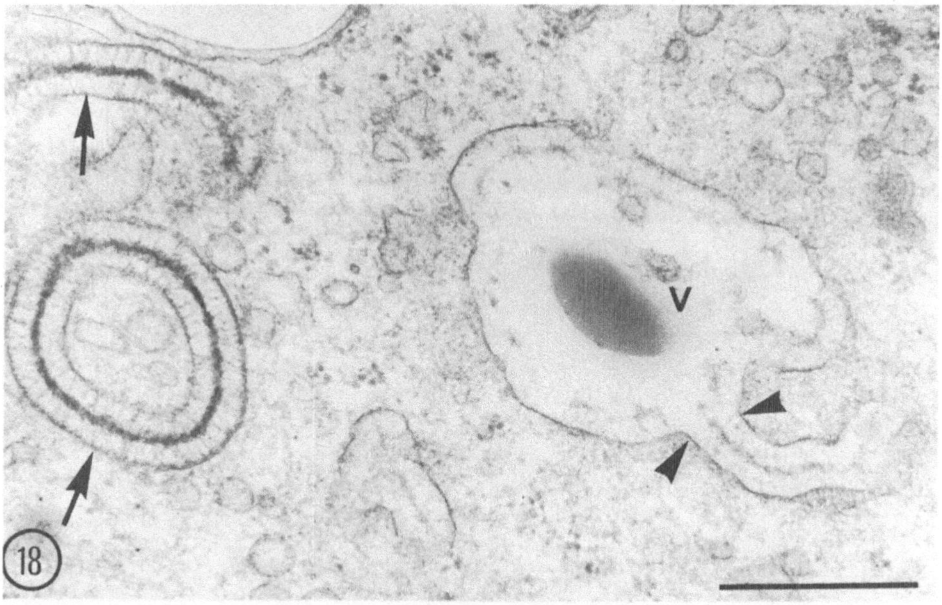

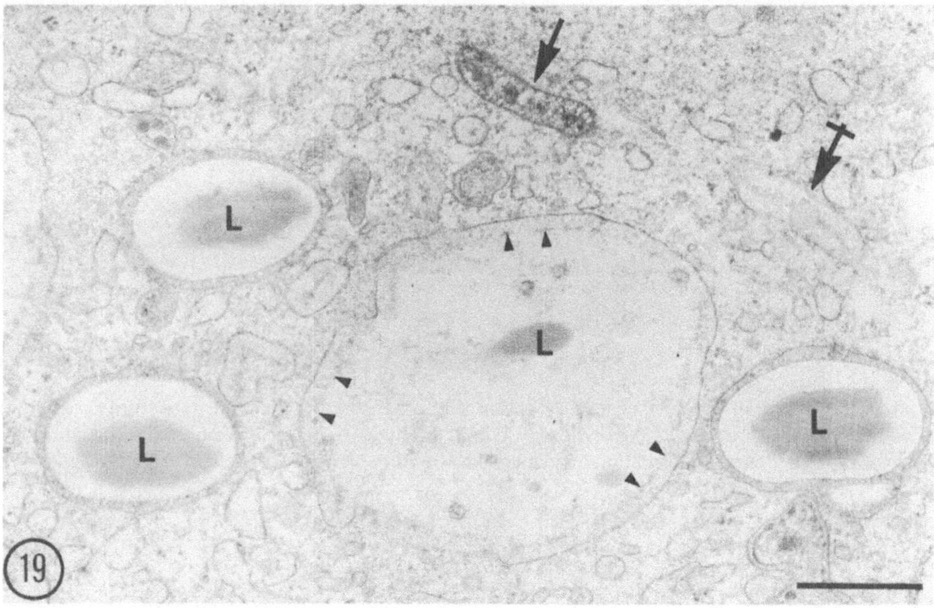

FIGURES 18, 19. Parts of Kupffer cells—from rats injected with latex—fixed by perfusion with glutaraldehyde plus tannic acid and postfixed with osmium. With this fixation procedure, structures still continuous with the extracellular space are strongly contrasted by the tannic acid and can therefore be differentiated from (normally contrasted) intracellular structures.

FIGURE 18. Two wormlike structures (arrows) are still continuous with the sinusoidal lumen, whereas a latex-containing vacuole (V) with a thick coat has been interiorized. Note continuity of thick coat in vacuole with that of a wormlike structure (at arrowheads). Bar: 0.5 μm.

FIGURE 19. Two lipoprotein-containing wormlike structures, one of which (arrow) is continuous with the extracellular space, whereas the other (crossed arrow) has been interiorized, thus showing that the wormlike structures are endocytic. Only one of the latex-containing vacuoles (L) has a thick coat (arrowheads). Bar: 0.5 μm.

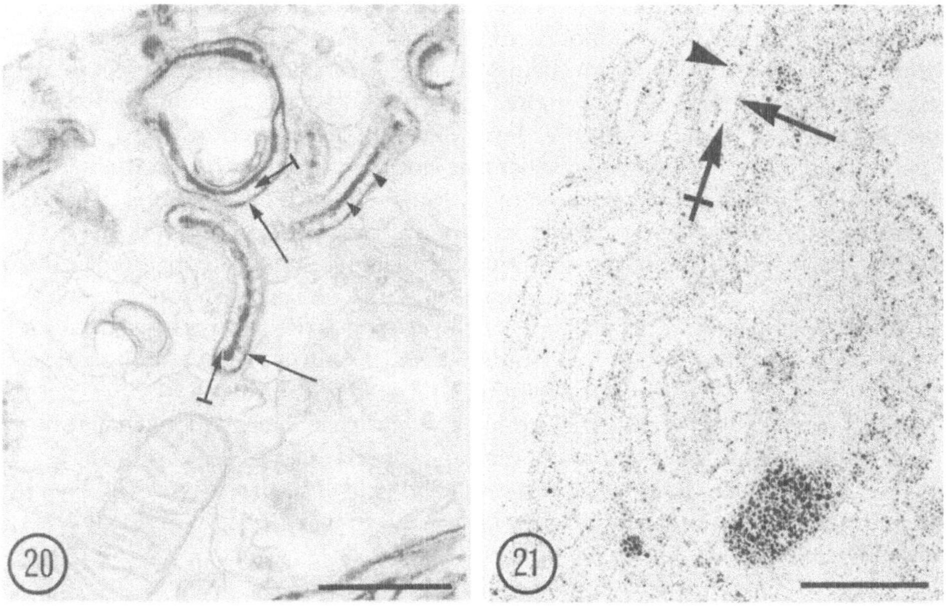

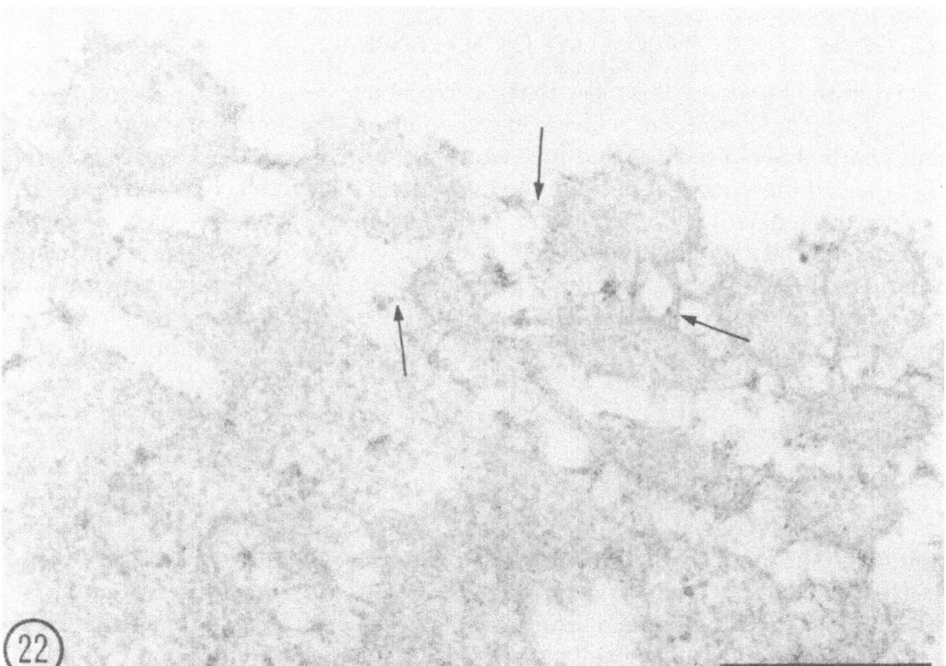

FIGURES 20, 21. Cytochemical features of the cell coat of Kupffer cells. Wormlike structures stained for the presence of carbohydrates by the Concanavalin A-peroxidase procedure (Figure 20), and by the periodic acid-silvermethenamine procedure (Figure 21). With both techniques, the thin coat (arrows) close to the plasma membrane is stained, whereas the outer layer of the thick coat is only occasionally faintly contrasted (crossed arrows). The midline of wormlike structures (arrowheads) is also stained. Bar: 0.5 μm.

FIGURE 22. Cell incubated for the demonstration of protein by the acrolein-thiocarbohydrazide-osmium procedure. The filaments of the outer layer of the thick coat (arrows) are contrasted. Bar: 0.5 μm.

types of cell (Brederoo and Daems, 1972). However, there was a quantitative difference with respect to the amount of cell coat and the number of wormlike structures of these cells, both being smaller in resident than in exudate peritoneal macrophages. Furthermore, the presence of cell coat on peritoneal macrophages proved to be highly dependent on the procedure used for the isolation and fixation of the cells, as well as, for the exudate cells, on the nature of the irritant. These observations were in agreement with observations in Kupffer cells, where wormlike structures were found more often in livers fixed by vascular perfusion than in those subjected to immersion fixation (Wisse and Daems, 1970a). For peritoneal macrophages, the use of direct fixation (i.e., without prior centrifugation and washing of the cell suspensions) was essential for the presence of these structures (Brederoo and Daems, 1972), although they were occasionally also found under other preparative conditions.

Most of the observations on wormlike structures have been made in Kupffer cells (Figures 13 and 18). As already mentioned, and as can be seen from Table 1, the presence of these structures is not restricted to liver macrophages. However, the findings strongly suggest that they only occur in macrophages and may thus be a morphological characteristic of this type of cell.

2.3. OCCURRENCE OF CELL COAT ON MACROPHAGES

From the foregoing it is clear that a cell coat as present on the macrophage cell surface is by definition, a layer on the outer plasma membrane with a constant width of about half the width of wormlike structures (Figure 10). This coat will be called the thick coat here. In the literature the term cell coat has also been used for the constant spacing between two adjacent macrophages, or between a macrophage and any other type of cell, or for the gap between a layer of particles sticking to the cell surface and the outer plasma membrane. But in these cases, the width of the cell coat often is less than half the width of a wormlike structure (Figure 15). Therefore, in the following this type of cell coat will be called the thin coat.

2.3.1. Thick Coat

Few morphological data are available concerning a thick coat on macrophages. The published findings are, to the best of our knowledge, summarized in Table 2. The thickness of the cell coat measured by different authors varies from 50 to 70 nm, which represents half the width of wormlike structures very well (see Table 1). The cell coat was found under different conditions in these studies. Occasionally, a fuzzy coating was present on a free cell surface of a Kupffer cell (Wisse, 1974a). In most instances, however, the cell coat was present between the outer plasma membrane and certain particles, e.g., chylomicrons (Röhlich and Törö, 1964), ribosomes (Pfeifer, 1970), colloidal silver (Emeis, 1976), and colloidal gold (Wisse, 1977a,b). For peritoneal macrophages a cell coat has been observed in cell clusters in both the guinea pig (Brederoo and Daems, 1972) and the mouse [Koerten, unpublished observations (1975)]. A thick coat has also been observed between red cell ghosts (Figure 14) and Kupffer cells (Emeis, 1976) and between fibrin and peritoneal macrophages (Figure 4) or Kupffer cells (Prose *et al.*, 1965: their Figures 2 and 5; Brederoo and Daems, 1972; Emeis and Lindeman, 1976).

2.3.2. Thin Coat

A thin coat has been found on macrophages under the same conditions as those described for the presence of a thick coat. The reported width of the gap between the plasma membrane of the macrophage and adherent particles or cells ranges from 5.5 to 20 nm (see Table 3). Our morphological experiments showed a subdivision of the thick cell coat into three layers (Figures 2 and 16), the middle one being thicker than the other two (Brederoo and Daems, 1972). The inner layer lying on the plasma membrane was about 17 nm thick, the middle layer about 26 nm, and the outer layer about 20 nm. The thickness of the layer lying on the outer leaflet of the plasma membrane agrees well with the thickness of the thin coat mentioned above. Unlike the other layer(s) of the thick coat, the thin coat contains glycoproteins, which can be specifically stained with, e.g., silver stains and Alcian blue (see Section 3).

2.4. THE EXISTENCE OF A CELL COAT ON MACROPHAGES

It is generally accepted that all animal cells possess a glycocalyx on the outer plasma membrane. This glycocalyx differs morphologically from cell type to cell type and is very pronounced on ameba and the microvilli of intestinal epithelial cells (Parsons and Subjeck, 1972). In view of the presence in macrophages of wormlike structures and their involvement in the uptake of foreign material, one would expect a pronounced cell coat on macrophages too. Such a cell coat, i.e., a layer on the outer plasma membrane with a uniform width of 50–70 nm (see Table 2), a cross-pattern of fine filaments (Törö *et al.*, 1962; Röhlich and Törö, 1964; Brederoo and Daems, 1972; Emeis and Wisse, 1975; Emeis, 1976), and subdivided into three layers (Brederoo and Daems, 1972; see also Table 3) has, however, never been described to cover the entire macrophage. In fact, all of the reports on macrophage cell coat concern fragments present under certain conditions, i.e., covered with foreign particulate material or cells (for references, see Table 2). Larger areas of cell coat occur particularly when macrophages are found in clusters [Brederoo and Daems, 1972; Koerten, unpublished observations (1975)]. These observations make it obvious, in our opinion, that a certain "protection" is needed for the observation of a cell coat on macrophages. Protection may also explain the presence of cell coat in invaginations of the plasma membrane (Figures 9, 10, and 17), wormlike structures (for references, see Table 1), and vacuoles (Figures 18 and 19) (Röhlich and Törö, 1964; Orci *et al.*, 1967; Matter *et al.*, 1968; Han *et al.*, 1970; Pfeifer, 1970; Ericsson *et al.*, 1971; Emeis and

TABLE 1. OCCURRENCE OF WORMLIKE STRUCTURES

Tissue	Cell type	Animal	Width[a] (nm)	Authors
Liver	Kupffer cell	Rat	100–150	Toro *et al.* (1962)
			*	Wiener *et al.* (1964)
			*	Melis and Orci (1967)
			*	Orci *et al.* (1967)
			*	Rouiller *et al.* (1967)
			*	Matter *et al.* (1968)
			*	Cotutiu and Ericsson (1970)
			*	Fahimi (1970)
			120	Pfeifer (1970)
			*	Wisse and Daems (1970a)
			*	Wisse and Daems (1970b)
			*	Cotutiu and Ericsson (1971)
			140	Emeis and Wisse (1971)
			*	Widmann *et al.* (1972)
			*	Ogawa *et al.* (1973)
			140	Wisse (1974a,b)
			*	Arborgh *et al.* (1974)
			140	Emeis and Wisse (1975)
			130	Emeis (1976)
			*	Emeis and Lindeman (1976)

		Rabbit	*	Orci *et al.* (1967)
			100	Horn *et al.* (1969)
		Mouse	*	Ishihara (1973)
		Bat	120	Tanuma (1978)
Lung	Kupffer cell (migrated)	Cat	90–100	Schneeberger-Keeley and Burger (1970)
	Macrophage	Calf	*	Rybicka *et al.* (1974)
Spleen	Macrophage	Mouse	*	Orci *et al.* (1967)
		Rabbit	*	Burke and Simon (1970)
Peritoneum	Resident Macrophage	Guinea pig	*	Brederoo and Daems (1972)
	Exudate Macrophage		125	Brederoo and Daems (1972)
	Exudate Macrophage	Mouse	140	Daems [unpublished observations (1978)]
Lymph node	Macrophage	Rat	*	Han *et al.* (1970)
	Reticular cell		*	Röhlich and Törö (1964)
	Reticulum cell		*	Hoefsmit (1975)
Subcutis	Macrophage	Rat	96–144	Katenkamp and Stiller (1975)

[a] Asterisk indicates that no values are given by the authors.

TABLE 2. OCCURRENCE OF CELL COAT ON MACROPHAGES

Tissue	Animal	Width[a] (nm)	Gap with	Authors
Liver	Rat	50–60	Ribosomes	Pfeifer (1970)
		70	Free surface	Wisse (1974a)
		70	Colloidal gold	Wisse (1977a,b)
		70	Alcian blue	Emeis and Wisse (1975)
		65	Red cell ghosts	Emeis (1976)
		*	Colloidal silver	Emeis (1976)
		60	Fibrin	Emeis and Lindeman (1976)
	Rabbit	*	Fibrin	Prose *et al.* (1965)
Peritoneum	Guinea pig	63	Macrophage or fibrin	Brederoo and Daems (1972)
	Mouse	61	Macrophage	Koerten [unpublished observations (1975)]
Lung	Calf	*	Macrophage	Rybicka *et al.* (1974)
Lymph node	Rat	50	Chylomicrons	Röhlich and Törö (1964)

[a] Asterisk indicates that no values are given by the authors.

TABLE 3. WIDTH OF THIN COAT ON MACROPHAGES

Tissue	Animal	Width[a] (nm)	Gap with	Authors
Liver	Rat	15–20	Carbon	Widmann *et al.* (1972)
		20	Thorotrast	Emeis and Wisse (1971); Wisse (1974b)
	Rabbit	20–30	Erythrocyte	Tamaru and Fujita (1978)
Peritoneum	Guinea pig	17	Layer of thick coat	Brederoo and Daems (1972)
		46	Macrophage (in labyrinth)	Daems [unpublished observations (1978)]
	Mouse	10	Ferritin	Hardy *et al.* (1976)
		30	Erythrocyte ghosts	Daems (this volume)
Blood	Man[b]	15–20	Macrophage	Sanel and Serpick (1970)
Bone marrow	Man[b]	20	Plasma cell	Blom (1973)
		11–27	Plasma cell	Blom (1977)
		30	Macrophage	Bainton and Golde (1978)
Leptomeninges	Dog	36	Macrophage	Merchant *et al.* (1977)
		40	Neutrophil	Merchant *et al.* (1977)
Subcutis	Rat	27	Macrophage	Van der Rhee *et al.* (1979)

[a] Values represent gap between adjacent cells: when the cells are of the same type, these values can be considered to represent about twice the width of the thin coat.
[b] Human data concern pathological or culture conditions.

Wisse, 1971, 1975; Wisse, 1974a, 1977a; Hoefsmit, 1975; Emeis, 1976). However, as already noticed by Röhlich and Törö (1964), the cell coat in cytoplasmic vacuoles disappears, although it may still be present in a connected wormlike structure (Figure 18). If protection is a prerequisite for the presence of cell coat, one would also expect to find cell coat in labyrinths (invaginations of the plasma membrane of resident peritoneal macrophages; see Brederoo and Daems, 1972). However, in guinea pig peritoneal macrophages cell coat was never observed in a labyrinth, and in the mouse a piece of thin cell coat was only found once (Daems, this volume) (Figure 11). It is, therefore clear that not only protection but also the involvement of the cell in active processes such as phagocytosis influence the presence of the cell coat (Brederoo and Daems, 1972; see also Figures 6 and 12). With respect to this point, the observations on cytoplasmic protrusions are contradictory: cell coat was sometimes seen on protrusions of exudate macrophages (Brederoo and Daems, 1972; see Figures 7 and 8), whereas none was found on hyaloplasmic pseudopodia (Figure 14) of Kupffer cells (Emeis, 1976).

It is also clear that adequate fixation is essential for the optimal preservation of cell coat exposed to the extracellular space or occurring in wormlike structures and vacuoles. For Kupffer cells this means the use of a perfusion fixation method (Wisse, 1974a; Emeis, 1976), whereas peritoneal macrophages require direct fixation, i.e., without prior centrifugation and washing of the cells (Brederoo and Daems, 1972). Furthermore, our experiments (Brederoo and Daems, 1972) have indicated that a certain preservation of the cell coat on exudate macrophages was obtained by the intraperitoneal injection of a 3.6% sodium chloride solution. Such macrophages showed more cell coat than those evoked by an isoosmotic (0.9%) saline solution. These findings were applied in further experiments in an attempt to find cell coat on still larger areas of the peritoneal macrophage. However, the exudate collected after the injection of erythrocyte ghosts, thrombin, colloidal silver, or Alcian blue into the peritoneal cavity containing a 3.6% saline-induced exudate, contained diminished rather than increased amounts of cell coat on the macrophages compared to the controls (De Ruyter *et al.*, 1976).

3. CYTOCHEMISTRY

3.1. CYTOCHEMICAL TECHNIQUES

Although the presence of a cell coat can be demonstrated morphologically with routine electron-microscopical procedures—especially on epithelial cells of, for instance, the intestine, gall bladder, and kidney—the visualization of cell-coat material usually requires the application of special cytochemical staining techniques (for reviews, see Martínez-Palomo, 1970; Rambourg, 1971; Luft, 1976).

These staining techniques can be subdivided into three broad categories: (a) techniques using polycationic compounds (stains for acidic groups); (b) techniques using lectins (carbohydrate stains); and (c) techniques based on the periodic acid–Schiff reaction (carbohydrate stains).

Of the other available techniques, only those used for the staining of macrophage cell coat will be mentioned below.

Category a: Electron-dense polycationic compounds can be used for the demonstration of negatively charged groups on the cell surface (for review, see Luft, 1976). Metal colloids (iron, thorium), Ruthenium red, Alcian blue, and cationized ferritin are generally used for this purpose. Because these compounds bind to the cell surface by electrostatic interaction, the specificity of the reaction can be enhanced by incubating at low pH (pH 1–3), so that only sulfate or carboxyl groups are still negatively charged.

Category b: Lectins are proteins that bind to specific saccharides of the cell surface in a reaction not unlike an antigen–antibody reaction (see e.g., the review by Brown and Hunt, 1978). For cytochemical purposes, a specific sugar moiety can be demonstrated by attaching a marker molecule to the sugar by means of the appropriate lectin. This can be done in a one-step reaction, using a lectin covalently bound to an electron-dense marker molecule (ferritin, iron dextran) or an enzyme (peroxidase). The method as originally published involved a two-step reaction: in the first step, the lectin binds to the cell surface, and, in the second, the complex is incubated with a marker substance (usually peroxidase) that binds to the attached lectin via its own carbohydrate moieties (Bernhard and Avrameas, 1971).

Category c: As for the (light-microscopic) periodic acid–Schiff (PAS) reaction, the first step of PAS-related electron-microscopic techniques makes use of periodic acid to oxidize adjacent glycol and α-amino alcohol groups, thus transforming them into aldehyde groups. These aldehydic groups are then visualized with silvermethenamine (Rambourg, 1967) or alkaline bismuth subnitrate (Ainsworth *et al.*, 1972). Another way to visualize the periodate-induced aldehydes is to condense them with thiocarbohydrazide, followed by silver proteinate or osmium vapors (Thiery, 1967). When applied to aldehyde-fixed tissue, these PAS-related techniques can be considered to demonstrate glycoproteins (and glycogen) specifically, provided the oxidation step is not unduly prolonged [oxidation by 5 mM periodic acid for 10 min is even thought to oxidize exclusively neuraminic acid (Weber *et al.*, 1975)]. For detailed discussions of PAS-related techniques, reference can be made to Rambourg (1967, 1971).

To increase the specificity of the above-mentioned cytochemical techniques, these procedures should preferably be used in conjunction with blocking procedures for specific chemical groups (e.g., methylation of carboxyl groups or acetylation of hydroxyl moieties) or with enzymatic digestion of cell surface components (e.g., by neuraminidase or trypsin).

3.2. CYTOCHEMISTRY OF THE CELL COAT ON MACROPHAGES

Histochemical studies have not shown any major differences in cell coat composition between macrophages and other cells, although the cell surface of macrophages is known to differ both morphologically and functionally from the surface of other cell types. As already mentioned, the thick coat of the macrophage is particularly difficult to preserve by routine fixation methods. Therefore, we will discuss histochemical observations on the thin coat before presenting the few data available for the thick coat.

3.2.1. Thin Coat

As shown in Table 4, many histochemical techniques have been used to analyze the chemical composition of the macrophage cell coat, often in conjunction with enzymatic degradation or chemical modification of surface components.

Despite the differences in the sources of the macrophages and animal species studied, the results have been remarkably consistent, the over-all conclusion being that the thin coat of the macrophage is composed of acidic glycoproteins whose charge is provided by trypsin-resistant sialic acid residues.

3.2.1a. Acidic Groups. All studies on acidic groups, detected by Ruthenium red, Alcian blue, colloidal iron, or cationized ferritin (see Table 4) have shown homogeneous, diffuse staining of the cell surface. No areas with decreased or increased staining intensity were observed. Methods giving an amorphous precipitate in (or on) the cell coat allow measurement of the thickness of the coat. Carr *et al.* (1970) found an average thickness of 12 nm (occasionally 25 nm), and Santer *et al.* (1973) described a denser inner layer of 9 nm within a coat with an over-all thickness of 22 nm. Emeis (1976) reported a thickness of 15 nm. In grazing sections of cultured macrophages, Ben-Ishay *et al.* (1975) found a crystalline pattern of bound cationized ferritin particles with a center-to-center spacing of 14–16 nm. Using the same technique, Hardy *et al.* (1976) observed differences in charge-density between macrophages from newborn and adult mice, the former having about twice as many acidic groups per surface area. Also, resident peritoneal macrophages and macrophages induced by oil or thioglycollate showed differences in labeling densities with cationized ferritin (Knyszynski *et al.*, 1978).

Methylation (which blocks carboxyl groups) or treatment with neuraminidase (which removes the highly negatively charged carboxyl groups of sialic acid residues) abolished the binding of positively charged stains completely (Carr *et al.*, 1970; Emeis and Wisse, 1971; Ackerman, 1972; Ben-Ishay *et al.*, 1975). Treatment of cells with trypsin (Carr *et al.*, 1970; Ben-Ishay *et al.*, 1975) generally had no effect.

The binding of cationic compounds to the macrophage surface therefore seems to be due to (trypsin-insensitive, neuraminidase-sensitive) carboxyl groups, presumably of sialic acid residues.

3.2.1b. Lectin Binding and Silver Staining. Concanavalin A (Con A) is still the most popular lectin for cell-surface studies. On macrophages, Con A—visualized either by postcoupling with peroxidase (Ackerman and Waksal 1974; Emeis, 1976; Parmley *et al.*, 1973) or by using the lectin bound to colloidal gold (Geoghegan and Ackerman, 1977)—binds diffusely and homogeneously (Figure 20). Wheat-germ agglutinin (Geoghegan and Ackerman, 1977) and

TABLE 4. HISTOCHEMICAL DATA ON MACROPHAGE CELL COAT

Animal and cell type	Methods used	Width[a]	Enzyme degradation or chemical modification	Authors
Rat, Kupffer cell	Thorotrast Alcian blue Silver staining	10-15 nm (thin coat)	Yes	Emeis and Wisse (1971)
	Silicotungstic acid	*	No	Pease and Peterson (1972)
	Protein staining	*	No	Emeis and Wisse (1975)
	Alcian blue Colloidal iron Concanavalin A Silver staining Protein staining	15 nm (thin coat) 65 nm (thick coat)	No	Emeis (1976)
Mouse, peritoneal macrophage	Colloidal iron Ruthenium red Thorotrast	8-16 nm	Yes	Carr *et al* (1970)
	Cationized Ferritin	*	Yes	Ben-Ishay *et al* (1975)
	Cationized Ferritin	zero nm (newborn) 10 nm (adult)	No	Hardy *et al* (1976)
	Cationized Ferritin	*	No	Knyszynski *et al* (1978)
Mouse, peritoneal macrophage	Concanavalin A Wheat-germ agglutinin	*	No	Geoghegan and Ackerman (1977)
Guinea pig; peritoneal macrophage	Os-ferrocyanide Alcian blue Colloidal iron Ruthenium red	Variable	Yes	Dvorak *et al* (1972)
	Os-ferrocyanide	Variable	No	Hammond *et al* (1978)
Mouse; spleen and thymus macrophages	Ruthenium red	19-22nm (inner layer 9 nm)	No	Santer *et al* (1973)
Human bone marrow; macrophage	Ruthenium red Thorotrast Pyroantimonate	*	Yes	Ackerman (1972)
	Concanavalin A	19-21 nm	No	Ackerman and Waksal (1974)
Human and rabbit blood; monocyte	Concanavalin A *Lens culinaris* agglutinin	*	Yes	Parmley *et al.* (1973)

[a] Asterisk indicates that no values are given by the authors

hemagglutinins A and B from *Lens culinaris* (Parmley *et al.*, 1973) gave the same binding pattern as Con A.

Con A binding sites on monocytes were blocked by prior acetylation of hydroxyl groups and by periodic acid oxidation, but not by methylation or neuraminidase treatment (Parmley *et al.*, 1973). Periodic acid–silvermethenamine or periodic acid–silverproteinate staining procedures gave a weak deposit lying parallel to the macrophage plasma membrane (Emeis, 1976; Emeis and Wisse, 1971) (Figure 21).

3.2.1c. Miscellaneous Techniques. Postfixation of macrophages with ferrocyanide-reduced osmium tetroxide (Dvorak *et al.*, 1972) produced linear and globular deposits on the plasma membrane of guinea pig peritoneal macrophages *in vitro* (similar deposits were found with Alcian blue and Ruthenium red staining). These deposits were absent when migration of these macrophages had been inhibited by substances obtained from sensitized lymphocytes (Dvorak *et al.*, 1972), whereas enhanced staining was found in the presence of *E*-aminocaproic acid, a compound that stimulates macrophage migration (Hammond *et al.*, 1978). The specificity of osmium–ferrocyanide staining has not been defined.

Pyroantimonate, which is used to demonstrate inorganic cations, gave no staining of human bone marrow macrophages, in contrast to other bone marrow cells (Ackerman, 1972). Pease and Peterson (1972) stained liver tissue (embedded in mixtures of glutaraldehyde and urea) with silicotungstic acid, and observed a thin, densely stained surface coat on the plasma membrane of Kupffer cells. The acrolein–Schiff method of Van Duijn (1961) for the demonstration of proteins was adapted for electron microscopy by the substitution of thiocarbohydrazide followed by osmium for the Schiff reagent (Emeis, 1976). With this method a thin layer (about 20 nm) on the surface of Kupffer cells was demonstrated (Figure 22).

3.2.1d. Thin Coat: Conclusions. The thin coat (thickness: 12–20 nm) of macrophages reacts positively when methods to demonstrate the presence of acidic groups (mainly carboxyl groups of sialic acid), neutral sugars, and protein are used; inorganic cations are not demonstrable. Therefore, this thin coat is presumably composed of acid glycoproteins containing (trypsin-insensitive) sialic acid.

3.2.2. Thick Coat

As discussed above (Section 2.3.1), the thick coat of macrophages can be subdivided morphologically into three layers (Brederoo and Daems, 1972). Cytochemically, the thick coat is composed of two layers: the thin coat close to the plasma membrane and an outer layer about 50 nm thick. This outer layer, unlike the thin coat, does not react to cytochemical stains such as Ruthenium red, colloidal iron, Alcian blue, Con A, and silver-staining procedures. The only staining procedure known to give a positive reaction in outer layer is the acrolein–TCH–osmium method for the demonstration of proteins (Emeis, 1976) (Figure 22). Occasionally, a faint positive reaction of carbohydrate stains was observed in the outer layer of the thick coat in wormlike structures. (The dense midline of wormlike structures reacted similarly to the thin coat; see Figures 20, 21, and 22.)

4. CORRELATION WITH FUNCTION

One of the most conspicuous functions of macrophages is the removal of foreign, effete, or damaged material via endocytosis. The first step in this uptake process is the attachment of the material to the macrophage surface. Study of the attachment phase of endocytosis might, therefore, provide useful information about the morphology and chemistry of the outermost boundary of the macrophage: its cell coat.

Many studies have been devoted to the presence and role of Fc and complement receptors on macrophages during endocytosis. The findings concerning the layers in the cell coat raise the question of the localization of these receptors. Ultrastructural studies (McKeever *et al.*, 1976) have suggested that the Fc receptor is localized in or below the thin coat. For an extensive review on the subject of these receptors, the reader is referred to the chapter by McKeever and Spicer in this volume. Our discussion will be limited to morphological observations on cell coat during attachment and data relevant to the histochemical observations cited above.

Most studies on attachment and endocytosis by macrophages have been performed *in vitro* with freshly isolated or cultured cells (usually from the peritoneal cavity). With respect to the study of the cell coat, this is unfortunate. As discussed above, morphological studies have shown that the outer layer of the thick coat is an extremely vulnerable structure. Moreover, there is no evidence that this outer layer occurs on cultured macrophages. Wormlike structures (the morphological sequel of the thick coat) have not been found in cultured Kupffer cells, although this type of cell is known to have these structures abundantly *in vivo* (Emeis and Planqué, 1976). Possibly, macrophages cannot repair lost cell coat. In any case, it is clear that attempts to correlate *in vitro* observations on attachment with observations made *in vivo* are hazardous.

Attachment to the thick cell coat is known to occur *in vivo,* since it has been shown for foreign material, e.g., erythrocyte ghosts (see Figure 14), colloidal silver, dextran sulfate, colloidal carbon (see Emeis, 1976), components of damaged cells (Pfeifer, 1970; Ericsson *et al.*, 1971; Wisse, 1974a), and fibrin (Prose *et al.*, 1965; Brederoo and Daems, 1972; De Ruyter *et al.*, 1976; Emeis and Lindeman, 1976). On the other hand, comparable material (aged erythrocytes, colloidal thorium, parts of damaged cells) can be found close to the plasma membrane, presumably attached to the thin coat.

Carbon (Figure 13), erythrocyte ghosts, and fibrin can be found attached to both thick and thin coats (Emeis and Lindeman, 1976). This probably explains why fibrin-like material persists on macrophages after repeated washing (Colvin and Dvorak, 1975). Thus, the chemical composition of the cell coat layers cannot be deduced from characteristics of the materials attaching to them. Moreover, there seems to be no correlation between particle size and the gap with the plasma membrane (Tables 2 and 3).

Of interest is the observation that pseudopods formed by Kupffer cells as a reaction to the presence of foreign material lack the thick cell coat (Figure 14; Emeis, 1976). It is not clear why these areas of the cell surface have no thick outer layer, especially because comparable cytoplasmic extensions in peritoneal macrophages occasionally show a thick coat (Figures 7 and 8). The interesting suggestion put forward by Saba (1978), i.e., that α_1-opsonic glycoprotein (fibronectin) might be present in the outer layer, was not substantiated by an immunohistochemical study (Stenman and Vaheri, 1978).

In vitro studies have shown that glycoproteins are involved in various macrophage functions (e.g., endocytosis, cell migration). Con A has also been shown light microscopically (using fluorescent Con A) to bind diffusely to the macrophage surface (Allen *et al.*, 1971), this binding being trypsin insensitive (Lutton, 1973). Binding of Con A to peritoneal macrophages induced pinocytosis, rounding of the cells, and vacuole formation (Allen *et al.*, 1971; Goldman, 1974a), and endocytosis caused the disappearance of about one-third of Con A binding sites from the surface (Lutton, 1973). Preincubation of macrophages with Con A reduced the binding of erythrocytes (Goldman, 1974b) and latex (Friend *et al.*, 1975) to these cells. On the other hand, Con A induced binding, but not ingestion, of unopsonized *Bacillus subtilis* (Allen *et al.*, 1971) and adhesion of tumor cells to peritoneal macrophages (Inoue *et al.*, 1972).

Neuraminidase-sensitive cell surface components are also involved in macrophage phagocytosis, as shown by the enhanced uptake of bacteria after neuraminidase treatment of macrophages (Knop *et al.*, 1978; Ögmundsdottir *et al.*, 1978), although this treatment had no effect on pinocytosis of albumin and ferritin (Lagunoff, 1971).

Further evidence for the involvement of cell-surface glycoproteins in endocytosis was provided by the finding of Ögmundsdottir *et al.* (1978) that treatment of cells with periodic acid or β-galactosidase reduced the binding of *Corynebacterium parvum*, and that pretreatment of *Escherichia coli* or *Salmonella typhi* with mannose also reduced binding.

Thus, the above-mentioned surface properties of macrophages involved in endocytosis (Con-A-binding; periodate, neuraminidase, and β-galactosidase sensitivity; and involvement of mannose) are closely correlated with the histochemical properties of the thin coat.

With respect to macrophage migration, Kumagai and Arai (1973) showed that Con A inhibits migration, and Remold (1973) that the presence of fucose on the macrophage surface is essential for the binding of migration inhibition factor to macrophages.

Morphological changes in the cell coat of macrophages *in vitro* during migration inhibition and enhanced migration have also been reported (Dvorak *et al.*, 1972; Hammond *et al.*, 1978). Presumably, the cell coat is also involved in migration processes, and changes under the influence of factors stimulating or inhibiting macrophage migration.

5. CONCLUSIONS

Morphological studies have shown the presence of a (50–70 nm) thick cell coat on macrophages. This cell coat, which requires special preparative precautions for intact preservation, is never found over the whole cell surface. Morphologically, this coat shows three layers. The thickness of the innermost layer (thin coat, in our material 17 nm) agrees well with the size of the gap found between attached material and the plasma membrane (see Table 3). Cytochemically, two layers with different staining properties can be distinguished, the inner layer containing glycoprotein, the outer layer only protein. The thickness of the cytochemically-defined thin coat (in our material 15 nm; see also Table 4) corresponds with the thickness of the morphologically-defined thin coat.

Because foreign material can be found attached to both the thick and the thin coat, the involvement of the cell coat of macrophages in the process of endocytosis is not well understood.

REFERENCES

Ackerman, G. A., 1972, Localization of pyroantimonate-precipitable cation and surface coat anionic binding sites in developing erythrocytic cells and macrophages in normal human bone marrow, *Z. Zellforsch. Mikrosk. Anat.* **134:**153.

Ackerman, G. A., and Waksal, S. D., 1974, Ultrastructural localization of Concanavalin A binding sites on the surface of differentiating hemopoietic cells, *Cell Tissue Res.* **150:**331.

Ainsworth, S. K., Ito, S., and Karnovsky, M. J., 1972, Alkaline bismuth reagent for high resolution ultrastructural demonstration of periodate-reactive sites, *J. Histochem. Cytochem.* **20:**995.

Allen, J. M., Cook, G.M.W., and Poole, A. R., 1971, Action of Concanavalin A on the attachment stage of phagocytosis by macrophages, *Exp. Cell Res.* **68:**466.

Arborgh, B., Berg, T., and Ericsson, J.L.E., 1974, Evaluation of methods for specific loading of Kupffer cell lysosomes with heavy colloidal particles, *Acta Pathol. Microbiol. Scand. A* **82:**747.

Bainton, D. F., and Golde, D. W., 1978, Differentiation of macrophages from normal human bone marrow in liquid culture, *J. Clin. Invest.* **61:**1555.

Ben-Ishay, Z., Reichert, F., and Gallily, R., 1975, Crystalline-like surface charge array of murine macrophages and lymphocytes: Visualization with cationized ferritin, *J. Ultrastruct. Res.* **53:**119.

Bennett, H. S., 1963, Morphological aspects of extracellular polysaccharides, *J. Histochem. Cytochem.* **11:**14.

Bennett, H. S., 1969, The cell surface: components and configurations, in: *Handbook of Molecular Cytology* (A. Lima-de-Faria, ed.), pp. 1216–1293, North-Holland, Amsterdam.

Bernhard, W., and Avrameas, S., 1971, Ultrastructural visualization of cellular carbohydrate components by means of Concanavalin A, *Exp. Cell Res.*, **64:**232.

Blom, J., 1973, The ultrastructure of contact zones between plasma cells and dendritic macrophages from patients with multiple myeloma, *Acta Pathol. Microbiol. Scand. A* **81:**734.

Blom, J. 1977, The ultrastructure of contact zones between plasma cells and macrophages in the bone marrow of patients with multiple myeloma, *Acta Pathol. Microbiol. Scand. A* **85:**345.

Brederoo, P., and Daems, W. T., 1972, Cell coat, worm-like structures, and labyrinths in guinea pig resident and exudate peritoneal macrophages, as demonstrated by an abbreviated fixation procedure for electron microscopy, *Z. Zellforsch. Mikrosk. Anat.* **126:**135.

Brown, J. C., and Hunt, R. C., 1978, Lectins, in: *International Review of Cytology* (G. H. Bourne, J. F. Danielli, and K. W. Jeon, eds.), Vol. 52, pp. 277–349, Academic Press, New York.

Burke, J. S., and Simon, G. T., 1970, Electron microscopy of the spleen. II. Phagocytosis of colloidal carbon, *Am. J. Pathol.* **58:**157.

Carr, I., Everson, G., Rankin, A., and Rutherford, J., 1970, The fine structure of the cell coat of the peritoneal macrophage and its role in the recognition of foreign material, *Z. Zellforsch. Mikrosk. Anat.* **105:**339.

Colvin, R. B., and Dvorak, H. F., 1975, Fibrinogen/fibrin on the surface of macrophages: detection, distribution, binding requirements, and possible role in macrophage adherence phenomena, *J. Exp. Med.* **142:**1377.

Cotutiu, C. C., and Ericsson, J.L.E., 1970, Two types of "microendocytosis" in Kupffer cells studied with electron opaque tracers, in: *Microscopie Electronique 1970* (P. Favard, ed.), Rés. 7ième Congr. Intern., Grenoble, Vol. 3, pp. 53–54, Soc. Franç. Microsc. Electr., Paris.

Cotutiu, C. C., and Ericsson, J.L.E., 1971, Characterization of "micropinocytosis vermiformis" in Kupffer cells with the aid of electron dense tracers, *J. Ultrastruct. Res.* **36:**512.

De Ruyter, A., Brederoo, P., Emeis, J. J., and Daems, W. T., 1976, Cell coat on guinea-pig peritoneal macrophages, *Ultramicroscopy* **2:**124.

Dvorak, A. M., Hammond, M. E., Dvorak, H. F., and Karnovsky, M. J., 1972, Loss of cell surface material from peritoneal exudate cells associated with lymphocyte-mediated inhibition of macrophage migration from capillary tubes, *Lab. Invest.* **27:**561.

Emeis, J. J., 1976, Morphologic and cytochemical heterogeneity of the cell coat of rat liver Kupffer cells, *J. Reticuloendothel. Soc.* **20:**31.

Emeis, J. J., and Lindeman, J., 1976, Rat liver macrophages will not phagocytose fibrin during disseminated intravascular coagulation, *Haemostasis* **5:**193.

Emeis, J. J., and Planqué, B., 1976, Heterogeneity of cells isolated from rat liver by pronase digestion: Ultrastructure, cytochemistry and cell culture, *J. Reticuloendothel. Soc.* **20:**11.

Emeis, J. J., and Wisse, E., 1971, Electron-microscopic cytochemistry of the cell coat of Kupffer cells in rat liver, in: *The Reticuloendothelial System and Immune Phenomena* (N. R. DiLuzio and K. Flemming, eds.), Proceedings of the Sixth International Meeting of the Reticuloendothelial Society, Freiburg, 1970, pp. 1–12, Plenum Press, New York.

Emeis, J. J., and Wisse, E., 1975, On the cell coat of rat liver Kupffer cells, in: *Mononuclear Phagocytes in Immunity, Infection, and Pathology* (R. van Furth, ed.), pp. 315–325, Blackwell, Oxford.

Ericsson, J.L.E., Cotutiu, C. C., and Glaumann, H., 1971, Heterophagocytosis of intravenously injected subcellular organelles by Kupffer cells, *J. Ultrastruct. Res.* **36:**518.

Fahimi, H. D., 1970, The fine structural localization of endogenous and exogenous peroxidase activity in Kupffer cells of rat liver, *J. Cell Biol.* **47:**247.

Friend, K., Ekstedt, R. D., and Duncan, J. L., 1975, Effect of Concanavalin A on phagocytosis by mouse peritoneal macrophages, *J. Reticuloendothel. Soc.* **17:**10.

Geoghegan, W. D., and Ackerman, G. A., 1977, Adsorbtion of horseradish peroxidase, ovomucoid and anti-immunoglobulin to colloidal gold for the indirect detection of Concanavalin A, wheat germ agglutinin and goat anti-human immunoglobulin on cell surfaces at the electron microscopic level: A new method, theory and application, *J. Histochem. Cytochem.* **25:**1187.

Goldman, R., 1974a, Induction of vacuolation in the mouse peritoneal macrophage by Concanavalin A, *FEBS Lett.* **46:**203.

Goldman, R., 1974b, Effect of Concanavalin A on phagocytosis by macrophages, *FEBS Lett.* **46:**209.

Hammond, M. E., Dvorak, A. M., Roblin, R. O., Morgan, E. S., and Dvorak, H. F., 1978, Effect of *E*-amino caproic acid on macrophage cell coat and migration inhibition, *J. Reticuloendothel. Soc.* **24:**63.

Han, S. S., Han, J. H., and Johnson, A. G., 1970, The fate of isologous, homologous, and heterologous ferritin molecules in the rat, *Am. J. Anat.* **129:**141.

Hardy, B., Skutelsky, E., Globerson, A., and Danon, D., 1976, Ultrastructural differences between macrophages of newborn and adult mice, *J. Reticuloendothel. Soc.* **19:**291.

Hausmann, E., 1975, Struktur und Zusammensetzung der Mucoidschicht verschiedener Amöbenarten, *Cytobiologie* **10:**285.

Hoefsmit, E.C.M., 1975, Mononuclear phagocytes, reticulum cells, and dendritic cells in lymphoid tissues, in: *Mononuclear Phagocytes in Immunity, Infection, and Pathology* (R. van Furth, ed.), pp. 129–146, Blackwell, Oxford.

Horn, R. G., Koenig, M. G., Goodman, J. S., and Collins, R. D., 1969, Phagocytosis of *Staphylococcus aureus* by hepatic reticuloendothelial cells. An ultrastructural study, *Lab. Invest.* **21:**406.

Inoue, M., Mori, M., Utsumi, K., and Seno, S., 1972, Role of Concanavalin A in tumor cell agglutination and adhesion to macrophages, *Gann* **63:**795.
Ishihara, T., 1973, Experimental amyloidosis using silver nitrate. Electron microscopic study on the relationship between silver granules, amyloid fibrils and reticuloendothelial system, *Acta Pathol. Jpn.* **23**:439.
Ito, S., 1974, Form and function of the glycocalyx on free cell surfaces, *Phil. Trans. Royal Soc. London B* **268**:55.
Katenkamp, D., and Stiller, D., 1975, Structural patterns and histological behavior of experimental sarcomas. 2. Ultrastructural cytology, *Exp. Pathol.* **11**:190.
Knop, J., Ax, W., Sedlacek, H. H., and Seiler, F. R., 1978, Effect of *Vibrio cholerae* neuraminidase on the phagocytosis of *E. coli* by macrophages *in vivo* and *in vitro*, *Immunology* **34:**555.
Knyszinski, A., Leibovich, S. J., Skutelski, E., and Danon, D., 1978, Macrophages from unstimulated, mineral oil and thioglycollate-stimulated mice: A comparative study of surface anionic sites and phagocytosis of "old" red blood cells, *J. Reticuloendothel. Soc.* **24:**205.
Kumagai, K., and Arai, S., 1973, Inhibition of macrophage migration by Concanavalin A, *J. Reticuloendothel. Soc.* **13:**507.
Lagunoff, D., 1971, Macrophage pinocytosis: the removal and resynthesis of a cell surface factor, *Proc. Soc. Exp. Biol. Med.* **138:**118.
Luft, J. H., 1976, The structure and properties of the cell surface coat, in: *International Review of Cytology* (G. H. Bourne, J. F. Danielli, and K. W. Jeon, eds.), Vol. 45, pp. 291–382, Academic Press, New York.
Lutton, J. D., 1973, The effect of phagocytosis and spreading on macrophage surface receptors for Concanavalin A, *J. Cell Biol.* **56:**611.
Martínez-Palomo, A., 1970, The surface coats of animal cells, in: *International Review of Cytology* (G. H. Bourne, J. F. Danielli, and K. W. Jeon, eds.), Vol. 29, pp. 29–75, Academic Press, New York.
Matter, A., Orci, L., Forsmann, W. G., and Rouiller, C., 1968, The stereological analysis of the fine structure of the "micropinocytosis vermiformis" in Kupffer cells of the rat, *J. Ultrastruct. Res.* **23:**272.
McKeever, P. E., Garvin, A. J., and Spicer, S. S., 1976, Immune complex receptors on cell surfaces. I. Ultrastructural demonstration on macrophages, *J. Histochem. Cytochem.* **24:**948.
Melis, M., and Orci, L., 1967, Sugli aspetti ultrastrutturali delle cellule di Kupffer nel ratto dopo epatectomia parziale, *Fegato* **13:**356.
Merchant, R. E., Olson, G. E., and Frank, N. L., 1977, Scanning and transmission electron microscopy of leptomeningeal free cells: cell interactions in response to challenge by *Bacillus* Calmette-Guerin, *J. Reticuloendothel. Soc.* **22:**199.
Ogawa, K., Minase, T., Yokoyama, S., and Onoé, T., 1973, An ultrastructural study of peroxidatic and phagocytic activities of two types of sinusoidal lining cells in rat liver, *Tohoku J. Exp. Med.* **111:**253.
Ögmundsdottir, H. M., Weir, D. M., and Marmion, B. P., 1978, Binding of microorganisms to the macrophage plasma membrane; effects of enzymes and periodate, *Br. J. Exp. Pathol.* **59:**1.
Orci, L., Pictet, R., and Rouiller, C., 1967, Image ultrastructurale de pinocytose dans la cellule de Kupffer du foie de rat, *J. Microscopie* **6:**413.
Parmley, R. T., Martin, B. J., and Spicer, S. S., 1973, Staining of blood cell surfaces with a lectin-horseradish peroxidase method, *J. Histochem. Cytochem.* **21:**912.
Parsons, D. F., and Subjeck, J. R., 1972, The morphology of the polysaccharide coat of mammalian cells, *Biochim. Biophys. Acta* **265:**85.
Pease, D. C., and Peterson, R. G., 1972, Polymerizable glutaraldehyde-urea mixtures as polar, water-containing embedding media, *J. Ultrastruct. Res.* **42:**133.
Pfiefer, U., 1970, Über Endocytose in Kupfferschen Sternzellen nach Parenchymschädigung durch 3/4-Teilhepatektomie, *Virchows Arch. B* **6:**263.
Prose, P. H., Lee, L., and Balk, S. D., 1965, Electron microscopic study of the phagocytic fibrin-clearing mechanism, *Am. J. Pathol.* **47:**403.
Rambourg, A., 1967, An improved silver methenamine technique for the detection of periodic acid-reactive complex carbohydrates with the electron microscope, *J. Histochem. Cytochem.* **15:**409.
Rambourg, A., 1971, Morphological and histochemical aspects of glycoproteins at the surface of

animal cells, in: *International Review of Cytology* (G. H. Bourne, J. F. Danielli, and K. W. Jeon, eds.), Vol. 31, pp. 57–114, Academic Press, New York.

Remold, H. G., 1973, Requirement for α-L-fucose on the macrophage membrane receptor of MIF, *J. Exp. Med.* **138:**1065.

Revel, J.-P., and Ito, S., 1967, The surface components of cells, in: *The Specificity of Cell Surfaces* (B. D. Davis and L. Warren, eds.), pp. 211–234, Prentice-Hall, Englewood Cliffs, N.J.

Röhlich, P., and Törö, I., 1964, Uptake of chylomicron particles by reticular cells of mesenterial lymph nodes of the rat, in: *Elecron Microscopy 1964* (M. Titlebach, ed.), Proceedings of the Third European Regional Conference, Prague, Vol. B, pp. 225–226, Czechoslovak Acad. Sci., Prague.

Rouiller, C., Pichet, R., Niculescu, P., and Orci, L., 1967, Les cellules reticulo-endotheliales du foie et l'espace de Disse, *Rev. Int. Hépatol.* **17:**827.

Rybicka, K., Daly, B.D.T., Migliore, J. J., and Norman, J. C., 1974, Intravascular macrophages in normal calf lung. An electron microscopic study, *Am. J. Anat.* **139:**353.

Saba, T. M., 1978, Humoral control of Kupffer cell function after injury, *Bull. Kupffer Cell Found.* **1:**12.

Sanel, F. T., and Serpick, A. A., 1970, Plasmalemmal and subsurface complexes in human leukemic cells: membrane bonding by zipperlike junctions, *Science* **168:**1458.

Santer, V., Cone, R. E., and Marchalonis, J. J., 1973, The glycoprotein surface coat on different classes of murine lymphocytes, *Exp. Cell Res.* **79:**404.

Schneeberger-Keeley, E. E., and Burger, E. J., 1970, Intravascular macrophages in cat lungs after open chest ventilation, *Lab. Invest.* **22:**361.

Stenman, S., and Vaheri, A., 1978, Distribution of a major connective tissue protein, fibronectin, in normal human tissues, *J. Exp. Med.* **147:**1054.

Tamaru, T., and Fujita, H., 1978, Electron-microscopic studies on Kupffer's stellate cells and sinusoidal endothelial cells in the liver of normal and experimental rabbits, *Anat. Embryol.* **154:**125.

Tanuma, Y., 1978, Fine structure of the Kupffer cell in the bat, with special reference to the worm-like bodies, *Arch. Histol. Jpn.* **41:**113.

Thiery, J. P., 1967, Mise en évidence des polysaccharides sur coupes fines en microscopie électronique, *J. Microscopie* **6:**987.

Tòro, J., Rusza, P., and Röhlich, P., 1962, Ultrastructue of early phagocytic stages in sinus endothelial and Kupffer cells of the liver, *Exp. Cell Res.* **26:**601.

van der Rhee, H. J., van der Burgh-de Winter, C. P. M., and Daems, W. T., 1979, The differentiation of monocytes into macrophages, epithelioid cells, and multinucleated giant cells in subcutaneous granulomas. I. Fine structure, *Cell Tissue Res.* **197:**355.

van Duijn, P., 1961, Acrolein-Schiff, a new staining method for proteins, *J. Histochem. Cytochem.* **9:**234.

Weber, P., Harrison, F. W., and Hof, L., 1975, The histochemical application of dansylhydrazine as a fluorescent labeling reagent for sialic acid in cellular glycoconjugates, *Histochemistry* **45:**271.

Widmann, J. -J., Cotran, R. S., and Fahimi, H. D., 1972, Mononuclear phagocytes (Kupffer cells) and endothelial cells. Identification of two functional cell types in rat liver sinusoids by endogenous peroxidase activity, *J. Cell Biol.* **52:**159.

Wiener, J., Spiro, D., and Margaretten, W., 1964, An electron microscopic study of reticuloendothelial system blockade, *Am. J. Pathol.* **45:**783.

Winzler, R. J., 1970, Carbohydrates in cell surfaces, in: *International Review of Cytology* (G. H. Bourne, J. F. Danielli, and K. W. Jeon, eds.), Vol. 29, pp. 77–125, Academic Press, New York.

Wisse, E., 1974a, Observations on the fine structure and peroxidase cytochemistry of normal rat liver Kupffer cells, *J. Ultrastruct. Res.* **46:**393.

Wisse, E., 1974b, Kupffer cell reactions in rat liver under various conditions as observed in the electron microscope, *J. Ultrastruct. Res.* **46:**499.

Wisse, E., 1977a, Ultrastructure and function of Kupffer cells and other sinusoidal cells in the liver, in: *Kupffer Cells and Other Liver Sinusoidal Cells* (E. Wisse and D. L. Knook, eds.), pp. 33–60, Elsevier/North-Holland, Amsterdam.

Wisse, E., 1977b, Ultrastructure and function of Kupffer cells and other sinusoidal cells in the liver, *Méd. Chir. Dig.* **6:**409.

Wisse, E., and Daems, W. T., 1970a, Fine structural study on the sinusoidal lining cells of rat liver, in: *Mononuclear Phagocytes* (R. van Furth, ed.), pp. 200–210, Blackwell, Oxford.
Wisse, E., and Daems, W. T., 1970b, Differences between endothelial and Kupffer cells in rat liver, in: *Microscopie Electronique 1970* (P. Favard, ed.), Rés. 7ième Congr. Intern., Grenoble, Vol. 3, pp. 57–58, Soc. Franç. Microsc. Electr., Paris.

5

Surface Receptors of Mononuclear Phagocytes

PAUL E. MCKEEVER
and SAMUEL S. SPICER

1. INTRODUCTION

The surface receptors of obvious importance to the macrophage's biologic activities are receptors for the Fc portion of IgG, complement receptors, and the "nonspecific" receptor referred to here as the "receptor for foreign substances." Descriptions of morphologic and functional parameters which govern the activities of these receptors constitute a major portion of this chapter. Drugs are mentioned where they specifically affect activity of one of these receptors; however, the discussion of macrophage receptors for drugs is beyond the scope of this review. Fewer morphologic data are available on recently defined receptors for insulin, fibrin, lacto-ferrin, chemotactic, lymphokine, carbohydrate, and lipid receptors. They are included for their biologic and potential clinical importance. The glycoprotein receptor provides important insight into foreign substance receptor activity.

The role of macrophage surface receptors in immune cell respones is becoming increasingly apparent. Recent morphologic studies of macrophage–lymphocyte interactions in immune responses to antigen provide an opportunity to appreciate correlations of structure and function in this aspect of immunity. Macrophage participation in immune responses to antigens is considered at length with emphasis on the functional significance of morphologic relationships. Recent reviews discuss in detail the genetic restrictions and antigenic specificities of these immune cell interactions (Rosenthal *et al.*, 1978;

PAUL E. MCKEEVER • Laboratory of Microbial Immunity, National Institute of Allergy and Infectious Diseases, Bethesda, Maryland 20205; and Surgical Neurology Branch, National Institute of Neurological and Communicative Disorders and Stroke, Bethesda, Maryland 20205. SAMUEL S. SPICER • Department of Pathology, Medical University of South Carolina, Charleston, South Carolina 29403.

Niederhuber, 1978; Rosenthal, 1978; Schwartz *et al.*, 1978; Thomas *et al.*, 1978; Beller *et al.*, 1978; Pierce and Kapp, 1978). Detailed reviews of antigen processing by macrophages are available (Cohn, 1968; Unanue, 1972, 1978). In addition, it is likely that surface receptors on macrophages are involved in many other immunologic phenomena including the migration of macrophage subpopulations carrying foreign material toward lymphoid tissue (McKeever and Balentine, 1978), release of soluble mediators of lymphocyte stimulation (Beller *et al.*, 1978; Opitz *et al.*, 1978), macrophage aggregation (Postlethwaite and Kang, 1976), development of tolerance (Diener *et al.*, 1976), and macrophage extravasation (Musson and Henson, 1979).

The importance of receptor isolation and structural analysis for understanding receptor function warranted their inclusion as separate sections. Selected examples where macrophage or monocyte receptors have potential value in the clinical evaluation of metabolic disorders in the treatment of neoplasia, or in understanding pathogenesis of disease, have been included in an additional section.

The morphologic relationship of microfilaments and microtubules are noted briefly under binding or endocytosis of the specific receptor. Comprehensive coverage of the excitatory and contractile aspects of phagocytic receptors are available (Stossel, 1975, 1976, 1977). Macrophage ectoenzymes have been reviewed elsewhere (Karnovsky *et al.*, 1975).

A few terms used widely throughout this chapter require definition. *Induced macrophages* or *exudate macrophages* are elicited from an animal by the prior injection of a foreign irritant, usually into the peritoneal cavity. Hydrocarbons, oils, complex carbohydrates, and viable or killed microorganisms are commonly employed irritants (Stuart *et al.*, 1970). The macrophages obtained a few days to weeks following induction are usually activated compared with uninduced macrophages from the same location. *Uninduced macrophages* are also referred to as *resident* or *resting macrophages.* An *activated macrophage* has been defined as a macrophage capable of increased microbicidal capacity (Ratzan *et al.*, 1972). The term is commonly used for a macrophage which differs from a resident macrophage by spontaneous synthesis and secretion of neutral proteases (Unkeless *et al.*, 1974; Vassalli and Reich, 1977; Werb and Gordon, 1975a, b), increased endocytosis (Edelson *et al.*, 1975; Edelson and Erbs, 1978; Bianco *et al.*, 1975), increased spreading and attachment (Carr and Carr, 1970; Bianco *et al.*, 1976; Edelson and Erbs, 1978; Burgaleta *et al.*, 1978), lower membrane 5′-nucleotidase activity (Edelson and Erbs, 1978; Edelson and Cohn, 1976), increased cytotoxicity (Melsom and Seljelid, 1973; Hibbs, 1976; Hanna *et al.*, 1976), greater ability to proliferate *in vitro* (Stewart *et al.*, 1975; Van der Zeijst *et al.*, 1978), increased lysosomal enzymes (Dannenberg, 1968), increased capacity to ingest complement-coated erythrocytes (Bianco *et al.*, 1975; Edelsen and Erbs, 1978), larger cell volume (McKeever *et al.*, 1976b; Burgaleta *et al.*, 1978; Territo *et al.*, 1976), augmented chemotactic activity (Burgaleta *et al.*, 1978; Poplack *et al.*, 1976), and increased glucose metabolism (Ratzan *et al.*, 1972; Stubbs *et al.*, 1973). Each of the various agents which activates macrophages does not necessarily produce all of the effects listed (Stubbs *et al.*, 1973; Melsom and Seljelid, 1973).

We use the term *mononuclear phagocyte* to represent both the circulating monocyte and the macrophage found in tissue (van Furth *et al.*, 1975). Receptors of IgG, complement, and foreign substances are together referred to as the classical macrophage receptors.

This chapter attempts an updated and thorough definition of surface receptors on mononuclear phagocytes. Important comparisons of receptors on resident macrophages, activated macrophages, and monocytes are included. Comparisons of receptors on mononuclear phagocytes with receptors on other types of immune cells, in general, are not included.

2. RECEPTORS FOR IMMUNOGLOBULIN G (Fc RECEPTORS)

2.1. BIOLOGIC ACTIVITY

Originally known as opsonins (Briscoe, 1908; Berry and Spies, 1949), some of these factors have been specifically related to antibodies (Vaughan, 1965), especially the immunoglobulin G fraction (IgG) (Jones, 1975; Lawrence *et al.*, 1975; Huber *et al.*, 1971), and complement (see also Section 3; Müller-Eberhard, 1968). IgG and immunoglobulin with bound complement components account for much of the known selectivity of macrophage adherence (Haranaka *et al.*, 1977). Immunoglobulin coating of bacteria has been shown to facilitate selective adherence and phagocytic uptake of bacteria and thus to enhance the bactericidal action of macrophages (Reynolds and Thompson, 1973; Reynolds, 1974). The role of antibody in mediating selective phagocytosis depends on linkage of a region of the antibody to a recognition site or receptor on the macrophage surface. Thus, the IgG molecule apparently binds immunospecifically to the microorganisms through the variable portion of its Fab region and to the macrophage surface via its Fc segment (C_H2 or C_H3 homology region) (Diamond *et al.*, 1979; Dorrington, 1976). The latter portion of the macromolecule does not combine with antigen and, independent of its immunospecificity, has a constant amino acid sequence within an immunoglobulin subclass (i.e., IgG). The role of surface receptors for immunoglobulin in macrophage–lymphocyte interactions is described in Section 5. Cytophilic antibody participation in contact sensitization reactions has been reported (Askenase and Hayden, 1974). Receptors for IgG are important in combating parasitic infestation (Shear *et al.*, 1979; Kassis *et al.*, 1979).

In addition to their function in the elimination of foreign agents, the surface immunoglobulin receptors (Fc receptors) apparently function in the physiologic removal of the host's aged erythrocytes by macrophages (Knyszynski *et al.*, 1977), analogous to their removal of noncellular endogenous components such as antigen–antibody complexes. Aging may expose previously masked, terminal carbohydrate groups on erythrocytes which are then opsonized by autologous IgG *in situ*, promoting phagocytosis by hepatic and splenic macrophages (Kay, 1975; Jancik and Schauer, 1978; Nicolson, 1973). The evidence for such an autologous monitoring system points to the possibility that a similar immunologic mechanism participates in pathologic circumstances where macrophages ingest autologous material (see Section 12).

The comparison of macrophage and lymphocyte receptors is beyond the scope of this chapter. Excellent reviews of lymphocyte receptors are available (Nussenzweig, 1974; Dickler, 1976).

The important role of the macrophage in phagocytic clearance of noxious elements entails surface binding of the target substance followed by its endocytic incorporation into the cell. For the macrophage to benefit rather than damage its host, the initial adherence to the target element obviously must be selective. Serum factors which coat the phagocytic target substances largely provide the basis for such selective binding of target moieties to the surface of macrophages (Rabinovitch, 1968; North, 1968).

2.2. TECHNIQUES FOR DEMONSTRATING Fc RECEPTORS

Fc receptors are mainly demonstrated by the erythrocyte rosetting assay wherein erythrocytes are coated with antibody to the red cell surface antigens and macrophages are then exposed to these sensitized erythrocytes (designated EA). The Fc-receptor activity is indicated by adherence of three or more erythrocytes on the macrophage surface in a rosette configuration. Detailed methods for the study of macrophage Fc receptors with indicator erythrocytes are available (Bianco, 1976). If the erythrocyte targets have been prelabeled with ^{51}Cr, they can be counted automatically (Kuhn and Cassida, 1978). At the ultrastructural level immune complex can be demonstrated with a peroxidase or ferritin label, the former used as the antigen in a soluble immune complex with specific antibody (McKeever *et al.*, 1976a; Papadimitriou, 1972) (Figure 1). Ultrastructural marker proteins coupled covalently to immunoglobulin are available commercially (Papadimitriou, 1972).

Techniques for determining whether target cells are bound or ingested (Chambers, 1973) are very useful in distinguishing rosetting from endocytosis. Simplest of these is osmotic lysis of bound erythrocytes. These techniques take advantage of the fact that the contents of a true endocytic vacuole are within the cell where they are inaccessible to reagents which do not penetrate cells.

Fluorescein-conjugated heat-aggregated immunoglobulin has been used to study lymphocyte receptors for IgG extensively (Dickler, 1974; Dickler and Sachs, 1974), and macrophage IgG receptors to a lesser extent (Schwartz *et al.*, 1976). Despite some problems encountered with retention of cytophilic properties for macrophages of less aggregated immunoglobulin after fluorescein conjugation (Thrasher *et al.*, 1975a,b), fluorescent reagents will probably increase in favor, since they are adaptable for use on machines which separate cells on the basis of differences in fluorescence (Schwartz *et al.*, 1976). Immunoglobulin bound to the macrophage surface can be detected indirectly with fluoresceinated anti-immunoglobulin antibodies (Thrasher *et al.*, 1975b). Target cells coated with fluorescent antibody will be bound by macrophages (Griffin *et al.*, 1976).

Receptors for IgG can be enumerated and their specificity and affinity as-

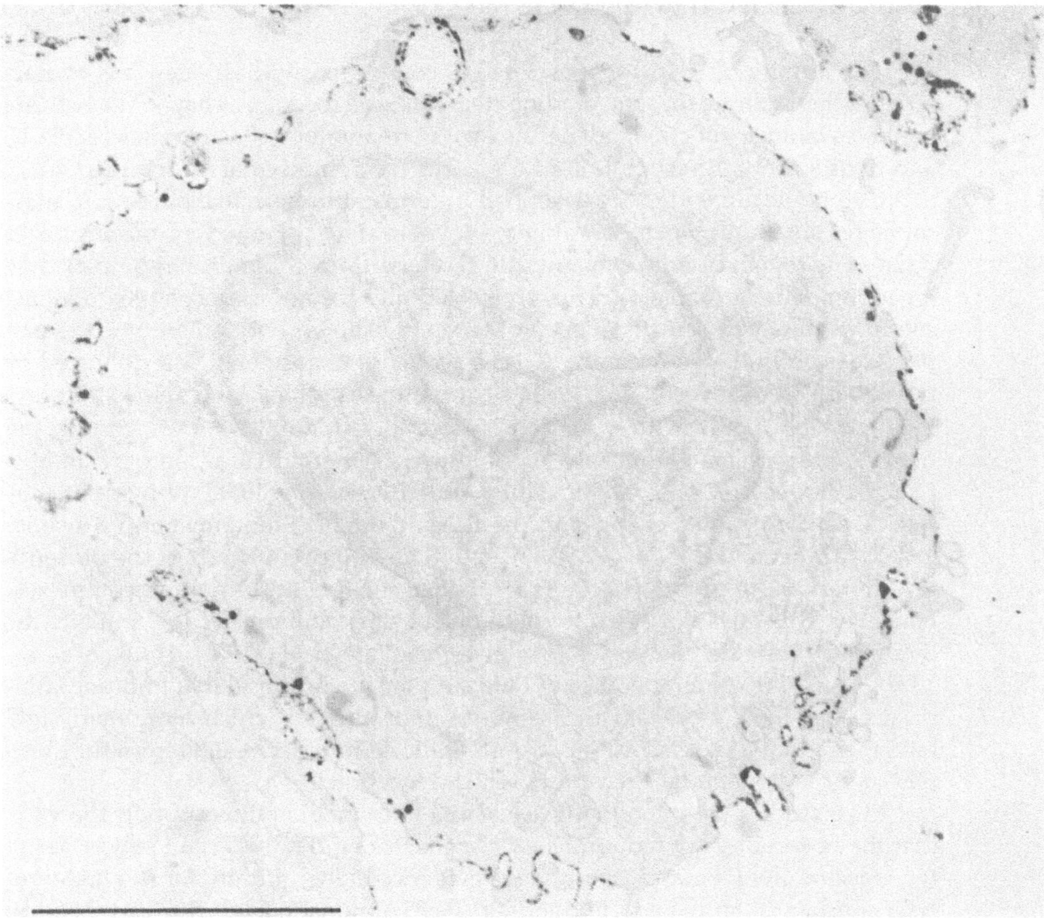

FIGURE 1. Ultrastructural morphology of macrophage receptors for IgG is demonstrated with electron dense diaminobenzidine reaction product localizing complexes of antibody to peroxidase and peroxidase. Focal deposits of reaction product have been enumerated on individual cellular profiles to estimate total number of stained receptors (Table 1). Rabbit peritoneal macrophage incubated at 4°C for 30 min with soluble peroxidase–rabbit-anti-peroxidase complexes, rinsed briefly, and stained with diaminobenzidine-peroxide (McKeever *et al.*, 1976a,b). Scale bar represents 3.8 μm.

sessed by binding and competitive inhibition of binding of radiolabeled immunoglobulins or antigen–antibody complexes. Macrophages are incubated in separate aliquots with radiolabled immunoglobulin plus an ascending series of concentrations of unlabeled protein. After incubation, the cells are washed and cell-bound radioactivity is assayed. Specificity is tested by using similar and different unlabeled proteins to determine whether only similar proteins inhibit binding. It is necessary to demonstrate saturation of receptor sites by increasing concentrations of immunoglobulin (Unkeless and Eisen, 1975; Segal and Hurwitz, 1977).

2.3. THE SPECIFICITY OF Fc RECEPTORS

The affinity of macrophages for IgG complexed with antigen has characteristics that indicate that the binding depends on a receptor. That is, the binding mediates biologic functions, depends on a finite number of active sites (Table 1), and occurs through recognition of a specific macromolecular structure.

Cytophilic antibody was described in terms of its selective binding to macrophages in the absence of antigen (Boyden, 1964). However, macrophages possess up to four orders of magnitude greater affinity for immune complex than for monomeric Ig (Table 1). This greater affinity for immune complex may account for the increased prevalence of receptors shown with an antiperoxidase–peroxidase complex as compared with a sequence of antiperoxidase followed by peroxidase (McKeever *et al.*, 1976a,b). Immune complexes combine more avidly, possibly because they proffer an additional number of Fc groups per molecular complex with each Fc binding an individual receptor (Phillips-Quagliata *et al.*, 1971; Segal and Titus, 1978; Benacerraf, 1968), or because they induce conformational changes in the Fc component (Shinomiya and Koyama, 1976; Thrasher and Cohen, 1971). The latter possibility could reflect the presence of different Fc receptors on the macrophage for IgG and immune complexes. However, this does not seem to be the case, since monomeric IgG inhibits the binding of immune complexes (Phillips-Quagliate *et al.*, 1971; Knutson *et al.*, 1977). Thus, the preponderance of data support the proposal that immune complexes bind more avidly to the same receptor sites to which monomeric IgG binds, and that increased affinity for an immune complex results from the summation of individual IgG receptors which bind to it.

Most studies with autologous cells and immunoglobulins support the view that the Fc receptors of mononuclear phagocytes bind only certain subclasses of IgG specifically. Human macrophages evidence higher affinity for the IgG_1 and IgG_3 subclasses (Huber and Fudenberg, 1968; Abramson *et al.*, 1970; Huber and Holm, 1975; Hay *et al.*, 1972; Okafor *et al.*, 1974), whereas mouse macrophages bind IgG_{2a} and IgG_{2b} specifically. The receptor on the mouse cell for IgG_{2a} may be additional to or separate from that for IgG_{2b} (Grey, *et al.*, 1976; Unkeless, 1977, 1979; Heusser *et al.*, 1977; Unkeless *et al.*, 1979; Walker, 1976). By varying the size of IgG aggregates, trypsinizing the cells, blocking with monoclonal anti-Fc receptor antibody, and isolating cloned macrophage neoplasms, studies have shown subtle differences in macrophage binding of IgG_{2a} and IgG_{2b}. However, inhibition studies suggest that IgG_{2a} binds at the same site with equal affinity as IgG_{2b} (Segal and Titus, 1975). Binding of IgG_1 by mouse macrophages appears somewhat equivocal but clearly IgG_3 is not bound by these cells (Unkeless and Eisen, 1975; Segal and Titus, 1978). Binding-inhibition assays where one class of Ig competes with another to occupy a binding site provide particularly strong evidence favoring the specificity of the classical Fc receptor for IgG subclasses rather than IgA and IgM (Unkeless and Eisen, 1975; Segal and Titus, 1978). Binding of mouse IgG subclasses has also been assessed by inhibition of antibody-coated-erythrocyte binding (Schevach *et al.*, 1972; Cline *et al.*, 1972). Receptors for IgG subclasses have not been visualized ultrastructurally, how-

TABLE 1. NUMBER, SPECIFICITY, AND AFFINITY OF MACROPHAGE RECEPTORS FOR HOMOLOGOUS IMMUNOGLOBULINS

Assay		Macrophage source	Immunoglobulin			State of immunoglobulin aggregation	Receptor number per cell	Association constant (M^{-1})	Reference
System	Conditions		Source	Classes examined	Classes binding				
Electron microscopic	4°C and 37°C	Resident and induced peritoneal	Rabbit anti-peroxidase	Ig and IgG	IgG	Soluble Ag–Ab complexes	1.2×10^5 resident; 2×10^5 induced	No data	McKeever *et al.* (1975)
^{125}I-labeled Ab or Ig	4°C and 37°C	Alveolar	Rabbit anti-benzyl penicilloyl-heptalysine	IgG	IgG	Soluble Ag–Ab complexes with monovalent, divalent, and oligovalent hapten	2×10^6	No data	Phillips-Quagliata, *et al.* (1971)
	37°C	Alveolar after stimulation with i.v. Freund's adjuvant	Rabbit anti-human serum albumin	IgG	IgG	Monomeric immunoglobulin	1.2×10^6 minimally stimulated; 2.2×10^6 heavily stimulated	7.6×10^5 minimally stimulated; 9×10^5 heavily stimulated	Arend and Mannik (1973)
	4, 24, and 37°C	Neoplasm $P388_{D1}$; resident and induced peritoneal	Mouse myelomas	IgG, IgG_{2A}, IgG_{2B}, IgA, IgM	IgG_{2A} strongly; IgG_{2B} weakly	Monomeric immunoglobulin	1×10^5 resident and neoplasm for IgG_{2A}; 4×10^5 stimulated for IgG_{2A}	2×10^7 at 37°C	Unkeless and Eisen (1975)

(Continued)

TABLE 1 (*Continued*)

Assay			Immunoglobulin						
System	Conditions	Macrophage source	Source	Classes examined	Classes binding	State of immunoglobulin aggregation	Receptor number per cell	Association constant (M^{-1})	Reference
	0°C	Neoplasm $P388_{D1}$	Mouse myelomas	IgG_1 IgG_{2A} IgG_{2B} IgG_3 IgA IgM	IgG_1 IgG_{2A} IgG_{2B}	Monomeric, dimeric, and trimeric IgG	3.9×10^5 for trimer of IgG_1, IgG_{2A}, and IgG_{2B}	5×10^5 for trimers of IgG_1, IgG_{2A}, and IgG_{2B} from inhibition studies	Segal and Hurwitz (1977); Segal and Titus (1978)
	4°C	Resident peritoneal	Normal rat	IgG	IgG	Multimeric IgG_9 to IgG_{74}	2.3×10^5 for IgG_9; 0.9×1 0.9×10^5 for IgG_{74}	$3\text{–}12 \times 10^8$	Knutson *et al.* (1977)
	20°C	Induced peritoneal	Normal guinea pig serum	IgG_2	IgG_2	Monomeric	2.5×10^6	1.5×10^6	Leslie and Cohen (1974)

ever, such localization, particularly employing double labeling techniques (Roth and Binder, 1978), with a different marker for each of two potential receptors, would provide needed information concerning the specificity and distribution of receptors for different subclasses of IgG. Such an approach could assess also the validity of the general assumption that the macrophage possesses a single type of Fc receptor with different affinities for monomeric IgG as compared with immune complexes rather than different receptors for these substances. Fc receptors with unusual affinity for either IgM or IgE have occasionally been reported (Lay and Nussenzweig, 1969; Rhodes, 1973; Capron *et al.*, 1975; Ehlenberger and Nussenzweig, 1977). The mouse macrophage receptor for IgG_{2b} binds the C_H2 domain of the molecule (Diamond *et al.*, 1979).

A portion of the binding of radiolabeled IgG to macrophages cannot be inhibited by large excesses of unlabeled IgG even when cells are incubated at 0°C (Segal and Hurwitz, 1977). It is not clear whether these results reflect a methodologic peculiarity, pinocytosis in the cold, or foreign substance receptors.

2.4. ABUNDANCE OF RECEPTORS

The problem of quantitating Fc receptors for monomeric IgG or immune complexes on the macrophage surface has been approached in different ways, including measurement of iodinated immunoglobulin bound to the cell and counting peroxidase-labeled immune complexes visualized ultrastructurally (McKeever *et al.*, 1976a,b) on macrophage profiles (Table 1). A macrophage appears to possess 10^5 to 10^6 receptors, allowing for variability according to the method employed and species examined.

Quantitation of Fc receptors for monomeric IgG or for immune complexes by electron microscopy (EM) encounters problems that are inherent in the properties of receptors. The affinity of monomeric IgG to macrophages is weak (Segal and Titus, 1978; Leslie and Cohen, 1974a); and the cytophilic IgG is susceptible to elution by repeated washing or displacement by unlabeled IgG (Benacerraf, 1968; Leslie and Cohen, 1974a,b).

Quantitation possibly errs in the direction of underestimating Fc receptors because certain factors have the effect of lowering the number of stainable Fc receptors. Monomeric IgG, for example, binds with lower affinity and tends to wash off during specimen preparation (McKeever *et al.*, 1976a,b; Papadimitriou, 1973). IgG complexed with antigen binds more avidly, producing better staining but introducing the risk of staining more than a single receptor. Guiding principles to optimize the staining of cellular receptors include washing briefly in small volumes and altering the concentration or conformational state of the target molecule in measured increments (Phillips-Quagliata *et al.*, 1971; Segal and Titus, 1978) as necessary to increase its avidity.

The abundance of receptors demonstrated on macrophages with antiperoxidase-peroxidase soluble complex greatly exceeds the number visualized on granulocytes which, in turn, have a higher prevalence of such receptors than lymphocytes (McKeever *et al.*, 1976b). Macrophages appear, then, to specialize to a greater extent than other cell types in phagocytosis of opsonized target elements.

A number of studies indicate that activated macrophages carry more IgG receptors per cell than resident or minimally stimulated peritoneal macrophages (Table 1) (Rhodes, 1975). The greater number of receptors in some cases reflects the larger size of macrophages activated by certain agents including mineral oil (McKeever *et al.*, 1976b) (Table 1).

2.5. ULTRASTRUCTURAL MORPHOLOGY, POSITION, AND DISTRIBUTION

Fine structural demonstration of macrophage Fc receptors provides clues to their molecular nature and biologic activity (Figure 1). IgG binds closely to the macrophage membrane (McKeever *et al.*, 1976a; Papadimitriou, 1973). Conceivably the 10-nm length of an immunoglobulin molecule (Valentine and Green, 1967; Feinstein and Rowe, 1965) would be expected to intervene between the Fc receptor in the plasmalemma and the antigen linked to the end of the immunoglobulin. Small molecular immune complexes offer advantages over sensitized erythrocytes or larger molecular markers artificially cross-linked to immunoglobulin not only in assessing the quantity and distribution of receptors but also in appraising the distance between plasmalemma and bound target substance (Bretton *et al.*, 1972).

The gap between an IgG-coated erythrocyte and the adjoined monocyte membrane measures 40 nm (Douglas and Huber, 1972). Actually the distance between a small antigen–antibody marker and the macrophage membrane measured on either small or large focal precipitates is less than 8 nm (Figure 2). This distance is surprisingly small and suggests either a very compact receptor molecule with receptor–Ig overlap or a receptor pocket within the macrophage membrane. It is smaller than the distance between the antigen marker alone and the macrophage membrane. Partial removal of the surface coat with enzyme (Bretton *et al.*, 1972) prior to ultrastructural receptor localization possibly could add information about the Fc receptor.

Fc receptors aggregate in the macrophage plasmalemma beneath large opsonized particles that become adherent on exposure at 37°C and microfilaments collect in the cytoplasm directly beneath these aggregated receptors (Griffin *et al.*, 1976; Douglas, 1978). The relative scarcity of these microfilaments associated with the binding and microendocytosis of soluble immune complexes (McKeever *et al.*, 1976b) (Figure 3) and their frequent accumulation near other large particles and surfaces suggests that most of them are important to the work of holding or translocating large particles.

At a level of resolution above 200 nm receptors for Ag–Ab complexes appear randomly distributed on macrophages in suspension (McKeever *et al.*, 1976a). Receptors are distributed less randomly on the exposed plasmalemma of macrophages bound to flat surfaces. Paradoxically, Ag–Ab complexes employing peroxidase (Figures 4 and 5) or hemocyanin as both antigen and marker preferentially stain the perinuclear plasmalemma of flattened cells as well as more

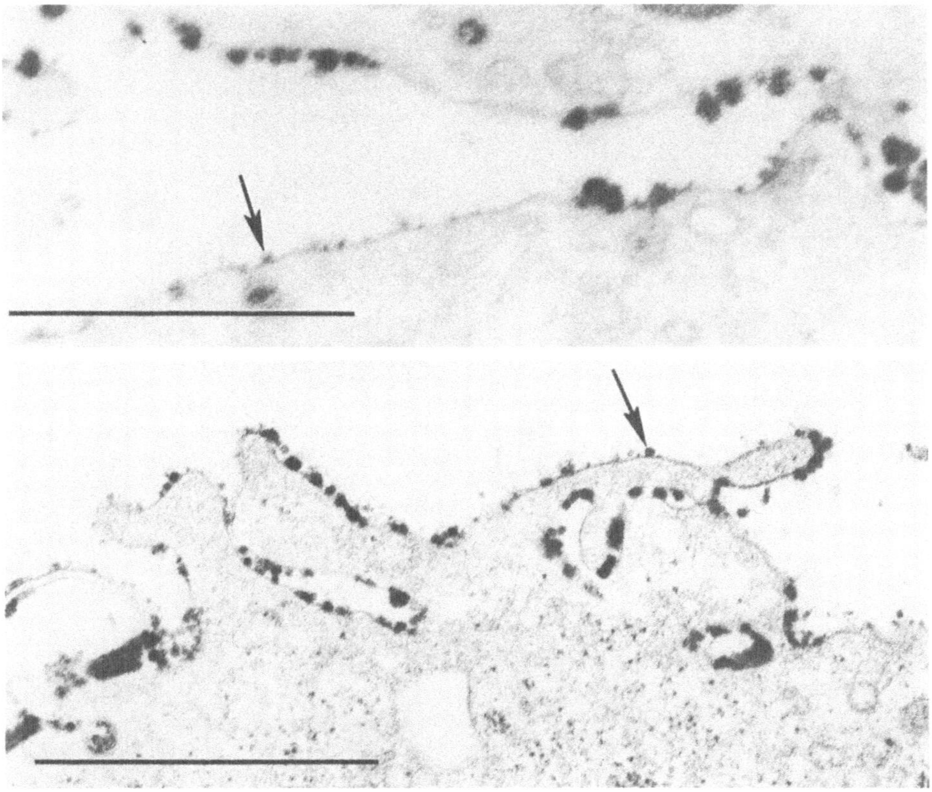

FIGURE 2. Higher magnifications of staining (arrows) for macrophage IgG receptors as for Figure 1. In the upper figure, the cell was incubated briefly in diaminobenzidine to minimize size of reaction product and facilitate estimation of the gap between antigen–antibody marker and membrane. Scale bars represent 0.75 μm (upper figure) and 1.1 μm (lower figure). Lower figure is reprinted from *The Journal of Histochemistry and Cytochemistry* **24:**951, with permission.

spherical cells (McKeever *et al.*, 1977; Kaplan *et al.*, 1975), whereas erythrocyte markers favor the more peripheral plasma membrane (Kaplan *et al.*, 1975; Tizard *et al.*, 1974) or show a random distribution (Kaplan, 1977; Munthe-Kaas *et al.*, 1976). The apparent differences observed with the molecular markers correlate with more active phagocytosis observed in the perinuclear region (McKeever *et al.*, 1977; Munthe-Kaas *et al.*, 1976). Perhaps phagocytosis of some erythrocytes over the perinuclear region renders these erythrocytes less visible by scanning electron microscopy.

At higher resolution, in the range below 200 nm, receptors for soluble Ag–Ab complexes appear discontinuous in distribution (McKeever *et al.*, 1976a, b). Their periodicity varies between 30 and 120 nm (Figures 1 and 2). Variability in the space between receptors possibly reflects a tendency for their aggregation within the plasmalemma. Periodicity of IgG receptors supports evidence from experiments demonstrating saturation of receptors with IgG or immune complex (McKeever *et al.*, 1976a,b; Segal and Titus, 1978) which indicates a finite number

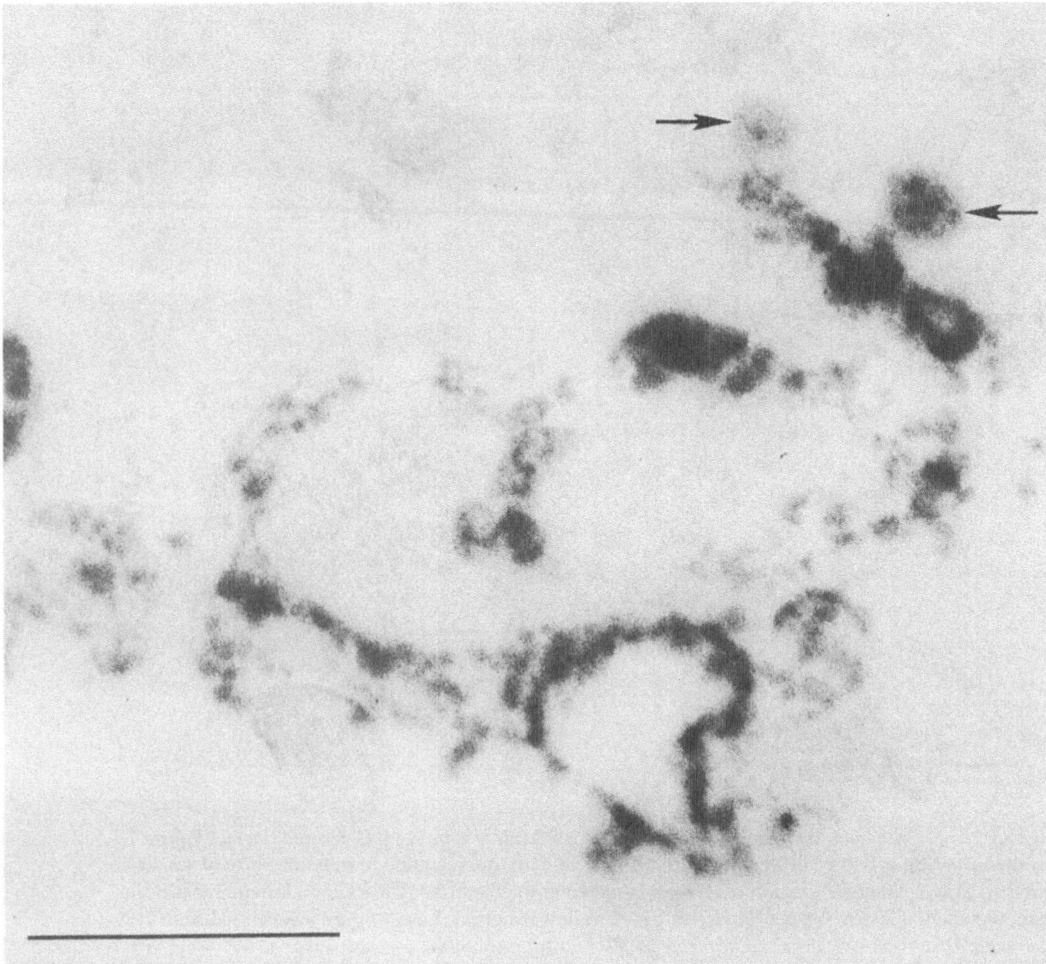

FIGURE 3. Adsorptive pinocytosis (arrows) via macrophage receptors for IgG is observed rarely at 4°C and similar to that seen commonly at more physiologic temperatures. Cells and reagents as for Figure 1. Scale bar represents 0.52 μm. Reprinted from *The American Journal of Pathology* **84**:450, with permission.

of IgG receptors. Similar immunocytochemical labeling of these receptors with Ab or Ag–Ab complexes of a variety of known molecular sizes (Segal and Titus, 1978; Knutson *et al.*, 1977) might determine the relative contribution of receptor aggregation to this periodicty.

Macrophage receptor activity for sheep erythrocytes coated with rabbit antibody against sheep erythrocytes (EA) is diminished in macrophages that have spread over glass coated with Ag–Ab complexes (Douglas, 1976; Rabinovitch *et al.*, 1975). Such exposure to a coating of immune complexes also decreases EA ingestion. The inhibited binding and uptake of EA suggests that exposure to the immune complexes either depletes receptors or inhibits the mechanism for their internalization. Trypsin resistant Fc receptors are inhibited to a greater degree than trypsin-sensitive (IgG_{2a}) receptors (Michl *et al.*, 1979). Exposure to immune

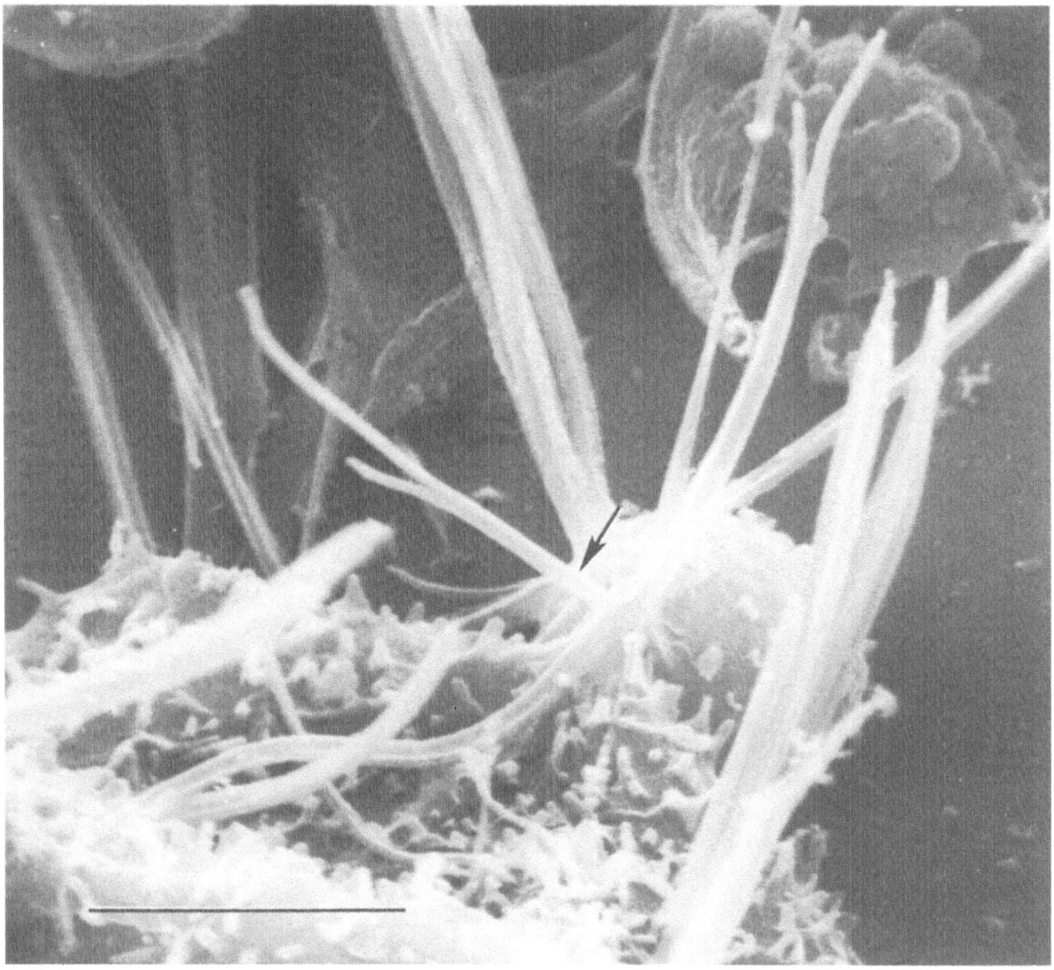

FIGURE 4. Surface topography of IgG receptors on rabbit alveolar macrophages is demonstrated with crystalline benzidine reaction product (arrow) localizing complexes of peroxidase and antibody to peroxidase. Cells incubated on coverslips at 4°C with soluble peroxidase–rabbit-anti-peroxidase complexes, rinsed and fixed briefly and incubated 15 min in benzidine-peroxide reagent. Scale bare represents 3.6 μm. Reprinted from *The Journal of Histochemistry and Cytochemistry* **25:**1066, with permission.

complexes on a flat surface alters macrophage secretion of fibrinolytic enzymes from plasminogen activator to elastase (Ragsdale and Arend, 1979). Spreading alone seems to affect receptors, since macrophages spread upon a flat coverslip reveal fewer receptors for HRP–anti-HRP complexes than rounded macrophages (McKeever *et al.*, 1977).

2.6. ENDOCYTOSIS AND REGENERATION OF RECEPTORS

Receptor activity is generally demonstrated by exposing the macrophage to sensitized target elements at lower than physiologic temperatures where bind-

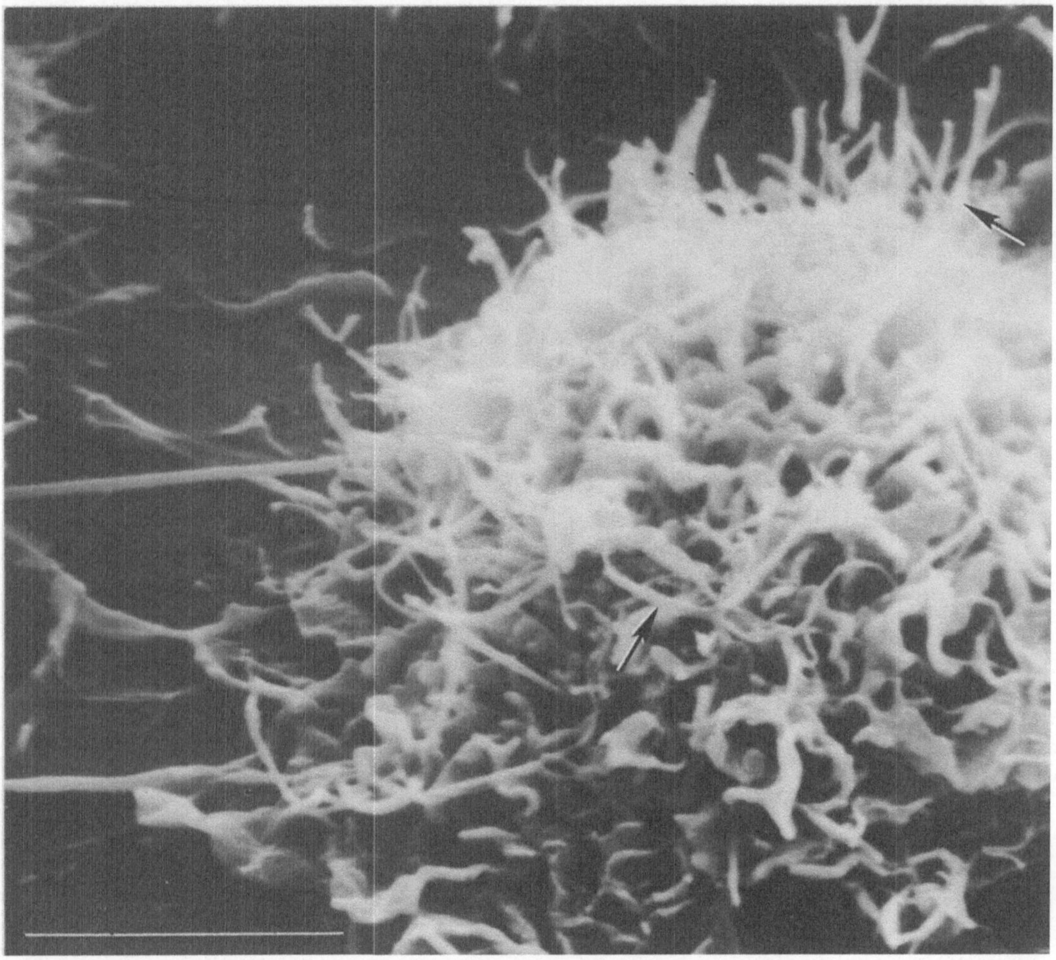

FIGURE 5. Macrophage receptors for IgG are localized with crystalline benzidine reaction product (arrows) on and between veils of plasma membrane covering the central portion of the cell. Receptors are scarce at the periphery. Cells and reagents as for Figure 4 except for 2.5 min decreased exposure to benzidine-peroxide in order to produce a smaller marker. Scale bar represents 5 μm. Reprinted from *The Journal of Histochemistry and Cytochemistry* **25:**1066, with permission.

ing occurs but energy-dependent uptake does not. When warmed, macrophages holding molecular immune complexes on their surface ingest them in small endocytic vesicles (Figure 3) by a process similar to pinocytosis except that the ingested molecules adhere to the membrane of the endosome rather than occupying its fluid phase (Steinman *et al.*, 1976). A similar process of interiorization of Concanavalin A which binds mannose and glucose residues on the macrophage surface is called adsorptive pinocytosis. Endocytized molecular immune complexes collect in large intracellular vacuoles (McKeever *et al.*, 1976b; Steinman and Cohn, 1972b) and smaller tubulovesicular structures (Figure 6). Degradation of the complexes can be demonstrated in hourly determinations

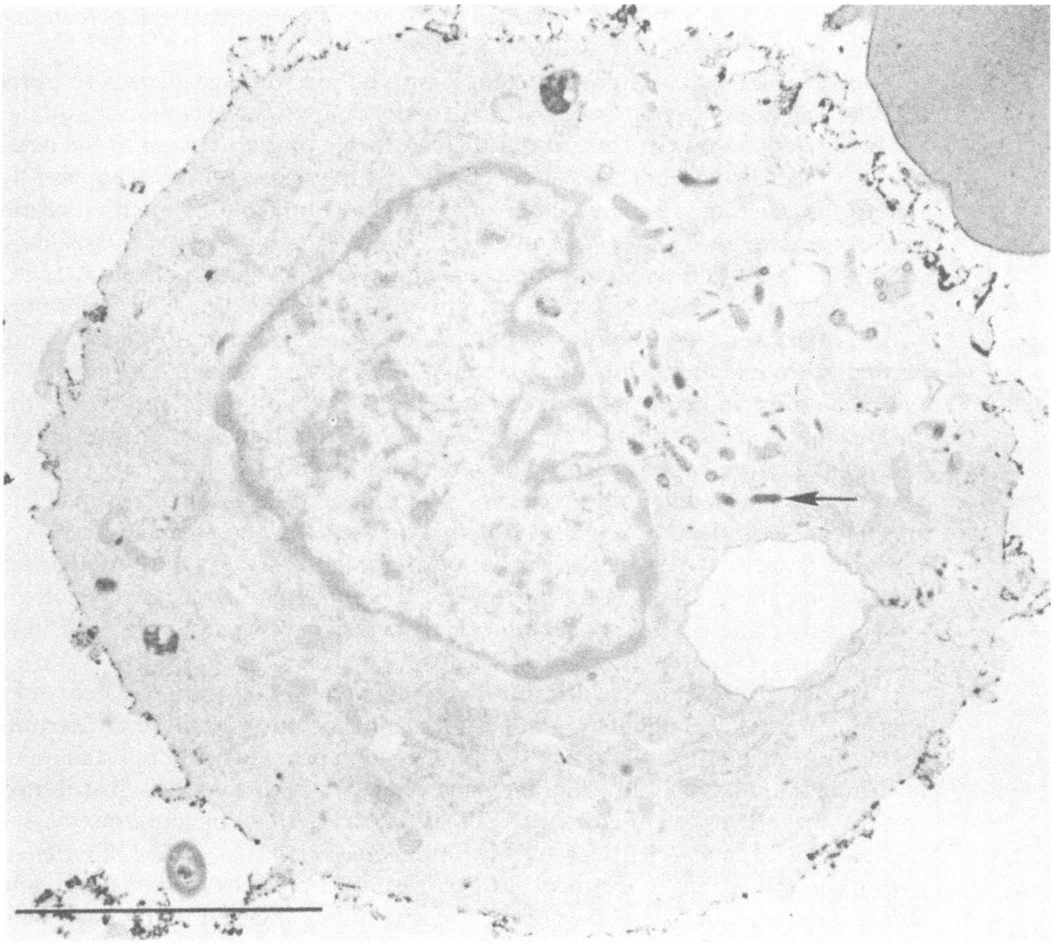

FIGURE 6. This macrophage had ingested a surface load of antibody to peroxidase complexed with peroxidase just prior to reincubation with the same antibody–antigen complex. Initial antibody–antigen complex stains within tubulovesicular structures in the cytoplasm (arrow). Receptors for IgG can still be observed on the macrophage surface. Rabbit peritoneal macrophage incubated at 4°C for 30 min with soluble peroxidase–rabbit-anti-peroxidase complexes, rinsed, warmed to 37°C for 30 min, reincubated with complexes, and stained with diaminobenzidine-peroxide. Scale bar represents 3.3 μm. Reprinted from Vol. 84, *The American Journal of Pathology*.

and after ingestion (Steinman and Cohn, 1972b; Shinomiya and Koyama, 1976), suggesting both structures function as phagocytic vacuoles. Uptake of antigen–antibody complexes can proceed at a rate up to 4000 times greater than for soluble antigen alone (Steinman and Cohn, 1972a,b; Feldman and Pollock, 1974). Complexes larger than one antigen linked to two antibodies are required for greatly accelerated interiorization by Kupffer cells (Arend and Mannik, 1971; Benacerraf *et al.*, 1959; Mannik *et al.*, 1971; Mannik and Arend, 1971). Mathematical constructs conform to these data (Van Oss *et al.*, 1974). The internalization of cytophilic antibody uncomplexed to antigen apparently differs in mechanism as well as rate since it is reported to entail formation of caps on a small percentage of macrophages at 37°C (Thrasher *et al.*, 1975).

Both the binding of immune complex to the macrophage surface receptor and the subsequent endocytosis which the loading of the receptor stimulates appear to depend on the integrity of the contractile protein system in the ectoplasm. Cytochalasin B impedes endocytic activity in various cell types apparently through interfering with the action of actin-like contractile filaments (Sannes and Spicer, 1978). This drug prevents binding of peroxidase–antiperoxidase complexes to rat peritoneal macrophages and also blocks internalization of the immune complex receptors loaded prior to exposure to the drug. The ionophore A23187 increases permeability of the plasmalemma to divalent cations and, perhaps through increased intracellular Ca^{2+}, appears similarly to affect the contractile protein filaments or to disrupt their connection to the receptor in the plasmalemma, as this agent also blocks binding of immune complex receptors and uptake of preloaded receptors (Sannes and Spicer, 1979).

At physiologic temperature target cells for phagocytosis, such as sensitized erythrocytes, are diffusely coated with antibodies to their surface antigens (Berken and Benacerraf, 1966). Redistribution of antibodies to target lymphocytes by allowing the antibodies to cap on the lymphocyte surface at physiologic temperature inhibits their subsequent phagocytosis by macrophages (Griffin *et al.*, 1976).

Blocking Fc receptors with anti-macrophage antibodies immediately after attachment of EIgG inhibits the subsequent phagocytosis of these antiserum-coated erythrocytes (Griffin *et al.*, 1975b). Ingestion proceeds only upon sequential circumferential binding of the macrophage receptors to IgG distributed around the target cell (Figure 7). Consistent with this concept, monocytes (LoBuglio *et al.*, 1967), Kupffer cells (Munthe-Kaas *et al.*, 1976) and stimulated peritoneal macrophages (Kaplan, 1977) reveal a cuplike extension of plasma membranes around IgG-coated erythrocytes during ingestion (compare with Sections 3.5 and 4.6).

Fetal calf serum promotes capping of antiserum-coated erythrocytes on macrophages (Romans *et al.*, 1976) presumably as a consequence of impairing the erythrophagocytosis that would be expected to ensue. We think this phenomenon may be attributable to a nonimmunospecific blockage of the Fc receptor or to interference with the work of endocytosis (see also Section 3.6).

Following the ingestion of small molecular Ag–Ab complexes, receptors loaded with peroxidase–antiperoxidase complexes again evidence binding of such complexes within 30 min of application of the initial surface load (Figure 6). This surprising phenomenon might be easily explained if the initial load of Ag–Ab complexes had not saturated the macrophage receptors. However, a tenfold increase in immune complexes applied to macrophages produced no increase in staining (McKeever *et al.*, 1976a). This restoration occurs at an order of magnitude faster than the cycloheximide-inhibitable regeneration which follows depletion of receptors by the phagocytosis of latex particles or formalin-treated erythrocytes (Schmidt and Douglas, 1972). Phagocytosis of latex particles causes a decrease in cell surface marker enzymes, presumably as a result of the

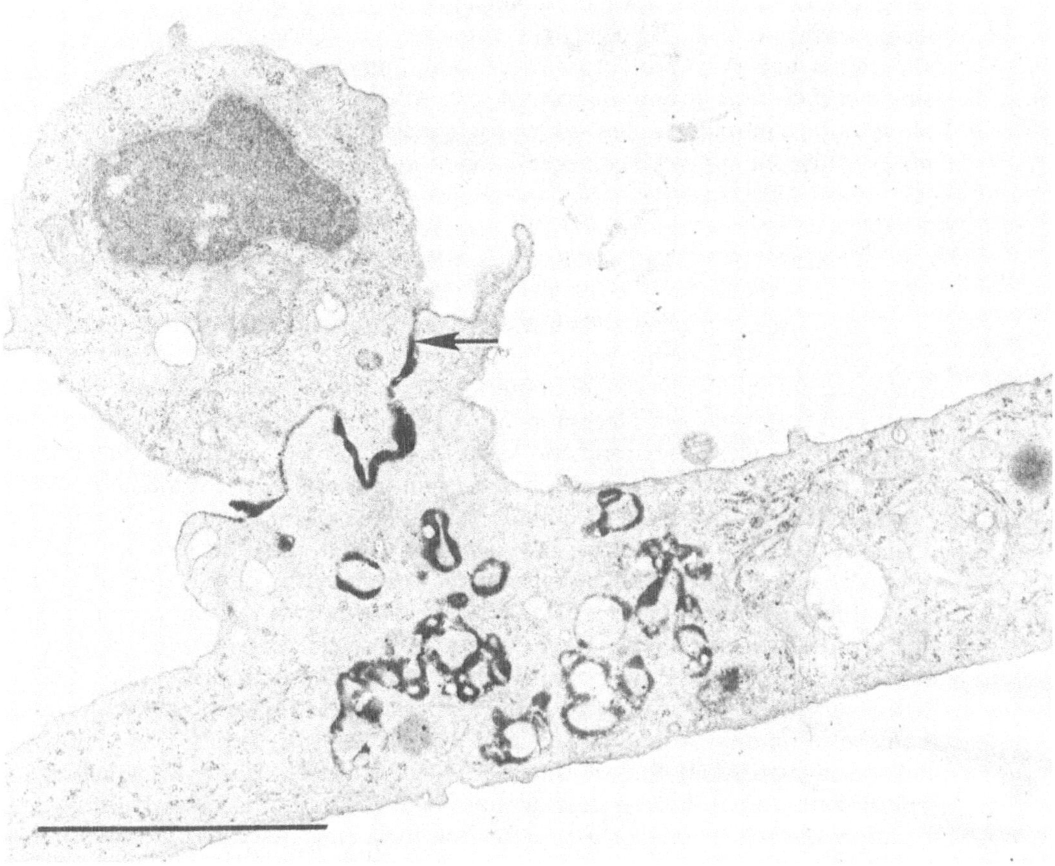

FIGURE 7. This mouse peritoneal macrophage has bound but at 37°C is unable to ingest a mouse B lymphocyte with IgG anti-immunoglobulin antibodies (arrow) capped on the B-cell surface. Prior to mixing these cells, lymphocytes were warmed to allow the antibody coating the lymphocyte to redistribute to one pole and a label for this antibody was introduced via peroxidase bound to a second anti-immunoglobulin reagent. Redistribution of the IgG has blocked ingestion of the lymphocyte via the macrophage IgG receptor, since the IgG no longer coats the entire surface of the target lymphocyte. Microfilaments have accumulated in the macrophage cytoplasm beneath the attached lymphocyte. Scale bar represents 1.5 μM. Reprinted from a section similar to that on p. 798, Vol. 144 of *The Journal of Experimental Medicine,* with permission of the journal and Drs. F. M. Griffin and S. C. Silverstein.

uptake of the enzyme into a phagolysosome where it is digested by the lysosomal enzymes (Werb and Cohn, 1972). Preloading with latex beads decreases Fc receptors less on alveolar macrophages than on blood monocytes (Daughaday and Douglas, 1976; Schmidt and Douglas, 1972).

Regeneration or selective conservation of receptors after a surface load of immune complexes may be a phenomenon related to size or total amount of the ingested material. A pulsed surface load of molecular Ab–Ag complexes does not interfere with macrophage receptor activity 30 min later. Continual phagocytosis of EIgG during 1 hr at 37°C seems to exhaust rat Kupffer cell receptors (Munthe-Kaas, 1976). The apparent difference in results may relate to a marginal pool of IgG receptors which is neither exposed to nor exhausted by a single surface load of immune complexes. Alternatively, adsorptive pinocytosis of immune complexes may allow for conservation of IgG receptors which phagocytosis of appreciably larger areas of membrane does not allow.

2.7. SENSITIVITY TO ENZYMES, IONS, AND DRUGS

The subject of trypsin-sensitive vs. trypsin-resistant Fc receptors has been reviewed recently (Dorrington, 1976; Silverstein *et al.*, 1977). Fc receptors which bind cytophilic antibodies as monomeric IgG molecules uncomplexed with antigen are trypsin sensitive. Receptors for Ag–Ab complexes are more resistant to such protease digestion (Arend and Mannik, 1972). However the fact that macrophages possess many times greater affinity for immune complex than for monomeric IgG (Table 1), may contribute to this difference in the measured receptor sensitivity to trypsin. Nevertheless, a number of studies have shown that Fc receptors assayed either with erythrocytes coated with complement-free antibody (EIgG) (Daughaday and Douglas, 1976; Howard and Benacerraf, 1966; Kossard and Nelson, 1968; Davey and Asherson, 1967), or with molecular immune complexes (Arend and Mannik, 1972; Steinman and Cohn, 1972b; McKeever *et al.*, 1976a) are more stable to trypsin and other proteolytic enzymes than are receptors for monomeric immunoglobulins of restricted IgG subclass and specificity (Unkeless and Eisen, 1975; Segal and Titus, 1978). These observations account in part for the clear demonstration of trypsin sensitivity on mouse macrophages where myeloma proteins are frequently used in binding studies compared with rabbit and guinea pig macrophages studied with polyclonal antibodies. Receptors regenerate following their removal with trypsin and this restoration is inhibited by cycloheximide (Unkeless and Eisen, 1975) indicating that restitution depends on protein synthesis. A number of the experiments on protease sensitivity were performed before the difference between Fc and complement receptors was appreciated. Investigations allowing a distinction between these receptors are covered elsewhere (Section 3.6).

The Fc receptor is also sensitive to digestion with phospholipase and apparently contains or depends for its activity on a lipid component. Phospholipase C for example has been found to inhibit Fc receptors (Howard and Benacerraf, 1966) and, in addition, to impair the capacity of the macrophage monolayers tested to adhere to glass (Davey and Asherson, 1967). Phospholipase A is said to inhibit cytophilic antibody receptors (Walker, 1976) while phospholipase D not. All of these phospholipases destroy lecithin and other phosphatides but vary in the cleavage products they liberate (Davey and Asherson, 1967; Wells and Hanahan, 1969; Ottolenghi, 1969; Kates and Sastry, 1969). Some of these cleavage products have detergent properties and further studies of their effects on Fc receptors might, therefore, be of value in determining whether the receptor inactivation is a direct result of the enzyme or is mediated by the cleavage product. Neuraminidase increases macrophage adherence to opsonized bacteria (Allen and Cook, 1970).

These receptors generally resist perturbation by common elemental ions, but they are inhibited, however, by oxidizing agents like periodate and nitrate (Davey and Asherson, 1967). Cytophilic antibody receptors are sensitive to agents which bind free sulfhydryl groups like iodoacetamide and *p*-chloromercuribenzoate (Howard and Benacerraf, 1966) and, therefore, appear to contain functional sulfhydryls.

Both polycations such as polylysine (Davey and Asherson, 1967) and polyanions as for example dextran sulfate (Wellek *et al.*, 1976a,b), inhibit Fc receptors, indicating that electropositive as well as negative forces influence receptor function. The influence of these large, charged molecules may be indirect through steric hindrance or stimulation of other activity in the macrophage plasma membrane. Preincubation of human monocytes with certain synthetic polynucleotides for 6 hr enhances their IgG receptors (Schmidt and Douglas, 1976b).

Corticosteroids are reported both to inhibit the binding of sensitized erythrocytes (EigG) to human monocytes (Schreiber *et al.*, 1975) and not to affect EIgG binding to these cells or guinea pig macrophages (Rinehardt *et al.*, 1974; Rinehardt *et al.*, 1975; Davey and Asherson, 1967). Steroids may need to be presented to monocytes in phospholipids to be effective. In the former studies, glucocorticoids proved effective at *in vivo* steroid levels, and mineralocorticoids also inhibited binding. Prostaglandins E_2, F_1, F_2, and A, enhance the phagocytosis of opsonized erythrocytes at physiologic concentrations and decrease it at higher concentrations (Razin *et al.*, 1978). Levamisole increases IgG receptor activity of human monocytes *in vitro* (Schmidt and Douglas, 1976a). Binding of EIgG alone is not affected by sodium azide (Howard and Benacerraf, 1966), but this agent does interfere with redistribution of bound EIgG upon warming to 21° or 37°C (Romans *et al.*, 1976).

Anti-macrophage, anti-erythrocyte and anti-lysosomal antibodies applied to the macrophage block subsequent attachment of EIgG to the macrophage and engulfment of the loaded Fc receptor. Such pretreatment has a smaller effect on the complement receptor (Holland *et al.*, 1972). These antibodies lack an effect on phagocytosis not mediated by immunoglobulin.

2.8. SUMMARY AND CONCLUSIONS

The receptor for IgG (Fc receptor) on the surface of phagocytes acting in concert with specific antibody plays a fundamental role in the selective binding and ingestion of antigens, microbes, and effete autologous erythrocytes. The receptor possesses relatively weak affinity for monomeric IgG and high affinity for immune complex. Summation of individual receptors identical to the monomeric receptor is the explanation of this high affinity most consistent with present data. The Fc receptor binds specific subclasses of IgG including the IgG_1

and IgG_3 subclasses in man and IgG_{2a}, IgG_{2b}, and possibly IgG_1 in the mouse. The number of receptors on the surface of a macrophage lies in the range of 10^5 or 10^6. The association constant for binding of immunoglobulins ranges between 10^6 and 10^7 M^{-1} and reflects a relatively low affinity. Increasing the size of the Ag–Ab immune complex greatly increases affinity but decreases the apparent number of receptors.

The ultrastructural topography of IgG receptors has recently been defined. A small gap less than 8 nm between the antigen marker and the macrophage membrane suggests a compact receptor pocket within the membrane. At relatively low magnification, receptors appear randomly distributed on macrophages in suspension and less randomly on macrophages forming monolayers. At high resolution, receptors for soluble Ag–Ab complexes are discontinuous showing a 30–120 nm periodicity.

The loading of Fc receptors with immune complex triggers endocytic uptake into the macrophage. Ingestion of IgG-coated cells requires sequential, circumferential binding of Fc receptors to IgG distributed around the perimeter of the cell. Endocytosis of bound Ag–Ab complexes resembles phagocytosis of EIgG after their initial attachment except that the loaded receptors become internalized by a microendocytic process into vesicles resembling pinocytic vesicles. The process resembles adsorptive pinocytosis. Microfilaments presumably composed of actin-like contractile protein appear to function in both the binding of immune complex to the macrophage surface receptor and the subsequent endocytosis of the loaded receptor. The relative scarcity of microfilaments associated with binding and microendocytosis of small Ag–Ab complexes and prominence of the filaments in cells engulfing EIgG correspond to the relative work performed and further associate the microfilaments with the function of binding and translocating endocytic target structures.

Regeneration or selective conservation of macrophage IgG receptors can be demonstrated soon after microendocytosis of a surface load of receptors saturated with small Ag–Ab complexes. Regeneration occurs considerably more slowly after bulk phagocytosis of latex and depends upon protein synthesis.

Studies which involve pretreatment of macrophages with various agents suggest the receptor possesses a functional sulfhydryl group and contains peptide sequences shielded from the extracellular environment. The receptor function appears to depend upon an intrinsic or closely associated phosphatide or is susceptible to hydrolytic cleavage products of phosphatides. Apparent receptor sensitivity to trypsin depends upon the subclass restriction and degree of aggregation of the immunoglobulins used in the assays.

3. COMPLEMENT RECEPTORS

3.1. BIOLOGIC ACTIVITY

As already noted (Section 2.1), complement factors function importantly in the selective elimination of invasive pathogens by macrophages. The complement system, formerly referred to as heat-labile opsonin consists of a group of polypeptides which are cleaved from parent macromolecules by protease activity in a series of steps.

Sequential adherence of the complement products up to the third complement component C3 on test particles generates a reactive site that has affinity for a macrophage receptor and, thus, renders the test particle maximally susceptible to ingestion (Gigli and Nelson, 1968). Numerous investigations implicate complement receptors in host defenses. The biological importance of this system is evidenced by patients with hereditary C3 deficiency or abnormal C3 metabolism. These patients, whose serum-opsonizing capacity is diminished (Alper *et al.*, 1970, 1972), are susceptible to recurrent bacterial infections which ordinarily can be controlled by phagocytes in the presence of normal serum (Ward and Enders, 1933). However, the precise deficit from lack of opsonization alone in these patients is obscured by the multiple functions of complement. The importance of complement or Fc receptors in enhancing killing by the polymorphonuclear phagocyte has also been demonstrated by *in vitro* experiments on phagocytic killing of bacteria which resist serum alone (Li *et al.*, 1963; Young, 1974).

It is possible that complement receptors regulate the inhibition of macrophage migration induced by endotoxin (Heilman, 1977). This influence could be effected by factors which are known to be generated by the complement and the blood coagulation system and may act upon the plasma membrane to induce macrophage activation (Bianco *et al.*, 1976).

Complement receptors are not unique to leukocytes. They have been detected on primate erythrocytes, nonprimate platelets, and epithelial cells of renal glomeruli in addition to B lymphocytes, polymorphonuclear leukocytes (PMN), macrophages, and monocytes (Nussenzweig, 1974; McConnell and Lachmann, 1977).

A link has been proposed between monocyte or lymphocyte complement receptors and histocompatibility antigens HLA-4a and -4b (Arnaiz-Villena and Festenstein, 1975), but this appears controversial at present (Ferreira *et al.*, 1976; McConnell and Lachmann, 1977).

3.2. TECHNIQUES FOR DETECTING AND QUANTITATING COMPLEMENT RECEPTORS

Complement receptor activity is assayed on macrophage monolayers by counting under the light microscope the number of appropriately sensitized erythrocytes (E) bound to a macrophage and averaging the number of E per macrophage. This number is multiplied by the percentage of macrophages with attached E, and all three numbers are generally tabulated (Griffin *et al.*, 1975a). Alternatively, the percentage of rosetting macrophages, i.e., those with three or more attached erythrocytes, may be tabulated (Daughaday and Douglas, 1976). A method using ^{51}Cr-labeled erythrocytes has also been described (Atkinson *et al.*, 1977). Detailed procedures for the sensitizing of erythrocytes, preparation of macrophages, and performances of these assays are available (Bianco, 1976).

Erythrocytes coated with IgM alone (EIgM) do not bind to macrophages in many systems, but when such red cells are incubated with complement components or with fresh serum, complement thereby fixes to the immune complex coating the cells, allowing them to bind to the macrophages. These sensitized erythrocytes which are designated EIgMC in this review, but are often abbreviated EAC, are used to test immune cells for complement receptor activity. Erythrocytes coated with IgG alone (EIgG) unlike EIgM, bind to macrophages via the Fc receptor. However, certain subclasses of IgG, depending on species, also bind complement. Accordingly, rosetting with EIgGC could reflect binding to either the Fc or the complement receptor of the macrophage. Therefore, some investigators (Mantovani *et al.*, 1972; Gigli and Nelson, 1968) have employed EIgG with and without added complement to differentiate Fc from complement-binding activity.

Indirect visualization of complement receptors by fluorescence microscopy is possible using fluorescein-labeled anti-complement antisera (Theofilopoulos *et al.*, 1974) or an immunoglobulin bridge (Ross and Polley, 1975) to localize complement components bound to cellular receptors.

A technique for assaying complement receptors utilizes target particles of antigen in oil droplets (Shurin and Stossel, 1978; Stossel, 1973). This assay accomodates a range of antigen concentrations and allows quantitation of receptor-mediated endocytosis. Quantitation of complement receptor sites could also be accomplished with radiolabeled (Stossel *et al.*, 1975) or ultrastructural probes like those employed for Fc receptors but these needed data have not been collected as yet. Unfortunately, the surface aggregation inherent in complement reactions and lability of the complement receptor hinder quantitation by direct binding assays.

3.3. COMPLEMENT RECEPTOR SPECIFICITY

Formal proof of specificity of macrophage receptors for complement is complicated by the rarity of direct binding and binding-inhibition assays (Section 2.3). Nevertheless, the available morphologic and functional methods for demonstrating complement receptors indicate specific binding of complement components. The complement reaction on target erythrocytes coated with IgM antibody (EIgM) begins with the sequential formation of C1, C4, C2, and C3. An IgM-coated erythrocyte which has bound the first four complement components may be designated EIgMC 1423. The EIgMC 1423 possesses maximal activity to form rosettes with macrophages and is more readily ingested by the phagocyte. Moreover, the C1 and C2 moieties can be removed from this complex and the resulting EIgMC 43 retains maximal activity for phagocytosis by guinea pig peritoneal exudate macrophages (Wellek *et al.*, 1976a; Gigli and Nelson, 1968). However, EIgMC 4 alone is not active (Wellek *et al.*, 1976a) and EIgM C142 has greatly decreased affinity for the complement receptor (Gigli and Nelson, 1968). The binding of sensitized erythrocytes to guinea pig macrophages apparently therefore requires C3 primarily. Human monocytes have receptor activity for C4

(Ross and Polley, 1975) in addition to C3 (Stossel, 1975; Ehlenberger and Nussenzweig, 1977).

Soluble C3 is enzymatically cleaved to C3a and C3b (nascent form) by earlier reacting complement components in the immune complex on the red cell surface (McConnell and Lachmann, 1977). The C3b attaches and remains fixed to the surface. A C3 inactivator can then cleave C3b into C3c and C3d (Ruddy and Austen, 1969) and, of these products, only the C3d component remains fixed on the target surface. The opsonically active fragment of C3 has been characterized as a molecule of 140,000 daltons (Stossel *et al.*, 1975). The preponderance of evidence favors receptor affinity for C3b. Recent studies have indicated that human monocytes and guinea pig peritoneal exudate macrophages possess receptor activity in addition for C3d (Ehlenberger and Nussenzweig, 1977; Wellek *et al.*, 1976a) but that human and guinea pig PMN and mouse peritoneal macrophages apparently lack this latter binding capacity (Ehlenberger and Nussenzweig, 1977; Griffin *et al.*, 1975a; Gigli and Nelson, 1968; Stossel *et al.*, 1975). The general concept that mononuclear phagocytes have receptors for both C3b and C3d while PMN possess exclusively the C3b receptors (Ehlenberger and Nussenzweig, 1977; Wellek *et al.*, 1976a) is contradicted by the lack of C3b affinity on mouse peritoneal macrophages (Griffin *et al.*, 1975a), and appears to be characteristic of some species. Purified complement is requisite for demonstration of the C3d receptor on human alveolar macrophages, as whole serum does not suffice (Reynolds *et al.*, 1975). This may explain the apparent lack of C3d binding by mouse macrophages since purified complement was not employed in the experiments on these cells. The failure of whole serum to serve in demonstration of C3d receptor on macrophages prompts speculation that serum may contain yet another factor which inactivates C3d and impairs its function of binding to complement receptors of mononuclear phagocytes. There is a study which finds C3b and C3d receptor activity on human neutrophils and which claims that all circulating phagocytic cells have similar membrane receptors for opsonized particles (Anwar and Kay, 1977). It is clear that close attention must be paid to animal species and method of complement purification in evaluating results.

A functional inhibition assay provided further evidence for the specificity of the complement receptor. In this assay system variably activated mouse peritoneal macrophages were allowed first to bind EIgMC and were then exposed to additional IgG-coated bacteria or latex beads to test accessibility and internalization of Fc receptors and foreign substance receptors. Both additional particles bound to the macrophages, but neither the IgG-coated bacteria nor the latex particles replaced the original EIgMC on the macrophage plasma membrane. The internalization of IgG-coated bacteria and latex particles on warming these macrophages contrasted with the failure of uptake of the EIgMC in this assay and provided functional evidence for a complement receptor distinct from the Fc and foreign-substance receptor (Griffin *et al.*, 1975a).

Direct binding assays of C3 have been applied to lymphoid neoplasms (Theofilopoulos *et al.*, 1974). They could be valuable in further defining the specificity of macrophage receptors for complement. Other molecular-binding assays utilize complement plus IgG which test cellular adherence to both complement and IgG receptors. A single study reports the assessment of complement-binding alone using commercial Fab'2 fragments of IgG (Cossman *et al.*, 1978). The manner in which these Fab'2 fragments fix complement is unclear.

The receptor for C3b on lymphocytes also binds C4b as demonstrated by cocapping of the two ligands (Ross and Polley, 1975). Macrophage C3b receptors may be similar in this regard, but the case is less clear (McConnell and Lachmann, 1977).

3.4. ULTRASTRUCTURAL MORPHOLOGY

The fine structural analysis of complement receptors is encumbered by the large size of the erythrocyte generally used to demonstrate the receptor. The area is not extensive where the macrophage membrane closely contacts the erythrocyte surface with its coat of IgM and complement (EIgMC). EIgMC, moreover, binds to macrophages with fewer regions of close contact than EIgG or E(Fab')2 (Griffin and Silverstein, 1974; Kaplan, 1977). This phenomenon may reflect the presence of fewer macrophage receptors for complement than for immune complex. However, the need for complement receptors may be less since fewer immunoglobulin molecules are required for binding erythrocytes to macrophages in the presence of complement than in its absence (Ehlenberger and Nussenzweig, 1977). These observations suggest that the macrophage surface contains a relatively small number of high-affinity receptors for complement. Proof of this consideration awaits better methods of analyzing the number and affinity of complement receptors (Section 2.3). The sensitized erythrocyte makes intimate contact with the monocyte as the intermembrane gap between EIgMC and monocyte measures approximately 40 nm (Douglas and Huber, 1972).

Receptors for EIgMC appear to be randomly distributed over the spread surface of endotoxin-induced peritoneal macrophages (Kaplan, 1977), and short-term cultured Kupffer cells (Munthe-Kaas *et al.*, 1976), except that the extreme periphery of these cells lacks affinity for EIgMC. Receptors for EIgG were evident randomly over these cells, however, including their periphery.

When rosettes formed in the cold are warmed to 37°C a zone of microfilaments collects in the macrophage cytoplasm beneath EIgMC (Kaplan, 1977). Such filaments, possibly composed of contractile actin-like protein, are thought to connect through hypothetical subplasmalemmal protein to the receptors and mediate endocytosis of the receptor-bound erythrocyte.

3.5. ENDOCYTOSIS AND REGENERATION OF RECEPTORS

At 37°C, nonactivated mononuclear cells including human blood monocytes (Ehlenberger and Nussenzweig, 1977) and uninduced mouse peritoneal macrophages bind but do not ingest complement-coated erythrocytes (EIgMC). However, induced mouse peritoneal macrophages both bind and ingest EIgMC (Bianco *et al.*, 1975; Morland and Kaplan, 1977). Accordingly, it appears that a major difference exists between nonactivated and activated macrophages in that only the latter are capable of phagocytosis via their complement receptor alone (Bianco *et al.*, 1975). This difference possibly explains past discrepancies in which phagocytosis of EIgMC failed in some systems (Mantovani *et al.*, 1972; Ehlenberger and Nussenzweig, 1977) but occurred in others (Kaplan, 1977; Wellek *et al.*, 1976a,b). Induced guinea pig peritoneal macrophages are also capable of ingesting complement-sensitized erythrocytes with the IgM removed (EC3) (Wellek *et al.*, 1976a). It would be interesting to determine whether the additional endocytic capability of activated macrophages for complement-coated erythrocytes is related to increased macrophage metabolism and ability to effect more cellular work or to a different relationship between complement receptors and subplasmalemmal molecules which mediate endocytosis. The endocytic capability of activated macrophages does apparently relate to the ability of the activated macrophage to redistribute its complement receptors on its plasma membrane (Michl *et al.*, 1979).

Complement receptors play a role also in enhancing endocytosis mediated by Fc receptors. Thus, PMN, monocytes (Ehlenberger and Nussenzweig, 1977), and peritoneal macrophages induced with starch gel (Wellek *et al.*, 1976b) ingest EIgG more efficiently in the presence of complement at limiting antibody concentrations. Apparently such enhancement of EIgG ingestion by complement is a property that characterizes phagocyte surface activity generally. Alternative experimental approaches suggest that complement bound to the Fc segments of IgG in immune complexes render them ineffective as ligands for macrophage Fc receptors, the binding occurring instead via complement receptors (Michl *et al.*, 1979).

Internalization of erythrocytes by Kupffer cells appears to differ for target cells bound by the complement receptor as compared with those bound by the Fc receptor. Thus, uptake of EIgG entails envelopment by extensions of macrophage membrane around the target cell. Scanning and transmission electron microscopy of phagocytosis of EIgMC by Kupffer cells (Munthe-Kaas *et al.*, 1976) and induced peritoneal macrophages (Kaplan, 1977), reveal that the target cell appears to sink directly into a crater or pit which forms on the macrophage surface. The regions where complement-coated erythrocyte and phagocyte membranes make close contact are uncommon except at the bottom of the phagocytic pit. This mode of ingestion differs from that by which Fc receptors become internalized (Section 2.6). Perhaps this difference depends on the stronger binding of complement than of the Fc portion of Ig so that less circumferential membrane approximation is necessary for endocytosis of EIgMC compared with EIgG (Figure 8). Complement and Fc receptors move independently of one another on the macrophage surface (Michl *et al.*, 1979).

In contrast to the aforementioned static morphologic impressions, a different view was obtained employing trypsinization during endocytosis. Trypsin treatment of activated macrophage–EIgMC complexes cleaved C3b and inhibited

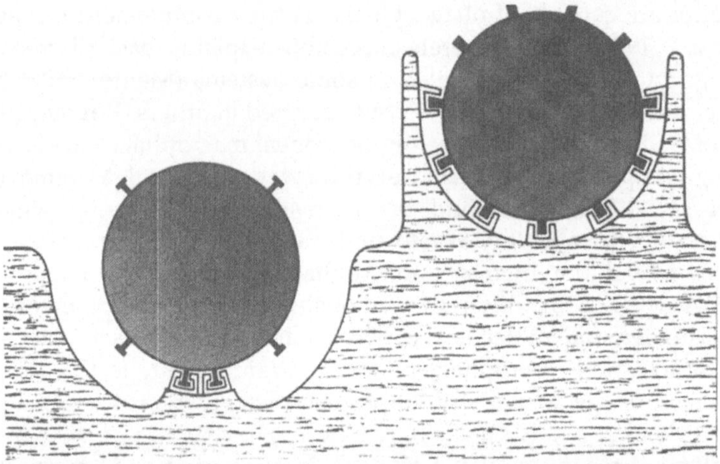

FIGURE 8. Authors' explanation for differences observed (Kaplan, 1977) between complement-mediated (solid T blocks) and Fc-mediated (solid rectangles) ingestion of target erythrocytes (gray-dotted circles) by the macrophage (streaked region). Receptors for complement (two white shapes) bind tightly to complement on the erythrocyte, allowing ingestion of the erythrocyte with minimal macrophage membrane (line across figure) and receptor contact with the target. Receptors for IgG (seven white shapes) bind less avidly to IgG-coated erythrocyte necessitating more macrophage membrane and receptor contact.

ingestion of the erythrocytes. This approach supports the concept of binding of additional surface complement molecules to macrophage receptors in a process referred to as circumferential sequential binding (Griffin *et al.*, 1975b). The latter experiments indicate that target moieties bind to the walls of the phagocytic pit or crater (Kaplan, 1977) rather than at the base alone. Full appreciation of the sequential events involved in macrophage ingestion of EIgMC would be enhanced by serial sections of specimens obtained at closely spaced time intervals.

Newborn calf serum triggers ingestion of EIgMC at 37°C (Mørland and Kaplan, 1977; Munthe-Kaas, 1976) by previously nonactivated macrophages. The mechanism whereby the serum effects such endocytosis is unexplained, but possibly relates to the known difference in endocytosis of EIgMC by activated compared with nonactivated macrophages. Newborn calf serum might contain factors which promote ingestion of EIgMC by stimulating aspects of macrophage activation.

After internalization of macrophage plasmalemma in the uptake of latex beads the remaining plasma membrane contains a functionally undepleted set of C3 receptors (Daughaday and Douglas, 1976), suggesting that the plasmalemma which provided the limiting membrane of bead-laden phagosomes: (1) equaled the plasma membrane in content of complement receptors and (2) became replaced by newly formed plasmalemma containing the usual abundance of complement receptors. This indication that plasmalemmal components are not sequestered during internalization of plasma membrane differs from other observations (Berlin and Oliver, 1978); selective concentration of Con A receptors in newly forming phagosome membranes is apparently a result of clustering of these receptors into patches from which endosomes formed.

3.6. SENSITIVITY TO ENZYMES, IONS, AND DRUGS

Complement receptors are labile to trypsin (Bianco *et al.*, 1975; Lay and Nussenzweig, 1968; Daughaday and Douglas, 1976; and Stossel, 1978) and were thought to differ from the Fc receptor in this respect until a trypsin-sensitive aspect of the Fc receptor was recognized (Section 2.7). A receptor mediating rapid complement-dependent phagocytosis has been distinguished from one mediating slow complement-dependent phagocytosis (Shurin and Stossel, 1978) in part on the basis of the greater susceptibility of the former to trypsin.

Rosette formation by EIgMC on mononuclear and polymorphonuclear phagocytes depends upon Mg^{2+} (Lay and Nussenzweig, 1968); binding to Fc receptors, on the other hand, differs from complement receptors in not requiring Mg^{2+} (Davey and Asherson, 1967). At the ingestion stage of endocytosis the internalization of complement-coated particles is stimulated by Ca^{2+}, Mg^{2+}, Mn^{2+}, and Co^{2+} (Stossel, 1973) probably through an effect of divalent cation on contractile proteins that carry out the work of engulfment. The polyanion, dextran sulfate, inhibits rosette formation and phagocytosis of EIgMC, EIgGC, and EIgG possibly by its own binding to the receptor (Wellek *et al.*, 1976b) or, alternatively, by interference with Mg^{2+} activation.

Corticosteroids inhibit the binding of EIgMC to human monocytes. Glucocorticoids were found to have this effect at levels that possibly occur *in vivo* and mineralocorticoids also afford evidence of inhibiting binding (Schreiber *et al.*, 1975). Levamisole increase C3 receptor activity of human monocytes *in vitro* (Schmidt and Douglas, 1976a).

Cytochalasins inhibit complement-coated erythrocytes from forming rosettes on alveolar macrophages (Atkinson *et al.*, 1977). Their order of effectiveness is cytochalasin A>E>D>B. Cytochalasin B inhibits ingestion of loaded complement receptors less effectively than it inhibits ingestion of the Fc receptors by Kupffer cells (Munthe-Kaas *et al.*, 1976). This inhibition, presumably dependent on antagonistic effect on contractile actin-type filaments, is consistent with the observation that microfilaments collect beneath both types of receptor at 37°C (Kaplan, 1977; Griffin *et al.*, 1976). Colchicine inhibits binding to and uptake of Fc and complement receptors to only a minor degree (Atkinson *et al.*, 1977; Munthe-Kaas *et al.*, 1976).

3.7. SUMMARY AND CONCLUSIONS

In addition to its conceivable role in cellular interactions, the complement receptor enhances the phagocytic ingestion of antigenic substances including infectious agents. Most of the data on complement receptors are derived from assays of rosetting of appropriately sensitized erythrocytes and observations of phagocytic ingestion of the rosetted cells. The binding activity of the complement receptor is primarily directed toward the third components of the complement system. Mononuclear phagocytes of man show evidence of receptor activity for both C3b and C3d, whereas mouse and possibly human polymorphonuclear leukocytes and perhaps mouse macrophages possess only the former. Complement receptors lie randomly distributed except at the extreme periphery of macrophages spread on a flat surface. Phagocytosis mediated by complement differs morphologically from that mediated by the Fc receptor in having fewer regions of close membrane contact and less extension of the macrophage plasma membrane around the target particle. These differences may reflect stronger binding of complement receptors and they await direct binding data for confirmation. Complement enhances binding and endocytosis of sensitized erythrocytes by activated macrophages but influences only the adherence and not the uptake of these red cells by nonactivated mononuclear phagocytes. Microfilaments function importantly in the binding and internalization steps of phagocytosis modulated by either receptor. The complement receptor is probably a membrane-bound macromolecule which requires Mg^{2+} for binding of complement and consists at least partly of protein. Fine structural markers for and biochemical isolation of the complement receptor would improve knowledge of its structure and biologic function.

4. RECEPTORS FOR FOREIGN SUBSTANCES (NONSPECIFIC RECEPTORS)

4.1. BIOLOGIC ACTIVITY

The macrophage exemplifies its given name in at least two ways. It is a large cell which ingests material. It is also a cell which ingests a large variety of material, for instance, asbestos (Spencer, 1969), talc (Goldner and Adams, 1977), silica (McKeever, 1976), tin (Spencer, 1969), polystyrene (latex) beads (Casley-Smith, 1969; Al-Ibrahim *et al.*, 1976), aluminum (Spencer, 1969; Michl *et al.*, 1976; Holland *et al.*, 1972; Polliack and Gordon, 1975), iridium (Gersten *et al.*, 1977), thorium dioxide (Thorotrast) (Casley-Smith, 1969; Goldner and Adams, 1977), more than one phase of carbon (Spencer, 1969; Casley and Smith, 1969; McKeever and Balentine, 1978), barium sulfate (Kronman *et al.*, 1977; Goldner and Adams, 1977), iron (Spencer, 1969; Hausmann, 1976), heterologous erythrocytes (Polliack and Gordon, 1975), glutaraldehyde-fixed heterologous erythrocytes (Holland *et al.*, 1972; Rabinovitch, 1976), extracted yeast cell walls (Zymosan) (Low, 1977; Michl *et al.*, 1976; Holland *et al.*, 1972; Walters *et al.*, 1976; Czop *et al.*, 1978), various fungi and bacteria (Wood, 1960; Casley-Smith, 1969; Walters *et al.*, 1975; Schroit and Gallily, 1977; McKeever *et al.*, 1978; Gee *et al.*, 1973), autologous injured tissue components (Powers and McKeever, 1976; Balentine *et al.*, 1974), cationized ferritin (Skutelsky and Hardy, 1976), horseradish peroxidase (McKeever, 1976; Steinman and Cohn, 1972a), keyhole limpet hemocyanin (Unanue, 1968), ferritin (Casley-Smith, 1969), various waxes and oils (Spencer, 1969), and numerous organic fibers (Spencer, 1969).

Macrophages ingest the majority of foreign substances which have been tested. From this response to a test sampling, it is reasonable to assume that the actual list of substances is much larger than any which could be compiled from the literature. The biologic importance of this highly diversified activity is most obvious within the lung where macrophages clear the alveolar respiratory surfaces of small particles (Heppleston, 1963; LaBelle and Brieger. 1960).

Nonspecific receptor is the term commonly used for defining the property of the macrophage surface that leads to the binding and ingestion of various foreign substances. The term exemplifies the major difficulty encountered in ascribing this activity to a receptor. Receptors are generally defined by their recognition of and interaction with a specific target component (Gove, 1967; Herbert and Wilkinson, 1977). As can be seen by the array of diverse materials ingested by macrophages, specificity of this receptor if present will be very difficult to determine. To avoid the ambiguity of the term *nonspecific receptor*, we prefer referring here to the hypothetical receptor(s) governing the binding and ingestion of such varied material as a *receptor for foreign substances*. It is uncertain whether one or more actual receptors with affinity for a wide range of chemically different components exist in the macrophage plasmalemma or rather that a glycocalyx or coat material mediates the binding. These cells possess, on whatever basis, a fairly unique capacity to bind foreign substances.

Innovations in the analysis of polypeptides have made it possible to estimate the number of polypeptide chains synthesized by one cell type. Cells examined reveal that 10^3 chains represents a major fraction of total cellular proteins 10^2 of which may be represented on the cellular surface (Ames and Nikaido, 1976; Jones, 1977). Should the macrophage contain similar numbers of different proteins, the existence of a different protein receptor for each foreign substance would be quite unlikely.

4.2. TECHNIQUES

Assays of macrophage activity for specific foreign substances can be found in the literature cited in Section 4.1. A major problem in interpreting data on foreign substance receptors is the wide variety of substances tested. Common techniques incubate the foreign target substance with macrophage monolayers for specific time intervals at specific temperatures. Excess target substance is rinsed away and the quantity of adherent and/or ingested foreign substance is assessed in a manner appropriate to the chosen substance. Target erythrocytes can be counted, for example, under the microscope (Rabinovitch, 1967), and smaller particulates can be assessed ultrastructurally (Casley-Smith, 1969; Feldman and Pollock, 1974) or biochemically (Feldman and Pollock, 1974). Simple and efficient assays of radiolabeled *Shigella* (Schroit and Gallifly, 1977), polystyrene particles (Al-Ibrahim *et al.*, 1976), and particulate ^{192}iridium

(Gersten *et al.*, 1977) exist. Two widely employed procedures for distinguishing between attachment and ingestion phases of phagocytosis entail incubating with aldehyde-fixed heterologous erythrocytes in different media for various times and at different temperatures (Rabinovitch, 1967). A chemical method for latex spheres is available (Gardner *et al.*, 1973).

Macrophage interactions with flat surfaces have been investigated by subjecting the adherent cells to various forces promoting their detachment. Centrifugal (McKeever and Gee, 1975) and fluid-shearing forces (McKeever and Gee, 1975; Weiss and Glaves, 1975) have served, for example, in determining the effects of drugs and certain macromolecules on adherence of macrophages to flat surfaces. Cells treated with the drug prior to surface contact were compared in these studies with untreated cells for their ability to establish adherence and, alternatively, cells already adhering to a surface were evaluated for their ability to maintain adherence against a detaching force.

A different but basically related assay has been developed to determine the effects of ionic strength, drugs, and protein upon macrophage spreading (Douglas, 1976; Rabinovitch and DeStefano, 1973). In another approach the forces required to detach macrophages from different surfaces by micromanipulation have been measured (Table 2). The unique capacity of macrophages to adhere to surfaces coated with microexudates has been utilized as the basis for a simple technique of obtaining purified isolates of monocytes (Ackerman and Douglas, 1978). A different type of assessment of the adhesive properties of the macrophage depends on measuring the contact angle between the cell surface and a precisely measured drop of isotonic saline by means of a telescope with crosshairs attached to a goniometer (Van Oss *et al.*, 1975; Van Oss and Gillman, 1972).

Interactions between macrophage and target surfaces can be predicted mathematically. Measurements of macrophage adherence obtained in this way agree with those yielded by other methods (Van Oss *et al.*, 1977; McKeever, 1974).

TABLE 2. FORCE REQUIRED TO REMOVE ALVEOLAR MACROPHAGES FROM VARIOUS MATERIALS[a]

Material	Number of observations	Mean force ($\times 10^{-2}$ dynes/cell)	±	Standard error
Glass	59	1.42[b]	±	0.16
Cellulose tetranitrate	22	0.85[c]	±	0.20
Polyethylene	16	0.63	±	0.12

[a] Rabbit alveolar macrophages in glucose phosphate buffer were incubated at 24°C for 15 min on a flat surface of the material to be tested. The material was then inverted creating a hanging drop on the stage of a microscope. A calibrated flexible glass microneedle removed the macrophages individually from the material and measured the force of detachment. (Reprinted from Vol. 16, p. 315, by permission of the *Journal of the Reticuloendothelial Society*.)
[b] Significant differences: 1,2 ($0.02 < p < 0.05$) and 1,3 ($0.01 < p < 0.02$).
[c] Not significant: 2,3.

4.3. SPECIFICITY

Macrophages exceed all other cells in the avidity with which they bind and internalize a variety of extrinsic substances or adhere to flat surfaces of glass or other composition. *Nonspecific receptor* is the term used with reservation for defining the property of the macrophage surface that leads to the binding and ingestion of various foreign substances (Silverstein *et al.*, 1977). The term exemplifies the major difficulty encountered in ascribing this activity to a receptor. Receptors are generally defined by their recognition of and interaction with a specific target component (Gove, 1967; Herbert and Wilkinson, 1977). Common tests of receptor specificity applied to receptors for foreign substances (Section 2) are often impractical because of the relationship of large particles to the cell surface (Rabinovitch, 1967, 1968) and the consequent problem of distinguishing between different receptors that are closely interspersed with particles larger than the periodic spacing of the receptors. To surmount this difficulty a competitive inhibition test of specificity has been applied to small foreign substances such as keyhole limpet hemocyanin either free in solution or bound to beads. Endocytosis of this protein by rat peritoneal exudate macrophages is partially inhibited nonspecifically by different serum proteins (Feldman and Pollock, 1974). The binding of cationized ferritin is inhibited by prior incubation with another positively charged macromolecule, poly-L-lysine (Skutelsky and Hardy, 1976). Another binding inhibition test was performed at 0°C to minimize ingestion of proteins bound by normal mouse spleen cells. Radioiodinated ribonuclease was bound by splenocytes even in the presence of a large excess of the unlabeled protein. Similar results were obtained with bovine serum albumin and cytochrome c (Segal and Hurwitz, 1977). These experiments suggest a lack of specificity in the physical interaction of cells and foreign protein. In contrast, other experiments discussed below indirectly suggest specific interactions between molecules in the membrane and foreign substances.

Foreign-substance receptors, i.e., the surface component mediating binding of foreign substances, can be distinguished, as a different entity, from the Fc receptor. This differentiation was accomplished by raising antiserum to cellular membranes and showing that they block the IgG (Fc) receptors but not the binding of foreign particles such as polystyrene, formalin-fixed erythrocytes, and yeast cell walls (Holland *et al.*, 1971).

The plasma membranes of macrophages exposed very briefly to cationized ferritin (CF) at physiologic temperature, bind this positively charged protein in clusters, leaving other membrane regions free of CF (Skutelsky and Hardy, 1976). Moreover, a second surface exposure to CF will not stain the clear regions. This distribution contrasts with the even distribution of ferritin particles observed after exposing fixed preparations of cells to CF. These distribution experiments suggest that surface molecules on the macrophage plasma membrane bind this cationic molecule and these sites are redistributed on unfixed cells. Reaction with CF represents perhaps a special case of foreign substance binding since it is positively charged and appears to be bound to the macrophage surface more avidly than naturally charged ferritin. Because many foreign substances bound by macrophages carry surface charge arrays, this difference is probably only a quantitative and not a qualitative distinction.

Macrophages, in any case, bind ionic substances more avidly than those lacking net electrostatic charge, thus evidencing a functional specialization (Table 2). This difference may reflect binding of specific receptors. Alternatively, a number of membrane surface molecules, some common to both surfaces, may bind. Ultrastructural cytochemical methods have demonstrated an anionic coat on the macrophage surface (Eguchi *et al.*, 1979). It is possible that this presumed acidic complex carbohydrate of the cell glycocalyx mediates the binding of cationized ferritin and could be assessed perhaps by comparison of the periodicty and distribution of reactive sites with two methods on fixed and unfixed cells.

At least three lines of indirect evidence point to receptors not requiring opsonins and delineate differences between the receptors for opsonized compared with nonopsonized targets. Thus, macrophage phagocytosis of nonopsonized latex or zymosan particles is not inhibited by 2-deoxyglucose in contrast to macrophage phagocytosis mediated by IgG and complement (Michl *et al.*, 1976). Macrophage ingestion of similar nonopsonized particles as noted earlier is not significantly reduced by anti-membrane antibodies, whereas their Fc receptor is blocked by these antibodies (Holland *et al.*, 1972). This property of distinguishing receptors for opsonized and nonopsonized particles may not be shared, however, by all anti-membrane antibodies (Unanue, 1968). Macrophages adhere to autologous erythrocytes altered by *in vitro* storage. Unlike the binding of aging erythrocytes *in vivo*, this adherence seems to be unaffected by differences in serum factors (Vaughan and Boyden, 1964).

Another indirect test of the specificity of various receptors for foreign substances depends on their differential lability to digestion. Monocyte ingestion of yeast cell wall (zymosan) particles is more sensitive to trypsin than is monocyte ingestion of latex (Czop *et al.*, 1978). Similarly the binding of glutaraldehyde-fixed erythrocytes and of immune complexes composed of the Fab2 portion of IgG and erythrocytes by macrophages is more sensitive to trypsin than binding of zymosan or latex particles (Steinman and Cohn, 1972b; Rabinovitch, 1968). The latter erythrocytes are not ingested as readily as latex (Griffin and Silverstein, 1974). They may, of course, represent a special case, since the erythrocytes are coated with a portion of IgG known to bind macrophages poorly.

The data summarized thus far are difficult to interpret because of differences in experimental design and in foreign target substances. They leave the impression that within the macrophage membrane, molecules containing protein or peptides (of low antigenicity) bind certain foreign substances. These membrane-associated molecules bind with low specificity detected only indirectly between broadly defined groups of foreign materials. Some of the binding probably depends on ionic forces. The few biochemical isolations of these and other receptors leaves their differentiation undetermined. Recognition of the different receptors is complicated by the possible effects that contaminating substances could have on binding of target substances in nonpurified systems as discussed next.

Undetected serum proteins conceivably could participate in macrophage ingestion of certain foreign materials, and undetected immunoglobulin and complement possibly play a role in phagocytosis of foreign substances via the Fc, complement, or fibrin receptor. These serum components could influence phagocytosis through their absorption on the surface of either the macrophage or the target entity. Macrophages are known, for example, to collect cytophilic antibodies from body fluids including heterologous serum, and small amounts of surface-associated proteins govern the surface adherence of cellular populations (Choi *et al.*, 1974). In addition, target particles such as aging erythrocytes accumulate autoantibodies which resist attempts at removal by washing and incubation in serum-free medium (Knyszynski *et al.*, 1977; Kay, 1975). Moreover, surfaces of inert particles including polystyrene possess capacity to bind nonspecifically to antigens alone (Wigzell *et al.*, 1972), immunoglobulins alone (Choi *et al.*, 1974), and immune complexes (Alexander and Henkart, 1976). Electron-dense material at polystyrene particle surfaces (Michl *et al.*, 1976; Reaven and Axline, 1973) may consist, at least in part, of absorbed protein. Zymosan particles, as another example, have been shown to fix complement when incubated in serum (Henson, 1971). *In vivo* observations of erythrophagocytosis during experimental low-grade intravascular coagulation are also hard to ignore (Prose *et al.*, 1965). It is difficult to devise a completely serum-free system to assay macrophage receptors for foreign substances in the absence of immunoglobulin and complement. Possibly, therefore, small quantities of absorbed serum factors contribute to a portion of macrophage activity in ingesting foreign substances. Indeed, since one complex system of immunologic recognition already exists in immunoglobulins and is available to macrophages via the IgG receptor, it is cumbersome to postulate yet another recognition system with function overlapping the first.

As discussed additionally in Sections 4.4 and 4.5, biomechanical factors may contribute significantly to macrophage-binding and ingestion of foreign substances. At physiologic temperature the macrophage surrounds and engulfs substances by means of its membranous extensions and preformed invaginations and tends to grip defects on the surface of large particles (Walters *et al.*, 1975; Polliack and Gordon, 1975; Emeis, 1976). The macrophage also has a propensity to spread onto surfaces of various types, thereby increasing contact with the target moiety. The rapid motility and plasticity of the macrophage membrane affords extensive contact which can greatly enhance even low-affinity binding. These activities may then facilitate endocytic capacity of macrophages in a nonspecific manner. Studies on the binding of foreign substances in the cold or in the presence of metabolic inhibitors minimize these potentially confusing factors by intefering with macrophage membrane mobility. The role of mobility in establishing contact and adherence is exemplified by experiments on the influence of metabolic inhibition on binding of macrophages to polystyrene surfaces. Pretreating the macrophages with potassium cyanide inhibited binding, but once they had contacted and adhered to the polystyrene, their adherence was much more resistance to cyanide treatment (McKeever and Gee, 1975).

The theoretical importance of macrophage-spreading in relation to its almost ubiquitous binding capability might be more commonly appreciated by considering the large membrane surface area which phagocytes like the macrophage present to a particle nearly its own size (Emeis, 1976) or to a flat surface (Fenn, 1922; Reaven and Axline, 1973). This membrane area may be more than 100-fold greater than that presented by an erythrocyte or lymphocyte (Gudat and Villiger, 1973). The difference in surface contact is like the difference between a surface contact with a fully inflated balloon as compared with one only partially inflated. The effect of lidocaine anesthetic on macrophages is interesting in this regard. Individual cells exposed to lidocaine first round up and then detach from their substrate (Rabinovitch and DeStefano, 1975, 1976).

Whereas most macrophages ingest only senescent autologous erythrocytes, alveolar macrophages ingest young and old erythrocytes when they are artifically introduced by intratracheal injection (Kay, 1975; Emeis, 1976; Collet and Petrik, 1971). Tissue location then appears to influence macrophage activity in avoiding the ingestion of healthy autologous cells as if they were foreign substances. Since the macrophages in all tissues are thought to originate as monocytes in bone marrow and to reside in the tissue after migration from peripheral blood, these differences reflect an influence of the microenvironment on the nature of the target cell or on the macrophage and its transformation from a blood-borne monocyte.

Determining the nature of a receptor for a foreign substance independent of other binding factors requires a system that is precisely defined in regard to serum factors, particle roughness, and macrophage source. A clean system in this sense might employ macrophages or their neoplastic counterparts that have been extensively cultured in serum-free medium and are tested with optically smooth particles or flat surfaces never exposed to serum. We have tried to simulate such a system by culturing the murine macrophage-like neoplasm $P388_{D1}$ on optically flat plastic dishes through multiple changes of serum-free medium. Factors which may affect macrophage adherence can then be added individually in this system to test their influence on the binding of the cell to the surface. A major difficulty with this approach is the inability of the $P388_{D1}$ line to thrive in culture in the absence of serum. Nevertheless, it is clear that many cells still adhere after multiple complete changes of serum-free medium over a 10-day period. Artificial roughening of the surface noticeably increases the adherence of $P388_{D1}$. This effect has been observed before in a less rigorously defined system using rabbit alveolar macrophages. Centrifugation clarified the difference in adherence to rough and smooth surfaces by subjecting all cells to a force tending to detach them from the surface (Table 3).

An alternative explanation of macrophage recognition of foreign substances is to consider the macrophage capable of ingesting any substance which it cannot specifically recognize and leave undisturbed. Such activity would perhaps be expected to occur at a basal level considerably below the rate of endocytosis stimulated by IgG and complement. Such a proposed explanation offers an advantage in not requiring recognition of each of the multitude of foreign substances which macrophages ingest. Instead, it would require a capacity to avoid ingesting autologous constituents which the cell commonly encounters *in situ*.

TABLE 3. EFFECT OF A ROUGH SURFACE ON ALVEOLAR MACROPHAGE ADHERENCE TO POLYSTYRENE[a]

Surface	Assay	Mean number cells, final/initial	Mean percentage adherent cells	±	Standard error	p value[b]
Rough	Centrifuge	40.1/53.4	75.1	±	6.0	0.025
Smooth	Centrifuge	33.7/57.1	59.0	±	11.9	
Rough	Uncentrifuged[c]	50.7/51.4	98.7	±	5.6	0.250
Smooth	Uncentrifuged	49.4/54.7	90.3	±	9.6	

[a] Rabbit alveolar macrophages in glucose phosphate buffer were allowed to attach to a petri dish with a grid for counting cells. The dish was rinsed, initial cell number counted, and the dish was centrifuged at 16,000*g*. The initial counting grid was then relocated and the final cell number counted. (Reprinted from Vol. 18, p. 225, by permission of the *Journal of the Reticuloendothelial Society*.)

[b] p values are derived from statistical comparison of experimental and control groups by the method for proportions.

[c] This control assay was performed in the same manner as the centrifuge assay but the centrifuge was not started.

Such a capability could be mediated by a limited number of receptors that recognize autologous components and, on contacting them, inhibit the cell's endocytic mechanisms. This recognition might be limited to tissue ground substances and components in basement membranes and the glycocalyx or plasmalemma of stromal or epithelial cells. Cell coat materials loosely attached or constantly shed from living cells might afford protection against phagocytosis of autologous tissue. This scheme is consistent with the peculiar abundance of macrophage processes *in situ*, revealed most convincingly by recent scanning electron microscopic studies (Orenstein and Shelton, 1976a). Indirect support for this rather reversed view of the foreign substance receptor may be found in the fact that removing sialic acid from the macrophage surface with neuraminidase increases macrophage ability to bind and ingest negatively charged particles (Weiss *et al.*, 1966; Weiss, 1973). This observation suggests that the negative charge on the macrophage surface impedes binding of macrophages to other cell types most of which contain acid complex carbohydrate in their glycocalyx. Examples of negatively charged molecules producing a barrier against the binding of proteins and particulates derive from controlled experiments in synthetic membrane technology. In addition, digesting a portion of the surface coat of mycoplasma, unfixed and fixed erythrocytes, and streptococci with proteolytic enzymes renders them more susceptible to phagocytosis (Jancik and Schauer, 1978; Jones, 1975; Wood, 1960), conceivably as a consequence of elimination of the constituents that macrophages normally recognize.

4.4. ULTRASTRUCTURAL MORPHOLOGY

Their fine structure holds clues to the manner in which macrophages interact with foreign substances. Distances between the electron-dense external edge of the macrophage membrane and its target substance have been measured in published electron micrographs (Table 4). The macrophage membrane was found to be closer to small objects than large. This intriguing generalization, based on differences collected from the literature, perhaps requires more definite validation which could be achieved by examining serial sections and utilizing a goniometer stage for aligning measurements at an angle normal to the plane of the membrane. Large cellular targets such as heterologous erythrocytes have their own surface coat or glycocalyx which very likely add to the gap between macrophage and erythrocyte. Presumably, such an extra coat is missing on inert particles, although some protein from serum or other sources may coat these particles. Steric hindrance by the macrophage's own glycocalyx must be considered as a probable contributor to the direct relation between particle size and proximity to the macrophage membrane. The macrophage contribution possibly is related to the finding that at least two gaps can be demonstrated between the cell membrane and the particle, including a large gap of 30–40 nm and intermittent smaller gaps (Reaven and Axline, 1973; Figure 9). The effects of fixation could influence the observed distance between macrophage membrane and a

TABLE 4. CLOSEST APPROACH OF MACROPHAGE AND TARGET FOREIGN SUBSTANCE VISUALIZED BY TRANSMISSION ELECTRON MICROSCOPY

	Incubation		Foreign substance			
Macrophage source	Site	Temperature	Type	Size	Apparent distance between macrophage membrane and foreign substance	Reference
NCS/PA mouse peritoneum (unstimulated)	*In vitro*	37°C	Polystyrene spheres	14,000 nm diameter	20–40 nm	Reaven and Axline (1973)
Rat liver	*In vivo*	37°C	Fixed heterologous erythrocytes	5,000 nm diameter	20–30 nm	Emeis (1976)
Albino rat liver	*In vivo*	37°C	Carbon particles	50–150 nm diameter	20–30 nm	Törö *et al.* (1962)
Lewis rat peritoneum (proteose peptone stimulated)	*In vitro*	4°C	Keyhole limpet hemocyanin	3×10^6 mol. wt. (15 nm diameter)[a]	20–25 nm	Feldman and Pollack (1974)
Mouse peritoneum (unstimulated); rabbit peritoneum (liquid paraffin stimulated)	*In vitro*	37°C	Heterologous ferritin	7×10^5 mol. wt.	10–15 nm	Casley-Smith (1969)
C3M/EB mouse peritoneum (thioglycollate stimulated)	*In vitro*	37°C	Cationized ferritin	7×10^5 mol. wt.	5–10 nm	Skutelsky and Hardy (1976)
New Zealand white rabbit peritoneum (minced oil induced and uninduced)	*In vitro*	4°C	Horseradish peroxidase	4×10^4 mol. wt.	5–15 nm (binding rarely encountered)	McKeever, 1976; Steinman *et al.* (1976); Stahl *et al.* (1978)

[a] Estimated from molecular weight assuming globularity.

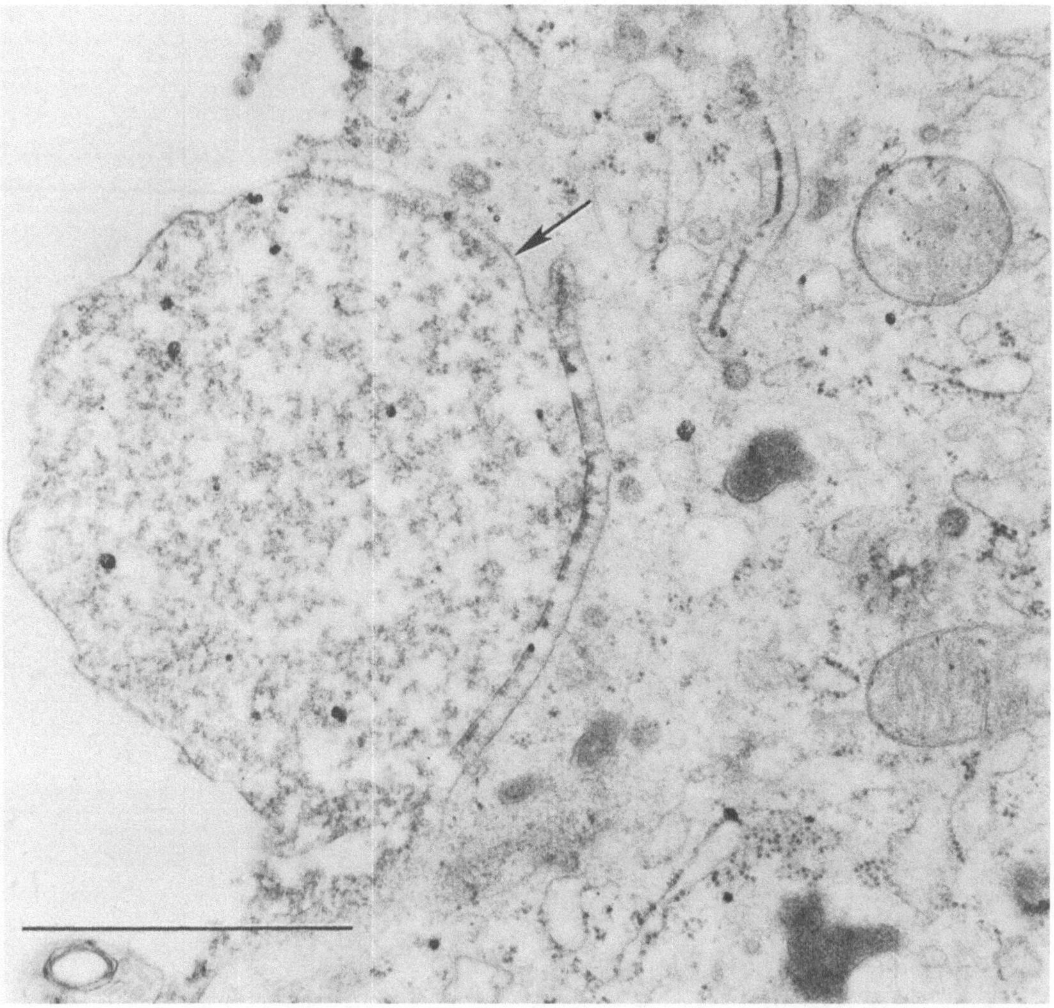

FIGURE 9. A Kupffer cell bound this tanned heterologous erythrocyte ghost. Its cell coat can be seen in contact with the erythrocyte where it forms a dense line. Contact is mainly at a level within the glycocalyx which is 80 nm from the plasma membrane of the Kupffer cell. Contact is more intimate in one region (arrow). Fixed *in vivo*, glutaraldehyde-tannic acid, scale bar represents 1.4 nm. Reprinted from *Journal of the Reticuloendothelial Society* **20**:37, with permission of the journal and Dr. J. J. Emeis.

target particle (Casley-Smith, 1969). The measurement of this distance would more directly reflect the living condition, therefore, if they were made on specimens that were rapidly frozen and processed by freeze substitution. Nonetheless, the differences observed in these distances under various conditions fall into a consistent pattern in which the fixation effect is apparently inconsequential.

An excessive positive charge on cationized ferritin brings it closer to the macrophage membrane (Table 4). It would not be surprising if this charge effect were a general phenomenon applicable to other cationic proteins and other negatively charged cell surfaces and inversely manifested by anionic macromolecules. Different levels within the cell surface coat (glycocalyx) are thought to have different densities (Pethica, 1961; Weiss, 1973). Mobility of these charged constituents within the membrane might facilitate their binding to highly charged foreign substances. The sialic acid outer coat on macrophages appears to protect negatively charged particles from ingestion (Weiss *et al.*, 1966), possibly through repulsive electrostatic forces. In the opposite sense the apparent penetration of cationized ferritin into the cellular coat to bind in close proximity to the macrophage membrane could reflect an electrostatic attraction between acidic groups of the plasmalemma and the electrostatic macromolecule.

Recent biochemical evidence strongly suggests that macrophages bind horseradish peroxidase via receptors for glycoproteins (Section 10). Nonetheless, this binding is difficult to demonstrate ultrastructurally (Steinman *et al.*, 1976; McKeever *et al.*, 1976a, b), perhaps due to low receptor affinity. Since peroxidase is not a large macromolecule, it appears that macrophages perhaps cannot bind foreign substances below a minimal size. This consideration derives support from experiments with hemocyanin in which the size of the carrier to which hemocyanin was bound strongly influenced its endocytosis by macrophages. Assuming that the macrophage has finite periodically spaced receptors which bind hemocyanin or other foreign substances with low affinity, the number of binding sites per target becomes crucial. Below a critical number of linkages between cell surface and target, as determined by the size of the foreign particle, binding may not be either demonstrable *in vitro* or biologically significant (Segal and Titus, 1978; Crothers and Metzger, 1972) (Section 2.3).

Although macrophage membrane binding of the small protein horseradish peroxidase is difficult to detect ultrastructurally (McKeever *et al.*, 1976a; Steinman *et al.*, 1976), by employing a system of rapid incubations in cold buffer, adherence of peroxidase can be demonstrated on a small portion of the plasma membrane (McKeever, 1976). Of interest in these infrequent areas is the gap between peroxidase and the macrophage membrane (Table 4). Despite some variability in the distance separating peroxidase from the plasmalemma, it is clear that the average distance between peroxidase and the macrophage membrane (Figure 10) is greater than the distance between the peroxidase in an immune complex and the membrane (Figure 2). The macrophage also appears to bind ferritin alone (Casley-Smith, 1969) at a slightly greater distance from its plasmalemma than ferritin-labeled IgG (Papadimitriou, 1973). These data suggest that the macrophage binds a foreign substance by a low-affinity force at a relatively distant level within its external surface coat. This level lies farther from the electron-dense lipid layer of the macrophage when the macrophage binds the protein alone than when it binds the protein via IgG and its Fc receptor.

4.5. TOPOGRAPHIC DISTRIBUTION

On a scale above 200 nm, macrophage receptors for foreign substances appear generally to be distributed at random on the cell surface (Skutelsky and

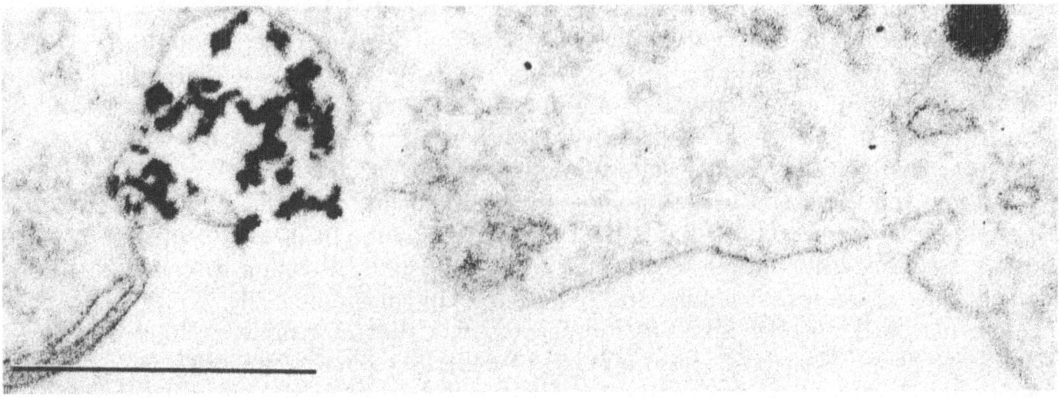

FIGURE 10. Evidence of macrophages directly binding peroxidase is uncommon. As visualized here in an indented portion of macrophage membrane, there is a gap of variable distance between the peroxidase and the macrophage membrane. Rabbit peritoneal macrophage incubated at 4°C for 30 min with horseradish peroxidase, rinsed briefly and stained with diaminobenzidine-peroxide. Scale bar represents 0.66 μm.

Hardy, 1976; Casley-Smith, 1969). However, foreign particles larger than 1 nm tend not to bind at the extreme periphery of macrophages attached to a flat surface (Kaplan *et al.*, 1975; Walters *et al.*, 1975). This topographic distribution resembles that of complement receptors (Section 3.4).

On a scale below 200 nm, macrophage receptors for foreign substances are discontinuous in distribution. Surface replicas of macrophages pulsed at 4°C with hemocyanin for 30 min have a tendency to form small clusters (Unanue and Calderon, 1975). Comparisons of this observation and of fixed and living macrophages binding cationized ferritin suggest that cellular surface activity partially restricted by low temperature (Unanue and Caleron, 1975) or brief incubation (Skutelsky and Hardy, 1976) contributes to clustering and periodicity of markers. Fixed macrophages stain more evenly than macrophages briefly incubated at physiologic temperature with cationized ferritin before fixation.

4.6. ENDOCYTOSIS

The binding of small proteins exemplified by peroxidase is rather weak despite the evidence that a receptor is involved (McKeever *et al.*, 1976a; Steinman *et al.*, 1976; Stahl *et al.*, 1978). Thus, a considerable amount of peroxidase may be ingested in fluid phase by pinocytosis. For larger macromolecules and particles that are internalized by either pinocytic or phagocytic mechanisms, binding to the plasmalemma precedes the action of ingestion.

Microcinematography (Hirsch, 1965), scanning (Walters *et al.*, 1975; Polliack and Gordon, 1975), and transmission electron microscopy of macrophages phagocytizing bacteria and particles of similar size reveal the initial attachment of the target and the formation of cell pockets and membranous extensions which accompany and appear to effect the ingestion process. During initial stages of ingestion, macrophages apparently pinch the membranes of fixed erythrocyte ghosts (Emeis, 1976). This pinching and close interaction between membrane and target particles resemble Fc-receptor- but not complement-receptor-mediated phagocytosis (Munthe-Kaas *et al.*, 1976; Kaplan, 1977).

Macrophage activity leading to ingestion of large particles is generally accompanied and enhanced by biomechanical interactions. The term *biomechanical factor* is used in reference to form changes which the macrophage and its membranes may undergo and which promote either total enveloping of a target particle or gripping of surface deformities on the target substance. Biomechanical factors probably represent the cellular equivalent of the grasp of a hand or the fitting of amalgam in the drilled tooth pocket and, as an example of the latter situation, can be considered to underlie the enhanced bacterial phagocytosis that results from roughness of the bacterial surface (Wood, 1960).

The plasmalemmal region of macrophages that are either active in the process of endocytosis or attached to flat surfaces contains microfilaments and microtubules (Reaven and Axline, 1973; Skutelsky and Hardy, 1976). These filaments probably consist of contractile protein and represent the cell structure that effects the motility essential to the cellular work of endocytosis.

4.7. REGENERATION OF FOREIGN SUBSTANCE RECEPTORS FOLLOWING ENDOCYTOSIS

A number of cellular events occurring after endocytosis relate to the degradation or recycling of macrophage membrane components (Silverstein *et al.*, 1977). Some of these events such as the engorgement of the cells, enzyme release by the cells into the cytoplasm from heterophagic bodies, or temporary depletion of plasma membrane constituents may decrease macrophage ability to ingest foreign substances (Weissmann *et al.*, 1971). Following extensive phagocytosis of polystyrene latex particles, the phagocytic and pinocytic activity of macrophages decreases to less than 20% of control activity (Werb and Cohn, 1972). The amount of macrophage plasma membrane and of a surface enzyme marker also decreases. After the initial phagocytosis ceases, endocytic activity and membrane constituents return to control levels over a 12-hr period. This return to control levels can be blocked with inhibitors of protein and RNA synthesis. It would be of interest particularly in regard to this requirement for protein synthesis to compare the recovery following phagocytosis of digestible and indigestible particles.

Cationized ferritin (CF) is a foreign substance with a high positive charge density. It is bound by more macrophage surface molecules having exposed anionic groups than ferritin alone (Casley-Smith, 1969; Skutelsky and Hardy, 1976). CF is bound avidly by living as well as fixed macrophages. Therefore, it is possible to trace the binding, ingestion, and regeneration of CF binding sites by interposing fixation and reexposure to CF at various times after initial exposure of living macrophages to CF (Skutelsky and Hardy, 1976). Live macrophages interiorize and shed most of a pulsed surface load of CF in 30 min while incubating in ferritin-free tissue culture medium. By 60 min after the initial CF pulse, regeneration of receptors for CF can be visualized on the upper part of adherent cells after fixation and reexposure to CF (Figure 11). The new receptors are distributed either continuously or discontinuously with large gaps of up to 3 nm on the apical portion of adherent cells. The lateral regions are free of label. The attached CF particles on the regenerated receptors are closer to the membrane and their density is higher than the ambient CF receptors on normal control macrophages. Regeneration of receptors for CF is complete after 3-hr incubation in ferritin-free medium.

The patches of regenerated receptors are consistent with the concept that groups of the receptors are inserted into the plasmalemma by some process,

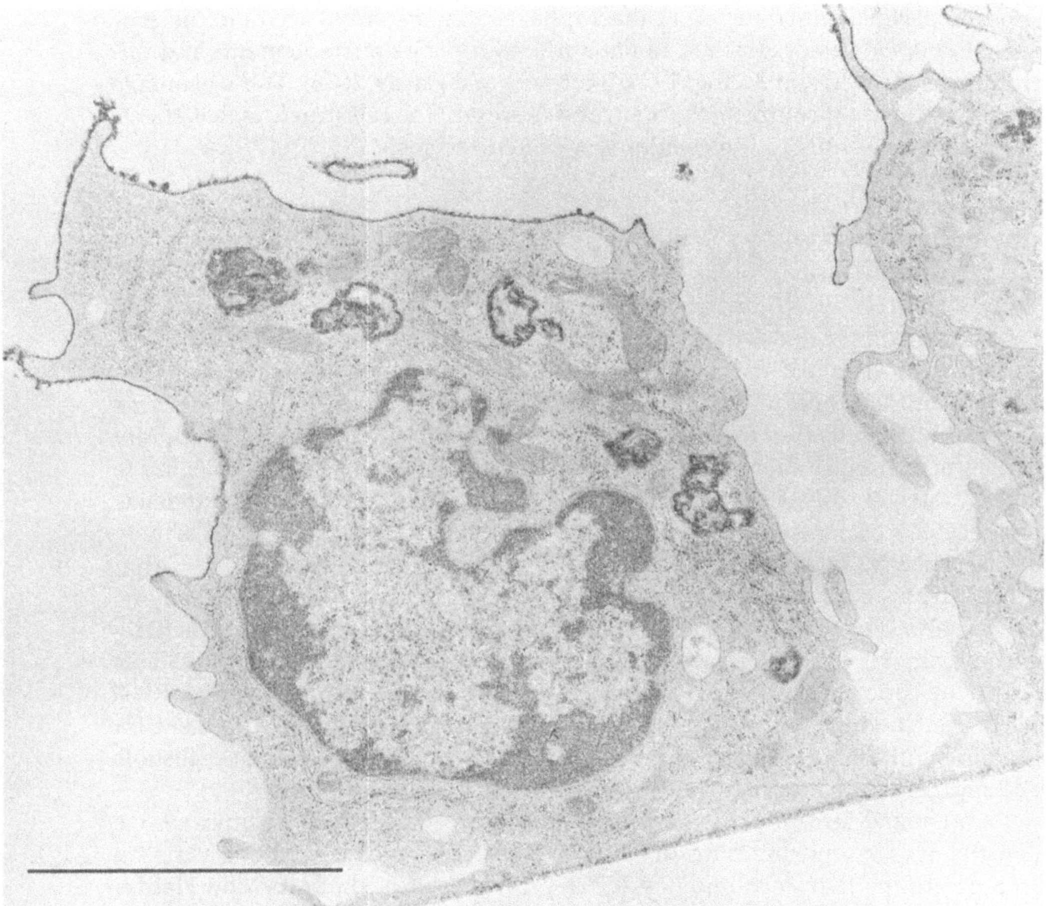

FIGURE 11. Rapid regeneration of surface receptors binding cationized ferritin occurs first at the apical plasma membrane of macrophages adhering to a flat surface. Almost no labelling of the lateral and basal membrane occurs. Mouse peritoneal macrophage incubated at 37°C with cationized ferritin, then for 60 min in ferritin-free culture medium, fixed with aldehydes, and again stained for cationized ferritin. Scale bar represents 2.2 μm. Reprinted from *Experimental Cell Research* **101**:342, with permission of the journal and Dr. E. Skutelsky.

perhaps fusion of vesicles. Comparison of CF ingestion and latex ingestion reveals that regeneration after CF ingestion is about four times faster (Skutelsky and Hardy, 1976; Werb and Cohn, 1972). The differences are analogous to differences in IgG receptor regeneration discussed earlier (Section 2.6). The faster regenerations may reflect receptor conservation in a marginal pool of receptors which is not depleted by a single surface load of CF or immune complexes. Studies testing CF or immune complex receptor regeneration in the presence of inhibitors of protein synthesis would be valuable.

4.8. SENSITIVITY OF THE RECEPTOR TO PROTEINS, IONS, AND DRUGS

Differential sensitivities of foreign substance receptors to trypsin provide indirect evidence for limited receptor specificity. Trypsin and neuraminidase sensitivities are discussed in Section 4.3. Migration inhibition factor renders adherent macrophages more resistant to displacement by shearing forces (Weiss and Glaves, 1975).

The separate examination of the attachment and ingestion phases of the phagocytosis of heterologous, glutaraldehyde-treated erythrocytes by macrophages allows observation of serum and ion effects on each phase (Rabinovitch, 1967). Whereas both attachment and ingestion are temperature dependent (Rabinovitch, 1967), only the ingestion phase requires the presence of newborn calf serum. The ingestion phase alone is affected by chelation and removal of divalent cations with EDTA, and this inhibition of ingestion by EDTA can be almost completely reversed by replacement of either calcium or magnesium cations. Dextran sulfate inhibits phagocytosis of bacteria and binding of cationized ferritin (Skutelsky and Hardy, 1976; Lilga, 1976).

Drugs, including aurothiomalate and sodium salicylate, have been reported to inhibit phagocytosis of *Candida albicans* by monocytes (Viken and Lamvik, 1976; Viken, 1976). Uptake of large foreign particles by peritoneal macrophages is prevented by inhibitors of glycolysis and mitochondrial oxidative phosphorylation (Casley-Smith, 1969) indicating requirement of glycolysis or oxidation-derived energy for the internalization process. The effect of these metabolic inhibitors on pinocytosis of small particles by peritoneal macrophages is less definite (Casley-Smith, 1969). Different effects on ingestion of small and large particles may reflect the greater cellular work required to ingest the large target moieties. Macrophages in different areas vary in energy dependence, those derived from pulmonary alveoli showing greater sensitivity to mitochondrial inhibitors than those of peritoneal origin (Oren *et al.*, 1963). The effect of 2-deoxy-D-glucose on internalization via foreign substance and Fc receptors of macrophages is discussed in Section 4.3.

A question remains, however, whether metabolic inhibitors affect the attachment phase of endocytosis. The distinct characteristics of this phase are partially obscured in assays which allow fixation to contribute to the macrophage binding of foreign particles (Casley-Smith, 1969). An assay system has been developed to eliminate fixation artifacts and to distinguish macrophage ability to establish adherence from its ability to maintain adherence. In this system, cyanide inhibits the former with little effect upon the latter (Table 5) suggesting an energy requirement for establishing attachment.

Macrophage phagocytosis of zymosan elevates their cyclic AMP response to PGE_1 and induces synthesis and release of PGE_1. It is suggested that phagocytosis-induced enhancement of PGE_1 sensitivity combined with a subsequent release of PGE_1 may regulate macrophage function under physiologic conditions (Gemsa *et al.*, 1978). Levamisole increases human monocyte phagocytosis of heat-killed staphylococci (Schmidt and Douglas, 1976).

4.9. SUMMARY AND CONCLUSIONS

The ability to endocytize a wide variety of foreign substances is a distinctive characteristic of the macrophage and one that is especially important to the function of the alveolar macrophage in pulmonary clearance. Assays of this macrophage function commonly involve incubating macrophage monolayers or suspensions with the target substance, and then counting bound or ingested targets. Attachment and ingestion phases of endocytosis can be distinguished. The attachment phase can be further differentiated into the preliminary establishment and the subsequent maintenance of adherence. The term *receptor for foreign substances* is used to avoid the ambiguity of the term *nonspecific receptor.* However, the degree to which specific receptors for foreign substances participate in the macrophage uptake of a wide variety of extrinsic materials is largely undetermined. Subtle metabolic and antigenic differences are apparent between immunoglobulin receptors and receptors for foreign substances. Receptors for certain groups of foreign substances differ from receptors for other substances in susceptibility to digestion by hydrolytic enzymes. This difference, and the rather

TABLE 5. EFFECT OF 10^{-2} M POTASSIUM CYANIDE ON ALVEOLAR MACROPHAGE ADHERENCE TO POLYSTYRENE[a]

Cyanide treatment	Mean number cells, final/initial	Mean percentage adherent cells	±	Standard error	p value[b]
Before cells contacted surface	0.5/86.4	0.6	±	0.2	0.005
None	92.7/97.6	95.0	±	15.9	
After cells contacted surface	81.5/100.9	80.8	±	12.3	0.250
None	102.8/102.9	99.9	±	18.3	

[a] Rabbit alveolar macrophages (5 × 10^3 cells) were incubated in glucose phosphate buffer over counting grids on polystyrene petri dishes and initial number of cells in a known area were counted. The dishes were covered and incubated to allow cells to attach. The dishes were then rinsed, the counting area relocated, and final number of adherent cells counted. Cyanide was added at appropriate times to experimental groups. (Reprinted from Vol. 18, p. 225, by permission of the *Journal of the Reticuloendothelial Society*.)

[b] p values are derived from statistical comparison of experimental and control groups by the method for proportions.

nonselective inhibition by serum proteins of macrophage uptake of some proteins such as hemocyanin, suggest a low level of receptor specificity. Macrophage receptors for cationized ferritin appear to redistribute on the membrane surface like discrete molecules. Serum proteins and membrane biomechanical factors may contribute to the activity observed in a wide variety of assays for binding and ingestion by macrophages.

Macrophages appear to bind ionized foreign substances and particularly cationic material more avidly and more closely to their plasmalemma than substances with less charge. Macrophages generally bind smaller substances closer to their plasmalemma, but there is probably a minimum target size below which the plasma membrane is incapable of perceptibly binding the component for ingestion. Proteins are bound closer to the plasmalemma via the IgG receptor than as free protein. The chemical basis for foreign substance adherence has only recently been approached with techniques capable of elucidating the chemical nature of the binding component of the plasmalemma of high-affinity target substances (Section 11.3).

Foreign substances tend to distribute randomly on the macrophage surface, but large particles avoid the peripheral membrane of flat macrophages. Bio-

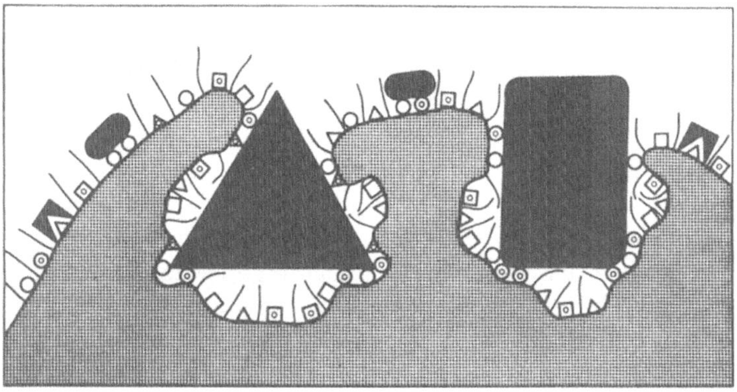

FIGURE 12. Model for macrophage interaction with foreign substances. The meandering line extending from lower left to upper right represents the macrophage membrane. Attached to the external surface of this membrane are small empty circles, triangles, and squares with and without central circles representing six different molecules which are intrinsic constitutents of the macrophage membrane. The cytoplasm below is stippled gray. Solid black shapes above represent extrinsic serum or foreign substances only one of which binds with high specific affinity to its receptor (open triangle) on the macrophage surface. The two circles and triangles with inner circle bind two different foreign substances (three solid rectangles with rounded edges and solid triangle) with overlapping specificity. Overlapping specificity is represented by circles binding both substances. They require more than a single surface molecule attachment to the foreign substance for any noticable attachment to occur. Lines protruding perpendicularly from the membrane represent extended surface molecules which do not bind extrinsic substances but electrostatically or sterically hinder the binding interactions. They keep portions of the macrophage membrane at a greater distance from large foreign particles (two large solid shapes). In contrast, smaller foreign substances intermingle between the extended surface molecules. The mechanical contribution is apparent where the flexible macrophage membrane has formed pits which hold the larger foreign particles.

mechanical interactions, close membrane–particle juxtaposition, gripping of surface deformities and creation of cups, pits, and veils around foreign substances, are specialized features mediating macrophage endocytosis of foreign substance. Microfilaments and microtubules collect beneath the plasmalemma.

Pinocytic and phagocytic activity and plasma membrane constituents decrease following extensive phagocytosis of latex spheres. The recovery of these features to prephagocytic control levels takes 12 hr and depends upon protein synthesis. Following endocytosis of a surface pulse of cationized ferritin, regeneration of its receptor takes 3 hr and occurs in patches at the apex of the adherent macrophage.

Preliminary to the biochemical isolation of receptors, indirect evidence provides clues to their identity and nature. Receptors for certain foreign substances are sensitive to proteolytic enzymes suggesting they are composed, at least in part, of peptide or protein. Certain receptors seem to be shielded by sialic acid or can be inhibited by polyanions. Metabolic energy and divalent cations are essential for phagocytosis to occur. Under certain conditions, metabolic energy is also necessary for the initial phase of macrophage attachment to a surface. One model which is consistent with major observations on macrophage interactions with foreign substances and provides a framework for further study is illustrated in Figure 12.

5. MACROPHAGE SURFACE RECEPTORS FOR LYMPHOID CELLS AND ANTIGEN

5.1. BIOLOGIC ACTIVITY

Thymus-derived lymphocytes (T cells) and bursal or bone marrow lymphocytes (B cells, plasma cell precursors) interact to generate antibody responses to complex antigens. Optimal antibody responses require the additional participation of adherent cells which resemble macrophages (Pierce *et al.*, 1974; Unanue, 1972; Mosier, 1967).

The section concentrates on the problems of visualizing these interactions and on the participation therein of receptors on the macrophage surface. Macrophages interact with antigen, T, and B cells in several ways to regulate the immune response to antigens. Interactions which probably involve macrophage-lymphocyte contacts include the presentation of antigen to T lymphocytes and B lymphocytes, modulation of the amount of antigen available to lymphocytes in different states of immunity to this antigen, and facilitation of T- and B-cell contacts (Pierce *et al.*, 1974; Werdelin *et al.*, 1974; Nielsen *et al.*, 1974; Bartfeld and Kelly, 1968; McIntyre *et al.*, 1973; Diener *et al.*, 1976). Nurturing both lymphocytes and plasma cells *in vitro* and macrophage secretion of lymphocyte-stimulating factors might also involve initial cell–cell interactions dependent upon macrophage receptors (Pierce *et al.*, 1974; Beller *et al.*, 1978).

Macrophages or a macrophage-like cellular subpopulation present antigen to lymphocytes. Both adherent and nonadherent splenic cell populations are required for the induction of antibody formation *in vitro* (Mosier, 1967). The majority of the adherent cells are macrophages; contributions by dendritic cells (Nossal *et al.*, 1968a,b) are also possible. Contributions by a small number of adherent B cells (Nathan *et al.*, 1977) are possible but not likely since B lymphocytes are generally radiosensitive and the adherent presenting cells are radioresistant (Roseman, 1969). The exact nature of the adherent presenting cells is under study (Schwartz *et al.*, 1978).

Both the primary immune response (the response of cells on first exposure to an antigen) (Mosier, 1969) and the secondary immune response (the response upon reexposure to an antigen observed in an animal primed weeks earlier by immunizing to the same antigen) (Unanue and Askonas, 1968) require the macrophage-like presenting cells (herafter called macrophages) to be in proximity to antigen and lymphocytes. Induction of the primary immune response *in vitro* measured by antibody production requires a distinct subpopulation of macrophages in the mouse (Mosier and Coppleson, 1968). Macrophage presentation of antigen is also important to the lymphocyte proliferative response (assay defined in Section 5.2). Peritoneal exudate cell populations depleted of macrophages respond at a relatively low level to antigen compared with populations which have been reconsituted with these adherent cells (Rosenstreich and Wilton, 1975). To promote a secondary immune response the lymphocyte population but not the macrophage population needs prior immunization with antigen (Rosenthal *et al.*, 1976).

Different lines of evidence indicate that to stimulate the immune response maximally the antigen must be associated with the macrophage surface and macrophage–lymphocyte contact must occur. The macrophage has surface-binding capacities which mediate this process, including receptors for foreign substances (Section 4), IgG (Section 2), and complement (Section 3). The first receptor is of anticipated importance in primary responses, the others in secondary responses. If macrophages are cultured for several hours following brief exposure to radioiodinated hemocyanin, a few molecules escape endocytosis and are retained on the plasma membrane of these macrophages (Unanue and Cerottini, 1970). The latter membrane-bound hemocyanin antigen is highly immunogenic, and this immunogenicity can be inhibited by treatment of these macrophages with trypsin or anti-hemocyanin antibody. A similar effect has been shown using heterologous gamma globulin (HGG) as antigen and anti-HGG to inhibit (Harris, 1965). In a different system, macrophages binding sheep red blood cells (SRBC) stimulate the ability of lymphoid cells to make specific anti-SRBC antibody unless they are treated with ammonium chloride to lyse the SRBC (Pierce, 1973; Leserman *et al.*, 1972). In contrast to surface-bound antigen, supernatants from cultures of antigen-exposed macrophages containing antigen and fragments of antigen in solution stimulate antibody responses relatively poorly (Pierce and Kapp, 1976). Macrophages and peritoneal exudate lymphocytes responding to *Listeria* infection stimulate thymidine uptake when cultured together for a day. However, if the two cell types are separated by a cell-impermeable membrane, thymidine uptake is diminished to near background levels (Beller *et al.*, 1978).

Evidence indicates that the macrophage retains enough information about antigens it encounters to boost the response of other immune cells to these antigens. The details of how the macrophage does this, in spite of the fact that it destroys much of the antigen it encounters (Steinman and Cohn, 1972b; Calderon and Unanue, 1974), are not clear. The immunogenic antigen seems to be bound to the macrophage surface, and it is possible that this binding brings together enough antigenic determinants to stimulate lymphocyte recognition (Feldmann, 1974; Feldmann and Nossal, 1972).

The simplest concept of macrophage activation of lymphocyte responses is that the macrophage binds antigen by a foreign substance receptor and thereby presents enough antigenic determinants to lymphocytes to stimulate a primary response. The participation of the foreign substance receptor is consistent with the observation that other proteins in excess will inhibit the primary *in vitro* immune response (Martineau and Johnson, 1978) (Section 5.3). The macrophage may also bind antigen by its foreign substance, IgG, or complement receptors to stimulate lymphocytes in the secondary response. Immune complex will suppress the immune response by its effect on the macrophages (Morgan and Tempelis, 1978). Indeed hapten-specific inhibition experiments suggest that macrophages from immune animals present antigen to lymphocytes by means of antibody attached to their IgG receptors. DNP on another protein carrier can inhibit macrophage presentation of DNP bound to albumin to lymphocytes from guinea pigs immune to DNP–albumin by competition for DNP–hapten-specific cytophilic antibody on the macrophage surface (Cohen *et al.*, 1973). However subsequent macrophage–lymphocyte interactions are more complicated than a simple proximity of macrophage with surface antigen to lymphocyte.

Primary and secondary responses require macrophage–lymphocyte histocompatibility in the I region (Thomas and Shevach, 1976; Niederhuber, 1978) and therefore involve Ia molecule (I region gene product) surface interactions. Subpopulations of macrophages appear to bind lymphocytes in immune interactions and bear surface Ia (Niederhuber, 1978; Schwartz *et al.*, 1976). Antigen-pulsed macrophage binding of antigen-primed T lymphocytes cannot be inhibited by the antigen in 5000-fold excess, soluble or bound to beads (Ben-Sasson *et al.*, 1977). These phenomena argue for a separate receptor in addition to IgG, complement, or foreign substance receptors important in macrophage–lymphocyte interactions with antigen. Indeed, removing only the Ia-bearing macrophage subpopulation will delete normal macrophage stimulation of the primary *in vitro* immune response. The Ia gene product associated with the I-J subregion of the mouse major histocompatibility complex is primarily responsible for that stimulation (Niederhuber, 1978). Thus, the causal relationship between Ia surface molecules and macrophage stimulation of the immune response is on a firmer basis than the relationship between the classic macrophage IgG, complement, and foreign substance receptors. The latter are present on all macrophages; this obviates testing for their activity by selective deletion experiments. Moreover, many obvious blocking experiments which could be performed on the classic receptors have not been done.

In addition to the experiments implicating macrophage surface factors in the immune response to antigen, there is evidence for a stable pool of intracellular antigen or immunogenic antigen fragments in the macrophage which persists for at least 3 days in culture after the macrophage was pulsed with antigen. If macrophages are trypsinized immediately after the antigen pulse, their ability to stimulate lymphocyte proliferation is decreased. However, if macrophages are trypsinized 3 hr after the antigen pulse their ability to stimulate lymphocytes is similar to untrypsinized 3-hr control macrophages (Ellner and Rosenthal, 1975). This has led to the proposal of two pathways of antigen presentation by macrophages to lymphocytes (Unanue, 1978).

5.2. TECHNIQUES

From the previous discussion it is clear that macrophages can modulate immunoglobulin production by their functional interactions with lymphocytes. Major techniques contributing to this insight include determining *in vitro* plaque formation and *in vivo* antibody production. The sections of this chapter describing the morphology of immune cell interactions with macrophages indicate the ubiquitous nature of cellular contacts with macrophages. Evidence previously noted indicates the importance of macrophage–lymphocyte contacts in stimulating at least one immune response: the lymphoid cell response to antigen followed by specific antibody production.

The major challenge, then, has been to develop methods to determine which macrophage–lymphocyte contacts regulate this immune response. Presumably this would involve a differentiation of stimulatory from inhibitory contacts and from nonessential contacts. The following paragraphs outline experiments which undertake to correlate the functional and morphologic activities of macrophages interacting with other immune cells.

Standard light microscopic (Yokomuro and Nozima, 1972; Smith and Goldman, 1970), microcinematographic (Berman, 1966), transmission electron microscopic (TEM) (Farr and Debruyn, 1975; Lipsky and Rosenthal, 1973; Schoenberg *et al.*, 1964), and scanning electron microscopic (SEM) (Orenstein and Shelton, 1976a,b; Fujita *et al.*, 1972) techniques have been employed to study interactions of macrophages and lymphoid cells of nonimmunized animals. Fixation of the specimen is followed by dehydration and either embedding for transmission electron microscopy or critical point drying for scanning electron microscopy. The difficulty of positive identification of macrophages and lymphocytes by SEM alone necessitated ingenious methods of viewing the same cell by light and scanning electron microscopy (Orenstein and Shelton, 1976; Alexander and Wetzel, 1975). The careful comparison of cellular contacts reveals that they are visible much less frequently *in situ* than *in vitro* (Orenstein and Shelton, 1976a,b). The authors suggest that more direct contact between the cells *in situ* may obscure lymphocyte–lymphocyte or lymphocyte–macrophage interactions because the cells lie atop one another and touch with relatively short processes (Figure 13). Whatever the explanation for differences noticed *in situ* and *in vitro*, it is clear that experimental alteration of the immediate cellular environment may introduce artifactual differences in macrophage–lymphocyte interactions in the absence of antigen administration. Such alteration must be carefully controlled in the design of *in vitro* experiments involving these cellular interactions.

Microcinematographic and interference or phase constrast techniques have the obvious advantage of viewing the dynamic interaction between these two cell types. Employed as screening procedures preliminary to ultrastructural techniques their value could be considerable.

Efforts to correlate functional activity in the immune response with morphology of the macrophage and lymphocyte cellular contacts have usually sought technical variations providing a common ground between functional and structural assays. Four variations have been particularly useful.

The first variation uses a fluorescent or radiolabeled antigen which is traced during its processing by macrophages and lymphocytes (Leduc *et al.*, 1955; Harris, 1965; Sulitzeanu *et al.*, 1971; Diener *et al.*, 1976; Unanue *et al.*, 1969). *Salmonella* flagellin or another antigen is radiolabeled with ^{125}I or ^{131}I (McConahey and Dixon, 1966). It is then introduced into a system appropriate for following interactions of immune cells. Cells or tissues are fixed and processed for autoradiography. Tritium-labeled antigen would afford somewhat more precise resolution than radioiodine for autoradiography (Diener and Paetkau, 1972).

The second variation compares interactions *in vitro* between cells from antigen-primed and -unprimed animals. The antigen used most successfully is tuberculin-purified protein derivative (Lipsky and Rosenthal, 1975b; Bartfeld and Kelly, 1968; Werdelin *et al.*, 1974), but other antigens have also been used (Hersh and Harris, 1968; Sharp and Burwell, 1960). The frequency and morphology of interactions between antigen-pulsed or -nonpulsed macrophages and antigen-primed lymph node lymphocytes is compared to interactions with unprimed lymphocytes. Since macrophage–lymphocyte binding occurs with and without the antigen, careful experimental design and interpretation are required to distinguish antigen-dependent and -independent binding.

The third variation utilizies the radiolabeled thymidine incorporation technique in combination with autoradiography. This determines the relationship of macrophages and other cells to those cells actively incorporating thymidine as they undergo blastogenic transformation (Mosier, 1969; Lipsky and Rosenthal, 1975b; Harris, 1965; Hanifin and Cline, 1970; Sulitzeanu *et al.*, 1971). Lymphocytes incorporate radiolabeled thymidine as they proliferate in response to an antigenic stimulus (Harris, 1965; Schwartz *et al.*, 1978).

The fourth variation combines morphologic examination of cells or clusters of cells with immunologic identification of specific antibody-producing cells or clusters of cells (Itoh *et al.*, 1975; McIntyre *et al.*, 1973; Saunders and Hammond, 1970; McIntyre and Pierce, 1973; Miller and Avrameas, 1971; Mosier, 1969; Straus, 1974; Avrameas and Leduc, 1970). Immunologic identification of specific antibody-producing cells *in vivo* is frequently accomplished by using a small marker protein such as peroxidase as the antigen (Avrameas and Leduc, 1970;

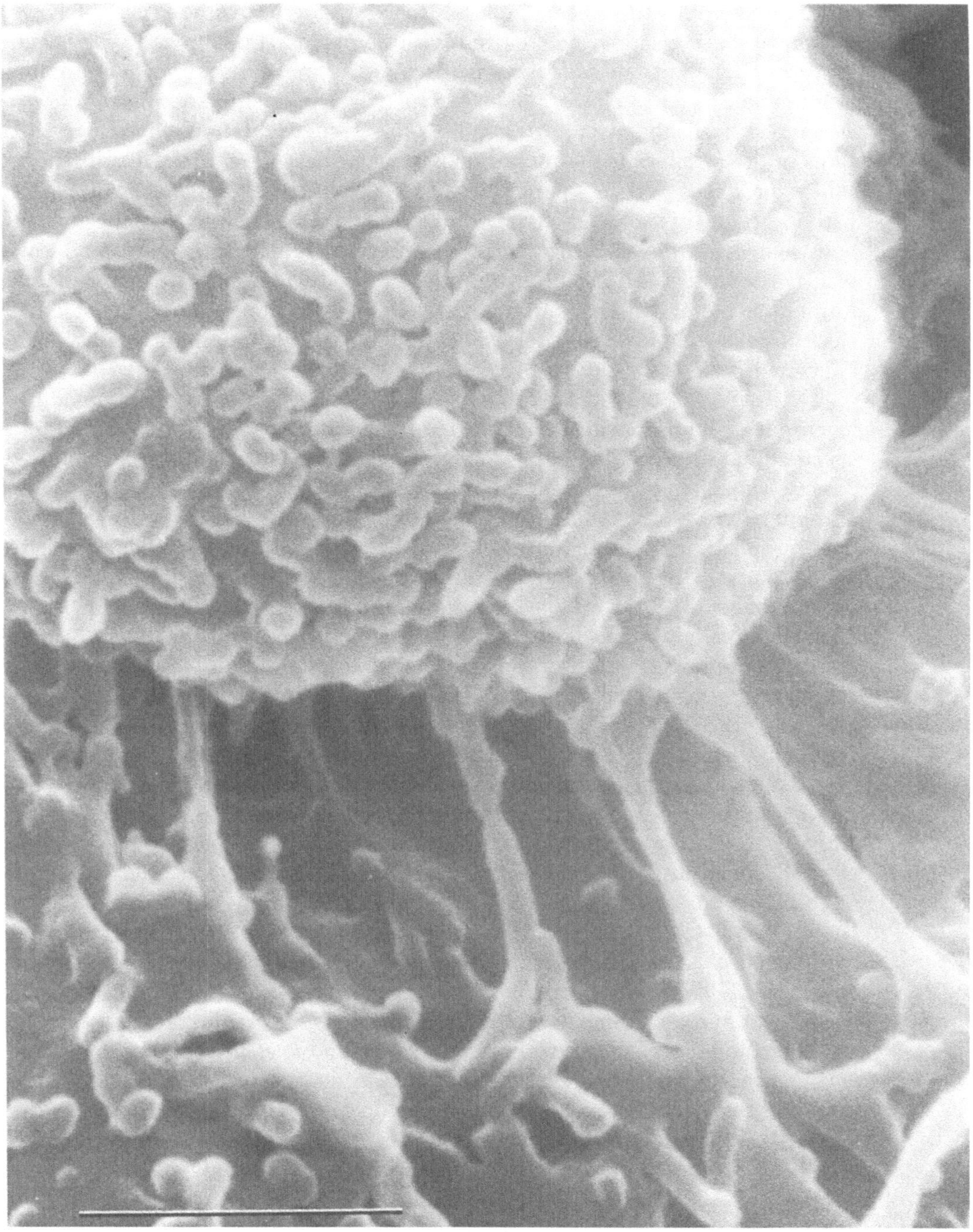

FIGURE 13. Scanning electron micrograph of a portion of macrophage membrane (below) interacting with an attached lymphocyte (above) in a nonimmunized mouse fixed *in situ* with buffered glutaraldehyde. Processes between the two cells make intimate contacts. From a milky spot of the peritoneum with light microscopic monitoring of cell types. Scale bar represents 1.3 μm. Reprinted from *Experimental and Molecular Pathology* **24**:422, with permission of the journal and Drs. J. M. Orenstein and E. Shelton.

Straus, 1974). At various time intervals following immunization regional lymph nodes can be excised, fixed, reacted with peroxidase on a frozen section, and processed for light and electron microscopy. In another system, immunologic examination of those clusters of cells which are producing specific antibody *in vitro* is accomplished by culturing mouse spleen cells with heterologous erythrocytes in the *in vitro* plaque formation assay (Mosier, 1969; Mishell and Dutton, 1967; Pierce *et al.*, 1974). After culture, spleen cells or cellular clusters are mixed with indicator erythrocytes in agar and spread onto a flat surface. Specific antibody will lyse the erythrocytes in the presence of complement and form a plaque. Specific anti-erythrocyte-antibody-producing cell clusters (plaque-forming clusters) can be examined directly by light microscopy or isolated for electron microscopic processing using a microinjection apparatus (McIntyre and Pierce, 1973).

The four described techniques correlating structure and function of macrophage–lymphocyte interactions provide the most useful information at specific time intervals following mixing of cells and antigen. The technique of directly visualizing macrophage–lymphocyte binding shows interaction within an hour after culture initiations (Lipsky and Rosenthal, 1975b). However, the binding this time occurs with or without the antigen to which the lymphocytes were primed. Only after 8 hr of culture does the initial, antigen-independent binding subside sufficiently so that antigen-dependent binding can be distinguished (Lipsky and Rosenthal, 1975b). At 20 hr of culture, clusters of more than six lymphocytes around a macrophage are seen about 90% of the time in association with specific antigen (Werdelin *et al.*, 1974). The technique of following radiolabeled antigen *in vivo* or *in vitro* is most useful during the earliest phases of the immune response when whole antigen may still be distinguishable from radiolabeled fragments and metabolites some of which may no longer be relevant to the immune response (Calderon and Unanue, 1974). Antigens which are very resistant to degradation can be followed longer. Earliest incorporation of radiolabeled thymidine by primed lymphocytes is detectable after 24 to 40 hr of culture with antigen-sensitized macrophages (Unanue and Calderon, 1975; Hanifin and Cline, 1970). Primary antibody production in macrophage–lymphocyte clusters can be seen at 24 hr (McIntyre *et al.*, 1973) to 96 hr (McIntyre and Pierce, 1973) *in vitro* by plaque-forming assay and 144 hr *in vivo* using a peroxidase marker antigen (Miller and Avrameas, 1971). Fluorescent markers can detect a primary response *in vivo* in 96 hr (Leduc *et al.*, 1955). The problem encountered in forcing these assay systems to detect very early interactions is that the specifically stimulated interactions do not occur at a sufficiently higher level than background until later. Thus, definitive plaque-forming or lymphocyte proliferative responses after antigen contact commonly require 96 hr to develop (Harris, 1965; McIntyre and Pierce, 1973). The temporal restrictions of these systems and the very small number of initial antigen-specific interactions impose difficulties in relating structure and function at early time intervals during the immune response. Antigen tracing studies are presently providing the best clues (Binz and Wigzell, 1975; Diener *et al.*, 1976). Such tracing might be enhanced in screening ability and resolution by using fluorescein or radiolabeled viral or heme-protein antigens with ultrastructurally characteristic features for dual light and electron microscopic analysis.

5.3. SPECIFICITY

At least three specific interactions are possible during macrophage associations with immune cells: specificity for the type of immune cell interacting, requirement for histocompatibility of the interacting cells, and specificity for the antigen involved in the cell–cell interaction. Specificity for antigen is included in Section 5.4.

Without antigen or prior immunization, macrophages have a certain specificity for interacting with the T cell. Heterologous thymocytes do not adhere well to guinea pig macrophages (Siegel, 1970). In the absence of antigen, syngeneic macrophages bind about twice as many thymocytes as lymph node lymphocytes and about thirty times as many thymocytes as erythrocytes. Other glass-adherent cells do not bind thymocytes as well as macrophages. Fibroblasts and polymorphonuclear leukocytes bind about one fifth as many thymocytes as peritoneal or splenic macrophages (Lipsky and Rosenthal, 1973).

The requirement for histocompatibility of macrophages and lymphocytes is different during the primary from that during the secondary immune response to antigen. Macrophages from a different mouse strain (allogeneic) plus antigen will stimulate unprimed lymphoid cells *in vitro* to produce antibody as measured by the plaque formation assay (Pierce and Kapp, 1978; Mosier, 1976). In contrast, certain allogeneic macrophages plus antigen have little or no stimulatory effect in the secondary immune response in guinea pigs (Thomas and Shevach, 1976; Rosenthal and Shevach, 1973) and in mice (Yano *et al.*, 1977). This genetic restriction on macrophage presentation of antigen is in the I region of the major histocompatibility complex.

The primary *in vitro* antibody response in the mouse is more sensitive to killing of Ia-positive macrophages (Niederhuber, 1978) than to macrophage–lymphocyte histocompatibility (Pierce and Kapp, 1978). Apparently the possession of Ia molecules by macrophages is more important to antigen presentation than the histocompatibility of the cells. Perhaps this system depends upon an Ia function other than cellular recognition.

5.4. MORPHOLOGY

Structural study and interpretation of the macrophage receptors which interact with lymphoid cells is complicated by the number of interactions which take place and the difficulty in distinguishing one type of interaction from others. Moreover, the initial specific binding sites of these cells with antigen and with each other may be quite small in number (Binz and Wigzell, 1975; Diener *et al.*, 1976). Nonetheless, as early as 8- to 20-hr differences in binding begin to appear between normal background activity and immunospecific reactions to complex antigens such as tuberculin-purified protein derivative (Lipsky and Rosenthal, 1975b; Werdelin *et al.*, 1974). Later immune reactions can be traced by the formation of specific antibody or by blastogenic transformation of lymphocytes. The method assessing specific antibody is perhaps preferable since it measures the actual end product of the immune response.

The optimal cell concentrations for inducing human primed peripheral blood lymphocyte proliferation in response to antigen are 3×10^6 to 4×10^6 lymphocytes and 0.2×10^6 to 0.6×10^6 macrophages per 3 ml Leighton tube culture (Hersh and Harris, 1968). This ratio of 5–20 lymphocytes per macrophage is comparable to the number of lymphocytes clustering around a macrophage during the first 24 hr of an immune response. Limiting dilution experiments predict a ratio of two lymphocytes per macrophage (Mosier and Coppleson, 1968). It should be recalled that most macrophages bind none or one lymphocyte in a secondary immune cell interaction with or without antigen (Werdelin *et al.*, 1974) and that in a primary immune response *in vitro* only 10% of cell clusters make specific antibody (Mosier, 1969). Interpretation of all strictly morphologic data is limited by the fact that subsets of both macrophage and lymphocyte populations are the only specifically interacting cells.

The following discussion will consider the temporal relationships, spatial relationships, membrane proximity, and the relation of macrophage–lymphocyte clusters to the antigen. Whenever possible, antigen-specific interactions will be compared with normal background interactions. The discussion proceeds from the earliest interactions studied through interactions which occur at progressively later time intervals following exposure or reexposure of lymphocytes and macrophages to antigen, macrophages bind and ingest most of the antigen (Harris, 1965; Sulitzeanu *et al.*, 1971; Nossal *et al.*, 1968). For example, radioiodinated *Salmonella* flagellar antigen injected into rat foot pads enters the draining popliteal lymph node. Some of the antigen passes briefly through the subcapsular sinus overlying the cortex and a small quantity of antigen can be detected adhering to sinus-lining cells for as long as one day. A moderate amount of antigen is ingested by sinus macrophages but much antigen escapes these cells, and during the first 4 hr of a primary response or 30 min of a secondary response, moves between lymphocytes separating the sinus from cortical follicles below. Antigen ultimately settles upon the surface of reticular cells within follicles. Although reticular cells are often considered as macrophages, they seem in fact to be much less phagocytic (Steinman and Cohn, 1975; Nossal *et al.*, 1968b). The bulk of the antigen in lymph node follicles remains for at least 3 weeks on the surface of dendritic reticular cells where some of it is also in contact with interdigitating lymphocyte processes. A small amount of antigen or a metabolic derivative appears to localize within lymphocytic nuclei. Other portions of the injected antigen enter the lymph node medulla during the first 4 hr after footpad injection. Most of this antigen is ingested by macrophages where radioactivity can be detected at least 3 weeks later in both cortex and medulla. In these regions macrophages, antigen, and lymphocytes lie in proximity, but no specific interaction was discerned (Nossal *et al.*, 1968a,b). These and other (Kaplan *et al.*, 1950; McDevitt *et al.*, 1966) elegant studies have provided a morphologic correlate for observing antigen processing *in vivo*. They are limited in demonstrating mainly sequential associations between antigen, macrophages, and lymphocytes. The situation is further complicated by the fact that in nonimmunized animals lymphocytes interact with other immune cells including macrophages (Figure 13) rather often. Combined quantitative approaches of defining interacting cellular populations with morphologic and biochemical correlation offer the best opportunity of understanding the role of macrophages in the immune response at present.

The long-recognized reticular cell (Stuart, 1975) which is difficult to distinguish from a macrophage or fibroblast by morphology alone is undergoing renewed scrutiny with modern techniques (Carr, 1975; Hoefsmit, 1975). Perhaps this new interest will help determine the relative importance of macrophage and reticular cells in the immune response.

In vitro morphologic studies offer advantages over *in vivo* experiments for distinguishing specific immune cell interactions to antigen, because of their capacity to separate the reactants of immune cell interaction. During the first 4 hr after exposure of primed or unprimed cells to antigen, conventional *in vitro* assays do not distinguish immunologic response to this antigen from background cellular activity. Antigen processing and cellular binding can be followed in this early phase of the immune response. Antigen processing studies during the earliest phases *in vitro* have concentrated on demonstrating the structural relationship of a single cell type with antigen. Thus, surface replicas of macrophages pulsed at 4°C with keyhole limpet hemocyanin for 30 min reveal hemocyanin distributed throughout the plasmalemma with a tendency to form small clusters (Unanue and Calderon, 1975). Although technical variations restrict precise morphologic comparisons, the clustering appears comparable to that of another foreign material, cationized ferritin (Section 4.3), and of immune complexes (Section 2.5). Subsequent culture of the hemocyanin-pulsed and washed macrophages for 3 hr produces a different surface distribution in which the majority of the hemocyanin molecules have been cleared from the plasmalemma, but isolated molecules remain (Unanue and Calderon, 1975). At 4 hr after a hemocyanin pulse, the surface hemocyanin can be demonstrated by the specifically sensitive autoradiographic method (Unanue *et al.*, 1969). About 3% of the originally bound radioiodinated hemocyanin can be detected on the macrophage by trypsinization 48 hr after an initial pulse, and smaller amounts persist at 72 hr. Whether the brief clustering that occurs before 3 hr with this antigen is important in gathering enough antigenic determinants to stimulate antigen recognition by lymphocytes (Feldmann, 1974; Feldmann and Nossal, 1972) encountering these macrophages is not clear. Possibly more protracted phenomena related to persistence of some antigen on the macrophage surface are also important to antigen recognition and resulting antibody response, since direct cell-to-cell contact for more than 2 days is required for proliferation of cells making specific antibody in the primary immune cells more than 10 hr before lymphocyte proliferation occurs in the secondary response (Harris, 1965).

Cells from mouse lymph nodes (Diener and Paetkau, 1972) and spleens (Diener *et al.*, 1976) bind polymerized *Salmonella* flagellin *in vitro*. Unlike the hemocyanin-binding cells these cells resemble lymphocytes in their sedimentation velocity profile and light microscopic appearance. Only 0.001–0.005% of the total cell population binds tritiated polymerized flagellin. The ability of these cells to cap small immunogenic but not large tolerogenic doses of flagellar antigen is interpreted to mean that capping is essential to immunospecific lymphoid proliferation. Moreover, other cells termed "A cells" and showing properties of macrophages are thought to interfere with the cells binding flagellar antigen. One caution must be raised in the interpretation of these otherwise impressive experiments. The molecular weight of the flagellin polymer is probably about 8 $\times 10^6$ and thus (Diener and Paetkau, 1972) in the vicinity of hemocyanin (See Section 4 and Table 4). Therefore polymerized flagellin is an atypical foreign substance in not being bound by macrophages (Diener *et al.*, 1976, 1973). However, before eliminating macrophages in this immune response the possibility should be excluded that rather than lymphocytes a small number of macrophages or reticular cells contaminating spleen and lymph node cell preparations are actually binding the polymerized flagellin. It would be useful to evaluate the endocytic and histochemical properties of these antigen-binding cells to identify them more precisely.

Between 5 and 36 hr after exposure of macrophages to antigen and to primed lymphocytes, macrophage–lymphocyte interactions become conspicuous. Studies examine the contact and binding between macrophages and lymph node lymphocytes from animals previously immunized to purified protein derivative of tuberculin (PPD). The presence of specific antigen enhanced the degree of macrophage–lymphocyte interaction. Macrophages exposed to PPD and washed were found to bind four times as many primed lymphocytes as unexposed macrophages after incubation for 20 hr at 37°C. The antigen-mediated macrophage–lymphocyte-binding interaction required syngeneic cells in contrast to the simple binding between unexposed macrophages and lymphocytes. This antigen effect was obscured at earlier time intervals because of binding of lymphocytes by macrophages not exposed to PPD but could be discerned at 8 hr and barely at 4 hr (Lipsky and Rosenthal, 1975b). At 20 hr of incubation, only cultures with macrophages exposed to PPD formed clusters with more than 10 lymphocytes attached to a macrophage (Werdelin *et al.*, 1974). The latter observation seems particularly important since it virtually eliminates the problem of contamination by non-antigen-specific clusters.

Clusters contained a macrophage adherent to the coverslip support, binding one central lymphocyte which in turn bound numerous peripheral lymphocytes. The macrophage plasmalemma was extensively folded. The membrane and cytoplasmic features of these macrophages suggest some degree of activation. They were probably influenced by the oil used to induce the peritoneal exudate (Werdelin *et al.*, 1974; Carr and Carr, 1970).

A relatively large spherical lymphocyte with one flat side was attached directly to this macrophage. This centrally located lymphocyte was completely covered with short microvilli. It is difficult to interpret the significance of this microvillous surface due to the variability of T- and B-lymphocyte surface properties (Polliack *et al.*, 1973; Perkins *et al.*, 1972; Chen *et al.*, 1972; Alexander and

Wetzel, 1975). The question of whether these central lymphocytes are T or B cells is important to understanding these immune cell clusters and could easily be resolved. Prominent euchromatin, enlarged nucleoli, and deep indentations were features of its nucleus. Its cytoplasm was notable for a prominent Golgi apparatus surrounded by numerous small vesicles. The nuclear and cytoplasmic features indicate an early stage of blast transformation (Nielsen *et al.*, 1974). The observation of blast transformation in this lymphocyte attached directly to the macrophage bears critically on the importance of these clusters. Blast transformation is considered to be closely associated with specific immune recognition. Only a portion of the available lymphocytes undergo blast transformation and this portion includes those directly attached to macrophages. It is therefore unlikely that cluster formation is the result of nonspecific macrophage activation by soluble migration inhibiting factors (MIF) alone. Extensive areas of plasma membrane contact occurred between this central lymphocyte and the macrophage beneath it. The membranes were 16 nm apart when running in parallel (Figure 14).

Peripheral lymphocytes in the cellular clusters had a generally smooth surface and attached to the central lymphocyte via their own structures resembling uropods. The uropod is a structure frequently encountered in the mixed leukocyte reaction (McFarland *et al.*, 1966; McFarland and Heilman, 1965) and the response to phytohemagglutinin (Biberfeld, 1971). In the immune response to PPD, the structures resembling uropods were 0.7–2.8 nm long and conical, except for a flat tip which contacted the central lymphocyte. The intermembranous gap was 16 nm. The subplasmalemma of both cells at their contact region, the uropod-like structure, and other regions of peripheral cytoplasm contained microfilaments. The foot appendages of peripheral lymphocytes (Nielsen *et al.*, 1974) resembled classic lymphocyte uropods in having microspikes, vacuoles, and occasional mitochondria (Rosenthal and Rosenstreich, 1974). Microspikes and organelles were relatively scarce. Uropod-bearing mouse lymphocytes carry T-lymphocyte markers (Rosenthal and Rosenstreich, 1974; Rosenstreich *et al.*, 1972; Matter *et al.*, 1972). These observations suggest that the lymphocytes bearing the uropods in response to PPD are T lymphocytes (Nielsen *et al.*, 1974). In contrast to central lymphocytes, these lymphocytes at the periphery of the cellular clusters did not possess the nuclear and cytoplasmic features of early blast transformation (Nielsen *et al.*, 1974). Recently, the interaction between peripheral T lymphocytes and the central lymphocyte has been found to be genetically unrestricted and not immunologically specific for the antigen which drives the macrophage–central lymphocyte interactions (Braendstrup *et al.*, 1979). Thus, despite the impressive display of uropods, it is quite likely that the macrophage–lymphocyte interaction has more central importance to the immune response than the lymphocyte–lymphocyte interactions.

In general, it appears that the lymphocytes intimately connected with macrophages are most frequently undergoing early blast transformation. Suitable labeling and tracing of the PPD antigen within these immune cell clusters might reveal important details in regard to antigen processing.

Macrophage–thymocyte clusters from nonimmunized guinea pigs unex-

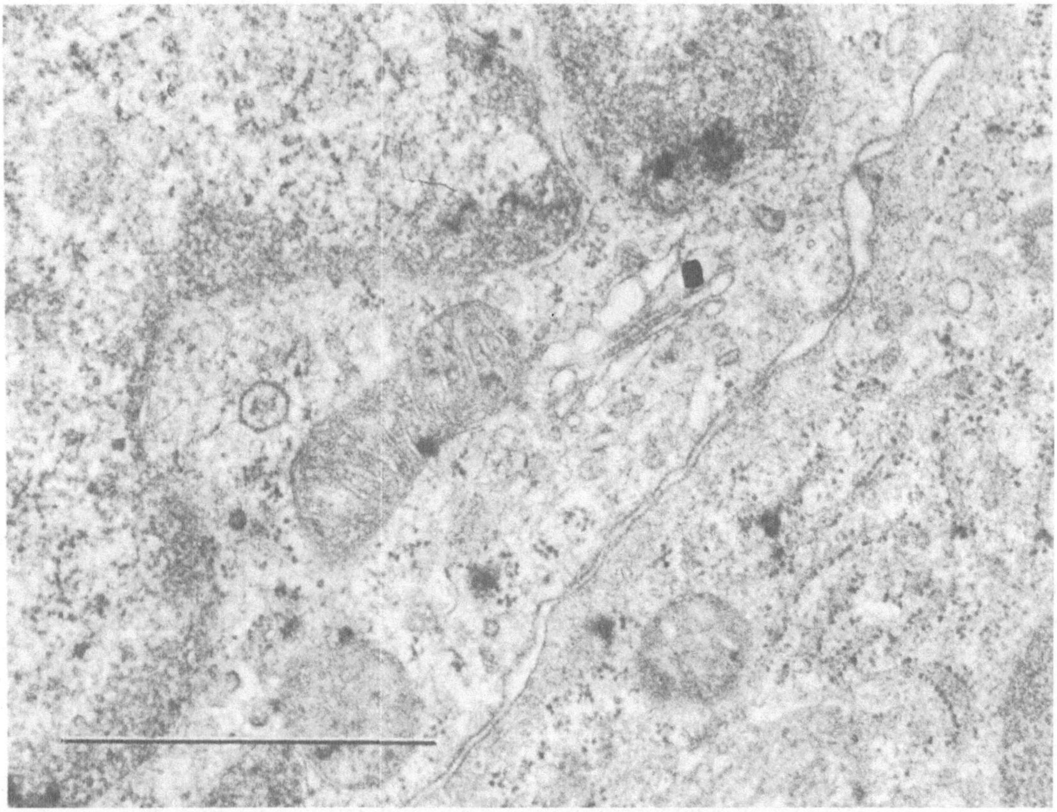

FIGURE 14. A portion of the intimate and extensive area of plasma membrane contact between a peritoneal macrophage (lower right) and a lymph node lymphocyte (upper left). These cells from previously immunized guinea pigs are the central portion of a cellular cluster which formed upon reexposure to purified protein derivative of tuberculin *in vitro* for 20 hr. The Golgi apparatus and nuclear cleft of the central lymphocyte face the macrophage. Microfilaments are visible mainly beneath the macrophage plasma membrane. Scale bar represents 0.95 μm. Reprinted from *The Journal of Experimental Medicine* **104:**1268, with permission of the journal and Dr. M. H. Nielsen.

posed to PPD differed from clusters obtained from guinea pigs that were immunized and reexposed to PPD in the following ways. Lymphocytes in the former animals show no signs of undergoing blast transformation and appear to bind directly to glass-adherent macrophages rather than through an intermediary transforming lymphocyte. Thus, these antigen-independent clusters have more macrophage–lymphocyte contacts and less lymphocyte–lymphocyte contacts than antigen-dependent clusters. Uropods are not seen in these clusters and the macrophage contains less phagocytized material. Fewer microfilaments occur beneath the macrophage plasmalemma binding the lymphocyte. These differences have been gleaned from study of published micrographs which differed with regard to incubation time *in vitro,* tissue source of lymphocytes, and culture serum (Nielsen *et al.,* 1974; Rosenthal *et al.,* 1978; Lipsky and Rosenthal, 1973).

After 36 hr of exposure of macrophages to antigen and to lymphocytes, blast transformation in the secondary response can be visualized by autoradiography, and antibody-producing cells or cell clusters can also be located in the primary or secondary response at this time. Light microscopic autoradiographic impressions (Hanifin and Cline, 1970; Sulitzeanu *et al.*, 1971) are in agreement with our impressions from routine electron microscopy (Nielsen *et al.*, 1974; Rosenthal *et al.*, 1978) that blast transformation occurs first in those lymphocytes in immediate contact with macrophages.

Studies with transmission electron microscopic autoradiography combined with antibody localization with the peroxidase immunocytochemical method have demonstrated that at a phase in the maturation of a lymphoid cell it incorporates thymidine and secretes antibody (Itoh *et al.*, 1975).

The primary *in vitro* response of mouse spleen cells to sheep red blood cells generates complex clusters of macrophages, lymphocytes undergoing blastogenesis, and plasma cells (Saunders and Hammond, 1970; McIntyre *et al.*, 1973; McIntyre and Pierce, 1973). Early cellular interactions similar to those just described with PPD occur, but their specificity for an immune response is uncertain because of the background hemolysis encountered in control cultures (McIntyre *et al.*, 1973). At 4 days about 10% of the cell clusters are hemolytic (Mosier, 1969) and these clusters can be isolated, analyzed by electron microscopy (Saunders and Hammond, 1970), and compared with nonhemolytic clusters (McIntyre *et al.*, 1973). The cellular relationships in these hemolytic clusters differ in several ways from those observed in the clusters which occur at 20 hr in a guinea pig secondary response to PPD. Plasma cells are present in the hemolytic clusters at this later time interval but small lymphocytes are scarce. Elongated cells and interdigitating cellular processes suggest complex intercellular relationships. These are difficult to interpret on transmission micrographs and invite scanning microscopic analysis. Both macrophages and, surprisingly, up to 20% of the lymphoid blast cells appear to contain remanents of erythrocytes, presumably sheep erythrocyte antigen. The latter are seen as early as 18 hr after culture initiation.

Antibody-producing cells in the medullary cords of lymph nodes and teased cell preparations have been studied during *in vivo* primary and secondary immune responses (Avrameas and Leduc, 1970; Miller and Avrameas, 1970; Itoh *et al.*, 1975). Six days and longer after primary immunization with the antigen markers peroxidase or alkaline phosphatase, many specific antibody-producing plasma cells are in contact with macrophages. Moreover, macrophages appear to be extensively phagocytizing plasma cells (Miller and Avrameas, 1973). Perhaps phagocytosis of certain plasma cell clones is an important regulatory mechanism resulting in the eventual production of antiserum of high specificity. On the other hand, to verify this observation, the distinction between phagocytosis and invagination of the macrophage seems to require examination of serial sections or use of ruthenium red or other tracers (Lipsky and Rosenthal, 1973).

Time-lapse cinematography has permitted demonstration of prolonged contacts between macrophages and lymphocytes cultured from spleens of rabbits immunized 1–2 weeks previously with horse serum. Lymphocytes were observed to move relatively rapidly around and over macrophages, a phenomenon called *peripolesis*. Control cultures of spleens did not show prolonged contacts (Sharp and Burwell, 1960).

Contacts between macrophages and lymphocytes in the absence of immunization provide a significant impediment to the complete analysis of the immune response. Cinematography has revealed a phenomenon very similar to peripolesis occurring in short term cultures of human leukocytes (Berman, 1966). Clusters of lymphocytes around macrophages occur also in mouse peritoneal milky spots (Orenstein and Shelton, 1976) and in lymphatic sinuses in lymph nodes (Farr and DeBruyn, 1975). The clusters, however, may reflect immune processes as they have been considered to be the *in vivo* response to naturally occurring antigenic stimulation, equivalent to cell reactions seen in the primary and secondary immune responses *in vitro*.

A contact zone with an intermembranous gap of 11–27 nm has been observed in patients with multiple myeloma (Blom, 1977). Micropinocytosis is reported to mediate bidirectional exchanges between macrophages and plasma cells (Thiery, 1962).

5.5. SENSITIVITY TO ENZYMES, IONS, AND DRUGS

The affinity of guinea pig macrophages for lymphocytes and the subsequent lymphocyte proliferation have been studied as features of an *in vitro* secondary immune response either by simultaneously mixing cells, the antigen PPD, and an inhibitor (antigen-independent clusters) or by mixing cells and antigen and then adding inhibitor at progressively later time intervals (antigen-dependent clusters). Alternatively, antigen-independent binding was assayed on cells never exposed to PPD. In this system the mediator of the macrophage binding is called a *macrophage receptor* and the mediator of the lymphocyte binding is called a *lymphocyte acceptor*. In general, the antigen-independent cluster is the more labile of the two and its binding is reversible (Rosenthal *et al.*, 1978). That is, the antigen-independent macrophage receptor is sensitive to trypsin and cytochalasin B. Antigen-independent binding is decreased by EDTA with partial reversal by adding either calcium or magnesium (Lipsky and Rosenthal, 1973, 1975a). Antigen-dependent clusters, or at least the central lymphocyte within the cluster which directly binds the macrophage (Section 5.4), is much less sensitive to these agents. The central lymphocyte in these clusters is antigen specific and requires a macrophage of the same genetic constitution at the major histocompatibility locus (Lipsky and Rosenthal, 1975b) to form an antigen-dependent cluster, in contrast to antigen-independent binding.

Trypsin and antisera have been used to define the biologic activity of the macrophage in antigen presentation to lymphocytes. These experiments are described in Section 5.1. The binding of antigen to macrophages not previously exposed to this antigen has the sensitivities described in Section 4.8.

Macrophage antigen-independent binding of lymphocytes is inhibited 75–95% by pretreating the macrophage with 20 mM sodium azide or sodium iodoacetate, or heating at 56°C for 1 hr. Binding is reduced 25% by 20 mM 2-deoxyglucose. Heat-killed thymocytes are still bound by macrophages (Lipsky and Rosenthal, 1973).

5.6. SUMMARY AND CONCLUSIONS

Macrophage presentation of antigen to lymphocytes is frequently measured by determining the effect of macrophages compared with antigen alone in enhancing lymphocyte proliferation or antibody formation. Morphologic techniques for studying macrophage–lymphocyte interactions are derived either from lymphocyte proliferation and antibody formation assays or from directly tracing antigen or observing cluster formation by immune cells. Their sensitivities are limited to specific time intervals during the immune response. Surface receptors for foreign substances and in some cases IgG and complement receptors are implicated in the early hours of macrophage processing of antigens, and are involved perhaps at later times for reticular cells *in vivo*. In large measure, studies which have traced the fate of antigen have demonstrated the effectiveness of one or more of these macrophage receptors in gathering most of the available antigen.

Other molecules on the macrophage surface are also involved in stimulating the immune response. In the mouse, Ia molecules associated with the I-J subregion of the major histocompatibility gene complex must be on macrophages to stimulate the immune response. Unlike the IgG, complement, and foreign substance receptors, Ia molecules occur on a subpopulation (15–65%) of macrophages. A receptor for primed lymphocytes has been defined on antigen-bearing macrophages *in vitro* in the guinea pig secondary immune response to purified protein derivative of tuberculin. This antigen-dependent binding requires histocompatibility of macrophage and lymphocyte, is antigen-specific, and is relatively stable to trypsin, chelators, and cytochalasin B. This cell interaction has been demonstrated ultrastructurally as an intimate macrophage–lymphocyte association which entailed extensive areas of plasma membrane contact at an intermembrane distance of 16 nm. The lymphocyte in these cell interactions appeared to be undergoing early blast transformation. Cells in the clusters involved in the secondary immune response include untransformed peripheral lymphocytes which attached only to the central transformed lymphocyte via their uropods. This specific macrophage–lymphocyte interaction can be recognized apart from background interactions 8–20 hr after mixing cells plus antigen, an early stage compared with lymphocyte proliferation or specific antibody production. It would be especially interesting should this model apply to the mouse and Ia molecules be involved in antigen-dependent cluster formation.

Short-term, reversible macrophage–lymphocyte binding occurs among guinea pig cells *in vitro* in the absence of antigen or of primed lymphocytes. While this binding has a certain specificity for cell type and animal species and requires living macrophages but not living thymocytes, it does not have the histocompatibility requirements of antigen-dependent binding and is less stable to trypsin, chelation, and cytochalasin B. These short-term clusters reveal differences in types of cells interacting and differences in the participating macrophages.

Hemolytic and nonhemolytic cell clusters formed during the primary *in vitro* response of mouse spleen cells to sheep erythrocytes have been examined by transmission electron microscopy at later time intervals. Hemolytic clusters are composed of macrophages, blast-transformed lymphocytes, and plasma cells. The blast cells and macrophages contain ingested erythrocytes. Nonhemolytic clusters have few or no plasma cells, but instead contain small lymphocytes.

Macrophage–lymphocyte interactions which in many ways mimic clusters formed *in vitro* can be found *in vivo* in unimmunized animals. These clusters may include immune cell responses to natural antigens and must be distinguished from experimental clusters and immune responses to the antigen of interest. Macrophage–lymphocyte interactions associated with secondary responses to antigen frequently involve more intimate membrane-to-membrane contact over a larger surface area than interactions among nonstimulated cells. Selection of the appropriate washing procedures may therefore contribute to enhanced separation of antigen specific macrophage–lymphocyte interactions from other interactions.

We envision macrophages interacting with complex antigens and then with lymphocytes to stimulate immune responses within, but not restricted to, lymph nodes, spleen, and inflammatory exudates. Antigen alone (primary response) or complexed with antibody and complement (secondary response) is initially bound by the appropriate IgG, complement, or foreign substance receptors and extensively degraded within the macrophage. A small percentage of antigen or fragments of antigen either remains on the macrophage and reticular cell surface or is returned to the surface for presentation to lymphocytes. Intimate and extensive plasma membrane contact of macrophage and lymphocyte is necessary for stimulation of lymphoid proliferation. The macrophage must bear Ia surface molecules to stimulate antibody production by lymphocytes and plasma cells. It is possible that Ia surface antigen is important to the intimate contact of macrophage and lymphocyte membranes. It would be interesting to know whether reticular cells are Ia-positive. During this contact, perhaps a special association of lymphocyte surface molecules and antigenic determinants is facilitated by the extensive capability of the macrophage to mobilize membrane or to cleave macromolecules. This hypothesis should be examined with complete appraisal of the contact zone with extracellular surface markers and freeze-fracture studies of these interacting cells and surface markers. Sequential monitoring with T- and B-cell markers of the lymphocytes exhibiting antigen-specific contact with macrophages needs to be accomplished. Immune cell clusters at later time intervals include antibody producing cells of B-cell lineage.

6. FIBRIN RECEPTORS

6.1. BIOLOGIC ACTIVITY

Mononuclear phagocytes which have direct access to the bloodstream, primarily the circulating monocytes, Kupffer cells, and splenic macrophages, play a critical role in the clearance of circulating fibrin. The clearance of fibrin complexes by the RES removes them from the circulation, thus preventing deleterious deposition of intravascular fibrin (Lee, 1962). The importance of this system was demonstrated in the rabbit where intravascular coagulation experimentally produced by thrombin injections did not result in massive deposition of fibrin unless the RES was blocked (Lee, 1962). Reticuloendothelial blockade following extensive injection of colloidal particles results in a temporary decrease in the ability of phagocytes to remove material from the circulation (Stuart, 1970; Benacerraf and Sebestyen, 1957).

This impairment apparently represents an *in vivo* counterpart of the *in vitro* receptor unresponsiveness which follows ingestion of large quantities of latex particles (Sections 2.6 and 4.7).

It has also been suggested that fibrin binding to the macrophage surface promotes intercellular adherence in the macrophage disappearance reaction (Colvin and Dvorak, 1975), a reaction characterized by the adherence of macrophages to each other and to peritoneal serosal cells after an antigenic challenge of a sensitized guinea pig (Nelson and North, 1965). This interpretation of the disappearance reaction was based on the similar macrophage binding of fibrin and disappearance reaction encountered after treatment with anticoagulants and thrombin (Nelson, 1965; Jókay and Karczag, 1973).

6.2. SPECIFICITY

The binding of macrophages to soluble fibrin has been studied with radioiodinated fibrin. Cells were incubated with [^{125}I]fibrin for 45 min under conditions inhibiting their metabolism. These cells were subsequently centrifuged and washed and the cell-associated radioactivity was determined by gamma counter. Specificity was demonstrated by the inhibition of [^{125}I]fibrin-binding by nonradioactive fibrin but not by immune complexes or IgG. The binding was presumably of high affinity, since only limited reversibility of binding occurred upon prolonged incubation. Removal of [^{125}I]fibrin by trypsin and uptake in the presence of metabolic inhibitors suggested that binding occurred on surface receptors. Soluble fibrin existed in these experiments as a macromolecular complex of undetermined size, so that each site was considered to bind a polymer containing several fibrin molecules. Since 6.9×10^6 molecules of [^{125}I]fibrin were bound per cell, the number of receptor sites per cell was estimated at less than 6.9×10^6. Binding of fibrin by PMN resembled that of macrophages, whereas lymphocytes, fibroblasts, and erythrocytes bound ten times less fibrin (Sherman and Lee, 1977).

6.3. MORPHOLOGY

Stained with fluorescent antiserum to fibrinogen, resident and exudate guinea pig peritoneal macrophages reveal a speckled, random distribution of fluorescent patches on a loose meshwork of fibrils that surrounded the cells like a net. The latter distribution has been interpreted as indicating fibrin formation on the cell surface. Notably the pulmonary alveolar macrophages lacked this surface fibrin net (Colvin and Dvorak, 1975).

Electron microscopy of rabbit liver following intravenous infusion of thrombin or an endotoxin injection reveals hypertrophied Kupffer cells containing fibrin within intracytoplasmic vacuoles. Fibrin was identified by its regular cross banding with periodicity slightly less than the usual 22–23 nm (Prose *et al.*, 1965).

In our experience fibrin can be demonstrated binding to the plasmalemma of Kupffer cells and peritoneal macrophages with membrane-to-fibrin gaps of less than 35 nm (Prose *et al.*, 1965; Emeis and Lindeman, 1976; Dvorak *et al.*, 1978). The exact membrane-to-fibrin distance cannot be measured, since the periphery of the fibrin microaggregate blends into the cellular coat. Fibrin microaggregates were not evenly distributed along the portion of the Kupffer cell membrane exposed to the lumen of the sinusoid (Prose *et al.*, 1965). Large regions of this membrane bound fibrin at intervals with a 100–300 nm periodicity. These distances appeared probably to be governed by the appreciable size of the fibrin aggregates. Other regions of the sinusoidal membrane appeared to bind less fibrin as did the membrane bordering the space of Disse.

6.4. ENDOCYTOSIS

Contrary to previous observations on rabbit macrophages (Prose *et al.*, 1965; Sherman *et al.*, 1975), rat liver macrophages have been found not to phagocytize fibrin during intravascular coagulation (Emeis and Lindeman, 1976). This discrepancy can conceivably be attributed to the degree of intravascular coagulation that was experimentally induced or the difference in sampling intervals examined following thrombin. Alternatively, a true difference in the manner in which these two animal species dispose of fibrin may exist.

6.5. SENSITIVITY TO ENZYMES AND IONS

Macrophage binding of fibrin is not affected by prior treatment of cells with trypsin or plasmin. The receptor requires calcium but not magnesium cations for fibrin binding (Colvin and Dvorak, 1975; Sherman and Lee, 1977).

7. RECEPTORS FOR INSULIN

Insulin receptors on cell surfaces are central to the biologic activity of this hormone in numerous organs and tissues including liver (Soll *et al.*, 1975), adipose tissue (Kahn *et al.*, 1977), muscle (Kahn and Roth, 1975), and thymus (Soll *et al.*, 1974). Among the peripheral blood leukocytes, the monocyte carries the preponderant receptor activity for insulin (Schwartz *et al.*, 1975). This receptor has many features in common with the insulin receptors on adipocytes

(Olefsky, 1976) and hepatocytes (Gavin *et al.*, 1972, 1974) including specificity, sensitivity, binding kinetics, and alterations in pathologic states such as diabetes and obesity. Thus, the insulin receptor of moncytes is not a specific feature confined to this cell type or to phagocytes but rather reflects the insulin-binding capability common to many metabolically active cells. The importance of the monocyte receptor for insulin is primarily clinical, since these cells are so accessible through venipuncture.

Insulin receptors are assayed by the binding of radioiodinated insulin to cells and specific binding inhibition. Cells are incubated with labeled insulin in separate aliquots in the absence and presence of a graded series of increasing amounts of unlabeled insulin. Resultant binding curves reflect specific inhibition from increasing quantities of unlabeled insulin but not from unrelated proteins and polypeptides. Moreover, the relative affinities of insulin analogs for receptors closely parallels the biologic activity of these analogs (Gavin *et al.*, 1972, 1974; Kahn *et al.*, 1977). Although monocytes exhibit greater receptor activity for insulin (Schwartz *et al.*, 1975), the presence of lymphocytes within circulating leukocyte populations does not alter the clinical usefulness of the data obtained from normal and diabetic patients (Olefsky and Reaven, 1976b).

Normal monocytes of human volunteers in the basal, fed state carry 15,000–28,000 insulin-receptor sites per cell (Bar *et al.*, 1976). Early estimates of 2200–12,000 sites per cell may have been low because of contamination with minimal and nonbinding lymphocytes (Gavin *et al.*, 1972, 1974; Olefsky and Reaven, 1974). Binding affinity is relatively high, around 10^8 or 10^9 M^{-1} (Bar *et al.*, 1976). In contrast, thymic lymphocytes have only about 300 high-affinity receptors in laboratory animals (Soll *et al.*, 1974).

Insulin receptors on human peripheral blood leukocytes have been localized by autoradiography of ^{125}I-labeled insulin. All of the cells with overlying silver grains could be morphologically classified as large mononuclear cells, and over 85% could be positively identified as monocytes. In preparations preincubated with latex particles, ninety percent of the cells binding insulin disclosed ingested latex. Remaining cells binding insulin could not be distinguished from large lymphocytes by the light microscopic criteria employed. Fourteen percent of the latex-ingesting cells did not reveal affinity for insulin. Thus, the majority of ^{125}I-labeled insulin was bound by monocytes identified as phagocytic by morphologic features and the functional criterion of latex ingestion. In addition, subpopulations may exist consisting of monocytes lacking insulin receptors and of nonphagocytic, large mononuclear cells with insulin receptors (Schwartz *et al.*, 1975).

Metabolic inhibitors including 3 mM sodium azide and 0.1 mM iodoacetamide either alone or combined did not prevent binding of ^{125}I-labeled insulin, nor did they significantly affect cell viability. Moreover phagocytosis of latex particles did not affect the macrophage's insulin-binding capability (Schwartz *et al.*, 1975).

Mouse splenic macrophages and the mouse neoplasm $P388_{D1}$ whose cells resemble macrophages both possess receptors for insulin which are similar to those on human monocytes (Kahn and Roth, 1975; Bar *et al.*, 1977; Koren *et al.*,

1975). Insulin at physiologic concentrations inhibits the ability of these cell populations to lyse antibody-coated erythrocytes (Bar *et al.*, 1977).

In summary, since human monocytes possess receptors for insulin similar to functional receptors in metabolically active tissues, peripheral blood samples provide a convenient test site to study patients with metabolic diseases or insulin resistance. Mouse macrophages and their neoplastic counterpart also have insulin receptors. The binding of radioiodinated insulin to monocytes is specific and of relatively high affinity ($10^8 - 10^9\ M^{-1}$). A monocyte has about 22,000 receptor sites for insulin. Receptors have been localized primarily on monocytes by light microscopic autoradiography, which also suggests subpopulations of mononuclear phagocytes without receptors and nonphagocytic cells with receptors. Single addition of metabolic inhibitors or latex particles did not inhibit binding. The changes of macrophage insulin receptors in relation to metabolic diseases are discussed in Section 12.

8. RECEPTORS FOR LACTOFERRIN

The participation of macrophages in iron metabolism has long been recognized (Fillet *et al.*, 1974; Stuart, 1970). Lactoferrin is a glycoprotein which reversibly binds two ferric ions and has a particular affinity for the reticuloendothelial system (Van Snick *et al.*, 1974) but not for reticulocytes (Van Snick *et al.*, 1975). We refer to the molecule bound with iron as lactoferrin and without iron as apolactoferrin. Lactoferrin is probably involved in iron metabolism and may be the principal effector molecule in the blockade of iron release from the RES occurring in the hyposideremia of inflammation (Van Snick and Masson, 1976). Inflammatory sequestration of iron is potentially bacteriostatic (Masson *et al.*, 1966). The binding of radiolabeled human lactoferrin to resident mouse peritoneal macrophages occurs via surface receptors which number 20×10^6 per cell. The affinity constant of these receptors is $0.9 \times 10^6\ M^{-1}$. Apolactoferrin has been shown to inhibit binding of lactoferrin less well than lactoferrin itself suggesting steric changes in the molecules that influence its binding. The relatively high proportion of positive charges on lactoferrin may partly account for binding, as removal of sialic acid from lactoferrin increased its binding tendency. However, cytochrome c which possesses a greater net positive charge did not inhibit lactoferrin binding (Van Snick and Masson, 1976). Removal of the sialic acid from lactoferrin may have altered the steric structure of the molecule and hence its affinity for the receptor. Various proteins unrelated to transferrin do not significantly inhibit the binding of lactoferrin by mouse peritoneal cells (Van Snick and Masson, 1976).

Lactoferrin has been localized by fluorescent immunostaining (Masson *et al.*, 1966; Masson *et al.*, 1969). Thus far, in the phagocytic cells, lactoferrin has been demonstrated within polymorphonuclear leukocytes and their myeloid precursors as early as the promyelocyte state (Masson *et al.*, 1969). Lactoferrin is associated with the lysozyme-rich fraction of human (Leffell and Spitznagel,

1972) and rabbit (Baggiolini *et al.*, 1970) PMN granules, equivalent to secondary granules in the rabbit.

9. RECEPTORS FOR CHEMOTACTIC FACTORS AND LYMPHOKINES

A number of naturally occuring biologic mediators cause leukocytes to migrate (Wilkinson, 1974). These include polypeptides derived from complement (Snyderman *et al.*, 1968; Bianco *et al.*, 1979; DeSouza and Nes, 1968), denatured proteins (Wilkinson, 1974), oxidized fatty acids (Turner *et al.*, 1975), and other substances (Isturiz *et al.*, 1978; Ward, 1968). Such migration into regions of inflammation is important to host defenses against microbes and neoplasms. Chemotactic factors induce monocytes and PMN to increase their adherence to foreign substances (O'Flaherty *et al.*, 1978).

Primary, definitive studies of surface receptors for a chemotactic peptide have focused on the PMN rather than the mononuclear leukocyte (Williams *et al.*, 1977; Aswanikumar *et al.*, 1977). The binding of tritiated *N*-formylmethionyl-leucyl-phenylalanine to human PMN is of high affinity with an association constant of 9×10^7 M^{-1} at 37°C. These receptors are relatively sparse, amounting to about 2000 per cell. Mononuclear leukocytes (81% lymphocytes and 19% monocytes) have 29% as much binding as the PMN preparation. Much of this binding can be accounted for by monocytes, since removal of cells adhering to nylon mesh reduced binding to 11% (Williams *et al.*, 1977). The binding of monocytes to the chemotactic substance may soon be assessed since methods of purifying monocytes are now available (Ackerman and Douglas, 1978). Morphologic assessment of monocyte binding to *N*-formylmethionyl-leucyl-phenyl-alanine is also conceivable by autoradiography similar to that applied to antigen and insulin receptors (Schwartz *et al.*, 1975).

Lymphocytes stimulated with appropriate exposures to antigen produce soluble factors with diverse biologic activities. Among these factors are some which activate macrophages (MAF), and others which inhibit macrophage migration out of capillary tubes (MIF) (Remold *et al.*, 1970; Postlethwaite *et al.*, 1976; Postlethwaite and Kang, 1976; David, 1966). MIF and MAF molecules from the guinea pig are similar in many respects including their molecular masses of 35,000–70,000 daltons by isopycnic density gradient centrifugation, and sensitivity to neuraminidase (Nathan *et al.*, 1973; Remold *et al.*, 1970; Bennet and Bloom, 1968).

Evidence that MIF interacts with the macrophage surface is indirect but substantial. Macrophages incubated with trypsin no longer respond to MIF, but regenerate responsiveness 24 hr after trypsin removal (David and David, 1972; David *et al.*, 1964). Trypsinized macrophages differ from the untreated cells in not removing MIF from supernatants (Leu *et al.*, 1972). Treatment of macrophages with chemical (Remold, 1977) or protein inhibitors of esterase increases their responsiveness to MIF (Remold, 1974; Remold and Rosenberg, 1975). Neither the chemical employed nor the protein could enter the cytoplasm in an active form, and their actions were therefore inferred to affect an esterase at the macrophage surface (Remold, 1974, 1977).

An essential part of the macrophage receptor for MIF is γ-L-fucose. Macrophages incubated with γ-L-fucosidase no longer responded to MIF while this enzyme had no effect on MIF alone (Remold, 1973). Of a number tested only this sugar reversibly abolished the activity of guinea pig MIF on macrophages. The recent discovery of a cloned macrophage-like cell line which responds to MIF may allow further characterization of the MIF receptor (Newman *et al.*, 1979).

Inhibition of macrophage migration and cellular aggregation are associated with the loss of guinea pig peritoneal exudate cell coat material. These observations were made with transmission electron microscopy using a ferrocyanide-reduced osmium tetroxide stain to visualize the cell coat (Dvorak *et al.*, 1972). The interrelationship of MIF and the foreign substance receptor is cited in Section 4.8.

10. RECEPTORS FOR GLYCOPROTEINS, POLYSACCHARIDES, AND LIPOPROTEINS

Rat alveolar macrophages possess specific receptors for glycoproteins and glycoconjugates that have terminal glucose, mannose, or *N*-acetylglucosamine (Stahl *et al.*, 1978) residues. This specific lectin-like affinity seems likely to explain macrophage binding of a subgroup of foreign substances with the appropriate terminal sugars. Zymosan, for example, contains both polyglucose (glucan) and polymannose (mannan) in association with protein and lipid (Northcote and Horne, 1952; Pillemer and Ecker, 1941). Specific types of pneumococci and certain other bacteria contain glucose residues either terminal or easily cleaved to a terminal position on their outer wall (Larm and Lindberg, 1976; How *et al.*, 1964). *N*-Acetyglucosamine residues comprise backbone and terminal positions in the cell walls of bacteria including *Staphylococcus aureus* (Strominger *et al.*, 1967; Lehninger, 1975). Certain streptococci contain glucose in the polysaccharides and glycolipids that are responsible for their serologic activity (Prescott *et al.*, 1971; King *et al.*, 1971), while *Mycoplasma pneumoniae* uses its own receptors for neuraminic acid to infect cells (Sobeslavsky *et al.*, 1968). Thus, macrophage glycoprotein receptors provide a clue to specific ligand-binding interactions which may account for a portion of the biologic agents susceptible to macrophage foreign substance receptors (Section 4). These receptors may be important in clearing some microorganisms from the lung and other exposed tissues before specific humoral immunity is generated. They may also aid in antigen presentation to lymphocytes or in the clearance of dangerous or effete autologous enzymes similar to an analogous hepatocyte receptor system (Stahl *et al.*, 1976, 1978).

The binding of glycoproteins, synthetic glycoconjugates, simple sugars, and albumin has been assayed by radiolabel binding and binding-inhibition studies. Alveolar macrophages were incubated with radiolabeled glycoprotein and various concentrations of unlabeled mannan or glycoprotein competitor. After the incubation, cells were centrifuged, washed, and assayed for radioactivity. The binding of radiolabeled glycoproteins and glycoconjugates having terminal glucoses, mannoses, or *N*-acetylglucosamines met minimal criteria of being mediated by a receptor in that it was saturable and had definable specificity. Mannan and various glycoproteins including horseradish peroxidase competitively inhibited binding of these radiolabeled glycoproteins. On the other hand, proteins lacking the appropriate sugars or with the sugar enzymatically removed did not inhibit binding (Stahl *et al.*, 1978).

Macrophage membranes are rich in free cholesterol. The majority of this cholesterol is derived from *in vitro* exchange with serum, and not from macrophage biosynthetic activity. Serum α-lipoproteins are the major heterologous proteins participating in the exchange with the plasma membranes of mouse peritoneal macrophages. This exchange has been measured by exposing macrophages to serum with tritium-labeled cholesterol equilibrated into the serum lipoproteins. Uptake of cholesterol was assessed at various time intervals by washing and solubilizing the cells for liquid scintillation counting. Washout of radiolabeled cholesterol, washing the cells, and reincubating in unlabeled medium for various times before washing, solubilizing, and counting. Thirty percent of cell cholesterol was exchanged per hour. Rapidly and slowly exchanging compartments were identified (Werb and Cohn, 1971). The rapidly exchanging compartment was trypsin-sensitive with a 7-hr duration of protein synthesis required to recover the normal exchange rate. This suggested that a plasma membrane receptor was involved in positioning the lipoproteins for exchange. Extensive phagocytosis or pinocytosis increased the size of the slowly exchanging compartment, suggesting that the slow exchange took place in intracellular membranes as, for example, those delimiting the lysosomes. Compared with uninduced macrophages, thioglycollate-induced mouse peritoneal macrophages and rabbit alveolar macrophages displayed increased cholesterol content of their slowly exchanging cholesterol pool. This latter increase was attributed to the increased number of secondary lysosomes in the activated macrophages (Werb and Cohn, 1971).

The disappearance from the circulation of *in vivo*-labeled β-lipoproteins has been followed after intravenous injection into rats. The half-life of circulating β-lipoprotein is 7.5 hr. Stimulation of the RES with repeated daily intramuscular injections of *Mycobacterium butyricum* increased the rate of clearance of radiolabel, thus implicating a macrophage receptor for lipoprotein in the clearance mechanism. Blocking the RES with intravascular injections of colloidal carbon decreased the clearance rate in further support of this implication (Monnier *et al.*, 1974).

11. ISOLATION OF SURFACE RECEPTORS

11.1. SEPARATION OF INTACT PLASMA MEMBRANES WITH RECEPTORS

Efforts have been made to isolate and characterize various macrophage receptors but it has proven difficult to separate the plasma membranes from macrophages by conventional homogenization and centrifugation in good yields. Brief swelling in water and fixation facilitates the isolation of macrophage membranes by Dounce homogenization and sucrose gradient centrifugation (Nachman *et al.*, 1971a). Some receptor activity is lost during this purification possibly at the brief fixation step. A method has been described for purifying BCG-induced rabbit alveolar macrophage plasma membranes to a 15-fold increase in specific activity. Quite possibly due to its state of activation, this cell type is reported to lack the classic membrane marker, 5′-nucleotidase (Edelson and Erbs, 1978; Wang *et al.*, 1976; DePierre and Karnovsky, 1973). The cells have also been labeled biosynthetically to investigate the insertion of proteins into membranes and their turnover. Insertion differed from one protein band to another, while radioactivity disappeared at the same rate. The results suggested regulated or modulated synthesis and bulk removal of plasma membrane proteins, possibly by ingestion (Nachman *et al.*, 1971b).

Membranes have been purified from phagosomes by choosing target material for ingestion which has an appropriate density to facilitate separation from other cellular constituents (Nachman *et al.*, 1971a; Stossel *et al.*, 1972). A problem with this technique is the exposure of surface molecules to lysosomal degradation during separation (Nachman *et al.*, 1971b; Werb and Cohn, 1972).

A recent method for isolating erythrocyte membranes seems applicable to macrophages that have been pretreated with inhibitors of ingestion or phagolysosomal interaction (Bhisey and Freed, 1975; Sbarra and Karnovsky, 1959). With this method, cells are allowed to adhere to charged beads of greater size than the cells until they cover the beads. The resulting aggregates are sonicated and the membranes which remain on the beads are easily removed free of cytoplasmic components (Jacobson and Branton, 1977).

Another method which has promise for purifying macrophage plasma membranes with intact receptors entails isolation of peripheral material shed from the cells. Macrophages and their neoplastic counterparts have a tendency to form peripheral vesicles surrounded by plasmalemma and enclosing either cytoplasm or a clear space (Franke *et al.*, 1971; McKeever *et al.*, 1976b; Merchant and Low, 1977; Skinnider and Ghadially, 1977; Daems and Brederoo, 1973) as seen by transmission electron microscopy (Figure 15). Both structures possess receptors for IgG (Scott and Rosenthal, 1977; McKeever *et al.*, 1976b). The possibility that these structures are related *in vivo* to the process of clasmatocytosis (Donné, 1884; Shields, 1972) has been suggested (McKeever *et al.*, 1976b). Formaldehyde, disulfide bond reducing agents, heating, cytochalasin B plus colchicine, and sonication will accelerate vesicle formation or shedding (Scott and Rosenthal, 1977; Davies and Stossel, 1977; Bhisey and Freed, 1975). Certain vesicles isolated from peritoneal exudate macrophages contain cytoplasm, form rosettes with thymic lumphocytes, agglutinate with lentil lectin, Con A, phytohemagglutinin and wheat germ agglutinin, and possess receptors for IgG and complement (Figure 16). Vesicles or peripheral hyalin blebs isolated from pulmonary macrophages are enriched in myosin- and actin-binding protein and increase their metabolism of ^{14}C-labeled glucose when incubated with latex in similar fashion to normal macrophages (Davies and Stossel, 1977). Further

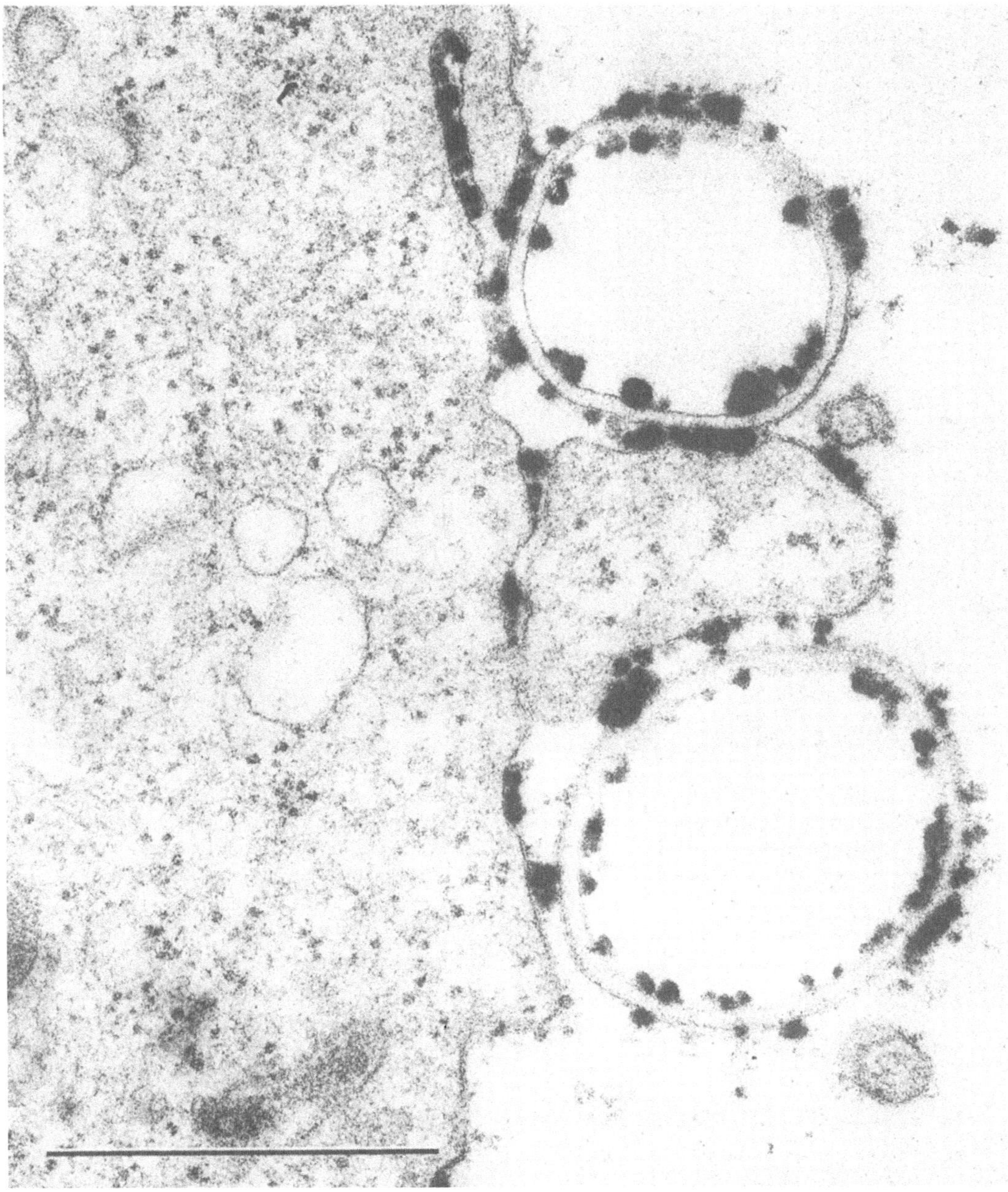

FIGURE 15. Vesicles or surface blebs of the macrophage plasma membrane have a double membrane and enclose a clear space. These membranes retain receptor activity for IgG demonstrated by their retention of peroxidase–anti-peroxidase complexes subsequently stained with diaminobenzidine-peroxide. Cells and incubation as for Figure 1. Scale bar represents 0.71 μm. Reprinted from *The American Journal of Pathology* **84**:450, with permission.

FIGURE 16. Scanning electron micrograph of a guinea pig peritoneal macrophage containing a cell surface bleb which is in the process of being shed. From a macrophage monolayer incubated in membrane vesiculant solution containing 25 mM formaldehyde and a 2 mM dithiothreitol. Scale bar represents 6.7 μm. Reprinted from *The Journal of Immunology* **119:**146, with permission of the journal and Dr. R. E. Scott.

purification on focusing sucrose density gradients yielded a separation of membrane vesicles from much of the cytoplasmic component. Formation of vesicles in response to noxious agents is a phenomenon which occurs among many cell types in addition to macrophages (Belkin and Hardy, 1961).

11.2. RECEPTORS FOR IMMUNOGLOBULIN G (Fc RECEPTORS)

Major advances have recently occurred in the isolation of Fc receptors for IgG from macrophages and macrophage-like neoplastic cell lines. Following extraction in nonionic detergent, the detergent was removed allowing the remaining receptors to be assayed for biologic activity. The assay involved inhibition of the binding of aggregated ^{125}I-labeled IgG_2 to cells. Receptors retained activity after removal from the membrane. The activity was associated with a 20 S macromolecule as measured by buoyant density and velocity sedimentation, but with detergent present the material was 7–9 S in size (Anderson and Grey, 1977).

In another approach macrophage Fc-binding protein has been found to bind specifically to affinity columns containing monomeric mouse IgG_{2a} or monomeric human IgG_1 and has been eluted from these columns. However, receptor activity bound specifically to a column containing aggregated IgG_{2b} could not be eluted. Perhaps the stronger binding to aggregated IgG and immune complexes so clearly manifested *in vivo* is still present in these lysates. The lactoperoxidase ^{125}I-labeled surface proteins of a neoplasm closely resembling macrophages ($P388_{D1}$) were extracted in nonionic detergent. The molecules which bound to IgG fell mainly in the 57,000-dalton range with minor 28,000- and 24,000-dalton components as measured by polyacrylamide gel electrophoresis in sodium dodecyl sulfate ionic detergent. The mobilities of these isolated components changed little on reduction, suggesting they represented single polypeptide chains. Using the monomeric IgG_{2a} subclass to purify the receptor may have resulted in the isolation of a receptor different from those isolated from lymphoid cells or tumors by binding to immune complexes or aggregated immunoglobulins.

The isolation of macrophage receptors for IgG represents a major advance in receptor biochemistry as most of the previous accomplishments have concerned isolation of IgG receptors from lymphocytes or lymphomas (Rask *et al.*, 1975; Cooper and Sambray, 1976; Molenaar *et al.*, 1977; Premkumar-Reddy *et al.*, 1976; Uhr and Vitetta, 1974; Andersson *et al.*, 1974; Wernet, 1976; Deacon and Ebringer, 1977). Some of the latter have resulted from fortuitous precipitations of molecules called MAID (membrane-associated immunoglobulin-detaining protein) (Premkumar *et al.*, 1975a,b) during the isolation of cellular immunoglobulins. Macrophage IgG receptors differ in molecular weight from these lymphoid IgG receptors which, in turn, differ among themselves. As methodologic problems such as proteolytic degradation during isolation (Loube *et al.*, 1978) become better controlled, it should be possible to determine whether: (1) the macrophage and lymphocyte IgG receptors are actually different molecules, (2) macrophage receptors for IgG_{2a} and IgG_{2b} are different molecules [These seem to have slightly different sensitivity or binding characteristics in some hands (Unkeless, 1977; Heusser *et al.*, 1977; Walker, 1976), but not in others (Segal and Titus, 1978).], (3) receptors for immune complexes differ from receptors for monomeric immunoglobulin, (4) molecular associations occur between Ia antigens and Fc receptors (Dickler and Sachs, 1974), and (5) the isoelectric points and amino acid sequences differ for various receptors for IgG.

11.3. RECEPTORS FOR FOREIGN SUBSTANCES (NONSPECIFIC RECEPTORS)

Using the macrophage-like neoplasm $P388_{D1}$ to obtain a pure population of adherent cells, one of us (PEM) has been inquiring into the nature of the receptor for foreign substances. The mouse tumor $P388_{D1}$ possesses common features of macrophages including the capacity to adhere and phagocytize, receptors for IgG and complement, and nonspecific esterase activity (Koren *et al.*, 1975). It also functions as an effector cell in antibody-dependent cell-mediated cytolysis (Bar *et al.*, 1977), bears insulin receptors (Bar *et al.*, 1977), and synthesizes lysozyme (Ralph *et al.*, 1976).

The initial questions and approaches entertained in this investigation have been straightforward. Is macrophage adherence a phenomenon attributable to one or more particular proteins synthesized by the cell? On the basis of a constructed model (Figure 12), are the molecules contributing to macrophage adherence heterogeneous or homogeneous? In these studies cells were labeled biosynthetically with tritiated leucine and lysine. Cellular lysates prepared with nonionic detergents were incubated with colloidal gold. The labeled proteins adhering to the colloidal gold were prepared for polyacrylamide gel electrophoresis in SDS. Figure 17 shows the migration pattern of proteins which have a strong tendency to bind colloidal gold. The findings suggests that gold affinity exists in a heterogeneous group of proteins which vary in their side chains and migrate in the range of 100,000 to above 200,000 daltons. The molecular weights of these molecules differ from those of the majority of isolated receptors for IgG.

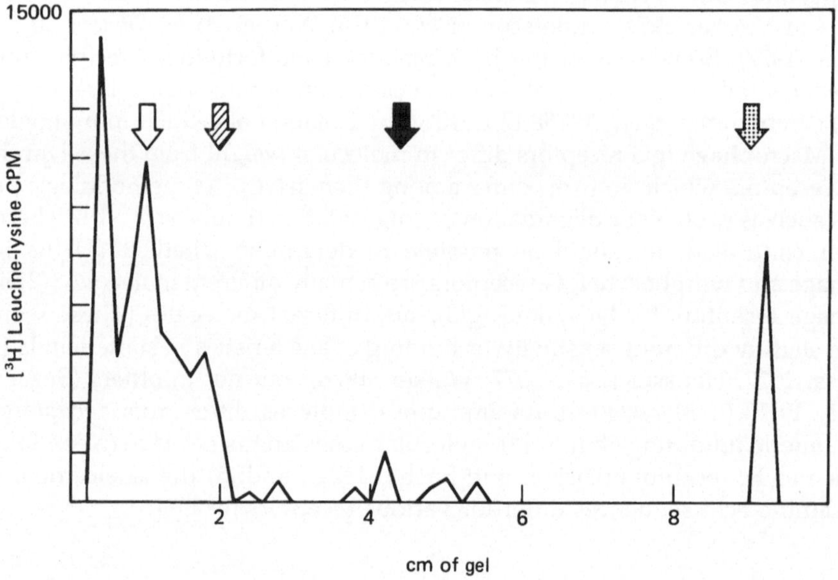

FIGURE 17. Gel electropherogram of protein-containing macromolecules having a tendency to bind colloidal gold. The counts per minute from an equivalent aliquot of molecules in the cellular lysate before incubating with gold have been subtracted from the counts per minute of proteins adherent to the gold. Marker proteins have the following molecular weights: 150,000 (clear arrow), 110,000 (hatched arrow), 50,000 (solid arrow), 23,000 (stippled arrow). Phenol red marker dye migrated 9.4 cm into the gel.

11.4. SUMMARY AND CONCLUSIONS

Conventional methods of isolating plasma membranes or phagolysosomal membranes have been applied to macrophages, but carry the risk of loss of receptor activity during isolation because of abundance of macrophage enzymes. The suitability of these methods for purification of certain receptors requires further evaluation. The induction and subsequent separation of membrane-coated vesicles from macrophages offers an alternative approach which appears to separate functional receptors from most degradative elements.

Direct extraction of cytoplasmic lysates of macrophages and neoplasms resembling macrophages into nonionic detergent with subsequent detergent removal has retained enough soluble IgG_{2b} receptor activity to be measured in a cellular binding inhibition assay. IgG receptor activity was associated with a 20 S macromolecule which was reduced to 7–9 S in the presence of detergent. Recently, surface receptors for IgG_{2a} from a histiocytic neoplasm labeled with ^{125}I have been isolated by affinity chromatography. Their molecular weights in sodium dodecyl sulfate are 57,000 with minor components of 28,000 and 24,000. These pioneering efforts in isolating macrophage receptors for IgG may eventually lead to chemical characterization of individual types of IgG receptors and answer a number of biological questions about these receptors.

Preliminary inquiries into the nature of macrophage receptors for a foreign substance used the histiocytic neoplasm $P388_{D1}$ and colloidal gold. Biosynthetically labeled protein-containing molecules which have a particular tendency to bind colloidal gold have heterogeneous molecular weights in sodium dodecyl sulfate from 100,000 to above 200,000 with minor components of less than 20,000.

12. SURFACE RECEPTORS OF MONONUCLEAR PHAGOCYTES IN DISEASE

The recognized importance of opsonization in the capacity of macrophages to protect the host against microbial disease is included in sections on the biologic activity and function of IgG and complement receptors. Macrophage participation in the lesions of pneumoconioses and the role of macrophages in protecting extremely exposed organs like the lung from foreign material are mentioned in the section on macrophage receptors for foreign substances. Selected additional examples where macrophage or monocyte receptors appear to be important in clinical evaluation or treatment of disease are discussed in this section.

Monocyte receptors for insulin have been investigated, for example, in insulin-related and other metabolic diseases, in part because the accessibility of human monocytes offers an advantage for such studies. Monocytes of obese patients with clinical insulin resistance have been found to contain half the number of receptors per cell that monocytes of nonresistant obese or normal patients contain (Bar *et al.*, 1976; Olefsky, 1976). After the insulin-resistant, obese patients fasted for 48-72 hr, their monocyte receptor affinity for insulin increased fivefold without change in receptor concentration. Chronic diet control increased the number of insulin receptors per monocyte to normal levels (Bar *et al.*, 1976). In all patients, total monocyte receptor concentration was inversely related to circulating insulin levels (Bar *et al.*, 1976; Olefsky, 1976). Thus, it appears that diet affects the affinity and number of insulin receptors per monocyte by determining the level of insulin to which the cells are exposed (Bar *et al.*, 1976; Olefsky, 1976; Soll *et al.*, 1975). Cells exposed in culture chronically to prolonged elevations of insulin also exhibit decreased insulin receptors (Gavin *et al.*, 1974). The simplest explanation for these data is a negative feedback loop whereby insulin inhibits its own receptor. Mononuclear leukocytes from untreated diabetics with fasting hyperglycemia and elevated fasting plasma insulin possess about half the normal number of insulin receptor sites per cell (Olefsky and Reaven, 1974, 1976b). Control patients with chronic hyperglycemia secondary to pancreatitis and with a normal or low fasting plasma insulin show normal insulin binding by monocytes. Plasma from the diabetic patients had no effect on a standard assay of insulin binding. Despite the diabetic propensity toward low plasma insulin levels following glucose challenge, search for a unifying hypothesis leads to the question whether the elevated fasting plasma insulin in these patients significantly affected the receptor activity of their leukocytes drawn after an overnight fast (Olefsky and Reaven, 1974). Treatment with the oral hypoglycemic agent chlorpropamide is associated with a return of mononuclear leukocytes of diabetic patients toward normal insulin binding (Olefsky and Reaven, 1976a).

Certain patients with acanthosis nigricans (Winkelmann *et al.*, 1960) have a syndrome which includes in addition to skin lesions, endocrinopathies, and neoplasms, autoimmune disease and extreme resistance to both exogenous and endogenous insulin. Insulin receptor affinity is decreased in the circulatory monocytes of these patients and also in patients with ataxia telangiectasia. The serum or the IgG or IgM serum fractions from these patients will block the standardized test of binding of insulin to monocytes. These patients appear therefore to have anti-insulin receptor antibodies which decrease receptor affinity and responsiveness to insulin (Bar *et al.*, 1978; Kahn *et al.*, 1976, 1977; Dukor *et al.*, 1970; Fu *et al.*, 1974). Receptor number is decreased and affinity increased in acromegaly (Muggeo *et al.*, 1979).

Neoplastic cells of human leukemias and lymphomas have been tested for their surface array of IgG, complement, and foreign substance receptors in addition to their content of surface markers for B and T lymphocytes (Nussenzweig, 1974; Catovsky *et al.*, 1974; Stein and Kaiserling, 1974). From this effort a more precise distinction between neoplasms of macrophage and lymphocyte origin has been achieved.

Macrophage participation in the host's response to neoplasia has been the subject of reviews and symposia (Fink, 1976; Alexander, 1976; Mathe *et al.*, 1976), and the morphologic aspects of macrophage–neoplastic cell interactions require consideration here. Certain agents which activate macrophages, including viable *Mycobacterium bovis* strain BCG (Hersh *et al.*, 1977; Zbar and Tanaka,

1971; Zbar *et al.*, 1972), and the predominantly B (1→3)-linked polyglucose present in zymosan (Di Luzio *et al.*, 1976; Mansell and DiLuzio, 1976), will cause tumor regression following direct injection into the tumor. This regression *in vivo* is associated with infiltration of the tumor by macrophages and contact between macrophages and neoplastic cells as well as development of tumor necrosis (Mansell and Di Luzio, 1976). The tumor regression frequently results in prolonged host survival, but is not a panacea (Hersh *et al.*, 1977).

In an effort to define more critically the relationship between macrophages and neoplastic cells, *in vitro* studies have been undertaken. Many studies utilize syngeneic guinea pig strains to eliminate the interference from histoincompatibility. Results generally indicate that before peritoneal macrophages can kill syngeneic neoplastic cells the macrophages must be activated *in vivo*, which can be accomplished by injection of parasites (Hibbs *et al.*, 1972a,b; Krahenbuhl and Remington, 1974), or certain bacteria (Hanna *et al.*, 1976; Ghaffar *et al.*, 1974; Scott, 1975), or by repetitive injection of bacterial lipopolysaccharides, nucleotides, and polysaccharides (Kaplan *et al.*, 1974; Doe and Henson; Alexander and Evans, 1971).

The morphology of the response of activated macrophages to neoplasia has been studied sequentially by light and electron microscopy *in vivo* and by interference time-lapse cinematography and transmission electron microscopy *in vitro* (Hanna *et al.*, 1976). Inbred strain-2 guinea pigs and the transplantable syngeneic line-10 hepatocarcinoma were used. *Mycobacterium bovis* was injected into these tumors causing tumor regression and elimination of regional lymph node metastases. Between 4 and 16 days after BCG injection, activated macrophages and occasional lymphocytes touched and formed rosettes around degenerating neoplastic cells in the tumor and regional lymph nodes. Regions of adhesion between the macrophage and neoplastic cell membranes and apparent intercellular bridges occurred. Small dense bodies, probably lysosomes, accumulated in the macrophage cytoplasm in the region of cell surface contact. When the hepatocarcinoma target cells were overlayed with cells from the regional lymph nodes of BCG–tumor-cured guinea pigs for 90 min at 37°C and then washed to remove nonadherent lymphocytes and cultured for 5 days, macrophages and a subpopulation of lymphocytes remained adherent to the neoplastic cells. Interference cinematography revealed the lymph node cells pulsating at cytoplasmic blebs which represented their region of attachment to neoplastic cells (Hanna *et al.*, 1976). Electron microscopy of the guinea pig system at this 90-min incubation time revealed an apparent destabilization and discontinuity of macrophage–neoplastic cell membranes at their point of contact. After longer incubation intervals, macrophages apparently project microcytoplasmic folds and perhaps their own lysosomes into the neoplastic cells (Hanna *et al.*, 1976).

In the systems previously described it is clear that membrane interactions are associated with the response of activated macrophages to neoplastic cells. It is not clear whether these membrane interactions are mediated by one of the classical macrophage surface receptors described in previous sections. However, in a syngeneic mouse adenocarcinoma model, neoplastic cells from immune mice appear to be coated with IgG which is predominantly an Fc-receptor-binding subclass (Haskill and Fett, 1976). Macrophages are cytotoxic for these neoplastic cells, but not for the neoplastic cells from immunologically compromised mice. Moreover, sera from tumor-bearing mice can be used to protect normal mice against a subsequent challenge with this tumor. It is possible that in this system cytotoxicity is mediated by a surface IgG receptor of the macrophage both *in vitro* and *in vivo*. Another mononuclear cell type which is nonadherent and bears Fc receptors (K cell) has been implicated in cytotoxicity for virus-infected Chang liver cells (Melewicz *et al.*, 1977).

Cytophilic antibodies to the acetylcholine receptor have been implicated in the pathogenesis of experimental autoimmune myasthenia gravis in rabbits (Martinez *et al.*, 1977). In rats with a similar experimental disease, mononuclear cells including macrophages invade the motor end-plate region and are associated with postsynaptic degeneration (Engel *et al.*, 1976). Although the inciting factors in this and several other models of autoimmune diseases seems to depend on lymphocytes, the resulting tissue destruction by phagocytes is mediated in part by specific antibody through the macrophage Fc receptor (Coates and Lennon, 1973; Allt, 1975). Such phagocytic destruction through their receptors for IgG and complement is probably a final common pathway for numerous diseases involving antibodies and immune complexes (Cochrane, 1968; McPhaul and Dixon, 1970).

A benign, frequently disfiguring condition known as sinus histiocytosis with massive lymphadenopathy is of particular interest to pathologists because it must be differentiated from a malignant process (Rosai and Dorfman, 1972; Dorfman and Warnke, 1974; Azoury and Reed, 1966). An outstanding feature of this disease of unknown etiology is the presence of lymphocytes and occasionally other cells of hematopoietic origin within the cytoplasm of macrophages which crowd the sinuses of lymph nodes. One explanation for this phenomenon is the phagocytosis of opsonized blood cells mediated by macrophage IgG receptors (Lennert *et al.*, 1972). This concept is particularly attractive in view of the elevated serum IgG levels in a high percentage of these cases (Rosai and Dorfman, 1972).

Monocytes from patients with the chronic granulomatous diseases tuberculosis, sarcoidosis, and regional enteritis have increased IgG receptor activity as measured by rosetting and phagocytosis of antibody-coated erythrocytes. The latter two diseases are also associated with increased monocyte C3 receptor activity (Douglas *et al.*, 1976; Schmidt and Douglas, 1977). Macrophage activation is expected in these diseases. Activated macrophages have increased IgG receptors per cell (Table 1) probably because of their larger cellular size (Territo *et al.*, 1976; McKeever *et al.*, 1976a,b; Burgaleta *et al.*, 1978). Thus, the results are probably secondary to macrophage activation. It is of interest that the activation occurs while these mononuclear phagocytes are still monocytes in the blood stream and before they have arrived within the affected tissues. We interpret these observations as evidence for feedback of information from diseased peripheral tissue to the macrophage precursors in the blood or marrow. Macrophages of rats responding to a subacute exposure to quartz dust (Miller and

Kagan, 1977) and asbestos (Miller and Kagan, 1976) also have an increase in IgG receptors. In contrast to monocytes, macrophages located within chronic granulomas eventually show diminished receptor capability (Mariano *et al.*, 1976). Perhaps the intense hydrolytic enzyme activity within granulomas (McKeever *et al.*, 1978) is deleterious to their receptors. Diminished Fc-receptor mediated phagocytosis in murine malaria is, in contrast, ascribed to the high level of circulating immune complexes in advanced stages of the disease (Shear *et al.*, 1979).

ACKNOWLEDGMENTS. We thank Richard Asofsky, Thelma Gaither, Donald Mosier, Joost Oppenheim, Alan Rosenthal, Philip Sannes, and David Segal for advice on individual sections. We thank Fran Cameron, Dot Smith, and Andrea Taylor for their valuable editorial assistance.

REFERENCES

Abramson, N., Gelfand, E. W., Jandl, J. H., and Rosen, F. S., 1970, The interaction of human monocytes and red cells, *J. Exp. Med.* **132:**1207.

Ackerman, S. K., and Douglas, S. D., 1978, Purification of human monocytes on microexudate-coated surfaces, *J. Immunol.* **120:**1372.

Alexander, E., and Henkart, P., 1976, The adherence of human Fc receptor-bearing lymphocytes to antigen-antibody complexes. II. Morphologic alterations induced by the substrate, *J. Exp. Med.* **143:**329.

Alexander, E. L., and Wetzel, B., 1975, Human lymphocytes: Similarity of B and T cell surface morphology, *Science* **188:**732.

Alexander, P., 1976, The functions of the macrophage in malignant disease, *Ann. Rev. Med.* **27:**207.

Alexander, P., and Evans, R., 1971, Endotoxin and double stranded RNA render macrophages cytotoxic, *Nature (New Biol.)* **232:**76.

Al-Ibrahim, M. S., Chandra, R., Kishore, R., Valentine, F. T., and Lawrence, H. S., 1976, A micromethod for evaluating the phagocytic activity of human macrophages by ingestion of radiolabelled polystyrene particles, *J. Immunol. Methods* **10:**207.

Allen, J. M., and Cook, G. M. W., 1970, A study of the attachment phase of phagocytosis by murine macrophages, *Exp. Cell Res.* **59:**105.

Allt, G., 1975, The node of Ranvier in experimental allergic neuritis; an electron microscope study, *J. Neurocytol.* **4:**63.

Alper, C. A., Abramson, N., Johnston, Jr., R. B., Jandl, J. H., and Rosen, F. S., 1970, Studies *in vivo* and *in vitro* on an abnormality in the metabolism of C3 in a patient with increased susceptibility to infection, *J. Clin. Invest.* **49:**1975.

Alper, C. A., Colten, H. R., Rosen, F. S., Rabson, A. R., Macnab, G. M., and Gear, J. S. S., 1972, Homozygous deficiency of C3 in a patient with repeated infections, *Lancet* **2:**1179.

Ames, G. F-L., and Nikaido, K., 1976, Two-dimensional gel electrophoresis of membrane proteins, *Biochemistry* **15:**616.

Anderson, C. L., and Grey, H. M., 1977, Solubilization and partial characterization of cell membrane Fc receptors, *J. Immunol.* **118:**819.

Andersson, J., Lafleur, L., and Melchers, F., 1974, IgG in bone marrow-derived lymphocytes. Synthesis, surface deposition, turnover and carbohydrate composition in unstimulated mouse B cells, *Eur. J. Immunol.* **4:**170.

Anwar, A. R. E., and Kay, A. B., 1977, Membrane receptors for IgG and complement (C4, C3b and C3d) on human eosinophils and neutrophils and their relation to eosinophilia, *J. Immunol.* **119:**976.

Arend, W. P., and Mannik, M., 1971, Studies on antigen-antibody complexes. II. Quantification of tissue uptake of soluble complexes in normal and complement-depleted rabbits, *J. Immunol.* **107:**63.

Arend, W. P., and Mannik, M., 1972, *In vitro* adherence of soluble immune complexes to macrophages, *J. Exp. Med.* **136:**514.

Arend, W. P., and Mannik, M., 1973, The macrophage receptor for IgG: Number and affinity of binding sites, *J. Immunol.* **110:**1455.

Arnaiz-Villena, A., and Festenstein, H., 1975, 4a(W4) and 4b(W6) human histocompatibility antigens are specifically associated with complement receptors, *Nature* **258:**732.

Askenase, P. W., and Hayden, B. J., 1974, Cytophilic antibodies in mice contact-sensitized with oxazolone. Immunochemical characterization and preferential binding to trypsin-sensitive macrophage receptors, *Immunology* **27:**563.

Aswanikumar, S., Corcoran, B., Schiffmann, E., Day, A. R., Freer, R. J., Showell, H. J., Becker, E. L., and Pert, C. B., 1977, Demonstration of a receptor on rabbit neutrophils for chemotactic peptides, *Biochem. Biophys. Res. Commun.* **74:**810.

Atkinson, J. P., Michael, J. M., Chaplin, Jr., H., and Parker, C. W., 1977, Modulation of macrophage C3b receptor function by cytochalasin-sensitive structures, *J. Immunol.* **118:**1292.

Avrameas, S., and Leduc, E. H., 1970, Detection of simultaneous antibody synthesis in plasma cells and specialized lymphocytes in rabbit lymph nodes, *J. Exp. Med.* **131:**1137.

Azoury, F. J., and Reed, R. J., 1966, Histiocytosis. Report of an unusual case, *N Engl. J. Med.* **274:**928.

Baggiolini, M., de Duve, C., Masson, P. L., and Heremans, J. F., 1970, Association of lactoferrin with specific granules in rabbit heterophil leukocytes, *J. Exp. Med.* **131:**559.

Balentine, J. D., McKeever, P. E., and Anderson, C. T., 1974, Ultrastructural pathology of central nervous system oxygen toxicity, in: *Proceedings of the Fifth International Hyperbaric Congress* (W. G. Trapp, E. W. Banister, A. J. Davidson, and P. A. Trapp, eds.), pp. 109–115, Vol. 1, Simon Fraser University Press, Burnaby 2, British Columbia.

Bar, R. S., Gorden, P., Roth, J., Kahn, C. R., and de Meyts, P., 1976, Fluctuations in the affinity and concentration of insulin receptors on circulating monocytes of obese patients. Effects of starvation, refeeding, and dieting, *J. Clin. Invest.* **58:**1123.

Bar, R. S., Kahn, C. R., and Karen, H. S., 1977, Insulin inhibition of antibody-dependent cytoxicity and insulin receptors in macrophages, *Nature* **265:**632.

Bar, R. S., Levis, W. R., Rechler, M. M., Harrison, L. C., Siebert, C., Podskalny, J., Roth, J., and Muggeo, M., 1978, Extreme insulin resistance in ataxia telangiectasia: Defect in affinity of insulin receptors, *N. Engl. J. Med.* **298:**1164.

Bartfeld, H., and Kelly, R., 1968, Mediation of delayed hypersensitivity by peripheral blood lymphocytes *in vitro* and by their products *in vivo* and *in vitro:* Morphology of *in vitro* lymphocyte-macrophage interaction, *J. Immunol.* **100:**1000.

Belkin, M., and Hardy, W. G., 1961, Relation between water permeability and integrity of sulfhydryl groups in malignant and normal cells, *J. Biophys. Biochem. Cytol.* **9:**733.

Beller, D. I., Farr, A. G., and Unanue, E. R., 1978, Regulation of lymphocyte proliferation and differentiation by macrophages, *Fed. Proc.* **37:**91.

Benacerraf, B., 1968, Cytophilic immunoglobulins and delayed hypersensitivity, *Fed. Proc.* **27:**46.

Benacerraf, B., and Sebestyen, M. M., 1957, Effect of bacterial endotoxins on the reticuloendothelial system, *Fed. Proc.* **16:**860.

Benacerraf, B., Sebestyen, M., and Cooper, N. S., 1959, The clearance of antigen–antibody complexes from the blood by the reticulo-endothelial system, *J. Immunol.* **82:**131.

Bennett, B., and Bloom, B. R., 1968, Reactions *in vivo* and *in vitro* produced by a soluble substance associated with delayed-type hypersensitivity, *Proc. Natl. Acad. Sci.* **59:**756.

Ben-Sasson, S. Z., Lipscomb, M. F., Tucker, T. F., and Uhr, J. W., 1977, Specific binding of T lymphocytes to macrophages. II. Role of macrophage-associated antigen, *J. Immunol.* **119:**1493.

Berken, A., and Benacerraf, B., 1966, Properties of antibodies cytophilic for macrophages, *J. Exp. Med.* **123:**119.

Berlin, R. D., and Oliver, J. M., 1978, Analogous ultrastructure and surface properties during capping and phagocytosis in leukocytes, *J. Cell Biol.* **77:**789.

Berman, L., 1966, Lymphocytes and macrophages *in vitro.* Their activities in relation to functions of small lymphocytes, *Lab. Invest.* **15:**1084.

Berry, L. J., and Spies, T. D., 1949, Phagocytosis, *Medicine* **28**:239.

Bhisey, A. N., and Freed, J. J., 1975, Remnant motility of macrophages treated with cytochalasin B in the presence of colchicine, *Exp. Cell Res.* **95**:376.

Bianco, C., 1976, Methods for the study of macrophage Fc and C3 receptors, in: *In Vitro Methods in Cell-Mediated and Tumor Immunity* (B. R. Bloom, and J. R. David, eds.), pp. 407–415, Academic Press, New York.

Bianco, C., Griffin, F. M., and Silverstein, S. C., 1975, Studies of the macrophage complement receptor. Alteration of receptor function upon macrophage activation, *J. Exp. Med.* **141**:1278.

Bianco, C., Eden, A., and Cohn, Z. A., 1976, The induction of macrophage spreading: Role of coagulation factors and the complement system, *J. Exp. Med.* **144**:1531.

Bianco, C., Gotze, O., and Cohn, Z. A., 1979, Regulation of macrophage migration by products of the complement system, *Proc. Natl. Acad. Sci. USA* **76**:888.

Biberfeld, P., 1971, Uropod formation in phytohaemagglutinin (PHA) stimulated lymphocytes, *Exp. Cell Res.* **66**:433.

Binz, H., and Wigzell, H., 1975, Shared idiotypic determinants on B and T lymphocytes reactive against the same antigenic determinants. II. Determination of frequency and characteristics of idiotypic T and B lymphocytes in normal rats using direct visualization, *J. Exp. Med.* **142**:1218.

Blom, J., 1977, The ultrastructure of contact zones between plasma cells and macrophages in the bone marrow of patients with multiple myeloma, *Acta. Pathol. Microbiol. Scand. A* **85**:345.

Boyden, S. V., 1964, Cytophilic antibody in guinea-pigs with delayed-type hypersensitivity, *Immunology* **7**:474.

Braendstrup, O., Werdelin, O., Shevach, E. M., and Rosenthal, A. S., 1979, Macrophage-lymphocyte clusters in the immune response to soluble protein antigen *in vitro*. VII. Genetically restricted and nonrestricted physical interactions, *J. Immunol.* **122**:1608.

Bretton, R., Ternynck, T., and Avrameas, S., 1972, Comparison of peroxidase and ferritin labelling of cell surface antigens, *Exp. Cell Res.* **71**:145.

Briscoe, J. C., 1908, An experimental investigation of the phagocytic action of the alveolar cells of the lung, *J. Pathol. Bacteriol.* **12**:66.

Burgaleta, C., Territo, M. C., Quan, S. G., and Golde, D. W., 1978, Glucan-activated macrophages: Functional characteristics and surface morphology, *J. Reticuloendothel. Soc.* **23**:195.

Calderon, J., and Unanue, E. R., 1974, The release of antigen molecules from macrophages: characterization of the phenomena, *J. Immunol.* **112**:1804.

Capron, A., Dessaint, J-P., Capron, M., and Bazin, H., 1975, Specific IgE antibodies in immune adherence of normal macrophages to *Schistosoma mansoni* schistosomules, *Nature* **253**:474.

Carr, I., 1975, The reticular cell, the reticulum cell, and the macrophage, in: *Mononuclear Phagocytes in Immunity, Infection, and Pathology* (R. van Furth, ed.), pp. 119–127, Blackwell, London.

Carr, K., and Carr, I., 1970, How cells settle on glass: a study by light and scanning electron microscopy of some properties of normal and stimulated macrophages, *Z. Zellforsch. Mikrosk. Anat.* **105**:234.

Casley-Smith, J. R., 1969, Endocytosis: The different energy requirements for the uptake of particles by small and large vesicles into peritoneal macrophages, *J. Microsc.* **90**:15.

Catovsky, D., Pettit, J. E., Galetto, J., Okos, A., and Galton, D. A. G., 1974, The B-lymphocyte nature of the hairy cell of leukaemic reticuloendotheliosis, *Br. J. Haematol.* **26**:29.

Chambers, V. C., 1973, The use of ruthenium red in an electron microscope study of cytophagocytosis, *J. Cell Biol.* **57**:874.

Chen, L-T., Eden, A., Nussenzweig, V., and Weiss, L., 1972, Electron microscopic study of the lymphocytes capable of binding antigen-antibody-complement complexes, *Cell. Immunol.* **4**:279.

Choi, T. K., Sleight, D. R., and Nisonoff, A., 1974, General method for isolation and recovery of B cells bearing specific receptors, *J. Exp. Med.* **139**:761.

Cline, M. J., Warner, N. L., and Metcalf, D., 1972, Identification of the bone marrow colony mononuclear phagocyte as a macrophage, *Blood* **39**:326.

Coates, A. S., and Lennon, V. A., 1973, Lymphocytes binding basic protein of myelin; cytophilic serum antibody and effect of adjuvant, *Immunology* **24**:425.

Cochrane, C. G., 1968, Immunologic tissue injury mediated by neutrophilic leukocytes, *Adv. Immunol.* **9**:97.

Cohen, B. E., Rosenthal, A. S., and Paul, W. E., 1973, Antigen-macrophage interaction. I. Hapten-specific inhibition of antigen interaction with macrophages from immune animals, *J. Immunol.* **111**:811.

Cohn, Z. A., 1968, The structure and function of monocytes and macrophages, *Adv. Immunol.* **9**:163.

Collet, A. J., and Petrik, P., 1971, Electron microscopic study of the *in vivo* erythrophagocytosis by alveolar macrophages of the cat. I. Early period: Hemolysis, *Z. Zellforsch. Mikrosk. Anat.* **116**:464.

Colvin, R. B., and Dvorak, H. F., 1975, Fibrinogen/fibrin on the surface of macrophages: detection, distribution, binding requirements, and possible role in macrophage adherence phenomena, *J. Exp. Med.* **142**:1377.

Cooper, S. M., and Sambray, Y., 1976, Isolation of a murine leukemia Fc receptor by selective release induced by surface redistribution, *J. Immunol.* **117**:511.

Cossman, J., Glorioso, J. C., and Adler, R., 1978, Complement receptors: Specific detection by molecular complexes, *J. Immunol. Methods* **19**:227.

Crothers, D. M., and Metzger, H., 1972, The influence of polyvalency on the binding properties of antibodies, *Immunochemistry* **9**:341.

Czop, J. K., Fearon, D. T., and Austen, K. F., 1978, Opsonin-independent phagocytosis of activators of the alternative complement pathway by human monocytes, *J. Immunol.* **120**:1132.

Daems, W. T., and Brederoo, P., 1973, Electron microscopical studies on the structure, phagocytic properties, and peroxidatic activity of resident and exudate peritoneal macrophages in the guinea pig, *Z. Zellforsch. Mikrosk. Anat.* **144**:247.

Dannenberg, Jr., A. M., 1968, Cellular hypersensitivity and cellular immunity in the pathogenesis of tuberculosis: specificity, systemic and local nature, and associated macrophage enzymes, *Bacteriol. Rev.* **32**:85.

da Silva, W. D., Eisele, J. W., and Lepow, I. H., 1967, Complement as a mediator of inflammation. III. Purification of the activity with anaphylatoxin properties generated by interaction of the first four components of complement and its identification as a cleavage product of C'3, *J. Exp. Med.* **126**:1027.

Daughaday, C. C., and Douglas, S. D., 1976, Membrane receptors on rabbit and human pulmonary alveolar macrophages, *J. Reticuloendothel. Soc.* **19**:37.

Davey, M. J., and Asherson, G. L., 1967, Cytophilic antibody. I. Nature of the macrophage receptor, *Immunology* **12**:13.

David, J. R., 1966, Delayed hypersensitivity *in vitro:* Its mediation by cell-free substances formed by lymphoid cell-antigen interaction, *Proc. Natl. Acad. Sci. USA* **56**:72.

David, J. R., and David, R. A., 1972, Cellular hypersensitivity and immunity: Inhibition of macrophage migration and the lymphocyte mediators, in: *Progress in Allergy* (P. Kallos, B. H. Waksman, and A. de Weck, eds.), p. 300, Karger, Basel.

David, J. R., Lawrence, H. S., and Thomas, L., 1964, The *in vitro* densitization of sensitive cells by trypsin, *J. Exp. Med.* **120**:1189.

Davies, W. A., and Stossel, T. P., 1977, Peripheral hyaline blebs (podosomes) of macrophages, *J. Cell Biol.* **75**:941.

Deacon, N. J., and Ebringer, A., 1977, Fucose incorporation into oocyte-synthesized rat immunoglobulins, *FEBS Lett.* **79**:191.

DePierre, J. W., and Karnovsky, M. L., 1973, Plasma membranes of mammalian cells. A review of methods for their characterization and isolation, *J. Cell Biol.* **56**:275.

DeSouza, N. J., and Nes, W. R., 1968, Chemotactic and anaphylatoxic fragment cleaved from the fifth component of guinea pig complement, *Science* **162**:361.

Diamond, B., Birshtein, B. K., and Scharff, M. D., 1979, Site of binding of mouse IgG_{2b} to the Fc receptor on mouse macrophages, *J. Exp. Med.* **150**:721.

Dickler, H. B., 1974, Studies of the human lymphocyte receptor for heat-aggregated or antigen-complexed immunoglobulin, *J. Exp. Med.* **140**:508.

Dickler, H. B., 1976, Lymphocyte receptors for immunoglobulin, *Adv. Immunol.* **24**:167.

Dickler, H. B., and Sachs, D. H., 1974, Evidence for identity or close association of the Fc receptor of B lymphocytes and alloantigens determined by the *Ir* region of the *H-2* complex, *J. Exp. Med.* **140**:779.

Diener, E., and Paetkau, V. H., 1972, Antigen recognition; early surface-receptor phenomena induced by binding of a tritium-labeled antigen, *Proc. Natl. Acad. Sci. USA* **69**:2364.

Diener, E., Kraft, N., and Armstrong, W. D., 1973, Antigen recognition. I. Immunological significance of antigen-binding cells, *Cell. Immunol.* **6:**80.

Diener, E., Kraft, N., Lee, K.-C., and Shiozaw a, C., 1976, Antigen recognition. IV. Discrimination by antigen-binding immunocompetent B cells between immunity and tolerance is determined by adherent cells, *J. Exp. Med.* **143:**805.

Di Luzio, N. R., McNamee, R., Jones, E., Cook, J. A., and Hoffman, E. O., 1976, The employment of glucan and glucan activated macrophages in the enhancement of host resistance to malignancies in experiemental animals, in: *The Macrophage in Neoplasia* (M. A. Fink, ed.), pp. 181–198, Academic Press, New York.

Doe, W. F., and Henson, P. M., 1979, Macrophage stimulation by bacterial lipopolysaccharides. III. Selective unresponsiveness of C3H/HeJ macrophages to the lipid A differentiation signal, *J. Immunol.* **123:**2304.

Donné, A., 1884, *Course de Microscopie* (J. B. Balliere, ed.), pp. 550–560.

Dorfman, R. F., and Warnke, R., 1974, Lymphadenopathy simulating the malignant lymphomas, *Hum. Pathol.* **5:**519.

Dorrington, K. J., 1976, Properties of the Fc receptor on macrophages and monocytes, *Immunol. Commun.* **5:**263.

Douglas, S. D., 1976, Human monocyte spreading *in vitro*—inducers and effects on Fc and C3 receptors, *Cell. Immunol.* **21:**344.

Douglas, S. D., 1978, Alternatives in intramembrane particle distribution during interaction of erythrocyte-bound ligands with immunoprotein receptors, *J. Immunol.* **120:**151.

Douglas, S. D., and Huber, H., 1972, Electron microscopic studies of human monocyte and lymphocyte interaction with immunoglobulin- and complement-coated erythrocytes, *Exp. Cell Res.* **70:**161.

Douglas, S. D., Daughaday, C. C., Schmidt, M. E., and Siltzbach, L. E., 1976, Kinetics of monocyte receptor activity for immunoproteins in patients with sarcoidosis, *Ann. N.Y. Acad. Sci.* **278:**190.

Dukor, P., Bianco, C., and Nussenzweig, V., 1970, Tissue localization of lymphocytes bearing a membrane receptor for antigen-antibody-complement complexes, *Proc. Natl. Acad. Sci. USA* **67:**991.

Dvorak, A. M., Hammond, M. E., Dvorak, H. F., and Karnovsky, M. J., 1972, Loss of cell surface material from peritoneal exudate cells associated with lymphocyte-mediated inhibition of macrophage migration from capillary tubes, *Lab. Invest.* **27:**561.

Dvorak, A. M., Connell, A. B., Proppe, K., and Dvorak, H. F., 1978, Immunologic rejection of mammary adenocarcinoma (TA3-St) in C57BL/6 mice: Participation of neutrophils and activated macrophages with fibrin formation, *J. Immunol.* **120:**1240.

Edelson, P. J., and Cohn, Z. A., 1976, 5′-nucleotidase activity of mouse peritoneal macrophages. I. Synthesis and degradation in resident and inflammatory populations, *J. Exp. Med.* **144:**1581.

Edelson, P. J., and Erbs, C., 1978, Biochemical and functional characteristics of the plasma membrane of macrophages from BCG-infected mice, *J. Immunol.* **120:**1532.

Edelson, P. J., Zwiebel, R., and Cohn, Z. A., 1975, The pinocytic rate of activated macrophages, *J. Exp. Med.*, **142:**1150.

Eguchi, M., Sannes, P. L., and Spicer, S. S., 1979, Alterations in ultrastructural cytochemistry of complex carbohydrates of rat peritoneal macrophages during phagocytosis, *J. Reticuloendothel. Soc.* **25:**207.

Ehlenberger, A. G., and Nussenzweig, V., 1977, The role of membrane receptors for C3b and C3d in phagocytosis, *J. Exp. Med.* **145:**357.

Ellner, J. J., and Rosenthal, A. S., 1975, Quantitative and immunologic aspects of handling of 2,4-dinitrophenyl guinea pig albumin by macrophages, *J. Immunol.* **114:**1563.

Emeis, J. J., 1976, Morphologic and cytochemical heterogeneity of the cell coat of rat liver Kupffer cells, *J. Reticuloendothel. Soc.* **20:**31.

Emeis, J. J., and Lindeman, J., 1976, Rat liver macrophages will not phagocytose fibrin during disseminated intravascular coagulation, *Haemostasis* **5:**193.

Engel, A. G., Tsujihata, M., Lindstrom, J., and Lennon, V. A., 1976, The motor end-plate in myasthenia gravis and in experimental autoimmune myasthenia gravis. A quantitative ultrastructural study, *Ann. N.Y. Acad. Sci. USA* **274:**60.

Farr, A. G., and De Bruyn, P. H., 1975, Macrophage-lymphocyte clusters in lymph nodes: A possible substrate for cellular interactions in the immune response, *Am. J. Anat.* **144:**209.

Feinstein, A., and Rowe, A. J., 1965, Molecular mechanism of formation of an antigen-antibody complex, *Nature* **205:**147.

Feldmann, M., 1974, Immunogenicity *in vitro:* Structural correlation, in: *Contemporary Topics in Molecular Immunology* (G. L. Ada, ed.), pp. 57–79, Plenum Press, New York.

Feldmann, M., and Nossal, G. J. V., 1972, Tolerance, enhancement and the regulation of interactions between T cells, B cells and macrophages, *Transplant Rev.* **13:**3.

Feldman, J. D., and Pollock, E. M., 1974, Endocytosis by macrophages of altered soluble protein. The effect of binding to particulate surfaces and of IgM and IgG antibody, *J. Immunol.* **113:**329.

Fenn, W. O., 1922, The adhesiveness of leucocytes to solid surfaces, *J. Gen. Physiol.* **5:**143.

Ferreira, A., Fotino, M., and Nussenzweig, V., 1976, Relationship between 4a and 4b HLA-determined specificities and C3 receptors of leukocyte membrane, *Eur. J. Immunol.* **6:**832.

Fillet, G., Cook, J. D., and Finch, C. A., 1974, Storage iron kinetics. VII. A biological model for reticuloendothelial iron transport, *J. Clin. Invest.* **53:**1527.

Fink, M. A., 1976, *The Macrophage in Neoplasia* (M. A. Fink, ed.), Academic Press, New York.

Franke, W. W., Kartenbeck, J., Zentgraf, H., Scheer, U., and Falk, H., 1971, Membrane-to-membrane cross-bridges, *J. Cell Biol.* **51:**881.

Fu, S. M., Winchester, R. J., Rai, K. R., and Kunkel, H. G., 1974, Hairy cell leukemia: Proliferation of a cell with phagocytic and B-lymphocyte properties, *Scand. J. Immunol.* **3:**847.

Fujita, T., Miyoshi, M., and Murakami, T., 1972, Scanning electron microscope observation of the dog mesenteric lymph node, *Z. Zellforsch.* **133:**147.

Gardner, D. E., Graham, J. A., Miller, F. J., Illing, J. W., and Coffin, D. L., 1973, Technique for differentiating particles that are cell-associated or ingested by macrophages, *Appl. Microbiol.* **25:**471.

Gavin, III, J. R., Roth, J., Jen, P., and Freychet, P., 1972. Insulin receptors in human circulating cells and fibroblasts, *Proc. Natl. Acad. Sci. USA* **69:**747.

Gavin, III, J. R., Roth, J., Neville, Jr., D. M., De Meyts, P., and Buell, D. N., 1974. Insulin-dependent regulation of insulin receptor concentrations: A direct demonstration in cell culture, *Proc.Natl. Acad. Sci. USA* **71:**84.

Gee, J. B. L., Sachs, F. L., McKeever, P. E., Douglas, J. S., and Malawista, S. E., 1973, Phagocytosis by alveolar macrophages: Pharmacologic features, *Chest* **63:**20.

Gemsa, D., Seitz, M., Kramer, W., Till, G., and Resch, K., 1978, The effects of phagocytosis, dextran sulfate, and cell damage on PGE_1 sensitivity and PGE_1 production of macrophages, *J. Immunol.* **120:**1187.

Gersten, D. M., Fogler, W. E., and Fidler, I. J., 1977, Quantitative analysis of macrophage phagocytosis by uptake of particulate ^{192}iridium, *J. Immunol. Methods* **17:**349.

Ghaffar, A., Cullen, R. T., Dunbar, N., and Woodruff, M. F. A., 1974, Anti-tumour effect *in vitro* of lymphocytes and macrophages from mice treated with *Corynebacterium parvum, Br. J. Cancer* **29:**199.

Gigli, I., and Nelson, Jr., R. A., 1968, Complement dependent immune phagocytosis I. Requirements for C'1, C'4, C'2, C'3, *Exp. Cell Res.* **51:**45.

Goldner, R. D., and Adams, D. O., 1977, The structure of mononuclear phagocytes differentiating *in vivo, Am. J. Pathol.* **89:**335.

Gove, P. B. (ed.), 1967, *Webster's Seventh New Collegiate Dictionary,* G. and C. Merriam Company, Springfield.

Grey, H. M., Anderson, C. L., Heusser, C. H., Borthistle, B. K., Von Eschen, K. B., and Chiller, J. M., 1976, Structural and functional heterogeneity of F_c receptors, *Cold Spring Harbor Symp. Quant. Biol.* **41:**315.

Griffin, Jr., F. M., and Silverstein, S. C., 1974, Segmental response of the macrophage plasma membrane to a phagocytic stimulus, *J. Exp. Med.* **139:**323.

Griffin, Jr., F. M., and Silverstein, S. C., 1975, Discrimination by the macrophage during the ingestion phase of phagocytosis, in: *Mononuclear Phagocytes in Immunity, Infection, and Pathology* (R. van Furth, ed.), pp. 283–286, Blackwell, Oxford.

Griffin, Jr., F. M., Bianco, C., and Silverstein, S. C., 1975a, Characterization of the macrophage

receptor for complement and demonstration of its functional independence from the receptor for the F_c portion of immunoglobulin G, *J. Exp. Med.* **141:**1269.
Griffin, Jr., F. M., Griffin, J. A., Leider, J. E., and Silverstein, S. C., 1975b, Studies on the mechanism of phagocytosis. I. Requirements for circumferential attachment of particle-bound ligands to specific receptors on the macrophage plasma membrane, *J. Exp. Med.* **142:**1263.
Griffin, Jr., F. M., Griffin, J. A., and Silverstein, S. C., 1976, Studies on the mechanism of phagocytosis. II. The interaction of macrophages with anti-immunoglobulin IgG-coated bone marrow-derived lymphocytes, *J. Exp. Med.* **144:**788.
Gudat, F. G., and Villiger, W., 1973, A scanning and transmission electron microscope study of antigen binding sites on rosette-forming cells, *J. Exp. Med.* **137:**483.
Hanifin, J. M., and Cline, M. J., 1970, Human monocytes and macrophages. Interaction with antigen and lymphocytes, *J. Cell Biol.* **46:**97.
Hanna, Jr., M. G., Bucana, C., Hobbs, B., and Fidler, I. J., 1976, Morphologic aspects of tumor cell cytotoxicity by effector cells of the macrophage-histiocyte compartment: *In vitro* and *in vivo* studies in BCG-mediated tumor regression, in: *The Macrophage in Neoplasia* (M. A. Fink, ed.), pp. 113–133, Academic Press, New York.
Haranaka, K., Matsuo, M., and Mashimo, K., 1977, The enhancement of phagocytosis and intracellular killing of *Pseudomonas aeruginosa* and its common antigen (OEP) coated latex particles by mouse spleen macrophages to which anti-OEP-IgG and gentamicin have been added, *Jpn. J. Exp. Med.* **47:**35.
Harris, G., 1965, Studies of the mechanism of antigen stimulation of DNA synthesis in rabbit spleen cultures, *Immunology* **9:**529.
Haskill, J. S., and Fett, J. W., 1976, Possible evidence for antibody-dependent macrophage-mediated cytotoxicity directed against murine adenocarcinoma cells *in vivo*, *J. Immunol.* **117:**1992.
Hausmann, K., Wulfhekel, U., Düllmann, J., and Kuse, R., 1976, Iron storage in macrophages and endothelial cells. Histochemistry, ultrastructure, and clinical significance, *Blut* **32:**289.
Hay, F. C., Torrigiani, G., and Roitt, I. M., 1972, The binding of human IgG subclasses to human monocytes, *Eur. J. Immunol.* **2:**257.
Heilman, D. H., 1977, Regulation of endotoxin-induced inhibition of macrophage migration by fresh serum, *Infect. Immun.* **17:**371.
Henson, P. M., 1971, The immunologic release of constituents from neutrophil leukocytes. I. The role of antibody and complement on nonphagocytosable surfaces or phagocytosable particles, *J. Immunol.* **107:**1535.
Heppleston, A. G., 1963, The disposal of inhaled particulate matter; a unifying hypothesis, *Am. J. Pathol.* **42:**119.
Herbert, W. J., and Wilkinson, P. C., 1977, *A Dictionary of Immunology*, J. B. Lippincott, Philadelphia.
Hersh, E. M., and Harris, J. E., 1968, Macrophage-lymphocyte interaction in the antigen-induced blastogenic response of human peripheral blood leukocytes, *J. Immunol.* **100:**1184.
Hersh, E. M., Gutterman, J. U., and Mavligit, G. M., 1977, BCG as adjuvant immunotherapy for neoplasia, *Ann. Rev. Med.* **28:**489.
Heusser, C. H., Anderson, C. L., and Grey, H. M., 1977, Receptors for IgG: Subclass specificity of receptors on different mouse cell types and the definition of two distinct receptors on a macrophage cell line, *J. Exp. Med.* **145:**1316.
Hibbs, Jr., J. B., 1976, The macrophage as a tumoricidal effector cell: A review of *in vivo* and *in vitro* studies on the mechanism of the activated macrophage nonspecific cytotoxic reaction, in: *The Macrophage in Neoplasia* (M. A. Fink, ed.), pp. 83–111, Academic Press, New York.
Hibbs, Jr., J. B., Lambert, Jr., L. H., and Remington, J. S., 1972a, Possible role of macrophage mediated nonspecific cytotoxicity in tumor resistance, *Nature, New Biol.* **235:**48.
Hibbs, Jr., J. B., Lambert, Jr., L. H., and Remington, J. S., 1972b, Control of carcinogenesis: A possible role for the activated macrophage, *Science* **177:**998.
Hirsch, J. G., 1965, Phagocytosis, in: *Annual Review of Microbiology* (C. E. Clifton, ed., S. Raffel and M. P. Starr, assoc. eds.), pp. 339–350, Annual Reviews, Palo Alto.
Hoefsmit, E. C. M., 1975, Mononuclear phagocytes, reticulum cells, and dendritic cells in lymphoid tissues, in: *Mononuclear Phagocytes in Immunity, Infection, and Pathology* (R. van Furth, ed.), pp. 129–146, Blackwell, Oxford.

Holland, P., Holland, N. H., and Cohn, Z. A., 1972, The selective inhibition of macrophage phagocytic receptors by anti-membrane antibodies, *J. Exp. Med.* **135:**458.

How, M. J., Brimacombe, J. S., and Stacey, M., 1964, The pneumococcal polysaccharides, in: *Advances in Carbohydrate Chemistry* (M. L. Wolfrom, ed., R. S. Tipson, assoc. ed.), pp. 303–358, Academic Press, New York.

Howard, J. G., and Benacerraf, B., 1966, Properties of macrophage receptors for cytophilic antibodies, *Br. J. Exp. Pathol.* **47:**193.

Huber, H., and Fudenberg, H. H., 1968, Receptor sites of human monocytes for IgG, *Int. Arch. Allergy Appl. Immunol.* **34:**18.

Huber, H., and Holm, G., 1975, Surface receptors of mononuclear phagocytes: effect of immune complexes on *in vitro* function in human monocytes, in: *Mononuclear Phagocytes in Immunity, Infection, and Pathology* (R. van Furth, ed.), pp. 291–301, Blackwell, Oxford.

Huber, H., Douglas, S. D., Nusbacher, J., Kochwa, S., and Rosenfield, R. E., 1971, Subclass specificity for human monocyte receptor sites, *Nature (London)* **229:**419.

Isturiz, M. A., Sandberg, A. L., Schiffmann, E., Wahl, S. M., and Notkins, A. L., 1978, Chemotactic antibody, *Science* **200:**554.

Itoh, G., Hirabayashi, N., and Kurashina, M., 1975, Electron microscopic autoradiographic and electron microscopic immunohistochemical studies on the anti-HPO antibody-producing cells, *Exp. Cell Res.* **95:**287.

Jacobson, B. S., and Branton, D., 1977, Plasma membrane: Rapid isolation and exposure of the cytoplasmic surface by use of positively charged beads, *Science* **195:**302.

Jancik, J. M., and Schauer, R., 1978, Sequestration of neuraminidase-treated erythrocytes. Studies on its topographic, morphologic and immunologic aspects, *Cell Tissue Res.* **186:**209.

Jókay, I., and Karczag, E., 1973, Thrombin-induced macrophage disappearance reaction in mice, *Experientia (Basel)* **29:**334.

Jones, P. P., 1977, Analysis of H-2 and Ia molecules by two-dimensional gel electrophoresis, *J. Exp. Med.* **146:**1261.

Jones, T. C., 1975, Attachment and ingestion phases of phagocytosis, in: *Mononuclear Phagocytes in Immunity, Infection and Pathology* (R. van Furth, ed.), pp. 269–282, Blackwell, Oxford.

Kahn, C. R., and Roth, J., 1975, Cell membrane receptors for polypeptide hormones, *Am. J. Clin. Pathol.* **63:**656.

Kahn, C. R., Flier, J. S., Bar, R. S., Archer, J. A., Gorden, P., Martin, M. M., and Roth, J., 1976, The syndromes of insulin resistance and acanthosis nigricans. Insulin-receptor disorders in man, *N. Engl. J. Med.* **294:**739.

Kahn, C. R., Megyesi, K., Bar, R. S., Eastman, R. C., and Flier, J. S., 1977, Receptors for peptide hormones. New insights into the pathophysiology of disease states in man, *Ann. Intern. Med.* **86:**205.

Kaplan, A. M., Morahan, P. S., and Regelson, W., 1974, Induction of macrophage-mediated tumor-cell cytotoxicity by pyran copolymer, *J. Natl. Cancer Inst.* **52:**1919.

Kaplan, G., 1977, Differences in the mode of phagocytosis with F_c and C_3 receptors in macrophages, *Scand. J. Immunol.* **6:**797.

Kaplan, G., Gaudernack, G., and Seljelid, R., 1975, Localization of receptors and early events of phagocytosis in the macrophage, *Exp. Cell Res.* **95:**365.

Kaplan, M. H., Coons, A. H., and Deane, H. W., 1950, Localization of antigen in tissue cells. III. Cellular distribution of pneumococcal polysaccharides types II and III in the mouse, *J. Exp. Med.* **91:**15.

Karnovsky, M. L., Lazdins, J., Drath, D., and Harper, A., 1975, Biochemical characteristics of activated macrophages, *Ann. N.Y. Acad. Sci.* **256:**266.

Karnovsky, M. L., Drath, D., and Lazdins, J., 1976, Biochemical aspects of the function of the reticulo-endothelial system, in: *Biochemistry and Physiology of the RES* (S. M. Reichard and M. R. Escobar, eds.), pp. 121–129, Plenum Press, New York.

Kassis, A. I., Aikawa, M., and Mahmoud, A. F., 1979, Mouse antibody-dependent eosinophil and macrophage adherence and damage to schistosomula of *Schistosoma mansoni*, *J. Immunol.* **122:**398.

Kates, M., and Sastry, P. S., 1969, Phospholipase D, in: *Methods in Enzymology, Lipids* (J. M. Lowenstein, ed.), pp. 197–209, Academic Press, New York.

Kay, M. M. B., 1975, Mechanism of removal of senescent cells by human macrophages *in situ, Proc. Natl. Acad. Sci.* **72:**3521.
King, J. R., Prescott, B., and Caldes, G., 1971, Differences in the carbohydrate content of *Streptococcus faecium* F24 and its stable L form, in: *Bacteriological Proceedings* (W. A. Wood, H. Gooder, and L. Levintow, eds.), p. 43, Abstracts-American Society for Microbiology, Minneapolis.
Knutson, D. W., Kijlstra, A., and van Es, L. A., 1977, Association and dissociation of aggregated IgG from rat peritoneal macrophages, *J. Exp. Med.* **145:**1368.
Knyszynski, A., Lebovich, S. J., and Danon, D., 1977, Phagocytosis of "Old" red blood cells by macrophages from syngeneic mice *in vitro, Exp. Hematol.* **5:**480.
Koren, H. S., Handwerger, B. S., and Wunderlich, J. R., 1975, Identification of macrophage-like characteristics in a cultured murine tumor line, *J. Immunol.* **114:**894.
Kossard, S., and Nelson, D. S., 1968, Studies on cytophilic antibodies. IV. The effects of proteolytic enzymes (trypsin and papain) on the attachment to macrophages of cytophilic antibodies, *Aust. J. Exp. Biol. Med. Sci.* **46:**63.
Krahenbuhl, J. L., and Remington, J. S., 1974, The role of activated macrophages in specific and nonspecific cytostasis of tumor cells, *J. Immunol.* **113:**507.
Kronman, J. H., Goldman, M., Lin, P. S., Goldman, L. B., and Kliment, C., 1977, Evaluation of intracytoplasmic particles in histiocytes after endodontic therapy with a hydrophilic plastic, *J. Dent. Res.* **56:**795.
Kuhn, R. E., and Cassida, G. W., 1978, An indirect, quantifiable assay for cytophilic antibody, *J. Immunol. Methods* **19:**387.
LaBelle, C. W., and Brieger, H., 1960, The fate of inhaled particles in the early post-exposure period. II. The role of pulmonary phagocytosis. *Arch. Environ. Health.* **1:**423.
Larm, O., and Lindberg, B., 1976, The pneumococcal polysaccharides: A re-examination, in: *Advances in Carbohydrate Chemistry and Biochemistry* (R. S. Tipson and D. Horton, eds.), pp. 295–322, Academic Press, New York.
Lawrence, D. A., Weigle, W. O., and Spiegelberg, H. L., 1975, Immunoglobulins cytophilic for human lymphocytes, monocytes, and neutrophils, *J. Clin. Invest.* **55:**368.
Lay, W. H., and Nussenzweig, V., 1968, Receptors for complement on leukocytes, *J. Exp. Med.* **128:**991.
Lay, W. H., and Nussenzweig, V., 1969, Ca^{++}-dependent binding of antigen-19S antibody complexes to macrophages, *J. Immunol.* **102:**1172.
Leduc, E. H., Coons, A. H., and Connolly, J. M., 1955, Studies on antibody production. II. The primary and secondary responses in the popliteal lymph node of the rabbit, *J. Exp. Med.* **102:**61.
Lee, L., 1962, Reticuloendothelial clearance of circulating fibrin in the pathogenesis of the generalized Schwartzman reaction, *J. Exp. Med.* **115:**1065.
Leffell, M. S., and Spitznagel, J. K., 1972, Association of lactoferrin with lysozyme in granules of human polymorphonuclear leukocytes, *Infect. Immun.* **6:**761.
Lehninger, A. L., 1975, *Biochemistry,* Worth, New York.
Lennert, K., Niedorf, H. R., and Blümcke, S., Hardmeier, T., 1972, Lymphadenitis with massive hemophagocytic sinus histiocytosis, *Virch. Arch. B* **10**:14.
Leserman, L. D., Cosenza, H., and Roseman, J. M., 1972, Cell interactions in antibody formation *in vitro.* II. The interaction of the third cell and antigen, *J. Immunol.* **109:**587.
Leslie, R. G. Q., and Cohen, S., 1974a, Cytophilic activity of IgG_2 from sera of unimmunized guinea-pigs, *Immunology* **27:**577.
Leslie, R. G. Q., and Cohen, S., 1974b, Cytophilic activity of IgG_2 from sera of guinea-pigs immunized with bovine γ-globulin, *Immunology* **27:**589.
Leu, R. W., Eddeston, A. L. W. F., Hadden, J. W., and Good, R. A., 1972, Mechanism of action of migration inhibitory factor (MIF). I. Evidence for a receptor for MIF present on the peritoneal macrophage but not on the alveolar macrophage, *J. Exp. Med.* **136:**589.
Li, I. W., Mudd, S., and Kapral, F. A., 1963, Dissociation of phagocytosis and intracellular killing of *Staphylococcus aureus* by human blood leukocytes, *J. Immunol.* **90:**804.
Lilga, J., 1976, Phagocytosis in rabbit heterophils and early myelocytes and effects of sulfated acid mucopolysaccharides on activity of heterophil granule components, Master's Thesis, Medical University of South Carolina, Charleston.

Lipscomb, M. F., Ben-Sasson, S. Z., and Uhr, J. W., 1977, Specific binding of T lymphocyte to macrophages. I. Kinetics of binding, *J. Immunol.* **118:**1748.

Lipsky, P. E., and Rosenthal, A. S., 1973, Macrophage-lymphocyte interaction. I. Characteristics of the antigen-independent-binding of guinea pig thymocytes and lymphocytes to syngeneic macrophages, *J. Exp. Med.* **138:**900.

Lipsky, P. E., and Rosenthal, A. S., 1975a, Macrophage-lymphocyte interaction: Antigen-independent binding of guinea pig lymph node lymphocytes by macrophages, *J. Immunol.* **115:**440.

Lipsky, P. E., and Rosenthal, A. S., 1975b, Macrophage-lymphocyte interaction: II. Antigen-mediated physical interactions between immune guinea pig lymph node lymphocytes and syngeneic macrophages, *J. Exp. Med.* **141:**138.

LoBuglio, A. F., Cotran, R. S., and Jandl, J. H., 1967, Red cells coated with immunoglobulin G: Binding and sphering by mononuclear cells in man, *Science* **158:**1582.

Loube, S. R., McNabb, T. C., and Dorrington, K. J., 1978, Isolation of an Fc_γ-binding protein from the cell membrane of a macrophage-like cell line ($P338D_1$) after detergent solubilization, *J. Immunol.* **120:**709.

Low, R. B., 1977, Macromolecule synthesis by alveolar macrophages: Response to a phagocytic load, *J. Reticuloendothel. Soc.* **22:**99.

Luk, S. C., Nopajaroonsri, C., and Simon, G. T., 1973, The architecture of the normal lymph node and hemolymph node, *Lab. Invest.* **29:**258.

Mannik, M., and Arend, W. P., 1971, Fate of preformed immune complexes in rabbits and rhesus monkeys, *J. Exp. Med.* **134:**19s.

Mannik, M., Arend, W. P., Hall, A. P., and Gilliland, B. C., 1971, Studies on antigen-antibody complexes. I. Elimination of soluble complexes from rabbit circulation, *J. Exp. Med.* **133:**713.

Mansell, P. W. A., and Di Luzio, N. R., 1976, The *in vivo* destruction of human tumor by glucan activated macrophages, in: *The Macrophage in Neoplasia* (M. A. Fink, ed.), pp. 227–243, Academic Press, New York.

Mantovani, B., Rabinovitch, M., and Nussenzweig, V., 1972, Phagocytosis of immune complexes by macrophages. Different roles of the macrophage receptor sites for complement (C_3) and for immunoglobulin (IgG), *J. Exp. Med.* **135:**780.

Mariano, M., Nikitin, T., and Malucelli, B. E., 1976, Immunological and nonimmunological phagocytosis by inflammatory macrophages, epithelioid cells and macrophage polykaryons from foreign body granulomata, *J. Pathol.* **120:**151.

Martineau, R. S., and Johnson, J. S., 1978, Normal mouse serum immunosuppressive activity: Action on adherent cells, *J. Immunol.* **120:**1550.

Martinez, R. D., Tarrab-Hazdai, R., Aharonov, A., and Fuchs, S., 1977, Cytophilic antibodies in experimental autoimmune myasthenia gravis, *J. Immunol.* **118:**17.

Masson, P. L., Heremans, J. F., Prignot, J. J., and Waters, G., 1966, Immunohistochemical localisation and bacteriostatic properties of an iron-binding protein from bronchial mucus, *Thorax* **21:**538.

Masson, P. L., Heremans, J. F., and Schonne, E., 1969, Lactoferrin, an iron-binding protein in neutrophilic leukocytes, *J. Exp. Med.* **130:**643.

Mathé, G., Florentin, I., and Simmler, M.-C., 1976, Lymphocytes, macrophages, and cancer, in: *Recent Results in Cancer Research* (P. R. Genéve, ed.), Springer Verlag, Berlin.

Matter, A., Lisowska-Bernstein, B., Ryser, J. E., Lamelin, J.-P., and Vassalli, P., 1972, Mouse thymus-independent and thymus-derived lymphoid cells. II. Ultrastructural studies, *J. Exp. Med.* **136:**1008.

McConahey, P. J., and Dixon, F. J., 1966, A method of trace iodination of proteins for immunologic studies, *Int. Arch. Allergy Appl. Immunol.* **29:**185.

McConnell, I., and Lachmann, P. J., 1977, Complement receptors and cell associated complement components, in: *Immunology of Receptors* (B. Cinader, ed.), p. 111, Marcel Dekker, New York.

McDevitt, H. O., Askonas, B. A., Humphrey, J. H., Schechter, I., and Sela, M., 1966, The localization of antigen in relation to specific antibody-producing cells. Use of a synthetic polypeptide [(T,G)-A–L] labelled with iodine-125, *Immunology* **11:**337.

McFarland, W., and Heilman, D. H., 1965, Lymphocyte foot appendage: Its role in lymphocyte function and in immunological reactions, *Nature* **205:**887.
McFarland, W., Heilman, D. H., and Moorhead, J. F., 1966, Functional anatomy of the lymphocyte in immunological reactions *in vitro, J. Exp. Med.* **124:**851.
McIntyre, J. A., and Pierce, C. W., 1973, Immune responses *in vitro.* VIII. Analysis of cell clusters, *J. Immunol.* **111:**512.
McIntyre, J. A., La Via, M. F., Prater, T. F. K., and Niblack, G. D., 1973, Studies of the immune response *in vitro.* I. Ultrastructural examination of cell types and cluster formation and functional evaluation of clusters, *Lab. Invest.* **29:**703.
McKeever, P. E., 1974, Methods to study pulmonary alveolar macrophage adherence: Micromanipulation and quantitation, *J. Reticuloendothel. Soc.* **16:**313.
McKeever, P. E., 1976, Normal and pathological macrophage function *in vitro, in vivo* and in the nervous system, Dissertation, Medical University of South Carolina, Charleston.
McKeever, P. E., and Balentine, J. D., 1978, Macrophage migration through the brain parenchyma to the perivascular space following particle ingestion, *Am. J. Pathol.* **93:**191.
McKeever, P. E., Gee, J. B. L., 1975, Methods to study pulmonary alveolar macrophage adherence: Rinsing and centrifugation, *J. Reticuloendothel. Soc.* **18:**221.
McKeever, P. E., Garvin, A. J., and Spicer, S. S., 1976a, Immune complex receptors on cell surfaces. I. Ultrastructural demonstration on macrophages, *J. Histochem. Cytochem.* **24**:948 and 1212.
McKeever, P. E., Garvin, A. J., Hardin, D. H., and Spicer, S. S., 1976b, Immune complex receptors on cell surfaces. II. Cytochemical evaluation of their abundance on different immune cells: distribution, uptake, and regeneration, *Am. J. Pathol.* **84:**437.
McKeever, P. E., Spicer, S. S., Brissie, N. T., and Garvin, A. J., 1977, Immune complex receptors on cell surfaces. III. Topography of macrophage receptors demonstrated by new scanning electron microscopic peroxidase marker, *J. Histochem. Cytochem.* **25:**1063.
McKeever, P. E., Walsh, G. P., Storrs, E. E., and Balentine, J. D., 1978, Electron microscopy of enzymes in leprous and uninfected armadillo macrophages: Peroxidase positive cells lack bacilli, *Am. J. Trop. Med. Hyg.* **27:**1019.
McPhaul, J. J., and Dixon, F. J., 1970, Characterization of human antiglomerular basement membrane antibodies eluted from glomerulonephritic kidneys, *J. Clin. Invest.* **49:**308.
Melewicz, F. M., Shore, S. L., Ades, E. W., and Phillips, D. J., 1977, The mononuclear cell in human blood which mediates antibody-dependent cellular cytotoxicity to virus-infected target cells. II. Identification as a K cell, *J. Immunol.* **118:**567.
Melsom, H., and Seljelid, R., 1973, The cytotoxic effect of mouse macrophages on syngeneic and allogeneic erythrocytes, *J. Exp. Med.* **137:**807.
Merchant, R. E., and Low, F. N., 1977, Identification of challenged subarachnoid free cells, *Am. J. Anat.* **148:**143.
Michl, J., Ohlbaum, D. J., and Silverstein, S. C., 1976, 2-Deoxyglucose selectively inhibits F_c and complement receptor-mediated phagocytosis in mouse peritoneal macrophages. I. Description of the inhibitory effect, *J. Exp. Med.* **144:**1465.
Michl, J., Pieczonka, M. M., Unkeless, J. C., and Silverstein, S. C., 1979, Effects of immobilized immune complexes on F_c- and complement-receptor function in resident and thioglycollate-elicited mouse peritoneal macrophages, *J. Exp. Med.* **150:**607.
Miller, H. R. P., and Avrameas, S., 1970, Association between macrophages and specific antibody producing cells, *Nature, New Biol.* **229:**184.
Miller, K., and Kagan, E., 1976, The *in vivo* effects of asbestos on macrophage membrane structure and population characteristics of macrophages: A scanning electron microscope study, *J. Reticuloendothel. Soc.* **20:**159.
Miller, K., and Kagan, E., 1977, The *in vivo* effects of quartz on alveolar macrophage membrane topography and on the characteristics of the intrapulmonary cell population, *J. Reticuloendothel. Soc.* **21:**307.
Mishell, R. I., and Dutton, R. W., 1967, Immunization of dissociated spleen cell cultures from normal mice, *J. Exp. Med.* **126:**423.
Molenaar, J. L., van Galen, M., Hannema, A. J., Zeijlemaker, W., and Pondman, K. W., 1977,

Spontaneous release of F_c receptor-like material from human lymphoblastoid cell lines, *Eur. J. Immunol.* **7**:230.

Monnier, G., Jacotot, B., and Beaumont, J. L., 1974, Effects of factors influencing the activity of the reticuloendothelial system on the half-life of labeled β-lipoproteins. Preliminary report, *J. Med.* **5**:217.

Morgan, E. L., and Tempelis, C. H., 1978, The requirement for the F_c portion of antibody in antigen-antibody complex-mediated suppression, *J. Immunol.* **120**:1669.

Mørland, B., and Kaplan, G., 1977, Macrophage activation *in vivo* and *in vitro, Exp. Cell Res.* **108**:279.

Mosier, D. E., 1967, A requirement for two cell types for antibody formation *in vitro, Science* **158**:1573.

Mosier, D. E., 1969, Cell interactions in the primary immune response *in vitro:* A requirement for specific cell clusters, *J. Exp. Med.* **129**:351.

Mosier, D. E., 1976, The role of macrophages in the specific determination of immunogenicity and tolerogenicity, in: *Immunobiology of the Macrophage* (D. S. Nelson, ed.), pp. 35–44, Academic Press, New York.

Mosier, D. E., and Coppleson, L. W., 1968, A three-cell interaction required for the induction of the primary immune response *in vitro, Proc. Natl. Acad. Sci USA* **61**:542.

Muggeo, M., Bar, R. S., Roth, J., Kahn, C. R., and Gorden, P., 1979, The insulin resistance of acromegaly: Evidence for two alterations in the insulin receptor on circulating monocytes, *J. Clin. Endocrinol. Metab.* **48**:17.

Müller-Eberhard, H. J., 1968, Chemistry and reaction mechanisms of complement, *Adv. Immunol.* **8**:1.

Munthe-Kaas, A. C., 1976, Phagocytosis in rat Kupffer cells *in vitro, Exp. Cell Res.* **99**:319.

Munthe-Kaas, A. C., Kaplan, G., and Seljelid, R., 1976, On the mechanism of internalization of opsonized particles by rat Kupffer cells *in vitro, Exp. Cell Res.* **103**:201.

Musson, R. A., and Henson, P. M., 1979, Humoral and formed elements of blood modulate the response of peripheral blood monocytes. I. Plasma and serum inhibit and platelets enhance monocyte adherence, *J. Immunol.* **122**:2026.

Nachman, R. L., Ferris, B., and Hirsch, J. G., 1971a, Macrophage plasma membranes. I. Isolation and studies on protein components, *J. Exp. Med.* **133**:785.

Nachman, R. L., Ferris, B., and Hirsch, J. G., 1971b, Macrophage plasma membrane. II. Studies on synthesis and turnover of protein constituents, *J. Exp. Med.* **133**:807.

Nathan, C. F., Remold, H. G., and David, J. R., 1973, Characterization of a lymphocyte factor which alters macrophage functions, *J. Exp. Med.* **137**:275.

Nathan, C. F., Asofsky, R., and Terry, W. D., 1977, Characterization of the nonphagocytic adherent cell from the peritoneal cavity of normal and BCG-treated mice, *J. Immunol.* **118**:1612.

Nelson, D. S., 1965, The effects of anticoagulants and other drugs on cellular and cutaneous reactions to antigen in guinea pigs with delayed-type hypersensitivity, *Immunology* **9**:219.

Nelson, D. S., and North, R. J., 1965, The fate of peritoneal macrophages after the injection of antigen into guinea pigs with delayed-type hypersensitivity, *Lab. Invest.* **14**:89.

Newman, W., Diamond, B., Flomenberg, P., Scharff, M. D., and Bloom, B. R., 1979, Response of a continuous macrophage-like cell line to MIF, *J. Immunol.* **123**:2292.

Nicolson, G. L., 1973, Neuraminidase "unmasking" and failure of trypsin to "unmask" β-D-galactose-like sites on erythrocyte, lymphoma, and normal and virus-transformed fibroblast cell membranes, *J. Natl. Cancer Inst.* **50**:1443.

Niederhuber, J. E., 1978, The role of I region gene products in macrophage-T lymphocyte interaction, in: *Immunological Reviews* pp. 3–52, Munksgaard, Copenhagen.

Nielsen, M. H., Jensen, H., Braendstrup, Werdelin, O., 1974, Macrophage–lymphocyte clusters in the immune response to soluble protein antigen *in vitro*. II. Ultrastructure of clusters formed during the early response, *J. Exp. Med.* **140**:1260.

North, R. J., 1968, The uptake of particulate antigens, *J. Reticuloendothel. Soc.* **5**:203.

Northcote, D. H., and Horne, R. W., 1952, The chemical composition and structure of the yeast cell wall, *Biochem. J.* **51**:232.

Nossal, G. J. V., Abbot, A., and Mitchell, J., 1968a, Antigens in immunity. XIV. Electron microscopic radioautographic studies of antigen capture in the lymph node medulla, *J. Exp. Med.* **127**:263.

Nossal, G. J. V., Abbot, A., Mitchell, J., and Lummus, Z., 1968b, Antigens in immunity. XV. Ultrastructural features of antigen capture in primary and secondary lymphoid follicles, *J. Exp. Med.* **127:**277.

Nussenzweig, V., 1974, Receptors for immune complexes on lymphocytes, in: *Advances in Immunology* (F. J. Dixon and H. G. Kunkel, eds.), pp. 217–254, Academic Press, New York.

O'Flaherty, J. T., Kreutzer, D. L., and Ward, P. A., 1978, Chemotactic factor influences on the aggregation, swelling, and foreign surface adhesiveness of human leukocytes, *Am. J. Pathol.* **90:**537.

Okafor, G. O., Turner, M. W., and Hay, F. C., 1974, Localization of monocyte binding site of human immunoglobulin G, *Nature* **248:**228.

Olefsky, J. M., 1976, Decreased insulin binding to adipocytes and circulating monocytes from obese subjects, *J. Clin. Invest.* **57:**1165.

Olefsky, J. M., and Reaven, G. M., 1974, Decreased insulin binding to lymphocytes from diabetic subjects, *J. Clin. Invest.* **54:**1323.

Olefsky, J. M., and Reaven, G. M., 1976a, Effects of sulfonylurea therapy on insulin binding to mononuclear leukocytes of diabetic patients, *Am. J. Med.* **60:**89.

Olefsky, J. M., and Reaven, G. M., 1976b, Insulin binding to monocytes and total mononuclear leukocytes from normal and diabetic patients, *J. Clin. Endocrinol. Metab.* **43:**226.

Opitz, H. G., Lemke, H., and Hewlett, G., 1978, Activation of T-cells by a macrophage or 2-mercaptoethanol. Activated serum factor is essential for induction of a primary immune response to heterologous red cells *in vitro, Immunol. Rev.* **40:**53.

Oren, R., Farnham, A. E., Saito, K., Milofsky, E., and Karnovsky, M. L., 1963, Metabolic patterns in three types of phagocytizing cells, *J. Cell Biol.* **17:**487.

Orenstein, J. M., and Shelton, E., 1976a, Surface topography and interactions between mouse peritoneal cells allowed to settle on an artificial substrate: observations by scanning electron microscopy, *Exp. Molec. Pathol.* **24:**201.

Orenstein, J. M., and Shelton, E., 1976b, Surface topography of leukocytes *in situ:* Cells of mouse peritoneal milky spots, *Exp. Molec. Pathol.* **24:**415.

Ottolenghi, A. C., 1969, Phospholipase C, in: *Methods in Enzymology*, Vol. XIV, *Lipids* (J. M. Lowenstein, ed.), pp. 188–197, Academic Press, New York.

Papadimitriou, J. M., 1973, Detection of macrophage receptors for heterologous IgG by scanning and transmission electron microscopy, *J. Pathol.* **110:**213.

Perkins, W. D., Karnovsky, M. J., and Unanue, E. R., 1972, An ultrastructural study of lymphocytes with surface-bound immunoglobulin, *J. Exp. Med.* **135:**267.

Pethica, B. A., 1961, The physical chemistry of cell adhesion, *Exp. Cell Res.* **123:**140.

Phillips-Quagliata, J. M., Levine, B. B., Quagliata, F., and Uhr, J. W., 1971, Mechanisms underlying binding of immune complexes to macrophages, *J. Exp. Med.* **133:**589.

Pierce, C. W., 1973, Immune responses *in vitro.* VI. Cell interactions in the development of primary IgM, IgG and IgA plaque-forming cell responses *in vitro, Cell. Immunol.* **9:**453.

Pierce, C. W., and Kapp, J. A., 1976, The role of macrophages in antibody responses *in vitro,* in: *Immunobiology of the Macrophage* (D. S. Nelson, ed.), pp. 1–33, Academic Press, New York.

Pierce, C. W., and Kapp, J. A., 1978, Functions of macrophages in antibody responses *in vitro, Fed. Proc.* **37:**86.

Pierce, C. W., Kapp, J. A., Wood, D. D., and Benacerraf, B., 1974, Immune responses *in vitro.* X. Functions of macrophages, *J. Immunol.* **112:**1181.

Pillemer, L., and Ecker, E. E., 1941, Anticomplementary factor in fresh yeast, *J. Biol. Chem.* **137:**139.

Plackett, P., Marmion, B. P., Shaw, E. J., and Lemcke, R. M., 1969, Immunochemical analysis of *Mycoplasma pneumoniae.* 3. Separation and chemical identification of serologically active lipids, *Aust. J. Exp. Biol.* **47:**171.

Polliack, A., and Gordon, S., 1975, Scanning electron microscopy of murine macrophages. Surface characteristics during maturation, activation, and phagocytosis, *Lab. Invest.* **33:**469.

Polliack, A., Lampen, N., Clarkson, B. D., de Harven, E., Bentwich, Z., Siegal, F. P., and Kunkel, H. G., 1973, Identification of human B and T lymphocytes by scanning electron microscopy, *J. Exp. Med.* **138:**607.

Poplack, D. G., Sher, N. A., Chaparas, S. D., and Blaese, R. M., 1976, The effect of *Mycobacterium bovis (Bacillus Calmette-Guerin)* on macrophage random migration, chemotaxis, and pinocytosis, *Cancer Res.* **36:**1233.

Postlethwaite, A. E., and Kang, A. H., 1976, Guinea pig lymphocyte-derived macrophage aggregation factor: Its separation from macrophage migration inhibitory factor, *J. Immunol.* **117:**1651.

Postlethwaite, A. E., Townes, A. S., and Kang, A. H., 1976, Characterization of macrophage migration inhibitory factor activity produced *in vivo* by a cell-mediated immune reaction in the guinea pig, *J. Immunol.* **117:**1716.

Powers, J. M., and McKeever, P. E., 1976, Central pontine myelinolysis: An ultrastructural and biochemical study, *J. Neurol. Sci.* **29:**65.

Premkumar, E., Singer, P. A., and Williamson, A. R., 1975a, A human lymphoid cell line secreting immunoglobulin G and retaining immunoglobulin M in the plasma membrane, *Cell* **5:**87.

Premkumar, E., Potter, M., Singer, P. A., and Sklar, M. D., 1975b, Synthesis, surface deposition and secretion of immunoglobulins by Abelson virus-transformed lymphosarcoma cell lines, *Cell* **6:**149.

Premkumar-Reddy, E., Price, P. J., Chung, K.-C., and Sarma, P. S., 1976, Continuous culturing of murine splenic B-lymphocytes: Synthesis and surface deposition of IgM and putative IgD molecules, *Cell* **8:**397.

Prescott, B., King, J. R., Caldes, G., Whitt, R. S., and Cole, R. M., 1971, Serologic reactions of lipid fractions of *streptococci,* in: *Bacteriological Proceedings* (W. A. Wood, H. Gooder, and L. Levintow, eds.), p. 74, American Society for Microbiology, Minneapolis.

Prose, P. H., Lee, L., and Balk, S. D., 1965, Electron microscopic study of the phagocytic fibrin-clearing mechanism, *Am. J. Pathol.* **47:**403.

Rabinovitch, M., 1967, The dissociation of the attachment and ingestion phases of phagocytosis by macrophages, *Exp. Cell Res.* **46:**19.

Rabinovitch, M., 1968, Effect of antiserum on the attachment of modified erythrocytes to normal or to trypsinized macrophages, *Proc. Soc. Exp. Biol. Med.* **127:**351.

Rabinovitch, M., and DeStefano, M. J., 1973, Macrophage spreading *in vitro.* I. Inducers of spreading, *Exp. Cell Res.* **77:**323.

Rabinovitch, M., and DeStefano, M. J., 1975, Use of the local anesthetic lidocaine for cell harvesting and subcultivation, *In Vitro* **12:**379.

Rabinovitch, M., and DeStefano, M. J., 1976, Cell shape changes induced by cationic anesthetics, *J. Exp. Med.* **143:**290.

Rabinovitch, M., Manejias, R. E., and Nussenzweig, V., 1975, Selective phagocytic paralysis induced by immobilized immune complexes, *J. Exp. Med.* **142:**827.

Ragsdale, C. G., and Arend, W. P., 1979, Neutral protease secretion by human monocytes. Effect of surface-bound immune complexes, *J. Exp. Med.* **149:**954.

Ralph, P., Moore, M. A. S., and Nilsson, K., 1976. Lysozyme synthesis by established human and murine histiocytic lymphoma cell lines, *J. Exp. Med.* **143:**1528.

Rask, L., Klareskog, L., Ostberg, L., and Peterson, P. A., 1975, Isolation and properties of a murine spleen cell F_c receptor, *Nature* **257:**231.

Ratzan, K. R., Musher, D. M., Keusch, G. T., and Weinstein, L., 1972, Correlation of increased metabolic activity, resistance to infection, enhanced phagocytosis, and inhibition of bacterial growth by macrophages from *Listeria-* and BCG-infected mice, *Infect. Immun.* **5:**499.

Razin, E., Bauminger, S., and Globerson, A., 1978, Effect of prostaglandins on phagocytosis of sheep erythrocytes by mouse peritoneal macrophages, *J. Reticuloendothel. Soc.* **23:**237.

Reaven, E. P., and Axline, S. G., 1973, Subplasmalemmal microfilaments and microtubules in resting and phagocytizing cultivated macrophages, *J. Cell Biol.* **59:**12.

Remold, H. G., 1973, Requirement for α-L-fucose on the macrophage membrane receptor for MIF, *J. Exp. Med.* **138:**1065.

Remold, H. G., 1974, The enhancement of MIF activity by inhibition of macrophage associated esterases, *J. Immunol.* **112:**1571.

Remold, H. G., 1977, Chemical treatment of macrophages increases their responsiveness to migration inhibitory factor (MIF), *J. Immunol.* **118:**1.

Remold, H. G., Katz, A. B., Haber, E., and David, J. R., 1970, Studies on migration inhibitory factor

(MIF): Recovery of MIF activity after purification by gel filtration and disc electrophoresis, *Cell Immunol.* **1:**133.
Remold, H. G., and Rosenberg, R. D., 1975, Enhancement of migration inhibitory factor activity by plasma esterase inhibitors, *J. Biol. Chem.* **250:**6608.
Reynolds, H. Y., 1974, Pulmonary host defenses in rabbits after immunization with *Pseudomonas* antigens: The interaction of bacteria, antibodies, macrophages, and lymphocytes, *J. Infect. Dis.* **130:**S-134.
Reynolds, H. Y., and Thompson, R. E., 1973, Pulmonary host defenses. II. Interaction of respiratory antibodies with *Pseudomonas aeruginosa* and alveolar macrophages, *J. Immunol.* **111:**369.
Reynolds, H. Y., Atkinson, J. P., Newball, H. H., and Frank, M. M., 1975, Receptors for immunoglobulin and complement on human alveolar macrophages, *J. Immunol.* **114:**1813.
Rhodes, J., 1973, Receptor for monomeric IgM on guinea-pig splenic macrophages, *Nature* **243:**527.
Rhodes, J., 1975, Macrophage heterogeneity in receptor activity: The activation of macrophage F_c receptor function *in vivo* and *in vitro*, *J. Immunol.* **114:**976.
Rinehardt, J. J., Balcerzak, S. P., Sagone, A. L., and LoBuglio, A. F., 1974, Effects of corticosteroids on human monocyte function, *J. Clin. Invest.* **54:**1337.
Rinehardt, J. J., Sagone, A. L., Balcerzak, S. P., Ackerman, G. A., and LoBuglio, A. F., 1975, Effects of corticosteroid therapy on human monocyte function, *N. Engl. J. Med.* **292:**236.
Romans, D. G., Pinteric, L., Falk, R. E., and Dorrington, K. J., 1976, Redistribution of the F_c receptor on human blood monocytes and peritoneal macrophages induced by immunoglobulin G-sensitized erythrocytes, *J. Immunol.* **116:**1473.
Rosai, J., and Dorfman, R. F., 1972, Sinus histiocytosis with massive lymphadenopathy: a pseudolymphomatous benign disorder, *Cancer* **30:**1174.
Roseman, J., 1969, X-ray resistant cell required for the induction of *in vitro* antibody formation, *Science* **165:**1125.
Rosenstreich, D. L., and Wilton, J. M., 1975, The mechanism of action of macrophages in the activation of T-lymphocytes *in vitro* by antigens and mitogens, in: *Immune Recognition* (A. S. Rosenthal, ed.), p. 113, Academic Press, New York.
Rosenstreich, D. L., Shevach, E., Green, I., and Rosenthal, A. S., 1972, The uropod-bearing lymphocyte of the guinea pig. Evidence for thymic origin. *J. Exp. Med.* **135:**1037.
Rosenthal, A. S., 1978, Determinant selection and macrophage function in genetic control of the immune response, in: *Immunological Reviews*, Vol. 40, pp. 136–152, Munksgaard, Copenhagen.
Rosenthal, A. S., and Rosenstreich, D. L., 1974, The lymphocyte uropod: A specialized surface site for immunologic recognition, in: *Biomembranes*, Vol. 5 (L. A. Manson, ed.), pp. 1–284, Plenum Press, New York.
Rosenthal, A. S., and Shevach, E. M., 1973, Function of macrophages in antigen recognition by guinea pig T lymphocytes. I. Requirement for histocompatible macrophages and lymphocytes, *J. Exp. Med.* **138:**1194.
Rosenthal, A. S., Lipsky, P. E., and Shevach, E. M., 1975, Macrophage–lymphocyte interaction and antigen recognition, *Fed. Proc.* **34:**1743.
Rosenthal, A. S., Blake, J. T., Ellner, J. J., Greineder, D. K., and Lipsky, P. E., 1976, Macrophage function in antigen recognition by T lymphocytes, in: *Immunobiology of the Macrophage* (D. S. Nelson, ed.), pp. 131–160, Academic Press, New York.
Rosenthal, A. S., Barcinski, M. A., and Rosenwasser, L. J., 1978, Function of macrophages in genetic control of immune responsiveness, *Fed. Proc.* **37:**79.
Ross, G. D., and Polley, M. J., 1975, Specificity of human lymphocyte complement receptors, *J. Exp. Med.* **141:**1163.
Roth, J., and Binder, M., 1978, Colloidal gold, ferritin and peroxidase as markers for electron microscopic double labeling lectin techniques, *J. Histochem. Cytochem.* **26:**163.
Ruddy, S., and Austen, K. F., 1969, C3 inactivator of man. I. Hemolytic measurement by the inactivation of cell-bound C3, *J. Immunol.* **102:**533.
Sannes, P. L., and Spicer, S. S., 1979, Inhibitory effects of cytochalasin B and the ionophores A23187 and X537A on binding and uptake of immune complex by alveolar macrophages, *J. Reticuloendothel. Soc.* **26:**317.

Saunders, G. C., and Hammond, W. S., 1970, Ultrastructural analysis of hemolysin-forming cell clusters. I. Preliminary observations, *J. Immunol.* **105:**1299.

Sbarra, A. J., and Karnovsky, M. L., 1959, The biochemical basis of phagocytosis. I. Metabolic changes during the ingestion of particles by polymorphonuclear leukocytes, *J. Biol. Chem.* **234:**1355.

Schmidt, M. E., and Douglas, S. D., 1972, Disappearance and recovery of human monocyte IgG receptor activity after phagocytosis, *J. Immunol.* **109:**914.

Schmidt, M. E., and Douglas, S. D., 1976a, Effects of levamisole on human monocyte function and immunoprotein receptors, *Clin. Immunol. Immunopathol.* **6:**299.

Schmidt, M. E., and Douglas, S. D. 1976b, Effects of synthetic single- and multistranded polynucleotides on human monocyte IgG receptor activity *in vitro, Proc. Soc. Exp. Biol. Med.* **151:**376.

Schmidt, M. E., and Douglas, S. D., 1977, Monocyte IgG receptor activity, dynamics, and modulation—normal individuals and patients with granulomatous diseases, *J. Lab. Clin. Med.* **89**:332.

Schoenberg, M. D., Mumaw, V. R., Moore, R. D., and Weisberger, A. S., 1964, Cytoplasmic interaction between macrophages and lymphocytic cells in antibody synthesis, *Science* **143:**964.

Schreiber, A. D., Parsons, J., McDermott, P., and Cooper, R. A., 1975, Effect of corticosteroids on the human monocyte IgG and complement receptors, *J. Clin. Invest.* **56:**1189.

Schroit, A. J., and Gallily, R., 1977, Quantitative *in vitro* phagocytic rate measurements, *J. Immunol. Methods* **17:**123.

Schwartz, R. H., Bianco, A. R., Handwerger, B. S., and Kahn, C. R., 1975, Demonstration that monocytes rather than lymphocytes are the insulin-binding cells in preparations of human peripheral blood mononuclear leukocytes: Implications for studies of insulin-resistant states in man, *Proc. Natl. Acad. Sci. USA* **72:**474.

Schwartz, R. H., Dickler, H. B., Sachs, D. H., and Schwartz, B. D., 1976, Studies of Ia antigens on murine peritoneal macrophages, *Scand. J. Immunol.* **5:**731.

Schwartz, R. H., Yano, A., and Paul, W. E., 1978, Interaction between antigen-presenting cells and primed T lymphocytes: An assessment of Ir gene expression in the antigen-presenting cell, in: *Immunological Reviews,* Vol. 40, pp. 153–180, Munksgaard, Copenhagen.

Scott, M. T., 1975, *In vivo* cortisone sensitivity of nonspecific antitumor activity of *Corynebacterium parvum*-activated mouse peritoneal macrophages, *J. Natl. Cancer Inst.* **54:**789.

Scott, R. E., and Rosenthal, A. S., 1977, Isolation of receptor-bearing plasma membrane vesicles from guinea pig macrophages, *J. Immunol.* **119:**143.

Segal, D. M., and Hurwitz, E., 1977, Binding of affinity cross-linked oligomers of IgG to cells bearing F_c receptors, *J. Immunol.* **118:**1338.

Segal, D. M., and Titus, J. A., 1978, The subclass specificity for the binding of murine myeloma proteins to macrophage and lymphocyte cell lines and to normal spleen cells, *J. Immunol.* **120:**1395.

Sharp, J. A., and Burwell, R. G., 1960, Interaction ("peripolesis") of macrophages and lymphocytes after skin homografting or challenge with soluble antigens, *Nature* **188:**474.

Shear, H. L., Nussenzweig, R. S., and Bianco, C., 1979, Immune phagocytosis in murine malaria, *J. Exp. Med.* **149:**1288.

Sherman, L. A., and Lee, J., 1977, Specific binding of soluble fibrin to macrophages, *J. Exp. Med.* **145:**76.

Sherman, L. A., Harwig, S., and Lee, J., 1975, *In vitro* formation and *in vivo* clearance of fibrinogen: Fibrin complexes, *J. Lab. Clin. Med.* **86:**100.

Shevach, E., Herberman, R., Liberman, R., Frank, M. M., and Green, I., 1972, Receptors for immunoglobulin and complement on mouse leukemias and lymphomas, *J. Immunol.* **108:**325.

Shields, J. W., 1972, *The Trophic Function of Lymphoid Elements,* Charles C. Thomas, Springfield.

Shinomiya, T., and Koyama, J., 1976, *In vitro* uptake and digestion of immune complexes containing guinea-pig IgG1 and IgG2 antibodies by macrophages, *Immunology* **30:**267.

Shurin, S. B., and Stossel, T. P., 1978, Complement (C3)-activated phagocytosis by lung macrophages, *J. Immunol.* **120:**1305.

Siegel, I., 1970, Natural and antibody-induced adherence of guinea pig phagocytic cells to autologous and heterologous thymocytes, *J. Immunol.* **105:**879.

Silverstein, S. C., Steinman, R. M., and Cohn, Z. A., 1977, Endocytosis, *Annu. Rev. Biochem.* **46**:669.
Skinnider, L. F., and Ghadially, F. N., 1977, Ultrastructure of cell surface abnormalities in neoplastic histiocytes, *Br. J. Cancer* **35**:657.
Skutelsky, E., and Hardy, B., 1976, Regeneration of plasmalemma and surface properties in macrophages, *Exp. Cell Res.* **101**:337.
Smith, C. W., and Goldman, A. S., 1970, Interactions of lymphocytes and macrophages from human colostrum: Characteristics of the interacting lymphocyte, *J. Reticuloendothel. Soc.* **8**:91.
Snyderman, R., Gewurz, H., and Mergenhagen, S. E., 1968, Interactions of the complement system with endotoxic lipopolysaccharide, *J. Exp. Med.* **128**:259.
Sobeslavsky, O., Prescott, B., and Chanock, R. M., 1968, Adsorption of *Mycoplasma pneumoniae* to neuraminic acid receptors of various cells and possible role in virulence, *J. Bacteriol.* **96**:695.
Soll, A. H., Goldfine, I. D., Roth, J., Kahn, C. R., and Neville, Jr., D. M., 1974, Thymic lymphocytes in obese (ob/ob) mice. A mirror of the insulin receptor defect in liver and fat, *J. Biol. Chem.* **249**:4127.
Soll, A. H., Kahn, C. R., Neville, Jr., D. M., and Roth, J., 1975, Insulin receptor deficiency in genetic and acquired obesity, *J. Clin. Invest.* **56**:769.
Spencer, H., 1969, *Pathology of the Lung (Excluding Pulmonary Tuberculosis)*, 2nd ed., Pergamon Press, Oxford.
Stahl, P., Six, H., Rodman, J. S., Schlesinger, P., Tulsient, D. R. P., and Touster, O., 1976, Evidence for specific recognition sites mediating clearance of lysosomal enzymes *in vivo*, *Proc. Natl. Acad. Sci. USA* **73**:4045.
Stahl, P., Rodman, J. S., Miller, M. J., and Schlesinger, P. H., 1978, Evidence for receptor-mediator binding of glycoproteins, glycoconjugates, and lysosomal glycosidases by alveolar macrophages, *Proc. Natl. Acad. Sci. USA* **75**:1399.
Stein, H., and Kaiserling, E., 1974, Surface immunoglobulins and lymphocyte-specific surface antigens on leukaemic reticuloendotheliosis cells, *Clin. Exp. Immunol.* **18**:63.
Steinman, R. M., and Cohn, Z. A., 1972a, The interaction of soluble horseradish peroxidase with mouse peritoneal macrophages *in vitro*, *J. Cell Biol.* **55**:186.
Steinman, R. M., and Cohn, Z. A., 1972b, The interaction of particulate horseradish peroxidase (HRP)-anti HRP immune complexes with mouse peritoneal macrophages *in vitro*, *J. Cell Biol.* **55**:616.
Steinman, R. M., and Cohn, Z. A., 1975, Dendritic cells, reticular cells, and macrophages, in: *Mononuclear Phagocytes in Immunity, Infection, and Pathology* (R. van Furth, ed.), pp. 95–107, Blackwell, Oxford.
Steinman, R. M., Brodie, S. E., and Cohn, Z. A., 1976, Membrane flow during pinocytosis. A stereologic analysis, *J. Cell Biol.* **68**:665.
Stewart, C. C., Lin, H.-S., and Adles, C., 1975, Proliferation and colony-forming ability of peritoneal exudate cells in liquid culture, *J. Exp. Med.* **141**:1114.
Stossel, T. P., 1973, Quantitative studies of phagocytosis. Kinetic effects of cations and heat-labile opsonin, *J. Cell Biol.* **58**:346.
Stossel, T. P., 1975, Phagocytosis: Recognition and ingestion, *Semin. Hematol.* **12**:83.
Stossel, T. P., 1976, The mechanism of phagocytosis, *J. Reticuloendothel. Soc.* **19**:237.
Stossel, T. P., 1977, Contractile proteins in phagocytosis: An example of cell surface-to-cytoplasm communication, *Fed. Proc.* **36**:2181.
Stossel, T. P., Mason, R. J., Pollard, T. D., and Vaughan, M., 1972, Isolation and properties of phagocytic vesicles. II. Alveolar macrophages, *J. Clin. Invest.* **51**:604.
Stossel, T. P., Field, R. J., Gitlin, J. D., Alper, C. A., and Rosen, F. S., 1975, The opsonic fragment of the third component of human complement (C3), *J. Exp. Med.* **141**:1329.
Straus, W., 1974, Immunocytochemical observations on hypersensitivity skin reactions to horseradish peroxidase and to antihorseradish peroxidase γ-globulin, *J. Histochem. Cytochem.* **22**:303.
Strominger, J. L., Izaki, K., Matsuhashi, M., and Tipper, D. J., 1967, Peptidoglycan transpeptidase and D-alanine carboxypeptidase: Penicillin-sensitive enzymatic reactions, *Fed. Proc.* **26**:9.
Stuart, A. E., 1970, *The Reticuloendothelial System*, Livingstone, Edinburgh.
Stuart, A., 1975, Perspectives on the reticulum cell and fibre networks, in: *Mononuclear Phagocytes in Immunity, Infection, and Pathology* (R. van Furth, ed.), pp. 111–118, Blackwell, Oxford.

Stubbs, M., Kühner, A. V., Glass, E. A., David, J. R., and Karnovsky, M. L., 1973, Metabolic and functional studies on activated mouse macrophages, *J. Exp. Med.* **137:**537.

Sulitzeanu, D., Kleinman, R., Benezra, D., and Gery, I., 1971, Cellular interactions and the secondary response *in vitro, Nature, New Biol.* **229:**254.

Territo, M. C., Golde, D. W., and Cline, M. J., 1976, Macrophage activation and function, in: *Manual of Clinical Immunology* (N. R. Rose and H. Friedman, eds.), pp. 142–158, American Society for Microbiology, Washington, D.C.

Theofilopoulos, A. N., Bokisch, V. A., and Dixon, F. J., 1974, Receptor for soluble C3 and C3b on human lymphoblastoid (Raji) cells. Properties and biological significance, *J. Exp. Med.* **139:**696.

Thiéry, J.-P., 1962, Etude au microscope électronique de l'îlot plasmocytaire, *J. Microscopie* **1:**275.

Thomas, D. W., and Shevach, E. M., 1976, Nature of the antigenic complex recognized by T lymphocytes. I. Analysis with an *in vitro* primary response to soluble protein antigens, *J. Exp. Med.* **144:**1263.

Thomas, D. W., Clement, L., and Shevach, E. M., 1978, T lymphocyte stimulation by hapten-conjugated macrophages: A model system for the study of immunocompetent cell interactions, in: *Immunological Reviews,* Vol. 40, pp. 181–203, Munksgaard, Copenhagen.

Thrasher, S. G., and Cohen, S., 1971, Studies of the mechanism of binding of chemically modified cytophilic antibodies to macrophages, *J. Immunol.* **107:**672.

Thrasher, S. G., Bigazzi, P. E., Yoshida, T., and Cohen, S., 1975a, The effect of fluorescein conjugation on F_c-dependent properties of rabbit antibody, *J. Immunol.* **114:**762.

Thrasher, S. G., Bigazzi, P. E., Yoshida, T., and Cohen, S., 1975b, Distribution of cytophilic and anti-macrophage antibody on the macrophage surface, *Immunol. Commun.* **4:**219.

Tizard, I. R., Holmes, W. L., and Parappally, N. P., 1974, Phagocytosis of sheep erythrocytes by macrophages: A study of the attachment phase by scanning electron microscopy, *J. Reticuloendothel. Soc.* **15:**225.

Törö, I., Ruzsa, P., and Röhlich, P., 1962, Ultrastructure of early phagocytic stages in sinus endothelial and Kupffer cells of the liver, *Exp. Cell Res.* **26:**601.

Turner, S. R., Campbell, J. A., and Lynn, W. S., 1975, Polymorphonuclear leukocyte chemotaxis toward oxidized lipid components of cell membranes, *J. Exp. Med.* **141:**1437.

Uhr, J. W., and Vitetta, E. S., 1974, Cell surface immunoglobulin. VIII. Synthesis, secretion and cell surface expression of immunoglobulin in murine thoracic duct lymphocytes, *J. Exp. Med.* **139:**1013.

Unanue, E. R., 1968, Properties and some uses of anti-macrophage antibodies, *Nature* **218:**36.

Unanue, E. R., 1972, The regulatory role of macrophages in antigenic stimulation, *Adv. Immunol.* **15:**95.

Unanue, E. R., 1978, The regulation of lymphocyte functions by the macrophage, in: *Immunological Review,* Vol. 40, pp. 227–255, Munksgaard, Copenhagen.

Unanue, E. R., and Askonas, B. A., 1968, Persistence of immunogenicity of antigen after uptake by macrophages, *J. Exp. Med.* **127:**915.

Unanue, E. R., and Calderon, J., 1975, Evaluation of the role of macrophages in immune induction, *Fed. Proc.* **34:**1737.

Unanue, E. R., and Cerottini, J.-C., 1970, The immunogenicity of antigen bound to the plasma membrane of macrophages, *J. Exp. Med.* **131:**711.

Unanue, E. R., Dixon, F. S., 1967, Experimental glomerulonephritis: Immunological events and pathologic mechanisms, *Adv. Immunol.* **6:**1.

Unanue, E. R., Cerottini, J.-C., and Bedford, M., 1969, Persistence of antigen on the surface of macrophages, *Nature* **222:**1193.

Unkeless, J. C., 1977, The presence of two F_c receptors on mouse macrophages: Evidence from a variant cell line and differential trypsin sensitivity, *J. Exp. Med.* **145:**931.

Unkeless, J. C., 1979, Characterization of a monoclonal antibody directed against mouse macrophage and lymphocyte F_c receptors, *J. Exp. Med.* **150:**580.

Unkeless, J. C., and Eisen, H. N., 1975, Binding of monomeric immunoglobulins to F_c receptors of mouse macrophages, *J. Exp. Med.* **142:**1520.

Unkeless, J. C., Gordon, S., and Reich, E., 1974, Secretion of plasminogen activator by stimulated macrophages, *J. Exp. Med.* **139:**834.

Unkeless, J. C., Kaplan, G., Plutner, H., and Cohn, Z. A., 1979, F_c-receptor variants of a mouse macrophage cell line, *Proc. Natl. Acad. Sci. USA* **76:**1400.

Valentine, R. C., and Green, N. M., 1967, Electron microscopy of an antibody-hapten complex, *J. Mol. Biol.* **27:**615.

van der Zeijst, B. A. M., Stewart, C. C., and Schlesinger, S., 1978, Proliferative capacity of mouse peritoneal macrophages *in vitro, J. Exp. Med.* **147:**1253.

van Furth, R., Langevoort, H. L., and Schaberg, A., 1975, Mononuclear phagocytes in human pathology—proposal for an approach to improved classification, in: *Mononuclear Phagocytes in Immunity, Infection, and Pathology* (R. van Furth, ed.), pp. 1–15, Blackwell, Oxford.

van Oss, C. J., and Gillman, C. F., 1972, Phagocytosis as a surface phenomenon. I. Contact angles and phagocytosis of non-opsonized bacteria, *J. Reticuloendothel. Soc.* **12:**283.

van Oss, C. J., Gillman, C. F., and Neumann, A. W., 1974, Phagocytosis as a surface phenomenon. IV. The minimum size and composition of antigen-antibody complexes that can become phagocytized, *Immunol. Commun.* **3:**77.

van Oss, C. J., Gillman, C. F., and Neumann, A. W., 1975, *Phagocytic Engulfment and Cell Adhesiveness,* Marcel Dekker, New York.

van Oss, C. J., Good, R. J., Neumann, A. W., Wieser, J. D., and Rosenberg, P., 1977, Comparison between attachment and detachment approaches to the quantitative study of cell adhesion to low-energy solids, *J. Colloid Interface Sci.* **59:**505.

van Snick, J. L., and Masson, P. L., 1976, The binding of human lactoferrin to mouse peritoneal cells, *J. Exp. Med.* **144:**1568.

van Snick, J. P., Masson, P. L., and Heremans, J. F., 1974, The involvement of lactoferrin in the hyposideremia of acute inflammation, *J. Exp. Med.* **140:**1068.

van Snick, J., Masson, P. L., and Heremans, J. F., 1975, The affinity of lactoferrin for the reticuloendothelial system (RES) as the molecular basis for the hyposideraemia of inflammation, in: *Proteins of Iron Storage and Transport in Biochemistry and Medicine* (R. R. Crichton, ed.), pp. 433–449, North Holland, Amsterdam.

Vassalli, J.-D., and Reich, E., 1977, Macrophage plasminogen activator: Induction by products of activated lymphoid cells, *J. Exp. Med.* **145:**429.

Vaughan, R. B., 1965, The discriminative behavior of rabbit phagocytes, *Br. J. Exp. Pathol.* **46:**71.

Vaughan, R. B., and Boyden, S. V., 1964, Interactions of macrophages and erythrocytes, *Immunology* **7:**118.

Viken, K. E., 1976, Effect of sodium salicylate on the function of cultured human mononuclear cells, *Acta Path. Microbiol. Scand. C* **84:**465.

Viken, K. E., and Lamvik, J. O., 1976, Effect of aurothiomalate on human mononuclear blood cells cultured *in vitro, Acta Path. Microbiol. Scand. C* **84:**419.

Walker, W. S., 1976, Separate F_c receptors for immunoglobulins IgG2a and IgG2b on an established cell line of mouse macrophages, *J. Immunol.* **116:**911.

Walters, M. N.-I., Papadimitriou, J. M., and Robertson, T. A., 1976, The surface morphology of the phagocytosis of micro-organisms by peritoneal macrophages, *J. Pathol.* **118:**221.

Wang, P., Shirley, P. S., DeChatelet, L. R., McCall, C. E., and Waite, B. M., 1976, Purification of plasma membrane from BCG-induced rabbit alveolar macrophages, *J. Reticuloendothel. Soc.* **19:**333.

Ward, H. K., and Enders, J. F., 1933, An analysis of the opsonic and tropic action of normal and immune sera based on experiments with the pneumococcus, *J. Exp. Med.* **57:**527.

Ward, P. A., 1968, Chemotaxis of mononuclear cells, *J. Exp. Med.* **128:**1201.

Weir, D. M., 1973, *Handbook of Experimental Immunology,* Blackwell, Oxford.

Weiss, L., 1973, Neuraminidase, sialic acids, and cell interactions, *J. Natl. Cancer Inst.* **50:**3.

Weiss, L., and Glaves, D., 1975, Effects of migration inhibiting factor(s) on the *in vitro* detachment of macrophages, *J. Immunol.* **115:**1362.

Weiss, L., Mayhew, E., Ulrich, K., 1966, The effect of neuraminidase on the phagocytic process in human monocytes, *Lab. Invest.* **15:**1304.

Weissmann, G., Dukor, P., and Zurier, R. B., 1971, Effect of cyclic AMP on release of lysosomal enzymes from phagocytes, *Nature, New Biol.* **231:**131.

Wellek, B., Hahn, H. H., and Opferkuch, W., 1975, Evidence for macrophage C3d-receptor active in phagocytosis, *J. Immunol.* **114:**1643.

Wellek, B., Hahn, H., and Opferkuch, W., 1976a, Opsonizing activities of IgG, IgM antibodies and the C3b inactivator-cleaved third component of complement in macrophage phagocytosis, *Agents Actions* **6:**260.

Wellek, B., Hahn, H., and Opferkuch, W., 1976b, Quantitative contributions of IgG, IgM and C3 to erythrophagocytosis and rosette formation by peritoneal macrophages, and anti-opsonin activity of dextran sulfate 500, *Eur. J. Immunol.* **5:**378.

Wells, M. A., and Hanahan, D. J., 1969, Phospholipase A from *Crotalus adamanteus* venom, in: *Methods in Enzymology* (Lowenstein, J. M., ed.), pp. 178–182, Academic Press, New York.

Werb, Z., and Cohn, Z. A., 1971a, Cholesterol metabolism in the macrophage. I. The regulation of cholesterol exchange, *J. Exp. Med.* **134:**1545.

Werb, Z., and Cohn, Z. A., 1971b. Cholesterol metabolism in the macrophage. II. Alteration of subcellular exchangeable cholesterol compartments and exchange in other cell types, *J. Exp. Med.* **134:**1570.

Werb, Z., and Cohn, Z. A., 1972, Plasma membrane synthesis in the macrophage following phagocytosis of polystyrene latex particles, *J. Biol. Chem.* **247:**2439.

Werb, Z., and Gordon, S., 1975a, Secretion of a specific collagenase by stimulated macrophages, *J. Exp. Med.* **142:**346.

Werb, Z., and Gordon, S., 1975b, Elastase secretion by stimulated macrophages. Characterization and regulation, *J. Exp. Med.* **142:**361.

Werdelin, O., Braendstrup, O., and Pedersen, E., 1974, Macrophage-lymphocyte clusters in the immune response to soluble protein antigen *in vitro.* I. Roles of lymphocytes and macrophages in cluster formation, *J. Exp. Med.* **140:**1245.

Wernet, P., 1976, Human Ia-type alloantigens: Methods of detection, aspects of chemistry and biology, markers for disease states, *Transplant Rev.* **30:**271.

Wigzell, H., Sundqvist, K. G., and Yoshida, T. O., 1972, Separation of cells according to surface antigens by the use of antibody-coated columns. Fractionation of cells carrying immunoglobulins and blood group antigen, *Scand. J. Immunol.* **1:**75.

Wilkinson, P. C., 1974, *Chemotaxis and Inflammation,* Churchill Livingstone, Edinburgh.

Williams, L. T., Snyderman, R., Pike, M. C., and Lefkowitz, R. J., 1977, Specific receptor sites for chemotactic peptides on human polymorphonuclear leukocytes, *Proc. Natl. Acad. Sci. USA* **74:**1204.

Winkelmann, R. K., Scheen, Jr., S. R., and Underdahl, L. O., 1960, Acanthosis nigricans and endocrine disease, *J. Am. Med. Assoc.* **174:**1145.

Wood, Jr., W. B., 1960, Phagocytosis, with particular reference to encapsulated bacteria, *Bacteriol. Rev.* **24:**41.

Yano, A., Schwartz, R. H., and Paul, W. E., 1977, Antigen presentation in the murine T-lymphocyte proliferative response. I. Requirement for genetic identity at the major histocompatibility complex, *J. Exp. Med.* **146:**828.

Yokomuro, K., and Nozima, T., 1972, Bridge formation between mouse peritoneal macrophage-macrophage and macrophage-lymph node cells and the influence of various chemicals, *J. Reticuloendothel. Soc.* **11:**579.

Young, L. S., 1974, The host: humoral and cellular factors. Role of antibody in infections due to *Pseudomonas aeruginosa, J. Infect. Dis.* **130:**S111.

Zbar, B., and Tanaka, T., 1971, Immunotherapy of cancer: Regression of tumors after intralesional injection of living *Mycobacterium bovis, Science* **172:**271.

Zbar, B., Bernstein, I. D., Bartlett, G. L., Hanna, Jr., M. G., and Rapp, H. J., 1972, Immunotherapy of cancer: Regression of intradermal tumors and prevention of growth of lymph node metastases after intralesional injection of living *Mycobacterium bovis, J. Natl. Cancer Inst.* **49:**119.

6

Scanning Electron Microscopy of Macrophages

K. E. CARR

1. INTRODUCTION: PREPARATION TECHNIQUES

The purpose of this chapter is to describe the contribution made by scanning electron microscopy (SEM) to the study of macrophages. This technique, used in the visualization of surfaces at a resolution well beyond that of the light microscope, has cast interesting new light on the structure and reactions of the macrophage cell surface. It is best used not alone but in conjunction with other techniques; presently, however, an attempt is being made to collect the information gained by this technique. References will, therefore, be made only where necessary to studies which do not involve SEM. If such a review seems superficial, it should be remembered that SEM, by definition, is a study of surfaces.

The biological specimen studied in the SEM has been fixed, dehydrated, and coated with heavy metal; the end result is a complex artefact. It is, however, with proper preparative techniques, an artefact consistent to the same extent as the artefact of transmission electron microscopy (TEM); preparative techniques are, therefore, important.

Tissue preparation for SEM (Boyde and Wood, 1969; Carr, 1971; Johari, 1968–1980) involves fixation in glutaraldehyde, formaldehyde, or osmium tetroxide, dehydration through a series of increasing concentrations of a drying agent, such as ethanol or acetone, with a final drying process.

Originally, air-drying was used, but recently improved results have been obtained with freeze-drying or critical-point-drying. The dried specimen is then given a conducting coat of metal *in vacuo,* or impregnated with osmium to avoid build-up of unwanted charge on the specimen surface.

These general techniques are used for the preparation of macrophages, monocytes, and other related cell types. There are particular problems in the

K. E. CARR • Department of Anatomy, University of Glasgow, Glasgow, G12 Scotland, United Kingdom.

collection, incubation, and drying of isolated cells: it is recommended that such cells be fixed with glutaraldehyde prior to collection without aspiration (Alexander *et al.*, 1976).

2. PHAGOCYTIC CELLS

2.1. MORPHOLOGY OF MACROPHAGES, MONOCYTES, EPITHELIOID CELLS, AND GIANT CELLS

2.1.1. Macrophages

The first SEM description of peritoneal exudates by Carr *et al.* (1969) presented the general morphology of the macrophage, using the standard methods of specimen preparation of that time, including air-drying. Specimen preparation techniques have changed greatly since then, particularly with regard to dehydration; these early studies must, therefore, be regarded with some reserve. Nevertheless, although critical-point-drying is now the accepted standard method, it is still possible to draw some worthwhile conclusions from earlier studies using other techniques.

While much has been made of the artefacts produced by air-drying, there has been surprisingly little comment on the possible artefactual alterations of membrane contours which might be introduced during other phases of the experimental manipulation of the macrophage. If valid comparisons are to be made with previous studies in the literature, standardization of the conditions under which macrophages are maintained before fixation is just as important as standardization of the dehydration process. For example, macrophages studied after prolonged culture (Güttner, 1975) may well not be directly comparable with cells which have undergone fewer manipulations prior to fixation.

From this early work, and from subsequent studies, we can draw up a list of the features by which the macrophage can be recognized under the scanning electron microscope. It must be emphasized that criteria such as these, based only upon surface detail, are in themselves inconclusive identification; TEM or some functional marker is required for final confirmation. In general, the macrophage is recognized by SEM as a large cell with a variety of membrane specializations. The cell may be round or flattened and may have foot processes of variable extent, depending upon its functional state. Many terms are used to describe the surface projections of macrophages. For instance, *Lamellipodia* are extended flaplike processes; *pseudopodia* are broad footlike processes; *filopodia* are thin threadlike processes; flanges, veils, undulating membranes, and microvilli are self-explanatory. The term *villi* is often inappropriately used to describe projections of various kinds from the surface of the cell. The surface ultrastructure of the macrophage is distinguished by varying combinations of these different cell processes.

The following reports on macrophage surface morphology have been reviewed in chronological order, beginning with the earliest papers and progressing to current work. In this way, allowance can more easily be made by the reader for the progressive changes in specimen preparation techniques.

Following Carr *et al.* (1969), Warfel and Elberg (1970) examined rabbit peritoneal macrophages after 18–24 hr in culture. The macrophage was described as having a variety of shapes, with surface processes, pseudopodia, and undulating membranes. Lysosome shaped structures were seen under the cell membrane and the macrophages were attached to the glass substrate by extensions of the membrane.

Mouse peritoneal macrophages have been described as being 6–15 μm in size varying in shape from compact and spherical to flattened and irregular (Papadimitriou *et al.*, 1973). Ridges and flaps were seen on the surface. Veil-like processes and lamellipodia were present, with thin filiform extensions protruding from the latter. Papadimitriou (1973) studied macrophage receptors for heterologous IgG using mouse peritoneal macrophages and epithelioid and giant cells. Both groups were challenged with a 2% suspension of sheep erythrocytes sensitized with rabbit anti-sheep erythrocyte immunoglobulin and incubated for 1 hr at 37°C. The erythrocytes were attached to macrophages in numbers varying from 3–25 per macrophage. They were more numerous at the lamellipodia, and the central paranuclear region was relatively free of erythrocytes. The distorted erythrocytes were angulated at the point of attachment.

Most experiments described have dealt with *in vitro* studies. Macrophages have also been identified as one of the cell types present on the surface of milky spots of mouse omentum and mesentery (Orenstein and Shelton, 1976b). The macrophages were round, discoid, or irregularly shaped and had ridges and folds on their surface. The flattened macrophages retained their ridged surface, and never had the "fried egg" appearance seen in macrophages settled on glass. Cell contacts through tapered processes were also seen *in vitro*. Albrecht *et al.* (1976) studied unstimulated mouse peritoneal macrophages cultured for 24 hrs before glutaraldehyde fixation and dehydration by air-drying from organic solvents or by critical-point-drying. The solvents used included water, xylene, benzene, isopentanol, acetone, ethanol, hexane, and ethyl ether from either ethanol or acetone as the final solvent. The critical-point-dried specimens showed many ridges, ruffles, and processes which were not well preserved by the other methods, although these techniques were satisfactory in assessing cell-to-cell relationships.

Wormlike processes (vermipodia) were described on the surface of histiocytes from histiocytic malignancies (Ghadially and Skinnider, 1976). It was claimed that they were different from uropodia, which were broad-based and tapering, and pseudopodia, which were blunt projections on the surface.

McKeever *et al.* (1977) assessed immune complex receptors on macrophages using horse radish peroxidase on rabbit lung alveolar cells. Receptors were visualized as sites of attachment of thin marker crystals and were most common on cytoplasmic veils, pseudopodia, and in the perinuclear region in minimally spread cells. Further spreading reduced the number of receptor sites stained, supporting the general view that macrophage spreading and endocytic activity are inversely related to contact with immune complexes (see also Section 5).

2.1.2. Monocytes

Monocytes from blood were seen as round cells with deeply folded membranes and cytoplasmic veils particularly prominent at the leading edge (Bessis and Boisfleury, 1971). The veils were similar to those described by Warfel and Elberg (1970) for macrophages.

Normal human blood cells were incubated for 1 hr and identified by light microscopy after Giemsa staining (Wetzel *et al.*, 1973). The same areas, critical-point-dried, were then examined with SEM and it was stressed that there was overlap in the surface morphology between the different types of leukocyte. The use of parallel light and scanning microscopy was therefore recommended.

Blood monocytes were estimated to be 5–7 μm in diameter with a ruffled membrane, and were similar to the histiocyte or unstimulated macrophages (Dantchev, 1976).

Polliack (1977) in his book on normal, transformed, and leukemic leukocytes summarized the findings of several groups on the optimum method for the preparation of isolated cells. Controversy had existed as to whether aspiration filtration was advisable. Alexander *et al.* (1976) finally showed that the purported difference between T and B lymphocyte surface morphology was an artefact created by the aspiration filtration. Polliack (1977) therefore recommended following the method of Alexander *et al.* (1976). This involved fixing the cells in suspension without aspiration prior to collection on a substrate. Critical-point-drying was used. With these techniques monocytes are seen as large cells, with ruffled membranes and surface ridges, able to attach and spread on glass and to phagocytize particles (Polliack, 1977). Leukemic monocytes are shown to be different in their structure from leukemic cells of the lymphocyte line.

2.1.3. Epithelioid Cells

Epithelioid cells resulting from subcutaneous implantation of cellophane strips were 50–100 μm in diameter and closely opposed to each other (Papadimitriou *et al.*, 1973). They lacked the peripheral lamellipodia of the giant cells although some of them had small apronlike equivalent structures. Filiform processes were common, in some cases apposed to adjacent cells.

Epithelioid cells were studied using sheep erythrocytes sensitized with rabbit anti-sheep erythrocyte immunoglobulin. There was less adherence than was present with macrophages, no doubt because of their poor phagocytic capacity (Papadimitriou *et al.*, 1973).

2.1.4. Giant Cells

Giant cells were described by Papadimitriou *et al.* (1973) as being 80–200 μm in diameter, with a raised central portion and a broad, thin peripheral region with lamellipodia. Ridges and folds occurred on the surface and pits 0.05–0.2 μm were seen between the ridges. There was a lack of adherence in giant cells similar to that of epithelioid cells, reflecting poor phagocytic ability (Papadimitriou, 1973).

The surface of BCG (bacillus Calmette-Guerin)-induced multinucleate giant cells of rabbit alveolar macrophage origin was studied by Warfel (1978). The cells had four areas: a central region covered with membranous veils, a transitional area having few veils, a peripheral area with pits and ridges, and terminal edges which were smooth or had filopodia. In general the surface was similar to that of macrophages.

2.1.5. Differentiation

Differentiation was investigated by studying bone marrow cells cultured from pathogen-free mice. In the mononuclear phagocyte colonies produced, three cell types were seen: monoblasts, promonocytes, and macrophages (Van Furth and Fedorko, 1976). The differences between the monoblast, the most immature of the series, and the immature granulocytes were proof that the mononuclear phagocytic series developed from an independent cell line.

The differentiation of monocytes into macrophages using human venous blood was studied over a period of 4–7 days: the cells were compared with stimulated rat peritoneal macrophages (Parakkal *et al.*, 1974). Monocytes were defined as the cells seen on the coverslips the day the cells were isolated from blood while macrophages were the cells maintained in culture for more than 7 days. The monocytes were smaller than macrophages: they were 12–18 μm in size and showed either ruffled membranes or lamellipodia, filopodia, and microvilli. The monocytes were too small to engulf Aminex particles easily dealt with by macrophages (Figure 1). After phagocytosis fewer ruffles were seen.

The relationship of macrophages to giant cells was proved by examining mouse macrophages on coverslips inserted subcutaneously into mice (Mariano and Spector, 1974). After 3 days, young macrophages newly arrived from the circulation fused with *in situ* macrophages to form multinucleate giant cells.

2.2. MORPHOLOGY OF SPREAD AND SETTLED CELLS

Early work on settled mouse peritoneal macrophages (Carr and Carr, 1970) showed that there was initial protrusion of fine processes followed by the formation of cytoplasmic veils, until finally the cell was markedly flattened. Full extension occurred more quickly in cells stimulated by glycerol trioleate. Blood-derived monocytes showed a similar settling process with the formation of veils and eventual flattening of the cells (Bessis and Boisfleury, 1971). The attachment stage of the settling process was studied in experiments on alveolar macrophages settled for 20 min on glass or millipore filters (Leake *et al.*, 1975). The macrophages on glass had more tendency to spread, whereas those on millipore filters remained rounded and resembled those *in situ* in the lungs, with the usual ruffled surfaces. Mouse peritoneal macrophages were incubated for 30–120 min

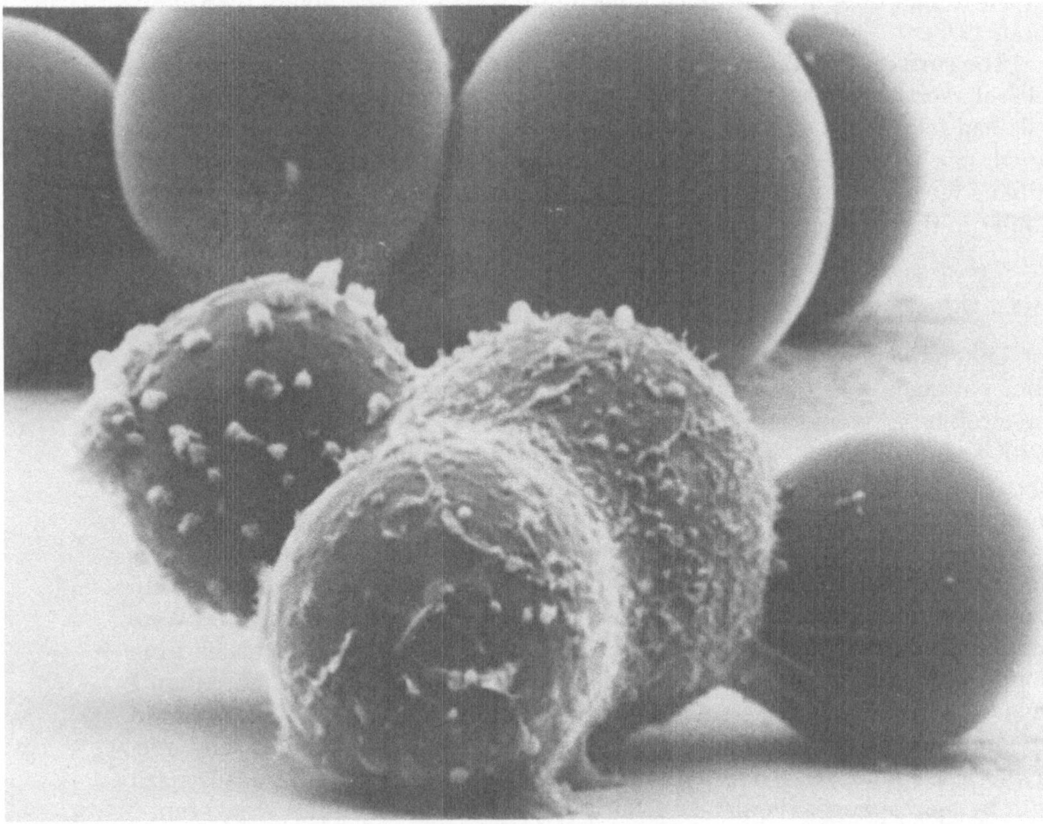

FIGURE 1. A macrophage which has engulfed three Aminex beads has formed a preengulfment attachment to a fourth bead. The cytoplasmic membranes are stretched taut, leaving no ruffled membranes. Irregular microvilli dot the surface. The macrophage is attached to the glass by small pedicles. ×2600. Micrograph courtesy of Dr. P. Parakkal (Parakkal, P., Pinto, J., and Hanifin, J. M., 1974, *J. Ultrastruc. Res.* **48**:216).

and then cultured for 24–48 hr: an increase in the flattening of the cells with culture time was observed, with great variation in the shape and degree of spreading (Polliack and Gordon, 1975). Slender filopodia were seen and ridges and fine pits appeared on the surface. This ridged appearance was confirmed in mouse peritoneal cells on artificial substrates (Orenstein and Shelton, 1976a): it became less obvious as the macrophages settled on millipore filters for longer periods of time. Mouse peritoneal exudate cells were examined after exposure to membrane vesiculant (Scott and Maercklein, 1977). It was found that the spreading varied with the culture medium. The first stage involved little sign of the extension of pseudopodia, the second stage showed small focal pseudopodia, while the third stage involved extensive spreading (Figure 2). With respect to media there was maximum spreading in buffers containing manganese or dithiothreitol. The difference was stressed between macrophages in suspension, where surface blebbing on exposure to vesiculant was occasionally seen, and

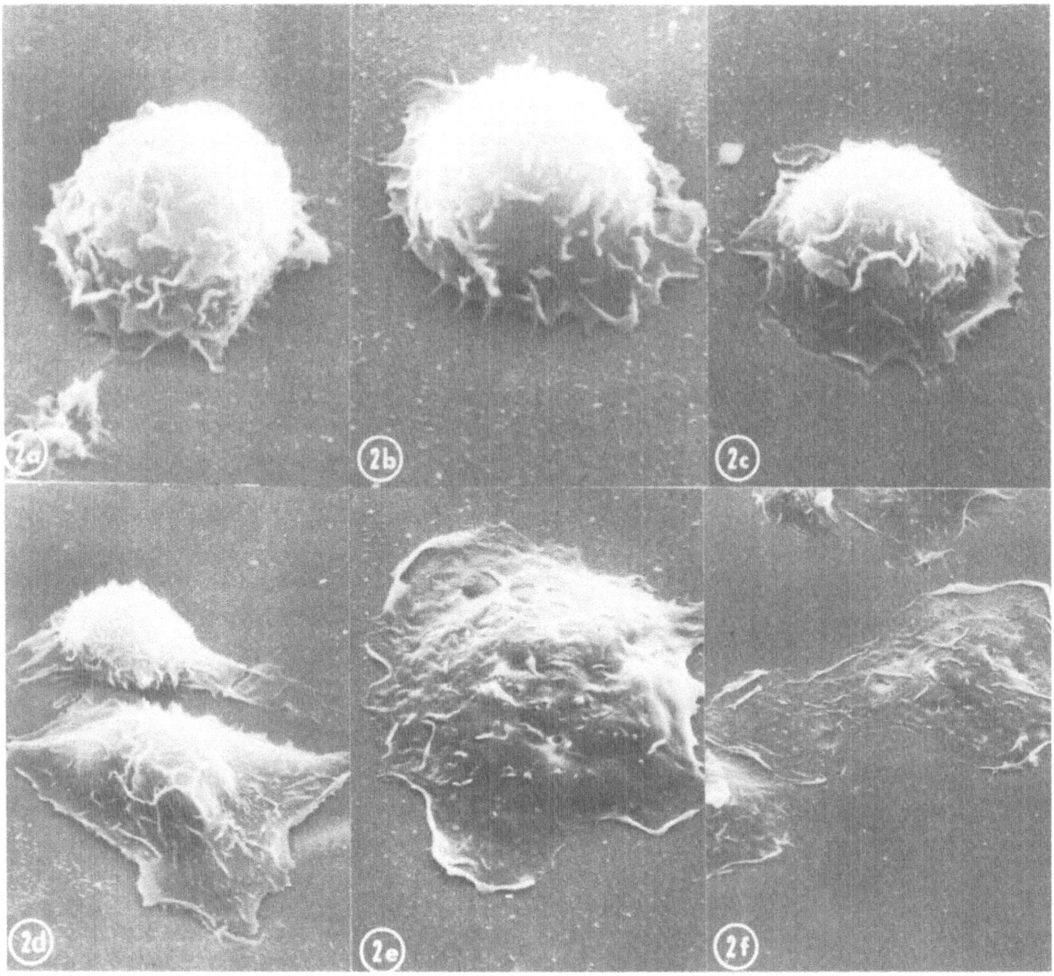

FIGURE 2. Scanning electron micrographs of macrophages incubated in SIM buffer and showing slight (2a) to extensive (2d) spreading. Macrophages incubated in SIM containing 0.1 mM Mn^{2+} (2e) or 2 mM DTT (2f) showed more extensive spreading with loss of most cell surface projections. Magnifications: (2a) ×4000; (2b) ×4800; (2c) ×4000; (2d) ×2400; (2e) ×1600; (2f) ×4000. Reduced 15% for reproduction. Micrographs courtesy of Dr. R. E. Scott and Dr. P. B. Maercklein. (Scott, R. E., and Maercklein, P. B., 1977, *Lab. Invest.* **37**:430.)

spread macrophages, where significant quantities of cell membrane vesicles were shed.

2.3. MORPHOLOGY OF STIMULATED CELLS

Normal macrophages were shown to have smooth or ridged surfaces, while cells stimulated with glycerol trioleate had flangelike processes, small ridges, and fingerlike projections (Carr *et al.*, 1969). Full extension occurred sooner in stimulated cells (Carr and Carr, 1970). Clarke *et al.* (1971) studied preparations of peritoneal macrophages and lymph node lymphocytes stimulated with phytohemagglutinin or antigen. It is of interest that they found no difference at a resolution of 20 nm between air-dried and critical-point-dried specimens. In the stimulated preparations, the two cell types developed long processes which could form intercellular bridges. Stimulated and unstimulated unfixed freeze-dried mouse macrophages were compared by Albrecht *et al.* (1972). Unstimulated cells had various shapes with footlike processes and ridges on the surface. Macrophages stimulated by intraperitoneal injection of light oil were larger, with a greater variety of membranous processes over the surface. Many small lateral projections were seen in addition to foot processes. Stimulated macrophages were described as being larger and flatter than unstimulated cells (Papadimitriou *et al.*, 1973). Thioglycollate- and endotoxin-stimulated macrophages were also more spread and flat than unstimulated ones, with extremely long processes, broad veils, and prominent rolled edges (Polliack and Gordon, 1975). There was more variation in the shape of the stimulated cells and ruffles, and ridges and filopodia were seen, the filopodia sometimes forming intercellular bridges. Activated macrophages were reported by Dantchev (1976) to be particularly adherent to glass.

2.4. MORPHOLOGY AND EXOCYTOSIS

This aspect of the function of macrophages has received little attention from scanning microscopists despite the possible importance of secretion in the functioning of the cell. While exocytosis and secretion are not identical, only exocytosis can be identified by SEM. Rabbit lung macrophages contained granules either near the ends of fine processes or on the cytoplasmic veil surrounding the macrophage (Thomas and Vilain, 1970). These granules were presumably secretory or lysosomal in nature and occurred singly or in groups, and were globular or polyhedral in shape and 0.2–0.7 μm in size (Thomas *et al.*, 1970). Mouse peritoneal macrophages showed a strong extracellular cytotoxic activity towards syngeneic erythrocytes, involving lysis of the target cell mediated by an extremely labile macrophage cytolytic factor (Melsom *et al.*, 1974). The target cells probably stuck first on to small projections on the macrophage surface, possibly receptor sites, and then lysed.

The reaction of BCG-exposed macrophages with tumor cells was examined by Bucana *et al.* (1976). The tumor cells were guinea pig line-10-hepatocarcinoma cells and the effector cells were peritoneal exudate cells from syngeneic strain-2-guinea pigs cured of a line-10-tumor by BCG injection. The BCG-sensitized macrophages spread more than normal ones and were predominantly spread in contact with tumor cells, whereas the normal macrophages were spherical when in contact. The results indicated that the cell-killing by activated macrophages was nonphagocytic; it was suggested that the response initiated by recognition resulted in extracellular release of lysosomal material which then affected the target cells.

3. PHAGOCYTIC CELLS IN ORGAN SYSTEMS

3.1. LYMPHORETICULAR SYSTEM

3.1.1. Lymph Node

Much of the attention directed to the study of macrophages in the lymph node has been aimed at identifying these cells and differentiating them from other cell types present. No intermediate cell types between macrophages and reticulum cells were found in a study of mesenteric lymph nodes from dogs (Fujita *et al.*, 1972).

Macrophages were numerous in the sinuses and there were few in the pulp. They were large and round (10–15 μm in diameter) densely covered with clubbed cytoplasmic processes and had few long, slender processes, best preserved by critical-point-drying. As well as stressing the fact that the macrophages unequivocally differed from the reticulum cells, Fujita *et al.* (1972) commented on the impropriety of the term *free macrophages* to describe those present in the sinuses. *Round-shaped* was preferred, since the cells were fixed to the reticulum by reticulum cell processes and had remained so despite relatively thorough perfusion by Ringer's solution and fixative. Macrophages were described in rat paraaortic and mesenteric lymph nodes and hemolymph nodes situated cephalic to the left renal artery (Luk *et al.*, 1973). In the lymph nodes the macrophages were recognized by their large size and the presence of pseudopodia and an irregular surface; cytoplasmic processes were seen anchoring them, in the sinuses, to the trabeculae. In the hemolymph node, phagocytosis of erythrocytes was seen, the erythrocytes being caught by micropseudopodia. Rosettes of erythrocytes were formed around the macrophages which hung like grapes from the trabeculae.

The theme of identification of macrophages in guinea pig popliteal lymph nodes was followed further in a study of teased-out lymph node cells allowed to settle on glass for 4 hrs at 37°C (Tizard and Holmes, 1975). A careful study was made of the cell populations so obtained from spleen, Peyer's patches, thymus, and peritoneal washings as well as in mesenteric and inguinal lymph nodes. Three cell types adhered to glass and had obvious cytoplasmic processes: these were further examined for their similarity to macrophages. Type I cells were rounded with a diameter of 12–16 μm, and had membranous ruffles and cytoplasmic processes. Type II cells had spread extensively and were generally smooth with some ruffles. Type III cells were spherical and smooth with a diameter of 8–12 μm (smaller than Type I) and had some fine branching processes. Type III cells were present only in lymph node and Peyer's patches, and had surface receptors for cytophilic antibodies, complement, and immunoglobulin. The Type III cell was, therefore, identified as distinct from macrophages and the term dendritic cell was chosen as the most suitable. Types I and II by the same evaluation were both macrophages, indicating that not only were there differences between macrophages and dendritic (or reticular) cells, but the surface structure of the macrophage in the lymphoreticular system was also widely variable.

3.1.2. Spleen

In the study of splenic macrophages, interest centers on their identification and also on their reaction with erythrocytes. In dogs, large spherical cells were described, 10–15 μm in diameter, with fringelike processes or undulating surfaces (Miyoshi and Fujita, 1971). In rats two types of cells in the cordal spaces were identified as macrophages, on the basis of their similarity to macrophages described by Carr (1970) and also on the basis of parallel TEM. The first type, covered with round-headed projections, was shown by transmission electron microscopy to have ample cytoplasm and lysosomes containing finely granular material. The second type which had thin foliate surface processes, contained large, dense granules and abundant cell debris. Cytoplasmic threads from adjacent reticular cells were seen to converge on the macrophages. The macrophages were one of three separate cell lines present, the other two being sinus-lining cells and cordal reticular cells. A general study of the structure of the spleen included illustrations of macrophages showing many surface folds and processes (Weiss, 1974). No gradations between macrophages and the cordal reticular cells were seen in a study (Fujita, 1974) using intense osmium impregnation (Murakami, 1974) and freeze-fracturing. Macrophages were identified by their microvillous projections and were attached to or projected into the sinus lumina. They sometimes pushed microprojections from the perisinal spaces into the sinus. In the Billroth or splenic cords macrophages showed a variation in shape not dissimilar to that described for rats by Miyoshi and Fujita (1971) in that, although usually large in size and covered with bubblelike or spinous processes, some macrophages were irregular in shape with pseudopodia, while others were flattened. Large vacuolar inclusions were seen in the macrophages fractured during preparation and it was thought these were phagosomes. Lymphocytes were sometimes seen close to macrophages or touching them with cytoplasmic processes. Species differences were studied in the spleen, using dog and rabbit as examples of a sinal spleen and cat as a nonsinal spleen (Song and Groom, 1974). Identification of the macrophages was made by comparison with light microscopy and by reference to the literature. Song and Groom (1974) described abnormally shaped erythrocytes adhering to the macrophages, forming clusters around them in some cases: it was suggested that this was associated with the maturation process of the erythrocytes rather than with their phagocytosis.

3.2. RESPIRATORY SYSTEM

The importance of alveolar macrophages was reviewed as a means of defense against inhaled pathogens, and the techniques of sampling and analysis used on macrophages obtained by pulmonary lavage were described by Voisin *et al.*, (1972a). The human alveolar macrophages studied had large bases and convoluted membranes. Later work by Voisin *et al.* (1977) described attempts to put alveolar macrophages *in vitro* in a situation similar to their environment in the lung. To this end, they were obtained by bronchopulmonary lavage and maintained on a filter in such a way that they obtained nutrients from the medium in the filter by capillary action but were in direct contact with the atmosphere above. These macrophages were very rounded and differed from those settled on glass, as was reported by Leake *et al.* (1975). Their membrane convolutions resembled those of macrophages *in situ*.

Brief reference was made to the presence of alveolar macrophages during the early stages of cell migration from a pulmonary fragment used to establish Type II epithelial cell cysts (Rosenbaum *et al.*, 1977).

Macrophages *in situ* were included in the cell types described by Kuhn and Finke (1972) in a study of the effects of different fixation procedures on the topography of pulmonary alveoli in mice, hamsters, and rats. The macrophages had a smooth surface towards air and pseudopodia spread over the alveolar surface. The alveolar macrophage was described in more detail in a study of rat, shrew, dog, and human material (Weibel *et al.*, 1976). In fetal rat lung the macrophages appeared in the last few days before birth and lay between the surfactant layer and type I cells. If the surface layer had been removed during preparation, the many filopodia of the macrophage were seen, attaching them to the epithelial surface. In discussion of the function of the alveolar macrophages, it was postulated that they might be involved in surfactant removal. There is an alteration in the properties of alveolar macrophages and dedifferentiation of specialized surface processes after exposure to smoke (Figures 3 and 4) or foreign material (Quan and Golde, 1975). Such lines of research are important in view of the key role respiratory macrophages may play in defense against inhaled particles.

Pleural macrophages were included in a study of stomas connecting the pleural cavity and lymphatics in the parietal pleura of rabbits and mice (Wang, 1975).

3.3. HEPATOBILIARY SYSTEM

A high yield of hepatic macrophages or Kupffer cells can be obtained by dispersing rat liver with collagenase perfusion, followed by selective pronase digestion of the parenchymal cells (Munthe-Kaas *et al.*, 1975). The cells were well spread, rounded or stellate, with small surface projections, and ruffling at the cell edges. They could not be distinguished from peritoneal macrophages on morphological grounds. Their phagocytic capacity was very high as measured in experiments using latex particles and although the Fc-receptor reading was not so high, it was postulated that not every Kupffer cell expresses this receptor at the same time.

There are several descriptions of Kupffer cells *in situ* in liver. In a report on guinea pig liver treated with the tannin–osmium impregnation method of Murakami (1974), it was claimed that the osmium impregnation decreased dehydration shrinkage so much that air drying was satisfactory (Itoshima *et al.*, 1974). Kupffer cells were described as lying mainly at the periphery of lobules

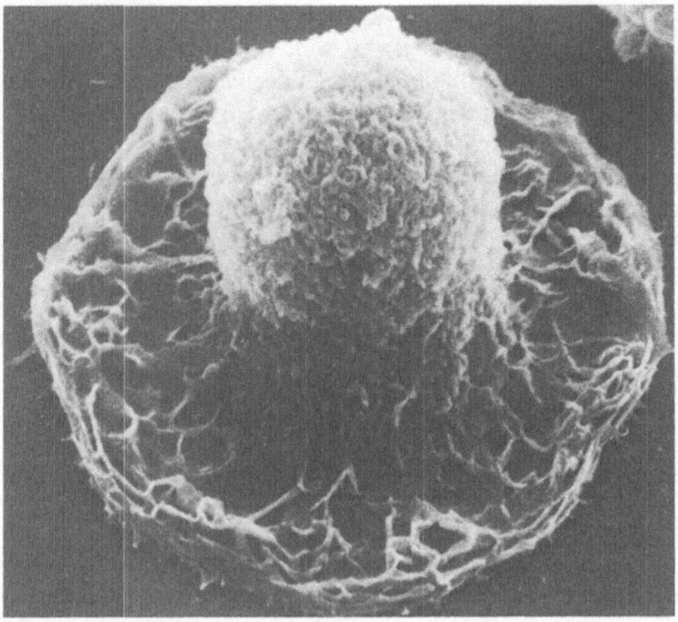

FIGURE 3. Typical macrophage from cigarette smoker. Cell surfaces are completely covered with very deep folds and few microvilli and blebs. The cell periphery is highly creased and has thickened lamellipodia. Settled for 10 min. ×2600. Micrograph courtesy of Dr. S. G. Quan and Dr. D. W. Golde. (Quan, S. G., and Golde, D. W., 1977, *Exp. Cell Res.* **109**:71.)

and protruding into the sinusoid lumen and extending processes in a stellate fashion, presumably with a filtering function. In the liver of albino rats the Kupffer cells extended processes to interrupt the continuity of the sinusoids and were different, at the time of fixation of the liver, from the fenestrated endothelial cells (Motta and Porter, 1974). It was suggested that the Kupffer cell in its role as a macrophage changed its position relative to the endothelial cells and altered the diameter of the sinusoid. This point was taken up by Motta (1975) who made the analogy to a sphincter in his description of the alteration of sinusoid diameter by macrophages in liver, again from albino rats. The macrophages were identified by correlation with TEM; the highly fenestrated endothelial cells differed completely from the rough-surfaced macrophages with their irregular short, surface processes. In addition the luminal surface in some places had a large number of short microvilli, microplicae, and invaginations, forming a kind of microlabyrinth. Clusters of red and white blood cells occasionally stick to the Kupffer cells and reduce the diameter of the sinusoids.

The difference between endothelial cells and Kupffer cells, both components of the sinusoidal wall, was again stressed in a study of human liver (Muto *et al.*, 1977). The endothelial cells were smooth and flat while the Kupffer cells had many microvilli, filopodia, and lamellipodia sometimes attached to endothelial cells. In contrast to the findings of Muto (1975) no intercellular gaps were found between the Kupffer cells and the endothelial cells.

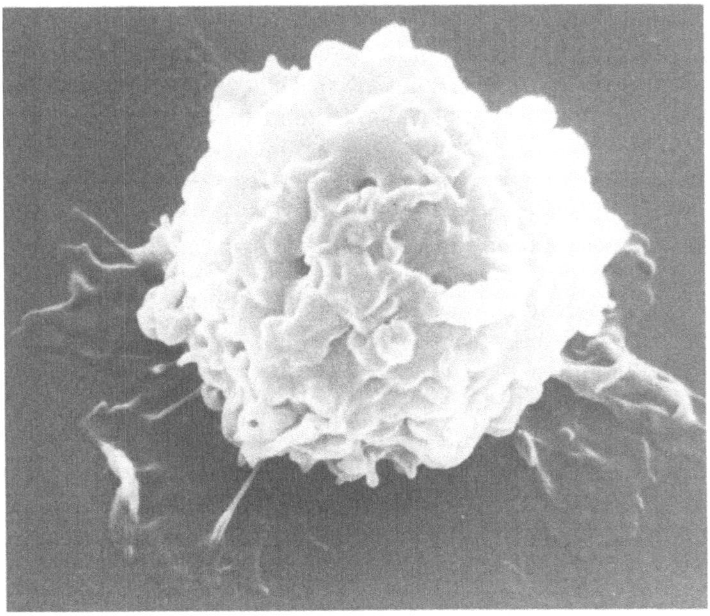

FIGURE 4. Alveolar macrophage from a nonsmoker settled for 10 min. The cell is rounded and attached to the substrate by several small and delicate lamellipodia. Surface has folds, but few blebs or microvilli. ×6100. Micrograph courtesy of Dr. S. G. Quan and Dr. D. W. Golde. (Quan, S. G., and Golde, D. W., 1977, *Exp. Cell Res.* **109**:71.)

On a few occasions the Kupffer cells were found in the sinusoidal lumen or within the space of Disse. The surface morphology of the Kupffer cell has been reviewed in detail by various authors in a text on the ultrastructure of the Kupffer cell (Wisse and Knook, 1977).

Takahashi (1977) studied amyloidosis of the liver produced in mice by subcutaneous injection of casein. Kupffer cells were identified by their attachment to sinusoidal walls and their surface ridges and ruffles. It was postulated that the macrophages were involved in the development of amyloidosis.

3.4. NERVOUS SYSTEM

There has been much interest in the study of the free cells present on the surfaces within the central nervous system.

In the first of a series of papers on the SEM of the subarachnoid space in the dog, free cells were seen in great numbers on all leptomeningeal surfaces at various spinal cord levels (Cloyd and Low, 1974). The similarity of the surface morphology of these cells to macrophages in other sites suggested a common identity. Cell shapes varied from spherical to elongated, from smooth to ruffled or wrinkled. Variations were seen within each cell and cytoplasmic processes or pseudopodia were prominent. Macrophage-like cells were also seen at spinal nerve exits (Malloy and Low, 1974). Many free cells were seen on the meningeal lining of the subarachnoid space: TEM was used to confirm the identity of these cells as macrophages (Allen and Low, 1975). They were found over most cranial leptomeningeal surfaces facing the subarachnoid space, being most numerous in the depths of the cerebral sulci and cerebellar laminae and along blood vessels. Their extreme pleomorphism was again noted. They were connected to the pial surface by microvilli which varied in length. The free cells were identified by TEM as macrophages, both in the subarachnoid space and in the choroid plexus (Malloy and Low, 1976). The cells were observed as phagocytic after introduction of horseradish peroxidase and red blood cells (Figures 5–8). Confirmation of this functional aspect came from work on animals after challenge. Free cells present on leptomeningeal sheaths were studied 12 days after intrathecal injection of BCG: TEM confirmed the identification of the three cell types recognized by SEM as macrophages, lymphoblasts, and neutrophils (Merchant and Low, 1977a,b). The free cell population increased ten-fold by comparison with the controls, 80% of these being macrophages; the stimulated cells were more pleomorphic and had rougher surfaces. The challenged macrophages often formed clusters and showed more surface blebbing, ruffles,

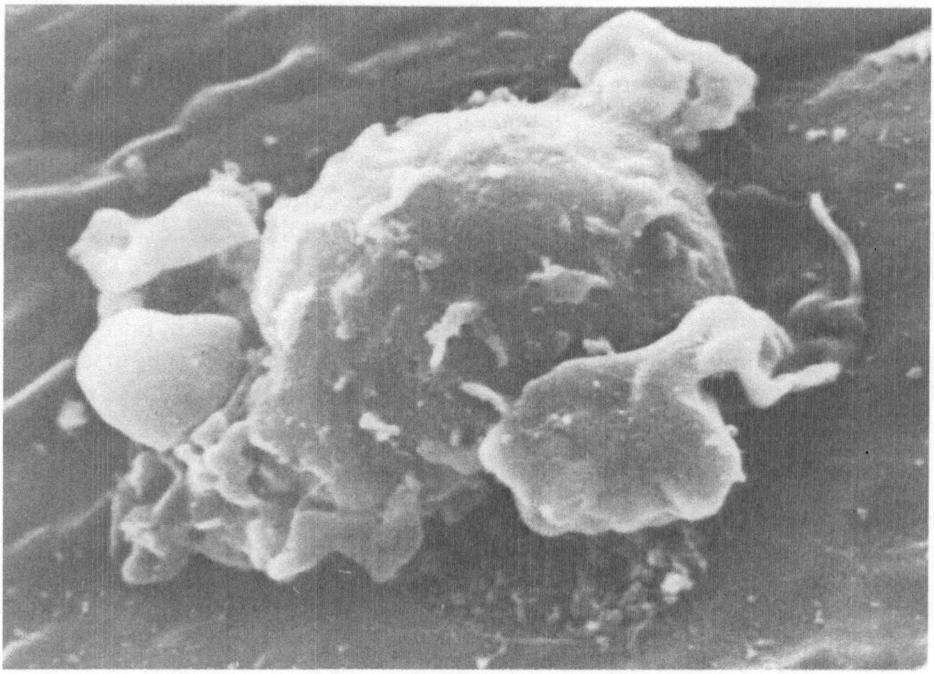

FIGURE 5. Scanning electron microscopy of a free cell on the pial surface following peroxidase injection. Free cells remain pleomorphic in character. Cells tend to be less smooth and microvillous projections are frequently found on their surfaces. A fenestration is present in the spinal cord pia. To demonstrate the phagocytic capacity of this cell the specimen was embedded and sections were obtained diagonally across the cell and fenestration. ×4800. Micrograph courtesy of Dr. J. J. Malloy and Dr. F. Low. (Malloy, J. J., and Low, F., 1976, *J. Comp. Neurol.* **167**:257.)

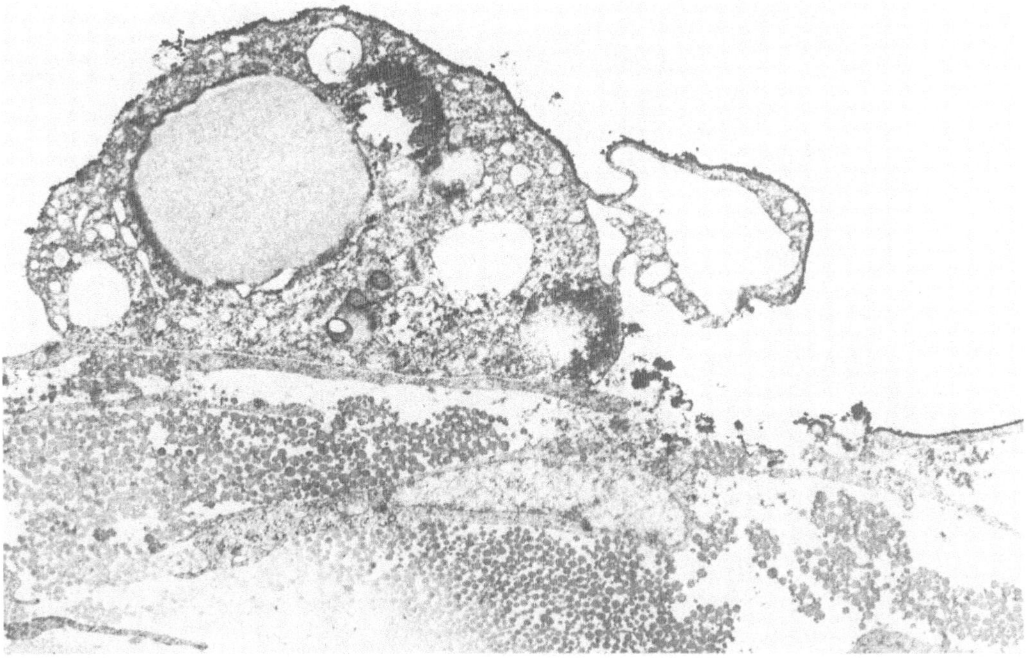

FIGURE 6. Transmission electron microscopy of the cell illustrated in Figure 5. Surface morphology clearly indicates that the cell in this figure is the same one as shown on the facing page. Vacuoles containing peroxidase reaction product are present. Numerous empty vacuoles and lipid inclusions are also situated in the cytoplasm. Pial cell processes in good morphological condition are interrupted to form a pial fenestration. This provides communication between the subarachnoid space and the pial connective tissue. The same fenestration is evident in Figure 5. ×11,000. Micrograph courtesy of Dr. J. J. Malloy and Dr. F. Low. (Malloy, J. J., and Low, F., 1976, *J. Comp. Neurol.* **167**:257.)

and microvilli than the controls. Cell clusters formed between the three free cell types 12 days after BCG challenge, with the formation of intercellular adhesions (Merchant *et al.*, 1977). This clustering of macrophages which increased with BCG challenge may be due to an increase in adhesiveness or phagocytic activity.

The effect of the phagocytic protozoon *Acanthamoeba culbertsoni* on the free cell population in the subarachnoid space of the spinal cord was studied in dogs killed 1 hr, 10 hr, and 100 hr after injection of the pathogen (Sarphie and Allen, 1977). At 1 hr after injection, macrophages were seen with smooth or ruffled membranes. Although there is an initial increase in the number of macrophages, by 100 hr the number has decreased again.

The next area examined by SEM lies within the ventricular spaces of the nervous system. The free cells of the choroid plexus were originally described by Kolmer (1921) and Ariens-Kappers (1953); the latter gave them the name of epiplexus cells. The suitability of the latter name has been discussed following recent thorough work with the scanning electron microscope. The free cells or Kolmer cells of the choroid plexus and the wall of the third ventricle of rats were

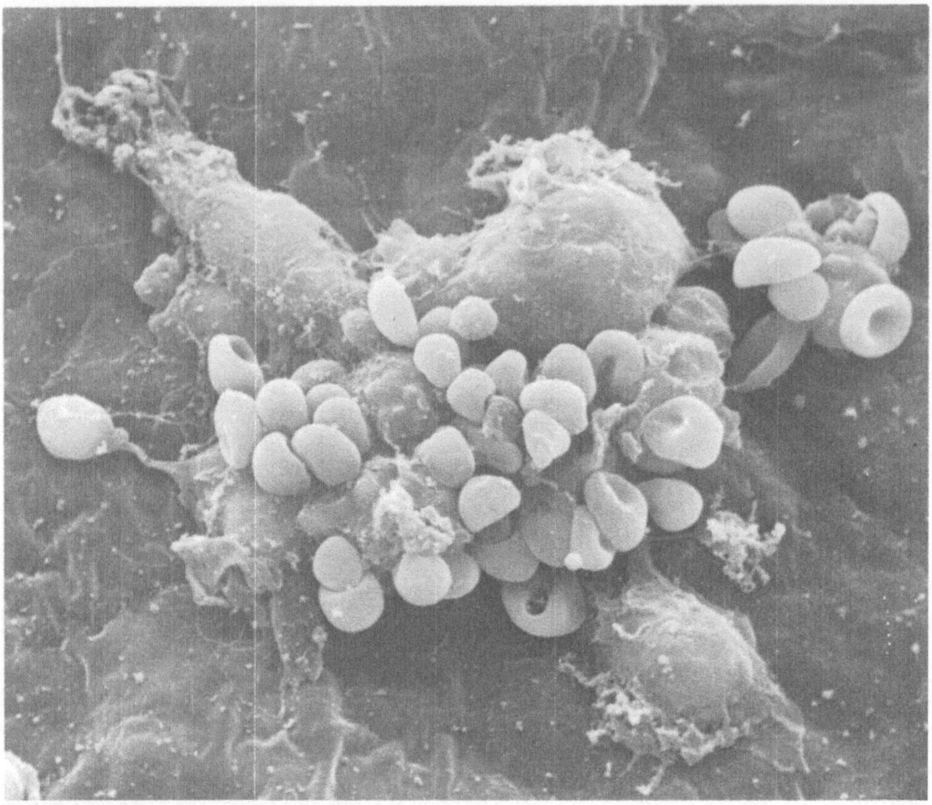

FIGURE 7. Extravasated red blood cells in the subarachnoid space. Following spinal puncture red blood cells may escape into the subarachnoid space. These are usually found attached to one or more free cells. Once attached the red cells tend to lose their biconcave shape and become more spherical. ×3200. Micrograph courtesy of Dr. J. J. Malloy and Dr. F. Low. (Malloy, J. J., and Low, F., 1976, *J. Comp. Neurol.* **167**:257.)

subdivided into two types (Hosoya and Fujita, 1973). Type I cells had a cell body which came together in a pointed fashion and had many radiating cytoplasmic processes extending over and among the microvilli of the underlying ependymal cells. Type II cells were much more flattened in shape, with pseudopodia running from the cell body, in the grooves between the underlying ependymal cell convexities. The numbers of each type varied and in some cases only the former type was found. Type I cells were thought to be in the resting phase, and Type II the moving phase. The Kolmer cells seem to be scavenger cells of the whole ventricular system; the term *epiplexus* is therefore inappropriate since it is too restrictive. The presence of a supraependymal population of cells was confirmed in a description of the ependymal surface of the temporal horn of the lateral ventricle of the dog (Allen and Low, 1973). Bleier (1971), in a study of tissue from the ependyma of the hypothalamic third ventricle in mice, described cells with pseudopodia. The fact that some of these were near mounds of debris implied a possible phagocytic role. Allen (1975) examined the inferior medullary velum

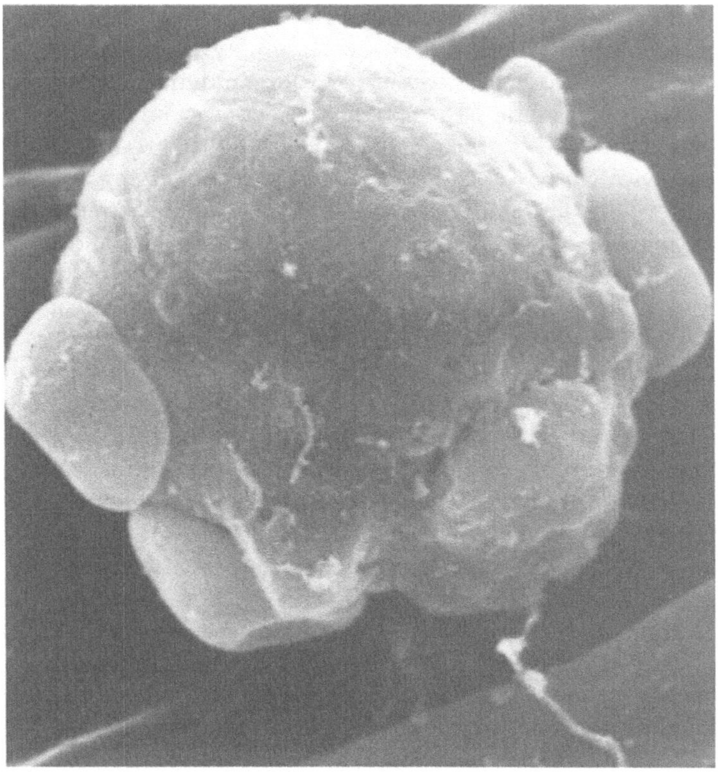

FIGURE 8. In this figure, three red blood cells are attached to a round smooth-surfaced free cell in the subarachnoid space. Two pseudopodlike extensions are present on the cell surface adjacent to two of the attached red blood cells. A fourth blood cell is almost covered by a membranous process from the plasmalemma of the free cell. ×4800. Micrograph courtesy of Dr. J. J. Malloy and Dr. F. Low. (Malloy, J. J., and Low, F., 1976, *J. Comp. Neurol.* **167**:257.)

and tela choroidea removed from the fourth ventricle. The Kolmer cells of the dog were not divisible into two types, but were smoother and very pleomorphic with variation in their cytoplasmic processes (Figures 9–12). They were present on the choroid plexus and the roof of the fourth ventricle: TEM was used to confirm their identity as macrophages (Allen, 1975). Bleier *et al.* (1975) used tissue from the hypothalamic ependyma of the tegu lizard to confirm the phagocytic properties of the free cells. In brain not exposed to latex beads the free cells were relatively smooth, with cytoplasmic extensions of various shapes lying over the ependymal surface. After exposure to latex beads, activated free cells were seen. These were rounder with a folded and microvillous surface. Latex beads were observed during ingestion producing lumpy, free cells. As a result, it was suggested that these free cells are resident phagocytes. Similarly, in a study of material from the median eminence of rhesus monkeys it was suggested that one of the two subpopulations of supraependymal cells was composed of histiocytes (Scott *et al.*, 1975). In the brain of a hydrocephalic mouse mutant there were many phagocytes on ependymal surfaces and the telencephalic choroid plexus in the first stage, while enlargement of the ventricular system took place (Borit and Sidman, 1972). Ependyma and choroid plexus of the lateral and fourth ventricles in hamsters were examined 7–49 days after injection of reovirus to produce hydrocephalus (Nielson and Gauger, 1974). Macrophages with warty cell bodies and long processes were seen as the hydrocephalus developed, but the reasons for the changes in the macrophage morphology were not clear. Ventricular walls from normal and hydrocephalic fetal rats were examined, the hydrocephaly being produced by the injection of 6-aminonicotinamide to the mothers on the thirteenth day of gestation (Chamberlain, 1974). The macrophages in untreated animals appeared less stimulated and rounder than those from the treated animals, whose macrophages had veil-like processes, folds, spherules, flanges, and long, thin processes, with periodic enlargements which made contact with the choroidal cells. There was an increase in the number of macrophages in the hydrocephalic rats. The macrophages in this fetal model were likened to the Type I Kolmer cells in the adult rat described by Hosoya and Fujita (1973). Excised pieces from the lateral ventricles of rats with hereditary hydrocephalus were examined: the number of neuron-like supraependymal cells was lower in hydrocephalic rats (Lindberg *et al.*, 1977). But on the other hand, the number of cells resembling Kolmer cells greatly increased, suggesting the presence of an ependymitis. Central nervous system macrophages were studied in an experimental cystic cavity produced by freeze-thawing (DeEstable-Puig *et al.*, 1976). Tissue was examined from normal animals and at intervals after irradiation. Free cells were identified as mac-

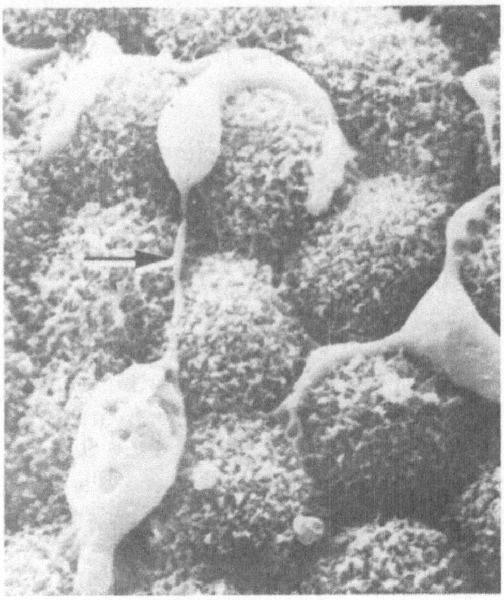

FIGURE 9. Supraependymal macrophages. Three morphologically different macrophages are illustrated here. A single thin delicate cytoplasmic process (arrow) and several larger ones are shown. These cells and processes have craterlike depressions on their surfaces that likely overlie internal vacuoles. ×2000. Micrograph courtesy of Dr. D. J. Allen. (Allen, D. J., 1975, *J. Comp. Neurol.* **161**:197.)

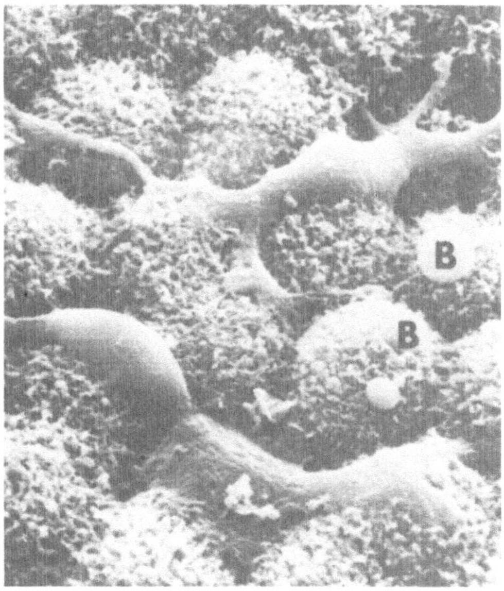

FIGURE 10. Flattened supraependymal macrophages. The processes of these macrophages tend to flatten out and blend with the ependymal cells. Several spherical blebs (B) appear on the free surfaces of several ependymal cells. ×2000. Micrograph courtesy of Dr. D. J. Allen. (Allen, D. J., 1975, *J. Comp. Neurol.* **161:**197.)

rophages by their size, their similarity to macrophages elsewhere and by correlation with light microscopy and TEM. Microvilli, blebs, and foliate structures were seen in the control animals, as were rounder, smoother cells. This latter type was chosen to calculate the surface area and volume of macrophages in

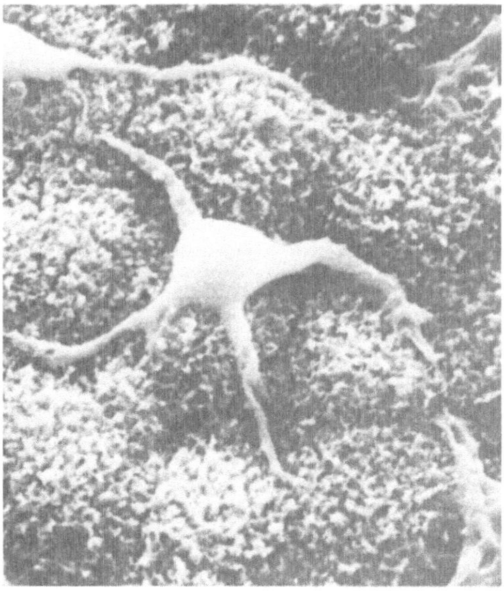

FIGURE 11. Star-shaped supraependymal macrophage. The macrophage in the center of the micrograph has a smooth oval cell body with four distinct cellular processes. ×2000. Micrograph courtesy of Dr. D. J. Allen. (Allen, D. J., 1975, *J. Comp. Neurol.* **161:**197.)

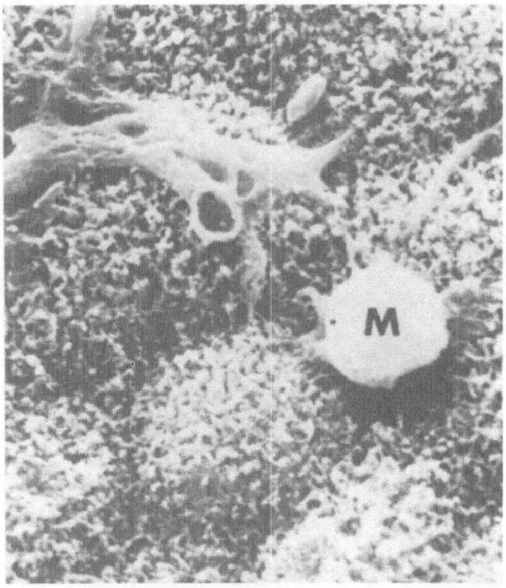

FIGURE 12. Rounded-up supraependymal macrophage (M). The rounded-up macrophage has many extremely fine filamentous processes. The processes on the left, which belong to another cell, are intimately associated with the ependymal surface and possess pits and craters. ×2000. Micrograph courtesy of Dr. D. J. Allen. (Allen, D. J., 1975, *J. Comp. Neurol.* **161**:197.)

control and irradiated specimens. The cells of the irradiated macrophages were rather larger in size.

Macrophages were identified by their surface morphology in guinea pig sympathetic ganglion cultures (Hill *et al.*, 1974), rat cerebellum cultures (Privat *et al.*, 1973 and Silberberg, 1975), meningeal granulations (Von Krahn and Richter, 1976) and guinea pig cochlear aqueducts and periotic ducts (Duckert, 1974). Migration of cells into the chicken corneal region was preceded at early stages by movement of macrophages between the corneal stroma and the lens (Bard *et al.*, 1975). The macrophages had rounded cell bodies and many surface processes. Ruffles were not present on the endothelial cells, despite their migratory behavior: it was concluded that the ruffles seen on the macrophages were therefore not necessarily connected with movement. Such macrophages with ruffles were also reported by Nelson and Revel (1975) in their study of chicken cornea.

4. PHAGOCYTOSIS

4.1. PHAGOCYTOSIS OF INERT PARTICLES

The phagocytosis of plutonium oxide was studied in an attempt to investigate the response of animals which had inhaled this toxic substance (Nolibe, 1973). A population of lung macrophages, from pulmonary lavage of rabbits, was exposed to plutonium oxide for periods varying from 30 min to 24 hrs. The unexposed macrophages were spherical with short pseudopodia and showed the usual veil after settling on glass. Some of the macrophages had a large pseudopod and were thought to be moving. After a 1-hr exposure to plutonium oxide they had an irregular veil, with thin, filamentous processes. After 3 and 6 hrs the veil had been replaced by thick processes, and after 24 hr of exposure 10% of the cells were missing and many more were very flat. Although there was little exact information on the phagocytosis of individual particles, macrophages were able to function normally when exposed to radioactive material. They spread less readily and some lysed. X-ray analysis and SEM were used on Epon-embedded 1-μm sections in a study of guinea pig lungs with experimental silicosis, (Morgenroth *et al.*, 1973). Inorganic material in the macrophages was analyzed and found to contain silicon.

A thorough investigation of different types of phagocytosis was given by Miller *et al.* (1978) in a study of rat macrophages exposed to asbestos and quartz particles. Crocidolite asbestos contained short and long forms. The short fibers were held by many cytoplasmic processes and were phagocytized quickly, since, after 20 min, some of the macrophages had the smooth surface expected of replete cells. The long fibers were attached to a pseudopod by thin processes and phagocytized end first. Glass fiber, though shorter and thicker than crocidolite, was phagocytized in a way similar to the short fibers of this form of asbestos. Quartz was treated differently by the macrophages. The particles were attached and quickly phagocytized by filopodia round the margins of the cell. *Staphylococcus aureus* was phagocytized in a similarly rapid way.

4.2. PHAGOCYTOSIS OF ORGANISMS

Rabbit peritoneal macrophages obtained approximately three weeks after subcutaneous injection of live *Brucella melitensis* were challenged with heat-killed brucellae and the process of phagocytosis was examined 2 hr later (Warfel *et al.*, 1973). The macrophages had undulating membranes and brucellae attached to the edges of these. The importance of stereopairs in seeing this process clearly was stressed (Figures 13 and 14). Mouse peritoneal macrophages were examined after incubation with *Bacillus cereus*, *Staphylococcus aureus*, brewer's yeast, and *Cryptococcus neoformans* (Walters *et al.*, 1976). The control macrophages were spherical, ridged, and had lamellipodia and filopodia. *Bacillus cereus* attached to and was phagocytized by the macrophages, and, as the number of organisms engulfed by any one macrophage increased, the ridges on the surface disappeared and the surface became smooth. The bacilli were finally engulfed along their short axes, with a cuff of macrophage membrane round them. *Staphylococcus aureus* attached to the lamellipodia or to the filopodia and the macrophage membrane did not become smooth. The yeast cells were covered with veil-like flaps while *Cryptococcus neoformans* showed very little sign of attachment to or phagocytosis by the macrophages. There was thus quite a variation in the response of the macrophages to the organisms.

Peritoneal macrophages from unstimulated mice were maintained for 24 hr and then exposed to *Mycoplasma pulmonis* (Jones *et al.*, 1977). The mycoplasmas attached to the macrophages, but no change was seen in the surface membrane.

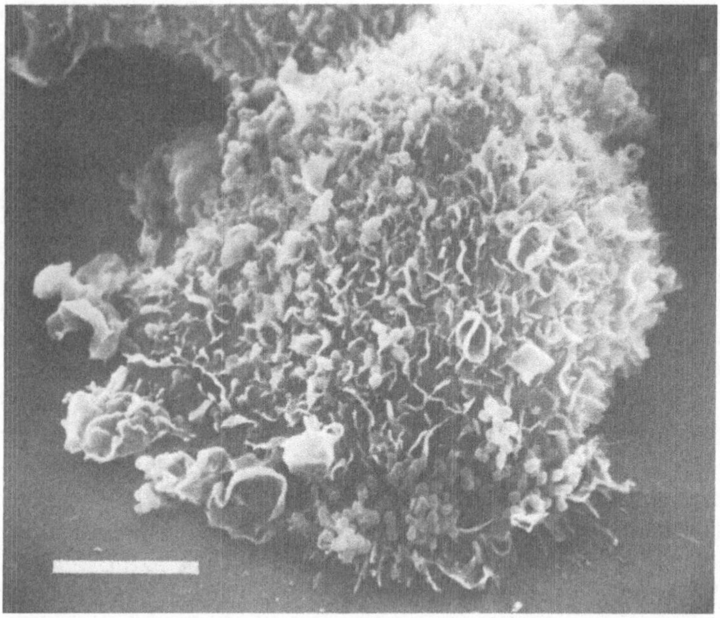

FIGURE 13. *Brucella melitensis* strain Rev. 1 attached to the undulating membranes of a macrophage derived from a homologous immune rabbit. ×3900. Micrograph courtesy of Dr. A. H. Warfel and Dr. T. L. Hayes. (Warfel, A. H., Elberg, S. S., and Hayes, T. L., 1973, *Infect. Immun.* **8**:665.)

However, when anti-mycoplasma antibody was added, there was an increase in membrane ruffling and folding and the macrophage became granular in appearance. Five minutes after the addition of antibody the mycoplasmas had been ingested by the macrophages and the cells looked engorged. The addition of trypsin instead of antibody induced slow ingestion of the mycoplasmas without the membrane activity described earlier. The reaction of guinea pig alveolar macrophages with *Mycoplasma pneumoniae* was followed, showing that the mycoplasmas could attach to the macrophage filopodia without initiating phagocytosis (Powell and Muse, 1977). Only after the addition of specific anti-*Mycoplasma pneumoniae* serum were the organisms engulfed by macrophage ruffled membranes (Figures 15–17). The macrophages after phagocytosis appeared highly spread with an area round them devoid of organisms.

Phagocytosis by rabbit lung macrophages of four different fungal spore types and inert particles of corresponding size was studied by Lundborg and Holma (1972). For *Aspergillus fumigatus* and for two strains of *Aspergillus flavus* the degree of phagocytosis was distinctly higher than that for the corresponding inert particles. This was not the case for *Rhizopus arrhizus*. The response of rabbit alveolar macrophages of two strains of *Nocardia asteroides* was described by Beaman (1977). The more virulent 14759 form produced aggregation of macrophages, in some cases producing tight cell contacts and in others producing multinucleate giant cells, which then destroyed the intracellular *Nocardia*. The

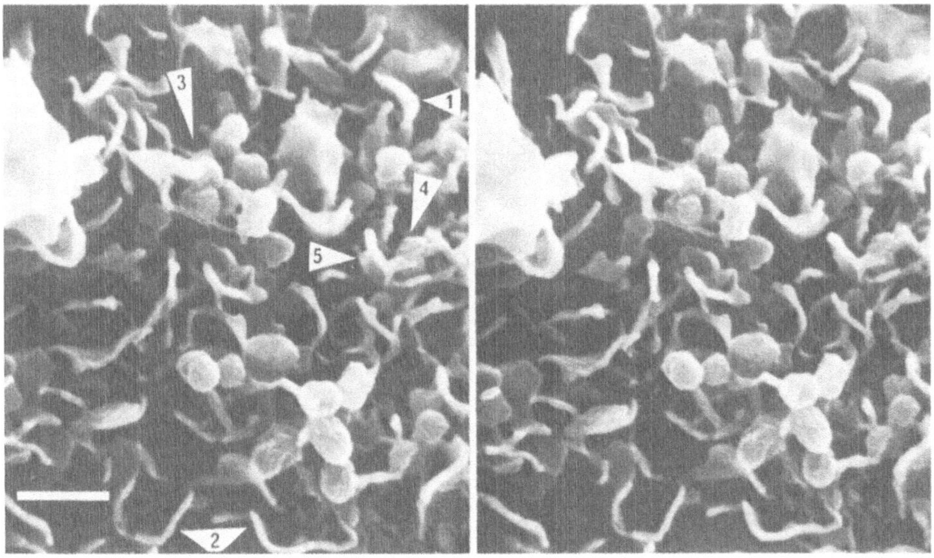

FIGURE 14. These two micrographs form a stereopair. The full details are best appreciated using a stereoviewer. The detailed surface morphology of the macrophage seen in Figure 13 is shown. Points of interest include a clump of brucellae attached to undulating membranes and located in crevices formed by these membranes (3). ×17,000. Micrograph courtesy of Dr. A. H. Warfel and Dr. T. L. Hayes. (Warfel, A. H., Elberg, S. S., and Hayes, T. L., 1973, *Infect, Immun.* **8**:665.)

macrophages that did not fuse seemed not to have the ability to carry out this destruction.

The response of rabbit alveolar macrophages to *Candida albicans* was examined by Arai *et al.* (1977). The reaction of the macrophages was compared to that of guinea pig neutrophils which killed the *Candida* cells presumably by the myeloperoxidase-halide-hydrogen peroxide system. The macrophages, lacking this system, engulfed the *Candida* cells; the *Candida* then proliferated within the macrophages and destroyed them.

There is some evidence that such organisms as the trophozoites of *Toxoplasma gondii* may actively penetrate some cells (e.g., L cells). In the light of this Klainer *et al.* (1973) studied the interaction between mouse peritoneal macrophages and the trophozoites of toxoplasma. Organisms were seen adhering to the surface of macrophages, trapped beneath their membranes or surrounded by pseudopods. This suggested that the macrophage phagocytized the toxoplasma rather than that the toxoplasma actively penetrated the macrophage.

The interaction between *Trichinella spiralis* larvae and peritoneal macrophages from normal and sensitized mice was described by Vernes *et al.* (1974). With the latter group only, the larvae stuck rapidly to the phagocytes thereby reducing their mobility but not being killed. The phagocytosis of *Trypanosoma gambiense* by rat macrophages was studied by Takayanagi *et al.* (1974) who deduced that the agglutination antigen is responsible for phagocytosis.

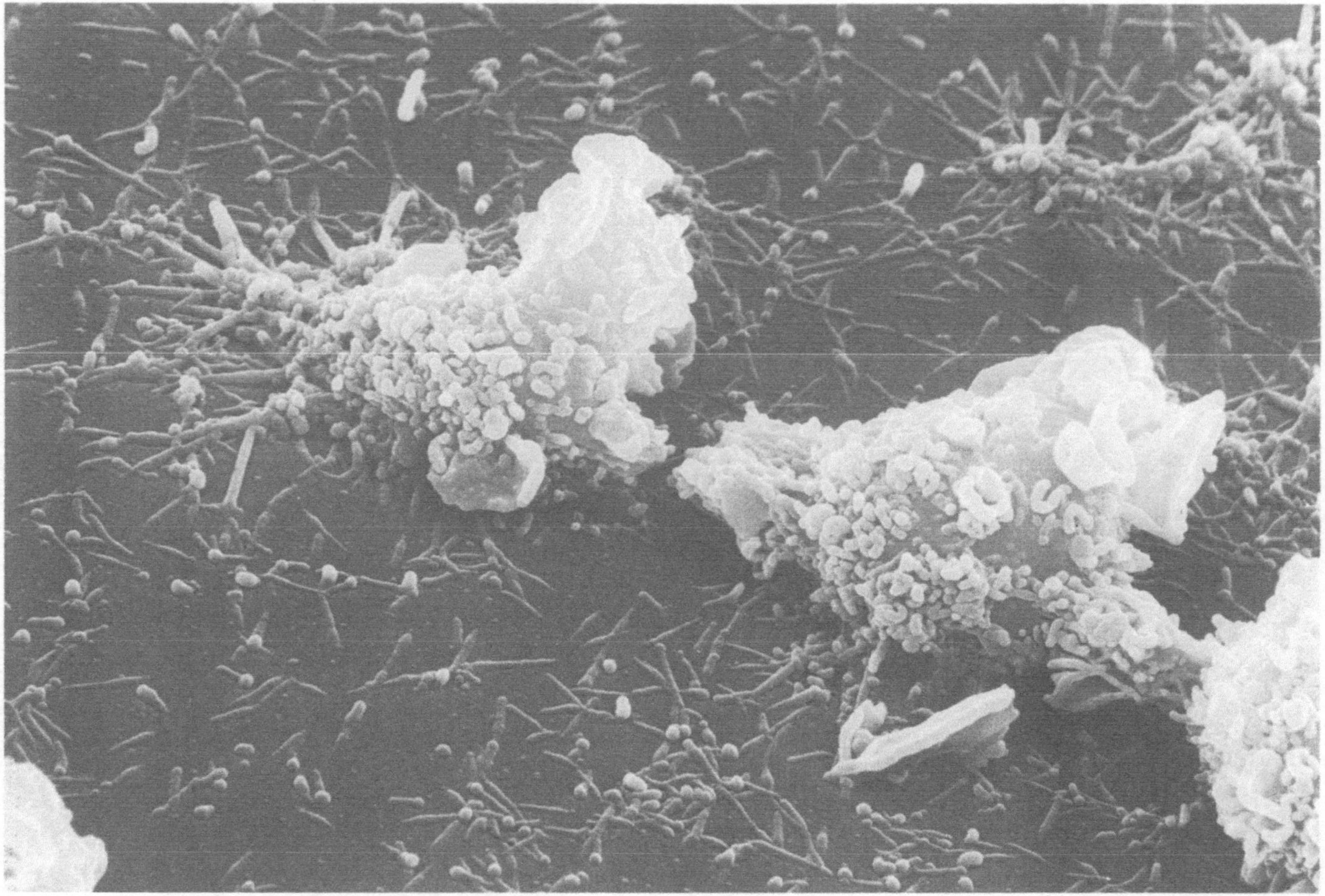

FIGURE 15. Interaction of *M. pneumoniae* and alveolar macrophages prior to addition of *M. pneumoniae* serum. Note the rounded macrophages surrounded by a lawn of filamentous mycoplasma many of which are clearly attaching to macrophage filopodia. ×6000. Reduced 17% for reproduction. Micrograph courtesy of Dr. D. A. Powell and Dr. K. A. Muse. (Powell, D. A., and Muse, K. A., 1977, *Lab. Invest.* **37**:535.)

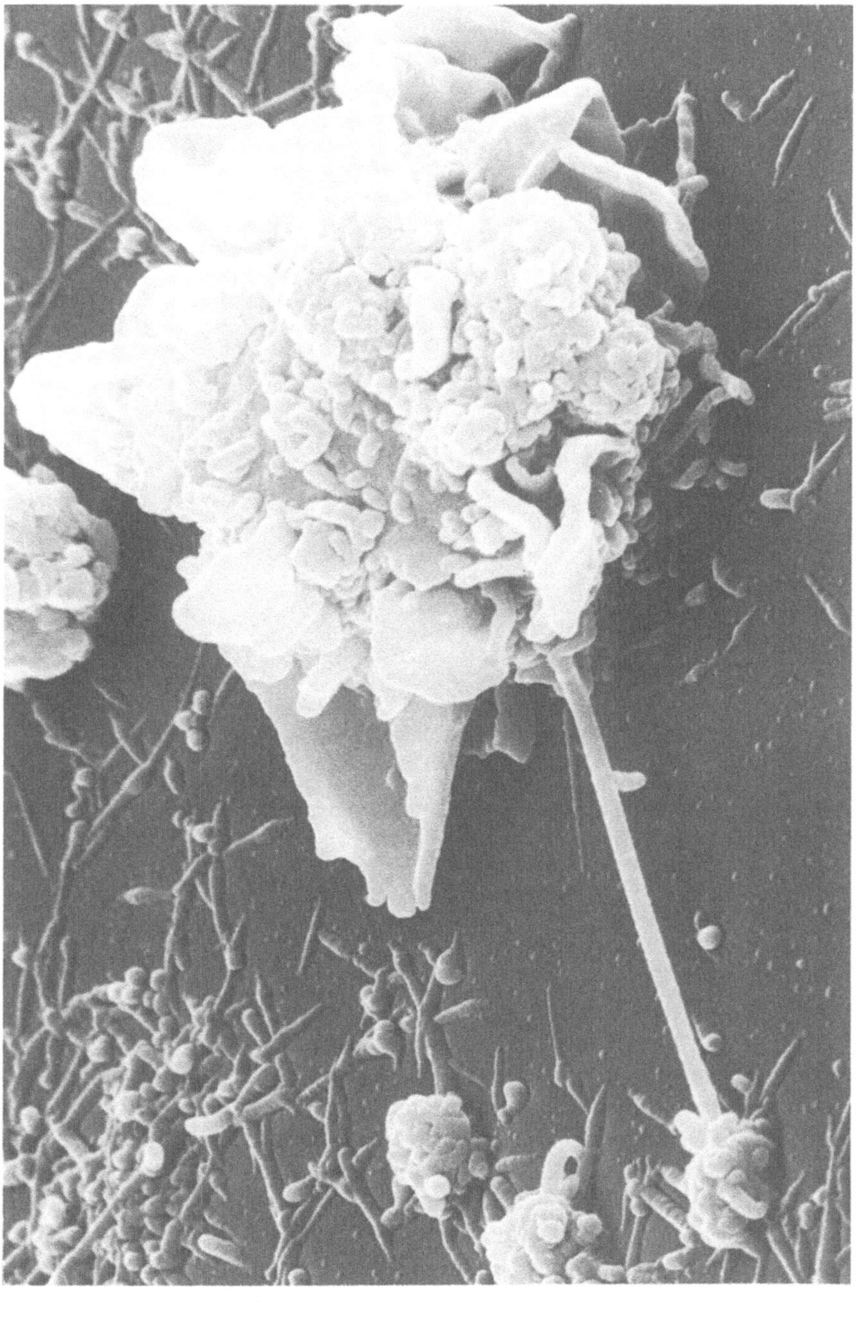

FIGURE 16. Interaction of *M. pneumoniae* and alveolar macrophages 10 min after addition of antiserum. Notable are the increases in macrophage ruffled membranes and a filopodial extension attaching to a microcolony of *M. pneumoniae*. ×11,000. Reduced 25% for reproduction. Micrograph courtesy of Dr. D. A. Powell and Dr. K. A. Muse. (Powell, D. A., and Muse, K. A., 1977, *Lab. Invest.* 37:535.)

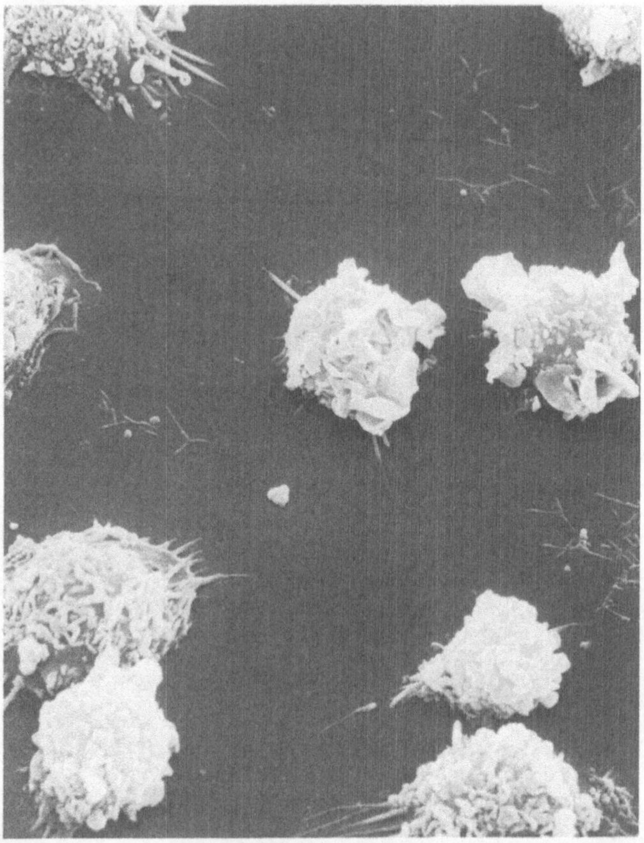

FIGURE 17. *M. pneumoniae*-infected alveolar macrophages incubated for 90 min with anti-*M. pneumoniae* serum. Note the presence of spread cells surrounded by extensive areas of cover slip cleared of *M. pneumoniae*. ×2300. Micrograph courtesy of Dr. D. A. Powell and Dr. K. A. Muse. (Powell, D. A., and Muse, K. A., 1977, *Lab. Invest.* **37**:535.)

4.3. PHAGOCYTOSIS OF MAMMALIAN CELLS

The attachment of erythrocytes to macrophages has been used as an indication of the surface activity of the macrophages. The phagocytosis of opsonized sheep erythrocytes after incubation for 90 min at 37°C was studied by Tizard and Holmes (1974). The stages of phagocytosis were described as attachment, enclosure by macrophage membrane, and centripetal movement of the ingested erythrocyte which was finally visible as a raised area on the surface. The erythrocyte became spherical after attachment but did not evidently fragment. Similar techniques were used on mouse peritoneal macrophages to study further the attachment phase by comparing different types of macrophage–erythrocyte rosettes (Tizard *et al.*, 1974). Expansion of the attachment area was fast with opsonized erythrocytes in the presence or absence of complement and led to total enclosure of the erythrocytes within minutes. On the other hand, if the process was mediated by mouse anti-sheep erythrocyte cytophilic antibodies the attachment area on the sensitized macrophages became only slightly bigger, perhaps because there was a low concentration of sites on the macrophage surface giving less stimulus for the generation of new membrane required for enclosure. Monocytes were identified in human blood as one of the cell types able to produce lysis of erythrocytes within 4 hr (Inglis *et al.*, 1976). Normal and sensitized mouse peritoneal macrophages were used in a study of membrane changes during phagocytosis of opsonized sheep erythrocytes (Orenstein and Shelton, 1977). *In vitro* experiments and *in vivo* studies on omentum, mesentery, and diaphragm were carried out. There were two methods of phagocytosis. Unflattened macrophages with highly folded membrane ruffles engulfed the erythrocytes using several pseudopodia, whereas flattened macrophages used a single cuplike or funnel-like pseudopod. The flat, smooth macrophages were themselves the result of spreading on a substrate or previous engulfing of erythrocytes. Both types were seen in phagocytosis using normal, sensitized or pyran-treated macrophages either *in vivo* or *in vitro*. Their reaction was observed with erythrocytes coated with antibody from mice, or with rabbit anti-sheep erythrocyte antibody. The membrane pattern was determined by the number of points of contact made between erythrocyte and macrophage.

The exact timing of phagocytosis of *Plasmodium berghei*-infected erythrocytes was worked out using preparations of mouse peritoneal macrophages from normal, immune, and infected animals (Seitz *et al.*, 1977). The rate of phagocytosis was calculated. The half-lives for intracellular elimination of the parasites were 27 min for "immune" cells, 57 min for "normal" cells, and 66 min for "infected" cells.

In the course of spermiophagocytosis in the epididymis of monkeys, fragments of disintegrated sperm were seen in macrophages, in some cases in the process of being engulfed by cytoplasmic flaps (Murakami *et al.*, 1977).

5. PATHOPHYSIOLOGY OF PHAGOCYTES

5.1. EFFECTS OF CHEMICAL, PHARMACOLOGICAL, AND PHYSICAL AGENTS ON PHAGOCYTOSIS

A study of the effects of cytochalasin B on the surface of peritoneal macrophages from pathogen free mice showed that the peripheral cytoplasm of affected macrophages was retracted, leading to rounding-up of the central part of the cell (Axline and Reaven, 1974). Extensive infoldings of the membrane gave the treated cells a wrinkled appearance and phagocytosis was inhibited. The effects of different metals on the viability of rabbit alveolar macrophages *in vitro* were studied by Waters *et al.* (1975). The concentration of salts was chosen to give two groups of low and high cell viability. Control cells had a viability range of 95–99% with many membrane processes visible on the cell after 20 hr of incubation at 37°C. Exposure to vanadate at 0.098 mM gave a viability of 43% with unusually smooth cells, although some had blebs or had broken up, while exposure to vanadate at 0.20 mM gave a viability of less than 10%, with many cells so damaged that only remnants of the cell membrane were seen. A similar drop in cell viability with surface blebbing and cell destruction was seen when low and high concentrations of nickel, manganese, and chromium were used. The macrophage damage included retraction of normal pseudopodia, appearance of blebs, smoothing of the surface membrane, and final disappearance of the cell architecture. There was a link between cell viability and surface damage except in the reaction to cadmium. Here exposure to high concentrations (0.067 mM) produced cell viability of less than 10%, but although there was an increase in the size of surface blebs, there was little or no cell lysis. This is highly significant since alveolar macrophages which remain intact in this way after ingestion of toxic substances could fulfill their protective role by being carried out of the lung, unlike those macrophages which undergo lysis and release their toxic material to do further damage.

In an attempt to link alterations in the functional state of phagocytic cells with morphology, mouse peritoneal macrophages were studied after subcutaneous administration of *N*-mustard-benzimidazole derivatives, chlorambucil and 1-barbituric acid derivatives, which reduce phagocytosis, and also after anti-metabolites 6-mercaptopurine and 5-fluorouracil (Güttner *et al.*, 1975). The spreading time was kept short (10 min at 37°C) a factor of importance when comparison is made between macrophages in different situations. The first group of chemicals made surface projections shorter, thicker, and fewer, whereas no change was seen in the surface structure after exposure to the second group; these results were in line with the effect on phagocytosis. The surface changes may have stemmed from a "stabilization" of the membrane, with resulting difficulty in the triggering of the attachment process of phagocytosis by immunogens of foreign material.

On the other hand there was an increase in phagocytosis of quartz by lung macrophages after prophylactic administration of polyvinylpyrrolidine-*N*-oxide (Arutyonov *et al.*, 1976). The macrophages of the treated rats were larger in size.

In a study of the *in vivo* effects of asbestos on peritoneal and alveolar macrophages, the cells were cultured for up to 24 hr: their viability, morphology, and surface receptor sites for IgG and C3 surface markers were assessed (Miller and Kagan, 1976). There was an increase in the number of IgG receptor sites, a greater ability to spread on glass, and more extensive cytoplasmic processes in the dusted macrophages. Mouse peritoneal macrophages were cultured in a medium containing colchicine or cytochalasin B (Takayama *et al.*, 1976). Ridges and pseudopodia were reduced in number and phagocytosis was decreased after exposure to cytochalasin B. On the other hand the cytoskeletal elements of the cell were sensitive to colchicine. The effects of zinc on rat peritoneal macrophages *in vitro* were studied by Chvapil *et al.* (1977). Oxygen consumption in the presence of yeast was used as a measure of the activity of the macrophages. Zinc had a reversible inhibitory effect on the macrophages, the inhibition being less marked with activated macrophages. The presence of magnesium ions was essential for this process while calcium ions reduced the effects of the zinc. Zinc-treated macrophages had lost the pleomorphism of the controls and showed only fine surface processes. It was thought that zinc could react directly with the macrophage cell membrane. The effects of quartz *in vivo* on rat alveolar macrophages were studied by Miller and Kagan (1977). Many of the macrophages exposed to quartz had bizarre shapes and there were fewer ruffles and more blebs. In some there was such a deterioration in the cell membrane that details of cellular morphology were difficult to make out. There was an increase in the number of IgG receptor sites. The changes described were seen in the large macrophages but not in the smaller macrophages present, perhaps because the latter population had more recently arrived.

The effects of silica treatment on the ability of mouse liver Kupffer cells to clear *Salmonella typhimurium* was described by Friedman and Moon (1977). The Kupffer cells of the treated livers were not so spread out as those from normal livers and were round and engorged with silica, with less ruffling of the cell membrane. The attachment of the Kupffer cells to the sinusoidal walls had become less strong. Silica treatment produces a drop in the rate of bacterial clearance with a greater susceptibility to gram-negative infections.

The effects on mouse peritoneal exudate and resident macrophages of high-molecular-weight levan were examined by Robertson *et al.* (1977). "Resident" cells were glass-adherent monocyte-like cells from peritoneal washings of normal mice, and "exudate" cells were similar cells from thioglycollate-stimulated mice. The resident macrophages were mostly rounded, while the exudate macrophages were generally flat. In both types, levan increased the amount of intercellular contact in the form of bridges and interdigitations and encouraged the arrangement of the cells in two layers. It was suggested that, by decreasing the surface tension of the cell surface, levan increased the tendency of the cells to adhere.

Mouse macrophages were examined after exposure to interferon (Schultz *et al.*, 1978). When settled on glass the treated cells were less flat than controls, appearing more compact with extensive pseudopodia. It was suggested that an increase in macrophage-mediated cytotoxicity may be the reason why interferon or interferon inducers make mice more resistant to experimental tumors. Mouse peritoneal macrophages were examined after injection of glucan, a partially purified derivative of zymosan and responsible for much of the reticuloendothelial-stimulating properties ascribed to zymosan (Burgaleta *et al.*, 1978). The treated macrophages had increased in size and had spread more rapidly and to a greater extent than controls. There was a 23% increase in their ability to adhere to glass by comparison with controls; there was also an increase in surface processes. In general, it was concluded that glucan was a potent activator of macrophages.

The effect of whole-body X-irradiation on mouse macrophages was studied by Geiger and Gallily (1974). Small holes were seen on the surface of the irradiated macrophages. Those exposed to 550 rads of X-irradiation were round, while macrophages exposed to 800 rads were elongated. Controls displayed no such uniformity. It was suggested that the surface changes were responsible for the drop in the ability of the macrophages to handle antigens following sublethal radiation.

5.2. PARTICIPATION OF MACROPHAGES IN THE IMMUNE RESPONSE

The destruction *in vitro* of HEp-2 by lymphoid cells from subjects immunized with HeLa antigen indicated that macrophages were produced during the response, these cells then killing the target cells by phagocytosis (Björklund *et al.*, 1972). Human alveolar macrophages from healthy lobes or segments obtained during operations for bronchial neoplasm were studied by Voisin *et al.*, (1972b, 1973). *In vitro* anti-lymphocytic globulin or serum (ALS) had a cytotoxic effect on the macrophages, with retraction of pseudopodia and rounding up of the cell. Cells with several nuclei were formed after exposure to low concentrations of ALS or in the absence of complement.

In a study of *in vitro* macrophage response to antigen challenge, guinea pig peritoneal exudates were exposed to migration inhibition factor (MIF) elaborated by lymphocytes in the presence of antigen to which they had previously been sensitized (Barber and Burkholder, 1974). Macrophages exposed to MIF behaved like activated cells in their surface structure and cell diameter. Many processes and ruffles were seen and the cells tended to adhere to each other and formed more clusters than did untreated cells. Macrophages were also seen to attach to lymphocytes by cytoplasmic bridges. The way in which macrophages clump with lymphocytes was further examined in a study of peritoneal exudate cells and autologous immune lymph node cells prepared from guinea pigs immunized with *Mycobacterium tuberculosis* (Nielsen *et al.*, 1974). Clusters formed after incubation for 20 hr in the presence of a purified derivative of tuberculin. The clusters had variable numbers of macrophages and lymphocytes. The reaction of lymphocytes to normal and stimulated macrophages indicated that there were more contacts between the two cell types when stimulated macrophages were involved (Favorskaya *et al.*, 1975). The role of macrophages in the cytotoxic killing of tumor cells *in vitro* was studied by Cichocki *et al.*, 1975). Thioglycollate-induced peritoneal mouse macrophages adhered to the target tumor cells by cytoplasmic bridges and some exocytosis of the granules of the contacting macrophages was seen. Phagocytosis was only seen as a process secondary to the degeneration of the target cell.

The reactions of macrophages exposed to stimuli other than physicochemical injury or immunological factors have been studied in many different situations, including wound healing and inflammation (Watters and Buck, 1973; Shekhter *et al.*, 1977; Noonan and Riddle, 1977).

6. CONCLUSIONS

The identification of macrophages by SEM is often a considerable problem, since scanning microscopy is analogous to braille, a technique for identification by surface contours. Too often cells are identified as macrophages by their surface ruffles alone; such identification is made more valid by parallel information on ultrastructure gained by TEM or by information on such functional properties as phagocytic ability or the presence of surface receptors.

The surface topography of macrophages as seen in many sites *in vivo*, and as seen *in vitro* is remarkably similar; the notable feature is the heavily ruffled membrane with veils, pseudopodia, and filopodia. Various appearances have been related to activity; for instance among brain ventricular macrophages a distinction is made between rounded resting cells, and flatter mobile cells. Similarly, in many other sites a round cell with few ruffles may be a resting cell, a flat cell may be motile or settled on a substrate after spreading, a cell with many ruffles may be engaged in or about to undertake phagocytosis, while a large, smooth, engorged cell may be replete after engulfing foreign material. Much of the work done in this area with cultured macrophages is, however, difficult to interpret because of lack of standardization of the methods of harvesting and culture and of the many different stimuli or activating agents used.

Macrophages have been identified with SEM using the above criteria in a wide variety of sites—lymph node, spleen, lung, liver sinusoid, and nervous system. Their surface characteristics in these sites distinguish them especially from endothelial cells.

The interaction between macrophages and their substrate has been explored and the formation of cytoplasmic processes and cell flattening described. There is surprisingly little information available on the topography of the interaction between macrophages and substrate, be it natural or artificial, beyond the descriptions of the formation of processes and flattening of the cell. Stimulated cells spread faster than resting cells, they have more surface membrane specializations and more readily form connections with other cells. Exocytic material can be identified external to the macrophage: the secreted material can be cytolytic or cytotoxic. Surface receptor sites have been identified by SEM. Such receptor sites on macrophages are common on peripheral veils and are reduced in number by spreading of the cells. Fewer receptor sites are seen in epithelioid and giant cells, in keeping with their less powerful phagocytic capacities.

The surface phenomena of phagocytosis have been illuminated by SEM, notably in studies of the phagocytosis of red blood cells. The progressive enlargement of the attachment area between red blood cell and macrophage has been described and two different forms of phagocytosis illustrated, one involving membrane ruffling and the other cuplike pseudopodia.

The expression on the face of the macrophage as seen by SEM undoubtedly betrays its intentions and recent activities.

ACKNOWLEDGMENTS. I wish to thank Drs. P. Parakkal, R. E. Scott, P. B. Maercklein, J. J. Malloy, F. Low, J. Allen, A. H. Warfel, T. L. Hayes, D. A. Powell, K. A. Muse, S. G. Quan, and D. W. Golde for providing illustrations, and also Academic Press, the American Society for Microbiology, The Wistar Institute Press, and the Williams and Wilkins Company for permission to reproduce the illustrations. I am grateful to Dr. P. G. Toner for help with the bibliography. I would also like to thank Dr. D. Whitlock for the use of the facilities of the Department of Anatomy, University of Colorado, Denver, during the completion of this chapter.

REFERENCES

Albrecht, R. M., Hindsill, R. D., Sandok, P. L., Mackenzie, A. P., and Sachs, I. B., 1972, A comparative study of the surface morphology of stimulated and unstimulated macrophages prepared without chemical fixation for scanning EM, *Exp. Cell Res.* **70:**230.

Albrecht, R. M., Rasmussen, D. H., Keller, C. S., and Hindsill, R. D., 1976, Preparation of cultured cells for SEM: Air drying from organic solvents, *J. Microsc.* **108:**21.

Alexander, E., Sanders, S., and Braylan, R., 1976, Purported difference between T and B cell surface morphology is an artefact, *Nature* **261:**239.

Allen, D. J., 1975, Scanning electron microscopy of epiplexus macrophages (Kolmer cells) in the dog, *J. Comp. Neurol.* **161:**197.

Allen, D. J., and Low, F. N., 1973, The ependymal surface of the lateral ventricle of the dog as revealed by scanning electron microscopy, *Am. J. Anat.* **137:**483.

Allen, D. J., and Low, F. N., 1975, Scanning electron microscopy of the subarachnoid space in the dog. III. Cranial levels, *J. Comp. Neurol.* **161:**515.

Arai, T., Mikami, Y., and Yokoyama, K., 1977, Phagocytosis of *Candida albicans* by rabbit alveolar macrophages and guinea pig neutrophils, *Sabouraudia* **15:**171.

Ariens-Kappers, J., 1953, Beitrag zur experimentellen Untersuchung von Funktion und Herkunft der Kolmerschen Zellen des Plexus choroideus beim Axolotl und Meerschweinchen, *Z, Anat. Entwicklungsgesch.* **117:**1.

Arutyonov, V. D., Kruglikov, G. G., Batsura, Y. D., and Fedorova, V. I., 1976, Morphofunctional state of macrophages of the lungs and their biological protection during phagocytosis of toxic elements, *Arkh. Patol.* **38:**16.

Axline, S. G., and Reaven, E. P., 1974, Inhibition of phagocytosis and plasma membrane mobility of the cultivated macrophage by cytochalasin B. Role of subplasmalemmal microfilaments, *J. Cell Biol.* **62:**647.

Barber, T. A., and Burkholder, P. M., 1974, Correlative SEM and TEM study of the *in vitro* macrophage response to antigen challenge, in: *Scanning Electron Microscopy 1974* (O. Johari, ed.), Illinois Institute of Technology Research Institute, Chicago.

Bard, J. B., Hay, E. D., and Meller, S. M., 1975, Formation of the endothelium of the avian cornea: A study of cell movement *in vivo, Dev. Biol.* **42:**334.

Barnhart, M. I., and Lusher, J. M., 1976, The human spleen as revealed by scanning electron microscopy, *Am. J. Hematol.* **1:**243.

Beaman, B. L., 1977, *In vitro* response of rabbit aveolar macrophages to infection with *Nocardia asteroides, Infect. Immun.* **15:**925.

Bessie, M., and Boisfleury, A. de, 1971, Les mouvements des leucocytes étudiés au microscope électronique à balayage, *Nouv. Rev. Fr. Hématol.* **11:**377.

Björklund, B., Björkland, V., Lundström, R., Eklund, G., Nilsson, L., and Gronneberg, R., 1972, Cytogenetics of the destruction of Hep-2 *in vitro* by lymphoid cells from subjects immunised with HeLa antigen and demonstration of a mechanism for biologic amplification of certain lymphoid cells by phase contrast, time-lapse cinemicrography, and scanning electron microscopy, *J. Reticuloendothel. Soc.* **11**:29.

Bleier, R., Albrecht, R., and Cruce, J. A., 1975, Supraependymal cells of hypothalamic third ventricle: Identification as resident phagocytes of the brain, *Science,* **189**:299.

Bleier, R., 1971, The relations of ependyma to neurons and capillaries in the hypothalamus: A Golgi-Cox study, *J. Comp. Neurol.* **142**:439.

Borit, A., and Sidman, R. L., 1972, New mutant mouse with communicating hydrocephalus and secondary aqueductal stenosis, *Acta Neuropathol. (Berl.)* **21**:316.

Boyde, A., and Wood, C., 1969, Preparation of animal tissues for surface-scanning electron microscopy, *J. Microsc.* **90**:221.

Braylan, R. C., Jaffe, E. S., Triche, T. J., Narba, K., Fowlkes, B. J., Metzger, H., Frank, M. M., Dolan, M. S., Yee, C. L., Green, I., and Berard, C. W., 1978, Structural and functional properties of the "hairy" cells of leukaemic reticuloendotheliosis, *Cancer* **41**:210.

Brody, A. R., and Craighead, J. E., 1975, Cytoplasmic inclusions in pulmonary macrophages of cigarette smokers, *Lab. Invest.* **32**:125.

Bucana, C., Hoyer, L. C., Hobbs, B., Breesman, S., McDaniel, M., and Hanna, M. G., 1976, Morphological evidence for the translocation of lysosomal organelles from cytotoxic macrophages into the cytoplasm of tumor target cells, *Cancer Res.* **36**:4444.

Burgaleta, C., Territo, M. C., Quan, S. G., and Golde, D. W., 1978, Glucan-activated macrophages: Functional characteristics and surface morphology, *J. Reticuloendothel. Soc.* **23**:195.

Carr, I., 1970, The fine structure of the mammalian lymphoreticular system, *Int. Rev. Cytol.* **27**:283.

Carr, I., Clarke, J. A., and Salsbury, A. J., 1969, The surface structure of mouse peritoneal cells—a study with the SEM, *J. Microsc.* **89**:105.

Carr, K. E., 1971, Applications of scanning electron microscopy in biology, *Int. Rev. Cytol.* **30**:183.

Carr, K., and Carr, I., 1970, How cells settle on glass: a study by light and scanning electron microscopy of some properties of normal and stimulated macrophages, *Z. Zellforsch. Mikrosk. Anat.* **105**:234.

Chamberlain, J. G., 1974, Scanning electron microscopy of epiplexus cells (macrophages) in the fetal rat brain, *Am. J. Anat.* **139**:443.

Chvapil, M., Stankova, L., Bernhard, D. S., Weldy, P. L., Carlson, E. C., and Campbell, J. B., 1977, Effect of zinc on peritoneal macrophages *in vitro, Infect. Immun.* **16**:367.

Cichocki, T., Zembala, M., and Ptak, W., 1975, The role of macrophages in the cytotoxic killing of tumor cells *in vitro*. II. The morphology and ultrastructure of target cell-macrophage interaction *in vitro, Pathologica* **67**:377.

Clarke, J. A., Salsbury, A. J., and Willoughby, D. A., 1971, Some scanning electron microscope observations on stimulated lymphocytes, *J. Pathol.* **104**:115.

Cloyd, M. W., and Low, F. N., 1974, Scanning electron microscopy of the subarachnoid space in the dog. I. Spinal cord levels, *J. Comp. Neurol.* **153**:325.

Dantchev, D., 1976, Revised terminology of the different mononuclear cells under scanning electron microscopy, *Recent Results Cancer Res.* **56**:8.

de Estable-Puig, R. F., de Slobodrian, M. L., and de Estable-Puig, J. F., 1976, SEM study of brain macrophages with quantitative data on cell surface areas and volumes, in: *Scanning Electron Microscopy 1976/II* (O. Johari and R. P. Becker, eds.), pp. 195–202, Illinois Institute of Technology Research Institute, Chicago.

Duckert, L., 1974, The morphology of the cochlear aqueduct and periotic duct of the guinea pig. A light and electron microscopic study, *Trans. Am. Acad. Ophthalmol. Otolaryngol.* **78**:ORL 21.

Favorskaya, Yu. N., Krymskyi, L. D., and Nestaiko, G. V., 1975, Surface structures of macrophages and lymphocytes in their interaction under conditions of antigenic stimulation as observed with the scanning electron microscope, *Arkh. Patol.* **37**:33.

Friedman, R. L., and Moon, R. J., 1977, Hepatic clearance of *salmonella typhimurium* in silica-treated mice, *Infect. Immun.* **16**:1005.

Fujita, T., 1974, A scanning electron microscope study of the human spleen, *Arch. Histol. Jpn.* **37**:187.

Fujita, T., Miyoshi, M., and Murakami, T., 1972, Scanning electron microscope observation of the dog mesenteric lymph node, *Z. Zellforsch. Mikrosk. Anat.* **133**:147.

Geiger, B., and Gallily, R., 1974, Surface morphology of irradiated macrophages, *J. Reticuloendothel. Soc.* **15**:274.

Ghadially, F. N., and Skinnider, L. F., 1976, Vermipodia—a new type of cell process, *Experientia* **32**:1061.

Güttner, J., Augsten, K., Bimber, R., and Lange, P., 1975, Modification of the surface structure of murine peritoneal macrophages following chemotherapy. A scanning electron microscopic study, *Exp. Pathol. (Jena)* **11**:209.

Hill, C. E., Chamley, J. H., and Burnstock, G., 1974, Cell surfaces and fibre relationships in sympathetic ganglion cultures: A scanning electron microscopic study, *J. Cell Sci.* **14**:657.

Hosoya, Y., and Fujita, T., 1973, Scanning electron microscope observation of intraventricular macrophages (Kolmer cells) in the rat brain, *Arch. Histol. Jpn.* **35**:133.

Inglis, J. R., Penhale, W. J., Irvine, W. J., and Williams, A. E., 1976, Scanning electron microscopy of antibody-dependent cell-mediated cytotoxicity against red cell monolayers, in: *Scanning Electron Microscopy 1976* (O. Johari, ed.), pp. 49–56, Illinois Institute of Technology Research Institute, Chicago.

Itoshima, T., Kobayashi, T., Shimada, Y., and Murakami, T., 1974, Fenestrated endothelium of the liver sinusoids of the guinea pig as revealed by scanning electron microscopy, *Arch. Histol. Jpn.* **37**:15.

Johari, O., 1968–1980, *Scanning Electron Microscopy 1968–1978*, Illinois Institute of Technology Research Institute, Chicago.

Jones, T. C., Minick, R., and Yang, L., 1977, Attachment and ingestion of mycoplasmas by mouse macrophages. II. Scanning electron microscopic observations, *Am. J. Pathol.* **87**:347.

Kay, M.M.B., and Kadin, M., 1975, Surface characteristics of Hodgkin's cells, *Lancet* **1**:748.

Klainer, A. S., Krahenbuhl, J. L., and Remington, J. S., 1973, Scanning electron microscopy of *Toxoplasma gondii*, *J. Gen. Microbiol.* **75**:111.

Kolmer, W., 1921, Über eine eigenartige Beziehung von Wanderzellen zu den Choroidealplexus des Gehirns der Wirbeltiere, *Anat. Anz.* **54**:15.

Krivinkova, H., Ponten, J., and Blondal, T., 1976, The diagnosis of cancer from body fluids. A comparison of cytology, DNA measurement, tissue culture, scanning and transmission microscopy, *Acta Pathol. Microbiol. Scand. A* **84**:455.

Kuhn, C., and Finke, E. H., 1972, The topography of the pulmonary alveolus: Scanning electron microscopy using different fixations, *J. Ultrastruct. Res.* **38**:161.

Leake, E. S., Wright, M. J., and Myrvik, Q. N., 1975, Differences in surface morphology of alveolar macrophages attached to glass and to millipore filters: A scanning electron microscope study, *J. Reticuloendothel. Soc.* **17**:370.

Lindberg, L. A., Vasenius, L., and Talanti, S., 1977, The surface fine structure of the ependymal lining of the lateral ventricle in rats with hereditary hydrocephalus, *Cell Tissue Res.* **179**:30.

Luk, S. C., Nopajaroonsri, C., and Simon, G. T., 1973, The architecture of the normal lymph node and hemolymph node. A scanning and transmission electron microscopic study, *Lab. Invest.* **29**:258.

Lundborg, M., and Holma, B., 1972, *In vitro* phagocytosis of fungal spores by rabbit lung macrophages, *Sabouraudia* **10**:152.

Malloy, J. J., and Low, F. N., 1974, Scanning electron microscopy of the subarachnoid space in the dog. II. Spinal nerve exits, *J. Comp. Neurol.* **157**:87.

Malloy, J. J., and Low, F. N., 1976, Scanning electron microscopy of the subarachnoid space in the dog. IV. Subarachnoid macrophages, *J. Comp. Neurol.* **167**:257.

Mariano, M., and Spector, W. G., 1974, The formation and properties of macrophage polykaryons (inflammatory giant cells), *J. Pathol.* **113**:1.

McKeever, P. E., Spicer, S. S., Brissie, N. T., and Garvin, A. J., 1977, Immune complex receptors on cell surfaces. III. Topography of macrophage receptors demonstrated by new scanning electron microscopic peroxidase marker, *J. Histochem. Cytochem.* **25**:1063.

Melsom, H., Kearny, G., Gruca, S., and Seljelid, R., 1974, Evidence for a cytolytic factor released by macrophages, *J. Exp. Med.* **140**:1085.

Merchant, R. E., and Low, F. N., 1977a, Identification of challenged subarachnoid free cells, *Am. J. Anat.* **148:**143.
Merchant, R. E., and Low, F. N., 1977b, Scanning electron microscopy of the subarachnoid space of the dog. V. Macrophages challenged by Bacillus Calmette-Guerin, *J. Comp. Neurol.* **172:**381.
Merchant, R. E., Olson, G. E., and Low, F. N., 1977, Scanning and transmission electron microscopy of leptomeningeal free cells: cell interaction in response to challenge by Bacillus Calmette-Guerin, *J. Reticuloendothel. Soc.* **22:**199.
Miller, K., and Kagan, E., 1976, The *in vivo* effects of asbestos on macrophage membrane structure and population characteristics of macrophages: a scanning electron microscope study, *J. Reticuloendothel. Soc.* **20:**159.
Miller, K., and Kagan, E., 1977, The *in vivo* effects of quartz on alveolar macrophage membrane topography and on the characteristics of the intrapulmonary cell population, *J. Reticuloendothel. Soc.* **21:**307.
Miller, K., Handfield, R. I., and Kagan, E., 1978, The effect of different mineral dusts on the mechanism of phagocytosis: A scanning electron microscope study, *Environ. Res.* **15:**139.
Miyoshi, M., and Fujita, T., 1971, Stereo-fine structure of the splenic red pulp. A combined scanning and transmission electron microscope study on dog and rat spleen, *Arch. Histol. Jpn.* **33:** 225.
Morgenroth, K., Blaschke, R., and Schlake, W., 1973, Energy dispersive x-ray analysis of semi-thin sections in the scanning transmission, *Beitr. Pathol.* **150:**406.
Motta, P., 1975, A scanning electron microscopic study of the rat liver sinusoid: Endothelial and Kupffer cells, *Cell Tissue Res.* **164:**371.
Motta, P., and Porter, K. R., 1974, Structure of rat liver sinusoids and associated tissue spaces as revealed by scanning electron microscopy, *Cell Tissue Res.* **148:**111.
Munthe-Kaas, A. C., Berg, T., Seglen, P. O., and Seljelid, R., 1975, Mass isolation and culture of rat Kupffer cells, *J. Exp. Med.* **141:**1.
Murakami, T., 1974, A revised tannin-osmium method for non-coated scanning electron microscope specimens, *Arch. Histol. Jpn.* **36:**189.
Murakami, M., Shimada, T., and Suefuji, K., 1977, Scanning electron microscopic observation of spermiophage cell within the lumen of the epididymal duct of the vasectomized Japanese monkey (*Macacus fuscatus*), *Experientia* **33:**1101.
Muto, M., 1975, A scanning electron microscopic study on endothelial cells and Kupffer cells in rat liver sinusoids, *Arch. Histol. Jpn.* **37:**369.
Muto, M., Nishi, M., and Fujita, T., 1977, Scanning electron microscopy of human liver sinusoids, *Arch. Histol. Jpn.* **40:**137.
Nelson, G. A., and Revel, J. P., 1975, Scanning electron microscopic study of cell movements in the corneal endothelium of the avian embryo, *Dev. Biol.* **42:**315.
Nielsen, S. L., and Gauger, G. E., 1974, Experimental hydrocephalus: Surface alterations of the lateral ventricle. Scanning electron microscopic studies, *Lab. Invest.* **30:**618.
Nielsen, M. H., Jensen, H., Braendstrup, O., and Werdelin, O., 1974, Macrophage–lymphocyte clusters in the immune response to soluble protein antigen *in vitro*. II. Ultrastructure of clusters formed during the early response, *J. Exp. Med.* **140:**1260.
Nolibe, D., 1973, Etude morphologique, par microcinématographie et microscopie électronique de balayage, de l'histiocyte alvéolaire après phagocytose, *in vitro*, d'oxyde de plutonium, *C. R. Acad. Sci (D) (Paris)* **276:**65.
Noonan, S. M., and Riddle, J. M., 1977, Dynamic surface activities of exudative leucocytes, in: *Scanning Electron Microscopy 1977* (O. Johari, ed.), pp. 53–58, Illinois Institute of Technology Research Institute, Chicago.
Orenstein, J. M., and Shelton, E., 1976a, Surface topography and interactions between mouse peritoneal cells allowed to settle on an artificial substrate: observations by scanning electron microscopy, *Exp. Mol. Pathol.* **24:**201.
Orenstein, J. M., and Shelton, E., 1976b, Surface topography of leukocytes *in situ:* Cells of mouse peritoneal milky spots, *Exp. Mol. Pathol.* **24:**415.
Orenstein, J. M., and Shelton, E., 1977, Membrane phenomena accompanying erythrophagocytosis. A scanning electron microscope study, *Lab. Invest.* **36:**363.

Papadimitriou, J. M., 1973, Detection of macrophage receptors for heterologous IgG by scanning and transmission electron microscopy, *J. Pathol.* **110:**213.

Papadimitriou, J. M., Finlay-Jones, J. M., and Walters, M.N-I., 1973, Surface characteristics of macrophages, epithelioid and giant cells using scanning electron microscopy, *Exp. Cell. Res.* **76:**353.

Parakkal, P., Pinto, J., and Hanifin, J. M., 1974, Surface morphology of human mononuclear phagocytes during maturation and phagocytosis, *J. Ultrastruct. Res.* **48:**216.

Polliack, A., 1977, *Normal, Transformed and Leukemic Leukocytes. A Scanning Electron Microscopy Atlas,* Springer Verlag, New York.

Polliack, A., and Gordon, S., 1975, Scanning electron microscopy of murine macrophages. Surface characteristics during maturation, activation and phagocytosis, *Lab. Invest.* **33:**469.

Powell, D. A., and Muse, K. A., 1977, Scanning electron microscopy of guinea pig alveolar macrophages: *In vitro* phagocytosis of *Mycoplasma pneumoniae, Lab. Invest.* **37:**535.

Privat, A., Drian, M. J. and Mandon, P., 1973, The outgrowth of rat cerebellum in organized culture, *Z. Zellforsch. Mikrosk. Anat.* **146:**45.

Quan, S. G., and Golde, D. W., 1977, Surface morphology of the human alveolar macrophage, *Exp. Cell Res.* **109:**71.

Robertson, T. A., Papadimitriou, J. M., Walters, M.N.-I., and Wolman, M., 1977, Effects of exposure of murine peritoneal exudate and resident macrophages to high molecular levan: A morphological study, *J. Pathol.* **123:**157.

Rosenbaum, R. M., Picciano, P., Kress, Y., and Wittner, M., 1977, Ultrastructure of *in vitro* type 2 epithelial cell cysts derived from adult rabbit lung cells, *Anat. Rec.* **188:**241.

Sarphie, T. G., and Allen, D. J., 1977, Scanning electron microscopy of Acanthamoeba culbertsoni as observed in the subarachnoid space, *Am. J. Clin. Pathol.* **68:**485.

Schultz, R. M., Chirigos, M. A., and Heine, U. I., 1978, Functional and morphologic characteristics of interferon-treated macrophages, *Cell Immunol.* **35:**84.

Scott, R. E., and Maercklein, P. B., 1977, Plasma membrane vesiculation: Correlation between macrophage spreading and the shedding of cell surface vesicles, *Lab. Invest.* **37:**430.

Scott, D. E., Krobisch-Dudley, G., Paull, W. K., Kozlowski, G. P., and Ribas, J., 1975, The primate median eminence. I. Correlative scanning–transmission electron microscopy, *Cell Tissue Res.* **162:**61.

Seitz, H. M., Weiblen, E. E., and Claviez, M., 1977, Light microscopic and scanning electron microscopic observations on phagocytosis of *Plasmodium berghei* infected erythrocytes by mouse macrophages, *Tropenmed. Parasitol.* **28:**481.

Shekhter, A. B., Berchenko, G. N., and Nikolaev, A. V., 1977, Macrophage–fibroblast interaction and its possible role in regulating collagen metabolism during wound healing, *Byull, Eksp. Biol. Med.* **83:**627.

Siegesmund, K. A., Funahashi, A., and Pintar, K., 1974, Identification of metals in lung from a patient with interstitial pneumonia, *Arch. Environ. Health* **28:**345.

Silberberg, D. H., 1975, Scanning electron microscopy of organotypic rat cerebellum cultures, *J. Neuropathol. Exp. Neurol.* **34:**189.

Song, S. H., and Groom, A. C., 1974, Scanning electron microscopic study of the splenic red pulp in relation to the sequestration of immature and abnormal red cells, *J. Morphol.* **144:**439.

Takahashi, M., 1977, Pathological study on amyloidosis. Scanning electron microscopic observation of amyloid-laden mouse liver, *Acta Pathol. Jpn.* **27:**809.

Takayama, H., Katsumoto, T., and Takagi, A., 1976, Electron microscopic studies on macrophages. I. Effects of colchicine and cytochalasin B on the cell cortex as revealed by scanning electron microscopy, *J. Electron Microsc. (Tokyo)* **25:**75.

Takayanagi, T., Nakatke, Y., and Enriquez, G. L., 1974, *Trypanosoma gambiense:* Phagocytosis *in vitro, Exp. Parasitol.* **36:**106.

Takenaga, A., Matsuda, M., Horai, T., Ikegami, H., and Hattori, S., 1977, Scanning electron microscopy in the study of lung cancer. New technique of comparative studies on the same lung cancer cells by light microscopy and scanning electron microscopy, *Acta Cytol. (Baltimore)* **21:**90.

Thomas, J. A., and Vilain, C., 1970, Cellules histiocytaires des alvéoles du poumon en culture. Analyse cytochimique de la sécrétion, *C. R. Acad. Sci (D) (Paris)* **271:**2011.

Thomas, J. A., Vilain, C., and Delage, G., 1970, Cellules histiocytaires des alvéoles du poumon. Evolution extracellulaire de la sécrétion *in vitro:* Comparaison biochimique avec le surfactant, *C. R. Acad. Sci (D) (Paris)* **271:**2357.
Tizard, I. R., and Holmes, W. L., 1974, Phagocytosis of sheep erythrocytes by macrophages. Some observations under the scanning electron microscope, *J. Reticuloendothel. Soc.* **15:**132.
Tizard, I. R., and Holmes, W. L., 1975, The dendritic cells of the guinea pig popliteal lymph node: Identification and classification of cells observed by scanning electron microscopy, *J. Reticuloendothel. Soc.* **17:**333.
Tizard, I. R., Holmes, W. L., and Parappally, N. P., 1974, Phagocytosis of sheep erythrocytes by macrophages: A study of the attachment phase by scanning electron microscopy, *J. Reticuloendothel. Soc.* **15:**225.
Underwood, J.C.E., 1976, An ultrastructural analysis of lymphoreticular cell interactions in primary cultures of human non-lymphoid neoplasms and lymphomas, *J. Pathol.* **120:**75.
van Furth, R., and Fedorko, M. E., 1976, Ultrastructure of mouse mononuclear phagocytes in bone marrow colonies grown *in vitro, Lab. Invest.* **34:**440.
Vernes, A., Poulain, D., Prensier, G., Deblock, S., and Biguet, J., 1974, Trichinose expérimentale, III, Action *in vitro* des cellules péritonéales sensibilisées sur les larves musculaires de premier stade. Etude préliminaire comparative en microscopie optique et électronique à transmission et à balayage, *Biomedicine Express (Paris)* **21:**140.
Voisin, C., Aerts, C., Tonnel, A. B., and Petitprez, A., 1972a, Infections broncho-pulmonaires et déficit de la fonction phagocytaire alvéolaire. Le macrophage alvéolaire moyens d'étude et rôle dans la défense de l'appareil respiratoire contre l'infection, *Rev. Tuberc. Pneumol. (Paris)* **36:**357.
Voisin, C., Aerts, C., Tonnel, A. B., Petitprez, A., and Tacquet, A., 1973, Action du sérum anti-lymphocytaire sur les macrophages alvéolaires humains en survie *in vitro, J. Urol. Nephrol. (Paris)* **79:**735.
Voisin, C., Aerts, C., Jakubczak, E., and Tonnel, A. B., 1977, La culture cellulaire en phase gazeuse. Un nouveau modèle expérimental d'étude *in vitro* des activités des macrophages alvéolaires, *Bull. Eur. Physiopathol. Respir.* **13:**69.
Voisin, C., Tonnel, A. B., Lelièvre, G., Aerts, C., and Petitprez, A., 1972b, Traitements immunodépresseurs et défense phagocytaire alvéolaire, *Rev. Tuberc. Pneumol. (Paris)* **36:**853.
von Krahn, V. and Richter, I. E., 1976, The structure of meningeal granules based on light and scanning electron microscopic studies, *Anat. Anz.* **140:**118.
Walters, M.N-I., Papadimitriou, J. M., and Robertson, T. A., 1976, The surface morphology of the phagocytosis of micro-organisms by peritoneal macrophages, *J. Pathol.* **118:**221.
Wang, N. S., 1975, The preformed stomas connecting the pleural cavity and the lymphatics in the parietal pleura, *Am. Rev. Respir. Dis.* **111:**12.
Warfel, A. H., 1978, Macrophage fusion and multinucleated giant cell formation, surface morphology, *Exp. Mol. Pathol.* **28:**163.
Warfel, A. H., and Elberg, S. S., 1970, Macrophage membranes viewed through a scanning electron microscope, *Science* **170:**446.
Warfel, A. H., Elberg, S. S., and Hayes, T. L., 1973, Surface morphology of peritoneal microphages during the attachment of *Brucella melitensis, Infect. Immun.* **8:**665.
Waters, M. D., Gardner, D. E., Aranyi, C., and Coffin, D. L., 1975, Metal toxicity for rabbit alveolar macrophages *in vitro, Environ. Res.* **9:**32.
Watters, W. B., and Buck, R. C., 1973, Mitotic activity of peritoneum in contact with a regenerating area of peitoneum, *Virch. Arch. (Zellpathol.)* **13:**48.
Weibel, E. R., Gehr, P., Haies, D., Gil, J., and Bachofen, M., 1976, The cell population of the normal lung, in: *Lung Cells in Disease* (A. Bouhuys, ed.), pp. 3–16, North-Holland, Amsterdam.
Weiss, L., 1974, A scanning electron microscopic study of the spleen, *Blood* **43:**665.
Wetzel, B., Erickson, B. W., and Levis, W. R., 1973, The need for positive identification of leukocytes examined by S.E.M., in: *Scanning Electron Microscopy 1973* (O. Johari, ed.), pp. 535–541, Illinois Institute of Technology Research Institute, Chicago.
Wisse, E., Knook, D. L., 1977, *Kupffer Cells and Other Liver Sinusoidal Cells,* Elsevier/North Holland, Amsterdam.

7

Monocytes

SAMUEL K. ACKERMAN
and STEVEN D. DOUGLAS

1. INTRODUCTION

Since modern study of mammalian phagocytes began with Metchnikoff in the 19th century, classification of these cells with respect to structure-function relationships has been controversial. Relatively recent studies of the ontogeny, kinetics, and function of phagocytic cells in animals has led to the concept of the mononuclear phagocyte system (MPS) (van Furth *et al.*, 1975); this is considered to be the most useful conceptual framework for these cells. The mononuclear phagocyte system as presently defined consists of bone marrow promonocytes, circulating blood monocytes, and both mobile and fixed tissue macrophages. Vascular endothelium, reticulum cells, and dendritic cells of lymphoid germinal centers are not included, although in the past terms such as *reticuloendothelial system* have denoted these and mononuclear phagocytes (MNP) collectively. The justification for the present classification of cells of the mononuclear phagocyte system includes numerous lines of investigation, notably: (1) Most tissue macrophages, as identified morphologically in tissue section, appear to have several important functional characteristics in common, particularly, pronounced phagocytic ability *in vivo* and *in vitro* and adhesiveness to glass or plastic surfaces *in vitro:* and (2) kinetic studies identify a bone marrow cell as precursor to the monocyte, and the circulating monocyte as the precursor of most, though not all, tissue macrophages.

The functions of mononuclear phagocytes include the following:

1. Phagocytosis and digestion of microorganisms and tissue debris.
2. Secretion of inflammatory mediators and regulators.

SAMUEL K. ACKERMAN • Department of Medicine, University of Minnesota Medical School, Minneapolis, Minnesota 55455. STEVEN D. DOUGLAS • Departments of Medicine and Microbiology, University of Minnesota Medical School, Minneapolis, Minnesota 55455. Work supported in the authors' laboratory under grants from the National Institutes of Health, AI 12478, the National Leukemia Association, Inc., the Kroc Foundation, and March of Dimes-Birth Defects Foundation (6-246); SKA is supported by a National Institutes of Health Clinical Investigator AM 00642.

3. Interaction with antigen and lymphocytes in the generation of the immune response.
4. Extracellular killing, as of some tumor cells.
5. Other functions specific for macrophages of particular tissues.

Most investigations of MNP function have utilized tissue macrophages, both for technical reasons and also in the belief that these are the most important functional element in the MPS; morphology of macrophages is considered in the remainder of this volume. This chapter will describe the morphology of the precursor of these cells, the blood monocyte, and attempt to illustrate several structure–function relationships as they pertain to the maturing phagocyte. We will briefly describe the major organelle systems of the monocyte and then attempt to correlate physiologic alterations of morphology with biologic function. The morphology of the mononuclear phagocyte has been informatively investigated by many techniques. Light microscopic techniques have included histologic and cytochemical stains, phase contrast, Nomarski, and reflection contrast (Ploem, 1975) optics. Electron microscopic techniques include scanning and transmission microscopy and more recently the freeze-fracture and freeze-etch procedures (Douglas, 1978). Each method has its own applications and will be emphasized where appropriate in the sections that follow.

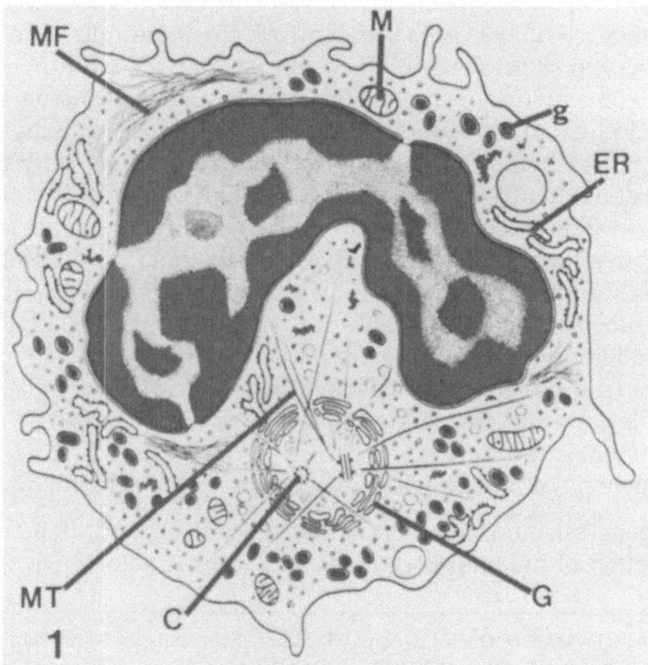

FIGURE 1. Schematic drawing of a blood monocyte. G, Golgi zone; c, centriole; g, granule; ER, endoplasmic reticulum; m, mitochondria; mf, microfilaments; mt, microtubules. Adapted from Bessis, M. (1977). *Blood Smears Reinterpreted*, Springer International, p. 159.

2. MORPHOLOGY OF BLOOD MONOCYTES

On Wright-stained blood smears, the human peripheral blood monocyte is 12–15 μm in diameter with an eccentrically placed horseshoe-shaped nucleus. The nuclear chromatin pattern has been called "raked" because of its fine-stranded appearance. The cytoplasm stains grayish-blue and contains many fine, pink or purple granules which are heterogeneous in size and shape. Although larger azurophilic granules may also be seen, monocyte granules do not, in general, form two distinct morphologic populations as do those of the neutrophil. Figure 1 shows a schematic drawing of a blood monocyte.

Phase contrast examination of monocytes (living or fixed) after isolation from peripheral blood (Zuckerman *et al.*, 1979) following short *in vitro* culture shows several prominent features. Since monocyte morphology is extremely sensitive to small changes in external milieu, culture conditions for such studies should be precisely controlled; small fluctuations of pH or of temperature, for example, may alter monocyte morphology dramatically. Figure 2 shows a human monocyte purified from peripheral blood (Ackerman and Douglas, 1978) and cultured for 1 hr at 37°C on a glass coverslip in Dulbecco's modified Eagle's medium with 10% fetal calf serum and 10% horse serum at pH 7.4 (Zuckerman *et al.*, 1979). Nearly all such cells are firmly adherent; they vary in shape from rounded to flattened. The cytoplasm contains many phase-dense granules (0.2–0.5 μm in diameter) and some cells contain clear vacuoles, 0.5–1.0 μm in size. In

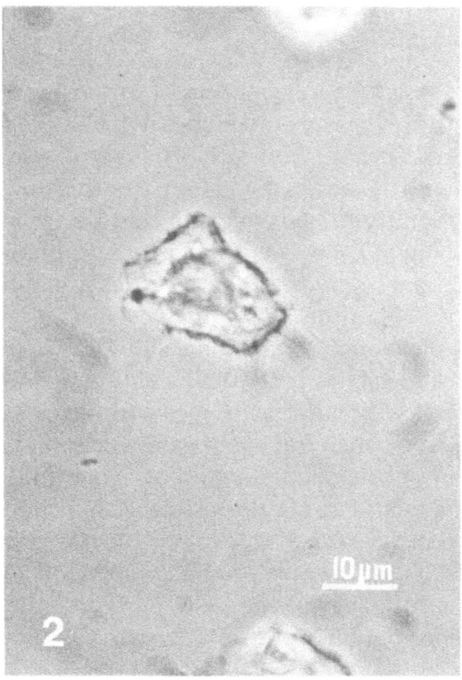

FIGURE 2. Human blood monocyte after *in vitro* culture for 1 hr. Note phase-dense thickening at the peripheral region of the cell, signifying ruffling of the membrane. Phase contrast micrograph; original magnification ×400.

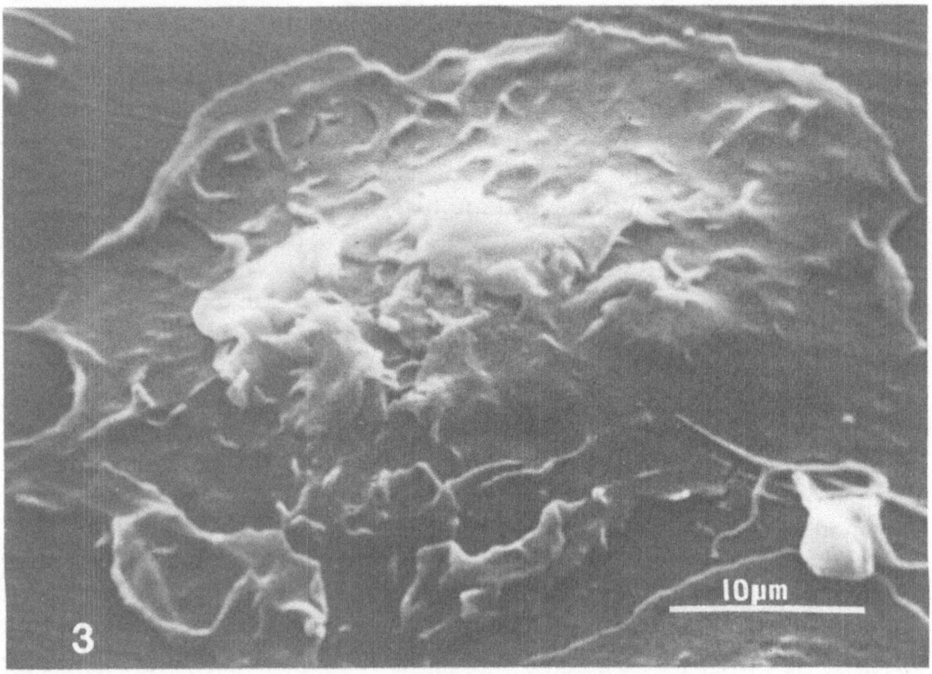

FIGURE 3. Blood monocyte cultured 1 hr *in vitro*. The cell is flattened against the substratum and possesses prominent surface ruffles. Scanning electron micrograph, ×2750.

some cells mitochondria form a small juxtanuclear rosette. Most striking is the ruffled plasma membrane which is visible as prominent phase-dense folds at the cell surface. Some cells have a dense thickening at the edge of the cytoplasm with microextensions on the thickened edge. The surface morphology of the monocyte, however, can best be appreciated in scanning electron micrographs. Figure 3 shows a human monocyte under conditions similar to those described above. Very prominent ruffles and small surface blebs are apparent.

The presence of extensive ruffling of the monocyte plasma membrane is considered of functional significance in several respects. First, the monocyte is a cell which is both mobile and phagocytic. These functions require physical contact with surfaces, either supportive or cellular. While the molecular details of the biochemical events which determine cell contact are totally unknown (as is, in fact, a good definition of *contact* in this context), it is believed that reduction in radius of curvature of the cell surface (as in ruffles or microvilli) may reduce repulsive forces due to cell surface charge and thus facilitate contact between the cell and the substratum or between two cells (Van Oss *et al.*, 1975). Secondly, redundancy of the cell membrane may provide reserve membrane required for locomotion and for phagocytosis.

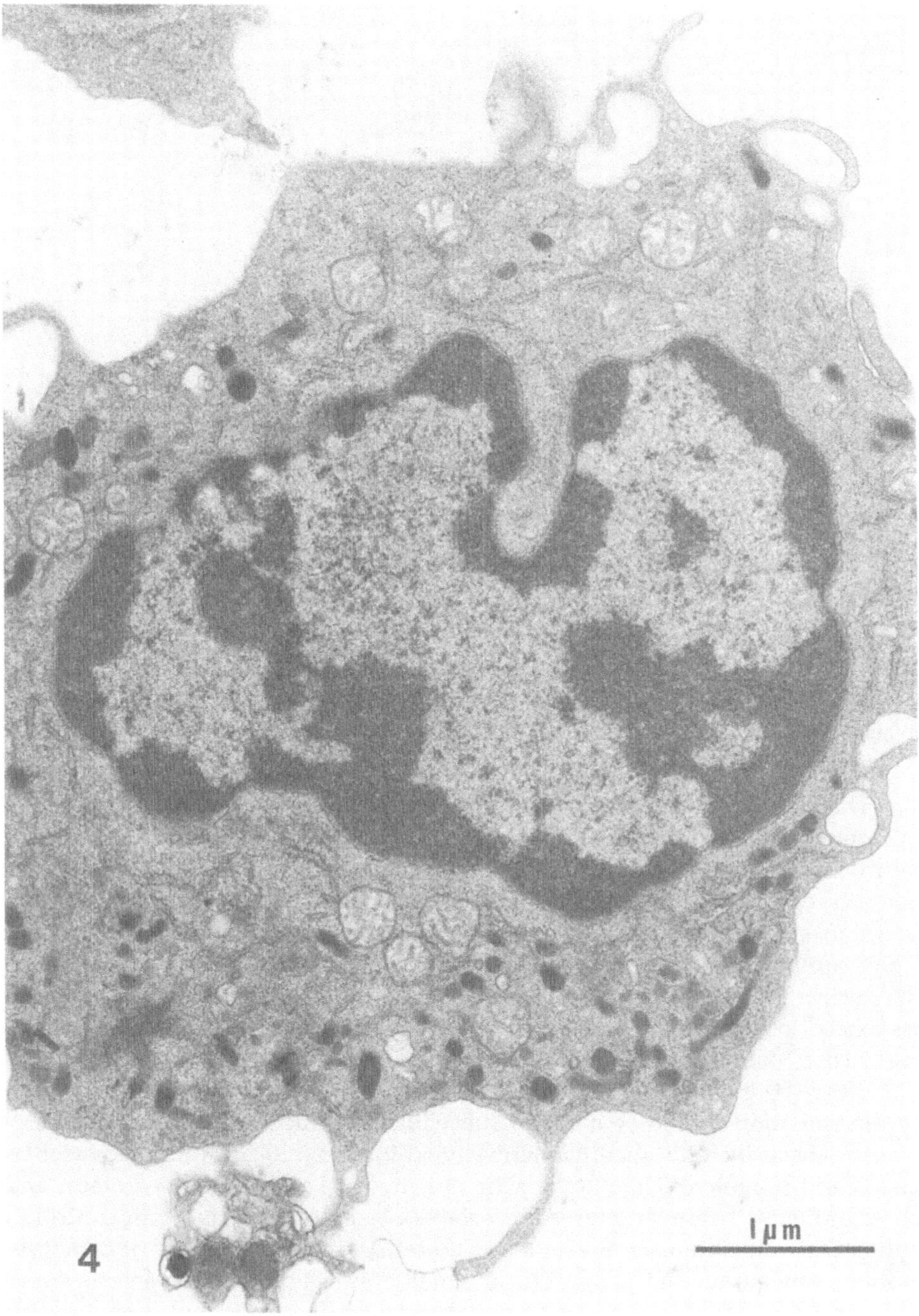

FIGURE 4. TEM of normal human blood monocyte. Note typical horseshoe-shaped nucleus, numerous round mitochondria, and multiple electron-dense granules represented. Prominent surface activity is evident in the numerous small microvilli. ×25,000.

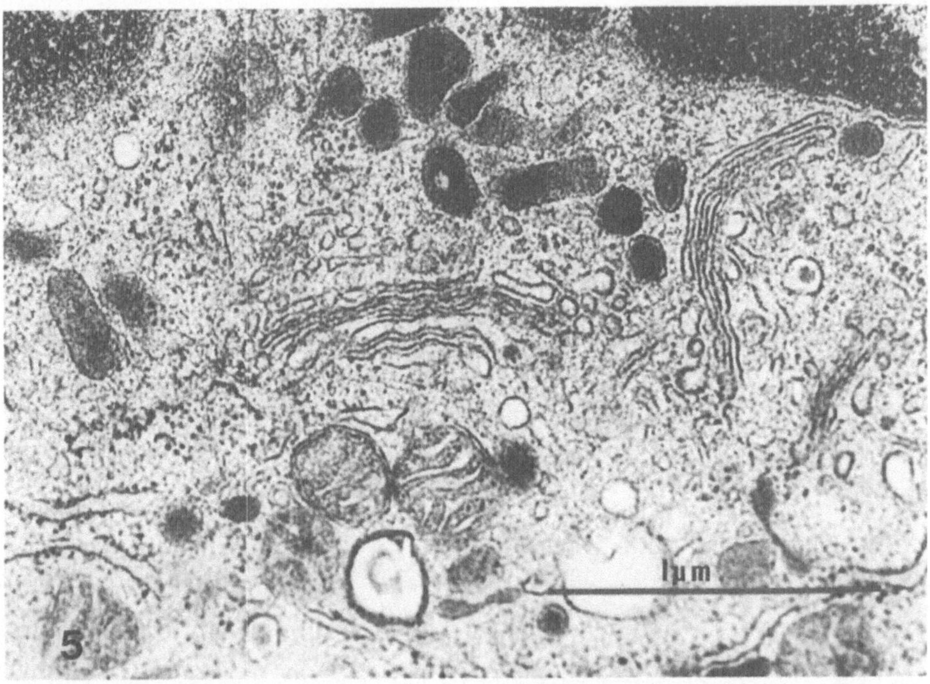

FIGURE 5. Human blood monocyte showing several stacks of typical Golgi apparatus with scattered rough endoplasmic reticulum, cytoplasmic granules, and mitochondria. ×45,500.

The electron microscopic appearance of human blood monocytes is shown in Figures 4–6. The nucleus often contains one or two small nucleoli surrounded by nucleolar-associated chromatin. The cytoplasm contains a relatively small quantity of endoplasmic reticulum and a variable quantity of ribosomes and polysomes. The mitochondria are usually numerous, small, and elongated. The Golgi complex is well-developed and is situated about the centrosome within the nuclear indentation. Centrioles and filamentous centriolar satellites are often visualized in this region. Numerous microvilli are apparent, again demonstrating "active" surface membranes.

Since the mononuclear phagocyte is a cell with prominent surface activity, specific mention should be made of subcellular contractile and cytoskeletal elements. These include microfilaments, microtubules, 10-nm intermediate filaments and nonmicrofilamentous actin and myosin, actin-binding protein, and other proteins. These structures are believed to be of importance in control of bulk plasma membrane motion and are, therefore, relevant to study of monocyte motility, spreading, and phagocytosis. Small bundles of fine filamentous structures are frequently seen in electron micrographs (Figure 7), but other specific morphologic observations on the blood monocyte are scant. In cultured macrophages collections of microfilaments may characteristically be seen underneath the plasma membrane near sites of cellular attachment to either sub-

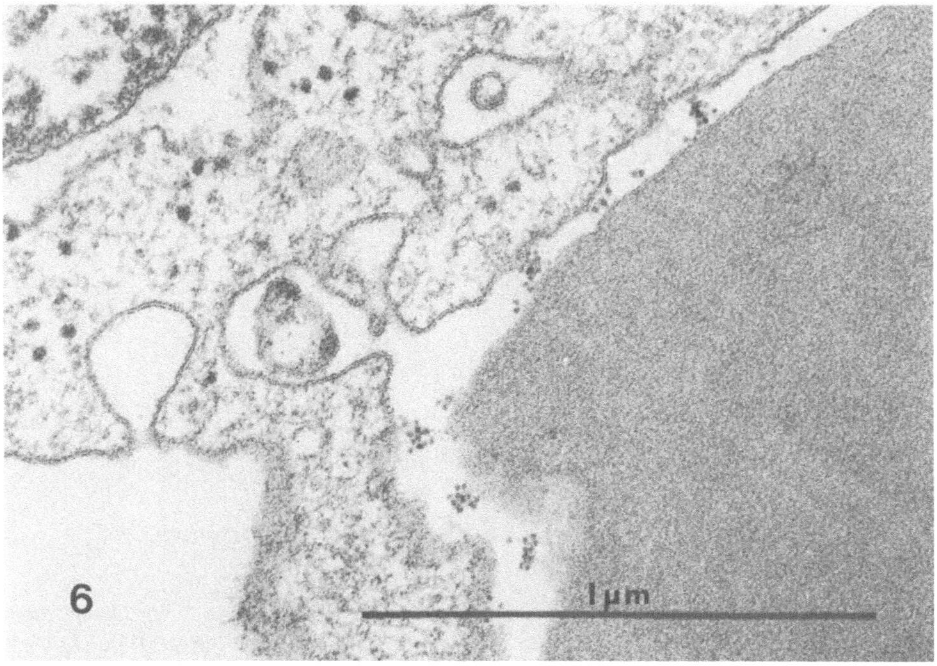

FIGURE 6. Human blood monocytes (upper left) interacting with antibody-coated erythrocyte (lower right). The monocyte is bounded by a typical trilaminar unit membrane and contains electron-dense glycogen granules. Ferritin-labeled antibody has been used to coat the red cell and this appears as small clusters of electron density between the two cells. Clustering of ferritin suggests that the interaction between monocyte and macrophages is occurring only along certain portions of membrane.

stratum or to phagocytizable particles (Reaven and Axline, 1973), as discussed elsewhere in this volume. Despite extensive biochemical characterization of macrophage contractile proteins (Hartwig and Stossel, 1975) there is, however, still no clear picture of the precise way in which the contractile mechanism powers motility and phagocytosis in intact cells.

Histochemical techniques have provided much information on the enzyme content of monocytes. Table 1 shows a summary of hydrolytic enzyme content as judged histochemically from monocytes, neutrophils, and lymphocytes (Braunsteiner and Schmalzl, 1970; Li *et al.*, 1973). In addition to these enzymes, monocytes give a weak but positive PAS reaction (for polysaccharides) and Sudan Black B reaction (for lipids).

Studies of the enzyme content of monocyte granules (Figures 7–9) have attempted to determine whether there are distinct granule populations as occur in the neutrophil and whether granule content can be related to monocyte ontogeny or functional status. Neutrophil granules for example consist of two populations separable by numerous criteria (Spitznagel, 1975): (1) The azurophilic (nonspecific, primary) granules are large (0.5–0.9 μm) and contain

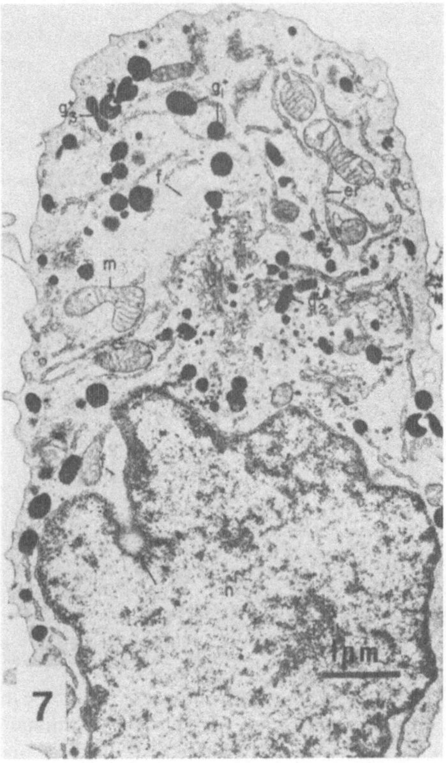

FIGURE 7. Promonocyte from human bone marrow, reacted for peroxidase. Reaction product is present in rough endoplasmic reticulum (er) and granules (g). Note also prominent microfilament bundles (f). (From Nichols and Bainton, 1973.)

acid hydrolases and peroxidase; they are modified primary lysosomes which fuse with phagosomes and perform intracellular digestion. (2) The specific or secondary granules are smaller (0.2–0.5 μm); in most species they contain alkaline phosphatase, lysozyme, and certain cationic proteins; their function is less well understood. Production of granules is completed when the neutrophil leaves the marrow and apparently does not resume during the several days of its subsequent life. In contrast, monocyte granules, although heterogeneous in size (0.3–0.6 μm), are not readily separable into populations by routine electron microscopic criteria except perhaps in the rat (van der Rhee *et al.*, 1977). Identification of monocyte granule populations has depended therefore on subcellular localization of monocyte enzymes by electron microscopic cytochemistry, which has been investigated by Nichols *et al.* (1971). In a series of studies on rabbit and human marrow promonocytes and blood monocytes, these workers have demonstrated that the granules do comprise two functionally distinct populations (Nichols *et al.*, 1971; Nichols and Bainton, 1973). One population contains the enzymes acid phosphatase, arylsulfatase, and, in the human (but not in the rabbit), also peroxidase; these granules are therefore modified primary lysosomes and analogous to the azurophil ("nonspecific") granules of the neutro-

TABLE 1. CYTOCHEMICAL REACTIONS OF LEUKOCYTE ENZYMES[a]

Chemical	Monocytes	Neutrophils	Lymphocytes
Acid phosphatase	++	+	+
β-Glucuronidase	++	+	0 to +
Sulfatase	+	+	0
N-Acetylglucosaminidase	+	+	0
Lysozyme	++	++	0
Naphthylamidase	++	+	0 to +
α-Naphthyl butyrate esterase	++	0 to +	0[b]
Naphthol AS-D chloroacetate esterase	0 to +	++	0
Peroxidase	+	++	0
Alkaline phosphatase	0	0 to +	0

[a] Modified from Braunsteiner and Schmalzl (1970) and Li *et al.* (1973).
[b] α-Naphthyl acetate esterase activity may appear in human T lymphocytes under certain conditions (Totterman *et al.*, 1977).

phil. It should be indicated that the azurophil granule population is heterogeneous with respect to cytochemical reactivity for peroxidase, acid phosphatase, and arylsulfatase (Nichols and Bainton, 1975). Moreover, it is becoming evident

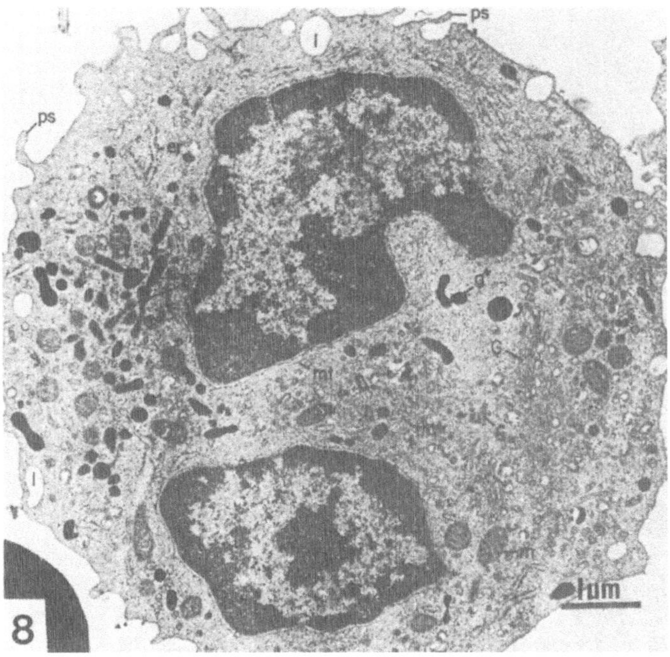

FIGURE 8. Monocyte from human peripheral blood, reacted for peroxidase. Reaction production is seen in only some of the granules and is absent from the ER. Abbreviations as in Figure 1. Note also prominent pseudopods (ps). (From Nichols and Bainton, 1973.)

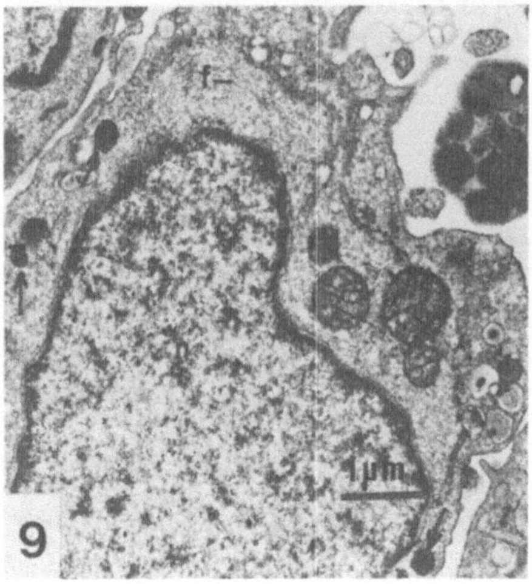

FIGURE 9. Promonocyte from human bone marrow, reacted for arylsulfatase. Arrows show dense reaction product in some, but not all, granules. (From Nichols and Bainton, 1973.)

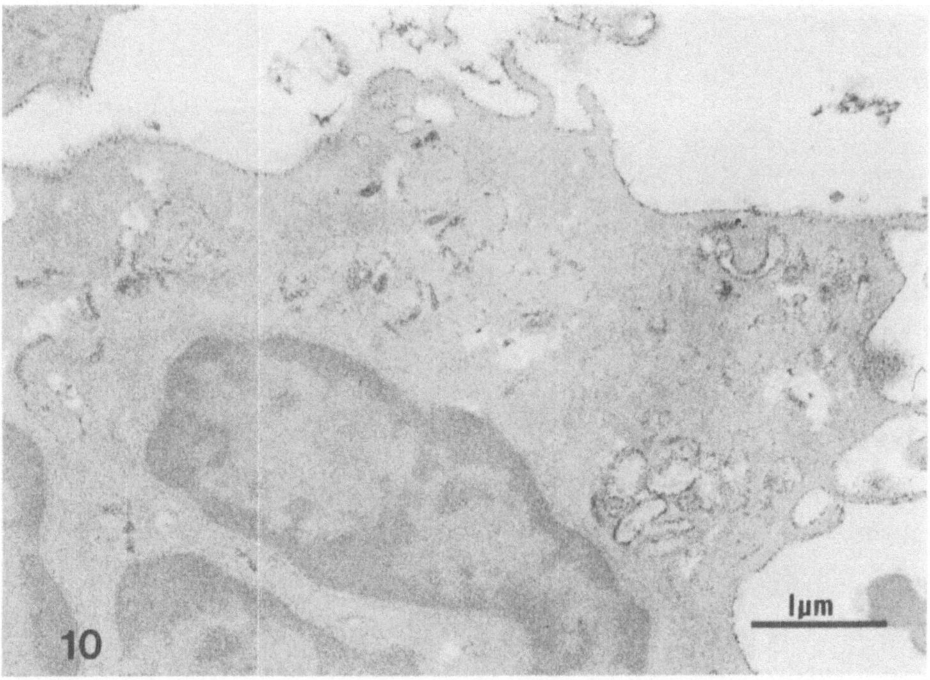

FIGURE 10. Blood monocyte, stained for acid complex carbohydrate. Staining may be seen as dark areas along plasma membrane and in about 10% of granules. (From Parmley *et al.*, 1978.)

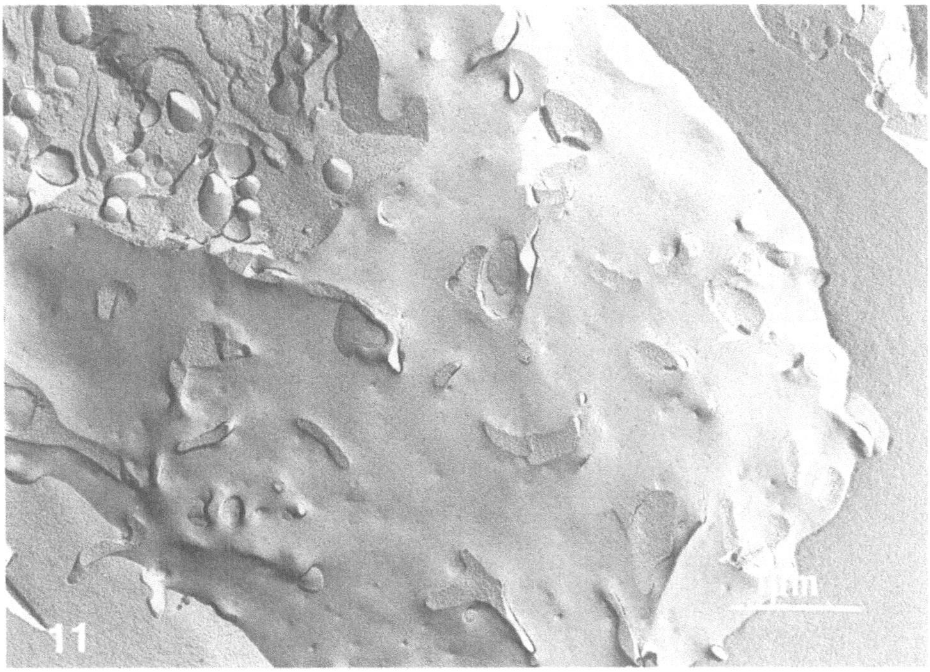

FIGURE 11. Human blood monocyte, freeze-fracture replica. Survey view; ×18,000.

that the vacuolar system is complex and that primary granules which are morphologically identical with other vesicles may be cytochemically identified as lysosomes. The content of the other population of monocyte granules is unknown; however, they have been shown to lack alkaline phosphatase (Nichols and Bainton, 1975), and hence are not strictly analogous to the neutrophil-specific granules. The function of the lysosomal type of granule is presumably digestive; that of the second population is not known.

Granule contents other than enzymes have also been investigated cytochemically in blood monocytes. Parmley *et al.* (1978) have recently reported that about 10% of granules in normal human blood monocytes stain with reagents which identify complex acid carbohydrates, or "acid mucosubstances" (Figure 10). Although the function of these substances is not known, they are also found in leukemic monocyte granules as well as in granules of normal neutrophils (Horn and Spicer, 1964).

The technique of freeze-fracture is beginning to be utilized for studies of MNPs. In this technique a cell suspension is first frozen then placed in a high vacuum chamber and struck with a blunt edge. A fracture is propagated through the frozen specimen and splits it. The utility of the procedure comes from the remarkable finding that when the fracture encounters a cell it tends to propagate along the interior of the plasma membrane and thus split the lipid bilayer in half.

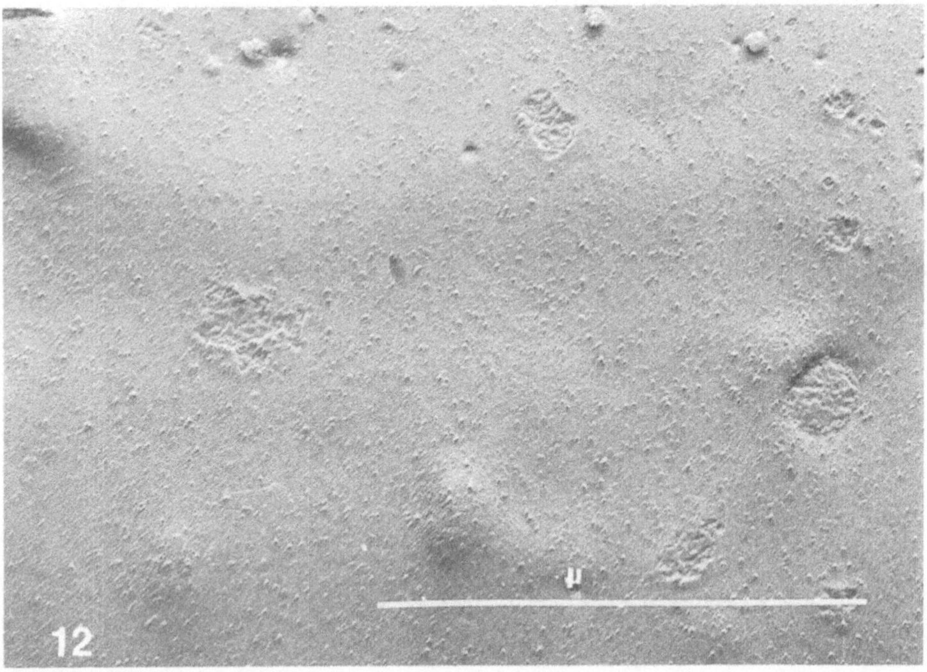

FIGURE 12. Human blood monocyte, freeze-fracture detail. Note many small particles whose normal distribution is random. These are believed to consist, at least in part, of intercalated membrane proteins. ×67,500.

After fracture, the specimen is coated with platinum, which is electron-dense, and viewed with conventional transmission electron microscopy (TEM). All cell types examined thus far by the freeze-fracture technique have shown intramembrane particles (IMP) as the predominant feature of the topography of the interior of the bilayer. Studies of the erythrocyte have shown that at least some particles may contain intercalated membrane proteins (Weinstein *et al.*, 1978), and this has been assumed to be the case with nucleated cells as well, although proof is lacking at present. The distribution of IMP are dramatically altered in a number of cell systems by physiologic stimuli, e.g., hormonal stimulation. Although the precise role of the IMP remains to be defined, they will undoubtedly be proven to be of great importance in many membrane-mediated intracellular events.

Figures 11–13 show a freeze-fracture replica of a human blood monocyte. The plasma membrane interior has been revealed and shows typical IMP in a normal, i.e., random, distribution. Also shown are the bases of many microvilli that were broken from the cell during fracturing. Recent studies from this laboratory have shown profound changes in the distribution of IMP on mononuclear phagocytes following binding of antibody-coated erythrocytes (Douglas, 1978). Since redistribution also occurs in some nonphagocytic Fc-receptor-bearing cells

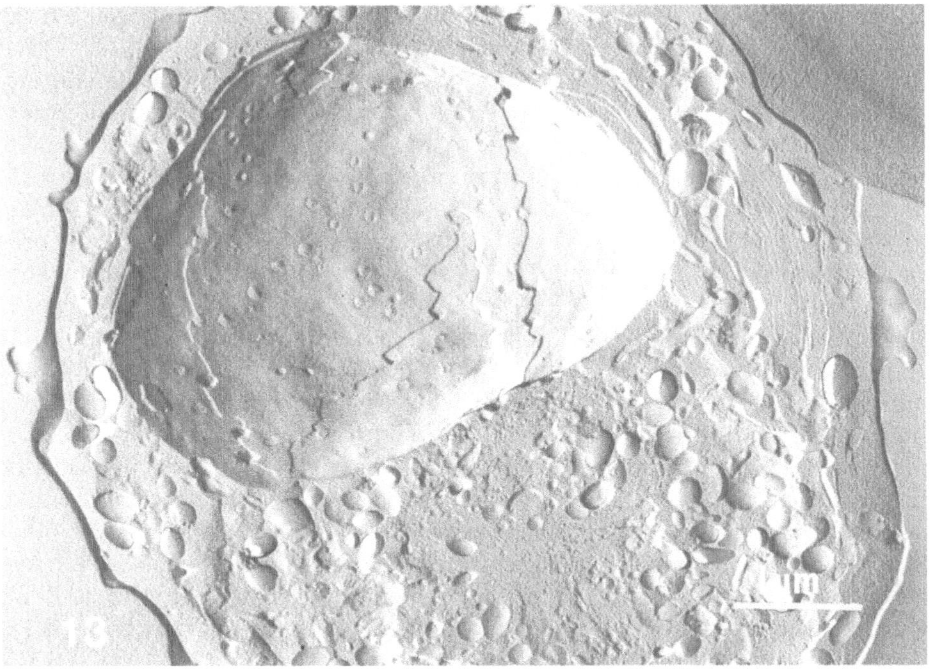

FIGURE 13. Human blood monocyte, freeze-fracture. In this instance, the fracture plane went through the cytoplasm instead of between membrane bilayers, producing a cross-fractured specimen. Numerous types of cytoplasmic organelles may be seen including granules, endoplasmic reticulum, and Golgi apparatus. The nucleus is not cross-fractured but rather cleaved along its bilayer thus appearing somewhat similar to Figure 11.

(Douglas, 1978), and after exposure to aggregated IgG (Douglas *et al.*, 1980), this alteration in IMP presumably reflects interaction with the Fc receptor.

3. FUNCTIONAL MORPHOLOGY OF BLOOD MONOCYTES

3.1. *IN VITRO* ATTACHMENT AND CULTURE

Attachment of monocytes to surfaces *in vitro* is studied with the expectation that information obtained will be of value in understanding the interaction of the MNP plasma membrane with physiologic structures *in vivo*. We will consider attachment of monocytes in the presence and absence of serum, and on several different surfaces.

When freshly isolated human peripheral blood monocytes are incubated at 37°C at pH 7.4 in 10% heat-inactivated serum (fetal calf or human) in a plastic culture vessel, the cells settle and within several minutes flatten onto the surface. At this time, cells are tenaciously adherent. With continued incubation (2–3 hr) the monocytes tend to round up. Some cells remain loosely attached to the substratum by a tuft of lamellapodia extending from one pole of the cell. Other cells remain firmly attached and begin active locomotion across the substratum. During locomotion across glass or plastic surfaces, the cells often assume a triangular shape with a flat leading edge and trailing tail. Occasionally the tail remains firmly adherent to the substratum during forward motion, resulting in elongation of the tail to lengths of 10–15 μm. When cells meet, they often remain adherent to one another, which may result in appearance of large clumps of cells during the first 24–48 hr of culture. These clumps often detach from the substratum.

Under optimal conditions (Zuckerman *et al.*, 1979) monocytes begin to show morphologic alterations suggestive of transition to macrophages within 6 hr of culture. Cells increase in size and begin to develop many phase-dense granulations and phase-lucent vacuoles. Cell growth is rapid the first week of culture, by which time virtually all cells have developed characteristic morphologic alterations. Cell shape after 7 days of culture ranges from round to fusiform; most nuclei have at least one nucleolus and the cytoplasm is often packed with phase-dense granulations (Figures 14 and 15). Occasional cells have large clear vacuoles, which, by Oil Red O staining, can be shown to be lipid. Multinucleated cells are present, usually near areas of prior cell clustering; these probably result from fusion of clumped monocytes.

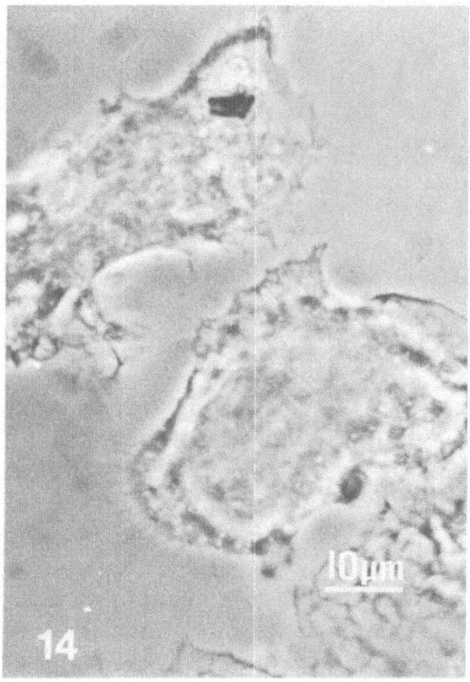

FIGURE 14. Human blood monocyte, 4-week culture. These cells are much larger than that shown in Figure 2. One pole of the central cell has an area with radiating microextensions, a common finding. Note phase-dense and phase-lucent granules. Membrane ruffles appear as dense areas along the cell periphery. Phase contrast micrograph; ×400.

In scanning electron microscopy (SEM), these cultured monocytes display a variety of plasma membrane configurations including ruffles (lamellipodia) and microextensions from the cell body. The latter specializations often appear to originate from one pole of the cell. Occasional knoblike specializations are seen along the length of the microextensions. The significance of these are unknown; however, Albrecht-Buehler and Goldman (1976) have reported similar structures on cultured fibroblasts; in these cells they move centripetally along the microextensions from substratum to cell body. Although most cultured monocytes continue to display extensively developed lamellipodia regardless of the overall shape of the cell, a few large, round cells show a much smoother surface.

Cultured monocytes interact with antibody-coated red cells much as do freshly isolated cells, i.e., mainly phagocytosis of IgG-coated particles (EA) and mainly rosetting with particles coated with IgM and complement (EAC) occur. A small number of cultured cells, usually large, round, giant cells, do not interact with either EA or EAC. The reason for this apparent loss of receptor activity is not known. In addition, we do not yet know whether phagocytosis of EAC will serve as a marker for activation of cultured human monocytes as it does in some mouse macrophage systems (Bianco *et al.*, 1975).

In contrast to the above, monocytes incubated over glass or plastic surfaces in tissue culture media that do not contain serum undergo a morphologically

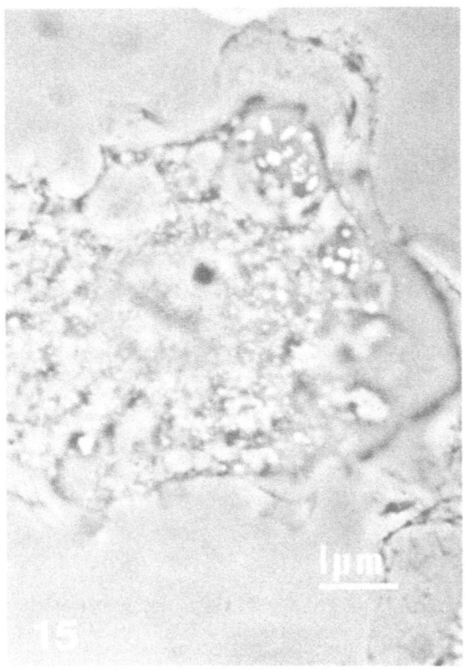

FIGURE 15. Human blood monocyte, 4-week culture. In addition to features mentioned above, note areas of organelle-free cytoplasm at periphery of cell, the so-called "ectoplasm." In other phagocytic and motile cells these areas are very rich in microfilaments which are thought to function in anchorage, phagocytosis, or motility. Phase contrast micrograph; ×400.

distinct interaction with the substratum called *spreading* (Rabinovitch and DeStefano, 1973a; Douglas, 1976). This phenomenon can be accentuated through the use of various inducers of spreading, which include hypotonicity of culture media, low pH, manganese ions, immune complexes, and the protease subtilisin (Rabinovitch and DeStefano, 1973b). When incubated under such appropriate conditions monocytes rapidly flatten onto glass surfaces. Ruffles over the cell body are partially lost as redundant membrane is utilized in extension of the plasma membrane over the substratum. The result is a cell with the appearance of a "fried egg," although other shapes occur as well. When spreading occurs on plain glass substrata, the process requires about 10–15 min for completion (Figure 16), after which monocytes begin to retract.

If human monocytes are allowed to spread on glass that has been previously coated with immune complexes, the morphology is somewhat different (Bumol and Douglas, 1977). Spreading occurs more rapidly and the monocytes spread very symmetrically; the spread cytoplasm is very round with little variability between cells. In addition, the border of the cells becomes phase-dense; SEM shows those edges to have a characteristic rolled appearance (Figure 17). The retraction phase of the process is also different on immune-complex-coated

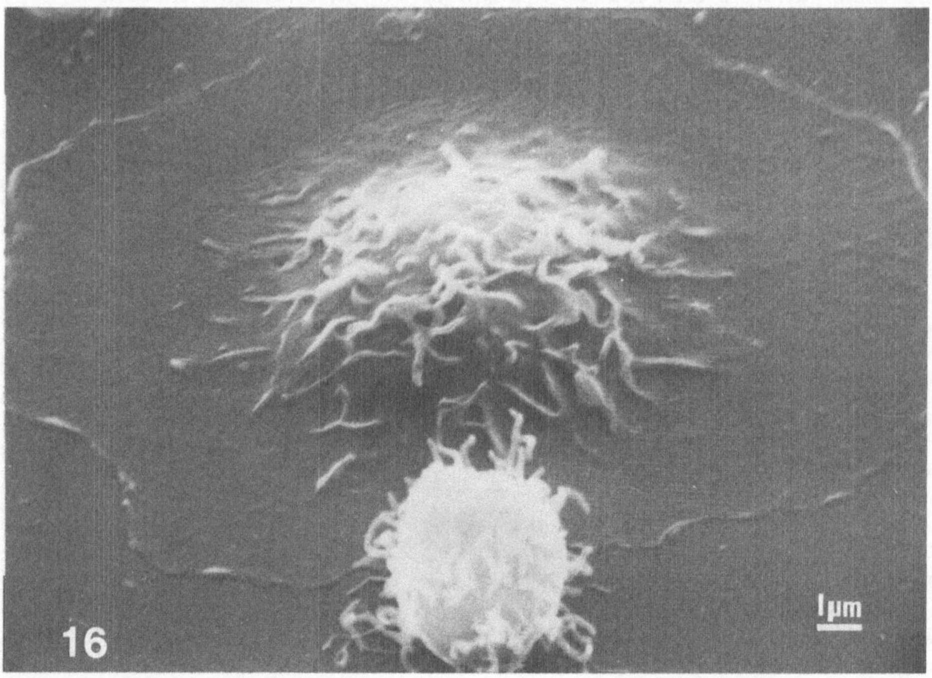

FIGURE 16. Human blood monocyte after 10 min of spreading on glass substratum. The cell has undergone extensive flattening onto the glass. In the foreground for comparison is an unspread lymphocyte. ×6250.

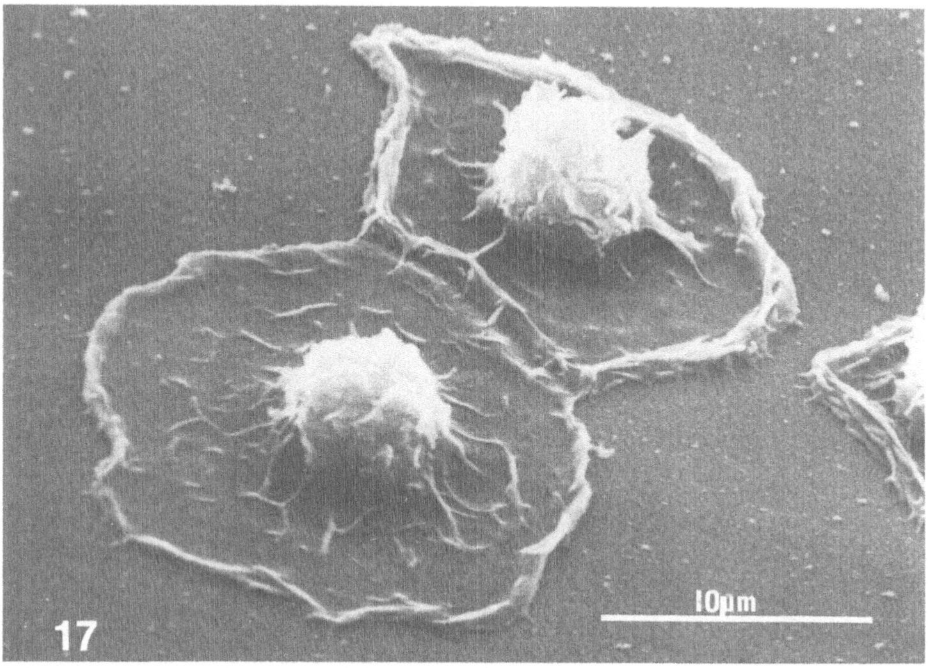

FIGURE 17. Human blood monocyte after 10 min spreading on glass precoated with BSA–anti-BSA immune complexes. Note that in comparison to Figure 16, cell margins are rolled and are beginning to show small microextensions. ×3750.

glass; retracing cells show the typical rolled edge but, in addition, many fine microextensions may be seen along the edge (Figure 18). Since spreading on immune complexes selectively blocks interaction of monocytes with IgG-coated erythrocytes (Douglas, 1976), the morphologic peculiarities of this process presumably reflect involvement of the Fc receptor. The precise way in which Fc-receptor occupancy results in those morphologic alterations, however, is not known.

We have recently observed that human blood monocytes have several distinctive characteristics if they are allowed to attach in the presence of serum to a plastic surface previously "conditioned" by other cells (Ackerman and Douglas, 1978). To demonstrate these effects, fibroblasts (e.g., BHK-21 cells) are grown to confluence in a tissue culture flask, and then treated with EDTA. Although the cells are completely removed by this procedure, they have been shown to leave on the substratum a complex mixture of macromolecules termed *microexudate*. As discussed further below, human monocytes adherent to such surfaces detach more easily than from untreated plastic, and it was therefore of interest to determine the presence of any morphologic peculiarities of such cells. Monocytes adherent onto microexudate-coated flasks, however, are quite similar in appearance to cells on uncoated flasks.

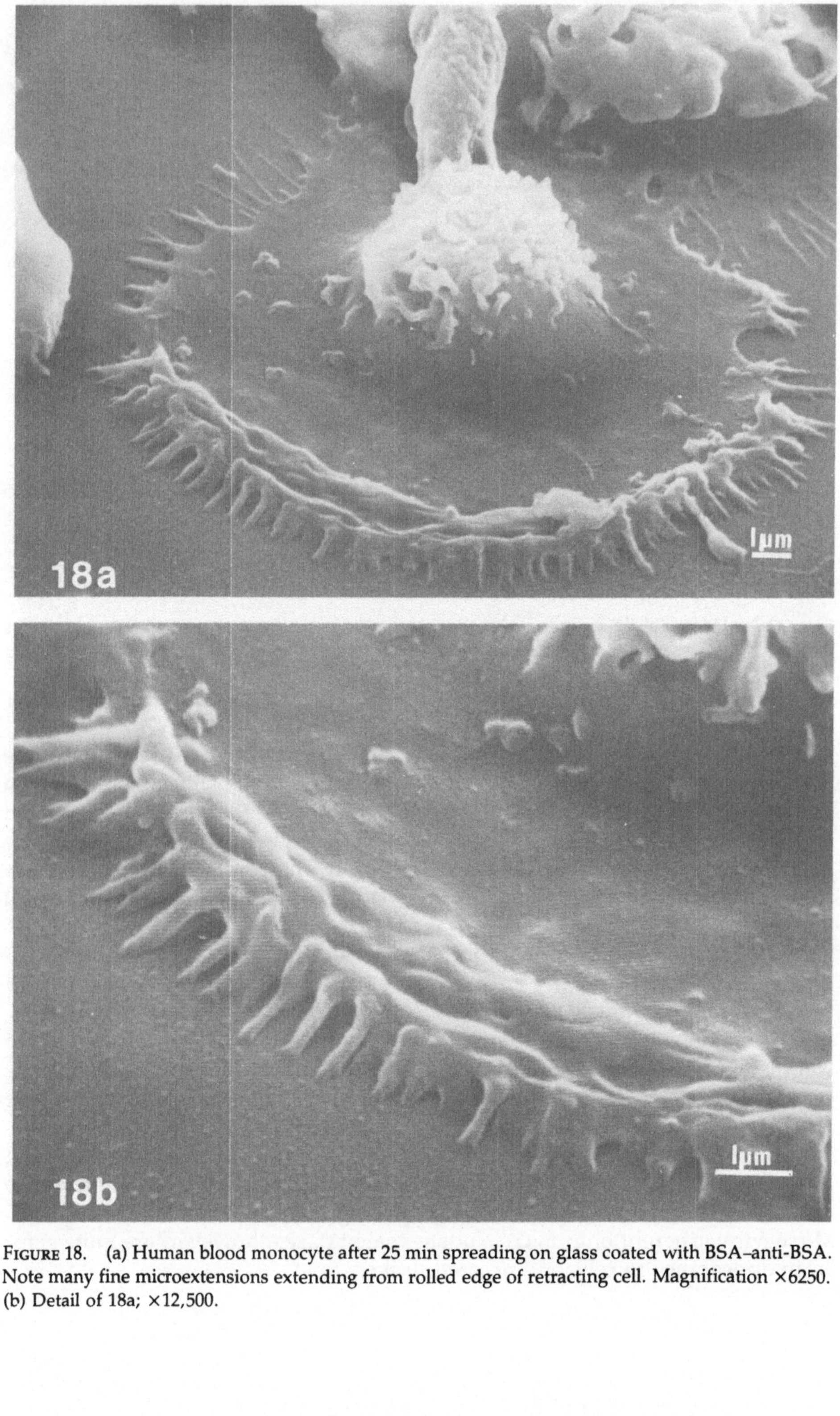

FIGURE 18. (a) Human blood monocyte after 25 min spreading on glass coated with BSA–anti-BSA. Note many fine microextensions extending from rolled edge of retracting cell. Magnification ×6250. (b) Detail of 18a; ×12,500.

Although there has been extensive morphologic, electron microscopic, and biochemical characterization of attachment of animal phagocytes to various surfaces for many years (Lucké *et al.*, 1933; Fenn, 1922), relatively little has been reported on the blood monocyte. In particular, studies on the roles of contractile and cytoskeletal structures in this process in blood monocytes are notably lacking, undoubtedly due, in part, to difficulties in procurement of large numbers of cells in acceptable purity. For a review of the neutrophil cytoskeleton, see Oliver (1978); for a review of the macrophage, see Hartwig *et al.* (1977) and this volume.

Recent investigations have revealed that monocytes may undergo characteristic metabolic alterations during adherence to a substratum. Johnston *et al.* (1976) have, for example, shown that human monocytes emit a burst of chemiluminescence upon adherence to glass-bound IgG, which presumably reflects a surface-stimulated respiratory burst similar to that seen in neutrophils and macrophages (Selvaraj and Sbarra, 1966) during phagocytosis. Another interesting consequence of monocyte adherence has recently been reported by Bodel *et al.* (1977) for rabbit, rat, and human blood monocytes. These workers have shown that during short-term *in vitro* culture, monocytes develop peroxidase reactivity within the rough endoplasmic reticulum (Bodel *et al.*, 1977) which is dependent on attachment to the culture surface. This showed that

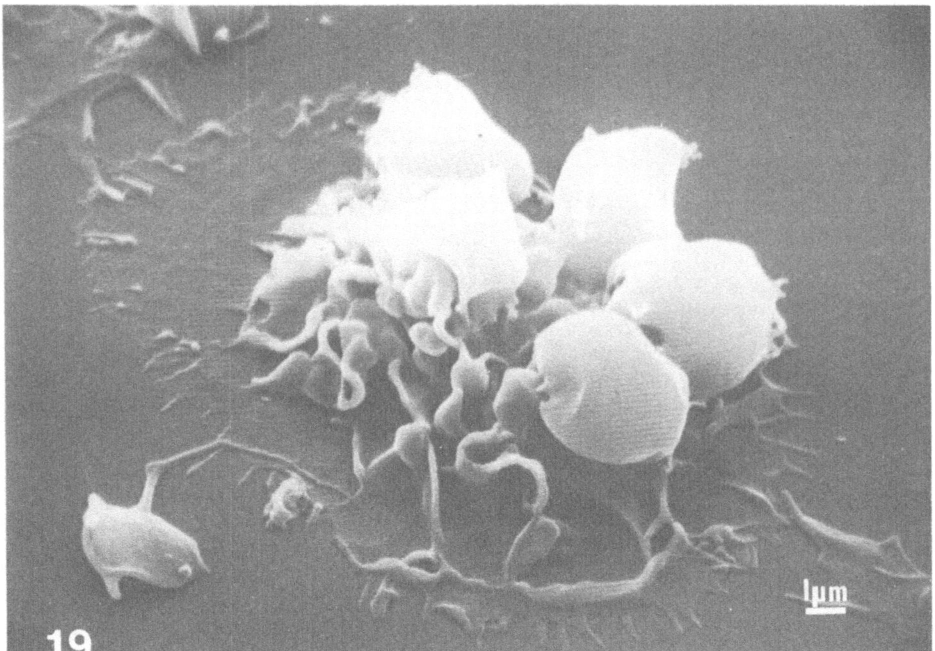

FIGURE 19. Human blood monocyte. Adherence of sheep erythrocyte coated with IgM and complement to the monocyte C3 receptor. Characteristically, adherence is preferentially over the cell body and there is no evidence of phagocytosis. ×2750.

peroxidase localization may vary with the functional state of the cell, a point of potential importance in consideration of monocyte ontogeny. Daems *et al.* (1976), for instance, have argued that although monocytes have clearly been shown to be the immediate predecessor of inflammatory or exudate macrophages, this may not be the case for resident macrophages. Different subcellular localization of peroxidase—endoplasmic reticular in the resident macrophage and lysosomal in the blood monocyte—has been used as evidence that the blood monocyte is not an immediate precursor of the resident macrophage (van der Rhee *et al.*, 1977; Daems and Koerten, 1978). However, the finding that peroxidase localization can vary with the functional state of the MNP reduces the force of this argument (Bodel *et al.*, 1977).

3.2. ADHERENCE AND PHAGOCYTOSIS OF PARTICLES (FIGURES 19–23)

Most mononuclear phagocytes, including human peripheral blood monocytes, have cell surface receptors for immunoglobulins and complement.

FIGURE 20. Human blood monocyte interacting with IgG-coated sheep erythrocytes. In contrast to Figure 19, note prominent phagocytosis as shown by a swollen, "lumpy" cell body. There is loss of surface ruffles as compared to Figure 19, indicating internalization of membrane during phagocytosis. Note also much smaller amount of plasma membrane in contact with the substratum than in Figure 19; phagocytosis of EA is a potent stimulus for monocyte detachment. ×2700.

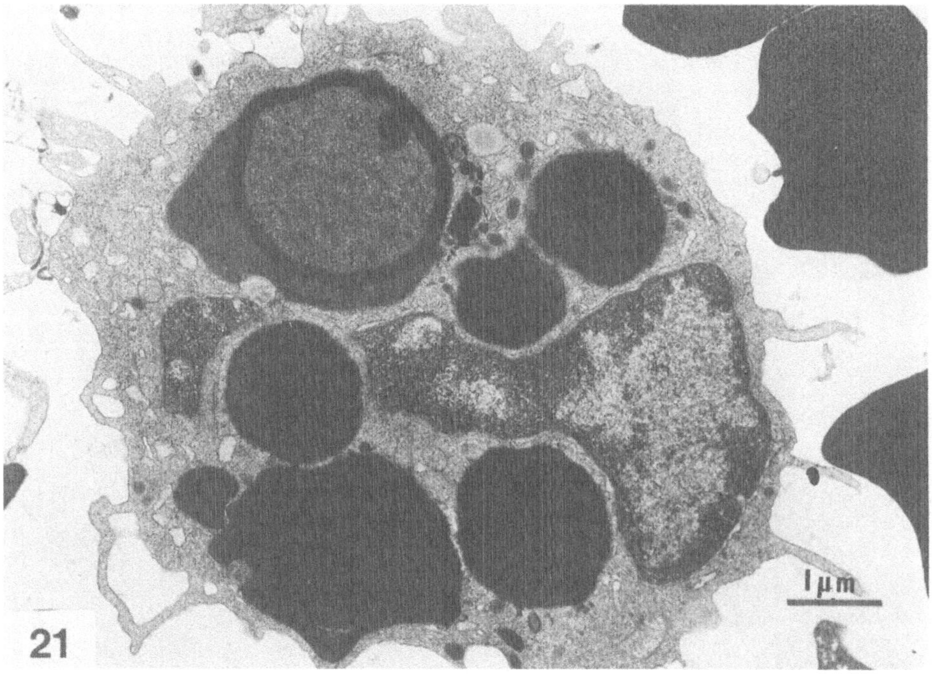

FIGURE 21. Human blood monocyte containing phagocytized EA. Note paucity of remaining granules; those that remain are close to large phagolysosomes with which they fuse following phagocytosis. ×13,000.

Human monocytes have receptors for the Fc portion of IgG, particularly of the IgG_1 and IgG_3 subclasses (Huber and Fudenberg, 1968; Huber *et al.*, 1971), as well as for the third component of complement, the C3 receptor (Huber *et al.*, 1968). When human monocytes are exposed to a particle, for example, sheep erythrocytes coated with IgM and complement, the red cells adhere via the monocyte C3 receptors. Adherence is predominantly to the lamellipodia, and phagocytosis of the erythrocyte is not usually seen. In contrast, when the erythrocyte is coated with IgG, red cells adhere and are rapidly phagocytized. During phagocytosis, lamellipodia may be seen surrounding the particles, then fusing to enclose it in a phagocytic vacuole or phagosome. Following this event, fusion of the phagosome with intracellular granules occurs and the erythrocyte is enzymatically digested. Similar events occur following ingestion of other particles such as *E. coli* or latex particles whose interaction with the monocyte does not depend on specific surface receptors.

It should be mentioned that phagocytosis is a cellular function not limited to organisms possessing a lymphoid system, but rather is phylogenetically ancient. Single celled protozoans such as amebae are avid phagocytes, and virtually every organism above this level has cells capable of active phagocytosis, which

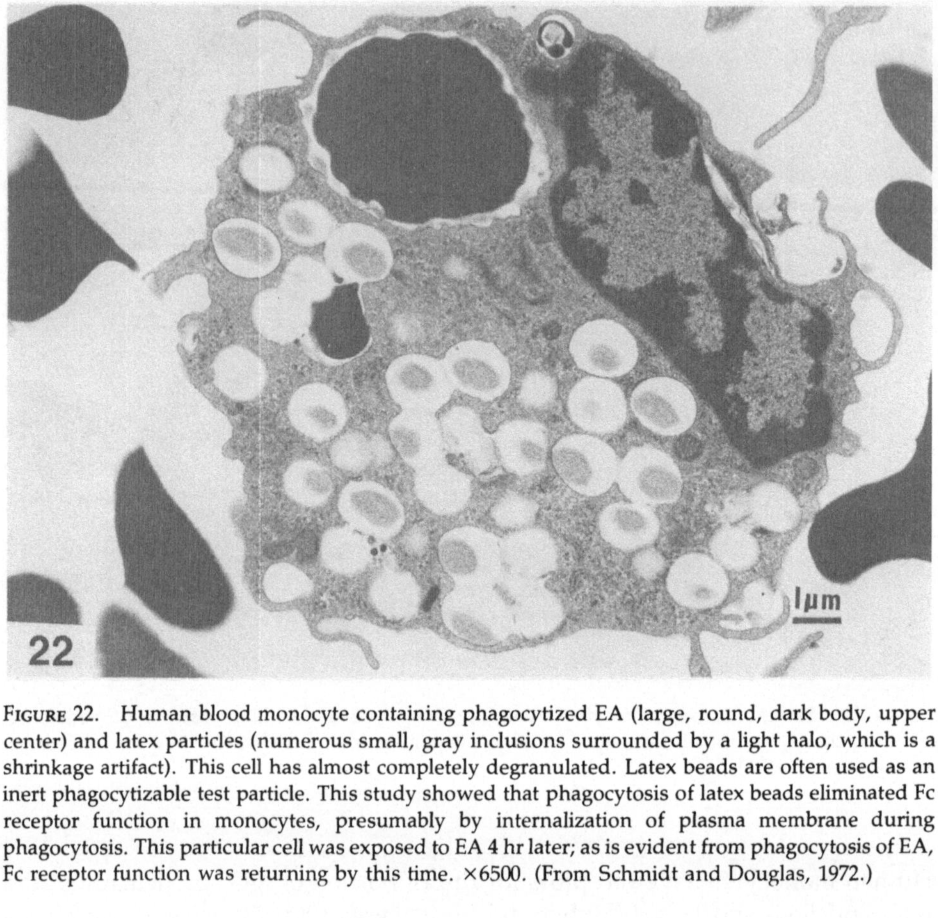

FIGURE 22. Human blood monocyte containing phagocytized EA (large, round, dark body, upper center) and latex particles (numerous small, gray inclusions surrounded by a light halo, which is a shrinkage artifact). This cell has almost completely degranulated. Latex beads are often used as an inert phagocytizable test particle. This study showed that phagocytosis of latex beads eliminated Fc receptor function in monocytes, presumably by internalization of plasma membrane during phagocytosis. This particular cell was exposed to EA 4 hr later; as is evident from phagocytosis of EA, Fc receptor function was returning by this time. ×6500. (From Schmidt and Douglas, 1972.)

frequently serves a protective function. Phagocytosis at all phylogenetic levels is characterized by at least a certain amount of discriminative ability. Thus, even at the mollusk level there is evidence of ability to distinguish between species (Cooper, 1976). The emergence of the lymphoid immune system at the level of the vertebrates may be partially seen as a mechanism to amplify, remember, and impart further specificity to these more primitive defense mechanisms. Figures 24–26 show several examples of invertebrate phagocytes.

3.3. DETACHMENT FROM *IN VITRO* SURFACES

In general blood monocytes attach to *in vitro* surfaces more rapidly and tenaciously than many other cells and they likewise resist detachment. Whereas trypsin readily detaches most fibroblasts from surfaces *in vitro,* it does not affect adhesion of mononuclear phagocytes. Chelators of divalent cations, such as

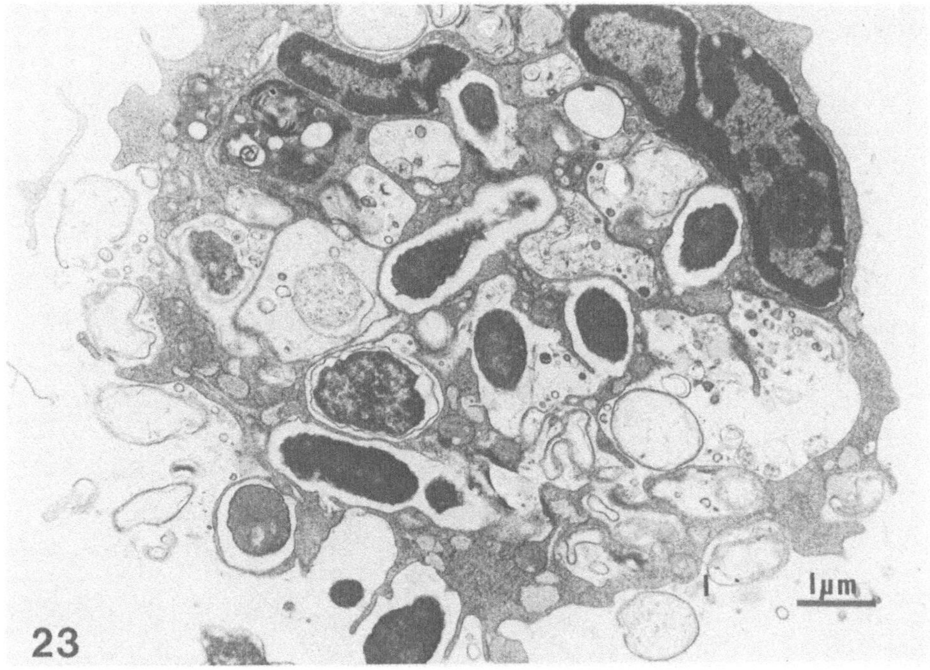

FIGURE 23. Human blood monocyte after exposure to *E. coli.* Note extensive phagocytosis with bacteria in all stages of intracellular digestion. ×11,250.

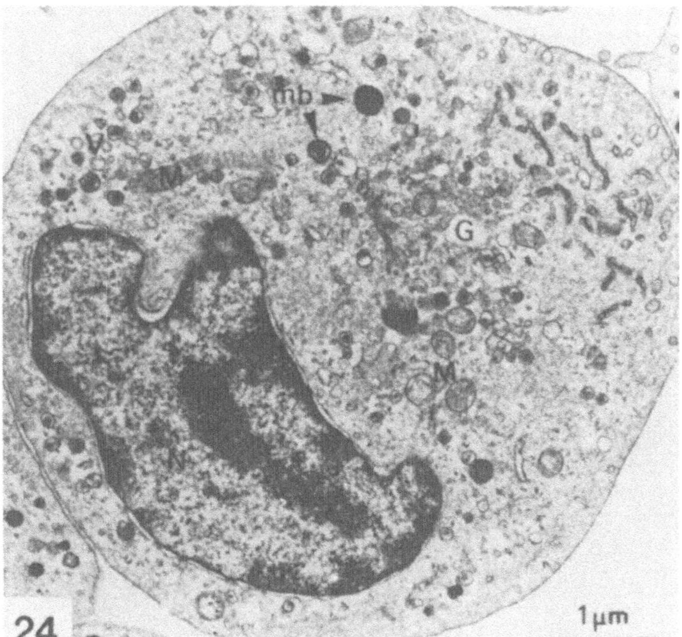

FIGURE 24. Blood monocyte from the hagfish, considered to be the most phylogenetically primitive vertebrate living. Note lysosomal vesicles (v), myelin bodies (mb), Golgi (G) and mitochondria (m). (From Linthicum, 1975.)

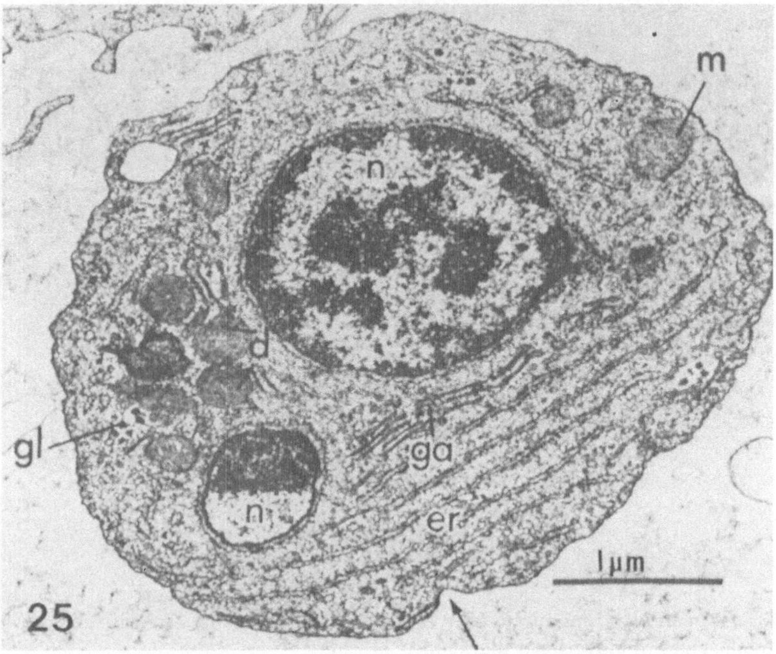

FIGURE 25. Circulating hemocyte from the chiton mollusk *Liolophura*. Ga, Golgi apparatus; n, nucleus; m, mitochondria; d, dense bodies of uncertain significance; er, endoplasmic reticulum. Arrow denotes pinocytic invagination. (From Killby *et al.*, 1973.)

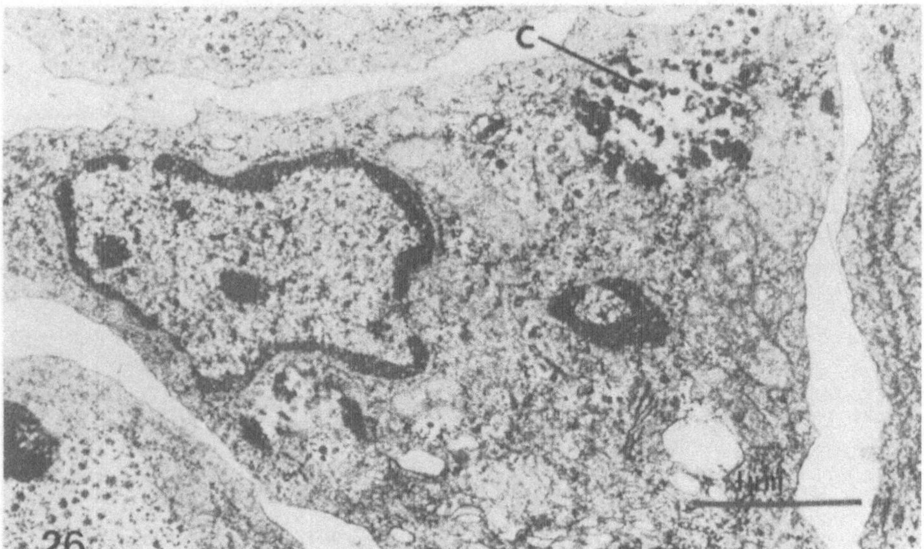

FIGURE 26. Another example of chiton hemocyte. This cell contains ingested carbon particles (c). (From Killby *et al.*, 1973.)

EDTA or EGTA, are also effective at detaching fibroblastlike cells, but relatively inactive with monocytes under usual conditions. It is clear, however, that during more physiologic processes the monocyte membrane may detach from the substratum, as during, for example, chemotaxis *in vitro*. We, therefore, have been studying characteristics of monocyte adherence on a variety of surfaces. As noted above, we have observed that human monocytes adhere to microexudate-coated surfaces and detach easily when treated with EDTA. SEM (Figure 27) of the detachment process shows that immediately following addition of EDTA the adherent membrane begins retraction toward the cell body. Within 10 min the cell remains attached only by a small tuft of lamellipodia under the cell body, which itself soon detaches from the surface. The retraction process is occasionally accompanied by formation of thin (200–400 Å) retraction fibrils which remain attached to the substratum and may break as the monocyte lifts from the surface. The mechanism of chelator-mediated detachment for any cell type is not known, but presumably involves disruption of the internal supporting cytoskeleton followed by collapse of the extended portions of the cell. Easier detachment of monocytes from microexudate-coated than from plastic surfaces is presumably due to decreased tenacity of adhesion to the former surface. SEM of monocytes with EDTA treatment shows qualitatively the same changes on plastic as on microexudate, except to a much lesser extent; many cells look essentially unchanged even 30 min after addition of chelator, though the few cells that do detach undergo the same morphologic changes (Figure 27).

We have also investigated chelator-mediated detachment of monocytes cultivated *in vitro* for periods up to 4 weeks. As noted above, by this time, cells are very large, from 40–100 μm, and many multinucleate. They appear as well-spread cells with large perinuclear collections of phase-lucent and phase-dense granules, and with a surrounding granule-free cytoplasmic rim. The susceptibility of these cells to EDTA-mediated detachment depends on numerous factors. Temperatures below 37°C favor rounding and detachment, whereas cells with greater membrane area in contact with the substratum resist detachment. Figure 28 shows EDTA-mediated detachment of human peripheral blood monocytes which had been cultured 4 weeks *in vitro*. As viewed by SEM the first stage in EDTA-induced detachment of cultured human blood monocytes is retraction of the peripheral cytoplasm toward the cell body. In contrast to the freshly isolated monocyte, retraction fibers in great numbers almost invariably appear during retraction, and the peripheral cytoplasm seems to recede along and between these fibrils. The cell body accumulates many blebs presumably reflecting acquisition of redundant membrane. During the latter stages of retraction (Figures 28 and 29) the cell body assumes a "mulberry" appearance and is attached to the substratum by radial retraction fibers, each angling down from the cell body to the surface. At the point of contact, the fiber becomes thinner and extends further from the cell body. These thin fibrils differ from the larger ones nearer the cell body in that they are all of uniform size (about 200 Å) and they are closely applied to the substratum along their entire length. Thus, these thin fibrils may contain cytoskeletal elements important in anchorage of the cultured monocytes to the surface.

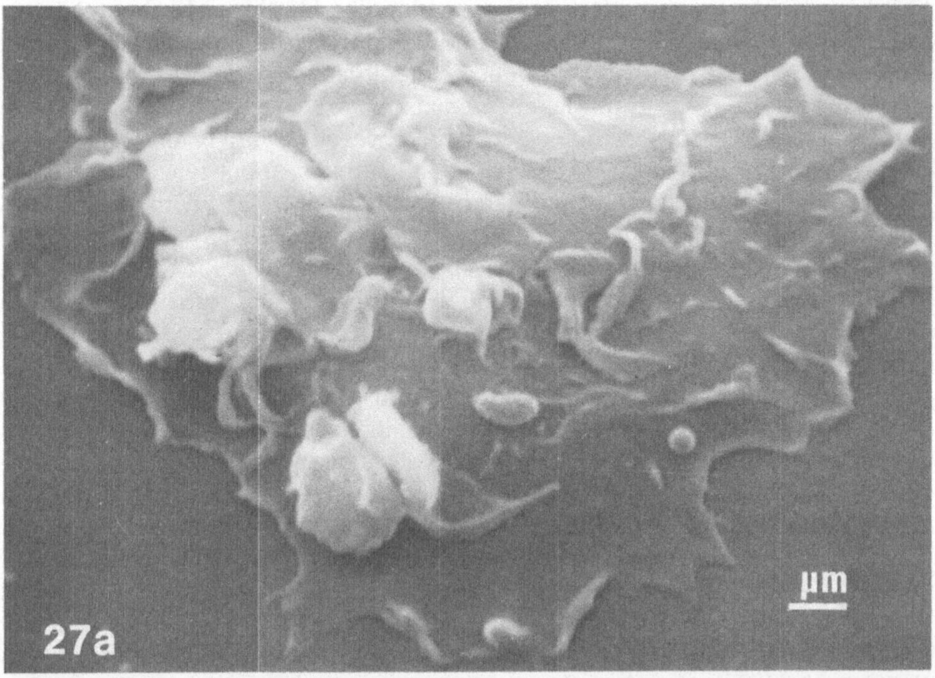

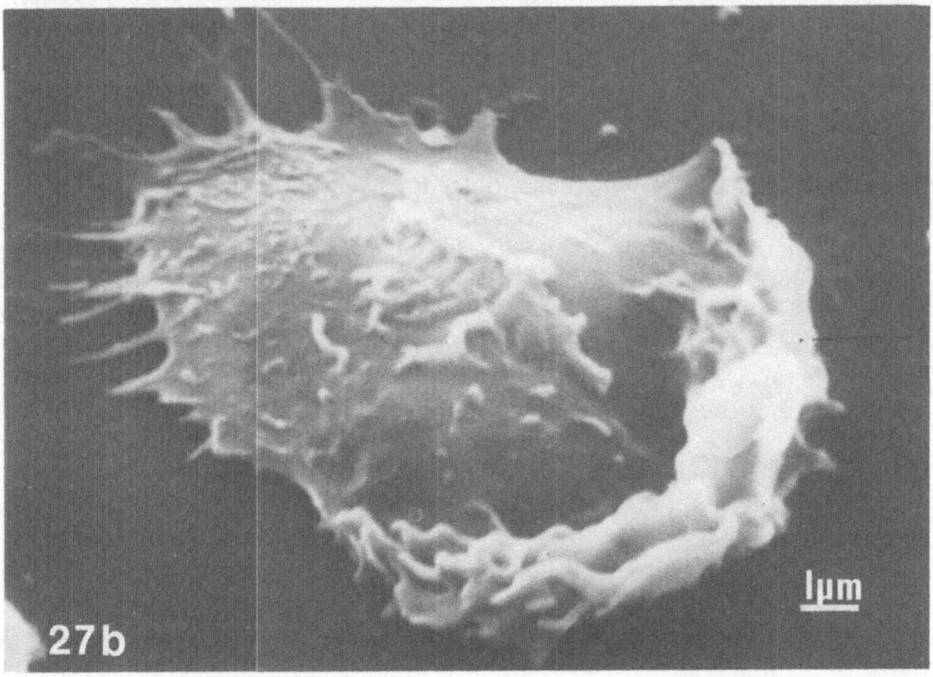

FIGURE 27. (a) Human blood monocyte after adherence to microexudate-coated flask for 1 hr. The cell is well-extended over the substratum and similar in appearance to cells on plain plastic surfaces, as shown in Figure 3. (b) Monocyte on microexudate after 1 min exposure to EDTA; cell has begun to

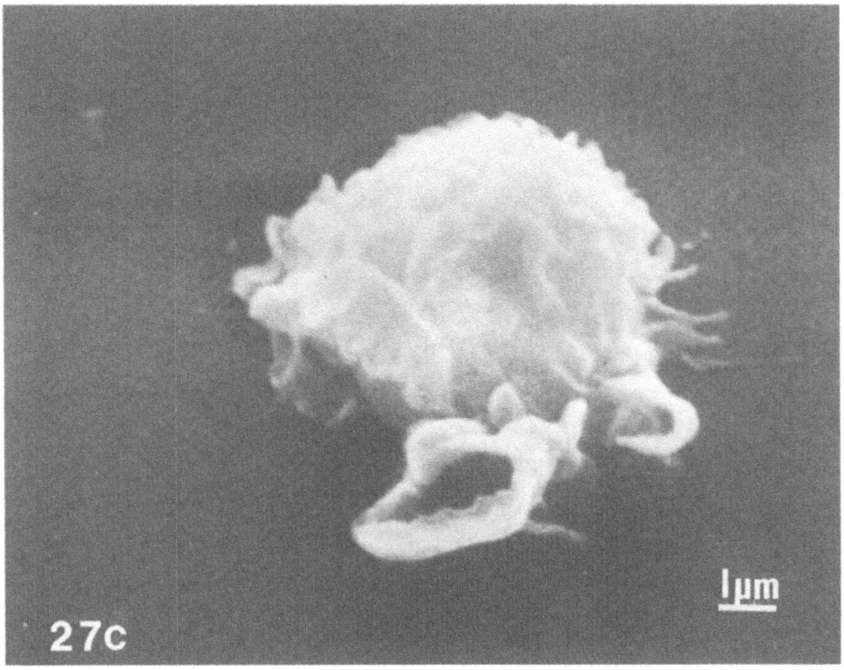

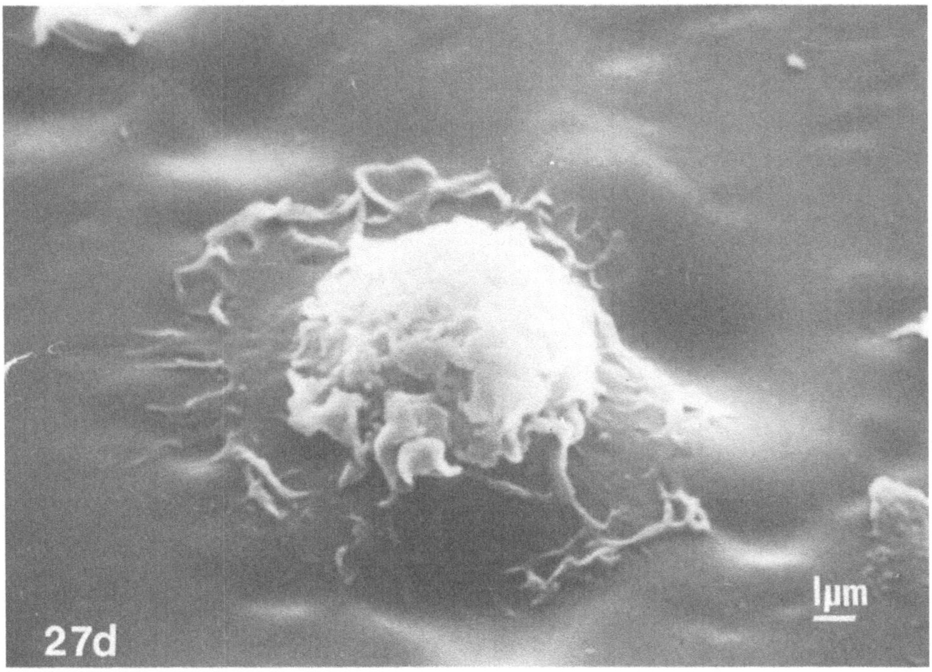

retract. (c) After 3 min exposure to EDTA; monocyte nearly totally rounded. (d) In contrast, monocyte plated on uncoated plastic remains extended even after 10 min in EDTA.

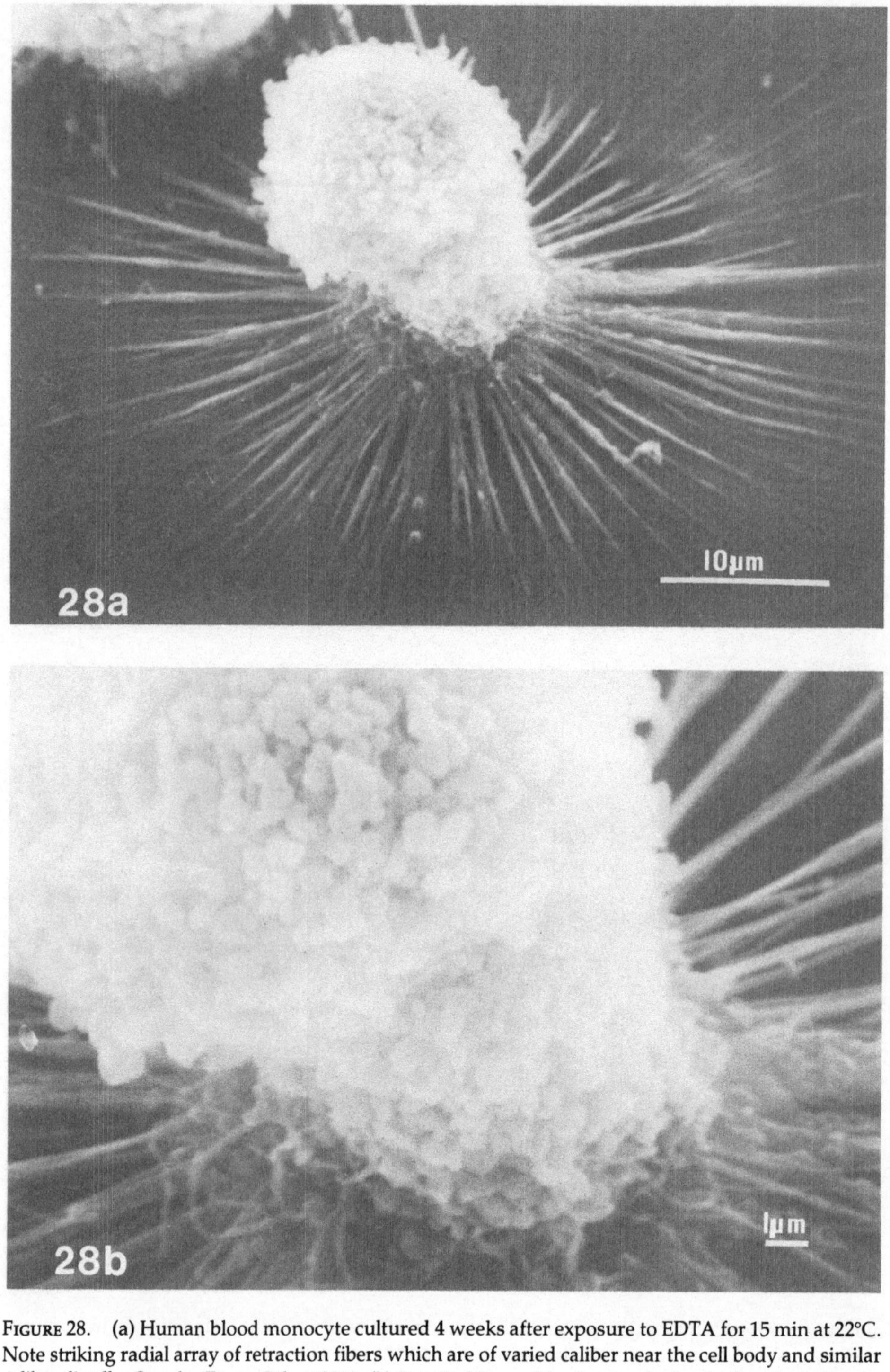

FIGURE 28. (a) Human blood monocyte cultured 4 weeks after exposure to EDTA for 15 min at 22°C. Note striking radial array of retraction fibers which are of varied caliber near the cell body and similar caliber distally. See also Figure 29b. ×2500. (b) Detail of Figure 28a showing body of cell with numerous "mulberry" blebs. ×6250.

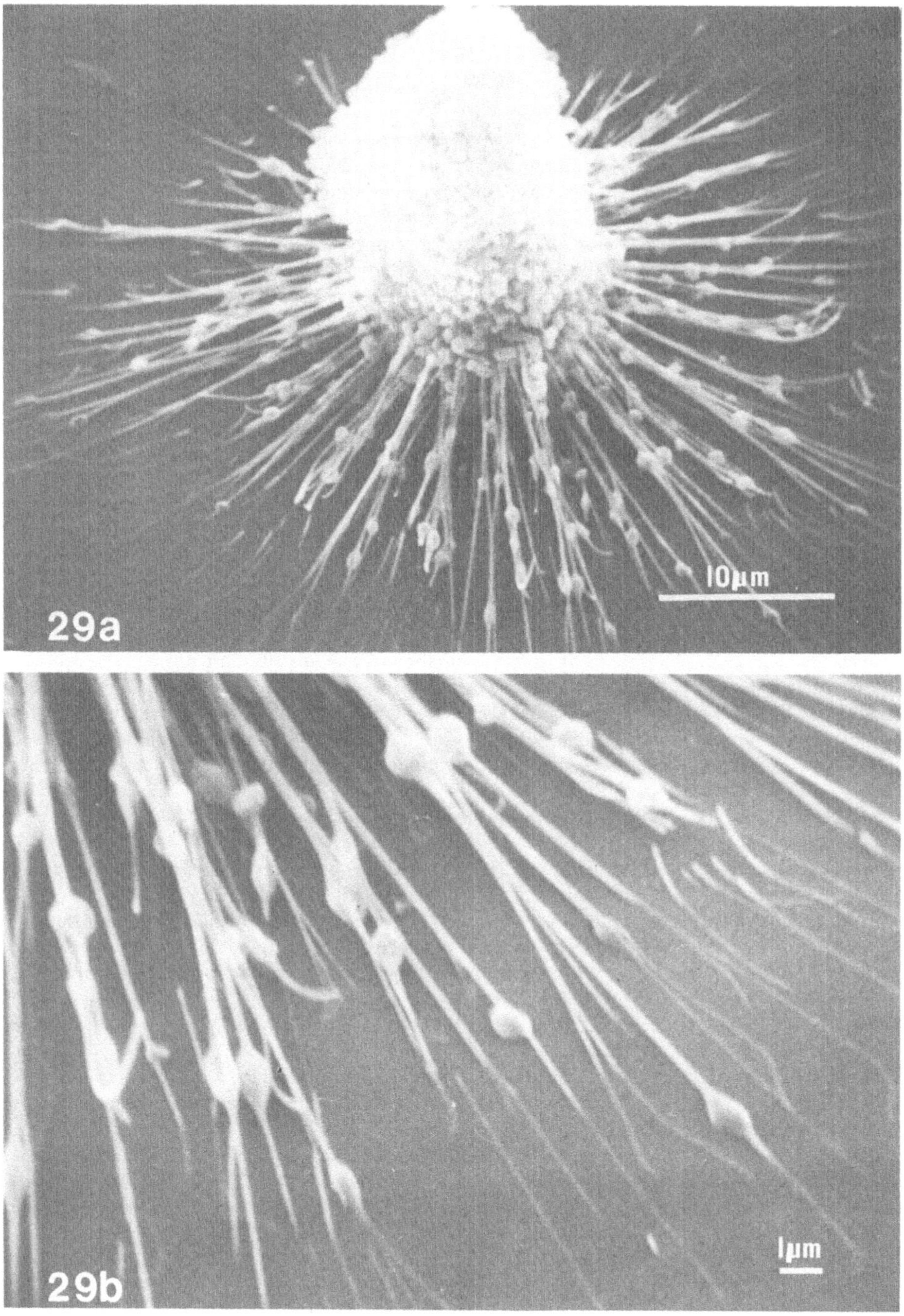

FIGURE 29. (a) Human blood monocyte, 4 week culture in EDTA for 15 min at 22°C. This cell is similar in appearance to that in Figure 28, except that the thicker fibrils here contain numerous nodules, which appear to contribute to the surface blebbing as the fibers are withdrawn to the cell body. ×2500. (b) Detail of Figure 29a. Note the thinner filaments (about 200 Å) are adherent to the substratum along their entire length. These may represent principal anchorage sites in the intact extended cell. ×6200.

REFERENCES

Ackerman, S. K., and Douglas, S. D., 1978, Purification of human monocytes on microexudate-coated surfaces, *J. Immunol.* **120:**1372.

Albrecht-Buehler, G., and Goldman, R. D., 1976, Microspike-mediated particle transport towards the cell body during early spreading of 3T3 cells, *Exp. Cell Res.* **97:**329.

Bessis, M., 1977, *Blood Smears Reinterpreted,* Springer International, p. 159.

Bianco, C., Griffin, F. M., Jr., and Silverstein, S. C., 1975, Studies of the macrophage complement receptor—alteration of receptor function upon macrophage activation, *J. Exp. Med.* **141:**1278.

Bodel, P. T., Nichols, B. A., and Bainton, D. F., 1977, Appearance of peroxidase reactivity within the rough ER of blood monocytes after surface adherence, *J. Exp. Med.* **145:**264.

Braunsteiner, H., and Schmalzl, F., 1970, Cytochemistry of monocytes and macrophages, in: *Mononuclear Phagocytes* (R. van Furth, ed.), pp. 62–80, Blackwell, London.

Bumol, T. F., and Douglas, S. D., 1977, Human monocyte cytoplasmic spread *in vitro*—early kinetics and scanning electron microscopy, *Cell. Immunol.* **34:**70.

Cooper, E. L., 1976, *Comparative Immunology,* Prentice Hall, Clinton, New Jersey.

Daems, W. T., and Koerten, H. K., 1978, Effects of various stimuli on the cellular composition of peritoneal exudates in the mouse, *Cell Tissue Res.* **190:**47.

Daems, W. T., Koerten, H. K., and Soranzo, M. R., 1976, Differences between monocyte derived and tissue macrophages, in: *The Reticuloendothelial System in Health and Disease: Functions and Characteristics* (S. M. Reichan, M. R. Escobar and H. Friedman, eds.), pp. 27–40, Plenum Press, New York.

Douglas, S. D., 1976, Human monocyte spreading *in vitro:* Inducers and effects on Fc and C3 receptors, *Cell. Immunol.* **21:**344.

Douglas, S. D., 1978, Alterations in intramembrane particle distribution during interaction of erythrocyte-bound ligands with immunoprotein receptors, *J. Immunol.* **120:**151.

Douglas, S. D., Zuckerman, S. H., and Cody, C. S., 1980, Alterations in intramembrane particle distribution during interaction of erythrocyte-bound ligands with immunoprotein receptors. II. Effect of the membrane mobility agent A_2C on immunologic and non-immunologic ligand binding, *J. Reticuloendothel. Soc.* (in press).

Fenn, W. O., 1922, The theoretical response of living cells to contact with solid bodies, *J. Gen. Physiol.* **4:**373.

Hartwig, J. H., and Stossel, T. P., 1975, Isolation and properties of actin, myosin and a new actin-binding protein in rabbit alveolar macrophages, *J. Biol. Chem.* **250:**5696.

Hartwig, J. H., Davies, W. A., and Stossel, T. P., 1977, Evidence for contractile protein translocation in macrophage spreading, phagocytosis and phagolysosome fermentation, *J. Cell Biol.* **75:**956.

Horn, R. G., and Spicer, S. S., 1964, Sulfated mucopolysaccharide and basic protein in certain granules of rabbit leukocytes, *J. Lab. Invest.* **13:**1.

Huber, H., and Fudenberg, H. H., 1968, Receptor sites of human monocytes for IgG, *Int. Arch. Allerg.* **34:**18.

Huber, H., Polley, M. G., Linscott, W. D., H. H. Fudenberg, and H. J. Müller-Eberhard, 1968, Human monocytes: Two distinct receptor sites for the third component of complement and for γG globulin, *Science* **162:**1281.

Huber, H., Douglas, S. D., Nusbacher, J., Kochwa, S., and Rosenfield, R. E., 1971, IgG subclass specificity of human monocyte receptor sites, *Nature* **229:**419.

Johnston, R. B., Lehmeyer, J. E., and Guthrie, L. A., 1976, Generation of superoxide and chemiluminescence by human monocytes, *J. Exp. Med.* **143:**1551.

Killby, V.A.A., Crichten, R., and Lafferty, K. J., 1973, Fine structure of phagocytic cells in the chiton *Liolophura gaimardi, Aust. J. Exp. Biol. Med. Sci.* **51:**373.

Li, C. Y., Lam, K. W., and Yam, L. T., 1973, Esterases in human leukocytes, *J. Histochem. Cytochem.* **21:**1.

Linthicum, D. S., 1975, Ultrastructure of hagfish leukocytes, *Adv. Exp. Med. Biol.* **64:**241.

Lucké, B., Strumia, M., Mudd, S., M. McCutcheon, and E.B.H. Mudd, 1933, On the comparative phagocytic activity of macrophages and polymorphonuclear leukocytes, *J. Immunol.* **24:**455.

Nichols, B. A., and Bainton, D. F., 1973, Differentiation of human monocytes in bone marrow and blood: Sequential formation of two granule populations, *Lab. Invest.* **29:**27.
Nichols, B. A., and Bainton, D. F., 1975, Ultrastructure and cytochemistry of mononuclear phagocytes, in: *Mononuclear Phagocytes in Immunity, Infection, and Pathology* (R. van Furth, ed.), pp. 17–56, Blackwell, London.
Nichols, B. A., Bainton, D. F., and Farquahr, M. G., 1971, Differentiation of monocytes. Origin, nature and fate of their azurophil granules, *J. Cell Biol.* **50:**498.
Oliver, J. M., 1978, Cell biology of leukocyte abnormalities—membrane and cytoskeletal function in normal and defective cells, *Am. J. Pathol.* **93:**221.
Parmley, R. T., Spicer, S. S., and O'Dell, R. F., 1978, Ultrastructural indentification of acid complex carbohydrate in cytoplasmic granules of normal and leukaemic human monocytes, *Br. J. Haematol.* **39:**33.
Ploem, J. S., 1975, Reflection contrast microscopy as a tool in investigations of the attachment of living cells to a glass surface, in: Mononuclear Phagocytes in Immunity, Infection, and Pathology (van Furth, R., ed.), pp. 405–422, Blackwell, London.
Rabinovitch, M., and DeStefano, M. J., 1973a, Macrophage spreading *in vitro.* I. Inducers of spreading, *Exp. Cell Res.* **77:**323.
Rabinovitch, M., and DeStefano, M. J., 1973b, Macrophage spreading *in vitro.* II. Manganese and other metals as inducers or as co-factors for inducing spreading, *Exp. Cell Res.* **79:**423.
Reaven, E. P., and Axline, S. G., 1973, Subplasmalemmal microfilaments and microtubules in resting and phagocytizing cultivated macrophages, *J. Cell Biol.* **59:**12.
Schmidt, M. E., and Douglas, S. D., 1972, Disappearance and recovery of human monocyte IgG receptor activity after phagocytosis, *J. Immunol.* **109:**914.
Selvaraj, R. J., and Sbarra, A. J., 1966, Relationship of glycolytic and oxidative metabolism to particle entry and destruction in phagocytosing cells, *Nature* **211:**1272.
Spitznagel, J. K., 1975, Advances in study of cytoplasmic granules of human polymorphonuclear leukocytes, in: *The Phagocytic Cell in Host Resistance* (J. A. Bellanti and D. H. Dayton, eds.), pp. 77–85, Raven Press, New York.
Totterman, T. H., Ranki, A., and Hayry, P., 1977, Expression of the acid alpha-naphthyl acetate esterase marker by activated and secondary lymphocytes in man, *Scand. J. Immunol.* **6:**305.
van der Rhee, H. J., de Winter, C.P.M., and Daems, W. T., 1977, Fine structure and peroxidatic activity of rat blood monocytes, *Cell Tissue Res.* **185:**1.
van Furth, R., Langevoort, H. L., and Schaberg, A., 1975, Mononuclear phagocytes in human pathology—proposal for an approach to improved classification, in: *Mononuclear Phagocytes in Immunity, Infection, and Pathology* (R. van Furth, ed.), pp. 1–16, Blackwell, London.
van Oss, C. J., Gillman, C. F., and Neuman, A. W., 1975, *Phagocytic Engulfment and Adhesiveness as Cellular Surface Phenomena,* Dekker, New York.
Weinstein, R. S., Khodadad, J. K., and Steck, T. L., 1978, Ultrastructural characterization of proteins at the natural surfaces of the red cell membrane, in: *The Red Cell* (G. J. Brewer, ed.), pp. 413–427, Alan Liss, New York.
Zuckerman, S. H., Ackerman, S. K., and Douglas, S. D., 1979, Long term peripheral blood monocyte cultures: Establishment, metabolism and morphology of primary human monocyte-macrophage cell culture, *Immunology* **38:**401–411.

8

Bone Marrow Macrophages

G. HUDSON and J. R. SHORTLAND

1. INTRODUCTION

Macrophages within the bone marrow constitute an important component of the mononuclear phagocytic system. Not only do they have proven phagocytic properties along with other macrophages, but, because of their unique juxtaposition to the developing blood cells, they are intimately concerned with hemopoiesis. Macrophage precursors constitute approximately 15% of the total macrophage population in bone marrow (Hirsch and Fedorko, 1970), and labeling experiments (Wickramasinghe and Hughes, 1978) indicate that only a small number, less than 1%, are active mitotically. Nevertheless, they are a dynamic population responding briskly to a variety of stimuli (van Furth and Cohn, 1968).

It is generally accepted that macrophages or phagocytic reticular cells are to be found in four situations within the bone marrow: within the erythroblastic islands; within the cords among other hemopoietic cells; applied to vascular sinuses; and, occasionally, free within the lumen of a sinus. Cells adjacent to the sinus extend pseudopodial projections through the endothelial cells into the vascular lumen.

The purpose of the chapter is to discuss the morphological aspects of the marrow macrophages and to relate them to their functional characteristics in the light of these extensive relationships within the marrow cavity.

2. GENERAL FEATURES

2.1. LIGHT MICROSCOPY

The macrophage within the bone marrow is a truly reticulated cell in that it sends out dendritic processes between the adjacent hemopoietic cells (Hudson

G. HUDSON • University Department of Hematology, Royal Hallamshire Hospital, Sheffield, S10 2JF, England. J. R. SHORTLAND • Department of Pathology, University of Sheffield, Sheffield, S10 2RX, England.

and Yoffey, 1963). This morphological interrelationship is not appreciated in marrow smear preparations (Bessis, 1973), where, due to the forces generated during aspiration and spreading, the reticular cells fragment. Sections of bone marrow reveal the reticular cells as large, mononuclear, rather faceted, cells surrounded by erythroblastic or myeloblastic cells. They are more easily identified after injection of a particulate material such as carbon, when the extent of the population becomes apparent (Hudson and Yoffey, 1963, 1968).

Their size increases with maturation (van Furth and Fedorko, 1976), mature forms measuring in the region of 15-17 μm diameter. In common with macrophages elsewhere, their size varies tremendously with the functional state. Metcalf and Wilson (1976) report an increase in size in response to the injection of endotoxin, and cells larger than 20 μm diameter are reported by Wickramasinghe and Hughes (1978) in thalassemia.

Earlier forms are rather smaller. The monoblast measures 8–10 μm diameter (van Furth and Fedorko, 1976). The cell progressively enlarges to the promonocyte state which measures 12–15 μm.

Phase contrast microscopy gives an image of the whole cell precluding the obvious sampling problem inherent in electron microscopic techniques.

2.2. ELECTRON MICROSCOPY

Basically two main methods of fixation have been used when studying the bone marrow: immersion and perfusion using glutaraldehyde and post-fixation with osmium tetroxide. Immersion into glutaraldehyde results in some distortion and disruption of the sinusoids, but except in those areas thus traumatized, the configuration of the macrophages and the overall architecture is well preserved. The perfusion technique has been used recently by Weiss (1976). As with most perfusion techniques, fixation is not uniform. In specimens taken from the perfused and well-fixed areas, Weiss has shown a gradation of fixation with a greater degree of cell separation around the venous sinusoid than around arteries.

The bone marrow macrophages (Figures 1 and 2) share common ultrastructural features with macrophages elsewhere (Hirsch and Fedorko, 1970). The external profile is characteristically irregular with dendritic processes bearing many irregular projections and indentations. The cytoplasm is dominated by vesicular structures and lysosomal granules.

The complement of cytoplasmic organelles varies in relation to the functional activity and state of maturity. The most primitive form, the monoblast, recently characterized in tissue culture by van Furth and Fedorko (1976), has relatively little cytoplasm. The cytoplasm contains abundant polyribosomes and relatively few granules. Scanning electron microscopy (SEM) has shown a paucity of cell projections, which, together with the insignificant number of lysosomes, correlates with the poor phagocytic activities of the monoblast. As the cell matures the number of polyribosomes decreases, while

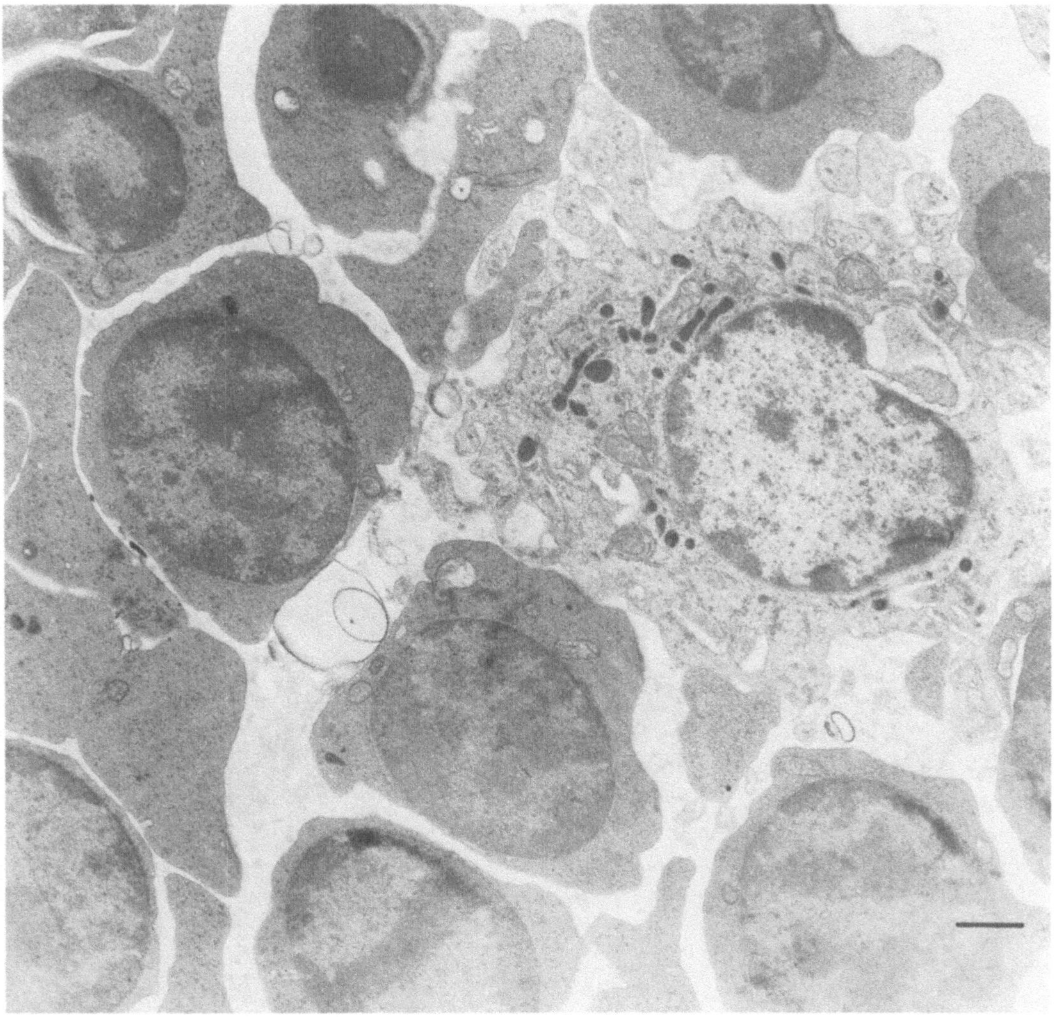

FIGURE 1. Central reticular cell in an erythroblastic island of normal monkey marrow. It has a reniform nucleus and abundant cytoplasm containing prominent lysosomes and vacuoles. The cell is surrounded by a ring of erythroblasts. The bar indicates 1 μm.

sparse fragments of rough endoplasmic reticulum (RER) appear together with small electron-dense granules.

The nucleus tends to be round, oval, or reniform in the precursor cells, taking on a more irregular profile with maturity and when stimulated. The nucleoplasm has a medium electron density with a prominent nucleolus, and heterochromatin that is grouped along the inner aspect of the nuclear membrane in glutaraldehyde-fixed cells. The nuclear membrane displays occasional nuclear pores with ribosomes adherent to the outer membrane. Daems and Brederoo

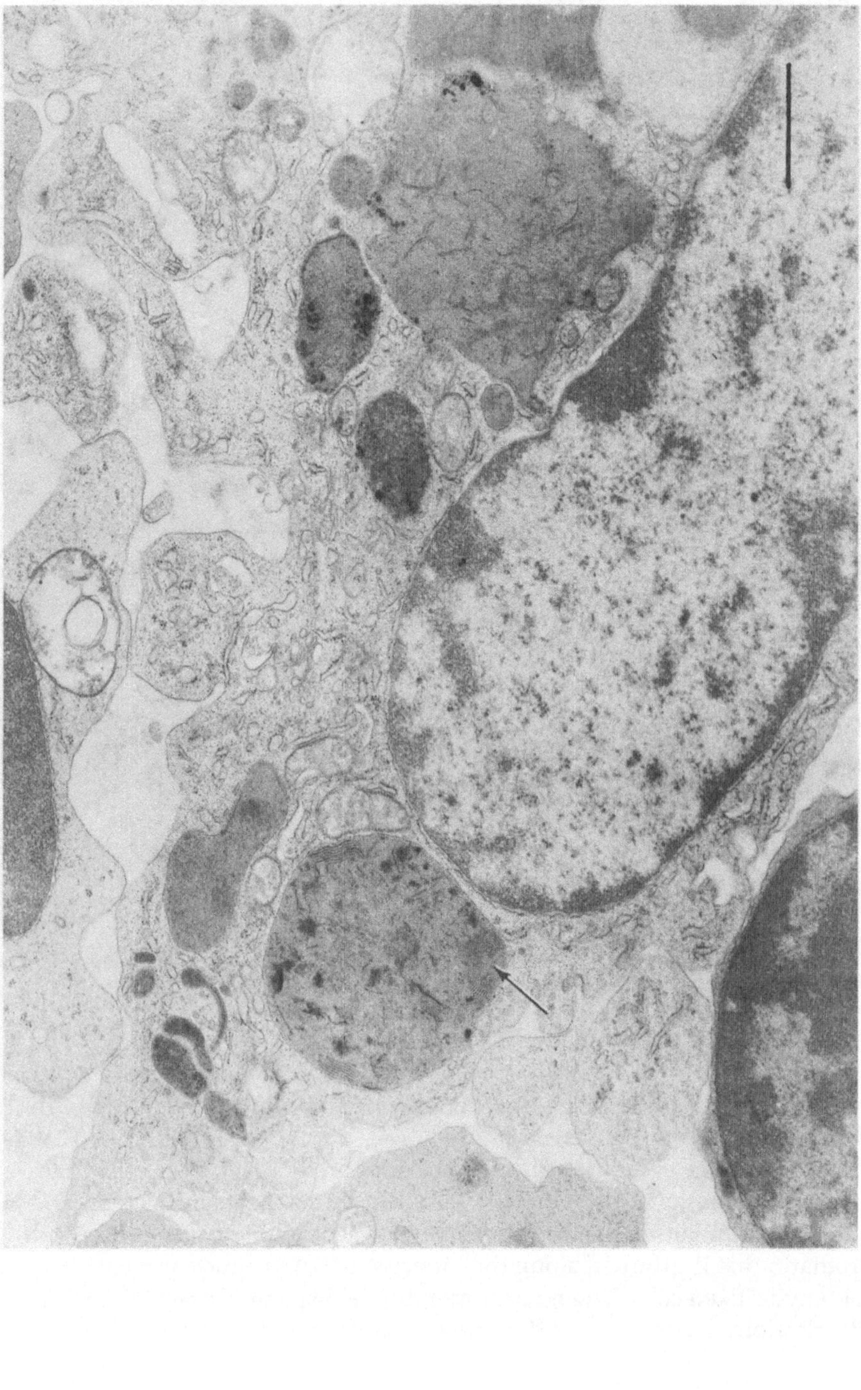

FIGURE 2. Higher power view of macrophage in normal monkey marrow. The cytoplasm shows several large phagosomes containing dense cellular remains and myelin-like material (arrow), as well as many small dense particles representing ferritin. The cell surface is irregular. The bar indicates 1 μm.

(1970) using a freeze-etching technique confirmed these features and added that the perinuclear cisterna is wider than envisaged by conventional chemical fixation. The significance of this observation is not yet clear.

The Golgi apparatus situated adjacent to the nucleus is prominent in the promonocyte stage, but is relatively inconspicuous in the mature marrow macrophage. The formation of intracytoplasmic vesicles from the Golgi cisternae was elegantly demonstrated by freeze-etching (Daems and Brederoo, 1970), and the relationship with the evolution of lysosomes originally proposed by Novikoff *et al.* (1964) is generally accepted. The detailed mechanism of transport of the lysosomal hydrolytic enzymes is still the subject of debate.

The cisternae of RER are relatively sparse and scattered throughout the cell interspersed with well-formed mitochondria. The cell sap contains a few ribosomes and abundant ferritin granules (Figure 2). Membrane-bound vesicles, vacuoles, and lysosomal granules dominate the cytoplasm. The lysosomal profiles vary considerably in size and shape. In the less mature cell the granules are small and of the primary or virgin type. With increasing maturity, the phagocytic functions become apparent, and not only does the number of lysosomes increase but also their complexity, due probably to fusion of secondary lysosomes. The majority of lysosomes in the mature macrophage contain ferritin and hemosiderin granules and are, thus, electron dense. In situations involving increased erythrophagocytosis, heavy iron loading results in true siderosome formation (Marton, 1973, 1975b). Depending on the level of phagocytic activity, fragments of nuclei and cells (Marton 1975a), and, eventually, residual myelin bodies are found. Wickramasinghe and Hughes (1978) have demonstrated the uptake of [^{3}H]leucine into the membrane region of the lysosome, implying the presence of newly synthesized protein which they postulated may come either from the cytoplasmic matrix or from fusion with primary lysosomes.

Interspersed within the cytoplasm are microfilamentous structures, often running in bundles, and possibly representing actin. These have been reported both in immature cells (Bentfeld *et al.*, 1977) and the processes of mature cells (Campbell, 1972).

The external contour of the cell is made irregular by the presence of many cell projections (Figures 2 and 3), a characteristic of the macrophage in general. These processes vary considerably in their morphology and are obviously related to phagocytic activity, since they are seen encompassing cells and fragments of cell debris. Watanabe (1964) originally described macrophage cell processes not only giving the endothelium an incomplete investment but occasionally extending through the endothelium into the sinusoidal lumen, and this has subsequently been confirmed (Hudson and Yoffey, 1968; Huang, 1971; Marton, 1971, 1973, 1975a; Ogawa, 1975; Wickramasinghe and Hughes, 1978). Serial sectioning indicates that these processes pass through rather than between endothelial cells (Tavassoli, 1977b). Marton (1975a) and others have demonstrated the important role of the bone marrow macrophage in the process of erythroclasis which will be discussed in greater detail later.

The freeze-etching studies by Daems and Brederoo (1970) have demonstrated that the microvillous projections seen in transmission electron micro-

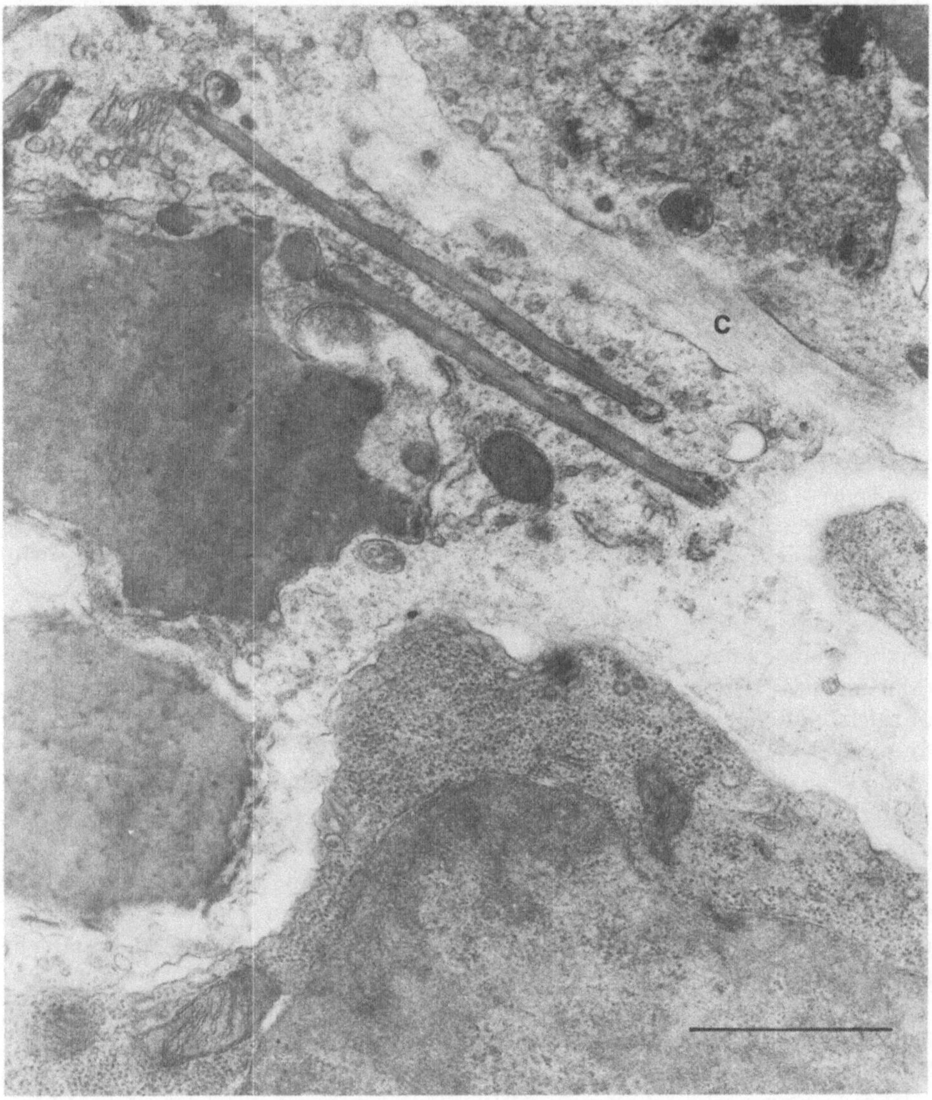

FIGURE 3. Macrophage projection in normal mouse bone marrow. It contains two strands of crystalloid material and some phagosomes. Its surface lies adjacent to fine collagen fibres (C). The bar indicates 1 μm.

scopic sections are not artifactual and in some instances are associated with invaginations of the cell membrane. Micropinocytic activity at the cell membrane reflects the functional state of the cell and ferritin granules are often seen in caveolae. However, the ferritin content is not as remarkable as the the micropinocytic vesicles of the erythroblast. The concept of rhopheocytosis introduced by Policard and Bessis (1958, 1962; Bessis and Breton-Gorius, 1962), as a means by which ferritin, having been processed by the central macrophage in the erythroblastic island, was transferred to the developing erythroblast, will be discussed below.

Special areas of contact between the macrophage and the adjacent erythroblasts in the erythroblastic island have been sought, but no such areas have been found by conventional means (Frisch *et al.*, 1976) or in freeze-etched material studied by Daems and Brederoo (1970), in which a close approximation only was observed. With improved fixation and the ability to tilt the specimen, the structure of both cell membranes can be seen to remain intact throughout. However, Tavassoli and Shaklai (1979) using freeze-fracture and lanthanum-tracer techniques reported the presence of small, adhering junctions between hemopoietic and stromal cells. Blom (1977) observed a modification of membranes of plasma cells at zones of contact with adjacent macrophages in patients with multiple myeloma.

In thalassemia, Wickramasinghe and Hughes (1978) have shown cell projections from the abnormal erythroblast into the adjacent macrophage which may be related to the monitoring function that the macrophage exerts on erythropoiesis. From the work of Bessis and co-workers (cited above), Marton (1975a), Weiss (1976), and others, it is clear that a close relationship exists between the macrophage and all developing hemopoietic cells. The importance of this relationship was substantiated by the work of Horland *et al.* (1977), with *in vitro* growth of erythroid colonies (see below).

In a combined transmission, scanning, and freeze-etched study following perfusion, Weiss (1976) demonstrated the cellular juxtaposition in three dimensions. He concluded that the macrophages are distinct from "fibrogenic" stromal cells which include the adventitial cells of the sinusoid; the relative paucity of lysosomes, the more abundant cisternae of rough endoplasmic reticulum, and filamentous structures are the ultrastructural features distinguishing this cell from the macrophage. In the literature, the terms *reticular cell* and *macrophage* are often used interchangeably, a point to which Weiss takes exception. Tavassoli (1977a) estimated that in the normal state, the adventitial reticular cells cover 29% of the external surface of the sinusoid. It is suggested that their processes have the property of retracting to expose the lining cell, because they possess microfilaments associated with cytoplasmic densities (Campbell, 1972) and, thus, may govern the release of erythrocytes into the blood (see below). The adventitial cells illustrated by Tavassoli send fine dendritic processes into the adjacent hemopoietic cords and lie in close approximation to the erythroblastic cells where they are indistinguishable from the processes of macrophages.

In unpublished ultrastructural studies in normal mice, the present authors have observed processes of macrophages lying adjacent to fine collagen fibrils (Figure 3). Wickramasinghe and Hughes (1978) also made this observation in human thalassemia. The proximity to the collagenous framework in the rat is stressed by Weiss (1976) as relevant to classifying the stromal reticular cell. Our findings suggest that this criterion would not always be tenable, if the view is accepted that histiocytic cells may form collagen and that their hydrolytic enzyme content and, presumably also, their lysosomal population, is inversely related to their fibrogenic activity (Mackenzie, 1970, 1975).

The thesis proposed by Weiss of a reticular stromal cell entirely separate from the macrophage line cannot be discounted. However, an alternative suggestion to explain the relatively minor morphological differences is that they are an expression of different functional specialization of the same cell type. Much more evidence will be required in order to clarify this situation and no attempt has been made to distinguish reticular cells and macrophages in the present chapter.

Cytochemically, macrophages contain a high level of hydrolytic enzymes which correlates with the abundant lysosomes and prominent Golgi apparatus seen in stimulated cells (Braunsteiner and Schmalzl, 1970). Peroxidase activity has been demonstrated ultrastructurally in rat monocytes by Daems *et al.* (1975) and more recently in the RER, Golgi apparatus, and lysosomal granules of bone marrow macrophages by Bentfeld *et al.* (1977). They noted that synthesis of peroxidase ceases at the promonocyte stage and thereby gives rise to a dual population of granules hitherto undescribed. This contrasts with the renewed synthesis of peroxidase by rat peritoneal cells (Robbins *et al.* 1971). A rather different dual population was noted in tissue culture bone marrow macrophages by van Furth and Fedorko (1976). Here two distinct populations of colonies were observed, composed of cells with either peroxidase-positive granules or peroxidase-negative granules. They comment, however, that some of the macrophages in "positive" colonies contain granules which are peroxidase-negative. Little is known of the chemical structure of the peroxidase-negative granules, and the evidence is difficult to evaluate.

3. PHAGOCYTIC FUNCTIONS

3.1. CLEARANCE MECHANISMS

The earliest descriptions of the role of the bone marrow in clearing injected particulate matter from the blood stream appear to have been those of Hoffman and Langerhans (1869), Hoyer (1869), and Ponfick (1869). These investigators gave intravenous injections of cinnabar into guinea pigs or rabbits and subsequently noted intracellular deposition of the material in cells of the bone marrow. The importance of the marrow in removing foreign particles from the blood was confirmed by numerous investigators including Cousin (1898), Kiyono (1914), Evans (1915), Nagao (1920, 1921), Wislocki (1921, 1924), Hashimoto (1936), Huggins and Noonan (1936), Patek and Bernick (1960), Hudson and Yoffey (1963, 1968), Hashimoto (1966), Carr (1968), Huang (1971), Bankston and De Bruyn (1974) and De Bruyn *et al.* (1975). In general terms, the marrow elements involved in the clearance mechanisms are the endothelial cells of sinusoids and veins and the macrophages, although an occasional neutrophil granulocyte may be seen to contain particles (Hudson and Yoffey, 1963). The latter authors (1963, 1968) injected a fine carbon suspension of about 25 nm particle size in guinea pigs and demonstrated the uptake of particles by parenchymal macrophages as early as 15 min, although the particles did not gain free

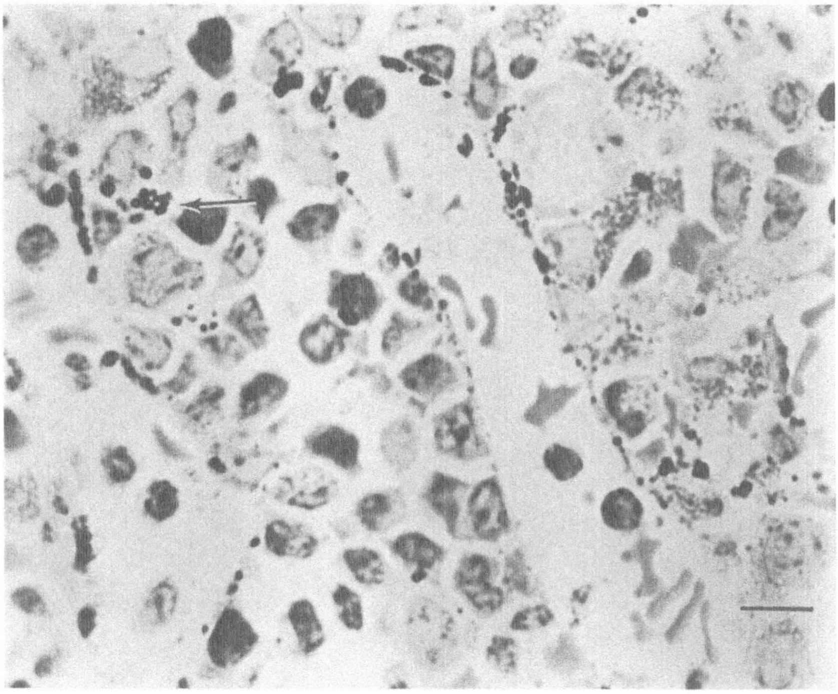

FIGURE 4. Light microscope picture of guinea pig marrow 90 min after intravenous carbon suspension. The endothelium of the sinusoids is speckled with carbon. A few collections of carbon are seen in a parenchymal macrophage (arrow). The bar indicates 10 μm.

access to the parenchymal spaces. At this stage the particles were found either in small cytoplasmic vesicles or in larger ones along with other inclusions, while at 14 days, cytoplasmic vesicles of up to 5 μm in diameter were seen, almost entirely filled with particles. In the early stages, particles were also seen to have been taken up by the sinusoidal endothelium but subsequently few particles were seen in this site. Carbon uptake is illustrated in Figures 4–8.

What are the mechanisms by which particulate matter from the blood stream is taken up and accumulated in the macrophages? Phagocytosis can clearly taken place when the cell surface of a macrophage comes into direct contact with particles, for example, when particulate matter is freely dispersed through the intercellular spaces of the parenchyma: this has, in fact, been described when suspensions of very small particles such as Thorotrast with a particle diameter of 10 nm are employed (Zamboni and Pease, 1961; Weiss, 1961). However, when larger particles, such as those of carbon are used, free particles are largely confined to the vascular bed and do not gain general access to the intercellular spaces.

As noted above, macrophages can sometimes be seen with processes extending through the walls of sinusoids into the lumen, and intracytoplasmic particles may be found within such processes after intravenous injection of particulate matter (Hudson and Yoffey, 1968; Huang 1971), as well as adhering

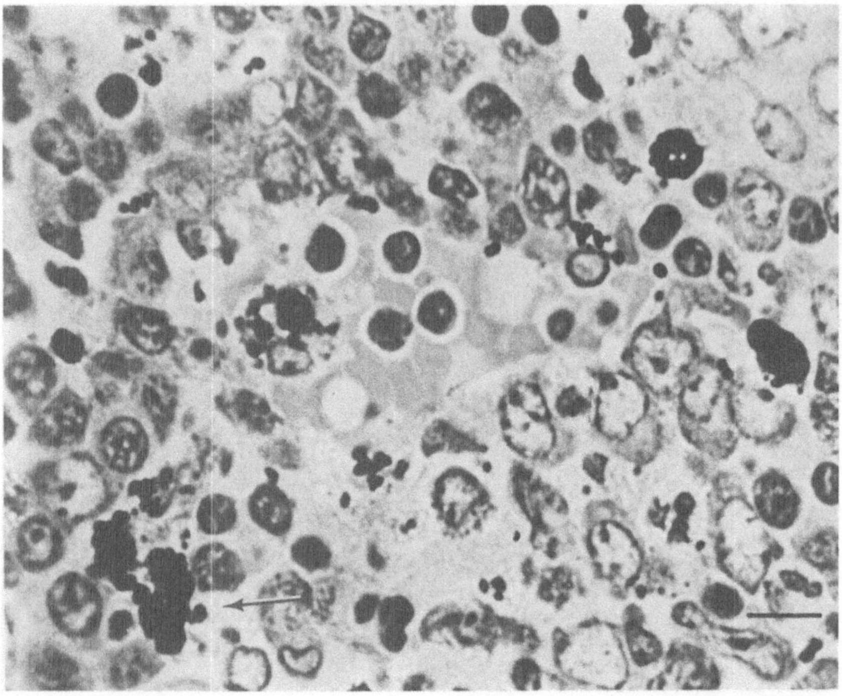

FIGURE 5. Guinea pig marrow 48 hr after intravenous carbon suspension, showing (left of center) an intrasinusoidal macrophage laden with particles. Other masses of carbon (arrow) indicate the position of a parenchymal macrophage. The bar indicates 10 μm.

to their surface or lying in indentations of the surface membrane, appearances which suggest that these intraluminal projections are active in phagocytosis. The projections are visualized by Marton (1975a) as part of an active motile process, being pushed into the lumen and retracted. At all events, the presence of these processes may provide a mechanism by which particles can be taken up from the blood stream and transferred into phagosomes of macrophages in the marrow parenchyma. It should also be noted that after intravenous injection of carbon suspension, particle-laden macrophages were seen lying wholly within the sinusoids (Hudson and Yoffey, 1963), an observation which raises the possibility that at least some of the macrophages with processes extending through the sinusoidal wall may be migrating either from marrow to blood or vice versa (Hudson and Yoffey, 1966; De Bruyn *et al.*, 1971). The possibility of migration of particle-laden macrophages was shown by the work of Patek and Bernick (1960), which indicated a massive movement from liver to lung with its peak at 3–4 days after intravenous injection of carbon in rabbits.

While particulate matter may, therefore, be cleared from the blood stream by the direct phagocytic activity of marrow macrophages, it is also necessary to consider possible mechanisms by which inert particles, initially taken up by sinusoidal cells, may later find their way into macrophages. Evidence that this occurs was provided by the observations of Nagao (1920) who noted that in

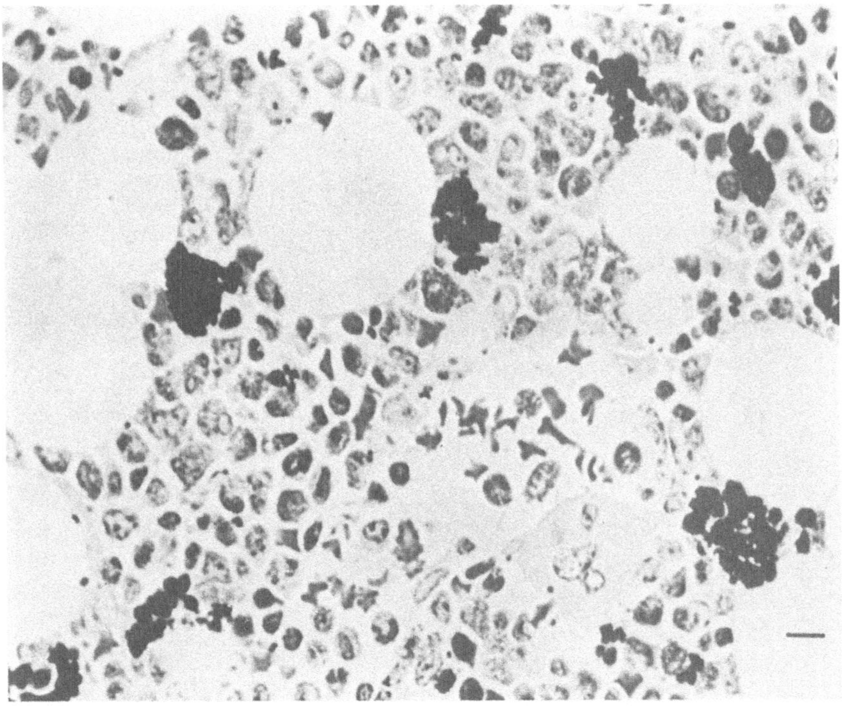

FIGURE 6. Guinea pig bone marrow 14 days after intravenous carbon suspension. Large masses of carbon, some closely related to fat vacuoles, indicate the position of laden macrophages. The bar indicates 10 μm.

guinea pigs and rabbits, endothelial cells of marrow sinusoids and veins gradually lose their ink particles, while larger masses of carbon accumulated in the parenchyma. Hudson and Yoffey (1968) reported that by 14 days particle-laden vesicles were only occasionally seen in sinusoidal and endothelial cells.

Three hypothetical mechanisms are perhaps worth considering, namely, intercellular transfer *in vivo,* transformation of endothelial cells into macrophages, and phagocytosis of degenerating particle-laden endothelial cells.

The first possibility is that particles are transported across the endothelial cell to the abluminal aspect where they are engulfed by investing macrophages (Hudson and Yoffey, 1963, 1968; Bankston and De Bruyn, 1974).

With regard to the second hypothesis, the early work of Weiss (1965, 1970) suggested that the walls of sinusoids are formed by cells which have the same fundamental properties as the macrophages or "phagocytic reticular cells" of the parenchyma and that the sinusoids and intersinal spaces might, from time to time, be interchangeable, depending on physiological requirements. In this interpretation, cells which lined the walls of sinusoids and helped in the clearance of particles from the blood, would later be identified as particle-containing macrophages in the marrow parenchyma. Although, at first sight, this hypothesis appears an attractive one, it would raise a number of difficulties. It has been observed that there are numerous rather loose junctional complexes between

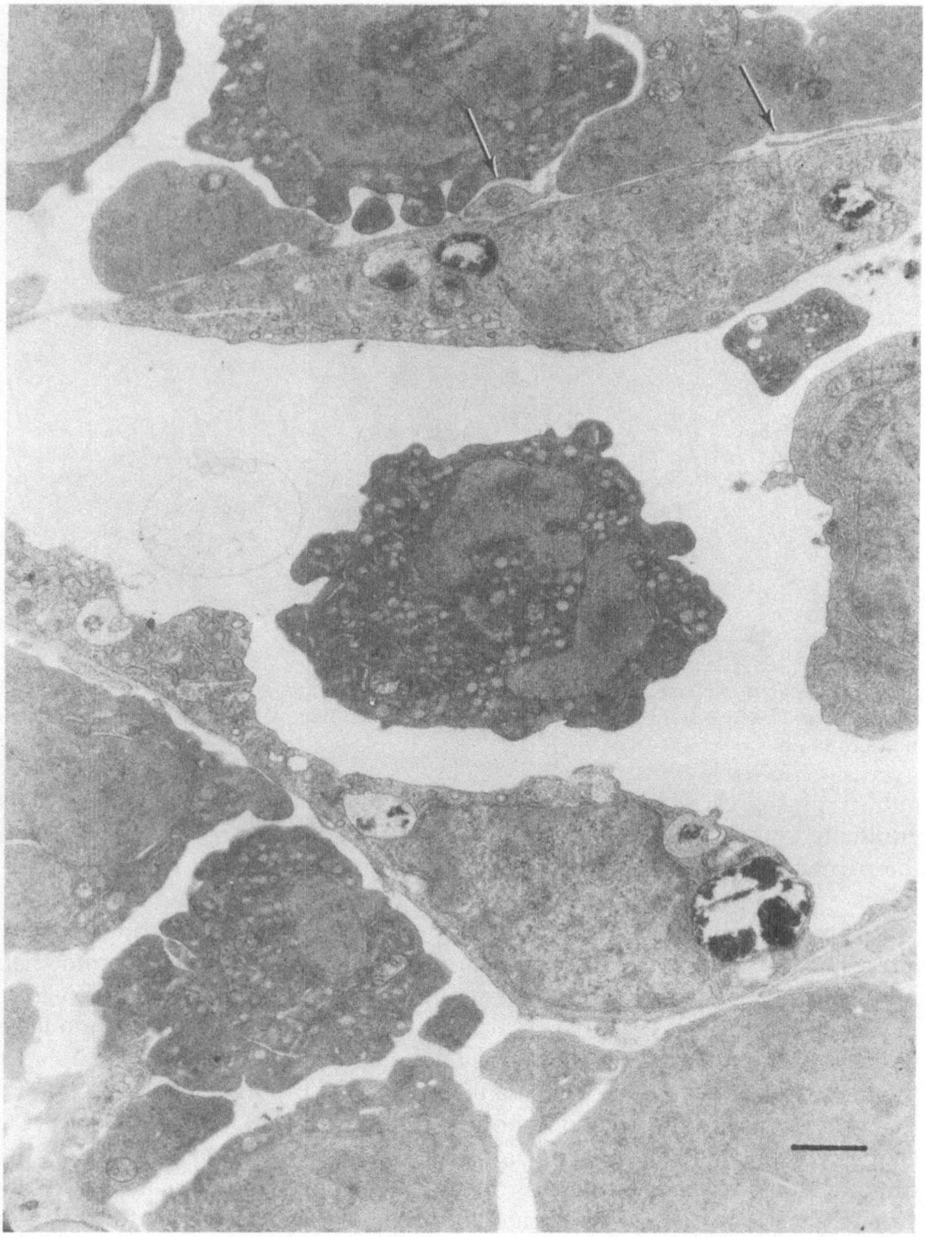

FIGURE 7. Electron micrograph of bone marrow of mouse 90 min after intravenous carbon suspension. The endothelium of a sinusoid shows a few vesicles containing carbon. On its abluminal aspect, the endothelium has an incomplete investment of attenuated macrophage processes (arrows) which do not contain carbon. The bar indicates 1 μm.

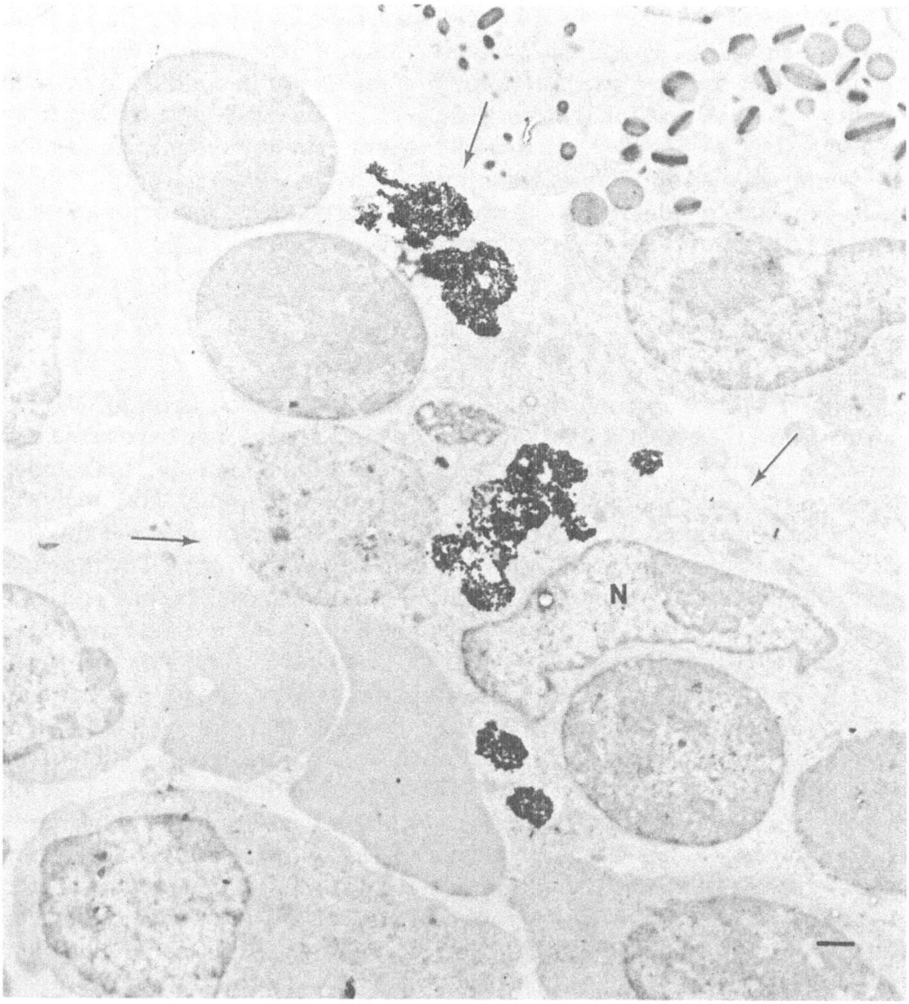

FIGURE 8. Guinea pig marrow 14 days after intravenous carbon suspension. A macrophage is shown containing several large accumulations of carbon, the profile of the largest having a diameter of over 3 μm. The macrophage nucleus is indicated at N and the cell boundaries by arrows. The cytoplasm shows several ingested nuclei and other inclusions. The bar indicates 1 μm. [This figure is reproduced from Hudson and Yoffey (1968) by kind permission of the Cambridge University Press.]

neighboring endothelial cells (Hudson and Yoffey, 1966), but junctions are only occasionally observed between the processes of macrophages or "adventitial cells" (Weiss, 1976, Tavassoli, 1977a). This, together with the special relationship of macrophage processes to the sinusoidal wall discussed above, suggests that both endothelial cell and reticular cell have a degree of differentiation which would preclude interchangeability. The *in vivo* studies of marrow sinusoids by Brånemark (1959) also lent no support for the idea that sinusoids and parenchymal spaces might be interchangeable.

With regard to the third possibility there is no evidence of significant degeneration of sinusoidal endothelial cells in clearance experiments using modern preparations of carbon particles (Halpern *et al.*, 1953), in contrast to the older experiments with India ink (Nagao, 1920, 1921). However, some cell death and replacement is presumably occurring all the time even in normal circumstances. Apparent examples of cell lysis were found by Weiss (1970) in control animals, and it seems safe to infer that some transfer of inert particles to macrophages will occur as cells die and are removed by phagocytosis.

3.2. ERYTHROPHAGOCYTOSIS AND ERYTHROCLASIS

While the bone marrow would not normally be called upon to deal with inert particles, the above studies of clearance mechanisms may have some bearing on the way that effete red cells, cell fragments, and other debris are removed from the circulation and digested. The early work of Miescher, (1957) who followed the destruction of labeled red cells in rabbits indicated that the bone marrow was the principal erythroclastic organ. This was confirmed by the experiments of von Ehrenstein and Lockner (1959) in which 54–74% of ^{59}Fe-labeled red cells cross-transfused in rabbits were broken down in the bone marrow, and the studies of Hughes-Jones (1961) with ^{51}Cr-labeled cells. Although the marrow had only a minor role in the rat in normal circumstances (Hughes-Jones and and Cheney, 1961) its clearance activities were increased markedly when it was depleted of blood-forming cells by hypertransfusion or protein deprivation (Keene and Jandl, 1965). Marton (1971, 1973, 1975a) found both light and electron microscopic evidence of erythrophagocytosis by marrow macrophages in the normal rat. He showed that this was markedly increased following transfusion of denatured erythrocytes or after splenectomy, and he drew attention to the role of the intrasinusoidal projections of macrophages as a means of removing effete red cells from the circulation. After injecting autologous red cells containing Heinz bodies into rabbits, Ogawa (1975) found that erythrophagocytosis by intrasinusoidal projections was prominent. Using the techniques of morphometry, Tavassoli (1977b) compared the number of erythrocytes in transit through the sinusoidal wall with the number being phagocytized intraluminally and concluded that 83% of circulating red cells of the rabbit may ultimately be removed by this mechanism.

The process of phagocytosis and digestion of an erythrocyte by a macrophage appears to be a very rapid one: in the living state, Bessis and Breton-Gorius (1962) found it took only about 10 min.

3.3. ROLE IN NORMAL AND INEFFECTIVE HEMOPOIESIS

While the function of marrow macrophages in breaking down red cells was described as long ago as 1925 by Peabody and Broun who reported a marked increase in megaloblastic anemia, their role in relation to blood cell formation was not suspected until the ultrastructure of erythroblastic islands had been described (Bessis, 1958). The "îlot érythroblastique" consists of a central reticular cell surrounded by one or more rings of erythroblasts. The central cell can be seen to extend cytoplasmic processes between and around the developing erythroblasts, and Bessis (1973) has suggested that these extensions are in continual movement encircling one erythroblast at one moment and another soon after. The arrangement of the erythroblasts indicates migration of the more mature forms towards the periphery of the island (Bessis and Breton-Gorius, 1962). Although the ultrastructural studies of Weiss (1965) in the rat indicated that erythropoiesis takes place in areas in close proximity to sinusoids, Mohandas and Prenant (1978) using the hypertransfused rat as a model, found that the islands are not spatially restricted to these sites, but occur over the entire marrow space. It is generally accepted that the central reticular cell is phagocytic and 1 hr after intravenous injection of carbon suspension in guinea pigs, Hudson and Yoffey (1963) found examples of these cells heavily laden with particles. The central reticular cell in fact appears indistinguishable from a macrophage. As Berman (1967) has pointed out, central reticular cells appear active in erythrocyte degradation in that they contain large masses of hemosiderin, scattered ferrogenous micelles, and other inclusions. They are also active in phagocytosis and destruction of the extruded nuclei of erythroblasts.

It has been been observed that the pyknotic nucleus extruded from the late normoblast as it transforms into an erythrocyte has attached to it a shred or rim of cytoplasm containing hemoglobin and perhaps a few organelles (Ralph, 1948). The ultrastructural observations of Skutelsky and Danon (1969) who stained cell surfaces with charged colloidal iron particles, showed that the "expulsed erythroid nuclei" were less negatively charged than any other cell in the bone marrow; this suggests that the reduced surface charge is one of the means used by the macrophage to identify these nuclei prior to phagocytosis. Although the intimate relationship of processes of the phagocytic cell to erythroblasts in the process of expelling their nuclei has been commented upon, there is no evidence that they are necessary for nuclear expulsion in these circumstances (Skutelsky and Danon, 1972). During primary hypoxia in the rat, Ben-Ishay and Yoffey (1971a) found central reticular cells which were larger than usual and contained a great number of inclusions; the increased numbers of erythroid nuclei to be ingested were thought to be a factor in this. During the "rebound" phase after discontinuing hypoxia, Ben-Ishay and Yoffey (1971b) found that the central reticular cells became smaller and that the erythroid cells of the island became replaced with small lymphocytes and transitional cells. In the early stages of hypoxia, transitional cells (presumed stem cells) were present in appreciable numbers and similar cells may occasionally be observed in the normal erythroblastic island of monkeys (Hudson, 1973).

The phagocytic function of the marrow macrophage is also important in relation to ineffective erythropoiesis. This is known to be marked in states where erythropoiesis is increased, but it is also present in the normal. Tavassoli (1974), for example, found that in normal rabbits, there was morphological evidence of phagocytosis of apparently intact nucleated red cells and reticulocytes in perisinal macrophages. Presumably, the processes of the central reticular cells exert a monitoring role to recognize and remove any defective cell. Ben-Ishay (1974) induced damage in marrow erythroblasts with an intraperitoneal injection of cytosine arabinoside in the rat and found numerous examples of phagocytosis of erythroblasts by central reticular cells, and similar results were obtained by Ben-Ishay and Yoffey (1974) following sublethal irradiation. Macrophage phagocytosis of erythroblasts at various stages of maturity has been reported in megaloblastic anemias (Goodman *et al.*, 1968), and increased numbers of erythroblastic islands were found in chronic myeloid leukemia by Sjögren (1976), presumably reflecting ineffective hemopoiesis.

Less attention has been given to the role of marrow macrophages in the phagocytosis of leukocytes. From scattered observations, it seems likely to have a similar significance to that discussed above for erythropoiesis in the removal of damaged, defective, or effete cells. In autoradiographic studies in normal guinea pigs injected with [^{3}H]thymidine, Osmond and Everett (1962) found phagocytosis of what appeared to be labeled small lymphocytes with a very short life-span, and some of the phagocytic cells were the central cells of erythroblastic islets. Examples of granulocytes engulfed by marrow macrophages may be seen in normal marrow, and Patt and Maloney (1964) suggested on the basis of their kinetic studies that granulocytic production may be death-controlled, i.e., adjusted by variations in the mortality of newly-formed cells. However, on the basis of current evidence, it is assumed that there is little or no marrow cell death in the granulocyte series in normal circumstances (Boggs, 1975). Increased phagocytosis of eosinophil granulocytes by marrow macrophages has been observed in guinea pigs sensitized to one foreign protein and then injected with a different antigen (Hudson, 1968a).

4. STORAGE FUNCTIONS

4.1. LIPID STORAGE

In studies of guinea pig bone marrow, after intravenous injection of carbon suspension (Hudson and Yoffey, 1968), macrophages containing large masses of particles were still present two years later. Although this was a somewhat artificial situation, the storage function of the mononuclear phagocytic cells is well authenticated. In normal marrow small fat vacuoles are occasionally encountered in macrophages along with other inclusions. Harris *et al.* (1963) found that fat storage and phagocytosis were commonly seen in "reticulum cells" of guinea pig bone marrow at an early stage in recovery from sublethal whole body irradiation, the stored fat being contained in large or small, rounded cytoplasmic vacuoles. On the basis of his extensive studies of normal rat bone marrow, Weiss (1976) concludes that fat cells occur adventitial to vascular sinuses and appear to be "reticular cells" which accumulate fat, and his conclusions were supported by Tavassoli (1976). Large fat cells or adipocytes are seen with increasing frequency in the "red" marrow with increasing age but rapidly disappear in response to a

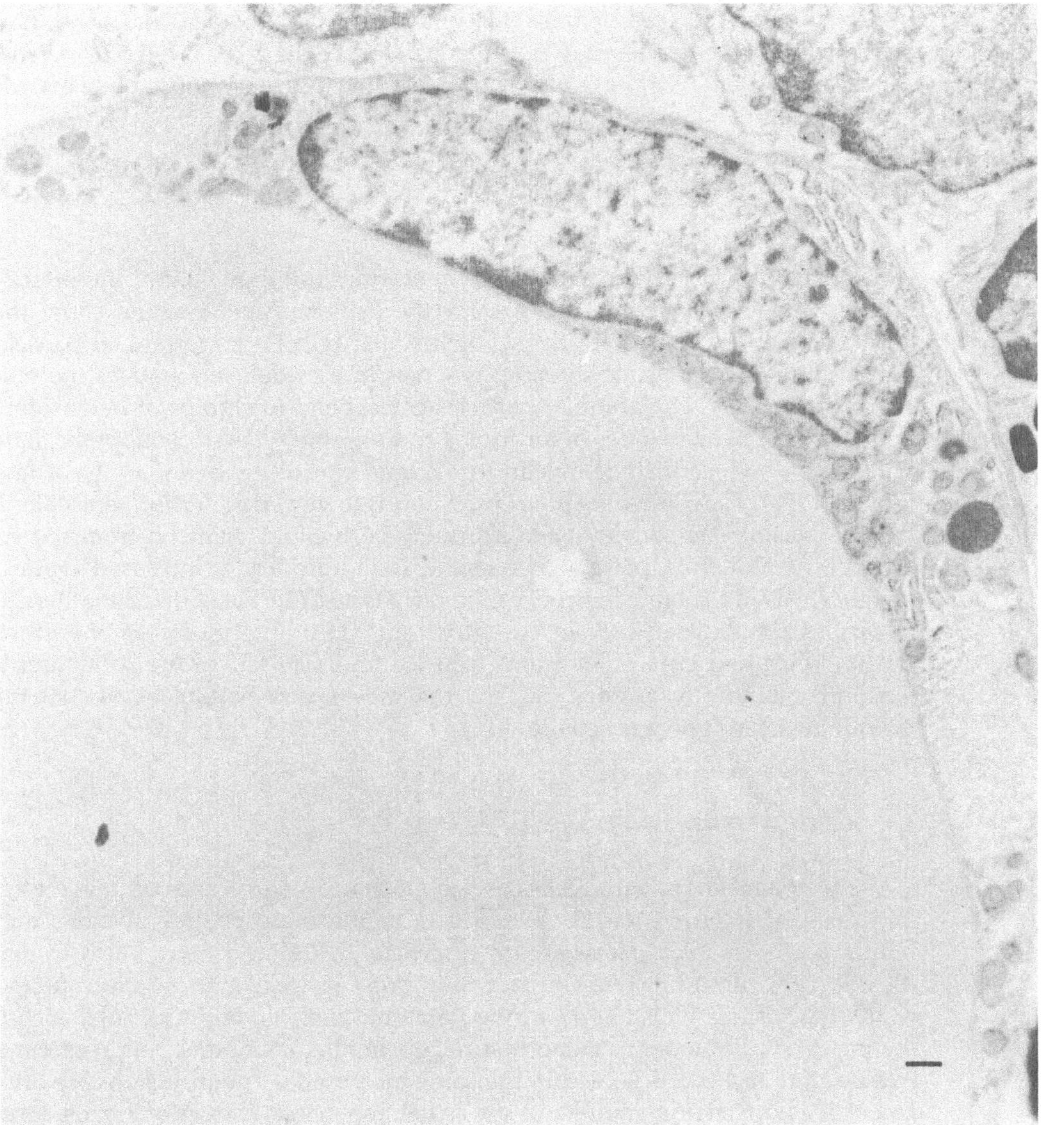

FIGURE 9. Section of a "fat cell" in normal guinea pig bone marrow showing a part of a large lipid inclusion vacuole. Smaller vacuoles, a few organelles, and a flattened nucleus may also be seen. The bar indicates 1 μm.

hemopoietic stimulus (Hudson, 1958). They have an eccentric, flattened nucleus and a thin rim of cytoplasm bordering the larger fat vacuole (Figure 9). Smaller lipid inclusions are also seen in the cytoplasm (Hudson, 1970; Tavassoli, 1976), and suggest that the large vacuole may be formed and added to by the confluence of smaller ones.

Zucker-Franklin *et al.* (1978) have shown that, in culture, monocytes can transform into fat cells, while still retaining their Fc receptors, and have suggested that monocytes may be the precursors of the fat cells of bone marrow. Allen and Dexter (1976) have also shown that in culture of mouse bone marrow giant fat cells can develop.

4.2. IRON STORAGE

The role of marrow macrophages in storing iron is of clinical importance. Routine staining of marrow smears for the Prussian blue reaction show that normally, iron is present in phagocytic reticular cells both in granular particles and in diffuse cytoplasmic staining, whereas in iron-deficiency states, no reaction may be seen. The granular particles correspond to clumps of hemosiderin which consist of aggregates of ferritin or complex mixtures of osmiophilic material, some of which include myelin forms, and constitute secondary lysosomes (Bessis, 1973). The diffuse staining corresponds to dispersed ferritin molecules in the cytoplasm. The source of the iron may be twofold, namely, from red cell breakdown and from plasma transferrin. Using artificially increased erythrophagocytosis in the bone marrow of the rat, Marton (1975b) showed that ferritin molecules accumulated both in the cytoplasm and in the lysosomes of erythroclastic "reticulum cells." Persistent iron loading also led to the formation of ferritin-containing lysosomes (or "siderosomes"), and he suggested that this may occur by autophagic activity.

4.3. CRYSTALLOID MATERIAL

The phagocytic reticular cells of the bone marrow sometimes contain crystalloid material (Figures 10–12). It is found in the macrophages of adult mice (Hudson, 1968b, 1969; Hudson and Shortland, 1974) and occurs also in man (Bessis, 1973). In adult mice of over 30 g body weight, marrow macrophages contain inclusions which appear long, straight, and slender. As a rule, each of the inclusions lies within a smooth-surfaced limiting membrane, but sometimes two or more inclusions lie within the same membrane. The inclusions are often associated with ferritin granules or dense heterogeneous masses of hemosiderin. In profiles of macrophages cut through their central region, the number of these inclusions varied from 6 to about 130 and measured up to 11.5 μm in length (Hudson, 1968b). The presence of these inclusions seemed related to the body weight and, by inference, to the age of the mice; very few were encountered in animals of less than 30 g (Hudson, 1969). Although most of these observations were made on mice of an inbred albino strain, similar inclusions were present in adult mice of the C57 Br strain; none were found in adult mice (germ-free) of the NMR1 strain which rarely weigh more than 30 g (Hudson and Shortland, 1974). Crystalloid material has also been reported as absent in mutant mice of the $S1/S1^d$ strain (Shaklai and Tavassoli, 1978). An incidental point of possible sig-

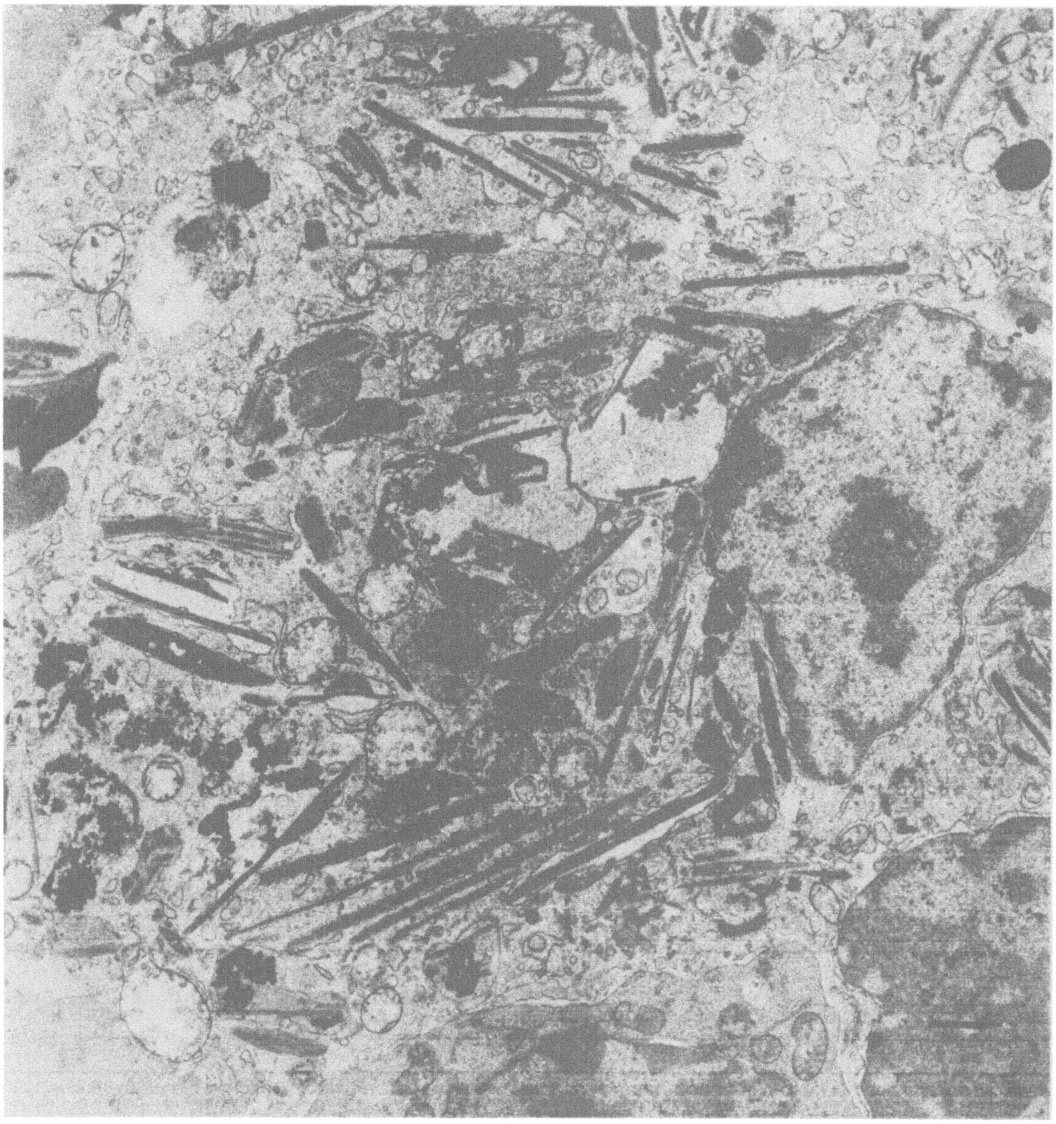

FIGURE 10. Macrophage in mouse marrow. The abundant cytoplasm contains a very large number of dense crystalloid inclusions, a few of which lie in large vacuoles. A cell process can be seen extending between two erythroblasts. The bar indicates 1 μm.

nificance is the relative paucity of fat vacuoles in the marrow of the mouse, i.e., it appears active.

The presence of crystalloid inclusions has been reported earlier by Ichikawa and Yoshioka (1960) who found them in the bone marrow macrophages of SL mice mice with myeloid leukemia but never in mice with lymphoid leukemia. Dalton *et al.* (1961) observed these inclusions in phagocytic cells of C57 B1 mice

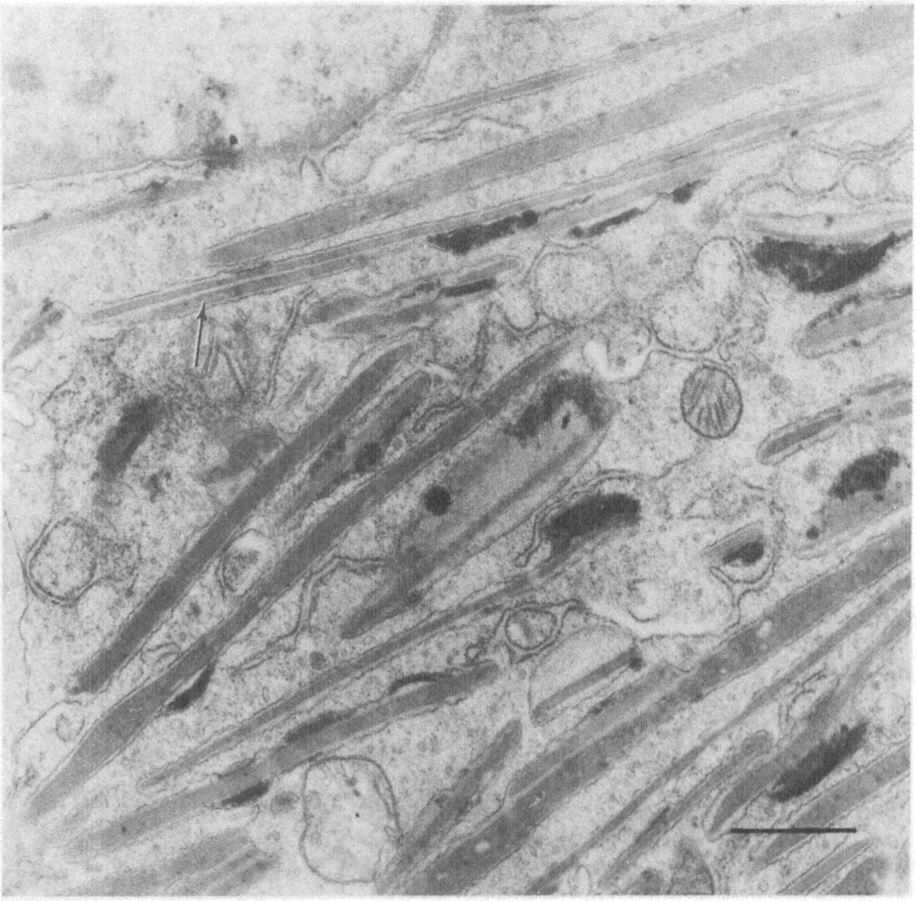

FIGURE 11. Higher power view of a mouse marrow macrophage. The crystalloid strands are enclosed in smooth-surfaced membranes; two or more inclusions are sometimes seen within the same membrane (arrow). Aggregates of heterogeneous dense material, representing hemosiderin, are associated with some of the crystalloids. The bar indicates 1 μm.

with radiation-induced leukemias, while Berman (1967) noted them in the marrow of normal mice of the C57 B1 strain. Crystalloid inclusions having a somewhat similar morphology to those described here have been reported in alveolar macrophages following *Toxocara canis* infestation in mice (Zyngier, 1976). Yang *et al.* (1979) while confirming that crystalloid-containing macrophages in the marrow generally increase in number with maturation and aging, also found a sharp decline in numbers after 12 months of age.

Although a few of the crystalloid inclusions may occasionally be observed lying within large phagosomes in the cytoplasm of marrow macrophages, there is little to suggest that crystalloid material is phagocytized as such. The inclusions are present in specific pathogen-free animals, and an exogenous origin from some common parasitic or infective process can be excluded (Hudson and

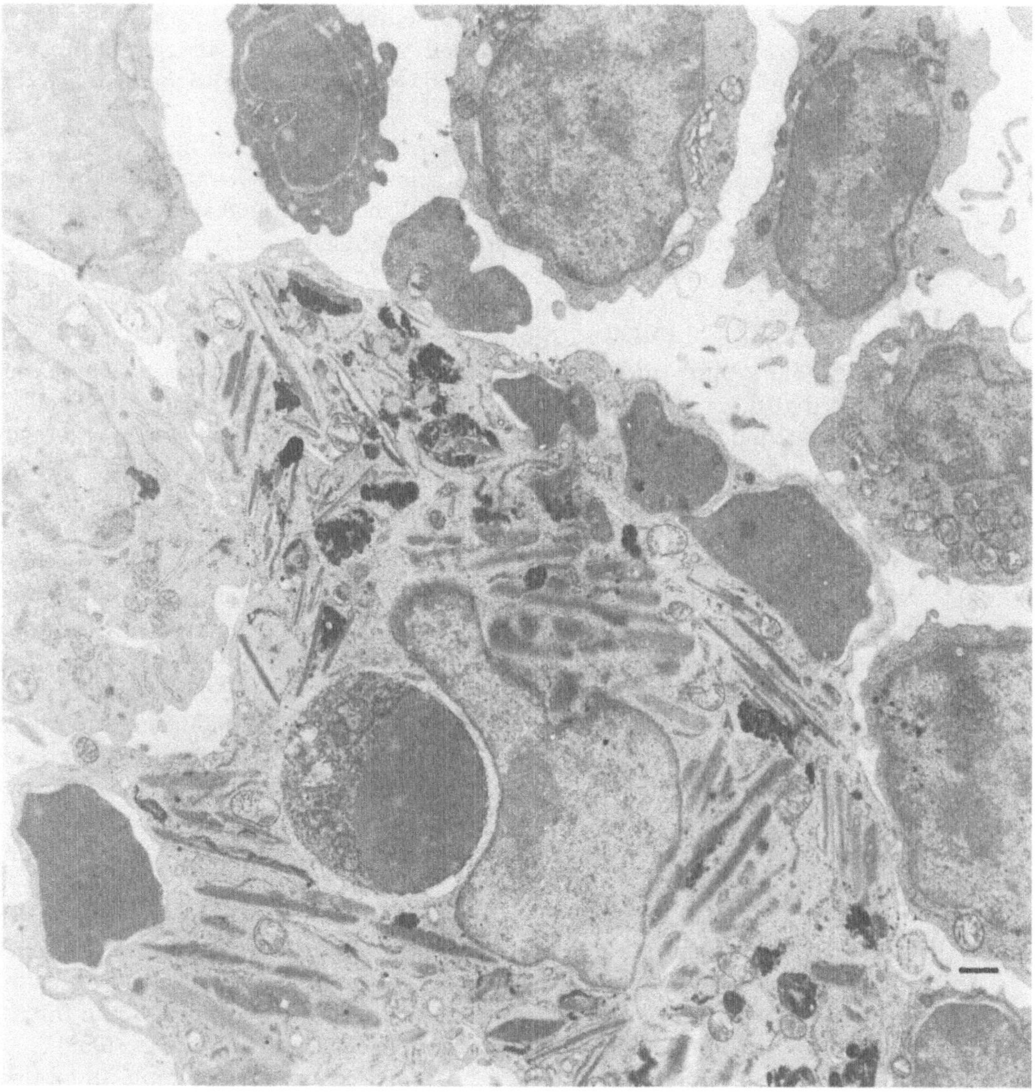

FIGURE 12. Macrophage in marrow of phenylhydrazine-injected mouse. Several ingested erythrocytes are present towards the periphery of the cell. The cytoplasm also contains abundant crystalloid inclusions and phagosomes. The bar indicates 1 μm.

Shortland, 1974). The assumption is, therefore, that it is either formed from material being broken down in the macrophage or else it is newly synthesized.

The suggestion that the deposition of crystalloid material was related to obesity was investigated by exposing a small group of adult mice to a restricted diet for 24 days, involving a loss of about 25% of their body weight. While there was evidence of a significant reduction in the number of crystalloid-containing macrophage profiles in sections of bone marrow (Hudson, 1969), the results are difficult to interpret. Starvation, for example, may itself produce a depression of erythropoiesis (Fruhman, 1966) as well as a decrease in the number of eosinophils and neutrophils (Fruhman and Gordon, 1955), and the results might be related to these.

The crystalloid material of mouse marrow macrophage is negative for Perl's reaction (Berman, 1967), and does not contain iron on X-ray dispersive microanalysis (Shortland and Hudson, unpublished findings). However its close association with masses of hemosiderin and aggregates of ferritin could be taken to imply its origin from ingested erythrocytes. In an attempt to substantiate this, Chin, Hudson, and Shortland (unpublished findings) observed the effects of a hemolytic episode. A single subcutaneous injection of phenylhydrazine was administered to adult mice of over 40 g body weight. Hematological examination showed that there was, on average, a 45% reduction in circulating red cells at Day 1, a reticulocyte response reaching a maximum of 24% of Day 6, and restoration to normal blood picture by Day 12. Study of marrow sections at intervals from Day 0 to Day 12 (five animals per observation), using the method described previously (Hudson, 1969), revealed no statistically significant differences in the number of crystalloid-containing macrophages per eyefield at any of the intervals studied. Thus, no evidence was found to suggest that either the intense initial red cell destruction and erythrophagocytosis, or the increased erythropoiesis which followed, had any effect on the inclusions.

The possibility that the crystalloid inclusions are related to the phagocytosis of granulocytes must also be considered, bearing in mind that in some species at least, these cell lines increase in number in the marrow with increasing body weight (Fand and Gordon, 1957). The possible association with myeloid leukemia in mice has already been mentioned (Dalton *et al.*, 1961). Rarely, a similar inclusion has been observed within the cytoplasm of either an eosinophil or a neutrophil granulocyte (Hudson, 1969). In this connection, however, it should be noted that when blocks were stained with phosphotungstic acid, the inclusions under discussion gave a different reaction from those of the crystalloid interna of the eosinophil granules. The former showed a densely reacting rim with an electron-lucent central area, while the eosinophil granule interna showed no reaction whatever with phosphotungstic acid (Hudson, 1969).

Material of similar morphology has been reported by Simon and Burke (1970) in the macrophages of normal rabbit spleen, associated with the phagocytosis of erythrocytes and leukocytes, while Bessis (1973) reported that large crystals may be found in the marrow macrophages of normal man as well as in a variety of pathological disorders. He suggested that they may be made up of "ceroid substance" giving a sea-blue color with May-Grünwald and Giemsa staining (Hartcroft and Porta, 1965). Elongated crystals of up to 20 μm in length and staining light-green with the above stain were found by Stavem *et al.* (1977) in some patients with myelomonocytic leukemia but not in patients with lymphoid leukemias; these crystals, however, appeared different in morphology from those of the mouse and did not show positive PAS-staining. We have not encountered crystalloid material in marrow macrophages of guinea pig, rat, horse, dog, or monkey during the course of other ultrastructural studies.

Although much remains to be learnt about the precise nature of the crystalloid material, it seems reasonable to think that where present, it represents a form of storage product derived from cellular breakdown.

5. INTERCELLULAR RELATIONSHIPS AND THE CONTROL OF HEMOPOIESIS

The intimate relationships of the phagocytic reticular cells not only with erythroblasts in the erythroblastic islands and the lining cells of the sinusoids, but with every other type of cell in the marrow space, has already been noted. Apart from their function in monitoring and culling defective or aged cells, the possibility that these extensive contacts have a significance in terms of providing a suitable microenvironment for hemopoiesis (Weiss, 1976), and even controlling it (Moore, 1976), is currently arousing some interest.

In relation to erythropoiesis, Bessis and his co-workers (Policard and Bessis, 1958, 1962; Bessis and Breton-Gorius, 1962) described a process which they termed rhopheocytosis, by which ferritin molecules appeared to be transferred from the central reticular cell to the developing erythroblasts. They suggested that, in this respect, the reticular cell was acting as a nurse cell, but noted that in iron-deficiency anemias the process of transfer still seemed to be present in the absence of ferritin. Tanaka *et al.* (1966) noted that ferritin-containing vesicles occurred in the membranes of two adjacent erythroblasts, demonstrating that the concept of rhopheocytosis could not be exclusive to transfer from the macrophage. The same point was subsequently made by Bessis (1973) who showed that with the newer techniques of fixation, the integrity of the membranes of cells was preserved and the idea of a direct transfer of ferritin from macrophage to erythroblast could no longer be accepted.

Weiss (1976) reviewed his own observations (discussed above) and those of other authors who had studied the induction and regeneration of hemopoiesis, and drew the inference that the reticular cells of the marrow, along with the blood vessels, have an important role in providing the requisite microenvironment for blood cell formation. A parallel conclusion drawn from *in vitro* study of cell cultures was made by Horland *et al.* (1977) and Goldstein (1975) who postulated that cellular interaction and the approximation of erythroblasts to macrophages is either necessary or facultative for the growth of an erythroid colony. The nature of the microenvironment provided by the marrow macrophages in physical and chemical terms is less certain. Macrophages have been shown to be a source of colony-stimulating factor (CSF), a factor essential for the *in vitro* formation of colonies from stem cells in a semisolid supporting medium (Metcalf and Moore, 1973; Moore, 1976), and there are indications that CSF may be a physiological regulator of leukopoiesis (Territo and Cline, 1975). The macrophage has also been found to produce prostaglandins and it has been suggested that these substances may exert an action antagonistic to CSF in mediating hemopoiesis (Moore, 1976). It may be noted in passing that the macrophage is probably the most important source of complement in reticuloendothelial tissues (Stecher, 1970) in addition to synthesizing interferon, pyrogens, and lysozyme.

With regard to red cell production, there has been speculation concerning the role of the macrophage in erythropoietin synthesis (Eisen, 1978). The observation of transitional cells in erythroblastic islands (Ben-Ishay and Yoffey, 1971a,b; Hudson, 1973) might be important here, leading one to speculate that the central reticular cell could have a role in committing stem cells to erythopoiesis.

The significance of the curious relationship of the processes of phagocytic reticular cells with the endothelium and lumen of the sinusoids has already been discussed in terms of phagocytosis and erythroclasis. There is also evidence that the relationship with the endothelial cells may reflect a delicate mechanism regulating the delivery of young cells into the circulation (Chamberlain *et al.*, 1975; Muto, 1976; Tavassoli, 1977a). Under experimental conditions, for example, after endotoxin administration in rats (Weiss, 1970), it has been found that the degree to which processes of reticular cells invest the sinusoidal endothelium is inversely proportional to the transmural passage of cells.

Although our knowledge of the significance of the intercellular relationships of the marrow macrophages is still largely based on inference, the present indications are that they will prove to be of importance in the control of hemopoiesis at all stages, from the stem cell stage to that of delivery of mature cells into the circulation.

6. ROLE IN SOME PATHOLOGICAL PROCESSES

The activity of marrow macrophages in various hematologic disorders has been referred to earlier, e.g., under ineffective hemopoiesis. Other pathological processes involving the bone marrow depend on the normal functional properties of the macrophage but their significance as a reflection of disease has been studied only infrequently.

6.1. INFECTIVE DISEASES

The ability of marrow macrophages to clear particulate matter from the blood stream finds its expression in some infective diseases where residual organisms may be detected in these cells. Brumpt (1940–1941) commented on the histiocytic changes in leprosy and noted that the diagnosis could be achieved by bone marrow examination. This is of particular interest in view of our lack of knowledge relating to immunological activities of marrow macrophages, although macrophages in general are important in the immune process. Unanue *et al.* (1969) and Unanue and Cerottini (1970) have shown that antigens become immunogenic by reacting with the surface of macrophages, and Huber and Fudenburg (1968) have demonstrated surface receptors which could be relevant in this context. The changes in macrophages are illustrated by the work of Turk

(1971) who studied leprosy and noted a distinct relationship between the histological picture and immune status. In tuberculoid leprosy there were giant macrophages associated with abundant lymphocytes, and bacilli were few in number. This was contrasted with lepromatous leprosy in which abundant histiocytes full of bacilli occur with few lymphocytes. Antibody production was predominant in patients with lepromatous leprosy compared with a marked cellular response in those with tuberculoid leprosy.

Tuberculosis and parasitic diseases, including leishmaniasis, malaria, and toxoplasmosis, can all be detected in the bone marrow macrophages (Bessis, 1973; Kitchens, 1977). The reactive histiocytes seen in patients with subacute infective endocarditis are reflected in the bone marrow (Bessis, 1973).

6.2. PHAGOCYTIC OVERLOADING

The phagocytic function of the marrow macrophage is involved in pathological processes where ingested material accumulates in the cytoplasm, and can be detected by bone marrow examination. Overloading of the macrophage with phagocytized material may result in a relative deficiency of digestive enzymes (Kitchens, 1977). An example is seen in thalassemia where iron turnover is increased. Enlarged macrophages containing secondary lysosomes have been noted (Zaino *et al.*, 1971; Bessis, 1973; Wickramasinghe and Bush, 1975), and erythrophagocytosis and lysosomes containing abundant iron are prominent. In this disease, Wickramasinghe and Hughes (1978) have demonstrated a peculiar structural relationship between the macrophages and the adjacent erythroblasts, whereby protrusions of erythroblast cytoplasm are pinched off and this may be important in eliminating alpha chains. Other examples of macrophage overloading are reviewed by Kitchens (1977).

The diagnostic use of Thorotrast, some three decades ago, resulted in retention of this alpha emitter by macrophages all over the body, the bone marrow included (Kitchens, 1977). Malignant change involving the RES following Thorotrast has been reviewed by Grampa (1971).

6.3. ENZYME DEFICIENCIES

A further group of pathological disorders relates to the digestive function of marrow macrophages. An absolute enzyme deficiency in macrophages has been found in conditions formerly referred to as the storage disorders ("thesauroses"). These conditions are now being characterized biochemically. Lysosomes in the macrophages of patients with Gaucher's disease lack the enzyme glucocerebrosidase. As a consequence, the phospholipid in red cell membranes cannot be degraded and glucose cerebroside accumulates within the macrophage cytoplasm giving rise to large (up to 100 μm), pale cells with characteristic striae. Ultrastructurally, the cytoplasm contains abundant curved, smooth-walled tubules 30–40 μm long (De Marsh and Kautz, 1957; Jordan, 1964;

Toujas *et al.*, 1966; Neimann *et al.*, 1968; Pennelli *et al.*, 1969), which display 6-nm striations and are thought to be aggregated glucocerebroside molecules (reviewed by Peters *et al.*, 1977).

In Niemann–Pick disease a deficiency of sphingomyelinase is involved leading to the accumulation of lipid which eventually totally replaces the cytoplasm. Electron microscopy reveals lipidic inclusions (Tanaka *et al.*, 1963) with prominent myelin forms in older cells. Hunter–Hurler disease results in lipid and mucopolysaccharide accumulating within the cytoplasm. The precise enzyme deficiency has not yet been characterized. Ultrastructurally, the inclusions are membrane-bound and contain granular material of medium electron density together with dense lysosomal structures (Callahan and Lorincz, 1966).

While the biochemical data is specific for each disease the morphological changes are by no means characteristic. Gaucher-like cells have been reported in a variety of leukemias (Witzleben *et al.*, 1970; Dosik *et al.*, 1972; Keyserlingk *et al.*, 1972) and in congenital dyserythropoietic anemia (van Dorpe *et al.*, 1973). Cells containing lipid granules resembling those in Niemann–Pick disease are found in "benign reticulopathies" (Hammarsten and Reis, 1957), essential hyperlipidemia (Tanaka *et al.*, 1963), reticulum cell sarcoma (Bessis, 1946), chronic myeloid leukemia (Lee and Ellis, 1970), and, rarely, in normal bone marrow (Bessis, 1973).

All these cells, due to some lipid fraction remaining in the cytoplasm during the processing, give a blue staining with Romanowsky stains. This has led to the concept of the "sea-blue histiocyte." It is possible that the condition of sea-blue histiocyte syndrome described by Silverstein *et al.* (1970) is a disease entity and has a familial incidence in some cases (Silverstein and Ellefson, 1972; Sawitsky *et al.*, 1972) but its identity has been questioned in a recent correlative histochemical and ultrastructural investigation by Long *et al.* (1977). It is clear that the staining reaction with Romanowsky stains is not pathognomic of any one condition.

6.4. MALIGNANT DISORDERS

Malignant change in mature macrophages is very rare, but neoplasms of histiocytic lineage do occur. Some of them involve the bone marrow, notably reticulum cell sarcoma, Hodgkin disease, the conditions grouped under the nebulous term histiocytosis X and histiocytic medullary reticulosis. The precise site of origin of these neoplasms is usually impossible to determine and the histogenesis is controversial. A detailed discussion, therefore, lies outside the scope of this chapter.

7. CONCLUSIONS

In terms of structure, the macrophages or phagocytic reticular cells of the bone marrow have many features in common with macrophages elsewhere. A distinctive feature is the intimate contact made by these cells with every other element in the bone marrow. Their processes partially invest the walls of sinusoids and, in places, penetrate the lumen; they have a central position in islands of erythroblasts and relate to developing leukocytes. The significance of these relationships lies in the functions of morrow macrophages. Phagocytosis and digestive functions are employed in the removal of effete red cells and particulate matter from the blood stream, in the disposal of extruded erythroid nuclei, and in the culling of defective cells during development. In association with their phagocytic and metabolic activities, they can act as storage cells, for example, in relation to inert particles, lipid, and iron; in certain circumstances, crystalloid material may be found in them. Their extensive relationships may enable the marrow macrophages to play an important role both in mediating hemopoiesis and in controlling the delivering of cells into the circulation. The phagocytic, digestive, and storage functions of normal marrow macrophages are highlighted by the changes in some pathological disorders involving the mononuclear phagocytic system.

Note Added in Proof. Since this chapter was written, a paper has been published by Hayhoe *et al.* (1979) on acquired lipidosis of human marrow macrophages which is relevant to the discussion in Sections 4 and 6.

REFERENCES

Allen, T. D., and Dexter, T. M., 1976, Cellular interrelationships during *in vitro* granulopoiesis, *Differentiation* **6**:i91.

Bankston, P. W., and De Bruyn, P. P. H., 1974, The permeability to carbon of the sinusoidal lining cells of the embryonic rat liver and rat bone marrow, *Am. J. Anat.* **141**:281.

Ben-Ishay, Z., 1974, Reticular cells in erythroid islands: Their erythrophagocytic function, *J. Reticuloendothal. Soc.* **16**:340.

Ben-Ishay, Z., and Yoffey, J. M., 1971a, Erythropoietic islands in rat bone marrow in different functional states. I. Changes in primary hypoxia, *Israel J. Med. Sci.* **7**:948.

Ben-Ishay, Z., and Yoffey, J. M., 1971b, Reticular cells of erythroid islands of rat bone marrow in hypoxia and rebound, *J. Reticuloendothel. Soc.* **10**:482.

Ben-Ishay, Z., and Yoffey, J. M., 1974, Ultrastructural studies of erythroblastic islands of rat bone marrow. III. Effect of sublethal irradiation, *Lab. Invest.* **30**:320.

Bentfeld, M. E., Nichols, B. A., and Bainton, D. F., 1977, Ultrastructural localization of peroxidase in leukocytes of rat bone marrow and blood, *Anat. Rec.* **187**:219.

Berman, I., 1967, The ultrastructure of erythroblastic islands and reticular cells in mouse bone marrow, *J. Ultrastruct. Res.* **17**:291.

Bessis, M., 1946, Contribution à l'étude cytologique des réticulosarcomes, *Sang* **17**:7.

Bessis, M., 1958, L'îlot érythroblastique, unité fonctionelle de la moelle osseuse, *Rev. Hématol.* **13**:8.

Bessis, M., 1973, *Living Blood Cells and Their Ultrastructure* (Translated by R. I. Weed), Springer Verlag, Berlin.

Bessis, M. C., and Breton-Gorius, J., 1962, Iron metabolism in the bone marrow as seen by electron microscopy: A critical review, *Blood* **19**:635.

Blom, J. 1977, The ultrastructure of macrophages found in contact with plasma cells in the bone marrow of patients with multiple myeloma, *Acta Pathol. Microbiol. Scand. A* **85**:335.

Boggs, D. R., 1975, Physiology of neutrophil proliferation, maturation and circulation, *Clin. Haematol.* **4**:535.

Braunsteiner, H., and Schmalzl, F., 1970, Cytochemistry of monocytes and macrophages, in: *Mononuclear Phagocytes* (R. van Furth, ed.), pp. 62–81, Blackwell, Oxford.

Brånemark, P-I, 1959, Vital microscopy of bone marrow in rabbit, *Scand, J. Clin. Lab. Invest. II Suppl.* **38**.

Brumpt, L. C., 1940–1941, La ponction de la moelle osseuse dans la lèpre; présence du bacille de Hansen et de la cellule écumeuse de Virchow, *Sang* **14**:403.

Callahan, W. P., and Lorincz, A. E., 1966, Hepatic ultrastructure in Hurler's syndrome, *Am. J. Pathol.* **48**:277.

Campbell, F. R., 1972, Ultrastructural studies of transmural migration of blood cells in the bone marrow of rats, mice and guinea pigs, *Am. J. Anat.* **135**:521.

Carr, I., 1968, Some aspects of fine structure of the reticuloendothelial system: The cells which clear colloids from the blood stream, *Z. Zellforsch. Mikrosk. Anat.* **89**:355.

Chamberlain, J. K., Weiss, L., and Weed, R. I., 1975, Bone marrow sinus cell packing: a determinant of cell release, *Blood* **46**:91.

Cousin, G., 1898, Notes biologiques sur l'endothélium vasculaire, *C. R. Soc. Biol. (Paris)* **5**:454.

Daems, W. T., and Brederoo, P., 1970, The fine structure of mononuclear phagocytes as revealed by freeze-etching, in: *Mononuclear Phagocytes* (R. van Furth, ed.), pp. 29–42, Blackwell, Oxford.

Daems, W. T., Wisse, E., Brederoo, P., and Emeis, J. J., 1975. Peroxidatic activity in monocytes and macrophages, in: *Mononuclear Phagocytes in Immunity, Infection, and Pathology* (R. van Furth, ed.), pp. 57–77, Blackwell, Oxford.

Dalton, A. J., Law, W. L., Moloney, J. B., and Manaker, R. A., 1961, An electron microscopic study of a series of murine lymphoid neoplasms. *J. Natl. Cancer Inst.* **27**:747.

De Bruyn, P. P. H., Michelson, S., and Thomas, T. B., 1971, The migration of blood cells of the bone marrow through the sinusoidal wall, *J. Morphol.* **133**:417.

De Bruyn, P. P. H., Michelson, S., and Becker, R. P., 1975, Endocytosis, transfer tubules and lysosomal activity in myeloid sinsusoidal endothelium, *J. Reticuloendothel. Soc.* **53**:133.

De Marsh, Q. B., and Kautz, J., 1957, The submicroscopic morphology of Gaucher cells, *Blood* **12**:324.

Dosik, H., Rosner, F., and Sawitsky, A., 1972, Acquired lipidosis: Gaucher-like cells and "blue cells" in chronic granulocytic leukemia, *Semin. Hematol.* **9**:309.

Eisen, H., 1978, Spectrin metabolism in normal and leukaemic erythroid tissue [quoted by Mohandas and Prenant (1978)].

Evans, H. M., 1915, The macrophages of mammals, *Am. J. Physiol.* **37**:243.

Fand, I., and Gordon, A. S., 1957, A quantitative study of the bone marrow in the guinea pig throughout life, *J. Morphol.* **100**:473.

Frisch, B., Lewis, S. M., and Swan. M., 1976, Intercellular contacts between erythroid precursors in the bone marrow in dyserythropoiesis, *Br. J. Haematol.* **33**:469.

Fruhman, G. J., 1966, Effects of starvation and refeeding on erythropoiesis in mice, *Z. Zellforsch. Mikrosk. Anat.* **75**:258.

Fruhman, G. J., and Gordon, A. S., 1955, A quantitative study of of adrenal influences upon the cellular elements of bone marrow, *Endocrinology* **57**:711.

Goldstein, K., 1975, Erythroblastic islands grown *in vitro, Blood* **46**:1023.

Goodman, J. R., Wallerstein, R. O., and Hall, S. G., 1968, The ultrastructure of bone marrow histiocytes in megaloblastic anaemia and the anaemia of infection, *Br. J. Haematol.* **14**:471.

Grampa, G., 1971, Radiation injury with particular reference to Thorotrast, in: *Pathology Annual* (S. C. Sommers, ed.), pp. 147–169, Butterworth, London.

Halpern, B. N., Benacerraf, B., and Biozzi, G., 1953, Quantitative study of the granulopectic activity of the reticulo-endothelial system. I. The effect of the ingredients present in India Ink and of substances affecting blood clotting *in vivo* on the fate of carbon particles administered intravenously in rats, mice and rabbits, *Br. J. Exp. Pathol.* **34**:426.

Hammarsten, G., and Reis, G. Von, 1957, A benign reticulopathy associated with lipophagic macrophages in the bone marrow occurring in late adult life, *Acta Med. Scand.* **158**:465.

Harris, P. F., Haigh, G., and Kugler, J. H., 1963, Observations on the accumulation of mononuclear cells and the activities of reticulum cells in bone marrow of guinea-pigs recovering from whole body gamma irradiation, *Acta Haematol.* **29**:166.

Hartroft, W. S., and Porta, E. A., 1965, Ceroid, *Am. J. Med. Sci.* **250**:324.

Hashimoto, M., 1936, Über das kapilläre Blutgefässystem des Kaninchenknochenmarks, *Trans. Soc. Pathol. Jpn.* **26**:300.

Hashimoto, M., 1966, Light-microscopic investigation of reticuloendothelial cells in the bone marrow, *Tohoku J. Exp. Med.* **89**:177.

Hayhoe, F. G. J., Flemans, F. J., and Cowling, D. C., 1979, Acquired lipidosis of marrow macrophages: Birefringent blue crystals and Gaucher-like cells, sea-blue histiocytes, and grey-green crystals, *J. Clin. Pathol.* **32**:420.
Hirsch, J. G., and Fedorko, M. E., 1970, Morphology of mouse mononuclear phagocytes, in: *Mononuclear Phagocytes* (R. van Furth, ed.), pp. 7–28, Blackwell, Oxford.
Hoffman, F. A., and Langerhans, P., 1869, Über den Verbleib des in die Circulation eingeführten Zinnobers, *Virchows Arch. A* **48**:303.
Horland, A. A., Wolman, S. R., Murphy, M. J., and Moore, M. A. S., 1977, Proliferation of erythroid colonies in semi-solid agar. *Br. J. Haematol.* **36**:495.
Hoyer, H. 1869, Zur Histologie des Knochenmarkes, *Zentralbl. Med. Wiss.* **7**:244, 257.
Huang, T-S., 1971, Passage of foreign particles through the sinusoidal wall of the rabbit bone marrow—an electron microscopic study, *Acta Pathol. Jpn.* **21**:349.
Huber, H., and Fudenberg, H. H., 1968, Receptor sites of human monocytes for IgG, *Int. Arch. Allergy Appl. Immunol.* **34**:18.
Hudson, G., 1958, Effect of hypoxia on bone marrow volume, *Br. J. Haematol.* **4**:239.
Hudson, G., 1968a, Quantitative study of eosinophil granulocytes, *Semin. Haematol.* **5**:166.
Hudson, G., 1968b, Crystalloid material in macrophages of mouse bone marrow, *Acta Anat.* **71**:100.
Hudson, G., 1969, Variations in the amount of crystalloid material in marrow macrophages in mice of different body weights, *Acta Anat.* **73**:136.
Hudson, G., 1970, Ultrastructural aspects of phagocytosis and storage in the bone marrow, Proceedings of the Thirteenth International Congress of Haematology, Munich (August 1970), p. 391.
Hudson, G. 1973, The LT compartment, in: *Haemopoietic Stem Cells* (G. E. W. Wolstenholme and M. O'Connor, eds.), Ciba Foundation Symposium 13, pp. 40–41, Associated Scientific Publishers, New York.
Hudson, G., and Shortland, J. R., 1974, Crystalloid material in marrow macrophages of specific pathogen-free mice, *Acta Anat.* **87**:404.
Hudson, G., and Yoffey, J. M., 1963, Reticuloendothelial cells in the bone marrow of the guinea pig, *J. Anat.* **97**:409.
Hudson, G., and Yoffey, J. M., 1966, The passage of lymphocytes through the sinusoidal endothelium of guinea-pig bone marrow, *Proc. Roy. Soc. B* **165**:486.
Hudson, G., and Yoffey, J. M., 1968, Ultrastructure of reticuloendothelial elements in guinea-pig bone marrow, *J. Anat.* **103**:515.
Huggins, C., and Noonan, W. J., 1936, An increase in reticuloendothelial cells in outlying bone marrow consequent upon a local increase in temperature, *J. Exp. Med.* **64**:275.
Hughes-Jones, N. C., 1961, The use of ^{51}Cr and ^{59}Fe as red cell labels to determine the fate of normal erythrocytes in the rabbit, *Clin. Sci.* **20**:315.
Hughes-Jones, N. C., and Cheney, B., 1961, The use of ^{51}Cr and ^{59}Fe as red cell labels to determine the fate of normal erythrocytes in the rat, *Clin. Sci.* **20**:323.
Huhn, D., 1966, Die Feinstruktur des Knochenmarks der Ratte bei Anwendung neuerer Aldehydfixationen, *Blut* **13**:291.
Ichikawa, Y., and Yoshioka, M., 1960, Electron microscopical and cytochemical observations on crystal formation in myeloid leukemia of SL and S strain mice, *Acta Haematol. Jpn.* **23**:669.
Jordan, S. W., 1964, Electron microscopy of Gaucher cells, *Exp. Mol. Pathol.* **3**:76.
Keene, W. R., and Jandl, J. H., 1965, The reticuloendothelial mass and sequestering function of rat bone marrow, *Blood* **26**:157.
Keyserlingk, D. G., Boll. I., and Albrecht, M., 1972, Elektronenmikroskopie und cytochemie der "Gaucher-zellen" bei chronischer myelose, *Klin. Wochenschr.* **50**:510.
Kitchens, C. S., 1977, Clinical observations of human bone marrow macrophages, *Medicine* **56**:503.
Kiyono, K., 1914, *Die Vitale Karminspeicherung*, Gustav Fischer, Jena.
Lee, R. E., and Ellis, L. D., 1970, "Gaucher cells" in chronic myelogenous leukemia, *Am. J. Pathol.* **59**:53a.
Long. R. G., Lake, B. D., Pettit, J. E., Scheuer, P. J., and Sherlock, S., 1977, Adult Niemann-Pick disease: Its relationship to the syndrome of the sea-blue histiocyte, *Am. J. Med.* **62**:627.
Mackenzie, D. H., 1970, *The Differential Diagnosis of Fibroblastic Disorders*, Blackwell, Oxford.

Mackenzie, D. H., 1975, Miscellaneous soft tissue sarcomas, in: *Recent Advances in Pathology*, Vol. 9 (C. V. Harrison and K. Weinbren eds.), p. 183, Churchill, London.
Marton, P. F., 1971, Erythrophagocytosis in the rat bone marrow following transfusion of heat-denatured erythrocytes, *Scand. J. Haematol.* **8**:328.
Marton, P. F., 1973, Erythrophagocytosis in the rat bone marrow following splenectomy, *Scand. J. Haematol.* **10**:81.
Marton, P. F., 1975a, Ultrastructural study of erythrophagocytosis in the rat bone marrow. I. Red cell engulfment by reticulum cells, *Scand. J. Haematol. Suppl.* **23**:1.
Marton, P. F., 1975b, Ultrastructural study of erythrophagocytosis in the rat bone marrow. II. Iron metabolism in reticulum cells following red cell digestion, *Scand. J. Haematol. Suppl.* **23**:27.
Metcalf, D., and Moore, M. A. S., 1973, Regulation of growth and differentiation in haemopoietic colonies growing in agar, in: *Haemopoietic Stem Cells* (G. E. W. Wolstenholme and M. O'Connor, eds.), Ciba Foundation Symposium 13, pp. 157–175, Associated Scientific Publishers, New York.
Metcalf, D., and Wilson, J. W., 1976, Endotoxin-induced size change in bone marrow progenitors of granulocytes and macrophages, *J. Cell Physiol.* **89**:381.
Miescher, P., 1957, The role of the reticulo-endothelial system in haematoclasia, in: *Physiopathology of the Reticulo-Endothelial System* (B. N. Halpern, B. Benacerraf, and J. F. Delafresnaye, eds.), pp. 147–171, Blackwell, Oxford.
Mohandas, N., and Prenant, M., 1978, Three dimensional model of bone marrow, *Blood* **51**:633.
Moore, M. A. S., 1976, Regulatory role of macrophages in hemopoiesis, *J. Reticuloendothel. Soc.* **20**:89.
Muto, M., 1976, A scanning and transmission electron microscopic study on rat bone marrow sinuses and transmural migration of blood cells, *Acta Histol. Jpn.* **39**:51.
Nagao, K, 1920, The fate of india ink injected into the blood. I. General observations, *J. Infect. Dis.* **27**:527.
Nagao, K., 1921, The fate of india ink injected into the blood, II. The formation of intracellular granules and their movements, *J. Infect. Dis.* **28**:294.
Neimann, N., Grignon, G., Gentin, G., Guedenet, J. C., and Vidailhet, M., 1968, Etude de la cellule de Gaucher en microscopie électronique, *Ann. Pediatr.* **44**:2393.
Novikoff, A. B., Essner, E., and Quintana, N., 1964, Golgi apparatus and lysosomes, *Fed. Proc.* **23**:1010.
Ogawa, T., 1975, An electron microscopic study of transmural migration of blood cells in the bone marrow, *Acta Haematol. Jpn.* **38**:656.
Osmond, D. G., and Everett, N. B., 1962, Nucleophagocytosis in the bone marrow, *Nature* **196**:488.
Patek, P. R., and Bernick, S. 1960, Time sequence studies of reticuloendothelial responses to foreign particles, *Anat. Rec.* **138**:27.
Patt, H. M., and Maloney, M. A., 1964, A model of granulocyte kinetics, *Ann. N. Y. Acad. Sci.* **113**:515.
Peabody, F. W., and Broun, G. O., 1925, Phagocytosis of erythrocytes in the bone marrow with special reference to pernicious anemia, *Am. J. Pathol.* **1**:169.
Pennelli, N., Scaravilli, F., and Zacchello, F., 1969, The morphogenesis of Gaucher cells investigated by electron microscopy, *Blood* **34**:331.
Peters, S. P., Lee, R. E., and Glew, R. H., 1977, Gaucher's disease, a review, *Medicine* **56**:425.
Policard, A., and Bessis, M., 1958, Sur un mode d'incorporation des macromolécules par la cellule, visible au microscope électronique: La rhophéocytose, *C. R. Acad. Sci.* **246**:3197.
Policard, A., and Bessis, M., 1962, Micropinocytosis and rhopheocytosis, *Nature* **194**:110.
Ponfick, E., 1869, Studien über die Schicksale körniger Farbstoffe im Organismus, *Virchows Arch. Pathol. Anat.* **48**:1.
Ralph, P. H., 1948, The occurrence and significance of motile erythrocytes in human blood and marrow in anemic states, *Blood* **3**:295.
Robbins, D., Fahimi, H. D., and Cotran, R. S., 1971, Fine structural cytochemical localization of peroxidase activity in rat peritoneal cells: Mononuclear cells, eosinophils, and mast cells, *J. Histochem. Cytochem.* **19**:571.
Sawitsky, A., Rosner, F., and Chodsky, S., 1972, The sea-blue histiocyte syndrome, a review: Genetic and biochemical studies, *Semin. Hematol.* **9**:285.

Shaklai, M., and Tavassoli, M., 1978, Structural analysis of hemopoiesis in S1/S1[d] anemic mice, *Am. J. Pathol.* **90**:633.
Silverstein, M. N., and Ellefson, R. D., 1972, The syndrome of the sea-blue histiocyte, *Semin. Hematol.* **9**:299.
Silverstein, M. N., Ellefson, R. D., and Ahern, E. J., 1970, The syndrome of the sea-blue histiocyte. *N. Engl. J. Med.* **282**:1.
Simon, G. T., and Burke, J. S., 1970, Electron microscopy of the spleen. III. Erythrophagocytosis, *Am. J. Pathol.* **53**:451.
Sjögren, U., 1976, Erythroblastic islands and extra-medullary erythropoiesis in chronic myeloid leukaemia, *Acta Haematol.* **55**:272.
Skutelsky, E., and Danon, D., 1969, Reduction of surface charge as an explanation of the recognition by macrophages of nuclei expelled from normoblasts, *J. Cell Biol.* **48**:8.
Skutelsky, E., and Danon, D., 1972, On the expulsion of the erythroid nucleus and its phagocytosis, *Anat. Rec.* **173**:123.
Stavem, P., Ly, B., and Bjørneklett, A., 1977, Light green crystals in May-Grünwald and Giemsa-stained bone marrow macrophages in patients with myeloid leukaemia, *Scand. J. Haematol.* **18**:67.
Stecher, V. J., 1970, Synthesis of proteins by mononuclear phagocytes, in: *Mononuclear Phagocytes* (R. van Furth, ed.), pp. 133–150, Blackwell, Oxford.
Tanaka, Y., Brecher, G., and Fredrickson, D. S., 1963, Cellules de la maladie de Niemann-Pick et de quelques autres lipidoses, *Nouv. Rev. Fr. Hématol.* **3**:5.
Tanaka, Y., Brecher, G., and Bull, B., 1966, Ferritin localisation on the erythroblast cell membrane and rhopheocytosis in hypersiderotic human bone marrows, *Blood* **28**:758.
Tavassoli, M., 1974, Bone marrow erythroclasia: The function of perisinal macrophages relative to the uptake of erythroid cells, *J. Reticuloendothel. Soc.* **15**:163.
Tavassoli, M., 1976, Ultrastructural development of bone marrow adipose cell, *Acta Anat.* **94**:65.
Tavassoli, M., 1977a, Adaptation of marrow sinus wall to fluctuation in the rate of cell delivery: Studies in rabbits after blood letting, *Br. J. Haematol.* **35**:25.
Tavassoli, M., 1977b, Intravascular phagocytosis in the rabbit bone marrow: A possible fate of normal senescent red cells, *Brit. J. Haematol.* **36**:323.
Tavassoli, M., and Shaklai, M., 1979, Junctional structures in hemopoiesis: A study of bone marrow using freeze-fracture and lanthanum impregnation techniques, *Br. J. Haematol.* **43**:235.
Territo, M. C., and Cline, M. J., 1975, Mononuclear phagocyte proliferation, maturation and function, *Clin. Haematol.* **4**:685.
Toujas, L., Juif, J. G., Cussac, Y., Porte, A., and Platt, E., 1966, Sur les modifications ultrastructurales du foie dans un cas de maladie de Gaucher, *Ann. Anat. Pathol.* **11**:101.
Turk, J. L., 1971, Granuloma formation in lymph nodes, *Proc. R. Soc. Med.* **64**:942.
Unanue, E. R., and Cerottini, J. C., 1970, The immunogenicity of antigen bound to the plasma membrane of macrophages, *J. Exp. Med.* **131**:711.
Unanue, E. R., Cerottini, J. C., and Bedford, M., 1969, Persistence of antigen on the surface of macrophages, *Nature* **222**:1193.
van Dorpe, A., Broeckaert-van Orshoven, A., Desmet, V., and Verwilghen, R. L., 1973, Gaucher-like cells and congenital dyserythropoietic anaemia II (HEMPAS), *Br. J. Haematol.* **25**:165.
van Furth, R., and Cohn, Z. A., 1968, The origin and kinetics of mononuclear phagocytes, *J. Exp. Med.* **128**:415.
van Furth, R., and Fedorko, M. E., 1976, Ultrastructure of mouse mononuclear phagocytes in bone marrow colonies grown *in vitro, Lab. Invest.* **34**:440.
von Ehrenstein, G., and Lockner, D., 1959, Physiologischer Erythrozytenabbau, *Acta Haematol.* **22**:129.
Watanabe, 1964, An electron microscopic observation on reticulum cell in the bone marrow, Proc. Fourth International Symposium on Reticulodothelial System, Otsu and Kyoto, May 1964, pp. 94–107.
Weiss, L, 1961, An electron microscopic study of the vascular sinuses of the bone marrow of the rabbit, *Johns Hopkins Hosp. Bull.* **108**:171.
Weiss, L., 1965, The structure of bone marrow. Functional interrelationships of vascular and

hematopoietic compartments in experimental hemolytic anemia. An electron microscopic study, *J. Morphol.* **17**:467.
Weiss, L., 1970, Transmural cellular passage in vascular sinuses of rat bone marrow, *Blood* **36**:189.
Weiss, L., 1976, The hematopoietic microenvironment of the bone marrow: An ultrastructural study of the stroma in rats, *Anat. Rec.* **186**:161.
Wickramasinghe, S. N., and Bush, V., 1975, Observations on the ultrastructure of erythropoietic cells and reticulum cells in the bone marrow of patients with homozygous β-thalassaemia, *Br. J. Haematol.* **30**:395.
Wickramasinghe, S. N., and Hughes, M., 1978, Some features of bone marrow macrophages in patients with homozygous β-thalassaemia, *Br. J. Haematol.* **38**:23.
Wislocki, G. B., 1921, Experimental observations on bone marrow, *Johns Hopkins Hosp. Bull.* **32**:132.
Wislocki, G. B., 1924, On the fate of carbon particles injected into the circulation with especial reference to the lungs, *Am. J. Anat.* **32**:423.
Witzleben, C. L., Drake, W. L., Sammon, J., and Mohabbat, O. M., 1970, Gaucher's cells in acute leukemia of childhood, *J. Pediatr.* **76**:129.
Yang, H-Y., Whest, G. M., and Nishimura, E. T., 1979, Age-related variations of paracrystalline inclusions in central reticular cells of mouse bone marrow, *Exp. Mol. Pathol.* **30**:303.
Zaino, E. C., Rossi, M. B., Pham, T. D., and Azar, H. A., 1971, Gaucher's cells in thalassemia, *Blood* **38**:457.
Zamboni, L., and Pease, D. C., 1961, The vascular bed of red bone marrow, *J. Ultrastruct. Res.* **5**:65.
Zucker-Franklin, D., Grusky, G., and Marcus, A., 1978, Transformation of monocytes into "fat" cells, Personal communication.
Zyngier, F. R., 1976, Cytoplasmic inclusions, *N. Engl. J. Med.* **295**:1483.

9

On the Fine Structure and Function of Rat Liver Kupffer Cells

E. WISSE

1. THE FINE STRUCTURE OF KUPFFER CELLS

In perfusion-fixed livers (Fahimi, 1967; Wisse, 1970) Kupffer cells are easily recognized on the basis of their fine structural characteristics. The shape of the cell is variable in such a way that it gives the impression of a cell which is either rearranging its internal contents, or is moving along the sinusoid. Cytoplasmic processes may radiate in all directions, the thicker ones giving the cell a stellate shape. This basic shape of the Kupffer cell can be observed light microscopically in thick (10–75 μm) Vibratome sections which were incubated for the demonstration of peroxidase, which specifically stains Kupffer cells (Fahimi, 1970; Widmann *et al.*, 1972; Wisse, 1974a); by using either bright or dark field illumination, Kupffer cells can be studied for their shape and distribution in the tissue. In these preparations, Kupffer cells are seen to be preferentially located in the periphery of the liver lobule (Figure 1).

Kupffer cells lie on endothelial cells or are embedded in the endothelial lining, and they may even partly invade the space of Disse. In all these circumstances the Kupffer cells maintain a close contact with endothelial cells. This attachment does not leave open gaps between the cells; specialized cell contacts such as desmosomes are, however, absent. Thick processes of Kupffer cells may traverse the sinusoidal lumen, which undoubtedly will influence the flow of red blood cells through the sinusoid involved. Also, the bulging of the body of the Kupffer cell into the sinusoid and the apparent swelling of the cells after heavy phagocytosis, might well influence local microcirculatory dynamics (Figure 2).

Kupffer cells are also seen in direct contact with the other sinusoidal elements like fat-storing cells, pit cells, and reticulin fibers. They occasionally make contact with almost all types of blood cells. None of these observations suggests

E. WISSE • Laboratory for Cell Biology and Histology, Free University of Brussels, 1090 Brussels-Jette, Belgium.

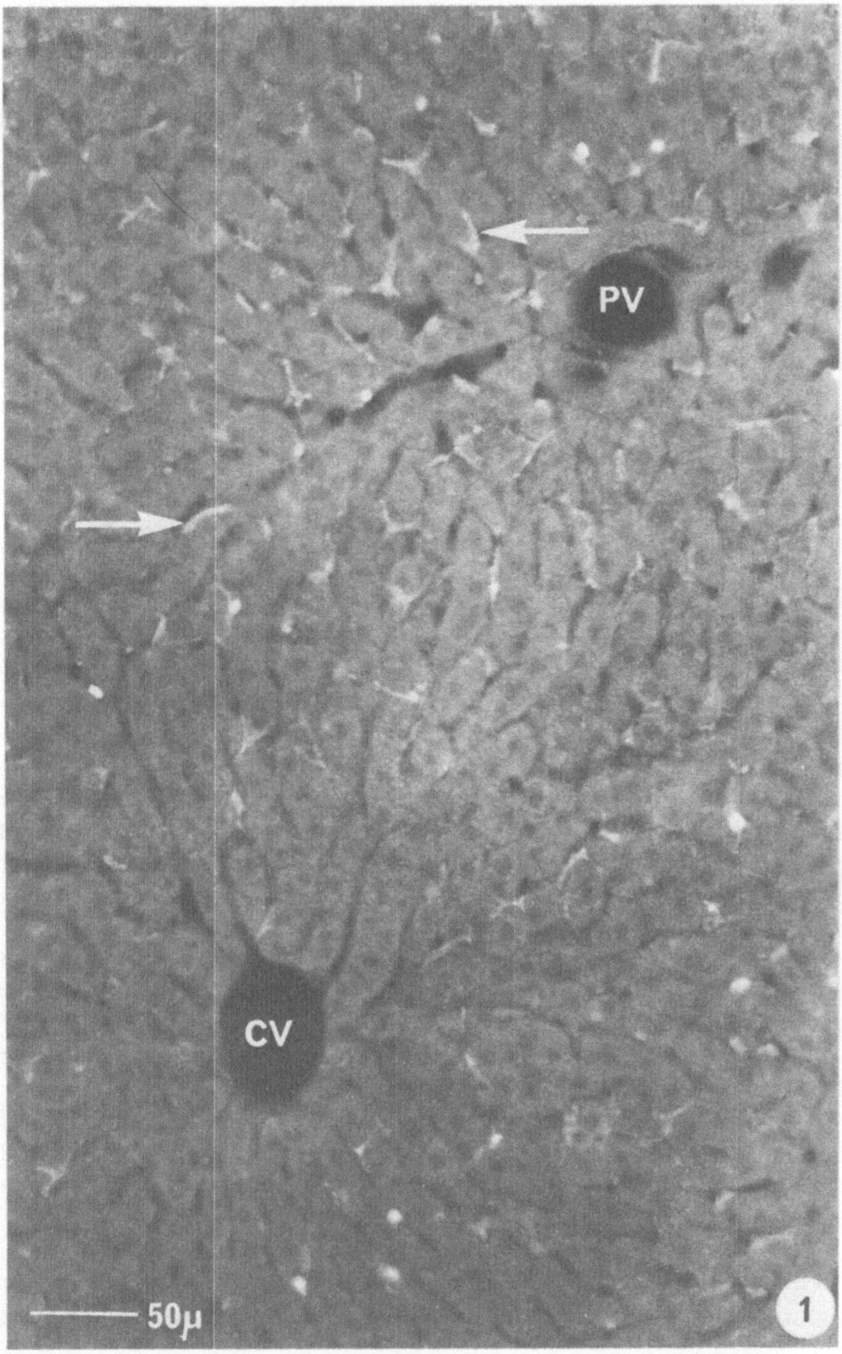

FIGURE 1. Dark field micrograph of a 20 μm Vibratome section of perfusion-fixed rat liver incubated for peroxidase. The light scattering properties of the reaction product in Kupffer cells can be used to study their shape and distribution within the tissue. Within the section, a branch of the portal vein (PV) and central vein (CV) can be recognized, as well as parenchymal cells and Kupffer cells (arrows).

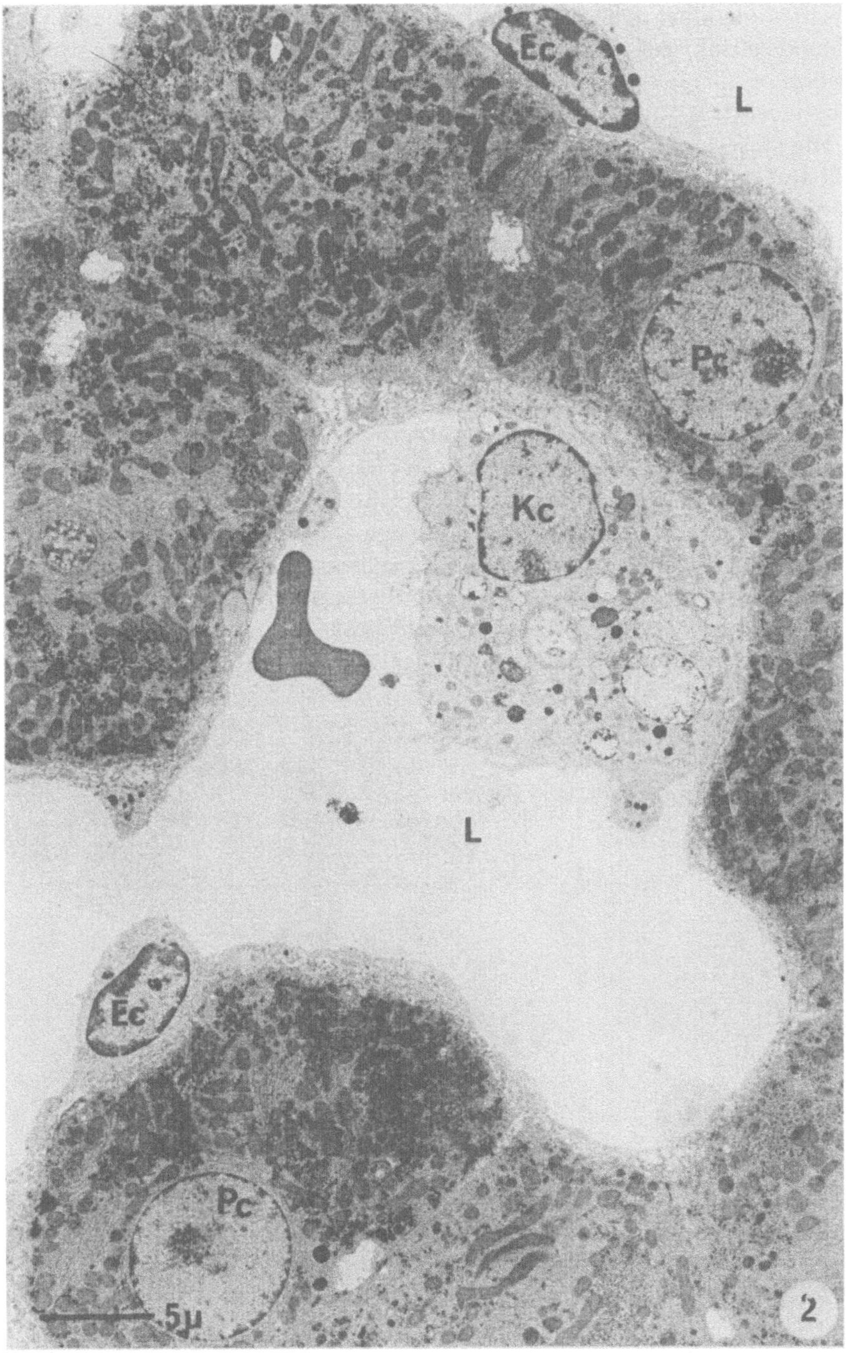

FIGURE 2. Electron micrograph at low magnification of rat liver, showing different types of cell. A Kupffer cell (Kc) and two endothelial cells (Ec) are seen at the border of the sinusoid. L, Sinusoidal lumen; Pc, parenchymal cell.

a specific functional relationship in normal rat liver, such as the phagocytosis of red blood cells (Bissell *et al.*, 1972) and other blood cells, the possible uptake and digestion of collagen fibers, or the presentation of antigen to lymphocytes. Contact between two Kupffer cells is only very seldom observed.

The surface of Kupffer cells, in normal, steady state conditions, bears microvilli and only occasionally lamellipodia. These surface specialization can best be studied either *in situ* (Figure 3) (Motta, 1975, 1977; Muto and Fujita, 1977; Satodate *et al.*, 1977; Vonnahme, 1977) or in cell cultures, with the scanning electron microscope. There seems to be a slight variation in the number of microvilli per unit cell surface; in normal animals, flat surfaces and also highly microvillous surfaces may be found. In animals in which the RES was stimulated by repeated injections of zymosan, the surfaces of Kupffer cells seem to be highly microvillous, although flat surfaces can still be found. The microvilli at the side of the Disse space are mostly intermingled with the microvilli of the parenchymal cells and this may, together with the contacts to endothelial and fact-storing cells, accomplish the attachment of the cells to the sinusoidal wall. There is never a basement membrane between a Kupffer cell and the space of Disse.

Specific organelles found in Kupffer cells are the so-called wormlike structures. These tubular profiles are apparently formed by tubular invaginations of the cell membrane. They consist, in longitudinal sections, of two parallel membranes about 140 nm apart, filled with faintly stained filamentous material giving

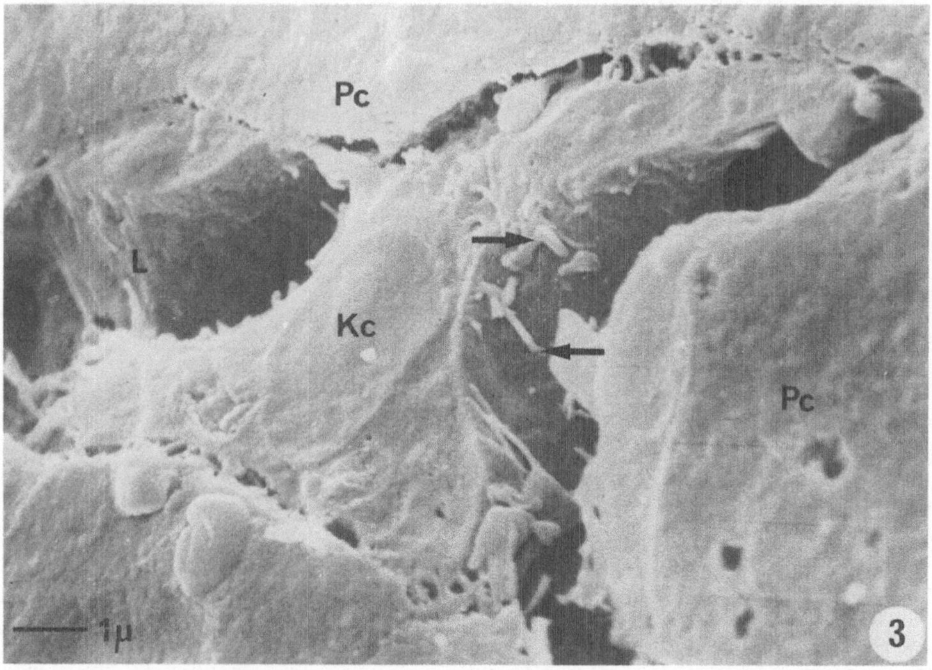

FIGURE 3. Scanning electron micrograph of a rat liver Kupffer cell (Kc), showing the shape and surface structure of the cell. A few microvilli (arrows) can be discerned. Two other sinusoidal cells can be observed. Parenchymal cells (Pc) lie peripherally.

rise to a dense midline running parallel and in the middle of the membranes. It is thought that the material inside the wormlike structures represents the thick, fuzzy coat present on the membranes. This coat (70 nm) is not seen after routine glutaraldehyde perfusion fixation, but is stabilized after plasma membrane invagination and probable fusion of the apical parts of the fuzzy coat molecules, which might cause the electron density of the microvilli. The plasma membrane character of the wormlike structures is further stressed by the pinching off of bristle-coated micropinocytic vesicles, a phenomenon which is also seen at the cell membrane. The function of wormlike structures is not completely clear. After injection of a number of small colloidal particles we have seen the presence of these particles within the wormlike structures at 10–30 min after injection. For this reason we consider the wormlike structures as one of the five described endocytic mechanisms which occur in sinusoidal cells (Wisse 1977a). One might object, however, that wormlike structures are in open contact with the sinusoidal lumen and that, for this reason, the wormlike structures are passively filled with particles, which are subsequently attached to the fuzzy coat in the same way as elsewhere on the sticky surface of the Kupffer cells. In this case, we might consider these structures as a reservoir of cell membrane material (Tanuma, 1978; Wisse, 1974b); this hypothesis is supported by the observation that after heavy phagocytosis, which is highly membrane consuming, wormlike structures are never found.

The further investigation of the functions of the thick, fuzzy coat is handicapped by the fact that routine glutaradehyde perfusion does not preserve this structural detail, which is assumed to play a crucial role in one of the primary functions of Kupffer cells, i.e., phagocytosis. Nevertheless, we believe this thick, fuzzy coat to be present *in vivo* for a number of reasons (Emeis, 1976; Emeis and Wisse, 1971, 1975; Wisse, 1974a):

1. The fuzzy coat is present in frozen sections of fresh liver, which are thawed in 1% osmium fixative.
2. The coat is also present after perfusion of the whole liver with 0.2% osmium.
3. Intravenous injections of a specially prepared gold colloid, with a particle size of 3.5 nm, apparently stabilizes the coat to glutaraldehyde perfusion by simply attaching to the end groups of the fuzzy coat molecules.
4. The fuzzy coat is preserved in wormlike structures as invaginations of the cell membrane.
5. In a special type of pinocytic vacuole, the fuzzy coat can be observed as an inner rim at the vacuolar membrane.

In cell cultures of isolated and purified Kupffer cells, the fuzzy coat is absent, and when special treatments as described above are applied, it can not be visualized. Also, wormlike structures and the fuzzy-coated pinocytic vacuoles are absent in cultured cells. These considerations seriously limit the comparability of *in vitro* and *in vivo* observations on Kupffer cell endocytosis, although such observations seem appropriate to study lysosomal digestion (Brouwer and Knook, 1977; Munthe-Kaas, 1977).

Earlier evidence indicated that Kupffer cells possessed at least four recognizable structural mechanisms for endocytosis (Wisse, 1977a). Three of them may be placed in the category of pinocytosis, one may be called phagocytosis in a strict sense. The pinocytic structures are described as: bristle-coated micropinocytosis, wormlike structures and (fuzzy-coated) pinocytosis (see 5 above). Phagocytosis occurs by the engulfment of relatively solid particles by sleevelike projections of hyaloplasmic cytoplasm. The pinocytic mechanisms seem to be mediated by predetermined or prefabricated structures, mainly taking up fluids, whereas the phagocytic mechanism adapts itself to the particle or cellular material to be taken up. Phagocytosis is clearly induced by the arrival of a particle at the cell surface, but pinocytosis in sinusoidal cells may be regarded as a steady-state process, since these structures, including endothelial pinocytosis, are seen in normal, untreated animals. It is not known what the turnover of these vesicles is or whether the four types of pinocytosis are subject to induced acceleration in certain circumstances. Another consequence of the presence of these structures in normal liver seems to be a contribution to the turnover of plasma constituents, since it might be expected that a certain quantity of extracellular fluid is taken up all the time.

Within the Kupffer cell cytoplasm, we may recognize a number of dense bodies and electron-lucent vacuoles. The dense bodies obviously differ in shape, diameter, and density. There can be composed a series of images showing transitions from relatively large, lucent vacuoles into dense bodies, which are lysosomes because they stain positively for acid phosphatase (Wisse *et al.*, 1974b; see also Sleyster and Knook, 1978). Both structures have a granular layer just underneath the membrane. In both lysosomes and lucent vacuoles this granular layer may simulate a second membrane.

Often more than one Golgi apparatus can be found in the cytoplasm. Bristle-coated vesicles, smaller than the pinocytic ones, pinch off at the periphery of the apparatus. In peroxidase-incubated specimens, in which the rough endoplasmic reticulum (RER), the nuclear envelope, and annulate lamellae are darkly stained, sometimes a close association of pieces of RER and one side of the Golgi apparatus can be seen. The RER shows continuity with cisternae of the annulate lamellae. No continuity of the RER and the nuclear envelope was observed. We did not observe smooth endoplasmic reticulum (SER) in connection with the RER. Free ribosomes are numerous. Further structures to be found in the Kupffer cell cytoplasm are centrioles, microfilaments, and microtubules. Mitochondria are present; they are smaller than those of the parenchymal cell. Kupffer cell cytoplasm apparently contains no glycogen, multivesicular bodies, fat droplets, or autophagic vacuoles.

Some fine structural components are characteristic for Kupffer cells, and distinguish them from other liver cells. These are wormlike structures and fuzzy coat, microvilli and pseudopodia at the cell surface, and annulate lamellae. The exclusive phagocytosis of particles larger and 0.1 μm and peroxidase staining can be added to this list. Kupffer cells characteristically show a positive reaction for peroxidase within the cisternae of the endoplasmic reticulum and the nuclear envelope (Figure 4). Peroxidase staining brings out an interesting difference

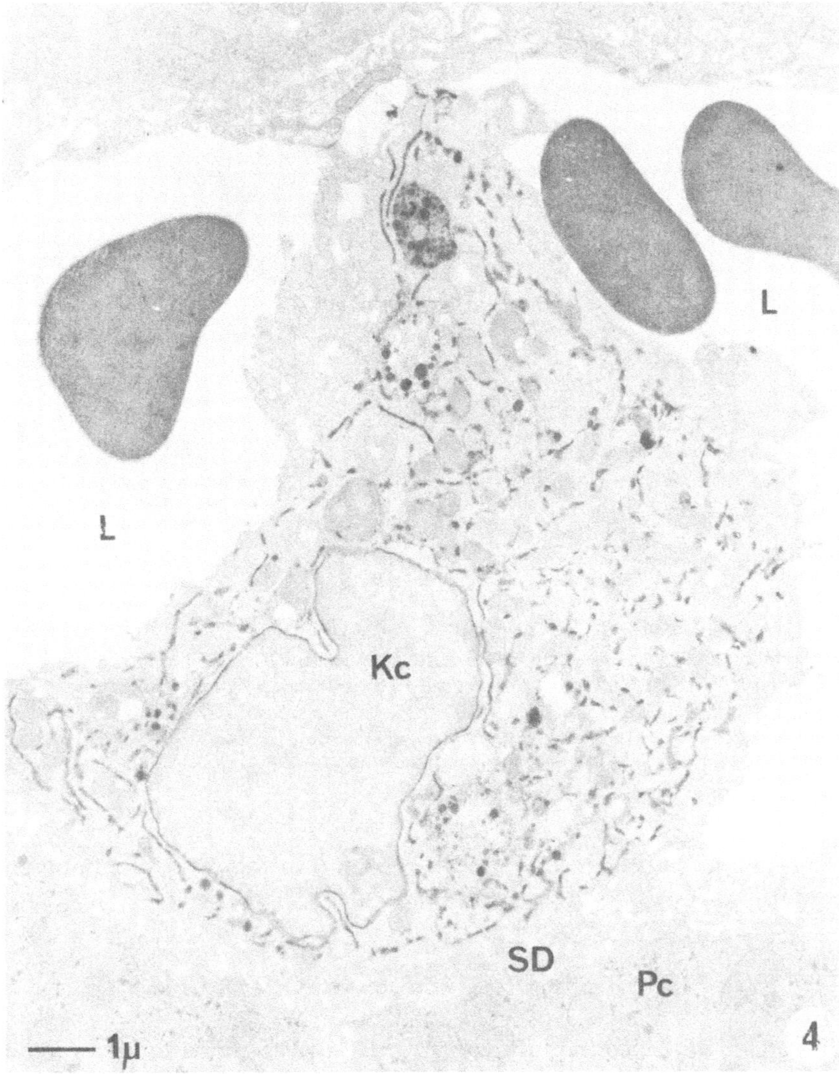

FIGURE 4. Kupffer cell (Kc) after incubation for peroxidase. An electron-dense reaction product can be observed in the cisternae of the RER and the nuclear envelope. Erythrocytes in the lumen (L) are also stained. SD, Space of Disse; PC, parenchymal cell.

between monocytes and Kupffer cells. After the intravenous injection of zymosan particles or bacteria, the monocyte shows fusion between peroxidase-containing granules and the phagosome; in the Kupffer cell, however, there is no evident fusion between the phagosome and organelles containing peroxidase (Figure 5). In the case of peroxidase, however, there is a problem for those interested in mouse liver, because Stöhr *et al.* (1978) recently demonstrated the presence of peroxidase in endothelial cells of mouse liver sinusoids. Van Berkel and Kruyt (1977) studied the biochemical characteristics of the Kupffer cell

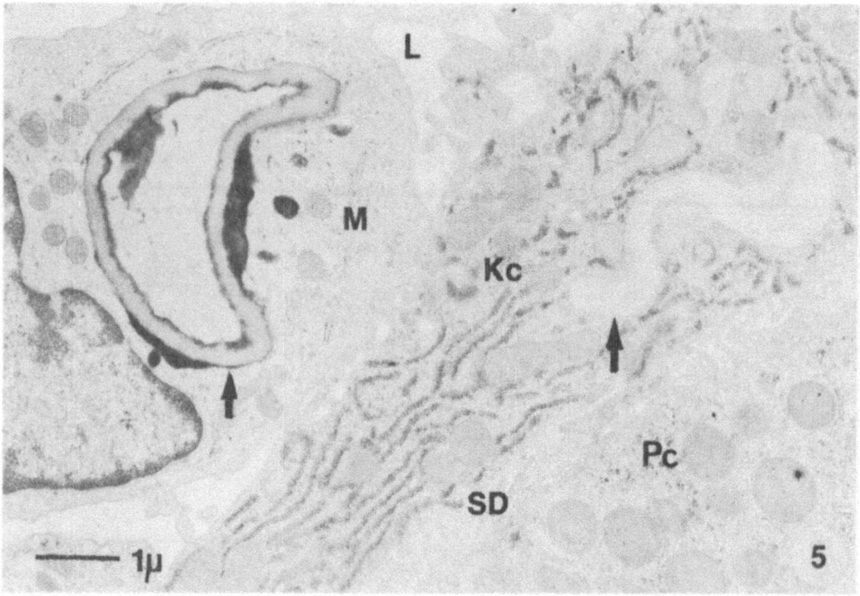

FIGURE 5. A Kupffer cell and a monocyte after incubation for peroxidase. In this case, the rat was injected with zymosan (yeast cell walls) in order to stimulate the RES. Both cells are seen to contain the zymosan particles (arrows). In the monocyte (M), the peroxidase-containing granules have fused with the phagosome, in the Kupffer cell (KC) there is no fusion of peroxidase with the vacuoles containing the zymosan. After the injection of bacteria the same reaction was observed (Wisse *et al.*, 1972). This reaction is one of the many basic differences between monocytes and Kupffer cells. L, Lumen; SD, space of Disse; Pc, parenchymal cell.

peroxidase, and cytochemical data can be found in the work of Fahimi (1970), Widmann *et al.* (1972), and Wisse (1974a).

2. THE FINE STRUCTURE OF OTHER SINUSOIDAL CELLS

For adequate recognition of Kupffer cells, it is essentiel to be aware of the presence of three other sinusoidal cell types, i.e., endothelial, fat-storing, and pit cells. A short description of these cells will be given.

2.1. THE ENDOTHELIAL CELL

Endothelial cells (Figure 6) have a streamlined shape, their perinuclear cytoplasm bulges slightly into the lumen, and their flat processes spread out to form the proper lining of the sinusoids. These endothelial processes are fenestrated, with a pore size of 0.1 μm (Wisse, 1970). The fenestrations are grouped together in so-called sieve plates. There is no basement membrane underneath the endothelial lining: the fenestrae form open connections between the lumen

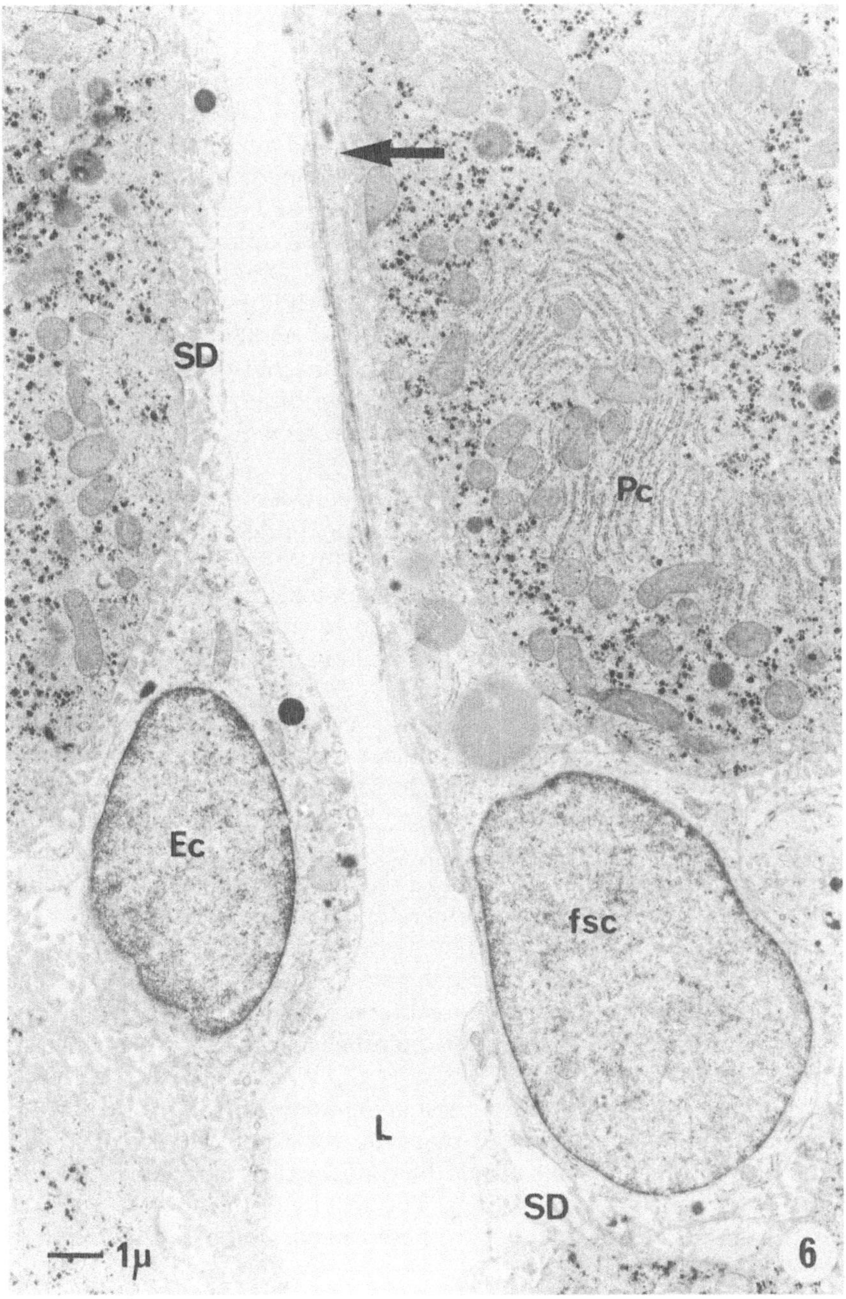

FIGURE 6. Electron micrograph of an endothelial cell and a fat-storing cell. The endothelial cell (Ec) can be recognized by the continuity with the fenestrated endothelial lining, the fat-storing cell (fsc) is situated in the space of Disse (SD) and contains fat droplets and long processes underlying the endothelial lining (arrow). L, Sinusoidal lumen; Pc, parenchymal cell.

and the space of Disse. Cell contacts between endothelial cells and between endothelial and other cells are simple.

Within the cytoplasm there are many vesicles, including bristle-coated micropinocytic (0.1 μm) and macropinocytic vesicles (0.7 μm) (Wisse, 1972), transfer tubules, small, coated vesicles of Golgi origin, and a number of vesicles whose origin and destination have not yet been determined. In addition, transitional stages between macropinocytic vacuoles and dense bodies (lysosomes) can be observed. This strong development of the vacuolar apparatus is ascribed to the continuous endocytic activity of endothelial cells. Other organelles include the Golgi apparatus, RER, SER, and centrioles. The RER may lie parallel to the cell membrane and show a loss of ribosomes on the membrane closely apposed to the cell membrane. The RER is occasionally connected to a special type of SER tubular network. The nucleus contains a sphaeridium, and transitional stages between this sphaeridium and a nucleolus have been observed. For further fine structural descriptions, see Wisse (1972).

We have never seen transitional stages between endothelial and Kupffer cells, or endothelial cells and any other cell type, under normal or experimental conditions (Wisse, 1972; Wisse and Daems, 1970). After partial hepatectomy and in neonatal and fetal livers we have observed mitoses in endothelial cells, which indicates that endothelial cells can give rise to their own offspring (see also Widmann and Fahimi, 1976). These observations have led us to conclude that endothelial cells are self-proliferating cells.

One of the main functions of endothelial cells can be found in shielding the parenchymal cells from coming into contact with blood corpuscles and other circulating particles. This shielding is, however, not complete. Fraser (1978), Fraser *et al.* (1978), and Naito and Wisse (1978) were able to demonstrate the filtering effect of the endothelial lining on the passage of chylomicrons. Van Dierendonck *et al.* (1979) demonstrated a reduction in the diameter of fenestrae under the influence of the hormones serotonin and noradrenalin.

Endothelial cells have two different pinocytic structures which contribute to the blood clearance of small colloidal particles (Wisse, 1972). Experimental lysosomal storage can be evoked by treating the animals with suramin (Buys *et al.*, 1978). The endocytic capacity of endothelial cells is further stressed by morphometric data (Blouin, 1977; Blouin *et al.*, 1977), showing that these cells might contribute about 15% of the plasma membranes, 15% of the lysosomes, and last, but not least, 45% of all pinocytic vesicles of the liver lobule. The presence of lysosomal enzymes could be confirmed in biochemical determinations and by cytochemical incubations. The activity of the enzyme arylsulfatase appeared to be three times higher in these cells as compared to Kupffer cells (Knook *et al.*, 1977).

2.2. THE FAT-STORING CELL

This third type of sinusoidal cell has been thoroughly described by others (Ito, 1973; Tanuma and Ito, 1978). The fat-storing cell (Figure 6) has a fixed localization underneath the endothelial lining within the space of Disse, and it is usually found in recesses between the parenchymal cells. The cytoplasm of these cells contains characteristic fat droplets whose number and diameter seem to vary between species and under different physiological conditions. The other organelles include multivesicular bodies, Golgi apparatus, small mitochondria, and characteristically swollen RER, giving the cell the appearance of a fibroblast. Centrioles are present, sometimes in the form of a basal body of a cilium projecting into the sinusoidal lumen. Characteristic features of fat-storing cells are long processes underlying the endothelial lining. These cytoplasmic extensions show numerous filaments and microtubules, and the close apposition to the endothelial lining also suggests a supportive or anchoring function. After the administration of vitamin A, Wake (1971) observed the proliferation of fat-storing cells and he also showed the vitamin to be present in the fat droplets. We have seen mitosis of fat-storing cells after partial hepatectomy, and Naito and Wisse (1977) also observed mitosis in fetal liver. Like other sinusoidal cells, the fat-storing cells have a capacity for self-proliferation. Fat-storing cells might play a role in the formation of collagen and resultant fibrosis and cirrhosis. If such is the case, this type of cell deserves thorough investigation.

2.3. THE PIT CELL

The fourth type of sinusoidal cell, the pit cell (Figure 7), has been described recently (Wisse *et al.*, 1976). We think this cell is a true inhabitant of rat liver sinusoids, because it is anchored to the sinusoidal wall and shows mitosis. The frequency of occurrence of this cell is low and seemingly variable. A preferential localization within the liver lobule could not be determined, due to the lack of specific staining methods at the light-microscopic level. This cell is found more often in sinusoidal cell isolates. Pit cells are nonphagocytic, although they possess many pseudopodia or flaps. These processes sometimes penetrate the endothelium and intermingle with microvilli on the surface of parenchymal cells. The pit cell has a strong polarity; all of its organelles are usually grouped at one side of the nucleus, while the other side shows electron-lucent hyaloplasm. The cytoplasm contains highly characteristic granules resembling those seen in endocrine cells occurring in the hypophysis, adrenals, and elsewhere. These characteristic granules sometimes show stages representing a transition to multivescular bodies. Normal organelles are also present, such as sparse RER, Golgi apparatus, and mitochondria. Outside the sinusoid the pit cells can be found within the tissue around portal vein branches and within granuloma-like cellular aggregates. Contacts are occasionally seen between pit cells and Kupffer cells which are either actively phagocytizing or swollen after heavy phagocytosis. The function and origin of pit cells are still unclear. We found a comparable cell in buffy-coat preparations made from rat blood, but we have only been able to find a few pit cells in semiperfusion-fixed human liver or buffy-coat preparations made from human blood.

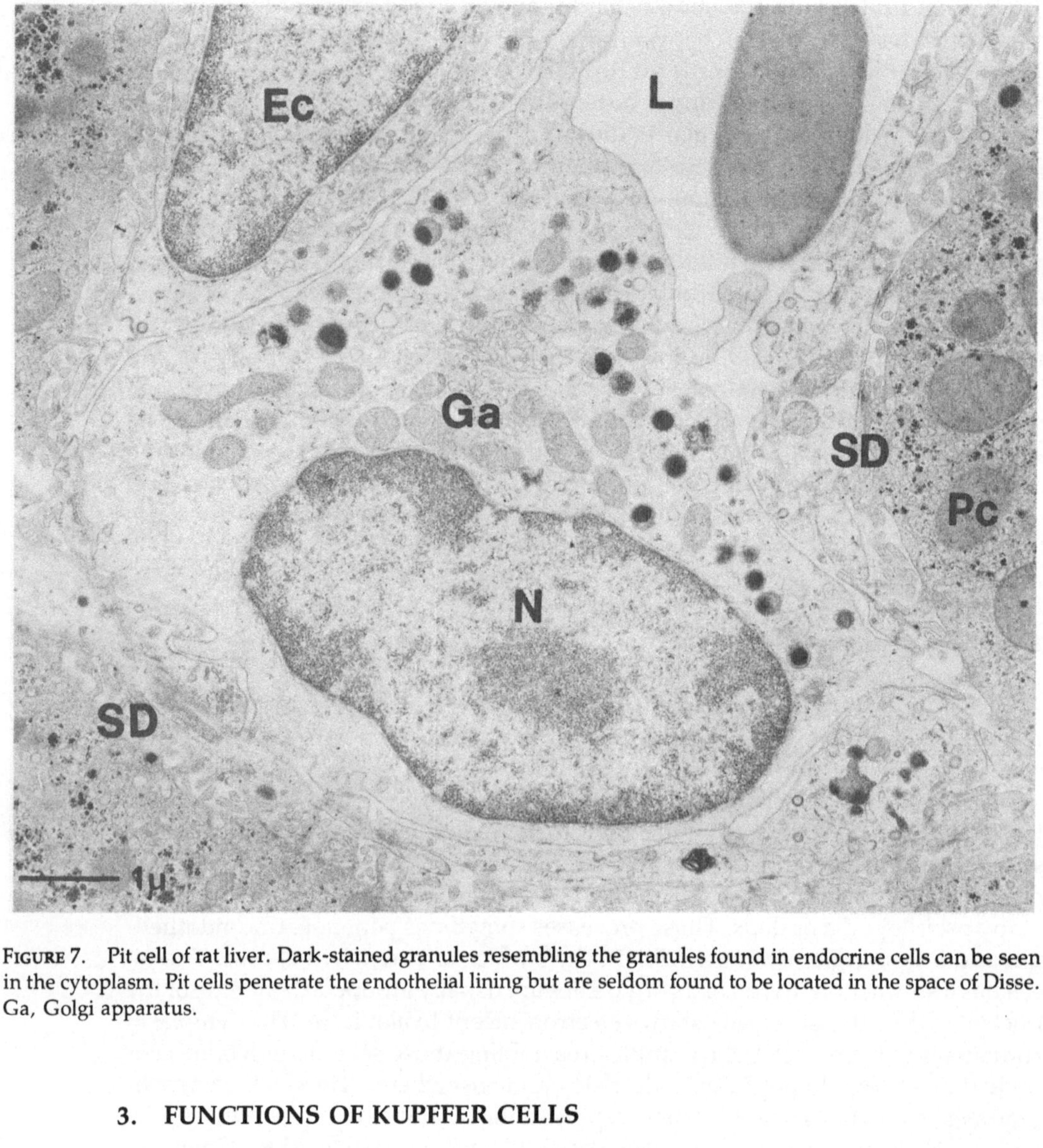

FIGURE 7. Pit cell of rat liver. Dark-stained granules resembling the granules found in endocrine cells can be seen in the cytoplasm. Pit cells penetrate the endothelial lining but are seldom found to be located in the space of Disse. Ga, Golgi apparatus.

3. FUNCTIONS OF KUPFFER CELLS

From the observations on ultrathin sections of normal rat liver, a detailed description of the fine structure of Kupffer cells can be composed. Recognition of Kupffer cells by using fine structural criteria can easily be performed, provided that, in individual cases, enough fine structural details are present. One of the most striking facts which can be derived from fine structural observations on Kupffer cells in normal adult rat liver, is the apparent lack of observations which directly suggest a specialized function. We know the Kupffer cell is a macrophage, which avidly phagocytizes injected test particles. We see the numerous lysosomes which comprise about 16% of the cell volume (Blouin, 1977), and which can digest almost all biological molecules (for references see Wisse and Knook, 1979), but in normal adult liver we can only very seldomly visualize the operation of this system. Nevertheless, there are specific functions described, which we will try to summarize in the following.

Firstly, Kupffer cells are seen to be highly active in early stages of liver development. Deimann and Fahimi (1977, 1978) and Naito and Wisse (1977) observed the presence of Kupffer cells in fetal rat liver, respectively, at 11 and 15 days of gestation. Kupffer cells at this stage show phagocytosis of extruded erythroblast nuclei and phagocytosis of erythrocytes at all stages of development, a situation which is not found in adults (Wisse, 1974a,b; see also Düllmann and Wulfhekel, 1978, and Hausmann *et al.*, 1976). They contain iron-loaded lysosomes. The position of the cell may not be restricted to sinusoids: they may also lie between developing erythroblasts or between parenchymal cells. An important observation in this material, is the fact that Kupffer cells show mitosis. This observation proves the self-proliferative capacity of Kupffer cells (Volkman, 1977; Widmann and Fahimi, 1975; Wisse, 1974b), a fact which is further supported by the occurrence of Kupffer cell mitosis in normal liver and in cell cultures (Clark and Pateman, 1978) and also after partial hepatectomy, RES stimulation, and RES blockade (Wisse, 1974b). Further evidence for Kupffer cell mitosis is given by the fact that they also seem to incorporate [^{3}H]thymidine (Crofton *et al.*, 1978; van Furth *et al.*, 1977; see also Wisse, 1978).

Secondly, a function of Kupffer cells or, more generally, of sinusoidal cells, can be ascribed to the clearance of endotoxin coming from the intestinal area and transported by the portal vein. There is substantial indirect evidence that systemic endotoxemia of intestinal origin is paralleled by Kupffer cell failure in patients with different liver diseases (Liehr and Grün, 1977). Further indirect evidence is provided by the observations by Berg and Midtvedt (1976), who demonstrated that the level of lysosomal hydrolase activity of sinusoidal cells is low in germ-free animals and about two times higher in conventional rats. Endotoxin seems to be a normal component of the portal vein blood in normal healthy animals or persons. Chromium-labeled endotoxin is taken up by the liver after intravenous injection (Noyes *et al.*, 1959; Zyldaszyk and Moon, 1976). Endotoxins induces mitoses in sinusoidal liver cells (Gospos *et al.*, 1977). Bacteria are avidly phagocytized by Kupffer cells (Friedman and Moon, 1977; Ruggiero *et al.*, 1977; Tamaru and Fujita, 1978; Wisse *et al.*, 1972). All these observations seem to indicate the involvement of sinusoidal cells in the clearance of endotoxins and bacteria. Further evidence, however, is needed to determine the cell type(s) and endocytic mechanism involved.

Thirdly, Kupffer cells are involved in a number of pathological reactions (Tanikawa and Ikejiri, 1977). and there are also a number of reports on various functions ascribed to Kupffer cells. Among these observations, the clearance of cellular debris after different traumas, such as burning, surgery, shock, and hemorrhage (Saba, 1978) is mentioned, also the phagocytosis of certain types of tumor cells (Roos and Dingemans, 1977), the reaction to virus (Gendrault *et al.*, 1977), including hepatitis virus (Sabesin and Koff, 1974), the production of pyrogen (Haeseler *et al.*, 1977) or lysosomal hydrolases (Loegering *et al.*, 1976), the secretion of erythropoietin (Gruber *et al.*, 1977; Peschle *et al.*, 1976) and colony stimulating factor (Joyce and Chervenick, 1975), collagenase (Fujiwara *et al.*, 1973), or complement (Littleton *et al.*, 1970).

It is obvious that methods for the isolation (Emeis and Planqué, 1976; Knook *et al.*, 1977; Mills and Zucker-Franklin, 1969; Munthe-Kaas *et al.*, 1975; Munthe-Kaas, 1977), purification (Knook *et al.*, 1977; Knook and Sleyster, 1976, 1977; Munthe-Kaas *et al.*, 1975; Munthe-Kaas, 1977; Sleyster *et al.*, 1977), culture (Munthe-Kaas *et al.*, 1975; Munthe-Kaas, 1977), or even coculture (Wanson, 1977) of Kupffer cells will provide new evidence in this area.

Emeis and Lindeman (1976) demonstrated a peculiar pattern of behavior of Kupffer cells in disseminated intravascular coagulation: fibrin attached to the Kupffer cells but was not phagocytized. Nevertheless, the fibrin disappeared, probably due to the action of fibrinolytic activity. Fibrin degradation products are taken up by sinusoidal cells (Fabre *et al.*, 1974; Sherman *et al.*, 1977). The strange point in this phenomenon is the paralysis of the Kupffer cell which is apparently unable to phagocytize the fibrin attached to the surface of the cell. This phenomenon induces a series of events, which, in normal circumstances leads to phagocytosis. To conclude this review, I will try to formulate the (hypothetical) series of events taking place before, during, and after phagocytosis in Kupffer cells.

4. SCHEME OF PHAGOCYTOSIS BY KUPFFER CELLS

Because phagocytosis by Kupffer cells seems to be one of its specialized and best-known functions, it might be of importance to summarize the basic reactions and to add a few hypothetical remarks.

Table 1 summarizes the series of events which might take place during particle phagocytosis by Kupffer cells. Firstly, intravenous particles will react with opsonins, such as were isolated, purified, and characterized recently as glycoproteins (3.2% neutral sugars, mol. wt. 800,000 at 4°C) (Blumenstock *et al.*, 1976), strongly reacting with foreign particles, and highly promoting Kupffer cell phagocytosis. It is supposed that the reaction with the particles changes the molecule in such a way, that it is immediately recognized when touching the fuzzy coated Kupffer cell surface. Emeis (1976) and Emeis and Wisse (1971, 1975) studied the nature of the fuzzy coat by cytochemical staining techniques, and one of their conclusions was that the fuzzy coat most probably consisted of protein. This means that the attachment reaction inducing phagocytosis is probably based on a protein–protein interaction. Specially prepared small colloidal gold particles (3.5 nm) and erythrocyte ghosts show, after attachment, a distance of about 70 nm to the cell membrane of the Kupffer cell; larger particles show only a 10–15 nm gap. This phenomenon might indicate contraction of the cell coat, which is supported by images of attached ghosts showing both distances simultaneously (Emeis, 1976). It was shown by Emeis (1976) that pseudopodia involved in phagocytosis lack the fuzzy coat, and, for this reason, we might assume that lateral displacement of cell coat molecules might occur during

TABLE 1. SCHEME OF EVENTS DURING PHAGOCYTOSIS BY KUPFFER CELLS

- Reaction of an intravenous particle with opsonin (α-2-macroglobulin)
- Recognition and attachment of the opsonin–particle complex to the fuzzy coat
- Contraction of the fuzzy coat, and possibly, lateral displacement
- Induction of phagocytosis by the rapid formation of hyaloplasm
- Formation of pseudopods or sleeve-like lamellipodia
- Engulfment of the particle(s); formation of a vacuole (phagosome)
- Transport of the phagosome through the cytoplasm
- Fusion with lysosome(s)
- Degradation of digestible substances within the lysosomes
- Defecation or distribution of indigestible product over daughter cells, or migration of loaded cells

formation of these pseudopodia. Munthe-Kaas *et al.* (1976) and Steer (1978) demonstrated the presence of specific receptors at the Kupffer cell membrane. These events at the outer cell surface apparently induce the formation of microfilamentous hyaloplasm from which all organelles are rapidly removed. The active deformation of this cytoplasmic area gives rise to engulfing lamellipodia (see also Munthe-Kaas *et al.*, 1976) resulting in the formation of a phagosome. These phagosomes are transported to meet the lysosomes and, normally, fusion would occur. Several bacteria are apparently able to avoid lysosomal fusion after phagocytosis by Kupffer cells (Mohelska *et al.*, 1975); also, opsonin or other factors are needed to establish lysosomal fusion (Horn *et al.*, 1969). In the case of RES blockade with colloidal particles, degradation of endocytized material is impossible. Nevertheless, at longer times (weeks) after this treatment, Kupffer cells show less abundant material in their cytoplasm, which might be explained by the observed mitosis, and which might also have been paralleled by Kupffer cell migration or exocytosis of material.

5. SUMMARY

From the descriptions given in this paper it may be concluded that apart from Kupffer cells, there are three other types of sinusoidal cells in the liver of the rat (Wisse, 1977a; Wisse, and Knook, 1979). These cell types differ in morphology, reaction to experimental conditions, and endocytic capacity.* Kupffer cells and endothelial cells possess five different endocytic mechanisms. Transitional stages between Kupffer cells and endothelial cells, and between Kupffer cells and other (blood) cells, were not found. Because mitoses were observed in Kupffer cells under five different conditions, they are to be considered as self-proliferating cells.

Specific functions can be assigned to Kupffer cells. These cells are mac-

*For the terminology of liver sinusoidal cells, see Bulletin of the Kupffer Cell Foundation [p. 1 (1977) and pp. 23–24 (1978)].

rophages that react strongly to foreign particles. In normal adult rat liver, Kupffer cells and endothelial cells may play a role in the uptake of endotoxin. In disease or experimental circumstances, Kupffer cells are supposed to endocytize circulating tumor cells, bacteria, cellular debris, antigens, immune complexes, and fibrin degradation products. Further studies, using microscopic techniques, together with biochemical methods, isolation, purification, and culture of Kupffer cells and other sinusoidal cells are needed to reveal the exact role of the various types of sinusoidal cells in these circumstances.

REFERENCES

Berg, T., and Midvedt, T., 1976, The influence of lysosomal enzymes in rat Kupffer cells and hepatocytes, *Acta Pathol. Microbiol. Scand. A* **84**:415.

Bissel, D. M., Hammaker, L., and Schmid, R., 1972, Liver phagocytes: Identification of a subpopulation for erythrocyte catabolism, *J. Cell Biol.* **54**:107.

Blouin, A., 1977, Morphometry of liver sinusoidal cells, in: *Kupffer Cells and Other Liver Sinusoidal Cells* (E. Wisse and D. L. Knook, eds.), pp. 61–73, Elsevier, Amsterdam.

Blouin, A., Bolender, R. P., and Weibel, E. R., 1977, Distribution of organelles and membranes between hepatocytes and nonhepatocytes in the rat liver parenchyma, *J. Cell Biol.* **72**:441.

Blumenstock, F., Saba, T. M., Weber, P., and Cho, E., 1976, Purification and biochemical characterization of a macrophage stimulating α-2-globulin opsonic protein, *J. Reticuloendothel. Soc.* **19**:157.

Brouwer, A., and Knook, D. L., 1977, Quantitative determination of endocytosis and intracellular digestion by rat liver Kupffer cells *in vitro*, in: *Kupffer Cells and Other Liver Sinusoidal Cells* (E. Wisse and D. L. Knook, eds.), pp. 343–352, Elsevier, Amsterdam.

Buys, C. H. C. M., Bouma, J. M. W., Gruber, M., and Wisse, E., 1978, Induction of lysosomal storage by suramin, *Naunyn-Schmiedeberg's Arch. Pharmacol.* **304**:183.

Clark, J. M., and Pateman, J. A., 1978, Long-term culture of Chinese hamster Kupffer cell lines isolated by a primary cloning step, *Exp. Cell Res.* **112**:207.

Crofton, R. W., Diesselhof-den Dulk, M. M. C., and van Furth, R., 1978, The origin, kinetics, and characteristics of the Kupffer cells in the normal steady state, *J. Exp. Med.* **148**:1.

Deimann, W., and Fahimi, H. D., 1977, The ontogeny of mononuclear phagocytes in fetal rat liver using endogenous peroxidase as a marker, in: *Kupffer Cells and Other Liver Sinusoidal Cells* (E. Wisse and D. L. Knook, eds.), pp. 487–495, Elsevier, Amsterdam.

Deimann, W., and Fahimi, H. D., 1978, Peroxidase cytochemistry and ultrastructure of resident macrophages in fetal rat liver, *Dev. Biol.* **66**:43.

Düllmann, J., and Wulfhekel, U., 1978, Kupffer cells, sinusoid lining endothelia, and hepatocytes in internal iron exchange and its disorders, *Bull. Kupffer Cell Found.* **1**:41.

Emeis, J. J., 1976, Morphological and cytochemical heterogeneity of the cell coat of rat liver Kupffer cells, *J. Reticuloendothel. Soc.* **20**:31.

Emeis, J. J., and Lindeman, J., 1976, Rat liver macrophages will not phagocytose fibrin during disseminated intravascular coagulation, *Haemostatis* **5**:193.

Emeis, J. J., and Planqué, B., 1976, Heterogeneity of cells isolated from rat liver by pronase digestion: Ultrastructure, cytochemistry and cell culture, *J. Reticuloendothel. Soc.* **20**:11.

Emeis, J. J., and Wisse, E., 1971, Electron microscopic cytochemistry of the cell coat of Kupffer cells in rat liver, in: *Advances in Experimental Medicine and Biology*, Vol. 15, *The RES and Immune Phenomena* (N. R. DiLuzio and K. Fleming, eds.), pp. 1–12, Plenum Press, New York.

Emeis, J. J., and Wisse, E., 1975, On the cell coat of rat liver Kupffer cells, in: *Mononuclear Phagocytes in Immunity, Infection, and Pathology* (R. van Furth, ed.), pp. 315–325, Blackwell, Oxford.

Fabre, J., Delsol, G., Familiades, J., Tapie, C., Boneu, B., Bierme, R., 1974, Apport de la morphologie dans la mise en évidence des dérivés du fibrinogens, *Pathol. Biol.* **22**:53.

Fahimi, H. D., 1967, Perfusion and immersion fixation of rat liver with glutaraldehyde, *Lab. Invest.* **16**:736.

Fahimi, H. D., 1970, The fine structural localization of endogenous and exogenous peroxidase activity in Kupffer cells of rat liver, *J. Cell Biol.* **47**:247.

Fraser, R., 1978, Thoughts on the liver sieve, *Bull. Kupffer Cell Found.* **1**:46.

Fraser, R., Bosanquet, A. G., and Day, W. A., 1978, Filtration of chylomicrons by the liver may influence cholesterol metabolism and atherosclerosis, *Atherosclerosis* **29**:113.

Friedman, R. L., and Moon, R. J., 1977, Hepatic clearance of *Salmonella typhimurium* in silica-treated mice, *Infect. Immun.* **16**:1005.

Fujiwara, K., Sakai, T., Oda, T., and Igarashi, S., 1973, The presence of collagenase in Kupffer cells of the rat liver, *Biochem. Biophys. Res. Commun.* **54**:531.

Gendrault, J. L., Steffan, A. M., Bingen, A., and Kirn, A., 1977, Interaction of frog virus 3 (FV3) with sinusoidal cells, in: *Kupffer Cells and Other Liver Sinusoidal Cells* (E. Wisse and D. L. Knook, eds.), pp. 223–233, Elsevier, Amsterdam.

Gospos, C., Freudenberg, N., Bank, A., and Freudenberg, M. A., 1977, Effect of endotoxin-induced shock on the reticuloendothelial system. Phagocytic activity and DNA-synthesis of reticuloendothelial cells following endotoxin transport, *Beitr. Pathol.* **161**:100.

Gruber, D. F., Zucali, J. R., Mirand, E. A., and Mirand, E. A., 1977, Identification of erythropoietin producing cells in fetal mouse liver cultures, *Exp. Hematol.* **5**:392.

Haeseler, F., Bodel, P., and Atkins, E., 1977, Characteristics of pyrogen production by isolated rabbit Kupffer cells *in vitro, J. Reticuloendothel. Soc.* **22**:569.

Hausmann, K., Wulfhekel, U., Düllmann, J., and Kuse, R., 1976, Iron storage in macrophages and endothelial cells, *Blut* **32**:289.

Horn, R. G., Koenig, M. G., Goodman, J. S., and Collins, R. D., 1969, Phagocytosis of *S. aureus* by hepatic RES cells, *Lab. Invest.* **21**:406.

Ito, T., 1973, Recent advances in the study of the fine structure of the hepatic sinusoidal wall: a review, *Gunma Rep. Med. Sci.* **6**:119.

Joyce, R. A., and Chervenick, P. A., 1975, Stimulation of granulopoiesis by liver macrophages, *J. Lab. Clin. Med.* **86**:112.

Knook, D. L., and Sleyster, E. C., 1976, Separation of Kupffer and endothelial cells of the rat liver by centrifugal elutriation, *Exp. Cell Res.* **99**:444.

Knook, D. L., and Sleyster, E. C., 1977, Preparation and characterization of Kupffer cells from rat and mouse liver, in: *Kupffer Cells and Other Liver Sinusoidal Cells* (E. Wisse and D. L. Knook, eds.), pp. 273–288, Elsevier, Amsterdam.

Knook, D. L., Blansjaar, N., and Sleyster, E. C., 1977, Isolation and characterization of Kupffer and endothelial cells from the rat liver, *Exp. Cell. Res.* **109**:317.

Liehr, H., and Grün, M., 1977, Clinical aspects of Kupffer cell failure in liver diseases, in: *Kupffer Cells and Other Liver Sinusoidal Cells* (E. Wisse and D. L. Knook, eds.), pp. 427–436, Elsevier, Amsterdam.

Littleton, C., Kessler, D., and Burkholder, P. M., 1970, Cellular basis for synthesis of the fourth component of guinea-pig complement as determined by a haemolytic plaque technique, *Immunology* **18**:693.

Loegering, D. J., Kaplan, J. E., and Saba, T. M., 1976, Correlation of plasma lysosomal enzyme levels with hepatic RE function after trauma, *Proc. Soc. Exp. Biol. Med.* **152**:42.

Mills, D. M., and Zucker-Franklin, D., 1969, Electron microscopic study of isolated Kupffer cells, *Am. J. Pathol.* **54**:147.

Mohelska, H., Holuša, R., Kubin, M., and Smetana, K., 1975, Endocellular parasitism of *Mycobacterium avium* in rabbit liver, *Exp. Pathol.* **10**:122.

Motta, P., 1975, A scanning electron microscopic study of the rat liver sinusoid: Endothelial and Kupffer cells, *Cell Tissue Res.* **164**:371.

Motta, P., 1977, Kupffer cells as revealed by scanning electron microscopy, in: *Kupffer Cells and Other Liver Sinusoidal Cells* (E. Wisse and D. L. Knook, eds.), pp. 93–103, Elsevier, Amsterdam.

Munthe-Kaas, A. C., 1977, Endocytosis studies on cultured rat Kupffer cells, in: *Kupffer Cells and Other Liver Sinusoidal Cells* (E. Wisse and D. L. Knook, eds.), pp. 325–332, Elsevier, Amsterdam.

Munthe-Kaas, A. C., Berg, T., Seglen, P. O., and Seljelid, R., 1975, Mass isolation and culture of rat Kupffer cells, *J. Exp. Med.* **141**:1.

Munthe-Kaas, A. C., Kaplan, G., and Seljelid, R., 1976, On the mechanism of internalization of opsonized particles by rat Kupffer cells *in vitro*, *Exp. Cell Res.* **103**:201.

Muto, M., and Fujita, T., 1977, Phagocytic activities of the Kupffer cell: a scanning electron microscope study, in: *Kupffer Cells and Other Liver Sinusoidal Cells* (E. Wisse and D. L. Knook, eds.), pp. 109–121, Elsevier, Amsterdam.

Naito, M., and Wisse, E., 1977, Observations on the fine structure of sinusoidal cells in fetal and neonatal rat liver, in: *Kupffer Cells and Other Liver Sinusoidal Cells* (E. Wisse and D. L. Knook, eds.), pp. 497–505, Elsevier, Amsterdam.

Naito, M., and Wisse, E., 1978, Filtration effect of endothelial fenestrations on chylomicron transport in neonatal rat liver sinusoids, *Cell Tissue Res.* **190**:371.

Noyes, H. E., McInturf, C. R., and Blahuta, G. J., 1959, Studies on the distribution of *E. coli* endotoxin in mice, *Proc. Soc. Exp. Biol. Med.* **100**:65.

Peschle, C., Marone, G., Genovese, A., Magli, C., and Condorelli, M., 1976, Hepatic erythropoietin: Enhanced production in anephric rats with hyperplasia by Kupffer cells, *Br. J. Haematol.* **32**:105.

Roos, E., and Dingemans, K. P., 1977, Phagocytosis of tumor cells by Kupffer cells *in vivo* and in the perfused mouse liver, in: *Kupffer Cells and Other Liver Sinusoidal Cells* (E. Wisse and D. L. Knook, eds.), pp. 183–190, Elsevier, Amsterdam.

Ruggiero, G., Utili, R., and Andreana, A., 1977, Clearance of viable salmonella strains by the isolated, perfused rat liver: a study of serum and cellular factors involved and of the effect of treatments with carbon tetrachloride or *Salmonella enteritidis* lipopolysaccharide, *J. Reticuloendothel. Soc.* **21**:79.

Saba, T. M., 1978, Humoral control of Kupffer cell function after injury, *Bull. Kupffer Cell Found.* **1**:12.

Sabesin, S. M., and Koff, R. S., 1974, Pathogenesis of experimental viral hepatitis, *N. Engl. J. Med.* **25**:944.

Satodate, R., Sasou, S., Oikawa, K., Hatakeyama, N., and Katsura, S., 1977, Scanning electron microscopical studies on the Kupffer cell in phagocytic activity, in: *Kupffer Cells and Other Liver Sinusoidal Cells* (E. Wisse and D. L. Knook, eds.), pp. 121–131, Elsevier, Amsterdam.

Sherman, L. A., Lee, J., and Jacobson, A., 1977, Quantitation of the reticuloendothelial system clearance of soluble fibrin, *Br. J. Haematol.* **37**:231.

Sleyster, E. C., and Knook, D. L., 1978, Multiple forms of acid phosphatase in rat liver parenchymal, endothelial and Kupffer cells, *Arch. Biochem. Biophys.* **190**:756–761.

Sleyster, E. C., Westerhuis, F. G., and Knook,, D. L., 1977, The purification of non-parenchymal liver cell classes by centrifugal elutriation, in: *Kupffer Cells and Other Liver Sinusoidal Cells* (E. Wisse and D. L. Knook, eds.), pp. 289–298, Elsevier, Amsterdam.

Steer, C., 1978, Kupffer cells and glycoproteins: Does a recognition phenomenon exist? *Bull. Kupffer Cell Found.* **1**:26.

Stöhr, G., Deimann, W., and Fahimi, H. D. 1978, Peroxidase-positive endothelial cells in sinusoids of the mouse liver, *J. Histochem. Cytochem.* **26**:409.

Tamaru, T., and Fujita, H., 1978, Electron microscopic studies on Kupffer's stellate cells and sinusoidal endothelial cells in the liver of normal and experimental rabbits, *Anat. Embryol.* **154**:125.

Tanikawa, K., and Ikejiri, N., 1977, Fine structural alterations of the sinusoidal lining cells in various liver diseases, in: *Kupffer Cells and Other Liver Sinusoidal Cells* (E. Wisse and D. L. Knook, eds.), pp. 153–162, Elsevier, Amsterdam.

Tanuma, Y., 1978, Fine structure of the Kupffer cell in the bat, with special reference to the wormlike bodies, *Arch. Histol. Jpn.* **41**:113.

Tanuma, Y., and Ito, T., 1978, Electron microscope study on the hepatic sinusoidal wall and fat-storing cells in the bat, *Arch. Histol. Jpn.* **41**:1.

van Berkel, T. J. C., and Kruyt, J. K., 1977, Identity of peroxidatic activities in non-parenchymal rat liver cells in relation to parenchymal liver cells, in: *Kupffer Cells and Other Liver Sinusoidal Cells* (E. Wisse and D. L. Knook, eds.), pp. 307–314, Elsevier, Amsterdam.

van Dierendonck, J. H., van der Meulen, J., De Zanger, R. B., and Wisse, E. 1979, The influence of hormones on the diameter of the fenestrae in the endothelial lining of rat liver sinusoids: A SEM study, *J. Ultramicrosc.* **4**:149 (abstr.).

van Furth, R., Crofton, R. W., and Diesselhoff-den Dulk, M. M. C., 1977, The bone marrow origin of Kupffer cells, in: *Kupffer Cells and Other Liver Sinusoidal Cells* (E. Wisse and D. L. Knook, eds.), pp. 471–480, Elsevier, Amsterdam.

Volkman, A., 1977, The unsteady state of the Kupffer cell, in: *Kupffer Cells and Other Liver Sinusoidal Cells* (E. Wisse and D. L. Knook, eds.), pp. 459–471, Elsevier, Amsterdam.

Vonnahme, J., 1977, A scanning electron microscopic study of Kupffer cells in the monkey liver, in: *Kupffer Cells and Other Liver Sinusoidal Cells* (E. Wisse, and D. L. Knook, eds.), pp. 103–109, Elsevier, Amsterdam.

Wake, K. 1971, Sternzellen in the liver: perisinusoidal cells with special reference to storage of vitamin A, Am. J. Anat. **132**:429.

Wanson, J. C., Drochmans, P., Mosselmans, R., and Knook, D. L., 1977, Symbiotic culture of adult hepatocytes and sinuslining cells, in: *Kupffer Cells and Other Liver Sinusoidal Cells* (E. Wisse and D. L. Knook, eds.), pp. 141-150, Elsevier, Amsterdam.

Widmann, J. J., and Fahimi, H. D., 1975, Proliferation of mononuclear phagocytes (Kupffer cells) and endothelial cells in regenerating rat liver, *Am. J. Pathol.* **80**:349.

Widmann, J. J., and Fahimi, H. D., 1976, Proliferation of endothelial cells in estrogen-stimulated rat liver, *Lab Invest.* **34**:141.

Widmann, J. J., Cotran, R. S., and Fahimi, H. D., 1972, Mononuclear phagocytes (Kupffer cells) and endothelial cells. Identification of two functional cell types in rat liver sinusoids by endogenous peroxidase activity. *J. Cell Biol.* **52**:159.

Wisse, E., 1970, An electron microscopic study of the fenestrated endothelial lining of rat liver sinusoids, *J. Ultrastruct. Res.* **31**:125.

Wisse, E., 1972, An ultrastructural characterization of the endothelial cell in the rat liver sinusoid under normal and various experimental conditions as a contribution to the distinction between endothelial and Kupffer cells, *J. Ultrastruct. Res.* **38**:528.

Wisse, E., 1974a, Observations on the fine structure and peroxidase cytochemistry of normal rat liver Kupffer cells, *J. Ultrastruct. Res.* **46**:393.

Wisse, E., 1974b, Kupffer cell reactions in rat liver under various conditions as observed in the electron microscope, *J. Ultrastruct. Res.* **46**:499.

Wisse, E., 1977a, Ultrastructure and function of Kupffer cells and other sinusoidal cells in the liver, in: *Kupffer Cells and Other Liver Sinusoidal Cells* (E. Wisse and D. L. Knook, eds.), pp. 33–60, Elsevier, Amsterdam.

Wisse, E., 1977b, Ultrastructure and function of Kupffer cells and other sinusoidal cells in the liver, *Méd. Chir. Dig.* **6**:409.

Wisse, E., 1978, Kupffer cells and peritoneal macrophages are different types of cells, *Blood Cells* **4**:319.

Wisse, E., and Daems, W. T., 1970, Fine structural study on the sinusoidal lining cells of rat liver, in: *Mononuclear Phagocytes* (R. van Furth, ed.), pp. 200–210, Blackwell, Oxford.

Wisse, E., and Knook, D. L., 1979, A new approach to the study of liver function: The investigation of sinusoidal cells *Progress in Liver Diseases,* Vol. 6 (H. Popper and F. Schaffner, eds.), pp. 153–171, Grune & Stratton, New York.

Wisse, E., Giphart, M., van der Meulen, J., Roels, F., Emeis, J. J., and Daems, W. T., 1972, Peroxidatic activity in bacteria phagocytizing Kupffer cells, demonstrated by a new method: Perfusion incubation, *Proceedings of the Fifth European Congress on Electron Microscopy,* p. 272, Manchester.

Wisse, E., Emeis, J. J., and Daems, W. T., 1974a, Some fine structural considerations on the possible involvement of the liver RES in lysosomal storage diseases, with observations in a case of Morquio's disease, in: *Enzyme Therapy in Lysosomal Storage Diseases* (J. M. Tager, C. J. M. Hooghwinkel, and W. Th. Daems, eds.), pp. 95–110, North-Holland, Amsterdam.

Wisse, E., van der Meulen, J., Emeis, J. J., and Daems, W. T., 1974b, Enzyme cytochemical study of rat liver endothelial and Kupffer cells, *Proceedings of the Eighth International Congress on Electron Microscopy,* Canberra, **2**:408.

Wisse, E., van't Noordende, J. J., van der Meulen, J., and Daems, W. T., 1976, The pit cell: Description of a new type of cell occurring in rat liver sinusoids and peripheral blood, *Cell Tissue Res.* **173**:423.

Zlydaszyk, J. C., and Moon, R. J., 1976, Fate of ^{51}Cr-labeled lipopolysaccharide in tissue culture cells and livers of normal mice, *Infect. Immun.* **14**:100.

10

Interdigitating Cells

JAN E. VELDMAN and EDWIN KAISERLING

1. INTRODUCTION

After the discovery of the interdigitating cell (IDC) and the first description of its microenvironment and presumable auxiliary function in the T-cell territory of lymphoid tissue (Veldman, 1970; Veldman and Keuning, 1978; Veldman *et al.*, 1978a,b), its existence has been confirmed by various workers (Kaiserling and Lennert, 1974; Hoefsmit, 1975; Veerman and van Ewijk, 1975; Müller-Hermelink *et al.*, 1976; Racz *et al.*, 1977; Kaiserling, 1977, 1978; Goos *et al.*, 1976; Lennert *et al.*, 1978a; Waksman, 1978).

The IDCs are a specific and, probably, functionally important cell constituent of the T-cell regions throughout the lymphoid system of many, if not all, species, including man. Enzyme histochemical and submicroscopic analyses in human lymphomas have recently demonstrated the importance of the IDC as a "guiding" cell in various lesions of the lymphoid system (Kaiserling, 1977, 1978; Lennert *et al.*, 1978b). IDCs were originally defined by their electron-microscopic appearance in experimentally B-cell-deprived animals (Veldman, 1970; Veldman *et al.*, 1978b). It was named interdigitating cell on account of its typical arrangement of plasma membranes. In addition, the nucleus of the IDC appeared both in light and electron microscopy as a very characteristic, bizarre, lobulated form, well-distinguishable from other nonlymphoid elements.

We should like to mention that the IDC has been given other names. Some investigators use the term *interdigitating cell,* originally introduced by one of the authors (J. E. V.) and thus emphasize the most important morphological feature of the cell. Other investigators, however, especially Lennert and his group (see Lennert *et al.*, 1978a), which includes the other author (E. K.), apply the term *interdigitating reticulum cell.* One of the reasons for adding *reticulum* cell to *interdigitating* was to point out that the IDC is thought to be a stationary structural cell of lymphoreticular tissue.

To understand the role which IDCs might play in the microarchitecture of

---

JAN E. VELDMAN • Laboratory of Histophysiology and Experimental Pathology, Department of Otorhinolaryngology, University of Utrecht, 3500 CG Utrecht, The Netherlands. EDWIN KAISERLING • Department of Pathology, Christian-Albrecht University, 2300 Kiel, West Germany.

lymphoid tissue and in the dynamics of an immune response, experimental procedures have been developed which enable the artificial separation—both functionally and morphologically—of the antibody response and cellular immunity reaction. One of the main difficulties in investigating lymphoid tissue histophysiology is the simultaneous occurrence of the plasma cell, germinal center, and cellular immunity reactions following most antigenic stimuli, at least when the lymph node is the site of the immune response.

The possibility of isolating these processes experimentally through X-irradiation procedures has fulfilled an important condition for adequate structural analyses—with both light and electron microscope—of IDCs and their (micro)environment in lymphoid tissue.

Our fragmentary knowledge of IDCs from animal models has recently been extrapolated into human (non)neoplastic disorders (Kaiserling, 1977; Lennert *et al.*, 1978a). In this review, we shall make an attempt to indicate where experimental observations relate to human lymph node pathology, mainly malignant lymphomas other than Hodgkin disease.

2. THE INTERDIGITATING CELL IN THE IMMUNOHISTOPHYSIOLOGY OF LYMPHOID TISSUE

Microscopic analyses of the immune response in the lymph node revealed a characteristic, strictly TDA*-bound basic pattern of IDCs. Electron microscopy elucidated (Veldman, 1970; Veldman *et al.*, 1978b) specific cytoplasmic details after silver-staining procedures: a tubulovesicular (TV) complex was found. In addition, the complicated labyrinthine membrane-membrane contacts between IDCs themselves and IDCs and T cells were marked by a surface coating, presumably the substance produced and released through this TV complex (Figure 1). The significance of these aspects in relation to B- and T-cell homing and B- and T-cell immune responsiveness will be discussed. It is tempting to assume that the IDC is the *in vivo* stand-in as an "auxiliary" cell, interacting with T cell and antigen, or with both of these and B cells, as questioned recently again by Waksman (1978).

2.1. FUNCTIONAL HISTOLOGY AND SUBMICROSCOPIC CYTOLOGY OF THE LYMPH NODE

Already at low magnification, a lymph node section shows the well-known characteristic pattern of its lymphoid tissue: a compact *cortex* with a single row of lymphoid *follicles* at the outer border, and a more "loosely built" *medulla*. The width of the cortex may vary considerably in the various parts of the node. A stretch of broad cortex results from the presence of a *paracortical area,* a large

*Thymus-dependent area (Parrott *et al.*, 1966).

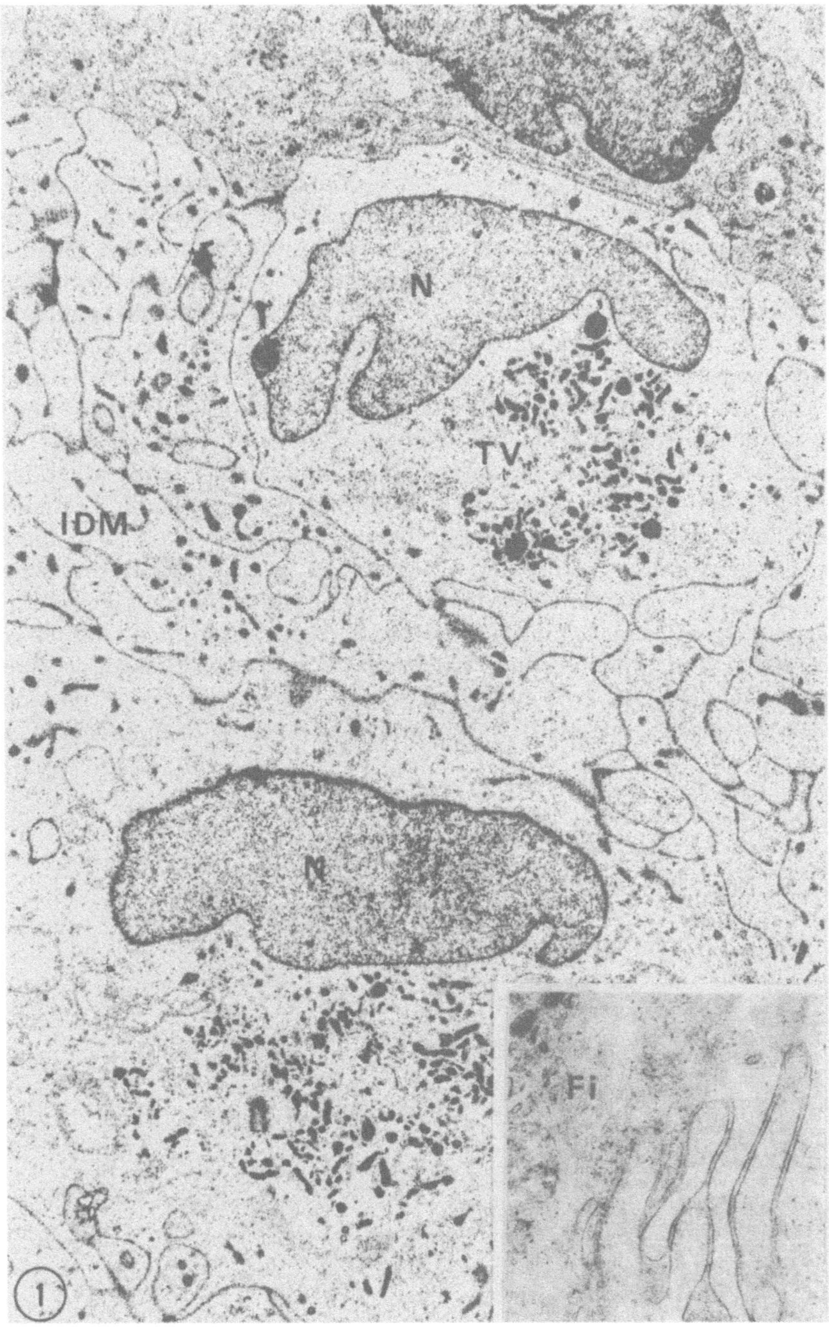

FIGURE 1. PSMA staining. Interdigitating cells. Bizarre nuclei (N), PSMA-positive substance in tubulovesicular complex (TV) and on interdigitating cell membranes (IDM) 2 days after skin allograft. ×9000. Inset: Ua–Pb staining. Details of IDC. Fi, Intracytoplasmic filaments. ×21,000. (From Veldman *et al.*, 1978b.)

egg-shaped mass of lymphocytic tissue protruding more or less deeply into the medulla. Between these, the cortex is a narrow band, hardly exceeding the diameter of the follicles. At the corticomedullary border the compact mass of cortical lymphoid tissue passes into the irregular, branching medullary cords (Figure 2).

The lymphoid tissue proper of the lymph node may be said to consist of (a) an *outer cortex* bearing the *primary* and *secondary lymphoid follicles,* (b) distinct *paracortical areas* approximated to the outer cortex, and (c) *medullary cords.* In the outer cortex, a characteristic *marginal zone* is present, apparently corresponding to the marginal zone of the splenic white pulp. It always consists of a few layers of lightly staining lymphoid cells which border the subcapsular sinus over the primary and secondary follicles and often between the follicles. The marginal zone cells are medium-sized lymphoid cells, which can be distinguished from small lymphocytes by a larger and lighter-staining nucleus with one or two distinct nucleoli and a larger amount of cytoplasm which is lightly to moderately pyroninophilic. They are seen best in methyl green-pyronine- or Giemsa-stained lymph node sections; usually they escape observation in hematoxylin and eosin-stained preparations. Also, electron microscopically, they constitute a quite characteristic class of cells, distinct from any type of small lymphocyte by nuclear and cytoplasmic features. The *open face*-type nucleus has a regular round or oval shape, and contains one or two small nucleoli clearly standing out in the "dispersed chromatin" type nucleoplasm. The nuclear envelope has marked numbers of nuclear pores with diaphragms. The cytoplasm contains one or two

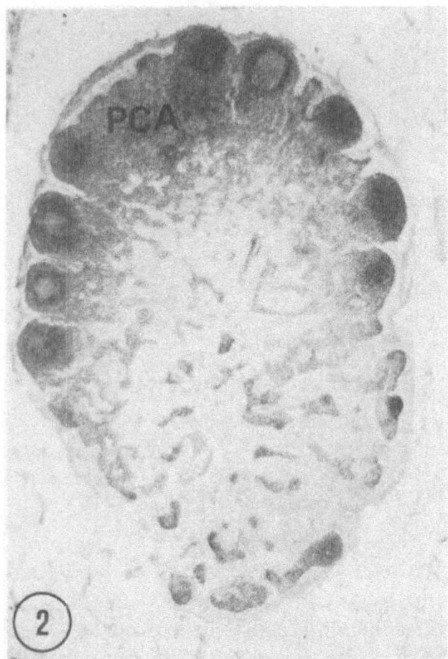

FIGURE 2. Normal lymph node. PCA, Paracortical area. ×20. (From Veldman *et al.,* 1978a.)

centrioles and numerous small, round mitochondria, often accumulated in one part of the cell. Mono- and polyribosomes are found in varying quantities. A few strands of rough-surfaced endoplasmic reticulum (RER) are always present, often connected with the nuclear envelope (Veldman 1970; Veldman *et al.*, 1978a). The marginal zone cells have been postulated to be a characteristic and essential cellular constitutent of those parts of lymphoid tissue that are potentially involved in antibody responses (Keuning *et al.*, 1963; Veldman, 1970; Nieuwenhuis and Keuning, 1974; Veldman *et al.*, 1978a).

A primary follicle in the lymph node outer cortex is essentially an aggregate of small lymphocytes with a cap of marginal zone cells. A secondary follicle consists of a follicular center surrounded by a lymphocyte corona, similarly, with a cap of marginal zone cells, the lymphocyte corona presumably corresponding to the primary follicle. The follicular center may be either indifferent, i.e., largely consisting of pale-staining reticular cells, or involved in a germinal center reaction. As a rule, even indifferent follicular centers have a basal segment showing some germinal center activity. Secondary follicles arise from primary ones through a germinal center reaction which, upon subsiding, leaves an indifferent, pale-staining follicular center. The follicles are maintained as such by a continuous supply of blood-borne, non-thymus-derived lymphocytes (Keuning and Bos, 1967; Keuning and van den Broek, 1968; Veldman, 1970; Nieuwenhuis *et al.*, 1970; Nieuwenhuis and Keuning, 1974; Veldman *et al.*, 1978a). Some of these subsequently transform into marginal zone cells. Submicroscopically, these marginal zone cell precursors have characteristic "condensed chromatin" type nuclei and a small quantity of cytoplasm. Apparently, the transformation of these lymphocytes into marginal zone cells consists of a marked change of the nucleus—including dispersing of chromatin and development of nucleoli—and cytoplasmic growth (Veldman, 1970; Veldman *et al.*, 1978a). The observations on plasmablast formation and plasma cell development leave little doubt as to the plasmablast precursor character of the marginal zone cell (see also Veldman *et al.*, 1980a). The transformation of lymphocytes into marginal zone cells seems to be an unconditional prerequisite for antibody-forming potential. Cytologically, the marginal zone cell would seem to be the immunocompetent "custodian" type of plasmablast precursor awaiting antigenic stimulation.

Lymphoid follicles should be considered the morphological signs of the above-mentioned particular type of lymphoid cell traffic of non-thymus-derived B lymphocytes. In the same way, the paracortical areas with their epithelioid venules are the homing sites of the recirculating T lymphocytes (Gowans and Knight, 1964; Veldman, 1970, 1977; Veldman and Keuning, 1978)—thymus-dependent areas (Waksman *et al.*, 1962; Parrott and de Sousa, 1966)—and, consequently, morphologically, represent that particular type of lymphoid cell traffic.

Taken together the lymph node can be considered as a continuously operating machinery, permanently kept ready for an immune response. Its structural organization at any given moment depends on a more or less basic pattern of a group of specialized reticular (?) cells (dendritic cells), blood vessels, and other nonlymphoid elements (interdigitating cells). In this basic structure, two types of lymphoid cell traffic precipitate distinct morphological—though essentially dynamic—entities.

Recent research added a third pathway, presumably with its own cell dynamics as well as cell constituents, an "alternative" route through the skin via afferent lymphatics to the lymph node (Kelly *et al.*, 1976; Silberberg-Sinakin *et al.*, 1978; Sainte-Marie, 1978). Both lymphoid cells and mononuclear phagocytic and nonphagocytic elements move to the lymph node by this route.

It is these structural dynamics which form the basis of immune responses induced by immunogens which have been carried along with the lymph in afferent lymph vessels.

2.1.1. The Immune Response in the Lymph Node: Plasma Cell, Germinal Center, and Cellular Immunity Reactions

Immune reactions in lymphoid tissue are considered to comprise three antigen-induced processes: the plasma cell reactions representing antibody formation, cellular immunity reactions leading to cell-mediated immunity, and the germinal center reactions. The latter process has been shown to be the source of "memory cells"—the plasma cell precursors of secondary response antibody formation (Wakefield and Thorbecke, 1968)—as well as the site of generation and amplification of antibody-forming cell precursors (marginal zone cells) in general (Nieuwenhuis and Keuning, 1974, Opstelten *et al.*, 1979). Their ultimate bone marrow origin is established beyond doubt (Balner and Dersjant, 1964; Mitchell and Miller, 1968; Miller and Mitchell, 1969). As already mentioned, simultaneous occurrence of these three reactions following the majority of antigenic stimuli is a main difficulty in investigating lymphoid tissue architecture and functional relationships in the different tissue compartments. Separation of these reactions—experimentally—has been an important tool for adequate structural analyses of plasma cell and cellular immunity reactions.

2.1.1a. Origin and Fate of B and T Cells. Neonatal thymectomy has been found to result in a long-lasting immunological unresponsiveness regarding transplantation immunity, delayed type sensitization, and so-called thymus-dependent antibody formation in various species. Adult thymectomy, on the other hand, has only negligible immediate effects on the immune response. However, severe impairment of immunological function could also be caused in adult animals if thymectomy was combined with potentially lethal X-irradiation. In rabbits, this was observed when adult thymectomy was combined with repeated sublethal X-irradiation at 14-day intervals (Keuning and van den Broek, 1968; Veldman, 1970; Keuning, 1972; Veldman *et al.*, 1978a). Histologically, totally lymphocyte-depleted thymus-dependent areas (TDAs) were seen in peripheral lymphoid tissue (Figure 3). Immunologically, the cellular immune capacity of these animals was abolished after this treatment. The outer cortex of the lymph node, including the follicles, had a quantitatively normal population of apparently non-thymus-derived small lymphocytes and marginal zone cells.

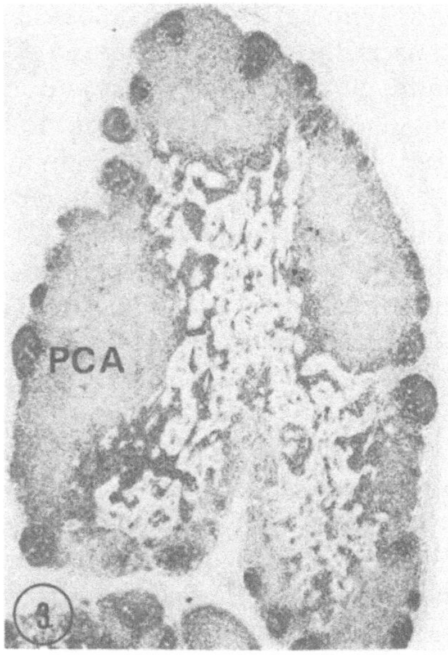

FIGURE 3. Axillary lymph node. T-cell-deprived paracortical area (PCA) with repleted outer cortex with primary follicles (B cells) 150 days after last sublethal total body X-irradiation in thymectomized rabbit. ×20. (From Veldman *et al.*, 1978a.)

In both animal models (Gutmann and Weissman, 1972) and in man (Veldman and Feltkamp-Vroom, unpublished) their B-cell nature could be demonstrated by means of membrane-bound Ig-staining techniques. This class of lymphoid cells, distinct from the thymus-dependent recirculating lymphocytes, presenting itself in conjunction with follicular structures and antibody-forming capacity, has been suggested by Nieuwenhuis and Keuning (1974) to originate mainly from germinal center reactions anywhere in the body. These authors proved by means of sophisticated isotope tracing experiments that germinal center reactions are sites of generation and amplification of antibody-forming cell precursors in general.

2.1.1b. Effects of X-Irradiation on the Immunohistophysiology of Lymphoid Tissue. The suppressive effects of X-irradiation on the immune response have been studied extensively. During the last decades, these effects have been analyzed in relation to lymphoid tissue damage. Following sublethal total body irradiation (450 R) in rabbits, the capacity to give a primary antibody response—histologically represented by plasma cell and germinal center reactions—against *Salmonella java* flagellar antigen was shown to be lost within 24 hr (Keuning *et al.*, 1964); it reappeared on the seventh day postirradiation.

This loss of antibody responsiveness histologically coincided with the destruction of follicular lymphocytes and marginal zone cells; its reappearance coincided with the reappearance of marginal zone cells in the lymph node outer cortex. A similar loss of immunological responsiveness has not been observed for cellular immunity reactions at sublethal irradiation levels (Salvin and Smith, 1959; Schipior and Maquire, 1966; Turk and Oort, 1969; Veldman, 1970; Veldman and Keuning, 1978). Histological analyses revealed strong immunoblast reactions in the paracortical areas of draining lymph nodes during the period of antibody unresponsiveness and in the absence of marginal zone cells and lymphoid follicles (van der Slikke and Keuning, 1964; Keuning, 1965; Turk and Oort, 1969; Veldman, 1970; Veldman and Keuning, 1978).

Higher doses of radiation (800–900 R) were found to suppress temporarily both types of immune response in rabbits (Uhr and Scharff, 1960). The same effect could, similarly, be obtained in rabbits through three or four times repeated sublethal total body irradiation (400–450 R) at seven-day intervals. Histologically, the whole lymphocyte population of the lymphoid tissue had been temporarily eradicated through this schedule of irradiation. Thus, sublethal X-irradiation (450 R) in rabbits may, for a limited period of time, dissociate the two main components of the immunological response of the lymphoid tissue: plasma cell and germinal center reaction on the one hand and TDA-bound cellular immunity reactions on the other hand.

Regarding the traffic of lymphoid cells towards peripheral lymphoid tissue, the following X-irradiation experiments on rabbits would seem to provide important information, which even present sophisticated techniques of cell-tracing by means of chromosome markers (Micklem *et al.*, 1968) and membrane-bound Ig (Gutman and Weissman, 1972) cannot elucidate.

Following *local irradiation of a lymph node* with a dose (750 R) which caused complete destruction of the existing lymphocytes, a repopulation with bloodborne lymphocytes was observed in two distinct areas within 12–24 hr (Figure 4)

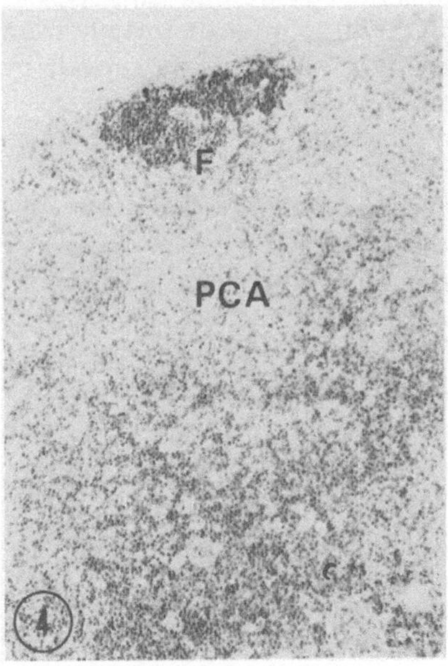

FIGURE 4. Twelve hours after local X-irradiation of popliteal lymph node in normal animal. Repleted outer cortical primary follicle (B cells) and paracortical area (T cells): F, Primary follicle; PCA, paracortical area. ×80. (From Veldman *et al.*, 1978a.)

(Bos, 1967; Veldman, 1970; Veldman *et al.*, 1978a). Firstly, an influx was observed into the midcortical regions—paracortical areas—by way of the cortical epithelioid venules. This influx corresponded with the influx of recirculating lymphocytes, as originally described by Gowans and Knight (1964). Secondly, a strictly localized repletion of the lymphoid follicles in the outer cortex was found with small, dark-staining lymphocytes. From these latter follicle-reconstituting small lymphocytes, a typical follicular cap of marginal zone cells developed in two or three days. Immunologically, the 24-hr-lasting influx into the locally irradiated lymph node was found to have restored the antibody-forming capacity to *Salmonella java* flagellar antigen in that node within 48 hr, with plasma cell formation and germinal center reactions as the morphological substrate.

The phenomenon of rapid and massive repletion of lymphoid follicles has not been observed in the recirculation experiments of Gowans (1959). However, in these experiments the cells used were thoracic duct lymphocytes which belong largely to the "long-lived" (mainly T-cell) category, whereas, in our experiments, the full complement of blood lymphocytes was available.

Local irradiation of lymph nodes in established T-cell-deprived animals gave additional information (Veldman 1970; Veldman *et al.*, 1978a). Whereas the damage of local irradiation of a lymph node in normal rabbits is overcome within 12–24 hr by an influx of blood-borne small lymphocytes in two distinct loci—the paracortical areas and the follicles of the outer cortex—the repletion with lymphocytes following local irradiation of a lymph node in T-cell-deprived animals was strictly limited to follicular structures (Figure 5). Histologically, the capacity

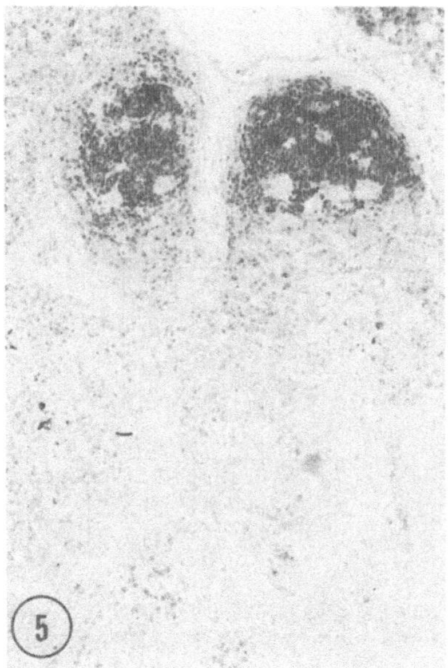

FIGURE 5. T-cell-deprived animal. Twenty-four hours after local X-irradiation of popliteal lymph node. Repleted primary follicles with blood-borne B lymphocytes. Paracortical area completely devoid of any repleting lymphoid cells. ×80. (From Veldman *et al.*, 1978a.)

to react to a subcutaneous antigenic challenge was restored in these animals. A plasma cell and (minimal) germinal center reaction occurred within the next days.

Neither X-irradiation treatment had any effect on the quantity or (sub)microscopic appearance of IDCs in the TDAs. Thymectomy or bone marrow depression did not seem to influence their presence or supposed function.

2.1.2. An X-Irradiation Model for Delineation of an "Isolated T-Cell System" (Figure 6)

The T-cell system has originally been described in "negative" terms as that part of the lymphoid system which is depleted after various thymectomy experiments (Waksman *et al.*, 1962; Parrott *et al.*, 1966). Later ^{3}H-labeling experiments—*in vitro* and *in vivo*—confirmed these observations in a more positive sense (Parrott and de Sousa, 1967; Weissman, 1967): labeled medium-sized and small lymphocytes—though in small numbers—appeared in circumscribed regions which corresponded to the thymus-dependent areas of spleen and

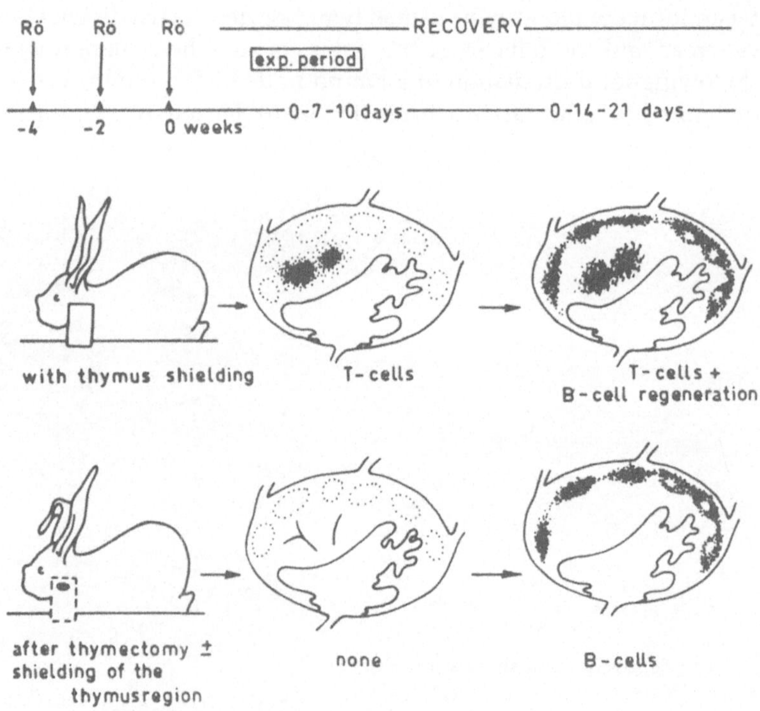

FIGURE 6. Diagram of lymph node repopulation with T and B cells after X-irradiation (×3) with thymus shielding or X-irradiation (×3) after thymectomy (with or without shielding of the thymus region) "Isolated T-cell system" for indicated experimental period

lymph nodes. Only recently have thymocytes been shown to make up the majority of lymphocytes in the paracortical areas of the lymph node (Veldman, 1970; Gutman and Weissman, 1972; Veldman, 1977, Veldman and Keuning, 1978). Since three times sublethal total body X-irradiation completely destroys the pool of immunocompetent cells for skin allograft rejection, and previous thymectomy precludes regeneration of this pool of lymphocytes, a rabbit thus treated—when given a few weeks for regeneration of its B-cell system—proved an experimental model for analysis of a plasma cell reaction in lymph nodes (Keuning and van den Broek, 1968; Nieuwenhuis *et al.*, 1970; Veldman, 1970; Veldman *et al.*, 1978a). In contrast, by maintaining three sublethal total body X-irradiations with two-week intervals, but with shielding of the thymus, it was possible to obtain a complementary experimental system. For a period of some 7–10 days following the last irradiation, any lymphocytes repopulating the peripheral lymphoid organs are recently thymus-derived; thus for 7–10 days an "isolated T-cell System" is achieved. Thereafter, the B-cell line reappears slowly in the non-TDAs of peripheral lymphoid tissue. In control experiments, the same procedure was followed, but, prior to irradiation with shielding of the thymus region, surgical thymectomy was performed.

2.1.2a. Histology of an "Isolated T-Cell" Animal Model. The thymus was markedly reduced in size 1–2 days following the third irradiation with thymus shielding in comparison with the thymus at the last irradiation time. As any notable destruction of thymocytes was not observed, the most probable

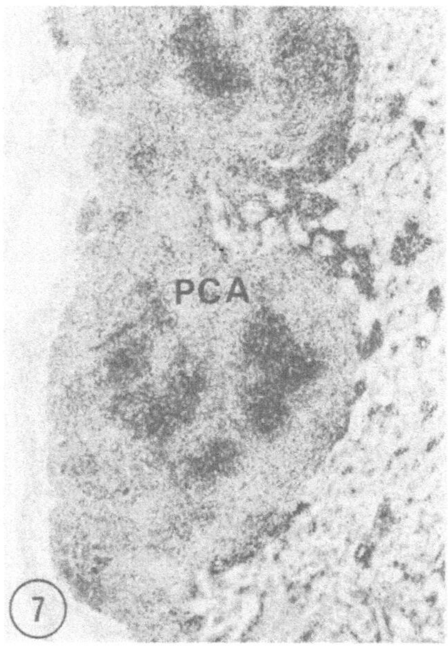

FIGURE 7. T-cell repleted paracortical area of popliteal lymph node, 7 days after last total body X-irradiation with thymus shielding. ×70. (From Veldman and Keuning, 1978.)

interpretation of this postirradiation decrease in size would seem to be a considerable outflow of cells towards the severely depleted peripheral lymphoid organs. Labeling experiments confirmed these observations. In *lymph nodes,* already 24 hr following the last irradiation, a marked patchy repletion of the paracortical areas with lymphocytes was observed. During the following days, this population of recently thymus-derived lymphocytes increased while leaving the outer cortical (B-cell) territory depleted till the seventh postirradiation day (Figure 7). A similar repopulation with thymocytes was observed in the *splenic* periarteriolar lymphocyte sheaths from 24 hr postirradiation onwards. The well-known depleted TDAs of T-cell-deprived animals were repopulated in this complementary system, while the non-TDAs remained depleted during a similar period, as mentioned above. The interfollicular areas of the *appendix, sacculus rotundus,* and *Peyer's patches*—depleted in the T-cell-deprived animals—24 hr after irradiation, had already started to repopulate with thymus-derived lymphocytes, demonstrating a similar complementary picture as the other peripheral lymphoid organs in these experiments (Veldman, 1970; Veldman and Keuning, 1978).

2.2. HISTOPHYSIOLOGY AND ELECTRON MICROSCOPY OF CELLULAR IMMUNITY REACTIONS IN THE LYMPH NODE

The conditions obtained in the X-irradiation model with thymus shielding represent an experimental system for proving thymocyte cell traffic and specific homing capacities as well as thymocyte immunocompetence in the kind of reactions studied. No interference with a normally present antibody response is possible during the experimental period; neither plasma cell nor germinal center reactivity are present during a testing period of ± 7 days post irradiation.

When rabbits were given a split-skin allograft 24 hr after the third irradiation with thymus shielding, already two days after grafting, extensive TDA-bound immunoblast activity was observed in draining lymph nodes. No outer cortical B-cell activity was present, in contrast to control nonirradiated animals. Allografts were completely rejected after ± 12 days. The rejection process itself varied between 3 and 9 days.

Contact sensitizer (oxazolone) and protein antigens (ferritin, horse-γ-globulin) elicited clear immunoblast reactions in the TDAs of the draining lymph nodes, first observed 2 days following antigen administration and maximal around 4–6 days. The most vigorous reaction was observed after application of a contact sensitizer. Autoradiographic studies revealed also a full capacity of these recently thymus-derived lymphocytes to react to antigenic challenge with a blast transformation having a final progeny of lymphocytes (Veldman, 1970; Veldman and Keuning, 1978). There was no antibody formation at the time of analysis (Keuning, 1972; Mulder, 1972). Cellular immunity reactions in the TDAs demonstrated a characteristic pattern: agglomerates of medium-sized lymphocytes and T blasts were seen surrounding IDCs (Figures 8–10). In lymph nodes obtained from these rabbits with an "isolated T-cell system," the first activity of cellular immunity (blast transformation) was observed among the thymus-derived cells lying between the interdigitating cell pattern. Figure 11 illustrates this situation at the ultrastructural level. Thymus-derived lymphocytes are shown embedded in a mosaic system of IDCs. The transitional elements and immunoblasts observed after antigenic stimulation by a skin allograft, contact sensitizer, or protein antigen were essentially similar in these experiments. Among these cells, two types could be clearly distinguished, one with a low and the other with a high electron density of the cytoplasmic matrix (Figure 12) (Veldman, 1970; Veldman *et al.*, 1978b; see also Veldman *et al.*, 1980b). This difference was also reflected in their packing and content of (poly)ribosomes: the electron-dense type contained more ribosomes than the electron-lucent variety. This very conspicuous difference was particularly clear following silver methenamine staining procedures. The difference in electron density—and optical density in the case of phase contrast microscopy—was seen not only in the cytoplasmic matrix, but also in that of nucleus and mitochondria. It could be hardly distinguished in conventional uranyl acetate-lead citrate stained sections. These different cell lines were always seen simultaneously and following each of the mentioned antigenic challenges, although there was a slight indication in the available material that the light variety appeared to predominate in the reactions to horse-γ-globulin (humoral immunity, T-cell dependent). In terms of these experiments, it would seem that the electron-dense cell line, predominantly present during the reaction to a contact sensitizer and skin allograft, represents the machinery of specificity and final expression of these types of cellular immunity.

2.2.1. Interdigitating Cells and T Cells. Their Microenvironment and Progeny in the Lymph Node

Submicroscopic analysis of cellular immunity reactions under the experimentally defined circumstances of exclusive T-cell presence and reactivity in the lymph node provided information on which two aspects deserve particular attention: (1) the presence of IDCs, exclusively in the T-cell territory of the lymph node with their intimate relationship with thymus-derived cells, and (2) the recognition of two T immunoblasts or, rather, cell lines. The IDCs are characterized by their most bizarre nuclei (Figures 1, 8, and 12). These nuclei have a lightly staining chromatin pattern, mostly diffuse, but also partly as a very thin layer associated with the nuclear membrane, and a small distinct nucleolus. The rather large mass of cytoplasm shows two main features: (1) the interdigitations in which the plasma membranes tightly adhere to each other and neighboring IDCs (Figures 1 and 13), as well as to the thymus-derived lymphocytes, and (2) the presence of an elaborate tubulovesicular (TV) system in the cytoplasm of these cells (Figures 1, 11, and 14). The contents of this TV system—a complex of vesicles and sacs—was intensively stained by periodic acid or chromic acid and silvermethenamine. However, no blackening was observed if periodic or chromic hydrolysis was omitted (Figure 15). Similarly, the perinuclear cisterna

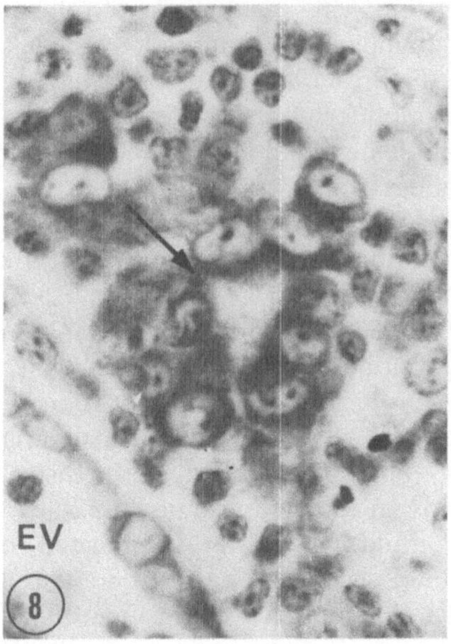

FIGURE 8. Cluster of T blasts around interdigitating cell (arrow) in paracortical area; popliteal lymph node, 4 days after last total body X-irradiation with thymus shielding (antigen: allogeneic T-cell homogenate) ×900. (From Veldman and Keuning, 1978.)

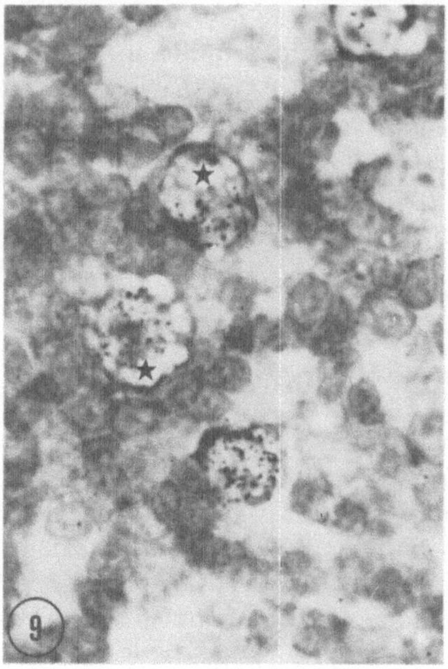

FIGURE 9. "Isolated T-cell system." Labeled T blasts 4 days after ferritin; popliteal lymph node, 1 hr after [^{3}H]thymidine. ×1200. (From Veldman and Keuning, 1978.)

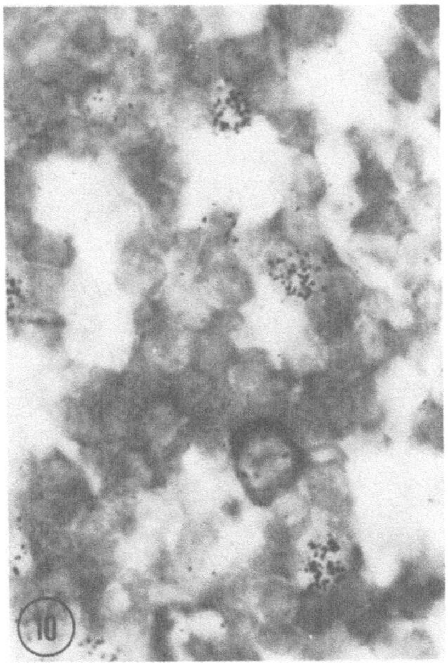

FIGURE 10. "Isolated T-cell system." Labeled small lymphocytes, 4 days after ferritin; popliteal lymph node, 1 day after [^{3}H]thymidine. ×1200. (From Veldman and Keuning, 1978.)

sometimes contains a PSMA-positive substance. The TV system appears to have some relation to the Golgi apparatus on one hand and with the interdigitating plasma membrane on the other. The Golgi vesicles sometimes contain silver-positive material with a gradient of increasing contrast towards the TV system (Figure 14). The presence of PSMA-positive material on the interdigitating plasma membrane would seem to represent a coating with the product of this TV system (Figures 11 and 16). Phagolysosomes are seen only occasionally in the IDC. In human lymphoid tissue, definite heterophagosomes have been demonstrated only in very rare instances. As far we know, heterophagosomes (melanosomes) have been found in human lymphoid tissue only in dermatopathic lymphadenitis (Rausch *et al.*, 1977). Mitochondria of IDCs are of a slender type. RER is sometimes observed in the form of a typical "whorl." A small number of mono- and polyribosomes and also bundles of intracytoplasmic filaments can be observed.

The TV system, the Golgi apparatus proper, and the whorls might together represent the so-called GERL (Novikoff *et al.*, 1966), i.e., the complex of Golgi-associated endoplasmic reticulum and lysosomes, responsible for synthesis and elaboration of the cell's product (Veldman *et al.*, 1978b). In view of the positive silver methenamine staining, the contents of this system and the plasma membrane coating seem to be a substance with a high carbohydrate moiety, presumably a glycoprotein. The production of this substance was not dependent on antigenic stimulation or related to the X-irradiation procedures, as shown by controls (Veldman, 1970).

According to light microscopic findings in imprints and cryostat sections of

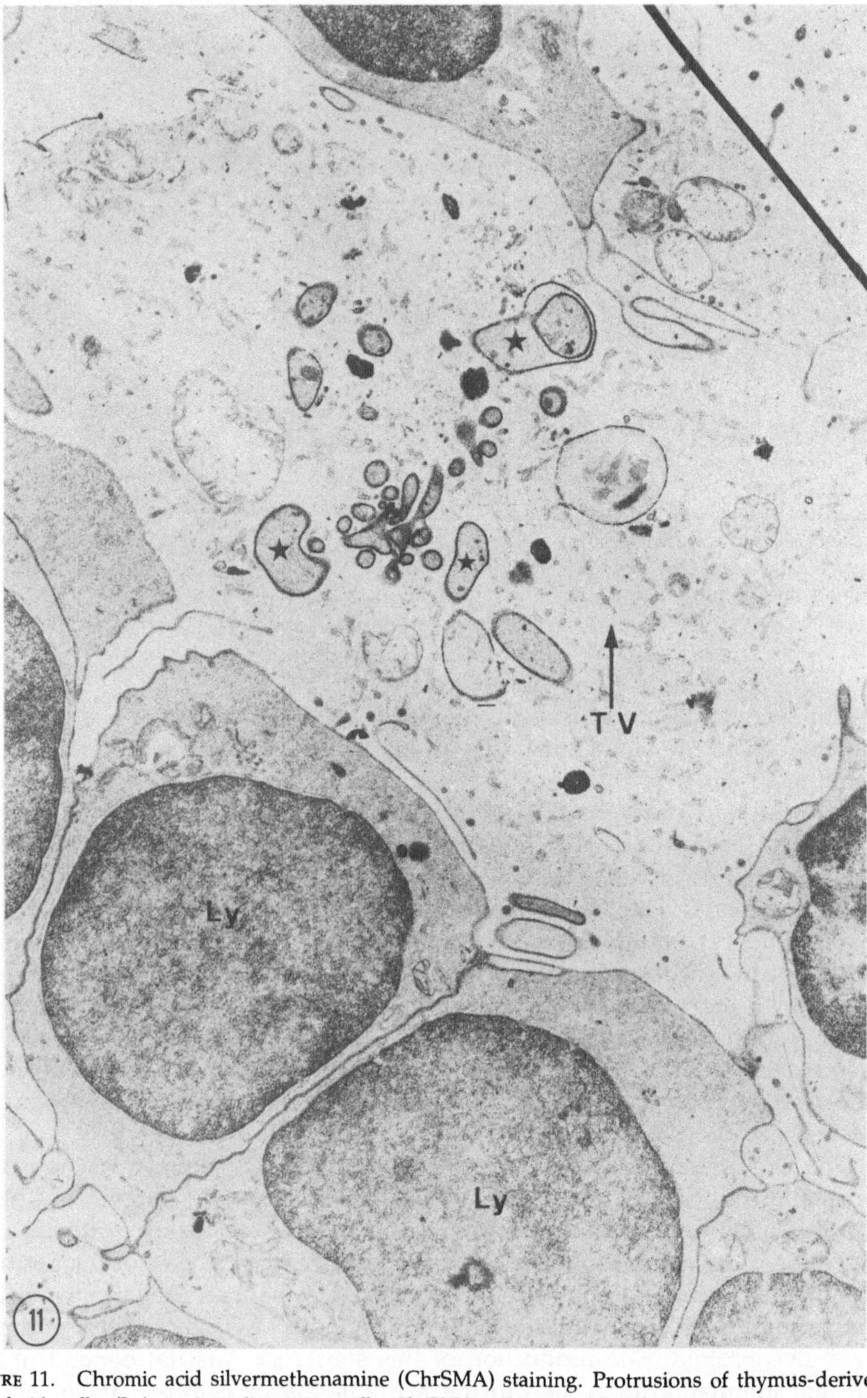

FIGURE 11. Chromic acid silvermethenamine (ChrSMA) staining. Protrusions of thymus-derived lymphoid cells (Ly) in interdigitating cell. ChrSMA-positive material covering fingerlike cell protrusions of thymus-derived cells (stars), piercing the body of IDC. Note ChrSMA-negative content of TV system, 4 days after ferritin. ×11,000. (From Veldman *et al.*, 1978b.)

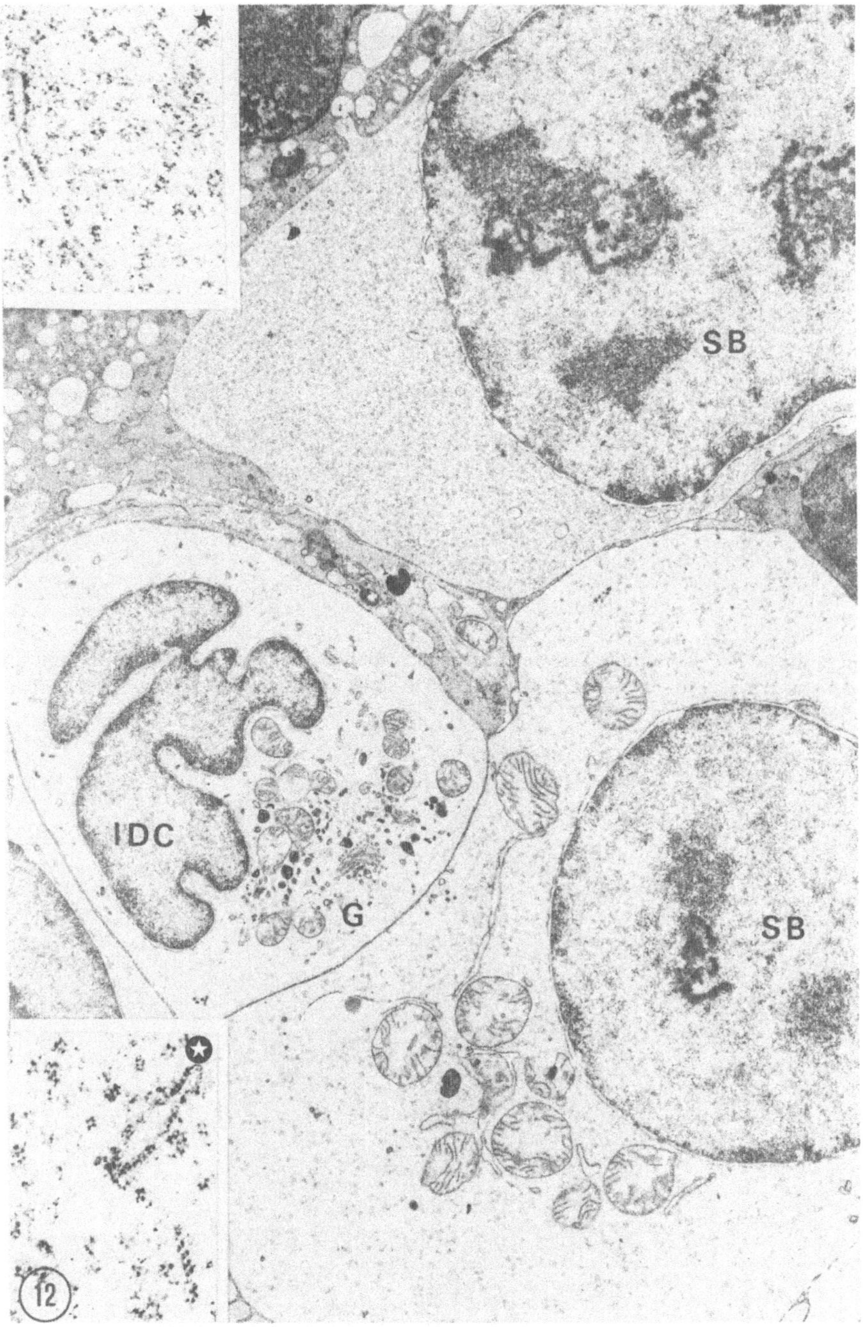

FIGURE 12. Periodic acid silvermethenamine staining. Two types of cellular immunity blasts (SB) in paracortical area. G, Golgi apparatus; IDC, interdigitating cell. Note difference in electron density and ultrastructure of both blast cells. Four days after oxazolone. ×6700. Insets: Ua–Pb staining. Note difference in packing of (poly-)ribosomes of electron-dense (black star) and electron-lucent (white star) cellular immunity blasts. ×10,000. (Inset Ua-Pb staining: ×29,000.) (From Veldman *et al.*, 1978b.)

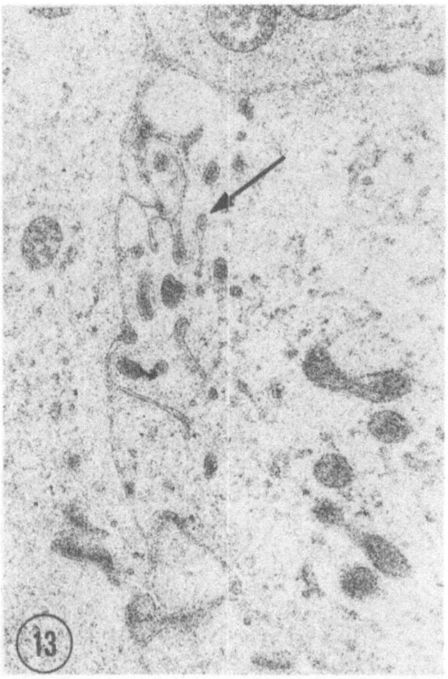

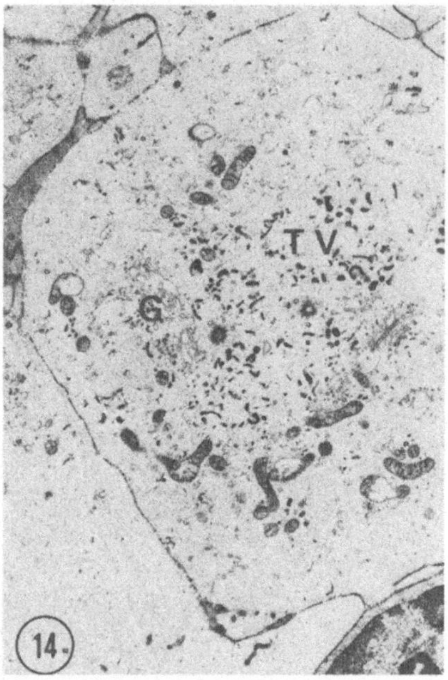

FIGURES 13 and 14. Continuity between intracellular tubulovesicular complex (TV) and cell surface. PSMA-positive deposits (arrow). See also relation to Golgi apparatus (G). Figure 13: ×15,000; Figure 14: ×7500. (From Veldman, 1970.)

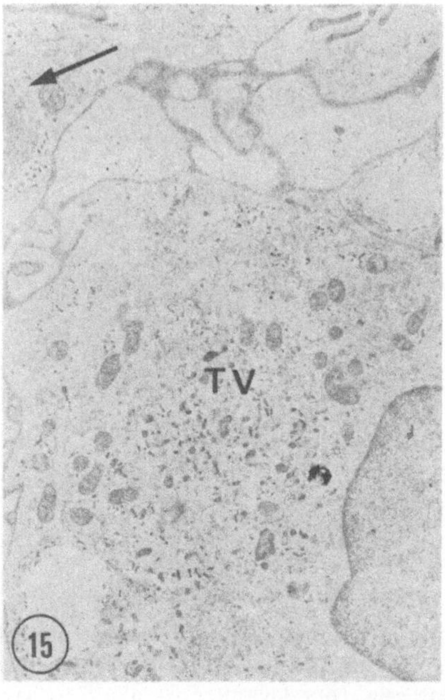

FIGURE 15. Control of Figure 14. AMA-staining without preoxidation. "Whorl" (arrow). ×8500. (From Veldman, 1970.)

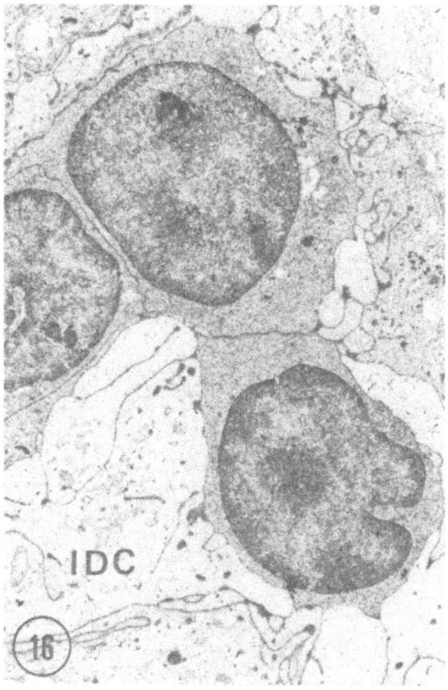

FIGURE 16. "Isolated T-cell system." IDC with thymus derived lymphocytes between membrane system. Axillary lymph node, 2 days after skin allograft. ×4500. (From Veldman, 1970.)

human lymph nodes, IDCs show weakly positive acid phosphatase and α-naphthyl acetate esterase activities (Kaiserling and Lennert, 1974; Lennert *et al.*, 1978a). The acid phosphatase reaction reveals a focal zone of activity, which is usually situated near the nucleus (Figure 17). The nonspecific esterase reaction is sometimes focal and paranuclear and sometimes diffusely positive. Another intracytoplasmic enzyme of the IDC is leucyl-β-naphthyl amidase (Elema and Poppema, 1978). IDCs are negative for alkaline phosphatase and naphthol-AS-D-chloroacetate esterase.

Mg^{2+}-dependent ATPase is a membrane enzyme that clearly distinguishes IDCs from other cells and particularly from outer cortical dendritic cells (Müller-Hermelink, 1974; Müller-Hermelink *et al.*, 1976; Rausch *et al.*, 1977, Lennert *et al.*, 1978a). The interdigitating plasma membrane exhibits a positive reaction for this enzyme, as seen in electron micrographs (Figure 18). The functional interpretation of these enzyme activities—convenient as additional cell markers—has not yet been clarified.

Thymus-derived lymphocytes are embedded in a mosaic system of IDCs. Fingerlike cell protrusions of lymphocytes sometimes pierce the cytoplasmic body of the IDC (Figure 11 and 16). High resolution electron microscopy occasionally even showed tight junctions between cytoplasmic extensions of IDCs and between lymphocytes and IDCs (Veerman and van Ewijk, 1975).

Müller-Hermelink *et al.* (1976) studied the ontogeny of IDCs in lymphoid tissue of human fetuses of 8–30 weeks gestational age. They came to the conclu-

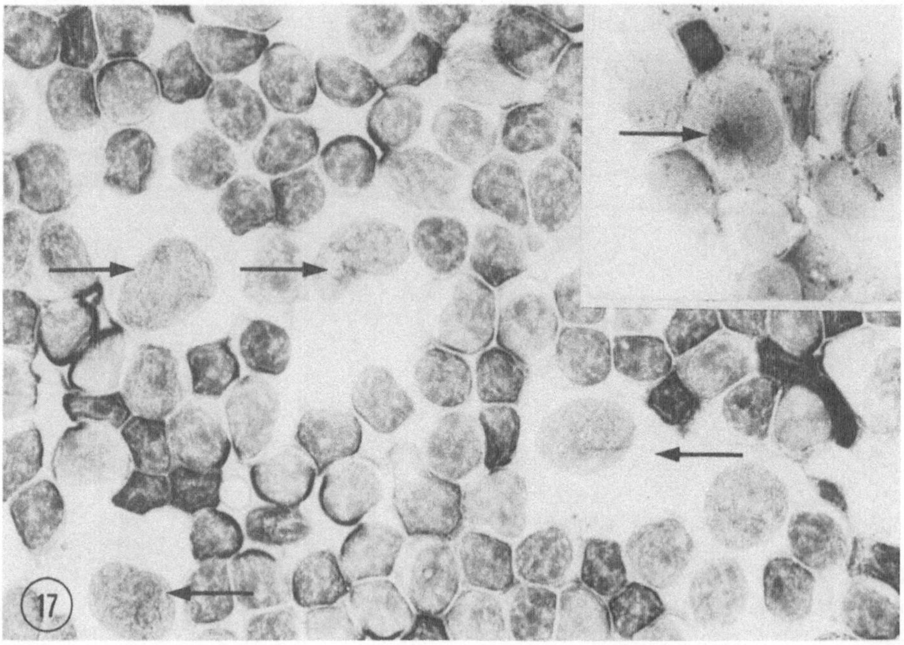

FIGURE 17. Interdigitating cells (arrows) and lymphocytes in an imprint. Human lymph node with chronic nonspecific lymphadenitis. Note abundant cytoplasm and oval or reniform nuclei of the IDC. Pappenheim. ×880. Inset: Acid phosphatase reaction of IDC; focal zone of slight enzyme activity near the nucleus (arrow). ×880.

sion that lymphocytes appeared in peripheral lymphoid tissue somewhat earlier than did IDCs. The appearance of typical T-cell regions is delayed until IDCs are present. Then also extramyeloid hemopoiesis is no longer found. The T-cell regions seem to be formed somewhat earlier than the B-cell territories in the lymph node (16 vs. 25–30 weeks gestational age).

The intimate contacts between the cytoplasmic extensions of thymus-derived lymphocytes and interdigitating cell membranes is reminiscent of the relationship between the outer cortical lymphocytes and marginal zone cells and the cytoplasmic processes of dendritic cells in the B-cell compartment. As a working hypothesis, it seems, therefore, logical to presume a functional relationship between the presence of IDCs and the homing and immunological reactivity of thymus-derived cells in the same areas. The first signs of cellular immune reactivity under defined immunohistophysiological conditions are observed around the IDCs. The membrane–membrane contact as realized in the IDC–T-cell contact might be particularly suited for adequate transfer of antigenic stimuli with further T-cell reactivity. Our knowledge in this respect, however, is still limited.

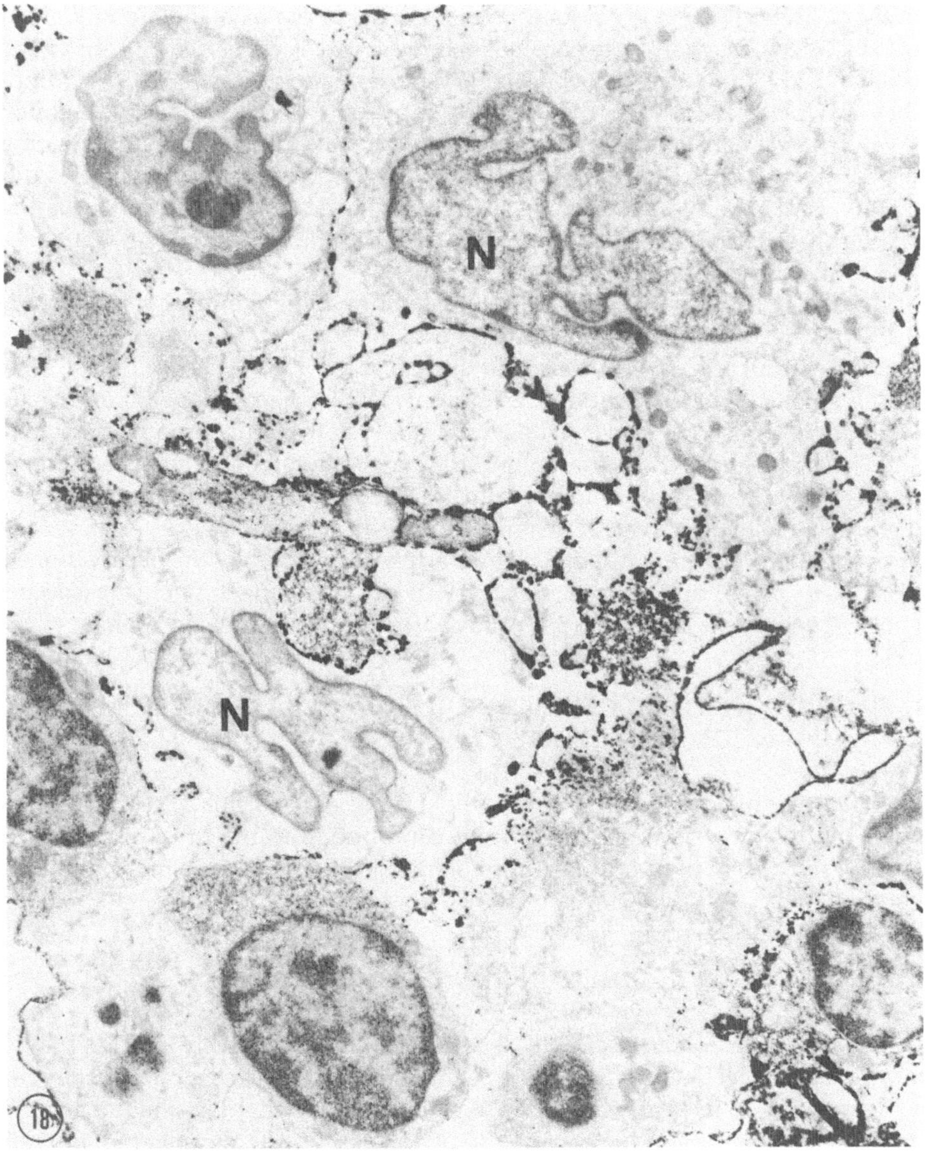

FIGURE 18. Two IDCs (N, nucleus). Human lymph node. Positive ATPase activity in the cell membrane. Unstained. ×9000. (Courtesy of Dr. H. K. Müller-Hermelink.) Reduced 12% for reproduction.

2.3. INTERDIGITATING CELLS IN OTHER T-CELL COMPARTMENTS

The concept that the IDC is an essential cell constituent in the T-cell microenvironment throughout lymphoid tissue is supported nowadays by several authors. The cells have been demonstrated in the periarteriolar lymphatic sheath of the spleen in rats (Veerman, 1974), mice (van Ewijk *et al.*, 1974), rabbits (Blijham, 1975), and human specimens (Heusermann *et al.*, 1974). In addition, they are found in human thymus of young children (Kaiserling and Lennert, 1974), in specimens of fetal origin (Müller-Hermelink *et al.*, 1976), and in guinea pigs (Töro and Oláh, 1966). In the thymus they are particularly localized in the medulla and inner cortex. Similar observations have been done in various animal species. T-cell proliferation is regularly seen around these IDCs (von Gaudecker and Müller-Hermelink, 1978).

In the TDA of the spleen a similar cluster formation of immunoblasts around IDCs has been observed after intravenous injection of sheep red blood cells (Blijham, 1975), as has been noticed in the reacting lymph node (Veldman, 1970). In the splenic reaction, we are dealing only with a T-cell-dependent primary antibody response, whereas in the lymph node, both cellular immune reactivity and T-cell-dependent humoral immunity may be involved in such as initiation process (Veldman, 1970; Veldman and Keuning, 1978). The integrity and intimate relationship between IDCs and T cells—in immunological terms the development of T immune reactivity—are further supported by the observations of Racz *et al.* (1977) and Myrvik *et al.* (1979), in analyzing bronchial-associated lymphoid tissue in BCG-induced cellular immunity reactions in rabbits.

3. INTERDIGITATING CELLS. A GUIDING CELL LINE IN HUMAN LYMPH NODE PATHOLOGY

The recognition of the IDC in lymphoid tissue as a specific cell constituent of the T-cell regions has been a significant contribution to the present concept of T-cell reactivity, both in normal lymphoid tissue and in various malignant lymphomas.

3.1. INTERDIGITATING CELLS IN NONNEOPLASTIC LESIONS

In human lymphoid tissue, IDCs are also found exclusively in TDAs. Thus, they often serve as important cytological markers for TDA-recognition. They can be demonstrated in nearly all types of reactive lymphadenitis. Enlargement of the TDAs is quite often due not only to an increase in the number of lymphoid cells but also to a significant augmentation of IDCs in the same regions. When reactivity encompasses both the T- and B-cell regions in the lymph node* it is generally impossible to distinguish between T and B cells, including their immunoblast progeny. However, when the IDC as an additional cytological marker is taken into consideration, immunoblasts that lie in the direct vicinity of these cells are generally of T-cell origin (Veldman, 1970; Lennert *et al.*, 1978b; Veldman and Keuning, 1978; Kaiserling, 1977, 1978). When the bias of the response in a lymph node is towards outer cortical B-cell reactivity (plasma cell reaction and

*In pathology: "polymorphic hyperplasis of the pulp" or "Bunte Pulpahyperplasie" (Lennert, 1961).

germinal center reaction) and T-cell reactivity is minimal, the number of IDCs is usually also small. It may also occur that T-cell reactivity (and IDC accumulation) predominates in the draining lymph node. This depends—as expected from our experimental analyses in rabbits—on the type of antigen, which elicits the immune response (Veldman, 1970; Veldman *et al.*, 1978a,b).

Dermatopathic lymphadenitis is a particular type of lymphadenitis: it is a benign reactive lymphadenopathy that develops in regional lymph nodes as a result of various types of chronically inflamed dermal lesions (Kaiserling and Lennert, 1974; Rausch *et al.*, 1977; Lennert, 1961; Lennert *et al.*, 1978a). The lymph node enlargement is mainly due to hyperplasia of the TDAs; in cytological terms, the number of nonlymphoid cells (IDCs) is particularly augmented (Figures 19 and 20). The local pool of lymphocytes is decreased. Immunoblasts are seen occasionally and mitotic figures are rare. IDCs are closely interwoven with their cytoplasmic extensions into a mosaic system throughout the depleted TDAs. Langerhans cells are also regularly found in these draining lymph nodes. This occurs as a special feature of dermatopathic lymphadenitis, in contrast to other types of nonspecific lymphadenitis, as revealed by ultrastructural analyses (Kaiserling, 1977). Although there are certain ultrastructural similarities between the Langerhans cell and the IDC, the Langerhans cell can be clearly distinguished from the IDC by the presence of Langerhans granules. Langerhans cells presumably migrate from the skin to the regional lymph node through afferent lymphatics, as suggested by their presence in the sinuses. This would be in

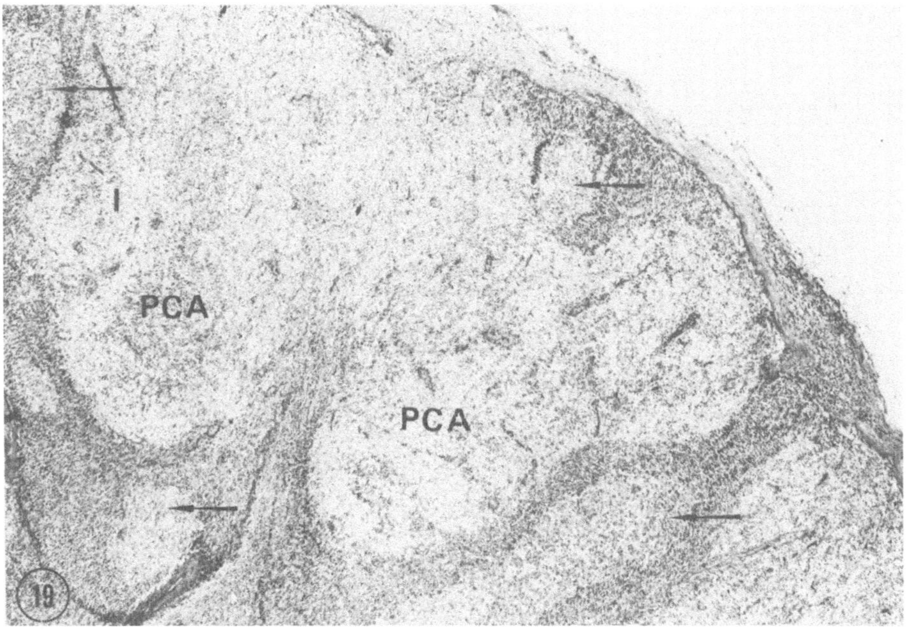

FIGURE 19. Human lymph node. Dermatopathic lymphadenitis. PCA, Paracortical areas; arrows, germinal centers. ×56.

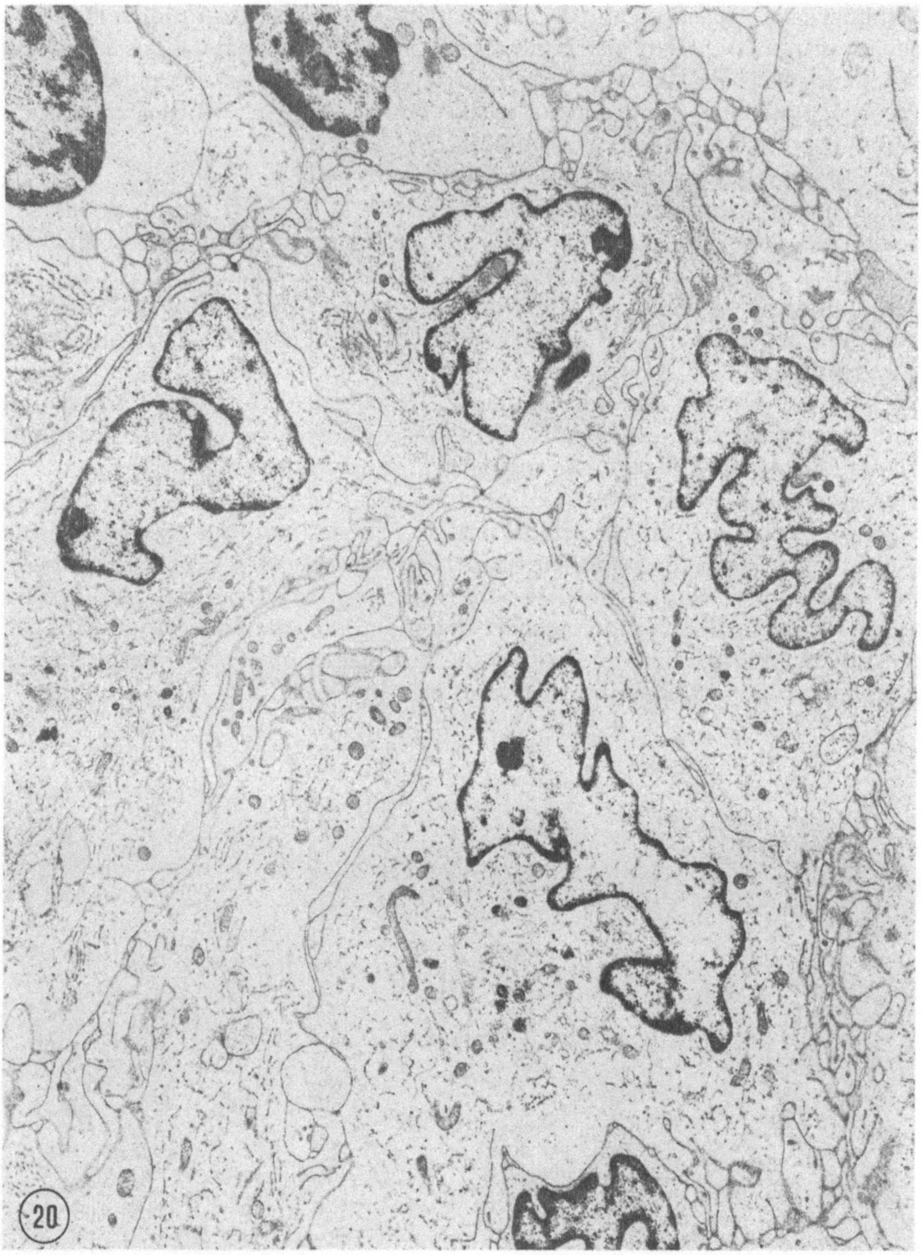

FIGURE 20. Human lymph node. Electron micrograph of PCA; note multiple interwoven IDCs with a few lymphocytes. ×9000.

accordance with experimental data of Kelly *et al.* (1976) and Silberberg-Sinakin *et al.* (1976, 1978) on Langerhans cell traffic. The discrepancy between the number of lymphoid cells and IDCs in dermatopathic lymphadenitis compared with nondermatopathic lymphadenitis is not yet understood. In general, it is still obscure whether the immune reaction is augmented, diminished, or even disturbed.

3.2. INTERDIGITATING CELLS IN MALIGNANT LYMPHOMAS

A systematic ultrastructural investigation of non-Hodgkin's lymphomas revealed that the IDC occurs in several lymphoma entities (Table 1). The IDCs are either actual components of the lymphoma, or they originate from completely or partially intact TDAs. In general, T-cell regions may be observed in all lymphoma entities. Malignant lymphomas in which IDCs are an integral part of the lymphoma are *T-zone lymphoma, T-CLL,* mycosis fungoides (M.F.),* and *Sézary's syndrome.* The T-zone lymphoma (Kaiserling, 1976, 1977; Lennert, 1976; Lennert *et al.*, 1978a) is a malignant lymphoma in which all characteristic components of a TDA occur: IDCs, epithelioid venules crowded with lymphocytes, occasionally so-called T-associated plasma cells (Lennert *et al.*, 1975, 1978b), and the tumor cells, by rosetting techniques to be characterized as T lymphocytes and T immunoblasts. At the ultrastructural level, these malignant cells lie between the cytoplasmic extensions of IDCs (Figure 21). The conditions in T-CLL are similar

*Chronic lymphocytic leukemia, T-cell type.

TABLE 1. OCCURRENCE OF INTERDIGITATING CELLS (IDC) AND DENDRITIC RETICULUM CELLS (DC) IN NON-HODGKIN'S LYMPHOMAS OF LOW-GRADE MALIGNANCY (KIEL CLASSIFICATION)[a]

Malignant lymphoma	IDC	DC
Lymphocytic		
Chronic lymphocytic leukemia, B-cell type	–[b]	–
Chronic lymphocytic leukemia, T-cell type	+[b]	–
Hairy-cell leukemia	–	–
Mycosis fungoides	++[b]	–
Sézary's syndrome	++	–
T-zone lymphoma	++	+
Lymphoplasmacytic/lymphoplasmacytoid (LP immunocytoma)	–	+
Plasmacytic (plasmacytoma)	–	–
Centrocytic	–	+
Centroblastic/centrocytic		
Follicular	++	+++[b]
Follicular and diffuse	++	+++
Diffuse	+	+++
With or without sclerosis	+/++	+++

[a] Gérard-Marchant *et al.* (1974); Kaiserling (1977); Lennert *et al.* (1978a).
[b] – = never, + = sometimes, ++ = often, +++ = as a rule.

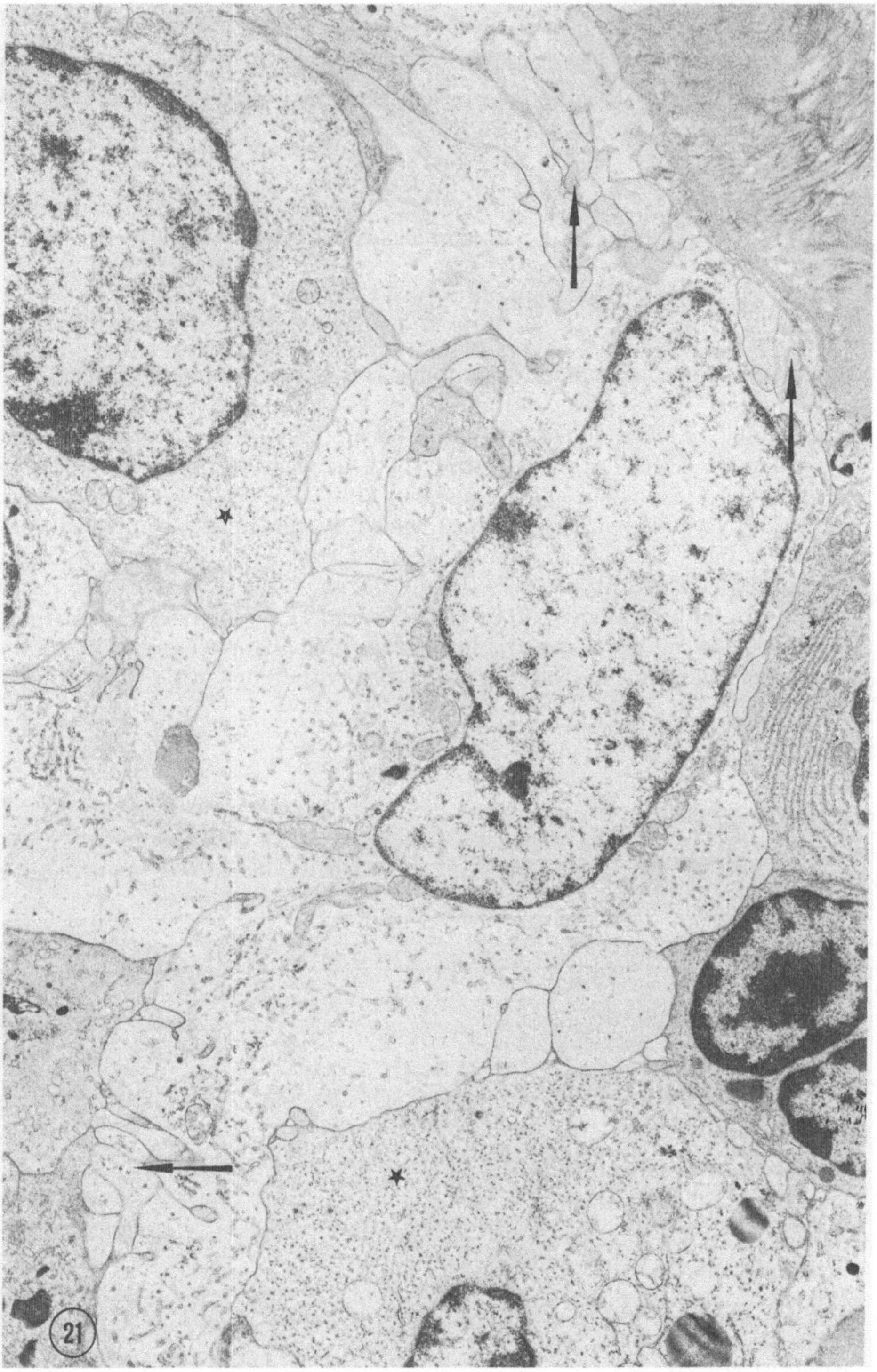

FIGURE 21. T-zone lymphoma. IDC protruding into lymphoid microenvironment (arrow). Stars, tumor cells. ×6600. (From Kaiserling, 1978.)

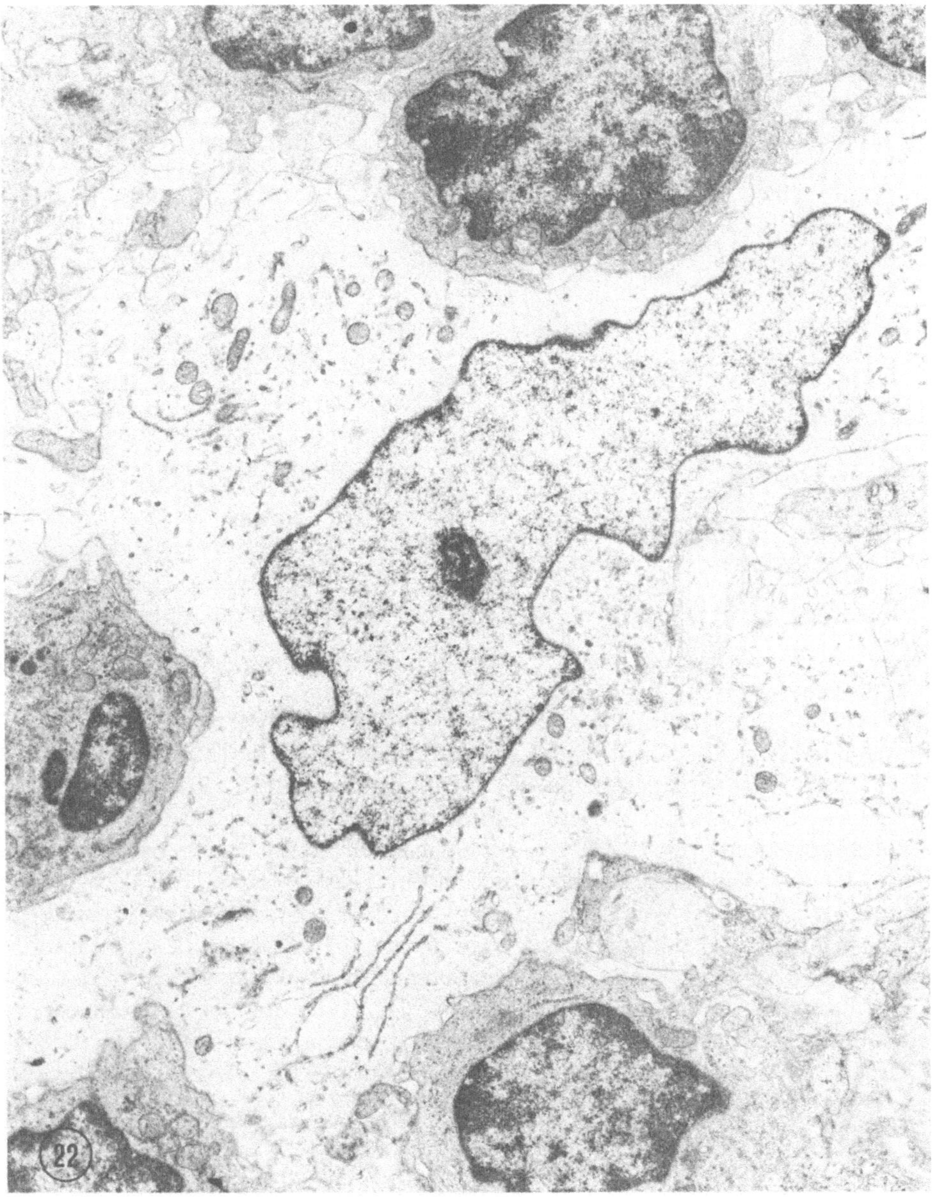

FIGURE 22. Mycosis fungoides. IDC in the dermis. Tumor cells (lymphocytes) are in close contact with IDC. ×7700. (From Kaiserling, 1978.)

to those in T-zone lymphoma. Lennert *et al.* (1978a) could demonstrate IDCs in tumor tissue of several cases at the light microscopic level. The IDC can be an essential component of a lymphoma, as suggested by the electron microscopic findings of Goos *et al.* (1976) in M.F. IDCs are seen among the tumor infiltrates in the dermis (Figure 22), whereas under physiological conditions, no IDCs are

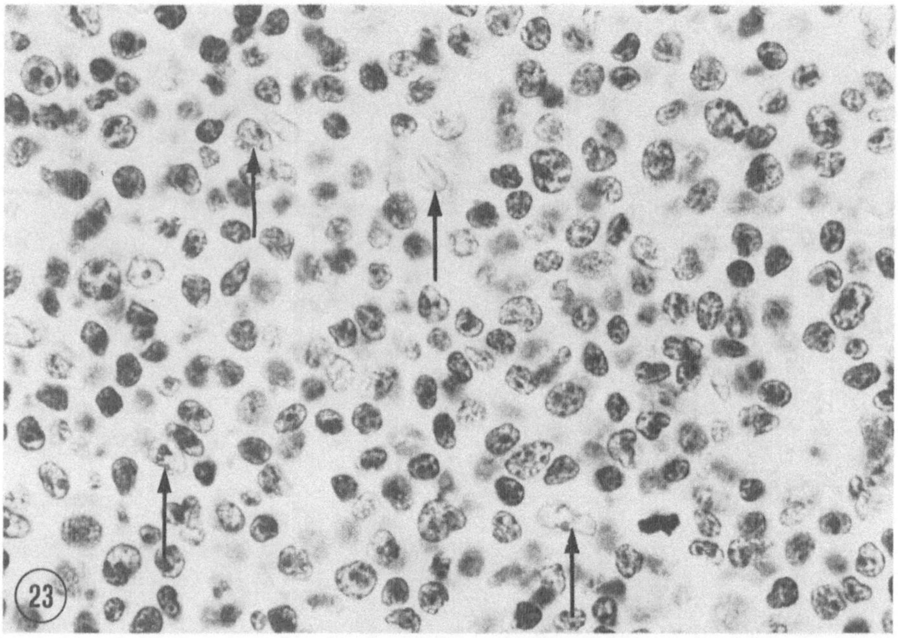

FIGURE 23. Mycosis fungoides. Lymph node IDCs (arrows) among polymorphic tumor cells Hematoxylin–eosin. ×880.

found in the skin. Furthermore, IDCs also occur in lymph nodes, infiltrated by the M.F. tumor cells (van Leeuwen *et al.*, 1976; Kaiserling, 1978). Their cytoplasmic processes extend between the surrounding tumor cells (Figure 23). Mitotic figures of the neoplastic lymphoid cells are seen occasionally in the vicinity of the IDCs. A particular case has recently been described by Lennert *et al.* (1978a): a monomorphic malignant proliferation of IDCs was observed in a patient—treated for M.F.—who developed large lymph node tumors. Atypical mitotic figures, infiltrative growth in the lymph node parenchyma and its surroundings, were seen. The subcapsular sinus was crowded with IDCs, in the manner of lymphangitis carcinomatosa. It is likely that we are dealing, in this case, with a sarcoma of IDCs. In Sézary's syndrome, the leukemic variant of M.F., IDCs have also been demonstrated (Kaiserling, 1978; Lennert *et al.*, 1978a).

All the malignant lymphomas, as mentioned above, in which IDCs are, very likely, essential components of the disease are *T-cell malignancies*. On the other hand, when dealing with a B-cell neoplasm—"malignant lymphoma, centroblastic/centrocytic"* of the so-called Kiel classification (Gérard-Marchant *et al.*, 1974), a germinal center tumor—only occasional IDCs occur in the interfollicular areas, if any at all. In these T regions only a few tumor cells are usually seen, whereas mainly T lymphocytes and epithelioid venules are present.

In T-cell lymphomas of *high-grade malignancy*—malignant lymphoma, lym-

*Also called: malignant lymphoma, FCC, cleaved (Lukes and Collins, 1975).

phoblastic, convoluted cell type, as originally described by Barkos and Lukes (1975) and T-immunoblastic lymphoma—IDCs are absent.*

Detailed investigations on the occurrence of IDCs in Hodgkin's disease have not yet been done. Preliminary studies by the Kiel group reveal that IDCs occasionally occur in Hodgkin's disease of the mixed cellularity type and the nodular sclerosis type (Lennert *et al.*, 1978a). There is sometimes a remarkable cytological relationship between these cells and the so-called lacunar cells and probably between IDCs and Sternberg-Reed cells as well. IDCs are also found in the lymphocyte-predominance type of Hodgkin's disease, especially the nodular subtype (nodular paragranuloma) (Lennert *et al.*, 1978a; Poppema *et al.*, 1979a,b). Whether IDCs in Hodgkin's disease are essential components of the malignancy itself is not known yet.

4. DISCUSSION

Since 1970, after the discovery and detailed description of the IDC in the TDA of lymph nodes (Veldman, 1970), there exists the same confusion in terminology as still remains in relation to its supposed analogue in the B-cell territory of lymphoid tissue. Namely, the dendritic (reticulum) cell or macrophage. As mentioned in Section 1, some authors call IDCs "reticulum cells." These authors chose to use a name that offers more than a mere description of the interdigitations and decided on "interdigitating reticulum cell" to point out that the cells are, at least to a large extent, stationary (?) and structural cells of lymphoid tissue.

Here, we prefer to use the term *interdigitating cells* as long as their properties are still in the process of further detailed definition and elucidation. There are still not enough arguments for or against placing them in the compartment of resident histiogenic macrophages (Daems *et al.*, 1976) or mononuclear phagocytes (van Furth *et al.*, 1972).

IDCs and Langerhans cells share many submicroscopic features (Rausch *et al.*, 1977). However, IDCs—as originally described by Veldman (1970)—lack the characteristic organelle of Langerhans cells, i.e., the submicroscopically identifiable granules (Birbeck *et al.*, 1961). Enzyme cytochemical patterns of both cell types are similar. Enzyme intensity and localization are identical (Kaiserling and Lennert, 1974; Müller-Hermelink *et al.*, 1976; Rausch *et al.*, 1977; Elema and Poppema, 1978).

From pure morphological studies, it is virtually unknown whether the IDC is a sessile cell constituent of the TDA or part of a mobile pool of nonlymphoid cells, specifically homing in the TDA of peripheral lymphoid tissue. Cannulation experiments on rabbits and pigs revealed, in the mobile pool, a population of so-called "veiled," nonlymphoid cells, present in the afferent lymphatics, which have a tendency to home in the TDAs of the regional lymph node (Balfour,

*IDCs have been demonstrated only in one case, at the EM level (Kaiserling, unpublished (1977)).

personal communication). Cinematographic studies by the same author demonstrated a different behaviour of these "veiled" cells than was observed in classical macrophages.

Whether this whole cell population also comprises precursors of IDCs is still unclear, although preliminary ultrastructural analyses and isotope cell-tracing experiments seem to support this concept. The cells observed in cinematography by Balfour do not have the visible "habits" of phagocytosis in contrast to macrophages under identical circumstances. This would be in accordance with the described properties of IDCs. These cells might also carry antigen on or in their cell membrane, as has been suggested for the Langerhans cell (Silberberg-Sinakin *et al.*, 1978).

Labeling experiments in guinea pigs by Dailey and Hunter (1974) indicate that lipid-rich antigens localize in a patchy distribution in the TDAs of the regional lymph node. Ultrastructural details are not known. It is tempting to assume that the antigen has also been trapped by the IDC on its cell membrane either in the periphery or directly in the lymph node.

Earlier experiments on cellular immunity have proved already that the skin route is not obligate for this induction process, since experimental allergic encephalitis (Horne and White, 1968), adjuvant arthritis (Newbould, 1964), and delayed-type sensitivity to contact sensitizers (Seeberg, 1951) could also be induced by direct intranodal application of the antigen.

The fact that, in guinea pigs and in rabbits, the migration of Langerhans cells (Silberberg-Sinakin *et al.*, 1976, 1978) and "veiled" cells (Balfour, personal communication) appears to be accelerated in the afferent lymphatics after immunological challenge at the dermal draining site—augmentation with a factor three to seven times the normal rate of migration—is noteworthy in relation to the above mentioned antigen-trapping phenomena. It is not yet clear whether this relationship also extends to the type of antigen used, e.g., a contact sensitizer vs. a notorious B-cell "challenger" such as a polysaccharide. It might be possible that there is a significant difference in migration rate under these circumstances. The lymph node pathology described for dermatopathic lymphadenitis might be a pathophysiological expression of such an extremely augmented migration rate from the inflammatory skin lesion toward the TDA of the regional lymph node [ecotaxopathy?; see deSousa (1978)].

It is still unclear whether IDCs are derived from monocytes. This possibility has been discussed occasionally in the literature (Veerman, 1974; Veerman and van Ewijk, 1975). The available data indicate, however, that IDCs are not, or are only rarely, capable of phagocytosis. In contrast, phagocytosis can be a significant feature of monocytes. At the ultrastructural level, however, similarities between IDCs and monocytes do exist.

It is not yet possible to give a definite answer to the question about the significance of IDCs in malignant lymphomas. To equate the occurrence of an IDC with a stroma cell might be correct, if one would like to express that fact that the IDCs are nonneoplastic components of the lymphoma. The term *stroma cell* would be an oversimplification, however, if one considers the specific functions that IDCs probably have in normal lymphoid tissue in immunologic reactions, and if one compares this term to the presence of fibroblasts, for example, in lymph node metastases of lymphoepithelial carcinoma (Schmincke's tumor), or of histiocytes in Burkitt's lymphoma. On the other hand, the observation that dendritic cells (DC) occur in various malignant lymphomas (Table 1) appears to be comparable to the occurrence of the IDC in malignant lymphomas. DCs are, as already mentioned by Veldman (1970), cells that presumably have a significance for the B-cell system similar to that which the IDC has for the T-cell system. Accordingly, DCs have been observed in malignant lymphomas exclusively of the B-cell type, especially in germinal center neoplasms. One regularly finds DCs in the lymphoma (Kojima, 1969; Lennert and Niedorf, 1969) that is called "centroblastic/centrocytic lymphoma" according to the Kiel classification

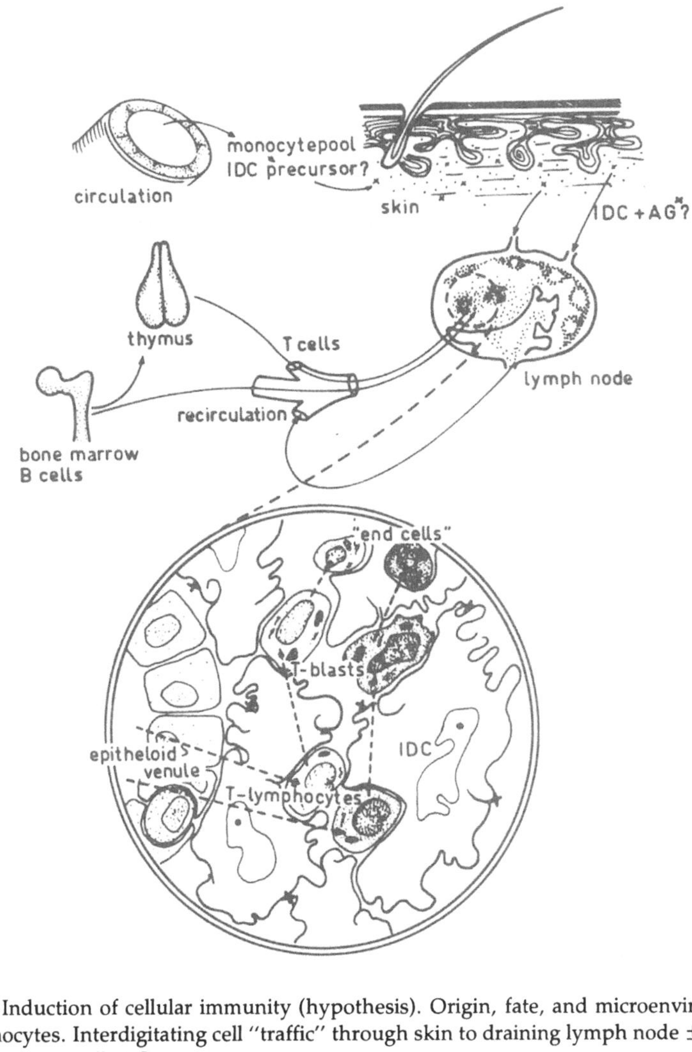

FIGURE 24. Induction of cellular immunity (hypothesis). Origin, fate, and microenvironment of T and B lymphocytes. Interdigitating cell "traffic" through skin to draining lymph node ± antigen (x). IDC, Interdigitating cell; AG, antigen.

(Gérard-Marchant *et al.*, 1974). Furthermore, they are sometimes seen in malignant lymphoma, centrocytic, which is a germinal center tumor with a diffuse growth pattern.

A number of experimental data is, however, of interest in this context. The results of Swartzendruber *et al.* (1967), Szakal and Hanna (1968), and Hanna *et al.* (1970) suggest the possibility that the DCs in malignant tumors are of pathogenetic significance. Hanna *et al.* (1970) have shown that the DCs in germinal centers of mice bind not only antigens, such as serum proteins or flagellin, but also oncogenic viruses (Rauscher leukemia viruses) to their surface projections. Swartzendruber *et al.* (1967), who found endogenous C virus particles in close contact with DCs in mice with a high rate of spontaneous leukemia, pointed out the significance DCs might have in the development of leukemia in mice. It is not known whether DCs or IDCs in human malignant lymphomas bind and concentrate oncogenic substances.

Noteworthy is that in T-cell lymphomas of high grade malignancy, IDCs are absent, whereas in the described T-cell lymphomas of low grade malignancy they appear to be an essential constituent of the disorder.

To conclude, the following scheme (Figure 24) summarizes the way we think the IDC fits into the dynamics of lymphoid cell traffic and lymph node immune reactivity. A lot of question marks illustrate our still limited knowledge, and indicate potential steps for future research. A monocyte origin—suggested by some authors—is still a speculation, although attractive as a working hypothesis. If it turns out that these cells originate from monocyte precursor cells, it occurs to us that that population might be a heterogeneous one, since the IDC does not behave like any "mononuclear phagocyte," but acts by its own rules, specifically in a liaison with its T-cell environment. Kinetic studies will be needed to elucidate the origin of the IDC, its behavior, and migration pattern, as well as its final fate in the various tissue compartments.

ACKNOWLEDGMENTS. We thank Professor F. J. Keuning (Groningen, The Netherlands) and Professor K. Lennert (Kiel, West Germany) for reviewing the manuscript. The assistance of Mr. H. R. A. Meiborg, Mr. G. van der Sleen and Miss M. Neubert is greatly appreciated. We thank Miss T. A. M. L. van Dun, Mrs. M. Soehring, and Miss D. P. M. J. Winkel for all their secretarial work.

REFERENCES

Balner, H., and Dersjant, H., 1964, Early lymphatic regeneration in thymectomized radiation chimaeras, *Nature* **204**:941.

Barcos, M. P., and Lukes, R. J. 1975, Malignant lymphoma of convoluted lymphocytes: A new entity of possible T-cell type, in: *Conflicts in Childhood Cancer. An Evaluation of Current Management,* Vol. **4** (L. F. Sinks and J. O. Godden, eds.), pp. 147–178, Liss, New York.

Birbeck, M. S., Breathnach, A. S., and Everall, J. D., 1961, An electron microscopic study of basal melanocytes and high level clear cells (Langerhans cells) in vitiligo, *J. Invest. Derm.* **37**:51.

Blijham, G. H., 1975, Histofysiologie van het helper T-cellen systeem in het konijn (Histophysiology of the helper-T-cell system in the rabbit), Ph.D. Thesis, Groningen, The Netherlands.
Bos, W. H., 1967, Recirculatie en transformatie van lymfocyten (Recirculation and transformation of lymphocytes), Ph.D. Thesis, Groningen, The Netherlands.
Daems, W. T., Koerten, H. K., and Soranzo, M. R., 1976, Differences between monocyte-derived and tissue macrophages, in: *Advances in Experimental Medicine and Biology,* Vol. 73A, *The Reticuloendothelial System in Health and Disease: Functions and Characteristics* (S. M. Reichard, M. R. Escobar, and H. Friedman, eds.), pp. 27–40, Plenum Press, New York.
Dailey, M. O., and Hunter, R. T., 1974, The role of lipid in the induction of hapten-specific delayed hypersensitivity and contact sensitivity, *J. Immunol.* **112**(4):1526.
deSousa, M., 1978, Ecotaxis, ecotaxopathy, and lymphoid malignancy: Terms, facts, and predictions, in: *The Immunopathology of Lymphoreticular Neoplasms* (J. J. Twomey and R. A. Good, eds.), pp. 325–359, Plenum Press, New York.
Elema, J. O., and Poppema, S., 1978, Infantile histiocytosis X (Letterer-Siwe disease). Investigations with enzyme-histochemical and sheep-erythrocyte rosetting techniques, *Cancer* **42:**555.
Gérard-Marchant, R., Hamlin, I., Lennert, K., Rilke, F., Stansfeld, A. G., and van Unnik, J. A. M., 1974, Classification of non-Hodgkin's lymphomas, *Lancet* **2**:406.
Goos, M., Kaiserling, E., and Lennert, K., 1976, Mycosis fungoides: Model for T-lymphocyte homing in the skin ? *Br. J. Dermatol.* **94**:221.
Gowans, J. L., 1959, The recirculation of lymphocytes from blood to lymph in the rat, *J. Physiol.* **146**:54.
Gowans, J. L., and Knight, E. J., 1964, The route of recirculation of lymphocytes in the rat, *Proc. Roy. Soc. London Ser. B* **159**:257.
Gutman, G., and Weissman, I. L., 1972, Lymphoid tissue architecture. Experimental analysis of the origin and distribution of T-cells and B-cells, *Immunology* **23**:465.
Hanna, Jr., M. G., Szakal, A. K., and Tyndall, R. L., 1970, Histoproliferative effect of Rauscher leukemia virus on lymphatic tissue: Histological and ultrastructural studies of germinal centers and their relation to leukemogenesis, *Cancer Res.* **30**:1748.
Heusermann, U., Stutte, H. J., and Müller-Hermelink, H. K., 1974, Interdigitating cells in the white pulp of human spleen, *Cell Tissue Res.* **153**:415.
Hoefsmit, E. C. M., 1975, Mononuclear phagocytes, reticulum cells and dendritic cells in lymphoid tissues, in: *Mononuclear Phagocytes in Immunity, Infection and Pathology* (R. van Furth, ed.), pp. 129–146, Blackwell, Oxford.
Horne, C. H. W., and White, R. G., 1968, Evaluation of the direct injection of antigen into a peripheral lymph node for the production of humoral and cell-mediated immunity in the guinea pig, *Immunology* **15**:65.
Kaiserling, E., 1976, Elektronenmikroskopische Befunde bei Non-Hodgkin-Lymphomen, in: *Hämatologie und Bluttransfusion,* Vol. 18, *Maligne Lymphome und monoklonale Gammopathien* (H. Löffler, ed.), pp. 185–198, Lehmanns, München.
Kaiserling, E., 1977, Non-Hodgkin-Lymphome, Ultrastruktur und Cytogenese, in: *Veröffentlichungen aus der Pathologie,* Vol. 105, Fischer, Stuttgart.
Kaiserling, E., 1978, Ultrastructure of Non-Hodgkin's lymphomas, in: *Malignant Lymphomas Other than Hodgkin's Disease* (K. Lennert, ed.), Springer Verlag, New York.
Kaiserling, E., and Lennert, K., 1974, Die interdigitierende Reticulumzelle im menschlichen Lymphknoten. Eine spezifische Zelle der thymusabhängigen Region, *Virchows Arch. B* **16**:51.
Kaiserling, E., Stein, H., and Müller-Hermelink, H. K., 1974, Interdigitating reticulum cells in the human thymus, *Cell Tissue Res.* **155:**47.
Kelly, R. H., Balfour, B. M., and Armstrong, J. A., 1976, Lymphborne Langerhans cells (abstract), *Am. J. Pathol.* **82.**
Keuning, F. J., 1965, Die Beantwortung eines homologen Gewebsstimulus (Homotransplantation) durch das lymphoreticuläre System, *Oncologia (Basel)* **19**:180.
Keuning, F. J., 1972, Dynamics of immunoglobulin forming cells and their precursors, in: *Immunoglobulins: Cell Bound Receptors and Humoral Antibodies* (R. E. Ballieux, M. Gruber, and H. G. Seyen, eds.), pp. 1–14, North-Holland, Amsterdam.

Keuning, F. J., and Bos, W. H., 1967, Regeneration pattern of lymphoid follicles in the rabbit spleen after sublethal X-irradiation, in: *Germinal Centers in Immune Responses* (H. Cottier, N. Odartchenko, R. Schindler, and C. Congdon, eds.), p. 250, Springer Verlag, Berlin.

Keuning, F. J., and van den Broek, A. A., 1968, Role of marginal zone cells of lymphoid follicles in the immune response of rabbits, *Exp. Haematol.* **17**:4.

Keuning, F. J., van der Meer, J., Nieuwenhuis, P., and Oudendijk, P., 1963, The histophysiology of the antibody responses and splenic plasma cell reactions in sublethally X-irradiated rabbits, *Lab. Invest.* **12**:156.

Keuning, F. J., Dijkhuis, A., and Dijkstra-van den Vliet, T. A., 1964, The effects of sublethal röntgen irradiation on the induction period of the antibody response in the rabbit, *Intl. J. Radiat. Biol.* **8**:279.

Kojima, M., 1969, Cytopathological study on lymph node, especially with reference to germinal center (in Japanese), *Trans. Soc. Pathol. Jpn.* **58**:3.

Lennert, K., 1961, *Lymphknoten, Diagnostik in Schnitt und Ausstrich. Cytologie und Lymphadenitis. Handbuch der speziellen Pathologischen Anatomie und Histologie,* I/3 A Springer, Berlin.

Lennert, K., 1976, Klassifikation und Morphologie der Non-Hodgkin-Lymphome, in: *Hämatologie und Bluttransfusion,* Vol. 18, *Maligne Lymphome und monoklonale Gammopathien* (H. Löffler, ed.), pp. 145–166, Lehmanns, Munchen.

Lennert, K., and Niedorf, H. R., 1969, Nachweis von desmosomal verknupften Reticulumzellen im follikulären Lymphom (Brill Symmers), *Vichows Arch. B* **4**:148.

Lennert, K., Kaiserling, E., and Müller-Hermelink, H. K., 1975, T-associated plasma cells, *Lancet* **1**:1031.

Lennert, K., in collaboration with H. Stein, N. Mohri, E. Kaiserling, H. K. Müller-Hermelink, 1978a, *Malignant Lymphomas Other than Hodgkin's Disease,* Springer Verlag, New York.

Lennert, K., Kaiserling, E., and Müller-Hermelink, H. K., 1978b, Malignant lymphomas: Models of differentiation and cooperation of lymphoreticular cells, in: *Differentiation of Normal and Neoplastic Hematopoietic Cells,* Book B, Cold Spring Harbor Conferences on Cell Proliferation, Vol. 5 (B. Clarkson, P. A. Marks, and J. E. Till, eds.), pp. 897–913, Cold Spring Harbor Laboratory, New York.

Lukes, R. J., and Collins, R.D., 1975, New approaches to the classification of the lymphomata, *Br. J. Cancer* **31** (suppl. II):1.

Micklem, H. S., Clarke, C. M., Evans, E. P., and Ford, C. E., 1968, Fate of chromosome-marked mouse bone marrow cells transfused into normal syngeneic recipients, *Transplantation* **6**:299.

Miller, J. F. A. P., and Mitchell, G. F., 1969, Thymus and antigen-reactive cells, in: *Transplantation Reviews,* Vol. 1, *Antigen Sensitive Cells, Their Source and Differentiation* (G. Möller, ed.), Munksgaard, Copenhagen.

Mitchell, G. F., and Miller, J. F. A. P., 1968, Cell to cell interaction in the immune response. II. The source of hemolysin-forming cells in irradiated mice given bone-marrow and thymus or thoracic duct lymphocytes, *J. Exp. Med.* **128**:821.

Mulder, N. H., 1972, Thymus-afhankelijkheid van de humorale immuun-reactie (Thymus dependency of the antibody response), Ph.D. Thesis, Groningen, The Netherlands.

Müller-Hermelink, H. K., 1974, Characterization of the B-cell and T-cell regions of human lymphatic tissue through enzyme histochemical demonstration of ATPase and 5'-nucleotidase activities, *Virchows Arch. B.* **16**:371.

Müller-Hermelink, H. K., Heusermann, U., Kaiserling, E., and Stutte, H. J., 1976, Human lymphatic microecology—specificity, characterization and ontogeny of different reticulum cells in the B-cell and T-cell regions, in: *Advances in Experimental Medicine and Biology,* Vol. 66, *Immune Reactivity of Lymphocytes* (M. Feldman, and A. Globerson, eds.), pp. 177–182, Plenum Press, New York.

Myrvik, Q. N., Racz, P., and Tenner-Racz, K., 1979, Light and electron microscopic characteristics of sinus reactions and cellular traffic in the hilar lymph node complex (HLNC) in rabbits undergoing a pulmonary CMI reaction, in: *Advances in Experimental Medicine and Biology,* Vol. 114, *Function and Structure of the Immune System* (W. Müller-Ruchholtz and H. K. Müller-Hermelink, eds.), p. 827, Plenum Press, New York.

Newbould, B. B., 1964, Lymphatic drainage and adjuvant induced arthritis in rats, *Br. J. Exp. Pathol.* **45**:375.

Nieuwenhuis, P., and Keuning, F. J., 1974, Germinal centres and the origin of the B-cell system. II. Germinal centres in the rabbit spleen and popliteal lymph node, *Immunology* **26**:509.

Nieuwenhuis, P., Veldman, J. E., van den Broek, A. A., and Keuning, F. J., 1970, Radiation induced dissociation of immune responsiveness in the rabbit; delineation of a thymus-derived and non-thymus derived lymphoid cell system, Proc. 4iéme Congrès International de Radiobiologie et de Physico-Chemie des Rayonnements, Evian, France Gordon and Breach, London.

Novikoff, A. B., Roheim, P. S., and Quintana, N., 1966, Changes in the rat liver cells induced by orotic acid feeding, *Lab. Invest.* **15**:27.

Opstelten, D., Van Der Heyden, D., Stikker, R., and Nieuwenhuis, P., 1979, Germinal centers and the B-cell system: B-cell differentiation in rabbit appendix germinal centers, in: *Advances in Experimental Medicine and Biology,* Vol. 114, *Function and Structure of the Immune System* (W. Müller-Ruchholtz and H. K. Müller-Hermelink, eds.), p. 125, Plenum Press, New York.

Parrott, D. M. V., and deSousa, M. A. B., 1966, Changes in the thymus-dependent areas of lymph nodes after immunological stimulation, *Nature* **212**:1316.

Parrott, D. M. V., and deSousa, M. A. B., 1967, The persistence of donor-derived cells in thymus grafts, lymph nodes and spleens of recipient mice, *Immunology* **13**:193.

Parrott, D. M. V., deSousa, M. A. B., and East, J, 1966, Thymus dependent areas in the lymphoid organs of neonatally thymectomized mice, *J. Exp. Med.* **123**:191.

Poppema, S., Kaiserling, E., and Lennert, K., 1979, Hodgkin's disease with lymphocytic predominance nodular type (nodular paragranuloma) and progressively transformed germinal centres—a cytohistological study, *Histopathology* **3**:259–308.

Poppema, S., Kaiserling, E., and Lennert, K., 1979, Nodular paragranuloma and progressively transformed germinal centres. Ultrastructural and immunohistologic findings, *Virchows Arch. B* **31**:211.

Racz, P., Tenner-Racz, K., Myrvik, Q. N., and Fainter, L. K., 1977, Functional architecture of bronchial associated lymphoid tissue and lymphoepithelium in pulmonary cell-mediated reactions in the rabbit, *J. Reticuloendothel. Soc.* **22**:59.

Rausch, E., Kaiserling, E., and Goos, M., 1977, Langerhans cells and interdigitating reticulum cells in the thymus-dependent region in human dermatopathic lymphadenitis, *Virchows Arch. B* **25**:327.

Sainte-Marie, G., 1978, Morphology and physiology of the structures of the lymph node related to lymphocyte traffic, *Z. Immun. Forsch.* **154**(4):359 (abstr.).

Salvin, S. B., and Smith, R. F., 1959, Delayed hypersensitivity in the development of circulating antibody. The effect of X-irradiation, *J. Exp. Med.* **109**:325.

Schipior, P., and Maquire, H. C., 1966, Resistance of the allergic contact dermatitis sensitization reaction to whole body X-ray in the guinea pig, *Int. Arch. Allergy Appl. Immunol.* **29**:447.

Seeberg, G., 1951, Eczematogenous sensitization via the lymphatic glands as compared with other routes, *Acta Derm. Venereol. (Stockholm)* **31**:592.

Silberberg-Sinakin, I., Thorbecke, G. J., Baer, R. L., Rosenthal, S. A., and Berezowsky, V., 1976, Antigen-bearing Langerhans cells in skin, dermal lymphatics and in lymph nodes, *Cell. Immunol.* **25**:137.

Silberberg-Sinakin, I., Baer, R. L., and Thorbecke, B., 1978, Langerhans cells. A review of their nature with emphasis on their immunologic functions, in: *Progress in Allergy,* Vol. 24 (P. Kallos, B. H. Waksman, and A. L. deWeck, eds.), pp. 268–294, Karger, Basel.

Swartzendruber, D. C., Il Ma, B., and Murphy, W. H., 1967, Localization of C-type virus particles in lymphoid germinal centers of C58 mice, *Proc. Soc. Exp. Biol. (N.Y.)* **126**:731.

Szakal, A. K., and Hanna Jr., M. G., 1968, The ultrastructure of antigen localization and virus-like particles in mouse spleen germinal centers, *Exp. Mol. Pathol.* **8**:75.

Törö, I., and Oláh, I., 1966, Electron microscopic study of guinea-pig thymus, *Acta Morphol. Acad. Sci. Hung.* **14**:75.

Turk, J. C., and Oort, J., 1969, Further studies on the relation between germinal centers and cell-mediated injury, in: *Advances in Experimental Medicine and Biology,* Vol. 5, *Lymphatic Tissue and*

Germinal Centers in the Immune Response (L. Fiore-Donati and M. G. Hanna, eds.), p. 317, Plenum Press, New York.

Uhr, J. W., and Scharff, M., 1960, Delayed hypersensitivity. V. The effect of X-irradiation on the development of delayed hypersensitivity and antibody formation, *J. Exp. Med.* **112**:65.

van der Slikke, L. B., and Keuning, F. J., 1964, Influence of sublethal X-irradiation on the survival time of skin homografts and the reaction of the lymphoid system, *Int. J. Rad. Biol.* **8**:279.

van Ewijk, W., Verzijden, J. H. M., van der Kwast, T. H., and Luijcx-Meijer, S. W. M., 1974, Reconstitution of the thymus dependent area in the spleen of lethally irradiated mice. A light- and electron microscopic study of the T-cell microenvironment, *Cell Tissue Res.* **149**:43.

van Furth, R., Cohn, Z. A., Hirsch, J. G., Humphrey, J. H., Spector, W. G., and Langevoort, H. L., 1972, The mononuclear phagocyte system: A new classification of macrophages, monocytes and their precursor cells. *Bull. W. H. O.* **46**:845.

van Leeuwen, A. W. F. M., Meyer, C. J. C. M., van Vloten, W. A., Scheffer, E., and de Man, J. C. H., 1976, Further evidence for the T-cell nature of the atypical mononuclear cells in mycosis fungoides, *Virchows Arch. B.* **21**:179.

Veerman, A. J. P., 1974, On the interdigitating cells in the thymus-dependent area of the rat spleen: A relation between the mononuclear phagocyte system and T-lymphocytes, *Cell Tissue Res.* **148**:247.

Veerman, A. J. P., and van Ewijk, W., 1975, White pulp compartment in the spleen of rats and mice. A light and electron microscopic study of lymphoid and non-lymphoid cell types in T- and B-areas, *Cell. Tissue Res.* **156**:417.

Veldman, J. E., 1970, Histophysiology and Electron Microscopy of the Immune Response, Part 1 and 2. Ph.D. Thesis, Groningen, The Netherlands.

Veldman, J. E., 1977, Progress in Immunology. III. Workshop No. 12. *Lymphoid Cell Traffic* (T. E. Mandel, C. Cheers, C. S. Hosking, J. F. C. Mckenzie, and G. J. V. Nossal, eds.), p. 755, Elsevier/North Holland, New York.

Veldman, J. E., and Keuning, F. J., 1978, Histophysiology of cellular immunity reactions in B-cell deprived rabbits. An X-irradiation model for delineation of an isolated T-cell system, *Virchows Arch. B* **28**:203.

Veldman, J. E., Keuning, F. J. and Molenaar I., 1978a, Site of initiation of the plasmacell reaction in the rabbit lymph node. Ultrastructural evidence for two distinct antibody forming cell precursors, *Virchows Arch. B* **28**:187.

Veldman, J. E., Molenaar, I., and Keuning, F. J., 1978b, Electron microscopy of cellular immunity reactions in B-cell deprived rabbits. Thymus derived antigen reactive cells, their microeniron-ment and progeny in the lymph node, *Virchows Arch. B* **28**:217.

Veldman, J. E., Nieuwenhuis. P., and Keuning, F. J., 1980a, A T-dependent plasma cell response: Part of a graft versus host and host versus graft reaction in the rabbit spleen, *Virchows Arch. B* **33** (in press).

Veldman, J. E., Nieuwenhuis, P., Molenaar, I., and Keuning, F. J., 1980b, The graft versus host reaction in the rabbit spleen. Ultrastructural evidence for two differentiated T-cell lines, *Virchows Arch. B* **33** (in press).

von Gaudecker, B., and Müller-Hermelink, H. K., 1979, Ontogenetic differentiation of epithelial and nonepithelial cells in the human thymus, in: *Advances in Experimental Medicine and Biology,* Vol. 114, *Function and Structure of the Immune System* (W. Müller-Ruchholtz and H. K. Müller Hermelink, eds.), pp. 19–23, Plenum Press, New York.

Wakefield, J. D., and Thorbecke, G. J., 1968, Relationship of germinal centers in lymphoid tissue to immunological memory. II. The detection of primed cells and their proliferation upon cell transfer to lethally irradiated syngeneic mice, *J. Exp. Med.* **128**:171.

Waksman, B. H., 1978, Introduction, in: *Progress in Allergy,* Vol. 24 (P. Kallos, B. H. Waksman, and A. L. deWeck, eds.), pp. 11–13, Karger, Basel.

Waksman, B. H., Arnason, B. G., and Jankovic, B. D., 1962, Role of the thymus in immune reactions in rats. III. Changes in the lymphoid organs of thymectomized rats, *J. Exp. Med.* **116**:187.

Weissman, J. L., 1967, Thymus cell migration, *J. Exp. Med.* **126**:191.

11

Lymph Node Macrophages

ELISABETH C. M. HOEFSMIT, ED W. A. KAMPERDIJK, HANS R. HENDRICKS, ROBERT H. J. BEELEN, and BRIGID M. BALFOUR

1. THE NORMAL LYMPH NODE

1.1. THE ANATOMICAL COMPARTMENTS OF THE LYMPH NODE AND THEIR SPECIALIZED IMMUNOLOGICAL FUNCTIONS

The lymph node filters tissue fluid which is conveyed to it by afferent lymph vessels. In most species, these vessels pierce the capsule on the convex surface of the bean-shaped organ and open into the subscapsular or marginal sinus which is continuous with the medullary sinus in the concavity or hilum of the organ (Figure 1). The lymph filters through the fine meshes of the reticuloendothelial tissue which supports the sinuses and the compact lymphatic tissue of the node (Figure 2). There are also large numbers of macrophages in the sinuses, and these cells clear the lymph by phagocytosis.

The lymph node is also a peripheral lymphoid organ which responds to antigenic stimulation by the production of specific antibodies, or the generation of effector cells. The lymphocytes include bone-marrow-derived B cells and thymus-derived T cells. In both populations, antigen-specific lymphocytes are triggered to proliferate, forming clones of B or T cells. In a humoral immune response, B cells generate plasma cells which synthesize immunoglobulins, and, in most cases, there is also T-cell proliferation, with the generation of helper and suppressor cells; in cell-mediated immunity, T-cell proliferation leads to the formation of specific effector T cells. These amplification and differentiation processes all occur within the reticuloendothelial framework of the node which also contains a mobile population of macrophages.

The cortex has a compact structure and is subdivided into an outer cortex

ELISABETH C. M. HOEFSMIT, ED W. A. KAMPERDIJK, HANS R. HENDRIKS, and ROBERT H. J. BEELEN • Department of Electron Microscopy, Medical Faculty, Free University, Amsterdam 1081 BT, The Netherlands. BRIGID M. BALFOUR • National Institute for Medical Research, London NW7 1AA, England.

comprising the follicles and the interfollicular areas, or marginal zone, and an inner cortex, or paracortex, which is continuous with the medulla. The medulla has a more open structure, being composed of a three-dimensional network of compact lymphatic tissue forming cordlike structures which cross the medullary sinus. The arteries supplying the node enter at the hilum and follow the cords to the cortex. Here they divide into arterioles which penetrate the inner and outer cortex and divide again to form capillaries. The nutrient vessels end in capillary loops just below the marginal sinus. At the venous end, these loops are collected into somewhat larger anastomosing vessels which open almost immediately into the high endothelial venules of the paracortex. In these venules, the endothelial cells are cuboidal or cylindrical in shape. These venules are connected to larger veins which reach the hilum again via the medullary cords.

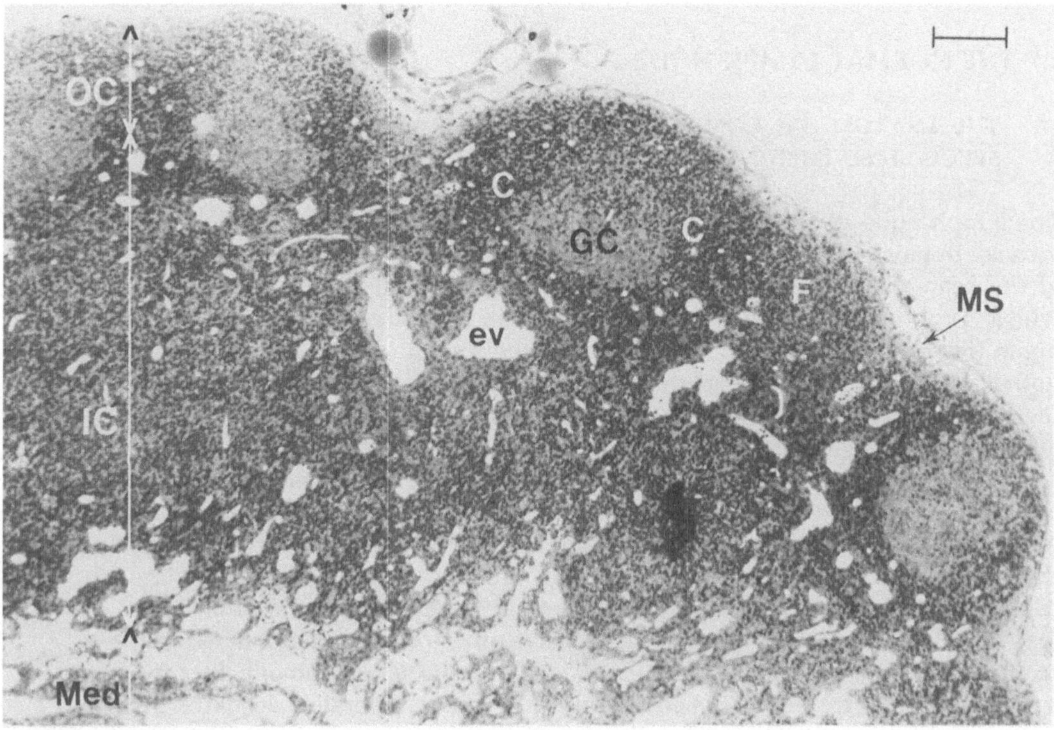

FIGURE 1. Semithin section of a normal rat lymph node (|—| = 100 μm). OC, Outer cortex; IC, inner cortex or paracortex; Med, medulla; C, corona or mantle layer; ev, high endothelial venule; GC, germinal center; F, follicle; MS, marginal sinus. Note: In order to obtain the figures for this chapter, normal lymph node structure was studied in rats and rabbits. Wistar rats weighing 200 g were obtained from the Central Institute for the Breeding of Laboratory Animals TNO, Zeist, The Netherlands. The rabbits were New Zealand White rabbits, weighing between 2.0–2.5 kg and Sandylops weighing about 2.5 kg, bred at the National Institute for Medical Research, London. A few experiments were also carried out on mice and guinea pigs, but no obvious differences were found in the fine structure of reticulum cells and macrophages in different species. All the rat lymph nodes were fixed by perfusion and processed for EM as described by Kamperdijk *et al.* (1978). The rabbit lymph nodes were usually fixed by immersion fixation in the fixative used for rat lymph nodes or in a mixture of glutaraldehyde–osmium tetroxide (Hirsch and Fedorko, 1968), and they were processed in the same manner as the rat lymph nodes.

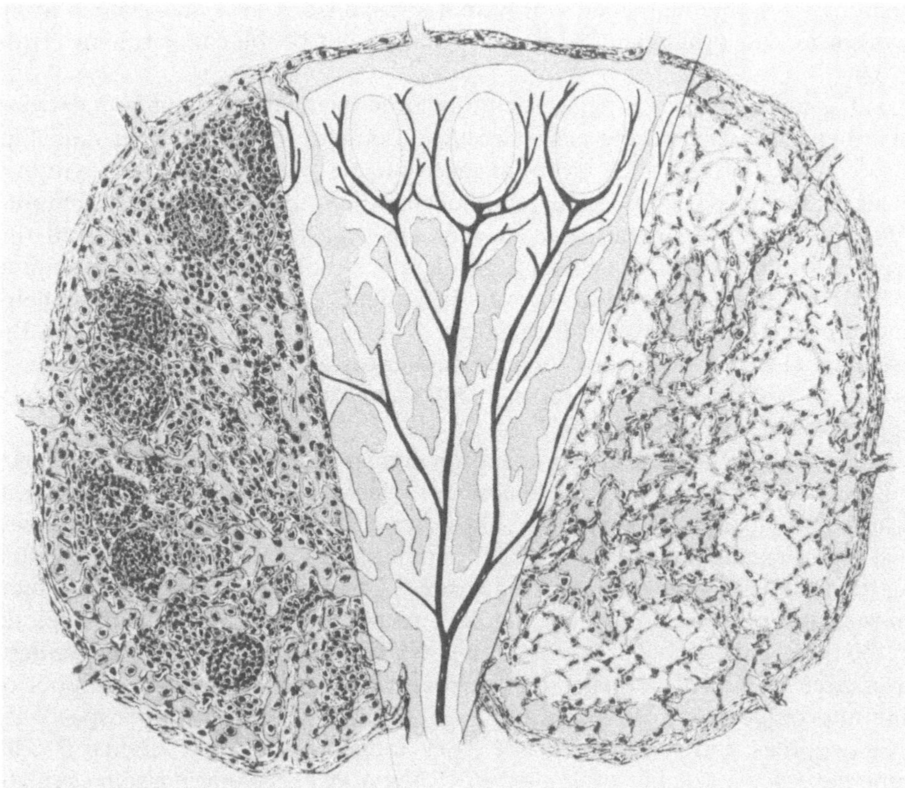

FIGURE 2. Schematic representation of the structural components of the lymph node (after Weiss, 1972). In the right hand segment, the reticulum of the node is shown, the reticular cells in nucleated outline and the fibres in stipple. The subcapsular sinus and the radial sinus are shaded. Afferent lymphatics penetrate the capsule emptying into the subcapsular sinus. Radial sinuses criss-crossed by the reticulum, run from the subcapsular or marginal sinus toward the hilus. The reticular meshwork is specialized to form the follicles and perifollicular zones of the cortex and medullary cords. In the central segment, the distribution of the veins is shown. The terminal twigs of the postcapillary venules originate about the cortical follicles as well as deep to the follicles. In the left hand segment, the reticulum together with the lymphocytes and other free cells is presented. The reticulum becomes masked by the crowds of lymphocytes.

It has been shown that the outer cortex, comprising the follicles and interfollicular areas or marginal zone, is populated by B lymphocytes (Howard *et al.*, 1972) and that the inner cortex is mainly a T-cell compartment (Parrott *et al.*, 1966; De Sousa *et al.*, 1969). Both populations enter the node by traversing the high endothelial venules in the paracortex (Gowans and Knight, 1964; Howard *et al.*, 1972; Nieuwenhuis and Ford, 1976). From this region, B cells migrate to the outer cortex but T cells remain in the paracortex. Both populations ultimately migrate to the medulla and leave the node via the efferent lymph vessels to rejoin the blood stream. During the development of a cell-mediated immune response or a thymus-dependent humoral response, T cells are retained in the paracortex and proliferate, leading to enlargement of the compartment (Turk, 1967).

The marginal zone is partly separated from the marginal sinus by a network of lining cells which belong to the reticuloendothelial framework. The reticulum also contains two or three layers of macrophages, interspersed with lymphocytes. In the deeper layers small lymphocytes and some medium-sized lymphocytes predominate. In this area, B cells are triggered to proliferate and differentiate into plasma cells which later migrate to the medullary cords (Keuning, 1972). The B-cell response to thymus-dependent antigens requires T-cell help, and the interaction between B cells and T helper cells probably occurs in the transitional zone between the inner and outer cortex, the area in which plasmablasts are first observed (Veldman, 1970). Later, large numbers of plasma cells appear in the medullary cords.

The primary follicles are compact conglomerates of B lymphocytes originating from the recirculating pool. Secondary follicles contain a germinal center in which lymphoblasts predominate. These structures are preferentially localized just above a small capillary vein which communicates almost immediately with a high endothelial venule in the paracortex. Germinal centers are covered by a crescent-shaped cap or corona containing small lymphocytes which also belong to the recirculating pool of B cells. The germinal centers are antigen-dependent structures in which virginal bone-marrow-derived B cells generate a pool of immunocompetent B cells with a wide spectrum of antigenic specificity (Nieuwenhuis and Keuning, 1974). After antigen stimulation virginal B cells generate a pool of highly antigen-specific "memory" cells which also recirculate (Thorbecke *et al.* 1964), and these cells proliferate and differentiate into plasma cells in response to a booster dose of the same antigen (Wakefield and Thorbecke, 1968).

It is clear that the lymph node contains several compartments, each with its own immunological function: the marginal zone is responsible for the induction of the humoral response, which culminates in the secretion of antibody by plasma cells in the medullary cords; the paracortex is responsible for the induction of cell-mediated immunity and the generation of T helper cells, and the germinal centers for the generation of immunocompetent B cells and "memory" cells (Keuning, 1972). The literature concerning lymph node function deals almost entirely with the functions of lymphocytes, the role of macrophages being largely neglected, probably because these cells are usually included in the same category as the supporting tissue.

1.2. IMMUNOLOGICAL FUNCTIONS OF MACROPHAGES: A REVIEW OF THE LITERATURE

Unicellular organisms destroy their enemies by endocytosis, followed by lysosomal digestion. Primitive multicellular organisms possess multifunctional blood cells which have phagocytic properties. These cells ingest bacteria and bacterial products but the reaction is nonspecific. Antigen-specific immunity evolved with the appearance of lymphocytes bearing specific receptors. T-cell-mediated immunity came first and was followed by the evolution of B cells capable of synthesizing specific antibody.

The function of macrophages is to digest material endocytized by pinocytosis or phagocytosis. These cells have a long life-span and they retain the ability to synthesize lysosomal enzymes throughout life. Moreover, the enzymatic composition of the lysosomes can be adapted to the requirements of the environment (Cohn and Wiener, 1963; Axline and Cohn, 1970). This means that the phagocytic properties of macrophages depend on the stage of development and on the microenvironment in which development occurs. Macrophages also have important functions in the regulation of the immune response (Unanue, 1972). Since they are mobile cells, they can concentrate antigens and transport them to distant sites, provided lysosomal digestion is not complete. Antigens may be preserved on the cell membrane or in intracellular compartments from which they may later be released. However, this release has not been unequivocally demonstrated (Unanue and Askonas, 1968; Steinman and Cohn, 1972).

Macrophages have Fc receptors and C3 receptors (Huber and Holm, 1975). If IgG-containing immune complexes are bound through the Fc receptor they are immediately endocytized, whereas IgM-containing complexes, bound through the C3 receptor are not immediately endocytized, at least by resident macrophages (Bianco *et al.*, 1975).

In vitro experiments have shown that macrophages or "adherent" cells are needed in almost all immunological reactions. The induction of DNA synthesis in mixed lymphocyte cultures (Greineder and Rosenthal, 1975) and the generation of cytotoxic T cells in allograft rejection (Wagner *et al.*, 1972) are macrophage dependent. The macrophage proved to be the stimulator cell in contact sensitivity (Thomas, 1977); sensitized T cells require macrophages in order to respond to challenge with the same antigen (Waldron *et al.*, 1973), and this response is restricted by the major histocompatibility complex (MHC) (Shevach, 1976). The T-cell response to mitogens also depends on the presence of macrophages (Rosenstreich *et al.*, 1976).

The majority, if not all, humoral immune responses are T-cell-dependent and the generation of T helper cells in response to stimulation by soluble antigens is dependent on the presence of a critical number of macrophages (Erb and Feldman, 1975a) with the same MHC restriction (Erb and Feldman, 1975b). The subsequent T-cell–B-Cell interaction requires macrophages (Shortman and Palmer, 1971; Feldmann, 1972; Askonas and Roelants, 1974). Even thymus-independent B-cell stimulation requires macrophages in order to prevent antigen overload (Mosier, 1976).

Macrophages secrete products which activate or inhibit lymphocyte functions (Nelson, 1976), and they themselves are influenced by lymphocyte products or lymphokines, such as migration inhibiting factor (MIF). They are also important effector cells in cellular and humoral immunity.

1.3. THE ULTRASTRUCTURE OF THE RETICULOENDOTHELIAL FRAMEWORK OF THE NODE AND OF THE MACROPHAGES

The ultrastructural organization of reticulum cells and endothelial cells is essentially different from that of macrophages (Hoefsmit, 1975) (Figure 3). The framework of the node is composed of reticulum cells, lining cells, and vascular endothelial cells which are stretched on a basement membrane or wrapped around collagen bundles, strands, or single fibers. The cells tend to be flattened and have a smooth surface. They sometimes overlap, and may be fixed to each other by desmosome-like structures or by specialized membrane contacts (Kamperdijk *et al.*, 1978) such as those described between vascular endothelial cells. Typical reticulum cells are "reticular" in shape, i.e., they have long, slender cell

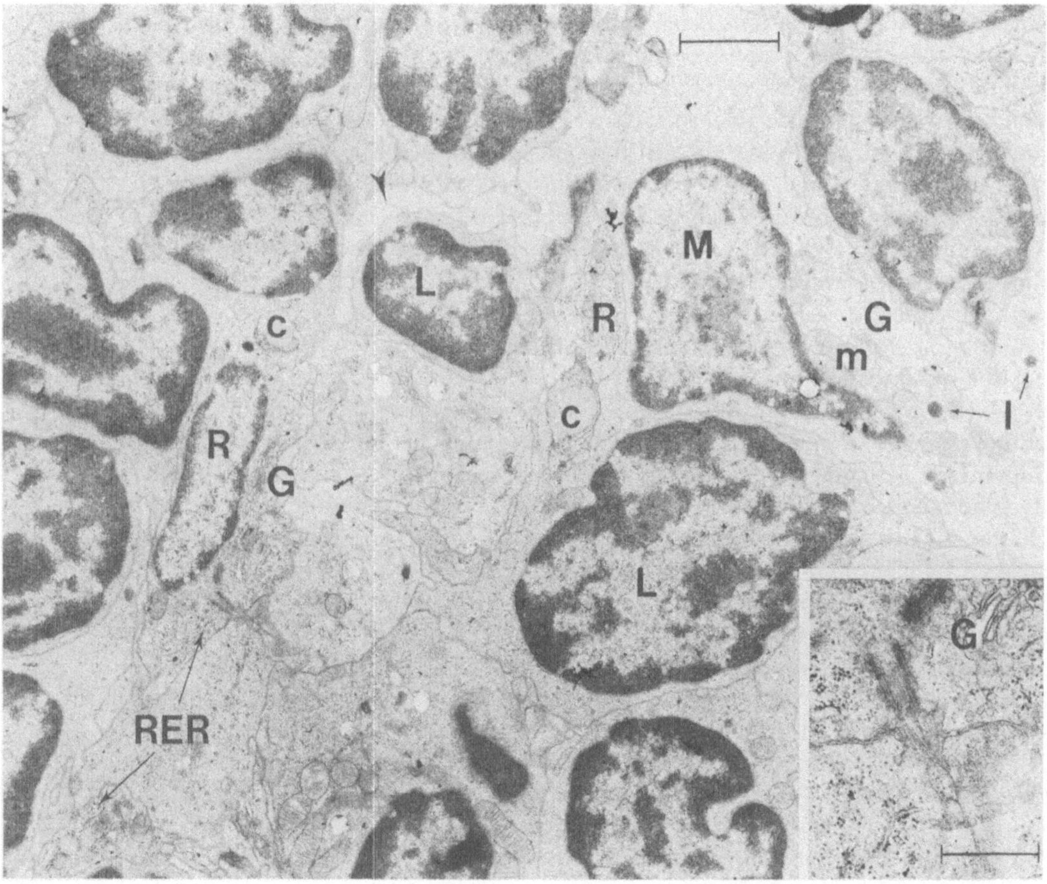

FIGURE 3. Reticulum cell and macrophage in paracortex of a normal rat lymph node. The reticulum has a cilium (inset, |–| = 1 μm) and envelops rather coarse bundles of collagen. The macrophage has a central Golgi area and lysosomes. The paracortex macrophages (IDC) have cell processes which interdigitate with surrounding lymphocytes (arrow head) (|–| = 2 μm). R, Reticulum cell; c, collagen; G, Golgi apparatus; RER, rough endoplasmic reticulum; L, lymphocyte; M, macrophage; m, mitochondrion; 1, phagolysosome. For experimental details, see the note in the caption for Figure 1.

processes which are connected to the cytoplasmic extensions of neighboring cells forming a cellular network between the fine fibers of the intercellular skeleton. They are sometimes "dendritic," i.e., with processes resembling the branches of a tree, or fibroblast-like cells encircling collagen fibers. In all these cells the nucleus occupies a central position and appears triangular or guadrangular in section; it is usually euchromatic with a distinct nucleolus. The nucleus is surrounded by a thin layer of cytoplasm, which extends into the cell processes. A variable amount of rough endoplasmic reticulum (RER) is aligned parallel to the cell membrane and moderate numbers of mitochondria are present between the strands of RER. Reticulum cells contain a well-developed Golgi system. Stacks of Golgi cisterns are present in the cell body and also in the cell processes. In both areas these organelles lie close to the cell membrane parallel with the surface. In some cells minute, smooth-surfaced vesicles are present in the region intervening between the so-called maturing face of the Golgi system and the cell membrane, and they may fuse with each other and with the cell membrane. Reticular cells sometimes contain a pair of centrioles close to the Golgi system and the cell membrane. Microtubules and microfilaments are usually present. The microfilaments lie close to that part of the plasma membrane which is in apposition with the intercellular skeleton. The microtubules are aligned between the Golgi system and the cell membrane and are also present in the cell processes. They form an intracellular skeleton and are probably involved in the transport of secretory vesicles towards the cell membrane (Ross, 1975). Reticulum cells sometimes contain a single dense body, but large phagolysosomes are never found. Some reticulum cells and lining cells also contain a cilium. In general, reticulum cells have the morphological characteristics of protein-synthesizing and -secreting cells and are apparently concerned in the production and maintenance of the intercellular skeleton, with which they actually constitute the three-dimensional network or reticulum of the node. They do not show any significant phagocytic activity. The function of the cilia in reticulum cells is not yet clear. Lauwerijns and Boussauw (1975) described rudimentary ciliary rootlets in vascular lymphatic endothelial cells in rabbit lung, and these structures could have played a role in the microcirculation of tissue fluid.

Macrophages are sometimes stretched over reticulum cells, but they are seldom in direct contact with the fibrous skeleton. They usually lack desmosome-like structures associated with membrane adherence. Macrophages may assume various shapes, but are mostly rounded with a ruffled membrane. The nucleus has an eccentric position. In the center of the cell there is a pair of centrioles from which many microtubules radiate. The centrosome is surrounded by varying numbers of Golgi stacks and a large number of coated vesicles measuring 50–70 nm in diameter. Variable numbers of large and small dense bodies, sometimes containing heterogeneous electron dense material, are aligned between the microtubules, the larger bodies in the periphery of the cell, and the smaller bodies towards the center. RER and mitochondria surround the nucleus and the central area. Microfilaments, 5–8 nm in diameter, are usually present, sometimes in coarse bundles surrounding the nucleus. Macrophages never contain a cilium. Large macrophages have a diameter of 25–30 μm. There are also smaller cells with a diameter of 8–10 μm. The large cells have a lower nucleocytoplasmic ratio, smaller numbers of polyribosomes, but more RER, Golgi cisterns, and lysosomes than small macrophages. Transitional forms between reticulum cells and macrophages have never been observed.

Macrophages are concerned with the production of lysosomes and lysosomal digestion. The centripetal movement of pinosomes and phagosomes and their conversion to secondary lysosomes in the central Golgi area was extensively studied by Cohn and Fedorko (1969) and Cohn (1970). The small, coated vesicles surrounding the Golgi cisterns were considered to be virginal lysosomes, although it was difficult to detect acid phosphatase in these vesicles by electron microscopic (EM) cytochemistry (Carr, 1968; Holtzman, 1976). In bone marrow promonocytes and monocytes, Nichols and Bainton (1975) described the maturation of larger virginal lysosomes, which gradually condensed to form small electron-dense storage granules. In exudate macrophages these storage granules were almost immediately used up for lysosomal digestion (Daems *et al.*, 1973; Nichols and Bainton, 1975) and there was no evidence that storage granules were manufactured by these cells though they were able to secrete various products (Gordon *et al.*, 1974). The selective release of the enzyme lysozyme could not be explained by "frustrated phagocytosis," i.e., fusion of virginal lysosomes with endocytic vesicles before they had been pinched off from the cell membrane (Henson, 1971). Microtubules and microfilaments are almost certainly involved in these macrophage functions, but their precise role is not yet known (Elsbach, 1977).

Ultrastructural studies on lymphatic tissue in different organs and in different species (Pictet *et al.*, 1969; Oláh *et al.*, 1975; Burke and Simon, 1970a,b; Carr, 1970, 1973; Weiss, 1972) have clearly distinguished "free macrophages" from the "reticular" cells composing the framework. The "reticular cells" were usually subdivided into phagocyte- and fibroblast-like, or undifferentiated, nonphagocytic reticular cells. Weiss (1972), who described the structure of the lymph node very clearly (Figure 2), distinguished two populations of phagocytic cells: one population of fixed macrophages or "phagocytic reticular cells" and another population of free macrophages, of which a proportion were bone-marrow-derived, mobile cells. These cells could, in turn, become sessile and contribute to the cellular reticulum. Thus, the position of the "fixed macrophages" was not clear. They could be fixed reticulum cells or bone-marrow-derived mobile cells. Later observations have shown that reticulum cells and macrophages are essentially different. Macrophages are sometimes reticular in shape and may adhere to the reticulum, but they do not manufacture collagen or sustain the intercellular skeleton. Reticulum cells are not phagocytic and transitional forms between reticulum cells and macrophages are never found. This classification is compatible with the concept of the MPS put forward by van Furth *et al.* (1975). In this concept, reticulum cells and macrophages are considered to be separate cell lines, the macrophages being mobile bone-marrow-derived cells, the reticuloendothelial tissue contributing the fixed framework into which lymphocytes and macrophages migrate.

1.4. RETICULUM CELLS IN THE VARIOUS COMPARTMENTS OF THE NODE

The reticuloendothelial framework of the various compartments of the node shows some degree of specialization. The outer cortex is traversed by trabecular extensions of the capsular connective tissue which are anchored to the reticular framework of the deeper layers in the interfollicular area. The connective tissue of the hilum is continuous with the rather coarse reticulum of the medullary cords and the paracortex. In the paracortex and medulla the reticulum cells resemble fibroblasts. They are stellate or fusiform cells closely apposed to bundles of collagen fibers. The cytoplasm contains many strands of RER and the Golgi system is correspondingly well-developed. In the marginal zone, primary follicles, and corona of the secondary follicles, the reticulum cells have long, slender processes which are apposed to fine collagen fibers. These cells contain some strands of RER and possess a rather extensive Golgi system (Hoefsmit *et al.*, 1980).

The reticulum cells in germinal centers are highly specialized cells (Figure 4), with multiple infoldings of the plasma membrane forming a labyrinth. This gives the cells their characteristic dendritic appearance. Dendritic reticulum cells (DRC) and the adjoining fibers are concentrated in a cap between the corona and the germinal center. Their processes extend into the germinal center forming an intricate web between the lymphoid cells. It is of interest that the upper half of the germinal center is predominantly occupied by medium-sized lymphocytes, whereas the immunoblasts and mitotic lymphocytes are concentrated in the basal half, toward the paracortex. Thus the germinal center forms a polarized structure in which DRC and mitotic lymphocytes are at opposite poles: there are no other reticulum cells present. DRC have an electron-lucent nucleus which may be deeply indented. A few strands of RER are present and the Golgi system fills the cytoplasm almost completely. Numerous smooth and coated vesicles are in contact with the cisterns of the Golgi system and the cell membrane, but no dense bodies are present, and microfilaments are rarely seen. The cells sometimes contain a cilium.

Germinal center reticulum cells have the same morphological characteristics as the cells described by Nossal *et al.* (1968) as "antigen-retaining dendritic reticular cells" since they trap labeled antigen–antibody complexes (IC) on the cell surface. In a critical review of the literature, Hanna and Hunter (1971) suggested that the retention of antigen in the microenvironment of the germinal center was essential for the continued differentiation of primary antibody-forming cells, and also for the expansion of the memory cell compartment (Wakefield and Thorbecke, 1968). The retention of antigen in germinal centers depends on the presence of circulating antibody (Humphrey and Frank, 1967), and in this location it probably plays a role in the amplification rather than the induction phase of the humoral immune response. The development of germinal centers and the retention of immune complexes are also dependent on the activation of complement (Papamichail *et al.*, 1975).

Cells with the morphological characteristics of the antigen-retaining cells described by Nossal *et al.* (1968) are only found in germinal centers and are not

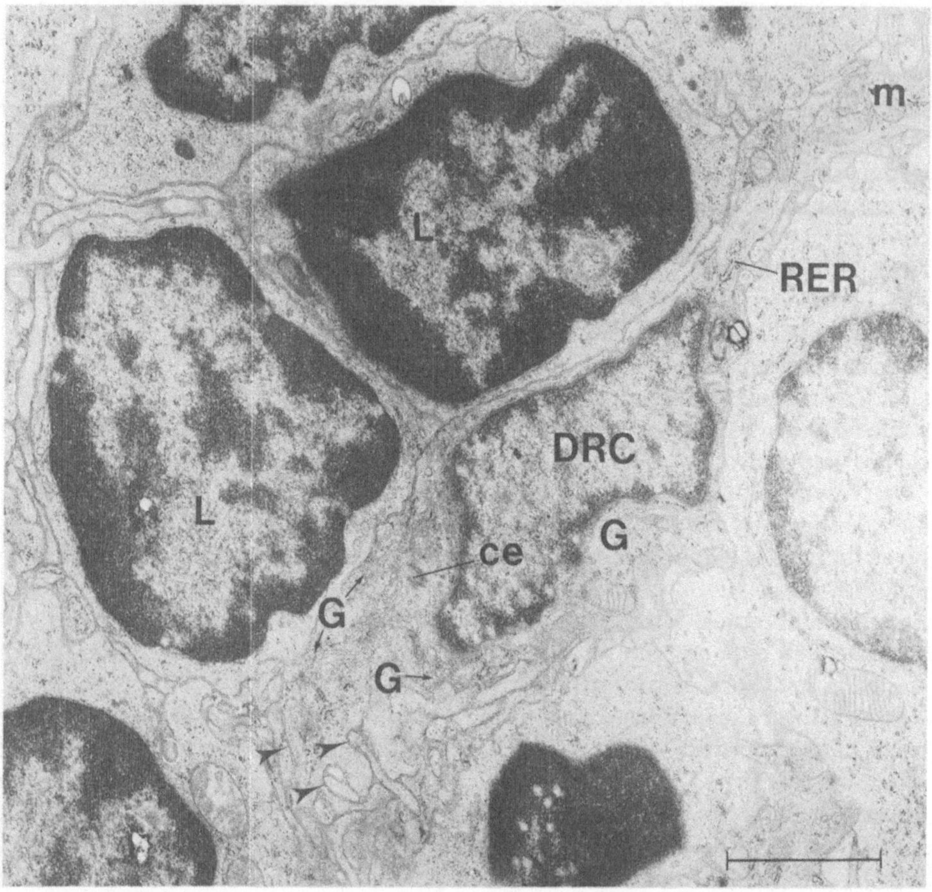

FIGURE 4. Dendritic reticulum cell (DRC) in germinal center of normal rabbit lymph node. The cell is characterized by the labyrinthine invaginations of the cell membrane (arrow heads) and contains only a few strands of RER, but many well-developed stacks of Golgi cisterns just beneath the cell membrane (arrow) (|—| = 2 μm). L, Lymphocyte; G, Golgi apparatus; ce, centriole; m, mitochondrion. For experimental details, see the note in the caption for Figure 1.

present throughout the outer cortex, as reported by Veldman (1970) and Veldman *et al.* (1978). These cells do not resemble macrophages nor do they endocytize and digest immune complexes, and should not, therefore, be described as "dendritic macrophages" (Carr, 1970, 1973; Stuart, 1975; White *et al.*, 1970). They also lack the machinery needed to migrate and are obviously not the "migratory antigen-retaining dendritic cells" described by White (*et al.*, 1970; White, 1975).

DRC are probably not involved in the initiation of the germinal center reaction. Ultrastructural studies have shown that the cells which are present in developing germinal centers are stimulated by the presence of immune complexes and only later develop the labyrinthine infoldings of the cell membrane which are characteristic of DRC (Szakal and Hanna, 1968; Hanna and Szakal, 1968; Sordat *et al.*, 1970).

It is generally believed that DRC trap antigen in the form of immune complexes by means of receptors for the Fc fragment of immunoglobulin and the third component of complement, C3. Herd and Ada (1969) demonstrated that the trapping of immune complexes required the presence of the Fc part of immunoglobulin and suggested that DRC had Fc receptors. However, the Fc fragment is also able to fix and activate the complement system, and Humphrey (1976) demonstrated that C3 activation was required for the retention of immune complexes on DRC, and that levan, a potent activator of C3 via the alternate pathway, was rapidly localized in germinal centers before antibody was present. He suggested that activation of C3 alone, was responsible for the localization of immune complexes in germinal centers. However, it is not certain that DRC possess complement receptors, since activated C3 and immune complexes can be easily eluted from germinal centers (Gajl-Peczalska *et al.*, 1969), and DRC do not undergo cytotoxic lysis after trapping immune complexes in the course of a primary immune response. The alternative view is that antigen–antibody complexes bind and activate the complement system in the microenvironment of germinal center reticulum cells, and this activation transforms them into DRC. The enormous development of the Golgi system is probably responsible for the hypertrophy of the plasma membrane, since the Golgi system is known to be concerned in the differentiation and glycosylation of the plasma membrane (Morré, 1977), and this hypertrophic membrane may be able to bind immunogenic (Van Rooijen, 1973) and nonimmunogenic materials (Cohen *et al.*, 1966) by a nonspecific mechanism.

The antigen-retaining dendritic cells have also been described as metallophilic reticular cells (Stuart, 1975), which are considered to be undifferentiated phagocytic reticular cells according to the RES classification (Marshall, 1956; White *et al.*, 1970). Eikelenboom (1978) combined silver staining of lymph nodes with the immunocytochemical demonstration of anti-horseradish peroxidase antibodies, and concluded that reticulum cells and macrophages both had the ability to bind immune complexes and to reduce silver, but reticulum cells did not possess phagocytic properties. In fact, silver reduction occurs in all glycoprotein-containing structures including the Golgi system, secretory vesicles, lysosomes, and the cell coat, and does not distinguish between different cell types.

1.5. MACROPHAGES IN THE VARIOUS COMPARTMENTS OF THE NODE

In previous sections we defined the different compartments of the lymph node (Section 1.1) and the general morphological characteristics of macrophages (Section 1.3). In this section we describe the specialized types of macrophages which are present in the various compartments.

1.5.1. Macrophages in the Marginal and Medullary Sinuses

In these communicating compartments the filter function of the lymph node predominates. The macrophages may be free in the sinus or apposed to the sinus reticulum cells or sinus lining cells. They are also in different stages of development, ranging from small monocyte-like cells measuring 8–12 μm to large macrophages with a diameter of 30 μm or more. The nucleocytoplasmic ratio of small macrophages is 1:2 whereas large macrophages may have ratios as low as 1:8.

Macrophages can be roughly subdivided into two groups, although their characteristics overlap. The first group consists of small cells with a ruffled membrane and fingerlike or flaplike cell processes. In these cells the irregularly formed nucleus is surrounded by bundles of microfilaments interspersed with mitochondria. The cell membrane is rather smooth, there is no evidence of active micropinocytosis, and large phagolysosomes are usually not present. The Golgi area is located in the nuclear hof and contains a variable number of small, coated vesicles measuring 50–70 nm, which are presumably virginal or primary lysosomes (Holtzman, 1976). However, this area may contain larger vesicles varying in diameter, but usually less than 1000 nm. The membrane of these vesicles is usually separated from the contents by a clear halo with a constant thickness of about 10 nm. These vesicles contain heterogeneous material of variable electron density and are considered to be secondary lysosomes or phagolysosomes (Cohn and Fedorko, 1965, 1969; Cohn *et al.*, 1966; Daems *et al.*, 1969).

The second type of macrophage is larger and has irregularly shaped, blunt-ended cell processes, and the nucleus is slightly indented or bilobate. It is euchromatic, and contains one or more well-developed nucleoli. The surrounding microfilaments are less prominent although some are always present throughout the cytoplasm. Mitochondria interspersed with numerous strands of RER surround the nucleus and the extensive Golgi area. The cells frequently have large phagolysosomes containing heterogeneous material of variable electron density and, sometimes, recognizable cellular debris.

1.5.2. Macrophages in the Outer Cortex

The outer cortex comprises the interfollicular area or marginal zone and the follicles. The primary follicles are conglomerates of small lymphocytes around small capillaries. In secondary follicles, a crescentic cap or mantle layer of small lymphocytes covers the germinal center on the peripheral side, that is the side from which antigen enters the cortex. In tangential histological and EM sections, the primary follicles cannot easily be distinguished from the mantle layer of secondary follicles. In fact, these two structures are ontogenetically the same and will be described together.

Both types of sinus macrophages are seen squeezing between the reticulum cells lining the inner wall of the marginal sinus, but the marginal zone is mainly populated by large, actively phagocytic cells (Figure 5). These cells may ingest all types of necrotic leukocytes; a single granulocyte, mast cell, lymphocyte, or, occasionally, a plasma cell can be found inside a phagolysosome (Figure 6). The nuclear chromatin appears to be rather resistant to lysosomal digestion since marginal zone macrophages sometimes contain one or two tingible bodies. Deeper in the marginal zone some macrophages contain larger numbers of incom-

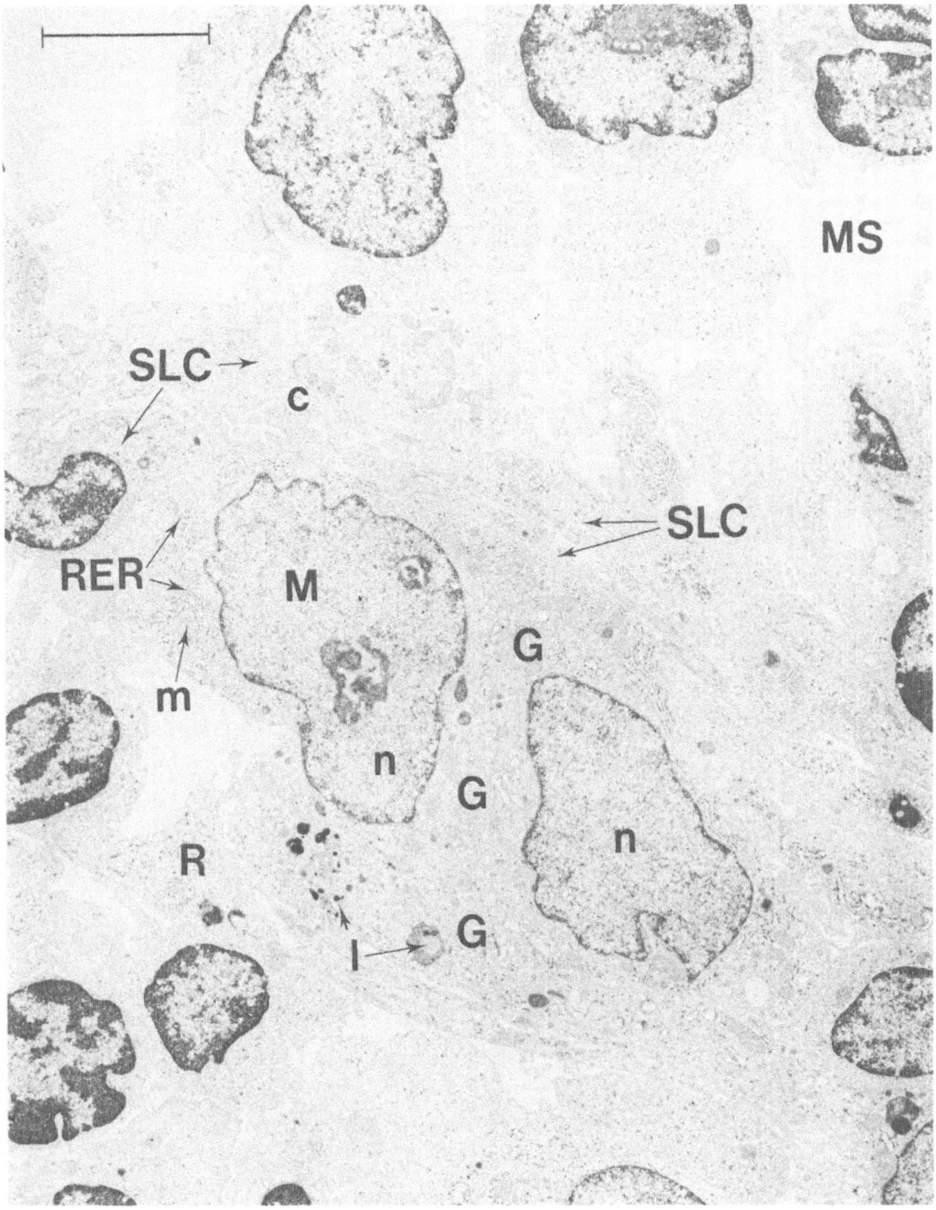

FIGURE 5. Macrophage in the marginal zone of a normal rabbit lymph node. The cell has a large euchromatic, bilobed nucleus with two nucleoles. The RER and the Golgi apparatus are well developed. The cytoplasm contains many phagolysosomes (|—| = 4 μm). SLC, Sinus-lining cell; m, mitochondrion; c, collagen; n, nucleus; MS, marginal sinus; R, reticulum cell. For experimental details, see the note in the caption for Figure 1.

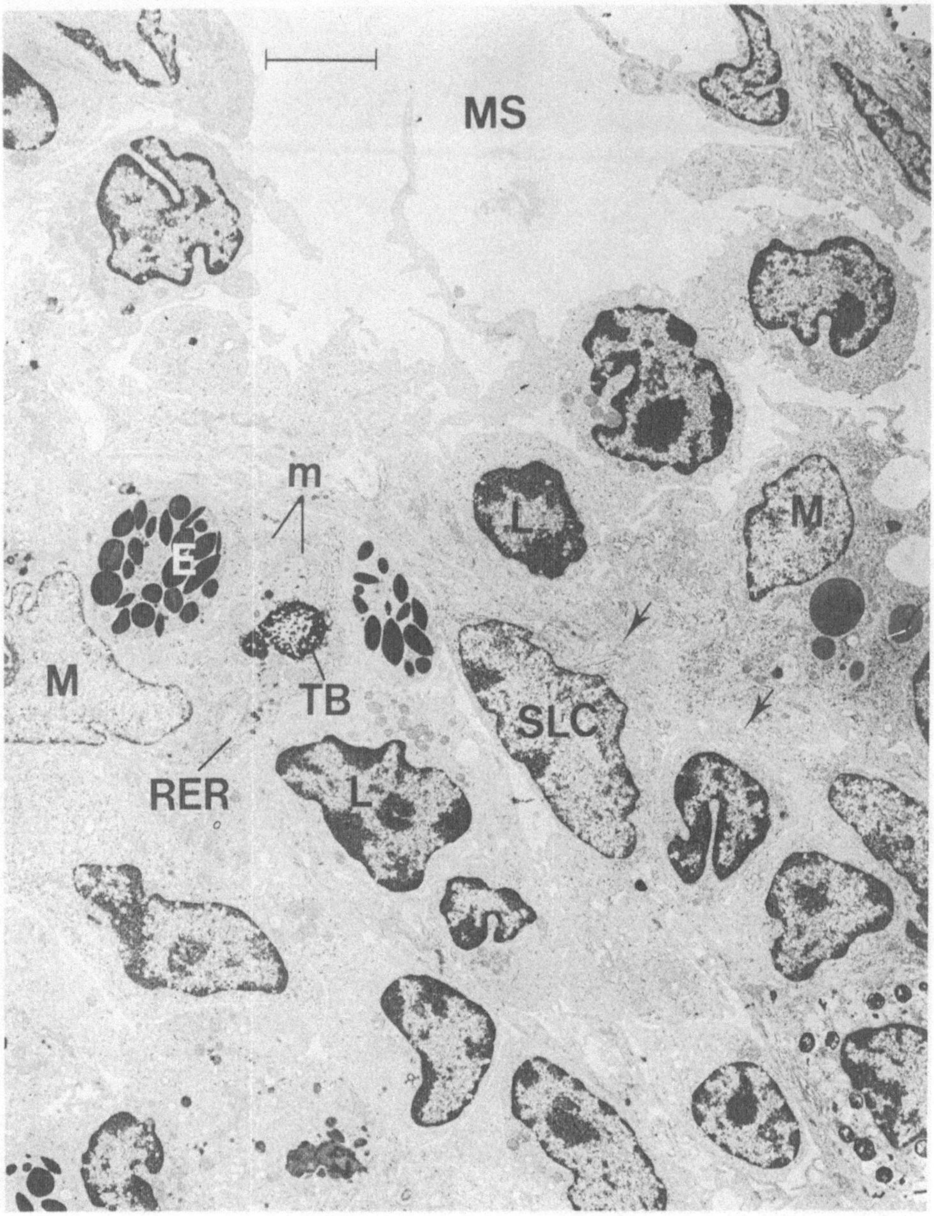

FIGURE 6. A macrophage (M) in the subcapsular sinus extends a pseudopod into the marginal zone through a pore between sinus-lining cells (indicated by arrows). Another macrophage in the marginal zone contains a tingible body (TB). This macrophage seems also to ingest an eosinophilic granulocyte (E) (|—| = 3 μm). RER, Rough endoplasmic reticulum; m, mitochondrion; L, lymphocyte; MS, marginal sinus. For experimental details, see the note in the caption for Figure 1.

pletely digested necrotic cells, and these macrophages resemble the tingible body macrophages (TBM) found in germinal centers. This means that TBM are are not restricted to germinal centers: they can be found in the marginal zone, in primary follicles, and in the mantle layer of secondary follicles, usually surrounding small capillaries. These perivascular macrophages contain large numbers of residual bodies with heterogeneous electron-dense contents or myelin figures in addition to phagolysosomes containing incompletely digested cellular material. Some phagocytic cells are themselves degenerating; they contain small amounts of RER, some of the Golgi cisterns are dilated, the lysosomal membranes are sometimes incomplete or absent, and the cytoplasm is electron lucent, probably due to an increased water content. In some macrophages, the cellular material inside phagolysosomes contains lysosomes, suggesting that phagocytosis of effete macrophages also occurs.

Normal lymph nodes sometimes contain germinal centers, with immunoblasts and some medium-sized and small lymphocytes, but mitotic figures are not usually present in these centers. The characteristic germinal center macrophage is the tingible body macrophage (Figure 7) (Flemming, 1885). At the outer border of the germinal center, the macrophages are preferentially localized around capillaries, as in all other parts of the outer cortex. Toward the base of the germinal center, the macrophages contain increased numbers of irregularly shaped residual bodies, lipid material, and myelin figures, frequently combined with incompletely digested cellular material. These cells have a smooth surface, they are more or less polygonal in shape, and they are adapted to the surrounding cells. The nucleus is usually heterochromatic and the cytoplasm electron lucent. Clearly, these cells are unable to synthesize the lysosomal enzymes needed to digest all the material which they have endocytized, and, as a result, they themselves have become degenerate. At the base of the germinal center aggregations of large effete macrophages surround the small venules (Figure 8), and these cells are filled with residual bodies and myelin figures.

One week after a footpad injection of carbon, the sinus-lining cells and reticulum cells in the outer cortex of the draining node contain small amounts of carbon located in micropinocytic vesicles, whereas the marginal zone and medullary macrophages contain large vacuoles filled with carbon. During the next two weeks, the number of carbon-containing macrophages diminishes but the amount of carbon per cell increases and these carbon-laden cells are found deeper in the cortex. After three weeks, all the TBM in germinal centers are filled with carbon, and there is also an accumulation of carbon-laden cells at the base of the germinal centers (Figure 9). In the outer cortex, the reticulum cells still contain a dust-like accumulation of carbon, whereas the marginal zone macrophages are free of carbon. Eight weeks after carbon administration, there is a dense accumulation of carbon-laden cells close to the vascular complex at the base of each germinal center. Some carbon-laden cells are also found in the paracortex, preferentially localized around high endothelial venules. In normal lymph nodes, solitary effete macrophages may occasionally be found in this area. The accumulation of carbon within individual cells is accompanied by a decrease in the total number of carbon-laden cells. This indicated that macrophages endocytize carbon, then die, and are themselves phagocytized by other macrophages.

These observations support the idea that reticulum cells and macrophages in the outer cortex are distinct cell lines. No transitional forms between these two cell types have ever been observed, though transitional forms between marginal zone macrophages and TBM are commonly found in normal lymph nodes, and probably represent different stages in the development of the large actively phagocytic type of macrophage.

The marginal zone macrophages contain many strands of RER, a well-developed Golgi complex, and large phagolysosomes and, sometimes, a necrotic leukocyte. The macrophages which lie deeper in the cortex contain increased numbers of necrotic cells and are preferentially localized around small cortical capillaries. They also appear to have ingested more necrotic cells than they are able to digest since they contain increased numbers of residual bodies and in-

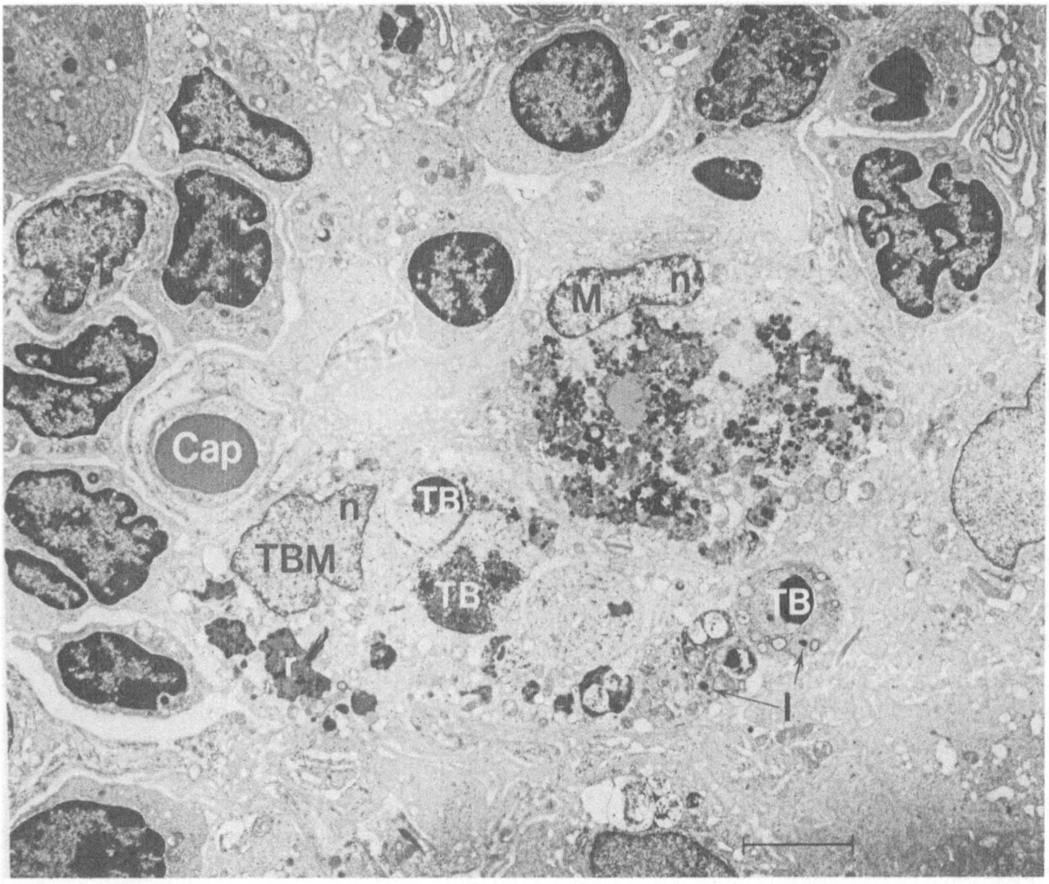

FIGURE 7. Tingible body macrophage (TBM) in a germinal center in a normal rabbit lymph node. Some ingested necrotic cells contain lysosomes themselves. Another macrophage contains numerous residual bodies (|—| = 3 μm). Cap, Capillary; n, nucleus; TB, tingible body; l, phagolysosome. For experimental details, see the note in the caption for Figure 1.

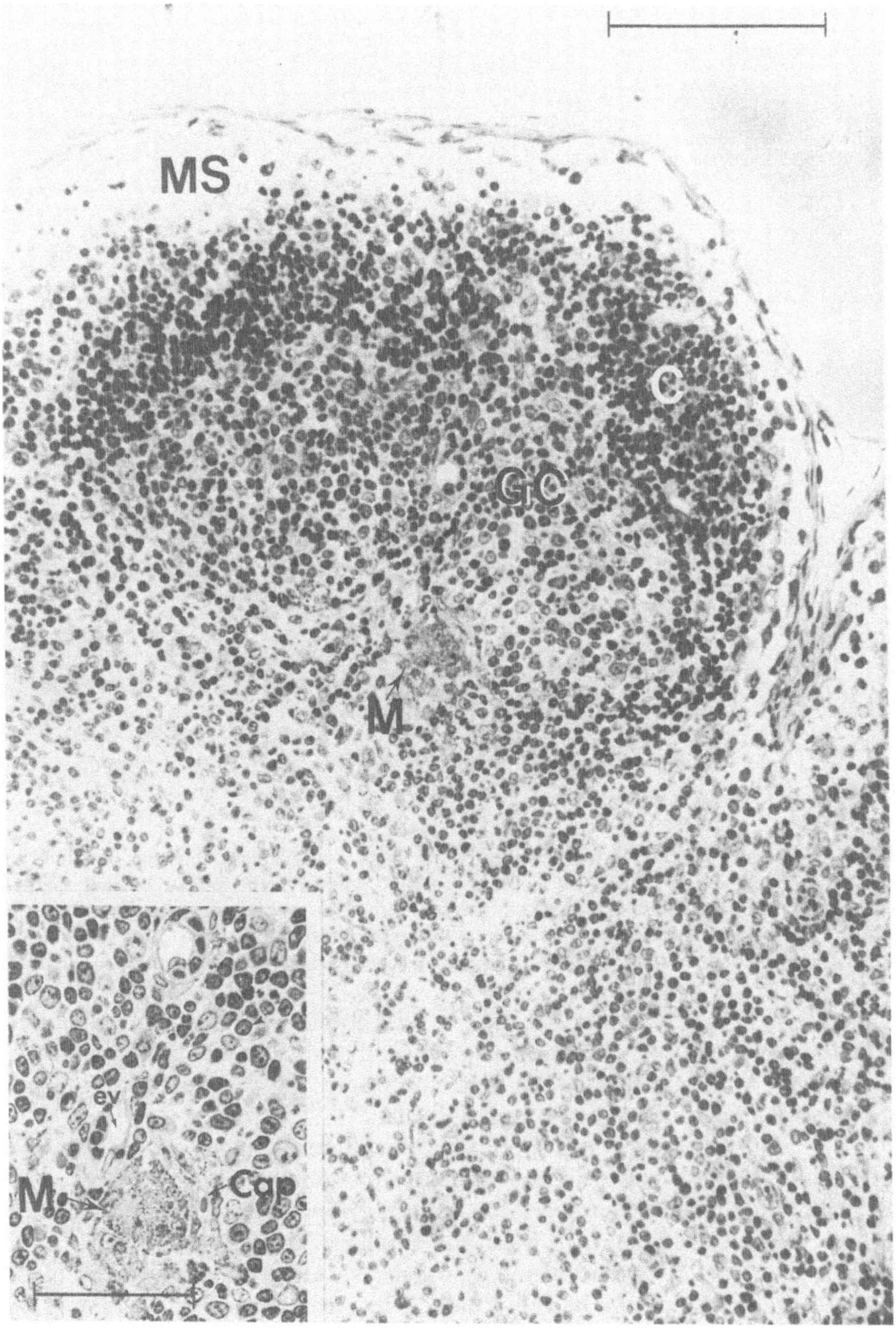

FIGURE 8. Normal rabbit lymph node, semithin section. At the base of the germinal center (GC) effete macrophages accumulate around small capillary (Cap) venules, near the point at which they merge into high endothelial venules (|—| = 100 μm, inset: |—| = 50 μm). MS, Marginal sinus; C, corona or mantle layer; ev, high endothelial venule. For experimental details, see the note in the caption for Figure 1.

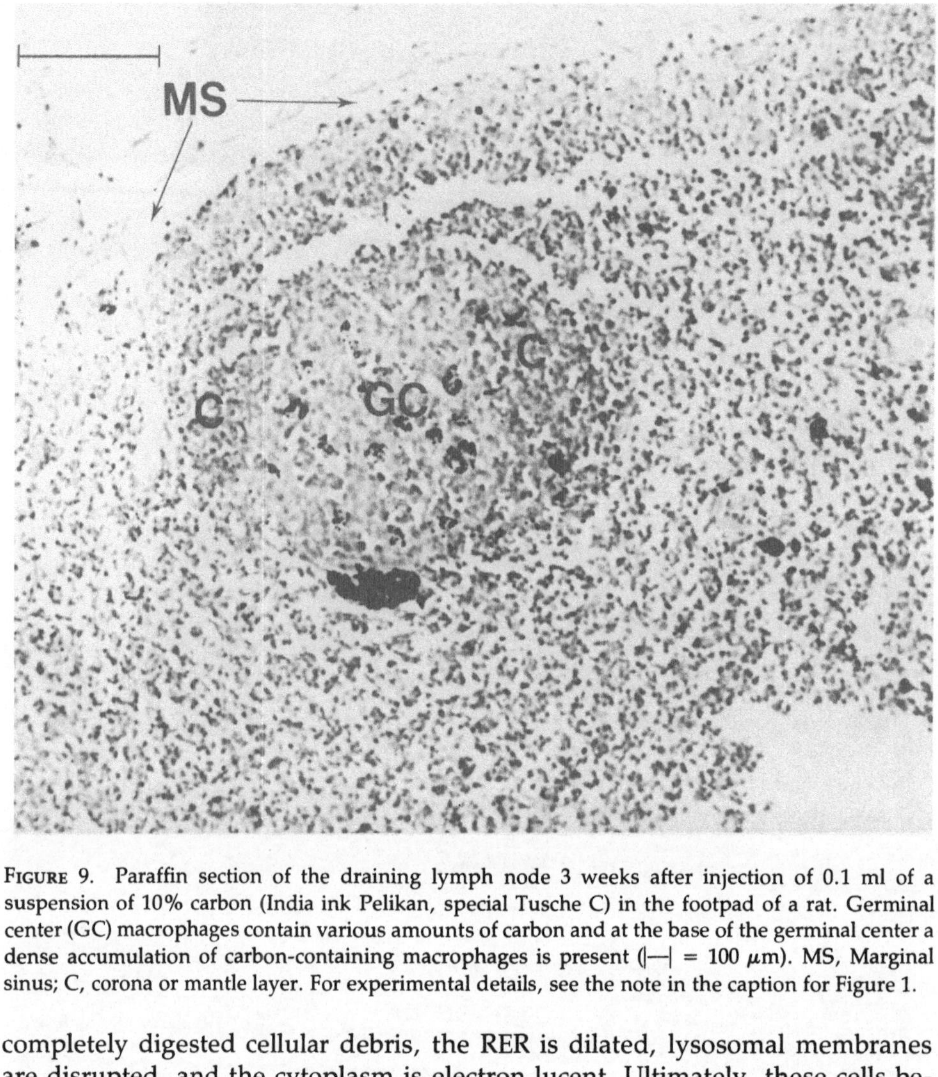

FIGURE 9. Paraffin section of the draining lymph node 3 weeks after injection of 0.1 ml of a suspension of 10% carbon (India ink Pelikan, special Tusche C) in the footpad of a rat. Germinal center (GC) macrophages contain various amounts of carbon and at the base of the germinal center a dense accumulation of carbon-containing macrophages is present (|—| = 100 μm). MS, Marginal sinus; C, corona or mantle layer. For experimental details, see the note in the caption for Figure 1.

completely digested cellular debris, the RER is dilated, lysosomal membranes are disrupted, and the cytoplasm is electron lucent. Ultimately, these cells become filled with residual bodies and contribute to the mass of effete macrophages which are found in germinal centers.

The distribution of macrophages in different stages of development suggests that these cells migrate from the marginal zone towards the small capillaries in the outer cortex. After entering the mantle layer of secondary follicles these cells migrate toward the capillary venules at the base of the germinal centers, and this movement probably results from the microcirculation of tissue fluid, rather than the active movement of the macrophages themselves. The pressure of the afferent lymph is relatively high as compared with the venous blood pressure (Drinker *et al.*, 1934); moreover, the actively phagocytic cells in the marginal zone and the TBM lack the microfilamentous equipment needed for active migration. Germinal centers have a predetermined location, which is related to the blood vascular system in the spleen (Smythyman *et al.*, 1978) and also in the lymph node (Kelly, 1975). The degenerate appearance of many TBM and the presence of cellular debris derived from phagocytic cells demonstrate that macrophages are themselves phagocytized when they become effete and their residual bodies are added to those of the surviving population. The results of the carbon experiments support this hypothesis.

TBM are not confined to germinal centers but can be found elsewhere in the outer cortex. Odartchenko *et al.* (1967) demonstrated that the lymphoid cells taken up by TBM become necrotic at the end of the period of DNA synthesis, just before the onset of mitosis. Swartzendruber and Congdon (1963) clearly demonstrated that TBM are nonselective and phagocytize lymphocytes, plasma cells, erythrocytes, and even granulocytes. Since erythrocytes and granulocytes are not normally present in germinal centers they must have been phagocytized elsewhere in the outer cortex and transported into the germinal center.

Nossal *et al.* (1968) used labeled flagellin to study antigen trapping in rat popliteal nodes. Almost immediately after administration, the antigen was found in macrophages in the marginal sinus and the outermost layers of the cortex. These authors suggested that the macrophages were monocyte-derived cells which migrated down into the germinal centers and there assumed their characteristic function and morphology. The amount of antigen taken up by the TBM in preexisting germinal centers was very variable. The label was found on or near the cell surface, in lysosomes, and also concentrated in myelin figures in residual bodies. However, little attention has been paid to these observations probably because the antigen-retaining dendritic cells were more heavily labeled. Also it was assumed that the TBM digest the antigen completely. Later it was proved that these animals had natural antibodies against the antigen used in these experiments.

White (1975) described the development of germinal centers in chicken spleen following stimulation with isotopically labeled soluble antigen. Development occurred in two phases, both of which were shown to depend on the presence of circulating immune complexes. In the first two days "dendritic" cells in the white pulp were labeled and migrated through the area beside the penicillary arteries to their point of origin from the central arteriole. The dendritic cells formed clusters with small lymphocytes and on the third day immunoblasts appeared in these foci and the characteristic antigen-retaining dendritic web of cells was first observed. We suggest that the migratory cells were macrophages, but the antigen-retaining cells in developed germinal centers were dendritic reticulum cells (DRC) as discussed above. It was found that the migratory cells were not able to transport labeled material into primordial germinal centers unless antibody was present. Han *et al.* (1969) studied the handling of isologous and heterologous ferritin and found no difference in the distribution of the two preparations on the first and second days after injection, but on the third day heterologous ferritin was ingested by macrophages and rapidly eliminated from the circulation in the form of antigen–antibody complexes. Humphrey (1978) found that acid polysaccharide–DNP conjugates were endocytized by various types of macrophages, and these conjugates were able to inhibit anti-DNP secondary responses, whereas neutral polysaccharide–DNP conjugates were only taken up by marginal zone and marginal sinus macrophages, and these conjugates were not able to inhibit the development of anti-DNP memory cells. Levan, a neutral polysaccharide, which activates the alternate pathway of complement, was rapidly localized in germinal centers. Thus, marginal zone macrophages may regulate the memory response. The macrophages which surround the capillary venules at the base of the germinal centers are in close apposition to the immunoblasts in the lower pole of the follicle, whereas the dendritic cells form a cap between the medium-sized lymphocytes in the upper pole and the mantle layer of small lymphocytes. These observations strongly suggest that macrophages are responsible for the germinal center reaction rather than dendritic reticulum cells.

1.5.3. Macrophages in the Inner Cortex

The interdigitating cell (IDC) is the characteristic type of macrophage in the thymus-dependent inner cortex or paracortex (Hoefsmit, 1975). These cells interdigitate with the surrounding lymphocytes (Figure 3) (Veldman, 1970), and their electron-lucent cytoplasm contrasts sharply with the surrounding lymphocytes which are usually more electron-dense than the lymphocytes in other parts of the paracortex. IDC have irregular, sometimes rather wide processes, which extend between the lymphocytes. The lymphocytes often insert narrow processes into invaginations of the IDC membrane. In immersion fixation, the typical pattern of cellular interaction may be lost and the IDC appear to have flaplike processes extending into the intercellular space (Hoefsmit *et al.*, 1980). IDC vary in size, but are usually large cells, 20–30 μm in diameter, excluding the cell processes (Figure 10). The nucleus is irregularly shaped and has an eccentric position in the cell, there is a small rim of heterochromatin in close apposition to the nuclear envelope, and a nucleolus is sometimes present. There is a centrosome in the center of the cell from which microtubules radiate. Bundles of microfilaments may surround the nucleus. The cytoplasm does not usually contain many free ribosomes or polyribosomes, and the mitochondria are small and variable in number. Some strands of RER surround the nucleus and the cytocenter. The RER may be somewhat dilated, and the membranes are irregularly studded with ribosomes. Strands of RER are sometimes concentrically arranged around a mitochondrium, forming a whorl.

The Golgi apparatus may be well-developed forming four to six stacks of cisterns radially arranged around the centrosome between the microtubules. In some cells, there are one or two well-developed stacks of Golgi cisterns in a radial array and other ill-defined Golgi stacks. Some of the cisterns are dilated and there may be a cluster of small vesicles occupying an area at least 1000 nm in diameter. Occasionally clusters of vesicles are assembled at one side of the cytocenter. These structures have been described as a "vesicular complex" by Veerman and Van Ewijk (1975) and Kamperdijk *et al.* (1978) and should be distinguished from the "tubulovesicular system" described by Veldman (1970).

Almost all the IDC present in normal lymph nodes contain large numbers of small lysosomes with heterogeneous electron-dense contents. These organelles are aligned between the microtubules and surround the cytocenter. They are irregularly shaped, often somewhat elongated, sometimes beaded, or even cylindrical. They frequently have a narrow zone of lower electron density forming a halo around the electron-dense contents. Large phagolysosomes are not usually observed, but may be present in the periphery of the cell. They contain moderately electron-dense material. After an injection of India ink, some IDC contain large vesicles filled with carbon and located in the peripheral part of the Golgi area (Figure 11).

In some IDC there are invaginations of the cell membrane into which lymphocytes insert narrow processes, and these tubular invaginations of the cell

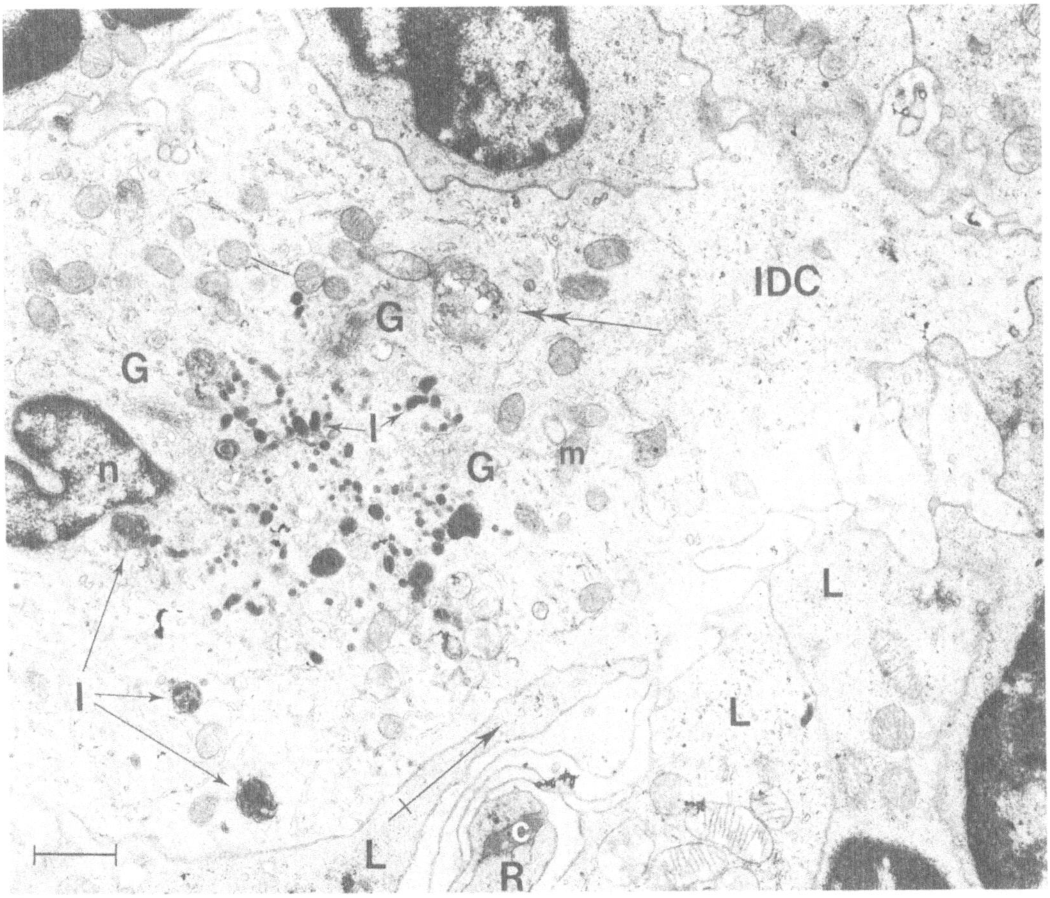

FIGURE 10. IDC in paracortex of normal rabbit lymph node. Adjacent lymphocyte inserts cell processes (barred arrow) into invaginations of the IDC cell membrane. Numerous small lysosomes, containing heterogeneous electron-dense material are present in the central area of the cell and larger phagolysosomes in the periphery. A vesicular complex (double arrow) is present in an eccentric position (|—| = 1 μm). G, Golgi apparatus; n, nucleus; m, mitochondrion; R, reticulum cell; c, collagen; L, lymphocyte; l, phagolysosome. For experimental details, see the note in the caption for Figure 1.

membrane can be followed in serial sections, appearing as tubules or rows of small vesicles arranged more or less perpendicular to the cell surface. Cilia have never been found in IDC.

In sections treated with silvermethenamine, the Golgi system, the vesicular complex, and smaller lysosomes are intensely reactive, whereas larger lysosomes and the cell coat, including tubular invaginations of the cell membrane, contain a moderate amount of silver precipitate.

In other experiments horseradish peroxidase (HRP) was used to delineate the extracellular space. The irregularity of the cell surface, including the tubular invaginations, was clearly seen (Figure 12). The nuclear envelope and the RER sometimes exhibited endogenous peroxidase-like activity (see below) but the Golgi system was invariably negative, and in the Golgi area HRP was only present in some small lysosomes.

The interpretation of the ultrastructure and immunological function of IDC

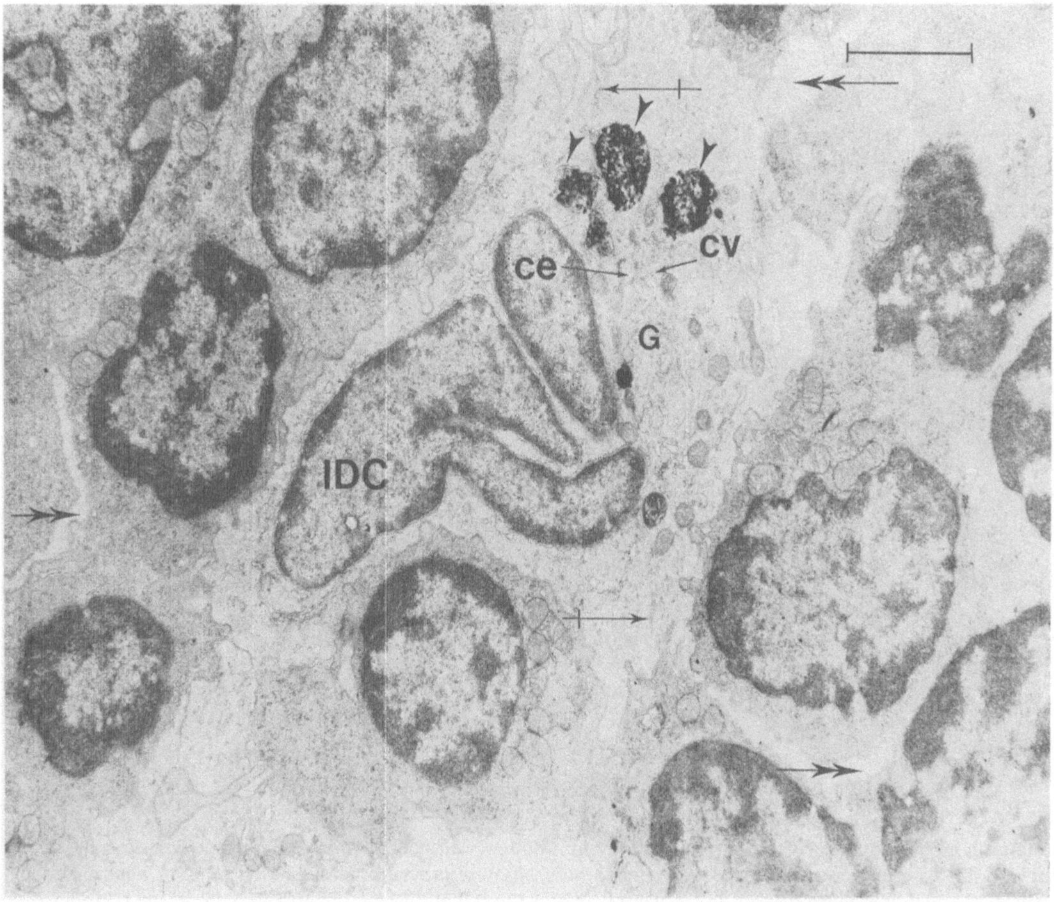

FIGURE 11. One week after injection of carbon in the rat footpad (for details see Figure 9.), some IDC contain large vesicles filled with carbon (arrow heads). The IDC extend cell processes between the adjacent lymphocytes (double arrows), and the cell membrane has many tubular invaginations (Barred arrows) (|—| = 2 μm). G, Golgi apparatus; ce, centriole; cv, coated vesicle. For experimental details, see the note in the caption for Figure 1.

is confused. The cell was first described by Veldman in 1970 in rabbit lymph nodes. He drew attention to the typical pattern of interaction between IDC and surrounding lymphocytes, and suggested that the lysosomal complex, together with the tubular invaginations of the cell membrane, was a secretory "tubulovesicular system." This system was considered to be responsible for the synthesis of glycoproteins, and for transporting them from the Golgi apparatus vai the tubulovesicular system to the cell membrane. Veldman did not detect lysosomal activity in these cells, and therefore classified them as nonphagocytic reticular cells. When treated with silvermethenamine (Rambourg and Leblond,

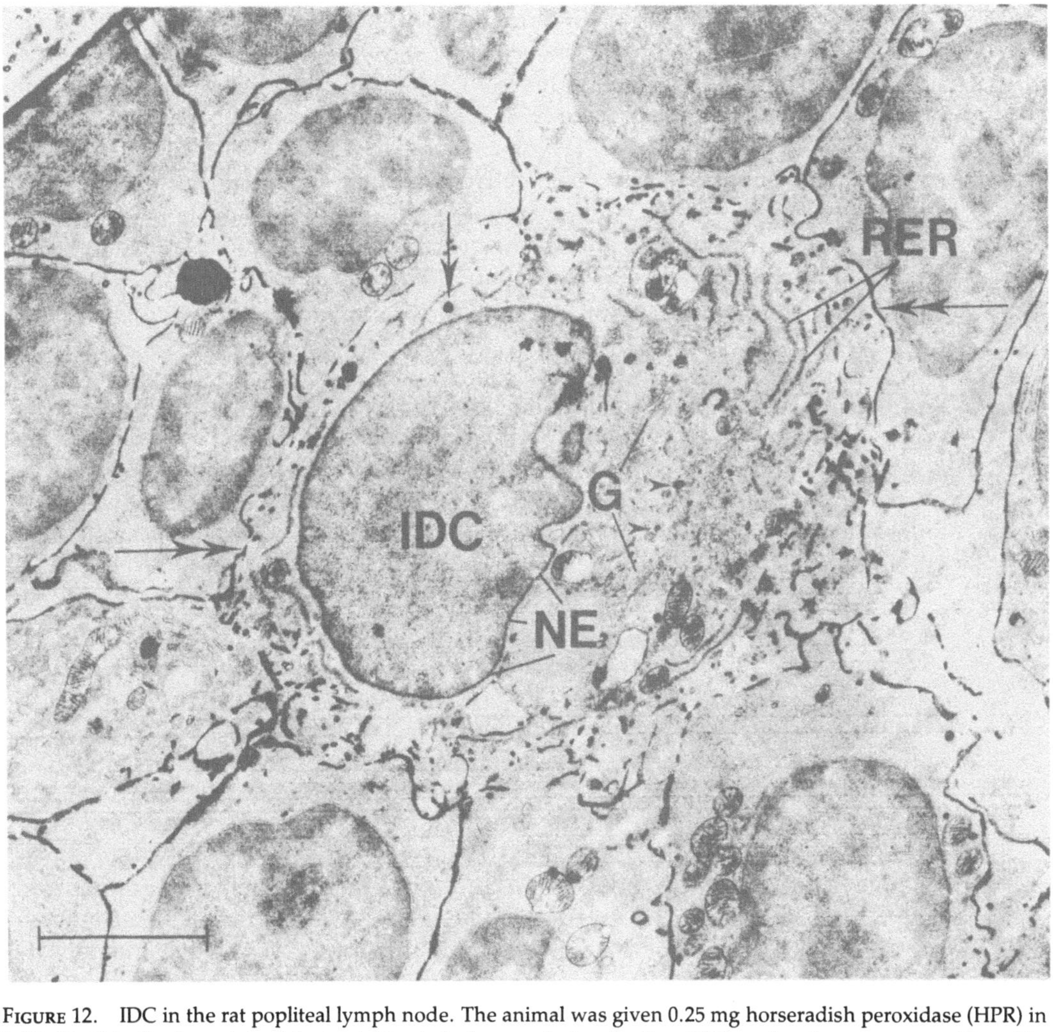

FIGURE 12. IDC in the rat popliteal lymph node. The animal was given 0.25 mg horseradish peroxidase (HPR) in 0.9% NaCl in the homolateral footpad 45 min before perfusion fixation. HPR and endogenous peroxidase-like activity was demonstrated with diaminobenzidine (Hoefsmit, 1975). HRP fills the intercellular space following the tubular invaginations of the cell membrane (double arrow) and pinocytic vesicles (arrow). Endogenous peroxidase-like activity is present in the NE and RER. The Golgi system is negative and there are some HRP-containing vesicles present in the Golgi area (arrow head) (|—| = 3 μm). For experimental details, see the note in the caption for Figure 1.

1967; Rambourg *et al.*, 1969), these organelles were covered by a silver precipitate. This observation was confirmed by Friess (1976). The tubulovesicular system found in IDC is not, however, characteristic of secretory cells. In the process of secretion "the condensing vacuoles," (Jamieson and Palade, 1971) derived from the Golgi system, decrease in diameter and increase in density, as they approach the cell membrane. In IDC, as in other types of macrophages the phagolysosomes become smaller as a result of lysosomal digestion when they move towards the central area of the cell. During this process, the nondigestible products are condensed to form electron-dense residual bodies, while the small molecules diffuse through the lysosomal membrane. Since these organelles all contain glycoproteins, a positive reaction with silvermethenamine is to be expected. Morphological observations clearly show that IDC have the characteristics of macrophages engaged in lysosomal digestion. The lysosomal complex in IDC was described by Veerman (1974), Veerman and Van Ewijk (1975), Hoefsmit (1975), and Kamperdijk *et al.* (1978).

IDC do not apparently phagocytize particulate material, but may "drink" substances like India ink. This material is not digestible, and persists in large vacuoles in the peripheral area of the cell. Horseradish peroxidase is an easily digestible protein (Steinman and Cohn, 1972), but can be found immediately after administration located in the tubular structures which communicate with the intercellular space and in some lysosomal vesicles.

The macrophages in the outer cortex, i.e., the actively phagocytic macrophages of the marginal zone and the TBM in germinal centers, seem to be concerned with phagocytosis of particulate matter, whereas IDC are more concerned with macropinocytosis. In this respect, the IDC resemble the smaller type of macrophage found in the marginal sinus. However, the IDC are much larger, more electron-lucent cells, and the larger they are, the more electron-lucent they become.

Veldman (1970) stressed the morphological and functional similarity between DRC in the B-cell area and IDC in the T-cell area of the lymph node. He suggested that the function of these two cell types is to present antigens to B and T lymphocytes; he also suggested that the spleen is incapable of giving a specific cellular immune response because it contains no IDC. However, IDC have been found in the thymus-dependent area in all peripheral lymphoid organs including the spleen and in the medulla of the thymus (Veerman, 1974; Veerman and Van Ewijk, 1975; Hoefsmit, 1975). According to Veerman (1974), Van Ewijk *et al.* (1974), Veerman and Van Ewijk (1975), and Friess (1976), IDC provide the microenvironment in which T cells home and proliferate. Lennert's group followed Veldman's classification. They described the IDC in the thymus-dependent area of the human lymph node (Kaiserling and Lennert, 1974), the spleen (Heusermann *et al.*, 1974), and the thymus (Kaiserling *et al.*, 1974) as nonphagocytic reticular cells. Recently, it has been shown that IDC in lymph nodes draining the skin may contain Birbeck granules (Kamperdijk *et al.*, 1978), and there is increasing evidence to suggest that macrophages which contain Birbeck granules play an important role in antigenic stimulation (Silberberg-Sinakin and Thorbecke, this volume; Hoefsmit *et al.*, 1979, 1980).

1.6. LYMPH NODE MACROPHAGES: A KINETIC SYSTEM OF LYMPH-BORNE OR BLOOD-BORNE MACROPHAGES

Following Veldman's hypothesis, Kaiserling (1977) presented a static model of antigen presentation by DRC and IDC.

The concept of the RES does not distinguish clearly between reticulum cells and macrophages. In this classification, bone-marrow-derived fixed macrophages are included in the cellular reticulum and, therefore, belong to the stationary elements or framework of the node in which lymphocytes reside for a certain time (Müller-Hermelink and Lennert, 1978).

However, it has been shown that the reticulum cells which manufacture and sustain the intercellular skeleton are essentially different from macrophages which have the morphological characteristics of mononuclear phagocytes (Hoefsmit, 1975). According to the concept of the MPS (van Furth *et al.*, 1972, 1975), the cellular reticulum together with the intercellular skeleton forms the framework in which lymphoid cells and macrophages home and interact.

In the preceding paragraphs the ultrastructure of the nonlymphoid population of the lymph node was described in detail. It was shown that the reticulum cells and macrophages in the various compartments are more or less specialized. The reticulum cells in the paracortex most resemble fibroblasts, forming a delicate reticulum throughout the area, except in the region of the germinal centers. In these structures, the antigen-retaining DRC are highly specialized reticulum cells with a hypertrophic cell membrane, forming an intricate web in which antigen is localized.

The macrophages in the outer cortex or B-cell compartment are strikingly different from those in the inner cortex or T-cell compartment. In the outer cortex, the macrophages are large, actively phagocytic cells in various stages of development. They appear to migrate from the marginal sinus through the marginal zone and the follicles towards the base of the germinal centers in the region of the capillary venules. In the T-cell compartment, the characteristic type of macrophage is the IDC.

Macrophages may enter the lymph node via the afferent lymph or via the blood vessels. The afferent lymph carries tissue macrophages, and in the next section it will be shown that macrophages in the inner cortex and the outer cortex are lymph-borne cells.

2. EVIDENCE SUGGESTING THAT MACROPHAGES IN NORMAL AND STIMULATED LYMPH NODES ARE MOBILE LYMPH-BORNE CELLS

2.1. EXUDATE AND RESIDENT MACROPHAGES IN PERIPHERAL CONNECTIVE TISSUE AND IN THE LYMPH NODE: ENDOGENOUS PEROXIDASE-LIKE ACTIVITY

Morphological and cytochemical studies have shown that exudate and resident peritoneal macrophages differ in a number of respects (Daems and Brederoo, 1971, 1973; Daems *et al.*, 1976). It is generally accepted that exudate macrophages are derived from blood monocytes which penetrate into the tissues by diapedesis. However, the origin of tissue macrophages is still in doubt since resident and exudate macrophages have different patterns of endogenous peroxidase-like activity. Hoefsmit (1975) and Beelen *et al.* (1978c), using Graham and Karnovsky's reagent (1966) to demonstrate the presence of endogenous peroxidase-like activity, were able to confirm these findings. In exudate macrophages, peroxidase-like activity can only be found in some of the virginal lysosomes or storage granules, whereas in resident macrophages, reaction products are located exclusively over the nuclear envelope and RER (Beelen *et al.*, 1978a). In acute inflammatory reactions in rats, some macrophages exhibit both patterns of activity. In these "exudate-resident" macrophages reaction products are located over the nuclear envelope, the RER, and virginal lysosomes, but not over the Golgi system. There are also macrophages without detectable peroxidase-like activity present in acute and chronic exudates (Beelen *et al.*, 1978b). Exudate macrophages expend their store of granules during lysosomal digestion (Daems *et al.*, 1973), and some cells may lose their granules before they have had time to manufacture further peroxidase-like enzymes in the RER. These macrophages, as well as monocytes and exudate macrophages, generate this enzyme *in vitro,* and the developing peroxidase-like activity has a resident pattern (Bodel *et al.*, 1978; Beelen *et al.*, 1978b, 1979). From these observations, it was concluded that the various patterns of peroxidase-like activity represent different stages in the development of mononuclear phagocytes.

In a previous publication (Hoefsmit, 1975), the endogenous peroxidase-like activity of lymph node macrophages was described. The macrophages in the marginal sinus and the marginal zone were invariably negative; some IDC showed a resident pattern (Figure 12), while others were negative; all TBM and almost all medullary macrophages had a resident pattern of peroxidase-like activity. At that time we concluded that the marginal zone macrophages were lymph-derived exudate macrophages, whereas the medullary macrophages were locally-derived resident macrophages. It now seems more probable that the marginal zone macrophages are recently-arrived cells and the IDC are a labile population, whereas the TBM and the medullary macrophages are the oldest inhabitants.

Milky spots are opaque patches located in the omentum. They consist of a cellular reticulum populated by macrophages and lymphocytes which is analogous to the reticuloendothelial tissue of the lymph node. Rats and mice were given an intraperitoneal (i.p.) injection of 10 ml of new-born calf serum. This provoked an acute peritonitis, with perivascular accumulations of neutrophilic leucocytes. After 12–48 hr the exudate became predominantly monocytic. Another group of mice were injected with BCG and developed cell-mediated immunity. These animals also received an intraperitoneal injection of new-born calf serum, and milky spots appeared in which the diapedesis of cells was much more vigorous than it was in normal animals, and there were accumulations of proliferating promonocytes and monocytes around blood vessels (Figure 13) (Hoefsmit and Smelt, 1978; Beelen *et al.*, 1978c). On the other hand, perivascular

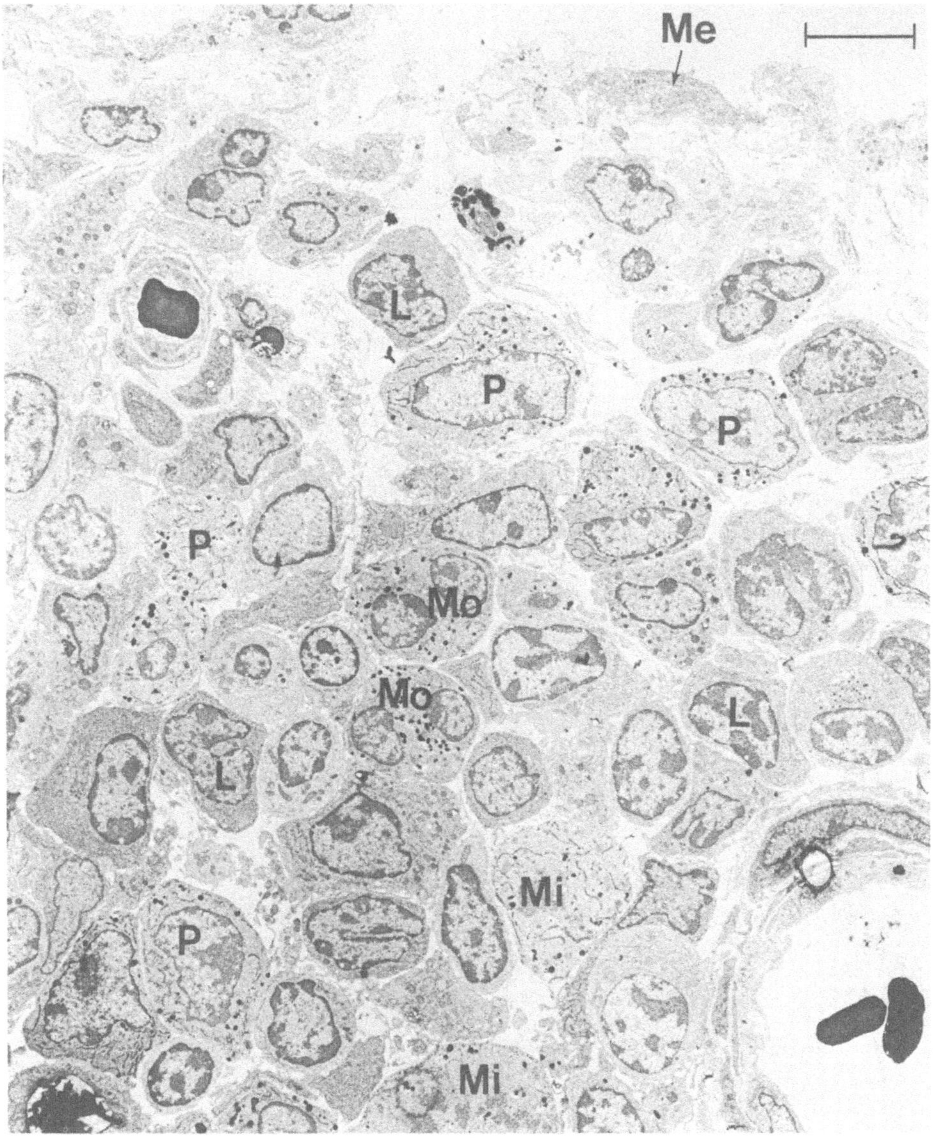

FIGURE 13. Milky spot of a mouse which had received 2.5 × 10^6 viable BCG intravenously (R.I.V., Zeist, The Netherlands), and which was further sensitized by intravenous inocculation of 10^7 BCG after 21 days. A peritoneal exudate was provoked by injection of 1 mg PPD (in phosphate buffer) intraperitoneally 3 days after the booster. Endogenous peroxidase is present in promonocytes (P), mitotic figures (Mi) of promonocytes, and monocytes (Mo), which are present perivascularly (|—| = 5 μm). L, Lymphocyte; Me, mesothelial cell.

accumulations of monocyte-like cells, proliferating promonocytes, and monocytes were never found in normal lymph nodes, or in lymph nodes in which a primary or secondary immune response was taking place (Kamperdijk, to be published). It appears that macrophages are not able to enter the nodes via

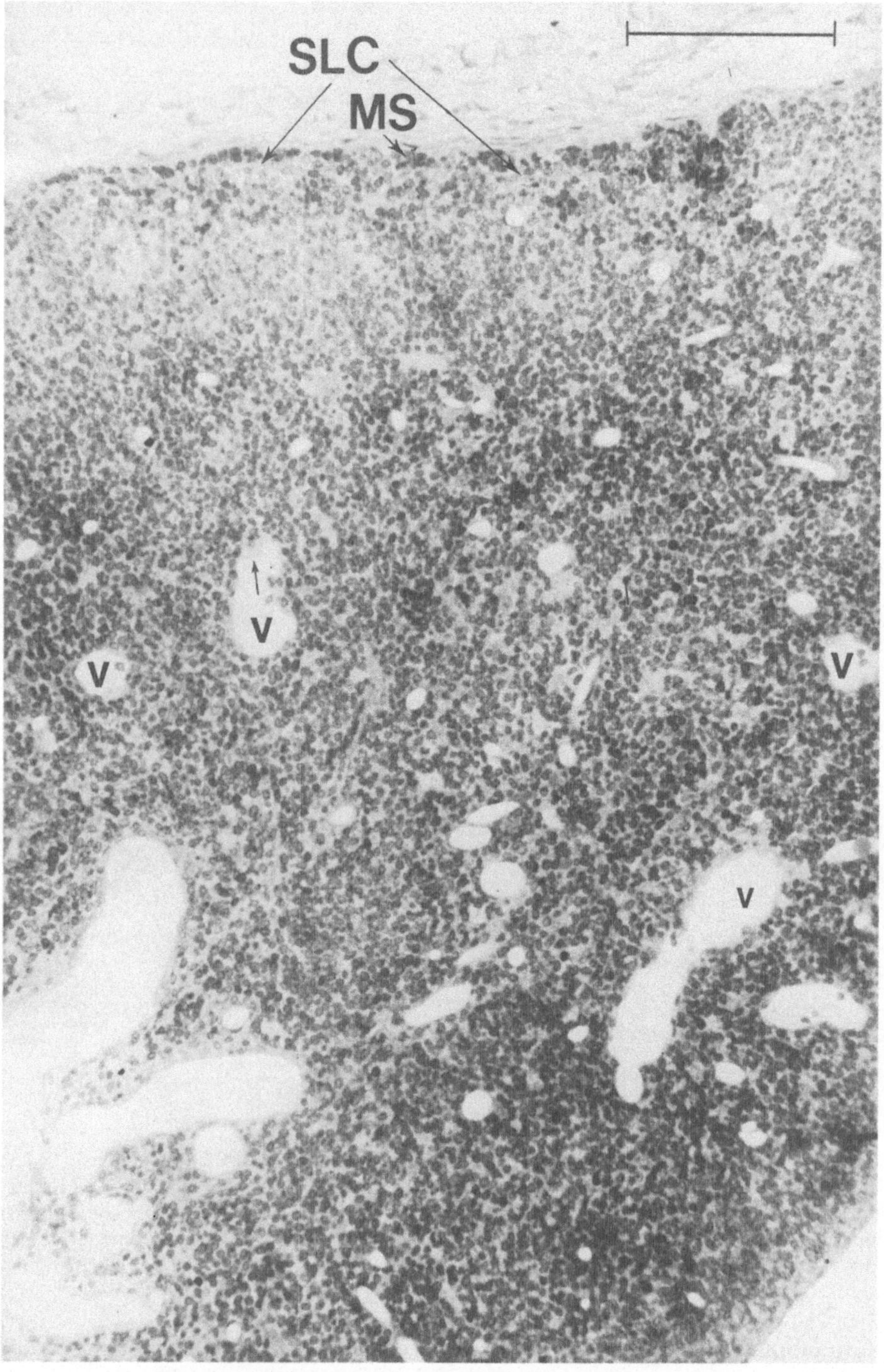

FIGURE 14. Rat lymph node deprived of afferent lymph for 6 weeks. The lymph node reticulum is populated with small lymphocytes; the macrophages have disappeared. Follicular structures are no longer present, and there are only some high endothelial cells lining the venules (arrow) (|—| = 100 μm). SLC, Sinus-lining cell; MS, marginal sinus; v, venule.

the blood vascular system, and must therefore arrive via the afferent lymph. This means that interruption of the afferent lymph supply should lead to depletion of the macrophages in the node. This experiment is described in the next section.

2.2. THE LYMPH NODE DEPRIVED OF THE AFFERENT LYMPH

There are very few studies which deal with the effects of interruption of the afferent lymph supply to a lymph node (Engeset and Nesheim, 1966; Kotani *et al.*, 1977). Osogoe (1969) obstructed the afferent and efferent lymphatics of a lymph node. He found that there were accumulations of lymphocytes in the sinuses, and that the cortex and medulla were packed with lymphocytes. However, communication between the obstructed lymphatic vessels and the node

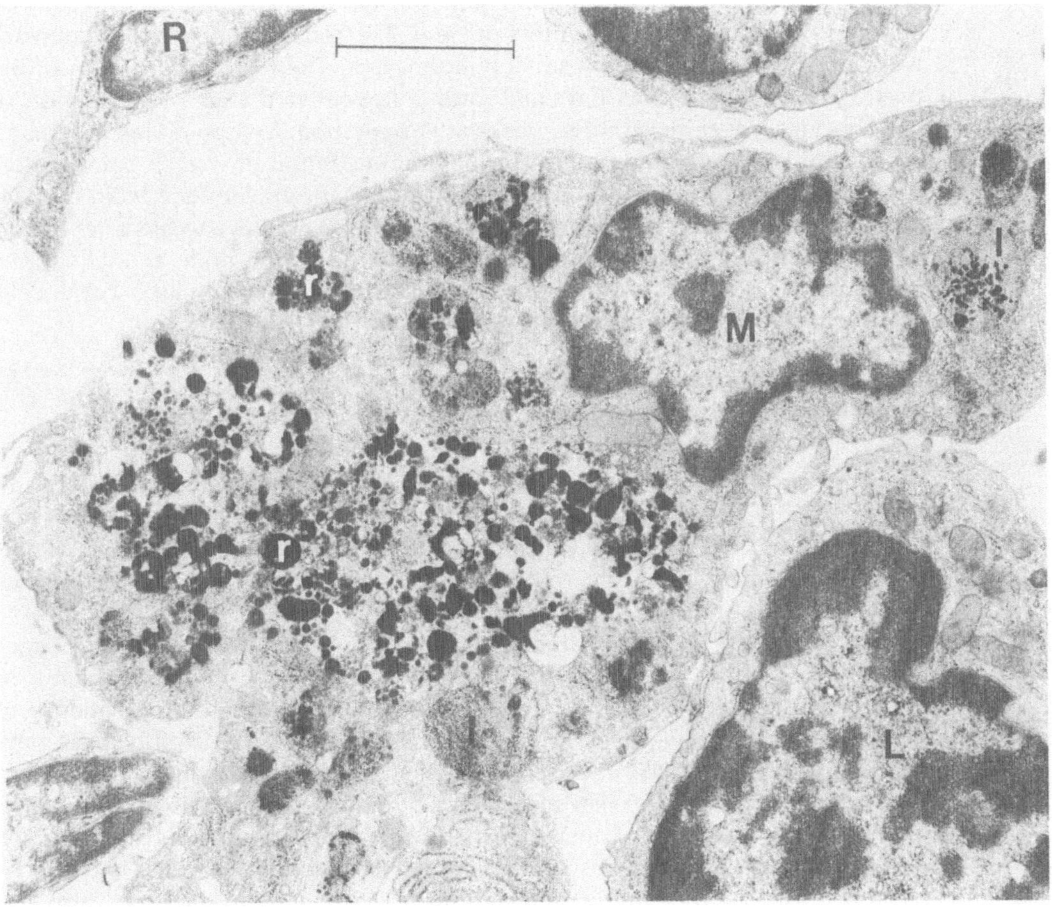

FIGURE 15. Macrophage in medulla of rat lymph node which was deprived of afferent lymph for 6 weeks. The macrophage contains many phagolysosomes and residual bodies (r) (|—| = 2 μm). R, Reticulum cell; L, lymphocyte. For experimental details, see the note in the caption for Figure 1.

was reestablished spontaneously 1–4 weeks after operation, and the structure of the node later returned to normal. After pilot studies (Hoefsmit, 1973), Hendriks (1978) succeeded in cutting off the afferent lymph supply to rat popliteal nodes for protracted periods. In these experiments, the efferent lymphatics and blood vessels supplying the node were left intact.

The most striking alteration in these lymph nodes was depletion of the macrophages in all compartments (Figure 14). The marginal sinus macrophages and the macrophages in the adjacent cortex disappeared within 1 week of operation, and, from that time onwards, there was a continuous decrease in the number of macrophages in the follicles and the paracortex. Six weeks after operation there were only a few macrophages left in the node, and these cells were in the medullary cords and sinuses. They contained increased numbers of large electron-dense phagolysosomes (Figure 15). It was concluded that lymph node macrophages are predominantly lymph-borne.

The immunological activity of these nodes was also completely abolished 8 weeks after operation. Lymphoblasts and plasma cells had almost disappeared, and there were no germinal centers present. The reticuloendothelial framework was populated exclusively by small lymphocytes. The high endothelium of the postcapillary venules was flattened, and it appeared that the recirculation of lymphocytes was reduced three weeks after operation. Lymph nodes which had been stimulated by paratyphoid vaccine administration one week before interruption of the afferent lymphatic supply were again practically devoid of macrophages, lymphoblasts, plasma cells, and germinal centers 6 weeks after operation. This implies that the normal structure of the lymph node and its various anatomical entities depends on physiological stimulation by lymph-borne cells and perhaps other afferent lymph constituents.

2.3. LYMPH-BORNE MACROPHAGES ENTERING THE LYMPH NODE

The cells present in normal afferent lymph were first studied by Yoffey and Drinker (1939), using cats and dogs. They found that 20% of the cells were monocytes and macrophages. These studies were extended by Hall and Morris (1962) using sheep, and the cellular composition of the efferent lymph was investigated by Smith *et al.* (1970) in the same species.

These studies showed that a small but continuous stream of macrophages entered the lymph node via the afferent lymph, but no macrophages or monocytes were found in the efferent lymph, nor was there any evidence to suggest that macrophages left the node via the blood vessels (Yoffey and Courtice, 1970). Morris (1968) and Morris *et al.* (1968) studied lymph-borne macrophages in the electron microscope, and noted that these cells had extensive pinocytic veils which they considered to be evidence of a high degree of surface activity. Smith *et al.* (1970) observed lymphocytes wandering around and attaching to the plasma membrane of macrophages (i.e., peripolesis); they also observed lymphocytes which appeared to be inside macrophages, without suffering any ill effects (i.e., emperipolesis).

Recently, it has been shown that the large mononuclear cells present in afferent skin lymph do not usually phagocytize particulate material. These cells have ceaselessly moving cytoplasmic veils, and may contain Birbeck granules (Kelly *et al.*, 1978; Hoefsmit *et al.*, 1978, 1979; Drexhage *et al.*, 1980), the characteristic organelle of the epidermal Langerhans cells (Birbeck *et al.*, 1961). Birbeck-granule-containing cells have also been found in the dermis and in dermal lymphatics at the site of contact allergic reactions (Silberberg *et al.*, 1974). The presence of Ia antigens (Stingl *et al.*, 1978) and Fc and C3 receptors (Stingl *et al.*, 1977) on Langerhans cells suggested that these cells play a role in immunological reactions (Silberberg-Sinakin *et al.*, 1978, and in this volume; Hoefsmit *et al.*, 1978, 1980).

The cellular composition of normal afferent skin lymph from rabbits, pigs, and man has been examined. Rabbits were found to have the highest proportion of veiled cells (Drexhage *et al.*, 1979). The total cell content of this lymph averages 18×10^4 cells per ml; 47% of the cells are lymphocytes, 3% granulocytes, and 49% large mononuclear cells, comprising 40% veiled cells and 9% nonveiled cells. It was estimated that 1×10^6–2×10^7 veiled cells entered the rabbit popliteal node each day (Kelly *et al.*, 1978). After warming to 37°C, the lymph-borne veiled cells become extremely active, constantly extending long boat-shaped veils, which then curve back until the free margin is in contact with the cell surface, some distance from the point of origin. Shortly afterwards the veil disappears, apparently incorporated in the cell body, and, at the same time, a large vacuole becomes visible inside the cell and then quickly vanishes. This behavior suggests that the main activity of these cells is to engulf large amounts of fluid. This is accomplished by sweeping movements of the veils leading to the formation of a vacuole. The movements are much more expansive than those carried out by the majority of peritoneal macrophages. During all this activity the cells remain stationary, but do not adhere to glass surfaces. Veiled cells do not usually ingest carbon particles, but they are sometimes seen in preformed aggregates with other lymph cells. After skin painting with dinitrofluorobenzene (DNFB), the veiled cells in lymph coming from the skin site, change their behavior, moving in a directional manner as if attracted toward other cells; the veils are larger and more fan-shaped, the cell body is constricted, and the coming and going of fluid vacuoles can no longer be observed. There are increased numbers of cell aggregates in the lymph, and almost all the aggregates contain a veiled cell, sometimes two or three (Drexhage *et al.*, 1980).

Eighteen percent of rabbit veiled cells form rosettes with antibody-coated bovine red cells, therefore they probably have Fc receptors, but the rosettes are small as compared with those formed by oil-induced peritoneal macrophages. Similar tests on pig cells show that the veiled cells all possess Fc receptors, but of rather weak affinity; they also possess C3 receptors. Pig veiled cells give very strong reactions for Ia-like antigens.

Acid phosphatase activity can be demonstrated in pig and rabbit veiled cells. The reaction is weak and diffusely distributed near the nuclear indentation, whereas in peritoneal exudate macrophages it is present in the form of discrete packages of intense activity. Nonspecific esterase similar to that found in monocytes and peritoneal macrophages can be demonstrated in pig and rabbit veiled cells. The ATPase activity of veiled cells is very strong, the entire plasma membrane, including the veils, giving a positive reaction, whereas in peritoneal macrophages the reaction is weak and distributed in the form of spots on the plasma membrane. This enzyme is absent in blood monocytes (Drexhage *et al.*, 1980).

Rabbit lymph-borne cells were studied in the electron microscope. In normal lymph, the population includes 9% of monocyte-like cells 7 μm in diameter. These cells have a nucleocytoplasmic ratio of 1:2. The nucleus is in an eccentric position, is reniform, and frequently contains a distinct nucleolus (Figure 16). There are many polyribosomes in the cytoplasm and some microfilaments in the

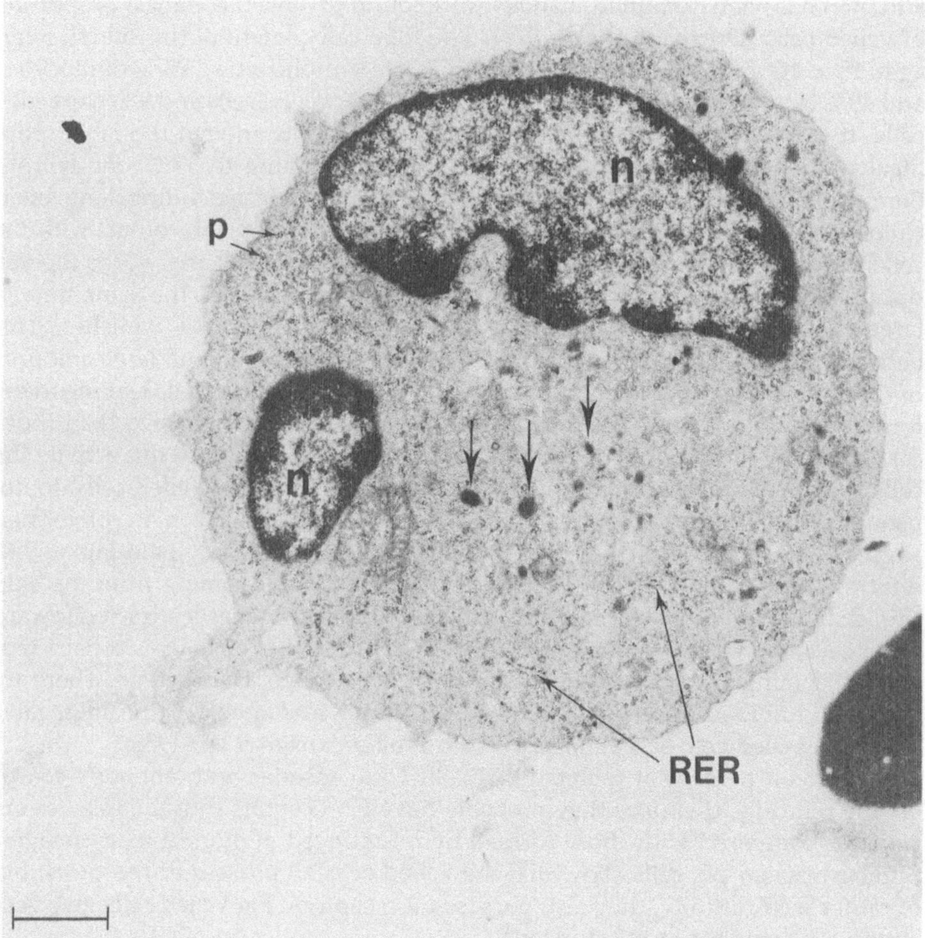

FIGURE 16. Monocyte from normal rabbit lymph. Vesicles containing homogeneous dense material (arrows) are present in the Golgi area (|—| = 1 μm). RER, Rough endoplasmic reticulum; p, polyribosomes; n, nucleus. For experimental details, see the note in the caption for Figure 1.

periphery of the cell. The central area contains small groups of Golgi vesicles and some coated vesicles. About 15% of these cells contain small vesicles with homogeneous electron-dense contents which resemble lysosomal storage granules in rabbit monocytes (Nichols and Bainton, 1975).

Veiled cells form the major component of the nonlymphoid mononuclear cell population in the afferent lymph. These cells (Figure 17) measure at least 8 μm in diameter and possess up to five extremely long cytoplasmic extensions or veils, some nearly half as long as the diameter of the cell body. The veils have a constant thickness of about 100 nm and do not contain any organelles. At the base of the veils there is often a row of electron-lucent vesicles also measuring 100 nm in diameter. The nucleocytoplasmic ratio of veiled cells is 1:3 or 1:4. The nucleus has an eccentric position, is reniform, with many indentations, and frequently contains a nucleolus. The karyoplasm is euchromatic with some condensation against the nuclear envelope. Veiled cells also contain heavy bundles of microfilaments, located near the nuclear indentations: mitochondria, polyribosomes, and some strands of RER are also enmeshed in a network of crisscrossing microfilaments. The central area of these cells is occupied by a cytocenter surrounded by large numbers of small smooth-surfaced vesicles with electron-lucent contents and some coated vesicles, but there are usually no distinct stacks of Golgi cisterns. Near the plasma membrane one or two large vacuoles are frequently seen containing a fluffy precipitate of low electron density. These vacuoles measure 0.5–4.0 μm in diameter; the outer wall is 100 nm thick and has the same structure as the veils themselves. At a somewhat deeper level towards the cell center, endocytic vacuoles of moderate size and irregular shape are often seen, containing a denser material, but phagolysosomes, containing electron-dense or recognizable cellular debris are usually absent. The endocytic vacuoles apparently condense their contents and become smaller as they move towards the central area of the cell. The diffuse acid phosphatase activity detected in this area with the light microscope indicates that lysosomal digestion is occurring, but the contents of the vesicles are apparently easily digestible since the cells usually lack residual bodies. The Golgi apparatus is well-developed in macrophages but ill-defined in veiled cells. The central complex of small electron-lucent vesicles may represent a site for the turnover of the cell membrane. Chains of small vesicles are often present at the base of the veils, and they may fuse together to form part of the plasma membrane of the veils. The enormous filamentous and mitochondrial equipment of veiled cells is obviously responsible for the vigorous movements of the veils. Seventeen percent of the veiled cells are somewhat larger, measuring up to 15 μm (Figure 18). These cells contain one or more large vacuoles near the cell surface, phagolysosomes with heterogeneous electron-dense contents, and, sometimes, small dense residual bodies.

In normal lymph about 5% of veiled cells contain Birbeck granules but only 0.5% also contain large phagolysosomes (Hoefsmit *et al.*, 1979). Birbeck granules are not found in any other cell type present in normal lymph.

The process of engulfing fluid in vacuoles was observed by Lewis in

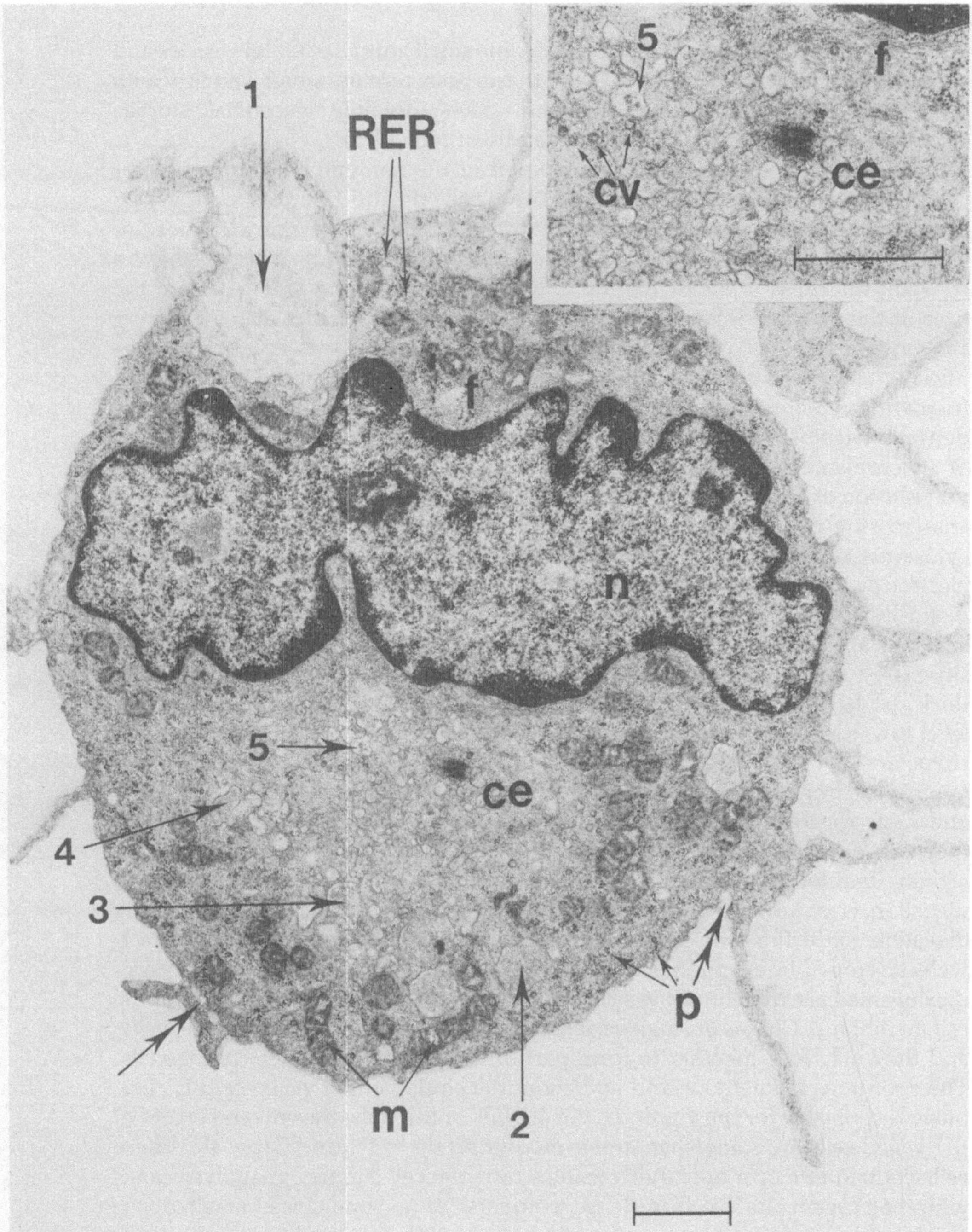

FIGURE 17. Veiled cell from normal rabbit lymph. Pinosomes are present in different stages of condensation and lysosomal digestion (arrows). Numerous vesicles surround the centriole, but no well-defined stacks of Golgi cisterns are present. There are smooth vesicles present at the bases of the veils (double arrows) (|—| = 1 μm). Inset: bundles of microfilaments near the nucleus and coated vesicles near the centriole (|—| = 1 μm). RER, Rough endoplasmic reticulum; f, filaments; m, mitochondrion; p, polyribosomes. For experimental details, see the note in the caption for Figure 1.

monocyte cultures and called by him "pinocytosis" (Lewis, 1931, 1937). It has also been observed in mouse peritoneal macrophages and extensively studied by Cohn (1966), Hirsch *et al.* (1968), and by Ehrenreich and Cohn (1968). It is a highly energy-dependent nonselective uptake of soluble material which can be induced *in vitro* by adding proteins to the medium (Cohn and Parks, 1967). It differs from "micropinocytosis," a process in which small drops of fluid are taken into invaginations of the cell membrane, forming tiny vesicles of about 100-nm diameter. Micropinocytosis may also be nonselective, and, in endothelial cells, it is nonenergy dependent, or it may be confined to membrane-bound proteins and is then low energy dependent.

The ultrastructure of veiled cells differs in several respects from that of peritoneal macrophages. Peritoneal cells have short ruffles, they contain small bundles of microfilaments, possess a well-developed Golgi apparatus, and variable numbers of lysosomes. They do not contain the central mass of small clear

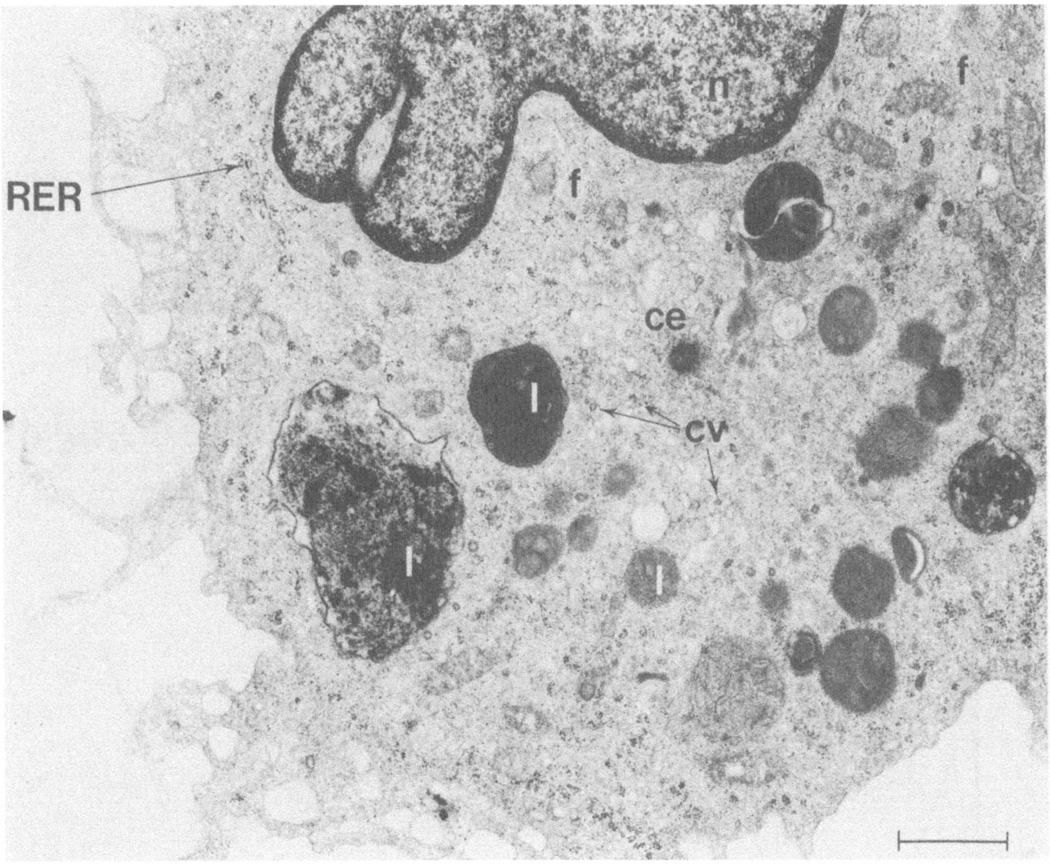

FIGURE 18. Veiled cell from normal rabbit lymph, containing many large phagolysosomes, which demonstrate phagocytosis of particulate material. Many small coated vesicles, presumably, primary lysosomes, are present in a central complex of smooth vesicles which surround the centrosome (|—| = 1 μm). RER, Rough endoplasmic reticulum; f, filaments; n, nucleus. For experimental details, see the note in the caption for Figure 1.

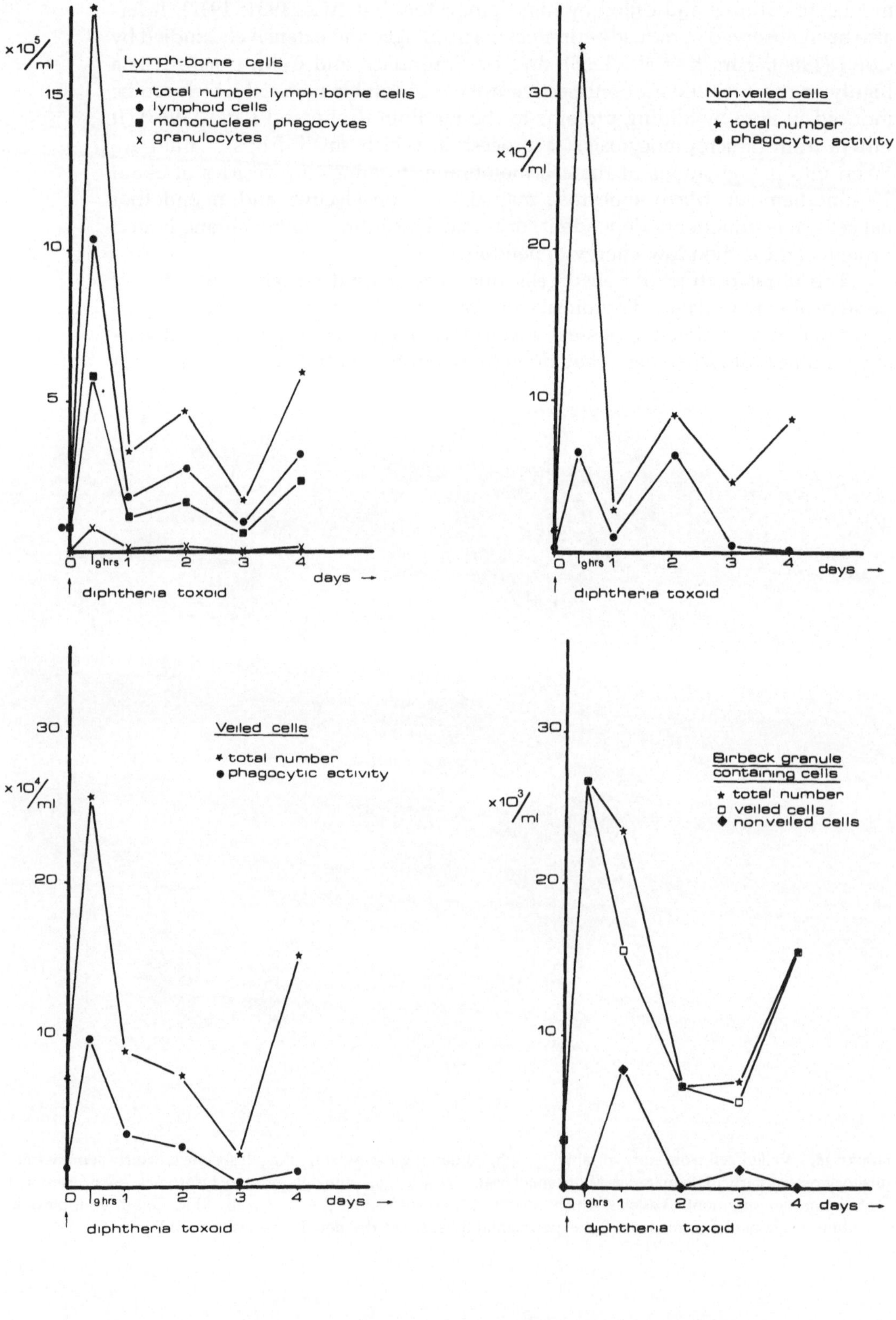
x10⁵/ml
Lymph-borne cells
total number lymph-borne cells
lymphoid cells
mononuclear phagocytes
granulocytes
15
10
5
0
9hrs
1
2
3
4
days
diphtheria toxoid
x10⁴/ml
Non veiled cells
total number
phagocytic activity
30
20
10
0
9hrs
1
2
3
4
days
diphtheria toxoid
x10⁴/ml
Veiled cells
total number
phagocytic activity
30
20
10
0
9hrs
1
2
3
4
days
diphtheria toxoid
x10³/ml
Birbeck granule containing cells
total number
veiled cells
nonveiled cells
30
20
10
0
9hrs
1
2
3
4
days
diphtheria toxoid

vesicles seen in veiled cells. The acid phosphatase activity of peritoneal cells often has a particulate distribution and the ATPase reaction in the plasma membrane is much weaker than it is in veiled cells. We propose that the term "macropinocyte" should be applied to veiled cells to indicate that they are especially adapted to engulf and digest large volumes of fluid, whereas other mononuclear phagocytes are adapted to endocytize and digest large amounts of particulate matter.

In order to study the effect of antigenic stimulation on lymph-borne macrophages rabbits were injected with diphtheria toxoid, 10 L.F. (flocculating units) in each hind footpad, and lymph coming from the site of injection was collected at intervals. The total number of cells per ml of lymph was determined, differential counts were carried out in the light microscope, and the cells were examined in the electron microscope.

The total number of cells per milliliter of lymph increases tenfold after 9 hr, and the number of mononuclear phagocytes is proportional to the total number of cells in the lymph (Figure 19a), but the numbers of veiled and nonveiled cells are divergent. The concentration of nonveiled cells increases 15-fold at 9 hr, mainly by the appearance of small monocyte-like cells, then decreases to the normal level at 24 hr, and there is a second six-fold increase at 48 hr (Figure 19b). At 9 hr, 20% of these cells contain endocytized particulate matter, and this proportion increases to a maximum of 70% at 48 hr. The veiled cells increase three-fold at 9 hr, by 24 hr, their number is again in the normal range (Figure 19c), and at this time an intermediate type of cell also appears in the lymph. These cells have short veils, folded back, and, apparently, adhering to the cell surface. The number of veiled cells continues to decrease until the third day when it is only ⅓ of the normal concentration. At 9 hr, the proportion of veiled cells engaged in active phagocytosis of particulate matter increased two-fold and remains at this level until the third day. There is also an increase in the number of veiled cells containing Birbeck granules (Figure 19d), and 24 hr after antigen administration these granules are also present in some nonveiled monocyte-like cells. However, the presence of granules seems to be unrelated to the phagocytic activity of veiled and nonveiled cells.

The presence of Birbeck granules in monocyte-like cells collected 24 hr after antigen administration suggests that there is a transition from monocytes to differentiated veiled cells. By contrast, the proportion of mononuclear phagocytes, engaged in active phagocytosis of particulate matter, is maximal on the second day. This means that the composition of the population of lymph-borne macrophages, which drain from the exudate in the antigen depot and arrive into the node, is not constant during the immune response.

FIGURE 19. Concentrations of nucleated cells in rabbit lymph drained away from the foodpad in which diphtheria toxoid was injected. (a) Differential counts of nucleated cells. (b) Nonveiled mononuclear phagocytes, including the cells which are phagocytizing particulate material. (c) Veiled mononuclear phagocytes, including the cells which are phagocytizing particulate material. (d) Cells containing Birbeck granules, subdivided into a population of veiled and a population of nonveiled cells. For experimental details, see text.

2.4. THE ANTIGEN-STIMULATED LYMPH NODE

The concept of the MPS was originally worked out for normal lymph nodes, but should also be applied to stimulated lymph nodes (Langevoort *et al.*, 1970), since mononuclear phagocytes may change their appearance after stimulation, and may also perform different immunological functions in the various compartments of the lymph node. There are still conflicting views as to the origin and function of macrophages including the IDC (Rausch *et al.*, 1977; Kamperdijk *et al.*, 1978).

Rats were stimulated with whole formol-killed paratyphoid microorganisms (0.1 ml of a suspension of 5×10^9 microorganisms/ml saline was injected into the foot-pad), in order to provoke a thymus-dependent, as well as a thymus-independent, humoral response. The draining lymph nodes were examined by electron microscopy after various time intervals ranging from 6 hr to 6 weeks. The secondary response was also studied in animals boosted 6 weeks after primary immunization.

In the primary response, specific antibody was first detected at 3 days and the maximum titer was found on the fourth day. The first lymphoblast appeared in the marginal zone and the paracortex 9 hr after antigen injection. After 2 days, the paracortex was enlarged and contained many lymphoblasts and mitotic lymphocytes. However, the proliferating cells did not seem to be grouped around macrophages or reticulum cells. After 3 days, the first plasmablasts appeared in the region of the corticomedullary junction, and, at 4 days, many plasma cells were present in the medullary cords. These morphological changes reflect the three phases of the humoral immune response.

In the induction phase, up to 24 hr after antigen injection, immunocompetent lymphocytes were stimulated, B cells in the marginal zone and T cells in the paracortex.

In the prolifereative phase, between the first and fourth day after stimulation, many immunoblasts and mitotic lymphoid cells were seen in the outer and inner cortex, and also many plasmablasts. It was assumed that T-cell–B-cell interaction also occurred in this phase.

In the memory phase from the fourth day onwards, the proliferation of lymphocytes in the marginal zone and paracortex was diminished, and the germinal center reaction became prominent. This reaction started with the appearance of TBM in association with a small number of lymphoblasts and one or two dendritic reticulum cells (DRC) in the region of a capillary venule at the base of a primary follicle. As the germinal center reaction developed, the reticulum cells in the primary follicle and marginal zone were incorporated within it and gradually transformed into the characteristic DRC. These cells contained a very large Golgi complex, which was probably responsible for the hypertrophy of the plasma membrane. The lymphocytes continued to proliferate at the base of the center, forcing the dendritic cells outwards until they formed a cap over the center.

During these phases of the immune response, the reticulum cells in the interfollicular area and paracortex did not show any striking morphological transformation. Transitional cells with the characteristics of reticulum cells and macrophages were not observed and there was no macrophage proliferation. The reticulum cells showed evidence of moderately active protein synthesis, as judged by the increase in RER and polyribosomes. The behavior of the macrophages was strikingly different. In each phase of the immune response there was a corresponding change in the morphology of the macrophages, and this occurred in the lymph node compartment most obviously involved in the reaction. The induction phase was characterized by penetration of newly arrived macrophages into the marginal zone and paracortex (Kamperdijk *et al.*, 1978). These macrophages were small cells with an indented or irregularly shaped euchromatic nucleus and a high nucleocytoplasmic ratio. The cytoplasm was moderately electron dense and contained some phagolysosomes and bundles of microfilaments. The cytocenter was surrounded by lysosomal vesicles and stacks of Golgi cisterns. In the paracortex these cells gradually transformed into IDC.

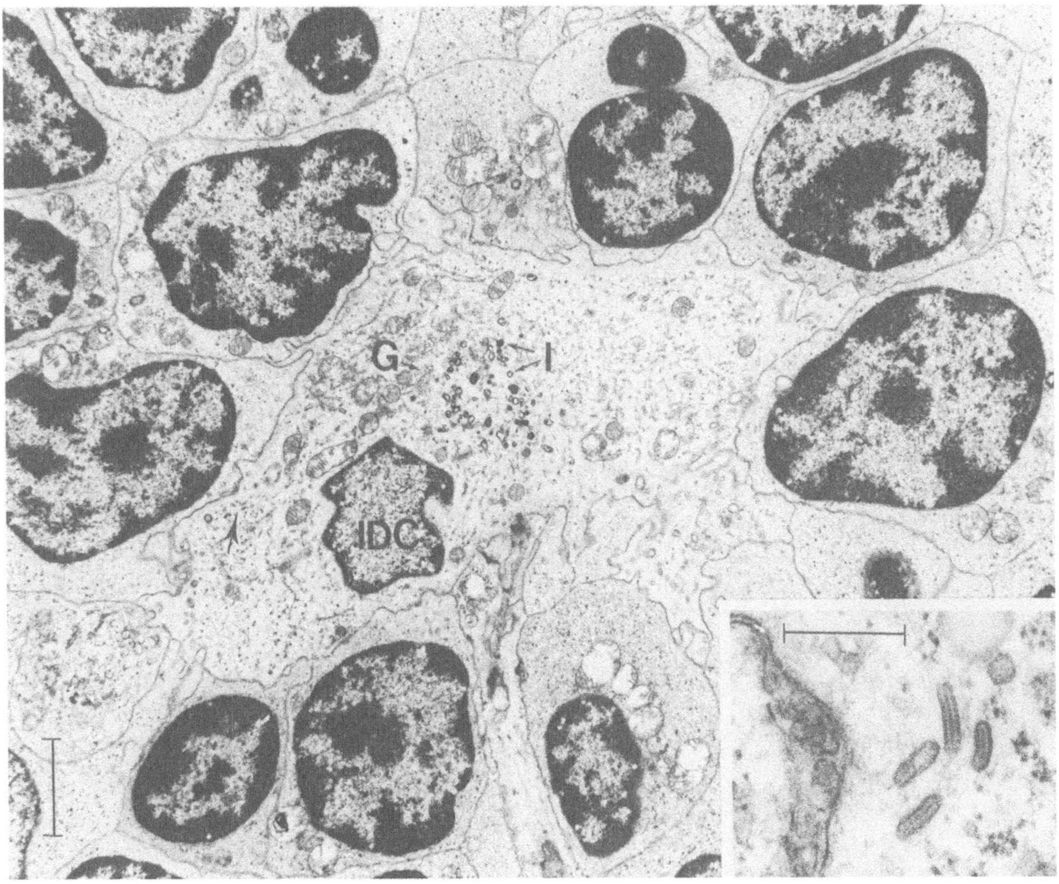

FIGURE 20. IDC in rat lymph node 1 day after stimulation with PTV. The cell contains many phagolysosomes in the central area and Birbeck granules (arrow) (|—| = 2 μm). Inset: detail of Birbeck granules (|—| = 400 μm). G, Golgi apparatus. For experimental details, see the note in the caption for Figure 1.

On the first day after antigenic stimulation, newly arrived macrophages, transitional forms, and mature IDC frequently contained Birbeck granules (Figure 20), whereas the large actively phagocytizing macrophages, which populated the marginal zone, never contained Birbeck granules. These organelles did not show any topographical relationship with the cell membrane or with other cell organelles. In the proliferative phase of the immune response, the number of mature IDC was decreased and the paracortex was mainly populated by transitional macrophages. No Birbeck granules could be found in any macrophage after the first 24 hr. In the memory phase of the immune response the proportion of newly arrived macrophages and transitional forms gradually decreased, and the paracortex was repopulated by mature IDC. Germinal center formation occurred at this stage; it was maximal on the 8th day after antigen administration when the centers contained the largest number of TBM.

In the secondary response the proliferative phase overlapped the induction phase, and the generation of memory cells started earlier. The transformation of

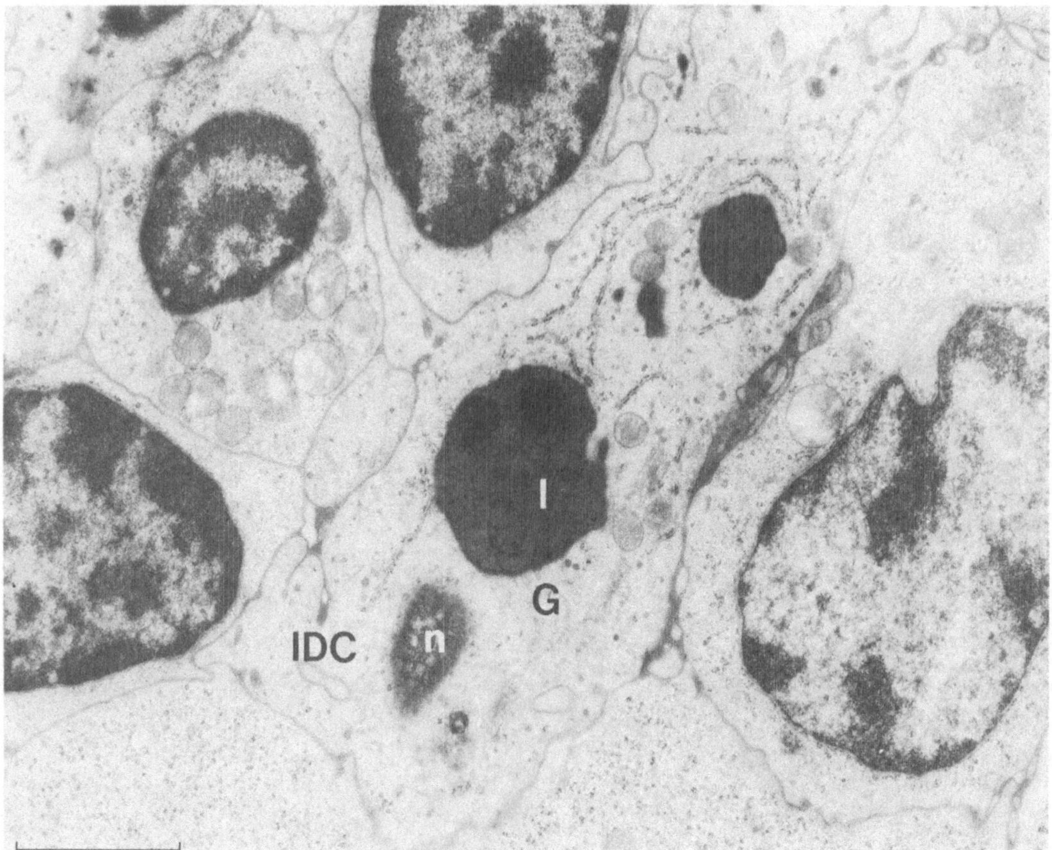

FIGURE 21. IDC with large phagolysosomes in rat lymph node 8 days after secondary immunization with PTV (|—| = 2 μm). G, Golgi apparatus; n, nucleus. For experimental details, see the note in the caption for Figure 1.

newly arrived macrophages into IDC was more rapid, and already visible one day after the booster injection. At three days, many well-developed IDC populated the paracortex. In the last phase of the response, the blast cell reaction in the paracortex gradually diminished and the number of IDC was much increased, and many of these cells were engaged in active phagocytosis (Figure 21). They frequently contained phagolysosomes with recognizable necrotic cellular contents, and later they formed clusters around damaged lymphocytes. These lymphocytes had electron-dense cytoplasm, they contained large amounts of glycogen, and their Golgi vesicles were dilated (Figure 22).

These observations illustrate the morphological changes which accompany the movement of lymph-borne macrophages through a stimulated lymph node. Changes in the cellular composition of rabbit lymph following antigenic stimulation, indicate that there is a shift in the population of lymph-borne macrophages

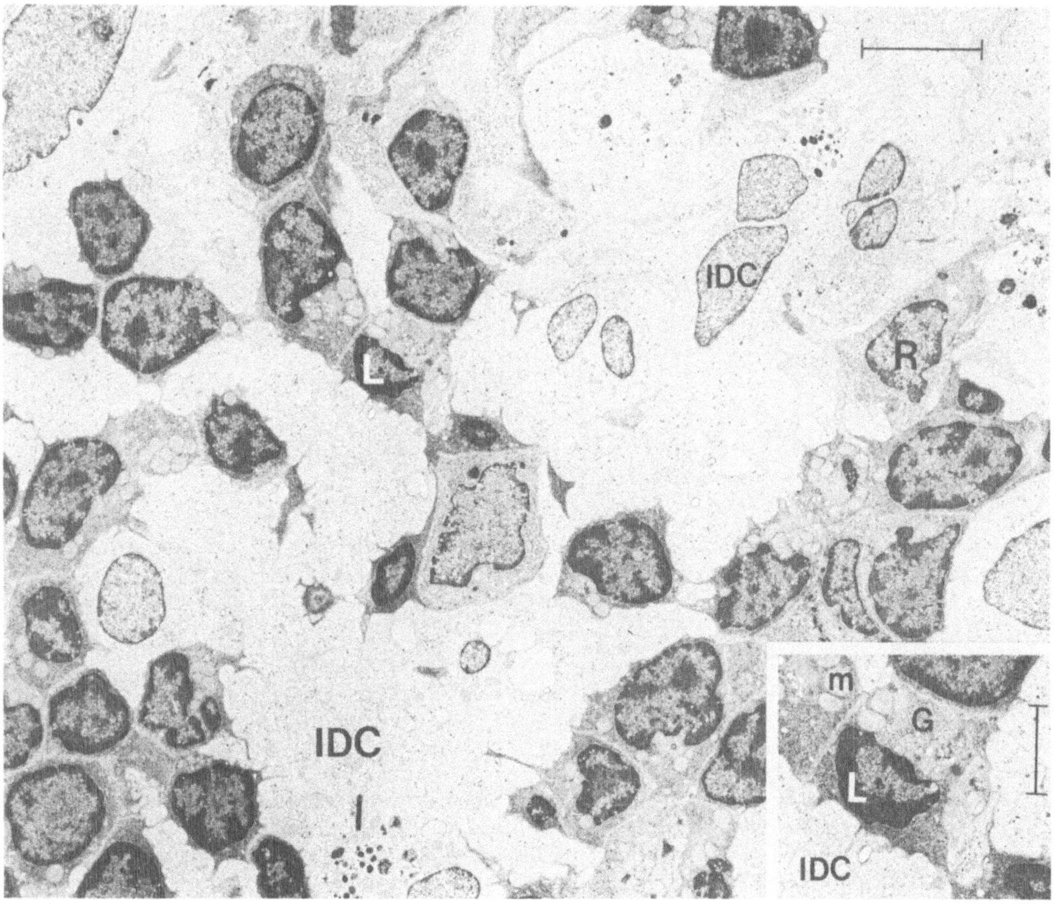

FIGURE 22. Paracortex of rat lymph node 3 weeks after the booster dose of PTV, clusters of IDC interdigitate with enclosed lymphocytes. The cytoplasm of the lymphocytes is filled with glycogen particles and the Golgi cisterns are dilated (Inset: |—| = 2 μm). Many IDC contain cellular debris in phagolysosomes (|—| = 5 μm). G, Golgi apparatus; m, mitochondrion; R, reticulum cell. For experimental details, see the note in the caption for Figure 1.

which could explain the sequence of events in stimulated lymph nodes (Figure 23). In the draining lymph node, cells containing Birbeck granules are found in the outer cortex 3 hr after an injection of diphtheria toxoid. During the next 24 hr, the paracortex becomes enlarged, mainly as a result of an increase in the number of cells with the characteristics of veiled cells. These findings are in agreement with the observations of Kelly *et al.* (1978). Thereafter, the number of lymph-borne veiled cells decreases but the content of nonveiled cells engaged in phagocytosis of particulate matter increases, reaching a maximum on the second day. The destination of these cells was not established, but it seems likely that the majority are held up in the outer cortex. Kelly *et al.* (1978) reported that on the third day after antigen injection, lymph-borne cells homed in the interfollicular cortex and failed to reach the paracortex.

We conclude that small macrophages with actively moving cells reach the paracortex during the first 24 hr after the antigenic stimulation, that is, during the induction phase of the immune response, at which time the first immuno-

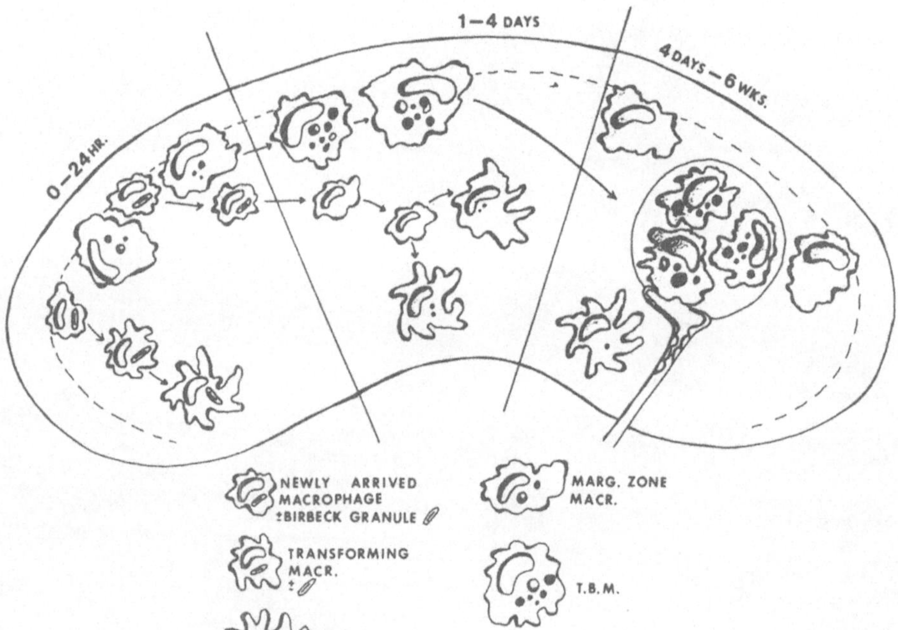

FIGURE 23. Schematic representation of the migratory pathways of newly arrived lymph-borne macrophages, following stimulation with antigens, which elicit a humoral immune response. 0–24 hr: In the induction phase of the immune response, small, newly arrived macrophages containing Birbeck granules penetrate the outer and inner cortex, and at the same time the first immunoblasts appear. 1–4 days: During the proliferation phase, the blast cell reaction increases. Increased numbers of tingible body macrophages are present in the outer cortex, and the macrophages which were newly arrived in the induction phase, transform into IDC. 4–8 days: During the induction of the memory phase, which was followed up to 6 weeks, the germinal center reaction is maximal and large numbers of TBM are present.

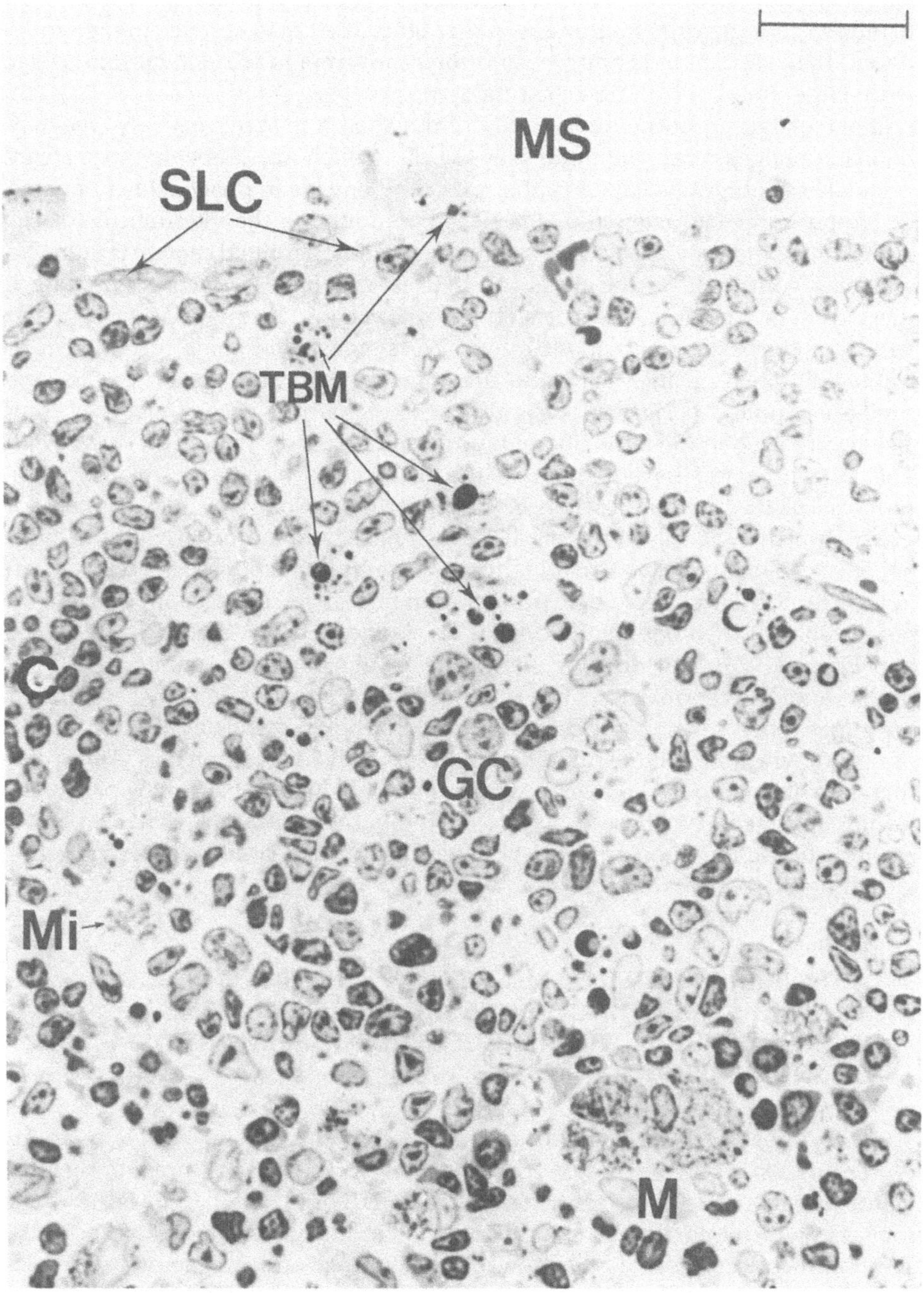

FIGURE 24. Germinal center in rabbit lymph node, 48 hr after injection of diphtheria toxoid in the footpad. TBM in the marginal sinus, marginal zone, and preexistent germinal center. Effete macrophages are located at the base of the germinal center (|—| = 40 μm). SLC, Sinus-lining cell; TBM, tingible body macrophage; C, corona or mantle layer; Mi, mitotic figure. For experimental details, see the note in the caption for Figure 1.

blasts appear in the outer and inner cortex. Veiled cells would seem to be well-equipped to trigger the immune response since they may contact antigen in the depot, they are rich in Ia antigen, and their movements bring them into contact with large numbers of recirculating lymphocytes. On the second day, the number of lymph-borne veiled cells diminishes but many actively phagocytic macrophages reach the node. These cells, which are probably more sticky, populate the outermost layers of the cortex and engage in phagocytosis of necrotic lymph-borne exudate cells. Many TBM are found in the interfollicular cortex at this time, and they are also present in preexisting germinal centers (Figure 24), and may contribute to the mass of effete macrophages which accumulate in these centers. The antigen associated with TBM could be involved in the reactivation of preexisting germinal centers. We suggest that some incoming macrophages transform into TBM and are transported to the region of the large capillary venules in primary follicles by the movement of tissue fluid; they could also be involved in the induction of new germinal centers (Hoefsmit *et al.*, 1980). The morphological development of the IDC seems also to be related to the progress of the immune response. Veiled cells predominate in the induction phase, transitional cells in the proliferative phase, and mature IDC in the memory phase. This indicates that in a humoral immune response, mature IDC are not concerned with lymphocyte proliferation or T–B-cell interaction, they are not responsible for the homing of T cells as was suggested by Van Ewijk *et al.* (1974) and Friess (1976), nor do they contribute to the microenvironment for T-cell proliferation as Veerman and Van Ewijk (1975) and Kaiserling (1977) concluded. We suggest that sensitized T cells produce lymphokines such as MIF (Sandok *et al.*, 1975), which inhibit the migration of newly arrived macrophages. These cells then transform into IDC. The increasing number of mature IDC in the paracortex during the memory phase is associated with the appearance of necrobiotic, electron-dense lymphocytes, which suggests that mature IDC are concerned in switching off T-cell proliferation.

3. A DYNAMIC MODEL OF MIGRATING LYMPH-BORNE MONONUCLEAR PHAGOCYTES AND RECIRCULATING LYMPHOCYTES IN THE RETICULOENDOTHELIAL STROMA OF THE LYMPH NODE

The three structural and functional components of the lymph node are represented in a scheme (Figure 25). This shows that the structure and function of the lymph node depends on a continuous supply of lymph-borne macrophages which migrate through the reticuloendothelial framework of the node. The framework consists of reticulum cells, including "antigen-retaining" dendritic cells (Nossal *et al.*, 1968) or DRC. These elements, together with the blood vessels, form the stroma of the lymph node. The macrophages probably induce the immune response in the various compartments of the node. Those which appear during the first 24 hr after antigen administration, may contain Birbeck granules and show macropinocytic activity rather than phagocytosis of particulate material. They are possible candidates for the triggering of B cells in the

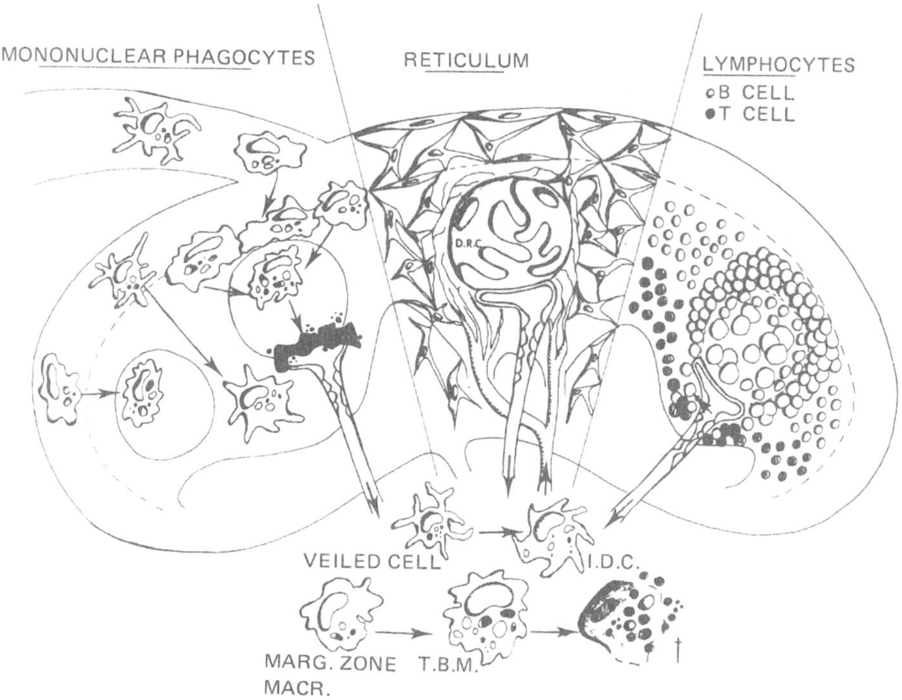

FIGURE 25. Schematic representation of the three structural and functional components of the lymph node. In the middle segment, the framework of the node is represented. It consists of reticulum cells, including the dendritic reticulum cells of the germinal centers and the walls of the small blood vessels. In the right hand segment, the recirculating B and T cells are shown. These cells populate the framework of the outer cortex and inner cortex, respectively. In the left hand segment, the lymph-borne mononuclear phagocytes are shown arriving in the lymph node. They may be actively moving veiled cells, which migrate into the paracortex, or actively phagocytic cells, which migrate through the outer cortex towards the capillary venules at the base of the germinal centers.

marginal zone and T cells in the paracortex before they home and transform into IDC. In the next phase of the response, actively phagocytic macrophages predominate in the lymph draining from the exudate and these macrophages probably populate the B-cell compartment. They enter the marginal zone, migrate through the outer cortex, and congregate around capillary venules in the deep layers of the interfollicular cortex, primary follicles, and mantle layer of secondary follicles; finally, they form a cluster of effete macrophages at the base of the germinal center. These cells may initiate the germinal center reaction.

This scheme represents a dynamic model of interaction between lymph-borne exudate macrophages and lymphocytes which enter the node via the high endothelial venules.

ACKNOWLEDGMENTS. The authors are greatly indebted to Prof. Dr. H. L. Langevoort for his stimulating discussions, to Mrs. I. C. Schadee-Eestermans, Mrs. D. M. Broekhuis-Fluitsma, and Mr. W. Benson for their skillful technical assistance, to Mr. C. den Daas for preparing the micrographs, and to Mrs. A. J. C. Steenvoorden-Bosma for typing the manuscript.

We also acknowledge the kind permission of Prentice-Hall, Inc., Englewood Cliffs, New Jersey, to reproduce Figure 2 from Weiss 1972, *The Cells and Tissues of the Immune System. Structure, Functions and Interactions;* Blackwell Scientific Publications, Ltd., Oxford, to reproduce Figure 12 from Hoefsmit (1975), Mononuclear phagocytes and reticulum cells in lymphoid tissues, in: *Mononuclear Phagocytes in Immunity, Infection and Pathology* (R. van Furth, ed.), Martinus Nijhoff Publishers B.V., The Hague, to reproduce Figure 23 and 25 from Hoefsmit *et al.* (1980), Lymph node macrophages and reticulum cells in the immune response, in: *Mononuclear Phagocytes, Functional Aspects* (R. van Furth, ed.).

REFERENCES

Askonas, B. A., and Roelants, G. E., 1974, Macrophages bearing haptencarrier molecules as foci inducers for T and B lymphocyte interaction, *Eur. J. Immunol.* **4**:1.

Axline, S. G., and Cohn, Z. A., 1970, *In vitro* induction of lysosomal enzymes by phagocytosis, *J. Exp. Med.* **131B**:1239.

Beelen, R. H. J., Broekhuis-Fluitsma, D. M., Korn, C., and Hoefsmit, E. C. M., 1978a, Identification of exudate-resident macrophages on the basis of peroxidatic activity, *J. Reticuloendothel. Soc.* **23**:103.

Beelen, R. H. J., Veer, M. van't, Fluitsma, D. M., and Hoefsmit, E. C. M., 1978b, Identification of different peroxidatic activity patterns in human macrophages *in vivo* and *in vitro*, *J. Reticuloendothel. Soc.* **24**:355.

Beelen, R. H. J., Broekhuis-Fluitsma, D. M., and Hoefsmit, E. C. M., 1978c, Peroxidatic activity in mononuclear phagocytes, *Ultramicroscopy* **3**:129.

Beelen, R. H. J., Fluitsma, D. M., J. W. M. Van der Meer, and Hoefsmit, E. C. M., 1979, Development of the different peroxidatic activity patterns in peritoneal macrophages *in vivo* and *in vitro*, *J. Reticuloendothel. Soc.* **25**:573.

Beelen, R. H. J., Fluitsma, C. M., and Hoefsmit, E. C. M., 1980, Development of exudate-resident macrophages on the basis of patterns of peroxidatic activity *in vivo* and *in vitro*, in: *Mononuclear Phagocytes, Functional Aspects* (R. van Furth, ed.), Martinus Nijhoff, The Hague (in press).

Bianco, S., Griffin, F. M., and Silverstein, S. M., 1975, Studies on the macrophage complement receptor. Alterations of receptor function upon macrophage activation, *J. Exp. Med.* **141**:1278.

Birbeck, M. S., Breathnach, S. A., and Everall, J. D., 1961, An electron microscopic study of basal melanocyte and high level clear cells (Langerhans cells) in vitiligo, *J. Invest. Dermatol.* **37**:51.

Bodel, P. T., Nichols, B. A., and Bainton, D. F., 1978, Differences in peroxidase localization of rabbit peritoneal macrophages after surface adherence, *Am. J. Pathol.* **91**:107.

Burke, J. S., and Simon, G. T., 1970a, Electron microscopy of the spleen. I. Anatomy and microcirculation, *Am. J. Pathol.* **58**:127.

Burke, J. S., and Simon, G. T., 1970b, Electron microscopy of the spleen. II. Phagocytosis of colloid carbon, *Am. J. Pathol.* **58**:157.

Carr, I., 1968, Lysosome formation and surface changes in stimulated peritoneal cells, *Z. Zellforsch. Mikrosk. Anat.* **89**:328.

Carr, I., 1970, The fine structure of the mammalian lymphoreticular system, *Int. Rev. Cytol.* **27**:283.

Carr, I., 1973, *The Macrophage, a Review of Ultrastructure and Function*, Academic Press, London.

Cohen, S., Vassali, P., Benacerraf, B., and McCluskey, R. T., 1966, The distribution of antigenic and non-antigenic compounds within the draining lymph node, *Lab. Invest.* **15**:1143.

Cohn, Z. A., 1966, The regulation of pinocytosis in mouse macrophages. I. Metabolic requirements as defined by the use of inhibitors, *J. Exp. Med.* **124**:557.

Cohn, Z. A., 1970, Lysosomes in mononuclear phagocytes, in: *Mononuclear Phagocytes* (R. van Furth ed.), pp. 50–58, Blackwell, Oxford.
Cohn, Z. A., and Benson, B., 1965, The differentiation of mononuclear phagocytes, Morphology, cytochemistry, biochemistry, *J. Exp. Med.* **121**:153.
Cohn, Z. A., and Fedorko, M. E., 1969, The formation and fate of lysosomes, in: *Lysosomes in Biology and Pathology* (J. T. Dingle and H. B. Fell., eds.), Vol. 1, pp. 43–62, North Holland, Amsterdam.
Cohn, Z. A., and Parks, E., 1967, The regulation of pinocytosis in mouse macrophages. II. Factors inducing vesicle formation, *J. Exp. Med.* **125**:213.
Cohn, Z. A., and Wiener, E., 1963, The particulate hydrolases of macrophages, I. Comparative enzymology, isolation and properties, *J. Exp. Med.* **118**:991.
Cohn, Z. A., Hirsch, J. G. and Fedorko, M. E., 1966, *In vitro* differentiation of mononuclear phagocytes, IV. The ultrastructure of macrophage differentiation in the peritoneal cavity and in culture, *J. Exp. Med.* **123B**:747.
Daems, W. T., and Brederoo, P., 1971, The fine structure and peroxidatic activity of resident and exudate peritoneal macrophages in the guinea pig, in: *The Reticuloendothelial System and Immune Phenomena* (N. R. Diluzio and K. Flemming eds.), pp. 19–31, Plenum Press, New York.
Daems, W. T., and Brederoo, P., 1973, Electron microscopical studies on the structure, phagocytic properties and peroxidatic activity of resident and exudate macrophages in the guinea pig, *Z. Zellforsch. Mikrosk. Anat.* **114**:247.
Daems, W. T., Wisse, E., and Brederoo, P., 1969, Electron microscopy of the vacuolar apparatus, in: *Lysosomes in Biology and Pathology* (J. T. Dingle and H. B. Fell, eds.), Vol. 1, pp. 64–112, North Holland, Amsterdam.
Daems, W. T., Poelman, R. E., and Brederoo, P., 1973, Peroxidatic activity in resident macrophages and exudative monocytes of the guinea pig after ingestion of latex particles, *J. Histochem. Cytochem.* **21**:93.
Daems, W. T., Koerten, H. K., and Soranzo, M. R., 1976, Differences between monocyte-derived and tissue macrophages, in: *The Reticuloendothelial System in Health and Disease: Functions and Characteristics* (S. M. Reichard, M. R. Escobar, and H. Friedman, eds.), pp. 27–40, Plenum Press, New York.
De Sousa, M. A. B., Parrott, D. M. V., and Pantelouris, E. M., 1969, The lymphoid tissues in mice with congenital aplasia of the thymus, *Clin. Exp. Immunol.* **4**:637.
Drexhage, H. A., Mullink, H., Groot, J. de, Clarke, J., and Balfour, B. M., 1979, Large mononuclear veiled cells, resembling Langerhans cells, present in lymph draining from the skin, *Cell Tiss. Res.* **202**:407.
Drexhage, H. A., Lens, J. W., Cvetanov, J., Kamperdijk, E. W. A., Mullink, R., and Balfour, B. M., 1980, Structure and functional behavior of veiled cells resembling Langerhans cells, present in lymph draining from normal skin after application of the contact sensitizing agent dinitrofluorobenzene, in: *Mononuclear Phagocytes, Functional Aspects* (R. van Furth, ed.), Martinus Nijhoff, The Hague (in press).
Drinker, C. K., Field, M., and Ward, H. K., 1934, The filtering capacity of lymph nodes, *J. Exp. Med.* **59**:393.
Ehrenreich, B. A., and Cohn, Z. A., 1968, Fate of haemoglobin pinocytized by macrophages *in vitro*. *J. Cell Biol.* **38**:244.
Eikelenboom, P., 1978, Dendritic cells in the rat spleen follicles. A combined immuno- and enzyme histochemical study, *Cell Tiss. Res.* **190**:79.
Elsbach, P., 1977, Cell surface changes in phagocytosis,in: *Cell Surface Reviews*, Vol. 4, *The Synthesis, Assembly and Turnover of Cell Surface Components* (G. Poste and G. L. Nicolson, eds.), pp. 361–402, North Holland, Amsterdam.
Engeset, A., and Nesheim, A., 1966, Sinus lymphocytosis and lymph flow, *Acta Pathol. Microbiol. Scand.* **68**:181.
Erb, P., and Feldmann, M., 1975a, The role of macrophages in the generation of T-helper cells. I. The requirement for macrophages in helper cell induction and characteristics of the macrophage–T cell interaction, *Cell. Immunol.* **19**:356.
Erb, P., and Feldmann, M., 1975b, The role of macrophages in the generation of T-helper cells. II. The genetic control of the macrophage–T cell interaction for helper cell induction with soluble antigens, *J. Exp. Med.* **142**:460.

Feldmann, M., 1972, Cell interactions in the immune response in vitro. II. The requirement for macrophages in lymphoid cell collaboration, *J. Exp. Med.* **135**:1049.

Flemming, W., 1885, Studien über Regeneration der Gewebe, *Arch. Mikrosk. Anat.* **24**:50.

Friess, A., 1976, Interdigitating reticulum cells in the popliteal lymph node of the rat. An ultrastructural and cytochemical study, *Cell Tiss, Res.* **170**:43.

Gajl-Peczalska, K. J., Fish, A. J., Meuwissen, H. J., Frommel, D., and Good, R. A., 1969, Localization of immunological complexes fixing β_{1c} (C_3) in germinal centers of lymph nodes, *J. Exp. Med.* **130**:1367.

Gordon, S., Todd, J., and Cohn, Z. A., 1974, *In vitro* synthesis and secretion of lysozyme by mononuclear phagocytes, *J. Exp. Med.* **139**:1228.

Gowans, J. L., and Knight, E. S., 1964, The route of recirculation of lymphocytes in the rat, *Proc. Roy. Soc. B* **159**:257.

Graham, R. C., and Karnovsky, M. J., 1966, The early stages of absorption of injected horse-radish peroxidase in the proximal tubules of mouse kidney: Ultrastructural cytochemistry by a new technique, *J. Histochem. Cytochem.* **14**:291.

Greineder, D. K., and Rosenthal, A. S., 1975, Macrophage activation of allogeneic lymphocyte proliferation in the guinea pig mixed leucocyte culture, *J. Immunol.* **114**:1541.

Hall, J. G., and Morris, B., 1962, The output of cells in the lymph from the popliteal node of sheep, *Quart. J. Exp. Physiol.* **47**:360.

Han, S. S., Han, J. H., and Johnson, A. G., 1969, The immunologic specificity of ferritin uptake by macrophages in the rat, *Adv. Exp. Med. Biol.* **5**:167.

Hanna, M. G., Jr., and Hunter, R. L., 1971, Localization of antigen and immune complexes in lymphatic tissue, with special reference to germinal centres, in: *Morphological and Functional Aspects of Immunity*, (K. Lindahl-Kiessling, G. Alm, and M. G. Hanna, Jr., eds.), pp. 257–279, Plenum Press, New York.

Hanna, M. G., and Szakal, A. K., 1968, Localization of ^{125}I-labeled antigen in germinal centers of mouse spleen. Histologic and ultrastructural autoradiographic studies of the secondary immune reaction, *J. Immunol.* **101**:949.

Hendriks, H. R., 1978, Occlusion of the lymph flow to rat popliteal lymph nodes for protracted periods, *Z. Versuchstierk.* **20**:105.

Henson, P. M., 1971, The immunologic release of constituents from neutrophil leucocytes. II. Mechanisms of release during phagocytosis and adherence to non-phagocytosable surfaces, *J. Immunol.* **107**:1547.

Herd, Z. L., and Ada, G. L., 1969, Distribution of ^{125}I-immunoglobulins, IgG subunits and antigen–antibody complexes in rat lymph nodes, *Aust. J. Exp. Biol. Med. Sci.* **47**:73.

Heusermann, U., Stutte, H. J., and Müller Hermelink, H. K., 1974, Interdigitating cells in the white pulp of human spleen, *Cell Tiss. Res.* **153**:415.

Hirsch, J. G., and Fedorko, M. E., 1968, Ultrastructural characteristics of human leucocytes after simultaneous fixation with glutaraldehyde and osmium tetroxide and postfixation in uranyl acetate, *J. Cell Biol* **38**:615.

Hirsch, J. G., Fedorko, M. E., and Cohn, Z. A., 1968, Vesicle fusion and formation at the surface of pinocytic vacuoles in macrophages, *J. Cell Biol.* **38**:629.

Hoefsmit, E. C. M., 1973, The cytology of phagocytizing reticular cells as a starting point towards an experimental model: the subcutaneously sutured lymph node, *Acta Morphol. Med. Scand.* **11**:170.

Hoefsmit, E. C. M., 1975, Mononuclear phagocytes, reticulum cells and dendritic cells in lymphoid tissues, in: *Mononuclear Phagocytes in Immunity, Infection and Pathology* (R. van Furth, ed.), pp. 129–146, Blackwell, Oxford.

Hoefsmit, E. C. M., and Smelt, A. H. M., 1978, Milky spots and peritoneal exudate and resident macrophages in a state of cell mediated immunity, *Ultramicroscopy* **3**:135.

Hoefsmit, E. C. M., Balfour, B. M., Kamperdijk, E. W. A., and Cvetanov, J., 1978, Cells containing Birbeck granules in the lymph and the lymph node, *Z. Immun. Forsch.* **154**:321.

Hoefsmit, E. C. M., Balfour, B. M., Kamperdijk, E. W. A., and Cvetanov, J., 1979, Cells containing Birbeck granules in the lymph and the lymph node, in: *Lymphatic Tissues and Germinal Centres in Immune Reaction* (W. Mueller-Ruchholz and H. K. Müller-Hermelink, eds.), Plenum Press, New York.

Hoefsmit, E. C. M., Kamperdijk, E. W. A., and Balfour, B. M., 1980, Lymph node macrophages and reticulum cells in the immune response, in: *Mononuclear Phagocytes, Functional Aspects* (R. van Furth, ed.), Martinus Nijhoff, The Hague (in press).
Holtzman, E., 1976, Lysosomes: A Survey, Springer Verlag, Wien, New York.
Howard, J. G., Hunt, S. V., and Gowans, J. L., 1972, Identification of marrow-derived and thymus-derived small lymphocytes in the lymphoid tissue and thoracic duct lymph of normal rats, *J. Exp. Med.* **135**:200.
Huber, H., and Holm, G., 1975, Surface receptors of mononuclear phagocytes: Effect of immunocomplexes on *in vitro* function in human monocytes, in: *Mononuclear Phagocytes in Immunity, Infection and Pathology* (R. van Furth, ed.), pp. 291–301, Blackwell, Oxford, Edinburgh.
Humphrey, J. H., 1976, The still unsolved germinal centre mystery, in: *Immune Reactivity of Lymphocyte Development—Expression and Control* (M. Feldmann and A. Globerson, eds.), pp. 711–723, Plenum Press, New York.
Humphrey, J. H., 1978, Marginal zone and marginal sinus macrophages—a distinct population, *Z. Immun. Forsch.* **154**:297.
Humphrey, J. H., and Frank, M. M., 1967, The localization of non-microbial antigens in the draining lymph nodes of tolerant, normal and primed rabbits, *Immunology* **13**:87.
Jamieson, D., and Palade, G. E., 1971, Condensing vacuole conversion and zymogen granule discharge in pancreatic exocrine cells: Metabolic studies, *J. Cell Biol.* **48**:503.
Kaiserling, E., 1977, *Non-Hodgkin-Lymphomas,* Fischer Verlag, Stuttgart.
Kaiserling, E., and Lennert, K., 1974, Die interdigitierende Reticulumzellen in menschlichen Lymphknoten, *Virchows. Arch. B* **16**:51.
Kaiserling, E., Stein, H., and Müller-Hermelink, H. K., 1974, Interdigitating reticulum cells in the human thymus, *Cell Tiss. Res.* **155**:47.
Kamperdijk, E. W. A., Raaymakers, E. M., Leeuw, J. H. S. de, and Hoefsmit, E. C. M., 1978, Lymph node macrophages and reticulum cells in the immune response. I. The primary response to paratyphoid vaccine, *Cell Tiss. Res.* **192**:1.
Kelly, R. H., 1975, Functional anatomy of lymph nodes, *Int. Arch. Allergy Appl. Immunol.* **48**:836.
Kelly, R. H., Balfour, B. M., Armstrong, J. A., and Griffiths, S., 1978, Functional anatomy of lymph nodes. II. Peripheral lymph-borne mononuclear cells, *Anat. Rec.* **190**:5.
Keuning, F. J., 1972, Dynamics of immunoglobulin forming cells and their precursors, in: *Immunoglobulins* (R. E. Ballieux, M. Gruber, and H. G. Seijen, eds.), pp. 1–14, North-Holland, Amsterdam.
Kotani, M., Okada, K., Fujii, M., Tsuchiya, H., Ekino, S., and Fukuda, S., 1977, Lymph macrophages enter the germinal centre of lymph nodes of guinea pigs, *Acta Anat.* **99**:391.
Langevoort, H. L., Cohn, Z. A., Hirsch, J. G., Humphrey, J. H., Spector, W. G., and van Furth, R., 1970, The nomenclature of mononuclear phagocyte cells: a proposal for a new classification, in: *Mononuclear Phagocytes* (R. van Furth, ed.), pp. 1–6, Blackwell, Oxford.
Lauwerijns, J. M., and Boussauw, L., 1973, Striated filamentous bundles associated with centrioles in pulmonary lymphatic endothelial cells, *J. Ultrastruct. Res.* **42**:25.
Lewis, W. H., 1931, Pinocytosis, *Bull. Johns Hopkins Hosp.* **49**:17.
Lewis, W. H., 1937, Intake of whole drops of fluid by "ruffle cellular pseudopodia" in tissue culture, *Am. J. Cancer* **29**:666.
Marshall, A. H. E., 1956, *An Outline of the Cytology and Pathology of the Reticular Tissue,* Oliver and Boyd, Edinburgh.
Morré, D. J., 1977, The Golgi apparatus and membrane biogenesis, in: *The Synthesis, Assembly and Turnover of Cell Surface Components* (G. Poste and G. L. Nicolson, eds.), pp. 1–83, Elsevier-North-Holland, Amsterdam.
Morris, B., 1968, Migration intratissulaire des lymphocytes du mouton, *Nouv. Rév. Fr. Hématol.* **8**:525.
Morris, B., Moreno, G. D., and Bessis, M., 1968, Ultrastructure des cellules de la lymphe afférente aux ganglions périphériques avant et après stimulation antigénique, *Nouv. Rev. Fr. Hématol.* **8**:145.
Mosier, D. E., 1976, The role of macrophages in the specific determination of immunogenicity and tolerogenicity, in: *Immunobiology of the Macrophage* (D. S. Nelson, ed.), pp. 35–90, Academic Press, New York.

Müller-Hermelink, H. K., and Lennert, K., 1978, The cytologic, histologic and functional bases for a modern classification of lymphomas, in: *Handbuch der speziellen pathologischen Anatomie und Histologie* (E. Ueheniger, ed.), Vol. I/IIIB, pp. 1–71, Springer, Berlin.

Nelson, D. S., 1976, Nonspecific immunoregulation by macrophages and their products, in: *Immunobiology of the Macrophage* (D. S. Nelson, ed.), pp. 235–257, Academic Press, New York.

Nichols, B. A., and Bainton, D. F., 1975, Ultrastructure and cytochemistry of mononuclear phagocytes, in: *Mononuclear Phagocytes in Immunity, Infection and Pathology* (R. van Furth, ed.), pp. 235–257, Blackwell, Oxford.

Nieuwenhuis, P., and Ford, W. L., 1976, Comparative migration of B- and T-lymphocytes in the rat spleen and lymph nodes, *Cell. Immunol.* **23**:254.

Nieuwenhuis, P., and Keuning, F. J., 1974, Germinal centres and the origin of the B-cell system. II. Germinal centres in the rabbit spleen and popliteal lymph nodes, *Immunology* **26**:509.

Nossal, G. J. V., Abbot, A., Mitchell, J., and Lummus, Z., 1968, Antigens in immunity. Ultrastructural features of antigen capture in primary and secondary lymphoid follicles, *J. Exp. Med.* **127**:277.

Odartchenko, N., Lewerenz, M., Sordat, B., Roos, B., and Cottier, H., 1967, Kinetics of cellular death in germinal centres of mouse spleen, in: *Germinal Centers in Immune Responses* (H. Cottier, N. Odartchenko, R. Schindler, and C. C. Congdon, eds.), pp. 212–217, Springer Berlin.

Oláh, I., Röhlich, P., and Törö, I., 1975, *Ultrastructure of Lymphoid Organs. An Electron-Microscopic Atlas,* Masson, Paris.

Osogoe, B., 1969, Changes in the cellular architecture of a lymph node after blocking its lymphatic circulation, *J. Anat.* **104**:495.

Papamichail, M., Gutierrez, C., Embling, P., Johnson, P., Holborow, E. J., and Pepys, M. B., 1975, Complement dependence of localization of aggregated IgG in germinal centres, *Scand. J. Immunol.* **4**:343.

Parrott, D. M. V., De Sousa, M. A. B., and East, J., 1966, Thymus dependent areas in the lymphoid organs of neonatally thymectomised mice, *J. Exp. Med.* **123**:191.

Pictet, R., Orci, G., Forssmann, W. G., and Girardier, L., 1969, An electron microscope study of the perfusion-fixed spleen. I. The splenic circulation and the RES concept, *Z. Zellforsch. Mikrosk. Anat.* **96**:372.

Rambourg, A., and Leblond, C. P., 1967, Electron microscope observations on the carbohydrate rich cell coat present at the surface of cells in the rat, *J. Cell Biol.* **32**:27.

Rambourg, A., Hernandez, W., and Leblond, C. P., 1969, Detection of complex carbohydrates in the Golgi apparatus of rat cells, *J. Cell Biol.* **40**:395.

Rausch, E., Kaiserling, E., and Boos, M., 1977, Langerhans cells and interdigitating reticulum cells in the thymus-dependent region in human dermatopathic lymphadenitis, *Virchows Arch. B* **25**:327.

Rosenstreich, D. L., Farrer, J. J., and Dougherty, S., 1976, Absolute macrophage dependency of T-lymphocyte activations by mitogens, *J. Immunol.* **116**:131.

Ross, R., 1975, Connective tissue cells, cell proliferation and synthesis of extracellular matrix—a review, *Philos. Trans. R. Soc. London Ser. B* **271**:247.

Sandok, P. L., Hinsdill, R. D., and Albrecht, R. M., 1975, Alterations in mouse macrophage migration: A function of assay systems, lymphocyte activation product preparation and fractionation, *Infect. Immun.* **11**:1100.

Shevach, E. M., 1976, The role of the macrophage in the genetic control of the immune response, *Fed. Proc.* **35**:2048.

Shortman, K., and Palmer, J., 1971, The requirements for macrophages in the *in vitro* immune response, *Cell Immunol.* **2**:399.

Silberberg, I., Baer, R. L., and Rosenthal, S. A., 1974, Circulating Langerhans cells in a dermal vessel, *Acta Derm. Venereol.* **45**:81.

Silberberg-Sinakin, I., Baer, R. L., and Thorbecke, G. J., 1978, Langerhans cells. A review of their nature with emphasis on their immunological functions, *Prog. Allergy* **24**:268.

Smith, J. B., McIntosch, G. H., and Morris, B., 1970, The traffic of cells through tissues: A study of peripheral lymph in sheep, *J. Anat. (London)* **107**:87.

Smythyman, A. M., Carr, K., Forman, D., and White, R. G., 1978, Separation of germinal centres from chicken spleen, *Z. Immun. Forsch.* **154**:366.

Sordat, B., Sordat, M., Hess, M., Stoner, R. D., and Cottier, H., 1970, Specific antibody within lymphoid germinal centre cells of mice after primary immunization with horse radish peroxidase: A light and electron microscopic study, *J. Exp. Med.* **131**:77.

Steinman, R. M., and Cohn, Z. A., 1972, The interaction of soluble horse radish peroxidase with mouse peritoneal macrophages *in vitro, J. Cell Biol.* **55**:186.

Stingl, G., Wolff-Schreiber, E. C. H., Pickler, W. J., Gachnait, F., Knapp, W., and Wolff, K., 1977, Epidermal Langerhans cells bear F_c and C_3 receptors, *Nature (London)* **268**:245.

Stingl, G., Katz, S. J., Shevach, E. M., Rosenthal, A. S., and Green, I., 1978, Analogous functions of macrophages and Langerhans cells in the initiation of the immune response, *J. Invest. Dermatol.* **71**:59.

Stuarat, A., 1975, Perspectives on the reticulum cells and fibre networks, in: *Mononuclear Phagocytes in Immunity, Infection and Pathology* (R. van Furth, ed.), pp. 111–118, Blackwell, Oxford.

Swartzendruber, D. C., and Congdon, C. C., 1963, Electron microscope observations on tingible body macrophages in mouse spleen, *J. Cell Biol.* **19**:641.

Szakal, A. K., and Hanna, M. G., Jr., 1968, Electron microscopical studies of antigen localization and virus-like particles in mouse spleen germinal centres, *Exp. Haematol.* **15**:23.

Thomas, D. W., 1977, The role of the macrophage as the stimulator cell in contact sensitivity, *J. Immunol.* **118**:1677.

Thorbecke, G. J., Jacobson, E. B., and Asofsky, R., 1964, γ-Globulin and antibody formation *in vitro.* IV. The effect on the secondary response of x irradiation given at varying intervals after a primary injection of bovine-γ-globulin, *J. Immunol.* **92**:734.

Turk, J. L. 1967, Cytology of the induction of hypersensitivity, *Br. Med. Bull.* **23**:3.

Unanue, E. R., 1972, The regulatory role of macrophages in antigenic stimulation, *Adv. Immunol.* **15**:95.

Unanue, E. R., and Askonas, B. A., 1968, The immune response of mice to antigenic stimulation, *Immunology* **15**:95.

van Ewijk, W., Verzijden, J. H. M., Kwast, Th. M. van der, and Luycx-Meyer, S. W. M., 1974, Reconstitution of the thymus dependent area in the spleen of lethally irradiated mice. A light and electron microscopic study of the T-cell micro-environment, *Cell Tiss. Res.* **149**:43.

van Furth, R., Cohn, Z. A., Hirsch, J. G., Spector, W. G., and Langevoort, H. L., 1972, The mononuclear phagocyte system: A new classification of macrophages, monocytes and their precursor cells, *Bull. W. H. O.* **46**:845.

van Furth, R., Langevoort, H. L., and Schaberg, A., 1975, Mononuclear phagocytes in human pathology—proposal for an approach to improved classification, in: *Mononuclear Phagocytes in Immunity, Infection and Pathology* (R. van Furth, ed.), pp. 1–15, Blackwell, Oxford.

van Rooijen, N., 1973, Non-specific nature of antigen trapping in lymphoid follicles, *Immunology* **25**:853.

Veerman, A. J. P., 1974, On the interdigitating cells in the thymus-dependent area of the rat spleen: A relation between the mononuclear phagocyte system and T-lymphocytes, *Cell Tiss. Res.* **148**:247.

Veerman, A. J. P., van Ewijk, W., 1975, White pulp compartments in the spleen of rats and mice. A light and electron microscopic study of lymphoid and non-lymphoid cell types in T- and B areas, *Cell Tiss. Res.* **156**:427.

Veldman, J. E., 1970, Histophysiology and Electron Microscopy of the Immune Response, Thesis, Groningen.

Veldman, J. E., Keuning, E. J., Molenaar, J., 1978, Site of initiation of the plasma cell reaction in the rabbit lymph node. Ultrastructural evidence for two distinct antibody forming cell precursors, *Virchows Arch. B* **28**:187.

Wagner, H., Feldmann, M., Boyle, W., and Schrader, J. W., 1972, Cell mediated immune response *in vitro:* The requirements for macrophages in cytotoxic reactions against cell bound and subcellular alloantigens, *J. Exp. Med.* **136**:331.

Wakefield, J. D., and Thorbecke, G. J., 1968, Relationship of germinal centres in lymphoid tissue to immunological memory, *J. Exp. Med.* **128**:153.

Waldron, J. A., Hain, R. G., and Rosenthal, A. S., 1973, Antigen induced of Guinea pig lymphocytes *in vitro:* Obligatory role of macrophages in the recognition of antigen by immune T-lymphocytes, *J. Immunol.* **111**:58.

Weiss, L., 1972, *The Cells and Tissues of the Immune System. Structure, Functions, Interactions,* Prentice-Hall, Englewood Cliffs, N. J.

White, R. G., 1975, Immunological functions of lymphoreticular tissues, in: *Clinical Aspects of Immunology* (P. G. H. Gell, R. R. A. Coombs, and P. J. Lachmann, eds.), 3rd ed., pp. 411–447, Blackwell, Oxford.

White, R. G., French, I. V., and Stark, J. M., 1970, A study of the localization of a protein antigen in the chicken spleen and its relation to the formation of germinal centres, *J. Med. Microbiol.* **3**:65.

Yoffey, J. M., and Courtice, F. C., 1970, *Lymphatics, Lymph and the Lymphomyeloid Complex,* Academic Press, London.

Yoffey, J. M., and Drinker, C. K., 1939, The cell content of peripheral lymph and its bearing on the problem of the circulation of the lymphocyte, *Anat. Rec.* **73**:417.

12

Splenic Macrophages

GERARD T. SIMON

1. INTRODUCTION

In the spleen of an adult human being, 3 million red blood cells are destroyed *per second,* to say nothing of the numerous neutrophils, eosinophils, and platelets also destroyed in this organ. This gives an indication of the extraordinary activity of the macrophages in the splenic red pulp, which alone are responsible for all this cytolysis. Low magnification electron micrographs show that the spaces between two sinuses (Figure 1), which correspond to the Billroth cords, are almost entirely occupied by macrophages involved in the phagocytosis and destruction of blood cells. It is clear that the splenic red pulp should be considered a macrophagic organ.

The light-microscopic aspect of the spleen has been thoroughly investigated (Lubarsch, 1927; Von Herrath, 1958; Rappaport, 1970). Its ultrastructural aspect, however, is still not fully known. Few groups of researchers have been or are studying the fine structure of this organ. Moreover, there is no consensus on the nomenclature or even on the nature of the different types of cells present in the spleen. All studies made at the ultrastructural level are, however, unanimous that light microscopy does not possess sufficient resolving power to reveal the relationship between the cells of the cord and sinus wall, and the complexity of the organization of the red pulp. It is, for example, only through electron microscopy that one can observe that the basement membranes of the sinuses have large gaps which permit the free passage of circulating blood cells from the lumina of the sinuses into the cords, or that pseudopods of the macrophages in the cords protrude through these gaps into the lumina of the sinuses. Such observations invalidate almost entirely the conclusions drawn from light microscopy.

Unfortunately, the magnifications normally used in electron microscopy

GERARD T. SIMON • Department of Pathology, University of Toronto, Toronto, Canada. Present affiliation: Electron Microscopy Facility and Department of Pathology, McMaster University, Hamilton, Ontario L8S 4K1, Canada.

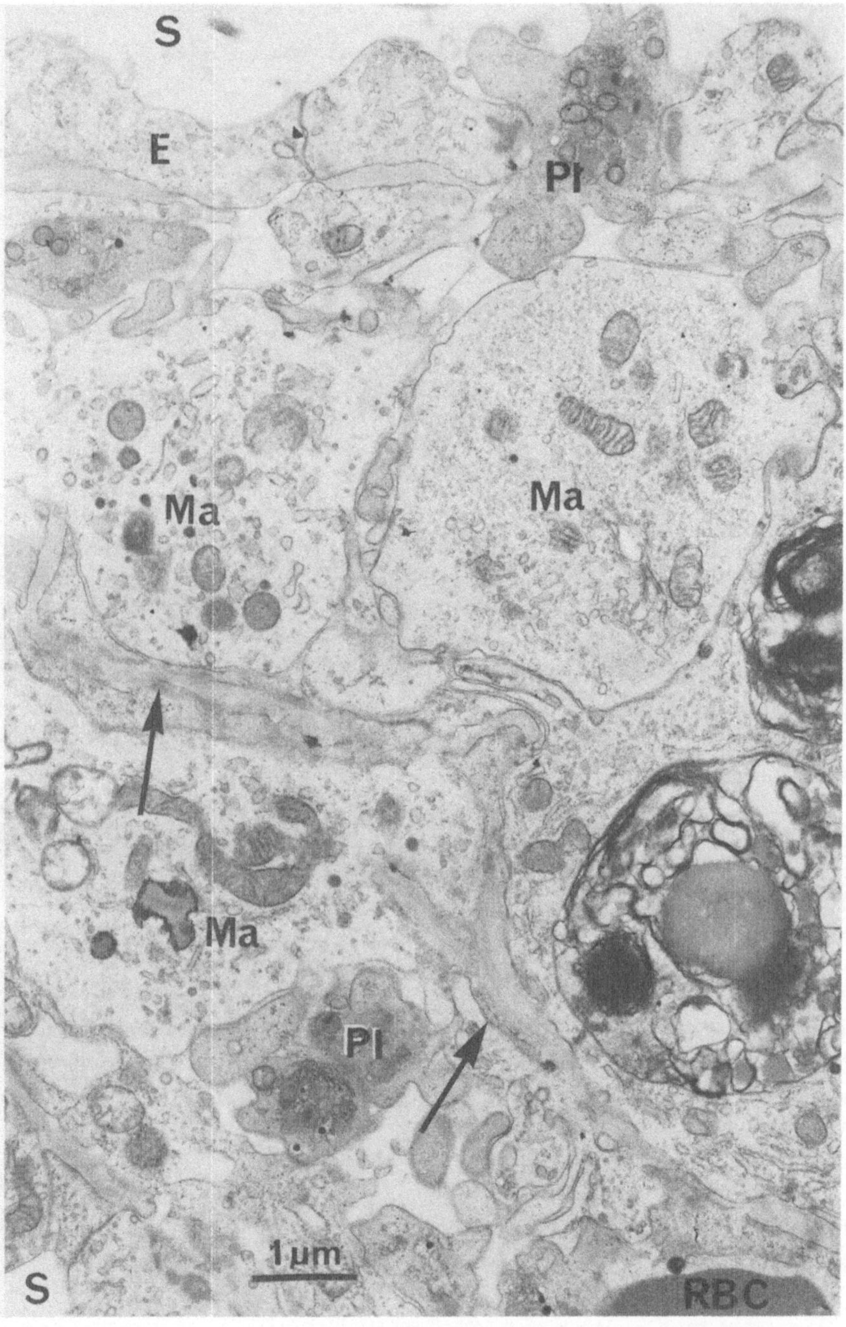

FIGURE 1. The cord between two small sinuses (S) contains mainly macrophages (Ma). The only other cellular component of the cords are prolongations of the fibroblasts (arrows) surrounding the collagenic network. E, Endothelium; Pl, platelets; RBC, red blood cell. (From Burke and Simon, 1970a.)

produce micrographs of small areas difficult to relate to the overall complex organization of the spleen. Many investigators have been misled by such micrographs and came to erroneous conclusions. It is, therefore, important in considering the function of the splenic macrophages to combine low and medium high magnification electron microscopy, which permits the establishment of the topographical relationships of these cells.

2. DEFINITION

One could define a macrophage as being a large cell containing in its cytoplasm, numerous mitochondria, prominent smooth endoplasmic reticulum, and a prominent Golgi apparatus, but whose main characteristic is the presence of several large phagolysosomes filled with various types of material. This description, accurate for active macrophages only, leads to the conclusion that cells of the red pulp which do not contain prominent phagolysosomes are not macrophages. In 1970, we followed this simplistic definition and labeled our micrographs accordingly. The cells without phagolysosomes were called reticulum cells. It now seems clear that these too are macrophages. Figure 2 of the present review which was labeled eight years ago had to be reassessed, and the cells classified as "reticulum cells" belong doubtlessly to the macrophagic series. This widening of the definition of macrophages is the result of our subsequent studies on lymph nodes, and particularly on the activities of macrophages under normal and several experimental conditions (Nopajaroonsri and Simon, 1971; Nopajaroonsri *et al.*, 1971, 1974; Simon *et al.*, 1975).

In the lymph nodes, circulating monocytes enter the parenchyma through the wall of the postcapillary venules and are transformed into macrophages, or may leave the lymph node unchanged. The different steps of differentiation are easily followed. It became evident that most of the cells in the spleen called reticulum cells in 1970 correspond to stages in the transformation of monocytes into macrophages. These images could also be produced if macrophages change back into monocytes. Although the transformation of some macrophages into monocytes is probable, no convincing data are available to prove it does occur. However, the constant extravasation of monocytes into the parenchyma of a lymph node does not cause an increase in the volume of the node. The transformation of these extravasated monocytes into macrophages is evident, but as the number of macrophages in the node does not increase, the inflow of monocytes must be matched by an equal outflow of macrophages. Presumably these macrophages are converted back into monocytes, and reenter the vessels. In the lymph node, such a traffic of monocytes is easy to visualize. In the spleen, it is more difficult to demonstrate. For example, Figure 3 shows a monocyte traversing a gap in the sinus wall. It is impossible to determine the direction of its passage. However, images such as this are frequently observed, and it is most probable that the spleen is also the site of a traffic of monocytes, which brings a constant renewal of part or all of its macrophagic cells.

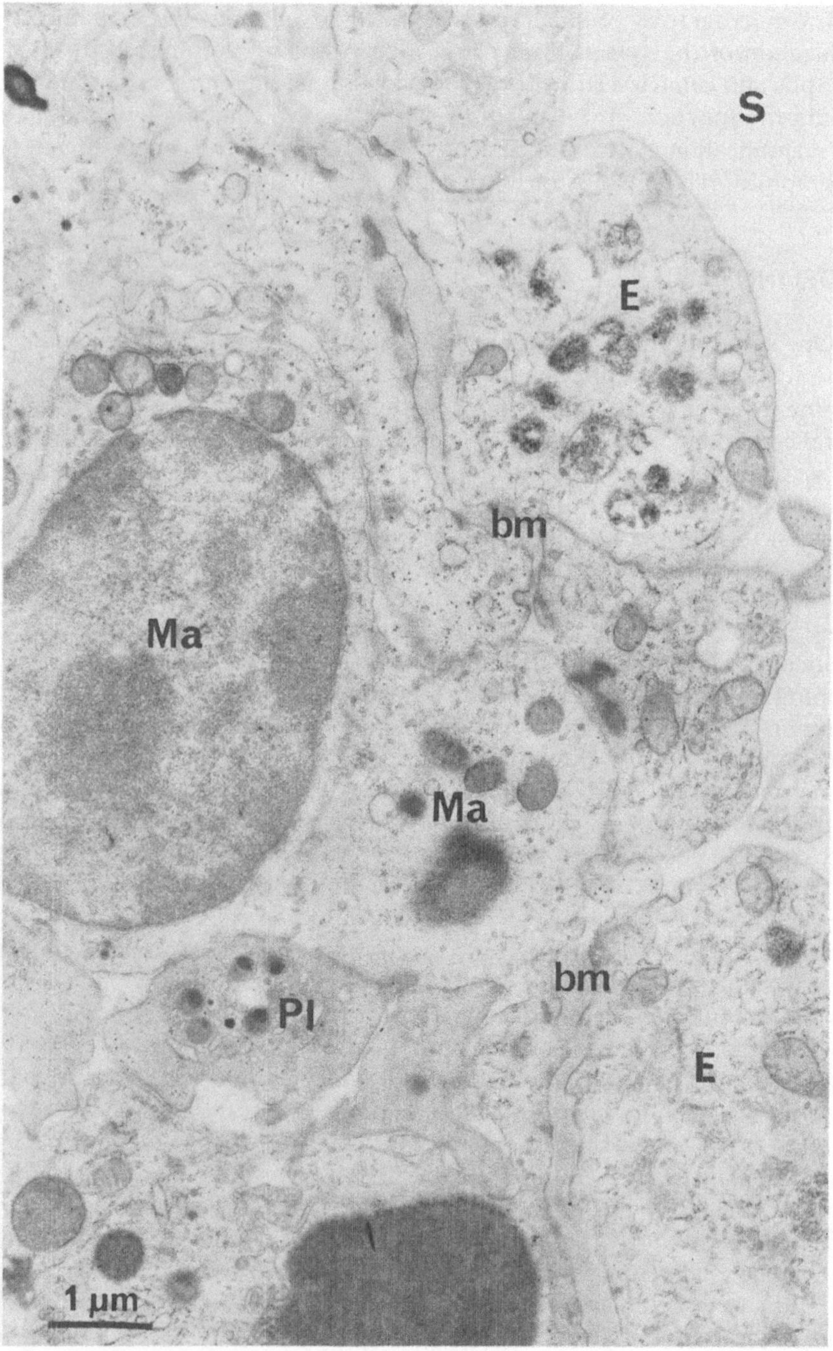

FIGURE 2. Close to a sinus (S) the cells in the cord do not exhibit any prominent lysosomes. Eight years ago they were labeled as "reticulum cells." There is no doubt today that they belong to the monocytic macrophagic line (Ma). bm, Basement membrane; Pl, platelet; E, endothelial cells.

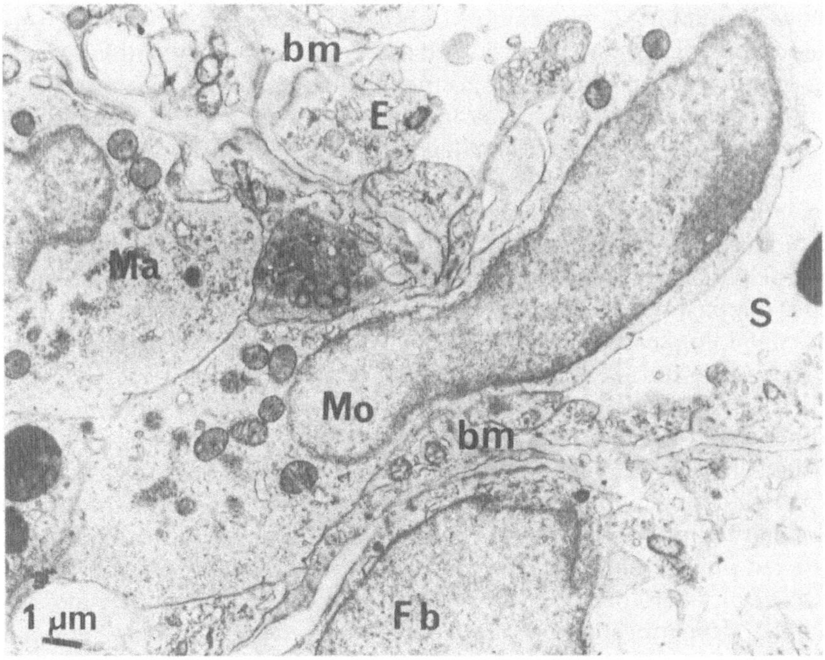

FIGURE 3. A monocyte (Mo) is seen passing the wall of a sinus (S) through a gap in the basement membrane (bm). It is impossible to assess the direction of its passage. Fb, Fibroblast; Ma, macrophage; E, endothelial cell. (From Burke and Simon, 1970a.)

3. DIFFERENT TYPES OF MACROPHAGES

Lung macrophages, Kupffer cells, foreign body giant cells, and osteoclasts are all of monocytic origin (Hancox, 1946; Fischman and Hay, 1962; Jee and Nolan, 1963; Sutton and Weiss, 1966; Brunstetter *et al.*, 1971; Gothlin and Ericsson, 1973; Walker, 1973; Kahn and Simmons, 1975; Simon and Luk, 1975; Gothlin and Ericsson, 1976; Luk *et al.*, unpublished). By extrapolation one should conclude that all macrophages are of the same origin, and, therefore, the question whether there is more than one type of macrophage in the spleen becomes irrelevant. But even if all macrophages are of the same origin, it appears that they behave differently, and have different functions.

3.1. RED PULP

Although no conclusive evidence is available one could assume that part of the red pulp macrophages belong to a trafficking pool, and can return to the monocytic form, and part belong to a nontrafficking pool of cells which remain in the spleen. This, however, is speculation. After intravenous injection of colloidal carbon, most of it is phagocytized by the splenic macrophages. This amorphous material cannot be degraded and is stored in large residual bodies.

Monocytes containing large amounts of colloidal carbon have never been seen in the blood, leading to the conclusion that the macrophages containing residual bodies cannot be retransformed into monocytes, and thus are included in the nontrafficking pool. This observation does not, however, indicate that these macrophages belong to a special category of cells which normally remain stationary in the cords. It is more probable that it is only when "storing" material, that macrophages lose their ability to dedifferentiate into monocytes. In the lymph nodes, carbon-loaded macrophages of the subcapsular sinuses reenter the parenchyma and appear to transfer the colloidal carbon to tissue macrophages (Nopajaroonsri and Simon, 1971). After several days or weeks, all the carbon-containing macrophages are located deep in the parenchyma close to the medulla, and do not show any phagocytic activity. Similar observations were made in the spleen, although a transfer of colloidal carbon from one macrophage to another could not be observed (Figure 4). However, although it has been established that macrophages containing undegradable substances cannot return to their monocytic form but remain in the parenchyma, the possibility remains that there are two different pools of macrophages in the Billroth cords.

The red pulp of the embryonic and fetal spleen has few macrophages (Von Herrath, 1958). Therefore, the Billroth cords have to be colonized by monocytes. It would be too simplistic to attribute to all the red pulp macrophages a bone marrow origin as has been suggested for Kupffer cells or pulmonary macrophages (Brunstetter *et al.*, 1971). The spleen is a very active hematopoietic

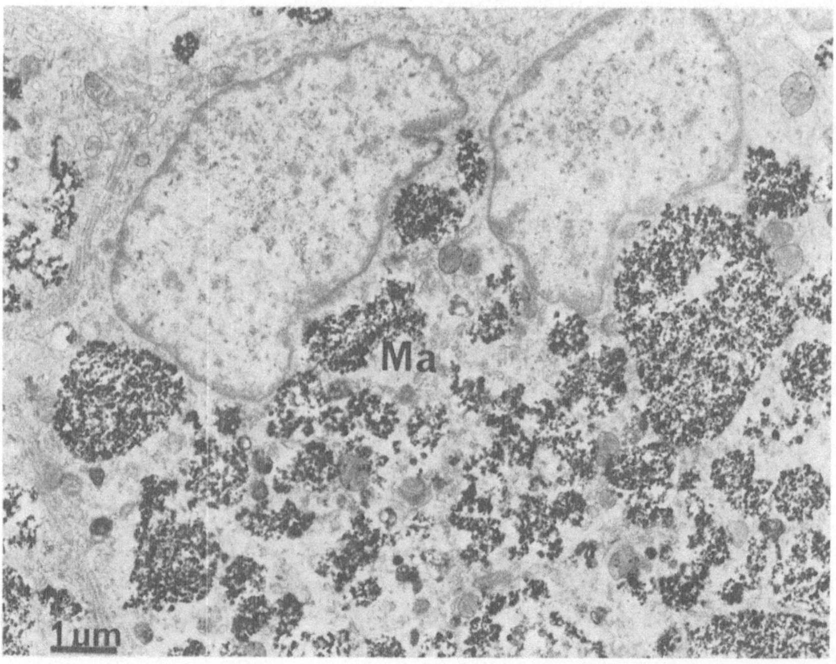

FIGURE 4. One week after the injection of colloidal carbon a macrophage (Ma) has accumulated a large amount of particulate matter. Note that the number of organelles in the cytoplasm is decreased in comparison to a phagocytizing macrophage. (From Burke and Simon, 1970b.)

organ during the fetal period, and, therefore, it is quite possible that some of the macrophages seen in the cords at birth are monocytes produced locally.

3.2. WHITE PULP

The macrophages seen in the germinal centers which develop in the white pulp as a result of a humoral immune reaction are different from those in the Billroth cords. The macrophages in the centers have the unique property of phagocytizing lymphocytes. Macrophages of monocytic origin in the red pulp of the spleen, in lymph nodes, in the liver, and in inflammatory reactions do not show any lymphophagocytic properties. To our knowledge, lymphophagocytosis is observed only in the germinal centers of the lymph nodes (Nopajaroonsri *et al.*, 1971), spleen, thymus (Hwang *et al.*, 1974), tonsils, and appendix (unpublished). In the tonsils, thymus, and appendix, there is reason to believe that these macrophages are of epithelial origin, although this is not universally accepted. In the germinal centers of spleen and lymph nodes it is most improbable that the macrophages originate from epithelial cells, although junctional complexes have been described between them.

3.3. OTHER CELLS WITH POSSIBLE MACROPHAGIC PROPERTIES

At the junction between the red and white pulp, and in the white pulp, but not related to germinal centers, elongated and possibly stellate cells take up and eventually completely degrade the antigenic material transported to the spleen by lymphocytes (see Section 11). Whether these cells should be categorized as macrophages is debatable.

4. NUMBER OF MACROPHAGES AND THE VOLUME THEY OCCUPY

It would be of interest to determine the number of macrophages present at a particular time in the spleen and the volume they occupy in order to assess their individual activities. von Herrath (1958) reports the volume occupied by different parts of the spleen and the differential distribution of splenic cells in several animal species as well as humans. It is unfortunate that these data based on light microscopic observations are obsolete. Only morphometric analysis (Weibel, 1969) on low magnification electron micrographs and the use of a unified terminology would provide this useful information. Such a study has never been done.

5. ULTRASTRUCTURAL ASPECT OF THE MACROPHAGES OF THE RED PULP

As mentioned, one should consider any trafficking monocytes undergoing transformation into a macrophagic cell as macrophages. This transformation begins when the monocytes pass from the sinuses into the cords of the red pulp. Circulating monocytes are round cells. Their cytoplasm contains a few, rather large mitochondria, a poorly developed rough endoplasmic reticulum (RER), a moderately extended smooth endoplasmic reticulum (SER), free ribosomes dispersed in patches and, when seen, a small Golgi apparatus. Pinocytic vesicles are almost absent. Phagolysosomes are rarely seen and seldom exceed 0.3–0.4 μm in diameter. The nucleus is round and slightly indented (Figure 5). The volume of a circulating monocyte is in the range of 80–100 μm^3. In contrast, the volume of a fully developed macrophage in a Billroth cord averages 1000 μm^3, a more than a tenfold increase. Studies on acute inflammation have shown that this transformation occurs in a very short period. In inflammatory conditions outside the spleen, monocytes are extravasated through the wall of post-capillary venules and capillaries, and there is no doubt as to the direction of their movement. Their transformation during the passage can be seen and this observation extrapolated for the spleen.

During the transformation of a monocyte into a macrophage, RER and Golgi apparatus increase (Figure 6). The extent of these modifications varies considerably according to the type of material to be phagocytized. In the normal spleen where erythroleucophagocytosis is the main function, the increase of these cytoplasmic organelles is moderate (Figure 6). After the intravenous injection of colloidal carbon, the proliferation of vacuoles and tubules is more extensive (Figure 7). In the fully developed macrophage, the number of mitochondria increases although their size is smaller than those observed in circulating

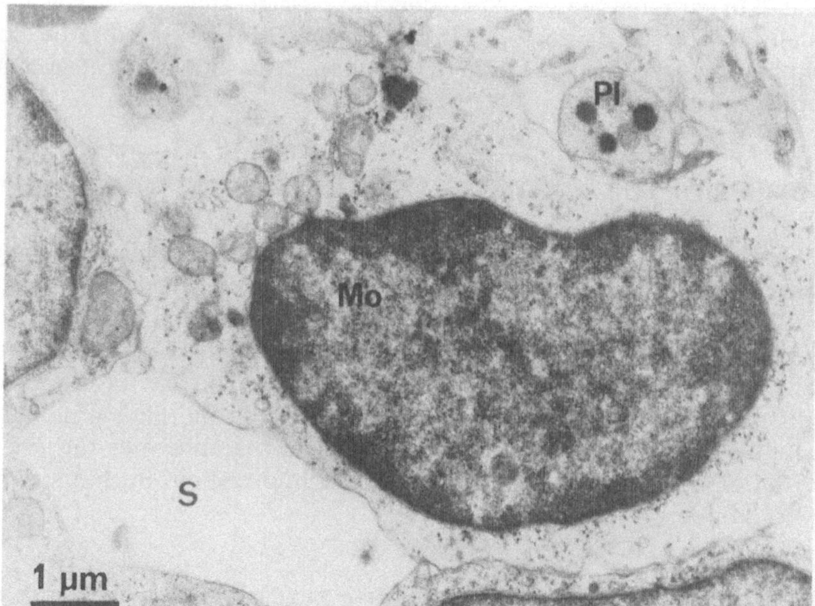

FIGURE 5. A circulating monocyte (Mo) in a sinus (S). The cytoplasm contains a few mitochondria and ribosomes dispersed in patches. Other organelles are not prominent. Pl, Platelet.

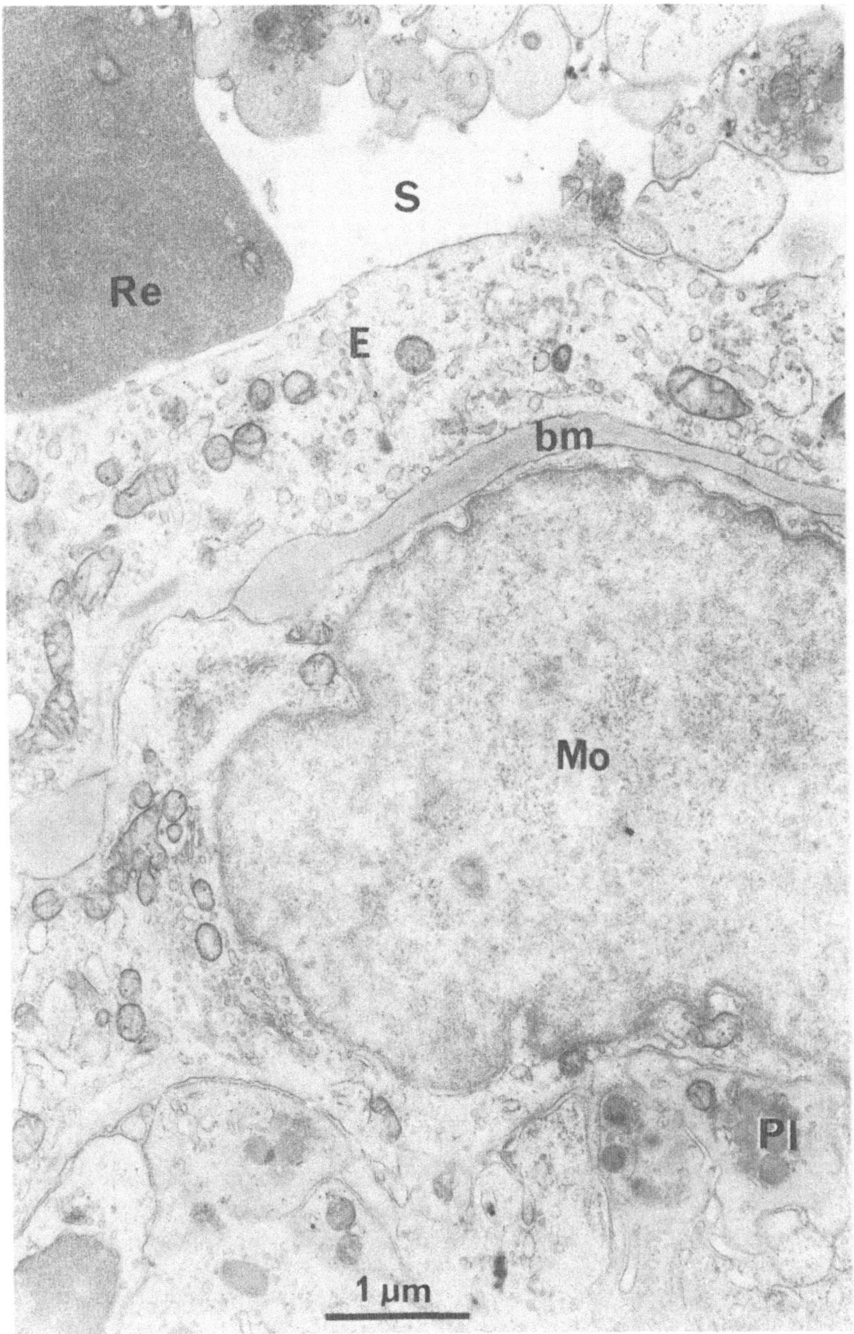

FIGURE 6. In a cord, a monocyte (Mo) is located directly below the basement membrane (bm). Compared to Figure 5 the number of organelles is considerably more. This aspect corresponds to transformation into a macrophage. S, Sinus; Re, reticulocyte; E, endothelial cell; Pl, platelet. (From Burke and Simon, 1970a.)

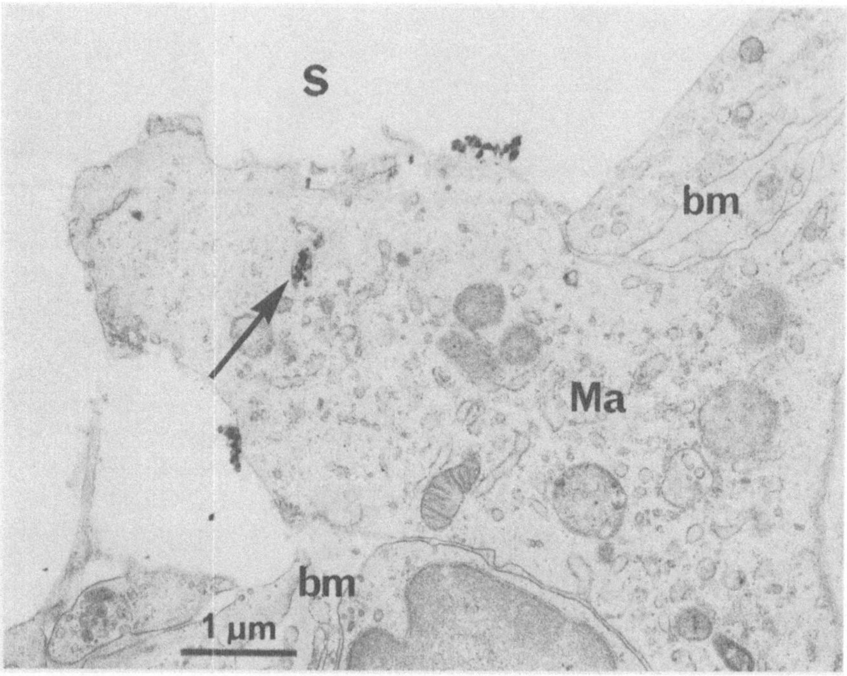

FIGURE 7. Twenty seconds after the injection of colloidal carbon the particles are already phagocytized (arrow) by a pseudopod of a macrophage (Ma) protruding into the sinus (S) wall. Note the prominence of the cytoplasmic organelles. bm, Basement membrane. (From Burke and Simon, 1970b.)

monocytes. The volume occupied by the Golgi apparatus can become very extensive.

When a macrophage is filled with undegradable material (Figure 4), the number of organelles decreases, and no phagocytic activity takes place.

6. ULTRASTRUCTURAL ASPECT OF PHAGOCYTOSIS IN THE RED PULP

Under normal conditions all circulating blood cells, with the exception of lymphocytes and monocytes, are phagocytized as whole cells and not fragmented prior to their total engulfment by the macrophages. Most of the circulating blood cells enter the Billroth cords through gaps in the sinus wall and are phagocytized inside the cords. Although free circulating cells can be seen in the cords, these cells usually meet the pseudopods of a macrophage in the gaps in the basement membrane of the sinuses. Less frequently, the macrophages extend pseudopods through the sinus wall (Figure 8) to grab circulating blood cells. In certain pathological conditions macrophages are said to have a "pitting function" (Wennberg and Weiss, 1968). These observations are controversial, since other authors (Rifkind, 1965) have not been able to confirm this fragmentation of cells before their engulfment by macrophages. The pitting function has

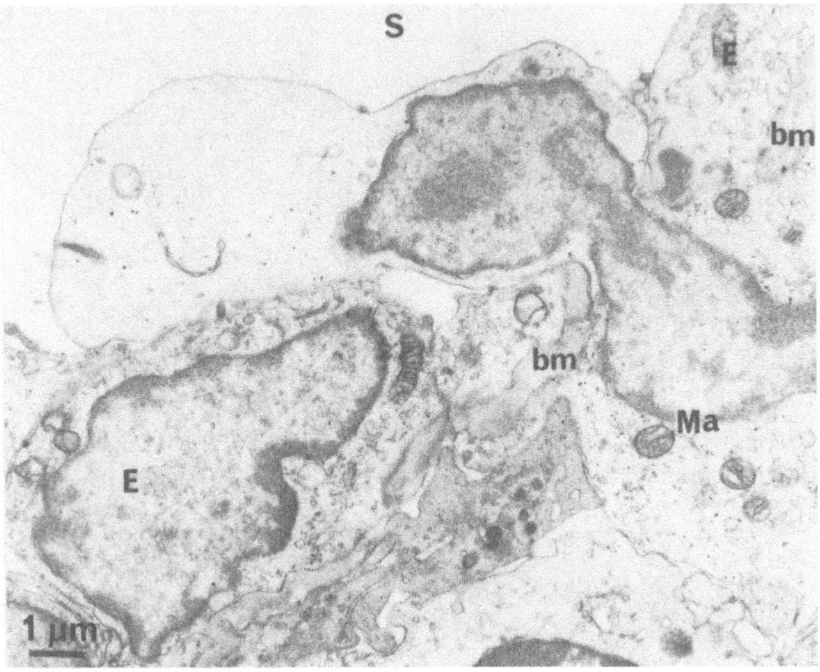

FIGURE 8. Normal rabbit spleen. A macrophage (Ma) extends a pseudopod into a sinus (S) through a gap in the basement membrane (bm). The presence of the nucleus in the gap is unusual. This picture could correspond to the passage of a monocyte transforming into a macrophage, or the reverse. E, Endothelial cell.

been put forward particularly in malaria. Schnitzer *et al.* (1972) show three pictures of this so-called pitting function, and none of them is really convincing. When penetrating into the cords, cells change their shape dramatically (Figure 9), into the most bizarre forms (Figure 10). In this micrograph there is no doubt that the fragment of a red blood cell belongs and is attached to its main body. The apparent interruption is due to the plane of sectioning. In other cases this relation is difficult or even impossible to establish. Only a reconstruction with serial sections can enable us to determine the true relation between two cell fragments.

When small foreign particles are injected intravenously, the number of pseudopods projected by the macrophages into the sinus increases considerably, and most of the circulating particles are phagocytized in the lumen of the sinuses (Figures 7, 11, 12, and 13).

7. ENDOTHELIUM-MACROPHAGE RELATIONSHIP

Ultrastructural studies have established that the basement membrane of the splenic sinuses is discontinuous. The sinuses are lined with stellate endothelial cells (Weiss, 1957, 1963). Tight junctions are almost absent between adjacent

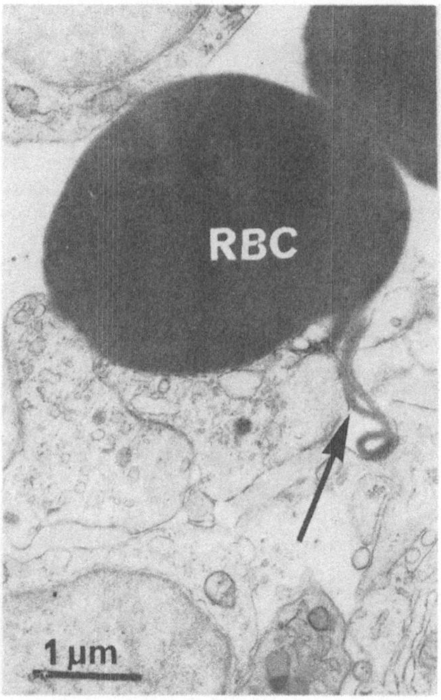

FIGURE 9. A red blood cell (RBC) is in the process of extravasation. Note the bizarre extension (arrow) of this cell in the gap of the sinus wall.

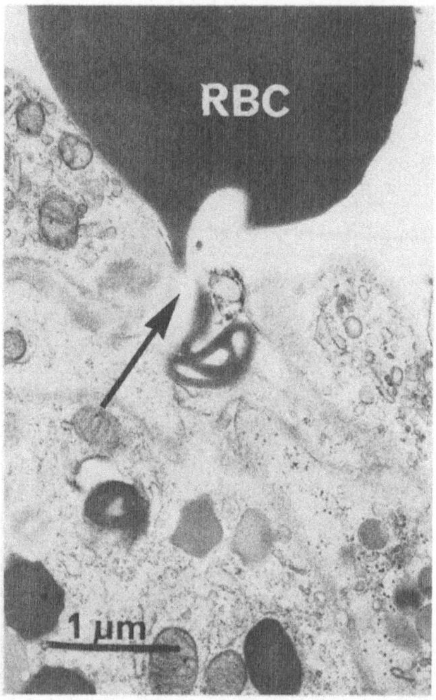

FIGURE 10. This picture shows the same bizarre extension as shown in Figure 9. The interruption (arrow), between the main body of the red blood cell (RBC) and the extension is artificial and is clearly due to a cutting angle.

extensions of these cells. The circulating blood cells and other particulate matter can, therefore, penetrate easily into the Billroth cords between the endothelial cells and through the openings in the basement membrane. Pseudopods of the cord macrophages also pass freely through these gaps, intercalating between two endothelial cells. These ultrastructural topographical observations could not be made with the light microscope. Knisely (1936a,b), for example, in the belief that the sinus wall was continuous and lined only by endothelial cells, made erroneous diagrams of the splenic circulation, leading to an incorrect conception of the destruction of red blood cells. These diagrams are unfortunately still reproduced in recent textbooks (Ham, 1969).

8. THE RETICULOENDOTHELIAL SYSTEM—A MYTH

The previous paragraph described the relationship between the lining endothelial cells and pseudopods of macrophages, indicating that the cellular lining of the sinus consists of a mosaic of these two types of cells. After intravenous injection of colloidal carbon, it is the pseudopods of the macrophages and not

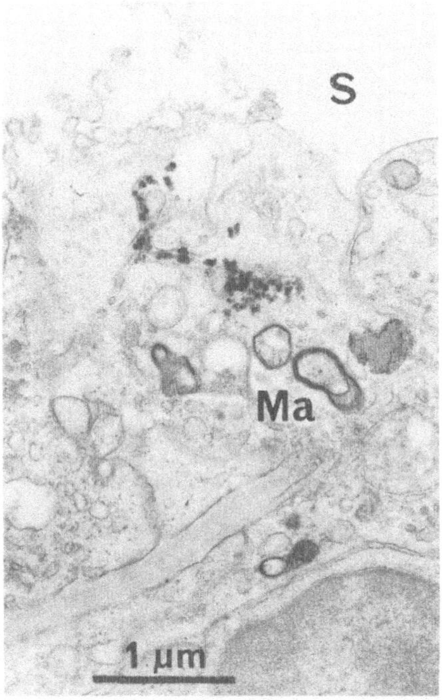

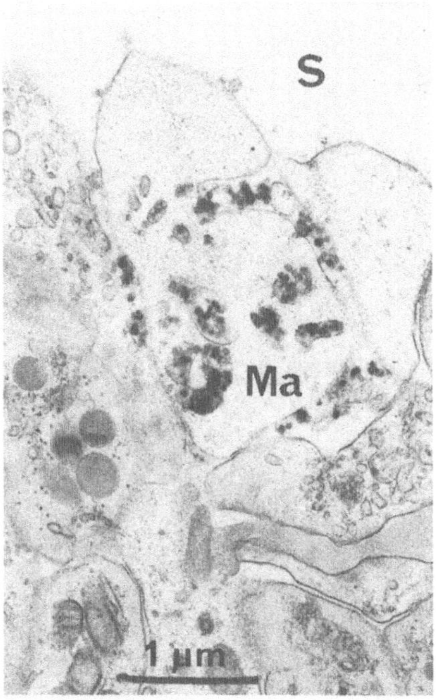

FIGURES 11 and 12. These two pictures show that the injected colloidal carbon is phagocytized in the sinus lumen (S) by pseudopods of macrophages (Ma). (From Burke and Simon, 1970b.)

the endothelial cells which phagocytize the foreign substance. In other organs such as liver, muscle (unpublished), lymph nodes (Nopajaroonsri and Simon, 1971) and bone marrow (Luk and Simon, 1974) the phagocytic activity of endothelial cells after the injection of carbon is identical to that of the splenic endothelial cells of the sinus. The phagocytic activity of endothelial cells increases only after the injection of a large dose of colloidal carbon or in case of high local concentration (Figure 14). When introducing the concept of the reticuloendothelial system, Aschoff (1924), like Knisely (1936a,b), could not see that the sinus walls were lined by two different types of cells. It is, therefore, understandable that after the injection of dyes he thought he saw great phagocytic activity in the splenic sinal lining cells. He actually depicted the phagocytic activity of the macrophages only. Ultrastructural studies have shown that in the bone marrow (Luk and Simon, 1974), the endothelial cells and macrophages have a relationship similar to that of the spleen. In the liver the Kupffer cells are intercalated with the endothelial cells (Aterman, 1963), and it is not surprising that Aschoff observed an increased phagocytic activity in the sinuses of these organs (Aschoff, 1924). His conception of a "reticuloendothelial system" was based on these observations. Today the term *reticuloendothelial system* is a misnomer and should be replaced by the more appropriate term of *macrophagic system*.

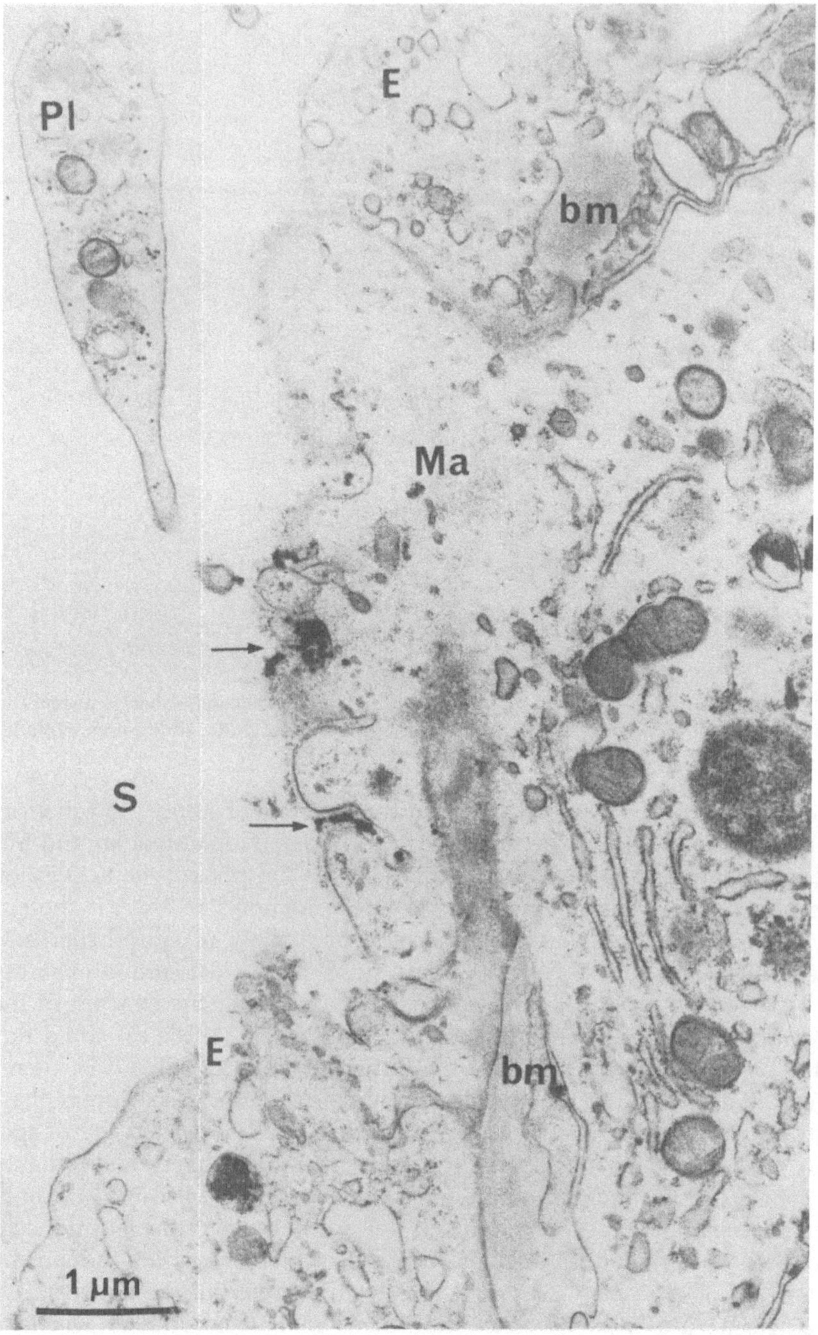

FIGURE 13. The pseudopod of a macrophage (Ma) extending through a gap in the basement membrane (bm) has topographically the same position as the endothelial cells (E). Note that the colloidal carbon is phagocytized by these pseudopods (arrows). S, Sinus; Pl, platelet. (From Burke and Simon, 1970b.)

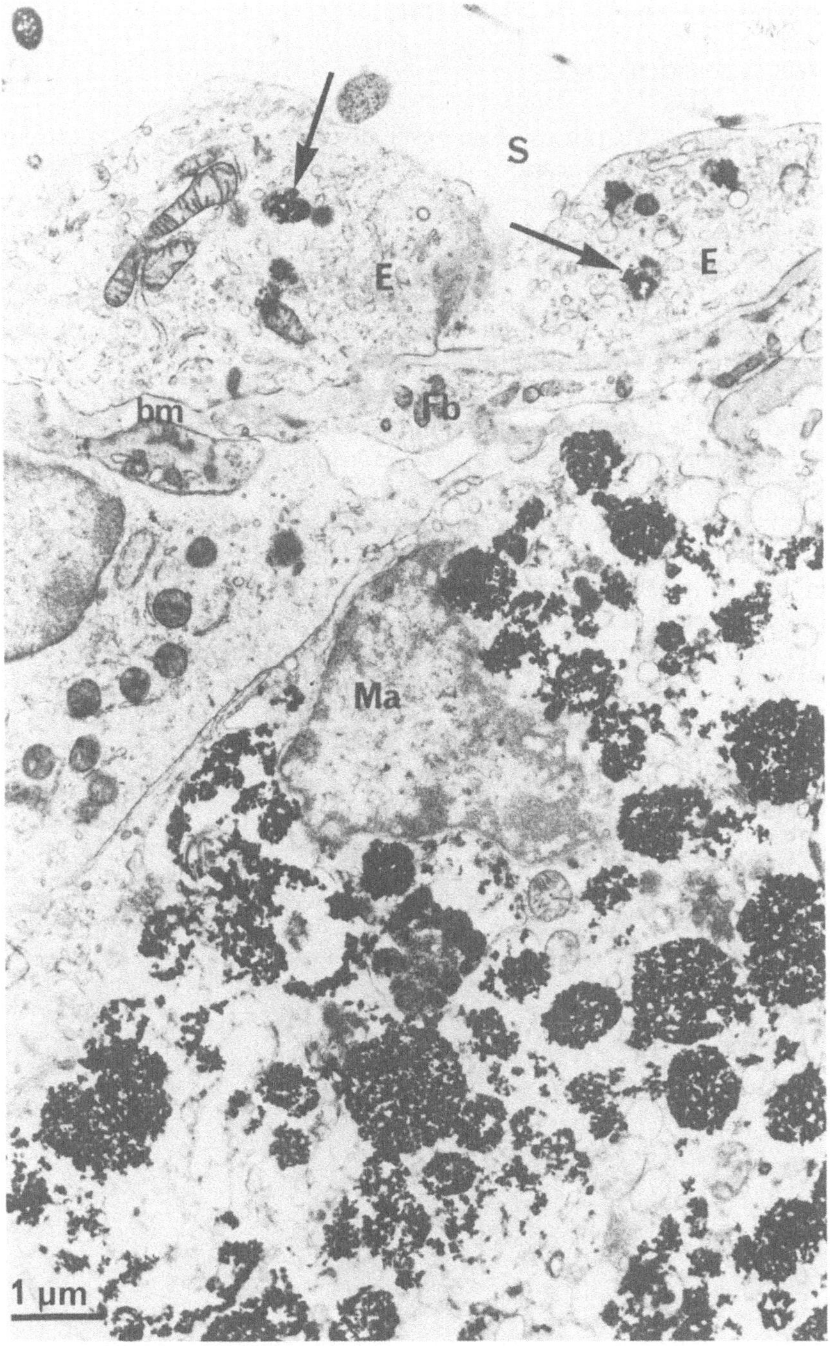

FIGURE 14. After a massive injection of colloidal carbon the endothelial cells (E) show phagosomes containing carbon particles (arrows). Below the basement membrane (bm), a cell prolongation belongs, possibly, to a fibroblast (Fb). In the cord, a macrophage (Ma) has accumulated a large amount of the injected material. (From Burke and Simon, 1970b.)

9. ERYTHROLEUKOPHAGOCYTOSIS

9.1. THE RED BLOOD CELL

The majority of the phagolysosomes in macrophages in the red pulp contain red blood cells in the process of degradation (Simon and Burke, 1970). The different steps in the destruction of the red blood cells can be followed from the engulfment of the red cells to their almost complete destruction (Figures 15 and 16). The ultrastructural manner of degradation varies from species to species. In the rabbit, the red blood cells are not fragmented inside the phagolysosomes, and during degradation numerous phospholipidic pseudomyelinic figures are formed (Simon and Burke, 1970). In the rat, the red blood cells are fragmented by a process of tunnelization, and fragmentation occurs before any modification of the red blood cell matrix is observed. During the subsequent degradation of the red blood cell in this animal, no pseudomyelinic figures are formed (Edwards and Simon, 1970).

Iron is released from heme in the early stages of red blood cell degradation and is almost immediately coupled to apoferritin, which is then detectable as ferritin by the electron microscope (Figure 17). Under normal conditions, haematoidin crystals (Figure 18) are rarely formed in rats or rabbits. Crystals do, however, occur in mice (see Hudson and Shortland, this volume).

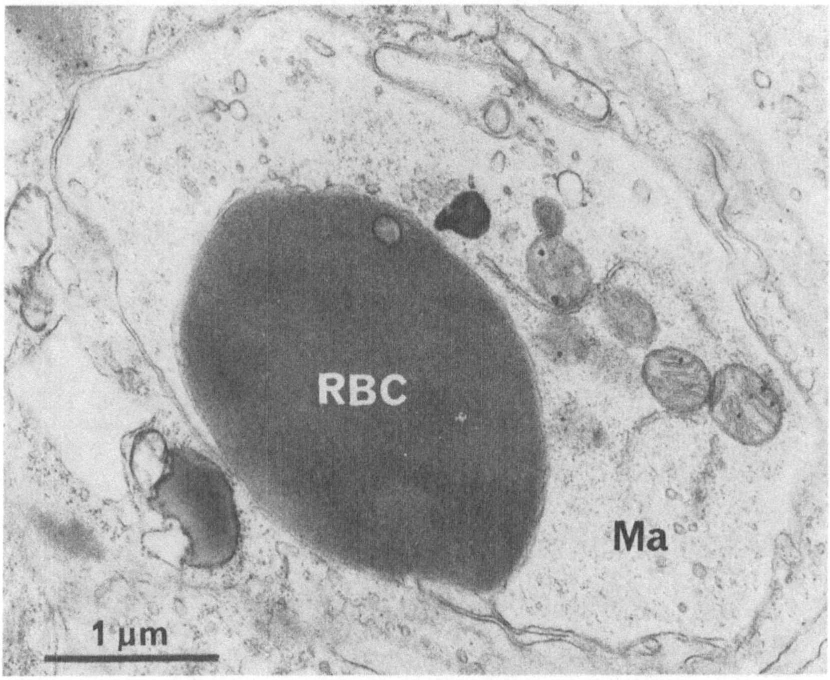

FIGURE 15. An extravasated red blood cell (RBC) is engulfed as a whole by a macrophage (Ma). (From Simon and Burke, 1970.)

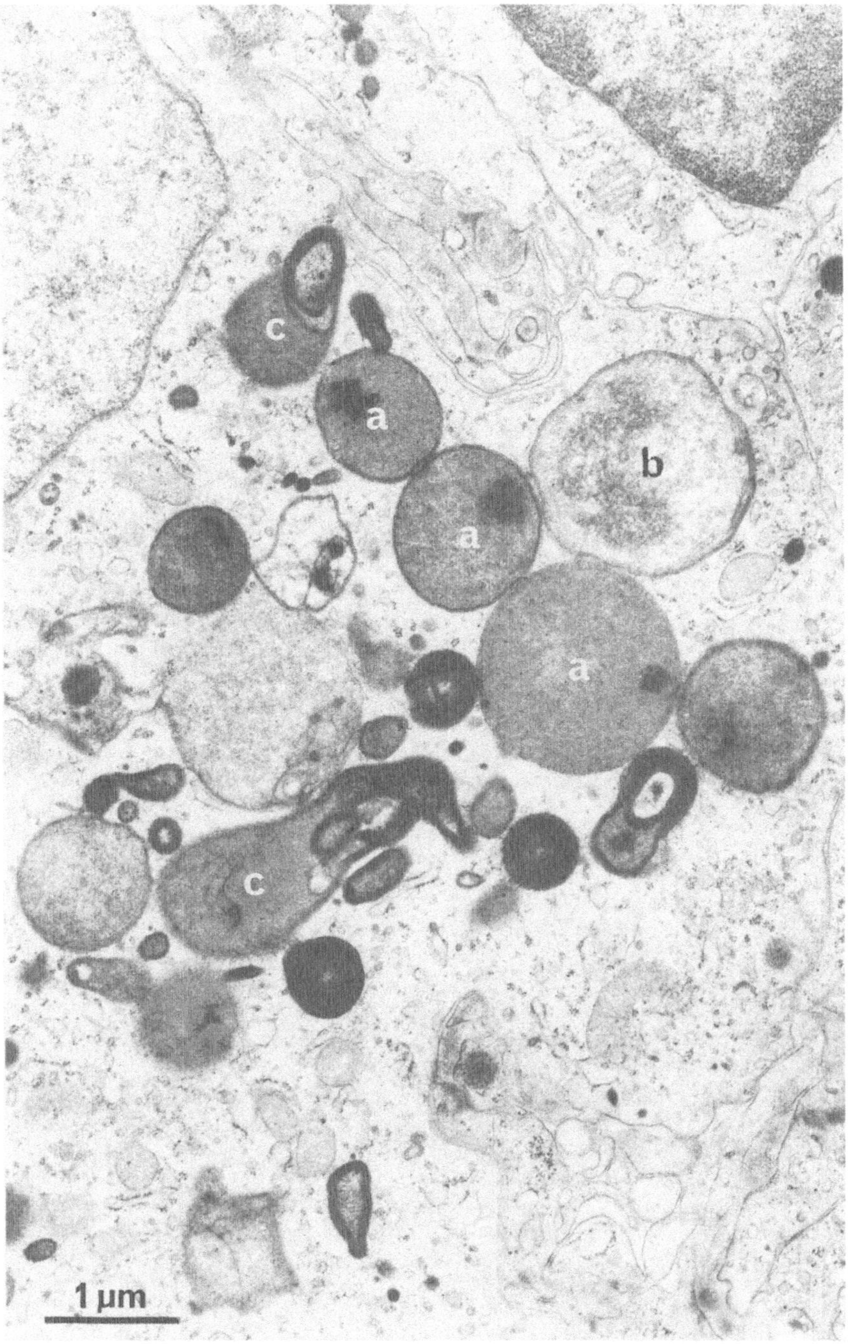

FIGURE 16. Macrophages in the cord contain several phagolysosomes degrading red blood cells. Different steps of degradation are seen. The vacuoles labeled "a" contain red blood cells with a still homogeneous granular matrix. In the vacuole "b" a red blood cell is at a more advanced step of degradation. The matrix is less homogeneous and exhibits a coarse granularity. In the vacuoles labeled "c" the matrix of the red blood cells is recondensed and pseudomyelinic figures are formed. (From Simon and Burke, 1970.)

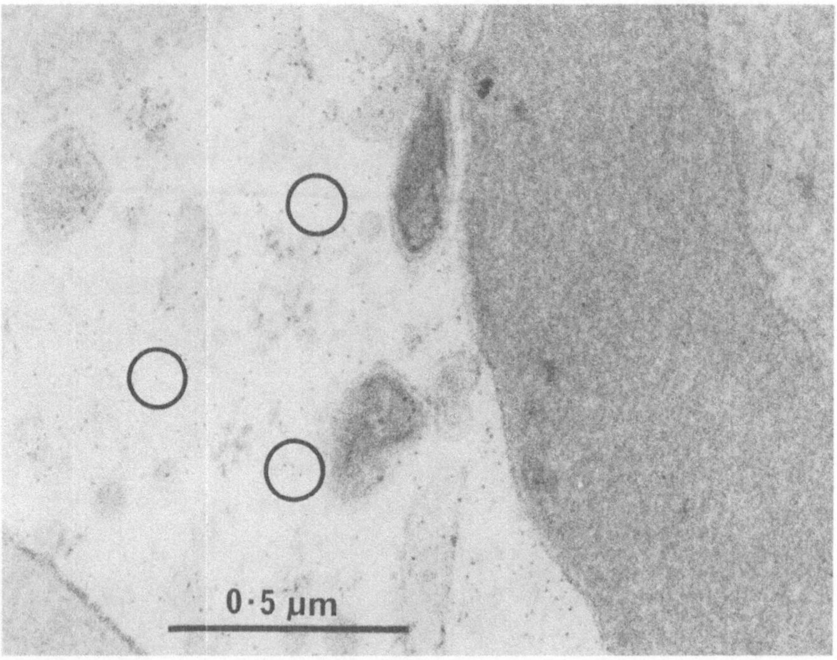

FIGURE 17. At an early stage of degradation, iron is released and is seen in the cytoplasm of the macrophage as ferritin (circles). Under normal conditions, no storage of ferritin is observed. (From Simon and Burke, 1970.)

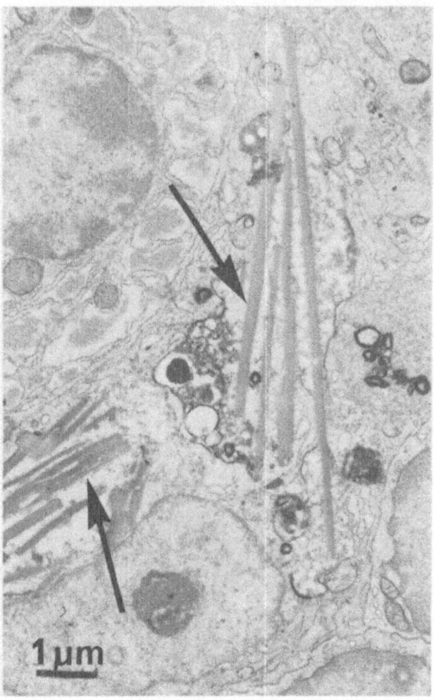

FIGURE 18. In a macrophage located at the junction of the red and white pulp hematoidin crystals (arrows) are seen in the cytoplasm. These crystals are rarely seen under normal conditions.

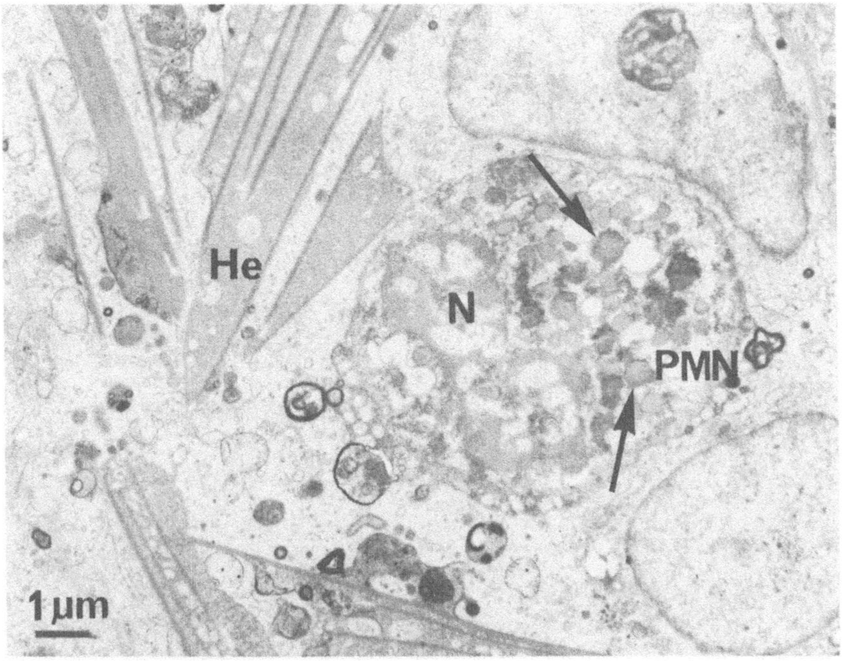

FIGURE 19. A polymorphonuclear neutrophil (PMN) is degraded in a phagolysosome of a macrophage. At this early stage of destruction, the nucleus (N) and the granules (arrows) are still recognizable. Note the presence in these macrophages of hematoidin crystals (HE). (From Simon and Burke, 1970.)

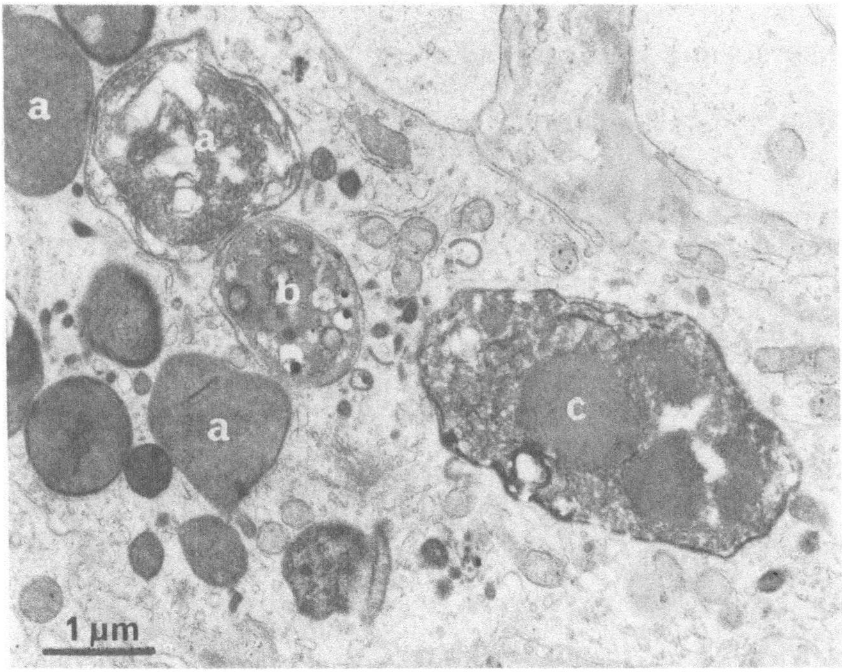

FIGURE 20. The cytoplasm of this macrophage contains phagolysosomes corresponding to the destruction of (a) red blood cells, (b) platelet, (c), polymorphonuclear neutrophil. (From Simon and Burke, 1970.)

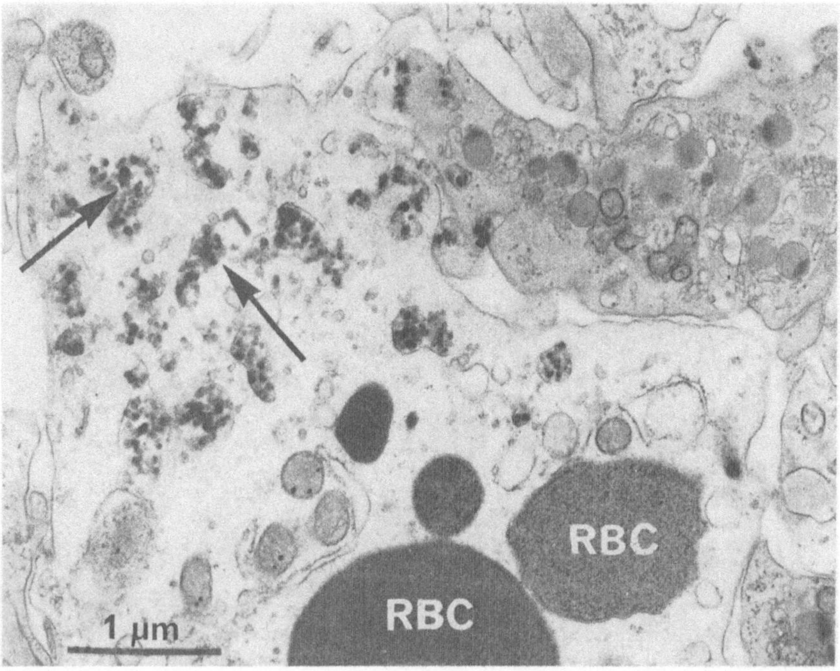

FIGURE 21. A macrophage which exhibits the destruction of red blood cells (RBC) also phagocytizes injected colloidal carbon (arrows). (From Burke and Simon, 1970b.)

9.2. NEUTROPHILS AND EOSINOPHILS

Neutrophils in the process of degradation are also seen in the macrophages of Billroth cords, though considerably less frequently than red blood cells (Figures 19 and 20). A macrophage containing a phagolysosome in which a polymorphonuclear neutrophil is being degraded can also contain red blood cells or platelets in the process of destruction, demonstrating that the macrophages are not in any way specialized to one type of phagocytosis. After injection of a colloidal substance these macrophages containing blood cells also engulf the foreign particulate matter (Figure 21).

Due to their limited number in the circulating blood, eosinophils in the process of degradation are very rarely seen in the macrophages, but they are found occasionally.

9.3. PLATELETS

The number of platelets in the spleen is particularly high (Ebbe, 1968). This can be observed even by light microscopy (Von Herrath, 1958). The platelets are often seen as aggregates which form plugs along the sinus wall. Although

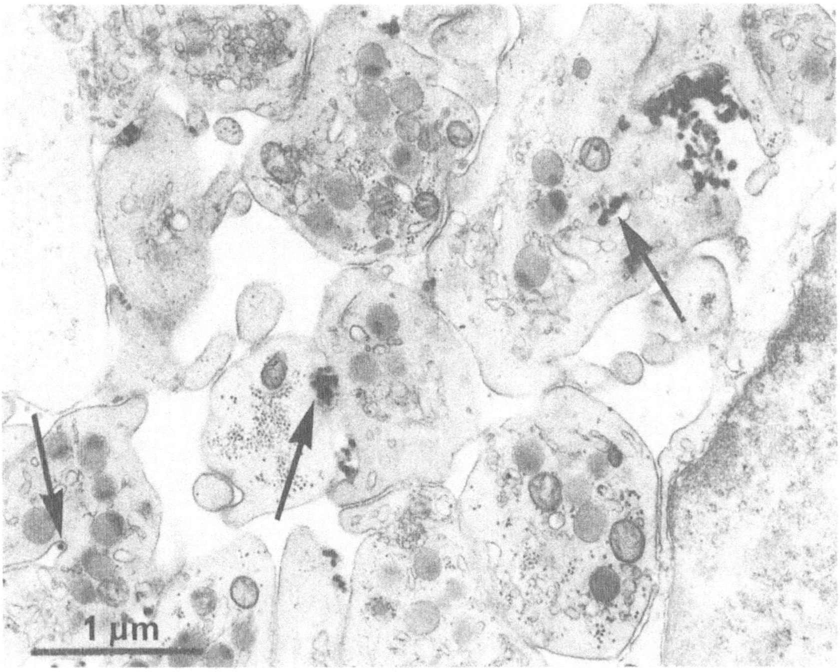

FIGURE 22. In the lumen of a sinus several platelets are phagocytizing injected colloidal carbon (arrows). It is remarkable that the platelets are not degranulated and that no fibrin filaments are formed. (From Nopajaroonsri *et al.*, 1974.)

aggregated, platelets do not degranulate and there is no formation of fibrin. This aggregation of platelets without coagulation is peculiar to the spleen. When large amounts of colloidal carbon are injected intravenously, white thrombi form in the bone marrow, lymph nodes, and other organs with aggregation of platelets and formation of fibrin. In the same experiments (Burke and Simon, 1970b) it is remarkable that no fibrin forms in the aggregates of platelets in the sinuses even though the platelets are actively involved in the phagocytosis of carbon (Figure 22). The reason for this difference is unknown.

The aggregated platelets are phagocytized, and not infrequently, several engulfed platelets are seen in a single segment of a macrophage. The platelets are each in its own phagolysosome, indicating that they are not engulfed as aggregated but individually (Figure 23).

9.4. PLASMA CELLS

At the junction of the white and red pulp, macrophages of the Billroth cord type phagocytize plasma cells (Figure 24). This observation is of particular importance because this type of macrophage is unable to phagocytize lymphocytes.

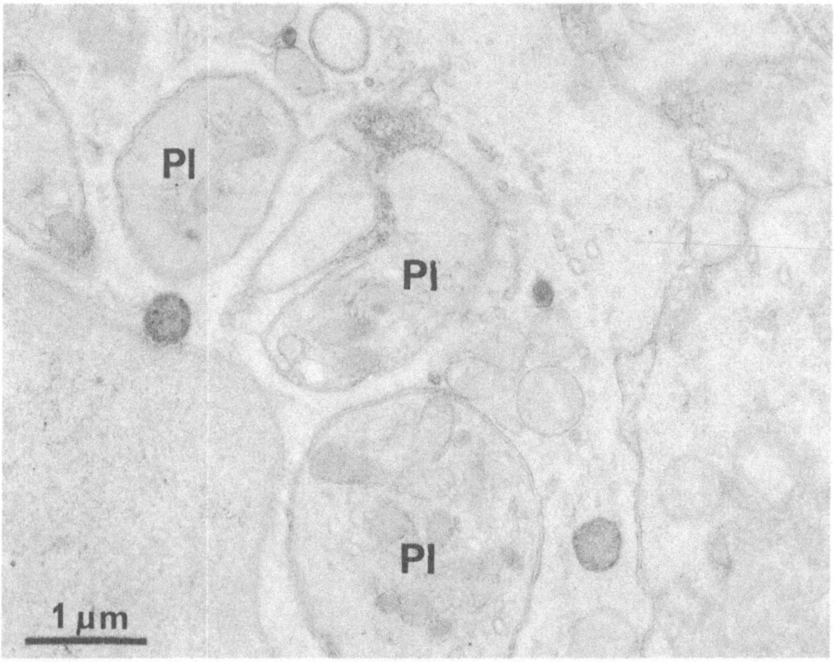

FIGURE 23. This macrophage shows several phagolysosomes each containing a platelet (Pl). Although the platelets are seen in aggregates close to the macrophages, they are, however, engulfed individually.

9.5. THE AGE OF THE PHAGOCYTIZED CELL

It has been repeatedly stated (Ham, 1969) that only aging circulating cells are phagocytized by the splenic macrophages. This is not so. It is not uncommon to see the phagocytosis of reticulocytes. Pictet *et al.* (1969a) emphasized the role of the parallel circulation in the phagocytic activity of the spleen. In the large sinuses, with their high flow rate, circulating cells are rarely engulfed by macrophages. In the parallel maze of smaller, interconnected sinuses, however, the great majority of circulating cells are phagocytized and destroyed. These authors (Pictet *et al.*, 1969a) concluded that erythroleukophagocytosis is a random phenomenon depending only on the passage of circulating cells either in large sinuses or the small ones. Our own experiments and observations confirm the above hypothesis, although it is possible that older cells, due to modifications of their plasma membrane and shape, preferentially take the capillaries leading to the network of small sinuses. Twenty seconds after the injection of colloidal carbon (Burke and Simon, 1970b), this substance is equally distributed in the large and the small sinuses. Phagocytosis is observed at this time in the small sinuses but only rarely in the large sinuses. It is only much later that some carbon is seen in macrophages of the cords lining the large sinuses.

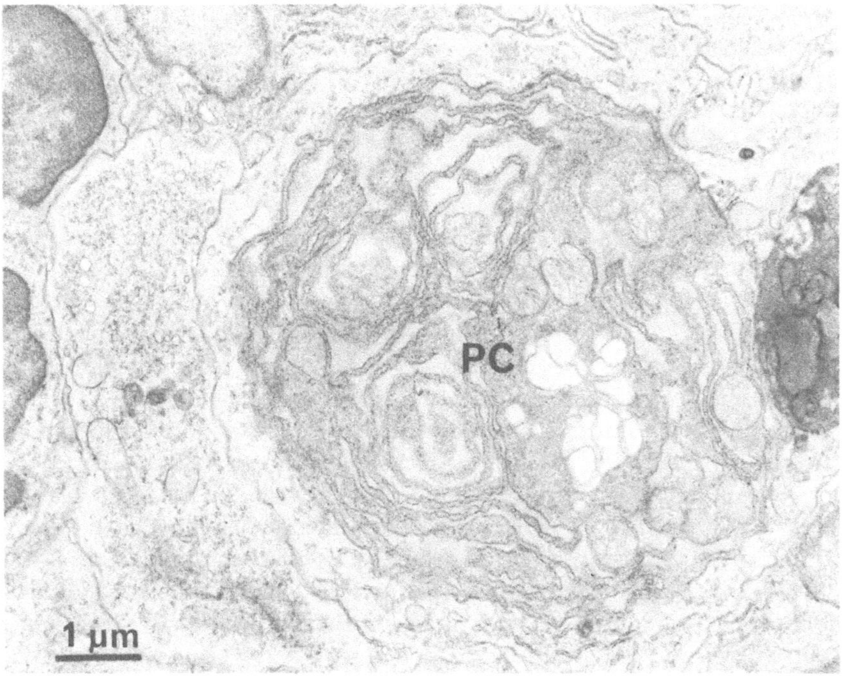

FIGURE 24. At the junction of the red and white pulp a macrophage shows in its phagolysosome a plasma cell (PC) in the process of degradation. It has to be emphasized that these macrophages are unable to phagocytize lymphocytes. (From Simon and Burke, 1970.)

10. STORAGE FUNCTION OF THE SPLEEN

The high content of iron in the spleen led to the conception that this organ has an important storage function (Von Herrath, 1958). This function, if existing, has to be related to macrophages because only these cells of all those in the spleen are capable of storing undegradable material for a long time. For example, amorphous material such as colloidal carbon is stored for ever in the macrophages of the red pulp. With most biological materials, such as antigens, storage is followed by slow degradation.

In experimental studies on iron absorption (Bédard *et al.*, 1974), it has been clearly demonstrated that iron not bound to protein is highly toxic. Normally, iron is first stored in the form of ferritin. In normal animals, the macrophages of the spleen do not exhibit vacuoles containing ferritin. The iron in the spleen is, therefore, directly derived from the destruction of red blood cells. In spite of the magnitude of the destruction of hemoglobin in the spleen, much of the iron released is returned immediately to circulation. Animals overloaded by intravenous injection of iron show a dramatic increase of hematoidin pseudocrystals long before ferritin accumulates in vacuoles in the macrophages (unpublished). A chronic overload with iron creates irreversible damage in the liver, like that in hemochromatosis.

11. MACROPHAGES IN IMMUNE RESPONSE

The spleen is an immunocompetent organ in the same way as is a lymph node. The three cellular components necessary for an immune response are present. These three types of cells are the macrophages, and the T and B lymphocytes, which meet at the limit of the red and white pulp.

The maze of sinuses which form the parallel slow-flowing circulation originate at the junction between the red and white pulp. Any circulating antigen trapped by macrophages in this location might well induce a primary immune response. This type of reaction can be illustrated by the modifications observed in the spleen in idiopathic thrombocytopenic purpura. In this condition the platelets are antigenic, and the splenic plasma cells produce anti-platelet antibodies (McMillan *et al.*, 1972, 1974). The platelets are phagocytized (Figure 25). They cannot be completely degraded, and accumulate in the macrophages (Figure 26). Eventually, the incompletely degraded platelets remain as residual bodies in the form of pseudomyelinic membranous structures (Figure 27). Interactions between B lymphocytes and these macrophages occurs, triggering the formation of anti-platelet antibodies.

The spleen also reacts, secondarily, to a primary immune response induced

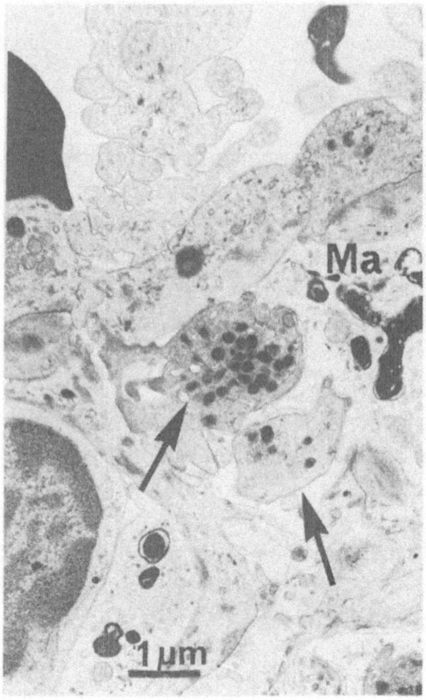

FIGURE 25. Human spleen in a case of idiopathic thrombocytopenic purpura. A macrophage (Ma) engulfs platelets (arrows) as whole cells.

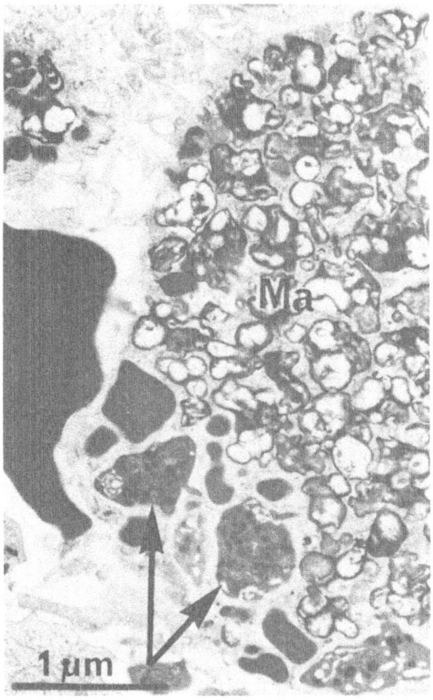

FIGURE 26. Human spleen in a case of idiopathic thrombocytopenic purpura. The engulfed platelets (arrows) are not completely degraded, resulting in an accumulation of residual bodies in the cytoplasm of the macrophage (Ma).

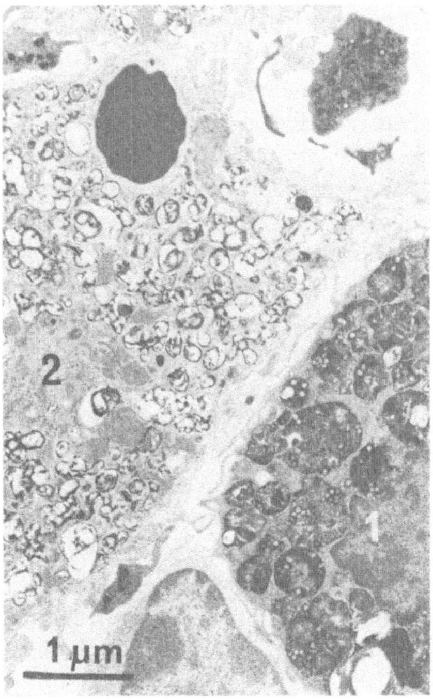

FIGURE 27. Human spleen in a case of idiopathic thrombocytopenic purpura. The cytoplasm of the macrophage (1) shows several phagolysosomes containing platelets in an early stage of degradation. The cytoplasm of the macrophage (2) exhibits a large number of residual bodies in the form of pseudomyelinic figures.

at the periphery. In a series of experiments, injecting autogeneic and xenogeneic red blood cells into the posterior part of the thigh of rats produced a primary immune response in the corresponding lymph node. Four days later, lymphocytes containing antigen leave the lymph node and colonize the spleen. On the fifth or the sixth day, germinal centers develop in the spleen, and exhibit macrophages of unknown origin which are capable of lymphophagocytosis. The lymphocytes which colonize the spleen undergo mitosis, eventually forming plasma cells. During this transformation, some of the lymphocytes transfer their antigenic inclusions to a cell which morphologically resembles neither the macrophages of the red pulp nor those of the germinal centers (Figure 28). These elongated, most probably stellate, cells, accumulate the antigenic material which is subsequently degradated. These mechanisms are far from being understood. Further studies should be made to determine whether such mechanisms are involved in all humoral responses.

12. CONCLUSIONS

In this article, the splenic macrophages have been studied in the context of their relationship with the ultrastructural topography of the spleen. It has been emphasized that electron microscopic observations have made most of the light microscopic studies obsolete.

In the red pulp, the origin, function, and behavior of the macrophages has been discussed. The possibility of a trafficking pool, implying the return to a monocytic form has been raised. The existence of a nontrafficking pool, possibly related to the presence of undegradable material in residual bodies of the macrophages, has also been postulated.

The mechanism of erythroleukophagocytosis has been considered. It has been postulated that the destruction of blood cells is essentially related to the peculiar circulation of the spleen, and not to the age of the blood cells. The pitting function of the macrophages has been questioned. All blood cells are phagocytized as a whole, and images which suggest such a function are mostly misinterpretations.

The macrophages of the red pulp are distinct from those found in germinal centers of the white pulp. The latter are involved in lymphophagocytosis, a property which the cord macrophages lack.

A possible special type of macrophage, located primarily at the junction of the white and red pulps has been described. This cell appears to be specialized in the ultimate degradation of antigenic material carried to the spleen, from the periphery, by lymphocytes. It is debatable whether this type of cell should be classified among splenic macrophages.

The storage function of the spleen which is directly related to the cord macrophages has been reassessed. For instance, the possible involvement of the spleen as a regulatory organ in iron metabolism is questioned. The amount of iron found in the spleen corresponds directly to its liberation during eryth-

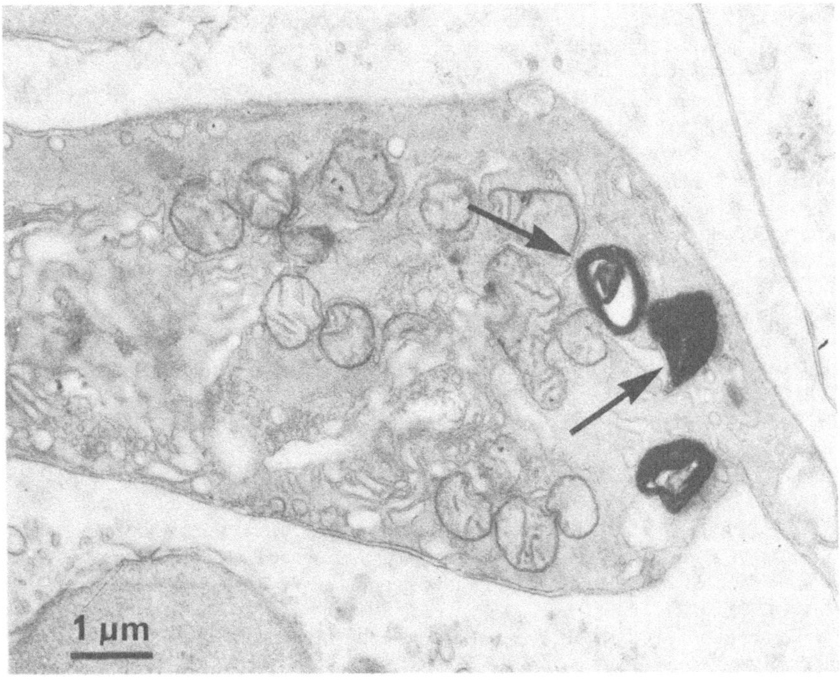

FIGURE 28. In the white pulp of a rat, 2 weeks after an intramuscular injection of xenogeneic red blood cells in the posterior part of the thigh, this type of cell is seen. The cytoplasm contains pseudomyelinic figures (arrows), identical to those seen in the lymphocytes of the corresponding lymph node in the first few days after the injection. This material still possesses antigenic properties. Pseudomyelinic figures are very rarely seen in the spleen of normal rats.

rophagocytosis. The liberated iron is released back into the blood circulation and, under normal conditions, no storage is observed in the spleen.

Finally, it is postulated that when antigenic material is injected intravenously, the pulp macrophages have an identical role in primary immune response to those of the lymph nodes when the antigen is administered intramuscularly.

ACKNOWLEDGMENTS. I wish to thank Mrs. Usha Bhargava for the literature research and for her photographic work. I also wish to thank Dr. A. C. Ritchie for his critical review of the manuscript.

REFERENCES

Aschoff, L., 1924, Das reticulo-endotheliale system, *Ergeb. Inn. Med. Kinderhelikd* **26**:1.

Aterman, K., 1963, The structure of the liver sinusoids and the sinusoidal cells, in: *The Liver*, Vol. I (C. Rouiller, ed.), pp. 61–136, Academic Press, New York.

Bédard, Y. C., Clarke, S., Pinkerton, P. H., and Simon, G. T., 1974, Effect of cycloheximide on iron absorption. "Toxicity of Iron," *Lab. Invest.* **30**:25.

Brunstetter, M. A., Hardie, J. A., Schiff, R., Lewis, J. R., and Cross, C. E., 1971, The origin of pulmonary alveolar macrophages. Studies of stem cells using the Es-2 marker of mice, *Arch. Intern. Med.* **127**:1064.

Burke, J. S., and Simon, G. T., 1970a, Electron microscopy of the spleen. I. Anatomy and microcirculation, *Am. J. Pathol.* **58**:127.

Burke, J. S ., and Simon, G. T., 1970b, Electron microscopy of the spleen. II. Phagocytosis of colloidal carbon, *Am. J. Pathol.* **58**:157.

Ebbe, S., 1968, Megakaryocytopoiesis and platelet turnover, *Ser. Haematol.* **1**(2):65.

Edwards, V. D., and Simon, G. T., 1970, Ultrastructural aspects of red cell destruction in the normal rat spleen, *J. Ultrastruct. Res.* **33**:187.

Fischman, D. A., and Hay, E. D., 1962, Origin of osteoclasts from mononuclear leucocytes in regenerating newt limbs, *Anat. Rec.* **143**:329.

Gothlin, G., and Ericsson, J. L. E., 1973, On the histogenesis of the cells in fracture callus. Electron microscopic autoradiographic observations in parabiotic rats and studies on labeled monocytes, *Virchows Arch. B* **12**:318.

Gothlin, G., and Ericsson, J. L. E., 1976, The osteoclast: Review of ultrastructure, origin, and structure–function relationship, *Clin. Orthop.* **120**:201.

Ham, A. W., 1969, *Histology,* 7th edn, Lippincott, Philadelphia.

Hancox, N. M., 1946, On occurrence *in vitro* of cells resembling osteoclasts, *J. Physiol.* **105**:66.

Hwang, W. S., Ho, T. Y., Luk, S. C., and Simon, G. T., 1974, Ultrastructure of the rat thymus—a transmission, scanning electron microscope, and morphometric study, *Lab. Invest.* **31**(5):473.

Jee, W. S. S., and Nolan, P. D., 1963, Origin of osteoclasts from the fusion of phagocytes, *Nature (London)* **200**:225.

Kahn, A. J., and Simmons, D. J., 1975, Investigation of cell lineage in bone using a chimaera of chick and quail embryonic tissue, *Nature* **258**(5533):325.

Kniseley, M. H., 1936a, Spleen studies. I. Microscopic observations of the circulatory system of living unstimulated mammalian spleens, *Anat. Rec.* **65**:23.

Kniseley, M. H., 1936b, Spleen studies. I. Microscopic observations of the circulatory system of living traumatized spleens and of dying spleens, *Anat. Rec.* **65**:131.

Lubarsch, O., 1927, Pathologische Anatomie der Milz, in: *Handbuch der Speziellen Pathologischen Anatomie und Histologie* (F. Henke and O. Lubarsch, eds.), pp. 373–774, Springer Verlag, Berlin.

Luk, S. C., and Simon, G. T., 1974, Phagocytosis of colloidal carbon and heterologous red blood cells in the bone marrow of rats and rabbits, *Am. J. Pathol.* **77**:423.

McMillan, R., Longmire, R. L., Yelenosky, R., Smith, R. S., and Craddock, C. G., 1972, Immunoglobulin synthesis *in vitro* by splenic tissue in idiopathic thrombocytopenic purpura, *N. Engl. J. Med.* **286**:681.

McMillan, R., Longmire, R. L., Yelenosky, R., Donnell, R. L., and Armstrong, S., 1974, The quantitation of platelet-binding IgG produced *in vitro* by spleens from patients with idiopathic thrombocytopenic purpura, *N. Engl. J. Med.* **29**:812.

Nopajaroonsri, C., and Simon, G. T., 1971, Phagocytosis of colloidal carbon in a lymph node, *Am. J. Pathol.* **65**:25.

Nopajaroonsri, C., Luk, S. C., and Simon, G. T., 1971, Ultrastructure of the normal lymph node, *Am. J. Pathol.* **65**:1.

Nopajaroonsri, C., Luk, S. C., and Simon, G. T., 1974, The passage of intravenously injected colloidal carbon into lymph node parenchyma, *Lab. Invest.* **30**:533.

Pictet, R., Orci, L., Forssman, W. G., Girardier, L., 1969a, An electron microscopic study of the perfusion fixed spleen. I. The splenic circulation and the RES concept. *Z. Zellforsch. Mikrosk. Anat.* **96**:372.

Pictet, R., Orci, L., Forssman, W. G., Girardier, L., 1969b, An electron microscopic study of the perfusion fixed spleen. II. Nurse cells and erythrophagocytosis, *Z. Zellforsch. Mikrosk. Anat.* **96**:400.

Rappaport, H., 1970, The pathologic anatomy of the splenic red pulp, in: *Die Milz* (K. Lennert and D. Harms, eds.), pp. 24–41, Springer Verlag, Berlin.

Rifkind, R. A., 1965, Heinz body anaemia: An ultrastructural study. II. Red cell sequestration and destruction, *Blood* **26**:433.

Schnitzer, B., Sodeman, T., Mead, M. L., 1972, Pitting function of the spleen in malaria: Ultrastructural observations, *Science* **177**:175.

Simon, G. T., and Burke, J. S., 1970, Electron microscopy of the spleen. III. Erythroleukophagocytosis, *Am. J. Pathol.* **58**(1):451.

Simon, G. T., and Luk, S. C., 1975, The origin of osteoclasts, *Proc. Micros. Soc. Can.* **2**:100.

Simon, G. T., Nopajaroonsri, C., Hwang, W. S., and Luk, S. C., 1975, Xenogeneic red blood cell degradation in a regional lymph node and dissemination of antigens by circulating lymphocytes, *Lab. Invest.* **33**:363.

Sutton, J. S., and Weiss, L., 1966, Transformation of monocytes in tissue culture into macrophages, epithelioid cells and multi-nucleated giant cells. An electron microscope study, *J. Cell Biol.* **28**:303.

Von Herrath, E., 1958, *Bau und Funktion der Normalen Milz,* Walter de Gruyter, Berlin.

Walker, D. G., 1973, Experimenta osteopetrosis, *Clin. Orthop.* **97**:158.

Weibel, G. R., 1969, Stereological principles for morphometry in electron microscopic cytology, *Int. Rev. Cytol.* **26**:235.

Weiss, L., 1957, A study of the structure of splenic sinuses in man and in the albino rat with the light microscope and the electron microscope, *J. Cell Biol.* **3**:599.

Weiss, L., 1963, The structure of intermediate vascular pathways in the spleen of rabbits, *Am. J. Anat.* **113**:51.

Wennberg, E., and Weiss, L., 1968, Splenic erythroclasia: An electron microscopic study of hemoglobin H disease, *Blood* **31**:778.

13

Lymphocyte – RES Interactions and Their Fine-Structural Correlates

PETER P. H. De BRUYN and ANDREW G. FARR

1. LYMPHOCYTE-RES INTERACTIONS

Although reciprocal functional relations of cells and tissues are an essential biological prerequisite for the maintenance of multicellular organisms, direct interactions of one cell type upon another, resulting in specific functional activities, have been subjected to systematic study only in the last decades. With the development of tissue culture methods enabling the *in vitro* maintenance of the functional state of cells, it has become possible to examine the direct consequences of such cell interactions. Studies of this nature have been particularly productive in the analysis of immunological phenomena involving the cellular interaction of lymphoid cells and reticuloendothelial components such as macrophages. The wealth of information that has resulted from such studies of cell interactions has not found an equally rewarding counterpart under *in vivo* conditions, and reports of cell interactions equivalent to those found *in vitro* are relatively scarce. The opposite is true for another form of lymphocyte–RES interaction, viz., the interaction of lymphocytes with the cells lining the sinuses of lymphatic and blood vascular channels in lymphopoietic and myelopoietic tissues. The passage of cells through this cellular interface is determined by an interaction between the migrating cell and the sinus-lining cell, and is part of the general phenomenon of selective transmural blood cell passage. This chapter will deal with the morphology of two aspects of lymphocyte–RES interaction: (1) The interaction between lymphocytes and macrophages occurring in the immune process and (2) the interactions between lymphocytes with the phagocytic sinus-lining cells of the RES and with the endothelial cells of postcapillary ven-

PETER P. H. De BRUYN • Department of Anatomy, University of Chicago, Chicago, Illinois 60637. ANDREW G. FARR • Department of Pathology, Harvard Medical School, Boston, Massachusetts. Present affiliation for AGF: National Jewish Hospital and Research Center, Denver, Colorado 80232.

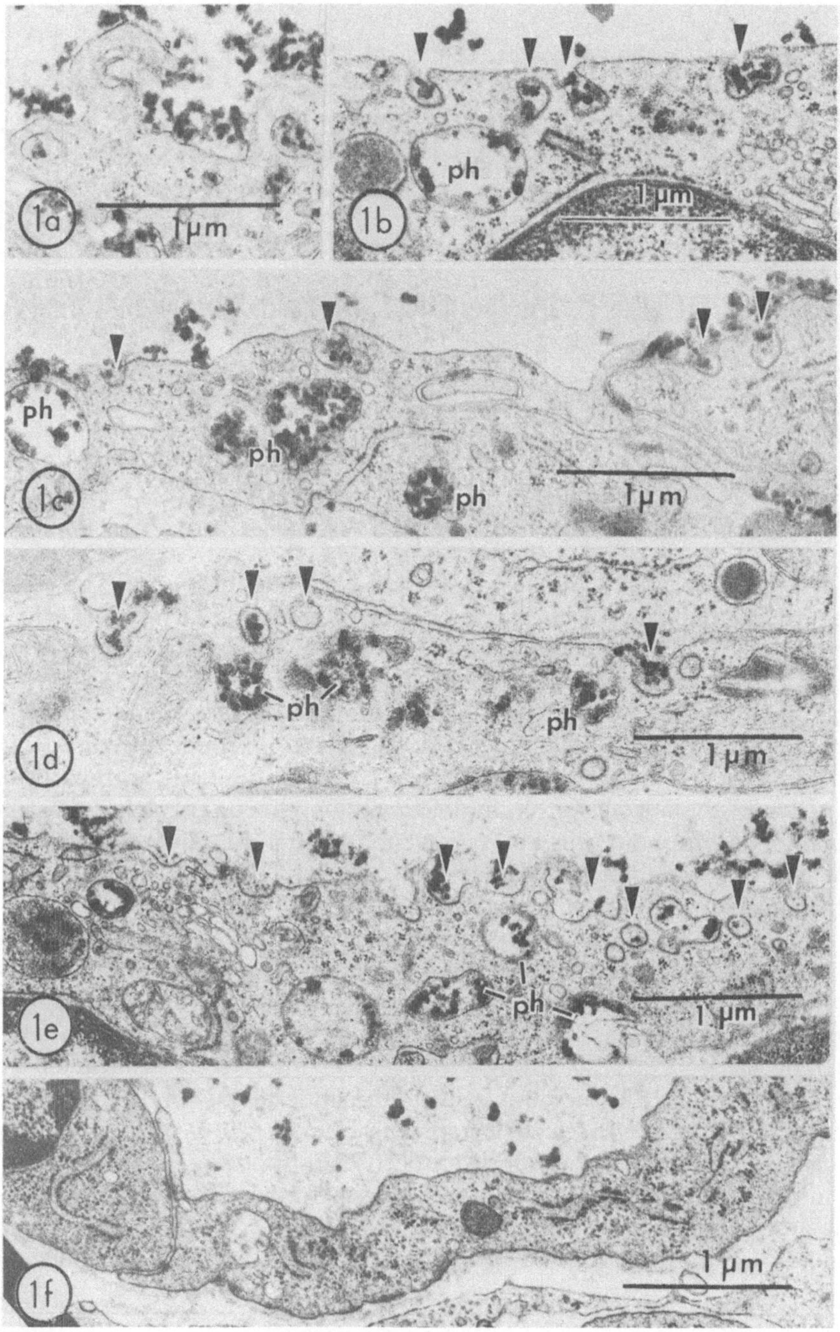

FIGURE 1. Phagocytosis of carbon particles by macrophage (Kupffer cell) and sinus-lining cells (a) Kupffer cell of rat liver phagocytizing large aggregates of carbon particles by means of filopodia (macrophagocytosis). Sinus-lining cells of rat liver (b), lymph node (c), spleen (d), and bone marrow (e) phagocytizing carbon particles by means of large, bristle-coated vesicles

ules in the lymph node. Such interactions occur in the transmural passage of lymphocytes. The postcapillary venules of lymphatic tissue appear to be an exception to the association of transmural cell traffic with elements of the RES. The transmural passage of lymphocytes from circulating blood back into the lymphatic parenchyma is, with respect to its pathway, as well as with respect to its selectivity, essentially identical to the transmural passage occurring through the endothelial cells of the sinuses of the blood forming organs (Marchesi and Gowans, 1964; Farr and De Bruyn, 1975; Cho and De Bruyn, 1979). In contrast to the latter, the endothelial cells of the postcapillary venules are not classified as belonging to the RES, because, under identical conditions, they are not capable of particulate uptake. However, as will be mentioned later, in special instances, they can become strikingly phagocytic.

The term *mononuclear phagocyte system,* as it is now defined, appears to exclude some or all of the cells lining the sinuses of the bone marrow, lymph node, spleen, and liver from the traditional concept of the RES. The definition of the MPS is based on origin, immunological properties, and quantitative phagocytic ability of this macrophage (Langevoort *et al.*, 1970; van Furth *et al.*, 1972). It is to be noted, however, that, although the macrophages, as identified by the term mononuclear phagocytes, take up larger particles and larger aggregates of particulate tracers than the sinus-lining cells, the latter have, nevertheless, considerable phagocytic ability (Figure 1; see also De Bruyn *et al.*, 1975; Farr *et al.*, 1980). A functional distinction between sinus-lining cells and vascular endothelia, which have been termed *non-RES endothelia* (Cotran, 1965), is useful. These non-RES vascular endothelia may take up small amounts of particulates, but only after systemic overloading by repeated administrations of high doses (Cotran, 1965; Cotran *et al.*, 1967). Lower dose levels of particulates, which consistently result in the uptake by macrophages and sinus-lining cells, do not cause phagocytic response in these non-RES endothelia (Figure 1). Furthermore, these endothelia lack the characteristic cellular organelles of the sinus endothelia for the uptake, intracellular transport and disposition of the endocytized particulates (De Bruyn *et al.*, 1975, 1977b). Prompted by these considerations, we propose, for the purposes of this chapter, to consider the RES as comprising: (1) macrophages in the sense defined by the term *mononuclear phagocytes,* and (2) sinus-lining cells.

2. ENDOCYTOSIS BY THE CELLS OF THE RES INTERACTING WITH LYMPHOCYTES

In the two instances of lymphocyte–RES interaction to be considered here, different cell types from the RES are involved, each exhibiting phagocytic properties, but each doing so in a different manner. The reticuloendothelial cells involved in immune interactions are considered to be macrophages (Unanue,

---

(microphagocytosis). (f) Endothelial cell of bone marrow capillary from the same section as depicted in (e)—no phagocytosis. ph, Phagosomes.

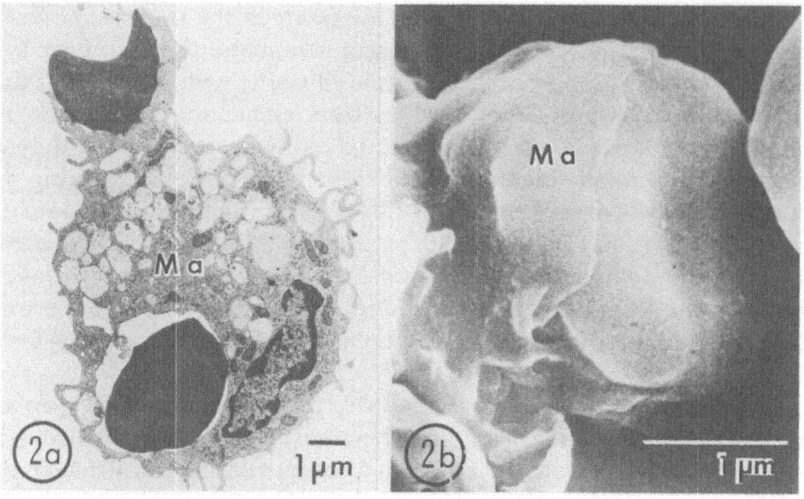

FIGURE 2. Phagocytosis of anti-sheep hemolysin-coated sheep erythrocytes by ubiquinone-8-stimulated guinea pig macrophages (Ma). (a) Transmission electron micrograph; (b) scanning electron micrograph. Note the cuplike extension of the macrophage cytoplasm over the erythrocyte. [From Biemesderfer *et al.* (1978), Changes in macrophage surface morphology and erythrophagocytosis induced by ubiquinone-8, in: *Scanning Electron Microscopy/II* (R. P. Becker and O. Johari, eds.), pp. 333–340, SEM Inc., AMF O'Hare, Illinois.]

1972). These cells have a wide distribution within the organism (reviewed by Carr, 1973). In addition to macrophage precursor cells circulating in the blood, macrophages are present in the parenchyma of myelopoietic and lymphopoietic tissue as well as in the loose connective tissue. Furthermore, they occur as Kupffer cells in the sinuses of the liver. Cytoplasmic extensions enable these cells to take up particles of a size as large as erythrocytes (Figure 2), or to engulf at once an accumulation of numerous small particles (Figure 1a). While these cytoplasmic extensions have the appearance of fingerlike pseudopods (filopodia) in sections (Figure 2a), they may also have the shape of a cup enclosing the particle being engulfed in its cavity (Figure 2b). The macrophages which, in this way, are capable of the uptake of relatively large particulates, have an additional endocytic apparatus: the large bristle-coated vesicles. These large bristle-coated vesicles are the sole endocytic apparatus of the lining cells of the sinuses of the liver, bone marrow, spleen, and lymph node. The vesicles are formed by invagination of the plasma membrane with the concomitant development of what, in cross sections, appear as small spikelike bristles at the cytoplasmic aspect of the plasmalemma. A simultaneous change in the random distribution of sialic acid at the luminal side of the cell surface occurs at this site (De Bruyn *et al.*, 1978). They pinch off from the plasmalemma and appear then as free large, bristle-coated vesicles in the cytoplasm. The diameter of the large, bristle-coated vesicles ranges from 1000–2100 Å (mean 1400 Å), and they should be distinguished from the small, bristle-coated vesicles occurring in the cytoplasm which have a diameter of 300–500 Å (mean 400 Å), and to which no endocytic function has been ascribed. The size of the large, bristle-coated vesicles determines the size of the particles ingested, and, consequently, uptake by these coated vesicles is restricted to particles of relatively small size. The large, coated vesicles may fuse with each other to form larger smooth-membraned storage vesicles (phagosomes) containing tracer particles, or they may fuse with smooth-membraned "transfer tubules," that convey the tracer to lysosomal bodies (De Bruyn *et al.*, 1975; Bankston, 1975). The latter process is prevalent when proteins are used as tracers. Considering the actual endocytic process of the cells of the RES, rather than the variable configurations of the intracellular vesicles resulting from this event (Wisse, 1977), two clearly defined mechanisms of phagocytosis occur in the cells of the RES: one by pseudopodial engulfment and one by the activity of the large, bristle-coated vesicles (Figures 1 and 2). For the purpose of this discussion, it is convenient to refer to these particular two different types of endocytosis as macrophagocytosis and microphagocytosis, a distinction based on the size of particles that these two endocytic components are capable of ingesting.*

The regulation of endocytosis is not well understood. In the case of the macrophage, endocytosis of material may be mediated by the association of extracellular material with two populations of receptors on the plasma membrane. One receptor (Fc receptor) binds the Fc portion of IgG antibodies and in doing so, causes these antibodies to display cytophilic properties (Berken and Benacerraf, 1966; Bianco *et al.*, 1975). The other receptor binds a complement cleavage product (C3b) which is bound by antigen–antibody complexes (Griffin *et al.*, 1975). Depending on the functional state of the macrophage, attachment of material to the macrophage by either of these two types of receptors may result in endocytosis of the bound material. Inflammatory exudate macrophages bind and ingest larger numbers of IgG-coated erythrocytes than do normal resident peritoneal macrophages. In addition to phagocytosis of these immunologically modified materials [termed *immunological phagocytosis* by Rabinovitch (1970)], macrophages are also capable of ingesting materials independently of serum-derived recognition factors. The manner in which this nonimmune phagocytosis is regulated remains unclear, and there have been few attempts to correlate the two different morphological forms of phagocytosis (here defined as macrophagocytosis and microphagocytosis) with the different mechanisms of phagocytosis (immunological or nonimmunological).

The cells lining the sinuses of the blood-forming organs and of the liver have vigorous phagocytic abilities (Hudson and Yoffey, 1963, 1968; De Bruyn *et al.*, 1975; Wisse, 1972, 1977), although they have only the microphagocytic, large, bristle-coated vesicles at their disposal for this process, and, thus, only small particulates and small aggregates of these particulates are sequestered in

*The pseudopodial activity of the macrophagocytic process is structurally much like the process described by W. H. Lewis in 1931 as pinocytosis, a term designed to describe the uptake of fluid through the engulfment of small, but light-microscopically visible droplets. Fawcett in 1965 drew attention to the indiscriminate use of the term pinocytosis which encompassed, then as now, a variety of surface activities, each with quite different cellular mechanisms, some of which clearly do not belong in the category of endocytic processes characteristic for the RES.

their cytoplasm (Figure 1). Since all the sinus-lining cells of the liver, bone marrow, spleen, and lymph node have microphagocytic properties, it is clear that the total phagocytic capacity by this means of ingestion is considerable.

Majno and Joris (1978) have drawn attention to the structural, functional, and metabolic differences between various endothelia, and they caution against general statements concerning "endothelium" without further specifications. In order to avoid confusion between the phagocytic sinus cells (often termed *endothelium*) and the non-RES vascular endothelium, we shall here continually define which type of endothelial cell is under discussion. For a more convenient distinction between the phagocytic endothelia lining sinuses and non-RES vascular endothelia, the older term *littoral cells* for the former is worthy to be considered for reinstatement.

The distinct structural differences between the two types of phagocytosis suggest functional differences. At present, nothing is known about any such differences related to this process other than particle size. In the macrophagocytic process of the macrophage, surface receptors for the Fc fragment of IgG complexes and for the third component of complement are known to have a role in the ingestive process. There is no information at present, whether or not there are similar receptors at the microphagocytic sites, i.e., the sites of large, coated vesicle formation at the surfaces of the sinus-lining cells and macrophages. It is of some interest that the macrophage, which is capable of both types of endocytosis, is a cell capable of interacting with lymphocytes, with subsequent immunological consequences. As far as is known, the sinus-lining cell which ingests only through microphagocytic action has no immunological potential.

3. MACROPHAGES AND SINUS-LINING CELLS

The internal complement of intracellular organelles in the cytoplasm of macrophages and sinus-lining cells corresponds to the differences in their functional characteristics. The macrophage, because of its ability to ingest large particulates and large aggregates of small particulates, is more richly endowed with lysosomes than the sinus-lining cells which are limited to the ingestion of small particulates and small aggregates of these particulates. As a consequence of the greater need for lysosomal enzymes, the macrophages have more ribosome-studded endoplasmic reticulum (RER) and more free polyribosomes than the sinus-lining cells. The larger content of cytoplasmic RNA is, however, not sufficient to impart a basophilia to the cytoplasm of the macrophages. The nucleus contains a rather substantial nucleolus. The microphagocytic sinus cells have fewer of these organelles related to protein synthesis. Phagocytosis is energy dependent (Sbarra and Karnovsky, 1959; Sastry and Hokin, 1966), and the higher energy requirements necessary for the ingestion and internal transport of the larger masses of particulates phagocytized by the macrophages as compared to the microphagocytic sinus-lining cells are reflected in a relatively larger number of mitochondria in the macrophages. The sinus-lining cells and the Kupffer cells contain transfer tubules. These tubules are derived from the Golgi complex and are acid-phosphatase- and phosphotungstic-acid-positive (De Bruyn *et al.*, 1975, 1977b).

4. LYMPHOCYTES AND PLASMA CELLS

At the time when the studies on the formation of blood cells centered almost exclusively around the question of whether the two cell lineages, lymphocytes and myeloid cells, originated from a common cell type or not, the term *stem cell* referred to a distinct morphological entity. Depending on whether one believed in a common ancestor cell or in two different ones, the stem cell was termed, respectively, hemocytoblast or lymphoblast and myeloblast. The functional studies of the last years have shown that a common hematopoietic stem cell exists in the various hematopoietic organs, but its structural configuration has not been clearly defined, or, at least, has not been generally accepted. Consequently, the term *stem cell* now has a somewhat different connotation than during the time when morphological studies on hematopoiesis were of sole concern. The terms *hemocytoblast, lymphoblast,* and *myeloblast* are now often replaced by the term *blast cell,* thus avoiding any commitment to either the monophyletic or the polyphyletic concept of hematopoiesis in areas of study where this is not of prime interest.

The small lymphocyte (Figure 3a) has a diameter of about 5–6 μm. There is a thin rim of cytoplasm, so that the nucleus, which contains a centrally located nucleolus, is relatively large. Clumps of condensed heterochromatin occur throughout the nuclear mass but are often seen in contact with the nuclear membrane. The nucleus frequently has a unilateral indentation and, in favorable sections, one may see a Golgi complex and centrioles at this site in the cytoplasm. A small complement of mitochondria is evenly distributed throughout the cytoplasm, which contains a sufficient number of ribosomes to give it a distinct basophilia in light microscopic preparations of lymphocytes stained with a basic (cationic) dye. Azurophil granules may occasionally be observed.

While once considered a homogeneous population of cells with unknown biological significance, lymphocytes have been shown in recent studies to be a heterogeneous population of cells with regard to developmental profile as well as immunological function. A review of these aspects of lymphocyte biology is outside the scope of this work, and for extensive consideration of this subject the reader is referred elsewhere (Katz, 1977). A brief discussion of some aspects of these topics is included here for background.

It has been possible to classify populations of lymphocytes on the basis of their developmental pathways: immature lymphocytes that differentiate within the microenvironment of the thymus (hereafter referred to as thymus-processed lymphocytes or T cells) and those which appear to undergo functional maturation without a thymic influence (subsequently referred to as B cells). These derive from the Bursa of Fabricius in fowl, and from an equivalent, possibly marrow, in mammals. In the course of this divergent differentiation, these subpopulations of lymphocytes acquire different functional capabilities, as well as

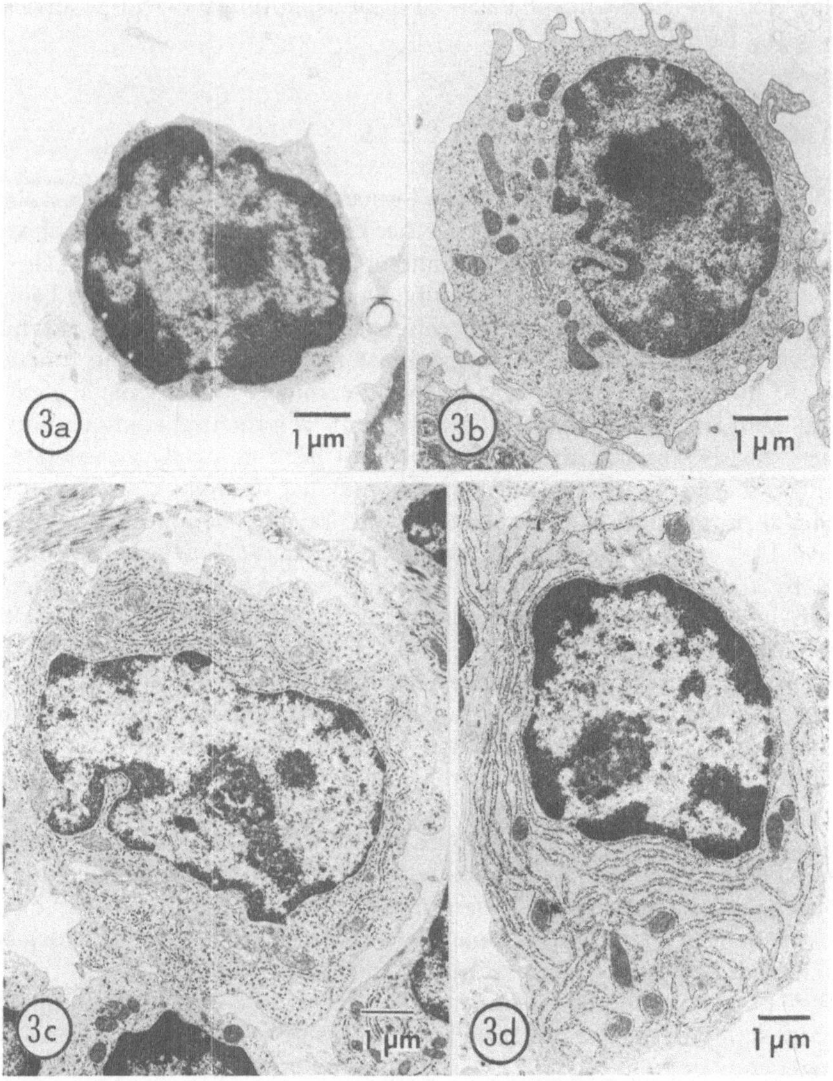

FIGURE 3. (a) Small lymphocyte; (b) lymphoblast; (c) plasmablast; (d) plasma cell. From sections of rat lymph node.

markers which allow their identification. Thus bone-marrow-derived lymphocytes may display different classes of immunoglobulin on their surface, as well as receptors for the third component of complement and for the Fc region of some classes of immunoglobulins. Upon receiving the proper antigenic stimulation, the B cell may then synthesize and secrete immunoglobulin (Katz, 1977; Möller, 1975a).

Thymus-processed lymphocytes, which may be identified on the basis of unique cell surface antigens, do not have surface immunoglobulin as an integral membrane component and do not synthesize immunoglobulins. For many antigens, subsets of T cells are required to provide an additional stimulus to the B cell for the production of antibody to occur (Katz, 1977; Katz and Benacerraf, 1972). Other subsets of T cells have also been identified which are capable of suppressing the production of antibody by B cells (Pierce and Kapp, 1976; Möller, 1975b). Although *in vitro* studies indicate that some B-cell functions may be affected by T-cell extracts, the mechanism of T-cell regulation of B-cell function is unclear, and it is not known whether direct cell contact between different cell types is required, or if soluble factors are involved *in vivo*. T cells are involved in a number of immunological processes in addition to the regulation of antibody formation by B cells. Delayed hypersensitivity (Oppenheim and Seeger, 1976), cell-mediated immunity to viral and bacterial infection (Blanden *et al.*, 1976), and the rejection of allografts and some tumors (Lohmann-Matthes, 1976), are all reflections of T-cell activity.

In addition to the cell surface antigens and receptors which allow one to distinguish between T and B lymphocytes, attempts have been made to identify these subpopulations of lymphocytes on the basis of cell surface morphology. Accordingly, B cells were reported to display a cell surface with increased numbers of microvilli and small projections while the T cell was thought to display a relatively smooth surface (Polliack *et al.*, 1973). Subsequent work has demonstrated that such differences in lymphocyte surface morphology result from the methods of preparation of the cells for scanning electron microscopy (SEM) and cannot be used to distinguish between thymus-processed and bone-marrow-derived lymphocytes (Alexander and Wetzel, 1975; Wetzel, 1976).

As mentioned above, after antigen stimulation under the proper conditions, the B cell may further differentiate into a cell capable of secreting large amounts of immunoglobulin. Electron microscopic studies of such antibody-producing cells reveal cells with a rather pleomorphic appearance, ranging from cells with the appearance of lymphoblasts (Figure 3b) with little apparent provision for protein secretion, to cells with extensive RER. This variable appearance of antibody-secreting cells is thought to represent different stages of development as the antigen-stimulated B cell develops into a mature antibody-secreting plasma cell (Figure 3d). The first indication of this series of changes is the accumulation of a few ribosome-studded cisternae of RER (plasmablast, Figure 3c). At this stage, the cell as well as the nucleus has increased in size, and a relatively large nucleolus gives evidence of the increased rRNA production. The cytoplasm continues to increase in mass and begins to acquire an extensively developed RER in the cytoplasmic interstices, in which there are a well-developed Golgi complex, a modest number of mitochondria, and some free ribosomes. The nucleus assumes an eccentric position as the cytoplasm develops and dense masses of heterochromatin are attached to the nuclear membrane. These masses of chromatin, together with a centrally located nucleolus, give the nucleus the appearance of a cartwheel which is most consistently evident in thick, light-microscopic sections. When stained with a basic dye, the cytoplasm shows a distinct basophilia as a result of the extensive presence of RNA. The fully developed plasma cell which contains immunoglobulins in the cisternae of the endoplasmic reticulum (Leduc *et al.*, 1969) now has all the appearance of a secreting cell.

Upon stimulation the small T lymphocyte also undergoes structural changes, and also develops into a cell with the appearance of a lymphoblast (Janossy *et al.*, 1972, Matter *et al.*. 1972). The cytoplasm and the nucleus increase in size and numerous polyribosomes appear in the cytoplasm, but there are only a few cisternae of RER. The nucleolus is relatively large, reflecting increased rRNA production. As the nucleus increases in size, it becomes more euchromatic and, consequently, lighter in appearance than that of the unstimulated small lymphocyte.

5. INTERACTION BETWEEN LYMPHOCYTES AND CELLS OF THE RES IN THE IMMUNE PROCESS

The interaction of lymphocytes with components of the RES is essential for the induction of many phenomena of the immune system. Various *in vitro* T-cell functions, such as antigen-dependent proliferation, mitogenic responsiveness, and generation of helper T-cell function for antibody responses have been shown to require a contribution from macrophages (Rosenstreich and Mizel, 1978; Schwartz *et al.*, 1978; Rosenthal, 1978). In a reciprocal fashion, a number of macrophage functions may be modified by interaction with immune T cells and the appropriate antigen. Following such interactions *in vitro*, the macrophage exhibits an increased basal rate of metabolic activity with increased rates of microendocytosis and [^{14}C]glucosamine incorporation (David and Remold, 1976). It also displays enhanced bacteriostatic and tumoricidal activity (Blanden *et al.*, 1976; Lohmann-Matthes, 1976). Under these conditions, there is also increased secretory activity by the macrophage resulting in the release of an array of enzymes such as lysozyme, plasminogen activator, elastase, and collagenase (Unanue, 1976). In addition to these materials, macrophages also release a number of lymphoregulatory molecules which can affect lymphocyte proliferation and differentiation (Beller *et al.*, 1978). By monitoring the production of these lymphoregulatory molecules in cultures of immune T cells and normal macrophages, it has been possible to gain some insight as to the nature of this lymphocyte–macrophage interaction (Farr *et al.*, 1977). First, the interaction of immune lymphocytes and macrophages resulting in the secretion of the lymphoregulatory materials requires that the lymphocyte and the macrophage be in direct physical contact. Separation of the lymphocyte and macrophage with a membrane impermeable to cells prevents the production of these mediators. Second, genetic analysis of this interaction revealed that the lymphocyte and macrophage must share regions of genetic homology for the interaction to occur. This genetic restriction is localized to a subregion of the major histocompatibility complex, the I region, which is known to contain genes which determine the ability of different mouse strains to respond to various antigens (Klein, 1975). In addition to these immune response genes, this subregion of the major histocompatibility complex also contains genes which code for cell surface antigens present on subpopulations of lymphocytes and macrophages which may be important structures for cell interactions (Niederhuber, 1978). It has been suggested that antigen recognition by T cells involves recognition of antigen in association with these cell surface antigens rather than as free antigen (Rosenthal, 1978). The manner in which macrophages take up potentially immunogenic material, process it, and then interact with lymphocytes remains a fascinating problem. Upon examining that aspect of macrophage physiology, one is faced with a paradoxical situation. On one hand, the macrophage is a powerfully catabolic cell which enzymatically degrades much of the ingested material, while, on the other hand, some of the material must be presented to the lymphocyte in such a way as to convey information regarding the structure of the antigen. There is evidence that a small but immunogenic amount of antigen may remain on the surface of the macrophage, from which it can be removed by treatment of the macrophage with proteolytic enzymes (Unanue, 1972).

The observed requirement for cell contact between the lymphocyte and macrophage in the secretion of lymphoregulatory materials is reflected morphologically in cell cultures generating these materials. When immune lymphocytes are cultured with normal macrophages in the presence of the appropriate antigen, there is a marked increase in the number of lymphocytes found associated with the macrophages. This association is antigen specific in that increased lymphocyte–macrophage association is not observed when antigen is not added to the culture (Farr and Unanue, unpublished observations). A similar requirement for T cell–macrophage interactions has been observed for the antigen-dependent proliferation of guinea pig lymphocytes as determined autoradiographically (Lipski and Rosenthal, 1975; Rosenthal *et al.*, 1976). Cell clusters consisting of one or a few central macrophages surrounded by small lymphocytes, lymphoblasts, and plasma cells in direct cell surface contact have been described in cultures of immune lymphocytes *in vitro* (Nielsen *et al.*, 1974; McFarland *et al.*, 1966; Sulitzeanu *et al.*, 1971), and have been held to be sites of interaction of these cell types involved in the production of antibody and mediators of delayed hypersensitivity (Lipski and Rosenthal, 1975; Bartfield and Kelly, 1968; Saunders and Hammond, 1970).

Similar cell clusters (Figures 4, 5, and 6) have been found strategically located in the sinuses of the lymphatic tissue (Farr and De Bruyn, 1975b). The sinuses, which are discrete vascular channels lined by reticular-lining cells, are spanned by trabecular structures consisting of bundles of collagenous fibrils covered by trabecular reticulum cells. The reticular cells have no filopodia and present a relatively smooth surface which displays a varying number of microphagocytic, large, bristle-coated vesicles. In this respect, the reticular cells lining the sinuses as well as those of the trabeculae, are similar to the sinus-lining cells of other hematopoietic organs and of the liver. Macrophages with both numerous filopodia and large, bristle-coated vesicles are in contact most often with the intraluminal trabeculae, but, occasionally, also with the sinus wall. Lymphocytes are often seen in contact with these macrophages but not with the trabecu-

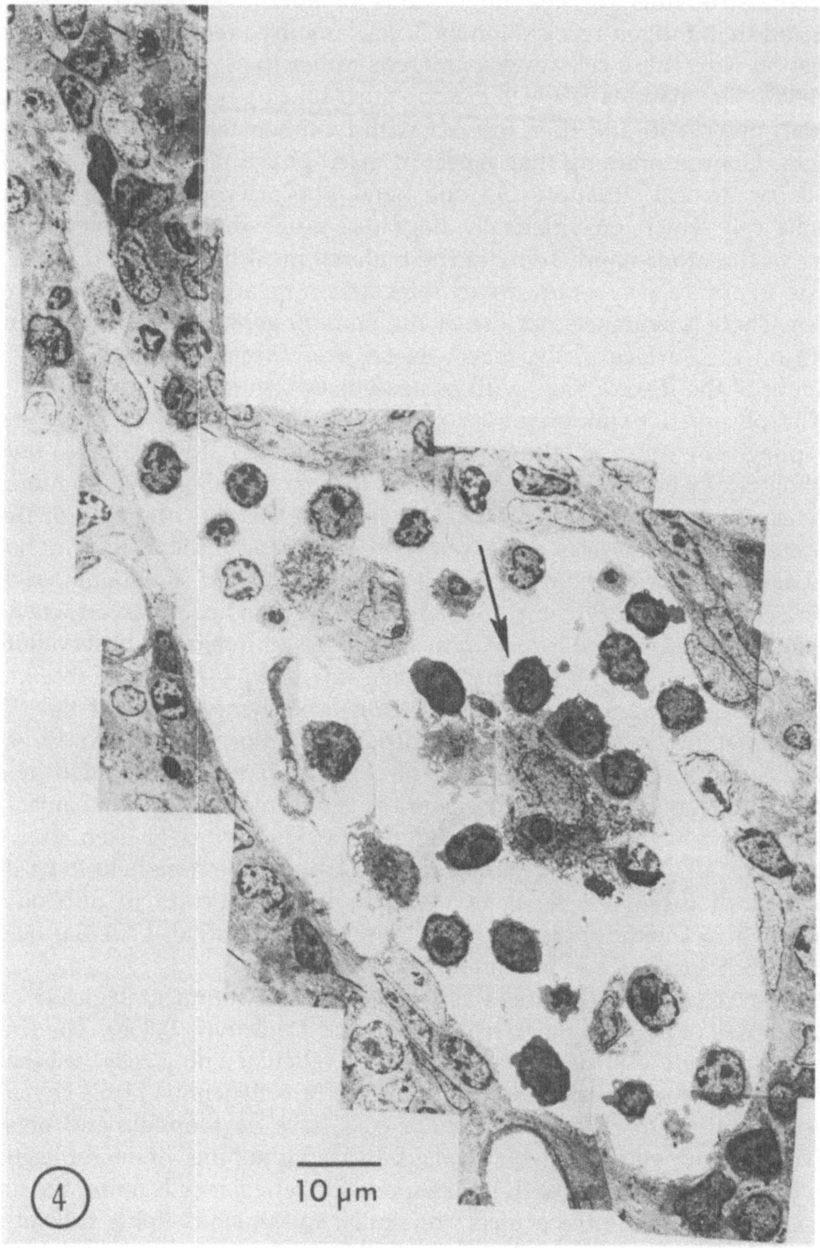

FIGURE 4. Sinus of a rat lymph node. There is one macrophage–lymphocyte cluster (arrow). The sinus wall is continuous. Composite of 14 electron micrographs.

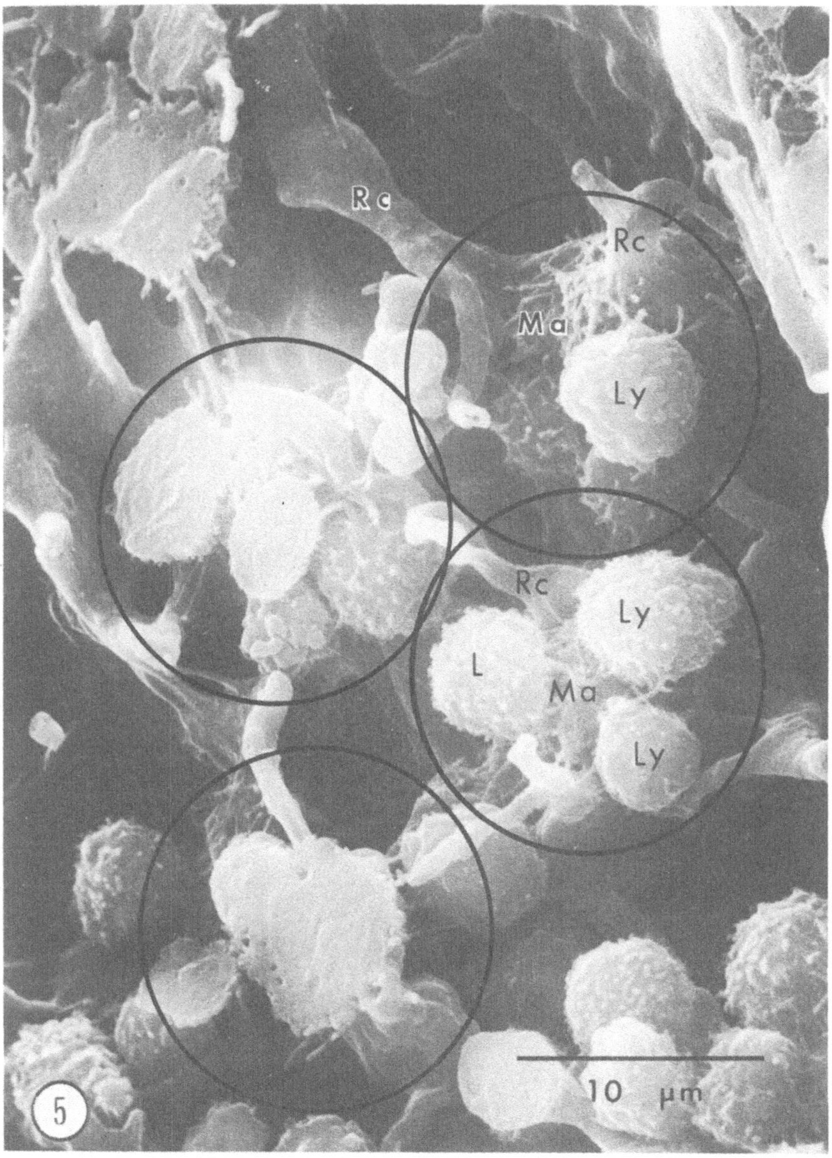

FIGURE 5. Scanning electron micrograph of the lumen of a sinus of a rat lymph node. There are four macrophage–lymphocyte clusters. Each cluster consists of lymphocytes (Ly) associated with a macrophage (Ma), which, in turn, is in contact with a reticular cell (Rc). The macrophages are identified by their filopodia, while the reticular cells have a smooth surface. The cells of two clusters are cut by the cryofracture. (Micrograph from R. P. Becker.)

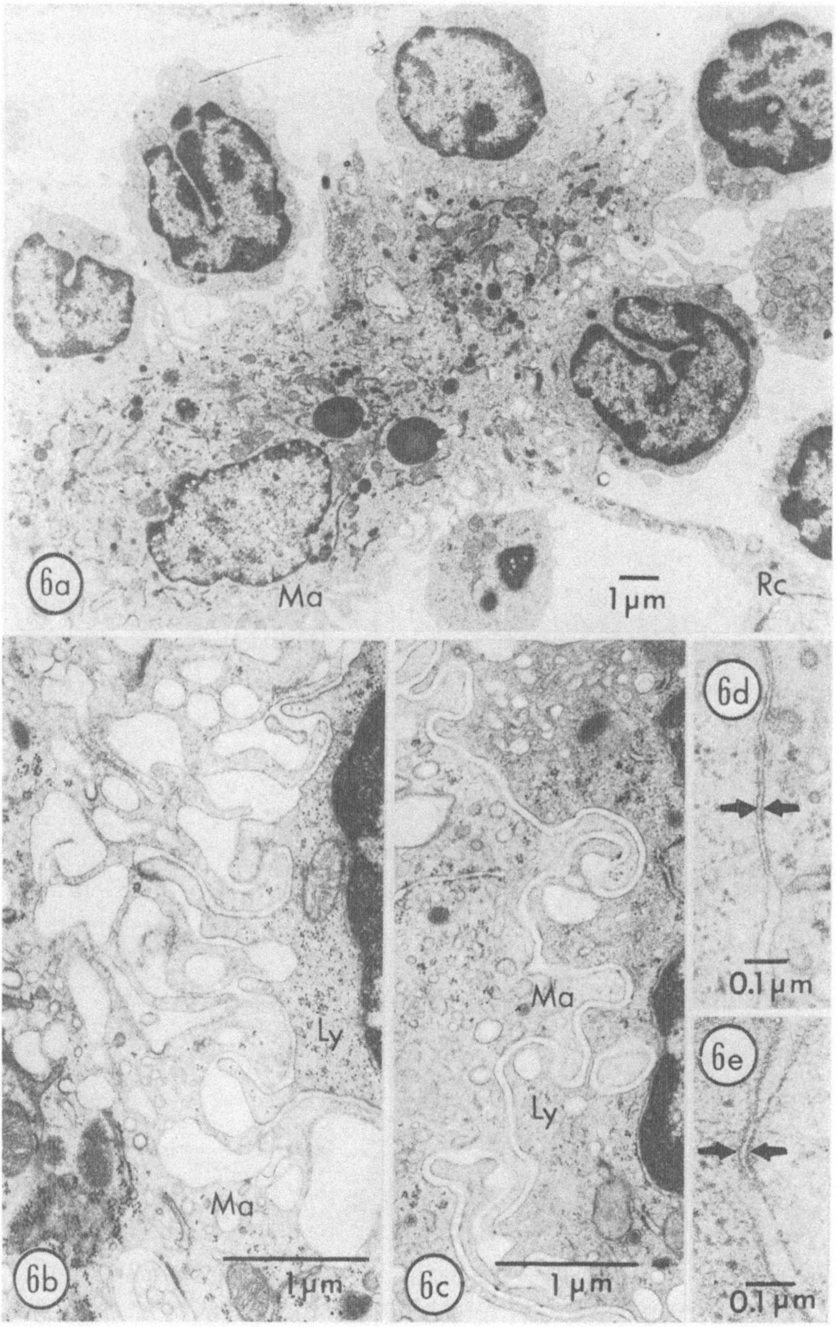

FIGURE 6. (a) Cell cluster in the sinus of a rat lymph node consisting of one central macrophage (Ma) with several associated lymphoid cells. The macrophage is attached to a reticular cell (Rc). (b, c) Sites of contact between macrophages (Ma) and lymphocytes of a cell cluster; (d, e) places where the distances between the plasmalemma of the macrophage (right) and lymphocyte (left) is reduced to 70–80 Å. (From Farr and De Bruyn, 1975, *Am. J. Anat.*, **144:** 209–232.)

lar reticulum cells (Figures 5 and 6a). These lymphoid cells range from small lymphocytes to those of the lymphoblastic type and cells with a sufficient number of cisternae of the RER to be termed plasmablasts. A number of morphological peculiarities occurring in these cell aggregates indicates that these cell associations in the lymphatic sinuses form definite structural units and are not simply the result of accidental groupings. At the site of contact between macrophage and reticular cell, there are characteristic paraplasmalemmal densities in the cytoplasm of the macrophage which cover large segments of the contact area. No such subplasmalemmal densities occur at the site of contact between macrophages and lymphocytes. In sections, the macrophages and lymphoid cells reveal an extensive contiguity between the two cell surfaces, presenting a variety of morphological configurations. Sometimes pseudopodial processes of the lymphoid cells deeply indent the cytoplasm of the macrophage, while in other instances the macrophage filopodia extensively intertwine with the pseudopodial processes of lymphocytes (Figure 6b). The contact sites may also be relatively smooth with the major portion of the total lymphocyte surface directly abutting a macrophage (Figure 6c). At the sites of contact, the distance between macrophages and lymphocytes generally ranges from 350–375 Å, but may, at clearly defined sites, be sharply reduced to 60–80 Å (Figure 6d,e). The grouping of lymphoid cells around a central macrophage results in increased contact between adhering lymphocytes, which would facilitate lymphocyte–lymphocyte interaction. The distinct configuration of these cell clusters consisting of cell types requisite for antibody synthesis and other immune phenomena, their strategic location within lymphoid tissue, and their similarity to the cell clusters associated with various immune processes *in vitro,* suggest that these clusters may be sites instrumental to such activities. The role of the reticular cells to which the central macrophage is attached, remains undetermined. The extent of these clusters or the number of associated lymphocytes cannot be adequately estimated in thin transmission electron microscopy (TEM) sections (Figure 4). Scanning electron micrographs, where large areas can be viewed in depth, reveal their frequency (Figure 5). These cell clusters have been found in normal lymphoid tissue undergoing only naturally occurring antigenic stimulation. It is to be noted, that antigen-independent cluster formation can also occur *in vitro* between T cells and macrophages (Lipsky and Rosenthal, 1973). Whether these cell clusters, observed *in vivo* under normal conditions, represent antigen-independent cell interactions or occur under conditions of natural antigenic stimulation, or both, remains to be determined. The presence of these intrasinal macrophage–lymphocyte clusters in lymphatic tissue has been confirmed by Friess (1977) in rats immunized with sheep red blood cells.

Associations between plasma cells and macrophages have been observed within the medullary cords of the lymph node by several investigators (Miller and Avrameas, 1971; Thiery, 1962), although, according to one report, their presence is a rare occurrence (Thiery, 1962). Utilizing an antigen which could be localized histochemically, close association has been observed between macrophages which had taken up the antigen and plasma cells secreting antibody directed against that antigen (Miller and Avrameas, 1971).

6. INTERACTION BETWEEN LYMPHOCYTES AND CELLS OF THE RES DURING TRANSMURAL CELL PASSAGE

The ability of a cell to enter or leave a vascular compartment, and to enter or leave the extravascular parenchyma by penetrating a continuous cellular interface, is a property which is, under normal conditions, peculiar to blood cells in hematopoietic organs. In the bone marrow, the entry of blood cells including lymphocytes from the extravascular site of formation into the circulation takes place through the continuous cellular lining of the sinuses and is selectively restricted to mature blood cells (De Bruyn *et al.*, 1966, 1971; Huhn, 1966; Huhn and Stich, 1969; Campbell, 1972; Leblond, 1973; Becker and De Bruyn, 1976; Muto, 1976). The transmural migration is through the cell body of the sinus cell and not at the sites of junction between the endothelial cells (Figure 7a,b). A temporary migration pore is formed in the body of the endothelial cell which closes again after the blood cell has entered the intravascular space. Erythrocytes, newly formed as a compensatory reaction to hemolytic destruction caused by erythroblastosis virus, appear to follow a similar transcellular pathway across

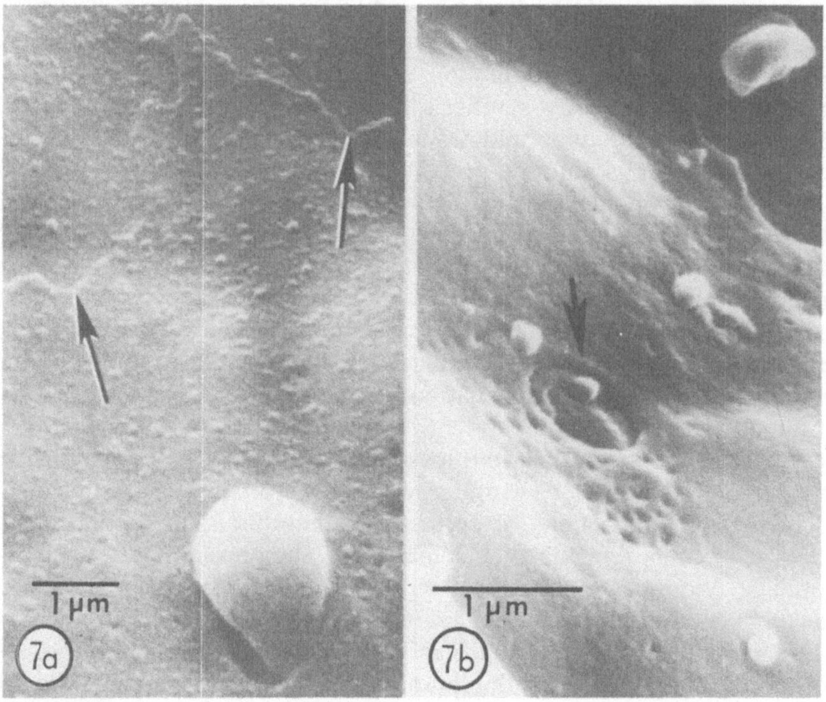

FIGURE 7. Scanning electron micrographs of the luminal aspect of rat bone marrow sinuses. (a) A blood cell is entering the lumen through the body of a sinus-lining cell. The arrows indicate the junctions of adjacent lining cells; (b) early stage of the entry of a blood cell into the circulation. The migration pore (arrow) is surrounded by fenestrae with diaphragms. (From Becker and De Bruyn, 1976, *Am. J. Anat.* **145**:183–206.)

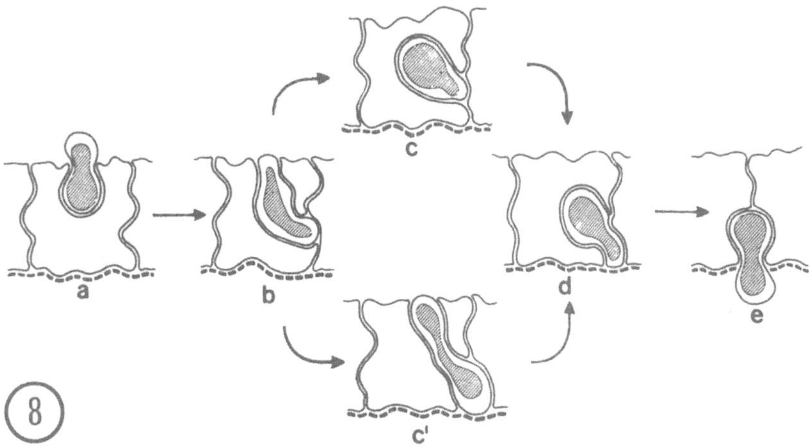

FIGURE 8. Diagram of the transcellular pathway of a lymphocyte through the endothelium of a postcapillary venule of a lymph node of a rat. (From Farr and De Bruyn, 1975, *Am. J. Anat.* **143:** 59–92.)

the sinus wall of the spleen (Cho and De Bruyn, 1975). It has been proposed that the transcellular pathway is related to the selectivity of transmural diapedesis and is based on an interaction of the mature blood cell and the sinus-lining cell (De Bruyn *et al.*, 1971). This selectivity is part of the mechanism regulating the quantitative and qualitative steady state of the peripheral blood pool.

The lymph node presents a more complex situation with regard to transmural passage of lymphocytes. In addition to diapedesis of lymphocytes through the endothelium of postcapillary venules, there is also extensive cell traffic occurring at the interface of the lymph node parenchyma and the lymphatic sinuses. Lymphocytes may enter lymph node parenchyma after arriving in the lymph node with the afferent lymph. Lymphocytes produced within the lymph node as well as those recirculating through the lymph node also leave the lymph node by crossing the wall of the lymphatic sinus. Examination of the structure of the lymphatic sinus wall in murine lymph nodes revealed a number of interesting features, relevant to the discussion here, regarding the interaction of blood cells with elements of the RES. The cells comprising the wall of the lymphatic sinus of the lymph node form a continuous layer (Figure 4), and there is no free communication between sinus lumen and lymphatic parenchyma (Clark, 1962). Thus, any cells traversing the lymphatic sinus wall must, in some way, penetrate this interface. SEM studies of lymphocytes in transmural passage through the sinus wall show that the lymphocyte pathway through the sinus wall is transcellular, suggesting a similar regulatory mechanism of transmural cell passage as occurs in the postcapillary venules (Farr *et al.*, 1980).

The passage of lymphocytes from the peripheral circulation into the lymphatic parenchyma through the postcapillary venules presents a special case. The lymphocytes enter the venule wall of penetrating the body of the endothelial cell and leave the endothelial cell body by entering the intervascular space between the endothelial cells, and from here gain access to the lymphatic parenchyma (Figure 8). The plasmalemma of both the migrating lymphocyte and of the endothelial cell retain their integrity during the transmural passage (Messier and Sainte-Marie, 1972). This essentially transcellular pathway has been resolved by determining, on the basis of contiguous serial sections, the location of the migrating lymphocyte in the endothelium at all stages of the transmural process (Farr and De Bruyn, 1975a), and by scanning electron microscope views of the luminal aspect of the sinus wall showing the entry of lymphocytes into the endothelium (Cho and De Bruyn, 1979). The transcellular phase of the pathway

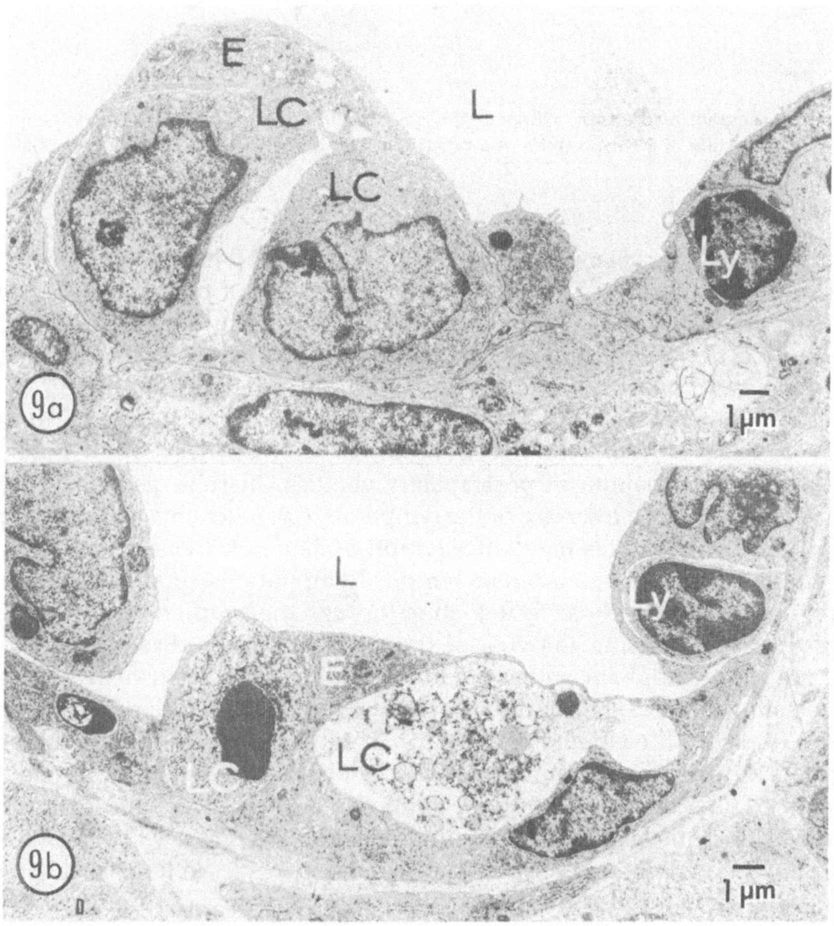

FIGURE 9. Ingestion of malignant myeloblasts (LC) in acute myelogenous leukemia (Shay) by the endothelial cells (E) of a postcapillary venule in the lymph node of a rat. (a) The malignant myeloblasts (LC) are still intact; (b) remnants of the malignant cells (LC) are enclosed by the endothelial cells (E). The transmural passage of lymphocytes (Ly) from the circulation into the extravascular lymphatic parenchyma continues unimpaired. L, Lumen of postcapillary venule. (From Cho and De Bruyn, 1978, *J. Natl. Cancer Inst.* **41**:185–195.)

of the lymphocyte through the endothelium of the postcapillary venule is a route which may be significant for the selective migration of only lymphocytes from the circulation into the lymphatic parenchyma (Marchesi and Gowans, 1964; Farr and De Bruyn, 1975a). In this respect, this selective process is essentially the same as the selective transcellular migration of mature blood cells through the sinus lining cells of the bone marrow. In contrast to the sinus cells of the bone marrow, the endothelial cells of the postcapillary venule are not considered to belong to the RES. It is noted, however, that under certain circumstances the endothelial cells of the postcapillary venule can display striking phagocytic abilities. A massive phagocytosis of malignant myeloblast-like cells in acute myelogenous leukemia (Shay) in rats has been observed in which the peripheral white blood cell count remained normal or exceptionally low (Cho and De Bruyn, 1978). This phagocytosis, which is followed by the destruction of the malignant cells, was effected by Kupffer cells in the liver, by macrophages in the spleen and bone marrow, and also by the endothelial cells of the postcapillary venules of the lymph nodes, which, at the same time, continue to be the locus of the return of normal lymphocytes from the circulation to the lymphatic parenchyma (Figure 9a,b). The reason for the curious difference in fate between normal lymphocytes and malignant cells is unknown. These observations, showing the destruction of malignant cells, are, in a sense, in agreement with the contention that the endothelial cells of the postcapillary venules may be, besides being part of the pathway for lymphatic recirculation, perhaps also a site of phagocytosis of defective lymphocytes (Wenk *et al.*, 1974). Experiments with inert particulate materials have, however, not been conclusive regarding the phagocytic ability of the endothelium of the postcapillary venules (Hummel, 1935; Mikata and Niki, 1971).

7. CELL CONTACT AND CELL SURFACE IN RES INTERACTIONS

It is likely that the interaction between the cells of the RES and lymphocytes is based on peculiarities of the cell surfaces. In the case of the lymphocyte–macrophage interaction in the immune response, gene products of the major histocompatibility complex and antigen receptors probably play a crucial role in regulating these cellular interactions. The selective affinity of recirculating lymphocytes and their interactions with the endothelium of postcapillary venules has been confirmed *in vitro* (Stamper and Woodruff, 1976, 1977). However the precise basis for this affinity, and for the selective transmural migration of lymphocytes and other blood cells through the lining cells of blood and lymph vessels of other hematopoietic organs, is unknown. There are, however, some indications that surface characteristics may be involved in the selectivity of this transmural passage. Interesting differences between cell surfaces of mature blood cells as compared to those of immature ones have been found. Mature blood cells have a lower concentration of Concanavalin A receptor sites than the corresponding immature forms (Ackerman and Waksal, 1974), while mature erythrocytes have a higher concentration of colloidal iron binding sites than their precursors (Skutelsky and Farquhar, 1976). Changes in the distribution of exposed sialic acid moieties at the abluminal face of myeloid sinus-lining cells at sites of transmural passage have been reported (De Bruyn, 1979). It remains to be determined whether the selective transmural passage is based on such, or related, changes of the cell surface.

In myelogenous leukemia, malignant myeloblasts in the extravascular marrow parenchyma enter the circulation through the cell body of the sinus-lining cells (Campbell, 1975; De Bruyn *et al.*, 1977a). These leukemic cells have, thus, acquired a property which, in the normal bone marrow, occurs only in mature blood cells, and which results in an interaction between the sinus-lining cell and the malignant cell, permitting the latter to perforate the vascular wall. The malignant cells from a subcutaneous tumor apparently acquire a similar attribute, as is evident from their transcellular pathway through non-RES endothelium (De Bruyn and Cho, 1978).

The complexity of the interaction between the reticuloendothelial cell and migrating blood cells is manifest from the following considerations regarding the selective transmural migration in hematopoietic organs. The first phase of this process is, in all likelihood, a discriminative recognition, selecting certain cell types for the transmural passage. It can be assumed that this recognition process requires a close approximation or perhaps direct cell contact. The absence of a fully constituted basal lamina at the sinuses of the bone marrow and at the sinuses of the lymphatic tissue could be significant in this respect, since its presence would preclude direct contact between the lining cell and the migrating blood cell. Regarding the close approximation of the two cell types, one should note that the cell surfaces are anionic. The repulsion resulting from the similar charge of the cell surfaces may be counteracted by molecules acting as specific ligands (Moscona, 1975). Also, evidence is now available for the presence of long range van der Waals–London forces (Parsegian and Gingell, 1973; Gingell and Fornes, 1976). An important theoretical discussion of the attractive and repulsive forces governing the interaction energy as a function of distance between two cells can be found in a paper by Weiss (1976).

Regarding the site of migration, Becker and De Bruyn (1976) and De Bruyn *et al.* (1977a) have observed that this place is regularly marked in the sinus-lining cells of the bone marrow by the presence of clusters of fenestrae with diaphragms (Figure 7b), sites which have been shown to be places permeable for blood-borne molecular tracers (Clementi and Palade, 1969; Simionescu *et al.*, 1972; Bankston and De Bruyn, 1974), One might speculate that the migrating cell may be attracted to these sites of diaphragmed fenestrae by a blood-borne factor. Alternatively, the diaphragmed fenestrae may simply be induced by the migrating blood cells as a precondition for diapedesis.

The formation of the migration pore which follows recognition occurs concomitantly with the transcellular progression of the migrating cell and probably is part of the selective interaction between the sinus cell and the migrating cell. Once the migration pore is formed, the transmural locomotory process should be considered as a separate event. In the case of the white blood cells, one might assume with reason that their ameboid properties produce the diapedetic movement. As far as erythrocytes, which have no locomotory properties of their own, are concerned, the movement through the lining cell may be the result of a *vis a tergo*. At this point, it is proper to consider the deformability of blood cells as a factor in their transmural passage. Measurements of cell deformability have been made, and the results have been interpreted to show that the deformability increases with the increase in the maturity of the blood cells involved. This has led to the suggestion that the increase in cell deformability plays a role in the egress of blood cells from the bone marrow (Lichtman and Weed, 1976; Lichtman *et al.*, 1978). Variations in deformability may have a role at that phase of the migratory event when a free communication between the extravascular and intravascular compartment has been established, and when a pore has been formed through which the cell has to squeeze.

8. CONCLUSIONS

It is of particular interest that both the immunological interaction as well as the interaction in transmural passage appear not to be related to the main functional characteristic of the RES, that of phagocytosis. The macrophage ingests and destroys antigen, but it is that small fraction of antigen which escapes the characteristic catabolic activity of the macrophage which has a role in the lymphocyte–macrophage interaction necessary for the generation of various T-cell functions as well as for the production of antibody. In a similar way, the typical microphagocytic function of the sinus-lining cells of the vessels in hemopoietic organs appears to have no relation to the function of these lining cells as a discriminating interface regulating the passage of blood cells. These two different processes, unrelated as they are to the traditional concept of the function of the RES as a phagocytic system, are, nevertheless, unique functional properties restricted to cells belonging to the RES.

REFERENCES

Ackerman, G. A., and Waksal, S. D., 1974, Ultrastructural localization of concanavalin A binding sites on the surface of differentiating hemopoietic cells, *Cell Tiss. Res.* **150**:331.

Alexander, E. L., and Wetzel, B., 1975, Human lymphocytes: Similarity of B- and T- cell surface morphology, *Science* **188**:732.

Bankston, P. W., 1975, Intracellular transport of carbon in endothelial cells of fetal rat liver sinusoids, *Anat. Rec.* **181**:306.

Bankston, P. W., and De Bruyn, P. P. H., 1974, The permeability to carbon of the sinusoidal lining cells of the embryonic rat liver and rat bone marrow, *Am. J. Anat.* **141**:281.

Bartfield, H., and Kelly, R., 1968, Mediation of delayed hypersensitivity by peripheral blood lymphocytes *in vitro* and by their products *in vivo* and *in vitro*, morphology of *in vitro* lymphocyte-macrophage interactions, *J. Immunol.* **100**:1000.

Becker, R. P., and De Bruyn, P. P. H., 1976, The transmural passage of blood cells into myeloid sinusoids and the entry of platelets into the sinusoidal circulation: A scanning electron microscopic investigation, *Am. J. Anat.* **145**:183.

Beller, D. I., Farr, A. G., and Unanue, E. R., 1978, Regulation of lymphocyte proliferation and differentiation by macrophages, *Fed. Proc.* **37**:91.

Berken, A., and Benacerraf, B., 1966, Properties of antibodies cytophilic for macrophages, *J. Exp. Med.* **123**:119.

Bianco, C., Griffin, F. M., and Silverstein, S., 1975, Studies of the macrophage complement receptor: Alteration of receptor function upon macrophage activation, *J. Exp. Med.* **141**:1278.

Biemesderfer, D., Block, L. H., Male, P., and Kashgarian, M., 1978, Changes in macrophage surface morphology and erythrophagocytosis induced by ubiquinone-8, in: *Scanning Electron Microscopy/II* (R. P. Becker and O. Johari, eds.), pp. 333–340, SEM Inc., Air Mail Facility, O'Hare, Illinois.

Blanden, R. V., Hapel, A. J., Doherty, P. C., and Zinkernagel, R. M., 1976, Lymphocyte-macrophage interactions and macrophage activation in the expression of antimicrobial immunity *in vivo*, in: *Immunobiology of the Macrophage* (D. S. Nelson, ed.), pp. 367–400, Academic Press, New York.

Campbell, F., 1972, Ultrastructural studies of transmural migration of blood cells in the bone marrow of rats, mice and guinea pigs, *Am. J. Anat.* **135**:521.

Campbell, F., 1975, Ultrastructure of the sinus wall of murine bone marrow in myelogenous leukemia, *Am. J. Anat.* **142**:319.

Carr, I., 1973, *The Macrophage*, Academic Press, London.

Cho, Y., and De Bruyn, P. P. H., 1975, Passage of red blood cells through the sinusoidal wall of the spleen, *Am. J. Anat.* 142:91.

Cho, Y., and De Bruyn, P. P. H., 1978, Destruction of circulating leukemia cells by phagocytosis in rats with myelogenous leukemia, *J. Natl. Cancer Inst.* **60**:185.

Cho, Y., and De Bruyn, P. P. H., 1979, The endothelial structure of the postcapillary venules of the lymph node and the passage of lymphocytes across the venule wall, *J. Ultrastruct. Res.* **69**:13.

Clark, S. L., 1962, The reticulum of lymph nodes in mice studied with the electron microscope, *Am. J. Anat.* **110**:217.

Clementi, F., and Palade, G. E., 1969, Intestinal capillaries. I. Permeability to peroxidase and ferritin, *J. Cell Biol.* **41**:33.

Cotran, R. S., 1965, Endothelial phagocytosis: An electronmicroscopic study, *Exp. Mol. Pathol.* **4**:217.

Cotran, R. S., Suter, E. R., and Majno, G., 1967, The use of colloidal carbon as a tracer for vascular injury, *Vasc. Dis.* **4**:107.

David, J. R., and Remold, H. G , 1976, Macrophage activation by lymphocyte mediators and studies on the interaction of macrophage inhibitory factor (MIF) with its target cell, in: *Immunobiology of the Macrophage* (D. S. Nelson, ed.), pp. 401–427, Academic Press, New York.

De Bruyn, P. P. H., 1979, The role of sialated glycoproteins in endocytosis, permeability and transmural passage in myeloid endothelium, *J. Histochem. Cytochem.* **27**:1174.

De Bruyn, P. P. H., and Cho, Y., 1978, The entry of metastatic malignant cells into the circulation from a subcutaneously growing myelogenous tumor, *J. Natl. Cancer Inst.* **62**:1221.

De Bruyn, P. P. H., Thomas, T. B., and Michelson, S., 1966, Fine structure of the vascular components of the guinea pig bone marrow, *Anat. Rec.* **154**:499.

De Bruyn, P. P. H., Michelson, S., and Thomas, T. B., 1971, The migration of blood cells of the bone marrow through the sinusoidal wall, *J. Morphol.* **133**:417.

De Bruyn, P. P. H., Michelson, S., and Becker, R. P., 1975, Endocytosis, transfer tubules, and lysosomal activity in myeloid sinusoidal endothelium, *J. Ultrastruct. Res.* **53**:133.

De Bruyn, P. P. H., Becker, R. P., and Michelson, S., 1977a, The transmural migration and release of blood cells in acute myelogenous leukemia, *Am. J. Anat.* **149**:247.

De Bruyn, P. P. H., Michelson, S., and Becker, R. P., 1977b, Phosphotungstic acid as a marker for the endocytic-lysosomal system (vacuolar apparatus) including transfer tubules of the lining cells of the sinusoids in the bone marrow and liver, *J. Ultrastruct. Res.* **58**:87.

De Bruyn, P. P. H., Michelson, S., and Becker, R. P., 1978, Nonrandom distribution of sialic acid over the cell surface of bristle coated endocytic vesicles of the sinusoidal endothelium cell, *J. Cell Biol.* **78**:379.

Farr, A. G., and De Bruyn, P. P. H., 1975a, The mode of lymphocyte migration through postcapillary venule endothelium in lymph node, *Am. J. Anat.* **143**:59.

Farr, A. G., and De Bruyn, P. P. H., 1975b, Macrophage-lymphocyte clusters in lymph nodes: A possible substrate for cellular interactions in the immune response, *Am. J. Anat.* **144**:209.

Farr, A. G., Dorf, M. E., and Unanue, E. R., 1977, Secretion of mediators following T lymphocyte-macrophage interaction is regulated by the major histocompatibility complex, *Proc. Natl. Acad. Sci. USA* **74**:3542.

Farr, A. G., Cho, Y., and De Bruyn, P. P. H., 1980, The structure of the sinus wall of the lymph node relative to its endocytic properties and transmural cell passage, *Am. J. Anat.* **157**:265.

Fawcett, D. W., 1965, Surface specializations of absorbing cells, *J. Histochem. Cytochem.* **13**:75.

Friess, A. E., 1977, Macrophage-lymphocyte cluster formation in the medullary sinus of lymph node after immunization with sheep red blood cells (SRBC), *Cell Tiss. Res.* **180**:505.

Gingell, D., and Fornés, J. A., 1976, Interaction of red blood cells with a polarized electrode. Evidence of long-range intermolecular forces, *Biophys. J.* **16**:1131.

Griffin, F. M., Bianco, C., and Silverstein, S., 1975, Characterization of the macrophage receptor for complement and demonstration of its functional independence from the receptor for the Fc portion of the immunoglobulin G, *J. Exp. Med.* **141**:1269.

Hudson, G., and Yoffey, J. M., 1963, Reticulo-endothelial cells in the bone marrow of the guinea-pig, *J. Anat.* **97**:409.

Hudson, G., and Yoffey, J. M., 1968, Ultrastructure of reticuloendothelial elements in guinea-pig bone marrow, *J. Anat.* **103**:515.

Huhn, D., 1966, Die Feinstruktur des Knochenmarks der Ratte bei Anwendung neuerer Aldehydfixationen, *Blut* **13**:291.

Huhn, D., and Stich, W., 1969, *Fine Structure of Blood and Bone Marrow*, Hafner, New York.

Hummel, K. P., 1935, The structure and development of the lymphatic tissue in the intestine of the albino rat, *Am. J. Anat.* **57**:351.

Janossy, G. M., Greaves, M. F., and Dourmashkin, R. R., 1972, Lymphocyte activation. IV. The ultrastructural pattern of the response of mouse T and B cells to mitogen stimulation *in vitro*, *Immunology* **24**:211.

Katz, D. H., 1977, *Lymphocyte Differentiation, Recognition, and Regulation*, Academic Press, New York.

Katz, D. H., and Benacerraf, B., 1972, The regulatory influence of activated T cells on B cell responses to antigen, *Adv. Immunol.* **15**:1.

Klein, J., 1975, *Biology of the Mouse Histocompatibility-2 Complex. Principles of Immunogenetics Applied to a Single System*, Springer Verlag, New York.

Langevoort, H. L., Cohn, Z. A., Hirsch, J. G., Humphrey, J. H., Spector, W. G., and van Furth, R., 1970, The nomenclature of mononuclear phagocyte cells: a proposal for a new classification, in: *Mononuclear Phagocytes* (R. van Furth, ed.), pp. 1–6, Blackwell, Oxford.

Leblond, P. F., 1973, Etude, au microscope électronique à balayage, de la migration des cellules sanguines à travers les parois des sinusoides spléniques et médullaires chez le rat, *Nouv. Rev. Fr. Hématol.* **13**:771.

Leduc, E. H., Avrameas, S., and Bouteille, M., 1969, Ultrastructural localization of antibody in differentiating plasma cells, *J. Exp. Med.* **127**:109–118.

Lewis, W. H., 1931, Pinocytosis, *Bull. Johns Hopkins Hosp.* **49**:17.

Lichtman, M. A., and Weed, R. I., 1976, Cellular deformability of normal and leukemic hematopoietic cells: a determinant of marrow and vascular egress, in: *Fundamental Aspects of Metastasis* (L. Weiss, ed.), pp. 319–325, North-Holland, Amsterdam.

Lichtman, M. A., Chamberlain, J. K., and Santillo, P. A., 1978, Factors thought to contribute to the regulation of egress of cells from marrow, in: *The Year in Hematology* (R. Silber, J. LoBue, and A. S. Gordon, eds.), pp. 243–279, Plenum Press, New York.

Lipski, P. E., and Rosenthal, A. S., 1973, Macrophage-lymphocyte interaction. I. Characteristics of the antigenic independent binding of guinea pig thymocytes and lymphocytes to syngeneic macrophages, *J. Exp. Med.* **138**:900.

Lipski, P. E., and Rosenthal, A. S., 1975, Macrophage-lymphocyte interaction. II. Antigen-mediated physical interactions between immune guinea pig lymph node lymphocytes and syngenic macrophages, *J. Exp. Med.* **141**:138.

Lohmann-Matthes, M. L., 1976, Induction of macrophage-mediated cytotoxicity, in: *Immunobiology of the Macrophage* (D. S. Nelson, ed.), pp. 463–486, Academic Press, New York.

Majno, G., and Joris, I., 1978, Endothelium 1977: A review, in: *The Thrombotic Process in*

Atherogenesis, Proceedings of a Symposium held in Reston, Virginia, October 16–19, 1977 (A. B. Chandler, K. Eurenius, G. C. McMillan, C. B. Nelson, C. J. Schwartz, and S. Wessler, eds.), Plenum Press, New York.

Marchesi, V. T., and Gowans, J. L., 1964, The migration of lymphocytes through the endothelium of venules in lymph nodes: An electron microscope study, *Proc. R. Soc. Ser. B* **159**:283.

Matter, A., Lisowska-Bernstein, B., Ryser, J. E., Lamelin, J. P., and Vassalli, P., 1972, Mouse thymus independent and thymus derived lymphoid cells. II. Ultrastructural studies, *J. Exp. Med.* **136**:1008.

McFarland, W., Heilman, D., and Moorhead, J., 1966, Functional anatomy of the lymphocyte in immunological reactions *in vitro*, *J. Exp. Med.* **124**:851.

Messier, P.-É., and Sainte-Marie, G., 1972, Location of lymphocytes in endothelium of postcapillary venules of rat lymph node, *Rev. Can. Biol.* **31**:231.

Mikata, A., and Niki, R., 1971, Permeability of postcapillary venules of the lymph node. An electron microscopic study, *Exp. Mol. Pathol.* **14**:289.

Miller, H. R. P., and Avrameas, A., 1971, Association between macrophages and specific antibody producing cells, *Nature New Biol.* **229**:184.

Möller, G. (ed.), 1975a, Subpopulation of B Lymphocytes, *Transplant Rev.*, Vol. 24, Munksgaard, Copenhagen.

Möller, G. (ed.), 1975b, Suppressor T Lymphocytes, *Transplant. Rev.*, Vol. 26, Munksgaard, Copenhagen.

Moscona, A. A., 1975, Surface specification of embryonic cells: Lectin receptors, cell recognition, and specific cell ligands, in: *The Cell Surface in Development* (A. A. Moscona, ed.), pp. 67–99, Wiley, New York.

Muto, M., 1976, A scanning and transmission electron microscopic study on rat bone marrow sinuses and transmural migration of blood cells, *Arch. Hist. Jpn.* **39**:51.

Niederhuber, J. E., 1978, The role of I region gene products in macrophage lymphocyte interaction, *Immunol. Rev.* **40**:28.

Nielsen, M. H., Jensen, H., Braendstrup, O., and Werdelin, O., 1974, Macrophage-lymphocyte clusters in the immune response to soluble protein antigen *in vitro*. II. Ultrastructure of clusters formed during the early response, *J. Exp. Med.* **140**:1260.

Oppenheim, J. J., and Seeger, R. C., 1976, The role of macrophages in the induction of cell-mediated immunity *in vitro*, in: *Immunobiology of the Macrophage* (D. S. Nelson, ed.), pp. 112–130. Academic Press, New York.

Parsegian, V. A., and Gingell, D., 1973, A physical force model of biological membrane interaction, in: *Recent Advances in Adhesion* (L. Lee, ed.), pp. 153–190, Gordon and Beech, London.

Pierce, C. W., and Kapp, J. A., 1976, Regulation of immune responses by suppressor T cells, *Contemp. Top. Immunobiol.* **5**:91.

Polliack, A., Lampen, N., Clarkson, B. D., DeHarven, E., Bentwich, Z., Siegal, F. P., and Kunkel, H. G., 1973, Identification of human B and T lymphocytes by scanning electron microscopy, *J. Exp. Med.* **138**:607.

Rabinovitch, M., 1970, Phagocytic recognition, in: *Mononuclear Phagocytes* (R. van Furth, ed.), pp. 299–313, Davis, Philadelphia.

Rosenstreich, D. L., and Mizel, S. B., 1978, The participation of macrophages and macrophage cell lines in the activation of T lymphocytes by mitogens, *Immunol. Rev.* **40**:102.

Rosenthal, A., 1978, Determinant selection and macrophage function in genetic control of the immune response, *Immunol. Rev.* **40**:136.

Rosenthal, A. S., Blake, J. T., Ellner, J. J., Greineder, D. K., and Lipsky, P. E., 1976, Macrophage function in antigen recognition by T lymphocytes, in: *Immunobiology of the Macrophage* (D. S. Nelson, ed.), pp. 131–161, Academic Press, New York.

Sastry, P. S., and Hokin, L. E., 1966, Studies on the role of phospholipids in phagocytosis, *J. Biol. Chem.* **241**:3354.

Saunders, G. C., and Hammond, W. S., 1970, Ultrastructural analysis of hemolysin forming cell clusters. I. Preliminary observations, *J. Immunol.* **105**:1299.

Sbarra, A. J., and Karnovsky, M. L., 1959, The biochemical basis of phagocytosis. I. Metabolic changes during the ingestion of particles by polymorphonuclear leukocytes, *J. Biol. Chem.* **234**:1355.

Schwartz, R. H., Yano, A., and Paul, W. E., 1978, Interaction between antigen presenting cells and primed T-cells, *Immunol. Rev.* **40**:153.

Simionescu, N., Simionescu, M., and Palade, G. E., 1972, Permeability of intestinal capillaries. Pathway followed by dextrans and glycogens, *J. Cell Biol.* **53**:365.

Skutelsky, E., and Farquhar, M. G., 1976, Variations in distribution of con A receptor sites and anionic groups during red blood cell differentiation in the rat, *J. Cell Biol.* **71**:218.

Stamper, H. B., Jr., and Woodruff, J. J., 1976, Lymphocyte homing into lymph nodes: *In vitro* demonstration of the selective affinity of recirculating lymphocytes for high-endothelial venules, *J. Exp. Med.* **144**:828.

Stamper, H. B., Jr., and Woodruff, J. J., 1977, An *in vitro* model of lymphocyte homing. I. Characterization of the interaction between thoracic duct lymphocytes and specialized high-endothelial venules of lymph nodes, *J. Immunol.* **119**:772.

Sulitzeanu, D., Kleinman, R., Benezra, D., and Gery, I., 1971, Cellular interactions and the secondary response *in vitro, Nature New Biol.* **229**:254.

Thiery, J.-P., 1962, Etude au microscope électronique de l'îlot plasmacytaire, *J. Microscopie* **1**:275.

Unanue, E. R., 1972, The regulatory role of macrophages in antigen stimulation, *Adv. Immunol.* **15**:95.

Unanue, E. R., 1976, Secretory function of mononuclear phagocytes, *Am. J. Pathol.* **83**:396.

van Furth, R., Cohn, Z. A., Hirsch, J. G., Spector, W. G., and Langevoort, H. L., 1972, The mononuclear phagocyte system: A new classification of macrophages, monocytes, and their precursor cells, *Bull. W.H.O.* **46**:845.

Weiss, L., 1976, Biophysical aspects of the metastatic cascade, in: *Fundamental Aspects of Metastasis* (L. Weiss, ed.), pp. 51–70, North-Holland, Amsterdam.

Wenk, E. J., Orlic, D., Reith, E. J., and Rhodin, J. A. G , 1974, The ultrastructure of mouse lymph node venules and the passage of lymphocytes across their walls, *J. Ultrastruct. Res.* **47**:214.

Wetzel, B., 1976, Cell kinesics: An interpretative review of the significance of cell surface form, in: *Scanning Electron Microscopy/1976/II* (O. Johari and R. P. Becker, eds.), pp. 137–144, IIT Research Institute, Chicago.

Wisse, E., 1972, An ultrastructural characterization of the endothelial cell in the rat liver sinusoid under normal and various experimental conditions, as a contribution to the distinction between endothelial and Kupffer cells, *J. Ultrastruct. Res.* **38**:528.

Wisse, E., 1977, Ultrastructure and function of Kupffer cells and other sinusoidal cells in the liver, in: *Kupffer Cells and Other Liver Sinusoidal Cells* (E. Wisse and D. L. Knook, eds.), pp. 33–60, Elsevier/North-Holland, Amsterdam.

14

Ultrastructure of Reticuloendothelial Clearance

K. DONALD

1. INTRODUCTION

Ultrastructural studies linked with functional studies of particle clearance indicate that many particle clearance studies in the literature which claim to measure phagocytosis do not do so. Instead, they appear to measure aggregation of particles in the blood, attachment to platelets, and attachment to the surface of macrophages, without sufficient time for phagocytosis to occur. This is because most clearance kinetics are reported over periods of 10 or 15 min following intravenous (i.v.) injection of the particles, during which time ultrastructural studies indicate the above processes are the major rate-controlling factors.

The tradition for conclusions based on morphological study of the RES linked with functional events appears early in the work of Elie Metchnikoff; the recent use of the electron microscope to extend observations to the subcellular level has produced a dimension not possible with the light microscope, and leads to the conclusions in the opening paragraph above.

Early, extensive light microscope studies of the clearance of carbon particles (Nagao, 1920, 1921) showed the distribution of large quantities of carbon in association with the reticuloendothelial (RE) cells of liver, spleen, and bone marrow. Intravascular deposits in many other organs were recorded. Nagao (1920, 1921) further described the localization of carbon particles within granules in the cell cytoplasm, and observed that the primary granules fused with one another or with other granules in the cells. He made passing reference to platelet aggregates, but attributed no functional significance to them. These early particle clearance studies were done with carbon particle preparations containing either no suspending agents or shellac as such an agent. Several workers (Elvidge, 1926; Nagao, 1920, 1921; Simpson, 1922; Tait and Elvidge, 1926; Wislocki, 1924; Wright, 1927) described the depletion of platelets following such injections.

K. DONALD • Department of Pathology, The University of Queensland Medical School, Herston, Queensland 4006, Australia.

Halpern *et al.* (1951) produced evidence of the toxicity of shellac, and subsequently performed extensive kinetic studies (Halpern *et al.*, 1953) with India ink C_{11}/1431a made by Günther Wagner. This contained fish gelatin as suspending agent and was less toxic (specifically, less fibrin deposition) than previous inks. The pattern of the carbon deposits seen with the light microscope followed the generally accepted pattern of the RES, but, without the availability of electron microscope examination, the precise localization of carbon in relation to cell membranes was not proven.

The clearance of such colloidal carbon suspensions from the bloodstream has usually been equated with the phagocytic function of RE cells (Halpern *et al.*, 1953), and forms the basis for a large volume of work concerning the supposed functions of the RES. Several workers have warned that the probable role played by platelets has been ignored in studies concerning the clearance of particles from the blood (Sanders *et al.*, 1951; Stehbens and Florey, 1960). Florey (1962) pointed out that much work had been done separately on the effects of circulating particles on either platelets or macrophages, and insufficient account was taken of one or the other in most studies. If, indeed, the commonly used methods of clearance of colloidal carbon do not provide a reliable measure of the rates of phagocytosis by the RES, then the large volume of work concerning this system becomes difficult to interpret.

The presence of transient thrombocytopenia following the injection of colloidal carbon as C_{11}/1431a, has been recorded (Cohen *et al.*, 1965; Salvidio and Crosby, 1960). Van Aken *et al.* (1968) suggested that substances which affect platelet number or function profoundly influence the clearance rate of carbon particles (C_{11}/1431a) injected intravenously. They demonstrated that adenosinenucleotides, which inhibit platelet aggregation, diminished the rate of carbon clearance from the blood and that platelet infusions accelerated carbon clearance. They suggested that platelet aggregation is an essential factor in the clearance of carbon from the bloodstream, and not an incidental associated event. The activation of fibrinolysis by urokinase or streptokinase reduced the carbon clearance rate, as did the injection of isolated fibrinogen degradation products (van Aken and Vreeken, 1969). The latter also caused disaggregation of platelets. They also found that inhibition of fibrinolysis by ϵ-aminocaproic acid accelerated the clearance of carbon from the blood and caused excess deposition of carbon particles in the lungs. They concluded that platelet aggregation and locally activated fibrinolysis are cofactors in the clearance of colloidal carbon. They did not provide any light or electron microscope evidence of the localization of carbon particles.

2. ULTRASTRUCTURE OF UNSTIMULATED CLEARANCE RATES

Recently, combined functional and ultrastructural studies have shown that the attachment of carbon particles to the surface membrane of platelets is a significant event in carbon clearance in normal rabbits (Donald, 1972a) (Figures 1 and 2). Further studies showed that raised or lowered carbon clearance rates

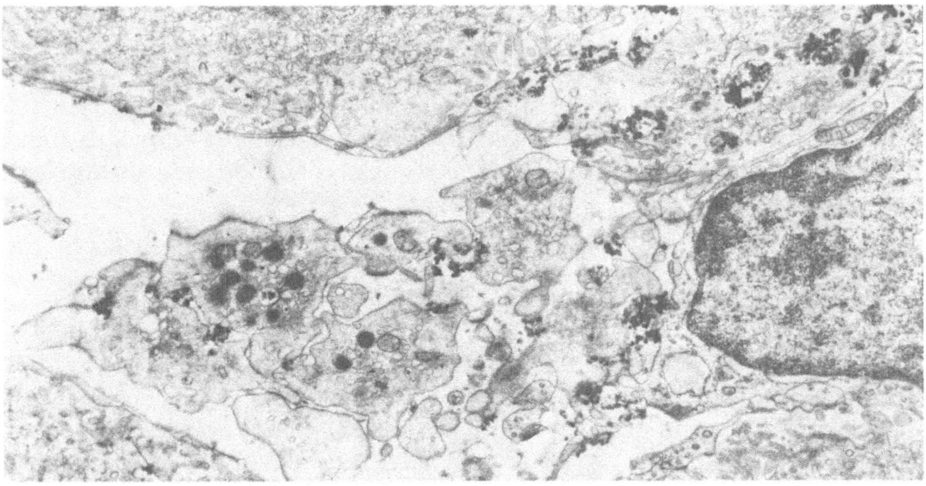

FIGURE 1. Platelet aggregates in a liver sinusoid of a rabbit given 8 mg/100 g body weight India ink alone. A small amount of carbon is attached to the surface membranes of the platelets. Some carbon is attached to the surface membrane of the sinusoidal lining cell, and some is present in membrane-bounded vesicles, which appear to be intracellular. Electron micrograph. Uranyl acetate and lead citrate. ×10,200.

may accompany increased or decreased levels of this attachment to platelets (Donald, 1972b; Donald and Pound, 1971; Donald and van't Hull, 1973; Tennent and Donald, 1976) (Figures 3, 4, 5, and 6).

Examination of the literature concerning kinetic studies, morphological studies, and hypotheses, presents evidence that a wide variety of variables affect carbon particle clearance rates, viz., concentration of the particles in the blood, intravascular coagulation and fibrinolysis, platelet aggregation and disaggrega-

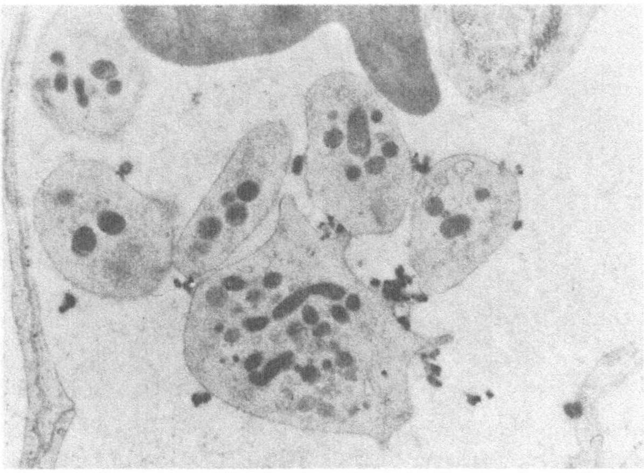

FIGURE 2. Electron micrograph of the lung of a rat given 1 mg/100 g body weight of colloidal carbon and killed 2 min later. A loose aggregate of platelets with small amounts of carbon attached to their surfaces is present in a capillary. There is no evidence of degranulation of the platelets. ×12,200.

tion, the number of phagocytic cells, phagocytic activity of each cell, blood flow through the RES, and loss of particles through endothelial gaps. The method has often been used to demonstrate supposed alterations in phagocytic function of RE cells without a systematic attempt to determine which of the various possible factors, singly or in combination, was responsible for changed rates of clearance in the particular experiments quoted. This requires combined ultrastructural and kinetic studies.

2.1. CARBON DOSE RESPONSE

The rates of carbon clearance at various dose levels are not always best represented by single order kinetics (Table 1, Figure 7). From Figure 7 it is apparent that, at doses of 16 mg and 2 mg per 100 g body weight, the rates of clearance in the first 8 min are faster than later. A detailed statistical method for

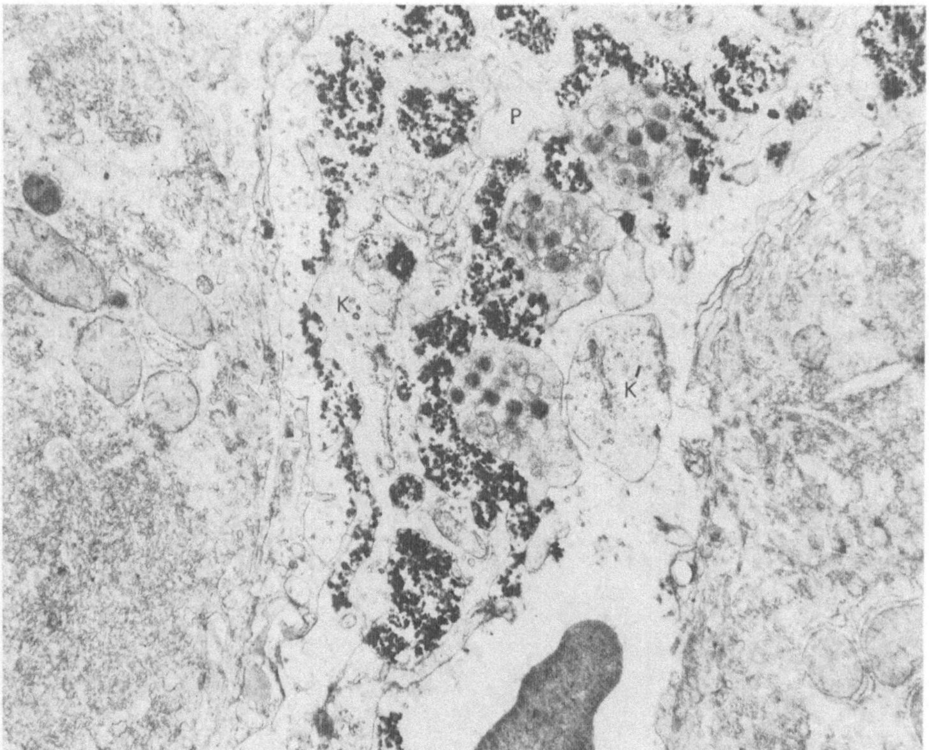

FIGURE 3. A large intrasinusoidal cytoplasmic process (K) arising from a Kupffer cell in a rabbit given India ink (8 mg/100 g body weight) 4 hr after tubercle lipid and killed 5 min after the carbon injection. A thick layer of carbon particles is attached to the surface membrane. The thick surface layer of carbon is accompanied by several attached platelets. These show large processes (p) and stages of degranulation. A cytoplasmic extension (K′) is cut in cross-section. Electron micrograph. Uranyl acetate and lead citrate. ×8400.

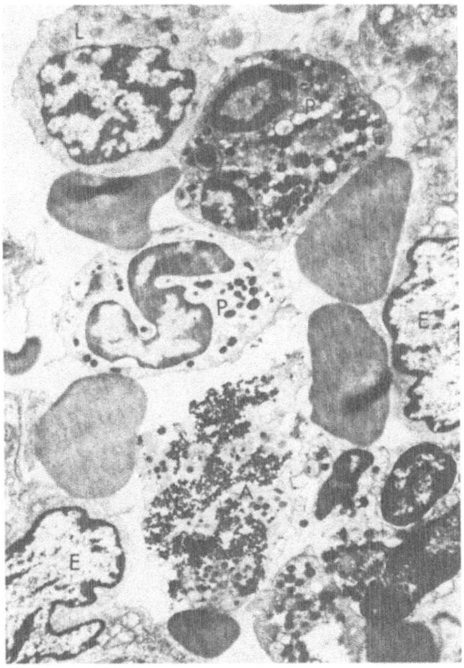

FIGURE 4. Rabbit treated as for Figure 3. An aggregate of platelets and carbon (A) in the lumen of a pulmonary alveolar capillary. No carbon is attached to polymorphonuclear leukocytes (P), lymphocytes (L), endothelial cells (E) or erythrocytes. Electron micrograph. Uranyl acetate and lead citrate. ×3250.

comparing such various slopes in carbon clearance studies has been published previously (Donald, 1972d). From Table 1 and Figure 7, it is also clear that the rates of clearance for various doses are not parallel for the first 8 min of clearance. Thereafter (i.e., from 8 to 64 min), the rates of clearance are parallel for

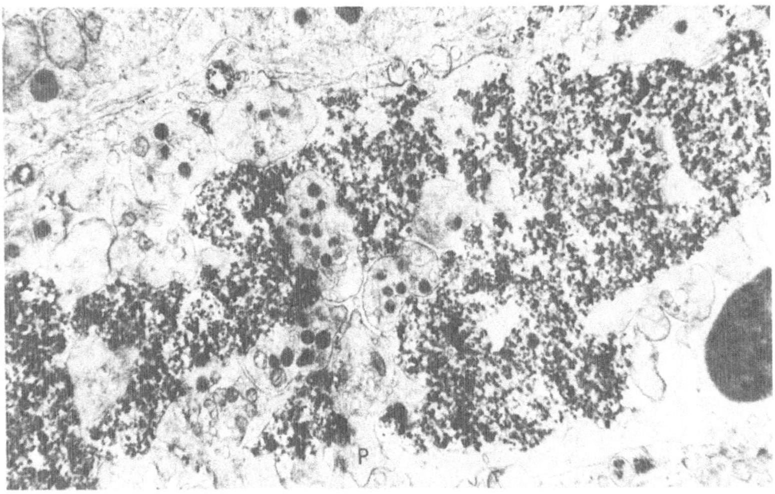

FIGURE 5. Rabbit treated as for Figures 3 and 4. An aggregate of platelets and carbon in a liver sinusoid. Degranulation and process formation (p) are evident in the platelets. Electron micrograph. Uranyl acetate and lead citrate. ×6400.

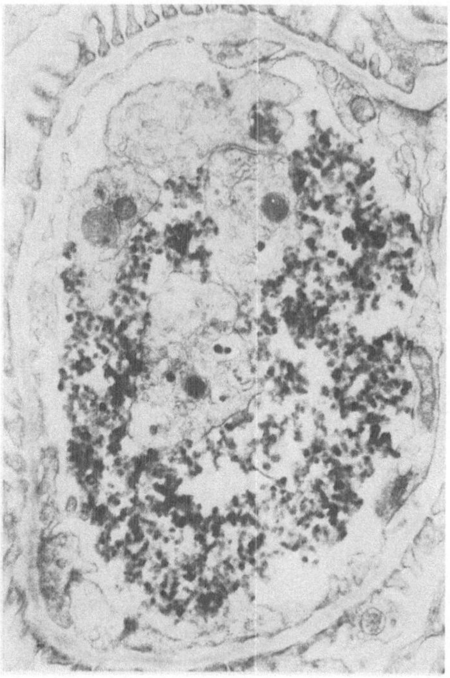

FIGURE 6. Rabbit treated as for figures 3, 4, and 5. An aggregate of platelets and carbon in a renal glomerular capillary. Degranulation and process formation are evident in the platelets. Electron micrograph. Uranyl acetate and lead citrate. ×15,200.

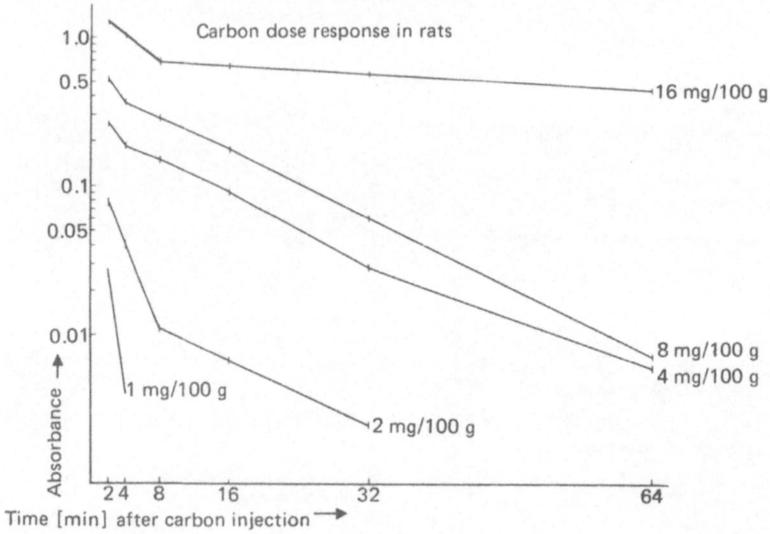

FIGURE 7. Semilogarithmic plot of absorbance against time in rats given various doses of colloidal carbon.

TABLE 1. CLEARANCE OF VARIOUS DOSES OF COLLOIDAL CARBON

Dose (mg/100 g body weight)	Absorbance					
	Time after injection (min)					
	2	4	8	16	32	64
16	1.29 ± 0.21	1.03 ± 0.18	0.68 ± 0.29	0.66 ± 0.30	0.58 ± 0.18	0.32 ± 0.12
8	0.50 ± 0.10	0.33 ± 0.06	0.28 ± 0.06	0.21 ± 0.08	0.07 ± 0.04	0.02 ± 0.01
4	0.25 ± 0.06	0.19 ± 0.06	0.15 ± 0.06	0.12 ± 0.06	0.03 ±0.02	0
2	0.08 ± 0.02	0.04 ± 0.01	0.014 ± 0.01	0.008 ± 0.008	0	0

doses of 2, 4, and 8 mg/100 g body weight. However, this "late" phase of clearance follows a different rate for animals given a dose of 16 mg/100 g body weight. Doses of 1 mg/100 g body weight are rapidly cleared and do not provide sufficient blood levels to produce significant results for analysis by the usual light absorption techniques. The presence of "early" and "late" phases of carbon clearance fits the morphological observations that different mechanisms appear dominant in the process at these times.

2.2. LIGHT MICROSCOPY

In animals killed at 2 min after the injection of any of the doses of colloidal carbon, considerable amounts of pigment are present in the vessels of the lungs. When animals are killed at 30 min after the injection, small deposits of carbon are found in the lungs of those given 8 mg/100 g body weight or less, but animals given 16 or 32 mg/100 g body weight still have considerable amounts in their lungs at this time. That is, the amount of carbon seen in the lungs depends on the dose and the time elapsed after injection of the particle suspension (Donald and Tennent, 1975).

2.3. ELECTRON MICROSCOPY

Electron microscope study demonstrates several mechanisms operating at all dose levels which may influence the rates of clearance as measured by

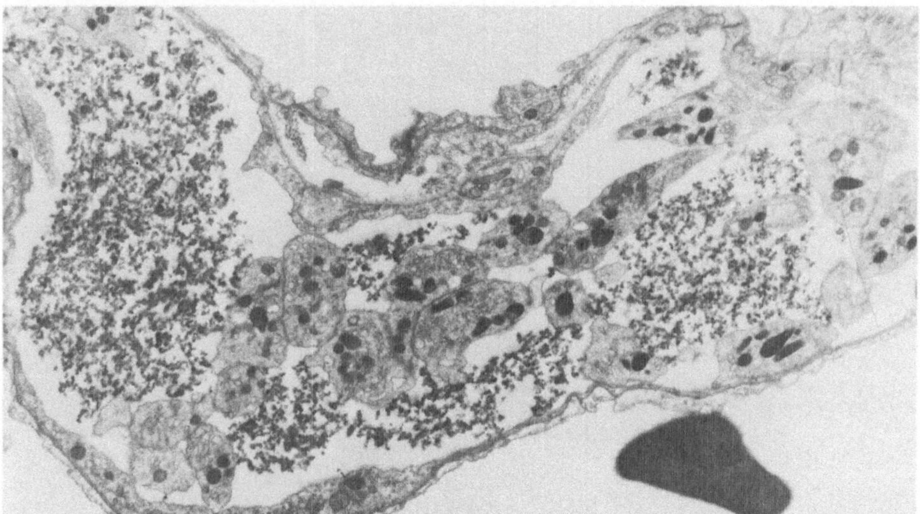

FIGURE 8. Electron micrograph of the lung of a rat given 8 mg/100 g body weight of colloidal carbon and killed 2 min later. Large amounts of colloidal carbon are aggregated in the plasma of a lung capillary and loose aggregates of platelets are associated with the carbon. Some of the platelets are showing pseudopod formation but there is no clear evidence of degranulation. ×8000.

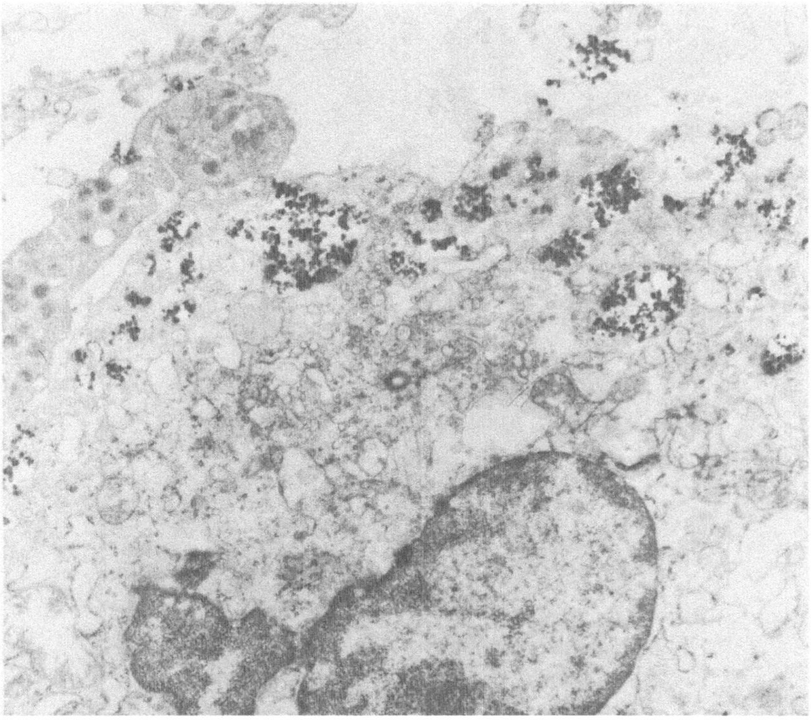

FIGURE 9. Electron micrograph of the liver of a rat given 1 mg/100 g body weight of colloidal carbon and killed 2 min later. A considerable amount of carbon is apparently in vacuoles in the cytoplasm of a Kupffer cell. A small amount of carbon is adherent to both the surface membrane of the cell and a small group of platelets adjacent to the cell. A small, free aggregate of carbon particles is present in the plasma adjacent to the cell membrane. ×9500.

the usual light absorption method. Aggregation within the bloodstream and attachment to platelets (Figure 8), attachment to macrophages without phagocytosis, and phagocytosis are all demonstrated (Figures 9, 10, and 11). The degrees to which these various processes are seen vary, depending upon the dose of colloidal carbon used and the time after the injection of carbon.

The amount of carbon phagocytized by individual Kupffer cells within 2 min of injection appears to remain constant for doses of 1, 4, 8, or 16 mg/100 g body weight (Figures 9, 10, and 11). As the dose increases, the number of Kupffer cells containing carbon appears to increase, both in tissue examined with the light microscope and the electron microscope. The amount of carbon attached to the surface membrane of Kupffer cells and nearby platelets is greater in animals given 8 mg/100 g body weight than in animals given 1 or 4 mg/100 g body weight (Figures 9, 10, and 11).

Platelet aggregates, with associated carbon particles, are present in large numbers in the lung vessels of animals given colloidal carbon and killed 2 min later. At doses from 1 to 4 mg/100 g body weight, these aggregates are loose; degranulation is not clearly established, and they are found mostly in capillaries (Figure 2). In animals given 8 or 16 mg/100 g body weight, large platelet–carbon

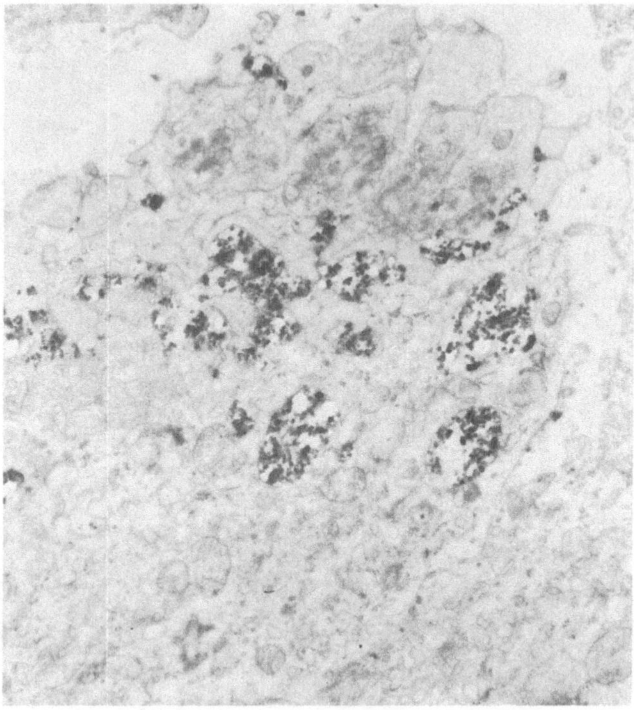

FIGURE 10. Electron micrograph of the liver of a rat given 4 mg/100 g body weight of colloidal carbon and killed 2 min later. A similar amount of carbon apears to be in vacuoles in the Kupffer cell to that seen in Figure 9, which is from a rat given only 1 mg/100 g body weight. Again, a small amount of carbon is attached to platelets adjacent to the cell membrane. ×9200.

aggregates are found in pulmonary venules (Figure 12), as well as capillaries. In doses below 8 mg/100 g body weight, platelet aggregates are rare in animals which survive 30 min after the injection. However, at doses of 16 and 32 mg/100 g body weight, these aggregates are massive (Figure 13), and many persist for 30 min in capillaries and sometimes in venules. At these higher doses, the platelet aggregates are dense with close adherence of platelets to one another, and degranulation is obvious (Figure 14).

The above studies show that, if results do not provide a sufficient number of estimations in the first 10 min after injection of the particles, then the early accelerated rates of clearance will not be demonstrated.

The doses of carbon used in the above studies were from 1 to 32 mg of colloidal carbon per 100 g body weight. Doses most frequently used in the literature to measure "reticuloendothelial phagocytosis" vary from 8 to 32 mg per 100 g body weight, which is the range in which the most extreme platelet changes occur, including a fall in number of circulating platelets, the formation of firm aggregates, and degranulation. The light microscope examination revealed a major degree of early sequestration in the lungs at these doses. Claims that lung sequestration plays no role in clearance rates are offset because the

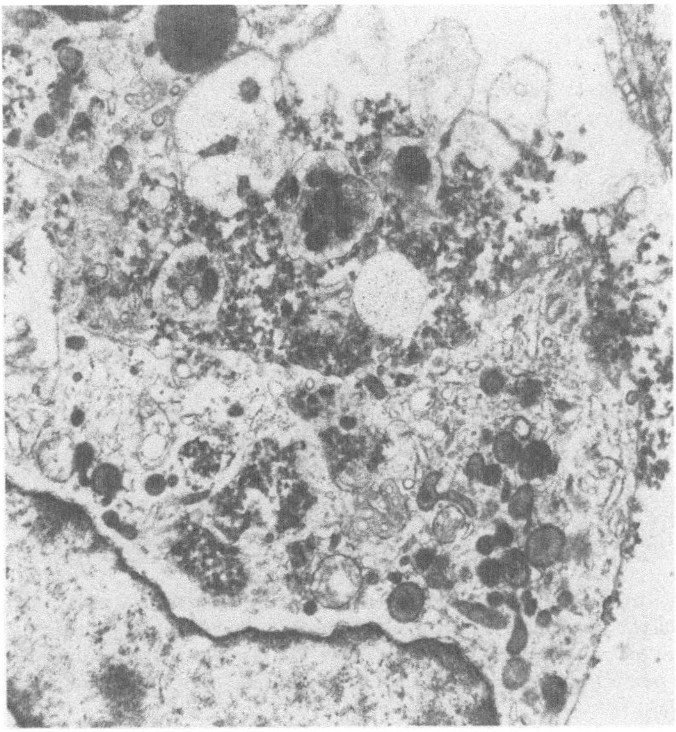

FIGURE 11. Electron micrograph of the liver of a rat given 8 mg/100 g body weight of colloidal carbon and killed 2 min later. A similar amount of carbon appears to be in vacuoles in the Kupffer cell to that seen in Figures 9 and 10. There is a thick layer of carbon particles attached to the cell surface and to adjacent platelets. ×8500.

animals in such studies are often killed late (i.e., after 20 or 30 min), when this lung sequestration has decreased, but the kinetics are usually reported over a period which includes this early major lung sequestration stage. We need further studies to determine how long this lung sequestration lasts, and we have, so far, seen it up to 5 min, but have not investigated it further. Doses of 1 or 2 mg per 100 g body weight are not used in studies in the literature because they are at the extreme of sensitivity of the light absorbance method of measurement used. However, it is at these levels that electron microscope study shows that phagocytosis in fact plays the major role in the early clearance, and platelet effects are least.

Experiments suggest that the rate of phagocytosis by Kupffer cells is not increased by increasing doses of carbon, but that the number of active Kupffer cells does increase with increasing doses (Donald and Tennent, 1975). However, the amount of carbon which may attach to the surface of macrophages and platelets increases with increasing dose, and previous work suggests that delay before phagocytosis of this material may be up to 30 min (Donald, 1972b). This means that, although the immediate phagocytic capacity of the RES may be exceeded, mechanisms still exist for the sequestration of circulating foreign par-

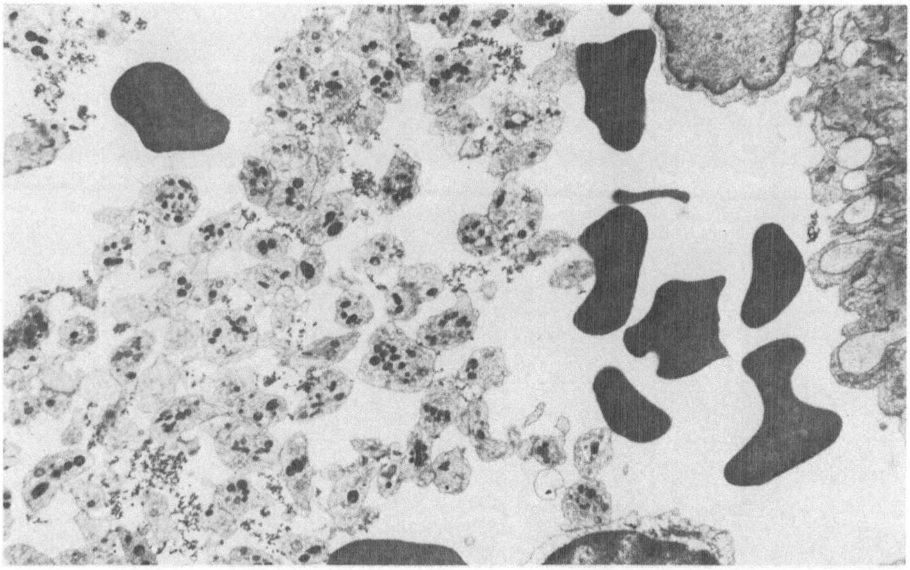

FIGURE 12. Electron micrograph of a lung venule of a rat given 8 mg/100 g body weight of colloidal carbon and killed 2 min later. A large, loose aggregate of platelets and colloidal carbon is present in the lumen of the venule. Some of the platelets show pseudopod formation and some appear to have reduced numbers of granules, but the morphological evidence of degranulation is equivocal. ×3600.

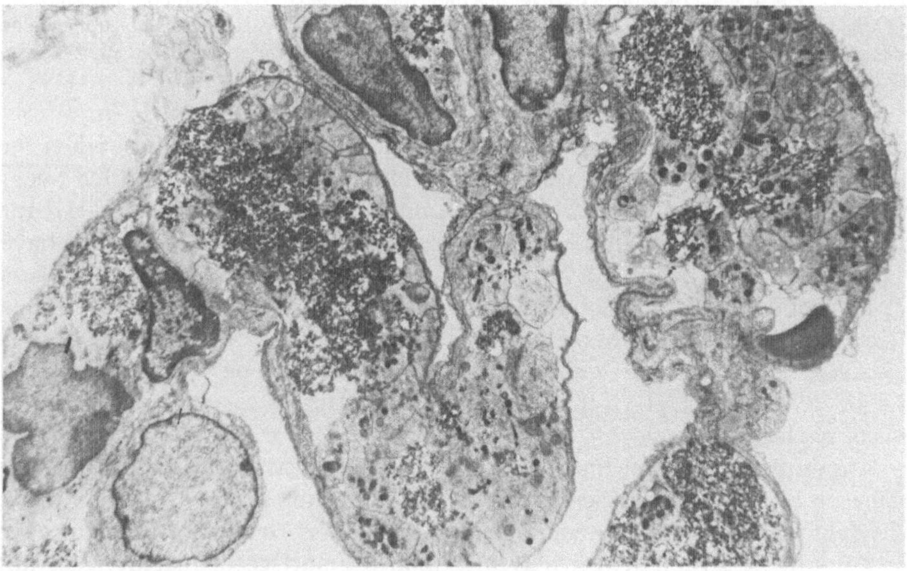

FIGURE 13. Electron micrograph of the lung of a rat given 16 mg/100 g body weight of colloidal carbon and killed 2 min later. Large, dense aggregates of carbon are present in capillaries along with aggregates of platelets in which individual platelets appear closely adherent to one another. Some of the platelets appear to have decreased numbers of granules. ×3600.

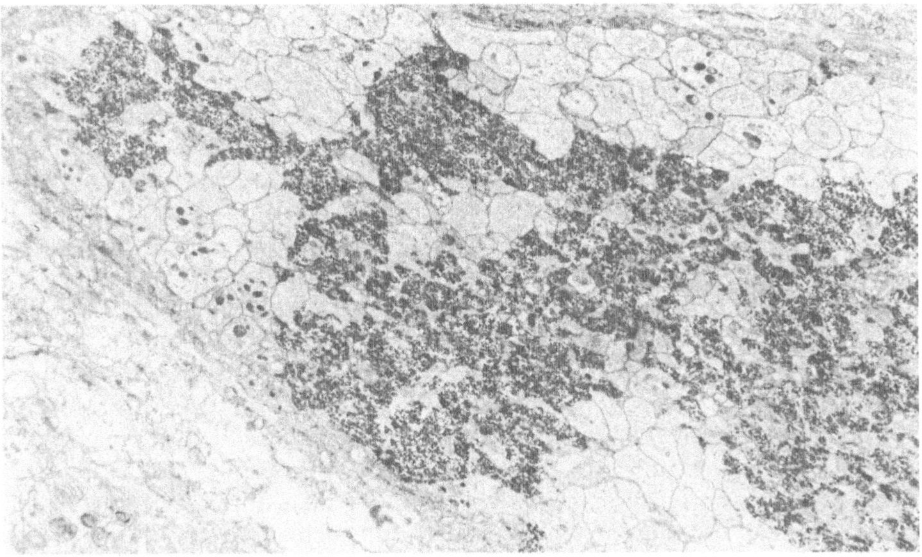

FIGURE 14. Electron micrograph of the lung of a rat given 32 mg/100 g body weight and killed 30 min later. A large, dense aggregate of carbon is present in a venule along with tightly packed platelets, many of which are devoid of granules. ×3600.

ticles. However, if the doses of particles are large, this system leads to degranulation of platelets, presumably with likely subsequent effects on coagulation. Nearly all the carbon clearance studies in the literature have been performed at these high dose levels where immediate phagocytic ability is exceeded and permanent platelet changes are initiated. It further follows that the controlling factors in such studies are likely to be the rate of regeneration of phagocytic ability by individual cells or the recruitment of new phagocytic cells, rather than the immediate rate of phagocytosis by individual cells. Such doses are very large in comparison with the numbers of circulating particles likely to be encountered by animals in physiological circumstances or even in disease states. Indeed, such doses initially render the animals black to casual inspection, and, at doses of 16 mg/100 g body weight and above, animals occasionally die soon after injection of the particles; the clearance rates at 16 mg/100 g body weight are also much slower than in those given 8 mg/100 g body weight or less, suggesting inability of the clearance processes to deal with such massive doses.

2.4. UPTAKE OF PARTICLES BY BLOOD PLATELETS

There is ample evidence in the literature of aggregation of platelets caused by a wide range of particles other than carbon (David-Ferreira, 1960, 1961; Glynn *et al.*, 1965; Haguenau *et al.*, 1964; Majno and Palade, 1961; Movat *et al.*, 1965; Schulz, 1961). Some authors have suggested that platelets phagocytize a variety of particles (Bloom, 1954; David-Ferreira, 1960, 1964; Haguenau *et al.*, 1964;

Majno and Palade, 1961; Movat *et al.*, 1965; Mustard and Packham, 1968; Schulz, 1961; Schulz and Wedell, 1962). However, White (1972) has shown that platelets do not phagocytize in the usual sense. Most of the latex particles which he found in platelets were lodged in channels of the open canalicular system, and these did not pinch off to form sealed phagocytic vacuoles, but remained open. This is part of the process of release of platelet factors which forms an integral part of the physiological response to circulating particles.

2.5. UPTAKE OF PARTICULATE MATTER BY ENDOTHELIAL CELLS

A number of authors have described phagocytosis of particles by "ordinary" endothelial cells in special circumstances or after repeated doses of particles (Cotran, 1965; Majno, 1965). Others have found evidence of such phagocytosis following single injections of "usual" doses of ferritin (Florey, 1967) and carbon (Carr, 1968). This probably contributes little to the rates of clearance of the usual doses of colloid, but there are again probably physiological consequences which form part of the total body response to circulating particles.

3. ULTRASTRUCTURE OF STIMULATED CLEARANCE RATES

Clear structural evidence exists that stimulation or suppression of several of the morphologically distinct processes previously mentioned may influence carbon clearance rates (Donald, 1972b,c,d; Donald and Pound, 1971; Donald and van't Hull, 1973; Tennent and Donald, 1976). Experiments claiming that a number of environmental agents produce increased or decreased phagocytosis of particles are often balanced in other parts of the literature by evidence that the same agents produce changes in the number or function of circulating platelets, e.g., zymosan (Dineen *et al.*, 1968), endotoxin (Des Prez *et al.*, 1961; Ream *et al.*, 1965; Shimamoto *et al.*, 1958), alcohol (Sullivan, 1971), tubercle bacillary lipid extracts (Donald and van't Hull, 1973), other bacteria (Clawson and White, 1971a,b), and irradiation (Ebbe and Stohlman, 1970).

3.1. TUBERCLE BACILLARY LIPID

Some lipid extracts of *Mycobacterium tuberculosis* augment various immunological phenomena when they are administered with a variety of antigens (Koga et al., 1969; Pound, 1955, 1958; Raffel, 1948; Raffel *et al.*, 1949; Raffel and Forney, 1948; White, 1967a,b). A conspicuous finding some days after injection of the tubercle lipid is hyperplasia of RE cells (O'Rourke and Pound, 1971; Pound, 1958). For this reason studies were made of the effect of the lipid on particle clearance processes.

Functional studies have shown that 4 hr after a single intravenous injection of 10 mg of the tubercle lipid there is a ten-fold increase in the rate of clearance of

TABLE 2. THE INFLUENCE OF INJECTION OF TUBERCLE BACILLARY LIPID ON CLEARANCE OF COLLOIDAL CARBON INJECTED 4 HR AFTER THE LIPID

Minutes after injection of carbon	Absorbance (mean value ± S.D.)
4	0.37 ± 0.30
8	0.26 ± 0.22
12	0.20 ± 0.2
16	0.19 ± 0.19
32	0.09 ± 0.1

colloidal carbon from the blood of rabbits (Donald and Pound, 1971) (Table 2, Figure 15). At the same time the distribution of carbon particles in the liver, as seen with the light microscope, shows conspicuous alterations as compared with that seen in control rabbits given India ink alone (Figures 16, 17, and 18). At this time, the carbon in the lipid-treated animals is deposited in much larger amounts at the periphery of the liver lobules than in control rabbits. In the lipid-treated rabbits these large peripheral deposits of carbon, with the almost complete absence of carbon rom the centrilobular areas, suggest almost total clearance of the blood in a single passage through the liver lobule. At the dose of carbon used, this does not occur in control animals, although the Kupffer cells at the periphery of lobules appear to take up more carbon than the more centrally placed sinusoidal lining cells. Also, in lipid-treated animals, carbon is constantly

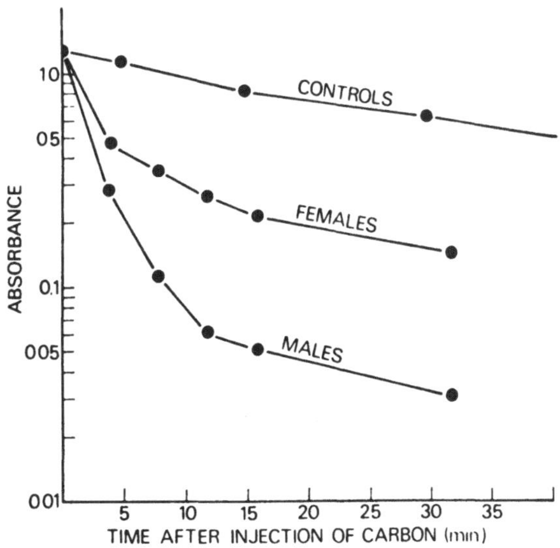

FIGURE 15. Absorbance in animals given carbon injection 4 hr after lipid.

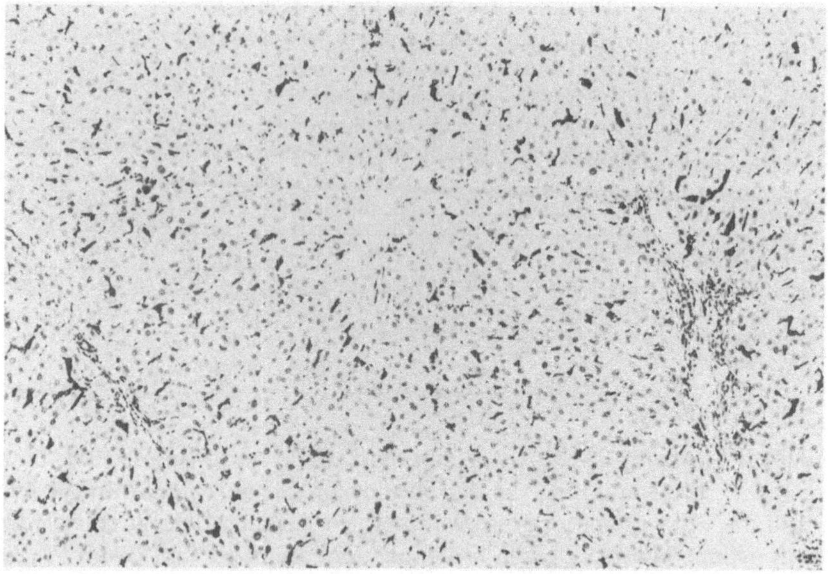

FIGURE 16. The distribution of carbon in the liver of normal rabbits given 8 mg/100 g body weight of colloidal carbon as India ink C_{11}/1431a and killed 35 min later. The carbon is distributed in aggregates throughout the lobule with larger collections in the peripheral zones. Hematoxylin and eosin. ×100.

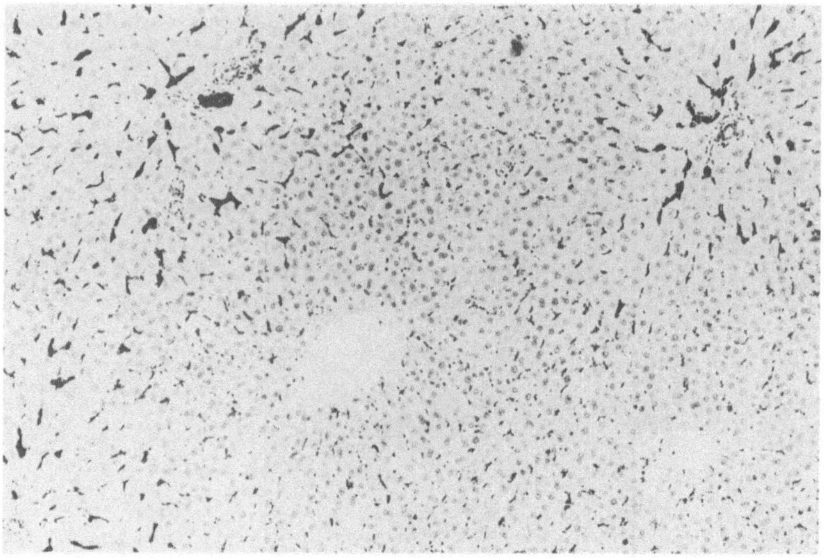

FIGURE 17. The distribution of carbon in the liver of rabbits given 8 mg/100 g body weight of colloidal carbon 4 hr after the injection of 10 mg of the tubercle lipid and killed 35 min after the carbon injection. There are large aggregates of carbon near the portal tracts, but little carbon has reached the center of the lobule. Hematoxylin and eosin. ×100.

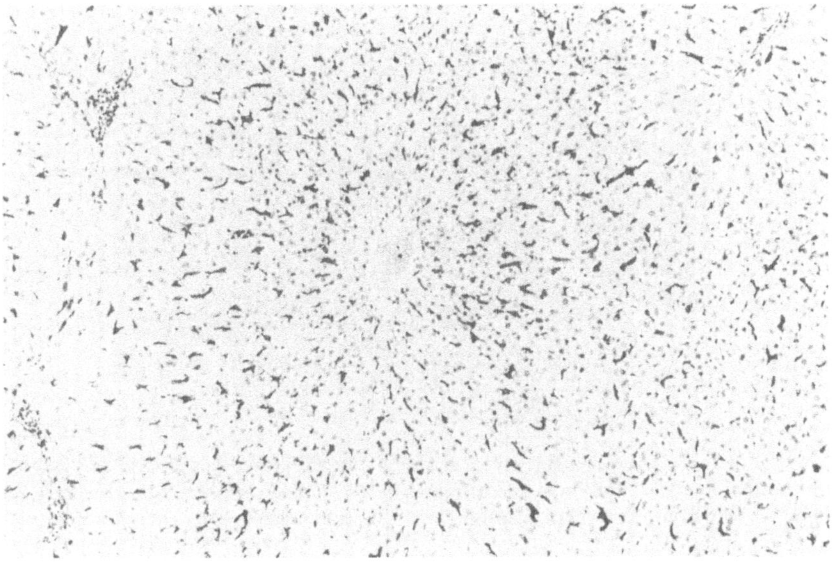

FIGURE 18. The distribution of carbon in the liver of rabbits given 8 mg/100 g body weight of colloidal carbon 36 hr after the injection of 10 mg of the tubercle and killed 35 min after the carbon injection. The aggregates of carbon particles are much more prominent in the central zones of the lobule than in control rabbits or those examined 4 hr after the lipid. Hematoxylin and eosin. ×100.

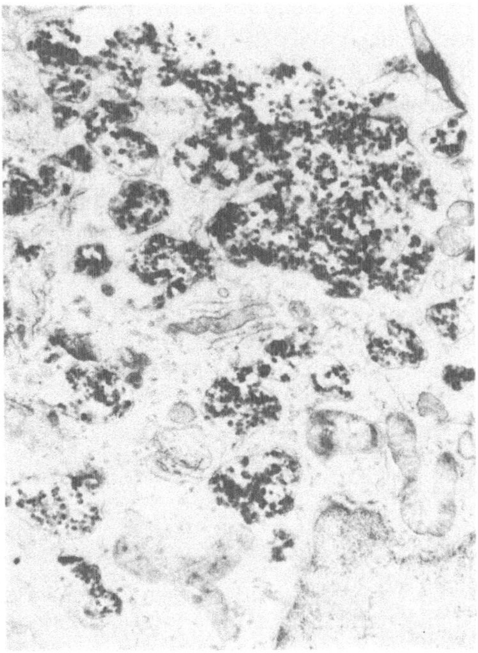

FIGURE 19. Surface membrane of a Kupffer cell from a rabbit treated as for Figures 3–6. A thick layer of carbon particles is present in many invaginations of the surface membrane. Some particles are present in membrane-bound vacuoles within the cytoplasm. Electron micrograph. Uranyl acetate and lead citrate. ×10,000.

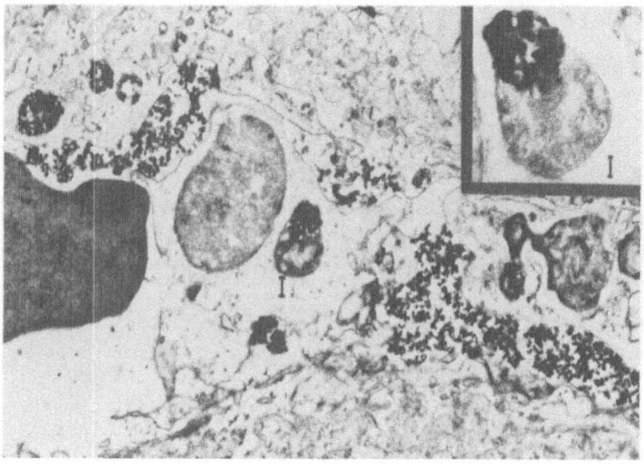

FIGURE 20. A Kupffer cell in a rabbit given India ink 4 hr after the tubercle lipid and killed 35 min after the carbon injection. Carbon is still attached to the surface membrane. Intracellular vacuoles containing carbon have fused with lysosomes (I). Electron micrograph. Uranyl acetate and lead citrate. ×8500. Inset: ×19,300.

found in some lung and renal glomerular capillaries. It is not possible with the light microscope to determine the exact location of the carbon at the cellular level.

Studies with the electron microscope show that the changed carbon distribution in the lipid-treated animals is associated with a very large increase in the adherence of the carbon particles to platelet and cell membranes (Figures 3–6). The surface membranes of macrophages show increased veil formation and endocytic activity in the lipid-treated animals (Figure 19). It is not clear whether this is a primary effect of the lipid, or a secondary effect that follows the increased platelet degranulation, and would be affected by the larger platelet aggregates in these animals. Carbon was not found attached to cells other than macrophages (Figure 4).

The work concerning functional studies of carbon clearance 4 hr after injection of this lipid showed a phase of very rapid clearance lasting 5–10 min after injection of the carbon, after which the rate slows down (Figure 15).

TABLE 3. INFLUENCE OF INJECTION OF TUBERCLE BACILLARY LIPID ON CLEARANCE OF COLLOIDAL CARBON INJECTED 30 MIN AND 36 HR AFTER THE LIPID

	Minutes after injection of the carbon (mean ± S.D.)			
	4	8	16	32
Absorbance 30 min after the lipid	1.16 ± 0.17	1.02 ± 0.08	0.91 ± 0.20	0.59 ± 0.21
Absorbance 36 hr after the lipid	0.94 ± 0.20	0.84 ± 0.19	0.49 ± 0.12	0.19 ± 0.20

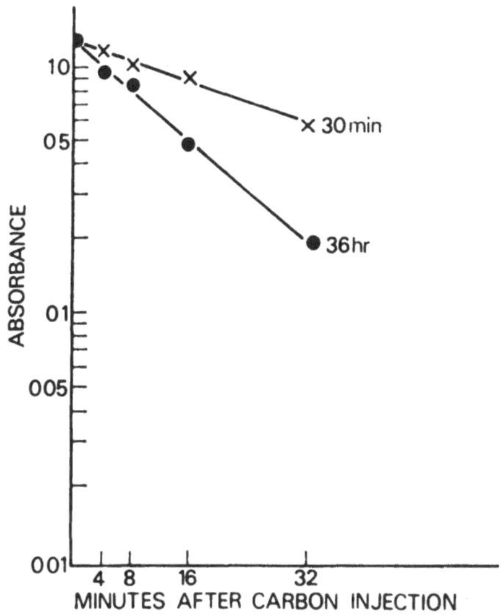

FIGURE 21. Absorbance in animals given carbon injection 30 min or 36 hr after lipid.

Some carbon was found still attached to the outside of the surface membrane of Kupffer cells 15 min after the circulating carbon levels had reached nonmeasurable values by light absorption techniques (Figure 20). This suggests that attachment of the particles to the cell surface is not always immediately followed by endocytosis.

In animals given the particle injections 36 hr after the lipid, clearance kinetics were of single order type (Table 3, Figure 21). There were now more deposits of carbon particles in each liver lobule, and they were evenly spread (Figure 18). Ultrastructural studies showed no increase in platelet attachment and normal phagocytosis. These findings suggest that an increased population of active phagocytic cells is responsible for the rapid clearance at this phase.

Thus the same substance appears to enhance different components of the clearance process at different times after its injection.

3.2. ZYMOSAN

The extensive literature concerning the biologic effects of zymosan (Pillemer and Ecker, 1941) involves three major areas, viz., phagocytosis (Benacerraf and Sebestyen, 1957), properdin-complement (Brade *et al.*, 1973), and platelets (Henson, 1970). It is likely that some *in vivo* experiments which have ascribed changes in clearance of various particles solely to RE macrophages have ignored a significant contribution by platelets. Further, the question of hyperphagocytosis vs. increased numbers of phagocytic cells is usually not explored.

Zymosan has long been known as a stimulant of particle clearance (Benacerraf and Sebestyen, 1957), and generally assumed to stimulate *phagocytosis*. Recent studies have examined the time relationships between the kinetics of stimu-

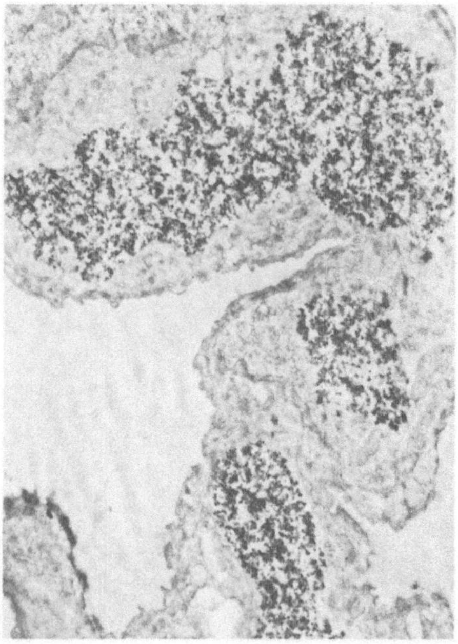

FIGURE 22. Electron micrograph of colloidal carbon in a lung capillary of an animal given carbon 2 hr after zymosan and killed 8 min after the injection of carbon. Large aggregates of carbon with few associated platelets (P) are present. ×3200.

lated clearance and the numbers of circulating platelets following a single i.v. injection of zymosan, and correlated these with detailed light and electron microscopic studies to determine the relative roles of platelets and macrophages (Tennent and Donald, 1976).

A single i.v. injection of zymosan is followed by a biphasic increase in the rate of clearance of colloidal carbon with peaks at 2 and 72 hr after the injection of zymosan. Circulating platelet levels are reduced at both these time points.

The combined light and electron microscopic studies confirm that a series of recognizable morphologic events contribute to particle clearance stimulated by zymosan. These processes are aggregation of carbon in the blood (Figure 22), attachment of carbon to platelets, attachment of carbon to macrophages, ingestion by macrophages, and variation in the number of active macrophages. Enhancement of any of these events produces increased rates of clearance as measured by the usual light absorption techniques. At 2 hr, the injected carbon appears to aggregate and to bind to platelets in greater than normal amounts, and to be cleared in the lungs. From 48–72 hr, the carbon aggregation and binding to platelets are less than in normal animals and clearance is largely by hyperactive Kupffer cells in the liver. That is, the late phase (72 hr) of increased rates of clearance following stimulation by zymosan is associated with inhibition of one or more stages of the normal clearance process (i.e., aggregation and platelet attachment).

The changes in distribution of carbon within organs and the kinetics of clearance provide evidence as to which clearance processes are stimulated or inhibited. Thus, lung sequestration and biphasic kinetics in the early phase of stimulated clearance correlate with above normal aggregation and binding to

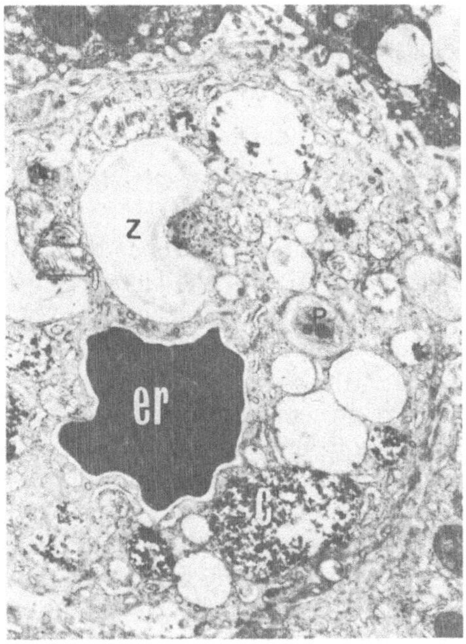

FIGURE 23. Electron micrograph of a Kupffer cell in an animal given colloidal carbon 48 hr after zymosan and killed 25 min after the colloidal carbon. Carbon (C), zymosan (Z), a platelet (P), and an erythrocyte (er) are all present in what appear to be vacuoles in the cytoplasm. ×4300.

platelets, whereas less than normal lung sequestration and single order clearance kinetics correlate with less than normal platelet attachment, but with both increased liver localization and increased phagocytosis. Hyperphagocytosis by individual Kupffer cells is prominent at this stage, as shown dramatically in Figure 23. In relation to platelet experiments, it should be noted that the same substance may produce either increased platelet aggregation by particles or decreased platelet aggregation by particles, depending solely on the time after injection.

3.3. OTHER STIMULANTS OF PARTICLE CLEARANCE

We have limited studies of the effects of endotoxin and *Corynebacterium parvum* upon carbon particle clearance. Endotoxin appears to produce dramatic increase of the platelet component, while *C. parvum* appears to have little effect on platelet clearance at any stage, but dramatic effects on the number and function of Kupffer cells. Endotoxin has a dramatic effect on particle aggregation with platelets (Donald and van't Hull, 1973).

3.4. CONCLUSIONS CONCERNING STIMULATED PARTICLE CLEARANCE STUDIES

From the effects of the stimulants discussed above, it seems unrealistic to consider merely increases of clearance rates without attempts to assess the actual processes involved. Because this is morphologically based and, therefore, difficult to quantitate and time-consuming does not mean that it should be ignored in defining the functions of the RES.

4. INTRAVASCULAR COAGULATION AND PARTICLE CLEARANCE

The rate of clearance has been shown to be influenced by substances that affect coagulation of the blood (Halpern *et al.*, 1951), and the hypothesis has been put forward that activation of blood coagulation mechanisms, particularly the deposition of fibrin, may be essential in the initiation of phagocytosis (Wilkins, 1971).

Ultrastructural studies have produced evidence that intravascular coagulation leads to the deposition of colloidal materials from the circulating blood in particular regions remote from the major RE organs (Donald *et al.*, 1973). Increased retention of carbon particles within renal glomerular capillaries has been demonstrated to occur after the administration of vasoactive drugs (Benacerraf *et al.*, 1959; Perez-Tamayo and Hernandez-Peon, 1953), following the injection of the provoking dose of bacterial endotoxin in the generalized Shwartzman reaction (McKay and Rowe, 1960), during development of renal cortical necrosis elicited by colchicine administration in the pregnant Syrian hamster (Galton, 1965), during the latent period of nephrotoxic serum nephritis 7–9 days following injection of nephrotoxic antiserum (Shibata and Mimura, 1970), and from 2–8 hr after the injection of tubercle bacillary lipid (Donald, 1972b,c).

Therefore, in the presence of disseminated intravascular coagulation, it is evident that the attachment of carbon particles to fibrin can contribute to their

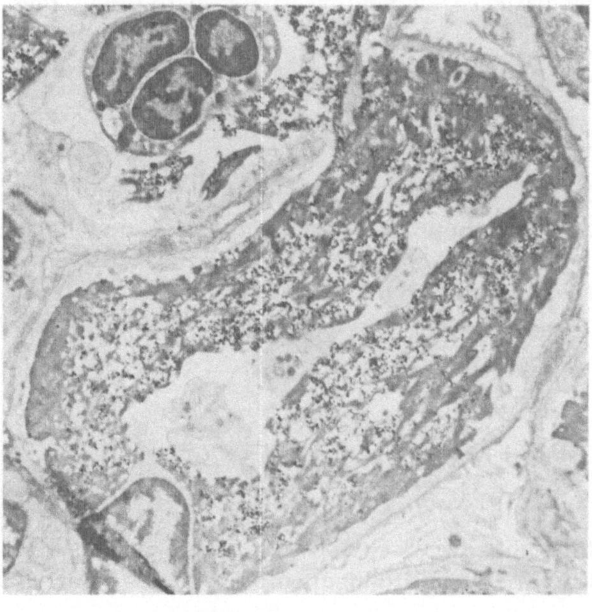

FIGURE 24. An electron micrograph from the kidney of a rabbit given an infusion of thrombin, epinephrine, and EACA, as well as 4 mg/100 g body weight of colloidal carbon simultaneously over a 2 hr period. Carbon particles are intermingled with the fibrin meshwork in the lumen of a glomerular capillary. ×5600.

clearance from the circulating blood (Figure 24). Furthermore, depending upon the localization of fibrin within the microcirculation, carbon deposition may occur in regions of the body remote from the phagocytic cells of the RES. Ultrastructural evidence suggests that this attachment of colloid carbon to fibrin is separate and distinct from its attachment to cell and platelet membranes.

The colloidal carbon in the biologic ink is maintained in suspension by a coating of fish gelatin, which is a foreign protein in rabbits, and may play a part in the attachment of the particles to the fibrin strands. It is likely that, in the

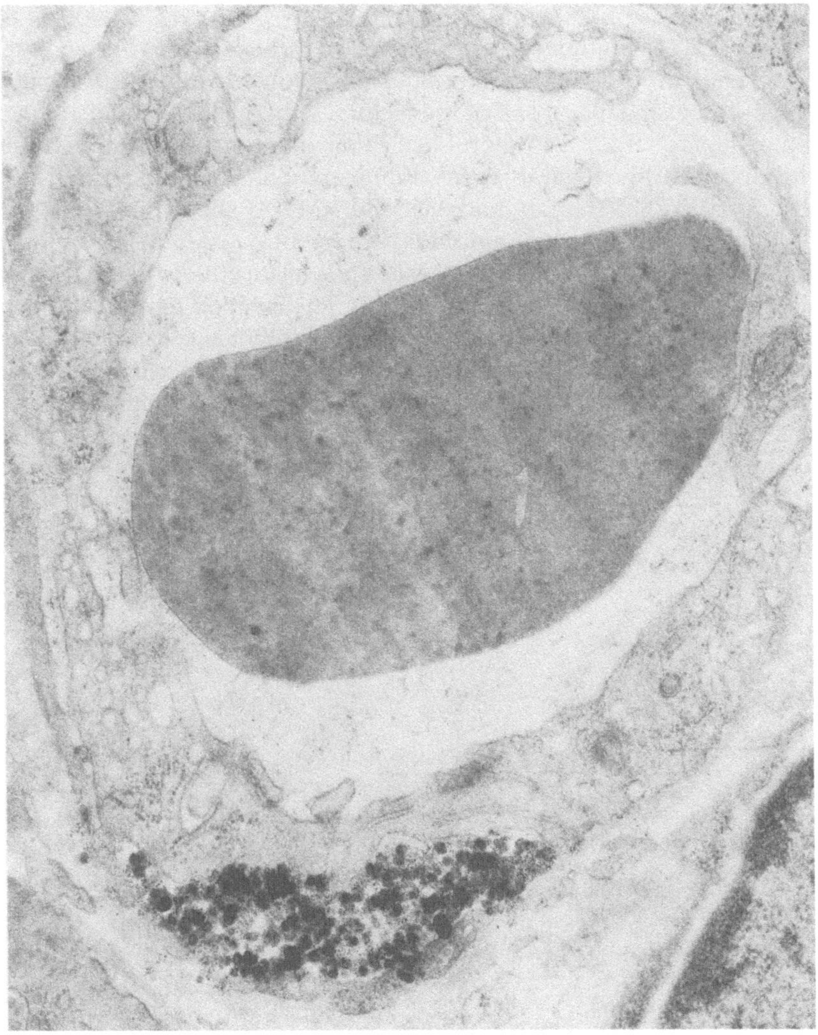

FIGURE 25. Electron micrograph of fenestrated capillary in small intestine of mouse injected with carbon and ferritin 35 min before killing. A large deposit of carbon and ferritin lies beneath a closed intercellular junction. Ferritin, but not carbon, is visible in the vascular lumen. ×28,600. (Supplied by Dr. J. Hurley.)

presence of intravascular coagulation, other circulating foreign particles and abnormal proteins might be trapped in capillaries in the kidney by the process demonstrated here. Such a mechanism might be implicated in the renal glomerular localization of antigen–antibody complexes, which are not immunologically cross-reactive with the kidney, and, therefore, might be relevant to the known importance of blood coagulation in the initiation of immune complex nephritis (Humair *et al.*, 1969).

5. EXTRAVASATION OF PARTICLES FROM BLOOD VESSELS

In discussing the ultrastructure of particle clearance, Carr (1968) pointed out the complexity of the cellular mechanisms involved, and included leakage through endothelial gaps as one of the factors. Others had pointed out (Casley-Smith and Reade, 1965; Hampton, 1958) that there was leakage between the endothelial cells of the liver after the administration of various colloids. Recent ultrastructural studies (Hurley and McCallum, 1974) show that large particles (carbon, ferritin, and mercuric sulphide) escape in considerable quantity from fenestrated and nonfenestrated small blood vessels in the intestinal mucosa of normal rats and mice (Figures 25, 26, and 27). The escaped particles form aggregates beneath the junctions between vascular endothelial cells, and leakage appears to be due to temporary opening of a minority of intercellular junctions. A small number of leaks was also described in the submaxillary gland.

This is likely to be a basically important physiological process, and may account for the escape of plasma protein into intestinal mucosa, submaxillary gland, and other normal tissues. The temporary opening of a minority of the intercellular junctions of the endothelium of small blood vessels may be the

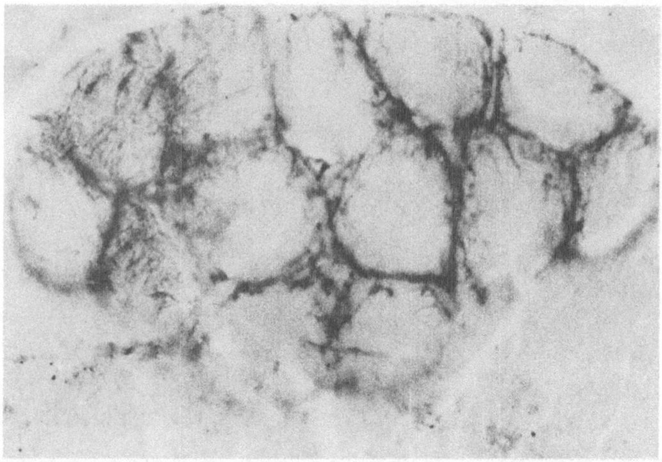

FIGURE 27. Serosal aspect of cleared specimen of small intestine of mouse given an injection of carbon and killed 1 hr later. Scattered small deposits of carbon in the mucosa are visible through the thin muscularis. × ≈30. (Supplied by Dr. Hurley.)

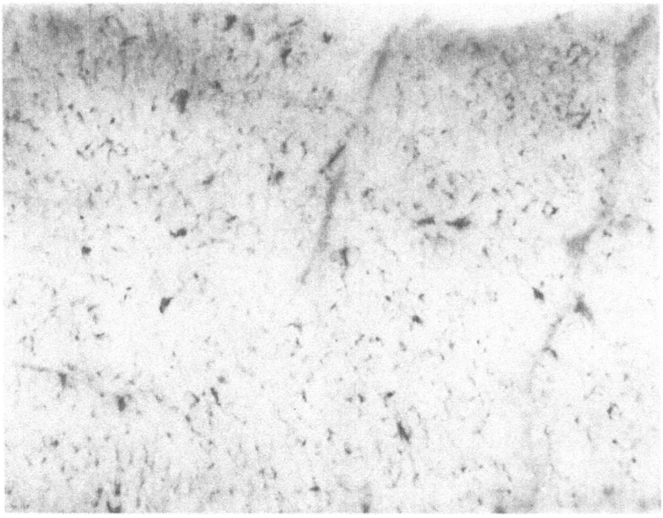

FIGURE 26. Cleared specimen of small intestine of rat given an injection of carbon and killed 1 hr later. Dense blackening of large venules within a Peyer's patch. × ≈10. (Supplied by Dr. J. Hurley.)

functional equivalent of the large pore system postulated by physiologists (Hurley and McCallum, 1974). It is doubtful that the process contributes a significant component to clearance rates as measured for relatively large unphysiological doses of particles, despite the fact that sufficient leakage to cause naked eye straining of intestine by carbon can be shown (Figures 26 and 27). However, in physiological conditions it clearly forms a component in the clearance of particulate matter from the circulation and its localization in tissues.

6. CLEARANCE OF BACTERIA AND NEOPLASTIC CELLS

There is convincing morphological evidence that bacteria interact with platelets (Clawson, 1973; Dudgeon and Goadby, 1931; Houlihan, 1947; Stehbens and Florey, 1960; Teale and Bach, 1920; Tocantins, 1938), and the clearance of microbes appears to involve the same visible processes as that of carbon particles (Clawson, 1973).

Aggregation of platelets, often with associated fibrin, appears to be a common feature in the entrapment of circulating tumor cells of many types (Figure 28) (Chew *et al.*, 1976; Chew and Wallace, 1976; Jones *et al.*, 1971; Warren, 1973; Wood, 1971). Some experimental tumor systems in which platelets are not attached to circulating tumor cells have been reported (Bishop and Donald, 1978; Dingemans, 1973; Locker *et al.*, 1970; Vlaeminck *et al.*, 1972). Although there is conflicting circumstantial evidence in the literature about the possible protective roles of platelets and coagulation in clinical infection and spread of neoplasms, the morphological evidence justifies the continued attempts to ascribe specific functional significance to the ultrastructural observations. Platelets have been

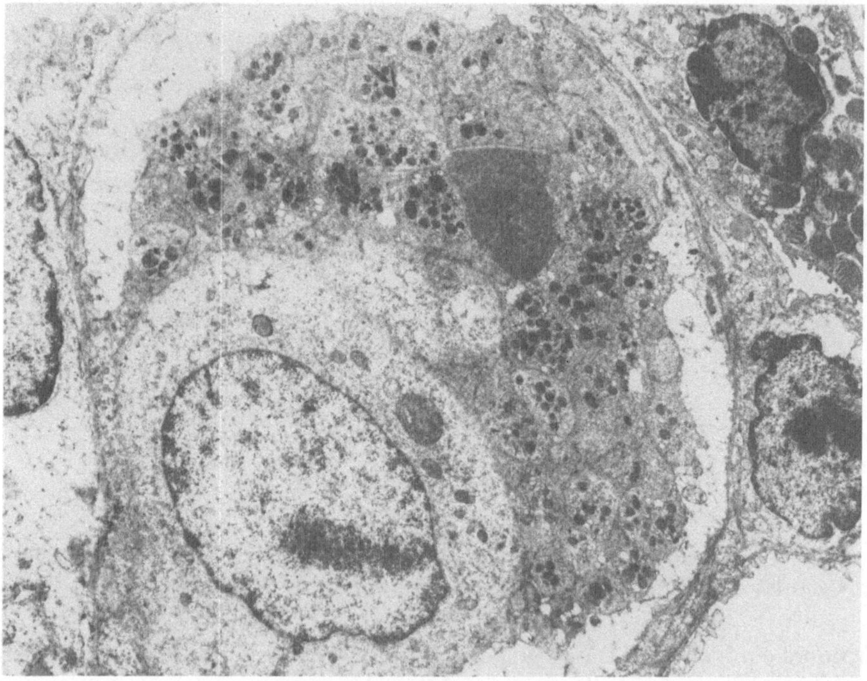

FIGURE 28. An electron micrograph of a mastocytoma tumor cell in a lung capillary 5 min after i.v. injection into a mouse. The tumor cell has a collar of platelets firmly attached to one side and is itself attached to the endothelial cell. ×6000.

shown to release physiologically active mediators after interaction with bacteria (Clawson *et al.*, 1975), and ultrastructural studies show platelet degranulation after interaction with tumor cells (Bishop and Donald, 1979; Chew *et al.*, 1976; Chew and Wallace, 1976; Jones *et al.*, 1971; Warren, 1973, 1976).

REFERENCES

Benacerraf, B., McCluskey, R. T., and Patras, D., 1959, Localization of colloidal substance in vascular endothelium: a mechanism of tissue damage. I. Factors causing the pathologic deposition of colloidal carbon, *Am. J. Pathol.* **35**:75.

Benacerraf, B., and Sebestyen, M. M., 1957, Effect of bacterial endotoxins on the reticuloendothelial system, *Fed. Proc.* **16**:860.

Bishop, C. J., and Donald, K. J., 1979, Non-immunological cell death of intravenously injected murine tumor cells, *Br. J. Exp. Pathol.* **60**:29.

Bloom, G., 1954, The disintegration of human blood platelets after taking up fine quartz particles, *Z. Zellforsch. Mikrosk. Anat.* **40**:222.

Brade, V., Lee, G. D., Nicholson, A., Shin, H. S., and Mayer, M M., 1973, The reaction of zymosan with the Properdin system in normal and C_4-deficient guinea pig serum, *J. Immunol.* **111**:1389.

Carr, I., 1968, Some aspects of the fine structure of the reticuloendothelial system: The cells which clear colloids from the blood stream, *Z. Zellforsch. Mikrosk. Anat.* **89**:355.

Casley-Smith, J. R., and Reade, P. C., 1965, An electron microscopic study of the uptake of foreign particles by the livers of foetal and adult rats, *Br. J. Exp. Pathol.* **46**:473.

Chew, E. C., and Wallace, A. C., 1976, Demonstration of fibrin in early stages of experimental metastases, *Cancer Res.* **36**:1904.
Chew, E. C., Josephson, R. L., and Wallace, A. C., 1976, Morphologic aspects of the arrest of circulating cancer cells, in: *Fundamental Aspects of Metastasis* (L. Weiss, ed.), pp. 121, North Holland, Amsterdam.
Clawson, C. C., 1973, Platelet interaction with bacteria. III. Ultrastructure, *Am. J. Pathol.* **70**:449.
Clawson, C. C., and White, J. G., 1971a, Platelet interaction with bacteria. I. Reaction phases and effects of inhibitors, *Am. J. Pathol.* **65**:367.
Clawson, C. C., and White, J. G., 1971b, Platelet interaction with bacteria. II. Fate of the bacteria, *Am. J. Pathol.* **65**:381.
Clawson, C. C., Rao, G. H. R., and White, J. G., 1975, Platelet interaction with bacteria. IV. Stimulation of the release reaction, *Am. J. Pathol.* **81**:411.
Cohen, P., Braunwald, J., and Gardner, F. H., 1965, Destruction of canine and rabbit platelets following intravenous administration of carbon particles or endotoxin, *J. Lab. Clin. Med.* **66**:263.
Cotran, R. S., 1965, Endothelial phagocytosis: an electron microscopic study, *Exp. Mol. Pathol.* **4**:217.
David-Ferreira, J. F., 1960, Démonstration du pouvoir phagocytaire des plaquettes sanguines chez le lapin, *Proc. Eur. Regional Conference on Electron Microscopy, Delft* **2**:917.
David-Ferreira, J. F., 1961, Sur la structure et le pouvoir phagocytaire des plaquettes sanguines, *Z. Zellforsch. Mikrosk. Anat.* **55**:89.
David-Ferreira, J. F., 1964, The blood platelet: Electron microscopic studies, *Int. Rev. Cytol.* **17**:99.
Des Prez, R. M., Horowitz, H. I., and Hook, E. W., 1961, Effects of bacterial endotoxin on rabbit platelets. II. Platelet aggregation and release of platelet factors *in vitro*, *J. Exp. Med.* **114**:857.
Dineen, J. J., Perillie, P. E., and Finch, S. C., 1968, Zymosan-induced thrombocytopenia in the rat, *J. Reticuloendothel. Soc.* **5**:161.
Dingemans, K. P., 1973, Behavior of intravenously injected malignant lymphoma cells. A morphologic study, *J. Natl. Cancer Inst.* **51**:1883.
Donald, K. J., 1972a, The mechanism of enhanced clearance of colloidal carbon from the blood of rabbits stimulated by a tubercle bacillary lipid, *J. Pathol.* **108**:97.
Donald, K. J., 1972b, Effects of a tubercle bacillary lipid on reticuloendothelial tissue, Ph.D. Thesis, University of Queensland.
Donald, K. J., 1972c, The role of platelets in the clearance from blood of colloidal carbon in rabbits: A light and electron microscope study, *Pathology* **4**:295.
Donald, K. J., 1972d, Elevated rates and altered characteristics of the clearance of carbon from the blood of rabbits with variations in environmental conditions, *J. Pathol.* **107**:73.
Donald, K. J., and Pound, A. W., 1971, The effect of a tubercle bacillary lipid on the clearance rate of colloidal carbon from the blood of rabbits, *Br. J. Exp. Pathol.* **52**:256.
Donald, K. J., and Tennent, R. J., 1975, The relative roles of platelets and macrophages in clearing particles from the blood: the value of carbon clearance as a measure of reticuloendothelial phagocytosis, *J. Pathol.* **117**:235.
Donald, K. J., and van't Hull, E., 1973, Platelets and carbon clearance rates in rabbits affected by a tubercle bacillary lipid or endotoxin, *Virchows Arch. B* **13**:169.
Donald, K. J., Whitaker, A. W., and Bunce, I. H., 1973, The mechanism of renal glomerular capillary retention of carbon in disseminated intravascular coagulation, *Am. J. Pathol.* **70**:245.
Dudgeon, L. S., Goadby, H. K., 1931, The examination of the tissues and some observations on the blood platelets of rabbits at intervals of five minutes, and later, after intravenous inoculations of *Staphylococcus aureus* and India ink, *J. Hyg.* **31**:247.
Ebbe, S., and Stohlman, F., Jr., 1970, Stimulation of thrombocytopoiesis in irradiated mice, *Blood* **35**:783.
Elvidge, A. R., 1926, Foreign particles, the reticuloendothelial system and anaemia, *J. Path. Bact.* **29**:325.
Florey, H., 1962, *General Pathology*, 3rd edn., Lloyd-Luke, London.
Florey, H., 1967, The uptake of particulate matter by endothelial cells, *Proc. R. Soc. London Ser. B* **166**:375.
Galton, M., 1965, Particle retention by the renal glomerular capillaries in the generalized Shwartzman reaction, *Proc. Soc. Exp. Biol. Med.* **119**:1139.

Glynn, M. F., Movat, H. Z., Murphy, E. A., and Mustard, J. F., 1965, Study of platelet adhesiveness and aggregation with latex particles, *J. Lab. Clin. Med.* **65**:179.

Haguenau, F., Hollman, K. H., Levy, J. P., and Boiron, M., 1964, Etude au microscope électronique des plaquettes sanguines dans les leucémies humaines, *J. Microsc.* **2**:529.

Halpern, B. N., Biozzi, G., Mene, F., and Benacerraf, B., 1951, Etude quantitative de l'activité granulopexique du système réticulo-endothélial par l'injection intraveineuse d'encre de Chine chez les diverses espèces animals; méthode d'étude quantitative de l'activité granulopexique du système réticulo-endothélial par l'injection intraveineuse de particules de carbone de dimensions connues, *Ann. Inst. Pasteur* **80**:582.

Halpern, B. N., Benacerraf, B., Biozzi, G., 1953, Quantitative study of the granulopectic activity of the reticuloendothelial system. I. The effect of the ingredients present in India ink and of substances affecting blood clotting *in vivo* on the fate of carbon particles administered intravenously in rats, mice and rabbits, *Br. J. Exp. Pathol.* **34**:426.

Hampton, J. C., 1958, An electron microscopic study of the hepatic uptake and excretion of submicroscopic particles injected into the blood stream, and into the bile duct, *Acta Anat. (Basel)* **82**:262.

Henson, P. M., 1970, Mechanisms of release of constituents from rabbit platelets by antigen-antibody complexes and complement I and II, *J. Immunol.* **108**:476.

Houlihan, R. B., 1947, Studies on the adhesion of human blood platelets and bacteria, *Blood (Special Issue)* **1**:142.

Humair, L., Potter, E. V., and Kwaan, H. C., 1969, The role of fibrinogen in renal diseases. I. Production of experimental lesions in mice, *J. Lab. Clin. Med.* **74**:60.

Hurley, J. V., and McCallum, N. E. W., 1974, The degree and functional significance of the escape of marker particles from small blood vessels with fenestrated endothelium, *J. Pathol.* **113**:183.

Jones, D. S., Wallace, A. C., and Fraser, E. E., 1971, Sequence of events in experimental metastases of Walker 256 tumor: Light, immunofluorescent, and electron microscopic observations, *J. Natl. Cancer Inst.* **46**:493.

Koga, T., Ishibashi, T., Sugiyama, K., and Tanaka, A., 1969, Immunological adjuvants. III. A preliminary report about the mode of action of mycobacterial adjuvants and further confirmation of adjuvant activity of acetylated Wax D, *Intl. Arch. Allergy Appl. Immunol.* **36**:233.

Levaditi, C., 1901, Et des organismes vaccinés contre le vibrion cholérique, *Ann. Inst. Pasteur* **15**:894.

Locker, J., Goldblatt, P. J., and Leighton, J., 1970, Ultrastructural features of invasion in chick embryo liver metastasis of Yoshida ascites hepatoma, *Cancer Res.* **30**:1632.

Majno, G., 1965, Ultrastructure of the vascular membrane, in: *Handbook of Physiology*, Section 2, *Circulation* (W. F. Hamilton and P. Dow, eds.), Vol. 3, pp. 2293–2375, Williams and Wilkins, Baltimore.

Majno, G., and Palade, G. E., 1961, Studies on inflammation. I. The effect of histamine and serotonin on vascular permeability: An electron microscopic study, *J. Biophys. Biochem. Cytol.* **11**:571.

McKay, D. G., and Rowe, F. J., 1960, The effect on the arterial vascular system of bacterial endotoxin in the generalized Shwartzman reaction, *Lab. Invest.* **9**:117.

Movat, H. Z., Weiser, W. J., Glynn, M. F., and Mustard, J. F., 1965, Platelet phagocytosis and aggregation, *J. Cell Biol.* **27**:531.

Mustard, J. F., and Packham, M. A., 1968, Platelet phagocytosis, *Ser. Haematol.* **1**:168.

Nagao, K., 1920, The fate of India ink injected into the blood. I. General observations, *J. Infect. Dis.* **27**:527.

Nagao, K., 1921, The fate of India ink injected into the blood. II. The formation of intracellular granules and their movements, *J. Infect. Dis.* **28**:294.

O'Rourke, G. E., and Pound, A. W., 1971, The effects of a tubercle bacillary lipid on lymphoreticular tissue and on the natural haemolysin to sheep red cells in the rabbit, *Pathology* **3**:107.

Perez-Tamayo, R., and Hernandez-Peon, R., 1953, Intrarenal vascular responses to sympathetic stimulation and to adrenalin, *J. Lab. Clin. Med.* **41**:871.

Pillemer, L., and Ecker, E. E., 1941, Anti complementary factor in fresh yeast, *J. Biol. Chem.* **137**:139.

Pound, A. W., 1955, The effect of a tubercle bacillary "wax" on the response of guinea-pigs to the injection of tubercle bacillary protein, *J. Path. Bact.* **70**:119.

Pound, A. W., 1958, The effect of a tubercle bacillary lipid on antibody production to ovalbumin and horse serum albumin in the guinea-pig, *J. Path. Bact.* **75**:55.

Raffel, S., 1948, The components of the tubercle bacillus responsible for the delayed type of "infectious" allergy, *J. Infect. Dis.* **82**:267.

Raffel, S., and Forney, J. E., 1948, The role of the "wax" of the tubercle bacillus in establishing delayed hypersensitivity. I. Hypersensitivity to a simple chemical substance, picryl chloride, *J. Exp. Med.* **88**:485.

Raffel, S., Arnaud, L. E., Dukes, C. D., and Huang, J. S., 1949, The role of the "wax" of the tubercle bacillus in establishing delayed hypersensitivity. II. Hypersensitivity to a protein antigen, egg albumin, *J. Exp. Med.* **90**:53.

Ream, V. J., Deykin, D., Gurewich, V., and Wessler, S., 1965, The aggregation of human platelets by bacterial endotoxin, *J. Lab. Clin. Med.* **66**:245.

Salvidio, E., and Crosby, W. H., 1960, Thrombocytopaenia after intravenous injection of India ink, *J. Lab. Clin. Med.* **56**:711.

Sanders, A. G., Florey, H. W., and Wells, A. Q., 1951, The behaviour of intravenously injected particles of carbon and micrococci in normal and tuberculous tissues, *Br. J. Exp. Pathol.* **32**:452.

Schulz, H., 1961, Über die Phagozytose von kolloidalem Silizium-Dioxyd durch Thrombozyten mit Bemerkungen zur submikroskopischen Struktur der Thrombozytenmembran, *Folia Haematol. (Frankfurt)* **5**:195.

Schulz, H., and Wedell, J., 1962, Elektronenmikroskopische Untersuchungen zur Frage der Fettphagozytose und des Fettransportes durch Thrombozyten, *Klin. Wochenschr.* **40**:1114.

Shibata, S., Mimura, N., 1970, Studies on latent period of nephrotoxic serum nephritis in rabbits: Production of bilateral renal cortical necrosis by a single injection of carbon particles in latent period, *Jpn. J. Exp. Med.* **40**:265.

Shimamoto, T., Yamazaki, H., Ohno, K., Uchida, H., Konishi, T., and Iwahara, S., 1958, Effect of bacterial endotoxins on platelets and release of serotonin (5-hydroxytryptamine). I. Bacterial endotoxin on the number of circulating platelets, *Proc. Jpn. Acad.* **34**:444.

Simpson, M. E., 1922, The experimental production of macrophages in the circulating blood, *J. Med. Res.* **43**:77.

Stehbens, W. E., and Florey, H. W., 1960, The behaviour of intravenously injected particles observed in chambers in rabbit ears, *Quart. J. Exp. Physiol.* **45**:252.

Sullivan, L. W., 1971, Effect of alcohol on platelet production, in: *Platelet Kinetics: Radio-isotopic, Cytological, Mathematical and Clinical Aspects* (J. M. Paulus, ed.), pp. 247–252, North Holland, Amsterdam.

Tait, J., and Elvidge, A. R., 1926, Effect upon platelets and on blood coagulation of injecting foreign particles into the bloodstream, *J. Physiol.* **62**:129.

Teale, F. H., and Bach, E., 1920, The factors leading to the removal of bacteria from the peripheral circulation by phagocytosis, *Proc. R. Soc. Med. (Sec. 3)* **13**:77.

Tennent, R. J., and Donald, K. J., 1976, The ultrastructure of platelets and macrophages in particle clearance stimulated by zymosan, *J. Reticuloendothel. Soc.* **19**:269.

Tocantins, L. M., 1938, The mammalian blood platelet in health and disease, *Medicine* **17**:155.

Van Aken, W. G., and Vreeken, J., 1969, Accumulation of macromolecular particles in the reticuloendothelial system (RES) mediated by platelet aggregation and disaggregation, *Thromb. Diath. Haemorrh.* **22**:496.

Van Aken, W. G., Goote, T. M., and Vreeken, J., 1968, Platelet aggregation: An intermediary mechanism in carbon clearance, *Scand. J. Haematol.* **5**:333.

Vlaeminck, M. N., Adenis, L., Mouton, Y., and DeMaille, A., 1972, Experimental study of the metastatic spread of tumor cells in embryonated chicken egg. Distribution, histology and ultrastructural aspects of tumorous focuses, *Int. J. Cancer* **10**:619.

Warren, B. A., 1973, Environment of the blood-borne tumor embolus adherent to vessel wall, *J. Med.* **4**:150.

Warren, B. A., 1976, Some aspects of blood borne tumour emboli associated with thrombosis, *Z. Krebsforsch.* **87**:1.

White, J. G., 1972, Uptake of latex particles by blood platelets. Phagocytosis or sequestration?, *Am. J. Pathol.* **69**:439.

White, R. G., 1967a, Concepts relating to the mode of action of adjuvants, *Symp. Ser. Immunobiol. Stand.* **6**:3.

White, R. G., 1967b, Characterisation of mycobacterial components of adjuvant mixtures, *Symp. Ser. Immunobiol. Stand.* **6**:49.

Wilkins, D. J., 1971, Hypothesis: A possible relationship between blood clotting and the recognition by the body of foreign particles. The reticuloendothelial system in immune phenomena, *Adv. Exp. Med. Biol.* **15**:77.

Wislocki, G. B., 1924, On the fate of carbon particles injected into the circulation with especial reference to the lungs, *Am. J. Anat.* **32**:423.

Wood, S., 1971, Mechanisms of establishment of tumour metastases, in: *Pathobiology Annual* (H. L. Joachim, ed.), pp. 281–308, Butterworths, London.

Wright, H. D., 1927, Experimental pneumococcal septicaemia and anti-pneumococcal immunity, *J. Path. Bact.* **30**:185.

15

The Langerhans Cell

INGA SILBERBERG-SINAKIN
and G. JEANETTE THORBECKE

1. HISTORY OF THE DISCOVERY OF LANGERHANS CELLS AND A REVIEW OF THE OLDER CONCEPTS OF THEIR NATURE

The Langerhans cell is named after its discoverer, Paul Langerhans (1868), who described Langerhans cells (LC) in the epidermis of sections of skin that had been stained with gold chloride. He had used this stain in attempts to demonstrate nerve fibers in the skin. The LC were dendritic in form and were located between basal cells and prickle cells. For several years they were thought by many to be nerve cells (Ferreira-Marques, 1951; Niebauer, 1956; Richter, 1956; Wiedmann, 1952). However, the dendritic shape of LC as well as their location in the epidermis suggested to other observers that LC were related to melanocytes, being either stem cells, melanoblasts, or inactive, effete, or degenerated melanocytes (Billingham and Medawar, 1953; Breathnach, 1963a; Fan and Hunter, 1958; Fan *et al.*, 1959; Masson, 1951; Zelickson, 1965, 1966). Still others considered them to be lymphocytes or derivatives of lymphocytes (Andrew, 1968; Billingham and Silvers, 1965; Kuwahara, 1971; Ranvier, 1875). Until 1961, all the data used to support the various theories regarding the derivation of LC were derived from tissue sections observed with the light microscope and stained by nonspecific methods. Birbeck *et al.* (1961) first described the ultrastructural features of these cells. This made it possible to reliably identify individual LC on the basis of the presence of the characteristic organelle, the LC granule (Birbeck granule). Breathnach and Wyllie (1967) made dermatologists aware of studies which showed LC to be present in lesions of histiocytosis X (Basset *et al.*, 1965; Basset and Nezelof, 1966; Basset and Turiaf, 1965). This caused investigators to review the older concepts of the derivation of LC and to think of the possibility

---

INGA SILBERBERG-SINAKIN • Department of Dermatology, New York University School of Medicine, New York, New York 10016. G. JEANETTE THORBECKE • Department of Pathology, New York University School of Medicine, New York, New York 10016. ISS is presently in private practice in Vineland, New Jersey.

that LC might be histiocytes (Hashimoto, 1971; Hashimoto and Tarnowski, 1968; Prunieras, 1969; Tarnowski and Hashimoto, 1967). Breathnach (1977a) summarized the problem of the origin of LC by remarking that the presence of LC within the epidermis had, perhaps, in the past, led to an overconcentration of effort to relate their prime functions to purely epidermal processes, while embryological evidence, suggesting a mesenchymal origin of LC, showed that their function was related to the skin as a whole, as will be discussed in detail below. A detailed description of the studies on LC, supporting the older views, is presented in Wolff's review of LC (1972). A more recent excellent and comprehensive review of LC is by Shelley and Juhlin (1978). Two other reviews which, like the present one, stress the immunological significance of LC are by Baer and Berman (1980) and Silberberg-Sinakin *et al.* (1978).

2. METHODS OF IDENTIFICATION OF LANGERHANS CELLS

In routine histological sections of skin stained with hematoxylin and eosin, LC cannot be identified. Sometimes they appear as high level, clear cells due to shrinkage of their cytoplasm. For this reason special stains have been used: gold chloride (Breathnach, 1965; Fan and Hunter, 1958; Ferreira-Marques, 1951; Langerhans, 1868; Zelickson and Mottaz, 1968b); gold sodium thiomalate (Juhlin and Shelley, 1977); osmium iodide (Breathnach and Goodwin, 1966; Mishima and Miller-Milinska, 1961); osmium zinc iodide (Niebauer *et al.*, 1969); quinone-imine dyes (Billingham and Medawar, 1953; Niebauer and Sekido, 1965); adenosine triphosphatase (Bell, 1967; Bradshaw *et al.*, 1963; Cormane and Kalsbeek, 1963; Im and Montagna, 1965; Ishikawa and Klingmüller, 1964; Jarrett and Riley, 1963; Mishima *et al.*, 1972; Mustakallio, 1962; Wolff, 1963, 1964a,b; Wolff and Winkelmann, 1967a); mannosidase (Elleder, 1975); paraphenylenediamine (Juhlin and Shelley, 1977); cobalt chloride (Juhlin and Shelley, 1977); nickel chloride (Juhlin and Shelley, 1977); L-dopa, dopamine and certain catecholamines, like noradrenalin (Falck *et al.*, 1976; Axelsson *et al.*, 1978; Sjöborg *et al.*, 1978); and fluorescent staining with anti-Ia (Klareskog *et al.*, 1977; Rowden *et al.*, 1977; Stingl *et al.*, 1978a,c). The following enzyme activities of LC are found: guinea pig LC have aminopeptidase (Wolff, 1964b); rat and mouse LC have nonspecific esterases (Campo-Aasen and Pearse, 1966; Jarrett and Riley, 1963); subhuman primate LC have alkaline phosphatase (Kechijian, 1965; Montagna and Ellis, 1959; Montagna and Yun, 1962a,b; Quevedo and Montagna, 1962; Yasuda *et al.*, 1961); sheep and bat LC have cholinesterase (Bourlond *et al.*, 1967; Lyne and Chase, 1966).

The various stains are useful for identification of LC, especially in the epidermis, but may not always be specific. The most reliable method for identification of LC at this time is by their electron microscopic features which make them distinguishable from other cells not only in the epidermis but in other locations such as the dermis and lymphoid tissue as well. They were first described by Birbeck *et al.* (1961) who listed them as follows: (1) Langerhans cell or Birbeck granules; (2) a relatively clear cytoplasm; (3) a lobulated nucleus, and (4)

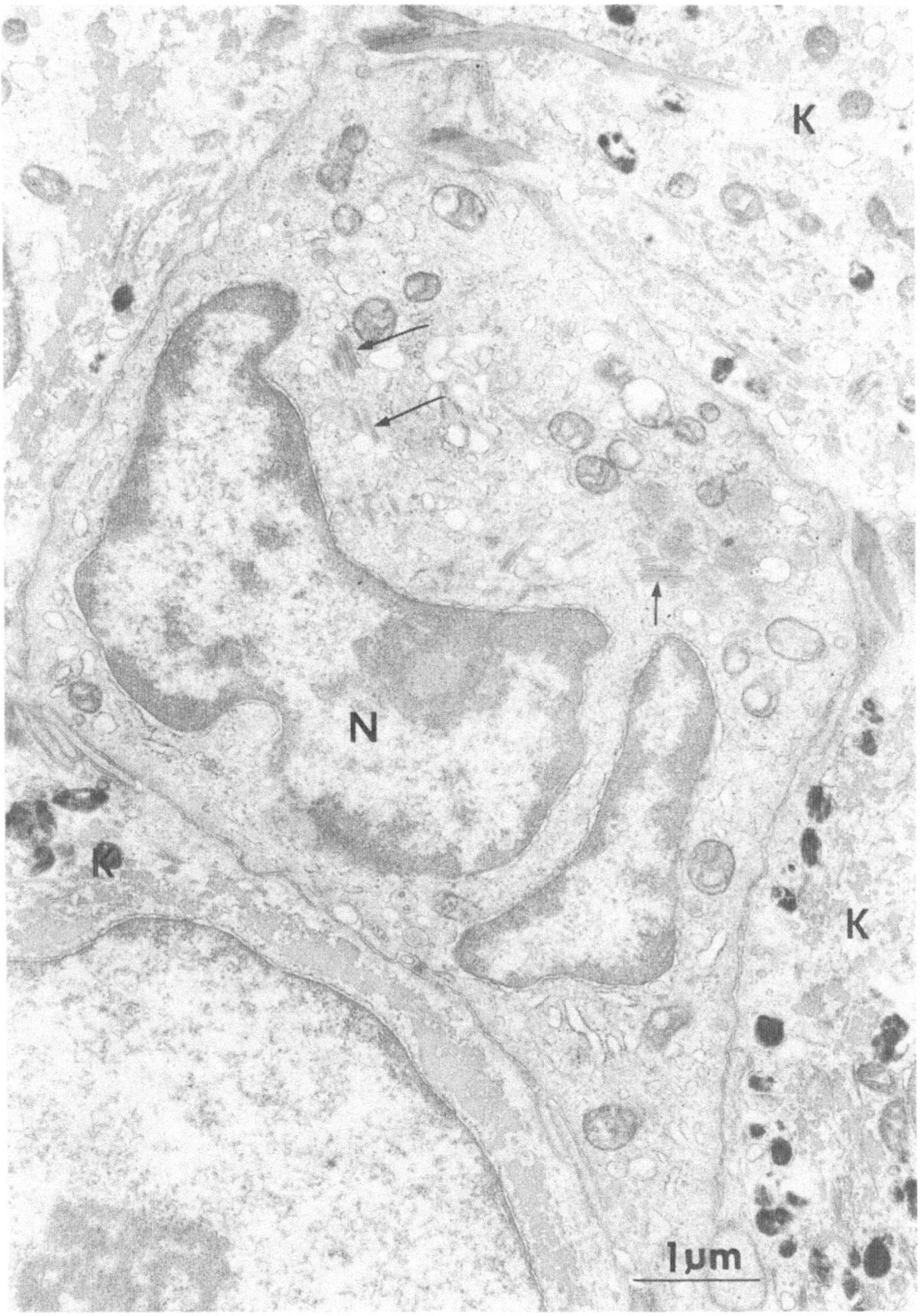

FIGURE 1. Ultrastructural appearance of a Langerhans cell in the suprabasal layer of the epidermis in a normal human subject. Shown are the lobulated nucleus (N), numerous Langerhans cell granules (arrows) with disklike or racket-shaped appearances; absence of tonofilaments, melanosomes, and desmosomes which are present in surrounding keratinocytes (K). Stained with uranyl acetate and lead citrate. ×17,180. (From Silberberg, 1971b, with permission.)

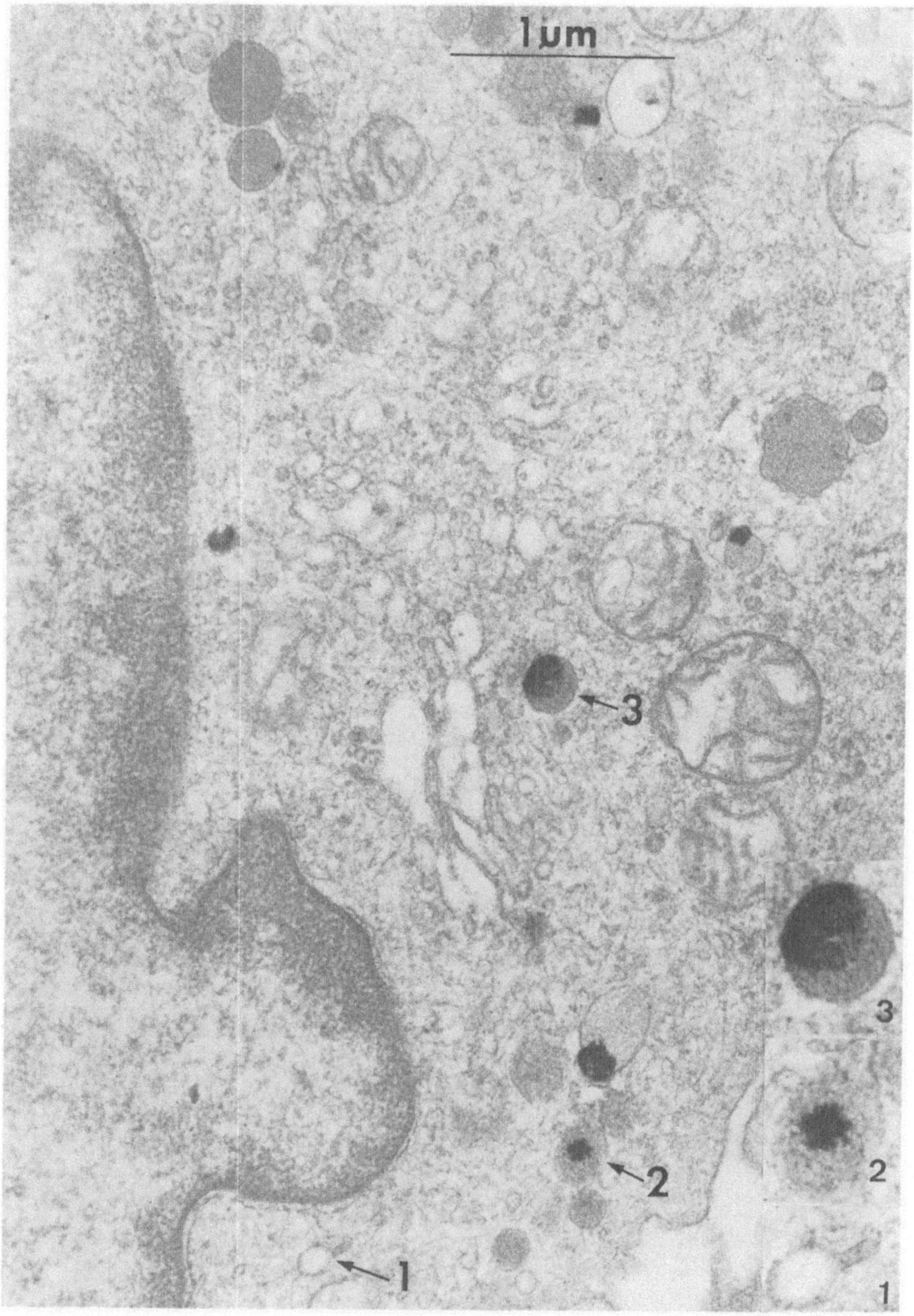

FIGURE 2. Langerhans cell with several lysosomes containing acid phosphatase reaction product (arrows 2, 3). A Langerhans cell granule is present in the cytoplasm (arrow 1). This skin specimen was taken 19 hr after 2,4-dinitro-1-chlorobenzene (DNCB) challenge in a guinea pig passively sensitized to DNCB. Stained with uranyl acetate and lead citrate. ×31,000. The insets show two lysosomes and a Langerhans cell granule. ×64,000. (From Silberberg *et al.*, 1975b, with permission.)

the absence of organelles which are characteristic for keratinocytes and melanocytes (i.e., tonofilaments, desmosomes, premelanosomes, and melanosomes) (Figure 1). However, LC may occasionally contain organelles which they have ingested, such as melanosomes. In addition to these characteristics, LC may share other organelles usually found in living cells. Under some conditions they have an active lysosomal system (Figure 2), osmiophilic dense bodies, microtubules, fine filaments, prominent channels of rough endoplasmic reticulum (RER), and dendrites (Olmos, 1975; Wolff, 1972; Silberberg-Sinakin *et al.*, 1977).

LC granules are the morphologic marker of the cell and, therefore, we will examine them in some detail. The LC granules cannot be seen by light microscopy. They may be stained with osmium zinc iodide (Niebauer *et al.*, 1969). Three dimensional models of individual granules in LC revealed that the granule is made up of a flattened or curved orthogonal net of particles and is bounded externally by a limiting membrane, which may be disk-shaped, cup-shaped, or a combination of both. The internal face of the membrane of the granule is lined by a leaflet of small particles spaced at 50–60 Å (Wolff, 1967b). Midway between the limiting membranes of the granule there is a linear density with a periodical striation which has been described as 50–60 Å (Wolff, 1967b; Zelickson, 1965) or 90 Å (Birbeck *et al.*, 1961; Sagebiel and Reed, 1968). Depending upon the axis along which the granules have been sectioned, their two-dimensional profiles may be "rod-shaped," "tennis-racket-shaped," or "flask-shaped" (Figures 1, 2, and 4–13) (Mishima, 1966; Zelickson, 1965; Zelickson, 1966). The length of the granules vary from 0.8 μm to greater than 2 μm (Tusques and Pradal, 1969). The shape of the granules, as determined by thin sectioning, has been confirmed by studies of freeze-fracture preparations (Breathnach *et al.*, 1973; Caputo *et al.*, 1976). The granules may be found anywhere in the cytoplasm of the cell. A common site for them is near the Golgi area; others show continuity with the outer cell membrane. The origin and significance of the granules is not known. Some investigators believe that the granule is formed as the result of an infolding of the cell membrane and regard the LC granules as a specialized form of phagocytosis, but without the classical phagocytic or pinocytic vacuoles (Hashimoto and Tarnowski, 1968). By contrast, others champion the view that the granules arise as vesicles in the Golgi area (Wolff, 1967b; Zelickson, 1965). These vesicles elaborate a rodlike extension which moves to the cell periphery, attaches to the membrane and opens to the extracellular space, presumably to discharge a secretion. And still others postulate that the granule may be formed in both ways (Cancilla, 1968).

It is important to note that although the LC granule (Birbeck granule) is the ultimate marker for LC, these granules may occasionally be taken up by classical macrophages, where they can be seen in phagocytic vacuoles (Silberberg *et al.*, 1974b; Silberberg-Sinakin *et al.*, 1977) (Figure 13) and possibly also by keratinocytes (Silberberg, 1971b). These possibilities must be considered before identifying the cells as LC. Therefore, the term LC, as used in this chapter, refers to any cell which contains cytoplasmic LC granules and which does not have the organelles typical of keratinocytes or melanocytes. Hoefsmit *et al.*, in Chapter 11 of this volume, describe a cell type in lymph nodes which has Langerhans cell granules (Birbeck granules) which they call an interdigitating reticulum cell and of which they point out the similarity to LC. Rausch *et al.* (1977) stress this similarity, but restrict the designation of interdigitating reticulum cell to cells without Langerhans cell granules in the lymph node. As will be shown in the following discussion, typical epidermal LC are so similar to the interdigitating reticulum cells with Langerhans cell granules that a distinction appears impossible.

3. DISTRIBUTION OF LANGERHANS CELLS

LC are a constant component of epidermis. In man there are about 460–1000 cells per mm^2 (Brown *et al.*, 1967), and in guinea pigs approximately 800–1000 cells per mm^2, except for the ears and paws which usually contain more (Wolff and Winkelmann, 1967b). Quantitative electron microscopic studies of human epidermis indicate LC occupy from 1.0 to 1.7% of the total volume of the epidermis (Zelickson and Mottaz, 1968a). However, more recent studies reveal LC make up about 3 to 8% of the epidermal cell population (Katz *et al.*, 1979). LC are found in skin appendages, such as the hair follicles, sebaceous glands (Breathnach, 1963b; Breathnach *et al.*, 1963; Jimbow *et al.*, 1969a; Wolff, 1963) and apocrine glands (Ito *et al.*, 1976). A few LC are seen in "normal" dermis (Hashimoto and Tarnowski, 1968; Kiistala and Mustakallio, 1968; Zelickson, 1965) and in small dermal lymphatics (Figures 3–5) under normal and abnormal conditions in guinea pigs (Silberberg *et al.*, 1974a; 1975a,b; Silberberg-Sinakin *et al.*, 1976), and man (Breathnach *et al.*, 1975; Breathnach, 1977a) as well as in major afferent lymphatics to lymph nodes (Hoefsmit *et al.*, 1978; Kelly *et al.*, 1978). They also have been described in mucous membranes of oral and pharyngeal tissue such as the tongue and tonsils (Hutchens *et al.*, 1971; Sagebiel *et al.*, 1971; Schroeder and Theilade, 1966; Waterhouse and Squier, 1967), esophagus (Al Yassin and Toner, 1976) as well as in gastric mucosa (Böck, 1974) and uterine cervix (Hackemann *et al.*, 1968; Younes *et al.*, 1968). Although LC have not been found in normal lung tissue of man, they have been described in diseased lung tissue (Basset *et al.*, 1976). Other sites where LC have been found are human, rabbit, and guinea pig lymph nodes (Gianotti *et al.*, 1968; Jimbow *et al.*, 1969b; Kondo, 1969; Silberberg *et al.*, 1975a) and rat thymus medulla (Hoefsmit and Gerver, 1975; Oláh *et al.*, 1968; van Haelst, 1969).

4. STUDIES OF THE ORIGIN AND TURNOVER OF LANGERHANS CELLS

Major approaches to the study of the origin of LC are summarized in Table 1. Our own studies led us to propose that the LC may be a highly differentiated end stage of a bone-marrow-derived cell line similar to, but separate from the more classical phagocytic macrophage cell line (Silberberg-Sinakin *et al.*, 1977). The finding that the bone marrow is one of the characteristic locations of histiocytosis X, for example, eosinophilic granuloma of the bone (Basset *et al.*, 1965), is of relevance in relation to a possible derivation of LC from bone marrow. In patients with monocytic leukemia, cells containing LC granules have

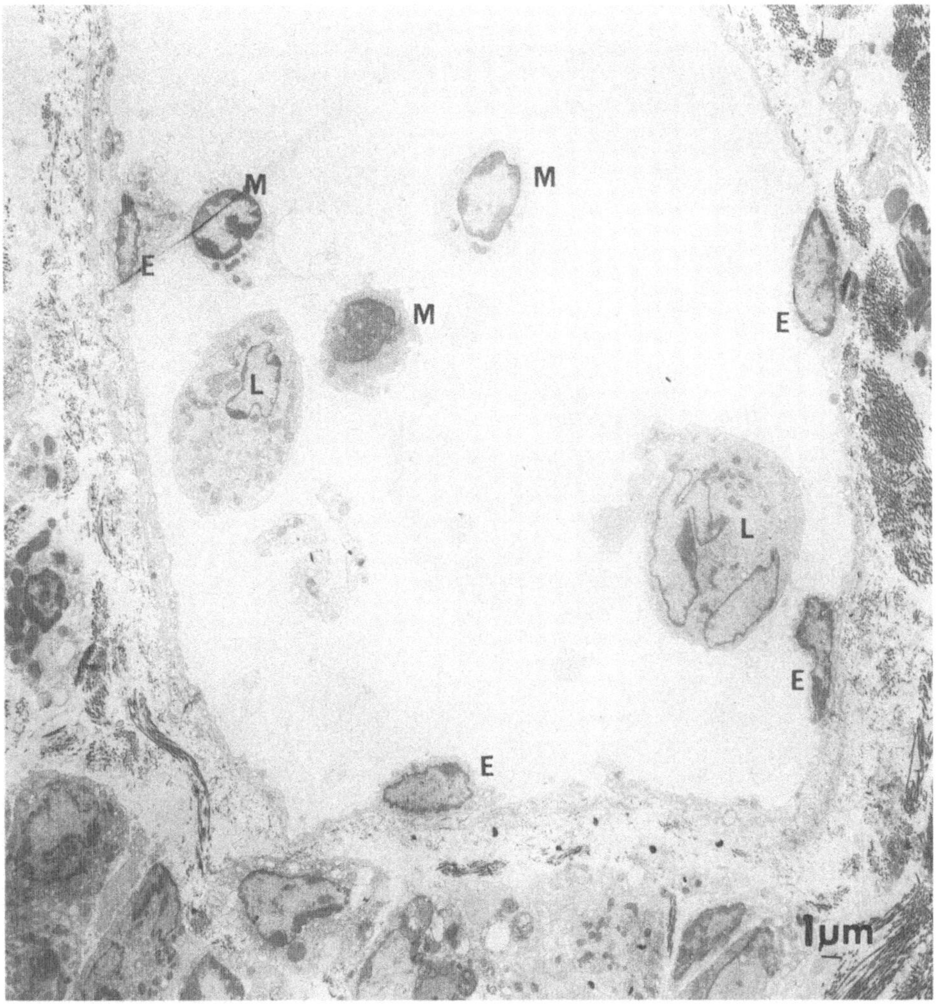

FIGURE 3. Part of the lumen of a vessel containing two Langerhans cells (L) and three mononuclear cells (M) which morphologically resemble lymphocytes. No red blood cells are seen. Four endothelial cells (E) are present. This vessel resembles a lymphatic. Stained with uranyl acetate and lead citrate. ×2500. (From Silberberg *et al.*, 1974a, with permission.)

also been found (Sanel and Serpick, 1970). This is particularly noteworthy since macrophages are derived from monocytes and monocytes are of bone marrow origin (van Furth and Thompson, 1971).

Recent studies by Katz and co-workers (1979) have made use of the fact that LC exhibit Ia antigens on their surface. They found that in murine chimeras, observed up to 11 weeks after injection of bone marrow from one strain into irradiated recipients of another (H-2-different) strain, a large percentage of LC in the epidermis were of donor origin. This type of result may indicate something about the turnover rate of the LC in normal skin as well as demonstrate the bone marrow origin of at least some LC (Katz *et al.*, 1979).

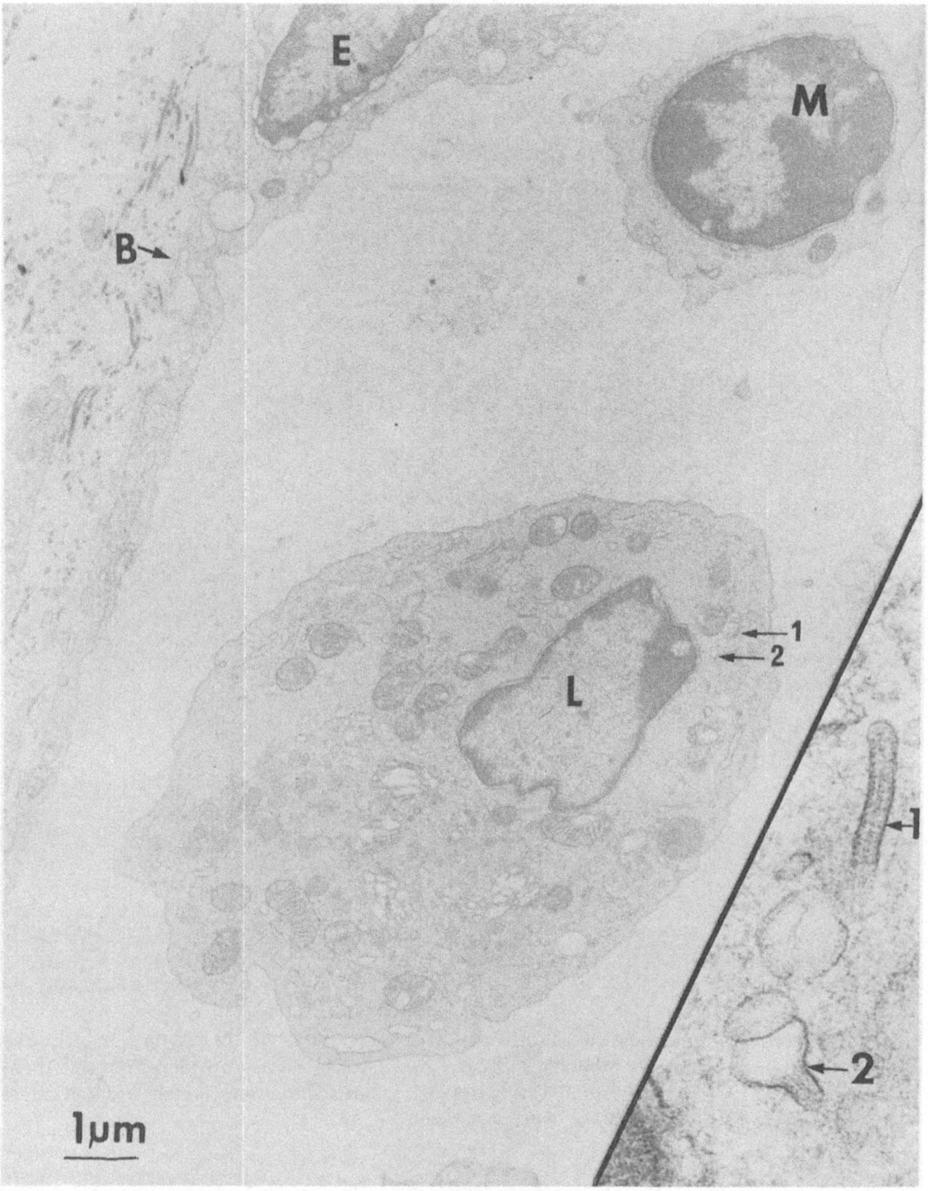

FIGURE 4. High power view of one of the Langerhans cells (L) shown in Figure 3. The arrows (marked 1 and 2) indicate the locations of Langerhans cell granules. A lymphocyte-like cell (M) and part of an endothelial cell (E) are also seen. The basement membrane (B) is thin and irregularly broken up. ×10,500. Langerhans cell granules are shown at higher magnification in the inset ×91,500. Stained with uranyl acetate and lead citrate. (From Silberberg *et al.*, 1974a, with permission.)

Autoradiographic studies of the mitotic activity of LC have shown that the cells are capable of incorporating thymidine into newly synthesized DNA and have suggested that the LC population is stable and self-replicating (Giacometti and Montagna, 1967; Schellander and Wolff, 1967). Occasional mitotic figures of

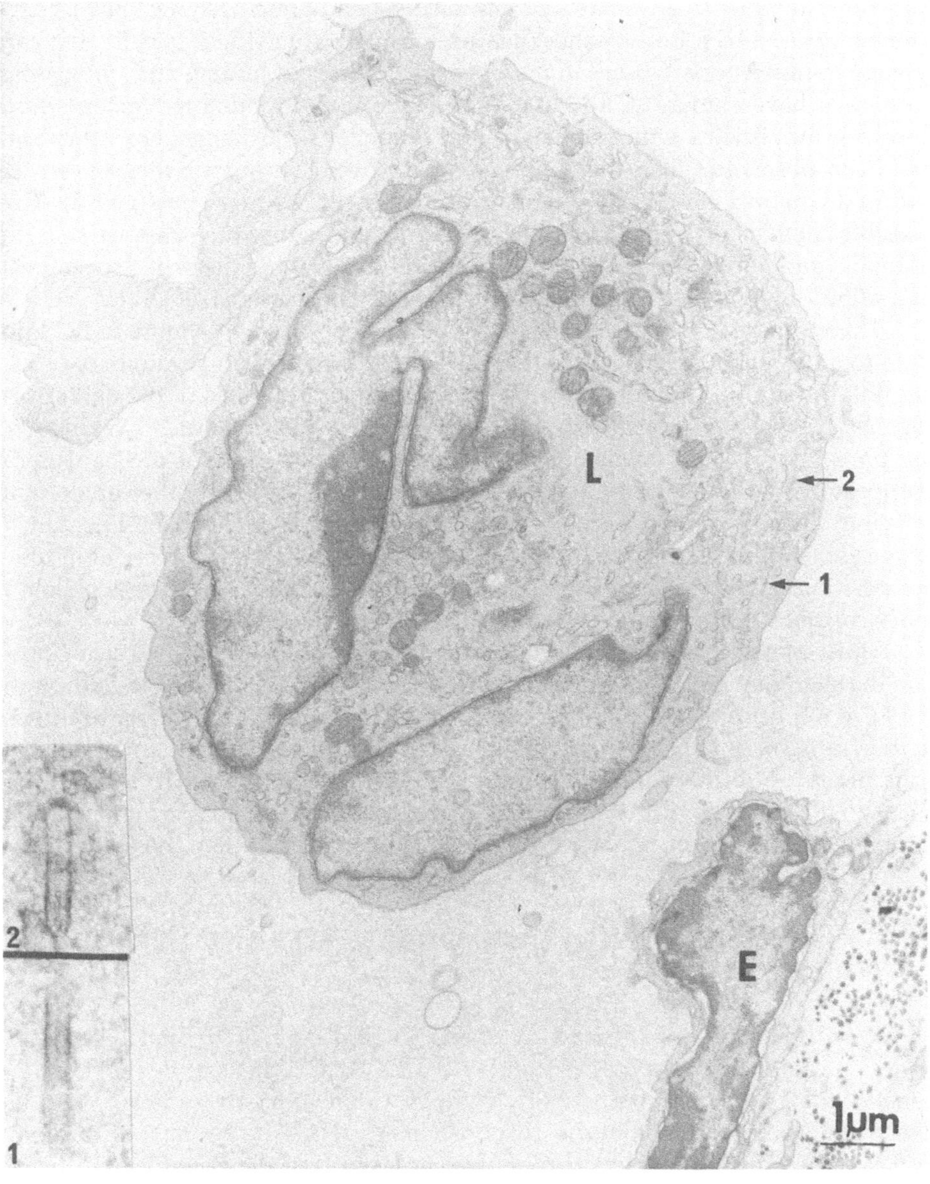

FIGURE 5. The other Langerhans cell (L) which was shown in Figure 3. The arrows (marked 1 and 2) indicate the positions of the Langerhans cell granules. Part of an endothelial cell (E) is seen. ×10,500. The Langerhans cell granules are shown at higher magnification in the inset. ×91,500. Stained with uranyl acetate and lead citrate. (From Silberberg *et al.*, 1974a, with permission.)

LC have been reported at the ultrastructural level (Gianotti and Caputo, 1969; Hashimoto and Tarnowski, 1968).

There is much evidence that LC are capable of movement. Ultrastructurally, they at times display a prominent network of microfilaments (Silberberg-Sinakin *et al.*, 1978; Wolff, 1972), occasionally with the dimensions of actin (Silberberg-

Sinakin *et al.*, 1977), as well as a prominent system of microtubules (Silberberg-Sinakin *et al.*, 1977). Both of these features usually are evident in cells that can engage in movement. Observations in tissue cultures of human and guinea pig epidermis have shown LC to have very active pseudopodial movements and dendritic movements with extensions and retractions. Strong contact inhibition has been observed, such that LC never pass over continuous monolayers of cultured keratinocytes (Cruickshank and Cooper, 1975; Fritsch *et al.*, 1973). The possible movement from epidermis to dermis and subsequent appearance in dermal lymphatics during immunologic reactions involving the skin also suggest migration of LC *in vivo* (Silberberg-Sinakin and Thorbecke, 1980).

The epidermal tape stripping method has been used to eliminate LC and then to study their reappearance (Lessard *et al.*, 1968). Normal numbers of LC are not found until the 15th day after injury, although the rest of the epidermis regenerates in 4 days. Tape stripping might be a good way to study LC kinetics, but preliminary results have been inconclusive with regard to the mechanism of reappearance of LC. During wound healing, LC migrate with other epidermal cells into the injured area. Some of the LC incorporate tritiated thymidine which suggests that they respond to injury both by migration and an increase of mitotic activity (Giacometti, 1969). There is, after a time, a tendency to reestablish a fairly constant LC to keratinocyte ratio (Lessard *et al.*, 1968).

Thus, in summary, it can be said that little is known about the turnover of LC, but that they most likely originate in the bone marrow. Therefore, although the LC is a constant resident of the normal epidermis with characteristic granules in its cytoplasm, it appears to originate in the bone marrow which, under normal conditions, up to now, does not show cells with LC granules. In Section 6 a survey will be made of the properties of a variety of cells that may qualify as possible relatives or precursors of the LC in locations other than the epidermis.

5. SURFACE RECEPTORS OF LANGERHANS CELLS AND METHODS OF ISOLATION

Recent observations show that human and guinea pig epidermal LC express Fc-IgG receptors, C3 receptors (Stingl *et al.*, 1977), and those antigens (Ia antigens) of the major histocompatibility complex (MHC) known to be involved in genetically determined immune responsiveness (HLA-D, human; IR-region, guinea pig and mouse). The presence of such cell surface components, which are also present on a variety of lymphoid cells and macrophages, suggest that LC may functionally be related to cells of the lymphoid and/or reticuloendothelial systems (Forsum *et al.*, 1978; Klareskog *et al.*, 1977; Rowden *et al.*, 1977; Stingl *et al.*, 1978a,c). At the ultrastructural level the Ia antigens were found on the cell surface and on dendrites of LC but not on the LC granules (Rowden, 1977).

Langerhans cells have now been successfully isolated from epidermis by exposing them to acrylamide beads coated with rabbit anti-Ia IgG. Density gradient centrifugation of beads produced a population of adherent cells which were identified as LC (Rowden *et al.*, 1978). Another method of isolating LC makes use of their ability to form rosettes with red blood cells coated with IgG (Stingl *et *al.*, 1977). Isolation of LC from dermis or lymphoid tissue through these techniques is impossible since many other cell types bear these receptors (Hammerling *et al.*, 1975; Schwartz *et al.*, 1976b; Shreffler and David, 1975).

6. COMPARISON OF LANGERHANS CELL PROPERTIES WITH THOSE OF SOME OTHER CELLS IN THE MACROPHAGE AND "RETICULUM" CELL FAMILIES

6.1. COMPARISON WITH MACROPHAGES

Table 2 lists some of the similarities and differences between LC and classical macrophages. It should be stressed that, although LC show several differences from "classical" macrophages, such as are found in the peritoneal cavity, the heterogeneity of such macrophages is beginning to be explored; for instance, only 8–15% of peritoneal cells have been found to bear Ia antigen (Schwartz *et al.*, 1976b). Considering the relationship between LC and macrophages, it seems possible that a subpopulation of the monocyte family which expresses Ia antigens might include the precursors of LC.

In any comparison with macrophages it might be important to consider the reports in the literature concerning the ability of these cells to take up substances from medium *in vitro* (Axelsson *et al.*, 1978; Falck *et al.*, 1976; Shelley and Juhlin, 1976, 1977), and, possibly, from their surroundings *in vivo* (Nagao *et al.*, 1976; Nordquist *et al.*, 1966; Sagebiel, 1972; Silberberg, 1971a,b; Silberberg *et al.*, 1969; Wolff, 1972; Wolff and Honigsmann, 1971; Wolff and Schreiner, 1970). Glycogen has been seen in LC but this may be present as a result of damage to the cell (Desmons *et al.*, 1977; Silberberg, 1971b). Although the mechanism of such uptake has not been studied in detail, two aspects require some attention. One is the observation by Shelley and Juhlin (1976, 1977a,b), who have shown that LC have selectively absorbed or removed the following compounds, each of which is a known potential cause of contact dermatitis: formaldehyde, glutaraldehyde, paraphenylenediamine, ethylenediamine, toluenediamine, nickel, cobalt, chromium, mercury, and gold. It is of interest that some of the same substances taken up by LC can also be used to stain them (Juhlin and Shelley, 1977). The other is the observation of the *in vitro* uptake of L-dopa, serotonin, and catecholamines (Axelsson *et al.*, 1978; Falck *et al.*, 1976).

The pinocytic or phagocytic* activities of these cells has also received attention. Pinocytosis of ferritin molecules is less in LC than in neighboring keratinocytes (Nordquist *et al.*, 1966; Sagebiel, 1972), and the same is true for thorotrast (Wolff and Honigsmann, 1971) and peroxidase (Wolff and Schreiner, 1970). Phagocytic activity *in vivo* which was studied with antigen–antibody complexes was much less than that of classical macrophages (Silberberg-Sinakin *et al.*, 1977).

*The uptake of particles greater than 1 μm in diameter has been defined as phagocytosis, while the uptake into pinocytoic vacuoles of smaller particles, as ferritin, is called pinocytosis (Rabinovitch, 1970).

The one property of paramount importance for the recently shown ability of LC to present antigen to T lymphocytes *in vitro* (Stingl *et al.*, 1978b) may be the binding of antigens to their surface (Shelley and Juhlin, 1976; Silberberg-Sinakin *et al.*, 1976), rather than pinocytic or phagocytic activities. Comparison with macrophages may be quite difficult at this stage, since those macrophages important in presenting antigen to lymphoid tissue are known to be Ia-positive (Schwartz *et al.*, 1976a), which identifies them as a subpopulation (Schwartz *et al.*, 1976b), but they have not been characterized with respect to their relative abilities to take up substances as compared to the majority of classical macrophages.

6.2. RELATIONSHIP OF LANGERHANS CELLS TO INDETERMINATE DENDRITIC CELLS IN SKIN

Early in the ultrastructural characterization of LC it was recognized that there were dendritic cells in the epidermis in which LC granules could not be seen. This may reflect the limited view one has with a high magnification study of the cytoplasm of a cell with a few LC granules. Nonetheless, this finding of dendritic cells containing neither LC granules nor melanosomes has given rise to the concept of indeterminate dendritic cells (Zelickson and Mottaz, 1968a). This cell has also been called Type 3 cell (Snell, 1965) or alpha dendritic cell (Mishima *et al.*, 1972). They appear to be more numerous in the basal than in suprabasal layers but are usually seen in lower numbers than LC. These cells are also found in oral mucous epithelium, in the lamina propria of oral mucous epithelium, and in the dermis of fetal and adult skin (Breathnach *et al.*, 1963; Breathnach, 1977a,b; Mishima *et al.*, 1972; Wolff, 1972) and in dermal lymphatics and marginal sinus of lymph nodes (Silberberg-Sinakin and Thorbecke, 1980). It has been shown that LC can accumulate rapidly, within 2–3 hr, at sites of certain contact allergic reactions (Silberberg *et al.*, 1975b; Silberberg-Sinakin *et al.*, 1978). There are three possible explanations for this observation: (1) pre-LC enter the skin rapidly from the circulation and differentiate locally; (2) indeterminate dendritic cells (functioning as "pre-LC") start forming Birbeck granules (Breathnach, 1977a; Silberberg *et al.*, 1975b; Wolff, 1972), and/or (3) LC become more readily detectable ultrastructurally because they contain more numerous granules. Breathnach (1977a) described a fetal dermal LC as well as the presence of many indeterminate cells fairly widely distributed in the dermis. Some of these cells were undergoing mitotic division. He postulated that this type of cell could have the latent capacity to produce LC granules to a variety of stimuli.

The presence of Ia antigen on these cells as well as on LC reemphasizes their similarity (Rowden *et al.*, 1979a) (Table 1). When comparing the LC to other dendritic cells and macrophages, however (Table 2), we have chosen not to include the whole indeterminate dendritic cell plus LC family, but to use only the typical epidermal LC with granules, in order to avoid confusion. It should be realized, however, that it is at present unclear whether some of the receptors, such as those for Fc and C3, as well as some of the enzyme-staining properties,

TABLE 1. SUMMARY STUDIES ON THE ORIGIN OF LANGERHANS CELLS

Tissue or cell system of origin considered	Experimental approach	Conclusion	Reference
Neural crest	Neural crest removed in the embryo	Neural crest excluded	Silvers (1957); Breathnach *et al.* (1968)
Thymus	Study of athymic mice	Thymus excluded	Hunter *et al.* (1976); Silberberg-Sinakin and Thorbecke (unpublished data)
Embryonic epidermis	Epidermal transplant of mouse to chicken	Ectodermal origin likely[a]	Reams and Thompkins (1973)
Keratinocytes	Detection of LC granules in fetal keratinocytes	Relation possible[b]	Bell (1969)
Melanocytes	Use of depigmenting agents	Melanocytes unrelated	Bleehan *et al.* (1968); Brun (1967); Pathak *et al.* (1966); Wolff (1967a)
Indeterminate dendritic cells	Detection of surface markers	Probably related	Rowden *et al.* (1979a)
	Quantitative relationships to LC during delayed hypersensitivity reactions	Probably related	Masutani (1974)
	Study of fetal skin	Probably related	Breathnach (1977a)
Monocytes	Detection of LC granules in monocytic leukemia cells	Possibly related	Sanel and Serpick (1970)
Bone marrow	Skin transplants Bone marrow chimeras	Bone marrow orgin likely	Katz *et al.* (1979)

[a] Only true if cells of mesenchymal origin are totally lacking in epidermis of 10.5-day-old mouse embryo.
[b] Bell (1969) points out that other interpretations are possible; this phenomenon is not frequently observed.

apply to indeterminate dendritic cells as well as to LC, since combined electron microscopy with many of the other techniques is not always readily possible, although results with mice suggest that the Fc receptor is present (Tamaki *et al.*, 1979).

6.3. RELATIONSHIP OF LANGERHANS CELLS TO INTERDIGITATING "RETICULUM" CELLS OF LYMPHOID ORGANS

6.3.1. Lymph Node

Interdigitating "reticulum" cells (RC) are found in the thymus-dependent, paracortical area of the lymph node (Eikelenboom, 1978b; van Ewijk *et al.*, 1974; Hoefsmit, 1975; Hoefsmit *et al.*, this volume; Kamperdijk *et al.*, 1978; Müller-Hermelink *et al.*, 1974; Rausch *et al.*, 1977; Veerman, 1974; Veldman, 1970). These cells, like LC, bear Ia antigens (Hoffmann-Fezer *et al.*, 1978), are ATPase positive, and resemble LC morphologically but the majority, like epidermal indeterminate dendritic cells, ultrastructurally lack LC granules (Table 2). LC granule-containing cells, looking exactly like LC, are present in normal lymph nodes (Kondo, 1969; Silberberg-Sinakin *et al.*, 1976; Vernon *et al.*, 1973), but are seen more frequently in lymph nodes draining an injection site of antigen (Kamperdijk *et al.*, 1978; Silberberg-Sinakin *et al.*, 1976). Cells looking exactly like indeterminate dendritic cells and LC have also been found in dermal lymphatics (Silberberg-Sinakin *et al.*, 1976; Silberberg-Sinakin and Thorbecke, 1980), afferent lymph vessels (Kelly *et al.*, 1978), and in lymph node sinuses (Silberberg-Sinakin *et al.*, 1976; Silberberg-Sinakin and Thorbecke, 1980), upon intradermal challenge with antigen. Kamperdijk and Hoefsmit (1978) have suggested that the paracortical interdigitating RC may be derived from the afferent lymph, since they disappear 6 weeks after ligation of all afferent lymph vessels (Hoefsmit *et al.*, 1978). Therefore, it appears that the interdigitating RC with or without typical LC granules may arrive in the paracortex of the lymph node from the skin. An alternate explanation for the transient increase in LC granule containing cells seen after local stimulation with antigen in paracortical areas could be that interdigitating RC form LC granules in response to stimuli (Kamperdijk *et al.*, 1978). These were only detected during the induction phase of the immune response for approximately one day, after which either the cells containing the LC granules, or just the LC granules themselves, disappeared (Kamperdijk *et al.*, 1978). It is at present difficult to differentiate between skin-derived LC migrating into the paracortex of the lymph node and interdigitating RC rapidly producing LC granules. The similarity between LC and paracortical interdigitating RC has also been stressed by Rausch *et al.* (1977), who suggest that these cells may have similar functions. More conclusive evidence, such as the ability to fix antigen on the cell surface of paracortical interdigitating RC needs to be obtained before such a functional relationship between LC and interdigitating RC can be put on a firm base. A more detailed description of interdigitating reticulum cells is found elsewhere in this volume (Hoefsmit *et al.*, this volume).

TABLE 2. COMPARISON OF LANGERHANS CELLS WITH OTHER DENDRITIC CELLS IN LYMPHOID ORGANS AND MACROPHAGES

Points of comparison	Langerhans cells[a]	Interdigitating reticulum cells[b,c,d]	Monocyte/macrophage[e]	Nossal's follicular DRC[f]
Morphological criteria				
Location	Squamous epithelia	Spleen (PALS) Thymus (medulla) Lymph node (paracortex)	Spleen red pulp Lymph node sinuses and medulla Peritoneal cavity	Mantle zone of lymphoid follicles
Cell shape[g]	Dendritic (Langerhans, 1868)	Dendritic	Irregular	Dendritic
Nuclear shape	Irregular	Irregular	Bean-shaped	Oval to quadrangular
Vesicles	+ to +++	+++	++ to +++	+++
Lysosomes	+ to ++ (Wolff, 1967a)	+ to ++	+++ (Nelson, 1969)	(+)[l]
RER	+ to ++	+ to ++	++	+
LC granules	+ to +++[h] (Birbeck *et al.*, 1961)	(+)[i] (Kamperdijk *et al.*, 1978)	–	–
Histochemistry				
ATPase	+ (Mishima *et al.*, 1972; Wolff and Winkelmann, 1967a)	+ (Müller-Hermelink *et al.*, 1974)	+ (North, 1966; Steinman and Cohn, 1973)	–; +[j] (Eikelenboom, 1978a; Müller-Hermelink *et al.*, 1974)
Aminopeptidase	–; +[k] (Wolff, 1964b)		+ (Wachsmuth, 1975)	
Nonspecific esterase	+ (Campo-Aasen and Pearse, 1966; Jarrett and Riley, 1963)	(+)[l] (Eikelenboom, 1978b)	++ (Fedorko, 1975)	+ (Eikelenboom, 1978b)

(*Continued*)

[a] Birbeck *et al.* (1961); Breathnach (1977a, 1980); Silberberg-Sinakin *et al.* (1978).

[b] Interdigitating "reticulum" cells (interdigitating RC) have first been given that name by Veldman (1970), because they show interlocking of cell membranes with each other and because of the presence of fingerlike protrusions into their cytoplasm extruding from the lymphocytes, which are trapped between their cytoplasm extensions.

[c] In the present discussion the word *reticulum* has been used for both the interdigitating RC and the follicular dendritic RC [called, respectively, IDC and DRC by Hoefsmit and co-workers (Hoefsmit *et al.*, this volume; Hoefsmit, 1975)] to help differentiate these cells in lymphoid tissue from the indeterminate dendritic cell in skin. The name does not imply that we assume a derivation from the reticulum for these cells, since another possibility is that they are bone-marrow-derived.

[d] Hoefsmit *et al.* (this volume); Oláh *et al.* (1968); Veerman (1974); Veldman (1970).

[e] Hirsch and Fedorko (1970); Steinman and Cohn (1973).

[f] Chen *et al.* (1978a); Hoefsmit (1975); Nossal *et al.* (1968).

[g] As is true of other cells, the morphology of LC and the appearance of the cell membrane [ruffled (Kelly *et al.*, 1978; Silberberg *et al.*, 1975b), invaginations and protrusions, number of dendrites, etc.] may be influenced by their location (epidermis, dermal lymphatics, marginal sinus of lymph node).

[h] In the table we are comparing Langerhans cells from squamous epithelia containing typical LC granules with other cell types. Although we regard the indeterminate dendritic cell from the epidermis as a closely related if not identical cell, poor or lacking in LC granules, we have not included them in the comparisons, since few of their properties have been specifically studied (see text).

[i] (+) indicates sometimes present, other times not. LC granules are seen in these cells most frequently in the thymus (Oláh *et al.*, 1968). When cells with this appearance show LC granules in the lymph node, it cannot be excluded that they have not just migrated into the lymph node and are epidermally derived, typical LC. The PALS interdigitating RC have no LC granules.

[j] + for rat; – for human.

[k] + in guinea pig; – in human.

[l] (+) sometimes present, other times not (within the same species).

TABLE 2 (*Continued*)

Points of comparison	Langerhans cells[a]	Interdigitating reticulum cells[b,c,d]	Monocyte/macrophage[e]	Nossal's follicular DRC[f]
5′-Nucleotidase		− (Müller-Hermelink *et al.*, 1974)	+ (Eikelenboom, 1978a)	+; −[m] (Eikelenboom, 1978a; Müller-Hermelink *et al.*, 1974)
Acid phosphatase	(+)[l] (Wolff, 1967a)	(+)[l] (Eikelenboom, 1978b)	+ to +++ (Eikelenboom, 1978a; Steinman and Cohn, 1973)	− (Eikelenboom, 1978b)
Peroxidase	(−)[n]	(+)[l] (Hoefsmit, 1975)	+ (Hoefsmit, 1975)	− (Hoefsmit, 1975)
Membrane receptors and antigens				
Fc (IgG)	+ (Stingl *et al.*, 1977)	(−)[o]	+ (Berken and Benacerraf, 1966; Huber *et al.*, 1968; Huber and Fudenberg, 1968; Rabinovitch, 1970)	+[p]
C3b[q]	+ (Gigli, 1980; Stingl *et al.*, 1977)	(−)[o]	+ (Lay and Nussenzweig, 1968)	and/or +[p]
Ia[r]	+ (Forsum *et al.*, 1978; Klareskog *et al.*, 1977; Rowden *et al.*, 1977; Stingl *et al.*, 1978a,c)	+ (Hoffman-Fezer *et al.*, 1978)	8–15% + (Schwartz *et al.*, 1976b; Shreffler and David, 1975)	
T-cell markers	− (Stingl *et al.*, 1977)		− (Unanue, 1972)	
Surface IgM	−		−	
Properties				
Apposition to lymphocytes	(+)[s] (Silberberg, 1971b, 1973; Silberberg *et al.*, 1974b)	+ (Veldman, 1970; Veerman, 1974; Veldman *et al.*, 1978)	+ (Gottlieb and Waldman, 1972; Lipsky and Rosenthal, 1975)	+ (Chen *et al.*, 1978a; Nossal *et al.*, 1968)
Glass adherence	+ (Nezelof *et al.*, 1973)		+ (Fedorko and Hirsch, 1970; Hirsch and Fedorko, 1970; Rabinovitch, 1970)	
	(+)[t] (Silberberg-Sinakin *et al.*, 1977)	(+)[t]	+++ (Fedorko, 1974; Rabinovitch, 1970; Steinman and Cohn, 1972)	

Pinocytosis	+ (Sagebiel, 1972; Wolff and Honigsmann, 1971)	+ (Hoefsmit, 1975)	+ (Cohn, 1970; van Furth *et al.*, 1975)	(+)[s] (Chen *et al.*, 1978a; Nossal *et al.*, 1968)
Antigen binding	+ (Silberberg-Sinakin *et al.*, 1976)		+ (Gottlieb and Waldman, 1972; Unanue, 1972)	+[t] (Chen *et al.*, 1978b; Cohen *et al.*, 1966; Nossal *et al.*, 1968)
Ability to present antigen to lymphoid cells	+ (Stingl *et al.*, 1978b)		+ (Thomas *et al.*, 1977a,b; Thomas and Shevach, 1976; Unanue, 1972)	
Induction of MLR	+ (Stingl *et al.*, 1978d)		+ (Rode and Gordon, 1974; Stingl *et al.*, 1978d)	
Radiation resistance	High (Silberberg-Sinakin *et al.*, 1976)	Medium[u] (van Ewijk *et al.*, 1974; Hoffman-Fezer *et al.*, 1978)	High (Askonas and Jarošková, 1970; Unanue, 1972)	Low (Nettesheim and Hammons, 1971; Nettesheim and Hanna, 1969)
Bone marrow origin	+ (Katz *et al.*, 1979)		+[v](Volkman and Gowans, 1968; van furth, 1970)	

[m] − for rat; + for human.

[n] Usually negative in normal LC (Silberberg-Sinakin, unpublished observations), but further studies of LC are needed to determine if under some conditions or stages of maturation they have positive peroxidase staining. Positive in LC granulomas (Basset, personal communication).

[o] Binding of EA or EAC to thymus-dependent regions in lymph node and spleen has not been observed in frozen tissue sections (Dukor *et al.*, 1970). This, however, does not exclude the presence of the Fc and C3 receptors on a minor population in these areas.

[p] Follicular dendritic "reticulum" cells (DRC) are known to bind the Fc but not the Fab fragments of IgG and, therefore, appear to have an Fc receptor (Herd and Ada, 1969). However, the absence of C3 in the serum has a negative influence on the localization of immune complexes in follicles (Papamichael *et al.*, 1975), which suggests a role of C3 and, therefore, of a C3 receptor on DRC as well. Thus, in the table, we have indicated the presence of one or the other or both of these receptors.

[q] Some of the cells have receptors for other determinants of C3, and among receptors for C3b and among Fc receptors there is variability in resistance to trypsin digestion (Gigli, 1980).

[r] While the comparison between cells at this time can best be made for the mouse, the equivalent of Ia antigen is also found on LC in guinea pigs (Stingl *et al.*, 1978c) and man (Forsum *et al.*, 1978; Stingl *et al.*, 1978a). In the mouse the products of at least two loci, both needed on the same macrophages for antigen presentation to T cells (Cowing *et al.*, 1978; Schwartz *et al.*, 1979), and called IA and IE/C respectively, have been identified on LC (Frelinger *et al.*, 1978; Tamaki *et al.*, 1979).

[s] Applicable only when lymphocytes are present in their vicinity, such as in contact dermatitis (Silberberg *et al.*, 1974b).

[t] Some so-called "sticky" antigens and nonantigenic substances may localize on DRC without antibody (Cohen *et al.*, 1966). This could be due to activation of C3 via the alternate pathway rather than to antigen binding as such.

[u] Although some decrease in interdigitating RC numbers may occur in the spleen of irradiated mice, thymocyte reconstituted animals have a fairly normal appearance (van Ewijk *et al.*, 1974). Thus, quantitation of the X-irradiation resistance of these cells is still needed, and the possibility that thymocytes reconstitute their presence also remains to be evaluated. However, the Ia + RC in the inner part of the periarteriolar sheath of the spleen appears to survive 900 rad whole-body irradiation in mice (Hoffman-Fezer *et al.*, 1978).

[v] Bone marrow origin has also been shown for the Ia + subpopulation of macrophages, which is needed for antigen presentation to T cells (Schwartz *et al.*, 1979).

[w] Reconstitution of antigen-binding function in irradiated animals on days 1 and 8 after irradiation is shown after spleen or peritoneal exudate, but not after bone marrow transfers (Nettesheim and Hammons, 1971).

6.3.2. Spleen

The thymus-dependent areas of the spleen, the inner portion of the periarteriolar lymphatic sheaths (inner PALS) contain a similar network of interdigitating RC as the paracortex of lymph nodes (Hoefsmit *et al.*, this volume; Veerman, 1974; Veldman, 1970), but cells containing typical LC granules have not yet been seen in this location (Hoefsmit, personal communication). Tight junctions (Veerman, 1974; Veldman, 1970) and gap junctions (Veerman, 1974) have been described between T cells and interdigitating RC as occurring with great regularity. In studies by van Ewijk *et al.* (1974), on lethally irradiated mice reconstituted with thymocytes, an intimate contact between the interdigitating RC of the PALS and the T lymphocytes was observed in the form of fingerlike protrusions, extending from the lymphocyte and protruding into the cytoplasm of the interdigitating RC. Electron-dense material, often seen as cross-bridges of about 120 Å in length, were present in the extracellular space of these contact regions.

6.3.3. Thymus

A close interrelationship between lymphocytes and mesenchymal interdigitating RC has also been described in the thymus, particularly in the medulla, and was interpreted as an integral part of thymocyte maturation and proliferation, since mitotic figures were seen to be more frequent in thymocytes around these glycoprotein-containing interdigitating RC than elsewhere in the thymus (Metcalf and Ishidate, 1962).

These mesenchymal RC in the thymus also interdigitate with the surrounding lymphocytes and have been described by Veldman (1970) as similar to interdigitating RC found in the thymus-dependent areas of the peripheral lymphoid organs, spleen, and lymph node. Hoefsmit and Gerver (1975) have suggested the presence of glycoprotein in lysosome-like structures of these cells. Typical LC granules are present in these cells in the rat (van Haelst, 1969; Oláh *et al.*, 1968). These cells are quite resistant to cortisone and therefore are more prominent after corticosteroid injection (van Haelst, 1969), but they are also present in normal rat thymus. It is, perhaps, no accident that LC granule-containing interdigitating RC are found in thymus which, like squamous epithelia, contain keratinizing epithelial elements.

Staining for the presence of Ia antigen on these mesenchymal interdigitating RC, as examined by fluorescence microscopy, has given inconclusive results in the thymus, because the epithelial reticulum cells which are also present throughout the thymus are strongly Ia-positive (Rouse *et al.*, 1979).

6.4. COMPARISON WITH NOSSAL'S FOLLICULAR DENDRITIC "RETICULUM" CELL

Table 2 also includes some of the known properties of the dendritic reticulum cells (DRC) of lymphoid follicles described by Nossal *et al.* (1968), Chen *et al.* (1978a), and Miller and Nossal (1964). These cells bear antigen or antigen–antibody complexes on their surface. They are located in the corona of lymphoid follicles, where antigen can be detected for long periods after injection (Lang and Ada, 1967).

Although these cells can bind some antigenic and nonantigenic "sticky" substances in the absence of antibody (Cohen *et al.*, 1966; Miller and Nossal, 1964), the presence of antibody of the IgG class greatly enhances antigen uptake in follicles (Balfour and Humphrey 1967; Humphrey and Frank, 1967; Nossal *et al.*, 1965). Conversely, the presence of antigen causes the localization of passively administered antibody on DRC (Pernis, 1967). Even though intact IgG localizes much better on DRC than F(ab)2 fragments (Johnson *et al.*, 1975), and Fc fragments much better than Fab (Herd and Ada, 1969), the presence of an Fc receptor on DRC has not been shown by direct means. Since C3 depletion greatly reduces intact IgG localization on DRC, the presence of a C3 receptor on DRC has been suggested (Papamichael *et al.*, 1975).

The follicular DRC are considered to represent a specialized form of the reticulum cells in the outer rather than in the inner portion of the splenic PALS and as such would be quite separate from the interdigitating RC. They show extensive membrane invaginations and protrusions which often enfold electron-dense material and viruses (Szakal and Hanna, 1968). The degree of membrane-folding may depend on the amount of stimulation by antigen or antigen–antibody complexes recently experienced (Veerman, 1974). Their morphological appearance is not very similar to the interdigitating RC or LC (Table 2), and when the membrane invaginations of the DRC are at a minimum its morphology approaches that of an ordinary reticulum cell. Their contact with neighboring lymphocytes is relatively intimate and involves B lymphocytes in the follicles rather than the T lymphocytes which are the neighbors of interdigitating RC.

It should be noted that in contrast to the antigen-presenting spleen cell (Schwartz *et al.*, 1978b), the interdigitating RC in T-dependent areas (see footnotes to Table 2) and the LC, the DRC in follicles have been shown to be X-irradiation sensitive (Hunter *et al.*, 1969; Nettesheim and Hammons, 1971; Nettesheim and Hanna, 1969; Williams, 1966) which, in combination with their morphological properties, tends to identify the DRC as a separate cell type. However, quantitation of X-irradiation sensitivity of all these cell types is still needed. Neither the studies by Hoefsmit and co-workers (Hoefsmit *et al.*, this volume; Kamperdijk *et al.*, 1978) nor our own observations (Silberberg-Sinakin *et al.*, 1976) have indicated the presence of LC in follicular areas of lymph node draining a site of antigen injection, although in both studies LC are seen in the paracortex.

6.5. COMPARISON WITH STEINMAN AND COHN'S DENDRITIC CELL

The cell described by Steinman and Cohn (Steinman and Cohn, 1973, 1974; Steinman *et al.*, 1974, 1975, 1979) as a dendritic cell primarily found in spleen, also present in lymph node and Peyer's patches, but absent everywhere else, has been characterized with respect to some of the properties listed in Table 2. With respect to cell shape and nuclear shape, the morphological properties of the cell are similar to the interdigitating reticulum cell, but the cell never shows LC granules. It may contain vesicles, lysosomes, and RER. The cell is negative for ATPase and peroxidase. It lacks all known cell surface receptors, but is positive for Ia antigens. It forms clusters with lymphocytes *in vitro* (Steinman and Cohn, 1975), is adherent to glass (Steinman and Cohn, 1973, 1974), is not phagocytic, but occasionally shows pinocytic activity (Steinman and Cohn, 1974), does not bind antigens (Steinman and Cohn, 1974), and has a low resistance to radiation (Steinman *et al.*, 1974). It is of bone marrow origin (Steinman *et al.*, 1974) and induces mixed lymphocyte reactions (Steinman and Witmer, 1978). A major problem with relating this cell to either the LC on the one hand, or to the follicular DRC on the other, is that they lack both C3 and Fc receptors.

6.6. SUMMARY

The combined properties of these cells suggest great similarity between the indeterminate dendritic cells + LC family in the epidermis and the interdigitating RC with or without LC granules in lymph node, spleen, and thymus. The resemblance of LC to classical monocytes and/or macrophages is less striking, but it should be realized that many of the properties listed for macrophages reflect the Ia-negative and probably highly phagocytic populations of macrophages rather than the Ia-positive form of adherent mononuclear cell which is capable of presenting antigen to T lymphocytes. Further studies are needed to evaluate the possible identity of the interdigitating RC with this Ia-positive subpopulation of macrophages. While the evidence available to date does not suggest that Nossal's follicular DRC and Steinman and Cohn's DC have properties identical to the interdigitating RC, the possibility that they all belong to one family of cells with variations in characteristics due to location and/or different stages of maturation cannot be excluded. Thus, LC might be the epidermal representative of a cell system that has a much more widespread distribution over lymphoid organs.

6.7. RELATIONSHIP TO LYMPHOMATOUS DENDRITIC CELLS

It would be of interest if tumors or lymphomas could be identified which consist of one of the dendritic cell types represented in Table 2. Histiocytosis X (Langerhans cell granuloma*) is a candidate (Basset *et al.*, 1965; Tarnowski and Hashimoto, 1967), although it is most likely that it represents a granulomatous reaction. There are reports in the literature of monocytic leukemias with cells containing LC granules (Sanel and Serpick, 1970). Hairy-cell lympho-

*The term *Langerhans cell granuloma* has been proposed by Dr. Pinkus to replace the name histiocytosis X (Pinkus, and Mehregan, 1980). Langerhans cell granulomatosis has been offered as a new name for eosinophilic granuloma (Lieberman, 1980).

ma cells also show LC granules (Padilla and Soloff, 1971); the cells are dendritic in appearance, phagocytic, and contain peroxidase, in addition to the B-cell characteristics of surface Ig (Reyes *et al.*, 1978). In recent years, we have studied transplantable reticulum cell sarcomas in SJL/J mice, lymphomas which contain at least 5% dendritic cells with the ultrastructural characteristics of the interdigitating reticulum cells, while a large percentage of cells have some B-cell properties (Carswell *et al.*, 1976). It is of interest that these Ia-positive cells (Ponzio *et al.*, 1977), like Steinman and Cohn's DC (Steinman and Witmer, 1978) stimulate a strong, mixed lymphocyte reaction in allogeneic T cells, and even more striking that they stimulate syngeneic T cells to proliferate (Lerman *et al.*, 1974). The relationship of these cells to the normal dendritic inhabitants of lymphoid tissue needs to be unraveled. A normally present Ia-positive, Fc-positive subpopulation of peritoneal exudate cells has recently been described which combines macrophage-like properties, such as nonspecific esterase activity and adherence, and B-cell properties, such as surface IgM (Nathan *et al.*, 1977). The occurrence of such apparently intermediate forms between cells of the monocyte and B-cell series may indicate the close relationship between these cell types and the difficulty of making rigid distinctions between cell types in general.

7. ROLE OF LANGERHANS CELLS IN IMMUNOLOGICAL PROCESSES

7.1. LANGERHANS CELLS IN DELAYED HYPERSENSITIVITY

As reviewed in Section 1, several authors have suggested that LC have possible immunologic functions. In this laboratory, we examined LC behavior in immune responses in man and guinea pigs. Initially we reported (Silberberg, 1971a) close apposition of lymphocyte-like cells to LC in man within 4–6 hr of topical application of mercury bichloride. This was found only at sites of positive allergic contact hypersensitivity reactions and not in contact primary irritant reactions (Figure 6). Such observations suggested to us that the possible role of LC in the pathogenesis of contact dermatitis needed more attention. Further experiments have provided strong suggestive evidence that LC are the peripheral representation of the cell population(s) which effectively present antigens to lymphocytes.

The following observations have been made:

a. Apposition of lymphocyte-like cells to LC in the epidermis and dermis (Figures 7 and 8) in allergic contact dermatitis (Silberberg, 1973; Silberberg *et al.*, 1974b).
b. LC in dermal lymphatics (Figures 3–5) and draining lymph nodes both in the marginal sinus and paracortical areas (Figures 9 and 10), with greater frequency in contact allergic reactions and after intradermal injections of immune complexes than in normal subjects (Silberberg *et al.*, 1975a,b; Silberberg-Sinakin *et al.*, 1976, 1977, 1978). Indeterminate dendritic cells are also seen in these locations (Silberberg-Sinakin and Thorbecke, 1980).

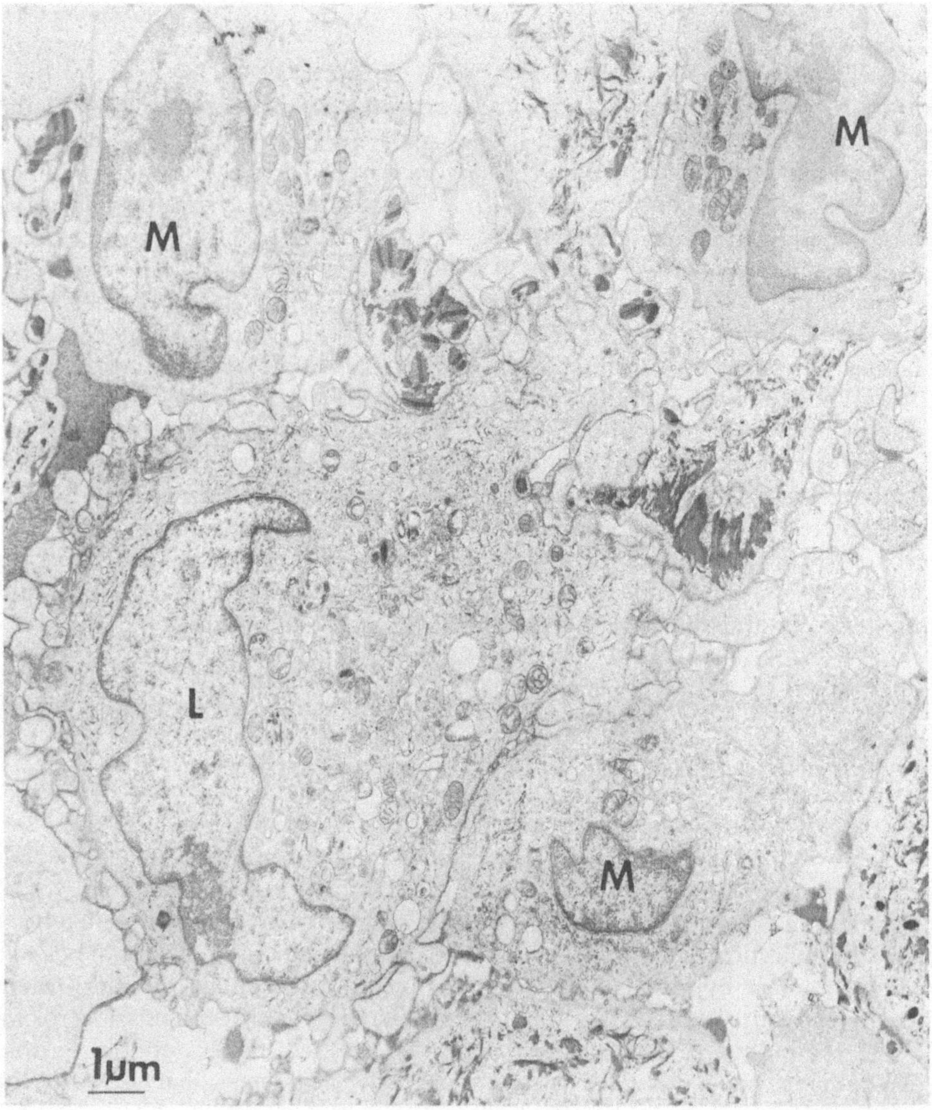

FIGURE 6. Langerhans cell (L) surrounded by three lymphocyte-like cells (M) in the epidermis, 6 hr after application of allergen topically on skin of an allergically sensitive human subject. Stained with uranyl acetate and lead citrate. ×7200. (From Silberberg, 1973, with permission.)

c. Ability of LC to pick up antigens such as ferritin and contact allergens (Shelley and Juhlin, 1977a,b) in skin and to carry these by way of dermal lymphatics to the regional lymph nodes (Figures 9 and 10) (Silberberg-Sinakin *et al.*, 1976).
d. Ability of antigen-bearing LC-enriched cell suspensions from the epidermis to replace antigen-bearing peritoneal cells in the *in vitro* induction of proliferation of primed T lymphocytes (Stingl *et al.*, 1978b).

Thus, LC represent a system of cells capable of migrating from the skin and carrying out a function analogous to macrophages. This relay system appears to

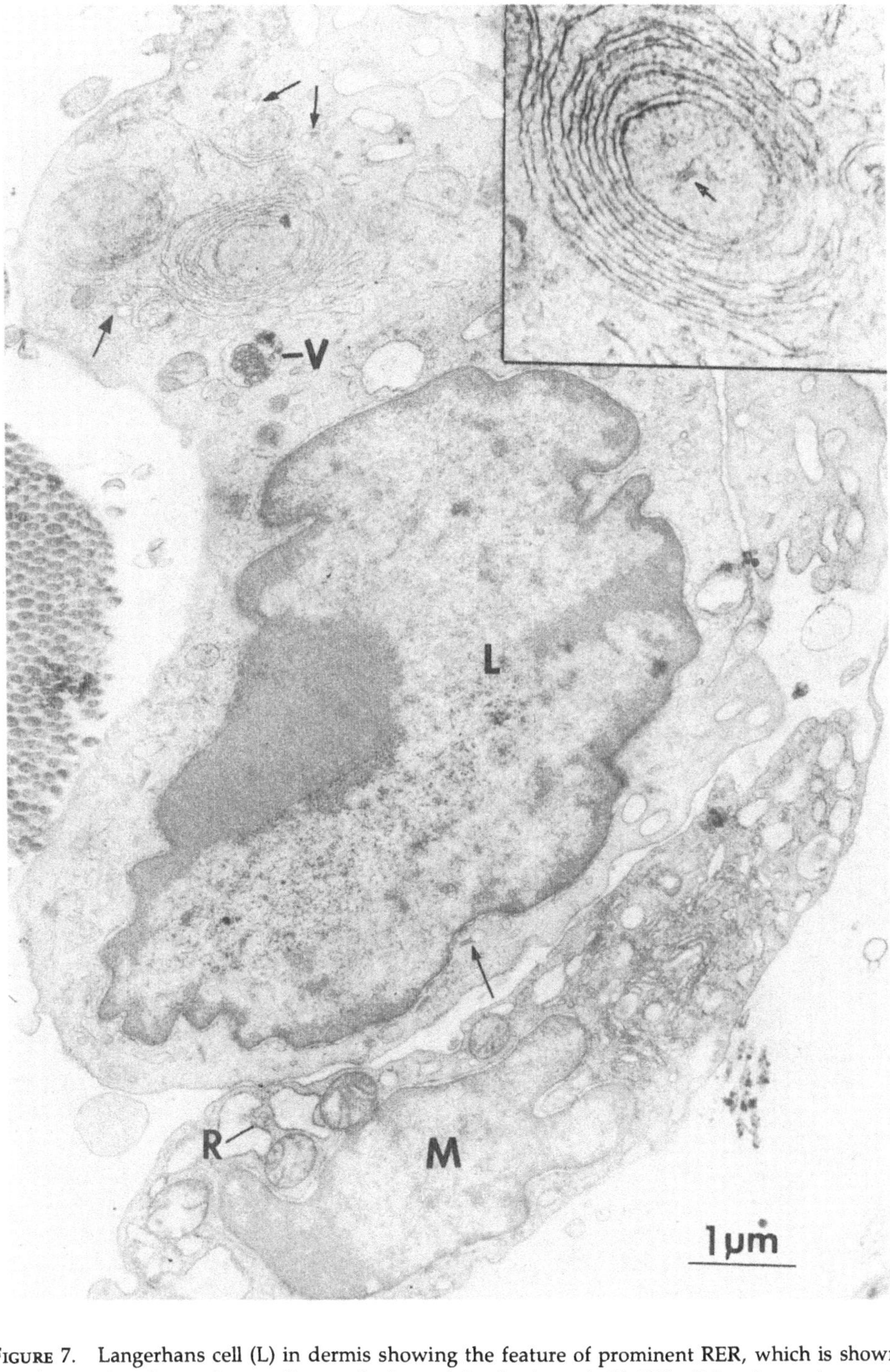

FIGURE 7. Langerhans cell (L) in dermis showing the feature of prominent RER, which is shown together with a Langerhans cell granule (arrow) in the inset (×32,500). Other Langerhans cell granules are shown at the arrows and a lysosome-like body (V) is present. A mononuclear cell (M), with several channels of RER (R) is seen in apposition to the Langerhans cell. This skin section was taken 3 hr after DNCB challenge in a guinea pig passively sensitized to DNCB. Stained with uranyl acetate and lead citrate. ×15,750. (From Silberberg *et al.*, 1975b, with permission.)

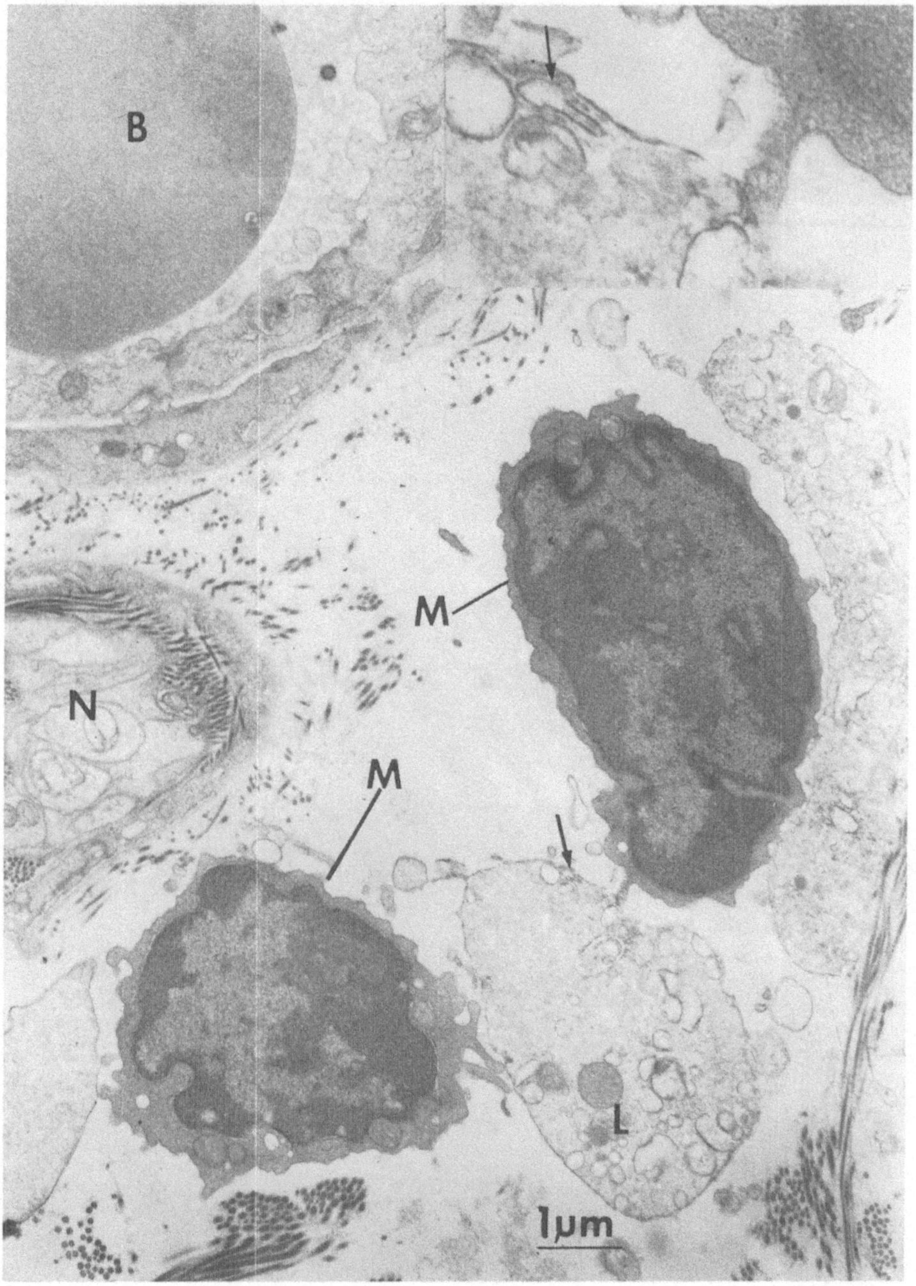

FIGURE 8. Part of a damaged Langerhans cell (L) in the dermis apposed by two mononuclear cells (M). These mononuclear cells resemble lymphocytes. A Langerhans cell granule is seen at the arrow. Part of a nerve (N) and blood vessel containing a red cell (B) are seen. Section taken 6 hr after DNCB challenge in a guinea pig passively sensitized to DNCB. Stained with uranyl acetate and lead citrate. ×12,250. The inset shows enlargement of Langerhans cell granule in the Langerhans cell. Part of the mononuclear cell is touching the Langerhans cell. ×45,500. (From Silberberg *et al.*, 1975b, with permission.)

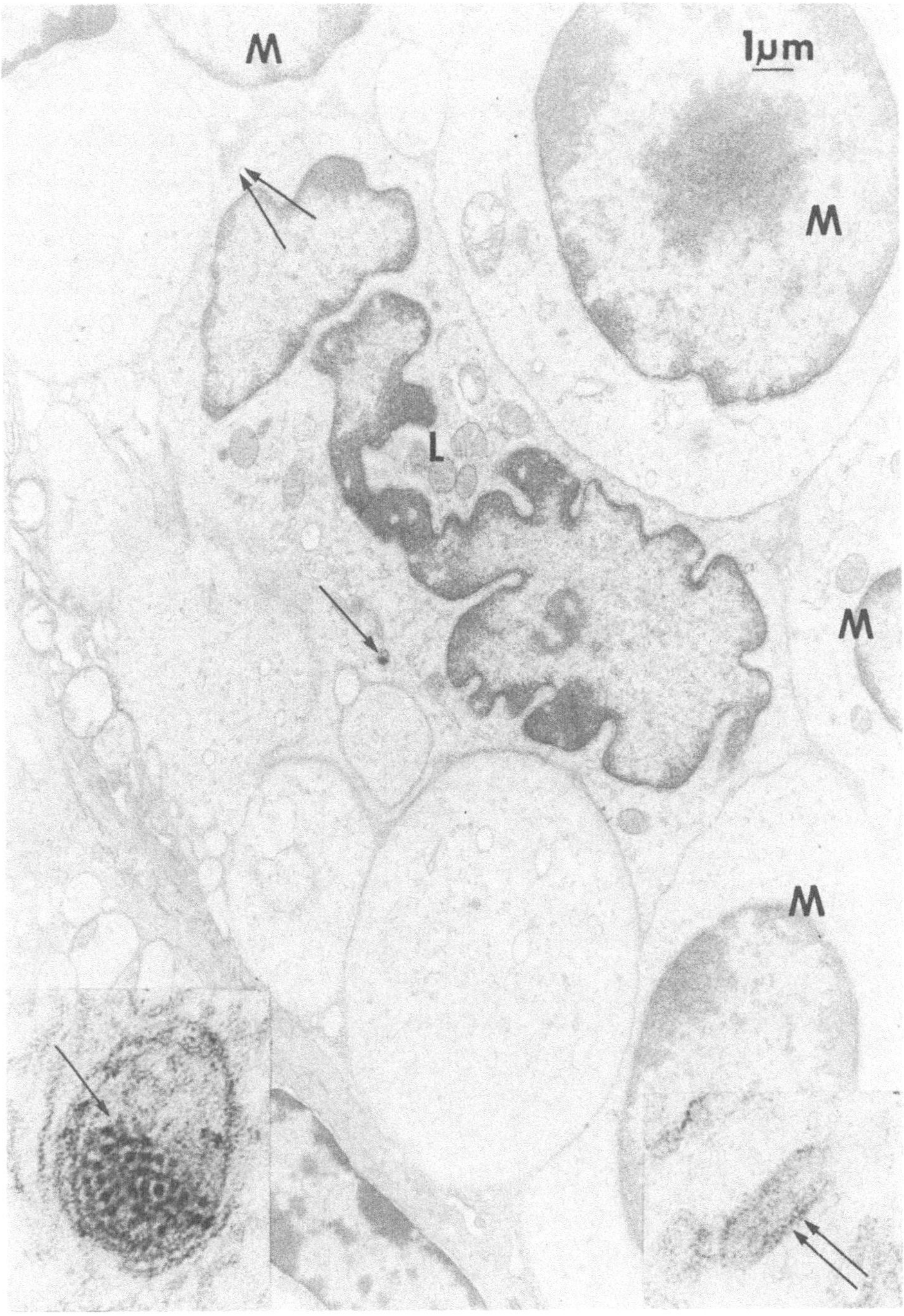

FIGURE 9. Langerhans cell (L) in the cortex of a popliteal lymph node 4 hr after intradermal challenge with 5 μg ferritin in a normal guinea pig. The single arrow denotes a membrane-bound aggregate of ferritin which is shown at higher magnification in inset. Double arrows denote site of a Langerhans cell granule which is shown at higher magnification in inset. Part of several lymphocyte-like cells (M) are seen touching the Langerhans cell. Section stained with uranyl acetate and lead citrate. ×6000. Both insets, ×240,000. (From Silberberg-Sinakin *et al.*, 1976, with permission.)

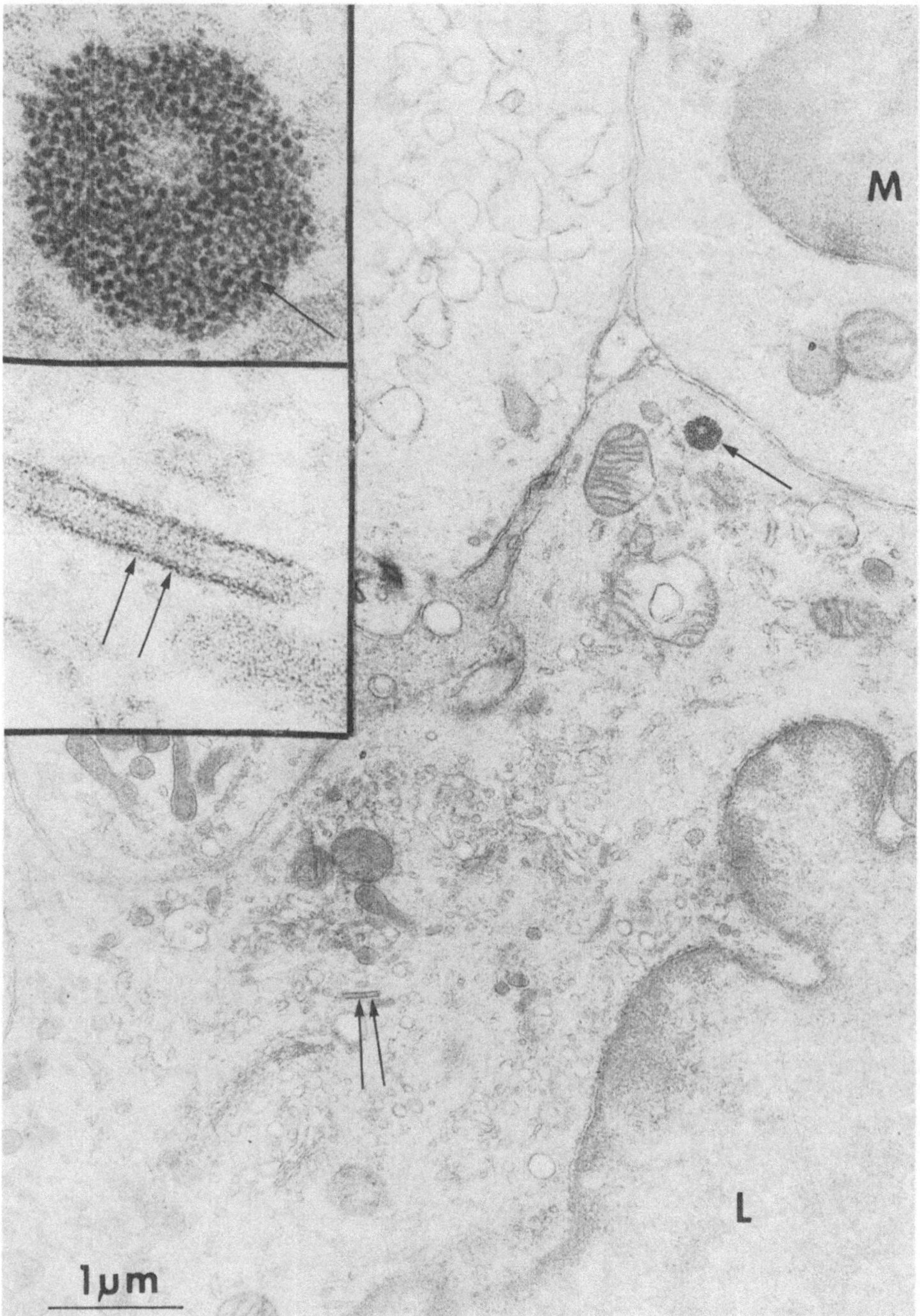

FIGURE 10. A Langerhans cell (L) in the cortex of a popliteal lymph node 24 hr after footpad inoculation with 300 μg ferritin in complete Freund's adjuvant. Part of one lymphocyte-like cell (M) is touching the Langerhans cell. The single arrow denotes the dense aggregate of ferritin in the cytoplasm and double arrows denote a Langerhans cell granule. These are shown at higher magnification in the insets. Section stained with uranyl acetate and lead citrate. ×18,750. Both insets, ×240,000. (From Silberberg-Sinakin *et al.*, 1976, with permission.)

be more active during immunological reactions in the skin (Silberberg-Sinakin *et al.*, 1976, 1978), and to involve antigen bearing cells. Ferritin-bearing guinea pig LC are apparently resistant to X-irradiation since they have been found to persist both in skin and lymph node for at least 24 hr after total body irradiation (800 rad).

On the basis of experimental data reviewed above, one may infer that, upon introduction of allergens or antigens on or in the skin, such substances may localize at the surface of and in LC (Silberberg-Sinakin *et al.*, 1976). LC can then present these substances to lymphocytes in the epidermis, dermis, dermal lymphatics, and lymph nodes (paracortical areas). If an animal has not been previously sensitized, exposure to antigen on LC may set in motion the induction of delayed hypersensitivity, which, in mice, is known to be a function of the Ly 1+, 2− subclass of T lymphocytes (Huber *et al.*, Vadas *et al.*, 1976). It has been demonstrated that, while hapten conjugates invariably need to be presented via macrophages in order to induce proliferation in T cells from contact-sensitized animals, direct hapten conjugates of macrophage-like cells can also induce such proliferation (Thomas *et al.*, 1977a). It is probable that one major mechanism of induction of contact sensitivity is via direct conjugation of sensitizing agents on LC locally in the epidermis. The hapten–LC complex would then serve as the inducing and eliciting antigen complex in contact dermatitis. However, formation of a hapten–keratinocyte complex, which then gets presented to lymphoid cells via LC, is an equally good possibility.

In contrast, exposure to a contact sensitizer via escape of the allergen from the skin directly into the general circulation (Macher and Chase, 1969), via ingestion (Chase, 1946), or via the intravenous (i.v.) route (Frei, 1928; Sulzberger, 1929), might lead to tolerance instead (Elson and Taylor, 1975; Fidler and Golub, 1973; Polak *et al.*, 1974). Exposure by these routes bypasses the LC system and thus appropriate presentation of the antigen by LC to the Ly 1+, 2− subclass of lymphocytes might not take place (Silberberg-Sinakin *et al.*, 1978). If antigen is presented separately from the correct Ia antigens, suppressor T-cell proliferation may result, rather than helper T-cell proliferation (Miller *et al.*, 1979). If Ia antigens on LC, or LC themselves, are destroyed this would interfere with the induction and elicitation of contact allergic reactions.

7.2. DAMAGE OF LANGERHANS CELLS DURING IMMUNE REACTIONS

Additional evidence suggests that LC undergo further consequences from local immune reactions:

a. Spotty damage to LC (Figure 8) and decrease in their number in the epidermis in contact allergic reaction sites within 6 hr after challenge (Silberberg, 1973; Silberberg *et al.*, 1974b, 1975b).
b. Damage to LC by intradermal injections of immune complexes requiring the presence of complement (C) (Silberberg-Sinakin *et al.*, 1977).

In an animal that has been previously sensitized, exposure to the antigen on LC may lead to blast transformation and proliferation of sensitized cells, proba-

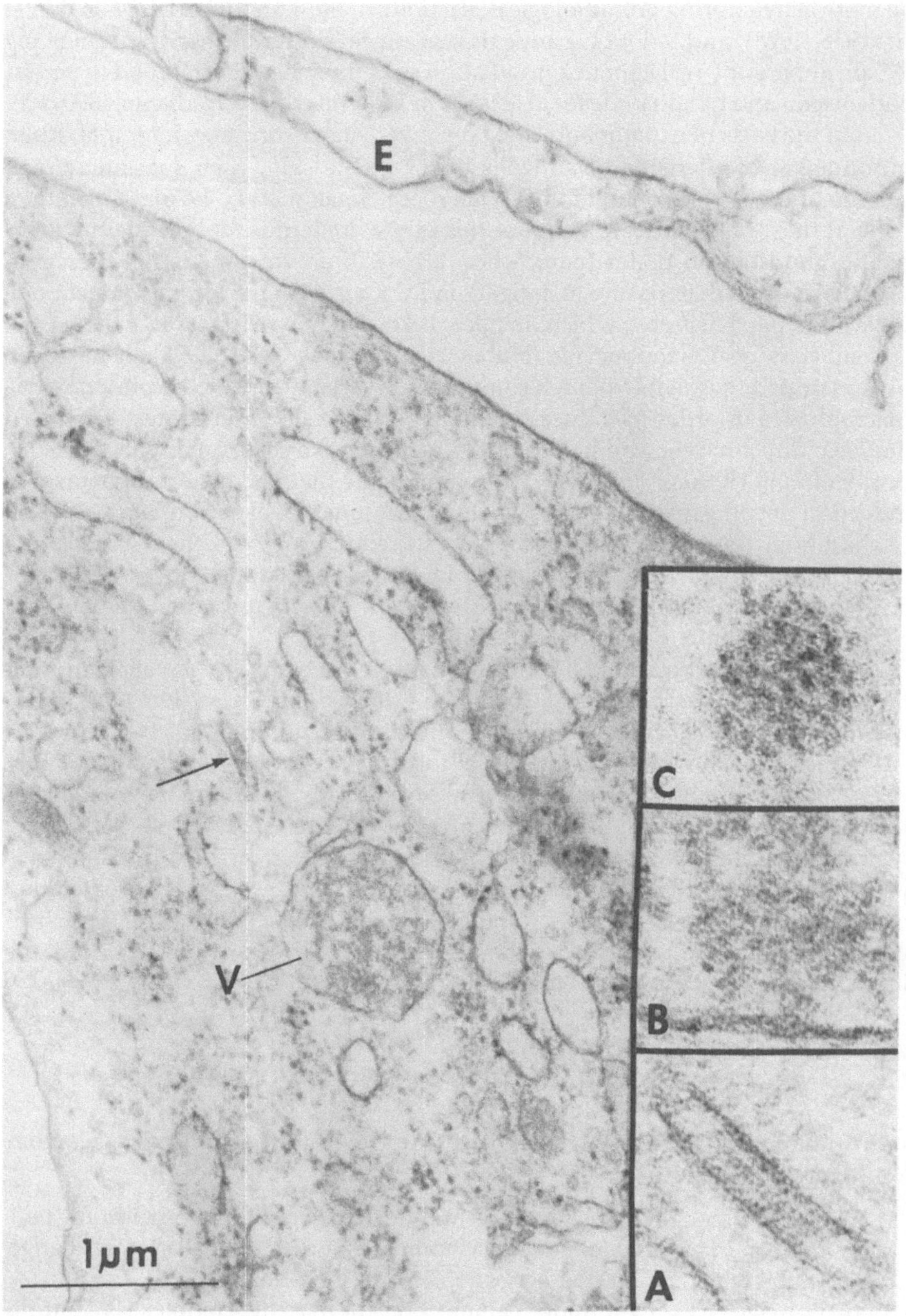

FIGURE 11. High-power magnification of part of a Langerhans cell in a dermal lymphatic of a cobra-venom-factor-treated guinea pig challenged with ferritin–antiferritin immune complexes 1.5 hr earlier. Part of an endothelial cell (E) is seen. A phagocytic vacuole containing material resembling ferritin–antiferritin is seen (V). Part of the vacuole shown within the box is enlarged in inset B. A

bly with the release of lymphokines. Whether lymphocytes, such as killer T cells, lymphokines, or antigen–antibody complexes plus complement cause the damage to LC observed during contact sensitivity reactions in the skin, is not clear (Silberberg-Sinakin *et al.*, 1977). The injured LC themselves, in turn, may release substances they contain, perhaps including enzymes from lysosomes, and thus cause further inflammatory changes in the surrounding epidermis and dermis. LC in this type of immunologic reaction play the role of an antigen carrier, elicitor, target cell, and inflammation-producing cell.

The events that occur in allergically sensitized skin after challenge with antigenic substances appear to depend on several factors, among them the route of exposure (e.g., contact, intradermal injection), the challenge dose, and immunologic state of the animal (e.g., cell-mediated and antibody-mediated hypersensitivity, presence of immune complexes, etc.). The presence of Fc receptors on LC may clearly mediate the binding of antigen–antibody complexes to the surface of LC. Even when substances are bound to LC via Fc and C3 receptors the phagocytic activities of LC are still limited as compared to those of classical macrophages (Silberberg-Sinakin *et al.*, 1977) (Figure 11).

If damage to LC is not severe, they can self-repair via autophagy (Figure 12) and circulate from skin to draining lymph nodes. When it is severe, they are likely to be lysed. Under such circumstances, remnants of LC, including LC granules, can be found in macrophages (Figure 13) (Silberberg-Sinakin *et al.*, 1977). The finding that immune complexes, in the presence of C, can damage LC suggests one possibility of how an acquired local deficiency in the LC system could be produced. In the presence of sufficient quantities of immune complexes and of C, damage to the LC system might be severe enough to interfere with the functions of the LC system in cell-mediated reactions. The finding that LC themselves are injured in some hypersensitivity reactions suggests that this phenomenon may be part of a homeostatic mechanism to prevent hyperstimulation of certain immune responses (Silberberg-Sinakin *et al.*, 1978).

7.3. ROLE OF LANGERHANS CELLS IN ANTIGEN PRESENTATION TO T CELLS AND IN DETERMINING GENETIC CONTROL OF CONTACT SENSITIVITY

Recent studies on the genetic control of the immune response show that one major factor determining the ability of animals to produce delayed hypersensitivity and T-helper activity to an antigen is the nature of the Ia molecules on their antigen-presenting macrophages (Rosenthal *et al.*, 1977; Thomas *et al.*,

---

Langerhans cell granule is seen at the arrow and is enlarged in inset A. The morphology of ferritin–antiferritin immune complexes is shown in inset B. As a control, in inset C, we show the morphology of ferritin alone. Although ferritin has a distinct morphology by which it can be identified, the ferritin in ferritin–antiferritin complex may not always be easy to identify. No vacuoles with contents of this morphology were seen in uninjected animals. The section is stained with uranyl acetate and lead citrate. ×28,500. Insets, ×228,000. (From Silberberg-Sinakin *et al.*, 1977, with permission.)

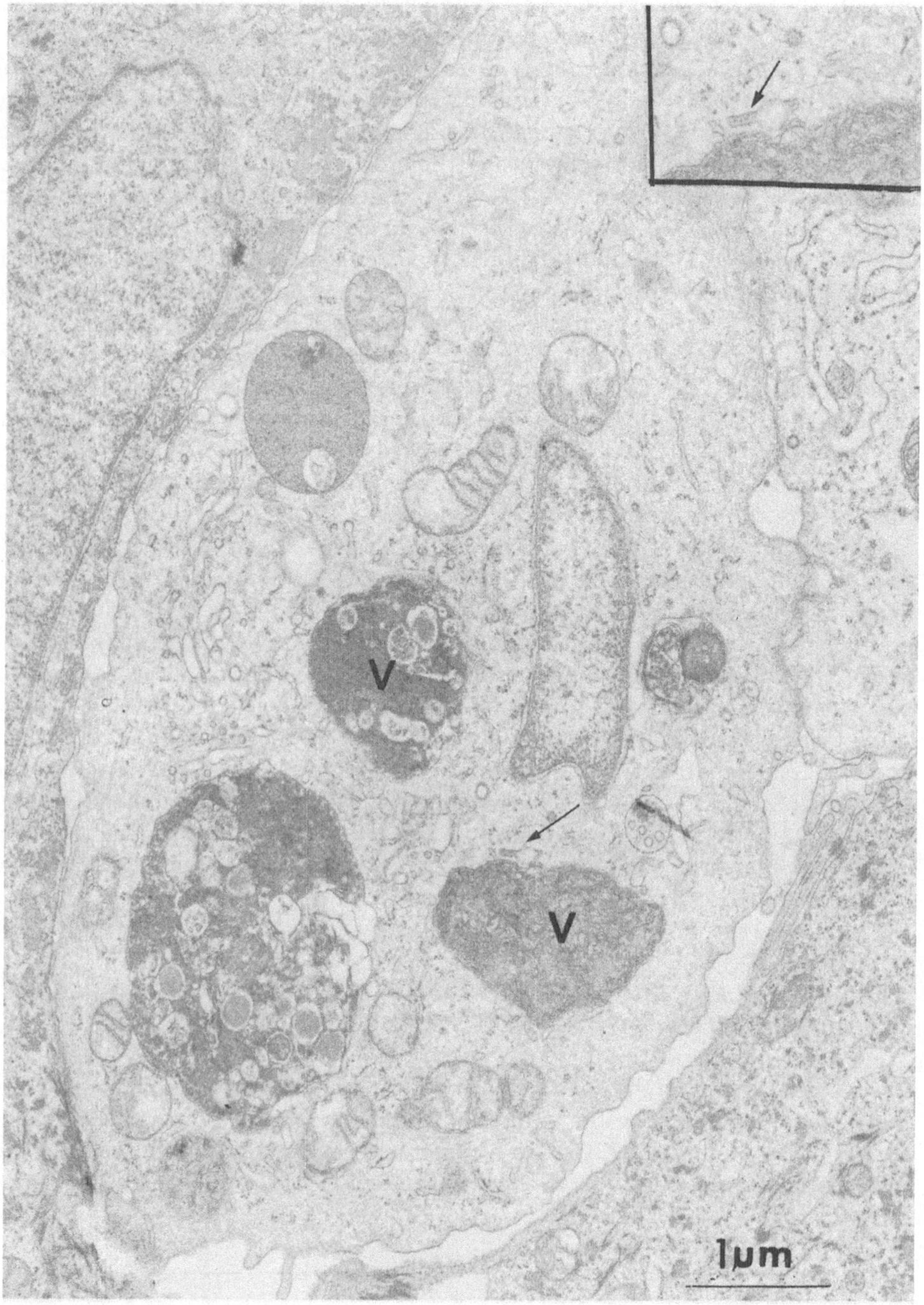

FIGURE 12. Contact allergic reaction at 48 hr. A cell morphologically compatible with a Langerhans cell is shown. A Langerhans cell granule is shown at the arrow and at higher magnification in the inset. Several phagocytic vacuoles (V) are present in the cytoplasm. Stained with uranyl acetate and lead citrate. ×19,500. Inset, ×32,500. (From Silberberg *et al.*, 1974b, with permission.)

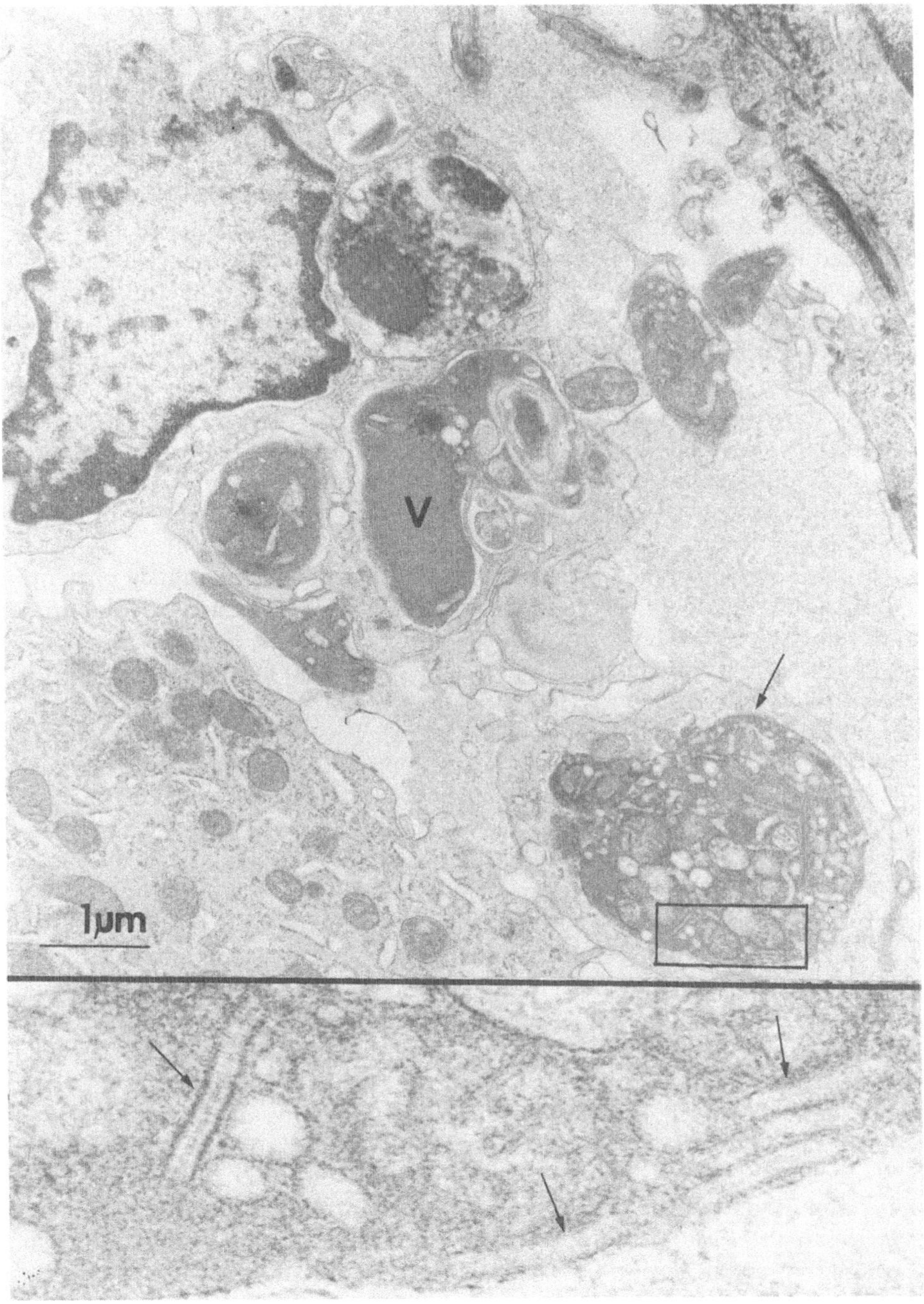

FIGURE 13. Contact allergic reaction at 48 hr. A macrophage is seen with several phagocytic vacuoles (V). Some of these (see arrow) contain Langerhans cell granules. The inset shows a higher magnification of the phagocytic vacuole containing Langerhans cell granules shown at the arrow. Stained with uranyl acetate and lead citrate. ×15,500. Inset, ×104,000. (From Silberberg *et al.*, 1974b, with permission.)

1977b). T cells from sensitized animals which proliferate upon a second exposure to antigen, both in mice (Pierce *et al.*, 1976) and in guinea pigs (Thomas and Shevach, 1976), recognize the antigen in conjunction with the Ia antigen of the presenting cell. They cannot respond when the second exposure is through a genetically different Ia-bearing cell or when anti-Ia is present in the medium (Ruhl and Shevach, 1975; Schwartz *et al.*, 1976a).

In vitro studies have recently confirmed similar observations for LC (Stingl *et al.*, 1978b). Antigen-primed T lymphocytes from inbred guinea pigs were incubated with syngeneic and allogeneic LC-enriched epidermal cells and with LC-deficient epidermal cells. T-cell activation was measured by an increase in [^{3}H]thymidine incorporation. Syngeneic, but not allogeneic, antigen-pulsed LC-enriched epidermal cells, induced a marked T-cell proliferative response similar in degree to that seen with antigen-pulsed syngeneic macrophages. It was also shown that LC-enriched epidermal cell suspensions were as effective as allogeneic macrophages in inducing stimulation in mixed leukocyte reactions. Both the antigen-presenting and mixed leukocyte-reaction-stimulating ability of LC-enriched epidermal cells could be abrogated by pretreatment with anti-Ia sera and complement.

In view of the fact that in contact allergy the major (only) antigen presentation is via the skin, it appears likely that LC with their Ia antigens are the prime contributors to the genetic control of this form of immune response.

Direct conjugation of hapten to Ia-positive adherent macrophages leads to a very effective stimulator cell for activation of hapten-specific T cells from contact-sensitive animals (Thomas *et al.*, 1977a). This suggests that hapten conjugations of the LC itself is a very effective method of sensitization.

The tendency of monocytes to make rosettes with lymphocytes *in vitro*, particularly that of antigen-bearing monocytes with specifically sensitized lymphocytes (Lipsky and Rosenthal, 1975), is reminiscent of the tendency of LC to become apposed to lymphocytes in the skin and lymph node during delayed hypersensitivity reactions (Silberberg *et al.*, 1975b). It seems possible that the intimate relationship between interdigitating RC and lymphocytes in the inner PALS and paracortex are representative of this same phenomenon. This could also be of importance for antigen presentation. Although antigen does not usually localize in large amounts in regions rich in interdigitating RC, it has been described that certain hydrophobic conjugation products of protein antigens, which are more effective than the unmodified proteins in inducing delayed hypersensitivity (Dailey *et al.*, 1977a), have an increased tendency to localize in the T-dependent areas (Daily *et al.*, 1977b). The ultrastructural localization of such compounds was not yet reported. Veldman and co-workers (Veldman, 1970; Veldman *et al.*, 1978) reported on electron microscopic observations of interdigitating RC in ferritin-injected rabbits, but they did not comment on the presence or absence of ferritin in the interdigitating RC–T-cell junctions. The LC-granule-containing cells we observed in paracortical areas of draining nodes 4–24 hr after ferritin challenge (Silberberg-Sinakin *et al.*, 1976) did exhibit ferritin on their surface and also in their cytoplasm.

7.4. PROBABLE ROLE OF LANGERHANS CELLS IN SKIN GRAFT IMMUNOGENICITY

Tissues presenting both Ia antigens and other major histocompatibility antigens are more effective in inducing graft rejection than those lacking Ia antigens (van Rood, 1977; Häyry, 1976). Any method which could selectively destroy LC in skin or damage Ia on LC might, therefore, be useful in promoting skin allograft acceptance. It is suggested that a study of substances with specific injurious effects on LC might be of value in this respect.

It has been noted in *in vitro* experiments that ultraviolet (UV) light destroys the ability of lymphoid cells to induce mixed lymphocyte reactions, and this has been related to the destruction of Ia antigen (Bach, 1976; Grillot-Courvalin *et al.*, 1977). It is conceivable that UV irradiation of skin might also decrease its Ia content and could thus be tried as an agent in the promotion of allograft acceptance.

7.5. POSSIBLE IMMUNOLOGICALLY NONSPECIFIC INFLUENCE ON LYMPHOID CELLS

The intimate relationship between monocytes or interdigitating RC and lymphocytes may have significance in addition to its possible role in the presentation of antigens. There is ample evidence that mitogen-induced proliferation of both T and B cells cannot be adequately performed *in vitro* without the presence of adherent monocytes (Rosenstreich and Oppenheim, 1976). Although 2-mercaptoethanol (2-ME) can replace monocytes partly in some of these functions, its effect is usually synergistic with that of macrophages (Umetsu *et al.*, 1979) or of macrophage supernatants (Nordin, 1978). It has, therefore, been postulated that monocytes play an important role in maintaining proper *in vitro* conditions for lymphocytes in an immunologically nonspecific manner, at least partly mediated via soluble products (lymphokines). Recent studies suggest that various lymphoma cells of both T- and B-cell origin also require monocytes for optimal *in vitro* growth (Nathan and Terry, 1975). In the case of a B-cell lymphoma of BALB/c mouse origin we found that Ia-antigen-bearing adherent cells were required for optimal growth, although they did not have to be of the same mouse strain origin as the lymphoma cell (Umetsu *et al.*, 1979). The results suggested strongly that cell-to-cell contact was important for lymphoma growth and that 2-ME could not replace this monocyte function. Steinman and Cohn have described a cluster-forming property of their dendritic cell with lymphoid cells, particularly with B lymphoblasts, which appears to help the proliferation and/or differentiation of these blasts *in vitro* (Steinman and Cohn, 1975).

The lymphokine, lymphocyte-activating factor (LAF), which stimulates T-cell proliferation is one of the better characterized products of macrophages (Gery and Waksman, 1972). Macrophages have also been found to promote thymocyte maturation *in vitro,* presumably via cell-to-cell contact (van den

Tweel and Walker, 1977). The presence of interdigitating RC with and without LC granules in thymus may well have an important role in the proliferation of thymocytes on the basis of comparable secretion products or cell-to-cell contact. In addition, if the LC represents an epidermal equivalent of one or more of these monocytes or dendritic "reticulum" cell types, it may be expected that they, too, will, at times, promote lymphoid cell proliferation of T-cell origin in the skin.

8. ROLE OF LANGERHANS CELLS IN NONIMMUNOLOGIC PROCESSES

8.1. POSSIBLE IMPORTANCE OF LANGERHANS CELLS FOR KERATINIZATION

A relationship has been suggested to exist between LC and the formation of the granular layer of the epidermis as well as with keratinocyte proliferation (Potten, 1975; Potten and Allen, 1975, 1976). Studies of the formation of the scale–interscale patterning in mouse tail epidermis revealed that LC appeared at precisely the same time as the granular layer (Schweizer, 1977; Schweizer and Marks, 1977). Similar observations have been made in mouse back skin (Reams and Tompkins, 1973; Weiss and Zelickson, 1975a,b) and in man (Breathnach and Wyllie, 1965; Hashimoto *et al.*, 1966). It has been proposed that LC are involved in the process of orthokeratinization (Jarrett and Spearman, 1964; Riley, 1966). In the orthokeratinized epidermis numerous hydrolytic enzymes are strongly concentrated in the region of the granular layer, forming a continuous band upon appropriate histochemical staining (Jarrett *et al.*, 1959; Jarrett and Spearman, 1964). These enzymes are thought to be responsible for the degradation of keratinocytes. Since similar enzymes can be demonstrated in LC (Wolff and Winkelmann, 1967c), it was proposed that these cells may contribute to the autolytic process of keratinocytes (Schweitzer and Marks, 1977). LC are reported to be greatly reduced in human skin diseases characterized by parakeratotic differentiation, i.e., chronic psoriasis and chronic eczema; whereas their number is increased in disorders with augmented orthokeratinization, i.e., lichen ruber and ichthyosis (Prunieras, 1969). LC have been shown to appear in the urinary bladder epithelium of vitamin-A-deficient rats where they are normally absent (Wong and Buck, 1971). It is known that vitamin A deficiency results in a conversion of nonkeratinized bladder epithelium to an orthokeratinized squamous epithelium (Hicks, 1968).

In some species LC do not have the same anatomical relationship in the epidermis as they do in the mouse (Mackenzie, 1975). However, because of the dendritic nature of LC, these cells may exert their influence over relatively diffusely distributed areas (Silberberg-Sinakin *et al.*, 1978), thus making precise anatomical position not a prerequisite for speculating that LC influence keratinocytes. Since LC can assume the morphologic appearance consistent with what has been described for macrophages during prostaglandin synthesis (Brune *et al.*, 1978; Silberberg-Sinakin and Thorbecke, 1980), it seems possible that a secreted product of LC could affect the surrounding keratinocytes. The possible secretory products, including prostaglandins, of LC are discussed in the following section. It has recently been shown that prostaglandin B_1 selectively affects morphology of epidermal mitochondria in developing chick skin in organ cultures, and also accelerates keratinization and differentiation (Kischer, 1973). Regarding LC as members of the monocyte–macrophage family does not exclude a role for LC in keratinization in view of the many varied properties that a monocyte-related cell system may have.

8.2. POSSIBLE SECRETORY FUNCTIONS OF LANGERHANS CELLS

The mononuclear phagocyte has been suggested to be the principal hemopoietic cell source of prostaglandin E (Kurland and Moore, 1977; Brune *et al.*, 1978). Therefore, if LC represent the mononuclear phagocyte system in the epidermis, it would not at all be unlikely that they might be an important source of prostaglandins in the skin.

The role of prostaglandins in producing erythema in skin has recently been reviewed (Goldyne, 1975). For instance, after exposure to UV light a measurable increase in the amount of prostaglandins occurs in skin (Greaves and Søndergaard, 1970). Prostaglandin activity in the skin was also found in 35 of 45 patients with contact dermatitis of 48-hr duration. This activity was due to both E and F prostaglandins (Goldyne *et al.*, 1973; Greaves *et al.*, 1971). The major portion of prostaglandin activity appears to be in the epidermis rather than in the dermis (Jouvenaz *et al.*, 1970; van Dorp, 1971). If LC secretory products are mediating the erythematous reaction to such irritants as UV light, it is of importance to review the qualitative and quantitative changes in these cells induced by exposure to UV. Electron microscopic studies of epidermis after exposure to UV light show that LC contain numerous electron-dense bodies [see Figure 6 in Nix *et al.* (1965)]. The authors believe the postirradiation ultrastructural changes are not degenerative in nature, but indicate accelerated metabolic activity. Kumakiri *et al.* (1977) also reported droplets interpreted to be lipids in LC and a widening of intercellular spaces after long-wave UV. After exposure to UV light, quantitative studies of LC show variable results. Fan *et al.* (1959) observed an increase of dopa-positive melanocytes and a reciprocal decrease of suprabasal gold-positive cells. In another study, guinea pigs were irradiated with UV light daily, and the cell population of LC and melanocytes were assessed quantitatively after the ninth through the 25th day. UV light produced no significant changes in LC population (Wolff and Winkelmann, 1967d). However, in still another study on the effects of UV light, a decrease of LC was observed 48 hr after a single exposure and two weeks after daily exposures (Zelickson and Mottaz, 1970). In a recent study in man (Juhlin and Shelley, 1979) of the effect of ultraviolet light (UVA) plus topically applied trioxsalen, 1–3 weeks of treatment to skin sites caused some of the LC to appear by ATPase-staining to be two to four times as large as normal and to possess long dendrites. After treatment with UVA alone or with topical psoralens alone no such changes were seen.

Since injection of prostaglandin into rat skin produces changes in epidermal cells other than LC, but not in LC (Lupulescu, 1977), any changes in LC produced by UV light exposure are unlikely to be secondary to prostaglandins released by other cells. Thus, considering all the evidence, the LC may be an excellent candidate as a source of prostaglandins in the skin. Since catecholamines are known to represent an initial trigger of a cycle of events leading to prostaglandin synthesis (Zurier, 1974), it seems possible that the affinity of catecholamines for LC, discussed earlier in this review, is related to this postulated prostaglandin secretion.

There are many other functions of macrophages and/or monocytes, such as production of various enzymes (Davies and Allison, 1976; Unanue, 1976), C factors (Thorbecke *et al.*, 1965; Stecher and Thorbecke, 1967), and lymphokines (Gery and Waksman, 1972; Unanue, 1976), for which it is not known whether an Ia-positive or Ia-negative subpopulation of macrophages is involved. It is therefore difficult to predict which of these are relevant for a comparison with LC. It should be stressed, however, that the ultrastructural features of LC certainly indicate a variable activity in protein synthesis and secretion. Identification of these products should be the subject of further experimentation.

9. LANGERHANS CELLS IN DISEASE

A variety of diseases have been described in which increases in LC, either in the locally affected area of the skin or in draining lymph nodes, were reported. Many of these diseases are infectious or allergic in nature, or else suspected of having one or both of these components. These have been summarized in previous reviews (Baer and Berman, 1980; Lisi, 1973; Shelley and Juhlin, 1978; Silberberg-Sinakin *et al.*, 1978; Wolff, 1972). It should be stressed, however, that many of these reports concern description of one stage of the disease development, and that these, being clinical studies, frequently lack appropriate controls.

It has previously been pointed out that the release of damaging substances from sensitized lymphocytes as well as LC, which are participating in the reaction during *allergic contact dermatitis,* could account for the localization of early lesions in the lower rather than upper parts of the epidermis and in the dermis (Baer and Berman, 1980; Silberberg-Sinakin *et al.*, 1978). In response to antigen fixed to the surface of LC, lymphoid cells could react by release of lymphokines or by development of cytotoxic responses, which in turn could damage the LC. Moreover, as was shown, immune complexes plus complement may damage LC. Further experiments are needed to determine the exact mechanism by which LC are damaged and to which degree this damage contributes to the epidermal changes such as edema, spongiosus, vesiculation, bulla formation, and urtication (Silberberg-Sinakin *et al.*, 1978).

Another disease in which LC are damaged and cytolytic changes are seen in surrounding keratinocytes is *pityriasis rosea.* In this condition, IgM antibodies to cytoplasmic components of skin cells are found in serum (Takaki and Miyazaki, 1976).

Mycosis fungoides is a chronic disease, which is characterized by infiltration of the skin with T cells. The evidence for this has been reviewed by Mackie *et al.*

(1976) and Nordqvist and Kinney (1976). It has been suggested that the lymphocytes in mycosis fungoides are specifically reactive lymphocytes which are present in the skin because of localized antigen persistence. In other words, mycosis fungoides might be a form of protracted contact dermatitis resulting from chronic exposure to contact allergens (Rowden and Lewis, 1976). Such a chronic antigenic stimulation could induce the formation of a malignant clone which although at first limited to the skin would later disseminate (Goos *et al.*, 1976; Goudie *et al.*, 1974; Tan *et al.*, 1974). In mycosis fungoides apposition of lymphocytes to LC was seen and LC showed some signs of cell damage (Rowden and Lewis, 1976; Rowden *et al.*, 1979b). The frequent association between LC and T-lymphoma cells in mycosis fungoides might involve a nutritive role of LC for T lymphocytes, supporting their proliferation and differentiation (see Section 7.5). This association may, of course be the counterpart in malignant disease of a similar supportive function that interdigitating reticulum cells may have on normal T lymphocytes in T-dependent areas of lymphoid tissue.

In *histiocytosis* X (Langerhans cell granuloma*) there is a striking accumulation of LC (Nezelof *et al.*, 1973). There is no satisfactory explanation for the accumulation of vast numbers of LC in the lesions of about 80% of cases of histiocytosis X. Possibilities which have been considered are: (1) histiocytosis X is a chronic inflammatory reaction to an as yet unknown agent; (2) it represents a form of immunologic response to an undefined antigen; or (3) it is a benign or malignant proliferative tumorlike process (Basset *et al.*, 1976). Membrane receptors on the surface of histiocytosis X cells for IgG and for immune complexes (C3) have been identified by direct immunofluorescence and by rosette formation. Histiocytosis X cells also avidly phagocytize human red blood cells and latex particles. These immunologic characteristics suggest that histiocytosis X represents a pathological increase of a variant of a macrophage subpopulation, the LC lineage (Nezelof *et al.*, 1977), or else an LC granuloma (Lieberman, 1980; Pinkus and Mehregan, 1980).

10. POSSIBLE CONSEQUENCES OF INTERFERENCE WITH LANGERHANS CELL FUNCTION

Inactivation of Ia antigens on LC or removal of LC from skin might be of importance in promoting skin allograft acceptance, since tissues presenting both Ia antigens and other major histocompatibility antigens are more effective in inducing graft rejection than those lacking Ia antigens (van Rood, 1977; Häyry, 1976). Similarly, inactivation of Ia antigens or of LC themselves might interfere with local induction of contact reactions, as is strongly suggested by results of Streilein *et al.* (1979). This was discussed elsewhere (Silberberg-Sinakin and Thorbecke, 1980). If LC have a nurturing influence on T-lymphoid cells in the skin, their destruction might result, at times, in clearing of lymphomatous skin infiltrates. Ultraviolet light has been used to treat a variety of skin diseases,

*See the footnote on p. 574.

including some cutaneous lymphomas such as early stages of mycosis fungoides, with beneficial effects.

For a long time it was thought that the only disease in which LC were involved was histiocytosis X (Langerhans cell granuloma*). It now becomes apparent that there may be more diseases in which these cells have pathogenetic influences as a result of their interaction with other cells (lymphocytes, keratinocytes). The effect of a variety of commonly used therapeutic modalities on LC morphology and cell membrane receptors needs, therefore, to be studied in detail in order to gain insight into methods of monitoring LC function.

11. SUMMARY

The recently discovered similarities between LC and a subpopulation of macrophages are reviewed. These include, among others, morphological criteria, cell surface properties, migrating properties from skin to lymph node, fixing of protein antigens on the cell surface, and role in presentation of antigens to primed T lymphocytes. The importance of the presence of a macrophage-like cell in the epidermis for immunological reactions involving the skin, particularly allergic contact dermatitis, is discussed. In addition, it is pointed out that the macrophage-like nature of LC may also be of importance for other immunologically nonspecific and even for nonimmunological functions of these cells. Separate sections of this review are devoted to the origin of LC, their relationship to other dendritic cells of epidermis and lymphoid tissue, and to diseases involving changes in LC. It is concluded that LC are best thought of as a specialized Ia+ subpopulation of the mononuclear phagocyte system, most closely related to interdigitating "reticulum" cells of T-dependent areas in lymphoid organs.

ACKNOWLEDGMENTS. Research conducted in the laboratory of G. J. T. and quoted in this review was supported by grant A I 03076 from the USPHS. Research conducted in the laboratory of I. S. and quoted in this review was supported by grant A I 11103 from the USPHS and by funds donated to the department of Dermatology by the Polly Annenberg Levee Charitable Trust and the Chernow Fund. Discussions with Dr. Martha E. Fedorko and Dr. Rudolf L. Baer were greatly appreciated.

REFERENCES

Al Yassin, T. M., and Toner, P. G., 1976, Langerhans cells in the human esophagus, *J. Anat.* **122**:435.

Andrew, W., 1968, Differentiation of migrant cells in the human epidermis and the non-specificity of the germ-layers, *Arch. Histol. Jpn.* **29**:211.

Askonas, B. A., and Jarošková, L., 1970, Antigen in macrophages and antibody induction, in: *Mononuclear Phagocytes* (R. van Furth, ed.), pp. 595–612, Davis, Philadelphia.

Axelsson, S., Elofsson, R., Falck, B., and Sjöborg, S., 1978, *In vitro* uptake of L-dopa and catecholamines into the epidermal Langerhans cells, *Acta Dermatovener.* **58,** *Suppl.* **79**:31.

---

*See the footnote on p. 574.

Baer, R. L., and Berman, B., 1980, The role of Langerhans cells in cutaneous immunologic reactions in: *Comprehensive Immunology* Vol. 7, *Immunodermotology* (B. Safai and R. A. Good, eds.), Plenum Press, New York.

Bach, F. H., 1976, Differential function of MHC LD and SD determinants, in: *Leukocyte Membrane Determinants Regulating Immune Reactivity* (V. P. Eysvogel, D. Roos, and W. P. Zeylemaker, eds.), pp. 753–755, Academic Press, New York.

Balfour, B. M., and Humphrey, J. H., 1967, Localization of γ-globulin and labelled antigen in germinal centers in relation to the immune response, in: *Germinal Centers in Immune Responses* (H. Cottier, N. Odartchenko, R. Schindler, and C. C. Congdon, eds.), pp. 80–85, Springer Verlag, Berlin.

Basset, F., and Nezelof, C., 1966, Présence en microscopie électronique de structures filamenteuses originales dans les lésions pulmonaires et osseuses de l'histicytose X. Etat actuel de la question, *Soc. Med. Hop.Paris* **117**:413.

Basset, F., and Turiaf, J., 1965, Identification par la microscopie électronique de particules de nature probablement virale dans les lésions granulomateuses d'une histiocytose "X" pulmonaire, *C. R. Acad. Sci.* **261**:3701.

Basset, F., Nezelof, C., Mallet, R., and Turiaf, J., 1965, Nouvelle mise en évidence par la microscopie électronique, de particules d'allure virale dans une séconde forme clinique de l'histiocytose X, le granulome éosinophilie de l'os, *C. R. Acad. Sci.* **261**:5719.

Basset, F., Soler, P., Wyllie, L., Mazin, F., and Turiaf, J., 1976, Langerhans' cells and lung interstitium, 7th International Congress on Sarcoidosis and Other Granulomatous Diseases, New York, *Ann. N.Y. Acad. Sci.* **278**:599.

Bell, M., 1967, The Langerhans cells in the skin of subhuman primates, in: *Advances in Biology of Skin.* Volume 8, *The Pigmentary System* (W. Montagna and F. Hu, eds.), pp. 115–134, Pergamon Press, New York.

Bell, M., 1969, Langerhans granules in fetal keratinocytes, *J. Cell Biol.* **41**:914.

Berken, A., and Benacerraf, B., 1966, Properties of antibodies cytophilic for macrophages, *J. Exp. Med.* **123**:119.

Billingham, R. E., and Medawar, P. B., 1953, A study of the branched cells of mammalian epidermis with special reference to the fate of their division products, *Phil. Trans. R. Soc. London Ser. B* **237**:151.

Billingham, R. E., and Silvers, W. K., 1965, Some unsolved problems in the biology of the skin, in: *Biology of the Skin and Hair Growth* (A. G. Lyne and B. F. Short, eds.), pp. 1–24, Angus and Robertson, Sydney.

Birbeck, M. S., Breathnach, A. S., and Everall, J. D., 1961, An electron microscopic study of basal melanocytes and high level clear cells (Langerhans cells) in vitiligo, *Invest. Dermatol.* **37**:51.

Bleehan, S. S., Pathak, M. A., Hori, Y., and Fitzpatrick, T. B., 1968, Depigmentation of skin with 4-isopropylcatechol, mercaptoamines, and other compounds, *J. Invest. Dermatol.* **50**:103.

Böck, P., 1974, Fine structure of Langerhans cells in the stratified epithelia of the esophagus and stomach of mice, *Z. Zellforsch Mikrosk. Anat.* **147**:237.

Bourlond, A., Wolff, K., and Winkelmann, R. K., 1967, Cholinesterase in melanocytes of the bat, *Nature* **214**:846.

Bradshaw, M., Wachstein, M., Spence, J., and Elias, J. M., 1963, Adenosine triphosphatase activity in melanocytes and epidermal cells of human skin, *J. Histochem. Cytochem.* **11**:465.

Breathnach, A. S., 1963a, A new concept of the relation between the Langerhans cell and the melanocyte, *J. Invest. Dermatol.* **40**:279.

Breathnach, A. S., 1963b, The distribution of Langerhans cells within the human follicle and some observations on its staining properties with gold chloride, *J. Anat.* **97**:73.

Breathnach, A. S., 1965, The cell of Langerhans, in: *International Review of Cytology* (G. H. Bourne and J. F. Danielli, eds.), Vol. 18, pp. 1–28, Academic Press, New York.

Breathnach, A. S., 1977a, Electron micrographs from a collection, *Clin. Exp. Dermatol.* **2**:1.

Breathnach A. S., 1977b, Variations in ultrastructural appearance of Langerhans cells of normal human epidermis, *Br. J. Dermatol.* **97** *Suppl.* **15**:14.

Breathnach, A. S., 1980, Overview of branched cells in epidermis, *J. Invest. Dermatol.* **75:** (in press).

Breathnach, A. S., and Goodwin, D., 1966, Electron microscopy of guinea pig epidermis stained by the osmium-iodide technique, *J. Anat.* **100**:159.

Breathnach, A. S., and Wyllie, L. M. A., 1965, Electron microscopy of melanocytes and Langerhans cells in human fetal epidermis at fourteen weeks, *J. Invest. Dermatol.* **44**:51.

Breathnach, A. S., and Wyllie, L. M. A., 1967, The problem of the Langerhans cell, in: *Advances in Biology of Skin*, Vol. 8, *The Pigmentary System* (W. Montagna and F. Hu, eds.), pp. 97–113, Pergamon Press, Oxford.

Breathnach, A. S., Birbeck, M. S. C., and Everall, J. D., 1963, Observations bearing on the relationship between Langerhans cells and melanocytes, *Ann. N.Y. Acad. Sci.* **100**:223.

Breathnach, A. S., Silvers, W. K., Smith, J., and Heyner, S., 1968, Langerhans cells in mouse skin experimentally deprived of its neural crest component, *J. Invest. Dermatol.* **50**:147.

Breathnach, A. S., Gross, M., Basset, F., and Nezelof, C., 1973, Freeze-fracture replication of X-granules in cells of cutaneous lesions of histiocytosis X (Letterer-Siwe disease), *Br. J. Dermatol.* **89**:571.

Breathnach, A. S., Nazzaro Porro, M., and Martin, B., 1975, Ultrastructure of skin in pityriasis versicolor, *Minerva Dermatol.* **110**:457.

Brown, J., Winkelmann, R. K., and Wolff, K., 1967, Langerhans cells in vitiligo: a quantitative study, *J. Invest. Dermatol.* **49**:386.

Brun, R., 1967, Effect of the ethyl ether of hydroquinone on pigmentation and on the cells of Langerhans, *Dermatologica* **134**:125.

Brune K., Glatt, M., Kalin, H., and Peskar, B. A., 1978, Pharmacological control of prostaglandin and thromboxane release from macrophages, *Nature* **274**:261.

Campo-Aasen, I., and Pearse, A. G. E., 1966, Enzimologia de la célula de Langerhans, *Med. Cutànea* **1**:35.

Cancilla, P. A., 1968, Demonstration of the Langerhans granule by lanthanum, *J. Cell Biol.* **38**:248.

Caputo, R., Peluchetti, D., and Monti, M., 1976, Freeze-fracture of Langerhans granules. A comparative study, *J. Invest. Dermatol.* **66**:297.

Carswell, E. A., Lerman, S. P., and Thorbecke, G. J., 1976, Properties of reticulum cell sarcomas in SJL/J mice. II. Fate of labelled tumor cells in normal and irradiated syngeneic mice, *Cell Immunol.* **23**:39.

Chase, M. W., 1946, Inhibition of experimental drug allergy by prior feeding of the sensitizing agent, *Proc. Soc. Exp. Biol. Med.* **61**:257.

Chen, L., Adams, J., and Steinman, R. M., 1978a, Anatomy of germinal centers in mouse spleen with special reference to "follicular dendritic cells," *J. Cell Biol.* **77**:148.

Chen, L., Frank, A., Adams, J., and Steinman, R., 1978b, Distribution of horseradish peroxidase (HRP)-anti-HRP immune complexes in mouse spleen with special reference to follicular dendritic cells, *J. Cell Biol.* **79**:184.

Cohen, S., Vassalli, P., Benacerraf, B., and McCluskey, R., 1966, The distribution of antigenic and non-antigenic compounds within draining lymph nodes, *Lab. Invest.* **15**:1143.

Cohn, Z. A., 1970, Endocytosis and intracellular digestion, in: *Mononuclear Phagocytes* (R. van Furth, ed.), pp. 121–129, Davis, Philadelphia.

Cormane, R. H., and Kalsbeek, G. L., 1963, ATP-hydrolysing enzymes in normal human skin, *Dermatologica* **127**:381.

Cowing, C., Pincus, S., Sachs, D., and Dickler, H., 1978, A subpopulation of adherent accessory cells bearing both I-A and I-E or C subregion antigens is required for antigen specific murine T lymphocyte proliferation, *J. Immunol.* **121**:1680.

Cruickshank, C. N. D., and Cooper, J. R., 1975, Langerhans cells in tissue cultures of guinea pig epidermal cells, *Br. J. Dermatol.* **92**:121.

Dailey, M., and Hunter, R., 1977a, Induction of cell-mediated immunity to chemically modified antigens in guinea pigs. I. Characterization of the immune response to lipid-conjugated protein antigens, *J. Immunol.* **118**:957.

Dailey, M., Post, W., and Hunter, R., 1977b, Induction of cell-mediated immunity to chemically modified antigens in guinea pigs. II. The interaction between lipid-conjugated antigens, macrophages and T lymphocytes, *J. Immunol.* **118**:963.

Davies, P., and Allison, A. C., 1976, Secretion of macrophage enzymes in relation to the pathogenesis of chronic inflammation, in: *Immunobiology of the Macrophage* (D. S. Nelson, ed.), pp. 427–461, Academic Press, New York.

Desmons, F., Breuillard, F., Thomas, P., Leonardelli, J., and Hildebrand, H. F., 1977, Multiple clear-cell acanthoma (Degos). Histochemical and ultrastructural study of two cases, *Int. J. Dermatol.* **16**:203.

Dukor, P., Bianco, C., and Nussenzweig, V., 1970, Tissue localization of lymphocytes bearing a membrane receptor for antigen-antibody-complement complexes, *Proc. Natl. Acad. Sci.* **67**:991.

Eikelenboom, P., 1978a, Dendritic cells in the rat spleen follicles: A combined immuno- and enzyme histochemical study, *Cell Tiss. Res.* **190**:79.

Eikelenboom, P., 1978b, Characterization of non-lymphoid cells in the white pulp of the mouse spleen, and *in vivo* and *in vitro* study, *Cell Tiss. Res.* **195**:445.

Elleder, M., 1975, Activity of alpha-D-mannosidase in human Langerhans epidermal cells, *Virchows Arch. B* **19**:93.

Elson, C. J., and Taylor, R. B., 1975, Permanent hapten-specific tolerance in B lymphocytes, *Immunology* **28**:543.

Falck, B., Agrup, G., Jacobsson, S., Rorsman, H., Rosengren, E., Sachner, K., and Ögren, M., 1976, Uptake of L-dopa and functionally related aromatic amino acids into the Langerhans cell, *J. Invest. Dermatol.* **66**:265.

Fan, J., and Hunter, R., 1958, Langerhans cells and the modified technic of gold impregnation by Ferreira-Marques, *J. Invest. Dermatol.* **31**:115.

Fan, J., Schoenfeld, R. J., and Hunter, R., 1959, A study of the epidermal clear cells with special reference to their relationship to the cells of Langerhans, *J. Invest. Dermatol.* **32**:445.

Fedorko, M. E., 1974, Loss of iron from mouse peritoneal macrophages *in vitro* after uptake of (^{55}Fe)ferritin and (^{55}Fe)ferritin rabbit antiferritin complexes, *J. Cell Biol.* **62**:802.

Fedorko, M. E., 1975, Morphologic and functional characteristics of bone marrow macrophages from imferon-treated mice, *Blood* **45**:435.

Fedorko, M. E., and Hirsch, J. G., 1970, Structure of monocytes and macrophages, *Sem. Hemat.* **7**:109.

Ferreira-Marques, J., 1951, Systema sensitivum intra-epidermicum. Die Langerhansschen zellen als receptoren des hellen schmerzes. Doloriceptores, *Arch. Dermatol. Syph.* **193**:191.

Fidler, J. M., and Golub, E. S., 1973, Immunologic tolerance to a hapten. I. Induction and maintenance of tolerance to trinitrophenyl with trinitrobenzene sulfonic acid, *J. Exp. Med.* **137**:42.

Forsum, U., Klareskog, L., Tjernlund, U. M., and Peterson, P. A., 1978, Significance of the expression of HLA-DR antigen on epidermal Langerhans cells, *Acta Derm.-Venereol.* **58**, Suppl. **79**:37.

Frei, W., 1928, Über willkürliche sensibilisierung gegen chemisch-definierte substanzen. I. Untersuchungen mit neosalvarsan am menschen, *Klin. Wochenschr.* **7**:539.

Frelinger, J., Wettstein, P., Frelinger, J., and Hood, L., 1978, Epidermal Ia molecules from the I-A and I-EC subregions of the mouse H-2 complex, *Immunogenetics* **6**:125.

Fritsch, P., Diem, E., and Hönigsmann, H., 1973, Langerhans cells in cell culture. Survival and identification. *Arch. Dermatol. Forsch.* **248**:123.

Gery, I., and Waksman, B. H., 1972, Potentiation of T-lymphocyte response to mitogens. II. The cellular source of potentiation of mediators, *J. Exp. Med.* **136**:143.

Giacometti, L., 1969, The healing of skin wounds in primates. III. Behavior of the cells of Langerhans, *J. Invest. Dermatol.* **53**:151.

Giacometti, L., and Montagna, W., 1967, Langerhans cells. Uptake of tritiated thymidine, *Science* **157**:439.

Gianotti, F., and Caputo, R., 1969, Skin ultrastructure in Hand-Schüller-Christian disease, *Arch. Dermatol.* **100**:342.

Gianotti, F., Caputo, R., and Ranzi, T., 1968, Ultrastructural study of giant cells and "Langerhans cell granules" in cutaneous lesions and lymph node and liver biopsies from four cases of subacute disseminated histiocytosis of Letterer-Siwe, *Arch. Klin. Exp. Dermatol.* **233**:238.

Gigli, I., 1980, Receptors for complement on Langerhans cells, *J. Invest. Dermatol.* (in press).

Goldyne, M. E., 1975, Prostaglandins and cutaneous inflammation, *J. Invest. Dermatol.* **64**:377.

Goldyne, M. E., Winkelmann, R. K., and Ryan, R. J., 1973, Prostaglandin activity in human cutaneous inflammation, detection by radioimmunoassay, *Prostaglandins* **4**:737.

Goos, M., Kaiserling, E., and Lennert, K., 1976, Mycosis fungoides: Model for T-lymphocyte homing to the skin?, *Br. J. Dermatol.* **94**:221.

Gottlieb, A. A., and Waldman, S. R., 1972, The multiple functions of macrophages in immunity, in: *Macrophages and Cellular Immunity* (A. I. Laskin and H. Lechevalier, eds.), pp. 13–44, Chemical Rubber Co., Cleveland.

Goudie, R. B., MacFarlane, P. S., and Lindsay, M. K., 1974, Homing of lymphocytes to non-lymphoid tissues, *Lancet* **1**:292.

Greaves, M. W., and Søndergaard, J., 1970, Pharmacologic agents released in ultraviolet inflammation studied by continuous skin perfusion, *J. Invest. Dermatol.* **54**:365.

Greaves, M. W., Søndergaard, J., and McDonald-Gibson, W., 1971, Recovery of prostaglandins in human cutaneous inflammation, *Br. Med. J.* **2**:258.

Grillot-Courvalin, C., Alter, B. J., and Bach, F. H., 1977, Antigenic requirements for the generation of secondary cytotoxicity, *J. Immunol.* **119**:1253.

Hackemann, M., Grupp, C., and Hill, K. R., 1968, The ultrastructure of normal squamous epithelium of the human cervix uteri, *J. Ultrastruct. Res.* **22**:443.

Hammerling, G. J., Mauve, G., Goldberg, E., and McDevitt, H. O., 1975, Tissue distribution of Ia antigens—Ia on spermatozoa, macrophages, and epidermal cells, *Immunogenetics* **1**:428.

Hashimoto, K., 1971, Langerhans cell granule, *Arch. Dermatol.* **104**:148.

Hashimoto, K., and Tarnowski, W. M., 1968, Some new aspects of the Langerhans cell, *Arch. Dermatol.* **97**:450.

Hashimoto, K., Gross, B. G., Di Bella, R. S., and Lever, W. F., 1966, The ultrastructure of the skin of human embryos. IV. The epidermis, *J. Invest. Dermatol.* **47**:317.

Häyry, P., 1976, Problems and prospects in surgical immunology, *Med. Biol.* **54**:1, 73.

Herd, Z., and Ada, G., 1969, Distribution of ^{125}I-immunoglobulins, IgG subunits and antigen-antibody complexes in rat lymph nodes, *Aust. J. Exp. Biol. Med. Sci.* **47**:73.

Hicks, R. M., 1968, Hyperplasia and cornification of the transitional epithelium in the vitamin A deficient rat, *Ultrastruc. Res.* **22**:206.

Hirsch, J. G., and Fedorko, M. E., 1970, Morphology of mononuclear phagocytes, in: *Mononuclear Phagocytes* (R. van Furth, ed.), pp. 7–28, Davis, Philadelphia.

Hoefsmit, E. C. M., 1975, Mononuclear phagocytes, reticulum cells, and dendritic cells in lymphoid tissues, in: *Mononuclear Phagocytes in Immunity, Infection and Pathology* (R. van Furth, ed.), pp. 129–146, Blackwell, Oxford.

Hoefsmit, E. C. M., and Gerver, J. A. M., 1975, Epithelial cells and macrophages in the normal thymus, in: *Biological Activity of Thymic Hormones* (D. W. van Bekkum and A. M. Kruisbeck, eds.), pp. 63–68, Kooyker, Rotterdam.

Hoefsmit, E. C. M., Balfour, B. M., Kamperdijk, E. W. A., and Cvetanov, J., 1978, Cells containing Birbeck granules in the lymph and the lymph node, *Z. Immunol. Forsch.* **154**:321.

Hoffmann-Fezer, G., Götze, D., Rodt, H., and Thierfelder, S., 1978, Immunohistochemical localization of xenogeneic antibodies against Ia^k lymphocytes on B cells and reticular cells, *Immunogenetics* **6**:367.

Huber, B., Devinsky, O., Gershon, R., and Cantor, H., 1976, Cell-mediated immunity: Delayed type hypersensitivity and cytotoxic responses are mediated by different T cell subclasses, *J. Exp. Med.* **143**:1424.

Huber, H., and Fudenberg, H. H., 1968, Receptor sites of human monocytes for IgG, *Intl. Arch. Allergy Appl. Immunol.* **34**:18.

Huber, H., Polley, M. J., Linscott, W. D., Fudenberg, H. H., and Muller-Eberhard, H. J., 1968, Human monocytes: Distinct receptor sites for the third component of complement and for immunoglobulin G, *Science* **162**:1281.

Humphrey, J. H., and Frank, M. M., 1967, The localization of non-microbial antigens in the draining lymph nodes of tolerant, normal and primed rabbits, *Immunology* **13**:87.

Hunter, J. A. A., Fairley, D. J., Priestley, G. C., and Cubie, H. A., 1976, Langerhans cells in the epidermis of athymic mice, *Br. J. Dermatol.* **94**:119.

Hunter, R. L., Wissler, R. W., and Fitch, F. W., 1969, Studies on the kinetics and radiation sensitivity of dendritic macrophages, in: *Lymphatic Tissue and Germinal Centers in Immune Responses* (L. Fiore-Donati and M. G. Hanna, Jr., eds.), pp. 101–111, Plenum Press, New York.

Hutchens, L. H., Sagebiel, R. W., and Clarke, M. A., 1971, Oral epithelial dendritic cells of the Rhesus monkey—histologic demonstration, fine structure and quantitative distribution, *J. Invest. Dermatol.* **56**:325.

Im, M. J. C., and Montagna, W., 1965, The skin of primates, XXVI. Specific and nonspecific phosphatases in the skin of the Rhesus monkey, *Am. J. Phys. Anthropol.* **23**:131.
Ishikawa, H., and Klingmüller, G., 1964, Phosphatdarstellung nach Adenosintriphosphatinkubation an dendritischen Zellelementen. Beitrag zur Histogenese der epidermalen neuralen Zellen und des Naevuszellnaevus, *Arch. Klin. Exp. Dermatol.* **220**:191.
Ito, K., Kawada, A., Sato, S., and Kukita, A., 1976, Langerhans cells in human apocrine ducts, *Arch. Dermatol. Res.* **256**:291.
Jarrett, A., and Riley, P. A., 1963, Esterase activity in dendritic cells, *Br. J. Dermatol.* **75**:79.
Jarrett, A., and Spearman, R. I. C., 1964, *Histochemistry of the Skin—Psoriasis,* English Universities Press, London.
Jarrett, A., Spearman, R. I., and Hardy, J. A., 1959, The histochemistry of keratinization, *Br. J. Dermatol.* **71**:277.
Jimbow, K., Sato, S., and Kukita, A., 1969a, Langerhans' cells of the normal human pilosebaceous system. An electron microscopic investigation, *J. Invest. Dermatol.* **52**:177.
Jimbow, K., Sato, S., and Kukita, A., 1969b, Cells containing Langerhans granules in human lymph nodes of dermatopathic lymphadenopathy, *J. Invest. Dermatol.* **53**:295.
Johnson, P. M., Papamichail, M., Gutierre, Z., and Holborow, E. J., 1975, Interaction of the hinge region of human immunoglobulin G with a murine lymphocyte membrane receptor, relevance to the problem of antiglobulin induction in rheumatoid arthritis, *Immunology* **28**:797.
Jouvenaz, G. H., Nugteren, D. H., Beerthuis, R. K., and van Dorp, D. A., 1970, A sensitive method for the determination of prostaglandins by gas chromatography with electron-capture detection, *Biochim. Biophys. Acta* **202**:231.
Juhlin, L., and Shelley, W. B., 1977, New staining techniques for the Langerhans cell, *Acta Derm.-Venereol.* **57**:289.
Juhlin, L., and Shelley, W. B., 1979, Giant Langerhans cells induced by psoralen and ultraviolet radiation, *Arch. Dermatol. Res.* **266**:311.
Kamperdijk, E. W. A., and Hoefsmit, E. C. M., 1978, Birbeck granules in lymph node macrophages, *Ultramicroscopy* **3**:137.
Kamperdijk, E. W. A., Raaymakers, E. M., de Leeuw, J. N. S., and Hoefsmit, E. C. M., 1978, Lymph node macrophages and reticulum cells in the immune response, the primary response to paratyphoid vaccine, *Cell Tiss. Res.* **192**:1.
Katz, S. I., Tamaki, K., and Sachs, D., 1979, Epidermal Langerhans cells are derived from cells originating in bone marrow, *Nature* **282**:324.
Kechijian, P., 1965, The skin of primates, XIX. The relationship between melanocytes and alkaline phosphatase-positive cells in the potto (*Perodicticus potto*), *Anat. Rec.* **152**:317.
Kelly, R. H., Balfour, B. M., Armstrong, J. A., and Griffiths, S., 1978, Functional anatomy of lymph nodes. II. Peripheral lymph-borne mononuclear cells, *Anat. Rec.* **190**:5.
Kiistala, U., and Mustakallio, K. K., 1968, The presence of Langerhans cells in human dermis with special reference to their potential mesenchymal origin, *Acta Derm.-Venereol.* **48**:155.
Kischer, C. W., 1973, The epidermal response of developing skin to prostaglandin Bl: Mitochondrial alterations obtained *in vitro, Exp. Cell Res.* **81**:393.
Klareskog, L., Tjernlund, U. M., Forsum, U., and Peterson, P. A., 1977, Epidermal Langerhans cells express Ia antigens, *Nature* **268**:248.
Kondo, Y., 1969, Macrophages containing Langerhans cell granules in normal lymph nodes of the rabbit, *Z. Zellforsch. Mikrosk. Anat.* **98**:506.
Kumakiri, M., Hashimoto, K., and Willis, I., 1977, Biologic changes due to long-wave ultraviolet irradiation on human skin: Ultrastructural study, *J. Invest. Dermatol.* **69**:392.
Kurland, J., and Moore, M. A. S., 1977, Regulatory role of the macrophage in normal and neoplastic hemopoiesis, in: *Experimental Hematology Today* (S. Baum and G. Ledney, eds.), pp. 51–62, Springer Verlag, New York.
Kuwahara, H., 1971, Fine structures of Langerhans cell with special references to their function, *Jpn. J. Dermatol. Ser. B* **80**:195.
Lang, P. G., and Ada, G. L., 1967, Antigen in tissues. IV. The effect of antibody on the retention and localization of antigen in rat lymph nodes, *Immunology* **13**:523.
Langerhans, P., 1868, Über die nerven der menschlichen haut, *Virchows Arch. B* **44**:325.
Lay, W., and Nussenzweig, V., 1968, Receptors for complement on leukocytes, *J. Exp. Med.* **128**:991.

Lerman, S. P., Chapman, J. M., Carswell, E. A., and Thorbecke, G. J., 1974, Properties of reticulum cell sarcomas in SJL/J mice. I. Proliferative response to tumor cells of T-derived lymphoid cells from normal mice, *Intl. J. Cancer* **14**:808.

Lessard, R. J., Wolff, K., Winkelmann, R. K., 1968, The disappearance and regeneration of Langerhans cells following epidermal injury, *J. Invest. Dermatol.* **50**:171.

Lieberman, P. H., 1980, Eosinophilic granuloma and related syndromes, *J. Invest. Dermatol.* **75**:(in press).

Lipsky, P., and Rosenthal, A., 1975, Macrophage–lymphocyte interaction: Antigen independent binding of guinea pig lymph node lymphocytes by macrophages, *J. Immunol.* **115**:440.

Lisi, P., 1973, Investigation on Langerhans cells in pathological human epidermis, *Acta Derm.-Venereol.* **53**:425.

Lupulescu, A. P., 1977, Cytologic and metabolic effects of prostaglandins on rat skin, *J. Invest. Dermatol.* **68**:138.

Lyne, A. G., and Chase, H. B., 1966, Branched cells in the epidermis of the sheep, *Nature* **209**:1357.

Macher, E., and Chase, M. W., 1969, Studies on the sensitization of animals with simple chemical compounds. XII. The influence of excision of allergenic depots on onset of delayed hypersensitivity and tolerance, *J. Exp. Med.* **129**:103.

Mackenzie, I. C., 1975, Ordered structure of the epidermis, *J. Invest. Dermatol.* **65**:45.

Mackie, R., Sless, F. R., Cochran, R., and de Sousa, M., 1976, Lymphocyte abnormalities in mycosis fungoides, *Br. J. Dermatol.* **94**:173.

Masson, P., 1951, My conception of cellular nevi, *Cancer* **4**:9.

Masutani, M., 1974, Electron microscopic studies on epidermal dendritic cells of hairless mice, *Jpn. J. Dermatol. A* **84**:249.

Metcalf, D., and Ishidate, M., 1962, PAS-positive reticulum cells in the thymus cortex of high and low leukemia strains of mice, *Aust. J. Exp. Biol.* **40**:57.

Miller, J. J., and Nossal, G. J. V., 1964, Antigens in immunity. VI. The phagocytic reticulum of lymph node follicles, *J. Exp. Med.* **120**:1075.

Miller, S. D., Wetzig, R. P., and Claman, H. N., 1979, The induction of cell mediated immunity and tolerance with protein antigens coupled to syngeneic lymphoid cells, *J. Exp. Med.* **149**:758.

Mishima, Y., 1966, Melanosomes in phagocytic vacuoles in Langerhans cells. Electron microscopy of keratin-stripped human epidermis, *J. Cell. Biol.* **30**:417.

Mishima, Y., and Miller-Milinska, A., 1961, Junctional and high-level dendritic cells revealed with osmium iodide reaction in human and animal epidermis under conditions of hyperpigmentation and depigmentation, *J. Invest. Dermatol.* **37**:107.

Mishima, Y., Kawasaki, H., and Pinkus, H., 1972, Dendritic cell dynamics in progressive depigmentations. Distinctive cytokinetics of α-dendritic cells revealed by electron microscopy, *Arch. Derm. Forsch.* **243**:67.

Montagna, W., and Ellis, R. A., 1959, The skin of primates, VII. The skin of the potto (*Perodicticus potto*), *Am. J. Phys. Anthrop.* **17**:137.

Montagna, W., and Yun, J. S., 1962a, The skin of primates, VII. The skin of the great bushbaby (*Galago crassicaudatus*), *Am. J. Phys. Anthropol.* **20**:149.

Montagna, W., and Yun, J. S., 1962b, The skin of primates, XIV. Further observations on *Perodicticus potto*, *Am. J. Phys. Anthropol.* **20**:341.

Müller-Hermelink, H., Heusermann, U., and Stutte, H., 1974, Enzyme histochemical observations on the localization and structure of the T-cell and B-cell regions in the human spleen, *Cell Tiss. Res.* **154**:167.

Mustakallio, K., 1962, Adenosinetriphosphatase activity in neural elements of human epidermis, *Exp. Cell Res.* **28**:448.

Nagao, S., Inaba, S., and Iijima, S., 1976, Langerhans cells at the sites of vaccinia virus inoculation, *Arch. Dermatol. Res.* **256**:23.

Nathan, C. F., and Terry, W. D., 1975, Differential stimulation of murine lymphoma growth *in vitro* by normal and BCG-activated macrophages, *J. Exp. Med.* **142**:887.

Nathan, C. F., Asofsky, R., and Terry, W. D., 1977, Characterization of the non-phagocytic adherent cell from the peritoneal cavity of normal and BCG-treated mice, *J. Immunol.* **118**:1612.

Nelson, D. S., 1969, Macrophages and immunity, in: *Frontiers of Biology*, Vol. 2 (A. Neuberger and E. L. Tatum, eds.), p. 187, North-Holland, Amsterdam.

Nettesheim, P., and Hammons, A., 1971, Recovery of the antigen retention mechanism after sublethal X-irradiation in mice receiving cells from various sources, *J. Immunol.* **107**:518.

Nettesheim, P., and Hanna, M. G., Jr., 1969, Radiosensitivity of the antigen-trapping mechanism and its relation to the suppression of immune response, in: *Lymphatic Tissue and Germinal Centers in Immune Response* (L. Fiore-Donati and M. G. Hanna, Jr., eds.), pp. 167–175, Plenum Press, New York.

Nezelof, C., Basset, F., and Rousseau, M. F., 1973, Histiocytosis X. Histogenetic arguments for a Langerhans cell origin, *Biomedicine* **18**:365.

Nezelof, C., Diebold, N., and Rousseau-Merck, M. F., 1977, Ig surface receptors and erythrophagocytic activity of histiocytosis X cells *in vitro*, *J. Pathol.* **122**:105.

Niebauer, G., 1956, Über die interstitiellen Zellen der Haut, *Hautarzt* **7**:123.

Niebauer, G., and Sekido, N., 1965, Über die Dendritenzellen der Epidermis. Eine Studie über die Langerhans-Zellen in der normalen and ekzematösen Haut des Meerschweinchens, *Arch. Klin. Exp. Dermatol.* **222**:23.

Niebauer, G., Krawczyk, W. S., Kidd, R. L., and Wilgram, G. F., 1969, Osmium zinc iodide reactive sites in the epidermal Langerhans cell, *J. Cell Biol.* **43**:80.

Nix, T. E., Jr., Nordquist, R. E., Scott, J. R., and Everett, M. A., 1965, Ultrastructural changes induced by ultraviolet light in human epidermis: Basal and spinous layers, *J. Invest. Dermatol.* **45**:51.

Nordin, A. A., 1978, The *in vitro* immune response to a T-independent antigen. I. The effect of macrophages and 2-mercaptoethanol, *Eur. J. Immunol.* **8**:776.

Nordqvist, B. C., and Kinney, J. P., 1976, T and B cells and cell-mediated immunity in mycosis fungoides, *Cancer* **37**:714.

Nordquist, R. E., Olson, R. L., and Everett, M. A., 1966, The transport, uptake and storage of ferritin in human epidermis, *Arch. Dermatol.* **94**:482.

North, R. J., 1966, The localization by electron microscopy of nucleoside phosphatase activity in guinea pig phagocytic cells, *J. Ultrastruct. Res.* **16**:83.

Nossal, G. J. V., Ada, G. L., Austin, C. M., and Pye, J., 1965, Antigens in immunity. VII. Localization of ^{125}I-labelled antigens in the secondary response, *Immunology* **9**:349.

Nossal, G. J. V., Abbott, A., Mitchell, J., and Lummus, Z., 1968, Antigens in immunity. XV. Ultrastructural features of antigen capture in primary and secondary lymphoid follicles, *J. Exp. Med.* **127**:277.

Oláh, I., Dunay, C., Röhlich, P., and Törö, I., 1968, A special type of cell in the medulla of the rat thymus, *Acta Biol. Hung.* **19**:97.

Olmos, L., 1975, Dense bodies and Langerhans granules after application of podophyllum resin, *Arch. Dermatol. Res.* **253**:277.

Padilla, F., and Soloff, B. L., 1971, Leukemic reticuloendotheliosis: Electron microscopy of circulating lymphoid cells, *Nouv. Rev. Fr. Hematol.* **11**:547.

Papamichael, M., Gutierrez, C., Embling, P., Johnson, P., Holborow, E., and Pepys, M., 1975, Complement dependence of localization of aggregated IgG in germinal centers, *Scand. J. Immunol.* **4**:343.

Pathak, M. A., Frenk, E., Szabo, G., and Fitzpatrick, T. B., 1966, Cutaneous depigmentation, *Clin. Res.* **14**:272.

Pernis, B., 1967, The immunoglobulins present in the germinal centers, in: *Germinal Centers in Immune Responses* (H. Cottier, N. Odartchenko, R. Schindler, and C. C. Congdon, eds.), pp. 112–119, Springer Verlag, Berlin.

Pierce, C. W., Kapp, J. A., and Benacerraf, B., 1976, Regulation by the H-2 gene complex of macrophage-lymphoid cell interactions in secondary antibody responses *in vitro*, *J. Exp. Med.* **144**:371.

Pinkus, H., and Mehregan, A. H., 1980, *A Guide to Dermatohistopathology*, 3rd ed., Appleton-Century-Crofts, New York (in press).

Polak, L., Polak, A., and Frey, J. R., 1974, Increased DNA synthesis *in vitro* in guinea pigs unresponsive to DNP-skin protein conjugate, *Immunology* **27**:115.

Ponzio, N. M., David, C. S., Shreffler, D. C., and Thorbecke, G. J., 1977, Properties of reticulum cell sarcomas in SJL/J mice. V. Nature of reticulum cell sarcoma surface antigen which induces proliferation of normal SJL/J cells, *J. Exp. Med.* **146**:132.

Potten, C. S., 1975, Epidermal cell production rate, *J. Invest. Dermatol.* **65**:488.

Potten, C. S., and Allen, T. D., 1975, The fine structure and cell kinetics of mouse epidermis after wounding, *J. Cell Sci.* **17**:413.

Potten, C. S., and Allen, T. D., 1976, A model implicating the Langerhans cell in keratinocyte proliferation control, *Differentiation* **5**:443.

Prunieras, M., 1969, Interactions between keratinocytes and dendritic cells, *J. Invest. Dermatol.* **52**:1.

Quevedo, W. C., and Montagna, W., 1962, A new system of melanocytes in the skin of the potto (*Perodicticus potto*), *Anat. Rec.* **144**:279.

Rabinovitch, M., 1970, Phagocytic recognition, in: *Mononuclear Phagocytes* (R. van Furth, ed.), pp. 299–315, Blackwell, Oxford.

Ranvier, L., 1875, *Traité Technique d'Histologie*, Savy Edit., Paris.

Rausch, E., Kaiserling, E., and Goos, M., 1977, Langerhans cells and interdigitating reticulum cells in the thymus-dependent region in human dermatopathic lymphadenitis, *Virchows Arch. B* **25**:327.

Reams, W. M., Jr., and Tompkins, S. P., 1973, A developmental study of murine epidermal Langerhans cells, *Dev. Biol.* **31**:114.

Reyes, F., Gourdin, M. F., Farcet, J. P., Dreyfus, B., and Breton-Gorius, J., 1978, Synthesis of a peroxidase activity by cells of hairy cell leukemia: A study by ultrastructural cytochemistry, *Blood* **52**:537.

Richter, R., 1956, Studien zur Neurohistologie der nervösen vegetativen Peripherie der Haut bei verschiedenen chronisch-infektiösen Granulomen, mit besonderer Berücksichtigung der Langerhansschen Zellen, *Arch. Klin. Exp. Dermatol.* **202**:466.

Riley, P. A., 1966, Esterase in epidermal dendritic cells of the mouse. A study of the histochemical properties and distribution of activity of the enzyme in relation to patterning in the tail, *Br. J. Dermatol.* **78**:388.

Rode, H., and Gordon, J., 1974, Macrophages in the mixed leukocyte culture reaction (MLC), *Cell. Immunol.* **13**:87.

Rosenstreich, D. L., and Oppenheim, J. J., 1976, The role of macrophages in the activation of T and B lymphocytes *in vitro*, in: *Immunobiology of the Macrophage* (D. S. Nelson, ed.), pp. 161–199, Academic Press, New York.

Rosenthal, A. S., Barcinski, M. A., and Blake, J. T., 1977, Determinant selection is a macrophage dependent immune response gene function, *Nature* **267**:156.

Rouse, R. V., van Ewijk, W., Jones, P. P., and Weissman, I., 1979, Expression of MHC antigens by mouse thymic dendritic cells, *J. Immunol.* **122**:2508.

Rowden, G., 1977, Immuno-electron microscopic studies of surface receptors and antigens of human Langerhans cells, *Br. J. Dermatol.* **97**:593.

Rowden, G., and Lewis, M. G., 1976, Langerhans cells: Involvement in the pathogenesis of mycosis fungoides, *Br. J. Dermatol.* **95**:665.

Rowden, G., Lewis, M. G., and Sullivan, A. K., 1977, Ia antigen expression on human epidermal Langerhans cells, *Nature* **268**:247.

Rowden, G., Phillips, T. M., and Lewis, M., 1978, Isolation studies of human Langerhans cells using anti-Ia antisera. Presented at Eastern Regional meeting of Society for Investigative Dermatology, April.

Rowden, G., Philips, T. M., and Lewis, M. G., 1979a, Ia antigen on indeterminant cells of the epidermis. Immunoelectron microscopic studies of surface antigen, *Br. J. Dermatol.* **100**:531.

Rowden, G., Phillips, T. M., Lewis, M. G., and Wilkinson, R. D., 1979b, Target role of Langerhans cells in mycosis fungoides: Transmission and immunoelectron microscopic studies, *J. Cutaneous Pathol.* **6**:364.

Ruhl, H., and Shevach, E. M., 1975, The effect of alloantisera on antigen-induced T cell proliferation, *J. Immunol.* **115**:1493.

Sagebiel, R. W., 1972, *In vivo* and *in vitro* uptake of ferritin by Langerhans cells of the epidermis, *J. Invest. Dermatol.* **58**:47.

Sagebiel, R. W., and Reed, T. H., 1968, Serial reconstruction of the characteristic granule of the Langerhans cell, *J. Cell Biol.* **36**:595.

Sagebiel, R. W., Hutchens, L. H., and Clarke, M. A., 1971, Non-keratinocytes of the oral mucosa: Melanocytes and Langerhans cells, in: *The Histology of Oral Mucosa* (C. A. Squier and J. Meyer, eds.), Thomas, Springfield.

Sanel, F. T., and Serpick, A. A., 1970, Plasmalemmal and subsurface complexes in human leukemia cells: Membrane bonding by zipper-like junctions, *Science* **168**:1458.

Schellander, F., and Wolff, K., 1967, Zur autoradiographischen Markierung von Langerhans Zellen mit ^{3}H-Thymidin, *Arch. Klin. Exp. Dermatol.* **230**:140.

Schroeder, H., and Theilade, J., 1966, Electron microscopy of normal human gingival epithelium, *J. Periodontal Res.* **1**:95.

Schwartz, R. H., David, C. S., Sachs, D. H., and Paul, W. E., 1976a, T lymphocyte-enriched murine peritoneal exudate cells. III. Inhibition of antigen-induced T lymphocyte proliferation with anti-Ia antisera, *J. Immunol.* **117**:531.

Schwartz, R. H., Dickler, H. B., Sachs, D. H., and Schwartz, B. D., 1976b, Studies of Ia antigens on murine peritoneal macrophages, *Scand. J. Immunol.* **5**:731.

Schwartz, R., David, C., Dorf, M., Benacerraf, B., and Paul, W. E., 1978a, Inhibition of dual Ir Gene-controlled T-lymphocyte proliferative response to poly(Glu^{56} Lys^{35} Phe^{9})^{n} with anti-Ia antisera directed against products of either I-A of I-C subregion, *Proc. Natl. Acad. Sci.* **75**:2387.

Schwartz, R., Yano, A., and Paul, W. E., 1978b, Interaction between antigen-presenting cells and primed T lymphocytes: an assessment of Ir gene expression in the antigen-presenting cell, *Immunol. Rev.* **40**:153.

Schwartz, R., Yano, A., Stimpfling, J., and Paul, W., 1979, Gene complementation in the T-lymphocyte proliferative response to poly(Glu^{55} Lys^{36} Phe^{9})^{n}, *J. Exp. Med.* **149**:40.

Schweizer, J., 1977, Langerhans cells and epidermal keratinization, *J. Invest. Dermatol.* **68**:250.

Schweizer, J., and Marks, F., 1977, A developmental study of the distribution and frequency of Langerhans cells in relation to formation of patterning in mouse tail epidermis, *J. Invest. Dermatol.* **69**:198.

Shelley, W. B., and Juhlin, L., 1976, Langerhans cells form a reticuloepithelial trap for external contact antigens, *Nature* **261**:46.

Shelley, W. B., and Juhlin, L., 1977a, Selective uptake of contact allergens by the Langerhans cell, *Arch. Dermatol.* **113**:187.

Shelley, W. B., and Juhlin, L., 1977b, A reticuloepithelial system: Cutaneous trap for antigens, in: *Transactions of the Association of American Physicians,* Vol. 89, pp. 245–253, Dornan, Collingdale.

Shelley, W. B., and Juhlin, L., 1978, The Langerhans cell: Its origin, nature and function, *Acta Derm.-Venereol.* **58**, *Suppl.* **79**:7.

Shreffler, D. C., and David, C. S., 1975, The H-2 major histocompatibility complex and the I immune region: genetic variation, function and organization, *Adv. Immunol.* **20**:125.

Silberberg, I., 1971a, Ultrastructural studies of Langerhans cells in contact sensitivity and primary irritant reactions to mercuric chloride, *Clin. Res.* **19**:715.

Silberberg, I., 1971b, Studies by electron microscopy of epidermis after topical application of mercuric chloride. Morphologic and histochemical findings in epidermal cells of human subjects who do not show allergic sensitivity or primary irritant reactions to mercuric chloride (0.1%), *J. Invest. Dermatol.* **56**:147.

Silberberg, I., 1973, Apposition of mononuclear cells to Langerhans cells in contact allergic reactions. An ultrastructural study, *Acta Derm.-Venereol.* **53**:1.

Silberberg, I., Prutkin, L., and Leider, M., 1969, Electron microscopic studies of transepidermal absorption of mercury. Histochemical methods for demonstration of electron densities in mercury-treated skin, *Arch. Environ. Health* **19**:7.

Silberberg, I., Baer, R. L., and Rosenthal, S. A., 1974a, Circulating Langerhans cells in a dermal vessel, *Acta Dermatovener.* **54**:81.

Silberberg, I., Baer, R. L., and Rosenthal, S. A., 1974b, The role of Langerhans cells in contact allergy. I. An ultrastructural study in actively induced contact dermatitis in guinea pigs, *Acta Derm.-Venereol.* **54**:321.

Silberberg, I., Baer, R. L., Rosenthal. S. A., Thorbecke, G. J., and Berezowsky, V., 1975a,

Langerhans cells in skin, lymphatic-like vessels and draining lymph nodes in allergic contact sensitivity, *Clin. Res.* **23**:231A.

Silberberg, I., Baer, R. L., Rosenthal, S. A., Thorbecke, G. J., and Berezowsky, V., 1975b, Dermal and intravascular Langerhans cells at sites of passively induced allergic contact sensitivity, *Cell. Immunol.* **18**:435.

Silberberg, I., Baer, R. L., and Rosenthal, S. A., 1976, The role of Langerhans cells in allergic contact hypersensitivity. A review of findings in man and guinea pigs, *J. Invest. Dermatol.* **66**:210.

Silberberg-Sinakin, I., and Thorbecke, G. J., 1970, Contact hypersensitivity and Langerhans cells, *J. Invest. Dermatol.* **75**:(in press).

Silberberg-Sinakin, I., Thorbecke, G. J., Baer, R. L., Rosenthal, S. A., and Berezowsky, V., 1976, Antigen-bearing Langerhans cells in skin, dermal lymphatics and in lymph nodes, *Cell. Immunol.* **25**:137.

Silberberg-Sinakin, I., Fedorko, M. E., Baer, R. L., Rosenthal, S. A., Berezowsky, V., and Thorbecke, G. J., 1977, Langerhans cells: target cells in immune complex reactions, *Cell. Immunol.* **32**:400.

Silberberg-Sinakin, I., Baer, R. L., and Thorbecke, G. H., 1978, Langerhans cells. A review of their nature with emphasis on their immunologic functions, in: *Progress in Allergy* (P. Kallos, B. H. Waksman, and A. L. de Weck, eds.), Vol. 24, pp. 268–294, Karger, Basel.

Silvers, W. K., 1957, A histochemical and experimental approach to determine the relationship between gold-impregnated dendritic cells and melanocytes, *Am. J. Anat.* **100**:225.

Sjöborg, S., Axelsson, S., Falck, B., Jacobsson, S., and Ringberg, A., 1978, A new method for the visualization of the epidermal Langerhans cell and its application on normal and allergic skin, *Acta Derm.-Venereol.* **58**, Suppl. **79**:23.

Snell, R. S., 1965, An electron microscopic study of the dendritic cells in the basal layer of guinea pig epidermis, *Z. Zellforsch. Mikrosk. Anat.* **66**:457.

Stecher, V. J., and Thorbecke, G. J., 1967, Sites of synthesis of serum proteins. III. Production of β1C, β1E, and transferrin by primate and rodent cell lines, *J. Immunol.* **99**:660.

Steinman, R. M., and Cohn, Z. A., 1972, The interaction of particulate horseradish peroxidase (HRP)-anti HRP immune complexes with mouse peritoneal macrophages *in vitro*, *J. Cell. Biol.* **55**:616.

Steinman, R. M., and Cohn, Z. A., 1973, Identification of a novel cell type in peripheral lymphoid organs of mice. I. Morphology, quantification, tissue distribution, *J. Exp. Med.* **137**:1142.

Steinman, R. M., and Cohn, Z. A., 1974, Identification of a novel cell type in peripheral lymphoid organs of mice. II. Functional properties *in vitro*, *J. Exp. Med.* **139**:380.

Steinman, R. M., and Cohn, Z. A., 1975, A novel adherent cell in mouse lymphoid organs, in: *Immune Recognition* (A. S. Rosenthal, ed.), pp. 571–587, Academic Press, New York.

Steinman, R. M., and Witmer, M., 1978, Lymphoid dendritic cells are potent stimulators of the primary mixed leukocyte reaction in mice, *Proc. Natl. Acad. Aci.* **75**:5132.

Steinman, R. M., Lustig, D., and Cohn, Z. A., 1974, Identification of a novel cell type in peripheral lymphoid organs of mice. III. Functional properties *in vivo*, *J. Exp. Med.* **139**:1431.

Steinman, R. M., Adams, J., and Cohn, Z. A., 1975, Identification of a novel cell type in peripheral lymphoid organs of mice. IV. Identification and distribution in mouse spleen, *J. Exp. Med.* **141**:804.

Steinman, R. M., Kaplan, G., Witmer, M., and Cohn, Z. A., 1979, Identification of a novel cell in the peripheral lymphoid organs of mice. V. Purification of spleen dendritic cells, new surface markers, and maintenance *in vivo*, *J. Exp. Med.* **149**:1.

Stingl, G., Wolff-Schreiner, E. C. H., Pichler, W. J., Gschnait, F., Knapp, W., and Wolff, K., 1977, Epidermal Langerhans cells bear Fc and C3 receptors, *Nature* **268**:245.

Stingl, G., Katz, S. I., Abelson, L. D., and Mann, D. L., 1978a, Immunofluorescent detection of human B cell alloantigens on S-Ig-positive lymphocytes and epidermal Langerhans cells, *J. Immunol.* **120**:661.

Stingl, G., Katz, S. I., Shevach, E. M., Rosenthal, A. S., and Green, I., 1978b, Analogous functions of macrophages and Langerhans cells in the initiation of the immune response, *J. Invest. Dermatol.* **71**:59.

Stingl, G., Katz, S. I., Shevach, E. M., Wolff-Schreiner, E., and Green, I., 1978c, Detection of Ia antigens on Langerhans cells in guinea pig skin, *J. Immunol.* **120**:570.

Stingl, G., Katz, S. A., Clement, L., Green, I., and Shevach, E., 1978d, Immunologic functions of Ia bearing epidermal Langerhans cells, *J. Immunol.* **121**:2005.
Streilein, J. W., Toews, G. T., and Bergstresser, P. R., 1979, Corneal allografts fail to express Ia antigens, *Nature* **282**:326.
Sulzberger, M. B., 1929, Hypersensitiveness to arsphenamine in guinea pigs. I. Experiments in prevention and in desensitization, *Arch. Derm. Syph.* **20**:669.
Szakal, A. K., and Hanna, M. G., 1968, The ultrastructure of antigen localization and virus-like particles in mouse spleen germinal centers, *Exp. Mol. Pathol.* **8**:75.
Takaki, Y., and Miyazaki, H., 1976, Cytolytic degeneration of keratinocytes adjacent to Langerhans cells in pityriasis rosea (Gibert), *Acta Derm.-Venereol.* **56**:99.
Tamaki, K., Stingl, G., Gullino, M., Sachs, D. H., and Katz, S. I., 1979, Ia antigens in mouse skin are predominantly expressed on Langerhans cells, *J. Immunol.* **123**:784.
Tan, R. S. H., Butterworth, C. M., McLaughlin, H., Malka, S., and Samman, P. D., 1974, Mycosis fungoides—a disease of antigen persistence, *Br. J. Dermatol.* **91**:607.
Tarnowski, W. M., and Hashimoto, K., 1967, Langerhans cell granules in histiocytosis X. The epidermal Langerhans' cell as a macrophage, *Arch. Dermatol.* **96**:298.
Thomas, D. W., and Shevach, E. M., 1976, Nature of the antigenic complex recognized by T lymphocytes. I. Analysis with an *in vitro* primary response to soluble protein antigens, *J. Exp. Med.* **144**:1263.
Thomas, D. W., Forni, G., Shevach, E. M., and Green, I., 1977a, The role of the macrophage as the stimulatory cell in contact sensitivity, *J. Immunol.* **118**:1677.
Thomas, D. W., Yamashita, U., and Shevach, E. M., 1977b, The role of Ia antigens in T cell activation, *Immunol. Rev.* **35**:97.
Thorbecke, G. J., Hochwald, G. M., van Furth, R., Muller-Eberhard, H. J., and Jacobson, E. B., 1965, Problems in determining the site of synthesis of complement components, in: *Ciba Foundation Symposium on Complement* (G. E. W. Wolstenholme and J. Knight, eds.), pp. 99–115, Little Brown, Boston.
Tusques, J., and Pradal, G., 1969, Analyse tridimensionnelle des inclusions recontrées dans les histiocytes de l'histiocytose "X," en microscopie électronique. Comparaison avec les inclusions des cellules de Langerhans, *J. Microscopie* **8**:113.
Umetsu, D. T., Lerman, S. P., and Thorbecke, G. J., 1979, Accessory cell requirements for lymphoma growth *in vitro* and in irradiated mice, *Cell. Immunol.* **42**:139.
Unanue, E. R., 1972, The regulatory role of macrophages in antigenic stimulation, in: *Advances in Immunology*, Vol. 15, (F. J. Dixon and H. Kunkel, eds.), pp. 95–165, Academic Press, New York.
Unanue, E. R., 1976, Secretory function of mononuclear phagocytes, *Am. J. Pathol.* **83**:395.
Vadas, M. A., Miller, J. F. A. P., McKenzie, I. F. C., Chism, S. E., Shen, F. W., Boyse, E. A., Gamble, J. R., and Whitelaw, A. M., 1976, Ly and Ia antigen phenotypes of T cells involved in delayed-type hypersensitivity and in suppression, *J. Exp. Med.* **144**:10.
van den Tweel, J. G., and Walker, W. S., 1977, Macrophage-induced thymic lymphocyte maturation, *Immunology* **33**:817.
van Dorp, D., 1971, Recent developments in the biosynthesis and the analyses of prostaglandins, *Ann. N.Y. Acad. Sci.* **180**:181.
van Ewijk, W., Verzijden, J., and van der Kwast, T. H., and Luijex-Meijer, S., 1974, Reconstitution of the thymus dependent area in the spleen of lethally irradiated mice, a light and electron microscopical study of the T-cell microenvironment, *Cell Tiss. Res.* **149**:43.
van Furth, R., 1970, The origin and turnover of promonocytes, monocytes and macrophages in normal mice, in: *Mononuclear Phagocytes* (R. van Furth, ed.), pp. 151–165, Davis, Philadelphia.
van Furth, R., and Thompson, J., 1971, Review of the origin of kinetics of the promonocytes, monocytes, and macrophages and a brief discussion of the mononuclear phagocyte system, *Ann. Inst. Pasteur* **120**:337.
van Furth, R., Langevoort, H. L., and Schaberg, A., 1975, Mononuclear phagocytes in human pathology—proposal for an approach to improved classification, in: *Mononuclear Phagocytes in Immunity, Infection and Pathology* (R. van Furth, ed.), pp. 1–15, Blackwell, Oxford.
van Haelst, U. J. G., 1969, Light and electron microscopic study of the normal and pathological thymus of the rat. III. A mesenchymal histiocytic type of cell, *Z. Zellforsch. Mikrosk. Anat.* **99**:198.

van Rood, J. J., 1977, The relevance of HLA-D in renal transplantation, *Excerpta Medica Intl. Congr. Series* No. 423, 107.

Veerman, A., 1974, On the interdigitating cells in the thymus-dependent area of the rat spleen: A relation between the mononuclear phagocyte system and T-lymphocytes, *Cell Tiss. Res.* **148**:247.

Veldman, J. E., 1970, Histophysiology and electron microscopy of the immune response. Ph.D. Thesis, Groningen, The Netherlands.

Veldman, J. E., Molenaar, I., and Keuning, F. J., 1978, Electron microscopy of cellular immunity reactions in B-cell deprived rabbits, *Virchows Arch. B* **28**:217.

Vernon, M. L., Fountain, L., Krebs, H. M., Horta-Barbosa, L., Fuccillo, D. A., and Sever, J. L., 1973, Birbeck granules (Langerhans' cell granules) in human lymph nodes, *Am. J. Clin. Pathol.* **60**:771.

Volkman, A., and Gowans, J. L., 1968, The origin of macrophages from bone marrow in the rat, *Br. J. Exp. Pathol.* **46**:62.

Wachsmuth, E. D., 1975, Aminopeptidase as a marker for macrophage differentiation, *Exp. Cell Res.* **96**:409.

Waterhouse, J. P., and Squier, C. A., 1967, The Langerhans cell in human gingival epithelium, *Arch. Oral Biol.* **12**:341.

Weiss, L. W., and Zelickson, A. S., 1975a, Embryology of the epidermis: ultrastructural aspects. II. Period of differentiation in the mouse with mammalian comparisons, *Acta Derm.-Venereol.* **55**:321.

Weiss, L. W., and Zelickson, A. S., 1975b, Embryology of the epidermis: Ultrastructural aspects. III. Maturation and primary appearance of dendritic cells in the mouse with mammalian comparisons, *Acta Derm.-Venereol.* **55**:431.

Wiedmann, A., 1952, Zur Frage der sogenannten Langerhans-Zellen der Haut, *Hautarzt* **3**:249.

Williams, G. M., 1966, Antigen localization in lymphopenic states. II. Further studies on whole body X-irradiation, *Immunology* **11**:475.

Wolff, K., 1963, Histologische Beobachtungen an der normalen menschlichen Haut bei der Durchführung fermenthistochemischer Untersuchungen mit Adenosintriphosphat als Substrat, *Arch. Klin, Exp. Dermatol.* **216**:1.

Wolff, K., 1964a, Über die Adenosintriphosphatataseaktivität der menschlichen Haut, *Arch. Klin. Exp. Dermatol.* **218**:254.

Wolff, K., 1964b, Zur Enzymaktivität in den Langerhansschen Zellen, *Arch. Klin. Exp. Dermatol.* **218**:446.

Wolff, K., 1967a, Die Langerhans-Zelle. Ergebnisse neuerer experimenteller Untersuchungen, I und II, *Arch. Klin. Exp. Dermatol.* **229**:54.

Wolff, K., 1967b, The fine structure of the Langerhans cell granule, *J. Cell. Biol.* **35**:466.

Wolff, K., 1972, The Langerhans cell, in: *Current Problems in Dermatology*, Vol. 4 (J. W. H. Mali, ed.), pp. 79–145, Karger, Basel.

Wolff, K., and Honigsman, H., 1971, Permeability of the epidermis and the phagocytic activity of keratinocytes. Ultrastructural studies with Thorotrast as a marker, *J. Ultrastruct. Res.* **36**:176.

Wolff, K., and Schreiner, E., 1970, Uptake, intracellular transport and degradation of exogenous protein by Langerhans cells. An electron microscopic cytochemical study using peroxidase as tracer substances, *J. Invest. Dermatol.* **54**:37.

Wolff, K., and Winkelmann, R. K., 1967a, Ultrastructural localization of nucleoside triphosphatase in Langerhans cells, *J. Invest. Dermatol.* **48**:50.

Wolff, K., and Winkelmann, R. K., 1967b, Quantitative studies on the Langerhans cell population of guinea pig epidermis, *J. Invest. Dermatol.* **48**:504.

Wolff, K., and Winkelmann, R. K., 1967c, Non-pigmentary enzymes of the melanocyte-Langerhans cell system, in: *Advances in Biology of Skin*, Vol. 8, *The Pigmentary System* (W. Montagna and F. Hu, eds.), pp. 135–167, Pergamon, New York.

Wolff, K., and Winkelmann, R. K., 1967d, The influence of ultraviolet light on the Langerhans cell population and its hydrolytic enzymes in guinea pigs, *J. Invest. Dermatol.* **48**:531.

Wong, Y. C., and Buck, R. C., 1971, Langerhans cells in epidermoid metaplasia, *J. Invest. Dermatol.* **56**:10.

Yasuda, K., Aoki, T., and Montagna, W., 1961, The skin of primates. IV. The skin of the lesser bushbaby (*Galago senegalensis*), *Am. J. Phys. Anthropol.* **19**:23.

Younes, M. S., Robertson, E. M., and Bencosne, S. A., 1968, Electron microscopic observations on Langerhans cells in the cervix, *Am. J. Obstet. Gynecol.* **102**:397.

Zelickson, A. S., 1965, The Langerhans Cell, *J. Invest. Dermatol.* **44**:201.

Zelickson, A. S., 1966, Granule formation in the Langerhans cell, *J. Invest. Dermatol.* **47**:498.

Zelickson, A. S., and Mottaz, J. H., 1968a, Epidermal dendritic cells. A quantitative study. *Arch. Dermatol.* **98**:652.

Zelickson, A. S., and Mottaz, J. H., 1968b, Localization of gold chloride and adenosine triphosphatase in human Langerhans cells, *J. Invest. Dermatol.* **51**:365.

Zelickson, A. S., and Mottaz, J. H., 1970, The effect of sunlight on human epidermis, *Arch. Dermatol.* **101**:312.

Zurier, R. B., 1974, Prostaglandins, in: *Mediators of Inflammation* (G. Weissmann, ed.), pp. 163–180, Plenum Press, New York.

16

Pulmonary Macrophages

DRUMMOND H. BOWDEN
and IAN Y. R. ADAMSON

1. INTRODUCTION

Pulmonary macrophages are essential components of respiratory defense. As outriders of the overall mononuclear phagocyte system of the body they occupy a special position because the lung, with some 90 m^2 exposed to the environment, is vulnerable to injury by all manner of airborne agents. This vast surface is protected by a primary filtration system in the nose and airways which effectively prevents the majority of particles larger than 3 μm in diameter from reaching the air sacs. Harmful gases are also adsorbed to a significant degree. Within the alveolus, however, it is the macrophage that scavenges and protects the thin epithelium covering the pulmonary capillaries.

The existence of free alveolar cells has been known for many years. At the turn of the century, pathologists recognized the phagocytic potential of these cells by the presence of dust and blood pigments within their cytoplasm. Similarly, a presumptive route of elimination was identified when macrophages were found in the sputum. Only recently has the relationship of these free phagocytes to other pulmonary cells and to the cells of Aschoff's RES been determined. The common origin of all cells of the mononuclear phagocytic system from the bone marrow is acknowledged and, in general terms, this relationship holds for the pulmonary macrophages. The differences between the majority of tissue macrophages and the alveolar phagocytes appear to be related to the distinctive functional milieu of the lung. In contrast to the phagocytes of the liver sinusoids and the peritoneum, the alveolar macrophages operate in an aerobic atmosphere. In addition, faced with a constant threat of an unexpectedly high load of inhaled particulates the system must respond quickly with an enormous increase in cellular output. It appears that these functions are supported by a labile population of interstitial macrophages, a cellular compartment between the cir-

DRUMMOND H. BOWDEN and IAN Y. R. ADAMSON • Department of Pathology, University of Manitoba, Winnipeg, Canada R3E 0W3.

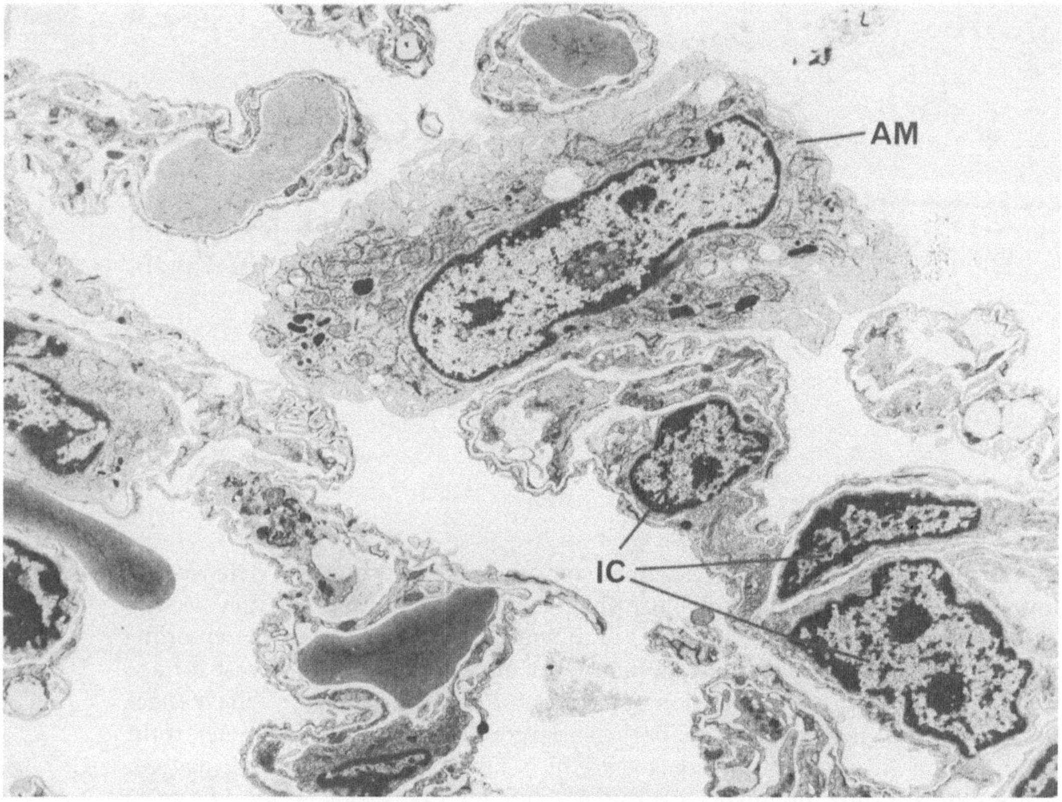

FIGURE 1. Alveolar macrophage (AM) lies free in the airspace; in the septa, several intersitial cells (IC) are seen.

culating blood monocytes and the free alveolar cells. In this compartment, cells may undergo biochemical adaptation prior to emigration into the air sacs and, in response to demand, adaptive proliferation of interstitial cells may supplement the normal population of alveolar macrophages. Thus, there are two components of the pulmonary macrophagic system, interstitial cells and free alveolar phagocytes (Figure 1).

2. STRUCTURE AND FUNCTION

The term *macrophage* embraces a diversity of cells but there are certain features, related to phagocytosis and to intracellular digestion, that are common to all macrophages whatever their origin. The restricted view of the alveolar macrophage obtained by the microscopic examination of lung sections provides only a fragmentary insight into the morphological variety of these cells. Detailed study of the morphology, metabolism, and function of alveolar macrophages is best performed on cells obtained by pulmonary lavage. Brain *et al.* (1977) have shown that in eight mammalian species studied, yields of 3–15 million cells per gram of lung may be obtained by this method.

The predominant morphological feature of these lavaged cells is their heterogeneity (Figure 2). This structural variety is dependent largely upon the age of the cell, the phase of its cellular activity, and the nature and quantity of material that has been phagocytized. Immature nonworking cells exhibit abundant smooth endoplasmic reticulum (SER), whereas cells that have engulfed material from the alveoli show a pronounced increase in rough endoplasmic reticulum (RER) and in the number of lysosomes. Lysosomal inclusions may include dust particles, bacteria, red cells, and cellular debris (Figure 2). Lamellar and lattice configurations observed within these cells are probably derived from the Type 2 cells as they spill their surfactant phospholipids on to the surface of the alveoli. The structural heterogeneity of the alveolar macrophage is a reflection of the metabolic and synthetic activities of the cell. Following the ingestion of particles such as carbon or iron oxide, there is, within 8 hr, a dramatic shift towards lysosomal production (Sorokin and Brain, 1975). In the unstimulated

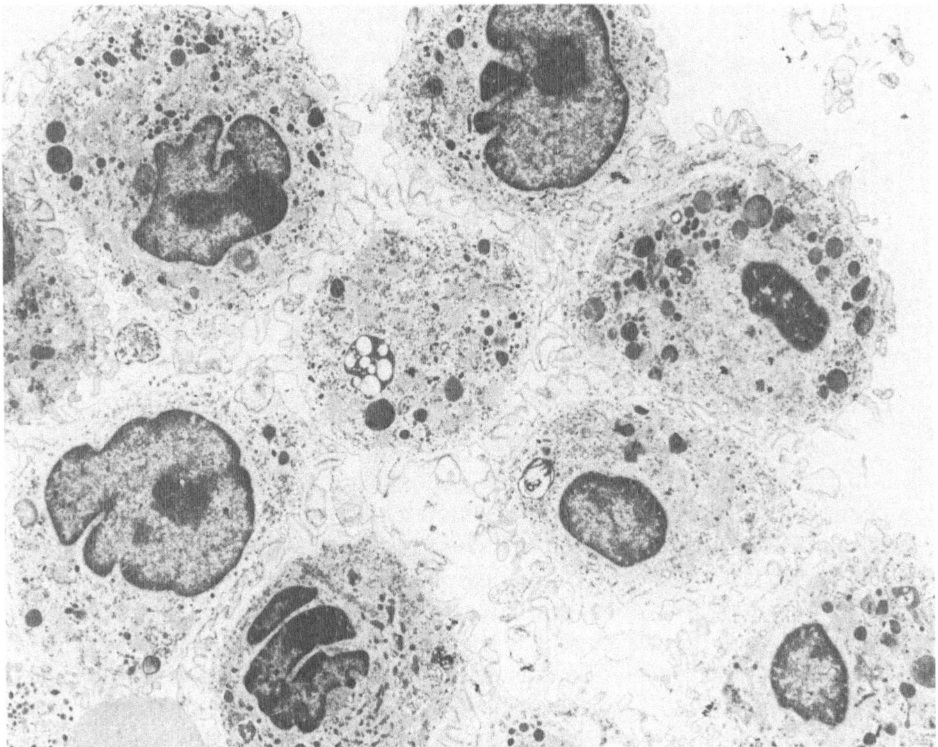

FIGURE 2. Alveolar macrophages lavaged from the lungs of a normal mouse. The cells are heterogeneous due to the variable number, size, and content of phagolysosomes.

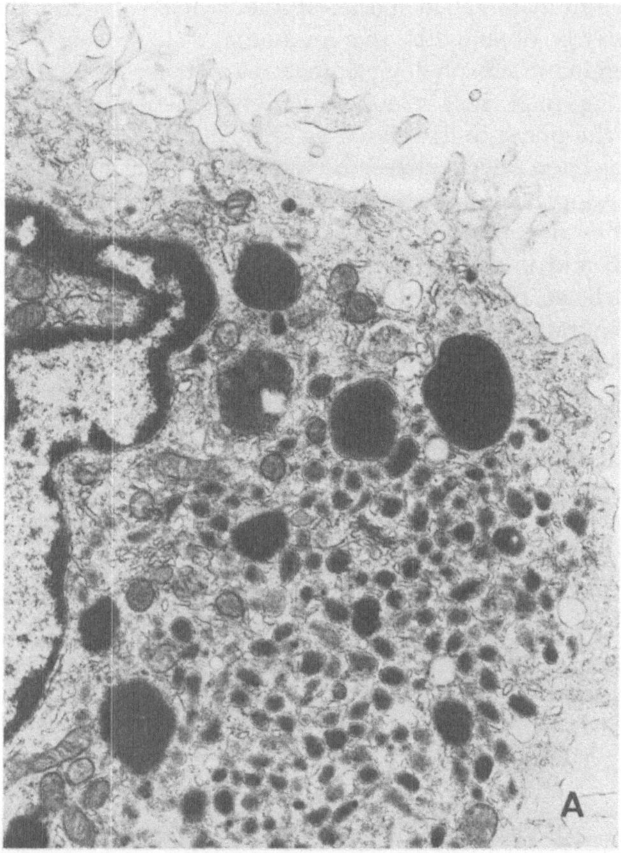

FIGURE 3. Cytoplasm of an alveolar macrophage. (A) Conventional preparation showing many membrane bound lysosomes. (B) Acid phosphatase preparation showing enzyme activity in the tubular network of lysosomes.

cell, acid phosphatase activity is almost exclusively perinuclear in distribution; after phagocytosis, the enzyme is more generally distributed throughout the cell.

Sorokin (1976) has described a branched system of smooth-surfaced tubules within the cytoplasm of the alveolar macrophage. This system, which appears to produce lysosomes, becomes particularly prominent within two hours of phagocytosis. Though not exclusive to the alveolar macrophage the tubular network is more readily detected in this cell than in macrophages from other tissues. A recent paper by Essner and Haimes (1977) describes a similar system in the alveolar macrophage of the beige mouse, a mutant considered to be homologous to human Chediak–Higashi syndrome. The authors emphasize the identity of the membranous array with GERL (Golgi apparatus, endoplasmic reticulum, lysosomes) which is thought to represent a specialized area of lysosome-producing endoplasmic reticulum (Novikoff *et al.*, 1971). The prominence of this system in the alveolar macrophage may be a reflection of the

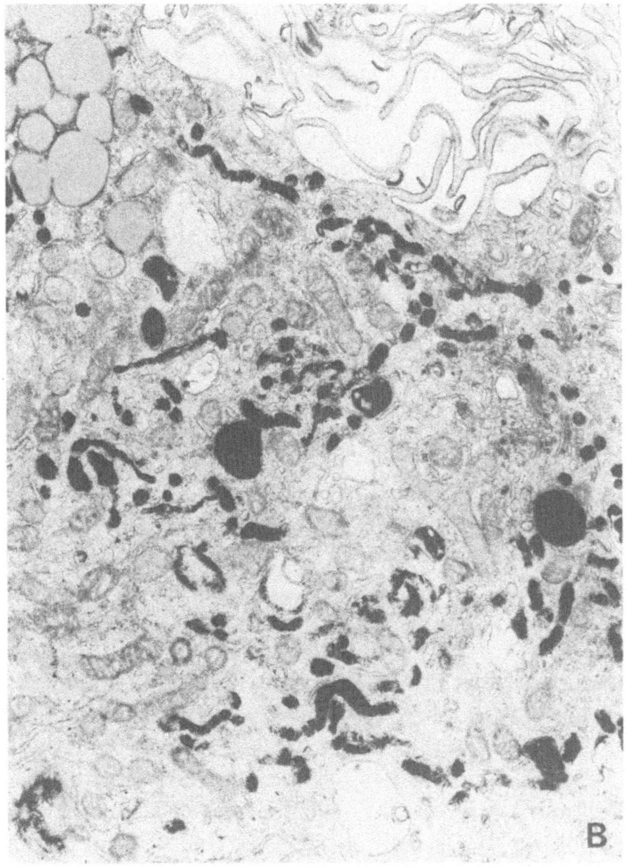

Figure 3 (*Continued*).

demands placed upon this cell for the rapid synthesis and packaging of acid hydrolases under load conditions (Figure 3).

The differences in enzyme activity between alveolar macrophages and phagocytes from other parts of the body appear to be related to the immediate and continuing functional demands placed upon the alveolar cell because of its environmental habitat. Alveolar macrophages contain considerably more acid phosphatase, lysozyme, and lipase than peritoneal macrophages (Cohn and Weiner, 1963a,b; Myrvik *et al.*, 1961). The importance of microbial phagocytosis to the initiation of enzyme synthesis in these cells has been emphasized by Myrvik and his associates. Enzyme activity in macrophages obtained from germ-free animals is much lower than that found in the cells of conventionally raised animals (Heise and Myvrik, 1966); similarly, the enzymatic content of the alveolar macrophage is much greater in the fall and winter than in spring and summer (Heise *et al.*, 1965). In addition to enzyme synthesis related predominantly to phagocytosis and lysosomal digestion, the alveolar macrophage, like its counterparts in other tissues, produces a variety of proteolytic enzymes, including elastases and collagenases, which are secreted externally (Werb and Gordon, 1975a,b).

The geographic situation of the alveolar macrophage, on the outer surface of the air–blood barrier has functional connotations also. The continuing process of phagocytosis and mucociliary clearance through the airways ensures the removal of inhaled particulates from the alveoli and maintains pulmonary sterility. Phagocytosis is also an important step in the induction of an immune response and, as the alveolar scavenger of inhaled antigens, the macrophage is uniquely situated for this initial manipulation. However, since these cells are eliminated through the bronchial passages and swallowed, it seems likely that most inhaled antigens are not available for interaction with lymphoid receptors within the interstitium of the lung. If this is so, the primary immunologic function of the alveolar macrophage may be to provide a filter for the protection of antibody-producing lymphoid cells against excess antigen which could induce immune paralysis.

3. THE INTERSTITIAL COMPARTMENT

It is widely held that the free alveolar macrophages, like the resident macrophages of tissues and serous cavities, are descendents of bone-marrow-derived blood monocytes (van Furth, 1970). Much of the evidence supporting this theory is based upon studies of monocytic egress during the inflammatory response (Volkman and Gowans, 1965). It is not known, however, if a similar mechanism operates under resting or basal conditions. Volkman (1976) has advanced persuasive evidence favoring a local origin for hepatic and peritoneal macrophages, and it seems likely that a similar mechanism pertains in the lung.

The free macrophages of the alveolar spaces, exposed directly to an aerobic and potentially hostile environment, differ in several ways from their counterparts in other organs and tissues. Whereas blood monocytes and tissue macrophages operate within an anaerobic milieu, the alveolar macrophage derives its energy aerobically. In addition, the defense of the air–blood barrier is dependent in no small measure upon the ability of the macrophagic system to respond to the demands of an increased inhalant load by a rapid outpouring of functioning cells.

The immediate proximity of the interstitial compartment of the lung to the air sacs suggests a relationship between the free alveolar phagocytes and the mononuclear cells of the interstitium. Casual microscopic examination of the lung suggests that the interstitium contains relatively few cells; quantitative studies have shown, however, that some 30% of lung cells reside in this compartment (Weibel *et al.*, 1976). Precise morphological and functional analysis of this population has not been accomplished, but there is little doubt that a significant proportion of interstitial cells are macrophages or precursors of macrophages. It has been postulated that the macrophage undergoes its biochemical adaptation to aerobic conditions in this compartment prior to its emigration into the air sacs (Bowden, 1971). Morphologic evidence favoring maturation within the interstitial compartment is scant indeed, since reliable markers capable of distinguishing macrophages from other mononuclear interstitial cells are not available. Schneeberger, quoted by Brain *et al.* (1977), has suggested that, in the guinea pig, interstitial macrophages may be distinguished by their peroxidase activity.

The concept of a compartment of interstitial cells which are the immediate antecedents of free alveolar macrophages is supported by observations in a number of clinical and experimental situations. In some viral pneumonias, proliferation of interstitial cells is a dominant feature and, in granulomatous conditions such as sarcoidosis, tuberculosis, and silicosis, and, experimentally, after the injection of Freund's adjuvant, proliferation of interstitial cells is accompanied by an efflux of free alveolar macrophages. Of particular interest in this context, is the observation that leukemic patients with no circulating monocytes are capable of maintaining a normal output of alveolar macrophages (Golde *et al.*, 1974). Observations such as this offer strong support to Volkman's tenet that under basal conditions the phagocytes of the lung are derived by division of resident macrophages with subsequent migration into the alveoli (Volkman, 1976).

Although these studies point to a strong kinship between interstitial cells and alveolar macrophages, none rule out the monocyte as a source of the free alveolar phagocyte either by direct migration or as an antecedent of the interstitial macrophage. If the monocyte is the immediate precursor of the alveolar macrophage, any reduction in the number of circulating monocytes should be followed within a short period by a fall in the output of alveolar cells. The relationship between the monocytes, the interstitial cells, and the free macrophages has been demonstrated in a study of cellular kinetics following whole-body irradiation (Bowden *et al.*, 1969). In this model, although the number of circulating monocytes falls significantly following irradiation, there is no concomitant or subsequent reduction in the output of alveolar macrophages. The number of macrophages recovered by lavage remains constant until the second week after irradiation, at which time a sharp peak of DNA synthesis is observed in the pulmonary interstitial cells. Within a few days, as the interstitial reaction subsides, an increase in the output of alveolar macrophages occurs. These observations strongly support the relationship of the interstitial cell to the macrophage, but they do not in any way negate the possibility of direct monocytic migration into the air sacs. It may be argued that, although the total mass of blood monocytes is reduced by irradiation, they are by no means eliminated; a continuing, though reduced, supply of monocytes is still present. This dilemma has been resolved by studying macrophage kinetics in a system which is completely free of circulating monocytes. Using cultured explants of murine lung, Bowden and Adamson (1972) have shown that, after an initial period of quiescence, the interstitial cells proliferate, an activity that is followed within 24 hr by the appearance of a new population of cells, identified by their ultrastructural features and phagocytic properties as macrophages (Figure 4). Cytokinetic studies using [^{3}H]thymidine confirm the interstitial origin of the free alveolar macrophages.

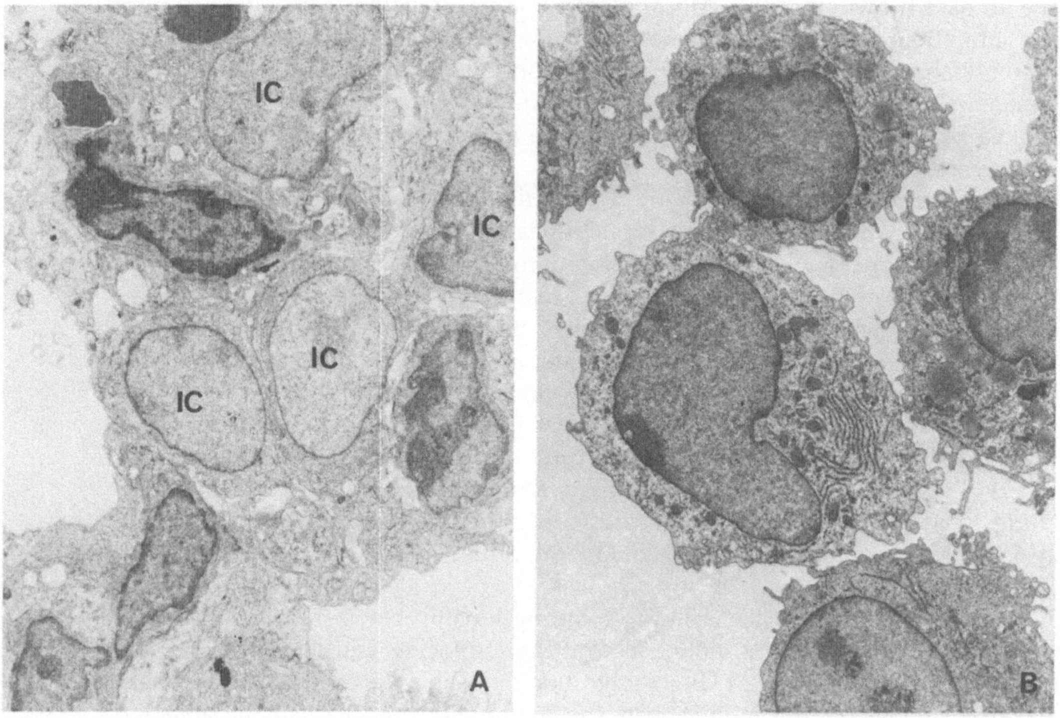

FIGURE 4. (A) Explant of mouse lung, cultured for 2 days, showing proliferation of interstitial cells (IC). (B) After 3 days of culture, a population of new macrophages are observed in the air spaces. (From Bowden, D. H., and Adamson, I. Y. R., 1972, The pulmonary interstitial cell as immediate precursor of the alveolar macrophage, *Am. J. Pathol.* **68**:521. With permission.)

It remains to be determined whether, under basal conditions, the population of free alveolar macrophages is maintained by mitosis of free cells or by continuing division and migration of interstitial cells. This question has been investigated in a modified organ culture system in which pulmonary explants, attached to cover glasses or cellophane, are maintained for periods up to two weeks (Bowden and Adamson, 1976). In this system, the explant becomes surrounded by a monolayer of migratory cells (Figure 5). These cells are identified as macrophages by their ability to phagocytize inert particles and bacteria, their adherence to glass and cellophane, and by their ultrastructural features (Figure 6). The free macrophages in the halo do not exhibit any mitotic activity, whereas the interstitial cells synthesize DNA and undergo mitosis.

The results of this experiment suggest that a continuing supply of alveolar macrophages is dependent upon the division and migration of interstitial cells, rather than the mitotic activity of the free macrophagic population. It would seem, therefore, that the free alveolar macrophage exists in the inactive G_0 phase of the mitotic cycle, an end-of-the-line cell destined to phagocytize and to be eliminated through the tracheobronchial tree. This may be the usual state of the free macrophages, but it is known that this cell may be induced to divide by the

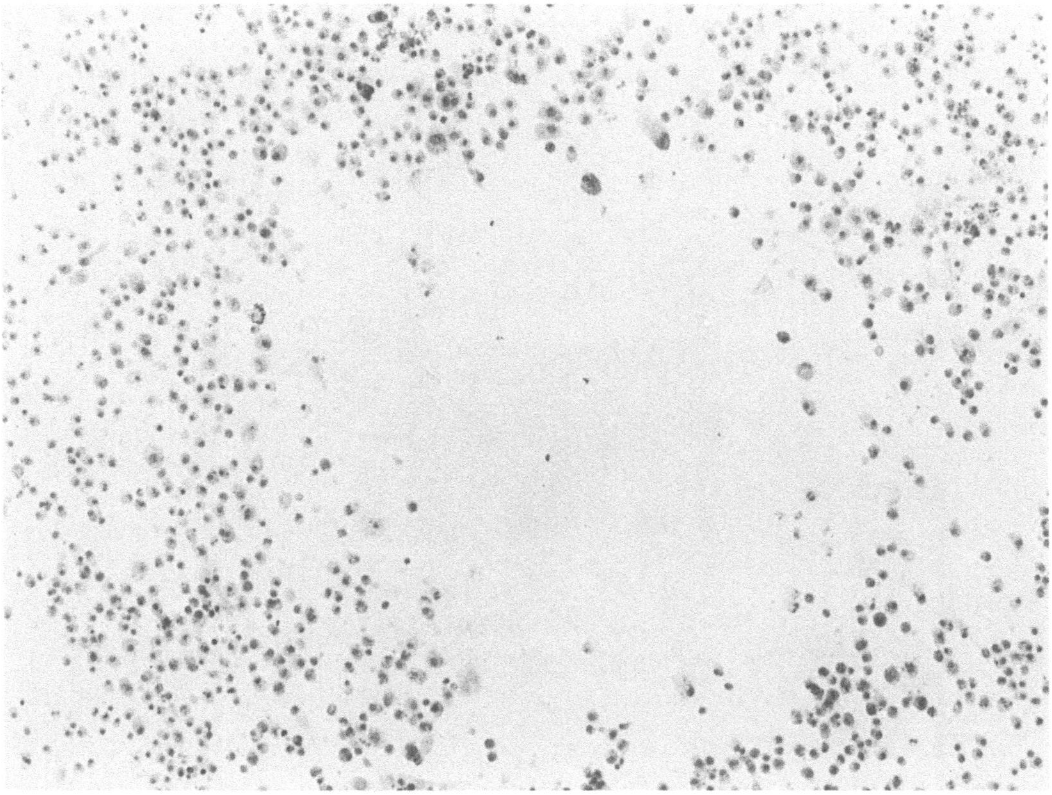

FIGURE 5. Light micrograph showing that, after 8 days of culture on a cover glass, a halo of macrophages migrates from a central explant, now removed.

addition of conditioned media and cell-free inflammatory exudates (Wynne *et al.*, 1975). Such an adaptive response suggests that, following alveolar injury, an inflammatory exudate of plasma proteins may induce *in situ* proliferation of free alveolar macrophages.

4. MACROPHAGES IN THE DEVELOPING LUNG

The conclusion that free alveolar macrophages are derived from precursors in the interstitium is further supported by studies of the developing lung. In the species that have been examined, macrophages are observed free in the terminal spaces only towards the end of gestation. For example, in the rat, which has a 22-day gestation period, no macrophages are seen in the developing airways before day 20. At this stage, when the rapid phase of epithelial cell division is over, DNA synthesis is predominant in interstitial cells; this mitotic activity coincides with the initial appearance of macrophages in the terminal sacs (Figure 7). Although it has been postulated that the alveolar macrophages develop their

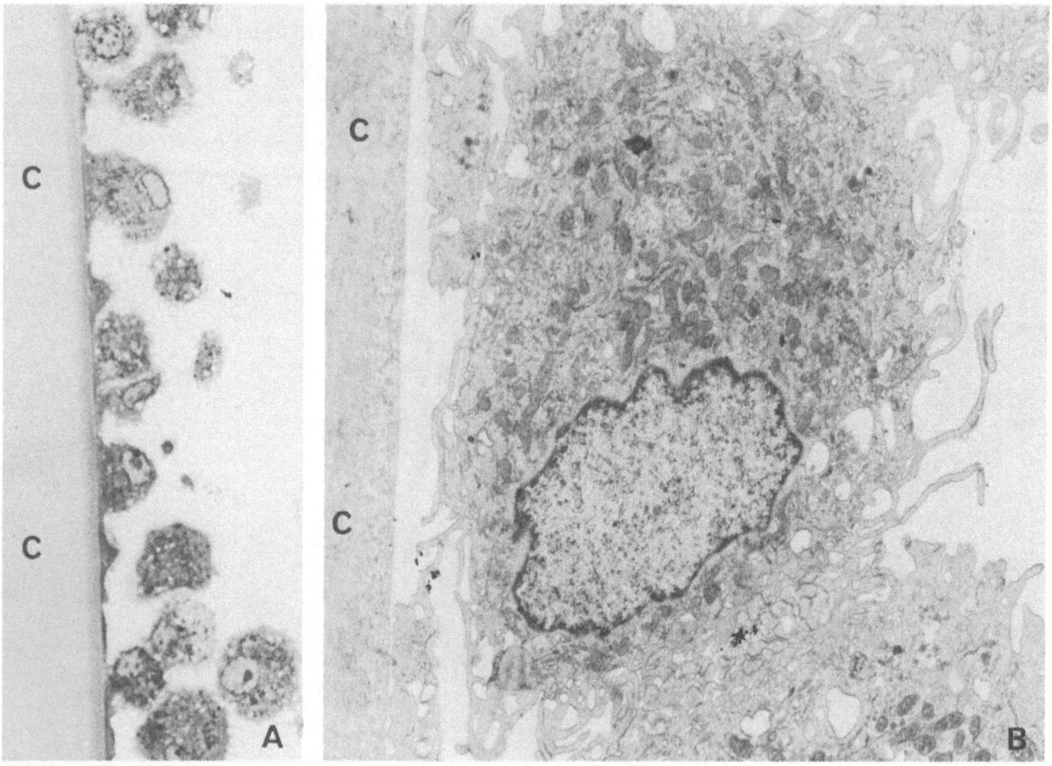

FIGURE 6. (A) Macrophages that migrated from an explant onto cellophane (C); 0.5-μm section. (B) Electron microscopy shows macrophages with numerous pseudopodia attached to the cellophane (C).

particular biochemical characteristics in an interstitial compartment before release into the air sacs, it appears that these fetal macrophages are not fully mature. When first observed at day 20, they have few pseudopodia and sparse cytoplasmic lysosomes. At this time epithelial cell differentiation is proceeding rapidly and many osmiophilic lamellar bodies of Type 2 cells are secreted into the fluid-filled sacs. In spite of the availability of material to be ingested, these young macrophages appear to be poor phagocytes (Figure 8). Subsequently, phagocytic activity increases and large quantities of lipid material are seen in the phagosomes. Similarly, in rabbits the prenatal alveolar macrophages resemble monocytes, but at birth they contain many myelin figures (Zeligs *et al.*, 1977). Additional evidence of functional immaturity at birth comes from studies on cellular motility of human monocytes; chemotactic activity of these cells is much lower in the newborn than in adults (Weston *et al.*, 1977).

The immediate postnatal period is characterized by a dramatic increase in the number of alveolar macrophages (Zeligs, 1977). This response appears to be related to the relatively high mitotic rate of interstitial cells as new alveoli develop (Adamson and Bowden, 1975). The ability to phagocytize and to digest bacteria also increases rapidly with age; in rabbits, adult levels of bactericidal

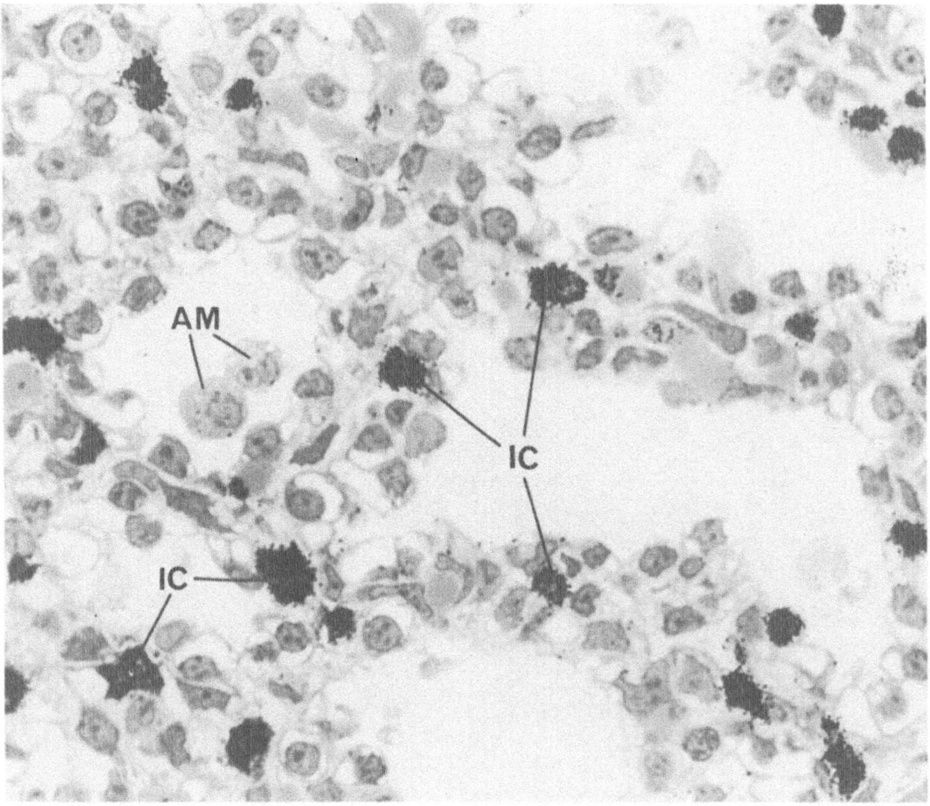

FIGURE 7. Autoradiograph (1-μm methacrylate section) of fetal rat lung at day 20 of gestation. Silver grains, indicative of [^{3}H]thymidine uptake, are over interstitial cells (IC), coincident with the initial appearance of free macrophages (AM) in the airways.

activity are achieved by four weeks (Nerurkar *et al.*, 1977). As macrophagic function develops, morphologic changes indicative of cellular maturation are observed. At 1–2 weeks postnatal, macrophages contain a well-developed Golgi system and RER; there is also a considerable increase in mitochondria and lysosomal bodies (Zeligs *et al.*, 1977). Further morphologic evidence of functional development has been obtained by cytochemistry; the acid phosphatase activity of 10-day-old rat macrophages is clearly increased over that of the prenatal cells (Figure 9) (Curle and Adamson, 1978).

These findings indicate that, in terms of cell number and functional immaturity, the poorly developed alveolar macrophage system at birth may contribute to the susceptibility of the newborn to infection. In this respect the fetal macrophage in its sterile environment resembles the macrophage of a germ-free animal which has not been stimulated to produce hydrolytic enzymes (Heise and Myrvik, 1966). Such enzymes are stimulated as required; for example, cells exposed to bacteria show increased synthesis of membranes and digestive enzymes (Cohn and Weiner, 1963a,b). It seems likely therefore that functional

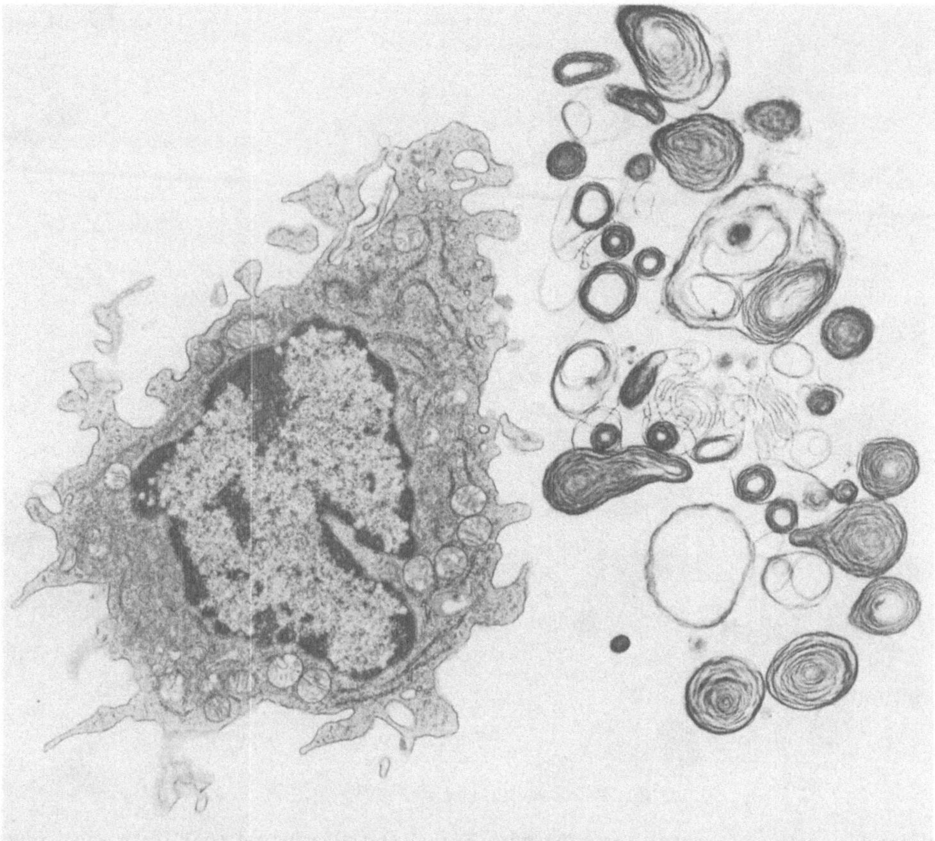

FIGURE 8. Macrophage of fetal rat lung. The cell has pseudopodia but no lysosomes. In spite of much lamellar debris in the airway, there is no phagocytosis at this time.

maturation and the increase in the number of macrophages found after birth are innate adaptive responses to the sudden change from sterile intrauterine environment to the external milieu of ambient air.

5. ADAPTIVE RESPONSES

Individual alveolar macrophages adapt by altering their internal structure and function. Thus, the heterogeneity of the macrophagic population in the lung may be attributed to the varying types of material handled by individual cells and to the varying lengths of time that the macrophages have spent in the alveoli. In this section the emphasis will be directed to the adaptive responses of the pulmonary macrophagic system as a whole.

This system usually responds to a heavy particulate load by increasing the output of free alveolar macrophages. These new cells might arise by: (a) division of preexisting macrophages in the alveoli; (b) direct migration of circulating

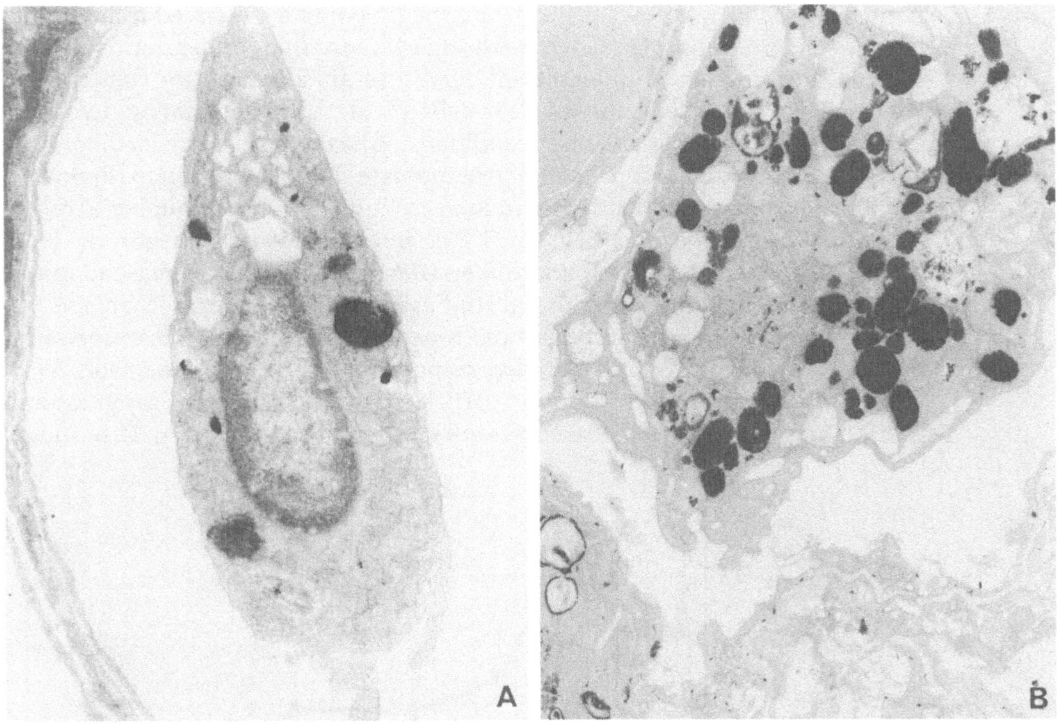

FIGURE 9. (A) Prenatal rat lung. Acid phosphatase content is low in the small immature alveolar macrophage. (B) Postnatal rat lung (day 10). Acid phosphatase activity is substantially increased in the alveolar macrophage (From Curle, D. C., and Adamson, I. Y. R., 1978, Retarded development of neonatal rat lung by maternal malnutrition, *J. Histochem. Cytochem.* **26**:401. With permission.)

blood monocytes; (c) migration of preexisting interstitial macrophages; (d) division of interstitial cells with subsequent migration of daughter cells to the alveoli, as already demonstrated in normal lung.

Division of free macrophages is unusual under normal circumstances. These cells are considered to be fully functional and in the inactive phase of the mitotic cycle. Alveolar macrophages are rarely labeled after a pulse of [^{3}H]thymidine and long-term culture of these cells under normal conditions does not result in any significant increase in cell number. Division of alveolar macrophages has been demonstrated after exposure to nitrogen dioxide (Evans *et al.*, 1973), and several investigators have succeeded in stimulating cultured macrophages to divide by adding "conditioned medium," usually consisting of factors produced from a line of actively dividing fibroblasts (Soderland and Naum, 1973). In addition, cell-free inflammatory exudates also promote mitosis in macrophages (Wynne *et al.*, 1975). There is no evidence, however, that phagocytic activity induces free macrophages to divide. For instance, it has been shown that the increased number of alveolar macrophages that follows the instillation of carbon *in vivo* is not accompanied by division of free cells (Bowden and Adamson, 1978).

The failure of alveolar macrophages to synthesize DNA in response to a particulate load suggests that the macrophagic exudate is derived from interstitial cells and/or monocytes. There is strong evidence that, under normal circumstances, the pulmonary interstitium serves as an intermediate compartment where marrow-derived mononuclear cells divide before migrating across the epithelium to the air sacs. Under conditions of heavy particulate loading, however, the origin of the supplemental macrophages has been disputed. Brain *et al.* (1977) have suggested that the rapid increase in macrophagic number is related to release of preexisting cells such as interstitial macrophages or blood monocytes, whereas Casarett and Milley (1964) attribute the increase to mitotic activity within the lung. More recent studies indicate that both mechanisms may be operative and that the apparent discrepancy merely represents observations made at different stages of the cellular responses (Bowden and Adamson, 1978). The instillation of 4 mg of carbon (particle size 300 Å) into the lungs of mice yields a massive increase in free macrophages within 12 hr, and by 24 hr the cell

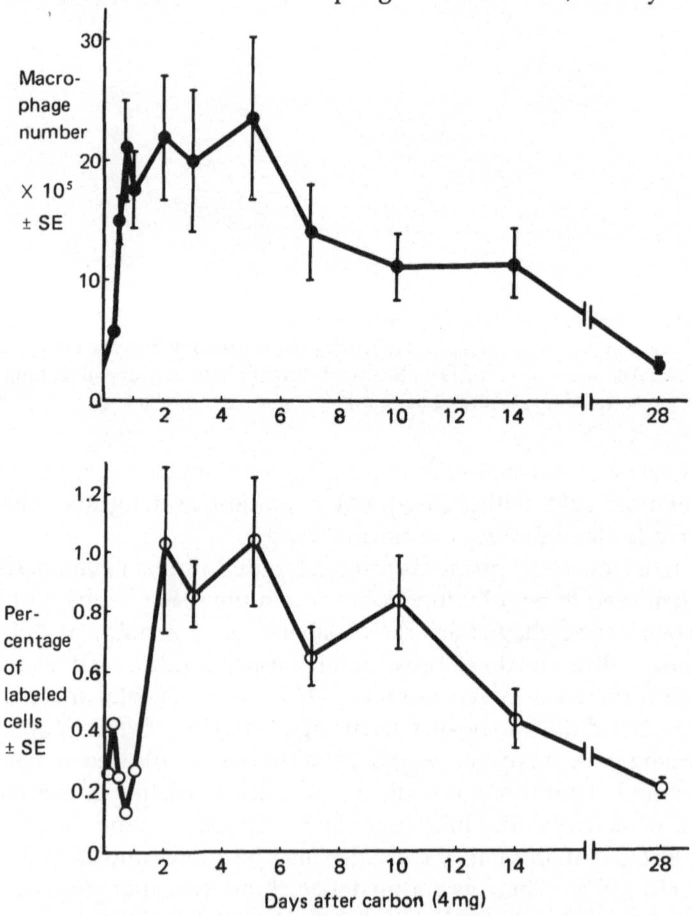

FIGURE 10. (Upper) The number of alveolar macrophages recovered from mouse lungs at intervals after carbon instillation. (Lower) Percentage of [^{3}H]thymidine-labeled cells in autoradiographs. DNA synthesis in the lung is not elevated at 1 day, but, from 2 days on, it correlates well with the increased number of macrophages.

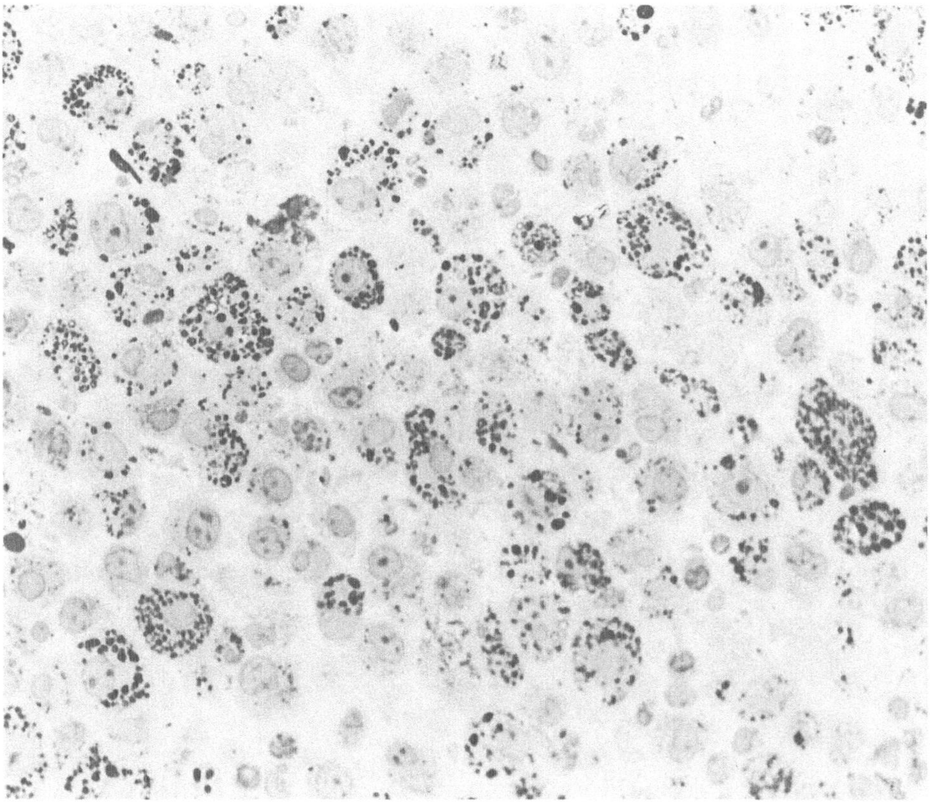

FIGURE 11. Cell pellet (0.5-μm section) of macrophages lavaged from mouse lung, 3 days after intratracheal carbon. Many cells are large and loaded with carbon; others are small, probably newly arrived in the alveoli.

number increases 10-fold (Figure 10). Within that period, DNA synthesis, as measured by the cellular incorporation of tritiated thymidine, remains at control levels. The initial sharp increase in macrophagic output is therefore unrelated to any cell division in the lung and appears to be the result of cellular migration. Further evidence for this has been obtained in a morphologic study which shows that the capillaries and small vessels of the lung become crowded by mononuclear cells within 6 hr of carbon administration (Adamson and Bowden, 1978). Cells are seen apparently migrating across the endothelium to reach the interstitium, and, by 24 hr, an increased number of interstitial mononuclear cells is observed. The newly arrived macrophages observed in the alveoli at 24 hr appear to be immature cells, bearing a close resemblance to monocytes; they are small, contain few cytoplasmic lysosomes and appear to be poor phagocytes. Within the alveoli, the variation in cellular size and configuration from these immature cells to large macrophages loaded with carbon accounts for the heterogeneity of cells lavaged from the lung (Figure 11).

In the period of 1–10 days after administering carbon, the vascular crowding by monocytes disappears rapidly, but the number of free macrophages remains

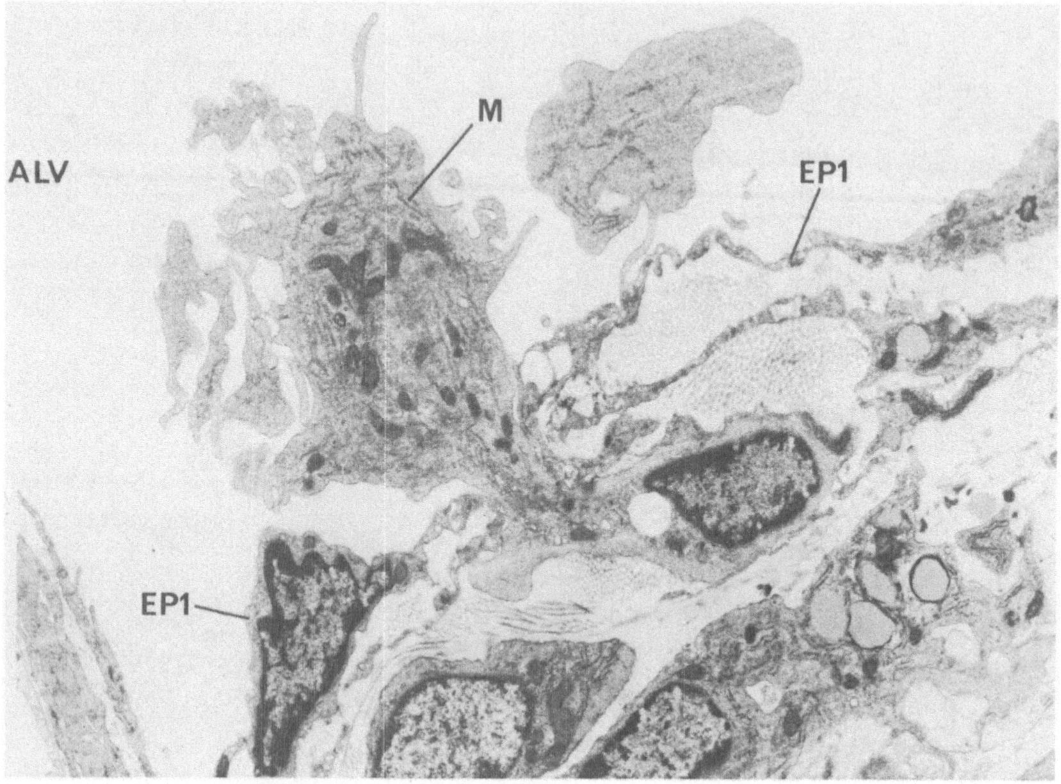

FIGURE 12. Macrophage (M) apparently crossing from the interstitium between Type 1 epithelial cells (EP1) to the alveolus (ALV). (From Adamson, I. Y. R., and Bowden, D. H. 1978, Adaptive responses of the pulmonary macrophagic system to carbon. II. Morphologic studies, *Lab Invest.* **38**:430. With permission.)

at a high level. Continued production of these cells correlates with an increase in [^{3}H]thymidine incorporation in the lung (Figure 10). Differential counts of labeled cells in autoradiographs reveal that the enhanced DNA synthesis is limited to the interstitial cells which are particularly prominent in perivascular locations. Elevated levels of DNA synthesis and numerous mitotic figures are seen in the interstitial cell population as long as the free macrophage number is increased. The results indicate that the response of the pulmonary macrophages to a large particulate load is biphasic. In the initial phase, new alveolar macrophages appear to arise by migration of blood monocytes through the interstitium to the alveoli with, perhaps, some release of preexisting interstitial macrophages. The second phase involves increased mitotic activity in the interstitial compartment where preexisting and/or newly arrived mononuclear cells divide and subsequently cross the epithelium into the alveolus (Figure 12).

The rapidity of the initial cellular efflux without cell division suggests that the monocytic migration is a response to a chemotactic stimulus. The nature of this stimulus is unknown; it could result from the interaction of particles with either free macrophages or, perhaps, the Type 1 alveolar epithelial cell, the site of transport of free particles to the interstitium. The secondary response, involving division of interstitial cells, suggests mitogenic stimulation at this site. The demonstration of free carbon in the interstitium at the exact time of mitosis suggests a causal relationship.

These changes are not specific to carbon since we have observed an identical sequence of cellular events following the instillation of polystyrene latex particles. It is known that the output of alveolar macrophages is influenced by factors such as total load, particle size, and chemical composition. Such variability of response may be explained by the relative predominance of either the monocytic or interstitial pathways of cellular recruitment. Our own observations suggest that the initial monocytic-related phase is transient whatever the load; it is the duration of the interstitial proliferative phase that determines the continuing adaptive outpouring of macrophages.

6. ABNORMAL RESPONSES

Macrophagic responses to the inhalation of foreign materials are largely adaptive in nature. Pathological changes may ensue if this adaptive capacity is overwhelmed, or if the protective function of these cells is perturbed in any way. In addition to its primary defensive role, it is possible that the pulmonary macrophage participates in the genesis of two major groups of disabling lung diseases, chronic pulmonary fibrosis and emphysema. In considering these inappropriate responses, we will examine the interactions of the macrophage with the fibroblast, and the role of the macrophage as a secretory cell.

6.1. THE PULMONARY MACROPHAGE AND THE FIBROBLAST

Chronic pulmonary injury may so interfere with reparative mechanisms that fibrosis ensues. Usually, the immediate cause is not known, but there are instances in which the macrophage is thought to play at least an intermediary role. Phagocytosis of immune complexes stimulates the synthesis and release of lysosomal enzymes which may provide the continuing stimulus to fibroblastic activity in conditions such as extrinsic fibrosing alveolitis (Cardella *et al.*, 1974). A more precise mechanism of intercellular communication has been elucidated in the response of the lung to silica. Particles of crystalline silica are extremely toxic to the macrophage, a toxicity that is dependent upon the chemical bonding of silicic acid with phospholipids and proteins of lysosomal membranes (Nash *et al.*, 1966). Massive intracellular discharge of lysosomal enzymes kills the macrophages (Figure 13), silica particles are released into the air sacs, and the sequence of phagocytosis, cell death, and release of toxic particles is repeated. This recurring process hinders mucociliary clearance and increases the likelihood of silica crystals penetrating the alveolar epithelium and gaining access to the pulmonary interstitium where they are engulfed by interstitial macrophages. The link between these early events and the development of fibrosis is thought to be related to some form of interaction between the particles and the macrophage.

Extracts of macrophages incubated with silica stimulate the production of hydroxyproline by cultured fibroblasts (Heppleston and Styles, 1967; Nourse *et al.*, 1975). Heppleston has suggested that death of the macrophage *per se* does not provide the stimulus for fibrogenesis, but that contact between the phagocyte and silica yields a fibrogenic principle. As an extension to this hypothesis, Spector (1974) has suggested that fibroblastic stimulation by injured macrophages may explain collagen deposition in a variety of the cryptogenic fibroses of the lung. However, the fibrotic response of the lung to asbestos suggests that the intermediary role of the macrophage may not be essential to fibroblastic stimulation. In asbestosis, the predominant location of fibrosis in the bronchiolar regions appears to be related to direct tissue injury by fibers that are too long to be phagocytized (Maroudas *et al.*, 1973). Short fibers, <20 μm in length, are

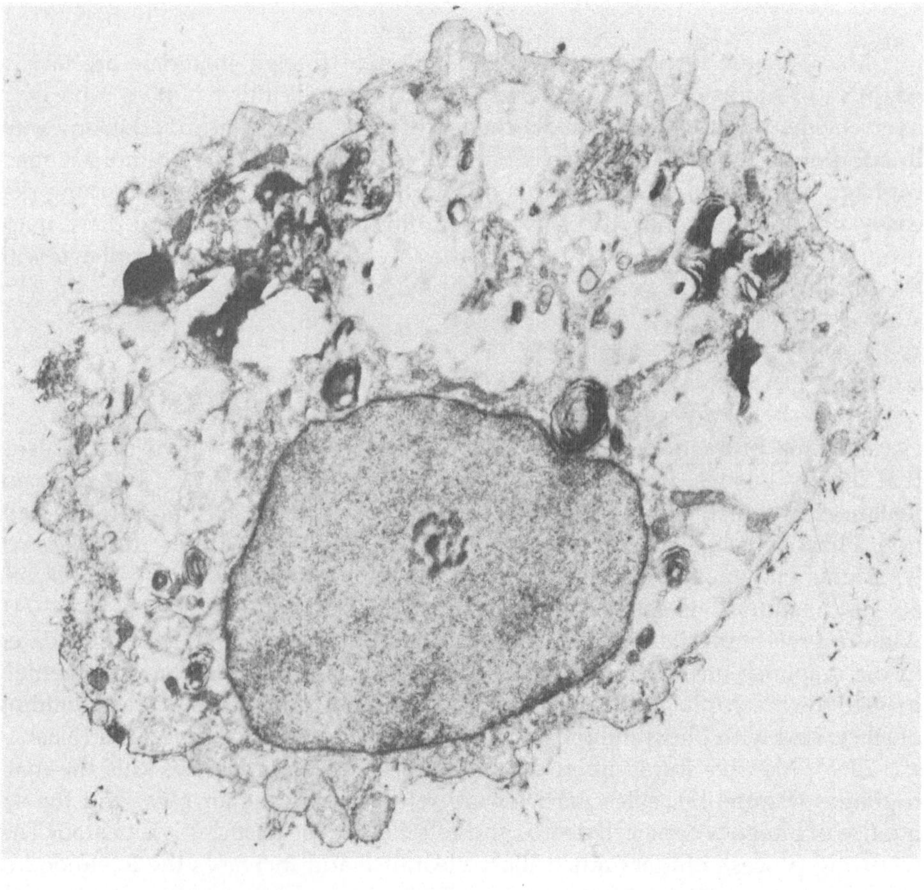

FIGURE 13. Alveolar macrophage after intratracheal silica. Lysosomal membranes have ruptured leading to cytoplasmic autodigestion.

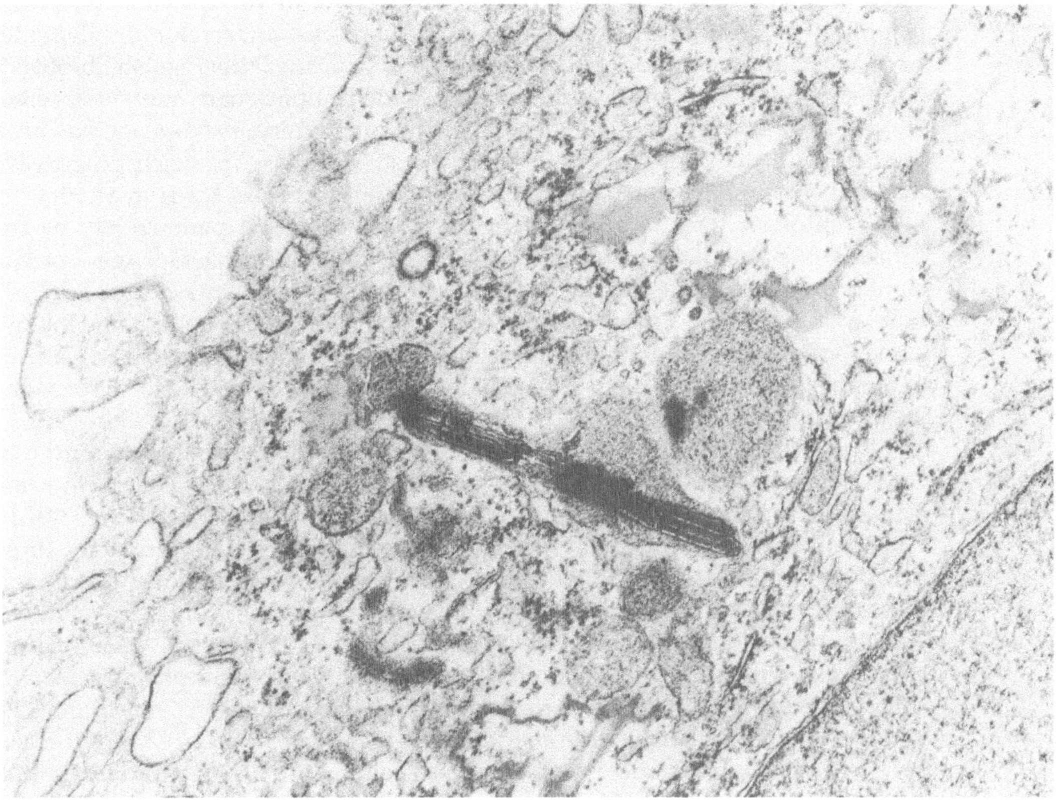

FIGURE 14. Alveolar macrophage after intratracheal asbestos (short fibers). Fibers are seen within a membrane-bound phagolysosome; there is no evidence of cell injury.

frequently seen in the phagosomes of macrophages (Figure 14), but in contrast to silica, these cells are not destroyed. This difference has been attributed to coating of the asbestos fibers by an iron-containing protein (Suzuki and Churg, 1969). These macrophages are cleared in the usual way and are not associated with fibrosis.

6.2. THE PULMONARY MACROPHAGE AS A SECRETORY CELL

The phagocytic and intrinsic digestive functions of macrophages in general are well recognized; the secretory function of these cells has only recently received some attention (Werb and Gordon, 1975a,b). It has been suggested that proteases secreted by macrophages may play a role in the lysis of collagen and the remodelling of scars (Spector, 1974). This idea is based upon the knowledge that, whereas the unstimulated or quiescent macrophage is concerned primarily with phagocytosis and endogenous digestion, the activated macrophage, identified morphologically as an epithelioid cell, is capable of exogenous secretion.

Proof of the postulate is not at hand, however, and the question is further complicated by the functional duality of the fibroblast as secretor and degrader of collagen, and by the lack of suitable markers which will distinguish fibroblasts and macrophages within the general population of pulmonary interstitial cells.

As a secretor of proteolytic enzymes the pulmonary macrophage has been implicated in the pathogenesis of emphysema, a disease in which progressive destruction and subsequent rearrangement of elastic fibers leads to an increasingly compliant lung. Similar lesions may be induced experimentally by the infusion of proteolytic enzymes into the lung and, since macrophages contain similar enzymes, it is tempting to suggest that their release may induce pulmonary autodigestion. Enzymatic secretion by the macrophage usually follows some form of extrinsic stimulus and, since cigarette smoking is a dominant factor in the etiology of emphysema, the response of the macrophage to tobacco smoke is of considerable interest.

The alveolar macrophages of cigarette smokers are decidedly different from those of nonsmokers. Bronchopulmonary lavage yields four times as many macrophages from smokers than are obtained from nonsmokers (Harris *et al.*, 1970). The smokers' cells are larger, their glucose utilization is increased about three times over control levels, and they exhibit distinctive morphologic features. Observed with the light microscope, smokers' macrophages contain pale brown sudanophilic and fluorescent cytoplasmic inclusions (Vassar *et al.*, 1960). Ultrastructurally, the inclusions are identified as membrane-bound phagolysosomes containing acid phosphatase; there is an increase of endoplasmic reticulum (ER) and the Golgi is unusually prominent (Figure 15), (Harris *et al.*, 1970; Pratt *et al.*, 1971). The phagolysosomes, which may be seen in both alveolar and interstitial macrophages, often contain inclusions consisting of kaolinite and aluminum silicate derived, probably, from inhaled tobacco smoke (Brody and Craighead, 1975). These changes probably represent the adaptive responses of the macrophagic delivery system and of individual cells to the increase in phagocytic load.

The functional significance of these morphologic changes is not known, an uncertainty that derives in part from conflicting results concerning the effects of tobacco smoke on particulate clearance and bactericidal activity in the lung (Bowden, 1973). The lung responds to any increase of particulate load with a boost in the production of macrophages. In this respect, the response to tobacco is not unusual and it is reversible. At the functional level, smokers' macrophages are competent phagocytes capable of digesting and killing bacteria (Harris *et al.*, 1970). As an adaptive outpouring of functionally active macrophages, the pulmonary response to cigarette smoking appears to be successful. But it is relevant to ask at this point whether the long-term cost of adaptation by the pulmonary macrophagic system may not compromise the defenses of the host and contribute to the development of chronic bronchitis.

The adapted macrophages with their secretory potential may also be involved in the destruction of terminal airways and air sacs, leading to emphysema. Proteases with elastase-like activity have been demonstrated in macrophages from various sites. White *et al.* (1977) have shown that an unstimulated

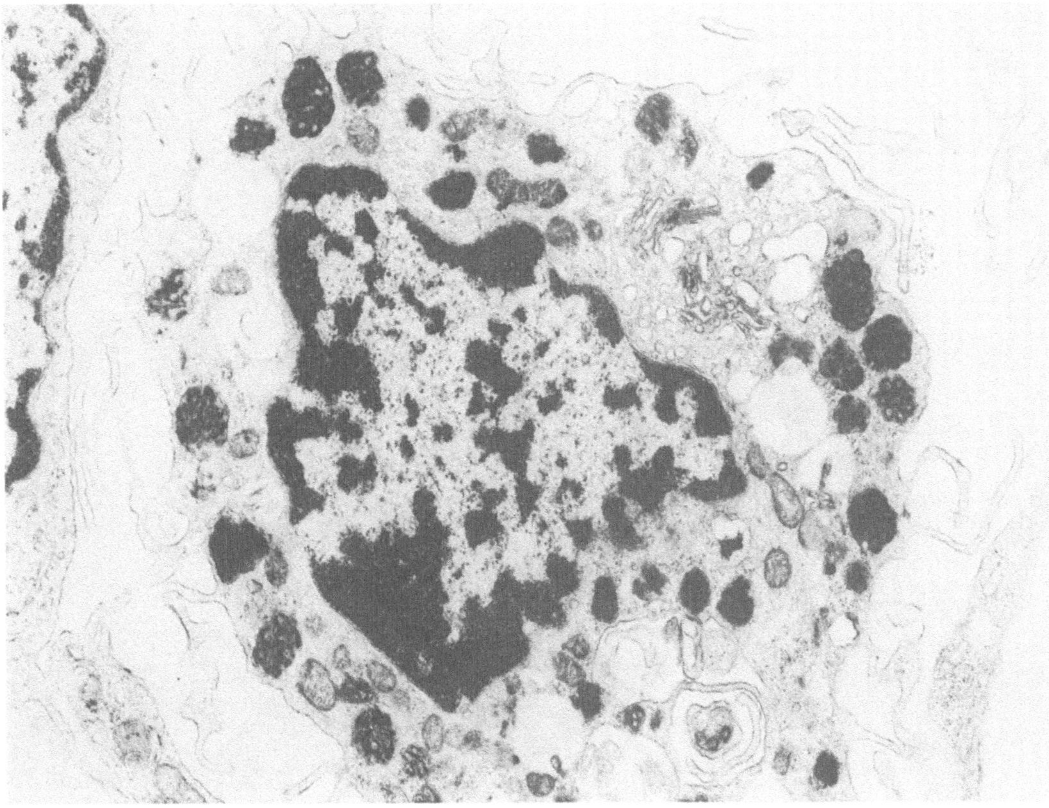

FIGURE 15. Alveolar macrophage from a biopsy of a "smoker's lung." The cell exhibits many pseudopodia, well developed Golgi, and many phagolysosomes with whorled inclusions.

alveolar macrophage produces more elastase than an activated peritoneal cell and, in smokers, alveolar macrophages exhibit greatly enhanced activity of this enzyme (Harris *et al.*, 1975). Whether these proteolytic enzymes are capable of digesting the tough collagen and elastic fibers of the lung *in vivo* is not known. It is difficult to comprehend how release of relatively low levels of digestive enzymes within the alveolar lumen can destroy, selectively, spiral elastic fibers located in the centrilobular interstitial tissues. On the other hand, it is conceivable that free particulate matter reaching the interstitium could induce secretory activity in interstitial macrophages located close to these elastic fibers. Current enthusiasm for a macrophagic role in destructive diseases of the lung should not obscure the protective role of this cell, since macrophages are known to bind and incorporate neutrophil-derived elastase (Campbell *et al.*, 1978). Neutrophil granulocytes, often observed in the interstitium and the lung washings of cigarette smokers, carry a large complement of neutral proteases. Phagocytosis of these cells (Figure 16) and neutralization of their enzymes may, in the final analysis, prove to be more important than the postulated role of the macrophage as a digester of elastic fibers (Blondin *et al.*, 1972).

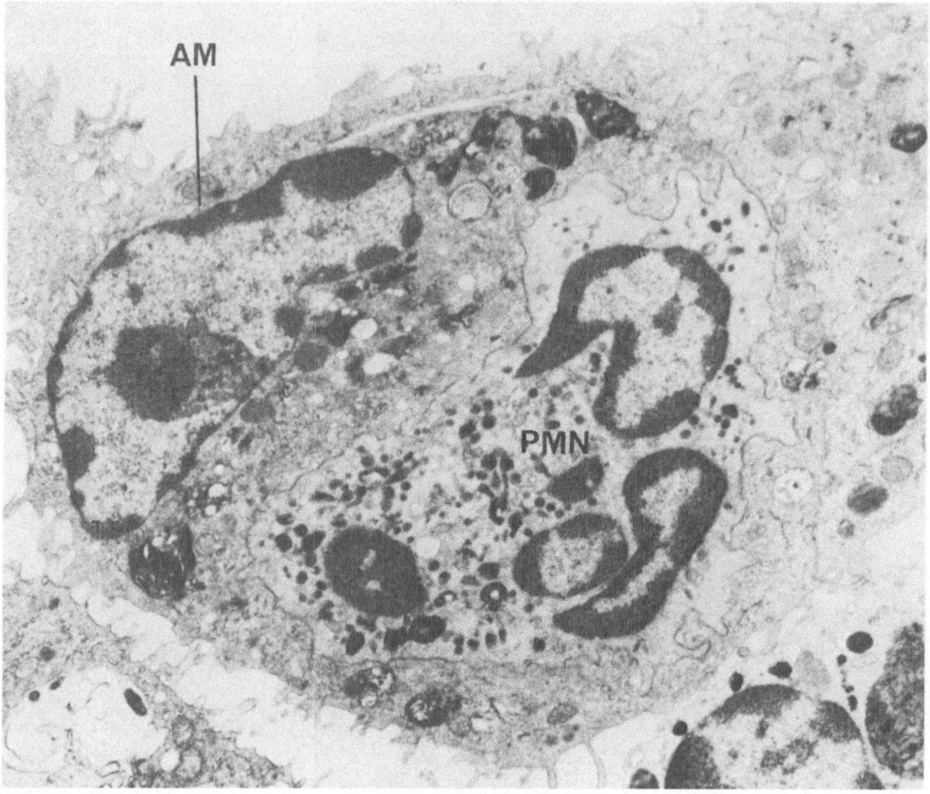

FIGURE 16. Phagocytosis of an intact polymorphonuclear leukocyte (PMN) by an alveolar macrophage (AM).

7. CLEARANCE

Alveolar macrophages exist as functioning cells in the air sacs, and their elimination from the lung with a phagocytized load is a normal event in their life cycle. The life span of the macrophage in the alveolus is not known. The difficulty of its determination is related partly to the low mitotic rate of the macrophagic system under basal conditions and the varying "normal" atmospheric conditions which may trigger an adaptive response. Under most conditions, the macrophage probably remains in the alveolus for about 2 weeks, but these cells are capable of long-term survival in the lung under conditions of impaired clearance; this is confirmed in tissue culture where cells are viable more than 100 days. The initial step in macrophagic clearance is the movement of cells from the air sacs to collect around the openings to terminal bronchioles. This motion is thought to be related to lung movement aided, perhaps, by the flow of an alveolar surface film (Kilburn, 1968). How macrophages are transported on to the cilia is not known, but once on the bronchial epithelium they are transported by mucociliary movement and swallowed. The major proportion of particles deposited in the alveoli are cleared by this route (Figure 17).

Evidence for macrophagic clearance by other routes is not so convincing. It has been postulated that free alveolar macrophages can recross the epithelium to the interstitium where they may proceed to lymphatics or to blood vessels. This pathway has never been proven despite the observation of particle-laden macrophages in the interstitium. Recently, it has been shown that, following the instillation of carbon to the lung, free particles cross the Type 1 alveolar epithelial cells to reach the interstitium where they are phagocytized by macrophages (Adamson and Bowden, 1978). These cells tend to aggregate around bronchioles and blood vessels where they may remain for a long time. Their persistence may be explained by the relative lack of mobility of peribronchial and perivascular tissue. Elimination of particulates may be accomplished eventually by migration of interstitial macrophages into the alveoli or into the lymphatics. Of particular interest is the rapidity with which a miner's black lung may be cleared of its interstitial pigment during an episode of pneumonia. Such fast clearance is

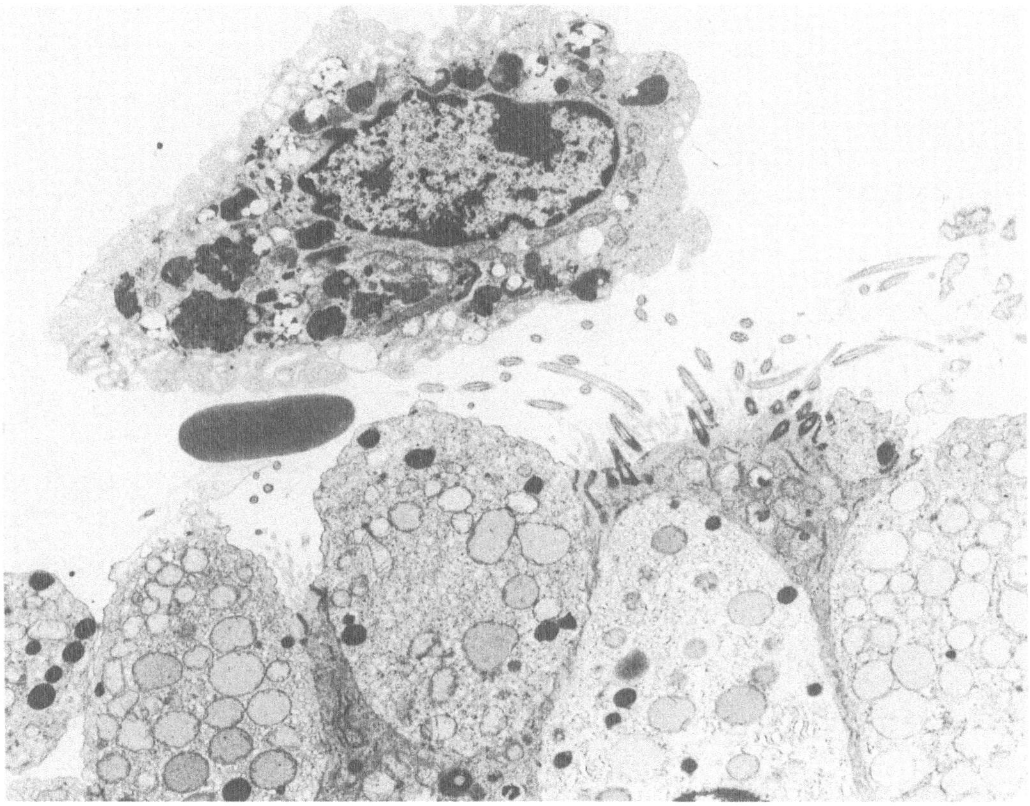

FIGURE 17. Surface of bronchiolar epithelium showing a heavily laden macrophage passing over ciliated and nonciliated epithelial cells.

dependent upon massive transport of pigment by resident and inflammatory macrophages to the alveoli and hence to the airways.

Inhaled particulates also reach the lymphatics. After administration of small particles such as proteins or carbon, there appears to be rapid passage of some particles across the Type 1 cells to the interstitium and hence to the lymphatics. Carbon is seen in macrophages lining the sinusoids of hilar lymph nodes within several hours of exposure and before phagocytosis of particles by interstitial cells occurs. This suggests that the early phase of lymphatic clearance involves transport of particles in the free state. Clearance of interstitial macrophages to the lymphatics appears to be a slower process. Particle-laden macrophages are not observed in lymph channels until several days after administration, and the amount of material removed by this route appears to be small compared to that eliminated by alveolar-bronchial clearance (Adamson and Bowden, 1978).

In certain circumstances, regular clearance mechanisms may be impaired. An agent such as silica destroys the macrophage, causing release of particles

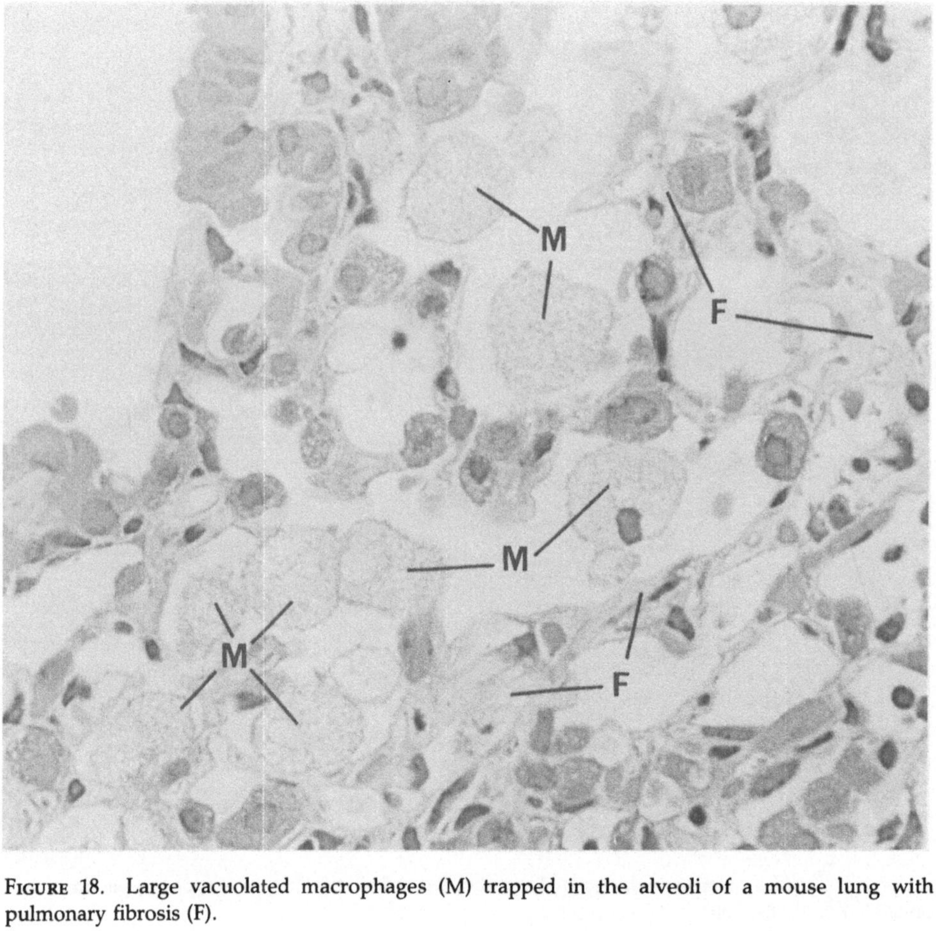

FIGURE 18. Large vacuolated macrophages (M) trapped in the alveoli of a mouse lung with pulmonary fibrosis (F).

which are engulfed by other macrophages. This process decreases clearance and facilitates transepithelial passage of the toxic material to the interstitium. Pollutant gases such as ozone and nitrogen dioxide, which selectively injure the respiratory bronchioles, cause necrosis at that site, thereby inhibiting the transfer of macrophages from alveolus to bronchus. Ciliostatic agents such as tobacco and alcohol may also inhibit clearance. Since alveolar movement is an important factor in macrophagic clearance, conditions such as interstitial edema and fibrosis, which reduce pulmonary compliance, are often associated with accumulations of macrophages, apparently trapped in distal air sacs. These cells appear viable, but usually are very large and the cytoplasm contains many vacuoles and debris, features characteristic of "old" macrophages (Figure 18).

In summary, the most important route for clearance of pulmonary macrophages is the mucociliary pathway. Clearance by lymphatics appears to be of minor significance, and there is no evidence that free alveolar macrophages recross the epithelium to reach the interstitium. Particulate matter that does reach the interstitium is transported across the epithelium in the free state and is phagocytized by interstitial macrophages.

REFERENCES

Adamson, I. Y. R., and Bowden, D. H., 1975, Derivation of type 1 epithelium from type 2 cells in the developing rat lung, *Lab. Invest.* **32:**736.

Adamson, I. Y. R., and Bowden, D. H., 1978, Adaptive responses of the pulmonary macrophagic system to carbon. II. Morphologic studies, *Lab. Invest.* **38:**430.

Blondin, J., Rosenberg, R., and Janoff, A., 1972, An inhibitor in human lung macrophages active against human neutrophil elastase, *Am. Rev. Resp. Dis.* **106:**477.

Bowden, D. H., 1971, The alveolar macrophage, in: *Current Topics in Pathology* (H. W. Altmann *et al.*, eds.), pp. 1–36, Springer Verlag, Heidelberg.

Bowden, D. H., 1973, The alveolar macrophage and its role in toxicology, *CRC Crit. Rev. Toxicol.* **2:**95.

Bowden, D. H., and Adamson, I. Y. R., 1972, The pulmonary interstitial cell as immediate precursor of the alveolar macrophage, *Am. J. Pathol.* **68:**521.

Bowden, D. H., and Adamson, I. Y. R., 1976, The alveolar macrophage delivery system. Kinetic studies in cultured explants of murine lung, *Am. J. Pathol.* **83:**123.

Bowden, D. H., and Adamson, I. Y. R., 1978, Adaptive responses of the pulmonary macrophagic system to carbon. I. Kinetic studies, *Lab. Invest.* **38:**422.

Bowden, D. H., Adamson, I. Y. R., Grantham, W. G., and Wyatt, J. P., 1969, Origin of the lung macrophage: Evidence derived from radiation injury, *Arch. Pathol.* **88:**540.

Brain, J. D., Godleski, J. G., and Sorokin, S. P., 1977, in: *Lung Biology in Health and Disease,* Vol. 5, *Respiratory Defense Mechanisms* (J. D. Brain, D. F. Proctor, and L. Reid, eds.), pp. 849–885, Dekker, New York.

Brody, A. R., Craighead, J. E., 1975, Cytoplasmic inclusions in pulmonary macrophages of cigarette smokers, *Lab. Invest.* **32:**125.

Campbell, E. J., White, R. R., Rodriguez, R. J., Senior, R. M., and Kuhn, C., 1978, Specific binding of human leukocyte elastase by human alveolar macrophages, *Am. Rev. Resp. Dis.* **117:**319.

Cardella, C. J., Davies, P., Allison, A. C., 1974, Immune complexes induce selective release of lysosomal hydrolases from macrophages, *Nature* **247:**46.

Casarett, L. J., and Milley, P. S., 1964, Alveolar reactivity following inhalation of particles, *Health Phys.* **10:**1003.

Cohn, Z. A., and Weiner, E., 1963a, The particulate hydrolases of macrophages, *J. Exp. Med.* **118**:991.

Cohn, Z. A., and Weiner, E., 1963b, The particulate hydrolases of macrophages. II. Biochemical and morphological response to particle ingestion, *J. Exp. Med.* **118**:1009.

Curle, D. C., and Adamson, I. Y. R., 1978, Retarded development of neonatal rat lung by maternal malnutrition, *J. Histochem. Cytochem.* **26**:401.

Essner, E., and Haimes, H., 1977, Ultrastructural study of GERL in beige mouse alveolar macrophages, *J. Cell Biol.* **75**:381.

Evans, M. J., Cabral, L. J., Stephens, R. J., and Freeman, G., 1973, Cell division of alveolar macrophages in rat lung following exposure to NO_2, *Am. J. Pathol.* **70**:199.

Golde, D. W., Finley, T. N., and Cline, M. J., 1974, The pulmonary macrophage in acute leukemia, *N. Engl. J. Med.* **290**:875.

Harris, J. O., Swenson, E. W., and Johnson, J. E., 1970, III. Human alveolar macrophages: Comparison of phagocytic ability, glucose utilization, and ultrastructure in smokers and nonsmokers, *J. Clin. Invest.* **49**:2086.

Harris, J. O., Olsen, G. N., Castle, J. R., and Maloney, A. S., 1975, Comparison of proteolytic enzyme activity in pulmonary alveolar macrophages and blood leukocytes in smokers and nonsmokers, *Am. Rev. Resp. Dis.* **111**:579.

Heise, E. R., and Myrvik, Q. N., 1966, Levels of lysosomal hydrolases in alveolar and peritoneal macrophages from conventional and germ-free rats, *Fed. Proc.* **25**:439.

Heise, E. R., Myrvik, Q. N., and Leake, E. S., 1965, Effects of *Bacillus* Calmette-Guerin on the levels of acid phosphatase, lysozyme and cathepsin in rabbit alveolar macrophages, *J. Immunol.* **95**:125.

Heppleston, A. G., and Styles, J. A., 1967, Activity of a macrophage factor in collagen formation by silica, *Nature* **214**:521.

Kilburn, K. H., 1968, A hypothesis for pulmonary clearance and its implications, *Am. Rev. Resp. Dis.* **98**:449.

Maroudas, N. G., O'Neill, C. H., and Stanton, M. F., 1973, Fibroblast anchorage in carcinogenesis by fibres, *Lancet* **1**:807.

Myrvik, Q. N., Leake, E. S., and Fariss, B., 1961, Lysozyme content of alveolar and peritoneal macrophages from the rabbit, *J. Immunol.* **86**:133.

Nash, T., Allison, A. C., and Harington, J. S., 1966, Physicochemical properties of silica in relation to its toxicity, *Nature* **210**:259.

Nerurkar, L. S., Zeligs, B. J., and Bellanti, J. A., 1977, Maturation of the rabbit alveolar macrophage during animal development. II. Biochemical and enzymatic studies, *Pediat. Res.* **11**:1202.

Nourse, L. D., Nourse, P. N., Botes, H., and Schwartz, H. M., 1975, The effects of macrophages isolated from the lungs of guinea pigs dusted with silica on collagen biosynthesis by guinea pig fibroblasts in cell culture, *Environ. Res.* **9**:115.

Novikoff, P. M., Novikoff, A. B., Quintana, N., and Hauw, J. J., 1971, Golgi apparatus, GERL, and lysosomes of neurons in rat dorsal root ganglia, studied by thick section and thin section cytochemistry, *J. Cell Biol.* **50**:859.

Pratt, S. A., Smith, M. H., Ladman, A. J., and Finley, T. N., 1971, The ultrastructure of alveolar macrophages from human cigarette smokers and nonsmokers, *Lab. Invest.* **24**:331.

Soderland, S. C., and Naum, Y., 1973, Growth of pulmonary alveolar macrophages *in vitro*, *Nature* **245**:150.

Sorokin, S. P., 1976, A labyrinthian precursor of primary lysosomes in alveolar macrophages, *J. Cell Biol.* **70**:38/a.

Sorokin, S. P., and Brain, J. D., 1975, Pathways of clearance in mouse lungs exposed to iron oxide aerosols, *Anat. Rec.* **181**:581.

Spector, W. G., 1974, Pulmonary fibrosis due to chemicals and particles, *Ann. N.Y. Acad. Sci.* **221**:309.

Suzuki, Y., and Churg, J., 1969, Structure and development of the asbestos body, *Am. J. Pathol.* **55**:79.

van Furth, R., 1970, The origin and turnover of promonocytes, monocytes and macrophages in normal mice, in: *Mononuclear Phagocytes* (R. van Furth, ed.), pp. 151–165, Blackwell, Oxford.

Vassar, P. S., Culling, C., and Saunders, A. M., 1960, Fluorescent histiocytes in sputum related to smoking, *Arch. Pathol.* **70:**649.

Volkman, A., 1976, Disparity in origin of mononuclear phagocyte populations, *J. Reticuloendothel. Soc.* **19:**249.

Volkman, A., and Gowans, J. L., 1965, The production of macrophages in the rat, *Br. J. Exp. Pathol.* **46:**50.

Weibel, E. R., Gehr, P., Haies, D., Gil, J., and Bachofen, M., 1976, Reaction of lung to injury, in: *Lung Cells in Disease* (A. Bouhuys, ed.), pp. 3–16, Elsevier/North-Holland, Amsterdam.

Werb, Z., and Gordon, S., 1975a, Secretion of a specific collagenase by stimulated macrophages, *J. Exp. Med.* **142:**346.

Werb, Z., and Gordon, S., 1975b, Elastase secretion by stimulated macrophages. Characterization and regulation, *J. Exp. Med.* **142:**361.

Weston, W. L., Carson, B. S., Barkin, R. M., Slater, G. D., Dustin, R. D., and Hecht, S. K., 1977, Monocyte-macrophage function in the newborn, *Am. J. Dis. Child.* **131:**1241.

White, R., Lin, H. S., Kuhn, C., 1977, Elastase secretion by peritoneal exudative and alveolar macrophages, *J. Exp. Med.* **146:**802.

Wynne, K. M., Spector, W. G., and Willoughby, D. A., 1975, Macrophage proliferation *in vitro* induced by exudates, *Nature* **253:**636.

Zeligs, B. J., Nerurkar, L. S., and Bellanti, J. A., 1977, Maturation of the rabbit alveolar macrophage during animal development. I. Perinatal influx into alveoli and ultrastructural differentiation, *Pediatr. Res.* **11:**197.

17

Microglia and Brain Macrophages

FRANK BALDWIN

1. MICROGLIA

1.1. MORPHOLOGY

More than 50 years ago it was shown that the neuropil of mammalian central nervous tissue possessed two distinct cellular elements in addition to neurons and astrocytes (Hortega, 1919a, 1920a,b); one of the cell populations was identified as oligodendrocytes and the other was characterized as microglia. The structure of microglia has since been elucidated at the light-microscopic level at different sites in central nervous tissue of various animals (Penfield, 1924, 1928, 1932; Gozzano, 1929; Hortega, 1932/1965; Dewulf, 1937; Glees, 1955; Hosokawa and Mannen, 1963; Oksche, 1968; Stensaas and Stensaas, 1968a,c; Cammermeyer, 1970; Ibrahim *et al.*, 1974; Baldwin, 1975a,b).

Recognition of the microglia at the light-microscopic level largely depends upon their demonstration by silver impregnation methods (Hortega, 1919a; Penfield, 1928). Although several types of microglia have been described (Schaffer, 1926; Hortega, 1932; Hosokawa, 1963), the two principal categories commonly recognized are bipolar (Figure 1) and multipolar (Figure 2) cells. They have in common a relatively small cell body from which arise two (Figure 1), three, and sometimes more (Figure 2), tortuous processes which branch dichotomously; the resultant processes are thinner and give off finer ones which wend their way between other cells in the neuropil but do not terminate on blood vessels.

There are frequently variations in thickness along the length of individual processes, small spines sometimes arise from them (Figure 2), but terminal filaments are only rarely apparent. Neither the location of the nucleus nor its structure are often visible in silver impregnations but it is elongated in bipolar microglia and sometimes triangular, or elongated, in multipolar microglia; the scanty cytoplasm is predominantly located at the cell poles from which the cytoplasmic processes arise. Thus the impression is of an argyrophilic, ramified cell whose

FRANK BALDWIN • Department of Veterinary Anatomy, Western College of Veterinary Medicine, University of Saskatchewan, Saskatoon, Saskatchewan, Canada 37N OWO.

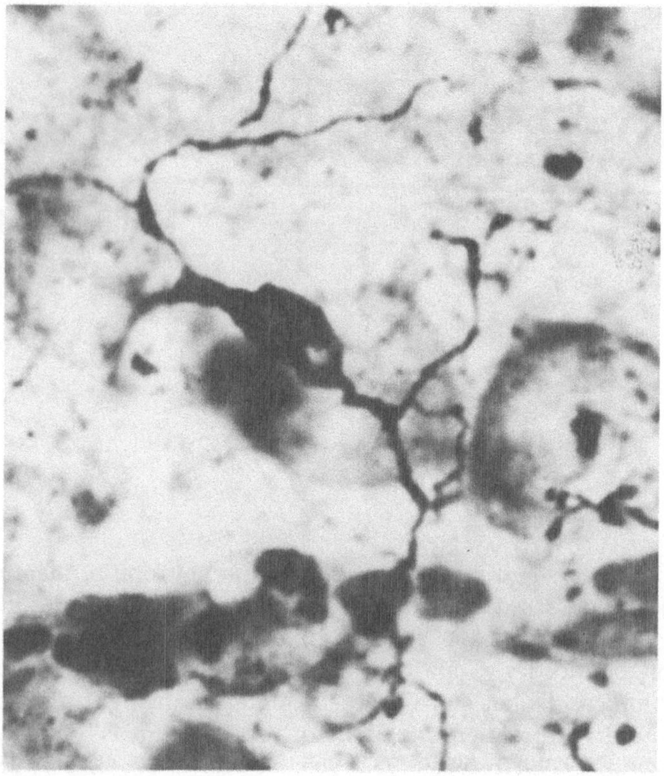

FIGURE 1. Bipolar microglia from cat cerebral cortex. Penfield's silver carbonate impregnation. ×2000. (From Baldwin, 1975b, courtesy of Blackwell Scientific Publications, Oxford.)

processes pass between other cells of the neuropil. The small size of the cell body when compared to most other cells in the neuropil does not provide a true indication of the span of microglia because measurements of microglia in thick sections indicate they may attain overall dimensions in excess of 50 μm. Variations in the argyrophilia of microglia have been reported, probably due to technical artifacts. Since silver impregnations designed to demonstrate microglia are only rarely specific, it is wise to identify as microglia only those cells strictly conforming to the above criteria.

The distribution of microglia in different areas of central nervous tissue, like their morphology, has received the attention of numerous investigators. Microglia lie usually, but not always, near neurons or blood vessels (Hortega, 1920a,b; Cajal, 1920; Hortega, 1932/1965; Penfield, 1932), and are more abundant in gray than in white matter (Penfield, 1924; Hortega, 1932; Rand and Courville, 1932; Schlote, 1959).

In white matter, microglial processes are often delicate with few branches (Schroeder, 1929; Cone and MacMillan, 1932; Hortega, 1932/1965; Dewulf, 1937), and their perikarya usually lie near blood vessels (Schroeder, 1929; Kamemura, 1934; Greenfield, 1963). Several different types of microglia have been described

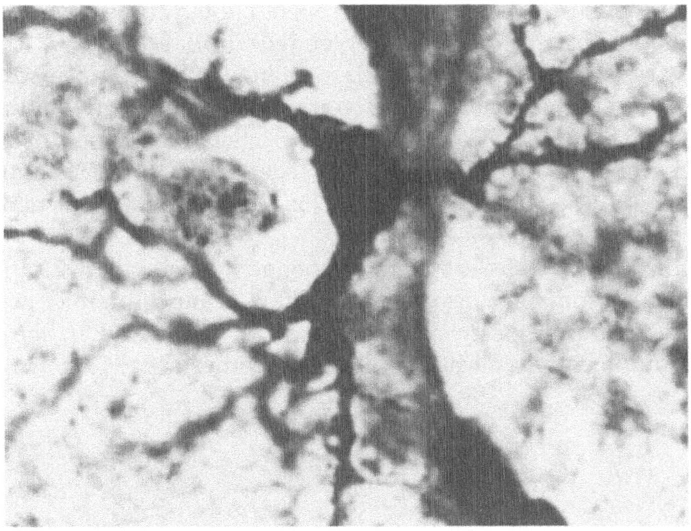

FIGURE 2. Multipolar microglia from cat cerebral cortex Penfield's silver carbonate impregnation. ×2000. (From Baldwin, 1975b, courtesy of Blackwell Scientific Publications, Oxford.)

(Schaffer, 1926; Hortega, 1932/1965; Hosokawa, 1963), but different types do not have specific sites. For instance, in cerebral cortex, opinions vary as to whether multipolar or bipolar microglia are more common (Hortega, 1920; Glees, 1955; Schaffer, 1926).

While the light-microscopic structure of microglia has been known for many years, it is only comparatively recently that the ultrastructure of microglia has been established. Early investigations were hampered by technical difficulties; initial suggestions that cells, with intensely electron-dense cytoplasm, were microglia (Luse, 1956; Schultz *et al.*, 1957; Maynard *et al.*, 1957; Mugnaini and Walberg, 1964), were untrue because such appearances are shown by damaged neurons (Gray, 1961, 1964; Field and Raine, 1964; Kruger, 1965; Magnaini, 1965) or poorly preserved or degenerate macroglia (Wendell-Smith *et al.*, 1966). For some time there was wide disagreement on the ultrastructural morphology of microglia, and some workers doubted the very existence of microglia (Malmfors, 1963; Kruger and Maxwell, 1966; Eager and Eager, 1966; King, 1968), furthermore, microglia were confused with oligodendrocytes (Kruger and Maxwell, 1966) and neurons (Mugnaini and Walberg, 1964) and/or with reactive cells (Luse, 1956; Blinzinger and Kreutzberg, 1968; Hollander *et al.*, 1969; Conradi, 1969a).

A number of different approaches have been used in attempting to define the ultrastructural appearance of microglia. For example, Blakemore (1969a) examined areas where microglia were present but oligodendrocytes were absent, Mori and Leblond (1969) examined preparations which had previously been silver-impregnated to demonstrate microglia; others were able to identify microglia in normal brain by their specific morphological characteristics (Baldwin *et al.*, 1969; Baldwin, 1972, 1975a,b) which were shown to be dissimilar to those generally accepted for macroglial cells (Wendell-Smith *et al.*, 1966). There is now widespread acceptance that certain cells as seen by electron microscopy (EM) in various sites, in many species, have a common ultrastructural pattern, and do correspond to the classical argyrophilic microglia (Blakemore, 1969a,b, 1975; Mori and Leblond, 1969; Blinzinger and Hager, 1964; Stensaas and Stensaas, 1968b; Baldwin, 1972, 1975a,b; Boya, 1975; Cragg, 1976). The ultrastructural characteristics are as follows.

Microglia (Figure 3) possess a small, elongated or irregularly shaped nucleus with clumps of chromatin unevenly distributed around the nuclear periphery; some clumps of varying size are also randomly scattered in the remaining karyoplasm in which a small nucleolus is sometimes seen.

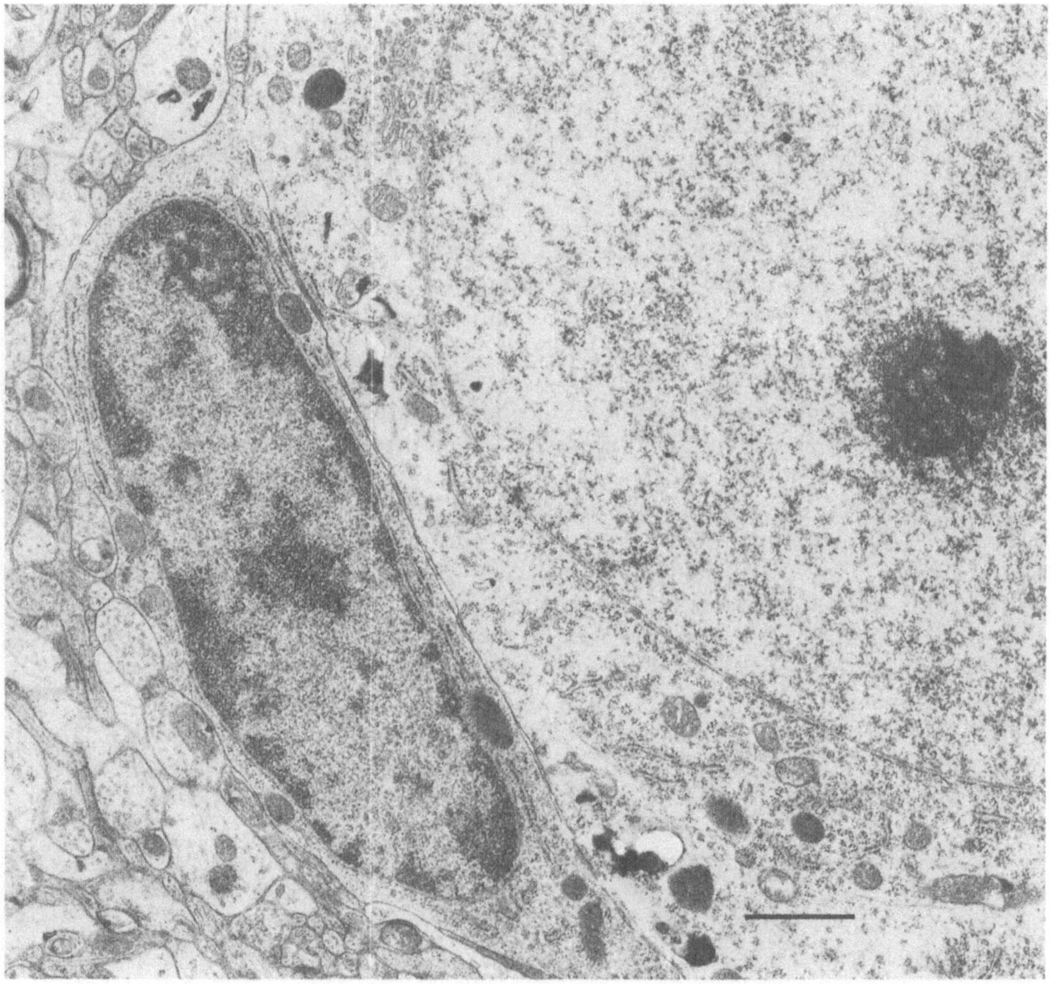

FIGURE 3. Perineuronal microglial cell. Rat cerebral cortex. Bar = 1 μm.

The cell body is fusiform or irregular in shape, with the major proportion of perikaryal cytoplasm distributed at the cell poles from which processes arise. The cytoplasm is of medium electron density and contains elongated cisternae of endoplasmic reticulum (ER) often oriented along the longitudinal axis of the cell. The cisternal contents are of similar or only slightly greater density than the surrounding cytoplasmic matrix; material of a like nature is also present between layers of the nuclear envelope. The outer surface of the cisternae of the ER and the outer nuclear membranes are irregularly studded with ribosomes which are more abundant in the former than in the latter case. Polyribosomes and some free ribosomes occur throughout the cytoplasm.

Mitochondria are randomly distributed, their profiles are small and rounded

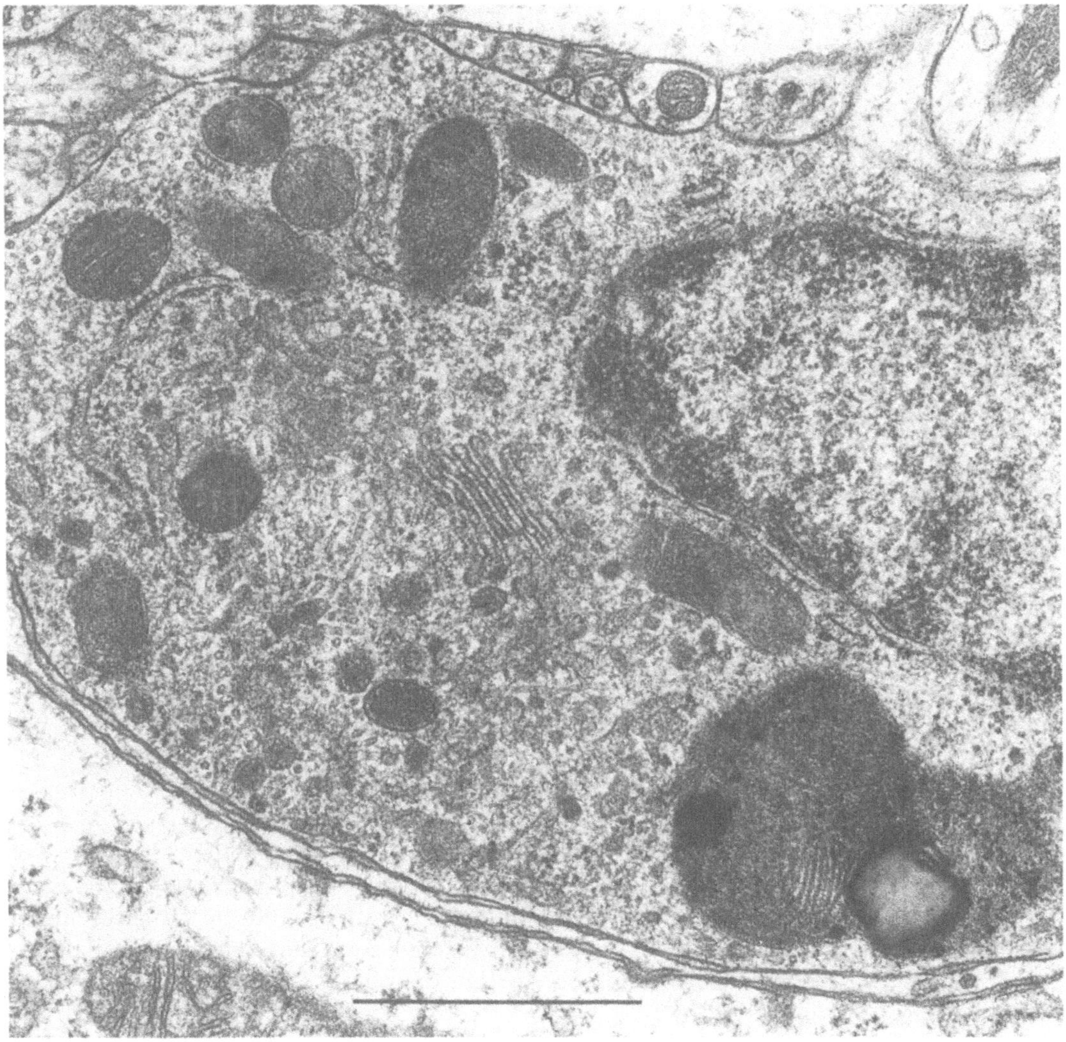

FIGURE 4. Polar cytoplasm of microglial cell. Rat cerebral cortex. Bar = 1 μm.

or oval (200–300 nm), and there are not usually many present in every section of microglial cell body. The mitochondria are not as numerous, and are usually, but not always, smaller than the majority of mitochondria seen in macroglia and neurons. Small smooth vesicles are of variable density; their contents are often more dense than those of nuclear envelope and ER cisternae; coated vesicles are sometimes seen.

Two types of membrane-bound bodies are often present (Figure 4), and their numbers vary in cells. One group, identified as primary lysosomes, are small (150–250 nm, sometimes 500 nm), rounded or oval bodies, with homogeneous finely granular electron-dense contents and have a clear area beneath the limiting membrane. Large (750 nm–1 μm or even larger), irregularly shaped membrane-bound bodies, with fine granular electron-dense contents and frequent additional electron-lucent and electron-dense components, have been identified as secondary lysosomes and probably contain much lipoprotein.

The processes which arise from the polar cytoplasm are often irregular in outline and pass between neuronal and macroglial elements. Microglial processes contain similar but fewer organelles than the perinuclear cytoplasm; rough endoplasmic reticulum (RER) and some microtubules are present, but filaments are not a common feature of microglial cytoplasm.

The establishment of the ultrastructural morphology of microglia in undamaged tissue has permitted their recognition in pathological situations where they were previously identified as M cells (Matthews, 1974), third neuroglial type (Vaughn *et al.*, 1970; Fernando, 1971), phagocytic oligodendrocytes (Maxwell and Kruger, 1966), or microglia (Blinzinger and Kreutzberg, 1968; Hollander *et al.*, 1969; Conradi, 1969a,b; Torvik and Skjorten, 1971).

1.2. ORIGIN OF MICROGLIA: PROSPECTIVE PRECURSORS

Despite much speculation over the past half century, it is not known what cells are the precursors of classical microglia; modern experimental techniques are casting much new light in this field. From past investigations, two principal contenders, "ameboid microglia" and monocytes, have emerged as possible precursors of classical microglia in physiological central nervous tissue; an ectodermal origin has arisen as a third possibility (Figure 5).

1.2.1. Ameboid Microglia

More than a century ago ameboid cells were described in the brains of young animals (Virchow, 1867); after the discovery of microglia such cells were considered to migrate from the meninges and from around the ventricles along myelinating tracts (Hortega, 1921). It was envisaged that morphological alterations accompanied their migration within the neuropil so that they developed pseudopodia and eventually became the branched, definitive forms which are classical microglia. Thus, for many years, these plump, fat laden cells, identified as "ameboid microglia" or amoeboid cells which are found in late fetal and early

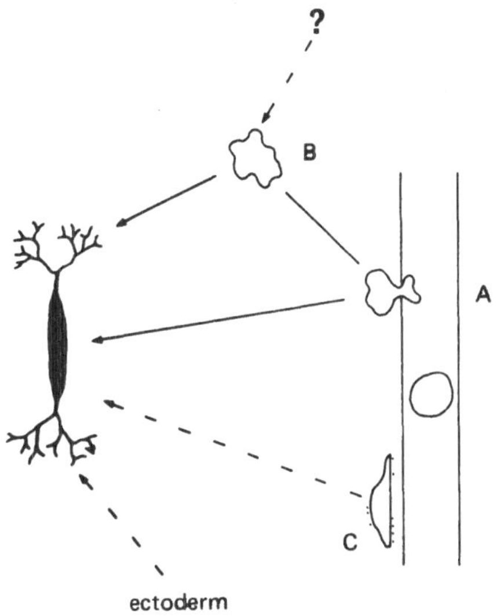

FIGURE 5. Diagrammatic representation of the origin of the microglial cell from (A) monocyte, (B) amoeboid cell (= amoeboid microglial cell, macrophage), (C) pericyte. It has been considered that microglia may also be derived from ectoderm. Broken lines indicate events for which conclusive evidence is not yet available.

neonate brain but not, reportedly, before vascularization (Hortega, 1921; Santha, 1932, 1933; Juba, 1933, 1934; Kershman, 1939; Field, 1955), have been suggested to be the precursors of classical microglia (Hortega, 1921, 1932/1965; Penfield, 1927, 1932; Pruijs, 1927; Gozzano, 1930, 1931; Rydberg, 1932; Juba, 1934; Kershman, 1939; Glees, 1955; Cammermeyer, 1970).

In recent times, ameboid microglial cells in the brains of neonates have been reinvestigated by means of light and electron microscopy (Stensaas and Reichert, 1971; Booz and Felsing, 1973; Schmitt, 1973; Ling, 1976), and there is now unequivocal evidence that such cells are normal features of prenatal and neonatal brain in a number of species. They are somewhat different from the classical microglia described in the brain of apparently healthy adult cats (Baldwin, 1972, 1973, 1975a; Baldwin *et al.*, 1969) and rats (Blakemore, 1969a,b; Mori and Leblond, 1969; Baldwin, 1975b).

In silver impregnations or in stained paraffin sections, ameboid microglia vary in structure from rounded, vacuolated cells to more highly vacuolated forms with branched processes (Kershman, 1939; Stensaas and Reichert, 1971); examination of suitably stained frozen sections reveal that many of the vacuoles contain fat; and histochemical tests establish the presence of acid phosphatase as well as other lysosomal enzymes in the cytoplasm (Ling, 1977).

With the electron microscope (Figure 6), the abundant cytoplasm is shown to be heavily laden with organelles and inclusions (Stensaas and Reichert, 1971; Booz and Felsing, 1973; Ling, 1976; Imamoto and Leblond, 1978). The nucleus is frequently round or oval and has coarse chromatin granules beneath its nuclear envelope. The most prominent cytoplasmic feature is the wealth of lipid droplets and the presence of vacuoles of varying density, some of whose contents may be flocculent; large numbers of primary and secondary lysosomes may be present and a variety of smooth, and some coated, vesicles are evident. A Golgi apparatus and lengths of RER are present, as well as smallish mitochondria of

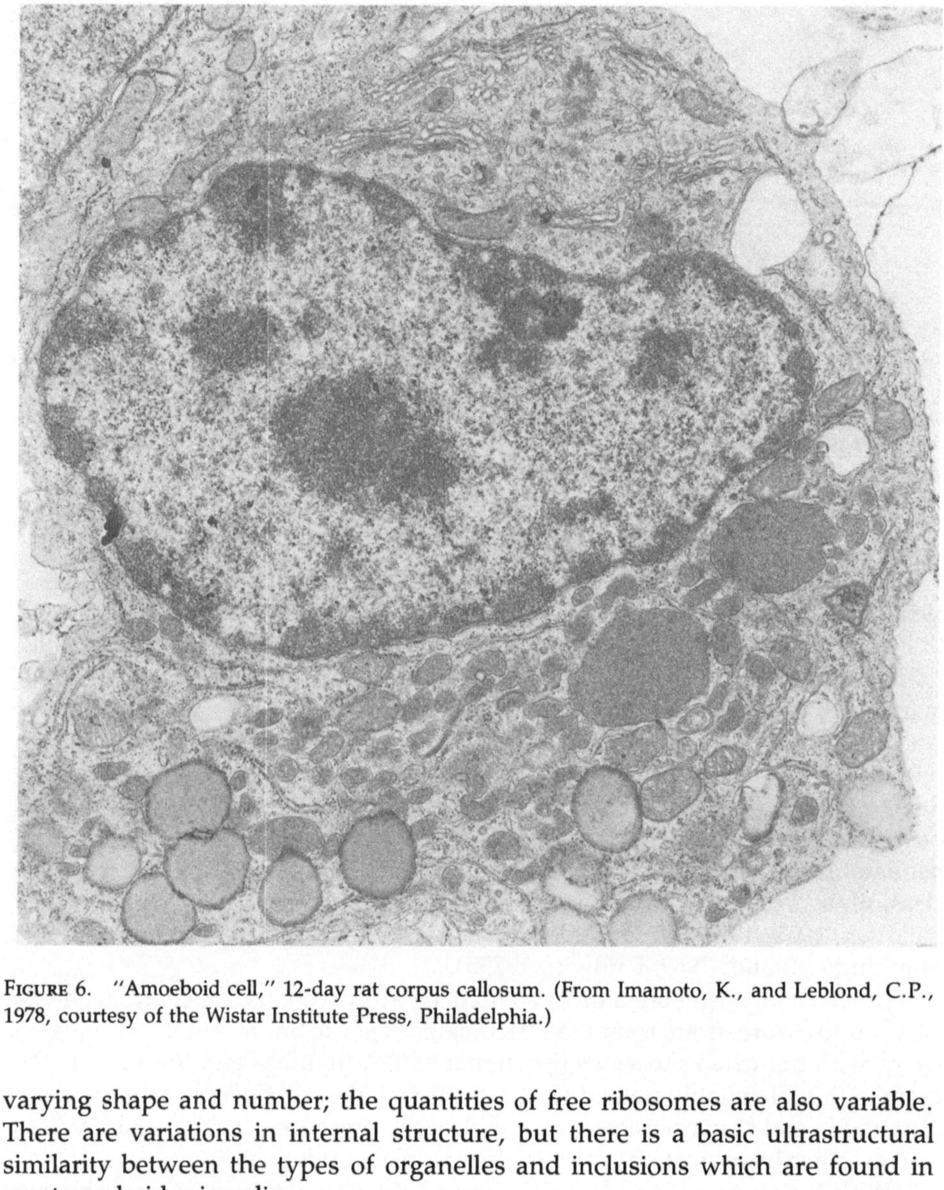

FIGURE 6. "Amoeboid cell," 12-day rat corpus callosum. (From Imamoto, K., and Leblond, C.P., 1978, courtesy of the Wistar Institute Press, Philadelphia.)

varying shape and number; the quantities of free ribosomes are also variable. There are variations in internal structure, but there is a basic ultrastructural similarity between the types of organelles and inclusions which are found in most ameboid microglia.

Fat-laden macrophages found in many pathological conditions are very similar to ameboid microglia; this can be seen by comparing the ultrastructure of ameboid microglia from the brain of young rabbits (Stensaas and Reichert, 1971) and rats (Ling, 1976; Das, 1976a,b), or kitten optic nerve (Baldwin, unpublished results) with that of macrophages in stab lesions (Baldwin, 1974, 1975a).

The role of ameboid microglia as precursors of classical microglia in development has recently received further attention; it has been suggested that ameboid microglia are phagocytic cells (Ling, 1976) or macrophages (Fraher and McDougall, 1975; Das, 1976a,b) which act as microglial precursors during development; furthermore, it has been concluded that macrophages as well as monocytes in adult stab lesions are microglial precursors (Imamoto and Leblond, 1977). However, it is not yet quite certain that ameboid microglia and macrophages are completely identical.

The loose usage of the term *microglia* to encompass not only classical and ameboid microglia but also to denote macrophages, i.e., rod cells (Nissl, 1899) and gitter cells (Nissl, 1904), found under pathological circumstances, is presently inadvisable and the term *microglia* should be used only for classical microglia, unless a prefix is added, e.g., reactive microglia, ameboid microglia (Baldwin, 1975a,b).

1.2.2. Monocytes

There is now some information to suggest that monocytes may be precursors of classical microglia (as opposed to reactive macrophages). Early investigations, which suggested that cells which migrate from around vessels to become microglia (Hortega, 1921), were followed by reports of an association between the first presence of microglia in developing brain and its vascularization (Hortega, 1921; Santha, 1932; Juba, 1933, 1934; Kershman, 1939; Field, 1955), thereby indicating the importance of the vasculature as a source of precursors and prompting some workers to postulate a monocyte origin for microglia (Santha and Juba, 1933).

The possibility that microglia may originate from blood monocytes has been extensively investigated in the past ten years despite considerable technical problems. For instance, it is difficult to identify small numbers of labeled cells, and it is difficult to interpret the results of combined silver impregnation and autoradiography. Despite these difficulties, early experiments, in which radioactively labeled bone marrow from donor animals was transferred to the circulation of recipients (Roessman and Friede, 1967, 1968), provided some indication that monocytes may migrate into undamaged adult brain as well as into brain under pathological circumstances. More recently, investigations, in which monocytes from animals with induced monocytosis, were labeled and transfused into recipients, showed that labeled monocytes migrated into healthy brain and took up locations identical to those classically occupied by microglia (Oehmichen, 1975a); such cells are probably microglia, although the definitive evidence that the labeled cells are classical microglia still eludes us because the entire morphology of the labeled cells is not represented.

The evidence for monocyte contributions to microglia has received further support (Das, 1976a; Imamoto and Leblond, 1977), and a combination of autoradiography and histochemistry has shown that labeled monocytes with high nonspecific esterase activity migrated into the neuropil, and were also present in vascular walls (Oehmichen *et al.*, 1973; Oehmichen, 1975a,b; Oehmichen and Haber, 1976).

1.2.3. Ectodermal Cells

Although the mesodermal origin of microglia has been in vogue for many years, there have always been some to whom an ectodermal origin for microglia has seemed feasible (Metz and Spatz, 1924; Urechia and Elekes, 1926; Creutzfeldt and Metz, 1926; Schaffer, 1926; Bergman, 1926; Jakob, 1927; Reynolds and Slater, 1928; Besta, 1929; Rydberg, 1932). Interest in the origin of microglia from neuroectodermal cells has been rekindled in recent times by new experimental approaches.

The subventricular neuroectodermal cells have once again come under close scrutiny; the presence of mitotically active subventricular neuroectodermal cells in adult rats (the subependymal plate) was described many years ago (Allen, 1912), and their occurrence in rats and other species confirmed on numerous occasions (Opalski, 1933; Kershman, 1938; Globus and Kuhlenbeck, 1944; Bryans, 1959; Smart, 1961; Altman, 1963); but while numerous workers have investigated the structure and role of subventricular cells as a source of neurons and macroglia (Kershman, 1939; Globus and Kuhlenbeck, 1944; Smart, 1961; Smart and Leblond, 1961; Noetzel and Siepmann, 1965; Altman and Das, 1964, 1966; Altman, 1966; Hassler, 1966; Hommes and Leblond, 1967; Privat, 1970; Stensaas and Gilson, 1972; Privat and Leblond, 1972; Paterson *et al.*, 1973; Privat, 1977), comparatively few have investigated the area as a source of microglia of neuroectodermal origin. Some time ago, it was suggested that microglia have a common ectodermal origin, both arising from the mitotic subventricular cells (Pruijs, 1927) which persist in infancy (Schwartz *et al.*, 1932; Opalski, 1933), and that subventricular cells are the source of microglia (Roussy *et al.*, 1930) and of the ameboid multipotential glia found in the subependymal matrix (Rydberg, 1932).

In more recent times the ectodermal origin of microglia was again suggested (Horstadius, 1950); light microscopy showed dividing subventricular cells, thought to be microglia (Lewis, 1968a), and this area was considered to be a general site for microglial production (Cammermeyer, 1965), and for replacement of microglia in the adult (Vaughn and Peters, 1968; Blakemore, 1969a), and thought to be the source of microglia, in addition to the pia and bloodstream (Lewis, 1968b). In optic nerve, microglia were shown to stem from neuroectodermal precursors during Wallerian degeneration (Vaughn, 1969; Vaughn *et al.*, 1970; Skoff, 1975); labeled microglia in autoradiographs were thought to originate from subventricular cells, although alternative origins could not be completely ruled out (Privat, 1970; Paterson *et al.*, 1973). It has been concluded that ameboid microglia in the normal brain of neonate rats are also derived from the proliferative subventricular cells (Ling, 1976) in common with macroglia (Altman, 1962; Paterson *et al.*, 1973), and that ameboid microglia and ramified microglia can transform into each other (Privat and Fulcrand, 1976), it has further been suggested that microglia are derived from the same progenitors as astrocytes and oligodendrocytes (Fujita and Kitamura, 1975).

Despite past and more recent experimentation, the precise contribution made by microglia of ectodermal origin to the overall microglia population is not known; nor are the relationships between microglia of possible ectodermal or mesodermal origin, nor their respective roles, fully understood. For example, it is not known whether microglia of ectodermal origin play a phagocytic role early in development, which predates and is independent of the potentially phagocytic cells of mesodermal origin; and it has not been determined whether microglia do in fact develop from an ectodermal cell line. Investigations currently taking place in this laboratory should throw some light on these topics.

2. BRAIN MACROPHAGES

2.1. ORIGIN OF MICROGLIA: PROSPECTIVE CONTRIBUTORS

2.1.1. Microglia and Monocytes

There has been diversity of opinion concerning the nature and origin of brain macrophages since they were first described (Gluge, 1841). Early investigators suggested brain macrophages originated from leukocytes (Gluge, 1841; Eichhorst, 1875; Ribbert, 1882; Kolbe, 1889; Stroebe, 1894; Schmaus and Sacki, 1901; Ziegler, 1901; Schmaus, 1904, Kaufmann, 1904), macroglia (Schwalbe, 1881, Friedmann, 1890; Nissl, 1899), both leukocytes and macroglial cells (Forster, 1908, Achúcarro, 1909, Marchand, 1909; Merzbacher, 1909), adventitial cells (Hatai, 1902, Da Fano, 1906; Alzheimer, 1904; Nissl, 1904; Devaux, 1908), adventitial cells and leukocytes (Minor, 1904; Farrar, 1908), and adventitial cells, leukocytes, and macroglial (Pick, 1904) and nervous elements (Popoff, 1875; Cerletti, 1903; Metschnikoff, 1913).

The discovery of the microglial cell in 1919 was followed by investigations of its role so that, from the examination of experimental wounds and other lesions, it was deduced that microglia substantially altered their morphology to become phagocytic cells (Figure 7); the alterations they underwent depended upon the prevailing pathological circumstances (Hortega, 1919b; 1920a,b; 1921; Penfield, 1925a,b). The phagocytic role of microglia and their importance as the precursor cells of the brain macrophages gained widespread acceptance and was further substantiated in the ensuing years (Hortega and Penfield, 1927; Carmichael, 1929; Russell, 1929; Hortega, 1932, 1939; Penfield, 1932; Glees, 1955; Hain, 1963; Fox and Inman, 1966; Polak and Haymaker, 1968; Feigin, 1969; Cammermeyer, 1970).

For many years, suggestions that there were alternative sources of brain macrophages received comparatively little serious consideration; however, the introduction of new experimental tools facilitated investigations which might clarify not only the structure and role of microglia but responses of other cells to brain damage, whose significance had long remained in the shadows. Since the early 1960s, there has been mounting evidence for the monocytic origin of many brain macrophages (Figure 7), migration from the general circulation was indicated by demonstrating that previously labeled mononuclear cells could be detected in autoradiographs of damaged brain (Adrian and Walker, 1962; Kosunen

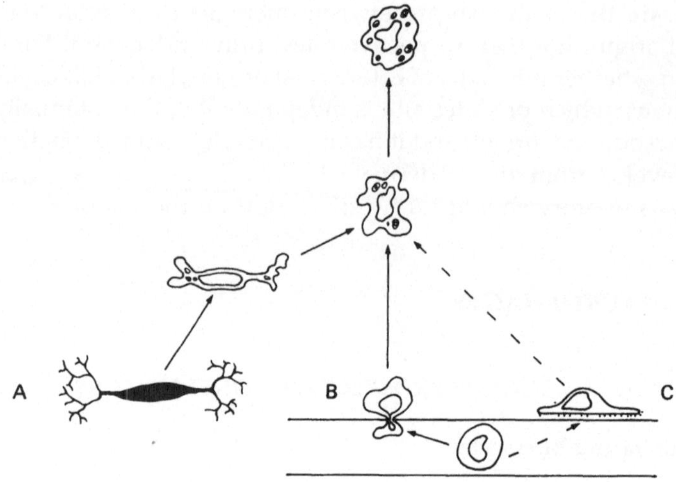

FIGURE 7. Origin of brain macrophages. Contributors: (A) microglia, (B) monocytes, and (C) pericytes. Broken lines represent incomplete evidence.

and Waksman, 1963; Konigsmark and Sidman, 1963; Huntington and Terry, 1966; Smith and Walker, 1967; Roessman and Friede, 1967, 1968; Adrian and Smothermon, 1970). Supporting, but nonautoradiographic, evidence of the contribution made to the brain macrophage population by circulating mononuclear cells was presented by Anderson *et al.* (1966), who labeled circulating mononuclear cells by repeated intraperitoneal injections of a carbon suspension (Hurley *et al.*, 1966); following cold injury of the brain carbon-labeled macrophages appeared in the lesion.

For a long time, knowledge of the role of microglia and contributors to brain macrophages was based upon the morphological interpretations of investigators who examined brain damaged by puncture (Hortega, 1919b; Hortega and Penfield, 1927; Penfield and Buckley, 1928). Many years later, light and ultrastructural examination of cell responses to similar stab wounds revealed that microglial activation was manifested by morphological alterations and accentuation of endocytosis (Baldwin, 1971b, 1972, 1973, 1975a). Microglia, however, were not the sole contributors to brain macrophages, because there was also participation by monocytes which were seen migrating from the circulation (Baldwin, 1975a; Imamoto and Leblond, 1977). The identifying characteristics of monocytic and microglial precursors of brain macrophages diminished with their advancing maturity, but, in addition to endocytic activities by cells of these origins, there were also perivascular cells which amassed products of degeneration (Baldwin, 1971a, 1975b). Similarities between perivascular cell morphology and their types of inclusions, and those of macrophages in the damaged neuropil, suggested a possible common identity. An ultrastructural study of thermal necrosis also provided evidence for the activation of microglia and microglial and monocyte contributions to brain macrophages (Blakemore, 1969a, 1972, 1975). Experiments which utilized the known monocyte and macrophage characteristics of adhering to the glasslike surfaces of polyester film (Melinex) implanted in the cerebral cortex of rats made it possible to examine the contribution of monocytes to brain macrophages and to examine other cellular responses in the neuropil (Baldwin, 1975b): the morphological and autoradiographic results showed that the major proportion of the adherent monolayer of macrophages, as well as the majority of those in the neuropil, were of monocyte origin, but also indicated activation of local microglia. Numerous other investigators have reported the monocytic origin of brain macrophages (Stenwig, 1972; Kitamura *et al.*, 1972; Murray and Walker, 1973; Matthews and Kruger, 1973) under different circumstances, for example, in electroshock (Cowan *et al.*, 1970) and experimental allergic neuritis (Astrom *et al.*, 1969; Robinson *et al.*, 1972).

In other studies, monocytes contributed to brain macrophages in brain stab wounds (Fujita and Kitamura, 1975) and in extensive lesions where there was increased vascular permeability (Torvik, 1975). Phagocytes of mononuclear cell origin were found in the degenerating optic nerve of the newt (Turner, 1975), but hematogenous cells contributed very few cells in degenerating rat optic nerve where phagocytes were principally derived from glial precursor cells (Skoff, 1975). The contribution of migrated monocytes to brain macrophages was further shown by autoradiographic and histochemical means (Oehmichen *et al.*, 1973; Oehmichen, 1975a, 1975b), although in Wallerian degeneration macrophages presumably came from local microglia (Oehmichen and Torvik, 1976). Mononuclear cells of vascular origin have also been shown to be macrophage contributors in degenerating retina (O'Steen and Karcioglu, 1974), and further investigations have corroborated the monocytic origin of macrophages in brain wounds (Imamoto and Leblond, 1977).

2.1.2. Morphology

The demonstration of microglia (Hortega, 1919a) opened the way for investigation of their reactions under a variety of pathological conditions (Hortega, 1919b, 1920a,b; Penfield, 1925a,b) and enabled recognition that the classical microglial cell could undergo alterations to become a rod cell or a large, fat- and inclusion-laden cell which became known by many different names (see Baldwin, 1975b), e.g., gitter cell and compound granular corpuscle. The changes undergone by microglia, the morphology of the resulting cell types, and the dependence of the level of alteration upon the prevailing pathological circumstances are well established in the literature of neuropathology. The classical literature has indicated that the formation of rod cells occurs by a retraction of microglial cell processes resulting in a long bipolar cell with an elongated nucleus. The cytoplasm may contain small vacuoles and fat droplets as well as small amounts of other inclusions; such cells occur under conditions of mild abnormality, particularly where microglial involvement takes place slowly.

The formation of vacuolated, fat- and inclusion-laden macrophages (i.e., gitter cells, compound granular corpuscles) from classical microglia, which occurs in more severe damage, and the suggested sequence of morphological events (Figure 7), have long been accepted. An increase in the perinuclear cyto-

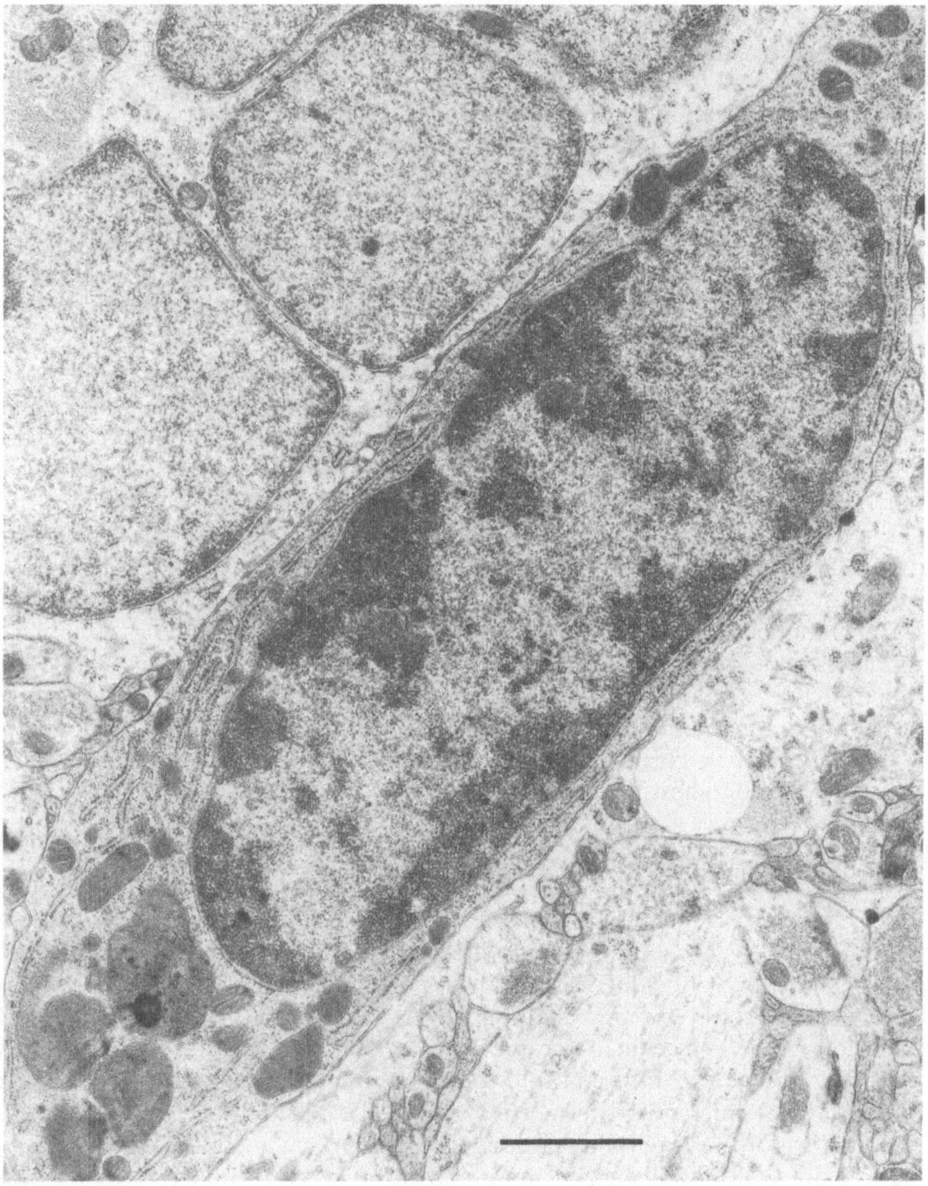

FIGURE 8. Reactive microglial cell in rat cortex damaged by Melinex implantation. Bar = 1 μm. (From Baldwin, F., 1975b, courtesy of Blackwell Scientific Publications, Oxford.)

plasm and a decrease in cytoplasmic processes is followed by a further shortening of processes, which become swollen at their bifurcations, and an increase in cytoplasmic vacuolation; a further gradual rise in perikaryal volume and almost complete retraction of processes, results in a roundish or lobulated cell with short pseudopodia and further increases in cytoplasmic volume. Distension with phagocytized inclusions, particularly lipid, culminates in a large, engorged macrophage.

In damaged brain, early ultrastructural alterations can be recognized in microglia identified by the same characteristics previously established for classical microglia in physiological brain; their identification is also assisted by their perineuronal or juxtavascular location.

Around cortical wounds, the first signs of microglial motivation or accentuation of activity are manifested by increased cytoplasmic vesiculation, and by

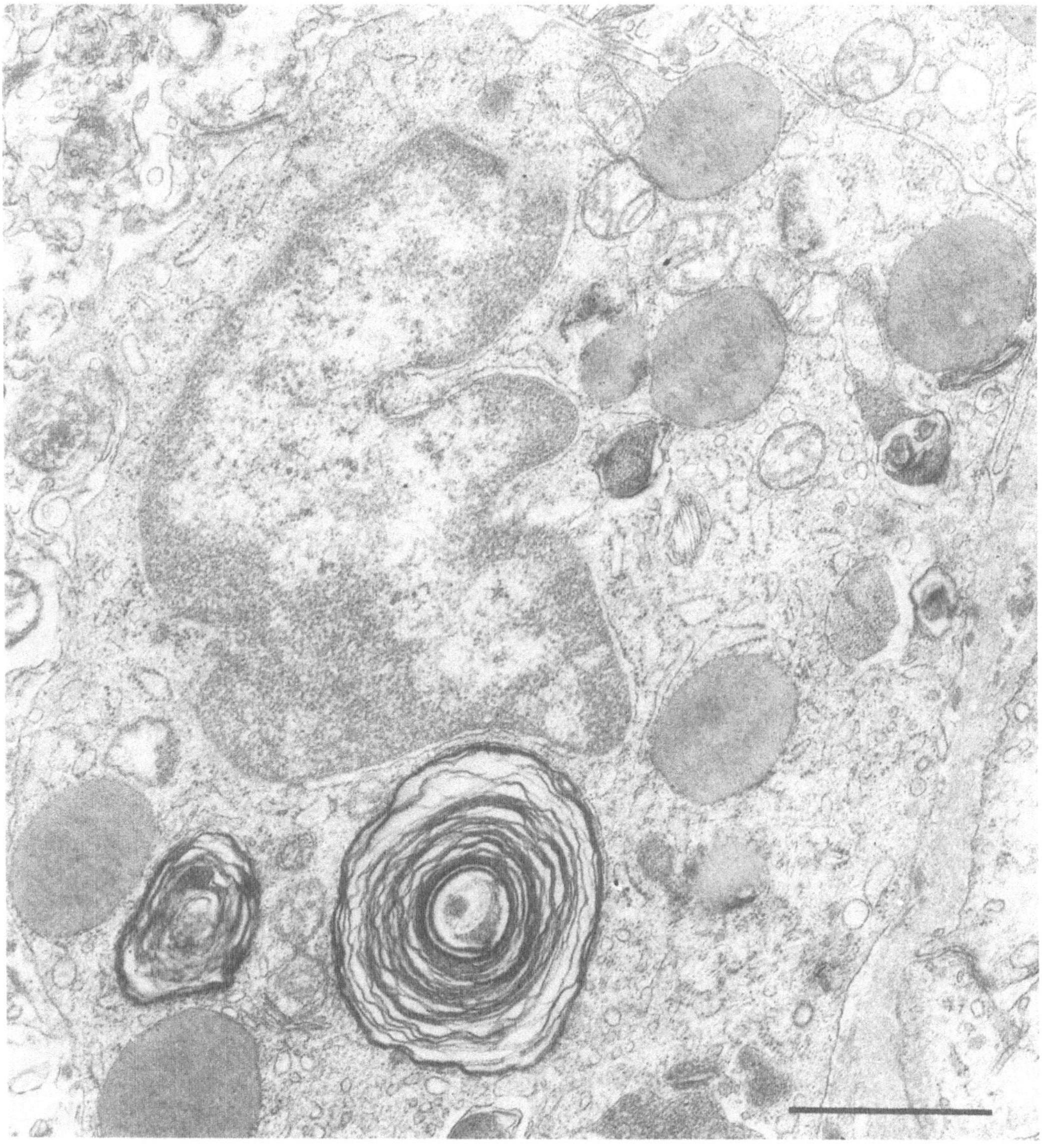

FIGURE 9. Macrophage. Stab-wounded cat cerebral cortex, 72 hr. Bar = 1 μm.

increasing amounts of inclusions and lysosomes (Figure 8); an increase in mitochondria, ribosomes, and ER, point to enhanced metabolic activity associated with the increasing activity of the cell. However, as activated microglia become more developed, their organelle complement changes, their overall morphology alters, and they progressively lose many of their former characteristics, so that a limit is reached beyond which they cannot be identified with certainty.

The contributions made by monocytes to brain macrophages have become more firmly established in the past 15 years, but while macrophages in general have separate ultrastructural features at different stages of development (Baldwin, 1975a), and while monocytes can be identified in vascular walls or after they enter the neuropil from the vasculature, once their endocytic activity markedly alters their morphology, macrophages of monocyte origin cannot be distinguished from those derived from microglia on morphological grounds alone (Blakemore, 1969b, 1972; Baldwin, 1975a,b).

The wide range of morphologies exhibited by brain macrophages as they

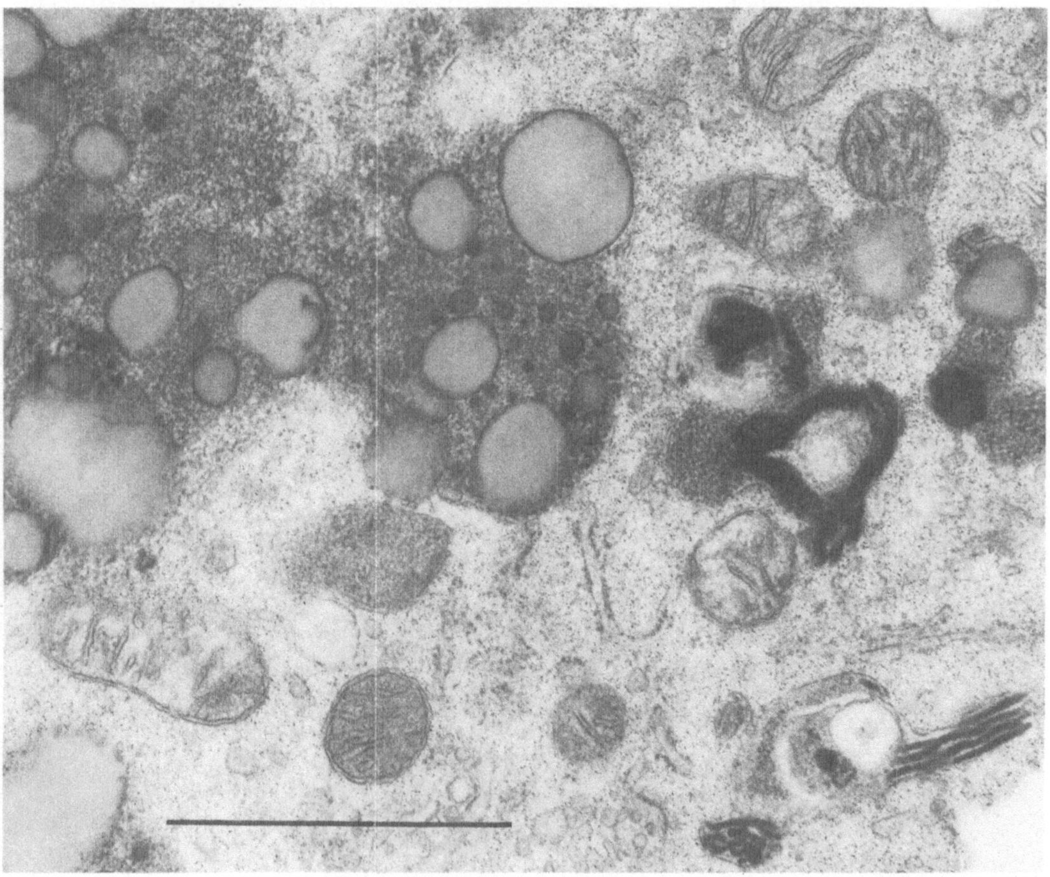

FIGURE 10. Macrophage. Stab-wounded cat cerebral cortex, 10 days. Bar = 1 μm.

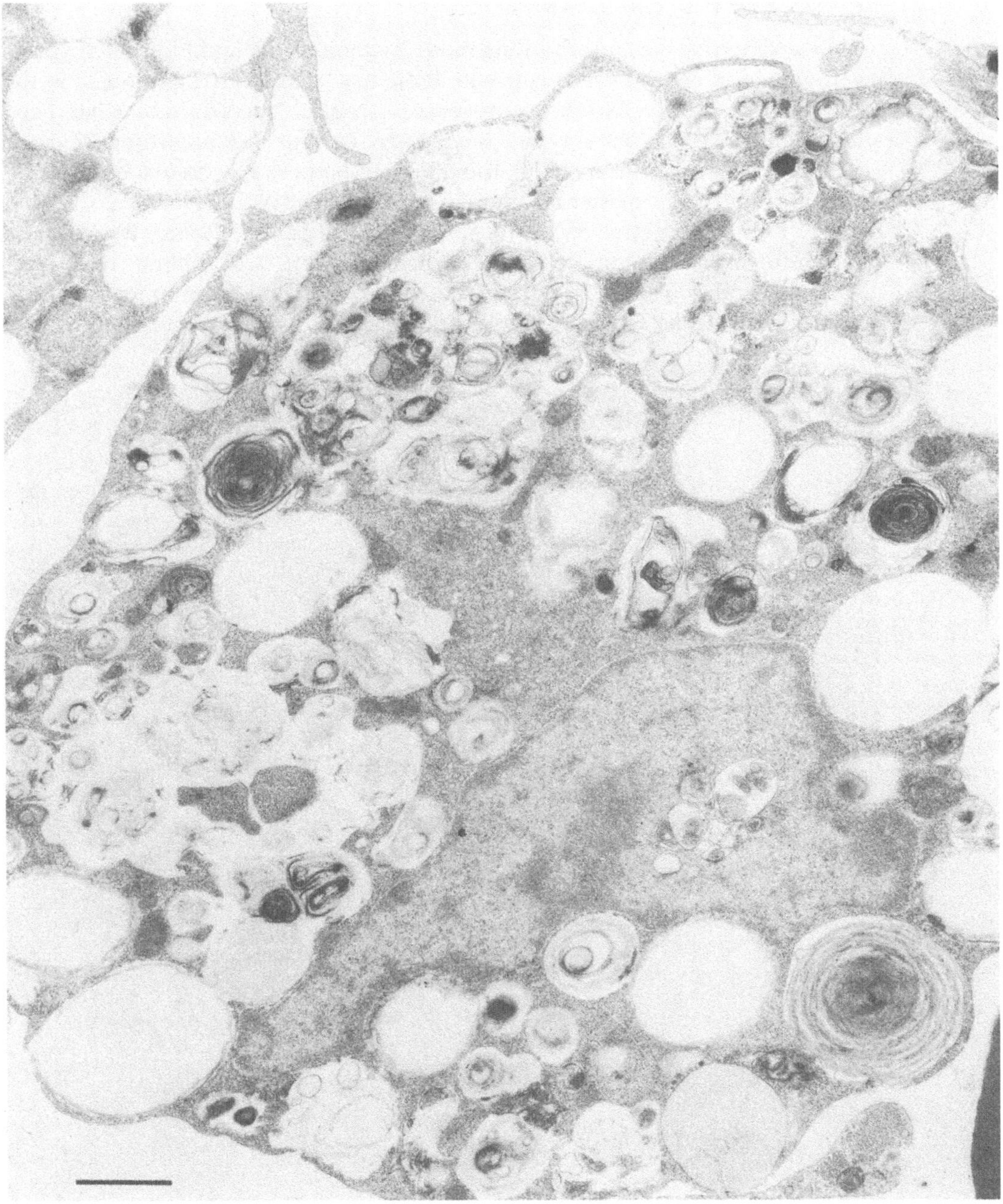

FIGURE 11. Large, heavily engorged macrophage. Stab-wounded cat cerebral cortex, 20 days. Bar = 1 μm.

become progressively altered by their endocytic activity culminates in macrophages whose cytoplasm is distended by a wealth of inclusions and organelles (Figures 9–11).

2.2. ROLE OF PERIVASCULAR CELLS

In addition to the contributions made by monocytes and local microglia to macrophages in central nervous tissue, there has been gathering interest in the role of perivascular cells whose presence in central nervous tissue has been widely documented. Two types have been categorized according to their granularity (Cammermeyer, 1970); the nongranular pericyte, occurs around vessels in gray and white matter of central nervous tissue (Wolff, 1963; Maxwell and Kruger, 1965; Cammermeyer, 1970; Jones, 1970; Baldwin, 1975a), is similar to pericytes described around vessels elsewhere (Fawcett, 1963; Majno, 1965; Rhodin, 1968), and is a long and arborized cell (Wolff, 1963); the other type, the granular pericyte, is found in similar perivascular locations and is more fre-

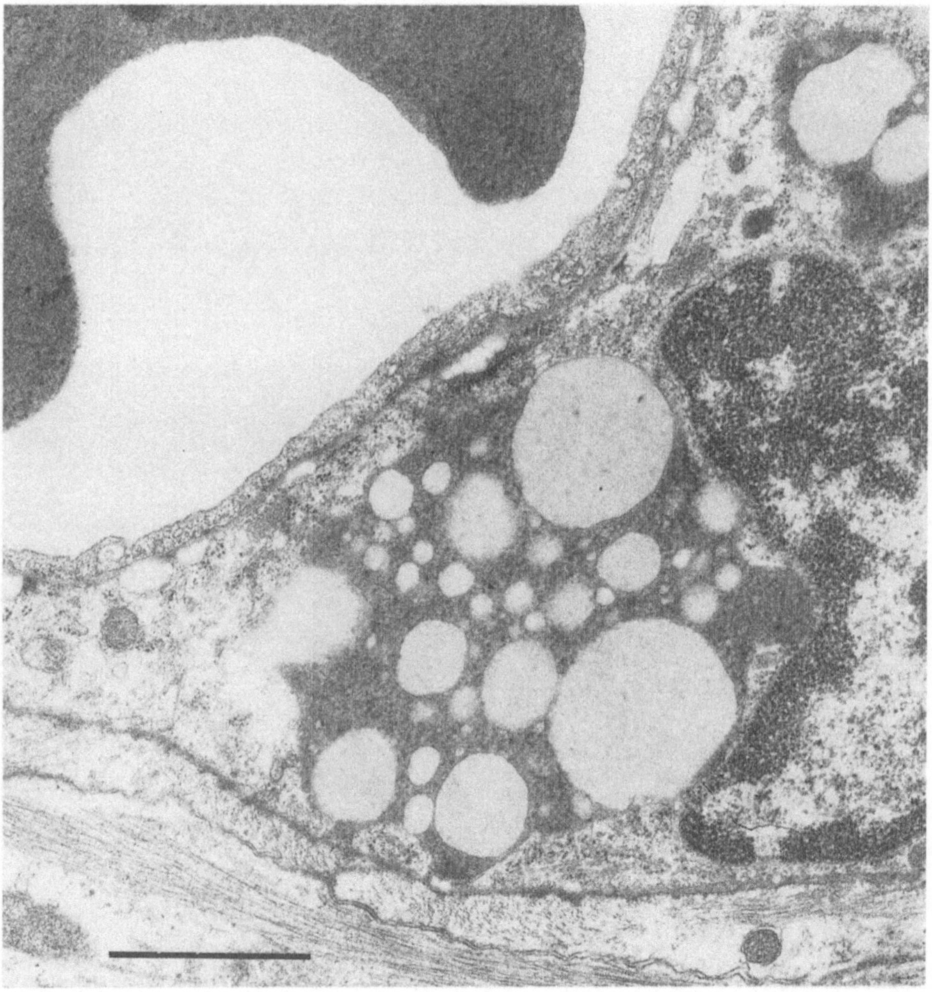

FIGURE 12. Perivascular cell (granular pericyte) showing vacuolated, granular lipofuscin. Cat cerebral cortex. Bar = 1 μm.

quently present in gray matter than in white. The properties of the granular pericyte have been widely documented (Dunning and Furth, 1935; Chessick, 1953; Gomori and Chessick, 1953; Barka and Anderson, 1962; Duckett and Pearse, 1965; Jones, 1970; Cammermeyer, 1970), and examinations of their structure and histochemistry in cat cerebral cortex (Baldwin, 1975a) have revealed that the membrane-bound vacuolated granular inclusions (Figure 12) are lipofuscin.

A number of functions have been generally ascribed to pericytes. For example, their phagocytic capability, in a variety of tissues, has been recognized for many years (Tschaschin, 1913; Vimtrup, 1922; Krogh and Vimtrup, 1932; Zimmerman, 1923; Bolsi, 1931; Lebowich, 1934; Worthington and Childs, 1959; Majno and Palade, 1961; Tani and Evans, 1965), and in neuropathology, their phagocytic role has been recorded in neoplasms (Nystrom, 1960; Torack, 1961), in Tay–Sachs disease (Terry and Weiss, 1963), in Krabbes disease (Andrews *et al.*, 1971), and following irradiation (McDonald, 1962; Maxwell and Kruger, 1965; Roizin and Schade, 1968); pericytes also incorporate tracers around needle tracks in brain (Selwood, 1970, 1971), and take up tracers following thermal injury (Cancilla *et al.*, 1970, 1972; Baker *et al.*, 1971). Several other roles for pericytes have been suggested; they have been implicated in the detection of substances in the interstitium (Rhodin, 1968), or in filtering substances passing between blood and tissue (Toro, 1942), they have also been suggested as regulators of the microcirculation (Tusques *et al.*, 1975), as well as participants in metabolic function (Cammermeyer, 1970; Tusques *et al.*, 1975), and may be responsible for the removal of lipid waste products (Scholz, 1957).

There is now considerable interest in the role of pericytes in central nervous tissue and their relationships to other cells. There has been increasing interest in their importance as a source of microglia (Cox, 1976; Hager, 1975) and of brain macrophages (Scholz, 1957; McDonald, 1962; Blinzinger and Hager, 1964; Maxwell and Kruger, 1965; Hager, 1968; Samorajski *et al.*, 1968; Mori and Leblond, 1969; Baron and Gallego, 1972; Matthews and Kruger, 1973; Hager, 1975), and various authors have referred to the similarities between microglia and pericytes. It has been suggested that pericytes are histiocytic cells corresponding to microglia (Baldwin *et al.*, 1969): "pericytal microglia" have been recognized (Mori and Leblond, 1969), and the microglial nature of pericytes has been described (Boya, 1975).

In addition to the endocytic roles played by microglia and monocytes in response to various pathological conditions, under some circumstances, there are also responses by perivascular cells. Taking cerebral stab wounding as an example, increased pinocytosis and cytoplasmic vesiculation are the earliest signs of activation shown by pericytes, but with advancing time a wide range of inclusions are present in cells compartmented within the basal lamina (Figure 13); some of the cells are evidently granular pericytes because of their lipofuscin inclusions. Whether all cells, contained within the vascular basal lamina, are all derived from pericytes is open to question, but the activation of pericytes in response to damage is certain. Perhaps debris-laden perivascular cells and macrophages in the neuropil are similar because pericytes, monocytes, and tissue

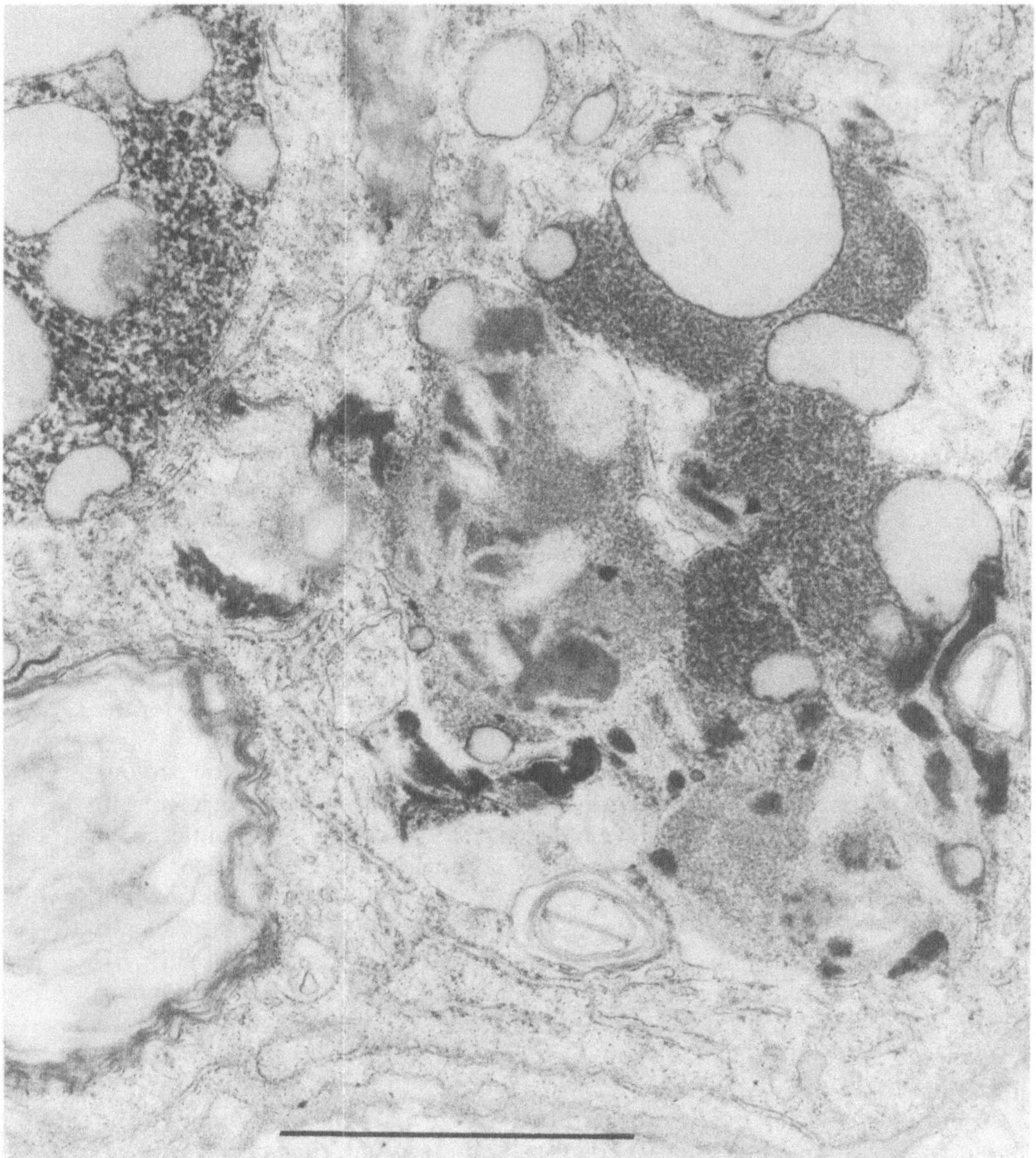

FIGURE 13. Perivascular cell cytoplasm showing a wide variety of inclusions; part of a vacuolated lipofuscin mass is at the upper left. Stab-wounded cat cerebral cortex, 10 days. Bar = 1 μm.

macrophages share a common origin (Spector and Willoughby, 1968). It has been shown, in a series of autoradiographic and enzyme histochemical experiments on healthy, as well as pathological brains, that monocytes from the general circulation take up locations in vessel walls where pericytes are usually found, thereby suggesting a monocyte origin for pericytes (Oehmichen *et al.*, 1973; Oehmichen, 1975a,b; Oehmichen and Huber, 1976).

It is not clear whether the inclusion-laden perivascular cells in stab lesions are cells which have collected their prodigious amounts of inclusions *in situ* or whether some are cells which have collected debris in the neuropil prior to moving into the perivascular space.

3. ENDOCYTIC ROLE OF ASTROCYTES

Many years ago it was suggested that astrocytes were capable of becoming phagocytic cells (Schwalbe, 1881; Friedmann, 1890; Nissl, 1899; Pick, 1904). More recently, the phagocytic ability of astrocytes has once again come under the scrutiny of experimentalists who have recognized inclusion-laden astrocytes and suggested their involvement in the removal of cellular debris; for example, in Wallerian degeneration (Vaughn and Pease, 1970; Vaughn *et al.*, 1970; Fernando, 1973; Torvik, 1975). In investigations of the cellular responses of optic nerve during degeneration, astrocytic cells accumulated a variety of cytoplasmic inclusions (Blunt and Baldwin, 1972, 1973), and were said to develop from macroglial precursors (Skoff, 1975). In other studies, astrocytes accumulated large lipid droplets following ultraviolet irradiation (Rubinstein *et al.*, 1971), as well as protein from the extracellular space in thermal wounds (Blakemore, 1971).

The importance of phagocytosis by astrocytes is as yet uncertain.

In developing nervous systems, several studies have implicated astrocytes, as well as macrophages, as the agents involved in phagocytosis (Levi-Montalcini, 1950, 1964; O'Connor and Wyttenbach, 1974) but the identity of the cell(s) responsible for removing products of degeneration is still in question; while it is quite probable that migrated monocytes and local microglia contribute to macrophages in developing brain, as they do in damaged brain, the precise identity of the responsible cells in early development, prior to vascularization, remains undecided. Monocytes are unlikely candidates if comparisons can be made between brain and other tissues where macrophages of monocyte origin are not present before vascularization (Anderson and Matthieson, 1966); microglia are also said to be absent from the brain before vascularization (Hortega, 1921). Perhaps in the absence of these contenders, it is possible that responsibility for the removal of cell debris in early development may rest with an alternative cell(s) of ectodermal origin. The identity of the cellular elements responsible is under investigation; whether astrocytes fulfill this purpose as one of their roles, remains to be seen.

4. DISCUSSION

There is now general agreement on the light and electron-microscopic characteristics by which classical microglia may be identified; there is also abundant evidence of endocytic activity by microglia, but it is equally clear that local microglia do not necessarily contribute all macrophages under all pathological circumstances, as was thought for many years.

The precise origin of classical microglia is still under investigation, and is a topic which has been approached recently with renewed vigor. Good evidence for the passage of monocytes from the circulation into the brain of healthy early neonates and, to a certain extent, older animals has begun to emerge. It has been shown that monocytes take up locations where microglia are classically found, and are contributors to the inclusion- and fat-laden cells known as ameboid microglia or ameboid cells, which have been recognized traditionally as the preferred microglial precursor; such cells bear strong resemblances to brain macrophages. Recent investigations implicate ameboid cells, in development, and macrophages, in damage, as contributors to the microglial population, although the events which accompany their transition, and the fate and possible function of their inclusions remain a mystery at present. Current knowledge of the precise origin and role of pericytes in central nervous tissue is somewhat sketchy, but there is evidence indicating that cells in a perivascular location are derived from monocytes; from the available evidence, it seems likely that pericytes have monocytes as their precursors. It appears possible that at least some pericytes are monocytes whose migration has not progressed beyond the vascular compartment and are "waiting in the wings" until called upon to perform by augmenting the activities of the microglial population when required.

While convincing evidence is mounting for a monocyte origin for microglia, the question of whether some microglia originate from additional ectodermal precursors has not yet been resolved; consequently the precise contribution by microglia of ectodermal origin to the overall population is unknown; nor are the relationships between microglia of possibly different origins or their roles understood. While there is no doubt about the role of monocytes as precursors of macrophages in damaged brain and evidence for their similar contributions in late development, the nature and origin of cells responsible for collecting cellular debris at earlier stages of development during intrauterine life and, particularly, before vascularization of brain, remain unresolved, but are currently being investigated in this laboratory.

The activation of microglia to endocytosis and their role as contributors of brain macrophages are well known, but, in addition, it is certain that monocytes are able to migrate from the vasculature and cross the blood–brain barrier and that they may contribute substantially to the macrophage population; the individual contributions made by local microglia or migrated monocytes depend upon the prevailing conditions. The precise requirements necessary for their participation await clarification. The mechanism which triggers activation of microglia is unknown, but it is, perhaps, possible to speculate, with the knowledge that pinocytosis is the earliest recognizible event in microglial activation in brain stab wounds, that uptake of some material by pinocytosis may be important in providing the necessary stimulus; the identity of such a factor or factors is unknown, but its local availability may account for the differences between the extent of reactivity of microglia in different areas around a wound. It may be that motivation can be provided by the availability of one, or a variety, of factors, whose presence depends upon the type of damage and dictates whether local microglia—monocytes, perivascular cells, perhaps some other cell types, for example, astrocytes, or perhaps a combination of these cells—are responsible for endocytic activity. Such a selective mechanism could account for the different participants under specific circumstances; for example, the previously noted roles played by microglia, monocytes, and perivascular elements following cortical puncture, microglia, monocytes, and astrocytes in thermal lesions, and astrocytes in Wallerian degeneration.

The evidence for the monocyte origin of pericytes which is now accumulating makes a close relationship between them, microglia, and brain macrophages seem more certain; it may be that nongranular pericytes replenish microglia when required in healthy tissue and act as a pool from which additional brain macrophages may be recruited in times of emergency, perhaps by proliferation, if the dividing cells seen in perivascular locations in wounds are indeed pericytes. The exact roles of the granular pericytes and origin of their cytoplasmic inclusions in undamaged brain await clarification; their contents may be accumulated locally, in which case there must, at some stage, be interruptions in the basal lamina for large particle ingestion to take place; on the other hand, they may have collected their inclusions in the neuropil before moving to a perivascular location (Figure 14). In cerebral stab lesions, granular pericytes within the basal lamina of the vascular compartment generally resemble macrophages in the neuropil. Numerous debris-laden macrophages are often present around vessels; those in the neuropil outside the basal lamina are undoubtedly macrophages and appear to have a common identity with those inside the basal lamina. The reasons for the increase in cells and the significance of their perivascular arrangement in wounds remains to be determined but, in part, may ultimately be associated with the fate of macrophages in lesions.

Generally speaking, there is little information about the fate of macrophages in inflammation, and almost nothing is known about how their decline is accomplished in central nervous tissue. It is evident that engorged macrophages may be present at the site of brain injury, months, or even years later, but the numbers present are far fewer than at earlier times.

In cerebral stab lesions the increase in number of macrophages, achieved by

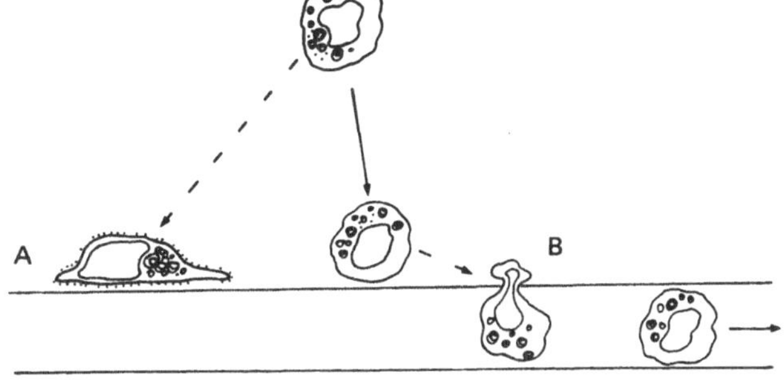

FIGURE 14. Fate of brain macrophages. (A) Migration of macrophages to perivascular locations where they may remain and be contained within the vascular compartment as granular pericytes surrounded by basal lamina. (B) Movement of macrophages into the vascular lumen. Broken lines indicate incomplete evidence.

monocyte migration and microglial activation and local proliferation and additional pericyte activity, reaches a peak by about 20 days postoperative, after which time a decline in macrophages in the neuropil is evident. Macrophages are found with increasing frequency around blood vessels and there is also a rise in the number of engorged cells contained within the basal lamina of the vascular wall. By 50 days postoperative, there are comparatively few macrophages left; those which remain are mostly enveloped by astrocyte processes of the glial scar, while a few are in vessel walls. In cerebral stab wounds many macrophages contain hemosiderin, derived from breakdown of phagocytized red cells; such macrophages whose cytoplasm also contains other debris, can be readily identified in the neuropil where many are arranged around vessels. Similar cells are also present in vessel walls; it is arguable that some of them may be pericytes, which have phagocytized red cells *in situ* to accumulate their iron contents. This would require the absence of the basal lamina between pericytes and neuropil at some stage for phagocytosis to take place, since at no stage was there any extracellular iron available for pinocytosis. On the other hand, if their inclusions were not derived from phagocytosis *in situ,* the cells must have moved into perivascular locations after picking up red cells and other debris in the neuropil, in which case they are migrated macrophages. Preliminary investigations into the fate of brain macrophages by the author show that peritoneal macrophages labeled with markers and placed in the brain as a small soft "pellet" gradually disperse and that many become located around vessels while others pass through vascular walls; some labeled cells appear to make their way to the ventricles, others are found in the choroid plexus. Thus far, while there is a little evidence for the role of the vasculature and the migration of macrophages out of brain wounds, there is no evidence for localized breakdown of macrophages which would account for their decline; the ways in which the reduction is achieved are open to speculation. It seems likely that the arrangement of macrophages around vessels in wounds is significant, and it appears feasible to suggest that migration out of the brain (Figure 14) is responsible for the reduced numbers of macrophages found with advancing time and that the brain blood vasculature, in the absence of lymphatics, provides a route, although other avenues, for example, the cerebrospinal fluid, where debris-laden macrophages are often found, may also be used.

There are a number of fascinating similarities between other tissues, where monocytes have been shown to be the precursors of both "fixed" and free macrophages (Goldman and Walker, 1963; Kosunen and Waksman, 1963; Volkman and Gowans, 1965a,b; Spector *et al.*, 1967; van Furth and Cohn, 1968; van Furth *et al.*, 1970; van Furth and Dulk, 1970; Langevoort *et al.*, 1970; Gordon and Cohn, 1973; and numerous others) and brain. For example, microglia are said not to appear in brain (Hortega, 1921, and others), or mononuclear phagocytes or their precursors in other organs (Anderson and Matthieson, 1966), until after vascularization of the respective tissues. It has been noted that microglial precursors elongate as migration proceeds, culminating in familiar argyrophilic microglia (Hortega, 1921; Kershman, 1939), and that there is a resemblance between them and similarily stained Kupffer cells (Hortega and Asua, 1921; Dunning and

Furth, 1935) and connective tissue macrophages (Juba, 1933) which are both of known monocytic origin.

In brain, the close relationship between surface membrane features of microglia, monocytes, reactive microglia, and brain macrophages (Oehmichen and Huber, 1976), the morphological similarities of brain macrophages, no matter whether derived from recently migrated monocytes or local microglia, and mounting autoradiographic evidence for their common origins (Imamato and Leblond, 1977, 1978) emphasizes that the role of the monocyte as a macrophage precursor is similar in brain as elsewhere; so that taking into account the role of the monocyte as a probable precursor of microglia, the known role of the monocyte as a contributor to macrophages in brain, and the established monocyte origins of macrophages in other tissues, it can be proposed that the microglia may occupy the same niche as tissue macrophages elsewhere, with which they have much in common, and that the contribution by monocytes in brain is identical to their contribution to tissue macrophages elsewhere. This would imply that, at times, after vascularization has taken place, microglial precursors (monocytes) migrate into the brain and assume the characteristically ramified morphology of microglia in an environment where cell ramification is the rule rather than the exception. It would be logical for microglia, as tissue macrophages, to take up locations where they could be easily called upon or where their phagocytic talents would be most required. Taking the perineuronal location of microglia in physiological brain as an example, it may be that the inclusions often found in perineuronal microglia are the ingested products of gradual degeneration of the adjacent neuron, and thus reflect its physiological state; the perivascular location of many microglia is comparable with the distribution of tissue macrophages elsewhere.

If monocytes are the precursors of microglia and brain macrophages, it follows that the situation is homologous with what takes place in other tissues. It is, therefore, justifiable to include microglia and brain macrophages within the MPS or RES, which operates in brain as it does elsewhere, although central nervous tissue is different in possessing additional cells, astrocytes, which may be phagocytic.

The evidence that monocytes migrate into brain and migrate into damaged brain in larger numbers, their activity as macrophages, their proliferative capacity, and their ability to persist for extended periods, indicates the presence of the three mechanisms for maintaining a macrophage population, recruitment, proliferation and longevity (Spector and Ryan, 1970). The inflammatory response might equally apply to brain as it does elsewhere; the notion of an endogenous population of microglia being the sole contributor of brain macrophages was bolstered by the concept of an impenetrable brain barrier, but it is now clear that it does not necessarily apply to cell migrations.

Further knowledge of brain macrophages and microglia will depend on such techniques as the study of isolated macrophages; it is now possible to obtain pure populations of macrophages from fetal and neonate brain (Baldwin, in preparation) by culturing cells from dissociated cerebral cortex and causing the preferential development of macrophages; by the use of suitable conditions of nutrition and pH, pure populations of brain macrophages are thus available for close scrutiny, which enables their nature, similarities, and differences to be determined and their relationships to members of the MPS in other tissues to be established; the results of preliminary investigations are heartening.

REFERENCES

Achúcarro, N., 1909, Cellules allongées et stäbchenzellen cellules neurologiques et cellules granuloadipeuse à la corne d'Ammon du lapin, *Trab. Lab. Invest. Biol.* **7**:201.

Adrian, G. N., and Smothermon, R. D., 1970, Leucocytic infiltration into the hypoglossal nucleus following injury to the hypoglossal nerve, *Anat. Rec.* **166**:99.

Adrian, E. N., and Walker, B. E., 1962, Incorporation of thymidine-H^3 by cells in normal and injured mouse spinal cord, *J. Neuropathol. Exp. Neurol.* **21**:597.

Allen, E., 1912, The cessation of mitosis in the central nervous system of the albino rat, *J. Comp. Neurol.* **22**:547.

Atlman, J., 1962, Autoradiographic study of degenerative and regenerative proliferation of neuroglia cells with tritiated thymidine. *Exp. Neurol.* **5**:302.

Altman, J., 1963, Autoradiographic investigation of cell proliferation in the brains of rats and cats, *Anat. Rec.* **145**:573.

Altman, J., 1966, Proliferation and migration of undifferentiated precursor cells in the rat during gliogenesis, *Exp. Neurol.* **16**:263.

Altman, J., and Das, G. D., 1964, Autoradiographic examination of the effects of enriched environment on the rate of glial multiplication in the adult rat brain, *Nature (London)* **204**:1161.

Altman, J., and Das, G. D., 1966, Autoradiographic and histological studies of postnatal neurogenesis, *J. Comp. Neurol.* **126**:337.

Alzheimer, A., 1904, Histologische Studien zur Differentialdiagnose der progressiven Paralyse, *Histopathol. Arb. Nissl-Alzheimer* **1**:18.

Anderson, H., and Matthieson, M. E., 1966, The histiocyte in human foetal tissues. Its morphology, cytochemistry, origin, function and fate, *Z. Zellforsch. Mikrosk. Anat.* **72**:193.

Anderson, R. McD., Arumugan, S., and Ryan, G. B., 1966, The origin of brain macrophages in the rat, *Proc. Austral. Assoc. Neurol.* **4**:113.

Andrews, J. M., Cancilla, P. A., Grippo, J., and Menkes, J. H., 1971, Globoid cell leukodystrophy (Krabbe's disease): Morphological and biochemical studies, *Neurology* **21**:337.

Astrom, N. E., Webster, H. de F., and Arnason, B. G., 1969, The initial lesion in experimental allergic neuritis, *J. Neuropathol. Exp. Neurol.* **28**:172.

Baker, R. N., Cancilla, P. A., Pollock, P. S., and Frommes, S. P., 1971, The movement of exogenous protein in experimental cerebral edema. An electron microscopic study after freeze injury, *J. Neuropathol. Exp. Neurol.* **30**:668.

Baldwin, F., 1971a, The structure of pericapillary cells in cerebral stab wounds of 10 days duration, *J. Anat.* **108**:204.

Baldwin, F., 1971b, The sequence of early cellular changes around cerebral stab wounds, *J. Anat.* **108**:204.

Baldwin, F., 1972, The identification of microglia and its response to injury, in: *Proceedings of the Electron Microscope Society of America* (C. J. Arceneaux, ed.), pp. 340–341, Claitor, Baton Rouge.

Baldwin, F., 1973, Ultrastructural characteristics of microglia and response to cortical puncture, *J. Anat.* **114**:154.

Baldwin, F., 1974, The nature and origin of endocytic cells in damaged brain, in: *Proceedings of the Eighth International Congress on Electron Microscopy* (J. N. Sanders and D. J. Goodchild, eds.), pp. 296–297, Australian Academy of Science, Canberra.

Baldwin, F., 1975a, The nature of the cerebral histiocyte and cellular responses to cerebral damage. Ph.D. thesis, University of New South Wales, Australia.

Baldwin, F., 1975b, The origin of macrophages in a simple cerebral lesion, in: *Mononuclear Phagocytes in Immunity, Infection and Pathology* (R. van Furth, ed.), pp. 241–257, Blackwell, Oxford.

Baldwin, F., Wendell-Smith, C. P., and Blunt, M. J., 1969, The nature of microglia, *J. Anat.* **104:**401.
Barka, T., and Anderson, P. J., 1962, Histochemical methods for acid phosphatase using hexazonium pararosanilin as coupler, *J. Histochem. Cytochem.* **10:**741.
Baron, M., and Gallego, A., 1972, The relation of the microglia with the pericytes in the cat cerebral cortex, *Z. Zellforsch. Mikrosk. Anat.* **128:**42.
Bergman, R., 1926, Über Hortegazellen und Neurophagie, *Nederlandsch. Tijdschr. Geneesk.* **70:**2064.
Besta, C., 1929, Dati sul reticulo periferico della cellula nervosa, sulla rete interstitiate diffusa e sulla loro probabile derivazione da particolari elementi cellulari, *Boll. Soc. Ital. Biol. Sper.* **3:**966.
Blakemore, W. F., 1969a, The ultrastructure of the subependymal plate in the rat. *J. Anat.* **104:**423.
Blakemore, W. F., 1969b, Reaction of neuroglia and microglia to injury, Ph.D. thesis, University of London.
Blakemore, W. F., 1971, The ultrastructural appearances of astrocytes following thermal lesions of the rat cortex, *J. Neurol. Sci.* **12:**319.
Blakemore, W. F., 1972, Microglial reaction following thermal necrosis of the cortex: An electron microscope study, *Acta Neuropathol.* **121:**11.
Blakemore, W. F., 1975, The ultrastructure of normal and reactive microglia, *Acta Neuropathol., Suppl.* **6:**273.
Blinzinger, K., and Hager, H., 1964, Elektronenmikroskopische Untersuchungen zur Feinstruktur runender und progressiver Microgliazellen im ZNS des Goldhamsters, *Prog. Brain Res.* **6:**99.
Blinzinger, K., and Kreutzberg, G., 1968, Displacement of synaptic terminals from regenerating motor neurons by microglial cells, *Z. Zellforsch. Mikrosk. Anat.* **85:**145.
Blunt, M. J., and Baldwin, F., 1972, Aspects of myelin-glial relationships in degeneration, *Proc. Aust. Physiol. Pharmacol. Soc.* **3:**191.
Blunt, M. J., and Baldwin, F., 1973, Glial reaction in degenerating optic nerve, *J. Anat.* **114.**
Bolsi, D., 1931, Ricerche sulla microglia e la oligodendroglia; colorazione vitale, *Riv. Patol. Nerv.* **37:**13.
Booz, K. H., and Felsing, T., 1973, Über ein transitorisches, perinatales, subependymales Zellsystem der weissen Ratte, *Z. Anat. Entwicklungsgesch.* **141:**275.
Boya, J., 1975, Contribution to the ultrastructural study of microglia in the cerebral cortex, *Acta Anat.* **92:**364.
Bryans, W. A., 1959, Mitotic activity in the brain of the adult rat, *Anat. Rec.* **133:**65.
Cajal, S. Ramon y, 1920, Algunas observaciones sobre la mesoglia de Robertson y Rio-Hortega, *Trab. Lab. Invest. Biol.* **18:**109.
Cammermeyer, J., 1965, The hypependymal microglial cell, *Z. Anat. Entwicklungsgesch.* **124:**543.
Cammermeyer, J., 1970, The life history of the microglial cell: A light microscopic study, in: *Neurosciences Research* (S. Ehrenpreis and O. C. Solnitzy, eds.), Vol. 3, pp. 43–129, Academic Press, New York.
Cancilla, P. A., Pollock, P. S., Baker, R. W., and Andrews, J. M., 1970, The fate of exogenous protein in the nervous system, in: *Sixth International Congress of Neuropathology,* pp. 1033–1034, Masson, Paris.
Cancilla, P. A., Baker, R. N., Pollock, P. S., and Frommes, S. P., 1972, The reaction of pericytes of the central nervous system to exogenous protein, *Lab. Invest.* **26:**376.
Carmichael, E. A., 1929, Microglia: An experimental study in rabbit after intracerebral injection of blood, *J. Neurol. Psychopathol.* **9:**209.
Cerletti, U., 1903, Sulla neuronofagia e sopra alcuni rapporti normali e patologici fra elementi nervosi ed elementi non nervosi, *Annali Ist. psichiat. R. Univ. Roma.* 91.
Chessick, R. D., 1953, Histochemical study of distribution of esterases, *J. Histochem.* **1:**471.
Cone, W., and MacMillan, J. A., 1932, The optic nerve and papilla, in: *Cytology and Cellular Pathology of the Nervous System,* Vol. 2 (W. Penfield, ed.), pp. 837–901, Hoeber, New York.
Conradi, S., 1969a, Ultrastructure and distribution of neuronal and glial elements on the motoneuron surface in the lumbosacral spinal cord of the adult cat, *Acta Physiol. Scand., Suppl.* **332:**5.
Conradi, S., 1969b, Ultrastructure of dorsal root location on lumbosacral mononeurons of the adult cat, as revealed by dorsal root section, *Acta Physiol. Scand., Suppl.* **332:**85.
Cowen, D., Geller, L. M., and Abner, W., 1970, Healing in the cerebral cortex of the infant rat after closed-head focal injury, *J. Neuropathol. Exp. Neurol.* **29:**21.

Cowan, W. M., 1973, Neuronal cell death as a regulative mechanism in the control of cell number in the nervous system, in: *Development and Aging in the Nervous System* (M. Rockstein, ed.), pp. 19–41, Academic Press, New York.

Cox, V. S., 1976, Ultrastructure of the axon reaction in the immature rat thalamus, *J. Neuropathol. Exp. Neurol.* **35:**191.

Cragg, B. G., 1976, Ultrastructural features of human cerebral cortex, *J. Anat.* **121:**331

Cruetzfeldt, H. G., and Metz, A., 1926, Über Gestalt and Tätigkeit der Hortegazellen bei pathologischen Vorgängen. *Z. ges. Neurol. Psychiat.* **106:**18.

Da Fano, C., 1906, Osservazioni sulla fine struttura della neuroglia, *Rev. Lab. Anat. Univ. Roma Altri Lab. Biol.* **12:**1.

Das, G. D., 1976a, Gitter cells and their relationship to macrophages in the developing cerebellum: An electron microscopic study, *Virchows Arch. B* **20:**299.

Das, G. D., 1976b, Resting and reactive macrophages in the developing cerebellum: An experimental ultrastructural study, *Virchows Arch. B* **20:**287.

Devaux, A., 1908, Etude histologique des foyers de nécrose de l'écorre cérébrale, *Histol. histopathol. Arb.* **1:**116.

Dewulf, A., 1937, La microglie normal chez le singe (*Macacus rhesus*), *J. Belge Neurol. Psychiat.* **37:**341.

Duckett, S., and Pearse, A. G. E., 1965, A pericapillary cell in the central nervous system of the human embryo and adult, *Acta Neuropathol.* **4:**442.

Dunning, H. S., and Furth, J., 1935, Studies on the relation between microglia, histiocytes and monocytes, *Am. J. Pathol.* **11:**6.

Eager, R. P., and Eager, P. R., 1966, Glial responses to degenerating cerebellar cortico-nuclear pathways in the cat, *Science* **153:**553.

Eichhorst, H., 1875, Über die Entwicklung des menschlichen Rückenmarks und seiner Formelemente, *Virchows Arch. A* **64:**425.

Farrar, C. B., 1908, On the phenomena of repair in the cerebral cortex. A study of mesodermal and ectodermal activities following the introduction of a foreign body, *Histol. histopathol. Arch. Grosshirnrinde, Jena* **2:**1.

Fawcett, D. W., 1963, Comparative observations on the fine structure of blood capillaries, in: *The Peripheral Blood Vessels* (J. L. Orbison and D. E. Smith, eds.), pp. 17–44, Williams and Wilkins, Baltimore.

Feigin, I., 1969, Mesenchymal tissues of the nervous system, *J. Neuropathol. Exp. Neurol.* **28:**6.

Fernando, D. A., 1971, A third glial cell seen in retrograde degeneration of the hypoglossal nerve, *Brain Res.* **27:**2.

Fernando, D. A., 1973, An electron microscopic study of the neuroglial reaction in Wallerian degeneration of the cortical spinal tract, *Acta. Anat.* **86:**459.

Field, E. J., 1955, Observations on the development of microglia together with a note on the influence of cortisone, *J. Anat.* **89:**201.

Field, E. J., and Raine, C. S., 1964, An electron microscopic study of scrapie in the mouse, *Acta Neuropathol.* **4:**200.

Forster, E., 1908, Experimentelle Beiträge zur Lehre der Phagocytose der Hirnrindenelemente, *Hist. Arb. Grosshirnrinde* **2:**173.

Fox, M. W., and Inman, O. R., 1966, Age differences in the cellular response to cerebral lesions in the dog and mouse, *Experientia* **22:**54.

Fraher, J. P., and McDougall, R. D., 1975, Macrophages related to leptomeninges and ventral nerve roots. An ultrastructural study, *J. Anat.* **120:**537.

Friedmann, M., 1890, Studien zur pathologischen Anatomie der akuten Encephalitis, *Arch. Psychiatr.* **21:**461.

Fujita, S., and Kitemura, T., 1975, Origin of brain macrophages and the nature of the so-called microglia, *Acta Neuropathol., Suppl.* **6:**291.

Glees, P., 1955, *Neuroglia,* Blackwell, Oxford.

Globus, J. H., and Kuhlenbeck, H., 1944, The subependymal cell plate (matrix) and its relationship to brain tumours of the ependymal type, *J. Neuropathol. Exp. Neurol.* **3:**1.

Glucksman, A., 1951, Cell deaths in normal vertebrate ontogeny, *Biol. Rev.* **26:**59.

Gluge, 1841, Experimentelles über Encephalitis, *Abh. Physiol. Pathol. Jena* **2:**13.
Goldman, A. S., and Walker, B. E., 1963, The origin of cells in the infiltrates found at the sites of foreign protein injection, *Lab. Invest.* **11:**808.
Gomori, G., and Chessick, R. D., 1953, Esterases and phosphatases of the brain. A histochemical study, *J. Neuropathol. Exp. Neurol.* **12:**387.
Gordon, S. F., and Cohn, Z. A., 1973, The macrophage, *Int. Rev. Cytol.* **36:**171.
Gozzano, M., 1929, Richerche sull'istogenesi della microglia, *Boll. Soc. Ital. Biol. Sper.* **4:**1028.
Gozzano, M., 1930, Quelques observations sur l'origine de la microglie, *Rev. Neurol. (Paris)* **1:**1024.
Gozzano, M., 1931, L'istogenesi della microglia, *Riv. Neurol.* **4:**225.
Gray, E. G., 1961, The granule cells, mossy synapses and Purkinje spine synapses of the cerebellum: Light and electron microscope observations, *J. Anat.* **95:**345.
Gray, E. G., 1964, Tissue of the central nervous system, in: *Electron Microscopic Anatomy* (S. M. Kurtz, ed.), pp. 369–370, Academic Press, New York.
Greenfield, J. G., 1963, *Neuropathology,* Edward Arnold, London.
Hager, H., 1968, Pathologie der Makro-und Mikroglia im elektronenmikroskopischen Bild, *Acta Neuropathol. (Suppl.)* **4:**86.
Hager, H., 1975, EM findings on the source of reactive microglia in the mammalian brain, *Acta Neuropathol. (Suppl.) (Berl.)* **6:**279.
Hain, R. F., 1963, Discussion in: Origin of gitter cells in the mouse brain (B. W. Konigsmark and R. L. Sidman), *J. Neuropathol. Exp. Neurol.* **22:**327.
Hamburger, V., 1975, Cell death in the development of the lateral motor column of the chick embryo, *J. Comp. Neurol.* **160:**535.
Hamburger, V., and Levi-Montalcini, R., 1949, Proliferation, differentiation and degeneration in the spinal ganglia of the chick embryo under normal and experimental conditions, *J. Exp. Zool.* **111:**457.
Hassler, O., 1966, Incorporation of tritiated thymidine into mouse brain after a simple dose of X-rays: An autoradiographic study, *J. Neuropathol. Exp. Neurol.* **25:**97.
Hatai, S., 1902, On the origin of neuroglia tissue from mesoblast, *J. Comp. Neruol.* **12:**291.
Holländer, H., Brodal, P., and Walberg, F., 1969, Electron microscopic observations on the structure of the pontine nuclei and the mode of termination of the corticopontine fibres. An experimental study in the cat, *Exp. Brain Res.* **7:**95.
Hommes, O. R., and Leblond, C. P., 1967, Mitotic divisions of neuroglia in the normal adult rat, *J. Comp. Neurol.* **129:**269.
Horstadius, R., 1950, *The Neural Crest. Its Properties and Derivation in the Light of Experimental Research,* Oxford University Press, London.
Hortega, P. Del Rio, 1919a, El tercer elemente de los centros nerviosos, *Bol. Soc. Esp. Biol.* **9:**69.
Hortega, P. Del Rio, 1919b, Poder fagocitario y movilidad de la microglia, *Bol. Soc. Esp. Biol.* **9:**154.
Hortega, P. Del Rio, 1920a, La microglia y su transformación en células en bastonicito y cuerpos, gránulo-adiposos, *Trab. Lab. Invest. Biol.* **18:**37.
Hortega, P. Del Rio, 1920b, La microglia y su transformación en células en bastonicito y cuerpos, gránulo-adiposos, *Arch. Neurobiol. (Madrid)* **1:**171.
Hortega, P. Del Rio, 1921, Histogénesis y evolución normal, éxodo y distribucion regional de la microglia, *Mem. Soc. Esp. Hist. Nat.* **11:**213.
Hortega, P. Del Rio, 1932/1965, Microglia, in: *Cytology and Cellular Pathology of the Nervous System,* Vol. 2 (W. Penfield, ed.), pp. 483–534, Hafner, New York.
Hortega, P. Del Rio, 1939, The microglia, *Lancet* **1:**1023.
Hortega, P. Del Rio, and Asua, F. J., 1921, Sobre la fagocitosis en los tumores y en otros processos patologicos, *Arch. Cardiol. Hematol.* **2:**161.
Hortega, P. Del Rio, and Penfield, W., 1927, Cerebral cicatrix. Reaction of neuroglia and microglia to brain wounds, *John Hopkins Hosp. Bull.* **41:**278.
Hosokawa, H., 1963, Microglia cells, in: *Morphology of Neuroglia* (J. Nakai, ed.), pp. 10–12, Igaku Shoin, Tokyo (Thomas, Springfield).
Hosokawa, H., and Mannen, H., 1963, Some aspects of the histology of neuroglia, in: *Morphology of Neuroglia* (J. Nakai, ed.), pp. 1–52, Igaku Shoin, Tokyo (Thomas, Springfield).

Huntington, H. W., and Terry, R. D., 1966, The origin of the reactive cells in cerebral stab wounds, *J. Neuropathol. Exp. Neurol.* **25:**646.
Hurley, J. V., Ryan, G. B., and Friedman, A., 1966, The mononuclear response to intrapleural injection in the rat, *J. Pathol. Bacteriol.* **91:**575.
Ibrahim, M. Z. M., Khreis, Y., and Koshayan, D. S., 1974, The histochemical identification of microglia, *J. Neurol. Sci.* **22:**211.
Imamoto, K., and Leblond, C. P., 1977, Presence of labelled monocytes, macrophages and microglia in a stab wound of the brain following an injection of bone marrow cells labelled with ^{3}H-thymidine into rats, *J. Comp. Neurol.* **174:**255.
Imamoto, K., and Leblond, C. P., 1978, Radioautographic investigations of gliogenesis in the corpus callosum of young rats. II. Origin of microglial cells, *J. Comp. Neurol.* **180:**139.
Jakob, A. M., 1927, *Normale und pathologische Anatomie und Histologie des Grosshirns,* Franz Dentiche, Leipzig.
Jones, E. G., 1970, On the mode of entry of blood vessels into the cerebral cortex, *J. Anat.* **106:**3.
Juba, A., 1933, Untersuchungen über die Entwicklung der Hortegaschen Mikroglia des Menschen, *Arch. Psychiatr. Nervenkr.* **101:**577.
Juba, A., 1934, Über die Entwicklung der Mikroglia mit besonderer Berücksichtigung der Cytogenese, *Z. Anat. Entwicklunggesch.* **103:**245.
Kamemura, T., 1934, Über die Entwicklung der Mikroglia sowie Einiges zur Frage der sog. embryonalen Fettkörnchenzellen, *Folia Psychiatr. Neurol. Jpn.* **1:**263.
Kaufmann, E., 1904, *Lehrbuch der speziellen pathologischen Anatomie,* Berlin.
Kershman, J., 1938, The medulloblast and the medulloblastoma, *Arch. Neurol. Psychiatry* **40:**337.
Kershman, J., 1939, Genesis of microglia in the human brain, *Arch. Neurol. Psychiatry* **31:**24.
King, J. S., 1968, A light and electron microscopic study of perineuronal glial cells and processes in the rabbit neocortex, *Anat. Rec.* **161:**111.
Langevoort, H. L., Cohn, Z. A., Hirsch, J. G., Humphrey, J. H., Spector, W. G., and van Furth, R., 1970, The nomenclature of mononuclear phagocytic cells, in: *Mononuclear Phagocytes* (R. van Furth, ed.), pp. 1–6, Blackwell, Oxford.
Lebowich, R. J., 1934, Phagocytic behaviour of interstitial cells of brain parenchyma of adult rabbit toward colloidal solutions and bacteria, *Arch. Pathol.* **18:**50.
Levi-Montalcini, R., 1950, The origin and development of the visceral system in the spinal cord of the chick embryo, *J. Morphol.* **86:**253.
Levi-Montalcini, R., 1964, Events in the developing nervous system, *Prog. Brain Res.* **4:**1.
Lewis, P. D., 1968a, A quantitative study of cell proliferation in the subependymal layer of the adult rat brain, *Exp. Neurol.* **20:**203.
Lewis, P. D., 1968b, The fate of the subependymal cell in the adult rat brain, with a note on the origin of microglia, *Brain* **91:**721.
Kitamura, T., Hattori, H., and Fujita, S., 1972, Autoradiographic studies on histogenesis of brain macrophages in the mouse, *J. Neuropathol. Exp. Neurol.* **31:**502.
Kolbe, C., 1889, Zur Kenntis der embolischem Gehirnerweichung, Inaugural Dissertation, Marburg.
Konigsmark, B. W., and Sidman, R. L., 1963, Origin of brain macrophages in the mouse, *J. Neuropathol. Exp. Neurol.* **22:**643.
Kosunen, T. V., and Waksman, B. H., 1963, Radio-autographic studies of experimental allergic encephalomyelitis (EAE) in rats, *J. Neuropathol. Exp. Neurol.* **22:**324.
Krogh, A., and Vimtrup, B., 1932, The capillaries, in: *Special Cytology* (E. V. Cowdry, ed.), 2nd edn., pp. 475–503, Hoeber, New York.
Kruger, L., 1965, The spectrum of oligodendrocytes in normal rat cerebrum, *Anat. Rec.* **151:**375.
Kruger, L., and Maxwell, D. S., 1966, E. M. of oligodendrocytes in normal rat cerebrum, *Am. J. Anat.* **118:**411.
Ling, E. A., 1976, Some aspects of ameboid microglia in the corpus callosum and neighbouring regions of neonatal rats, *J. Anat.* **121:**29.
Ling, E. A., 1977, Light and electron microscopic demonstration of some lysosomal enzymes in the ameboid microglia in neonatal rat brain, *J. Anat.* **123:**637.
Luse, S. A., 1956, Electron microscope observations of the central nervous system, *J. Biophys. Biochem. Cytol.* **2:**531.

Majno, G., 1965, Ultrastructure of the vascular membrane, in: *Handbook of Physiology*, Section 2, *Circulation*, Vol. 3 (W. F. Hamilton and P. Dow, eds.), pp. 2293–2375, American Physiological Society, Washington.

Majno, G., and Palade, G. E., 1961, Studies on inflammation. 1. The effect of histamine and serotonin on vascular permeability. An electron microscope study, *J. Biophys. Biochem. Cytol.* **11**:571.

Malmfors, T., 1963, Electron microscopic description of the glial cells in the nervous opticus in mice, *J. Ultrastruct. Res.* **8**:193.

Marchand, F., 1909, Herkunft des Körnchenzellen des Zentralnerven systems, *Beitr. pathol. Anat. allg. Pathol.* **45**:161.

Matthews, M. A., 1974, Microglia and reactive "M" cells of degenerating central nervous system: Does similar morphology and function imply a common origin? *Cell Tissue Res.* **148**:477.

Matthews, M. A., and Kruger, L., 1973, Electron microscopy of non-neuronal cellular changes accompanying neural degeneration in thalamic nuclei of the rabbit. 1. Reactive haematogenous and perivascular elements within the basal lamina, *J. Comp. Neurol.* **148**:285.

Maxwell, David S., and Kruger, L., 1965, Small blood vessels and the origin of phagocytes in the rat cerebral cortex following heavy particle irradiation, *Exp. Neurol.* **12**:33.

Maxwell, D. S., and Kruger, L., 1966, The reactive oligodendrocyte, *Am. J. Anat.* **118**:437.

Maynard, E. A., Schultz, R. L., and Pease, D. C., 1957, Electron microscopy of the vascular bed of rat cerebral cortex, *Am. J. Anat.* **100**:409.

McDonald, T. F., 1962, The formation of phagocytes from perivascular cells in the irradiated cerebral cortex of the rat as seen in the E. M., *Anat. Rec.* **142**:2.

Merzbacher, L., 1909, Untersuchungen über die Morphologie und Biologie der Abräumzellen, *Histol. histopath. Arb. Nissl-Alzheimer* **3**:1.

Metchnikoff, E., 1913, Die Lehre von den Phagocyten und deren experimentelle Grundlagen, in: *Handbuch der pathogenen Mikroorganismen* (W. Kalle and A. Wassermann, eds.), pp. 655–731, Fischer, Jena.

Metz, A., and Spatz, H., 1924, Die Hortegaschen Zellen, das sogenannte "dritte Element" und über ihre funktionelle Bedeutung, *Z. Neurol.* **89**:138.

Minor, L., 1904, Traumatische Erkrankungen des Rückenmarks, in: *Handbuch der pathologischen Anatomie des Nervensystems* (E. Flateau, L. Jacobsohn, and L. Minor, eds.), Vol. 2, pp. 1008–1058, Karger, Berlin.

Mori, S., and Leblond, C. P., 1969, Identification of microglia in light and electron microscopy, *J. Comp. Neurol.* **135**:57.

Mugnaini, E., 1965, "Dark cells" in electron micrographs from the central nervous system of vertebrates, *J. Ultrastruct. Res.* **12**:235.

Mugnaini, E., and Walberg, F., 1964, Ultrastructure of neuroglia, *Ergebn. Anat. Entwicklungsgesch.* **37**:194.

Murray, H. M., and Walker, B. G., 1973, Comparative study of astrocytes and mononuclear leukocytes reacting to brain trauma in mice, *Exp. Neurol.* **41**:290.

Nissl, F., 1899, Über einige Beziehungen zur Nervenzellen Erkrankungen und gliosen Erscheinungen bei verschiedenen Psychosen, *Arch. Psychiat.* **32**:656.

Nissl, F., 1904, Zur Histopathologie der paralytischen Rindenerkrankungen, *Histol. histopath. Arb. Nissl-Alzheimer* **1**:315.

Noetzel, H., and Siepmann, P., 1965, Autoradiographische Untersuchungen am Grosshirn neugeborener Ratten nach Injektion von Thymidine-H^3, *Dtsch. Z. Nervenheilkd.* **187**:637.

Nystrom, S., 1960, Pathological changes in blood vessels of human glioblastoma multiforma, *Acta Pathol. Microbiol. Scand.* **137** (Suppl. 2): 1.

O'Connor, T. M., and Wyttenbach, C. R., 1974, Cell death in the embryonic chick spinal cord, *J. Cell Biol.* **60**:448.

Oehmichen, M., 1975a, Monocytic origin of microglia cells, in: *Mononuclear Phagocytes in Immunity, Infection and Pathology* (R. van Furth, ed.), pp. 223–240, Blackwell, Oxford.

Oehmichen, M., 1975b, Experimental studies on kinetics and functions of mononuclear phagocytes of the central nervous system, *Acta. Neuropathol., Suppl.* **6**:285.

Oehmichen, M., and Huber, H., 1976, Reactive microglia with membrane features of mononuclear phagocytes, *J. Neuropathol. Exp. Neurol.* **35**:30.

Oehmichen, M., and Torvik, A., 1976, The origin of reactive cells in retrograde and Wallerian degeneration. Experiments with intravenous injection of ^{3}H-DFP-labelled macrophages, *Cell Tissue Res.* **173:**343.

Oehmichen, M., Gruninger, J., Saebisch, R., and Narita, Y., 1973, Mikroglia und Pericyten als Transformationsformen der Blut-Monocyten mit erhaltener Proliferationsfähigkeit. Experimentelle autoradiographische und enzymhistochemische Untersuchungen am normalen und geschädigten Kaninchen- und Rattengehirn, *Acta Neuropathol.* **23:**200.

Oksche, A., 1968, Die pränatale und vergleichende Entwicklungsgeschichte der Neuroglia, *Acta Neuropathol., Suppl.* **4:**4.

Opalski, A., 1933, Über lokale Unterschiede im Bau der Ventrikelwände beim Menschen, *Z. Neurol. Psychiatr.* **149:**221.

O'Steen, W. K., and Karcioglu, Z. A., 1974, Phagocytosis in the light-damaged albino rat eye: Light and electron microscopic study, *Am. J. Anat.* **139:**503.

Paterson, J. A., Privat, A., Ling, E. A., and Leblond, C. P., 1973, Investigation of glial cells in semithin sections. III. Transformation of subependymal cells into glial cells, as shown by radioautography after ^{3}H-thymidine injection into the lateral ventricle of the young rat, *J. Comp. Neurol.* **146:**277.

Penfield, W., 1924, Oligodendroglia and its relation to classical neuroglia, *Brain* **47:**430.

Penfield, W., 1925a, Microglia and the process of phagocytosis in gliomas, *Am. J. Pathol.* **1:**46.

Penfield, W., 1925b, Phagocytic activity of microglia in the central nervous system, *Proc. N.Y. Pathol. Soc.* **25:**77.

Penfield, W., 1927, The mechanism of cicatricial contraction of the brain, *Brain* **50:**499.

Penfield, W., 1928, A method of staining oligodendroglia and microglia (combined method), *Am. J. Pathol.* **4:**153.

Penfield, W., 1932, Neuroglia, normal and pathological, in: *Cytology and Cellular Pathology of the Nervous System,* Vol. 2 (W. Penfield ed.), pp. 421–479, Hoeber, New York.

Penfield, W., and Buckley, R. C., 1928, Punctures of the brain. The factors concerned in gliosis and cicatricial contraction, *Arch. Neurol. Psychiat.* **20:**1.

Pick, F., 1904, Compression; Myelitis; Rüchenmarksabscess, in: *Handbuch der pathologischen Anatomie des Nervensystem* (E. Flateau, L. Jacobsohn, and L. Minor, eds.), pp. 846–879, Karger, Berlin.

Polak, M., and Haymaker, W., 1968, Human central microglia and its pathologic changes, *Arch. Fund. Roux-Ocefa* **2:**249.

Popoff, L., 1875, Über Veränderungen im Gehirn bei Abdominaltyphus und traumatischer Entzündung, *Virch. Arch.* **63:**421.

Privat, A., 1970, Sur l'origine des divers types de microglie chez le rat, in: *Sixth International Congress of Neuropathology,* pp. 447–448, Masson, Paris.

Privat, A., 1977, The ependyma and subependymal layer of the young rat. A new contribution with freeze fracture, *Neuroscience* **2:**447.

Privat, A., and Fulcrand, J., 1976, Glial reactions in the optic nerve of the young rat after unilateral enucleation, *Anat. Rec.* **184:**505.

Privat, A., and Leblond, C. P., 1972, The subependymal layer and neighbouring region in the brain of the young rat, *J. Comp. Neurol.* **146:**277.

Pruijs, W. M., 1927, Über Mikroglia, ihre Herkunft, Funktion und ihr Verhältnis zu anderen Gliadementen, *Z. Neurol. Psychiat.* **108:**298.

Rand, C. W., and Courville, C. B., 1932, Histologic changes in the brain in cases of fatal injury to the head, *Arch. Neurol. Psychiatry.* **27:**605.

Reynolds, F. E., and Slater, J. K., 1928, A study of the structure and function of the interstitial tissue of the central nervous system, *Edinburgh Med. J.* **35:**49.

Rhodin, J. A. G., 1968, Ultrastructure of mammalian venous capillaries, venules, and small collecting veins, *J. Ultrastruct. Res.* **25:**452.

Ribbert, H., 1882, Über multiple Sclerose des Gehirns und Rückenmarks, *Virchows Arch.* **90:**243.

Robinson, H. C., Alt, G., and Evans, D. H. L., 1972, A study of the capacity of myelinated and unmyelinated nerves to induce experimental allergic neuritis, *Acta Neuropathol.* **21:**99.

Roessman, U., and Friede, R. L., 1967, Entry of labelled donor cells from the blood stream into the CNS, *J. Neuropathol. Exp. Neurol.* **26:**144.

Roessman, U., and Friede, R. L., 1968, Entry of labelled monocytic cells into the central nervous system, *Acta Neuropathol.* **10:**359.

Roizin, L., and Schade, J. P., 1968, Pathogenesis of x-irradiation effects in monkey cerebral cortex. V. Ultrastructural findings, *Brain Res.* **7:**87.

Roussy, G., L'Hermitte, J., and Oberling, C., 1930, La neuroglie et ses réactions pathologiques, *Revne Neurol.* **53:**878.

Rubinstein, L. J., Herman, M. M., Miquel, J., and Weibel, J., 1971, The short and long-term effects of ultraviolet irradiation on the exposed cat cerebrum, *J. Neurol. Sci.* **13:**333.

Russell, D. S., 1929, Intravital staining of microglia with trypan blue, *Am. J. Pathol.* **5:**451.

Rydberg, E., 1932, Cerebral injury in new-born children consequent on birth trauma with an inquiry into the normal and pathological anatomy of the neuroglia, *Acta Pathol. Microbiol. Scand., Suppl.* **10:**1.

Samorajski, T., Ordy, J. M., and Rady-Reimer, P., 1968, Lipofuscin pigment accumulation in the nervous system of aging mice, *Anat. Rec.* **160:**555.

Santha, K., 1932, Untersuchungen über die Entwicklung der Hortegaschen Mikroglia, *Arch. Psychiatr. Nervenkr.* **96:**36.

Santha, K., and Juba, A., 1933, Weitere Untersuchungen über die Entwicklung der Hortegaschen Mikroglia, *Arch. Psychiatr. Nervenkr.* **98:**59.

Schaffer, K., 1926, Über die Hortegaschen Mikroglia, *Z. Anat. Entwicklungsgesch.* **81:**715.

Schmaus, H., 1904, Akute myelitis, *Lubarsch-Ostertats Ergebn. Pathol.* **9:**1904.

Schmaus, H., and Sacki, S., 1901, *Vorlesungen über die pathologische Anatomie des Rückenmarks*, Bergmann, Wiesbaden.

Schmitt, D., 1973, Über glycoproteidhaltige amoeboide Zellen in embryonalen Huhnergehirn, *Z. Anat. Entwicklungsgesch.* **142:**341.

Schlote, W., 1959, Zur Gliaarchitektonik der Menschlichen Grosshirnrinde im Nissl-Bild, *Arch. Psychiatr. Nervenkr.* **199:**573.

Scholz, W., 1957, Erkrankungen des zentralen Nervensystems, in: *Handbuch der speziellen Pathologischen Anatomie und Histologie* (O. Lubarsch, F. Henke, and R. Rossle, eds.), Vol. 13, *Nervensystem* (W. Scholz, ed.), pp. 1–265, Springer Verlag, Berlin.

Schroeder, A. H., 1929, Die Gliaarchitektonik des menschlichen Kleinhirns, *J. Psychol. Neurol. Leipzig* **38:**234.

Schultz, R. L., Maynard, E. A., and Pease, D. C., 1957, Electron microscopy of neurons and neuroglia of cerebral cortex and corpus callosum, *Am. J. Anat.* **100:**369.

Schwalbe, G., 1881, *Lehrbuch der Neurologie*, Besold, Erlangen.

Schwarz, H., Goolker, P., and Globus, J. H., 1932, The normal histology of infants' brains, *Am. J. Dis. Child* **43:**889.

Selwood, L., 1970, Electron microscopy of the fate of exogenous ferritin in the feline visual cortex, *Z. Zellforsch. Mikrosk Anat.* **107:**6.

Selwood, L., 1971, Electron microscopy of endocytic elements in the cat cerebrum, *Acta Neuropathol.* **18:**299.

Skoff, R. P., 1975, The fine structure of pulse labelled (^{3}H-thymidine) cells in degenerating rat optic nerve, *J. Comp. Neurol.* **161:**595.

Skoff, R. P., Price, D. L., and Stocks, A., 1976, Electron microscopic autoradiographic studies of gliogenesis in rat optic nerve. 1. Cell proliferation. 2. Time of origin. *J. Comp. Neurol.* **189:**291.

Smart, I., and Leblond, C. P., 1961, Evidence for division and transformation of neuroglia cells in the mouse brain, as derived from radioautography after injection of thymidine-H^3, *J. Comp. Neurol.* **116:**349.

Smart, I., 1961, The sub-ependymal layer of the mouse brain and its cell production as shown by radioautography, after thymidine-H^3 injection, *J. Comp. Neurol.* **116:**325.

Smith, C. W., and Walker, B. E., 1967, Glial and lymphoid cell response to tumor implantation in mouse brain, *Texas Rep. Biol. Med.* **25:**585.

Spector, W. G., and Ryan, G. B., 1970, The mononuclear phagocyte in inflammation, in: *Mononuclear Phagocytes* (R. van Furth, ed.), pp. 219–232, Blackwell, Oxford.

Spector, W. G., and Willoughby, D. A., 1968, The origin of mononuclear cells in chronic inflammation and tubercular reactions in the rat, *J. Pathol. Bacteriol.* **96:**389.

Spector, W. G., Lykke, A. W. J., and Willoughby, D. A., 1967, A quantitative study of leucocyte emigration in chronic inflammatory granulomata, *J. Pathol. Bacteriol.* **93:**101.

Stensaas, L. J., and Gilson, B. C., 1972, Ependymal and subependymal cells in the caudate pallial junction in the lateral ventricle of the neonatal rabbit, *Z. Zellforsch. Mikrosk. Anat.* **132:**287.

Stensaas, L. J., and Reichert, W. H., 1971, Round and amoeboid microglial cells in the neo-natal rabbit brain, *Z. Zellforsch. Mikrosk. Anat.* **119:**2.

Stensaas, L. J., and Stensaas, S. S., 1968a, Astrocytic microglial cells, oligodendrocytes and microgliacytes in the spinal cord of the toad. I. Light microscopy, *Z. Zellforsch. Mikrosk. Anat.* **84:**473.

Stensaas, L. J., and Stensaas, S. S., 1968b, Astrocytic neuroglial cells, oligodendrocytes and microgliacytes in the spinal cord of the toad. II. Electron microscopy, *Z. Zellforsch. Mikrosk. Anat.* **86:**184.

Stensaas, L. J., and Stensaas, S. S., 1968c, Light microscopy of glial cells in the turtles and birds, *Z. Zellforsch. Mikrosk. Anat.* **91:**315.

Stenwig, A. E., 1972, The origin of brain macrophages in traumatic lesions, Wallerian degeneration and retrograde degeneration, *J. Neuropathol. Exp. Neurol.* **31:**686.

Tani, E., and Evans, J. P., 1965, Electron microscopic studies of cerebral swelling. I. Studies on the permeability of brain capillaries, using ferritin molecules as tracers, *Acta Neuropathol.* **4:**507.

Terry, R. D., and Weiss, M., 1963, Studies in Tay-Sachs disease. II. Ultrastructure of the cerebrum, *J. Neuropathol. Exp. Neurol.* **22:**18.

Torack, R. M., 1961, Ultrastructure of capillary reaction to brain tumors, *Arch. Neurol.* **5:**416.

Toro, I., 1942, Histologische Untersuchungen über die Beziehungen zwischen reticuloendothelialem System und Histaminwirkung, *Z. Mikrosk.-Anat. Forsch.* **52:**552.

Torvik, A., 1975, The relationship between microglia and brain macrophages, *Acta Neuropathol., Suppl.* **6:**297.

Torvik, A., and Skjorten, F., 1971, Electron microscopic observations on nerve cell regeneration and degeneration after axon lesions. II. Changes in the glial cells, *Acta Neuropathol.* **17:**265.

Torvik, A., and Soreide, A. J., 1975, The perineuronal glial reaction after axotomy, *Brain Res.* **95:**519.

Tschaschin, S., 1913, Über die "ruhenden Warderzellen" und ihre Beziehungen zu den anderen Zellformen des Bindegewebes und zu den Lymphocyten, *Folia Laemat. Archiv.* **17:**317.

Turner, J. E., 1975, Non glial phagocytes within the degenerating optic nerve of the newt (*Triturus viridescens*), *J. Exp. Zool.* **193:**87.

Tusques, J., Coudere, M., and George, Y., 1975, Microglia and pericytes of the human cerebral cortex, *Bull. Assoc. Anat.* **59:**535.

Urechia, G. I., and Elekes, N., 1926, Contribution a l'étude de la microglie, *Arch. Int. Neurol.* **45:**81.

van Furth, R., and Cohn, Z. A., 1968, The origin and kinetics of mononuclear phagocytes, *J. Exp. Med.* **128:**415.

van Furth, R., and Diesselhoff-Den Dulk, M. M. C., 1970, The kinetics of promonocytes and monocytes in the bone marrow, *J. Exp. Med.* **132:**813.

van Furth, R., Hirsch, J. G., and Fedorko, M. E., 1970, Morphology and peroxidase cytochemistry of mouse promonocytes, monocytes and macrophages, *J. Exp. Med.* **132:**794.

Vaughn, J. E., 1969, An electron microscopic analysis of gliogenesis in rat optic nerves, *Z. Zellforsch. Mikrosk. Anat.* **94:**293.

Vaughn, J. E., and Pease, D. C., 1970, Electron microscopic studies of Wallerian degeneration in rat optic nerves. II. Astrocytes, oligodendrocytes and adventitial cells, *J. Comp. Neurol.* **140:**207.

Vaughn, J. E., and Peters, A., 1968, A third neuroglial cell type. An electron microscope study, *J. Comp. Neurol.* **133:**269.

Vaughn, J. E., and Skoff, R. P., 1972, Neuroglia in experimentally altered central nervous system, in: *The Structure and Function of Nervous Tissue* (G. Bourne, ed.), pp. 39–72, Academic Press, New York.

Vaughn, J. E., Hinds, P. L., and Skoff, R. P., 1970, Electron microscopic studies of Wallerian degeneration in rat optic nerves, I. The multipotential glia. *J. Comp. Neurol.* **140:**175.

Vimtrup, B., 1922, Beiträge zur Anatomie der Capillaren: über contractile Elemente in der Gefässwand der Blut Capillaren, *Z. Anat. Entwicklungsgesch.* **65:**150.

Virchow, R. L. K., 1867, Zur pathologischen Anatomie des Gehirns: kongeniale Encephalitis und Myelitis, *Virchows Arch.* **38:**129.

Volkman, A., and Gowans, J. L., 1965a, The production of macrophages in the rat, *Br. J. Exp. Pathol.* **46:**50.

Volkman, A., and Gowans, J. L., 1965b, The origin of macrophages from bone marrow in the rat, *Br. J. Exp. Pathol.* **46:**62.

Wendell-Smith, C. P., Blunt, M. J., and Baldwin, F., 1966, The ultrastructural characterisation of macroglial cell types, *J. Comp. Neurol.* **127:**219.

Wolff, J., 1963, Contributions to the ultrastructure of the capillaries in the normal cerebral cortex, *Z. Zellforsch. Mikrosk. Anat.* **60:**409.

Worthington, W. C., and Childs, P. E., 1959, Granular adventitial cells of small cerebral vessels of mice, cats and humans, *Anat. Rec.* **133:**442.

Ziegler, E., 1901, *Lehrbuch der allgemeinen Pathologie und der pathologischen Anatomie für Ärtze und Studierende,* Vol. 1, Fischer, Jena.

Zimmerman, K. W., 1923, Der feinere Bau der Blutkapillaren, *Z. Anat. Entwicklungsgesch.* **68:**29.

18

Macrophages in Connective Tissues

The Granuloma Macrophage

IAN CARR

1. INTRODUCTION

This chapter describes the structure and function of macrophages as they are seen in the connective tissues of the body. These occur as normal residents of the connective tissues, and increase vastly by immigration of circulating monocytes in inflammation. The resident macrophages of the connective tissues probably have a relatively slow turnover and are found not only in the nonspecialized subcutaneous connective tissues, but in more specialized circumstances, like the connective tissues of the eye, the placenta, the testis, and the uterus. The latter fall outside the scope of this review.

1.1. THE CONNECTIVE TISSUE HISTIOCYTE

The connective tissue histiocyte in standard sections stained with hematoxylin and eosin (H and E) appears as a cell with a vesicular nucleus, often with nucleolus, and voluminous cytoplasm, often visible as elongated processes stretched in a rather stellate manner; or it may appear as a cell with a rather more basophilic and densely staining cytoplasm with visible "frilly" processes, though the latter may or may not be visible, and cannot be relied upon for identification. Such cells may be close to small blood vessels as pericytes. The cytoplasm stains positively but variably for acid phosphatase, β glucuronidase, and other lysosomal enzymes. The extent to which these cells are truly indigenous to the connective tissues and the extent to which they are part of a continu-

IAN CARR • Department of Pathology, University of Saskatchewan, Saskatoon, Saskatchewan, Canada S7N OWO.

ous process of monocyte immigration and cell turnover is uncertain and difficult to analyze because of the nature of the tissues. A number of early ultrastructural descriptions have been published (Kajikawa, 1964; Deane, 1964; Han and Avery, 1965). These give an account of macrophages of varying degrees of maturity with no specific cytological features. There is a lack of modern and detailed ultrastructural study of verifiable resident tissue histiocytes such as to make further detailed review unprofitable, except for comments on synovial macrophages and on the histochemical differences between resident histiocytes and monocyte derived macrophages.

Elsewhere in this volume (Chapter 3), Daems has described the differences in distribution between peroxidase distribution in resident peritoneal macrophages and in macrophages recently derived from monocytes. In line with this, it has recently been shown that, whereas peroxidase is present in the cytoplasmic granules of monocytes, it is found in the nuclear envelope and rough endoplasmic reticulum (RER) of the histiocytes of the lamina propria of the gut (Ogawa *et al.*, 1978).

1.2. THE SYNOVIAL MACROPHAGE

Phagocytic cells akin to macrophages are present in considerable numbers in the synovial tissues of joints. Two types of cells in the superficial layer of synovium were described by Barland *et al.* (1962)—Type A, highly phagocytic and possessed of numerous vacuoles and Golgi vesicles but little RER, and Type B, cells with well-developed RER and resembling a fibroblast and, possibly, able to manufacture components of synovial matrix or fluid. Extensive morphological study, using many forms of particulate matter from red cells to Thorotrast, has confirmed the phagocytic nature of Type A cells, shown the existence of intermediate forms, and led to the concept that "type A and B cells are not distinct and different races of cells with distinct and different functions, but merely cells whose morphology reflects the function they are performing at a given moment" (Ghadially, 1978). Fell and her colleagues (Fell *et al.*, 1976; Barratt *et al.*, 1977) have used pig synovium *in vivo* and in organ culture to demonstrate that synovial cells, like macrophages, have a highly variable form in appropriate culture conditions, are highly phagocytic, form opsonic rosettes with sensitized red blood cells, and may fuse to form giant cells. It seems that they are macrophages in all important senses. Their precise relation to other macrophages, and whether they are of ultimate marrow monocyte origin, is undetermined. This work confirms the view that synoviocytes belong to a single cell type, which assumes the A or B form according to local environmental conditions.

The important macrophage in connective tissues is the immigrant cell, derived from the blood monocyte, and of recent marrow origin; most of our information on tissue macrophages is derived from such cells because they are accessible to study. A detailed account will not be given of the evolution of the granulomatous inflammatory response (see Spector, 1969), nor of human granulomatous disease. It should be noted, however, that a mature granuloma is

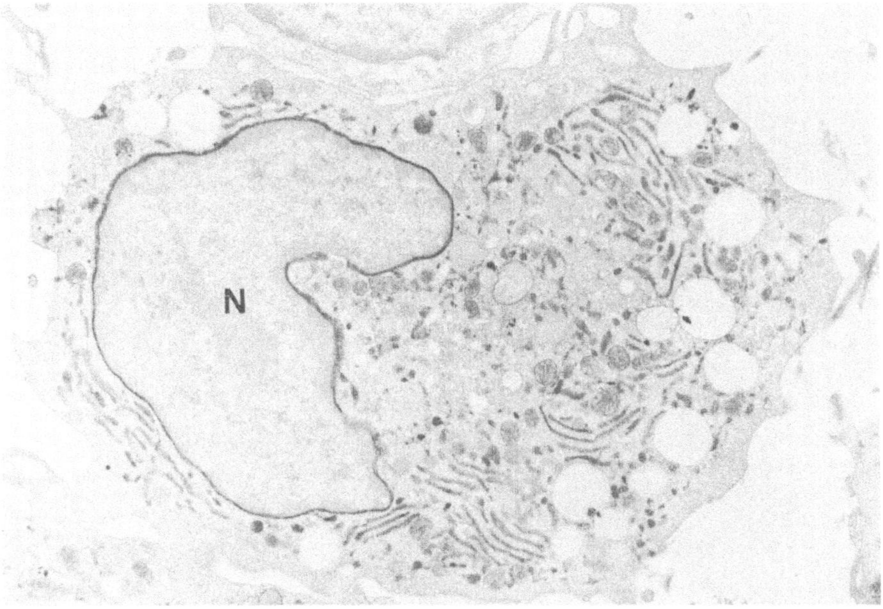

FIGURE 1. Slightly elongated resident peritoneal macrophage showing peroxidase activity in the rough endoplasmic reticulum and nuclear envelope. The granules are negative. ×7000. (Courtesy of Dr. H. J. van der Rhee.)

almost like an organized tissue, with a specialized vasculature like that of a lymph node and postcapillary venules through which lymphocytes migrate (Smith *et al.*, 1970). The present account is restricted to the structure of the component macrophages.

2. THE IMMIGRANT MACROPHAGE

The macrophages found in inflammatory lesions are recently derived from monocytes similar to those found in the circulation (see Chapter 7). When the monocyte enters the inflammatory lesion it matures into a macrophage. Cell proliferation may take place either at the level of an immature cell or of the mature macrophage. There is some, though not conclusive, evidence that mature macrophages may give rise to small cells once more (Spector and Lykke, 1966). Mature macrophages are usually immobilized in the lesion by factors released by sensitized lymphocytes, but small numbers of these cells reenter the draining lymphatics by passing between their endothelial cells by reverse diapedesis and then pass to the draining lymph node (Carr, 1977a). Their structure will, therefore, vary from that of a blood monocyte, through that of a mature mononuclear macrophage sometimes loaded with ingested bacilli or other material, to a multinucleate giant cell. The term epithelioid cell is often used to describe a mature macrophage-like cell, usually mononuclear, though sometimes, the term refers to multinucleate cells. It is variously used by different

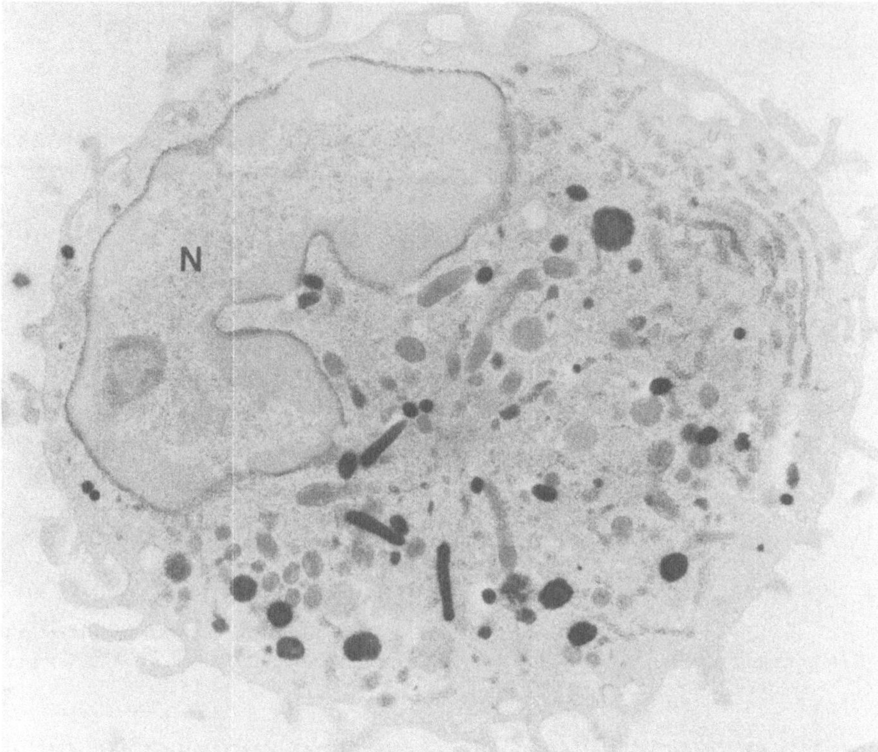

FIGURE 2. Peritoneal exudate macrophage showing peroxidase positive elongated primary and secondary granules and some peroxidase-negative macrophage granules. Weak peroxidase staining is also present in the rough endoplasmic reticulum and nuclear envelope. ×11,100. (Courtesy of Dr. H. J. van der Rhee.)

authors. The term will here be used to refer to a mature macrophage poorly phagocytic, which is probably in a secretory phase. A good account of the maturation sequence is given in an early and illuminating account by Gusek (1959).

It is usually assumed with a fair degree of accuracy that the macrophages found in an inflammatory lesion are similar whatever the site of the lesion. While this is probably true, there may be variations, dependent on the presence, to different degrees, of a local resident macrophage population, identifiable only by quite minor cytochemical differences. For instance, van der Rhee *et al.* (1979a,b,c) have studied in detail the inflammatory reactions induced by implanting Melinex at different sites in the subcutaneous tissues and in the peritoneal cavity. Striking differences occur in the reactions in the two sites (Figures 1, 2, and 3). Peroxidase activity is present if at all only in the granules of macrophages and multinucleate giant cells in the skin; on the other hand, in the peritoneum there are exudate macrophages with peroxidase activity restricted to granules, resident macrophages with peroxidase activity only in the RER and nuclear envelope, and macrophages with peroxidase activity in both granules,

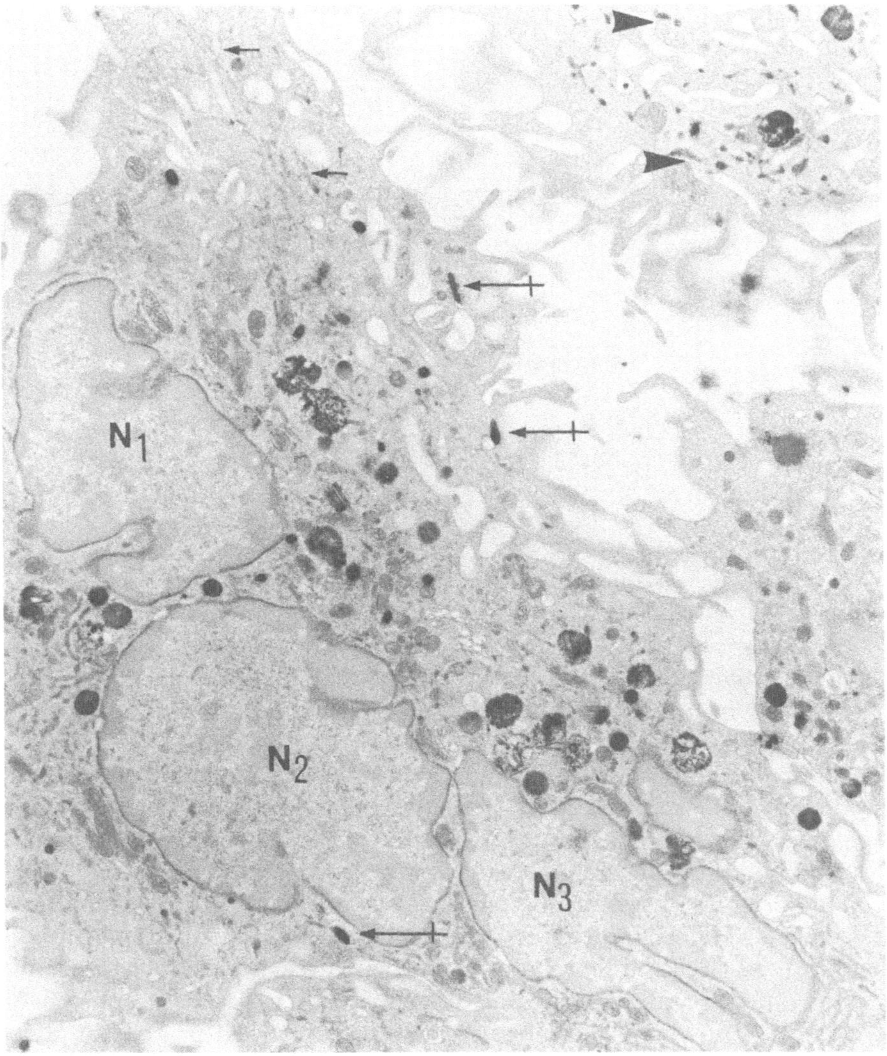

FIGURE 3. Peritoneal giant cell with 3 nuclei. Peroxidase activity is present in the rough endoplasmic reticulum (arrows) and nuclear envelope but is weaker than in a nearby resident macrophage (arrowheads). In addition to secondary lysosomes, several elongated peroxidase-positive granules can be seen (barred arrows). ×9000. (Courtesy of Dr. H. J. van der Rhee.)

RER, and nuclear envelope. Two kinds of giant cells occur, one with and one without peroxidase activity in the RER and nuclear envelope. The time relationships of the reaction led to the view that the appearance of peroxidase activity in RER and nuclear envelope of exudate macrophages and multinucleate giant cells is transient, and do not support the opinion that transitional forms exist between exudate and resident macrophages. The marrow origin of "resident" macrophages is, however, currently a topic of debate.

It is worth emphasizing in diagrammatic form (Figure 4), that macrophages

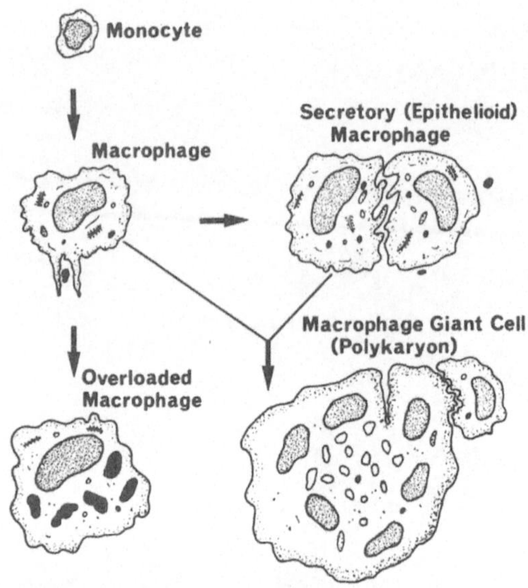

FIGURE 4. Diagram of evolution of macrophages in inflammatory lesions.

are found in different inflammatory lesions in all stages of transition between the monocyte, the mature macrophage, the macrophage loaded with phagocytized material, perhaps bacterial, perhaps mineral in origin, the macrophage with little evidence of phagocytic activity, but much evidence of secretory activity, often described as an epithelioid cell, and the multinucleate giant cell formed by macrophage fusion. A very similar sequence occurs when monocytes are cultured *in vitro* under appropriate conditions, as early described by Sutton and Weiss (1966). The modern morphologist, using transmission electron microscopy (TEM) and thin section technique, may overlook the fact that a thin section through one end of a nonspherical cell and that through the other end may look quite different.

The content of macrophages of different types varies from one type of granuloma to another. This point has been well-analyzed in a study of the ultrastructure of experimental granulomas induced by various substances (Papadimitriou and Spector, 1972). The macrophages in the lesion induced by carrageenan, which is indigestible and relatively nontoxic, are cells with a relatively low turnover, and show little evidence of secretory maturation but contain much phagocytized material. In the granuloma induced by *Bacillus pertussis* on the other hand there is a constant influx of young cells which mature, while in that induced by the tubercle bacillus (BCG), there are numerous large macrophages with numerous lysosomes and numerous epithelioid cells. Clearly, the nature of the granuloma, and the proportion of different types of macrophages in it, is dictated by whether the material inducing the granuloma is readily broken down by the macrophage and whether it is toxic to the macrophage, as well as by any possible question of the induction of immunity. The cellular environment of a granuloma is likely to be complex in many ways. For instance, monocytes enter the lesion, mature, phagocytize, and die, and their fragments or the bacteria which caused the lesion are phagocytized by a new group of macrophages. Mature macrophages containing tubercle bacilli rarely divide. The granuloma may act as a "trap" which can significantly deplete the bone marrow's supply of mononuclear cells. As would be expected, the turnover in a healing lesion is slower than in an active lesion (Dannenberg *et al.*, 1975; Tsuda *et al.*, 1976). Macrophages *in vitro* secrete a wide variety of substances which may well affect the maintenance of chronic inflammation (Allison and Davies, 1975). Granuloma macrophages have been shown histochemically to contain large amounts of various lysosomal enzymes, proteases, esterases, acid phosphatase, β-galactosidase, β-glucuronidase, lipase, lysozyme, deoxyribonuclease, and ribonuclease (Monis *et al.*, 1968; reviewed by Dannenberg, 1968).

3. THE MATURE MACROPHAGE

The ultrastructure of macrophages in inflammatory lesions has been studied by numerous authors (Gusek, 1959, 1964; Dumont and Sheldon, 1965; Galindo and Imaeda, 1966; Galindo *et al.*, 1969; Elias and Epstein, 1968; Jones Williams, 1972, Jones Williams *et al.*, 1970, 1972; James and Jones-Williams, 1974; Papadimitriou and Spector, 1972), and, in an excellent series of articles, by Adams (Adams, 1974–1976). The description that follows represents a consensus of these findings along with the author's own observations (Carr and Wright, 1979).

The mature macrophage in an inflammatory lesion is a large cell. In favorable H and E-stained sections it has a round, ovoid, or reniform nucleus, eosinophil cytoplasm which may contain visible inclusions, and an irregular surface; cytoplasmic processes may sometimes be seen. The cytoplasm stains positively for such lysosomal enzymes as acid phosphatase and β-glucuronidase.

A typical macrophage seen in an experimental animal granuloma is illustrated in Figure 5. It is an irregularly shaped cell 12–25 mμ in diameter. The nucleus is kidney shaped, circular, or ovoid in profile, with a peripheral fibrous lamina and nucleopores and often evident nucleolus. The margin of the cell is thrown into numerous processes, doubtless in movement or able to move, some of which are fingerlike and some flaplike (filopodia and lamellipodia), which do not interdigitate extensively with neighboring cells. After suitable staining there is a prominent cell coat. Between the processes are numerous invaginations of extracellular space some of which appear in section to be detached from the extracellular milieu. There are, in addition, some genuine vacuoles. Micropinocytic vacuoles, 60–80 nm in diameter, some "coated" with microfilaments, are present in varying numbers. Membranous profiles are prominent throughout the cytoplasm; the granular ER is prominent, sometimes localized in the periphery of the cytoplasm, sometimes more diffuse. In appropriate sections a well-developed Golgi apparatus is seen, composed of the usual channels and vacuoles. It may be on the concave surface of the nucleus, or be scattered more diffusely. Around it may radiate microtubules from the centriole and small

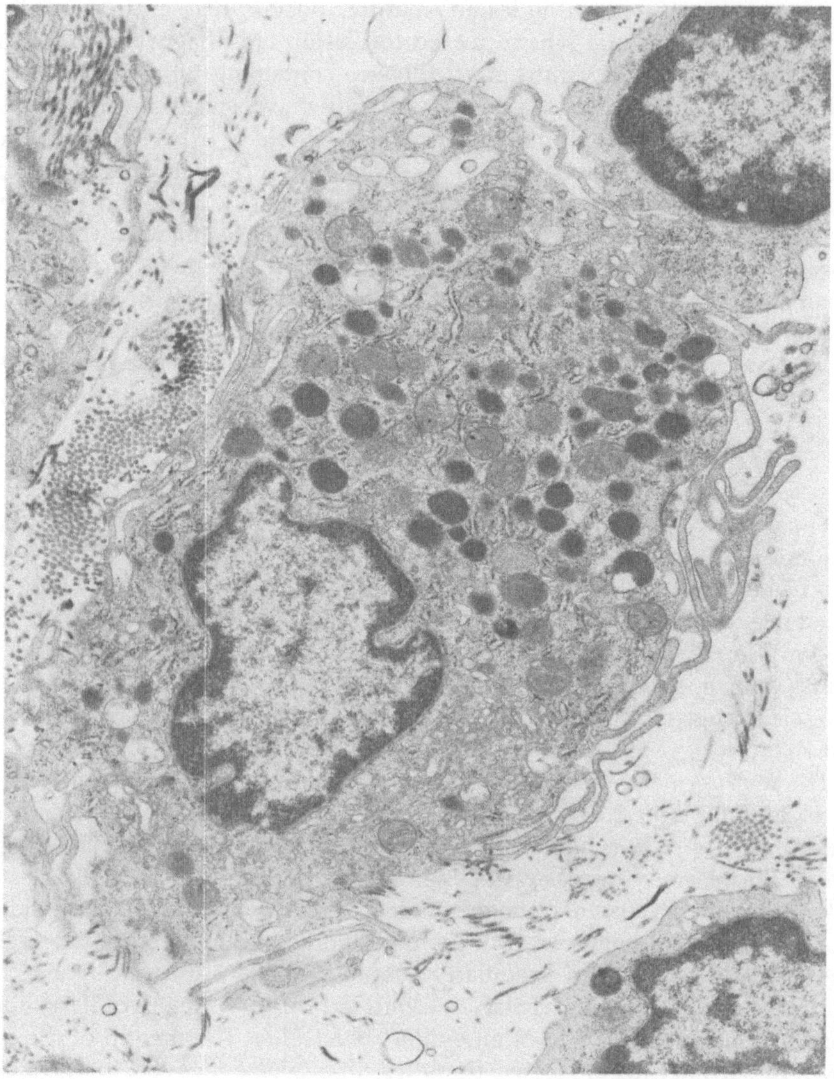

FIGURE 5. Macrophage from a rat granuloma. Diffuse profiles of granular ER are scattered throughout the cell. There are numerous dense bodies of various sizes, some of heterogeneous structure, vacuoles with less electron-dense material, and numerous mitochondria. There is a closely related adjacent lymphocyte. ×11,500.

channels or vacuoles filled with variably electron-dense material. Mitochondrial profiles are usually circular or ovoid, occasionally elongated, suggesting that mitochondria in these cells are usually short. Dense bodies of two major kinds are seen (Figure 6). Some are small and homogeneous with an electron-dense core, and usually have a less electron-dense area below the bounding membrane. They vary widely in shape from the spherical to elongated or dumbbell-shaped. These bodies (Figure 7) are characteristically bounded by a unit membrane, below which is an electron-lucent space. The core is finely granular; the granules are ~ 4 nm in diameter. Some, at least, of these bodies contain lysozyme. Probe microanalysis does not demonstrate iron in these bodies. At least, the smaller members of this group are presumably primary lysosomes (Figure 3). van der Rhee *et al.* (1979a,b,c) studied in great detail the evolution of monocytes into macrophages and giant cells. They noted several types of granules—primary and secondary granules, numerous in the Golgi area of monocytes and found near the cell membrane in maturing macrophages, and macrophage granules, round (~ 280 nm) with a halo between the limiting membrane and the granular matrix. The latter were essentially similar to those described by Carr (1968) in peritoneal macrophages.

van der Rhee *et al.* (1979a,b,c) considered in detail the various types of vesicles and granules found in macrophages. In their view, as seems reasonable, at least some bristle-coated vesicles, primary and secondary granules, and macrophage

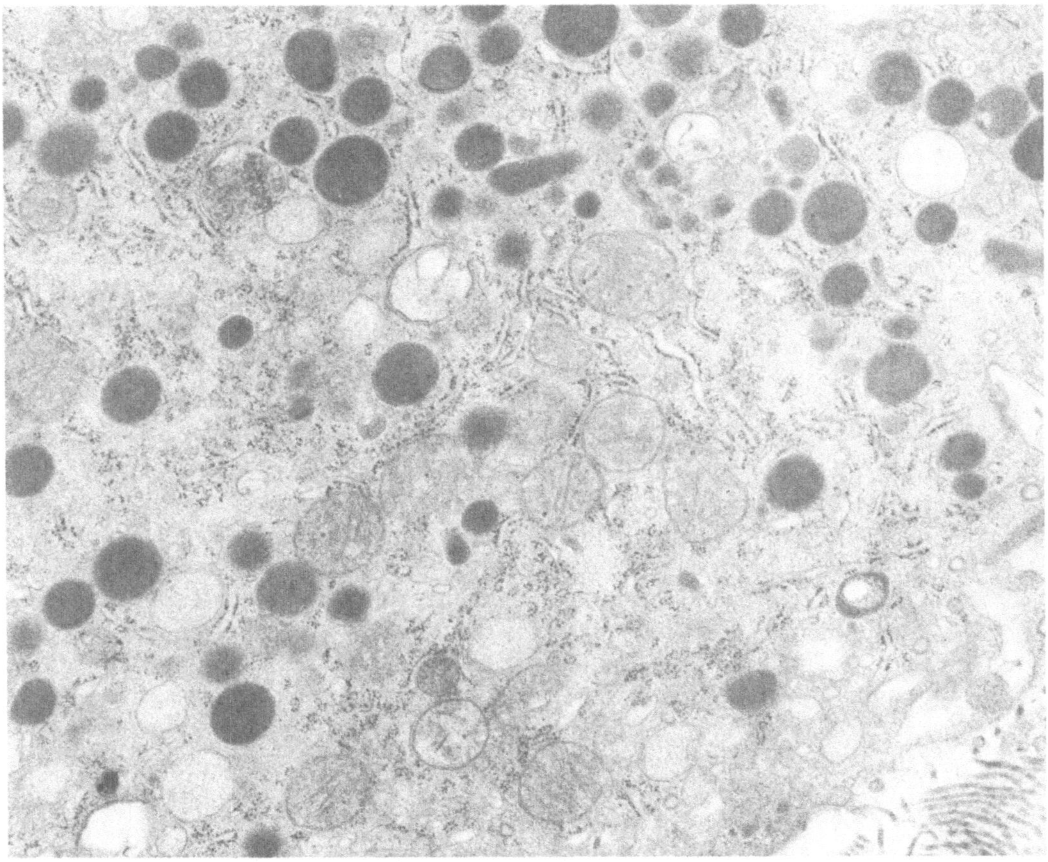

FIGURE 6. Detail of a macrophage similar to that seen in Figure 2. Dense bodies are of homogeneous structure and vary in size, one being elongated. Vacuoles with less electron-dense material are present. Near the edge of the cell lie several micropinocytic vesicles. ×22,000.

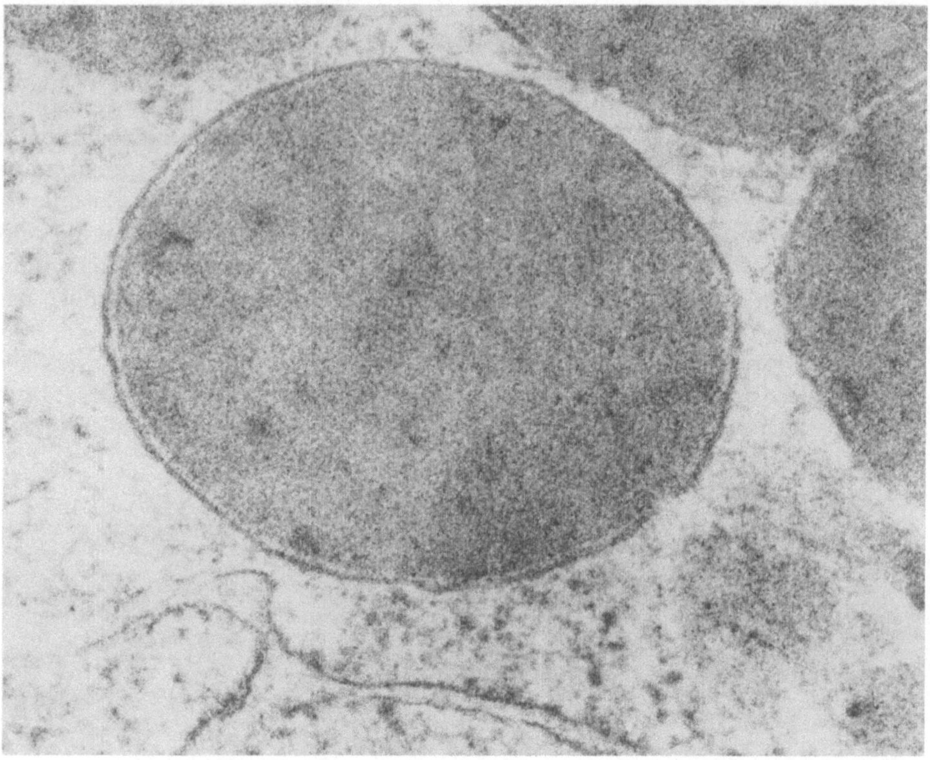

FIGURE 7. Detail of a dense granule from similar macrophage to that seen in Figure 3. Below the bounding unit membrane lies an electron-lucent zone. The core of the granule is composed of fine densities. ×120,000.

granules can, under various circumstances, and at various stages in cellular maturation, be primary lysosomes. Detailed cytochemical analysis shows differences in the nature and content of peroxidase between the different types of granules. While there is ample evidence that pinocytosis occurs freely *in vitro*, it is not clear to what extent these bodies contain pinocytized material *in vivo*. Other dense bodies are larger and more heterogeneous, containing electron-dense masses, myelin figures, and, often, ferritin. Recognizable phagocytized material may be present. There are also, in some cells, membrane-bound inclusions with a finely granular, much less electron-dense content. These will be discussed further below. As in many macrophages, microtubules and microfilaments are found in the cytoplasm. The latter may occur in bundles, and are of two sizes, 6–7 nm in diameter and 10 nm in diameter.

Such macrophages usually lie freely, independent of one another, or may be closely related to other macrophages or lymphocytes (Figure 5). They may interlock with adjacent cells by means of their cytoplasmic processes, and show poorly developed desmosomes. Such interlocking is stated by some authors to be indicative of the epithelioid state.

The mature macrophage seen in granulomatous lesions is phagocytically active. This has been studied in detail in an excellent series of articles on macrophages from pulmonary granulomas, induced by BCG in oil, by Leake, Myrvik, and their collaborators. As the granuloma develops, there is a parallel increase in lysozyme content of sera and cell extracts and of epithelioid cells with prominent ergastoplasm and, later, prominent intracellular granules (Leake and Myrvik, 1968). When such cells are allowed to interact with *Mycobacterium smeg-*

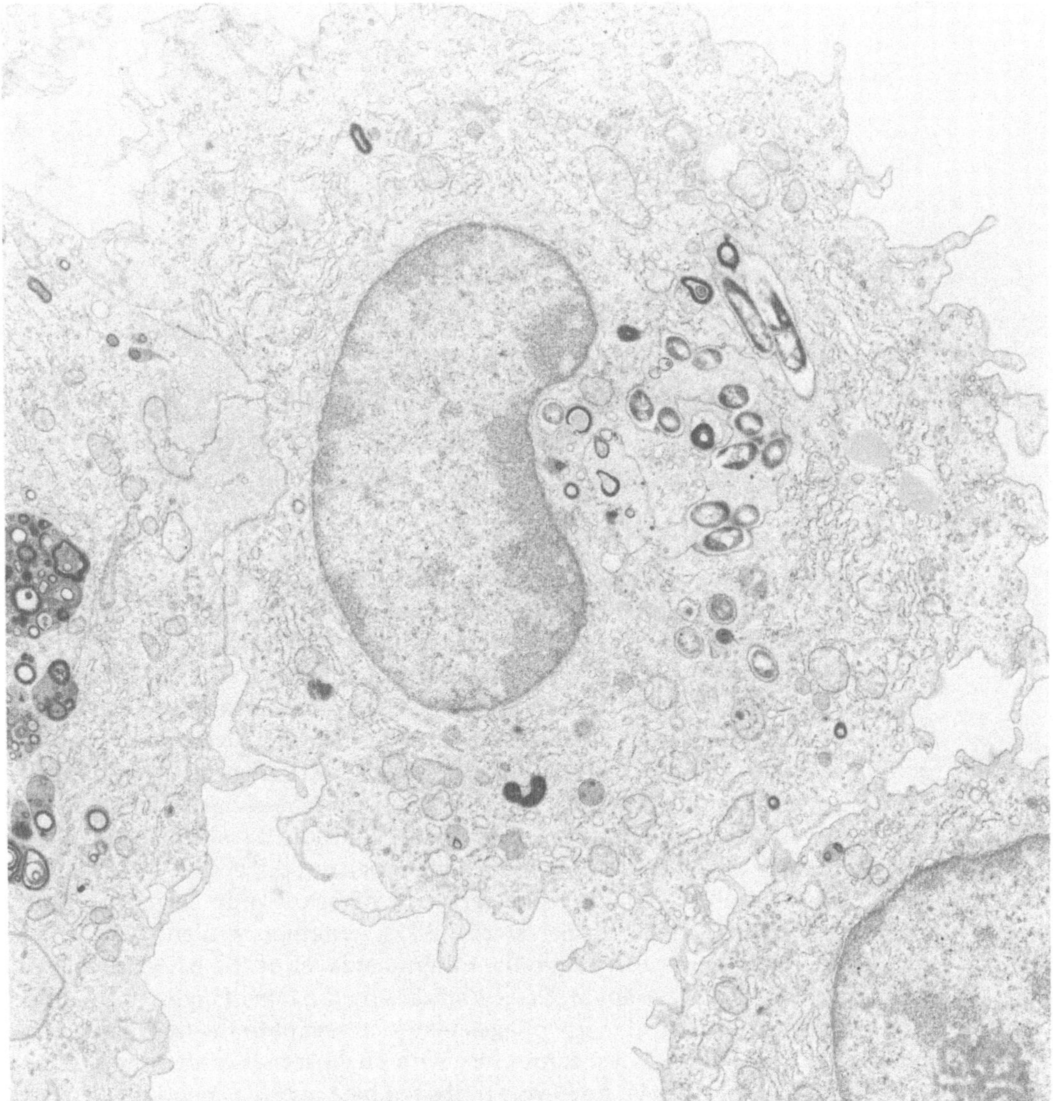

FIGURE 8. Alveolar macrophage from rabbit undergoing a pulmonary granulomatous reaction, showing numerous intracellular mycobacteria. ×7500. (Courtesy of Dr. E. S. Leake.)

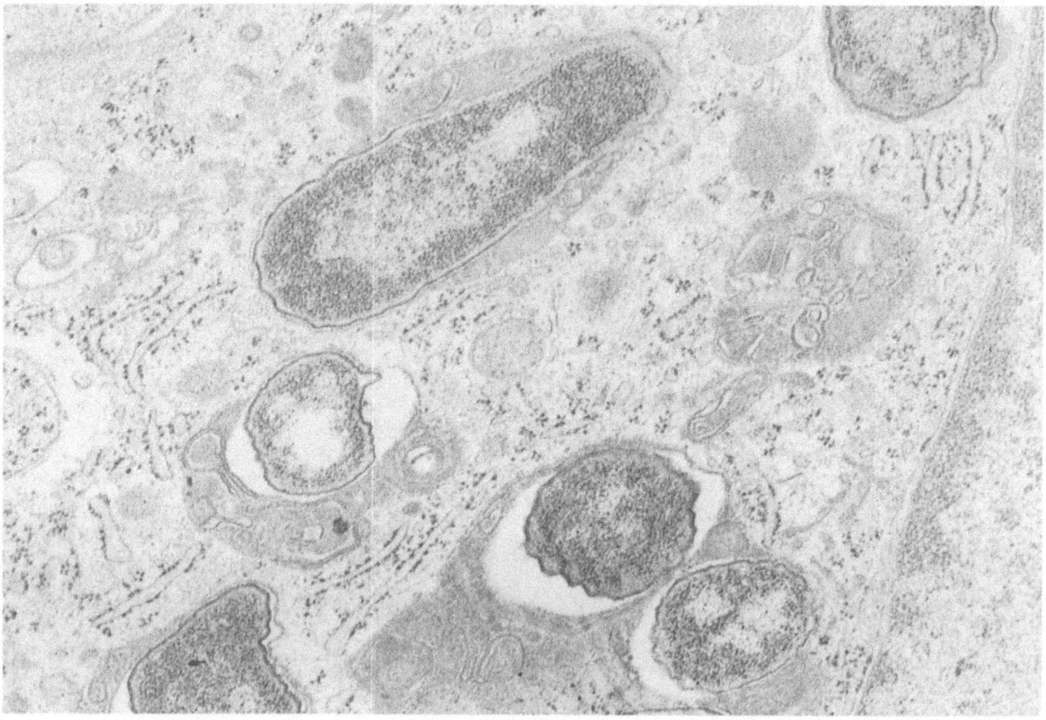

FIGURE 9. Similar cell showing electron dense material closely adherent to ingested *Salmonella*. ×27,000. (Courtesy of Dr. E. S. Leake.)

matis, small vesicles (50–70 nm) with an electron-dense core fuse with the phagosome, and larger (200–500 nm) bodies become adherent to the phagosome. The granulomatous cells did not appear to digest the organism at a faster rate than did normal alveolar macrophages (Leake and Myrvik, 1970). In a subsequent study of the breakdown of bacteria, small electron-dense granules were seen to fuse with the phagosome, and bacteria were seen to be surrounded by finely granular, electron-dense material (Leake *et al.*, 1971) (Figures 8 and 9). The electron-dense granules in these cells are often elongated or dumbbell-shaped and arranged in rosette fashion around the centriole in such fashion as to suggest that the granules forming the rosette represent distended Golgi saccules (Leake and Myrvik, 1972). Mycobacteria are digested only relatively slowly (Leake and Myrvik, 1966; Leake *et al.*, 1977), whether virulent or not. The phagosomal membrane adheres to the outermost layer of the bacterial cell wall to simulate a peribacillary space. *In vitro* studies on the fate of ingested mycobacteria in cultured peritoneal macrophages may be relevant to the fate of mycobacteria within a granuloma. Lysosomes fuse with phagosomes containing *Mycobacterium leprae*; the bacillus is, however, protected by a capsular peptidoglycolipid. Lysosomes do not, on the other hand, fuse with phagosomes containing viable *Mycobacterium tuberculosis*, but do fuse with phagosomes containing such organisms in a nonviable state. The inhibition of fusion may be mediated by a

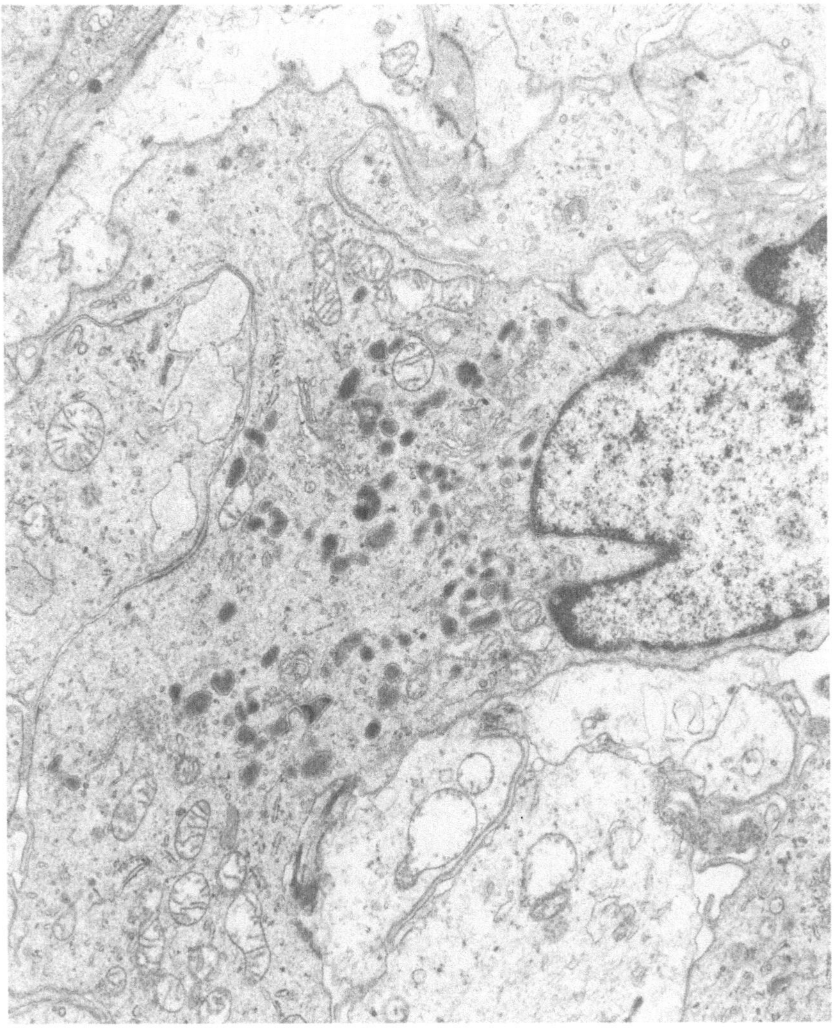

FIGURE 10. Macrophage from human sarcoid granuloma. The cell has a rather stellate shape and relates to adjacent cells by desmosomes. The cytoplasm contains numerous dense bodies, often elongated or hourglass-shaped. ×9500.

soluble product released by the mycobacterium (Armstrong and Hart, 1971, 1975; reviewed in Draper and Hart, 1975; Goren *et al.*, 1976).

The macrophages seen in human granulomatous lesions are very similar, though good recent ultrastructural accounts of such common lesions as human tuberculosis are scanty (Miller *et al.*, 1978). A typical macrophage as seen in a human sarcoid lesion is illustrated in Figure 10. The cell is highly irregular in shape, forming desmosomes with adjacent cells. The Golgi zone is diffuse and well-developed, and there are numerous dense bodies varying in profile from circular to elongated and dumbbell-shaped. Highly ordered paracrystalline

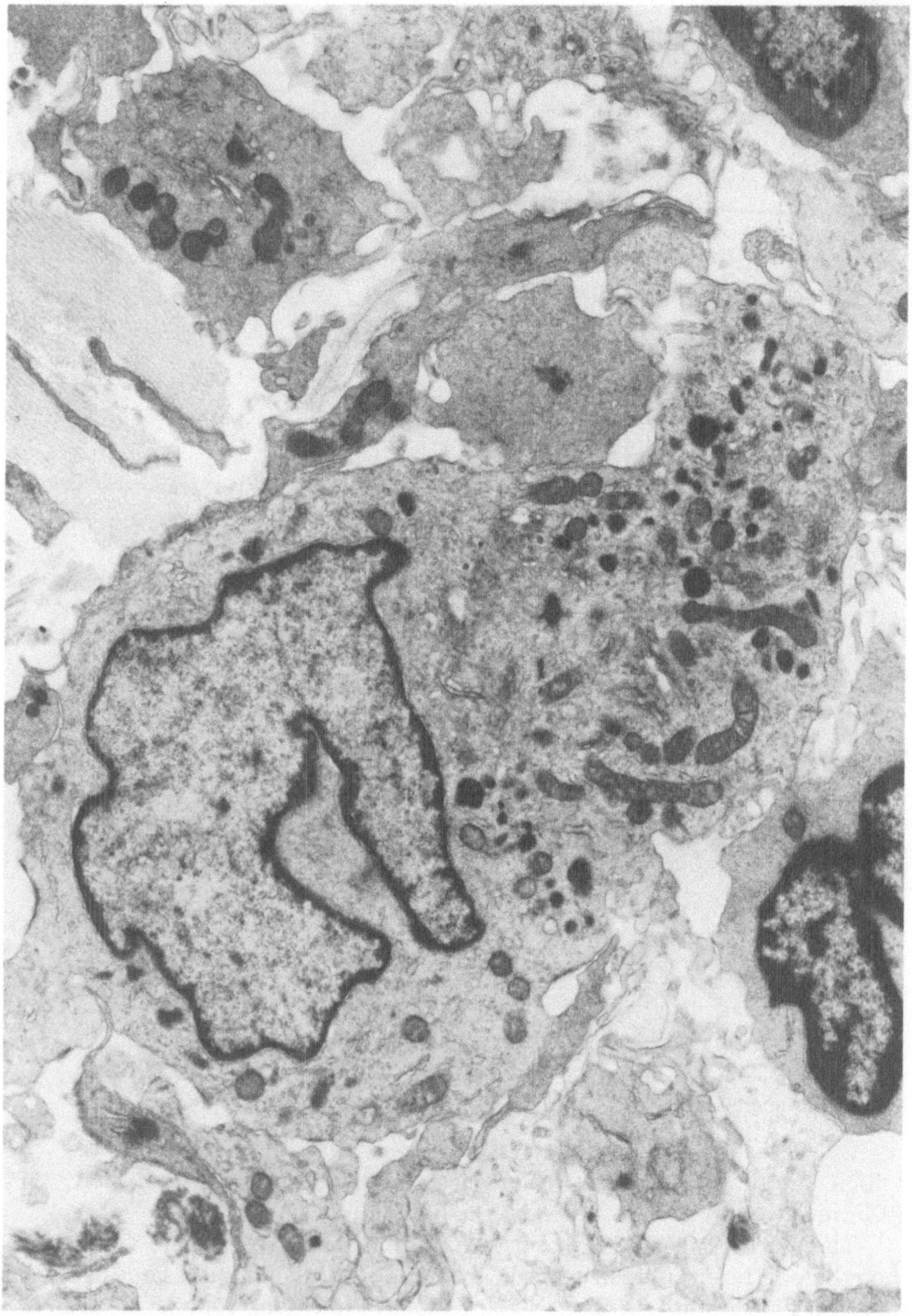

FIGURE 11. Macrophage from a human neoplasm. At one pole of the cell are clustered mitochondria and numerous dense bodies. A lymphocyte is closely adjacent. ×9000.

structures may be seen in some of these (Carr and Norris, 1977). The macrophages described by Miller *et al.* (1978) in human tuberculosis are similar, often showing dumbbell-shaped primary lysosomes. Transitional stages between monocyte and macrophage were identified, similar to those seen in experimental lesions.

Macrophages in tumors are in a rather special situation. It has been clearly shown that a mixed lymphoreticular cell response is present in many human and experimental tumors and macrophages are an important component of this response (Underwood and Carr, 1972; Carr and Underwood, 1974). It has been shown that macrophages may be cytotoxic in these situations (reviewed by Evans, 1977), and that macrophages may phagocytize apparently viable tumor cells in regressing experimental tumors (Carr *et al.*, 1974), but there is not yet solid evidence proving that macrophages play an important part in the defence against human neoplasms (reviewed by Carr, 1977b). A typical macrophage as seen in the infiltrate at the edge of a human neoplasm is illustrated in Figure 11.

4. THE LADEN MACROPHAGE

Under a number of situations, notably in granulomatous inflammation, macrophages may appear to be heavily laden with foreign material. This may be totally indigestible, as in the case of carbon in pulmonary macrophages, it may be a heavy load of digestible material in the process of destruction, or it may be material which would normally be digestible but is not destroyed at a normal rate because of enzymatic abnormality. Figure 12 illustrates an alveolar macrophage heavily loaded with ingested *Salmonella* in various stages of degradation and showing numerous secondary lysosomes probably representing progressive stages in breakdown.

An interesting light on the function of granuloma macrophages is cast by a consideration of the rare human disorder malakoplakia. This is an uncommon granuloma, usually of the urinary tract, but sometimes found elsewhere, The macrophages in this lesion contain large intracytoplasmic lysosomal inclusions laden with iron and calcium and known as Michaelis-Gutmann bodies. Bacteria such as *Escherichia coli* and *Klebsiella* species have been demonstrated within phagocytic vacuoles, and sequential stages between this and the Michaelis-Gutmann body have been demonstrated (Lou and Teplitz, 1974; McClurg *et al.*, 1973; Lewin *et al.*, 1974; reviewed by Abdou *et al.*, 1977). It has been postulated that there is impaired digestion of bacteria due to an intracellular abnormality probably in the phagolysosomes. It has recently been shown in a single case of this condition that the biochemistry and function of the circulating monocytes is abnormal. Many more monocytes than normal have large intralysosomal granules, the intracellular cyclic-GMP level is low, and the monocytes fail to release β-glucuronidase and are defective in their ability to inhibit the growth of *E. coli* in culture; cholinergic drugs correct the *in vitro* abnormality and are beneficial *in vivo* (Abdou *et al.*, 1977). It seems likely, therefore, that the mac-

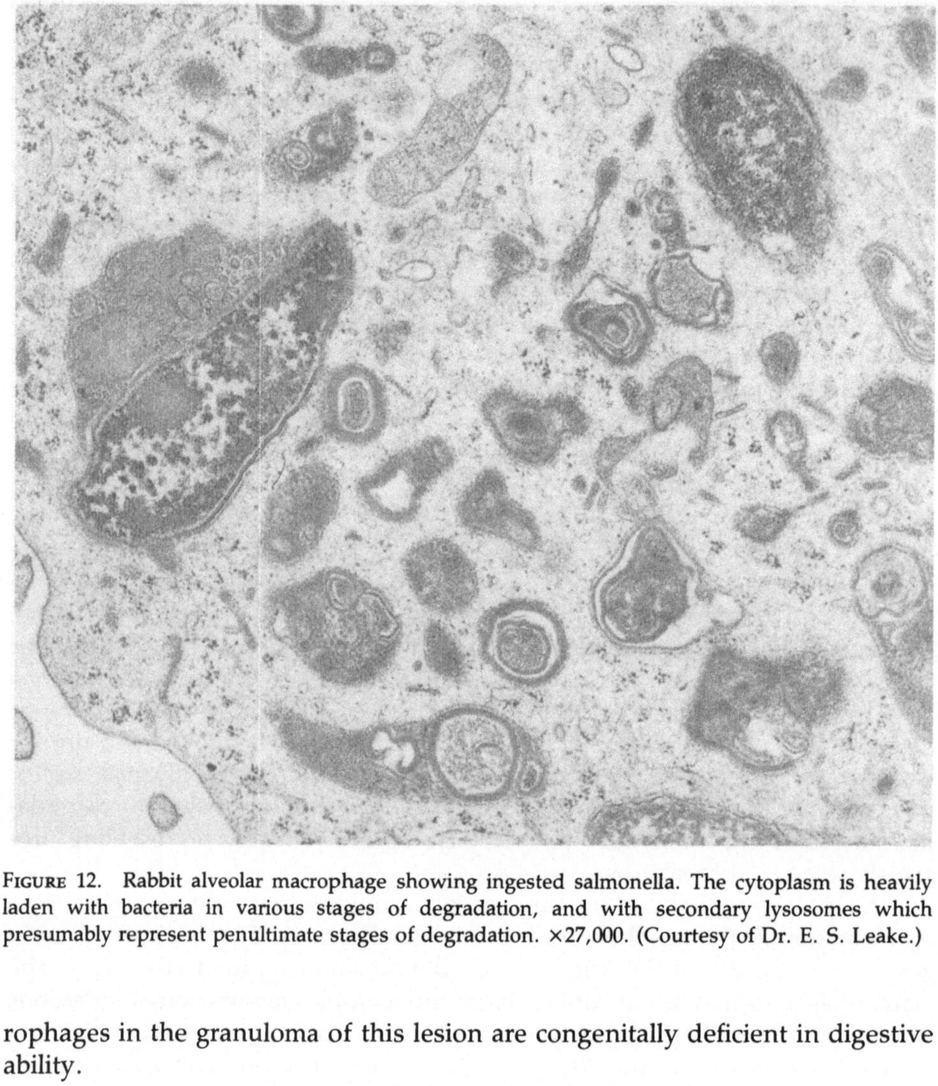

FIGURE 12. Rabbit alveolar macrophage showing ingested salmonella. The cytoplasm is heavily laden with bacteria in various stages of degradation, and with secondary lysosomes which presumably represent penultimate stages of degradation. ×27,000. (Courtesy of Dr. E. S. Leake.)

rophages in the granuloma of this lesion are congenitally deficient in digestive ability.

5. THE SECRETORY MACROPHAGE OR EPITHELIOID CELL

While epithelioid cells occur in experimental animal granulomas, they are best seen in human lesions. As Bernaudin *et al*. (1975) point out, epithelioid cells are found in a wide variety of human granulomas, notably those of tuberculosis, leprosy, sarcoidosis, and the granulomas induced by zirconium and beryllium. Descriptions of epithelioid cells in such lesions by Gusek (1964), Wanstrup and Christensen (1966), and Jones-Williams *et al.* (1972) are in broad general agreement. The epithelioid cell or secretory macrophage is large, with cytoplasm which is variably eosinophilic, and contains a wide variety of cytochemically demonstrable lysosomal enzymes; lysozyme is demonstrable in many epithelioid cells by immunocytochemical techniques. The epithelioid cells usually lie near the center of the sarcoid follicle (Soler and Basset, 1976). Ultrastructural examination shows appearances similar to those described above. The nucleus is often eccentric and the morphological features associated with secretion exaggerated. Thus, the granular ER may be very well-developed or there may be a particularly well-developed SER. Some of these cells contain numerous membrane-bound electron-dense granules, sometimes with an organized paracrystalline core (Carr and Norris, 1977) (Figure 13). Others contain numerous vacuoles of varying electron density. The cells described by Miller *et al.* (1978) as epithelioid, contained numerous large phagosomes, and interlocked laterally with their neighbors by means of cytoplasmic projections.

Jones-Williams and his colleagues pointed out that two kinds of epithelioid cells existed, one (A) with prominent granular ER, and the other (B) with prominent Golgi zone and many electron-lucent vesicles which contained mucoglycoprotein rather than acid phosphatase. Type A cells were more common in tuberculosis and Type B in berylliosis. The cell illustrated in Figure 14 is of the latter type, with many relatively electron-lucent vacuoles containing flocculent, electron-lucent material. Some of these vacuoles have a more electron-dense core (Figure 15). Bernaudin *et al.* (1975) illustrated one such vacuole in continuity with the extracellular space and suggested that this represents extrusion of secretory material. Transitional forms occur between these two cell types, or variations of the same cell type.

In such human granulomas as sarcoidosis, it is reasonably clear what is meant by an epithelioid cell, though even here it is not always easy to know whether to apply this term to a given cell. Some authors use the term to describe

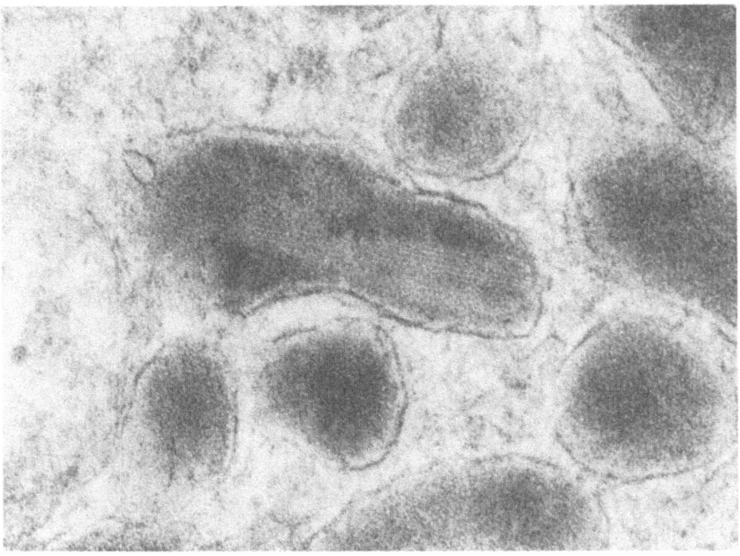

FIGURE 13. Dense granules from sarcoid macrophage showing ordered internal structure. ×80,500.

a cell which looks rather similar to a plasma cell, for example, Crawford and Hardwicke (1978); Turk *et al.* (1978) and others add confusion by talking about immature epithelioid cells. In a study of experimental pulmonary granulomas, Soler *et al.* (1975) point out that, while there are numerous cells in such lesions which resemble human epithelioid cells at the light microscope level, few of them resemble human epithelioid cells at the electron microscope (EM) level.

The most imaginative studies on the origin and nature of epithelioid cells have been those carried out by Spector and his colleagues on the cells which appear on the surface of cellophane strips implanted into the subcutaneous tissues of mice (Papadimitriou and Spector, 1971). It is clear from their sequential studies that epithelioid cells differentiated from monocytes by way of highly phagocytic macrophages. Under the conditions of these experiments epithelioid cells could not phagocytize bacteria, but could ingest colloidal gold or polystyrene particles by pinocytosis. One micrograph of a granule purportedly being excreted was shown. Macrophages which had completely digested staphylococci became epithelioid, as did cells which ingested and then extruded colloidal gold. Macrophages which ingested, but did not destroy, tubercle bacilli

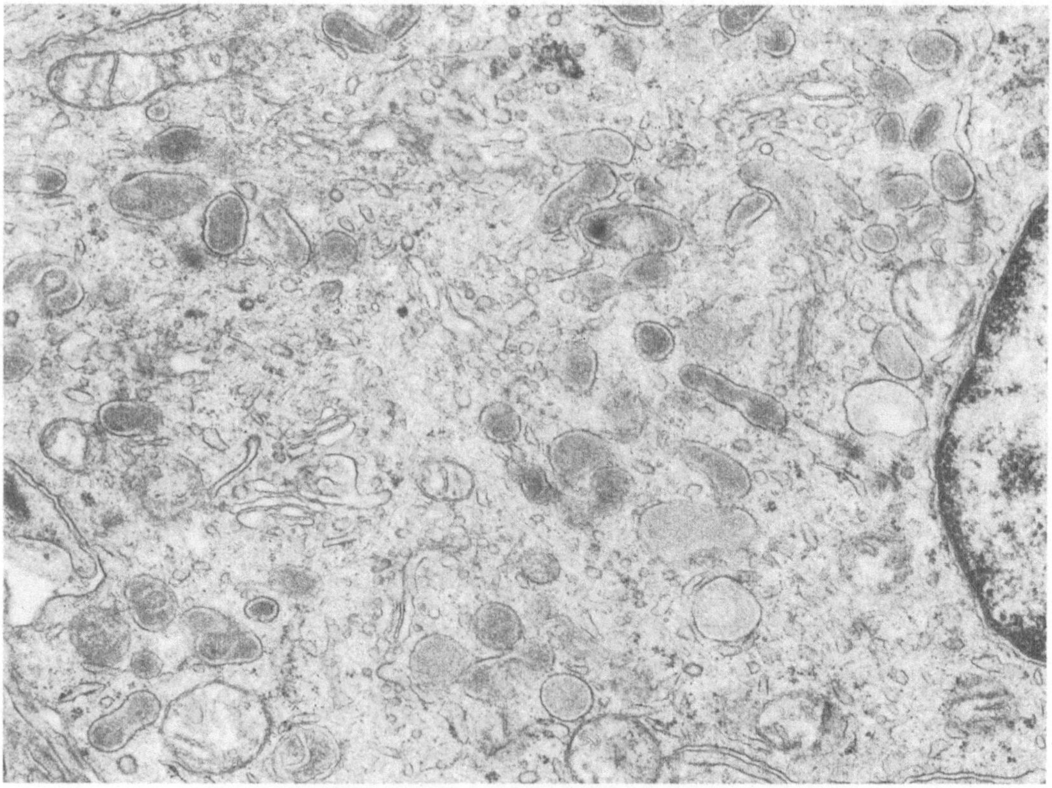

FIGURE 14. Human sarcoid macrophage showing numerous granules, predominantly electron lucent. Numerous small vesicles bud off the Golgi apparatus in two areas of the section ×27,500. Reduced 14% for reproduction.

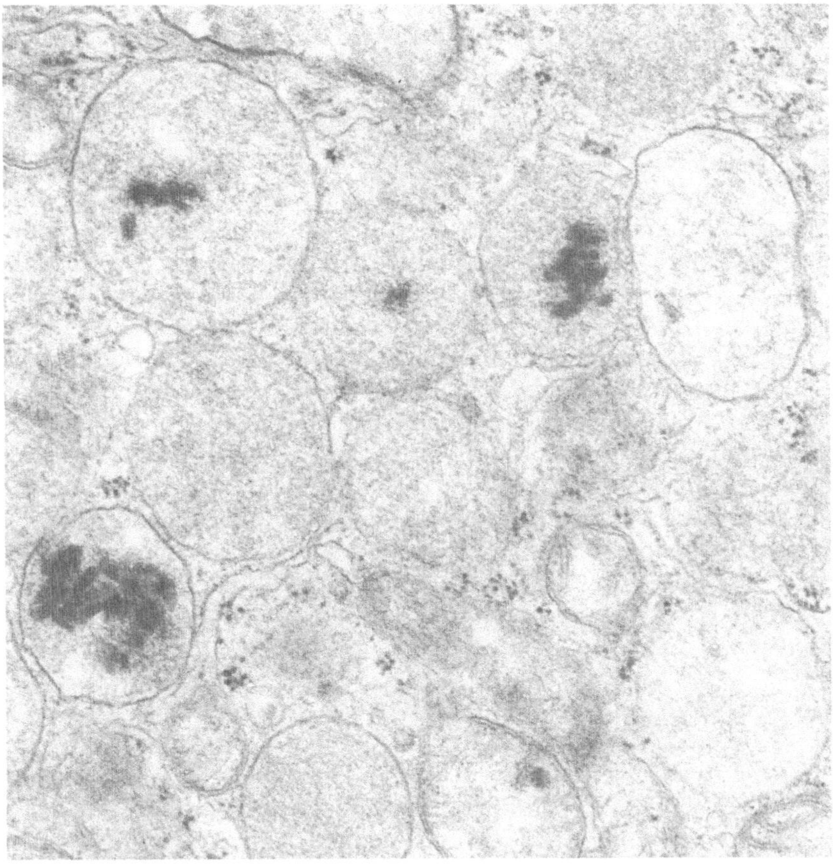

FIGURE 15. Detail of electron-lucent vacuoles from sarcoid macrophages, showing highly electron-dense core. ×42,900.

did not become epithelioid. Thymectomy or whole-body irradiation did not prevent epithelioid development. The conclusion was that epithelioid cells develop when there are macrophages in excess of those required for digestion, and that they form a barrier in a granuloma between the macrophages which contain bacteria, and the host. They may ingest, digest, and detoxify subparticulate material, or excrete it. They may, therefore, show "reciprocal changes in phagocytic and secretory activity." In a similar model studied in great detail by van der Rhee *et al.* (1979a,b,c), the granules in epithelioid cells were described as being distinctive, like macrophage granules, but slightly larger (~ 340 nm as opposed to ~ 280 nm), slightly more electron-dense, and with a less evident halo.

The epithelioid or secretory state may be accompanied by changes in surface receptors. When the surface receptors of macrophages, obtained by inserting glass coverslips into the connective tissues of mice, were studied for the presence of surface receptors, it was shown that most macrophages and giant cells expressed Fc, C, and nonspecific surface receptors, and that "epithelioid cells" did not, and were less actively phagocytic. Here, as in so many other studies, the criteria on which the use of the term "epithelioid" were based were not very strict (Mariano *et al.*, 1976). In granulomas associated with cellular immunity there is an initial increase in phagocytic ability, followed by a later diminution. This may facilitate the persistence of antigen (Mariano *et al.*, 1977).

There is no agreement as to whether cellular immunity is necessary for the development of the epithelioid cell state. Epithelioid cells can apparently develop in the slightly artificial situation of the subcutaneous coverslip or cellophane strip without cellular immunity (Papadimitriou and Spector, 1971; Spector and Mariano, 1975). On the other hand, studies on the granuloma induced by the injection of zirconium or beryllium salts into the skin showed clearly that epithelioid cells developed only in the presence of hypersensitivity (Epstein *et al.*, 1963; Elias and Epstein, 1968; Black and Epstein, 1974). A comparison of the granuloma induced by schistosome eggs with that induced by plastic beads, showed that the former was dependent on cellular immunity, being inhibited by immunosuppressives, while the latter was dependent on kinin formation. The sensitization is due to a soluble factor released by the schistosome eggs (Kellermeyer and Warren, 1970; Warren and Boros, 1975). Perhaps there are different types of epithelioid cells.

6. THE MACROPHAGE GIANT CELL OR MACROPHAGE POLYKARYON

Giant cells were classically identified in human granulomas such as tuberculosis and sarcoidosis. Their ultrastructure has been described by numerous authors (Gusek, 1959, 1964; Gusek and Naumann, 1959; Wanstrup and Christensen, 1966). Similar cells have been described in experimental animal lesions such as that of guinea pig tuberculosis.

The typical giant cell (Figure 16), as found in a granuloma, is composed of a mass of cytoplasm 100 μm or more in diameter. The nuclei are often arranged in a ring as seen in a section, and therefore, presumably, in a peripheral spheroidal shell around the cell center. Traditionally, two types of giant cell are distinguished; the foreign body giant cell whose nuclei are scattered throughout the cytoplasm, and the Langhans giant cell whose nuclei are arranged in a ring. This distinction is not always clear. The suggestion was made by Gusek and Naumann (1959) that Langhans cells form by a process akin to gastrulation. Considerable variations in giant cell structure are seen, those in sarcoidosis, for instance, have a more "epithelioid" appearance than those in tuberculosis. The edge of the cell (Figure 17) characteristically shows numerous processes, which may interdigitate with those of adjacent macrophages, very often young monocytoid macrophages. Desmosomes or hemidesmosomes may be evident. This probably represents a phase in the formation of a giant cell. These interdigitations may persist within the giant cell to various degrees, as clusters of elongated channels. There is often an area of granular ER lying peripheral to the

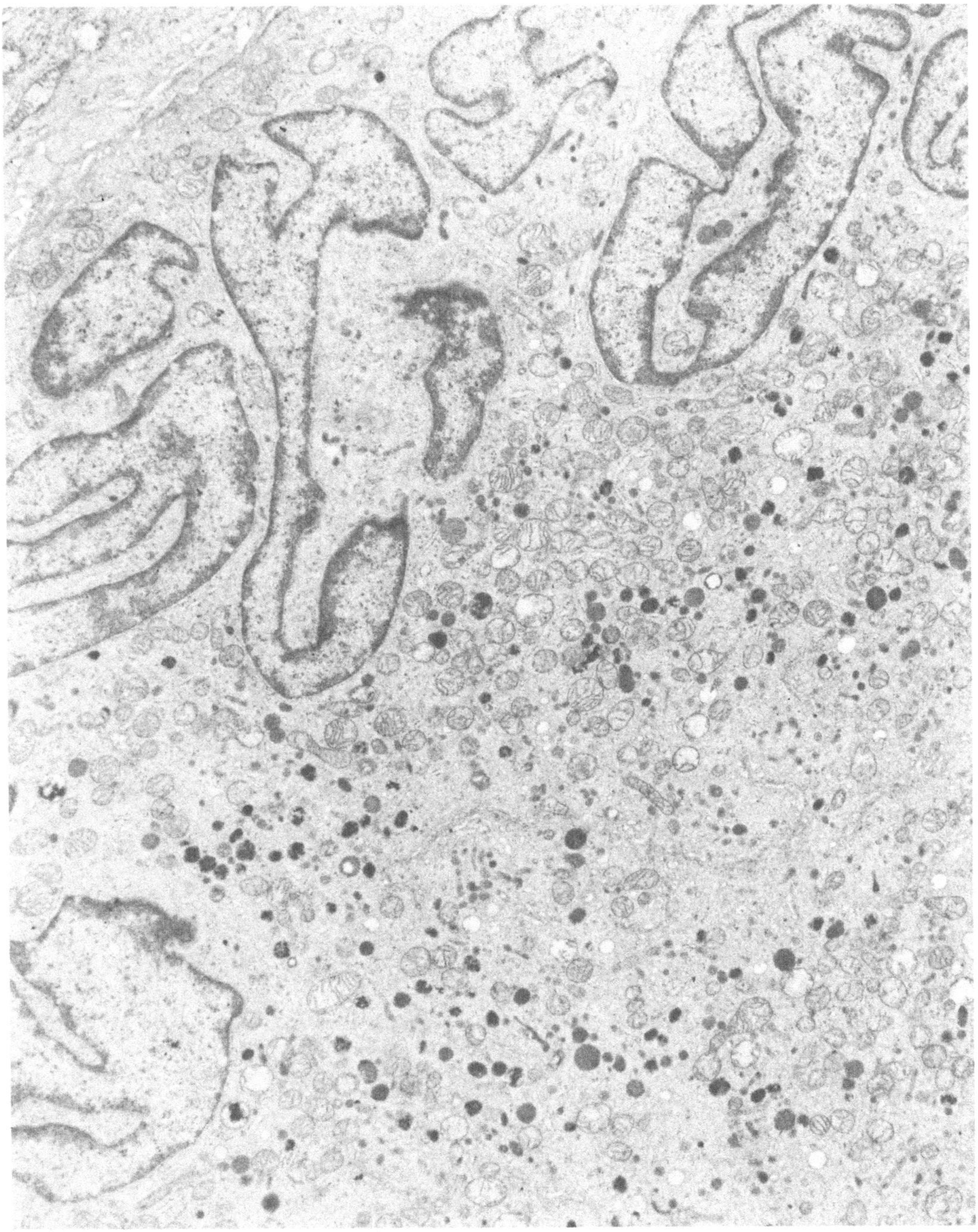

FIGURE 16. Part of a human sarcoid giant cell showing part of the peripheral ring of irregularly shaped nuclei surrounding the central part of the cell which contains numerous mitochondria, numerous lysosomal dense granules, and a Golgi zone. ×7000.

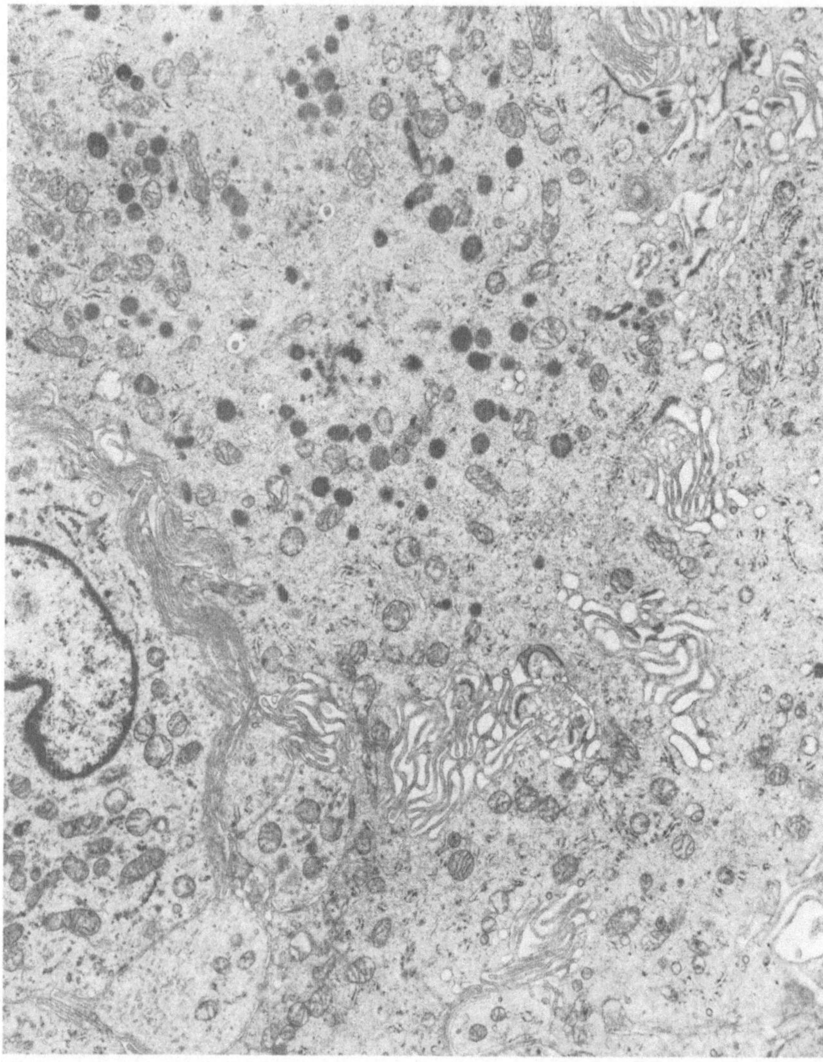

FIGURE 17. The edge of a sarcoid giant cell showing cell membrane interdigitating with that of an adjacent mononuclear macrophage. Desmosomes are evident at several points. Clusters of elongated channels dip deeply into the cytoplasm of the giant cell, probably representing the site of fusion of two cells. Many dense bodies lie in the cytoplasm ×7600.

nucleus, while within the ring of nuclei lie numerous membrane-bound structures similar to those seen in mononuclear macrophages. Some of these have an electron-dense core while others, usually larger, have a more electron-lucent internum (Figures 14 and 15). The center of the cell contains a very prominent Golgi zone and numerous membrane bound granules and vacuoles (Wanstrup and Christensen, 1966), many of which contain lysozyme (Lobo *et al.*, 1978). Star-shaped inclusions or "asteroid bodies" occur within giant cells in sarcoidosis (Azar and Lunardelli, 1969). These are largely composed of collagen—whether synthesized by the giant cell, or trapped, or phagocytized is not certain; they are often partly calcified (Figure 18). Other inclusions (Schaumann bodies) found in giant cells appear to be large calcified phagosomes containing the remnants of tubercle bacilli (Dumont and Sheldon, 1965). The prominent interdigitating cell processes found in giant cells in several human granulomas sup-

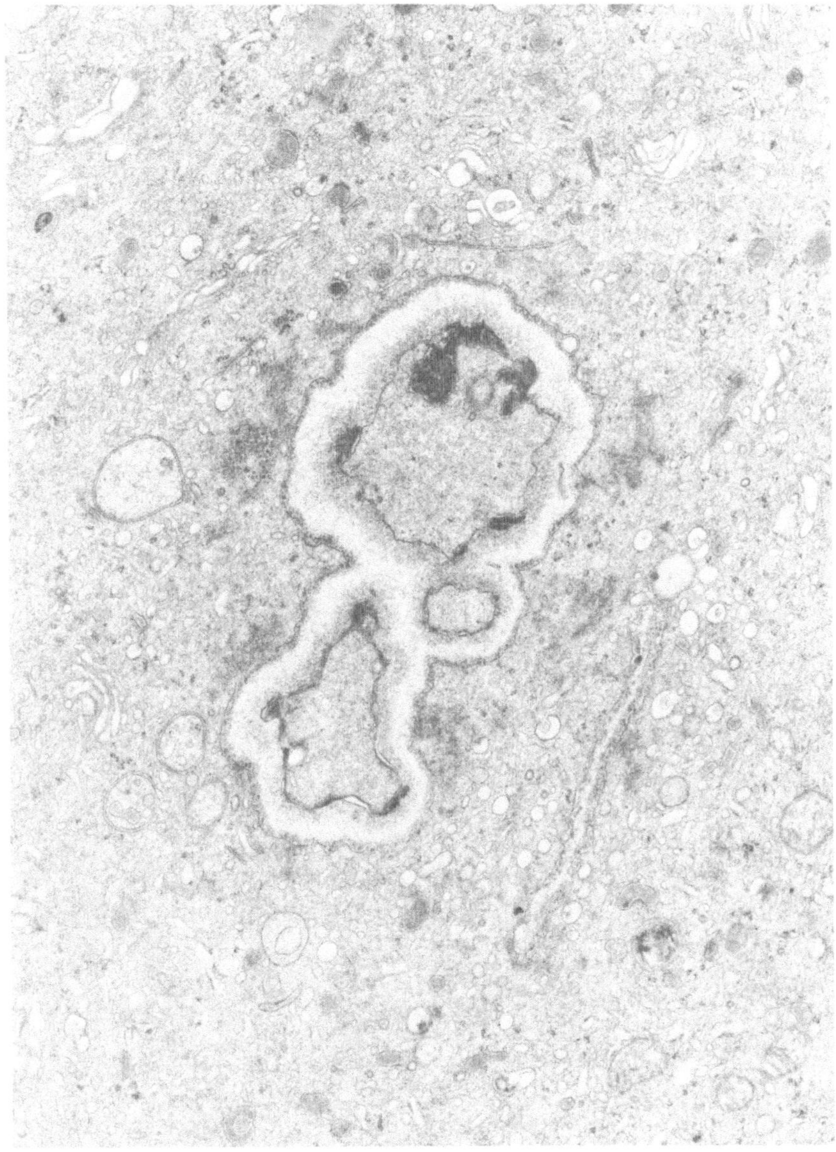

FIGURE 18. Large inclusion within a sarcoid giant cell composed apparently of invaginated basement-membrane-like material, partly calcified. ×12,285.

port the view that, like their animal counterparts, they form by cell fusion; this is seen, e.g., in sarcoidosis, tuberculosis, and pneumoconiosis (Policard *et al.*, 1965). Thymidine-labeling studies have clearly shown the origin of giant cells from circulating mononuclear cells (Gillman and Wright, 1966).

An extensive series of studies has been carried out on giant cells evoked on flat surfaces, either of material implanted into rodents, or in tissue culture, starting with the classic study of Mariano and Spector (1974), in which it was shown conclusively, by means of ultrastructural and cell-kinetic studies, that macrophage fusion was the most important process in giant cell formation. Young macrophages newly arrived on the site fused with older macrophages, many of which were already dividing and which showed a high proportion of chromosome breaks and gaps. While occasionally complete nuclear multiplication occurred within the cytoplasm, this was probably not a major factor in producing giant cells. Similarly, in the granuloma induced *in vivo* by combined injections of talc and prednisolone acetate, nuclear division does occur within giant cells but the chromosomal abnormalitites are so severe that it is unlikely that successful nuclear multiplication can occur (Dreher *et al.*, 1978). In coverslip preparations, two kinds of giant cells are distinguishable, foreign body giant

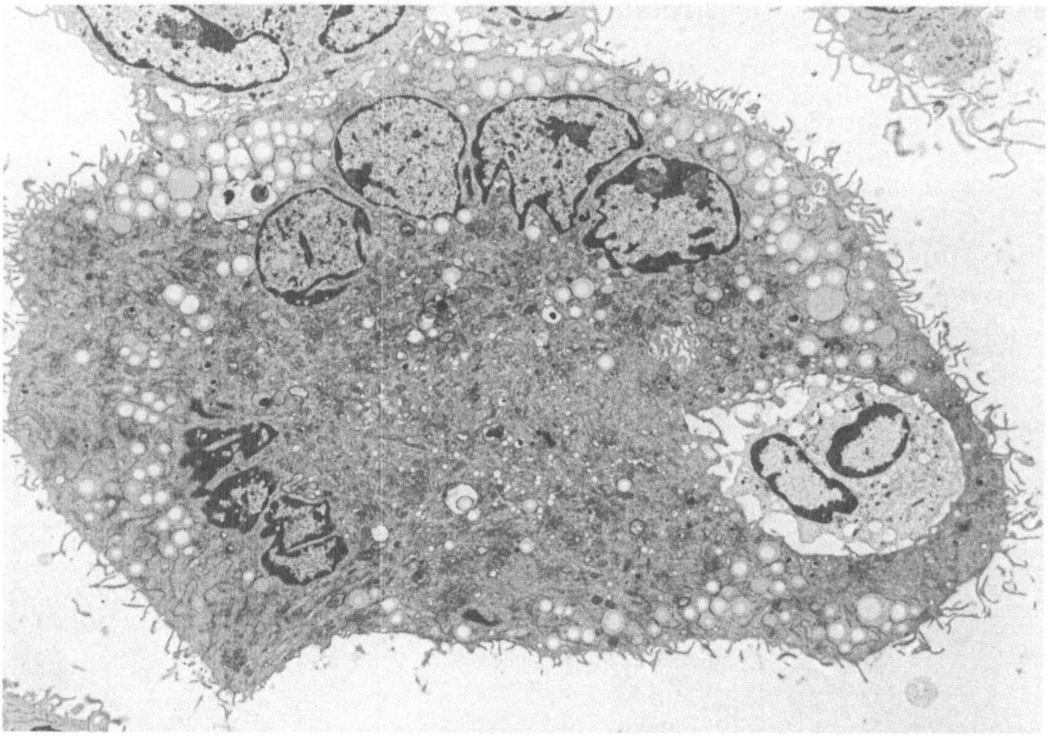

FIGURE 19. Multinucleate foreign body giant cell produced by the implantation of a plastic film into the dorsum of mice. The outline is irregular, showing numerous fine processes. Most of the organelles lie in the center of the syncytium. A small area of the vacuolar network is present. A mononuclear cell lies deeply indented within the syncytium. ×3600. Reduced 13% for reproduction. (Courtesy of Dr. J. M. Papadimitriou.)

cells with nuclei scattered throughout the cytoplasm and Langhans giant cells with a peripheral ring of nuclei and a central area positive for acid phosphatase and succinic dehydrogenase. Colchicine treatment inhibited the formation of the latter; their formation is probably, therefore, dependent on the formation of cytoplasmic microtubules. A detailed account of the ultrastructure of the murine foreign body giant cells which appeared on plastic film implants was given by Papadimitriou and Archer (1974). Phase contrast cinemicrography showed that these cells were motile, probing their environment by small slender processes like microvilli, and then moving by means of flaplike lamellipodia (Figures 19, 20, and 21). The cells vary in size from a surface area of 200 μm^2 to 0.25 mm^2 and contain a proportional number of nuclei from two to many hundreds. The nuclei arranged in a ring define a peripheral zone of cytoplasm with fine filaments (~ 4 nm), granular ER, ribosomes, fat, and glycogen. Within the ring of nuclei lies the cytocentrum with Golgi zone vacuoles and most, but not all, of the mitochondria. Most of the lysosomal dense bodies lie in this area and are of round, oval, semilunar, or hourglass shape; most are rather electron-dense, but some less so. It is in this area that most of the enzymes of oxidative phosphorylation,

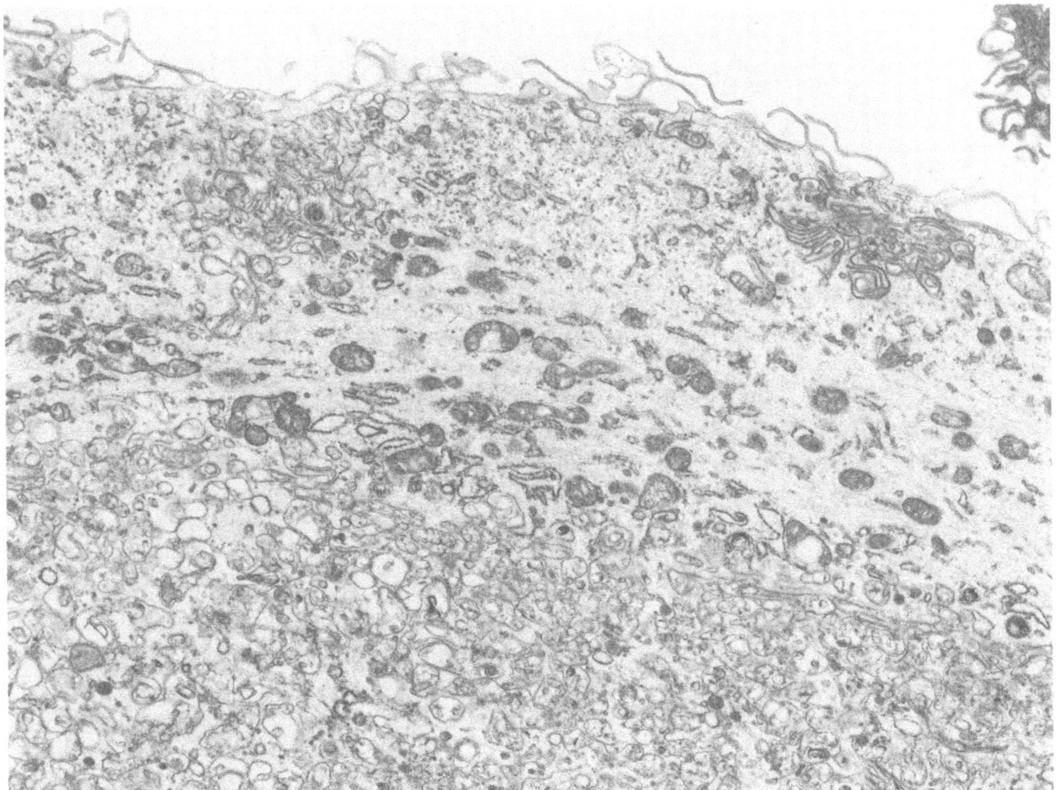

FIGURE 20. Peripheral cytoplasm of cell similar to that in Figure 16, showing cortical microfilaments. ×8700. (Courtesy of Dr. J. M. Papadimitriou.)

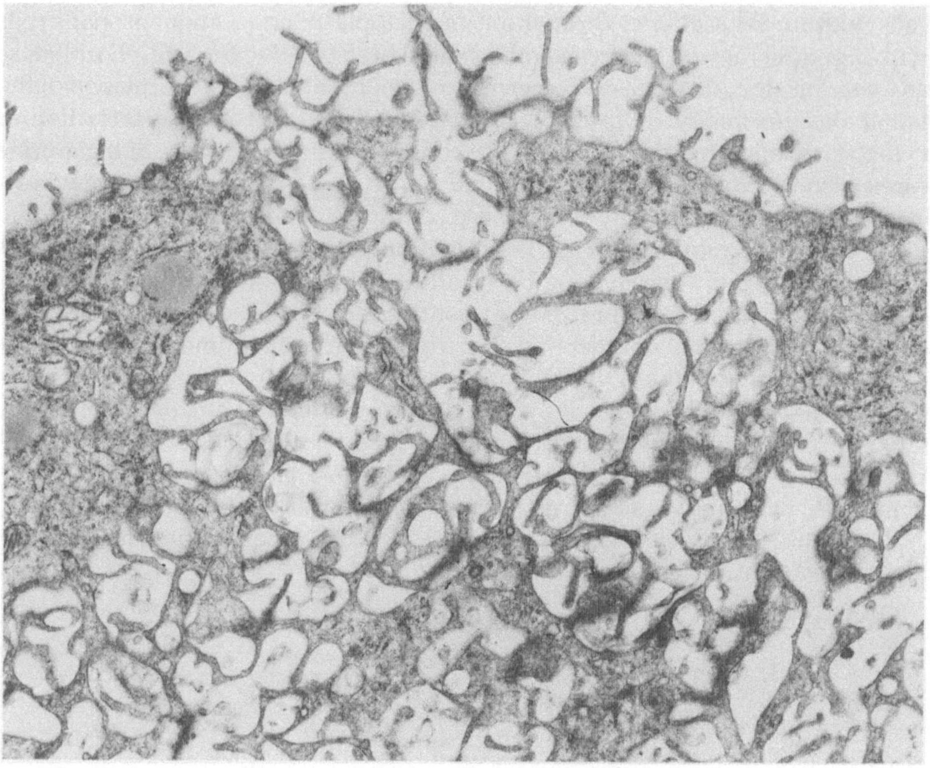

FIGURE 21. Irregular periphery of similar giant cell showing many cytoplasmic flaps and microvilli. There is apparent continuity between the extracellular space and the vacuolar network within the cell. ×13,050. (Courtesy of Dr. J. M. Papadimitriou.)

glycolysis, and the pentose shunt pathway have been demonstrated cytochemically.

The periphery of the cell shows numerous processes, some fingerlike microvilli and some flaplike lamellipodia, while the surface adjacent to the glass is flat. The cell membrane shows a well-marked glycocalyx, and the cortical cytoplasm shows numberous fine microfilaments. Peripherally, there are numerous vacuoles and vesicles 0.1–0.6 μm in diameter and, penetrating into the cytoplasm, a convoluted system of closely apposed smooth membranes, enclosing a space 6–10 nm in diameter; small particles (colloidal gold and Thorotrast) readily entered the vacuoles, while horseradish peroxidase, a smaller molecule, entered the narrower lamellar portion of the network. Acid phosphatase is found in the lysosomes and in the intracellular part of the membranous network.

The giant cells seen in similar subcutaneous granulomas were studied in detail by van der Rhee *et al.* (1979a,b,c). The structure varied with the number of contained nuclei. The smaller ("Langhans") giant cells are probably precursors of larger ("foreign body") giant cells. Smaller giant cells (2–10 nuclei) had peripheral nuclei. While there may be a few primary or secondary granules, most of the granules, which lie in the center of the cells are of macrophage or epithelioid type. Large giant cells (30 or more nuclei) have nuclei scattered throughout the cytoplasm, do not show the same clear central aggregation of granules, and have highly heterogeneous granules including elongated and beaded forms. Intermediate forms in size are seen but, in general, the larger the giant cell the more heterogeneous the granules. Morphologic evidence of cell fusion was rare, but fusing cells always had their Golgi areas adjacent to one another. This probably indicates that fusion is rapid.

Giant cells move more slowly (Papadimitriou and Kingston, 1977), and are less efficient phagocytes. Their efficiency is lower the larger the cell (Papadimitriou *et al.*, 1975; Chambers, 1977a). As they form, there is a rise in phospholipid content and glucose-6-phosphate dehydrogenase and a fall in succinic dehydrogenase and 5-nucleotidase in the cell population (Papadimitriou and Wyche, 1974). They probably selectively release lysosomal enzymes (Papadimitriou and Wee, 1976), which are plentiful in the center of the giant cell (Papadimitriou and Wyche, 1976). Studies with glutaraldehyde-fixed red cells, with antimacrophage antiserum, and with lectins make it likely that giant cell formation is related to simultaneous attempted phagocytosis by macrophages (Chambers, 1977b,c,d,e).

It has been shown that the interaction between sensitized lymphoid cells and antigen leads to the release of a macrophage fusion factor, falling into the group of lymphokines, which promotes macrophage fusion into giant cells (Galindo *et al.*, 1974; Parks and Weiser, 1975). Giant cell formation is evidently not thymus dependent, both in coverslip preparations (Papadimitriou, 1976) and in granulomas which appear on the injection of methyl cellulose into athymic nude mice (Machado and Lair, 1978), whereas the giant cells which develop in the graft-versus-host situation are of thymic origin (Bennett and Montes, 1973).

The phagocytic function of macrophages in granulomas is obvious, and will not be described here except to note that there are several human granulomas whose characteristic is the presence of macrophages stuffed with undigested or poorly digested microorganisms. It has, however, become evident from *in vitro* studies, that macrophages may secrete numerous substances into a tissue culture milieu. These include acid hydrolytic enzymes, plasminogen activator, cathepsin, elastase, collagenase, various cytolytic factors, and various substances which maintain the inflammatory reaction (see, e.g., Unanue, 1976, for review). Lysozyme is one of the best known of these; it has been shown that macrophages in many sites contain lysozyme as demonstrated immunocytochemically (Taylor and Burns, 1974), and that they release lysozyme *in vitro* (Cohn and Wiener, 1963). The release of a substance which lyses *Micrococcus lysodeikticus* has been recently and elegantly demonstrated by Biggar and Sturgess (1977). The secretory function of granuloma macrophages has been particularly clearly shown in relation to lysozyme in two situations, human and experimental. In human sarcoidosis and other granulomatous disorders there is often an elevation in serum lysozyme level (Pascual *et al.*, 1973; Perillie *et al.*, 1973; Falchuk *et al.*, 1975), and lysozyme is present in the giant cells (Taylor and

Burns, 1974; Klockars and Selroos, 1977). Lysozyme is demonstrable immunocytochemically at the EM level to be present in dense bodies; whether or not secretion of lysozyme is most active in giant cells, lysozyme is certainly present there in large amounts (Lobo *et al.*, 1978). Similarly, in an experimental rat granuloma, lysozyme is present in the granuloma macrophages and in the venous and lymphatic effluents, and there is a resultant elevation in serum lysozyme levels (Carr *et al.*, 1978). Again lysozyme can be identified at the EM level within the macrophage granules. The evidence would, therefore, seem to be quite good that *in vivo,* at least in the case of lysozyme, granuloma macrophages have a secretory function.

It seems likely that other biologically significant substances are secreted by granuloma macrophages *in vivo.* For instance, a glycoprotein is produced by a population of mononuclear cells within a granuloma; this has been shown to inhibit macrophage phagocytosis (Bole and Wright, 1976) and, therefore, could be of significance in the development of the epithelioid cell. Macrophages exposed to silica release a factor which stimulates collagen synthesis (Heppleston and Styles, 1967; Aho and Kulonen, 1977), though here it seems likely that the factor is released after death of the macrophage.

7. SUMMARY

The connective tissues of the body contain numerous connective tissue macrophages or histiocytes, of typical macrophage structure, but showing subtle differences in peroxidase cytochemistry which lead to the supposition that they form a population distinct from the macrophages which appear in inflammatory lesions. Relatively little is known about these resident histiocytes, with the exception of the macrophages of synovial tissues; these form a distinct subpopulation.

The most important and most numerous macrophages in the connective tissue are those which appear there in inflammatory lesions and are of recent marrow origin. These form a spectrum from maturing monocytes, to mature actively phagocytic macrophages, some heavily overloaded with ingested material, through secretory or epithelioid macrophages to macrophage giant cells, formed largely by cell fusion and probably secretory rather than phagocytic. At various stages of maturation, these cells contain membrane-bound secretory granules, differing in detailed structure with the phase of maturation. Apart from their obvious phagocytic function, it is likely that at one or other phase of their development macrophages have a secretory function which may modify profoundly the inflammatory process.

ACKNOWLEDGMENTS. I am grateful to Dr. Eva Leake for review of this manuscript, to Dr. Leake, Dr. J. M. Papadimitriou, and Dr. H. J. van der Rhee for illustrations, and to the publishers of *Cell and Tissue Research,* the *Journal of the Reticuloendothelial Society,* and the *Journal of Ultrastructure Research* for permission to reuse illustrations.

REFERENCES

Abdou, N. I., Napombejara, C., Sagawa, A., Ragland, C., Stechschulte, J., Nilsson, U., Gourley, W., Watanabe, I., Lindsey, N. J., and Allen, M. S., 1977, Malakoplakia: Evidence for monocyte lysosomal abnormality correctable by cholinergic agonist *in vitro* and *in vivo*, *N. Engl. J. Med.* **297**:1413.

Adams, D. O., 1974, The structure of mononuclear phagocytes differentiating *in vivo*. I. Sequential fine structural and histologic studies of the effect of *Bacillus* Calmette-Guérin (BCG), *Am. J. Pathol.* **76**:17.

Adams, D. O., 1975, The structure of mononuclear phagocytes differentiating in vivo. II. The effect of *Mycobacterium tuberculosis*, *Am. J. Pathol.* **80**:101.

Adams, D. O., 1976, The granulomatous inflammatory response. A review, *Am. J. Pathol.* **84**:164.

Aho, S., and Kulonen, E., 1977, Effect of silica-liberated macrophage factors on protein synthesis in cell-free systems, *Exp. Cell Res.* **104**:31.

Allison, A. C., and Davies, P., 1975, Increased biochemical and biological activities of mononuclear phagocytes exposed to various stimuli, with special reference to secretion of lysosomal enzymes, in: *Mononuclear Phagocytes in Immunity, Infection and Pathology* (R. van Furth, ed.), pp. 487–504, Blackwell, Oxford.

Armstrong, J. A., and Hart, D'Arcy P., 1971, Response of cultured macrophages to *Mycobacterium tuberculosis*, with observations on fusion of lysosomes with phagosomes, *J. Exp. Med.* **134**: 713.

Armstrong, J. A., and Hart, D'Arcy, P., 1975, Phagosome–lysosome interactions in cultured macrophages infected with virulent tubercle bacilli, *J. Exp. Med.* **142**:1.

Azar, H. A., and Lunardelli, E., 1969, Collagen nature of asteroid bodies of giant cells in sarcoidosis, *Am. J. Pathol.* **57**:81.

Barland, P., Novikoff, A. B., and Hamerman, D., 1962, Electron microscopy of the human synovial membrane, *J. Cell. Biol.* **14**:207.

Barratt, M. E. J., Fell, H. B., Coombs, R. R. A., and Glauert, A. M., 1977, The pig synovium. II. some properties of isolated intimal cells, *J. Anat.* **123**:47.

Bennett, M., and Montes, M., 1973, Graft-versus-host reactions in mice. 3. Epithelioid and multinucleated giant cells of thymic origin, *Am. J. Pathol.* **71**:119.

Bernaudin, J.-F., Soler, P., Basset, F., and Chretien, J., 1975, La cellule epithélioïde, *Pathol. Biol.* **23**:494.

Biggar, W. D., and Sturgess, J. M., 1977, Role of lysozyme in the microbicidal activity of rat alveolar macrophages, *Infect. Immun.* **16**:974.

Black, M. M., and Epstein, W. L., 1974, Formation of multinucleate giant cells in organized epithelioid cell granulomas, *Am. J. Pathol.* **74**:263.

Bole, G. G., and Wright, J. E., 1976, Biological properties of a granuloma glycoprotein that inhibits macrophage phagocytosis, *J. Lab. Clin. Med.* **87**:98.

Carr, I., 1968, Lysosome synthesis and surface changes in stimulated peritoneal cells, *Z. Zellforsch. Mikrosk. Anat.* **89**:328.

Carr, I., 1977a, The passage of macrophages across lymphatic walls by reverse diapedesis, *J. Reticuloendothel. Soc.* **6**:397.

Carr, I., 1977b, Macrophages in human cancer, in: *The Macrophage and Cancer* (K. James, B. McBride, and A. Stuart, eds.), pp. 364–374, Blackwell, Oxford.

Carr, I., and Norris, P., 1977, The fine structure of human macrophage granules in sarcoidosis, *J. Pathol.* **122**:29.

Carr, I., and Underwood, J. C. E., 1974, The ultrastructure of the local cellular reaction to neoplasia, *Int. Rev. Cytol.* **37**:329.

Carr, I., and Wright, J., 1979, The fine structure of macrophage granules in experimental granulomas in rodents, *J. Anat.* **128**:479.

Carr, I., Underwood, J. C. E., McGinty, F., Wood, P., 1974, The ultrastructure of the local lymphoreticular response to an experimental neoplasm, *J. Pathol.* **113**:75.

Carr, I., Carr, J., Lobo, A., and Malcolm, D., 1978, The secretion of lysozyme *in vivo*, by macrophages into lymph and blood in a rat granuloma, *J. Reticuloendothel. Soc.* **24**:41.

Chambers, T. J., 1977a, Studies on the phagocytic capacity of macrophage polykaryons, *J. Pathol.* **123**:65.
Chambers, T. J., 1977b, Fusion of macrophages following simultaneous attempted phagocytosis of glutaraldehyde-fixed red cells, *J. Pathol.* **122**:71.
Chambers, T. J., 1977c, The mechanism of fusion of hamster macrophages induced by anti-macrophage serum, *J. Pathol.* **122**:163.
Chambers, T. J., 1977d, Fusion of hamster macrophages induced by lectins, *J. Pathol.* **123**:53.
Chambers, T. J., 1977e, Failure of altered macrophage surface to lead to the formation of polykaryons, *J. Pathol.* **122**:185.
Cohn, Z. A., and Wiener, E., 1963, The particulate hydrolases of macrophages, *J. Exp. Med.* **118**:991.
Crawford, C. L., and Hardwicke, P. M. D., 1978, Ultrastructural features of epithelioid cell granulomas induced by intradermal injection of xenogeneic nerve tissue, *J. Pathol.* **125**:107.
Dannenberg, A. M., 1968, Cellular hypersensitivity and cellular immunity in the pathogenesis of tuberculosis: Specificity, systemic and local nature, and associated macrophage enzymes, *Bacteriol. Rev.* **32**:85.
Dannenberg, J., A. M., Ando, M., Shima, K., and Tsuda, T., 1975, Macrophage turnover and activation in tuberculous granulomata, in: *Mononuclear Phagocytes in Immunity, Infection and Pathology* (R. van Furth, ed.), pp. 959–978, Blackwell, London.
Deane, H. W., 1964, Some electron microscopic observations on the lamina propria of the gut, with comments on the close association of macrophages, plasma cells and eosinophils, *Anat. Rec.* **149**:453.
Draper, P., and Hart, P. D., 1975, Phagosomes, lysosomes and mycobacteria: Cellular and microbial aspects, in: *Mononuclear Phagocytes in Immunity, Infection and Pathology* (R. van Furth, ed.), pp. 575–589, Blackwell, London.
Dreher, R., Keller, H. U., Hess, M. W., Roos, B., and Cottier, H., 1978, Early appearance and mitotic activity of multinucleated giant cells in mice after combined injection of talc and prednisolone acetate, *Lab. Invest.* **38**:149.
Dumont, A., Sheldon, H., 1965, Changes in the fine structure of macrophages in experimentally produced tuberculous granulomas in hamsters, *Lab. Invest.* **14**:2034.
Elias, P. M., and Epstein, W. L., 1968, Ultrastructural observations on experimentally induced foreign body and organized epithelioid cell granulomas in man, *Am. J. Pathol.* **52**:1207.
Epstein, W. L., Skahen, J. R., and Krasnobrod, H., 1963, The organized epithelioid cell granuloma: Differentiation of allergic (zirconium) from colloidal (silica) types, *Am. J. Pathol.* **43**:391.
Evans, R., 1977, Macrophages in solid tumours, in: *The Macrophage and Cancer* (K. James, B. McBride, and A. Stuart, eds.), pp. 321–329, Blackwell, Oxford.
Falchuk, K. R., Perrotto, J. L., and Isselbacher, K. J., 1975, Serum lysozyme in Crohn's disease and ulcerative colitis, *N. Engl. J. Med.* **299**:395.
Fell, H. B., Glauert, A. M., Barratt, M. E. J., and Green, R., 1976, The pig synovium. I. The intact synovium *in vivo* and in organ culture, *J. Anat.* **122**:663.
Galindo, B., and Imaeda, T., 1966, Cellular response to Freund's adjuvant in the rabbit lung, *Lab. Invest.* **15**:1659.
Galindo, B., Imaeda, T., and Kanetsuna, F., 1969, Ultrastructural study of cellular events induced by cell walls and cytoplasm of mycobacteria, *J. Reticuloendothel. Soc.* **6**:59.
Galindo, B., Lazdins, J., and Castillo, R., 1974, Fusion of normal rabbit alveolar macrophage induced by supernatant fluids from BCG-sensitised lymph node cells after elicitation by antigen, *Infect. Immun.* **9**:212.
Ghadially, F. N., 1978, Fine structure of joints, in: *The Joints and Synovial Fluid,* Vol. 10 (L. Sokoloff, ed.), pp. 105–176, Academic Press, New York.
Gillman, T., and Wright, L. J., 1966, Probable *in vivo* origin of multi-nucleated giant cells from circulating mononuclears, *Nature* **209**:263.
Goren, M. B., Hart, D'Arcy, P., Young, M. R., and Armstrong, J. A., 1976, Prevention of phagosome-lysosome fusion in cultured macrophages by sulfatides of *Mycobacterium tuberculosis, Proc. Natl. Acad. Sci.* **73**:2510.
Gusek, W., 1959, Elektronenoptische Untersuchungen am Zellbild des Granulationsgewebes, *Semposio Della Societa Italiana e Della Societa Tedesca Di Patologia,* Miano, 6–7 October 1959.

Gusek, W., 1964, Histologische und vergleichende elektronenmikroskopische Untersuchungergebnisse zur Zytologie, Histogenese und Struktur des tuberkulosen und tuberkuloiden Granuloms, *Med. Welt.* **15**:1.

Gusek, W., and Naumann, P., 1959, Elektronenoptische Untersuchungen am tuberkulosen Granulationsgeweb, *Verh. dtsch. Ges. Pathol.* **43**:254.

Han, S. S., and Avery, J. K., 1965, The fine structure of intercellular substances and rounded cells in the incisor pulp of the guinea-pig, *Anat. Rec.* **151**:41.

Heppleston, A. G., and Styles, J. A., 1967, Activity of a macrophage factor in collagen formation by silica, *Nature* **214**:521.

James, E. M. V., Jones-Williams, W., 1974, Fine structure and histochemistry of epithelioid cells in sarcoidosis, *Thorax* **29**:115.

Jones-Williams, W., 1972, The fine structure of Kveim granulomas and review of the Kveim test, in: *Rapport du Symposium Européen de la Sarcoidose, Geneva* (Y. Gallopin, ed.), pp. 29–31, Review Swiss Praxis.

Jones-Williams, W., James, E. M. V., Erasmus, D. A., and Davies, T., 1970, The fine structure of sarcoid and tuberculous granulomas, *Postgrad. Med. J.* **46**:496.

Jones-Williams, W., Fry, E., and James, E. M. V., 1972, The fine structure of beryllium granulomas, *Acta Pathol. Microbiol. Scand. Sect. A* **223**:195.

Kajikawa, K., 1964, Electron microscopic studies on histiocytes, *Tohoku J. Exp. Med.* **81**:350.

Kellermeyer, R. W., and Warren, K. S., 1970, The role of chemical mediators in the inflammatory response induced by foreign bodies: Comparison with the schistosome egg granuloma, *J. Exp. Med.* **131**:21.

Klockars, M., and Selroos, O., 1977, Immunohistochemical demonstration of lysozyme in the lymph nodes and Kveim reaction papules in sarcoidosis, *Acta Pathol. Microbiol. Scand. Sect. A* **85**:169.

Leake, E. S., and Myrvik, Q. N., 1966, Digestive vacuole formation in alveolar macrophages after phagocytosis of *Mycobacterium smegmatis in vivo, J. Reticuloendothel. Soc.* **3**:83.

Leake, E. S., and Myrvik, Q. N., 1968, Changes in morphology and in lysozyme content of free alveolar cells after the intravenous injection of killed BCG in oil, *J. Reticuloendothel. Soc.* **5**:33.

Leake, E. S., and Myrvik, Q. N., 1970, Interaction of lysosome-like structures and phagosomes in normal and granulomatous alveolar macrophages, *J. Reticuloendothel. Soc.* **8**:407.

Leake, E. S., and Myrvik, Q. N., 1972, Rosette arrangement of electron-dense structures in granulomatous alveolar macrophages, *J. Reticuloendothel. Soc.* **12**:305.

Leake, E. S., Evans, D. G., and Myrvik, Q. N., 1971, Ultrastructural patterns of bacterial breakdown in normal and granulomatous rabbit alveolar macrophages, *J. Reticuloendothel. Soc.* **9**:174.

Leake, E. S., Ockers, J. R., and Myrvik, Q. N., 1977, *In vitro* interactions of the BCG and ravenel strains of *Mycobacterium bovis* with rabbit macrophages: Adherence of the phagosomal membrane to the bacterial cell wall and the problem of the peribacillary space, *J. Reticuloendothel. Soc.* **22**:129.

Lewin, K. J., Harrell, G. S., Lee, A. S., and Crowley, L. G., 1974, Malakoplakia: An electron microscopic study: Demonstration of bacilliform organisms in malacoplakic macrophages, *Gastroenterology* **66**:28.

Lobo, A., Carr, I., and Malcolm, D., 1978, The EM immunocytochemical demonstration of lysozyme in macrophage giant cells in sarcoidosis, *Experientia* **34**:1088.

Lou, T. Y., and Teplitz, C., 1974, Malakoplakia: Pathogenesis and ultrastructural morphogenesis. A problem of altered macrophage (phagolysosomal) response, *Human Pathol.* **5**:191.

Machado, E. A., and Lair, S. V., 1978, Giant multinucleate macrophages in methyl cellulose-stimulated athymic nude mice, *J. Reticuloendothel. Soc.* **23**:383.

Mariano, M., and Spector, W. G., 1974, The formation and properties of macrophage polykaryons (inflammatory giant cells), *J. Pathol.* **113**:1.

Mariano, M., Mikitin, T., and Malucelli, B. E., 1976, Immunological and non-immunological phagocytosis by inflammatory macrophages, epithelioid cells and macrophage polykaryons from foreign body granulomata, *J. Pathol.* **120**:151.

Mariano, M., Tamara, N., and Malucelli, B. E., 1977, Phagocytic potential of macrophages from within delayed hypersensitivity mediated granulomata, *J. Pathol.* **123**:27.

McClurg, F. V., D'Agostino, A. N., Martin, J. H., 1973, Ultrastructural demonstration of intracellular bacteria in three cases of malakoplakia of the bladder, *Am. J. Clin. Pathol.* **60**:780.

Miller, R. L., Krutchkoff, D. J., and Giammara, B. S., 1978, Human lingual tuberculosis, *Arch. Pathol. Lab. Med.* **102**:360.

Monis, B., Weinberg, T., and Spector, G. J., 1968, The carrageenan granuloma in the rat. A model for the study of the structure and function of macrophages, *Br. J. Exp. Pathol.* **49**:302.

Ogawa, T., Koerten, J. K., and Daems, W. T., 1978, Peroxidatic activity in monocytes and tissue macrophages of mice, *Cell Tissue Res.* **188**:361.

Papadimitriou, J. M., 1976, The influence of the thymus on multinucleate giant cell formation, *J. Pathol.* **118**:153.

Papadimitriou, J. M., and Archer, M., 1974, The morphology of murine foreign body multinucleate giant cells, *J. Ultrastruct. Res.* **49**:372.

Papadimitriou, J. M., and Kingston, K. J., 1977, The locomotory behaviour of the multinucleate giant cells of foreign body reactions, *J. Pathol.* **121**:27.

Papadimitriou, J. M., Robertson, T. A., and Walters, M. N. I., 1975, An analysis of the phagocytic potential of multinucleate foreign body giant cells, *Am. J. Pathol.* **78**:343.

Papadimitriou, J. M., and Spector, W. G., 1971, The origin, properties and fate of epithelioid cells, *J. Pathol.* **105**:187.

Papadimitriou, J. M., and Spector, W. G., 1972, The ultrastructure of high- and low-turnover inflammatory granulomata, *J. Pathol.* **106**:37.

Papadimitriou, J. M., and Wee, S. H., 1976, Selective release of lysosomal enzymes from cell populations containing multinucleate giant cells, *J. Pathol.* **120**:193.

Papadimitriou, J. M., and Wyche, P. A., 1974, An examination of murine foreign body giant cells using cytochemical techniques and thin layer chromatography, *J. Pathol.* **114**:75.

Papadimitriou, J. M., and Wyche, P. A., 1976, A biochemical profile of glass-adherent cell populations containing multinucleated foreign body giant cells, *J. Pathol.* **119**:239.

Parks, D. E., and Weiser, R. S., 1975, The role of phagocytosis and natural lymphokines in the fusion of alveolar macrophages to form Langhans giant cells, *J. Reticuloendothel. Soc.* **17**:219.

Pascual, R. S., Gee, J. B. L., and Finch, S. C., 1973, Usefulness of serum lysozyme measurements in diagnosis and evaluation of sarcoidosis, *N. Engl. J. Med.* **289**:1074.

Perillie, P. E., Khan, K., and Finch, S. C., 1973, Serum lysozyme in pulmonary tuberculosis, *Am. J. Med. Sci.* **265**:297.

Policard, A., Collet, A., Martin, J. C., and Reuet, C., 1965, Etude au microscope électronique sur les systèmes d'expansions foliacées présents dans les macrophages et les cellules epithélioïdes, *Z. Zellforsch. Mikrosk. Anat.* **66**:96.

Smith, J. B., McIntosh, and Morris, B., 1970, The migration of cells through chronically inflamed tissues, *J. Pathol.* **100**:21.

Soler, P., and Basset, F., 1976, Morphology and distribution of the cells of a sarcoid granuloma and ultrastructural study of serial sections, *Ann. N. Y. Acad. Sci.* **278**:147.

Soler, P., Bernaudin, J. F., and Basset, T. F., 1975, Ultrastructure of pulmonary granulomatosis induced in rats by intravenous complete Freund's adjuvant, *Virchows Arch. A* **368**:35.

Spector, W. G., 1969, The granulomatous inflammatory exudate, *Int. Rev. Exp. Pathol.* **8**:1.

Spector, W. G., and Lykke, A. W. J., 1966, The cellular evolution of inflammatory granuloma, *J. Pathol.* **92**:163.

Spector, W. G., and Mariano, M., 1975, Macrophage behaviour in experimental granulomas in: *Mononuclear Phagocytes in Immunity, Infection and Pathology* (R. van Furth, ed.), pp. 927–938, Blackwell, Oxford.

Sutton, J. S., and Weiss, L., 1966, Transformation of monocytes in tissue culture into macrophages, epithelioid cells and multinucleated giant cells, *J. Cell Biol.* **28**:303.

Taylor, C. R., and Burns, J., 1974, The demonstration of plasma cells and other immunoglobulin-containing cells in formalin-fixed, paraffin-embedded tissues using peroxidase-labelled antibody, *J. Clin. Pathol* **27**:14.

Tsuda, T., Dannenberg, A. M., Ando, M., Abbey, H., and Corrin, A., 1976, Mononuclear cell turnover in chronic inflammation, *Am. J. Pathol.* **83**:255.

Turk, J. L., Badenoch-Jones, P., and Parker, D., 1978, Ultrastructural observations on epithelioid cell granulomas induced by zirconium in the guinea pig, *J. Pathol.* **124**:45.

Unanue, E. R., 1976, Secretory function of mononuclear phagocytes. A review, *Am. J. Pathol.* **83**:396.

Underwood, J. C. E., and Carr, I., 1972, The ultrastructure of the lymphoreticular cells in non-lymphoid human neoplasms, *Virchows Arch. B* **12**:39.

van der Rhee, H. J., van der Burgh-deWinter, C. P. M., Tijssen, J. G. P., and Daems, W. T., 1979a, Comparative study on peroxidate activity in inflammatory cells on cutaneous and peritoneal implants, *Cell Tissue Res.* **197**:397.

van der Rhee, H. J., van der Burgh-deWinter, C. P. M., Daems, W. T., 1979b, The differentiation of monocytes into macrophages, epithelioid cells, and multinucleated giant cells in subcutaneous granulomas. I. Fine structure, *Cell Tissue Res.* **197**:355.

van der Rhee, H. J., van der Burgh-deWinter, C. P. M., Daems, W. T., 1979c The differentiation of monocytes into macrophages, epithelioid cells and multinucleated giant cells in subcutaneous granulomas. II. Peroxidatic activity, *Cell Tissue Res.* **197**:379.

Wanstrup, J., and Christensen, H. E., 1966, Sarcoidosis, *Acta Pathol. Microbiol. Scand.* **66**:169.

Warren, K. S., and Boros, D. L., 1975, The schistosome egg granuloma—a form of cell-mediated immunity, in: *Mononuclear Phagocytes in Immunity, Infection and Pathology* (R. van Furth, ed.), pp. 1015–1028, Blackwell, Oxford.

19

Osteoclasts
Ultrastructure and Functions

ULF LUCHT

1. INTRODUCTION

Osteoclasts were first described at the light-microscopic level by Robin (1849) who was able to distinguish these cells from megakaryocytes. Rollet (1870) observed similar cells at sites of enchondral ossification. Kölliker (1873) suggested that the cells resorb bone, and he gave them the name *osteoklast*, but in later literature the spelling osteoclast has been preferred. The ability of osteoclasts to resorb bone was confirmed much later in time-lapse studies of osteoclasts in tissue culture (Gaillard, 1955, 1959; Goldhaber, 1960). It was possible to see in these films that the matrix disappeared around the osteoclasts. In addition, Arnold and Jee (1957) showed by light-microscopic autoradiography, using plutonium, transport of this isotope from bone to the interior of osteoclasts. However, the cytological basis for the function of osteoclasts remained largely unknown although the structure of these cells, as observed in the light microscope, was described in numerous reports, and some enzymes were demonstrated by histochemistry (Hancox, 1972).

The first electron-microscopic study appeared in 1956 (Scott and Pease) and established that the so-called ruffled border belonged to the cell and not to the bone and consisted of cytoplasmic projections. Later reports have described other ultrastructural aspects which, together with biochemical, histochemical, and autoradiographic observations at the electron-microscopic level, have contributed to our knowledge of osteoclasts. The characteristics of osteoclasts have been reviewed several times in recent years (Vaes, 1969; Cameron, 1969, 1972; Vaugham, 1970; Rasmussen and Bordier, 1974; Göthlin and Ericsson, 1976; Holtrop and King, 1977).

In the present review, particular emphasis is placed on the recent

ULF LUCHT • Department of Cell Biology, Institute of Anatomy, University of Aarhus, DK-8000 Aarhus C, Denmark.

electron-microscopic studies, which provide a basis for the understanding of the function of the osteoclasts.

2. LIGHT MICROSCOPY

The osteoclast is a large, multinucleated cell which may exceed 100 μm in length. The number of nuclei varies from one or two in the smallest cells to more than 100 in the largest cells, but usually does not exceed 20. However, it should be noted that, in thin sections, a multinuclear cell may appear to have only one nucleus because of the plane of the section. Rodent osteoclasts tend to be smaller than those of the cat; human osteoclasts are of intermediate size, although there is overlap (Hancox, 1972).

Osteoclasts are mainly located on the endosteal bone surface, but are also present in Haversian systems and on the periosteal surface beneath the periost. However, the greatest number of osteoclasts always occurs in bone exposed to intensive remodelling, as, for example, in the metaphyses of growing bone or in bone which is in the process of fracture healing. The osteoclast is usually located in a bone cavity or it is bent around a bone trabeculum (Figure 1).

The appearance of the cell body varies. Sometimes it is more or less spherical, and sometimes it is flattened around the bone. It may be very irregular and show prominent cell processes. The nuclei are usually located together in the central part of the cell; they are more or less oval with one or two nucleoli.

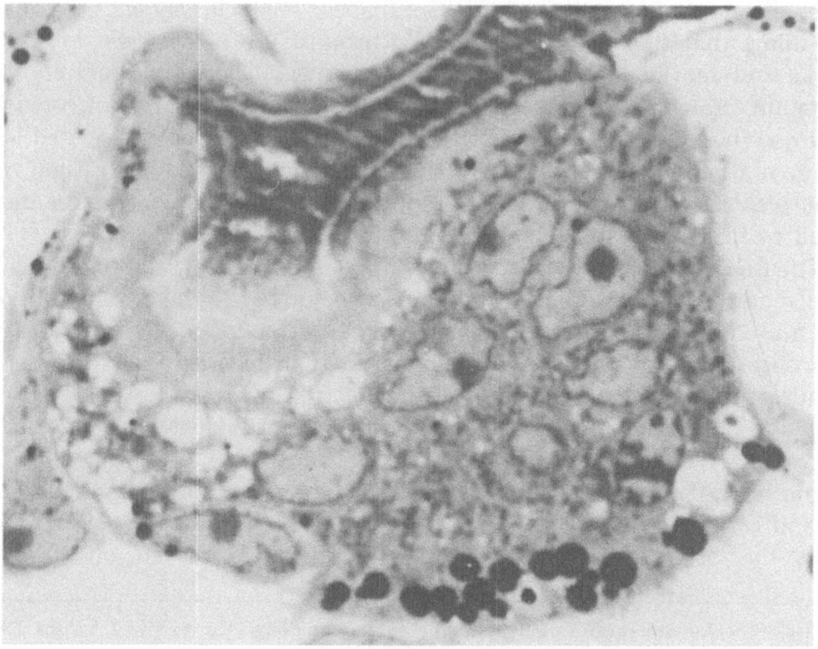

FIGURE 1. Light micrograph of multinucleated osteoclast from bone in tissue culture. The cell is located around a bone trabeculum, which is in process of resorption. ×2100.

Vacuoles may be seen in the cytoplasm close to the bone. The cell surface adjacent to the bone is different from the rest of the cell surface, but details cannot be studied by light microscopy. The cytoplasm is variable in appearance. In some cells it is palely stained and rather basophil, but in others it is more chromophil and tends toward eosinophilia (Hancox, 1972).

3. ELECTRON MICROSCOPY

The ultrastructure of osteoclasts has been studied both *in vivo* and *in vitro* cultures of the cells. Due to its mineral content, bone presents several technical difficulties, in particular, during sectioning, when processed for electron microscopy (EM). Therefore, much work has been done on growing bone, which is easier to section than adult bone.

3.1. GENERAL ULTRASTRUCTURE

The characteristic ultrastructural features of the osteoclast have been described by many authors (Dudley and Spiro, 1961; Gonzales and Karnovsky, 1961; Robinson and Cameron, 1964; Scott, 1967a; Fetter and Capen, 1971; Lucht, 1976). The typical osteoclast has a ruffled border and an adjacent transitional zone facing the bone, whereas the cell membrane away from the bone often shows some irregular microvilli (Figure 2). The nuclei appear more irregular in shape than noted by light microscopy. Cytoplasmic bodies and vacuoles are prominent, especially in the region behind the ruffled border. Mitochondria are numerous and may contain small granules in the matrix. A large Golgi complex is located adjacent to the nuclei. Free ribosomes are numerous and occur either as single ribosomes or as polyribosomes. The rough endoplasmic reticulum (RER) is rather sparse, and it is never the conspicuous element as in osteoblasts. Some cytoplasmic regions contain thin filaments and microtubules.

3.2. TYPES OF OSTEOCLASTS

In random sections, some osteoclasts appear devoid of a ruffled border, but the cells may contain a ruffled border at another level. Lucht (1972b), using interval sectioning of entire osteoclasts, demonstrated two cell types in the rat. Blocks were sectioned in such a way that ultrathin sections of these large cells were obtained at intervals of 2.5 μm.

The predominant type has a ruffled border and the other type has no ruffled border. The latter contains one or more nuclei and few vacuoles, but otherwise it has the same general ultrastructure. It does not have such a close relationship to the bone as the osteoclast with the ruffled border, and the bone beneath this cell type is not changed. Presumably, this is an inactive osteoclast corresponding to the inactive stage demonstrated *in vitro* in the light microscope by Gaillard

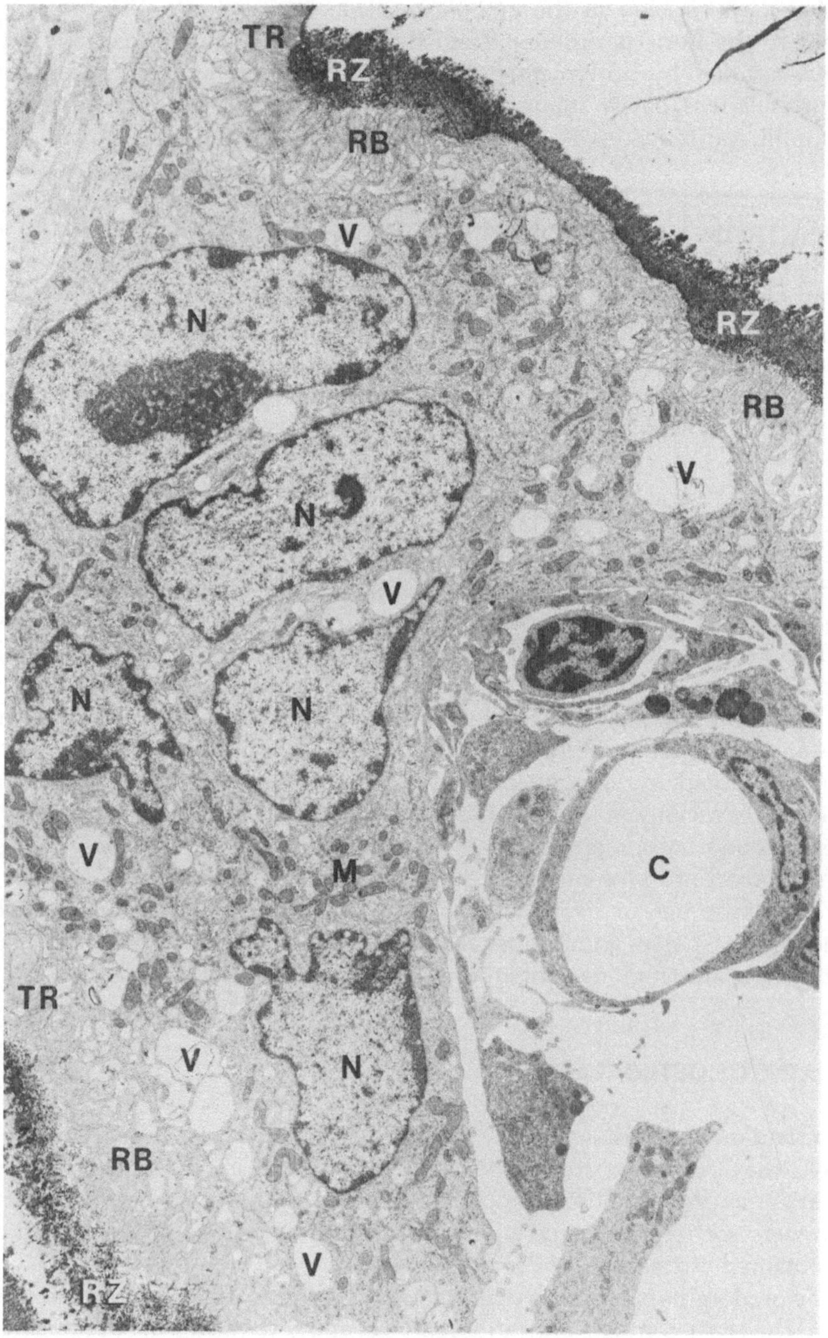

FIGURE 2. Multinucleated osteoclast with ruffled borders facing bone trabeculae. N, Nucleus; V, vacuoles; M, mitochondria; RB, ruffled border; TR, transitional zone; RZ, resorption zone; C, capillary. ×5900.

(1959), who observed that multinuclear cells for long periods were in close proximity to bone matrix until suddenly lysis started and continued for a limited period of time.

Cells that are ultrastructurally indistinguishable from osteoclasts and involved in the breakdown of calcified cartilage and dentine have, by some authors, been named chondroclasts (Schenk *et al.*, 1967; Thyberg *et al.*, 1975) or odontoclasts (Furseth, 1968; Freilich, 1971) and dentinoclasts (Nilsen, 1977). Knese (1972) suggests the osteoclasts be divided into mineraloclasts and chondroclasts, but convincing evidence for this classification is still lacking.

3.3. RELATIONSHIP BETWEEN OSTEOCLASTS AND BONE

The closest contact between osteoclasts and bone occurs at the level of the ruffled border (Figures 3 and 4). The bone beneath this border shows, in contrast to bone beyond this level, several changes indicative of resorption: the electron dense border of bone, or lamina limitans (Scherft, 1972), is absent; fragmented bone, isolated collagen fibrils, and numerous free bone crystals are present (Lucht, 1972b; Bonucci, 1974). Some collagen fibrils and some bone crystals are also located in the channels of the ruffled border (Hancox and Boothroyd, 1961;

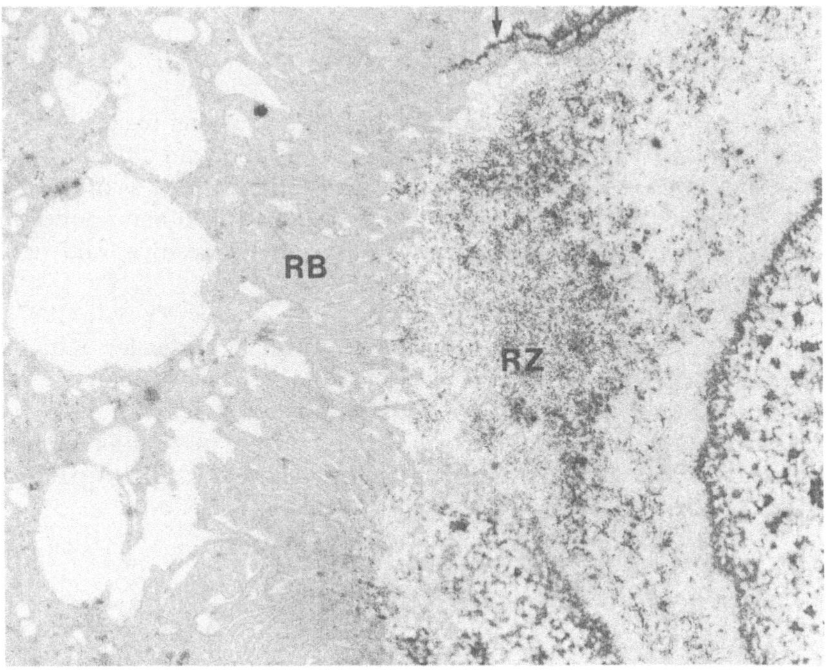

FIGURE 3. Ruffled border (RB) and adjacent resorption zone (RZ). The membrana limitans (→) is lacking in the resorption zone. ×9000.

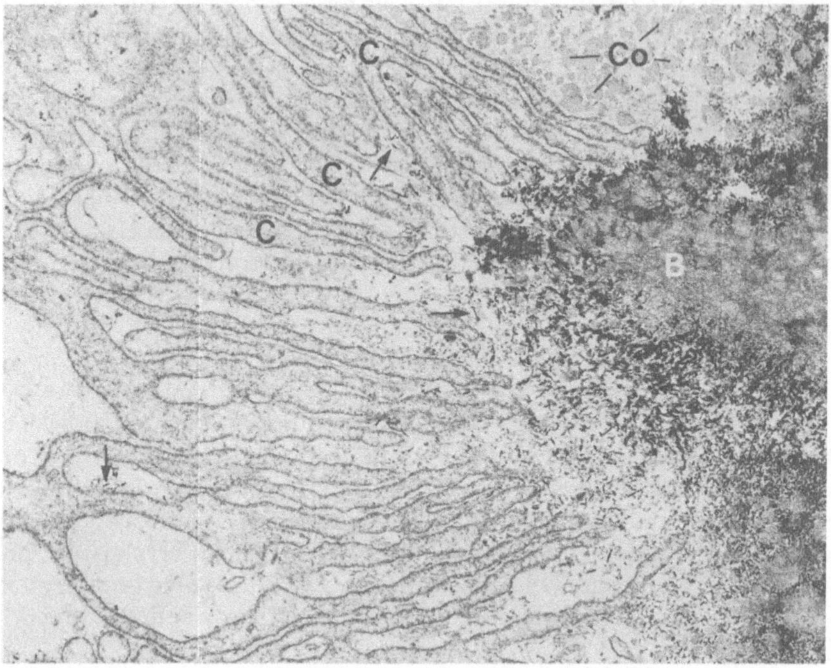

FIGURE 4. Resorption zone containing transversely cut collagen fibers (CO) and bone crystals (→). Some crystals and fibers are present between the cytoplasmic projections (C) of the ruffled border. B, Bone. ×28,000. (From Lucht, 1972b, by courtesy of publishers.)

Cameron, 1963; Lucht, 1972b; Bonucci, 1974). Conclusions regarding the frequency of bone crystals and the degree of mineralization are difficult, since the tissue has been processed in aqueous solutions and more or less of the mineral has been dissolved. Schenk *et al.*, (1967), in contrast to others, noticed bone crystals only when phosphate buffer was used in the fixative, and not with cacodylate buffer.

In a recent scanning electron microscope study Jones and Boyde (1977) have shown that osteoclasts have a rather complex form. In particular, remote portions of elongated and branched osteoclasts may be engaged in active resorption, while intermediate portions are not in intimate contact with the bone matrix surface or are even separated from the bone surface by an intervening layer of cells. The distance between osteoclasts and vessels varies, but parts of the osteoclasts are usually located close to a vessel (Lucht, 1972b).

3.4. RUFFLED BORDER AND TRANSITIONAL ZONE

The ruffled border consists of slender cytoplasmic projections, the shape of which varies from villuslike processes to folds (Figure 4). The projections are of the order 50–150 nm in thickness and normally several microns in length (Göthlin and Ericsson, 1976). The extracellular channels run toward the cell

FIGURE 5. The ruffled border forming a very complicated pattern. ×16,300.

interior where they often dilate. In some osteoclasts the ruffled border consists of closely packed projections while in others the projections are more separated. The projections may form a very complex pattern (Figure 5). The total area of the ruffled border varies from cell to cell, and some osteoclasts may have more than one ruffled border. The average area of the ruffled border occupied 3.4% of the cytoplasm in profiles of unstimulated osteoclasts sampled from metaphyses of rat long bones (King *et al.*, 1976). The ruffled border is surrounded by a 90-Å thick triple-layered membrane (Lucht, 1972c). Kallio *et al.* (1971) reported a coating of the membrane consisting of fine bristlelike structures, evenly spaced 200–250 Å apart and projected perpendicularly into the cytoplasm. The particles on the ruffled border were thought to represent sites of enzymatic activity; however, such structures have not been observed by others.

The ruffled border is commonly surrounded by a transitional zone (Figure 6) which has also been termed ectoplasmic layer (Göthlin and Ericsson, 1972; Scott, 1967a), granular zone (Cameron, 1963), clear zone (Schenk *et al.*, 1967), or contact zone (Malkani *et al.*, 1973). This zone consists of numerous thin filaments and is devoid of organelles, except that free ribosomes are sometimes present (Lucht, 1972b). Malkani *et al.* (1973) described parallel, dark bands in the transitional zone. These bands are directed perpendicularly to the bone surface, usually ending in short processes that extend into indentations of the bone surface. The filaments in the bands show affinity to heavy meromyosin and can, hence, be considered actinlike (King and Holtrop, 1975).

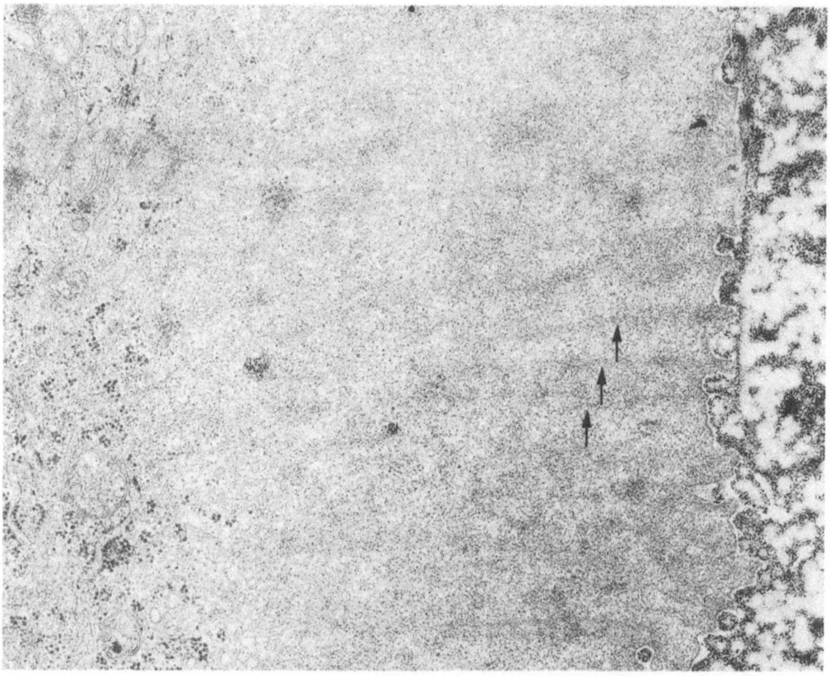

FIGURE 6. The transitional zone containing parallel dark bands (→). ×19,200.

The transitional zone has been suggested to function in the adhesion of the osteoclast to the bone (King and Holtrop, 1975) and to stabilize the cell during bone resorption (Lucht, 1972b). The material in the transitional zone cannot be distinguished from the material within the cytoplasmic extensions of the ruffled border (Kallio *et al.*, 1971); Malkani *et al.* (1973) have suggested that the transitional zone can progressively develop into a ruffled border.

3.5. CYTOPLASMIC BODIES

The cytoplasm of the osteoclast contains a population of membrane-limited cell organelles which may be referred to either as cytoplasmic bodies or vacuoles. They vary to some extent in their ultrastructure, but may be distinguished from mitochondria, endoplasmic reticulum (ER), and Golgi cisternae due to a greater thickness (about 90Å) of their limiting membrane (Lucht, 1972c). When these organelles appear devoid of stainable contents they are generally termed vacuoles, while those possessing an internal structure or a stainable matrix of varying density are referred to as cytoplasmic bodies. However, it should be emphasized that intermediate forms exist between vacuoles and bodies, and some membrane-limited structures appear only partially filled with matrix (Scott, 1967a; Lucht, 1972c).

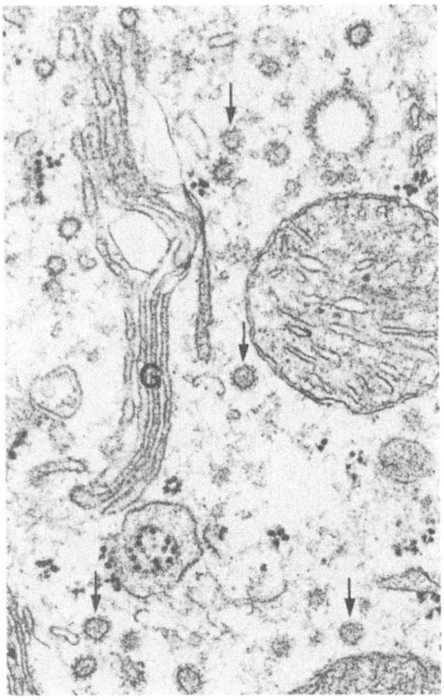

FIGURE 7. Golgi complex (G) with adjacent small cytoplasmic bodies of the coated type (→). ×49,000. (From Lucht, 1972c, by courtesy of the publishers.)

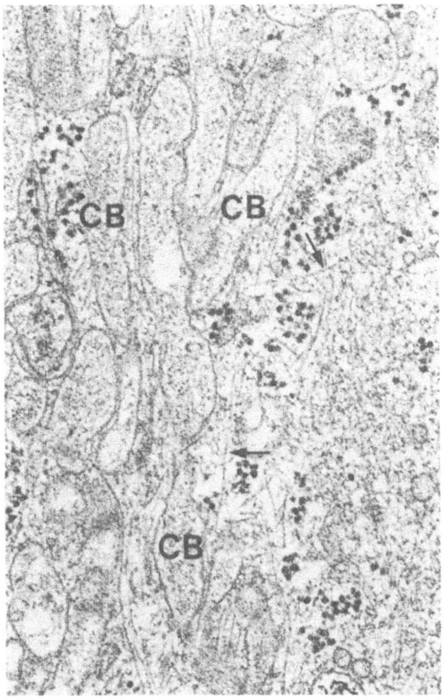

FIGURE 8. Elongated, light cytoplasmic bodies (CB) located close to the cell membrane. Thin filaments are present (→). ×50,400. (From Lucht, 1972c, by courtesy of the publishers.)

Many different names have been used for the cytoplasmic bodies in the osteoclast (Figures 7–10). They have been termed vesicles containing electron-opaque material (Schenk *et al.*, 1967), specific granules (Scott, 1967a), dense bodies (Freilich, 1971), dense granules (Fetter and Capen, 1971), acid-phosphatase-positive granules (Doty *et al.*, 1968), and lysosomes (Yaeger and Kraucunas, 1969; Lucht, 1971). Scott (1967a) divides the cytoplasmic bodies into a small solid type (0.2–0.5 μm) and a larger, paler vesicular type (0.5–3 μm). Lucht (1972c) describes four morphologically distinct groups: light cytoplasmic bodies, dense cytoplasmic bodies, coated cytoplasmic bodies, and cytoplasmic bodies with inclusions.

The light cytoplasmic bodies vary from about 0.02 to 3 μm in diameter and appear round, oval, or elongated (Figures 8 and 10). They represent the most frequent type of cytoplasmic bodies in the osteoclast, and are most frequently seen in the cytoplasm between the ruffled border and the nuclei. The smallest of these cytoplasmic bodies occur close to the Golgi complex, while elongated forms are usually located in regions near the cell membrane. Many cytoplasmic bodies are located close to the vacuoles and some are present in invaginations of the vacuole membrane.

A smaller number of cytoplasmic bodies show the same form and dimen-

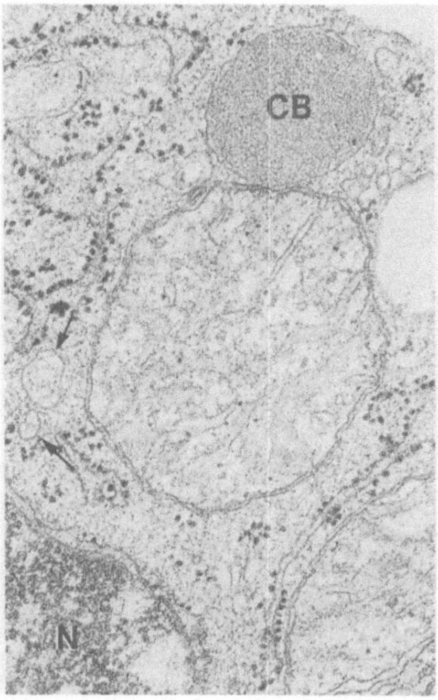

FIGURE 9. A dense cytoplasmic body (CB) and light cytoplasmic bodies (→) on the side of the nucleus opposite to that facing the ruffled border. ×50,400. (From Lucht, 1972c, by courtesy of the publishers.)

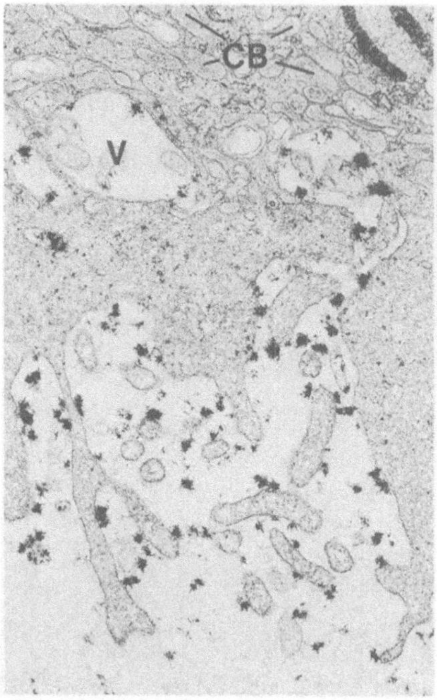

FIGURE 10. Colloidal lanthanum between the cytoplasmic projection of the ruffled border and in vacuole-like structures (V). Numerous light cytoplasmic bodies (CB) occur close to the ruffled border. ×25,200. (From Lucht, 1972c, by courtesy of the publishers.)

sion as light cytoplasmic bodies, but exhibit a matrix that appears more dense (Figure 9). These cytoplasmic bodies are scattered throughout the cytoplasm of the osteoclast.

Some small cytoplasmic bodies with a diameter up to 0.1 μm exhibit an external coat of radially arranged bristles (Figure 7). They contain a matrix similar to that of the light cytoplasmic bodies and are usually located close to the Golgi complex.

A few cytoplasmic bodies contain various forms of inclusion material. Usually the inclusions are not identifiable as organelles. However, occasionally it is possible to identify the inclusion material as modified mitochondria, which suggests that autophagy occurs in the osteoclast.

3.6. VACUOLES

The majority of vacuoles are located in the cytoplasm behind the ruffled border and, also, the largest vacuoles are present in this region (Figures 2 and 3).

They are usually more or less spherical. The diameter of the vacuoles varies from about 0.03 μm to 5 μm, the majority being between 0.1 and 2 μm (Lucht, 1972c). Some vacuoles behind the ruffled border contain bone crystals, and such crystals also occur in vacuoles deep in the cell interior. However, collagen fibers are not present in the vacuoles (Gonzales and Karnovsky, 1961; Knese and Knoop, 1961; Hancox and Boothroyd, 1963; Cameron, 1968; Lucht, 1972c; Bonucci, 1974).

The cell membrane away from the ruffled border shows coated invaginations (Lucht, 1972c), and the cytoplasm occasionally contains small, coated vacuoles with a diameter of up to 0.2 μm (Kallio *et al.*, 1971; Lucht, 1972c). Sometimes amorphous material is present in these vacuoles. These types of vacuoles may be specialized for endocytosis of protein, as suggested by Roth and Porter (1964) in another tissue.

In fixed tissue, colloidal lanthanum penetrates many of the vacuole-like structures behind the ruffled border (Figure 10), which demonstrates that they are connected to the extracellular space and thus, in reality, are channel expansions (Lucht, 1972c). The same connections have been demonstrated by serial sections (Dudley and Spiro, 1961).

3.7. GOLGI COMPLEX

The Golgi complex is located close to the nuclei and extends over a large part of each nuclear circumference (Figure 7). The Golgi complex is composed of stacks of 3–5 parallel cisternae and associated vacuoles, as well as small, coated and uncoated cytoplasmic bodies. The cisternae facing the nucleus are somewhat dilated but usually appear empty. Round or elongated vacuoles are often located close to these cisternae. The remaining cisternae are flat, although some are slightly dilated at their ends. These flat cisternae contain electron-dense material similar to that in the small cytoplasmic bodies adjacent to the Golgi cisternae. It may be suggested that the Golgi complex in the osteoclast, as in other cells, participates in the formation of lysosomes.

3.8. CENTRIOLES AND ASSOCIATE STRUCTURES

Centrioles are rarely observed, but are occasionally present in large numbers (Matthews *et al.*, 1967; Cameron, 1968; Lucht, 1973a). When two or more centrioles are observed they are always located close together in a centrosphere (Figure 11). The Golgi complex is present along part of the periphery of the centrosphere and sometimes surrounds the centrioles almost completely. Each centriole exhibits the same basic structure as that of centrioles in other cell types. The number of centriole "pairs" is suggested to correspond to the number of nuclei in the osteoclast (Matthews *et al.*, 1967), and it seems likely that the centrosphere of the osteoclast has been formed by a pooling of centrioles from precursor cells (Lucht, 1973a). There is no clear evidence of any specific functions of centrioles in osteoclasts.

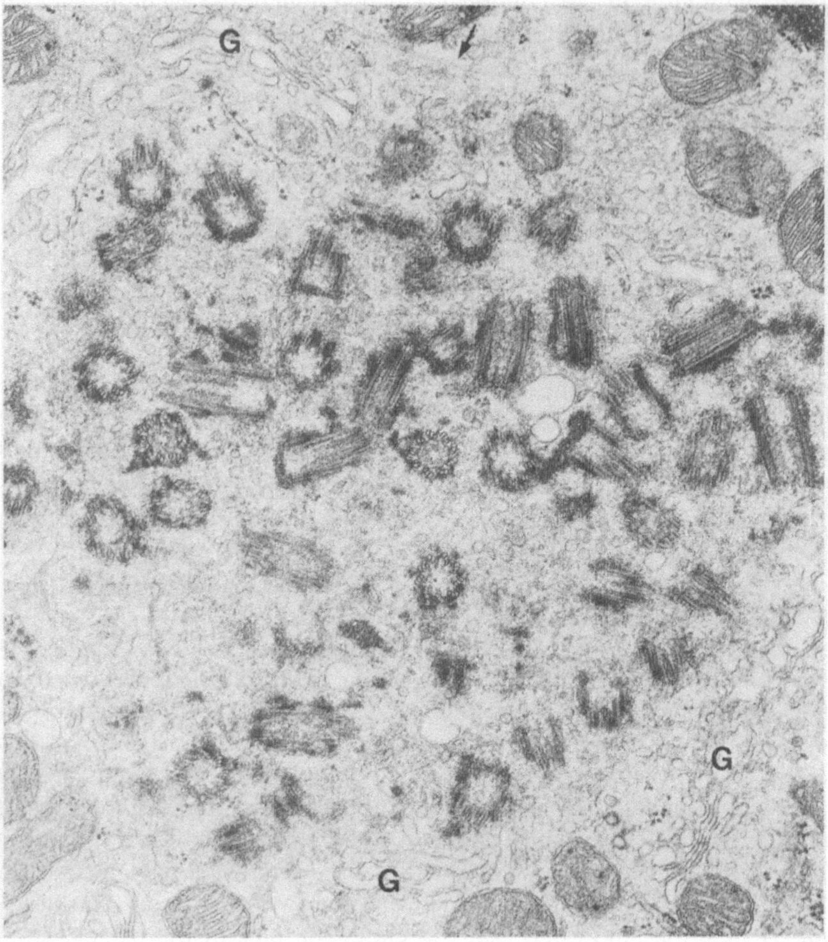

FIGURE 11. Approximately 40 centrioles are gathered in a centrosphere which is surrounded by the Golgi complex (G). →, Microtubule. ×33,100.

Microtubules, while present in all regions of the osteoclast, are sometimes especially numerous in regions close to centrioles (Cameron, 1969; Lucht, 1973a). When microtubules are disaggregated by colchicine, bone resorption induced in cultured bone by parathyroid hormone is inhibited; this suggests that microtubules are involved in the resorptive function of the osteoclast (Raisz *et al.*, 1973).

4. ENZYMES IN OSTEOCLASTS

The demonstration and localization of enzymes in the osteoclast has contributed to the understanding of the functions of this cell. As reviewed by Vaes (1969), light-microscopic histochemistry has revealed several acid hydrolases and other hydrolases such as aminopeptidase and ATPase. Succinic dehydrogenase (Burstone, 1960a; Walker, 1961; Takimoto *et al.*, 1966) and cytochrome oxidase (Burstone, 1960b) have been shown to be present in the osteoclast using histochemical techniques, and carbonic anhydrase by light-microscopic autoradiography (Gay and Mueller, 1974).

With the development of electron-microscopic histochemistry, it has been possible to determine the precise location of some enzymes and to relate them to the cell organelles and the cell membrane. Acid phosphatase, which has been studied most extensively, has been demonstrated in the Golgi complex (Doty *et al.*, 1968; Doty and Schofield, 1972; Lucht, 1971; Göthlin and Ericsson, 1971), inside the cytoplasmic bodies (Doty *et al.*, 1968; Doty and Schofield, 1972; Göthlin and Ericsson, 1971; Lucht, 1971; Thyberg *et al.*, 1975) and in the vacuoles (Göthlin and Ericsson, 1971; Lucht, 1971; Doty and Schofield, 1972; Thyberg *et al.*, 1975) of the osteoclast (Figure 12). Also, the small coated cytoplasmic bodies contain acid phosphatase (Doty and Schofield, 1972). Acid phosphatase is a lysosomal enzyme (de Duve, 1963; Novikoff, 1963), and the presence of this enzyme in the cytoplasmic bodies and vacuoles identifies these organelles as lysosomes. The small cytoplasmic bodies close to the Golgi complex may be interpreted as primary lysosomes and the coated type as early primary lysosomes. Some authors (Lucht, 1971; Doty and Schofield, 1972) have also demonstrated acid phosphatase in the channels of the ruffled border and in the extracellular resorption zone (Figure 13), while others did not observe the

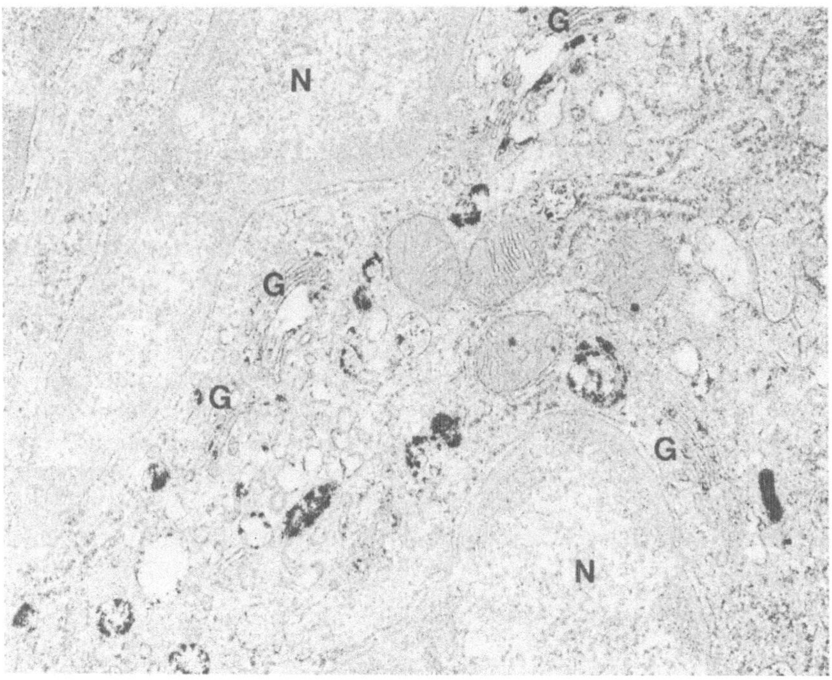

FIGURE 12. Osteoclast from bone tissue incubated for acid phosphatase. The reaction product is present in the Golgi complex (G) and in cytoplasmic bodies and vacuoles. N, Nucleus. ×44,800. (From Lucht, 1971, by courtesy of the publishers.)

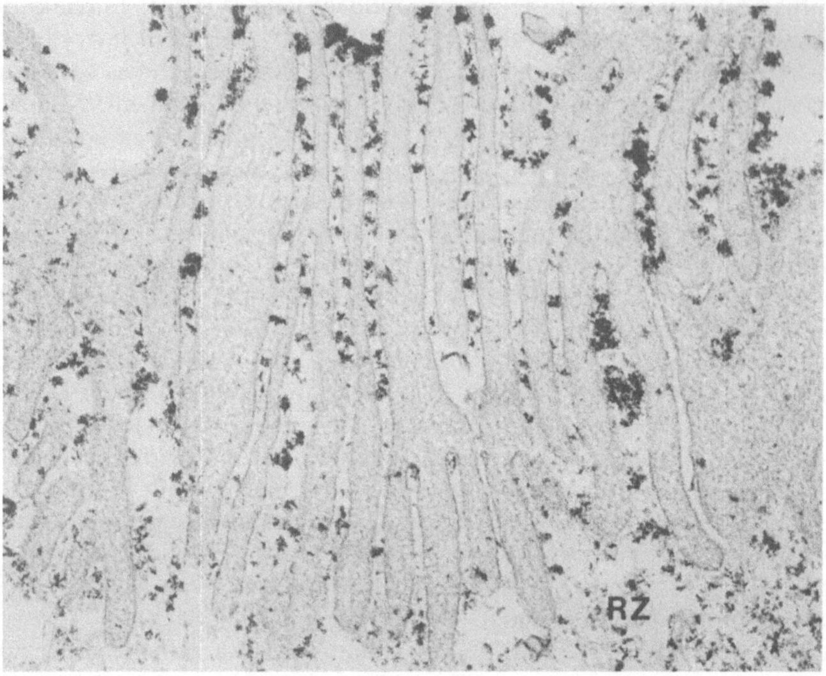

FIGURE 13. Acid phosphatase located in the channels of the ruffled border and the adjacent resorption zone (RZ). ×61,600. (From Lucht, 1971, by courtesy of the publishers.)

enzyme here (Göthlin and Ericsson, 1971; Thyberg *et al.*, 1975). This discrepancy may be caused by differences in species and techniques. Acid phosphatase has usually been visualized by the modified Gomori technique, and, before the incubation, the tissue needs to be decalcified (Lucht, 1971). The decalcification with EDTA can cause partial release of enzyme from lysosomes in unfixed tissue and to avoid nonspecific precipitations, the tissue has to be properly fixed before the incubation (Ericsson and Göthlin, 1971; Lucht, 1971). Aryl sulfatase is another lysosomal enzyme which has been found in the cytoplasmic bodies of the osteoclast (Schofield *et al.*, 1974; Thyberg *et al.*, 1975).

In addition, ATPase has been demonstrated at the electron-microscopic level having the same distribution as acid phosphatase (Doty *et al.*, 1968; Lucht, 1971; Göthlin and Ericsson, 1973b). Part of the ATPase in the ruffled border appeared Mg^{2+} dependent, but the function is still uncertain (Göthlin and Ericsson, 1973b).

Alkaline phosphatase which is present in osteoblasts appears absent from the osteoclast (Göthlin and Ericsson, 1973a; Thyberg *et al.*, 1975).

5. FUNCTIONS OF THE OSTEOCLAST

From time-lapse motion picture studies (Hancox and Boothroyd, 1961) it is known that the osteoclast is a motile cell which may resorb bone in one place and then move to another location in the tissue and start resorption again. As mentioned above, the resorption takes place at the region of the ruffled border and the resorption area is limited by the transitional zone. Schenk *et al.* (1967) suggested that the contact between the bone and the transitional zone was close enough to seal off the resorption zone so that enzymes secreted from the osteoclast could not escape. However, the electron microscopy (EM) always shows some space between the transitional zone and the bone, and peroxidase is seen in the channels of the ruffled border 5 min after intravenous injection (Lucht, 1972a), indicating that the resorption zone is not completely closed off.

In order for bone to be resorbed, both the mineral and the matrix components have to be dissolved. Bonucci (1974) suggests that the first component of bone to be attacked by the osteoclast is the ground substance and that this induces the release of crystals. In the next stage, the collagen fibrils lose their aggregation and are dissolved.

Although the resorption process is not yet understood in detail, there are strong indications that the osteoclast exerts some of its essential functions in bone resorption by means of exocytosis and endocytosis.

5.1. EXOCYTOSIS

The pH in the resorption zone has been suggested to be acid (Cretin, 1951; Neuman *et al.*, 1960), and some experiments (Walker, 1961, 1972; Vaes, 1970) have indicated that the osteoclast, by maintaining a high rate of citrate and lactate production, actively promotes decalcification of the bone.

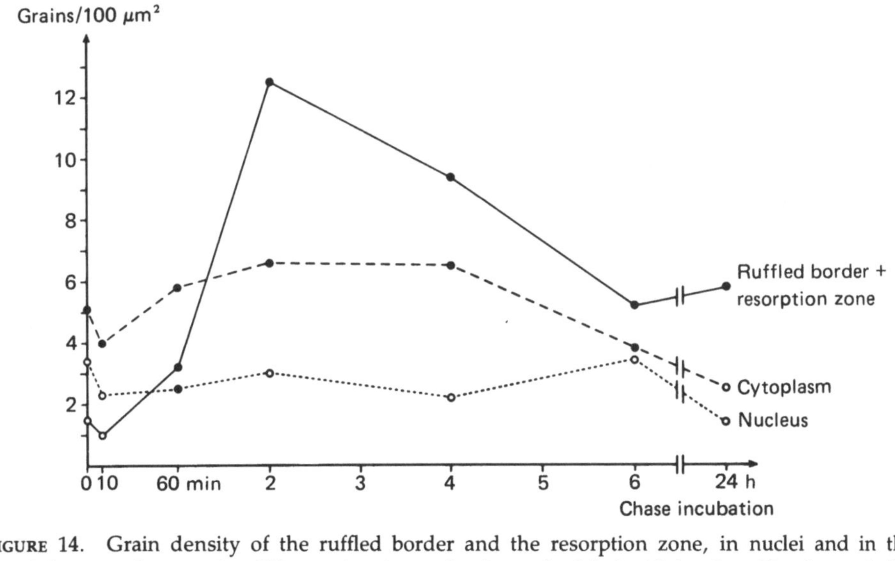

FIGURE 14. Grain density of the ruffled border and the resorption zone, in nuclei and in the remaining cytoplasm at the different time intervals after pulse label with leucine. The dense circles (●) indicate that the density is significantly higher than that of the background, while the open circles (○) indicates that the density is not significantly different from that of the background. (From Lucht and Nørgaard, 1976, by courtesy of the publishers.)

The presence of acid phosphatase in the extracellular zone beneath the ruffled border of the osteoclast (Lucht, 1971; Schofield *et al.*, 1974) suggests that lysosomes of the osteoclast release their enzymes by means of exocytosis into the extracellular space. This type of release is similar to that observed in other tissues, such as the pancreatic exocrine cells (Palade, 1959), where proteins enclosed in the zymogen granules are released by exocytosis. In the osteoclast, the occurrence of exocytosis of lysosomal enzymes is supported by an electron-microscopic autoradiographic observation of the turnover of tritiated leucine (Lucht and Nørgaard, 1976). A significant increase in the grain density over the ruffled border and the underlying resorption zone was demonstrated 2 hr post pulse, whereas the grain density of the remaining cytoplasm was relatively constant (Figure 14). This indicates a transport of newly synthesized protein from the osteoclast to the extracellular resorption zone. In addition, the biochemical studies of Vaes (1968) have shown a release of acid hydrolases and other lysosomal enzymes from bone in culture to the culture medium. The presence of lysosomal enzymes in the extracellular resorption zone suggests that these enzymes participate in the bone degradation.

The suggested role of lysosomal exocytosis in bone resorption is in agreement with observations on incisors-absent (*i.a.*) rats with osteopetrosis. The osteoclasts in these animals are not able to resorb bone (Marks, 1973; Schofield *et al.*, 1974). The cells did not contain a ruffled border and enzyme release could not be demonstrated.

5.2. ENDOCYTOSIS

Endocytosis of different tracers by the osteoclast has been demonstrated using peroxidase (Lucht, 1972a), thorium dioxide (Göthlin and Ericsson, 1972; Thyberg, 1975), and ferritin (Thyberg, 1975). Five minutes after intravenous injection of peroxidase the tracer was located around the osteoclast and in the space between the ruffled border and the bone. Inside the cell, peroxidase was located in some cytoplasmic vacuoles behind the ruffled border and along the cell membrane. Forty minutes after injection there was a large increase in the number of membrane-limited cytoplasmic structures containing reaction product, these being distributed throughout the cell, but with a high concentration adjacent to the ruffled border (Figure 15). This demonstrates an extensive endocytosis, in particular along the ruffled border.

It is possible that bone components such as collagen fibers and bone crystals can also be taken up by the osteoclast during bone resorption. However, the absence of collagen fibrils in cytoplasmic vacuoles suggests that collagen is degraded before uptake. For the degradation of collagen, a collagenase is necessary, and some studies (Sakamoto *et al.*, 1975) reveal that a collagenase is released into the medium during bone resorption *in vitro*. The enzyme has been localized to the cellular component of bone (Puzas and Brand, 1976), but the exact localization has not yet been demonstrated. Following labeling of bone collagen with [^{3}H]proline, it was possible to observe the removal of the label by osteoclasts, but the number of grains over the cells was insignificant (Irving and Heeley, 1970; Birkedal-Hansen, 1974).

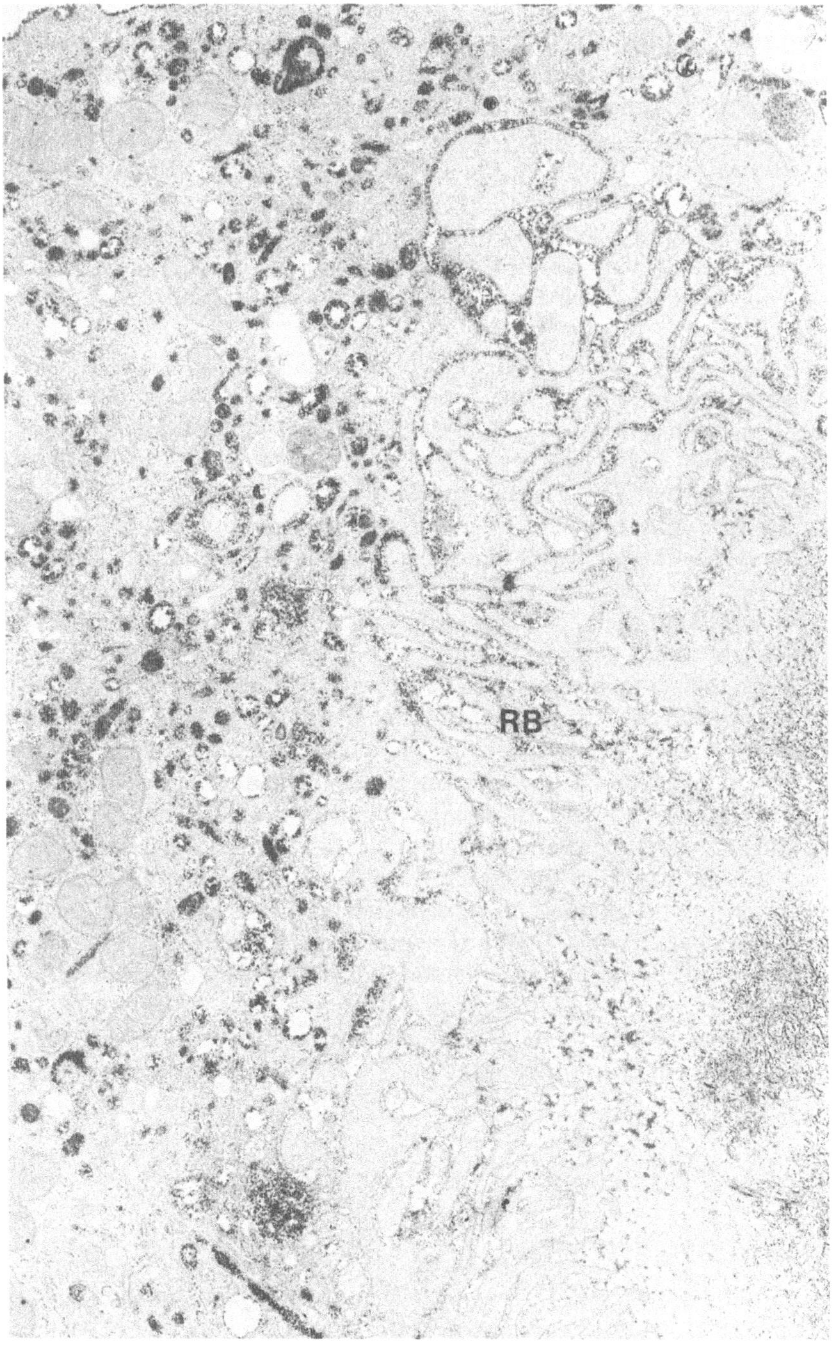

FIGURE 15. Osteoclast 40 min after intravenous injection of peroxidase. The reaction product is present in the channels of the ruffled border (RB) and in numerous cytoplasmic vacuoles and bodies in the area behind the ruffled border. ×16,500. (From Lucht, 1972a, by courtesy of the publishers.)

The presence of acid phosphatase in the vacuoles of the osteoclast (Doty and Schofield, 1968; Lucht, 1971) indicates that lysosomal enzymes participate in the intracellular degradation of material absorbed from the extracellular space. It appears likely that primary lysosomes, originating from the Golgi complex, fuse with the endocytosis vacuoles to form secondary lysosomes. The presence of residual bodies in the osteoclast (Scott, 1967a; Lucht, 1972c) may serve as an additional indication that internal digestion takes place.

It is interesting to note that in fracture callus the macrophages, which are ultrastructurally and histogenetically closely related to osteoclasts, contain collagenlike fibers in vacuoles deep in the cytoplasm (Göthlin and Ericsson, 1973d).

The presence of bone crystals in vacuolar profiles (Lucht, 1972c) indicates that endocytosis of mineral occurs. These vacuolar profiles could be considered true vacuoles because they did not contain the extracellular marker lanthanum. However, Göthlin and Ericsson (1972) did not notice crystals in vacuoles that had taken up thorium dioxide from the extracellular space.

The resorption by osteoclasts is always occurring along the ruffled border, and a recent study (Miller, 1977) indicated that osteoclasts in the Japanese quail can modulate their surface depending on the need for calcium mobilization. Thus, during the egg cycle, changes occur in the osteoclast cell surface. During egg shell calcification, osteoclasts with ruffled borders are closely apposed to bone surfaces; the cytoplasm is rich in vacuoles that contain mineral crystals and seem to derive from the ruffled border. At completion of egg shell calcification, osteoclasts with ruffled borders and vacuoles move away from the bone surface, although the osteoclast remains attached to the bone along the transitional zone. Associated with the disappearance of the ruffled border is the appearance of extensive interdigitated cell processes along the peripheral surface of the osteoclast away from the bone. These structures, which may seem as a reservoir of membrane, largely disappear when ruffled borders and associate structures reappear. It may, therefore, be suggested that the osteoclasts with and without ruffled border, as described in the rat, reflect two functinal stages of the same cell.

Recently vital osteoclasts have been isolated (Nelson and Bauer, 1977), and analysis of these cells may provide the basis for a further understanding of the cell function.

6. EFFECTS OF HORMONES ON THE OSTEOCLAST

6.1. PARATHYROID HORMONE

Parathyroid hormone (PTH) may induce an intensive bone resorption, both *in vivo* (Cameron *et al.*, 1967) and *in vitro* (Raisz, 1965). The number of osteoclasts rises after treatment with PTH (Toft and Talmage, 1960; Bingham *et al.*, 1969; Tatevossian, 1973; McGuire and Marks, 1974). Bingham *et al.* (1969) demonstrated by autoradiography that the rise of osteoclasts was preceded by an increase in the synthesis of RNA, protein, and mucoprotein, indicating an in-

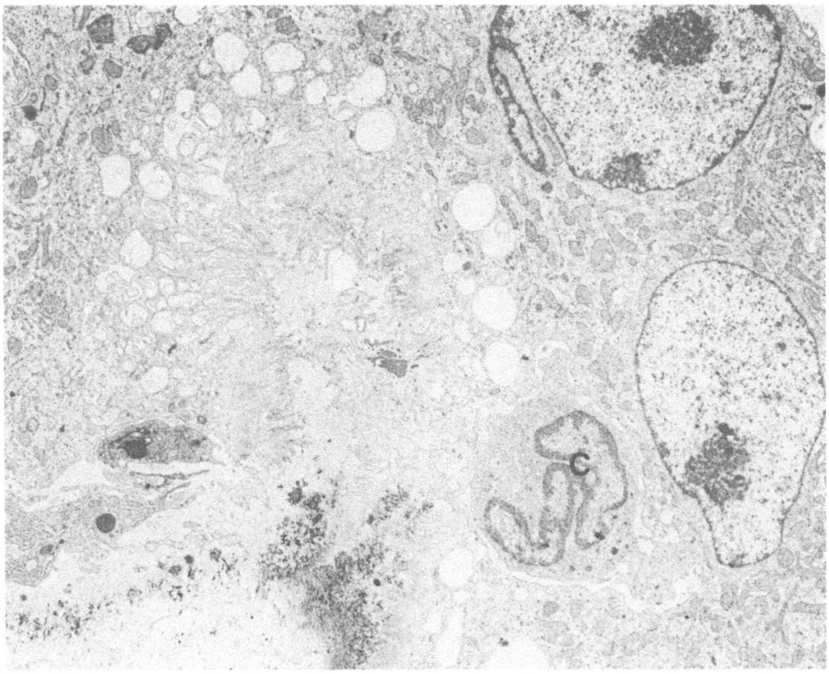

FIGURE 16. PTH-stimulated osteoclast with phagocytized cell (C). ×5400. (From Lucht and Maunsbach, 1973, by courtesy of the publishers.)

creased activity of the individual osteoclast; the increased cell activity correlated well with the rise in plasma calcium. In addition, the transmembrane potential of osteoclasts in culture changed rapidly in response to PTH and was followed over a longer period by an increased rate of RNA synthesis (Mears, 1971).

As judged from random light-microscopic sections, treatment with PTH increases the cell size of osteoclasts (McGuire and Marks, 1974; Lucht and Nørgaard, 1976). Lucht and Maunsbach (1973) found several ultrastructural changes in the osteoclasts after treatment of rats with PTH: The cytoplasm contained large vacuoles with phagocytized cells (Figure 16), some of which resembled osteoblasts or osteocytes; the vacuoles were interpreted as lysosomes because the engulfed cells often appeared partly digested and the vacuoles contained acid phosphatase. Small, coated cytoplasmic bodies, interpreted as early primary lysosomes, were numerous in the peripheral cytoplasm (Figure 17). Lipid droplets were also present in the cytoplasm. The membranes of the ER were fused in some places. Cytoplasmic regions with numerous free ribosomes were frequent. Large ring-shaped granules occurred in some mitochondria (Figure 18), and energy dispersive X-ray analysis of these granules provided evidence that they contained calcium and phosphorus. In some osteoclasts the mitochondria were enlarged, but not to the extent previously reported by Cameron *et al.* (1967). These findings are consistent with an increased activity of the osteoclasts, and, in particular, a stimulation of the lysosomal system with an increased

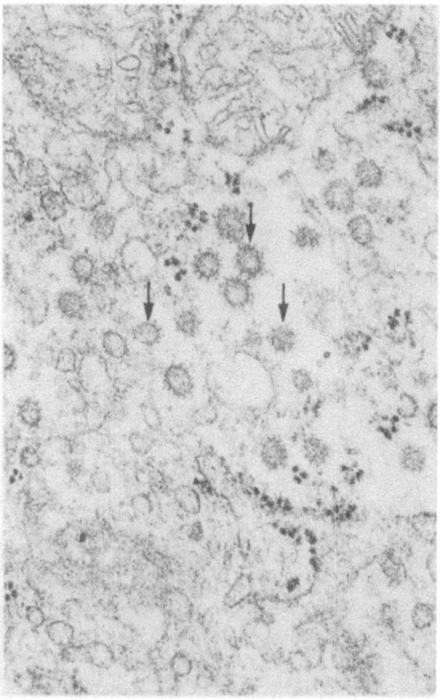

FIGURE 17. PTH-stimulated osteoclast containing numerous small, coated cytoplasmic bodies in the peripheral cytoplasm (→). ×49,900. (From Lucht and Maunsbach, 1973, by courtesy of the publishers.)

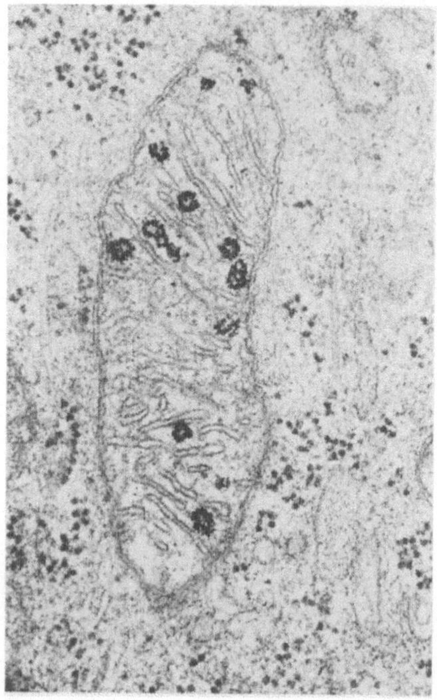

FIGURE 18. PTH-stimulated osteoclast. Large ring-shaped electron-dense granules are present in some mitochondria. ×56,000. (From Luch and Maunsbach, 1973, by courtesy of the publishers.)

production and transport of lysosomes to the cell membrane. This is in agreement with some *in vitro* studies (Tolnai, 1968; Vaes, 1968; Brown and Cameron, 1974) which have demonstrated a greater release of acid phosphatase to the culture medium when bone was treated with PTH. In a careful study on isolated osteoclasts, Walker (1972) found no significant change in the total activity of acid phosphatase compared to control cells. This finding is consistent with the possibility that the production of lysosomal enzymes is increased, but that the enzymes are rapidly released through exocytosis.

The presence of large calcium-containing granules in the mitochondria (Lucht and Maunsbach, 1973) may reflect an increase in the mitochondrial level of calcium during treatment with PTH. It has previously been shown that mitochondria have the ability to transport calcium (Lehninger, 1970; Selwyn *et al.*, 1970), and that kidney mitochondria accumulate calcium *in vitro* under the influence of PTH (Cohn *et al.*, 1967). Matthews (1970) observed that mitochondria located in the bone side of the osteoclast have several granules; mitochondria in the middle of the cell contain fewer granules and those on the vascular side have the least number. This intracellular gradient of mitochondrial granules is consistent with a calcium gradient and a transport of calcium from the resorbing surface through the osteoclast to the blood.

In quantitative studies, Holtrop *et al.* (1974) demonstrated that PTH increased the ruffled border as well as the transitional zone. When cultured bone from rats was treated with PTH for various periods, a significant increase in the cell and transitional zone sizes could be demonstrated after 3 hr, and in ruffled border size after 6 hr (King *et al.*, 1975); this correlated well with a significant increase in the release of calcium into the medium after 6 hr. In egg-laying Japanese quail, the osteoclasts were functionally inactive and ruffled borders adjacent to bone surfaces were rare during the period when medullary bone was not being resorbed (Miller, 1978). However, 20 min after the administration of PTH, over 70% of the osteoclast profiles had ruffled border adjacent to bone surfaces.

Some authors (Baud, 1966; Mills *et al.*, 1971; Freilich, 1971; Weisbrode *et al.*, 1974) did not demonstrate any ultrastructural changes in osteoclasts during treatment with PTH, but this may be explained by the different experimental conditions.

6.2. CALCITONIN

Several experimental studies have shown that calcitonin decreases bone resorption (Anast and Conaway, 1972). The number of osteoclasts has been reported to be reduced after treatment with calcitonin *in vitro* (Reynolds *et al.*, 1968) as well as *in vivo* (Foster *et al.*, 1966; Mills *et al.*, 1972). Evanson *et al.* (1967), on the contrary, found an increase in the number of osteoclasts after prolonged infusion of calcitonin in intact animals, but this can be explained as an effect of a secondary hyperparathyroidism.

Electron-microscopic studies show that osteoclasts are separated from bone and lose their ruffled border after treatment with calcitonin (Zichner, 1971; Kallio *et al.*, 1972; Mills *et al.*, 1972; Lucht, 1973b; Weisbrode and Capen, 1974). Intermediate forms between the typical ruffled border and the transitional zone occurred (Lucht, 1973b). The number of large vacuoles was decreased and the cell did not contain an accumulation of vacuoles behind the ruffled border; neither did the vacuoles contain bone crystals (Figure 19). This indicates a decreased endocytosis, but the calcitonin treatment still allows the osteoclasts to take up peroxidase (Lucht and Nørgaard, 1977). The calcitonin treatment increased the number of autophagic vacuoles (Lucht, 1973b) which shows an enhanced autophagocytosis, as in the liver after glucagon administration (Arstilla and Trump, 1968).

By electron-microscopic histochemistry no acid phosphatase could be demonstrated outside the osteoclast (Lucht, 1973b), which suggests that the exocytosis of lysosomal enzymes has been inhibited (Figure 20). This suggestion is consistent with *in vitro* experiments (Vaes, 1972) which showed that calcitonin reduces the release of lysosomal enzymes from the bone tissue to the medium.

In quantitative studies it was shown that both the size of the ruffled border and the transitional zone was decreased after treatment with calcitonin (Holtrop *et al.*, 1974). The frequency of the presence of ruffled borders and transitional zones in osteoclast profiles was determined in randomly selected sections of cultured bone pretreated with PTH. Four hours after the addition of calcitonin to

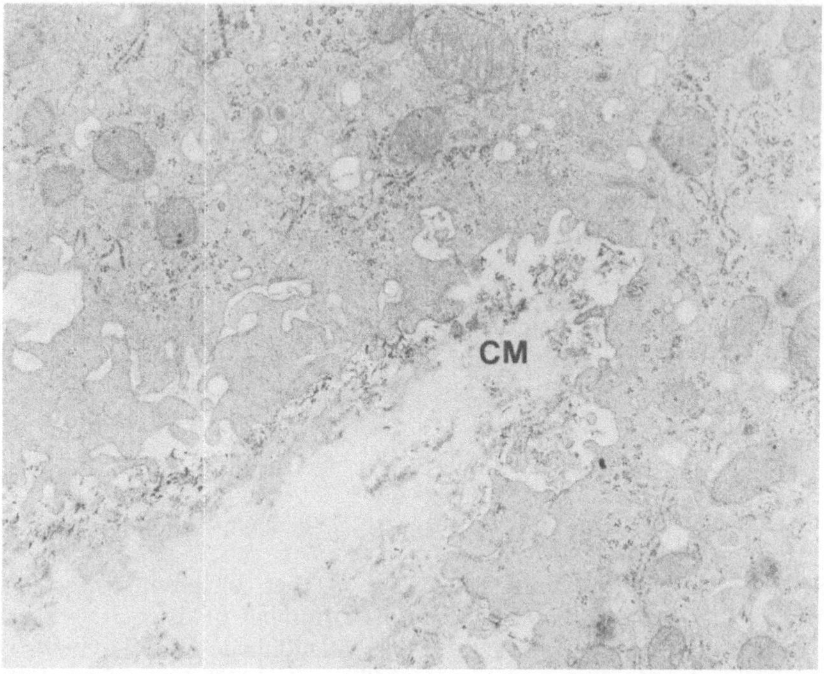

FIGURE 19. Calcitonin-affected osteoclast with an intermediate border adjacent to calcified matrix (CM), which has been undergoing resorption. ×19,200. (From Lucht, 1973b, by courtesy of the publishers.)

the medium the frequency of ruffled borders had decreased from 68% at the start to 10%.

These studies indicate that, in addition to a decrease in number, the individual osteoclast is transformed to an inactive stage by calcitonin. In rabbits, the decreased number of osteoclasts occurred within 15 min and was followed 15 min later by a fall in serum calcium (Mills *et al.*, 1972). Lucht (1974) showed that calcitonin treatment may cause a hypocalcemia without any significant variations in the osteoclast count. These experiments provide evidence that the hormone effect on the serum calcium primarily is mediated by a decreased cell activity.

It is quite interesting that osteoclasts seen in *i.a.* rats with osteopetrosis (Marks, 1973; Schofield *et al.*, 1974) are rather similar in structure to calcitonin-treated osteoclasts. However, it should be emphasized that the osteoclasts in *i.a.* rats with osteopetrosis cannot be transformed to an active stage after treatment with PTH (Marks, 1976).

6.3. VITAMIN D

It is well-known that vitamin D_3 is hydroxylated in the liver to produce 25-hydroxycholecalciferol (25-OHD_3) which is further transformed to two other metabolites, one being 1,25-dihydroxycholecalciferol (1,25-$(OH)_2D_3$). Both 25-

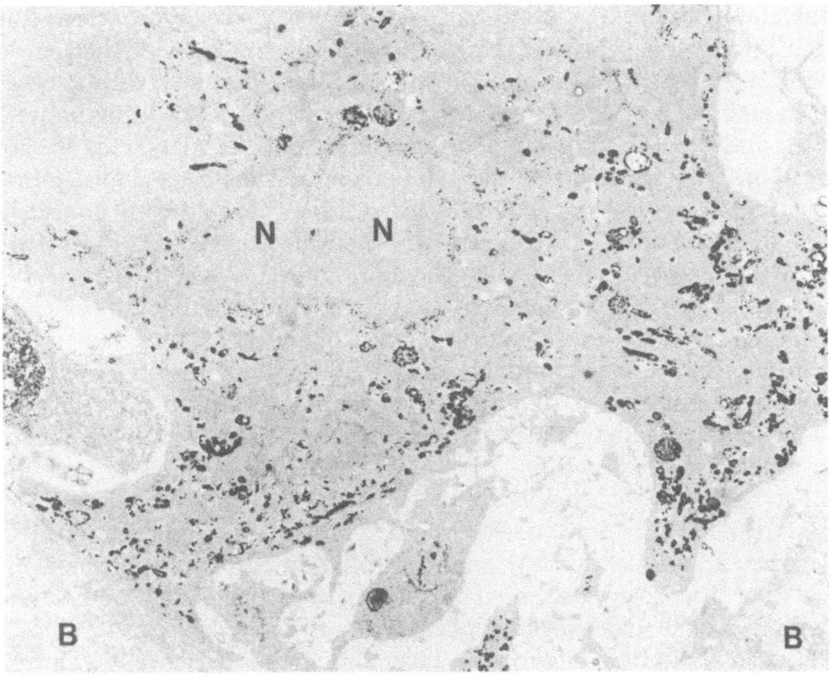

FIGURE 20. Calcitonin-affected osteoclast from bone tissue which has been incubated for acid phosphatase. Numerous lysosomes are present in the osteoclast, but no reaction product occurs in the extracellular space between the osteoclast and the bone. N, Nucleus; B, bone. ×6400.

OHD_3 and 1,25-$(OH)_2D_3$ are potent stimulators of bone resorption by osteoclasts in tissue culture (Raisz *et al.*, 1972; Reynolds *et al.*, 1973). The 1,25-metabolite was about 100 times more potent than 25-OHD_3, whereas vitamin D_3 itself is inactive *in vitro*. In thyroparathyroidectomized rats, vitamin D caused increased osteoclastic resorption (Baylink *et al.*, 1973; Weisbrode *et al.*, 1973). However, the ultrastructure of the osteoclasts was unchanged, and the phosphatase activity, localized as Ca-ATPase at pH 7.4, did not differ in sites of reactivity or intensity (Weisbrode *et al.*, 1973).

7. THE ORIGIN AND FATE OF OSTEOCLASTS

The origin of osteoclasts has for a long time been a subject for debate and the comprehensive literature concerning this problem was reviewed some years ago (Owen, 1970). It is generally accepted that the osteoclast is formed by fusion of precursor cells, but the nature of the last ones has been unclear. Some authors (Heller *et al.*, 1950; Kember, 1960; Tonna, 1960; Tonna and Cronkite, 1961; Young, 1962, 1964) have suggested that osteoblasts and osteoclasts derive from the same precursor cells. In contrast, Scott (1967b), in an autoradiographic study at the electron-microscopic level, found evidence for the concept that osteoclasts and osteoblasts derive from different osteoprogenitor cells. Thus, the population of proliferating osteoprogenitors was composed of two cell types which could be divided into preosteoblasts and preosteoclasts on the basis of their structure. Recognizable transitional forms between the two types were not observed. Jotereau and Duarin (1978) studied the developmental relationship between osteocytes and osteoclasts using the quail-chick nuclear marker in enchondral ossification. They showed that the hemopoietic and the osteogenic (comprising osteoblasts, osteocytes, and chondrocytes) cell lines have different embryological origins. The osteoclasts belonged to the hemopoietic cell line and were not derived from any cell type of the osteogenic line. This is consistent with light-microscopic autoradiographic experiments which indicate that the osteoclast derives from a mononuclear leukocyte (Fischman and Hay, 1962; Jee and Nolan, 1963; Buring, 1975). Göthlin and Ericsson (1973c) studied the origin of bone cells in fracture callus with two different techniques at the electron-microscopic level. The first series of experiments included electron-microscopic autoradiography using tritiated thymidine in parabiotic rats. In the other series of experiments, rats that had received exogenous thorium-dioxide-labeled mononuclear cells were studied by conventional EM. Both types of experiments indicated that osteoblasts and osteoclasts represent two different pathways of osteogenic differentiation. The osteoclasts seem to derive from phagocytic monocytes and macrophages while osteoblasts appear to take their origin from fibroblasts.

In order to induce heterotopic bone formation, Nilsen (1977) implanted dentine in the abdominal muscles of guinea pigs. He observed that monocytes and macrophages as well as osteoclasts (dentinoclasts) were active in the resorption of dentine; the monocytic cells dominated the resorption at an early stage, although later the osteoclasts increased. It was concluded that in this experimental system the monocytes and macrophages are important, not only as preosteoclasts, but also as dentine-resorbing cells.

All these experiments indicating an origin of osteoclasts from monocytes and macrophages are supported by experiments on *i.a.*, osteopetrotic rats (Marks, 1978a). The *i.a.* rat contains osteoclasts without a ruffled border. However, when these rats receive whole-body radiation and an injection of spleen cells from a normal littermate the osteopetrosis is cured, and normal osteoclasts with ruffled borders are present. This indicates that the mechanism of spleen cell cure for osteopetrosis in *i.a.* rats is a rapid remodelling of the skeleton produced by osteoclasts with ruffled borders deriving from spleen cells. The same results could be achieved by injection of cells from liver, thymus, and bone marrow, but not with cells from salivary gland, muscle, and brain (Marks, 1978b).

The fate of the osteoclast is rather obscure. In the previously mentioned studies, it was reported that PTH and calcitonin respectively increase and decrease the number of osteoclasts, which indicates that these hormones regulate the osteoclast population, depending on the need for bone resorption. In the same way as osteoclasts are formed by fusion of precursor cells and presumably grow by further incorporation of these, it may be suggested that the osteoclast in a reversible way can break down to precursor cells again. This would be consistent with the fact that no electron-microscopic reports have mentioned necrotic osteoclasts. In addition, Gaillard (1959), when studying osteoclasts in bone explants by microcinematography, was unable to detect any sign of osteoclast death. On the contrary, he observed their disappearance and replacement by "undifferentiated" cells.

8. SUMMARY AND CONCLUSIONS

Two types of osteoclasts occur in the rat, but these may be two functional stages of the same cell type as demonstrated in the Japanese quail. The first type or stage has no ruffled border, and is thought to be inactive with regard to bone resorption. The other, and more common type or stage, contains a ruffled border below which the bone shows changes indicative of resorption. The osteoclast contains a well-developed lysosomal system. Lysosomal enzymes are released by exocytosis during bone resorption to participate in the extracellular bone degradation. They also take part in the intracellular decomposition of endocytized bone components. In addition, there are indications that a collagenase is involved in the bone resorption, but there is still insufficiency in our knowledge of the resorption mechanism. However, isolation of vital osteoclasts may be a promising tool for further analysis of these cells.

Treatment of bone with parathyroid hormone induces an increased number of osteoclasts as well as an increased activity of the individual cell. Calcitonin, on the contrary, decreases the number of osteoclasts and the function of the individual cell.

Recent evidence indicates that osteoclasts derive from monocytes and macrophages. The last cells and osteoclasts are also rather closely related, both ultrastructurally and functionally.

REFERENCES

Anast, C. S., and Conoway, H. H., 1972, Calcitonin, *Clin. Orthopaed.* **84**:207.

Arnold, J. S., and Jee, W. S. S., 1957, Bone growth and osteoclastic activity as indicated by radioautographic distribution of plutonium, *Am. J. Anat.* **101**:367.

Arstila, A. U., and Trump, B. F., 1968, Studies on cellular autophagocytosis, *Am. J. Pathol.* **53**:587.

Baud, C. A., 1966, The fine structure of normal and parathormone-treated bone cells, in: *Fourth European Symposium on Calcified Tissues* (P. J. Gaillard, A. van den Hooff, and R. Steendijk, eds.), pp. 4–6, Excerpta Medica Foundation, Amsterdam.

Baylink, D., Sipe, J., Wergedal, J., and Whittemore, J., 1973, Vitamin D-enhanced osteocytic and osteoclastic bone resorption, *Am. J. Physiol.* **224**:1345.

Bingham, P. J., Brazell, I. A., and Owen, M., 1969, The effect of parathyroid extract on cellular activity and plasma calcium levels *in vivo*, *J. Endocrinol.* **45**:387.

Birkedal-Hansen, H., 1974, Osteoclastic resorption of ^{3}H-proline labelled bone, dentine and cementum in the rat, *Calcif. Tissue Res.* **15**:77.

Bonucci, E., 1974, The organic-inorganic relationships in bone matrix undergoing osteoclast resorption, *Calcif. Tissue Res.* **16**:13.

Brown R. M., and Cameron, D. A., 1974, Acid hydrolases and bone resorption in the remodelling phase of the development of bone fracture callus, *Pathology* **6**:53.

Buring, K., 1975, On the origin of cells in heterotopic bone formation, *Clin. Orthopaed.* **110**:293.

Burstone, M. S., 1960a, Histochemical demonstration of succinic dehydrogenase activity in osteoclasts, *Nature* **185**:866.

Burstone, M. S., 1960b, Histochemical demonstration of cytochrome oxidase activity in osteoclasts, *J. Histochem. Cytochem.* **8**:225.

Cameron, D. A., 1963, The fine structure of bone and calcified cartilage, *Clin. Orthopaed.* **26**:199.

Cameron, D. A., 1968, The Golgi apparatus in bone and cartilage cells, *Clin. Orthopaed.* **58**:191.

Cameron, D. A., 1969, The fine structure and function of bone cells, in: *The Biological Basis of Medicine* (E. E. Bittar and N. Bittar, eds.), Vol. 3, pp. 391–423, Academic Press, London.

Cameron, D. A., 1972, The ultrastructure of bone, in: *The Biochemistry and Physiology of Bone* (G. H. Bourne, ed.), Vol. 1, pp. 191–236, Academic Press, New York.

Cameron, D. A., Paschall, H. A., and Robinson, R. A., 1967, Changes in the fine structure of bone cells after the administration of parathyroid extract, *J. Cell Biol.* **33**:1.

Cohn, D. V., Bawdon, R., and Eller, G., 1967, The effect of parathyroid hormone *in vivo* on the accumulation of calcium and phosphate by kidney and on kidney mitochondrial function, *J. Biol. Chem.* **242**:1253.

Cretin, A., 1951, Contribution histochimique à l'étude de la construction et de la destruction osseuse, *Presse Méd.* **59**:1240.

de Duve, C., 1963, The lysosome, *Sci. Am.* **208**:64.

Doty, S. B., and Schofield, B. H., 1972, Electron microscopic localization of hydrolytic enzymes in osteoclasts, *Histochem. J.* **4**:245.

Doty, S. B., Schofield, H., and Robinson, R. A., 1968, The electron microscopic identification of acid phosphatase and adenosintriphosphatase in bone cells following parathyroid extract or thyrocalcitonin administration, in: *Parathyroid Hormone and Thyrocalcitonin (Calcitonin)* (R. V. Talmage and L. F. Bélanger, eds.), pp. 169–181, Excerpta Medica Foundation, Amsterdam.

Dudley, H. R., and Spiro, D., 1961, The fine structure of bone cells, *J. Biophys. Biochem. Cytol.* **11**:627.

Evanson, J. M., Garner, A., Holmes, A. M., Lumb, G. A., Stanbury, S. W., 1967, Interrelations between thyrocalcitonin and parathyroid hormone in rats, *Clin. Sci.* **32**:271.

Fetter, A. W., and Capen, C. C., 1971, The fine structure of bone in the nasal turbinates of young pigs, *Anat. Rec.* **171**:329.

Fischman, D. A., and Hay, E. D., 1962, Origin of osteoclasts from mononuclear leucocytes in regenerating newt limbs, *Anat. Rec.* **143**:329.

Foster, G. V., Doyle, F. H., Bordier, P., and Matrajt, H., 1966, Effect of thyrocalcitonin on bone, *Lancet* **2**:1428.

Freilich, L. S., 1971, Ultrastructure and acid phosphatase cytochemistry of odontoclasts: Effects of parthyroid extract, *J. Dental Res.* **50**:1047.

Furseth, R., 1968, The resorption processes of human deciduous teeth studied by light microscopy, microradiography and electron microscopy, *Arch. Oral Biol.* **13**:417.

Gaillard, P. J., 1955, Parathyroid gland tissue and bone *in vitro*, *Expt. Cell Res. (Suppl.)* **3**:154.

Gaillard, P. J., 1959, Parathyroid gland and bone *in vitro*, *Dev. Biol.* **1**:152.

Gay, C. V., and Mueller, W. J., 1974, Carbonic anhydrase and osteoclasts: Localization by labeled inhibitor autoradiography, *Science* **183**:432.

Goldhaber, P., 1960, Behavior of bone in tissue culture, in: *Calcification in Biological Systems* (R. F. Sognnaes, ed.), pp. 349–372, American Association for the Advancement of Science, Washington, D.C.

Gonzales, F., and Karnovsky, M. J., 1961, Electron microscopy of osteoclasts in healing fractures of rat bone, *J. Biophys. Biochem. Cytol.* **9**:299.

Göthlin, G., and Ericsson, J. L. E., 1971, Fine structural localization of acid phosphomonoesterase in the brush border region of osteoclasts, *Histochemie* **28**:337.

Göthlin, G., and Ericsson, J. L. E., 1972, Observations on the mode of uptake of thorium dioxide particles by osteoclasts in fracture callus, *Calcif. Tissue Res.* **10**:216.

Göthlin, G., and Ericsson, J. L. E., 1973a, Fine structural localization of alkaline phosphatase in the fracture callus of the rat, *Histochemie* **36**:225.

Göthlin, G., and Ericsson, J. L. E., 1973b, Studies on the ultrastructural localization of adenosine triphosphatase activity in fracture callus, *Histochemie* **35**:111.

Göthlin, G., and Ericsson, J. L. E., 1973c, On the histogenesis of the cells in fracture callus. Electron microscopic autoradiographic observations in parabiotic rats and studies on labeled monocytes, *Virchows Arch. B* **12**:318.

Göthlin, G., and Ericsson, J. L. E., 1973d, Electron microscopic studies on the uptake and storage of thorium dioxide molecules in different cell types of fracture callus, *Acta Pathol. Microbiol. Scand. A* **81**:523.

Göthlin, G., and Ericsson, J. L. E., 1976, The osteoclast, *Clin. Orthopaed.* **120**:201.

Hancox, N. M., 1972, The osteoclast, in: *The Biochemistry and Physiology of Bone* (G. H. Bourne, ed.), Vol. 1, pp. 46–67, Academic Press, New York.

Hancox, N. M., and Boothroyd, B., 1961, Motion picture and electron microscope studies on the embryonic avian osteoclast, *J. Biophys. Biochem. Cytol.* **11**:651.

Hancox, N. M., and Boothroyd, B., 1963, Structure-function relationships in the osteoclast, in: *Mechanisms of Hard Tissue Destruction* (R. F. Sognnaes, ed.), pp. 497–514, American Association for the Advancement of Science, Washington, D.C.

Heller, M., McLean, F. C., and Bloom, W., 1950, Cellular transformation in mammalian bones induced by parathyroid extract, *Am J. Anat.* **87**:315.

Holtrop, M., and King, G., 1977, The ultrastructure of the osteoclast and its functional implications, *Clin. Orthopaed.* **123**:177.

Holtrop, M. E., Raisz, L. G., and Simmons, H. A., 1974, The effect of parathyroid hormone, colchicine and calcitonin on the ultrastructure and the activity of osteoclasts in organ culture, *J. Cell Biol.* **60**:346.

Irving, J. T., and Heeley, J. D., 1970, Resorption of bone collagen by multinucleated cells, *Calcif. Tissue Res.* **6**:254.

Jee, W. S. S., and Nolan, P. D., 1963, Origin of osteoclasts from the fusion of phagocytes, *Nature (London)* **200**:225.

Jones, S. J., and Boyde, A., 1977, Some morphological observations on osteoclasts, *Cell Tissue Res.* **185**:387.

Jotereau, F. V., and Le Douarin, N. M., 1978, The developmental relationship between osteocytes and osteoclasts: A study using the quail-chick nuclear marker in enchondral ossification, *Dev. Biol.* **63**:253.

Kallio, D. M., Garant, P. R., and Minkin, C., 1971, Evidence of coated membranes in the ruffled border of the osteoclast, *J. Ultrastruct. Res.* **37**:169.

Kallio, D. M., Garant, P. R., and Minkin, C., 1972, Ultrastructural effects of calcitonin on osteoclasts in tissue culture, *J. Ultrastruct. Res.* **39**:205.

Kember, N. F., 1960, Cell division in enchondral ossification, *J. Bone Joint Surg.* **42B**:824.

King, G. J., and Holtrop, M. E., 1975, Demonstration of actin filaments in bone cells with heavy meromyosin, *J. Bone Joint Surg.* **57A**:578.

King, G. J., Holtrop, M. E., and Raisz, L. G., 1975, A quantitative study of the time course changes in the ultrastructure and activity of osteoclasts in bone stimulated by parathyroid hormone (PTH) in organ culture, *J. Bone Joint Surg.* **57A**:577.

King, G. J., Holtrop, M. E., and Reit, B., 1976, A quantitative study of the influence of parathyroid hormone on the ultrastructure and activity of osteoclasts in thyroparathyroidectomized rats, *Trans. 22nd Ann. Meeting Orthop. Res. Soc.* **1**:120.

Knese, K. H., 1972, Osteoklasten, Chondroklasten, Mineraloklasten, Kollagenoklasten, *Acta Anat.* **83**:275.

Knese, K. H., and Knoop, A. M., 1961, Elektronenmikroskopische Beobachtungen über die Zellen in der Eröffnungszone des Epiphysenknorpels, *Z. Zellforsch. Mikrosk. Anat.* **54**:1.

Kölliker, A., 1973, *Die normale Resorption des Knochengewebes*, Vogel, Leipzig.

Lehninger, A. L., 1970, Mitochondria and calcium ion transport, *Biochem, J.*, **119**:129.

Lucht, U., 1971, Acid phosphatase of osteoclasts demonstrated by electron microscopic histochemistry, *Histochemie* **28**:103.

Lucht, U., 1972a, Absorption of peroxidase by osteoclasts as studied by electron microscope histochemistry, *Histochemie* **29**:274.

Lucht, U., 1972b, Osteoclasts and their relationship to bone as studied by electron microscopy, *Z. Zellforsch. Mikrosk. Anat.* **135**:211.

Lucht, U., 1972c, Cytoplasmic vacuoles and bodies of the osteoclast: An electron microscope study, *Z. Zellforsch. Mikrosk. Anat.* **135**:229.

Lucht, U., 1973a, Electron Microscope observations of centrioles in osteoclasts, *Z. Anat. Entwicklungsgesch.* **140**:143.

Lucht, U., 1973b, Effects of calcitonin on osteoclasts *in vivo*, *Z. Zellforsch. Mikrosk. Anat.* **145**:75.

Lucht, U., 1974, Correlation between serum calcium and the number of osteoclasts following calcitonin treatment in rats, *International Research Communications System Med. Sci.* **2**:1357.

Lucht, U., 1976, Hormone induced changes in the lysosomal system of osteoclasts, in: *Proceedings of the Ninth Congress of the Nordic Society for Cell Biology* (F. Bierring, ed.), pp. 77–87, Odense University Press, Odense.

Lucht, U., and Maunsbach, A. B., 1973, Effects of parathyroid hormone on osteoclasts *in vivo*: An ultrastructural and histochemical study, *Z. Zellforsch. Mikroskop. Anat.* **141**:529.

Lucht, U., and Nørgaard, J. O., 1976, Export of protein from the osteoclast as studied by electron microscopic autoradiography, *Cell Tissue Res.* **168**:89.

Lucht, U., and Nørgaard, J. O., 1977, Uptake of peroxidase by calcitonin inhibited osteoclasts, *Histochemistry* **54**:143.

Malkani, K., Luxembourger, M. -M., and Rebel, A., 1973, Cytoplasmic modifications at the contact zone of osteoclasts and calcified tissue in the diaphyseal growing plate of foetal guinea-pig tibia, *Calcif. Tissue Res.* **11**:258.

Marks, S. C., 1973, Pathogenesis of osteopetrosis in the *ia* rat: Reduced bone resorption due to reduced osteoclast function, *Am. J. Anat.* **138**:165.

Marks, S. C., 1976, Tooth eruption and bone resorption: Experimental investigation of the *ia* (osteopetrotic) rat as a model for studying their relationships, *J. Oral Pathol.* **5**:149.

Marks, S. C., 1978a, Studies of the mechanism of spleen cell culture for osteopetrosis in *ia* rats: Appearance of osteoclasts with ruffled borders, *Am. J. Anat.* **151**:119.

Marks, S. C., 1978b, Studies of the cellular cure for osteopetrosis by transplanted cells. Specificity of the cell type in *ia* rats. *Am. J. Anat.* **151**:131.

Matthews, J. L., 1970, Ultrastructure of calcifying tissues, *Am. J. Anat.* **129**:451.

Matthews, J. L., Martin, J. H., and Race, G. J., 1967, Giant-cell centrioles, *Science* **155**:1423.

McGuire, J. L., and Marks, S. C., 1974, The effects of parathyroid hormone on bone cell structure and function, *Clin. Orthopaed.* **100**:392.

Mears, D. C., 1971, Effects of parathyroid hormone and thyrocalcitonin on the membrane potential of osteoclasts, *Endocrinology* **88**:1021.

Miller, S. C., 1977, Osteoclast cell-surface changes during the egg-laying cycle in Japanese quail, *J. Cell Biol.* **75**:104.

Miller, S. C., 1978, Rapid activation of the medullary bone osteoclast cell surface by parathyroid hormone, *J. Cell Biol.* **76**:615.

Mills, B. G., Holst, P., Haroutinian, A. M., and Bavetta, L. A., 1971, Bone cell response to serum calcium altering drugs, *Clin. Orthopaed.* **78**:56.

Mills, B. G., Haroutinian, A. M., Holst, P., Bordier, P. J., Tun-Chot, S., 1972, Ultrastructural and cellular changes at the costochondral junction following *in vivo* treatment with calcitonin or calcium chloride in the rabbit, in: *Endocrinology 1971* (S. Taylor, ed.), pp. 79–88, Heinemann, London.

Nelson, R. L., and Bauer, G. E., 1977, Isolation of osteoclasts by velocity sedimentation at unit gravity, *Calcif. Tissue Res.* **22**:303.

Neuman, W. F., Mulryan, B. J., and Martin, G. R., 1960, A chemical view of osteoclasis based on studies with yttrium, *Clin. Orthopaed.* **17**:124.

Nilsen, R., 1977, Electron microscopy of induced heterotopic bone formation in guinea pigs, *Arch. Oral Biol.* **22**:485.

Novikoff, A. B., 1963, Lysosomes in the physiology and pathology of cells: Contribution of staining methods, in: *Ciba Foundation Symposium on Lysosomes* (A. V. S. de Reuck and M. P. Cameron, eds.), pp. 36–77, Churchill, London.

Owen, M., 1970, The origin of bone cells, *Int. Rev. Cytol.* **28**:213.

Palade, G. E., 1959, Functional changes in the structure of cell components, in: *Subcellular Particles* (T. Hayashi, ed.), pp. 64–80, Ronald Press, New York.

Puzas, J. E., and Brand, J. S., 1976, Collagenolytic activity from isolated bone cells, *Biochim. Biophys. Acta* **429**:964.

Raisz, L. G., 1965, Bone resorption in tissue culture. Factors influencing the response to parathyroid hormone, *J. Clin. Invest.* **44**:103.

Raisz, L. G., Trummel, C. L., Holick, M. F., and DeLuca, H. F., 1972, 1,25-dihydroxycholecalciferol: A potent stimulater of bone resorption in tissue culture, *Science* **175**:768.

Raisz, L. G., Holtrop, M. E., and Simmons, H. A., 1973, Inhibition of bone resorption by colchicine in organ culture, *Endocrinology* **92**:556.
Rasmussen, H., and Bordier, P., 1974, *The Physiological and Cellular Basis of Metabolic Bone Disease*, Williams and Wilkins, Baltimore.
Reynolds, J. J., Dingle, J. T., Gudmundson, T. V., and MacIntyre, I., 1968, Bone resorption *in vitro* and its inhibition by calcitonin, in: *Calcitonin. Proceedings of the Symposium on Thyrocalcitonin and the C Cells* (S. Taylor, ed.), pp. 223–229, Heinemann, London.
Reynolds, J. J., Holick, M. F., and De Luca, H. F., 1973, The role of vitamin D metabolites in bone resorption, *Calcif. Tissue Res.* **12**:295.
Robin, M. C., 1849, Sur l'existence de deux espèces nouvelles d'éléments anatomiques qui se trouvent dans le canal médullaire des os, *C. R. Soc. Biol.* **1**:149.
Robinson, R. A., and Cameron, D. A., 1964, Bone, in: *Electron Microscopic Anatomy* (S. M. Kurtz, ed.), pp. 315–340, Academic Press, New York.
Rollet, A., 1870, in: *A Manual of Human and Comparative Histology* (S. Stricker, ed.), p. 95, New Sydenham Soc., London.
Roth, T. F., and Porter, K. R., 1964, Yolk protein uptake in the oocyte of the mosquito *Aedes Aegypti* L., *J. Cell Biol.* **20**:313.
Sakamoto, S., Sakamoto, M., Goldhaber, P., and Glimcher, M., 1975, Collagenase and bone resorption: Isolation of collagenase from culture medium containing serum after stimulation of bone resorption by addition of parathyroid hormone extract, *Biochem. Biophys. Res. Commun.* **63**:172.
Schenk, R. K., Spiro, D., and Wiener, J., 1967, Cartilage resorption in the tibial epiphyseal plate of growing rats. *J. Cell Biol.* **34**:275.
Scherft, J. P., 1972, The lamina limitans of the organic matrix of calcified cartilage and bone, *J. Ultrastruct. Res.* **38**:318.
Schofield, B. H., Levin, L. S., and Doty, S. B., 1974, Ultrastructure and lysosomal histochemistry of *ia* rat osteoclasts, *Calcif. Tissue Res.* **14**:153.
Scott, B. L., 1967a, The occurrence of specific cytoplasmic granules in the osteoclast, *J. Ultrastruct. Res.* **19**:417.
Scott, B. L., 1967b, Thymidine-^{3}H electron microscope radioautography of osteogenic cells in the fetal rat, *J. Cell Biol.* **35**:115.
Scott, B. L., and Pease, D. C., 1956, Electron microscopy of the epiphyseal apparatus, *Anat. Rec.* **126**:465.
Selwyn, M. J., Dawson, A. P., and Dunnett, S. J., 1970, Calcium transport in mitochondria, *FEBS Lett.* **10**:1.
Takimoto, K., Deguchi, T., and Mori, M., 1966, Histochemical detection of succinic dehydrogenase in osteoclasts following experimental tooth movement, *J. Dental Res.* **45**:1473.
Tatevossian, A., 1973, Effect of parathyroid extract on blood calcium and osteoclast count in mice, *Calcif. Tissue Res.* **11**:251.
Thyberg, J., 1975, Electron microscopic studies on the uptake of exogenous marker particles by different cell types in guinea pig metaphysis, *Cell Tissue Res.* **156**:301.
Thyberg, J., Nilsson, S., and Friberg, U., 1975, Electron microscopic and enzyme cytochemical studies on the guinea pig metaphysis with special reference to the lysosomal system of different cell types, *Cell Tissue Res.* **156**:273.
Toft, R. J., and Talmage, R. V., 1960, Quantitative relationship of osteoclasts to parathyroid function, *Proc. Soc. Exp. Biol. Med.* **103**:611.
Tolnai, S., 1968, Effects of parathyroid hormone on bone acid hydrolases in tissue culture, *Can. J. Physiol. Pharmacol.* **46**:261.
Tonna, E. A., 1960, Osteoclasts and the aging skeleton: A cytological, cytochemical and autoradiographic study. *Anat. Rec.* **137**:251.
Tonna, E. A., and Cronkite, E. P., 1961, Autoradiographic studies of cell proliferation in the periosteum of intact and fractured femora of mice utilizing DNA labeling with H^3-thymidine, *Proc. Soc. Exp. Biol. Med.* **107**:719.
Vaes, G., 1968, On the mechanisms of bone resorption. The action of parathyroid hormone on the excretion and synthesis of lysosomal enzymes and on the extracellular release of acid by bone cells, *J. Cell Biol.* **39**:676.

Vaes, G., 1969, Lysosomes and the cellular physiology of bone resorption, in: *Lysosomes in Biology and Pathology* (J. T. Dingle and H. B. Fell, eds.), pp. 217–253, North-Holland, Amsterdam.

Vaes, G., 1970, Enzymatic and other biochemical events accompanying bone resorption in tissue culture, *Calcif. Tissue Res. Suppl.* **4**:57.

Vaes, G., 1972, Inhibitory actions of calcitonin on resorbing bone explants in culture and on their release of lysosomal hydrolases, *J. Dental Res.* **51**:362.

Vaughan, J. M., 1970, *The Physiology of Bone*, Clarendon Press, Oxford.

Walker, D. G., 1961, Citric acid cycle in osteoblasts and osteoclasts, *Bull. Johns Hopkins Hosp.* **106**:80.

Walker, D. G., 1972, Enzymatic and electron microscopic analysis of isolated osteoclasts, *Calcif. Tissue Res.* **9**:296.

Weisbrode, S. E., and Capen, C. C., 1974, Ultrastructural evaluation of the effects of calcitonin on bone in thyroparathyroidectomized rats administered vitamin D, *Am. J. Pathol.* **77**:455.

Weisbrode, S. E., Capen, C. C., and Nagode, L. A., 1973, Fine structural and enzymatic evaluation of bone in thyroparathyroidectomized rats receiving various levels of vitamin D, *Lab. Invest.* **28**:29.

Weisbrode, S. E., Capen, C. C., and Nagode, L. A., 1974, Effects of parathyroid hormone on bone of thyroparathyroidectomized rats, *Am. J. Pathol.* **75**:529.

Yaeger, J. A., and Kraucunas, E., 1969, Fine structure of the resorptive cells in the teeth of frogs, *Anat. Rec.* **164**:1.

Young, R. W., 1962, Cell proliferation and specialization during endochondral osteogenesis in young rats, *J. Cell Biol.* **14**:357.

Young, R. W., 1964, Specialization of bone cells, in: *Bone Biodynamics* (H. M. Frost, ed.), pp. 117–189, Little, Brown, Boston.

Zichner, L., 1971, The effect of calcitonin on bone cells in young rats. An electron microscopic study, *Israel J. Med. Sci.* **7**:359.

20

Characteristics of Mononuclear Phagocytes in Culture

JOS W. M. VAN DER MEER

1. INTRODUCTION

Cell biologists have been interested in mononuclear phagocytes for many decades. Many investigators who cultured explants of various tissues, noticed that cells with pseudopods and phagocytic activity emerged from the explants (Lewis and Gey, 1923; Parker, 1932). These wandering cells were very similar in different tissues, and were considered to be macrophages (although confusion with fibroblasts often cannot be ruled out). In studies performed with buffy coats from avian blood, macrophages were shown to derive from monocytes (Carrel and Ebeling, 1926). These cultures also made it possible to obtain pure cultures of mononuclear phagocytes, because all other types of blood cell were eliminated by death. In 1933, cultures of mononuclear phagocytes on a glass surface to which these cells adhere, were described (Baker, 1933), and this property of mononuclear phagocytes is still being exploited for investigative purposes.

The proliferative capacity of mononuclear phagocytes *in vivo* and *in vitro* has proved to be a controversial subject (van Furth and Cohn, 1968; Volkman, 1976), which will be dealt with under the appropriate headings. Another source of controversy has been the transformation of mononuclear phagocytes into other cell types such as lymphocytes, fibroblasts, and other mesenchymal cells (van Furth *et al.*, 1972). Such transformation is denied by many investigators (see van Furth, 1978).

In this review studies on mononuclear phagocytes cultured *in vitro* and studies directly relevant to the subject are discussed. No attempt was made to cover the literature on mononuclear phagocytes.

JOS W. M. VAN DER MEER • Department of Infectious Diseases, University Hospital, 2333 AA Leiden, The Netherlands.

2. METHODS FOR THE CULTURE OF MONONUCLEAR PHAGOCYTES

2.1. LIQUID CULTURE

At present, most investigators who culture mononuclear phagocytes make use of their property to adhere to glass or plastic when grown in a liquid medium. This adherence provides separation from cell types that do not adhere firmly (e.g., lymphocytes).

After adherence to glass or plastic, normal mononuclear phagocytes spread gradually within 12–24 hr. When mice are previously injected intraperitoneally with a substance such as thioglycollate (see Section 3.1.2b), the peritoneal macrophages spread immediately after adherence *in vitro* (Werb and Cohn, 1971). Treatment of mice with interferon inducers and infection with, for example, *Listeria* or *Trypanosoma cruzi* enhance spreading (Rabinovitch *et al.*, 1977; North, 1969a; Hoff, 1975). Factors such a pH, temperature, divalent cations (especially Mg^{2+} and Mn^{2+}), and hydrogen ion concentrations are important for the process of spreading (Rabinovitch and DeStefano, 1973a,b). Spreading of unstimulated macrophages *in vitro* can be increased in the absence of serum by the addition of several reagents, such as ATP, dithiothreitol, and proteolytic enzymes (for a review see Rabinovitch, 1975). The contact phase of blood coagulation and complement activation (factor Bb, a large cleavage product of factor B) play a crucial role in macrophage spreading (Bianco *et al.*, 1976, Götze *et al.*, 1980).

The major disadvantage of culturing mononuclear phagocytes on a glass or plastic surface is the difficulty encountered in recovering the cells from the surface into suspension, which may be necessary for further investigations. Several methods have been recommended to recover adherent cells, such as repeatedly freezing and thawing (Cohn and Benson, 1965b), scraping with a rubber policeman at 4°C, treatment with trypsin or other enzymes, chelating agents (Paul, 1972), or cationic anesthetic agents (Rabinovitch and DeStefano, 1975), or a combination of these. However, these procedures damage the cells more or less, and recovery is far from complete (van der Meer *et al.*, 1978; Stewart, 1980). A number of different techniques have been devised to prevent attachment of cells to the surface, but these have been only partially successful for mononuclear phagocytes. Culturing of cells on surfaces coated with collagen or in a plasma clot and recovery of the cells by enzymatic digestion of the coating are currently used but have several disadvantages, such as ingestion of the coating substance, activation of the cells by the material on the surface, and interference from the coating material in microscopic studies. Furthermore, cells that normally do not adhere to the surface (e.g., lymphocytes) may adhere to the coated surface under such conditions.

Recently, we described methods for the culture of mononuclear phagocytes in suspension, making use of hydrophobic Teflon® film for cell support (van der Meer *et al.*, 1978, 1979b). Since mononuclear phagocytes attach very loosely or not at all to these films, the cells can be harvested easily without damage or functional impairment, and recovery is almost complete (> 75%). Moreover, since Teflon film is permeable for CO_2, O_2, and water vapor, it is possible to culture in closed Teflon bags, which minimizes the risk of infection.

Adherence of mononuclear phagocytes to these Teflon films can be induced by etching the films chemically (Munder *et al.*, 1971), or by coating with *l*-polylysine (Stulting and Berke, 1973).

2.2. CULTURE MEDIA AND SERA

A large variety of culture media containing amino acids, carbohydrates, vitamins, electrolytes, and buffers, are in use, but there are very few reports (Stecher and Thorbecke, 1967; Johnson *et al.*, 1977) of studies performed to compare the quality and efficacy of different media. To maintain mononuclear phagocytes *in vitro* in good condition, most investigators add serum to the culture medium. Various kinds of serum give satisfactory results. All batches of serum to be used, have to be tested for toxicity. Sera should be stored sterile at -20°C and, once thawed, can be kept at 4°C and should not be refrozen, since repeated freezing and thawing may lead to denaturation of proteins.

For cultures of mouse cells, for instance, newborn calf serum, fetal calf serum, horse serum, rabbit serum, guinea pig serum, and human serum have been used. Sera of some species are cytotoxic and can only be used after inactivation at 56°C. Cohn and Benson (1965b) showed that newborn calf serum stimulated spreading, pinocytosis, and hydrolase production of mouse macrophages *in vitro* to a greater extent than other sera tested. Johnson *et al.* (1977), who compared various sera for the culture of human monocytes, obtained optimal results with fresh autologous serum.

For some experiments mononuclear phagocytes must be cultured in the absence of serum. In that case, lactalbumin hydrolysate can be used as substitute, but more complex formulas for serum-free media are also available (Habeshaw, 1972; Guilbert and Iscove, 1976).

2.3. SEMISOLID MEDIA

For the culture of proliferating bone marrow cells in the presence of colony-stimulating factor (see Sections 3.2.1 and 4.1.1d)—provided by a feeder layer Dicke *et al.*, 1971), or added as a conditioned medium (Pluznik and Sachs, 1965; Bradley and Metcalf, 1966), or added in a purified form (Stanley and Heard, 1977)—many authors use semisolid media.

The method was originally described by Pluznik and Sachs (1965) and Bradley and Metcalf (1966), who used agar, and since these, many modifications of the original technique have been described. Methylcellulose is also used instead of agar (Worton *et al.*, 1969). However, both agar and methylcellulose hamper study of the morphology, cytochemistry, function, and proliferation of the cells in the colonies. Even with such modifications as the thin-layer agar techniques combined with agarase treatment (Dicke *et al.*, 1971) or a modified squash technique (Testa and Lord, 1970), characterization of the cells remains difficult. These difficulties can be overcome, however, by the use of a liquid culture system (Goud *et al.*, 1975). (For details on the proliferation of bone marrow mononuclear phagocytes, see Section 3.2.1c.)

2.4. INCUBATION

Most investigators culture mammalian macrophages at 37°C. Since almost all culture media contain bicarbonate buffers, the cultures are either gassed with a 5% CO_2 air mixture and closed, or placed in an incubator provided with the same gas mixture. The effect of different O_2 pressures on bone marrow cultures in agar has been investigated by Bradley *et al.* (1977). In their study 5% O_2 with 10% CO_2 gave the best results. Axline *et al.* (1980) studied the effects of different oxygen concentrations in culture media of macrophages on energy metabolism, phagocytosis, and pinocytosis.

3. CULTURES OF VARIOUS KINDS OF MONONUCLEAR PHAGOCYTES

For the discussion of the characteristics of the mononuclear phagocytes obtained from various sources and cultured *in vitro,* a division will be made according to the capacity of the cells to proliferate *in vitro.* It must be kept in mind here that without the addition of conditioning factors—for instance colony-stimulating factor (see Sections 2.3 and 4.1.1d)—little or no proliferation of mononuclear phagocytes is observed *in vitro.* Furthermore, the proliferative potential varies from species to species.

3.1. NONPROLIFERATING SYSTEMS

3.1.1. Blood Monocytes

Monocytes from human blood are relatively easy to obtain for the study of mononuclear phagocytes in man.

3.1.1a. Separation. Blood monocytes can be separated from other blood cells in various ways. The oldest method (Baker, 1933) makes use of the ability of mononuclear phagocytes to adhere to glass, and is still employed for separation (Einstein *et al.*, 1976). Other methods are separation over albumin or Ficoll–Hypaque gradients (Bennet and Cohn,1966; Boyum,1968; Steigbigel *et al.*, 1974; Johnson *et al.*, 1977). Van Furth *et al.* (1979) found a monocyte yield of 133% after Ficoll–Hypaque separation, which indicates that the differential counts done in blood smears represent an underestimation, and that the number of monocytes lost during the procedure is probably very low. This confirms the findings of Zucker-Franklin (1974). Lymphocytes are still present after these separation procedures, but most of them can be removed by washing after the monocytes become adherent. Alternatively, the continuous density gradient developed by Loos *et al.* (1976) can be used for this purpose, since the separation of monocytes from lymphocytes and other blood cells in this way gives about 30% monocyte yield with 80% purity.

After Ficoll–Hypaque passage, Sanderson *et al.* (1977) separated monocytes from lymphocytes by centrifugal elutriation (with a commercially available counterflow centrifuge). They obtained about 60% monocyte yield with 90% purity.

3.1.1b. *In Vitro* Culture. Human blood monocyte culture *in vitro* with current methods is associated with losses of up to 90% of the monocytes during the first 3 days. Johnson *et al.* (1977) found about 50% loss after a week of culture in Neumann–Tytell medium and fresh autologous serum. In Teflon culture bags, about 75% viable monocytes were found after a week of culture with Medium 199 containing newborn calf serum (Blussé van Oud Alblas *et al.*, 1978).

Monocytes cultured on glass or plastic surface can be maintained up to about 100 days (Einstein *et al.*,1976; Johnson *et al.*,1977), but exact data on the percentage of cells that are still alive are not available. Many giant cells are observed in such cultures on glass (see Section 3.1.6). Blood monocytes obtained from various kinds of experimental animals (e.g., mouse, rat, rabbit, horse, and birds) have been studied *in vitro* (Baker, 1933; Bennet and Cohn, 1966; van Furth and Cohn, 1968; Mangornkanok *et al.*, 1975).

3.1.2. Peritoneal Macrophages

Mononuclear phagocytes from the peritoneal cavity are the best-studied macrophages, because they are readily obtainable in many species. Since there are many differences between mononuclear phagocytes obtained by lavage of the peritoneal cavity in the normal steady state and those collected in the same way but after stimulation (usually by intraperitoneal injection of an inflammatory stimulus), these noninduced and induced mononuclear phagocytes will be dealt with separately.

3.1.2a. Noninduced Peritoneal Macrophages. These macrophages can be harvested in relatively large amounts from mice and rats. In other rodents, the number of peritoneal macrophages obtained under normal steady-state conditions is too low to make the use of untreated animals economical. Therefore, the number of cells must be increased by the use of agents that recruit monocytes from the blood (see Section 3.1.2b).

The peritoneal cavity of the mouse contains about 25% lymphocytes and a varying number of mast cells; both are lost by washing the cultures. Peritoneal cells can be maintained *in vitro* for many weeks. The [^{3}H]thymidine-labeling index of such cultures is very low (<3%) (van Furth and Cohn, 1968, North, 1969a). Van Furth *et al.* (1979) studied human peritoneal macrophages obtained via laparoscopy in women undergoing tubal coagulation for sterilization. During 6 hr of culture only 1.5% [^{3}H]thymidine labeling was found.

3.1.2b. Induced Peritoneal Macrophages. Many inducing agents—e.g., newborn calf serum, starch, mineral oil, thioglycollate, and BCG—are recommended in the literature. Injection of an inducing agent produces essentially three effects:

1. Exudate macrophages are recruited from the blood;
2. Macrophages, presumably both exudate macrophages and resident macrophages, are activated (i.e., their properties are altered); and
3. There is an influx of other cells, especially granulocytes during the first 24 hr.

The extent to which these effects occur is dependent on the agent used. Stuart *et al.* (1977) give a detailed description of various inducing agents, with special attention to their advantages and disadvantages (such as endocytosis of a particulate agent and clumping of the cells).

The influx of mononuclear phagocytes from the blood in response to various stimulants has been compared by Lin and Stewart (1974), Edelson *et al.* (1975), and van Waarde *et al.* (1977a).

The degree and kind of activation of macrophages is uncertain, as well as which cells are activated. Most investigators use either a limited number of stimuli and/or measure a limited number of effects. The question as to whether the macrophages recruited from the blood or the resident macrophages are activated, is neglected in many studies.

Some stimuli lead to a large influx of granulocytes into the peritoneal cavity, but the problem of these contaminating granulocytes can be circumvented by harvesting the exudate at least 24 hr after administration of the stimulus. Almost all lymphocytes can be removed by washing after adherence of the macrophages to the substratum.

3.1.3. Liver Macrophages (Kupffer Cells)

Reliable methods for the isolation and culture of liver macrophages based on enzymatic digestion of the parenchymal cells have been described by many authors (Garvey, 1961; Mills and Zucker-Franklin, 1969; Munthe-Kaas *et al.*, 1975; Emeis and Planqué, 1976; Crofton *et al.*, 1978). To eliminate blood monocytes, the liver is perfused prior to enzymatic digestion. In the mouse, this method yields on average 2.2×10^7 viable nonparenchymal cells comprising 0.9×10^7 liver macrophages (27%), the remainder being endothelial cells (34%), lymphocytes (20%), granulocytes (6%), and 10% unidentifiable cells (Crofton *et al.*, 1978). After culture for 48 to 72 hr, more than 90% of the adherent cells are Kupffer cells. Morphologically, the cells are heterogeneous in size and shape and tend to remain more rounded than peritoneal macrophages. The enzyme treatment may damage the membrane of macrophages, leading to the loss of Fc and complement receptors, but during culture these receptors reappear (Crofton *et al.*, 1978).

As seen *in vivo*, the uptake of tritiated thymidine by Kupffer cells *in vitro* lies below 1.5% during the normal steady state (North, 1969b; Volkman, 1976; Crofton *et al.*, 1978). As in the peritoneal cavity, certain stimulants (e.g., zymosan) can induce a strong influx of blood monocytes and a small number of immature mononuclear phagocytes (Diesselhoff-den Dulk *et al.*, 1979). For the separation of Kupffer cells from endothelial cells, centrifugal elutriation gives good results (Knook and Sleyster, 1976).

3.1.4. Lung Macrophages

Alveolar macrophages can be easily obtained by bronchial lavage not only in the experimental animal (Godleski and Brain, 1972; Weiden *et al.*, 1975) but also, since the development of the fiberoptic bronchoscope in man (Reynolds *et al.*, 1975). In the mouse, Blussé van Oud Alblas and van Furth (1979) compared the recovery of alveolar macrophages obtained with various lavage fluids, and found the highest yield when phosphate-buffered saline-containing EDTA (200 mg/liter) was used, i.e., on average 1.37×10^6 mononuclear phagocytes.

To study the macrophage in murine lung tissue, an enzyme digestion technique can be used (Blussé van Oud Alblas and van Furth, 1979). To eliminate monocytes in the pulmonary vascular tree, the lungs are perfused prior to alveolar lavage and enzyme digestion. Enzyme digestion gives on average 21.7×10^6 cells, of which about 0.58×10^6 are macrophages. The remainder are Type II alveolar epithelial cells, granulocytes, lymphocytes, and fibroblasts. Both alveolar macrophages and pulmonary tissue macrophages attach to glass in culture but exhibit little or no spreading. The adherent cells from the bronchial lavage fluid are exclusively macrophages, whereas cultures of the enzyme-digested lung tissue show a minority of granulocytes and fibroblasts among the adherent cells. These fibroblasts may overgrow the cultures when incubation is prolonged. Under normal culture conditions, pulmonary macrophages do not proliferate *in vitro;* proliferation has been obtained in the presence of various conditioned media (this will be discussed in Section 3.2.2).

3.1.5. Macrophages of the Spleen and Other Tissues

To obtain spleen macrophages, various mechanical methods are described in the literature, e.g., employing forceps, scissors, or sieves (Bennet, 1966; Stuart *et al.*, 1977; Steinman and Cohn, 1973). After mechanical disruption of the spleen, the cells are allowed to adhere to a glass or plastic surface. The yield of mononuclear phagocytes obtained with these techniques is rather low, however, and enzyme treatment seems preferable.

Steinman and Cohn (1973) used clostridial collagenase to prepare single-cell suspension from murine spleen. Of the total number of nucleated spleen cells, 10–50% are mononuclear phagocytes (Pierce, 1969; Steinman and Cohn, 1973). Good quantitative data on spleen macrophages are not available as yet.

Methods are also available for the culture of macrophages collected from lymph nodes, thymus, and tonsils (Stuart *et al.*, 1977), and glomeruli (Holdsworth *et al.*, 1978).

3.1.6. Macrophage Polykaryons and Granuloma Macrophages

According to most investigators, giant cells are formed by the fusion of macrophages, and the number of nuclei is a measure of the age of the cells thus formed (Sutton and Weiss,1966; Papadimitriou *et al.*,1973; Mariano and Spector, 1974; van der Rhee *et al.*, 1978). Such giant cells can be elicited in rodents by subcutaneous insertion of glass or plastic coverslips, to which mononuclear phagocytes adhere, transforming into epithelioid cells and heterokaryons. A granuloma evoked *in vivo* can be dissected and digested with collagenase, and the cell suspension thus obtained can be cultured *in vitro* (Bonney *et al.*, 1978).

In vitro cultures may show increasing numbers of macrophage polykaryons, especially after prolonged incubation. In cultures of human monocytes maintained up to 3 months *in vitro*, Johnston *et al.* (1977) found that 50–70% of the nuclei occurred in giant cells, which had up to 20 nuclei. In long-term cultures of human monocytes on a Teflon surface (up to 49 days), giant cells were only encountered occasionally (Blussé van Oud Alblas *et al.*, 1978), presumably due to differences in the substratum used.

3.2. PROLIFERATING SYSTEMS

3.2.1. Bone Marrow Cultures

The monoblast, a descendent of the (committed) hematopoietic stem cell, together with the promonocyte, which is produced by division of the monoblast, form the proliferating pool of the mononuclear phagocytes *in vivo*. These progenitor cells can also proliferate *in vitro* (Goud and van Furth, 1975).

3.2.1a. Promonocytes. These cells have been characterized in 6-hr bone marrow cultures (van Furth *et al.*, 1970; van Furth and Diesselhoff-den Dulk, 1970). Suspensions of cells from the bone marrow of the femurs of a mouse are incubated in Leighton tubes in culture medium containing newborn calf serum. After the cells have been allowed to attach to the coverslip, the cultures are washed, and reincubated.

Two types of cell adhere to the glass in these cultures, namely promonocytes and monocytes. On the basis of *in vitro* and *in vivo* labeling experiments with [^{3}H]thymidine and measurement of the relative amount of DNA, it was concluded that promonocytes are proliferating cells and monocytes are nondividing cells (van Furth and Diesselhoff-den Dulk, 1970).

3.2.1b. Bone Marrow Cultures in Semisolid Media. Bradley and Metcalf (1966) and Pluznik and Sachs (1965) were the first to describe methods for *in vitro* culture and proliferation of murine bone marrow cells. These authors used agar for cell support, and found that the presence of conditioning factors was necessary for proliferation. Colonies developed in these cultures, and were found to be composed of either granulocytes or macrophages; mixed colonies have also been described. Since it is difficult to perform adequate morphological, cytochemical, and functional studies on cells grown on semisolid media, these methods are not very suitable for the investigation of mononuclear phagocytes. Data on numbers of cells in these cultures are also difficult to obtain.

The most important conditioning factor in these cultures (at least for proliferation of precursor cells of the mononuclear phagocyte series in the mouse) is the colony-stimulating factor (CSF), a glycoprotein recently purified by Stanley and Heard (1977). This factor is secreted by mononuclear phagocytes and some other cell types (see Section 4.1.13). Most investigators still use colony-stimulating factor in an unpurified form and originating from many sources (e.g., pregnant mouse uterus extract, mouse lung, embryonic mouse fibroblasts, mouse L cells). For the human system, colony-stimulating factor, although produced by human leukocytes, is less active in promoting the growth of macrophage precursors.

3.2.1c. Bone Marrow in Liquid Culture. The liquid culture system described by Goud *et al.*(1975) provides a convenient method for morphological studies at both the light- and electron-microscopic levels, as well as for cytochemical and functional studies of murine bone marrow proliferating *in vitro*. Growth behavior, too, can be studied very precisely (Goud and van Furth, 1975). In this system two kinds of colony develop, mononuclear phagocyte colonies and a smaller number of granulocyte colonies (Goud *et al.*, 1975). Mixed colonies are never encountered. The structure of the granulocyte colonies is characterized by the close proximity of the cells, whereas the cells of mononuclear phagocyte colonies form a monolayer on the glass surface and show separation, the round cells lying centrally, the elongated cells swarming toward the periphery.

In the mononuclear phagocyte colonies the earliest precursor cell of the mononuclear phagocyte series, the monoblast, could be recognized and characterized. This actively proliferating cell has the typical features of a mononuclear phagocyte. Calculation showed that one monoblast has a progeny of more than 5×10^3 mononuclear phagocytes at day 14. Table 1 shows the characteristics of the monoblast and of the promonocyte and the macrophage, the other two types of cell present in mononuclear phagocyte colonies. Monocytes cannot be distinguished from macrophages in this system, since the cells beyond the promonocytes are stretched on the glass surface. Some properties of the monoblast (e.g., glass adherence, phagocytosis, and pinocytosis) are less well-developed than those of promonocytes and macrophages. The characteristics of the cells as well as the results of time-lapse studies indicate the sequence monoblast → promonocyte → macrophage, according to increasing maturity. A mathematical analysis of the data of the first four days of culture led Goud and van Furth (1975) to conclude that, *in vitro,* some of the dividing monoblasts and promonocytes replicate themselves, and that after division other monoblasts and promonocytes differentiate into promonocytes and macrophages, respectively. On the basis of the conclusion that the monoblast initiates the mononuclear phagocyte colony, the number of monoblasts present *in vivo* can be estimated to be about 1 per 1000 nucleated bone marrow cells (or 2.5×10^5 per mouse).

With this liquid culture technique, an attempt was made to culture murine bone marrow mononuclear phagocytes over a longer period of time (van der Meer *et al.*, 1980). In primary cultures, mononuclear phagocytes can be maintained for 3–4 weeks. After 2 weeks, the rate of proliferation declines, and after 3–4 weeks, the cultures deteriorate even if fresh medium is supplied. Cultured on a Teflon surface, the growth behavior of the cells was completely comparable to that on a glass surface. However, culturing on a Teflon surface permitted

TABLE 1. CHARACTERISTICS OF MONONUCLEAR PHAGOCYTES OF COLONIES GROWN IN LIQUID CULTURE[a]

	Mononuclear phagocytes		
	Monoblasts (%)	Promonocytes (%)	Macrophages (%)
Esterase[b]	91	90	93
Lysozyme	43	55	98
Fc receptors	94	99	100
C receptors	16	39	75
Phagocytosis	96	100	100
Pinocytosis	16	64	96

[a] From Goud *et al.* (1975).
[b] With α-naphthyl butyrate as substrate.

subculture of the cells at intervals without damage, and proliferation could be maintained over a very long period (> 100 days). Study of the characteristics of these proliferating cells showed that they are mononuclear phagocytes.

For the culture of proliferating *human* bone marrow mononuclear phagocytes, the system has not yet been very successful, first of all, because the human precursor cells adhere less well to glass, which hampers morphological studies. Secondly, a good conditioned medium for the proliferation of human bone marrow mononuclear phagocytes is not yet available.

In the rabbit, the mononuclear phagocytes from the bone marrow proliferate in the presence of a high concentration of fetal bovine serum and horse serum, even without conditioned medium; the precursor cells of the mononuclear phagocyte series in this species do not adhere well either (Hauser and Vaes, 1978).

A different kind of liquid culture system has been described by Cline and Sumner (1972). This system includes a modification of the *in vitro* diffusion chamber closed at one end with dialysis membrane, described by Marbrook (1967). The bone marrow cells are cultured on the dialysis membrane, under which a flask containing either conditioned medium or a feeder layer is placed. The results obtained with this system indicate that a blast cell precedes the promonocyte. This blast cell proved to differ from the monoblast described by Goud *et al.* (1975), being a strongly peroxidase-positive nonphagocytic cell with no Fc receptors on its surface. However, these cells were aspirated from suspension cultures, which means that the kind of colony they belonged to is unknown.

The modified Marbrook chamber has also been used for the culture of human bone marrow (Golde and Cline, 1973; Wilcox *et al.*, 1976; Bainton and Golde, 1978). In this system, a limited proliferation of mononuclear phagocytes occurs during the first 2 weeks of incubation. In an electron-microscopic study, Bainton and Golde (1978) found promonocytes, monocytes, and large macrophages in these cultures on the 2nd and 7th day of incubation, and large peroxidase-negative macrophages on day 14. Epithelioid cells and giant cells are also present.

Recently, Dexter *et al.* (1977) developed a culture system for murine cells in which not only mononuclear phagocytes and granulocytes, but also pluripotential hematopoietic stem cells (CFU-s) and committed stem cells (CFU-c) thrive and proliferate for several months. The first step in this system is the establishment of a very peculiar feeder layer of adherent cells from the bone marrow over a period of 3 weeks at 33°C. During this period at least half of the nonadherent cells and medium are removed and replaced by fresh medium every week. It seems to be essential for this system that very large (diameter > 50 μm) adherent cells with many fat globules appear in the feeder layer (Allen and Dexter, 1976). The exact nature of these cells is unknown. At the end of the three-week period, freshly isolated bone marrow cells are added to the cultures before incubation at 37°C. Again, removal of cells and medium is performed each week and fresh medium supplied. The cells removed in this way can be tested in a spleen colony assay (CFU-s, Till and McCullough, 1961) and in semisolid cultures (CFU-c). The choice of serum batches for this system is very critical (Williams *et al.*, 1977).

3.2.2. Proliferating Mononuclear Phagocytes from Other Tissues

Since the report by Virolainen and Defendi (1967), various studies have shown *in vitro* proliferation in liquid and in semisolid cultures of mononuclear phagocytes obtained from other tissues besides the bone marrow of rodents. Table 2 gives data from these studies.

A number of conditions have to be fulfilled before optimal growth of these colonies can be obtained:

1. The mouse strain selected must be a high responder in this respect (Lin *et al.*, 1978).
2. Under certain conditions (peritoneal cavity, pleural cavity) a strong inflammatory stimulus must be given before cells are collected. [For the efficacy of various stimulants with respect to colony formation, see Lin (1974).]
3. The source and batch of both colony-stimulating factor and sera are very important (Stewart, 1979).

Using Swiss mice and embryonic mouse fibroblast conditioned medium in Dulbecco's medium with 20% horse serum, a combination which gives us excellent results for bone marrow culture (Goud *et al.*, 1975; van der Meer *et al.*, 1980), we did not obtain numbers of colonies comparable to those of Lin and Stewart when we used cells from the stimulated peritoneal cavity, peripheral blood, or alveolar washings. Only suspensions of liver macrophages gave a variable amount of colonies (see Table 2; van der Meer *et al.*, 1976).

The cell type that initiates such non-bone-marrow colonies is still uncertain. In analogy with the bone marrow, monoblasts could be the colony-forming cells. There are indications in support of this possibility. Firstly, with bone marrow cells, colony formation does not occur after monoblasts are washed away and the remaining cells are promonocytes and monocytes (Goud *et al.*, 1975). Sec-

TABLE 2. MAXIMAL NUMBER OF MONONUCLEAR PHAGOCYTE COLONIES

Source[a]	Conditioned medium	Colonies per 1000 nucleated cells: Semisolid medium	Colonies per 1000 nucleated cells: Liquid culture	References
Peritoneal cavity				
Normal steady state	LCM[c]	0.01	3	Lin *et al.* (1973, 1974);
Thioglycollate-induced cells	LCM	90	200	van der Zeyst *et al.* (1978)
Pleural cavity				
Normal steady state	LCM	2	No data	Chu and Lin (1976)
Thioglycollate-induced cells	LCM	30	30	
Alveolar cells[b]	LCM	50	70	Lin *et al.* (1975)
Spleen	LCM	1	50	Stewart (1980)
Lymph node	PMUE[d]	0.03	No data	MacVittie and McCarthy (1977)
Thymus	PMUE	0.02	No data	
Liver macrophage suspension	EMFCM[e]	0.1	0.03	van der Meer *et al.* (1976)
Peripheral blood	LCM	4	40	Lin (1977)

[a] Murine cells, unless stated otherwise.
[b] Hamster cells.
[c] L cell conditioned medium.
[d] Pregnant mouse uterus extract.
[e] Embryonic mouse fibroblast conditioned medium.

ondly, a very strong inflammatory stimulus, one which might attract even monoblasts to the site of inflammation, is obligatory for colony growth of, e.g., peritoneal cells. On the other hand, cells obtained by alveolar lavage in the hamster or from the mouse spleen or liver, form mononuclear phagocyte colonies without an identifiable inflammatory stimulus. However, polluted air might stimulate the alveolar cells, and endotoxemia in the portal vein might stimulate the liver.

According to Stewart's view, the cell that initiates the colony is a mononuclear phagocyte recruited from the blood, and he calls these cells "epithelioid" macrophages, a name that can cause great confusion. These cells are approximately 10 μm in diameter and have a central nucleus and radially arranged organelles. He believes that these cells do not proliferate *in vivo,* whereas *in vitro* division occurs in the presence of colony-stimulating factor (Stewart, 1979).

3.2.3. Macrophage Cell Lines

In recent years a series of macrophage cell lines have been established, which facilitate the study of mononuclear phagocytes. Most of these cell lines have been obtained from the mouse and are the result of transformation by virus or originate from spontaneous tumors. A small number of macrophage cell lines from other species—e.g., in rabbit and man—have also been established (Mangornkanok *et al.,* 1975; Ralph, 1980). Studies on number of these cell lines in many laboratories have shown a wide divergence in characteristics and functions among them, which promotes the study of the mechanisms, regulation, and interrelations of properties also occurring in normal or activated macrophages.

It is of interest here that certain properties are inducible in some of these cell lines but not in others. A number of functions can be inhibited too, e.g., by glucocorticosteroids (Ralph *et al.,* 1978; Werb *et al.,* 1978a; Hamilton *et al.,* 1978). Proliferation of a number of cell lines was inhibited by agents such as lipopolysaccharides, BCG, and zymosan (Ralph *et al.,* 1977; Ralph and Nakoinz, 1977a). Under continuous culture the properties of these cell lines may change due to differentiation and selection.

Most of the lines are relatively easy to maintain *in vitro,* and, because they are generally less adherent than normal macrophages, replating can be done without many problems. Some data on the most extensively studied murine cell lines, namely WEHI-3, J774, P338 D_1, PU 5-1.8, and RAW 264, are given in Table 3; for more information, the references mentioned in the table and the review articles of Defendi (1976) and Ralph (1980) are recommended.

4. PROPERTIES OF MONONUCLEAR PHAGOCYTES IN CULTURE

At present, we have no single marker permitting distinction between mononuclear phagocytes and other cells. Therefore, a number of characteristics,

TABLE 3. PROPERTIES OF A NUMBER OF MURINE MACROPHAGE CELL LINES[a]

	$WEHI_3$	J774	$P388D_1$	PU5-1.8	RAW 264	References
Doubling time	24 hr	27 hr		20 hr		Van Furth and Elzenga-Claasen (1978)[e]
Adherence	+	+	+	+	+	
Esterase[b] activity	±	+	+			Van Furth and Elzenga-Claasen (1978)[e]
Peroxidatic activity						
Granules	±	±	−			Van Furth and Elzenga-Claasen (1978)[e]
NE, RER[c]	±	+	−			
Lysozyme production	+	+	+	+	+	Ralph *et al.* (1976); Werb *et al.* (1978a); Hamilton *et al.* (1978)
Production of other lysosomal enzymes		+(+)	+			Snyderman *et al.* (1977); Bloom *et al.* (1978)
Secretion of plasminogen activator			+		+(+)	Werb *et al.* (1978a); Hamilton *et al.* (1978)
Collagenase secretion			(+)			Werb *et al.* (1978a)
Elastase secretion			+			Werb *et al.* (1978a)
Production of colony-stimulating factor	+	−	−	(+)	−	Ralph *et al.* (1976); Ralph and Nakoinz (1977a)
Endogenous pyrogen production	+	+(+)	+	+(+)	+	Bodel (1978)
Prostaglandin production	+(+)	+(+)				Kurland and Bochman (1978)
Receptors						
Fc	+	+	+	+	+	Unkeless and Eisen (1975); Ralph *et al.* (1977a,b); Diamond *et al.* (1978)

C	±	+	±	+	+	Ralph *et al.* (1977a,b); Van Furth and Elzenga-Claasen (1978)[e]
Pinocytosis: dextran sulfate	+	+		+		Van Furth and Elzenga-Claasen (1978)[e]
Endocytosis						
Zymosan	+	+	+	+	+	Ralph and Nakoinz (1977a,b)
Latex	+	+	+	+		Ralph and Nakoinz (1977a); Snyderman *et al.* (1978); Muschel *et al.* (1977)
IgG-coated erythrocytes	+	+	+	+	+	Ralph and Nakoinz (1977b); Muschel *et al.* (1977); Diamond *et al.* (1978); Van Furth and Elzenga-Claasen (1978)[e]
Bacteria (opsonized)	+	+	+	±		Van Furth and Elzenga-Claasen (1978)[e]
Superoxide anion production	−	+	+	+		Johnston *et al.* (1978)
ADCC[d]						
Erythrocytes	±	±	+	+	+	Koren *et al.* (1975); Ralph and Nakoinz (1977b)
Tumor cells	−	−	−	+	+	Ralph and Nakoinz (1977b)
Lymphocyte stimulation	(+)	(+)	(+)	(+)		Lachman *et al.* (1977a); Meltzer and Oppenheim (1977)
Chemotactic response		+	−			Snyderman *et al.* (1978)

[a] ± weakly positive; (+) inducible by exogenous agents; + positive; +(+) stronger positive after exogenous stimulation.
[b] α-Naphthyl butyrate as substrate.
[c] Nuclear envelope and RER.
[d] Antibody-dependent cellular cytotoxicity.
[e] Unpublished observations.

enzymatic markers and functional or ultrastructural properties, must be used for this purpose.

4.1. BIOCHEMICAL MARKERS OF MONONUCLEAR PHAGOCYTES

4.1.1. Nonspecific Esterase Activity

According to many authors (Braunsteiner and Schmalzl, 1970; Ornstein *et al.*, 1976; van Furth, 1976; Stuart *et al.*, 1977), a very discriminative cytochemical marker for mononuclear phagocytes is nonspecific esterase activity. Commonly used substrates are α-naphthyl butyrate or α-naphtyl acetate. Mononuclear phagocytes of many species show strong cytoplasmic staining with these substrates, whereas granulocytes are negative. T lymphocytes have one or few distinct positive dots in the cytoplasm, whereas B lymphocytes are negative (Mueller *et al.*, 1975; Knowles *et al.*, 1978). Fibroblasts, endothelial cells, and epithelial cells may be weakly positive. The esterase activity of mononuclear phagocytes is inhibited by sodium fluoride. During culture, the intensity of esterase activity of murine mononuclear phagocytes increases (van Furth, 1976). Wiener and Levanon (1968), using β-naphthyl butyrate as a substrate at pH 6.5, demonstrated secretion of nonspecific esterase by cultured mouse peritoneal macrophages.

4.1.2. Peroxidatic Activity

The number of granules with peroxidatic activity is smaller in mononuclear phagocytes than in granulocytes. At the light-microscopic level, peroxidatic activity localized in the granules is found in monoblasts, promonocytes, monocytes, and exudate macrophages (van der Meer *et al.*, 1979a; van Furth *et al.*, 1970; Daems *et al.*, 1975). Few positive granules are present in monoblasts, but more occur in the promonocytes and a smaller number again in monocytes and exudate macrophages. During *in vitro* culture of promonocytes, monocytes, and exudate macrophages, the number of peroxidase-positive granules drops considerably (van Furth *et al.*, 1970).

During certain stages of development and/or activation, peroxidatic activity can also be observed electron microscopically in the rough endoplasmic reticulum (RER), the nuclear envelope, and the Golgi apparatus. These patterns of peroxidatic activity are shown in Figure 1.

The different patterns of distribution of peroxidatic activity in monocytes and exudate macrophages on the one hand and in resident macrophages on the other, has been used as evidence that the blood monocyte is not the precursor of resident macrophages (Daems *et al.*, 1975). However, recent studies have demonstrated the transition from exudate macrophages to resident macrophages *in vivo* (Beelen *et al.*, 1978a). Moreover, in many species, mononuclear phagocytes with the peroxidatic activity pattern of resident macrophages were obtained from blood monocytes *in vitro* (Bodel *et al.*, 1977, 1978; Beelen *et al.*, 1978b). In

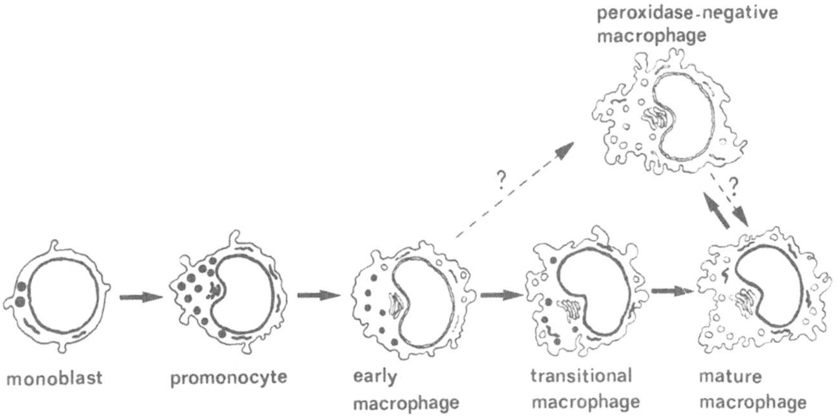

FIGURE 1. Peroxidatic activity patterns of mononuclear phagocytes in bone marrow cultures.

liquid cultures of mouse bone marrow (see Section 3.2.1c), not only monoblasts and promonocytes, but also four types of macrophage were recognized, namely early macrophages (with the peroxidatic activity pattern of blood monocytes or exudate macrophages), transitional macrophages [peroxidatic activity pattern of exudate-resident macrophages (Beelen *et al.*, 1978a)], mature macrophages [identical to resident macrophages (Daems *et al.*, 1975)], and peroxidase-negative macrophages (van der Meer *et al.*, 1979a). Figure 1 shows the sequence of these mononuclear phagocytes.

The identity of the enzyme responsible for peroxidatic activity is unknown. The cytochemical methods currently used are not specific for peroxidases, but also stain cytochromes, hemoglobin, and catalase. Depending on the fixation time, pH, and H_2O_2 concentration applied, the reactivity of these proteins varies.

4.1.3. Lysozyme

Lysozyme activity in mononuclear phagocytes can be shown at the cellular level by immunofluorescence (Glynn and Parkman, 1964; Schmalzl and Braunsteiner, 1970) or by demonstration of lysis of *Micrococcus Lysodeikticus* over the individual cells (Syren and Raeste, 1971). Mononuclear phagocytes are positive for this lysosomal enzyme during all of the developmental stages (Goud *et al.*, 1975; see Table 1). Granulocytes are strongly positive too, whereas lymphocytes and fibroblasts are negative (Braunsteiner and Schmalzl, 1970; Goud *et al.*, 1975). During *in vitro* cultivation of mononuclear phagocytes, Heise and Myrvik (1967) detected lysozyme in the supernatant. Gordon *et al.* (1974a) and McClelland and van Furth (1975) demonstrated continuous production and secretion of lysozyme by monocytes and macrophages during culture. The secretion is independent of cell stimulation and phagocytosis (Gordon *et al.*, 1974a). *In vitro* proliferating mononuclear phagocytes from murine bone marrow secrete an increasing amount of lysozyme (van der Meer *et al.*, 1977). Mature granulocytes are unable to synthesize new lysozyme *in vitro* (McClelland and van Furth, 1975).

4.1.4. Other Lysosomal Enzymes

Staining for lysosomal enzymes such as acid phosphatase and β-glucuronidase (Barka and Anderson, 1962; Lorbacher *et al.*, 1967) does not make it possible to distinguish mononuclear phagocytes from other cells such as granulocytes, lymphocytes, and fibroblasts. Various investigators have shown synthesis and release of various lysosomal hydrolases, such as *N*-acetyl-β-glucosaminidase, β-galactosidase, β-glucuronidase, and β-mannosidase by mouse macrophages during culture (Cohn and Benson, 1965a; Gordon *et al.*, 1974a,b; Schnyder and Baggiolini, 1980). A high concentration of serum (Cohn and Benson, 1965a) or endocytosis of material digestible by macrophages produce an increase in lysosomal hydrolases not seen for indigestible material (Axline and Cohn, 1970; Allison and Davis, 1975; Schnyder and Baggiolini, 1980).

4.1.5. 5′-Nucleotidase

The activity of 5′-nucleotidase, an enzyme present in the plasma membrane of mononuclear phagocytes, is high in resident macrophages and is reduced in inflammatory mononuclear phagocytes in the mouse (Edelson and Cohn, 1976a). It has been suggested that the decreased activity of this enzyme in inflammatory macrophages is due to increased degradation.

During culture of mouse macrophages, the enzyme activity increases (Edelson and Cohn, 1976b). In human monocytes, Johnson *et al.* (1977) observed a more than tenfold increase in the activity of this enzyme over the first 2 days of culture in the presence of fresh autologous serum. No further increase in activity was found with longer cultivation. When human monocytes were cultured with less than 5% serum or without serum, 5′-nucleotidase activity decreased. After endocytosis, there was a transient decrease in the activity of this enzyme in both murine macrophages and human monocytes (Edelson and Cohn, 1976b; Johnson *et al.*, 1977).

4.1.6. Alkaline Phosphodiesterase I

The activity of alkaline phosphodiesterase I, a membrane-bound enzyme, is somewhat higher in thioglycollate-activated peritoneal macrophages than in unstimulated macrophages in the mouse (Edelson and Erbs, 1978). Very little change is observed during culture.

4.1.7. Aminopeptidase

Aminopeptidase is a plasma membrane-bound enzyme occurring in macrophages; granulocytes and lymphocytes are negative, brush-border membranes of the proximal tubule and small intestine are positive (Wachsmuth and Torhorst, 1974). In peritoneal exudates elicited by thioglycollate injection, the enzyme activity of the macrophages increased from the 2nd to 6th day after injection (Wachsmuth, 1975). In bone marrow cultures with conditioned medium, Wachsmuth and Staber (1977) found that the aminopeptidase activity increased from the 5th to the 13th day and then decreased. It is not clear from their data which cells among the bone marrow mononuclear phagocytes have the highest aminopeptidase activity.

4.1.8. Plasminogen Activator

Thioglycollate-induced peritoneal macrophages produce and secrete—immediately after harvesting, as well as during culture—a neutral proteinase that splits plasminogen into the fibrinolytic enzyme plasmin (Unkeless *et al.*, 1974).

The plasminogen activator is barely detectable in culture fluids of unstimulated macrophages, and very little enzyme is stored in the macrophage.

Gordon *et al.* (1974b) have shown that the synthesis and secretion of plasminogen activator require a second trigger with an endocytizable material after a preparatory stimulus. Macrophages activated by products of stimulated T lymphocytes secrete high levels of plasminogen activator (Nogueira *et al.*, 1977; Vassali and Reich, 1977; Klimetzek and Sorg, 1977; Gordon and Cohn, 1978). Transformed fibroblasts also secrete plasminogen activator (Unkeless *et al.*, 1974).

4.1.9. Collagenase

Macrophages from various species produce and secrete collagenase, i.e., a specific neutral metalloproteinase that cleaves native collagen (Salthouse *et al.*, 1972; Fujuwara *et al.*, 1973; Robertson *et al.*, 1973; Wahl *et al.*, 1974; Werb and Gordon, 1975a). The enzyme is not stored in the macrophage. Cultured resident macrophages secrete a minimal amount of this enzyme, whereas thioglycollate-induced macrophages secrete a large amount at a constant rate for more than 7 days in culture (Werb and Gordon, 1975a). Secretion of the enzyme during culture is stimulated by phagocytosis of latex particles, pinocytosis of dextran sulphate (Werb and Gordon, 1975a), or exposure *in vitro* to endotoxin or to lymphokines (Wahl *et al.*, 1974, 1975a). Thus, secretion of this enzyme is regulated in a way similar to that for the other neutral proteinases (see Sections 4.1.8, 4.1.10, and 4.1.11).

Other cell types, i.e., fibroblasts (Werb and Burleigh, 1974) and granulocytes (Lazarus *et al.*, 1972; Ohlsson and Olson, 1973) also have collagenase activity.

4.1.10. Elastase

This enzyme is a serine proteinase which lyses insoluble elastin fibrils. As for plasminogen activator and collagenase, Werb and Gordon (1975b) showed that the secretion of this enzyme by murine macrophages is regulated by the state of activation of the macrophage: after thioglycollate stimulation, elastase secretion is increased about 30-fold; during culture in serum-free medium an increasing amount of elastase is secreted, and latex phagocytosis also increases the elastase secretion to some extent. Only a small amount of elastase is stored within the cells. White *et al.* (1977) found that murine alveolar macrophages in the steady state secrete similar amounts of elastase as peritoneal macrophages stimulated by thioglycollate *in vivo* and latex particles *in vitro*. Human monocytes in culture secrete an enzyme that breaks down elastin; this enzyme resembles the corresponding murine enzyme (Werb and Gordon, 1975b).

The elastase secreted by macrophages differs from the pancreatic and lysosomal elastases by a stronger substrate specificity. The sensitivity to certain inhibitors also differs (Werb and Gordon, 1975b).

4.1.11. Other Natural Proteinases

Apart from plasminogen activator, collagenase, and elastase, macrophages *in vitro* produce and release other neutral proteinases that degrade proteoglycans in cartilage (Hauser and Vaes, 1978), basic protein in myelin (Cammer *et al.*, 1978), and proteins as azo casein and gelatin (Gordon *et al.*, 1976).

4.1.12. Procoagulant Activity

During *in vitro* culture of monocytes a procoagulant activity identical to tissue thromboplastin activity is produced (van Ginkel *et al.*, 1977). Various stimuli such as antigen–antibody complexes, mitogens, and endotoxin (Rothburger *et al.*, 1976; Lerner *et al.*, 1971; Rickles *et al.*, 1973; Rivers *et al.*, 1975), and also adherence to glass and endocytosis of bacteria, induce generation of this procoagulant activity (van Ginkel *et al.*, 1977, 1979).

4.1.13. Secretion of Other Substances by Mononuclear Phagocytes

Many kinds of substances besides enzymes are produced by macrophages in culture, but it is not known whether the individual mononuclear phagocyte produces one, several, or all substances at the same time. Various *complement* factors such as C1 (Stecher, 1970; Lai A Fat and van Furth, 1975), C2 (Wyatt *et al.*, 1972; Einstein *et al.*, 1976), C3 (Lai A Fat and van Furth, 1975; Brade *et al.*, 1977), C4 (Littleton *et al.*, 1970; Wyatt *et al.*, 1972), factor B (Bentley *et al.*, 1976), and factor D (Brade *et al.*, 1977) are produced by macrophages. *Transferrin* (Stecher, 1970) and *transcobalamin II* (Rachmilewitz and Schlesinger, 1977) are macrophage products.

Endogenous pyrogen is released *in vitro* by all kinds of mononuclear phagocytes, for instance human blood monocytes (Bodel, 1974), lung macrophages (Atkins *et al.*, 1967), and liver macrophages (Haeseler *et al.*, 1977), after a variety of stimuli (e.g., endotoxin, endocytosis). Spontaneous release of endogenous pyrogen does not occur during culture.

Recently, production of *angiotensin-converting enzyme* by mononuclear phagocytes in culture was demonstrated especially after exposure to dexamethasone (Friedland *et al.*, 1977, 1978).

Interferon is synthesized and released by noninfected macrophages; virus-infected macrophages produce more interferon (Smith and Wagner, 1967).

Several factors that regulate proliferation of mononuclear phagocyte precursors and other cells are produced by macrophages, and include the following.

Colony-stimulating factor (CSF), which acts on the mononuclear phagocyte precursors and granulocyte precursors, is a glycoprotein with a molecular weight of 70,000 (Stanley and Heard, 1977), and is identical to the macrophage growth factor (Stanley *et al.*, 1976). Mononuclear phagocytes in culture produce CSF (Golde *et al.*, 1972) as do, for instance, mouse embryonic fibroblasts (Goud, 1975), mouse L cells (Austin *et al.*, 1971), and some of the established macrophage cell lines (see Section 3.2.3).

The *factor increasing monocytopoiesis* (FIM) is a protein of mol. wt. about 20,000 that enhances the proliferation of promonocytes and also of monoblasts (van Waarde *et al.*, 1977a,b). This factor is present during inflammation, and is probably released by macrophages at the site of inflammation (production by other cells at the site of inflammation has not been completely ruled out). Production of this factor *in vitro* has not been shown yet.

Prostaglandins are produced by mouse macrophages and human monocytes *in vitro* (Bray *et al.*,1974; Humes *et al.*,1977; Glatt *et al.*,1977; Kurland and Bockman,1978). The production was shown to be enhanced by endotoxin or by endocytosis *in vitro*. Among many other actions, prostaglandins inhibit proliferation of precursor cells of the granulocyte and the mononuclear phagocyte series (Kurland and Moore,1977).

Macrophages in culture synthesize and release a factor that inhibits DNA synthesis and proliferation of other cells, especially lymphoid cells (Nelson, 1973; Waldman and Gottlieb, 1973; Calderon *et al.*, 1974). This factor proved to be *thymidine* (Opitz *et al.*, 1975; Stadecker *et al.*, 1977); the secretion is due to a lack of thymidine kinase activity in these cultured macrophages (Stadecker and Unanue, 1979).

Macrophages activated *in vitro* or *in vivo* produce and release *arginase*, which deprives the medium of arginine, and, thus, these macrophages also interfere with other cells [e.g., tumor cells (Kung *et al.*, 1977; Currie, 1978)].

Proliferation of fibroblasts (Leibovitch and Ross, 1976) and of endothelial cells (Polverini *et al.*, 1977) are stimulated by macrophage factors.

Other factors originating from macrophages and playing a role in the interaction with lymphocytes are briefly discussed in Section 4.2.5.

4.2. FUNCTIONAL PROPERTIES OF MONONUCLEAR PHAGOCYTES

4.2.1. Receptors of Mononuclear Phagocytes

In recent years, a number of receptors of mononuclear phagocytes have been described. Of these, the most widely studied are the receptors for the Fc fragment of IgG (see Silverstein *et al.*, 1977) and for the third component of the complement system (C3b, perhaps also C3d) (see Bianco and Nussenzweig, 1977) on the cell surface. Fc receptor on human monocytes has specificity for the subclasses IgG_1 and IgG_3 (Huber and Fudenberg, 1968). Recently, Unkeless (1977) described two distinct types of Fc receptor on mouse mononuclear phagocytes, one resistant to trypsin that mediates the binding and ingestion of IgG-coated particles and soluble antigen–antibody complexes, and the other sensitive to trypsin and binding monomeric mouse IgG_{2a}. The complement receptor is also trypsin sensitive (Lay and Nussenzweig, 1968). After pronase treatment, the number of complement receptors also decreases strongly, but during *in vitro* culture these receptors reappear (Crofton *et al.*, 1978; Blussé van Oud Alblas and van Furth, 1979). Complement receptors are resynthesized *in vitro* 6 hr after trypsinization (Bianco and Nussenzweig,1977).

In all stages of development, mononuclear phagocytes carry Fc and C receptors on their surface, and the number of cells carrying C3 receptors increases in the sequence monoblast–promonocyte–macrophage; whereas the number of cells with Fc receptors is more than 90% in all stages of development (Goud *et al.*, 1975; see Table 1). With longer cultivation *in vitro* (14 days), the percentage of cells in bone marrow cultures that carry C3 receptors increased (van der Meer *et al.*, 1980). This seems to be in contradiction with the findings of Rabinovitch and DeStefano (1973c), i.e., that peritoneal macrophages lose their C receptors after 24 hr when cultured in serum-free medium and after a week when cultured in medium containing 10% serum. As already mentioned in Section 4.2.11, only activated macrophages ingest via the C3b receptor (Bianco *et al.*, 1975).

Membrane receptors for nonopsonized particles [e.g., latex, effete cells, zymosan (Rabinovitch,1975; Griffin *et al.*,1975; Michl *et al.*,1976)] are present on the macrophage. The significance of receptors that mediate ingestion of activators of the alternative complement pathway (Czop *et al.*, 1978) and the lectin-like receptors (Weir and Ögmundsdottir 1980) is not yet clear. Other receptors of the macrophage are the receptors for lymphokines (Remold,1973: Fox *et al.*,1974) for such hormones as insulin (Schwartz *et al.*,1975), and glucocorticosteroids (Werb *et al.*, 1978b). The effect of *in vitro* culture on these receptors has not been established yet.

4.2.2. Endocytosis

The main function of mononuclear phagocytes is endocytosis, a function they share with granulocytes. Endocytosis includes pinocytosis, which is divided into macropinocytosis (the uptake of particles or fluid in vacuoles of diamter 0.1–1 μm) land micropinocytosis (vacuoles smaller than 0.1 μm diameter), and phagocytosis (i.e., uptake of particles with a diameter of about 1 μm or larger (Cohn, 1970; Allison and Davies,1975).

4.2.2a. Pinocytosis. Pinocytosis of serum constituents occurs normally during culturing of mononuclear phagocytes, and can be observed as the formation of vacuoles in the cytoplasm. Several molecular species stimulate pinocytosis, anionic molecules being more effective than neutral or cationic molecules (Cohn, 1970). For the study of pinocytosis by cells in culture, a non-degradable material such as dextran sulfate (mol. wt. 500,000) is very convenient, but other degradable substances [e.g., horseradish peroxidase (Steinman and Cohn, 1972)] can also serve (for a review, see Silverstein *et al.*, 1977). Within 24 hr of incubation with dextran sulfate, more than 90% of the macrophages have pinocytized (van Furth, 1978).

Promonocytes and monoblasts have lower pinocytic activity (Goud *et al.*, 1975). The uptake of horseradish peroxidase by peritoneal macrophages remained constant during 3 days of incubation (Steinman and Cohn, 1972). Macrophages activated *in vivo* by thioglycollate pinocytize at a higher rate than unstimulated cells (Edelson *et al.*, 1975).

4.2.2b. Phagocytosis. Phagocytosis can be shown for many kinds of particle—e.g., polystyrene latex beads, zymosan, bacteria, and yeast—with macrophages either in a monolayer (Michell *et al.*, 1969) or in suspension (Cohn and Morse, 1959; Mackaness, 1970; van Furth *et al.*, 1978). If particles are opsonized by IgG or complement, uptake is more efficient and a high percentage of the macrophages can be seen under the microscope to have phagocytized. Uptake of opsonized sheep erythrocytes via the Fc receptor is one of the best markers available at present to distinguish mononuclear phagocytes from mesenchymal and lymphoid cells. Nonstimulated peritoneal macrophages only ingest opsonized particles via the Fc receptors, whereas activated macrophages endocytize via their complement receptors as well (Bianco *et al.*, 1975; van Furth *et al.*,1979). During *in vitro* culture, the phagocytic capacity of resident peritoneal macrophages for opsonized bacteria increases (Baughn and Bonventre, 1975).

Endocytic activity increases in the sequence monoblasts–promonocytes–macrophages (Goud *et al.*, 1975; see Table 1).

4.2.3. Intracellular Killing

Intracellular killing of micro-organisms can be measured *in vitro* independent of the rate of phagocytosis (van Furth *et al.*, 1978). The most reliable way to perform these experiments with cultured cells is with the cells in suspension (e.g., by culturing them on a Teflon surface, see Section 2.1). The capacity of cultured cells from the unstimulated peritoneal cavity to kill staphylococci diminishes during culture (Baughn and Bonventre, 1975; van der Meer *et al.*, 1978). Proliferating bone marrow mononuclear phagocytes are capable of killing bacteria after 7 and 14 days of culture (van der Meer *et al.*, 1980).

For efficient killing of obligate and facultative intracellular microorganisms (e.g., *Listeria, Mycobacteria, Toxoplasma, Leishmania, Trypanosoma*), macrophages have to be activated either *in vivo* or *in vitro* (Mackaness, 1964; Hirsch *et al.*, 1974; Anderson and Remington, 1974; Mauel and Behin, 1974; Nogueira *et al.*, 1977). Activated macrophages also kill staphylococci, streptococci, and Gram-negative microorganisms more efficiently than unactivated macrophages do.

The biochemical events that lead to killing of microorganisms within granulocytes, such as oxidative metabolism and iodination (for a review, see Babior, 1978), are presumably important in mononuclear phagocytes as well. However, the relative role of oxidative metabolism and the contribution of lysosomal enzymes after formation of the phagolysosome are not precisely known for the latter type of cell (for reviews see Johnston, 1978; Goren, 1977).

With respect to oxygen metabolism, marked differences between resident macrophages and activated macrophages have been found (Karnovsky *et al.*, 1975; Nathan and Root, 1977; Johnston, 1978). For example, macrophages activated either *in vivo* or *in vitro* release an increased amount of O_2^- and H_2O_2 after phagocytosis (Johnston, 1978). Interestingly, less O_2 seems to be produced at a high than at lower densities of macrophages in culture. Human monocytes cultured for 3 days released more O_2^- after phagocytosis or surface pertubation than those cultured for one day (Johnston, 1978). This seems difficult to reconcile with the above-mentioned decreased microbicidal capacity of mononuclear phagocytes during culture.

4.2.4. Tumoricidal Effect

Mononuclear phagocytes play an important role in the defence against tumor cells. A number of investigators have studied the capacity of mononuclear phagocytes to kill tumor cells *in vitro* (Weaver, 1958; Granger and Weiser, 1966; Bennet, 1965; Evans, 1975; Lohman-Matthes and Fischer, 1975; Remington *et al.*, 1975). Normal peritoneal macrophages obtained from untreated animals are not capable of killing tumor cells *in vitro,* whereas macrophages activated either *in vivo*—by immunization or in a nonspecific way (e.g., by *Toxoplasma* or BCG infection)—or *in vitro* (e.g., with endotoxin), are cytotoxic for tumor cells, predominantly via a nonphagocytic mechanism. Activated macrophages do not destroy normal embryonic fibroblasts (Remington *et al.*, 1975).

Several mechanisms probably act together in the killing of tumor cells by macrophages, such as antibody-dependent cellular cytotoxicity and extracellular killing by products of oxygen metabolism (Nathan, 1979), the complement cleavage product C3a (Ferluga *et al.*,1978), lysosomal enzymes, and other factors released by macrophages (see, e.g., Section 4.1.13).

4.2.5. Lymphocyte Interaction with Macrophages in Culture

Macrophages, play an essential role in the complicated processes of T- and B-lymphocyte responses.

It has been shown that a specific lymphocyte–antigen interaction results in nonspecific macrophage activation, for example, in the handling of facultative intracellular microorganisms (Mackaness, 1964). For example, the addition of sensitized lymphocytes and antigens to macrophage cultures leads to enhanced bacteriostatic activity against *Listeria monocytogenes* (Sheagren *et al.*, 1975). When soluble factors produced by sensitized T lymphocytes—called MIF-rich fractions (Remold *et al.*, 1972)—are added to macrophages in culture, a variety of changes take place in these macrophages. Increased adherence and spreading, increased phagocytosis of certain particles, increased bacteriostasis of, for example, listeria and mycobacteria, and increased tumor-cell killing, as well as cytochemical and metabolic alterations have been reported (Nathan *et al.*,1971; Godal *et al.*, 1971; Fowles *et al.*, 1973; Piessens *et al.*, 1974; Turk and Poulter, 1975; Lazdens *et al.*,

1978). In contrast with experiments where sensitized lymphocytes and antigens are added directly to macrophages the changes become apparent after 2 days of culture in the presence of MIF-rich fractions (for a discussion of this phenomenon, see Sheagren *et al.*, 1975, and David, 1975). Macrophages have been shown to be essential in all kinds of assays of immune response *in vitro* (see below). Rosenthal and Shevach demonstrated the importance of the macrophage in the genetic control of the macrophage–T-cell interaction for the induction of an antigen-specific immune response *in vitro* (Rosenthal and Shevach, 1973; Shevach and Rosenthal, 1973). The I region products of the major histocompatibility complex were shown to be crucial in this interaction. A number of soluble mediators present in the supernatants of macrophage cultures, are able—at least to some extent—to replace macrophages in antibody production *in vitro* (Hoffman and Dutton, 1971), in the mixed lymphocyte reaction (Bach *et al.*, 1970), the enhancing of the viability of lymphocytes in culture (Chen and Hirsch, 1972), the generation of T-helper lymphocytes *in vitro* (Erb and Feldmann, 1975a,b), T-lymphocyte stimulation (Gery and Waksman, 1972; Calderon *et al.*, 1975), and lymphokine production (Wahl *et al.*, 1975b). Some of these effects are attributed to factors released by macrophages *in vitro*, such as the lymphocyte-activating factor (LAF) [mol. wt. 13,000 (Lachman *et al.*, 1977b)], nonspecific macrophage factors (NMF) (Erb and Feldmann, 1975b), and the genetically related macrophage factor (GRMF) (Erb *et al.*, 1976). The last of these factors is a complex comprising the I region-associated antigen of the H-2 complex (Ia) and immunogenic fragments with a molecular weight of 50,000–60,000 (for a review, see Erb *et al.*, 1980).

Some substances with an inhibitory effect on lymphoid cells are discussed in Section 4.1.13.

4.2.6. Chemokinesis and Chemotaxis

General methods for the study of chemokinesis (nondirectional movement of cells) and chemotaxis (directional movement of cells) are described in the literature. Most studies have dealt with granulocytes or with *non*cultured mononuclear phagocytes. Blussé van Oud Alblas *et al.* (1978), who used mononuclear phagocytes cultured on a Teflon surface (van der Meer *et al.*, 1978, 1979b), found that chemotaxis and chemokinesis can be studied very satisfactorily with the method described by Wilkinson (1974). This method permits the use of a relatively small number of cells (about 2×10^5) and a relatively small volume of attractant. The results are scored according to the leading front method (Zigmond and Hirsch, 1973). Blussé van Oud Alblas *et al.* (1978) found that nondirectional movement of human monocytes is maintained during at least 10 days of culture, but directional movement disappears.

5. CONCLUDING REMARKS

Studies on mononuclear phagocytes *in vitro* have deepened our insight into the role played by these cells in health and disease. The use of primary cultures or macrophage cell lines makes it possible to obtain pure populations of mononuclear phagocytes, which is crucial for many studies on inflammation and in immunology, where the contribution of the macrophage must be assessed. However, for a proper understanding of the role of the mononuclear phagocyte, it is also essential to determine the stage of development, as well as the state of activation of these cells.

As will be clear from this review, the number of markers available for assessment of the stage of development and the state of activation is increasing. Future research will have to establish which are the best markers in this respect. At present, the term activation of macrophages is poorly defined and loosely applied. The spectrum of activation has not been established yet and the underlying mechanisms are poorly understood.

ACKNOWLEDGMENTS. I want to express gratitude to Ralph van Furth, Martina Diesselhoff-den Dulk, and Evert Hensen for their critical remarks and to Juul Noomen for typing the manuscript.

REFERENCES

Allen, T. D., and Dexter, T. M., 1976, Cellular interrelationships during *in vitro* granulocytopoiesis, *Differentiation* **6**:191.

Allison, A. C., and Davies, P., 1975, Increased biochemical and biological activities of mononuclear phagocytes exposed to various stimuli, with special reference to secretion of lysosomal enzymes, in: *Mononuclear Phagocytes in Immunity, Infection and Pathology* (R. van Furth, ed.), pp. 487–504, Blackwell, Oxford.

Anderson, S. E., and Remington, J. S., 1974, Effect of normal and activated human macrophages on *Toxoplasma gondii*, *J. Exp. Med.* **139**:1154.

Atkins, E., Bodel, P., and Francis, L., 1967, Release of endogenous pyrogen *in vitro* from rabbit mononuclear cells, *J. Exp. Med.* **126**:357.

Austin, P. E., McCulloch, E. A., and Till, J. E., 1971, Characterization of the factor in L-cell conditioned medium capable of stimulating colony formation by mouse marrow cells in culture, *J. Cell Physiol.* **77**:121.

Axline, S. G., and Cohn, Z. A., 1970, *In vitro* induction of lysosomal enzymes by phagocytosis, *J. Exp. Med.* **131**:1239.

Axline, S. G., Simon, L. M., Robin, E. D., and Pesanti, E. L., 1980, Effects of oxygen on metabolic patterns of endocytosis of macrophages, in: *Mononuclear Phagocytes - Functional Aspects* (R. van Furth, ed.), pp. 1247–1270, Martinus Nijhoff, The Hague.

Babior, B. M., 1978, Oxygen dependent microbial killing by phagocytes. I, *N. Engl. J. Med.* **298**:659.

Bach, F. H., Alter, B. J., Solliday, S., Zoschke, D. C., and Janis, M., 1970, Lymphocyte reactivity *in vitro*. II. Soluble reconstituting factor permitting response of purified lymphocytes, *Cell. Immunol.* **1**:2194.

Bainton, D. F., and Golde, D. W., 1978, Differentiation of macrophages from normal human bone marrow in liquid culture: Electron microscopy and cytochemistry, *J. Clin. Invest.* **61**:1555.

Baker, L. E., 1933, The culture of monocytes in fluid medium, *J. Exp. Med.* **58**:575.

Barka, T., and Anderson, P. J., 1962, Histochemical methods for acid phosphatase using hexazonium pararosanilin as coupler, *J. Histochem. Cytochem.* **10**:741.

Baughn, R. E., and Bonventre, P. F., 1975, Phagocytosis and intracellular killing of *Staphylococcus aureus* by normal mouse peritoneal macrophages, *Infect. Immun.* **12**:346.

Beelen, R. H. J., Broekhuis-Fluitsma, D. M., Kom, C., and Hoefsmit, E. C. M., 1978a, Identification of exudate-resident macrophages on the basis of peroxidatic activity, *J. Reticuloendothel. Soc.* **23**:103.

Beelen, R. H. J., van't Veer, M., Fluitsma, D. M., and Hoefsmit, E. C. M., 1978b, Identification of different peroxidatic activity patterns in human macrophages *in vivo* and *in vitro*, *J. Reticuloendothel. Soc.* **24**:355.

Bennet, B., 1965, Specific suppression of tumor growth by isolated peritoneal macrophages from immunized mice, *J. Immunol.* **95**:656.

Bennet, B., 1966, Isolation and cultivation *in vitro* of macrophages from various sources in the mouse, *Am. J. Pathol.* **48**:165.

Bennet, W. E., and Cohn, Z. A., 1966, The isolation and selected properties of blood monocytes, *J. Exp. Med.* **123**:145.

Bentley, C., Bitter-Suermann, D., Hadding, U., and Brade, V., 1976, *In vitro* synthesis of factor B of the alternative pathway of complement activation by mouse peritoneal macrophages, *Eur. J. Immunol.* **6**:393.

Bianco, C., and Nussenzweig, V., 1977, Complement receptors, in: *Contemporary Topics in Molecular Immunology* (R. R. Porter and G. L. Ada, eds.), pp. 145–175, Plenum Press, New York.

Bianco, C., Griffin, F. M., Jr., and Silverstein, S. C., 1975, Studies on the macrophage complement receptor. Alteration of receptor function upon macrophage activation, *J. Exp. Med.* **141**:1278.

Bianco, C., Eden, E., and Cohn, Z. A., 1976, The induction of macrophage spreading: role of coagulation factors and the complement system, *J. Exp. Med.* **144**:1531.

Bissel, D. M., Hammaker, L., and Schmid, R., 1972, Liver sinusoidal cells. Identification of a subpopulation for erythrocyte catabolism, *J. Cell Biol.* **54**:107.

Bloom, B. R., Diamond, B., Muschel, R., Rosen, N., Schnede, J., Dimiani, G., Rosen, O., and Scharff, M., 1978, Genetic approaches to the mechanisms of macrophage function, *Fed. Proc.* **37**:2765.

Blussé van Oud Alblas, A., and van Furth, R., 1979, The origin, kinetics and characteristics of pulmonary macrophages in the normal steady state, *J. Exp. Med.* **149**:1504.

Blussé van Oud Alblas, A., van der Meer, J. W. M., van de Gevel, J. S., and Matze-van der Lans, A., 1978, unpublished observations.

Bodel, P., 1974, Studies on the mechanisms of endogenous pyrogen production. III. Human blood monocytes, *J. Exp. Med.* **140**:954.

Bodel, P. T., 1978, Spontaneous pyrogen production by mouse histiocytic and myelomonocytic tumor cell lines *in vitro*, *J. Exp. Med.* **147**:1503.

Bodel, P. T., Nichols, B. A., and Bainton, D. F., 1977, Appearance of peroxidase reactivity within the rough endoplasmic reticulum of blood monocytes after surface adherence, *J. Exp. Med.* **145**:264.

Bodel, P. T., Nichols, B. A., and Bainton, D. F., 1978, Differences in peroxidase localisation of rabbit peritoneal macrophages after surface adherence, *Am. J. Pathol.* **91**:107.

Bonney, R. J., Gery, I., Lin, T-Y, Meyenhofer, M. F., Acevedo, W., and Davies, P., 1978, Mononuclear phagocytes from carrageenan-induced granulomas. Isolation, cultivation and characterization. *J. Exp. Med.* **148**:261.

Böyum, A., 1968, Isolation of mononuclear cells and granulocytes from human blood, *Scand. J. Clin. Lab. Invest.* **21** (Suppl. 97):77.

Bradley, T. R., and Metcalf, D., 1966, The growth of mouse bone marrow cells *in vitro*, *Austral. J. Exp. Biol. Med. Sci.* **44**:287.

Bradley, T. R., Hodgson, G. S., and Rosendaal, M., 1977, The effect of different oxygen tensions in haemopoietic cell cultures, *Exp. Hematol.* **5** (Suppl. 40).

Braunsteiner, H., and Schmalzl, F., 1970, Cytochemistry of monocytes and macrophages, in: *Mononuclear phagocytes* (R. van Furth, ed.), pp. 62–81, Blackwell, Oxford.

Bray, M. A., Gordon, D., and Morley, J., 1974, Role of prostaglandins in reactions of cellular immunity, *Br. J. Pharmacol.* **52**:453.

Calderon, J., Williams, R. T., and Unanue, E. R., 1974, An inhibitor of cell proliferation released by cultures of macrophages, *Proc. Natl. Acad. Sci. USA* **71**:4273.

Calderon, J., Kiely, J. M., Lefko, J. L., and Unanue, E. R., 1975, The modulation of lymphocyte functions by molecules secreted by macrophages. I. Description and partial biochemical analysis, *J. Exp. Med.* **142**:151.

Cammer, W., Bloom, B. R., Norton, W. T., and Gordon, S., 1978, Degradation of basic protein in myelin by proteases secreted by activated macrophages: A possible mechanism of inflammatory demyelination, *Proc. Natl. Acad. Sci. USA* **75**:1554.

Chen, C., and Hirsch, J. G., 1972, The effects of mercapto-ethanol and of peritoneal macrophages on the antibody-forming capacity of nonadherent mouse spleen cells *in vitro, J. Exp. Med.* **136**:604.

Chu, J., and Lin, H. S., 1976, Induction of macrophage colony forming cells in pleural cavity, *J. Reticuloendothel. Soc.* **20**:299.

Cline, M. J., and Sumner, M. A., 1972, Morphology, proliferation and some functional characteristics of the early cells of the mouse macrophage series, *Blood* **40**:62.

Cohn, Z. A., 1970, Endocytosis and intracellular digestion, in: *Mononuclear phagocytes* (R. van Furth, ed.), pp. 121–129, Blackwell, Oxford.

Cohn, Z. A., and Benson, B., 1965a, The differentiation of mononuclear phagocytes: Morphology, cytochemistry and biochemistry, *J. Exp. Med.* **121**:153.

Cohn, Z. A., and Benson, B., 1965b, The *in vitro* differentiation of mononuclear phagocytes. II. The influence of serum on granule formation, hydrolase production and pinocytosis, *J. Exp. Med.* **121**:835.

Cohn, Z. A., and Morse, S. I., 1959, Interactions between rabbit polymorphonuclear leucocytes and staphylococci, *J. Exp. Med.* **110**:419.

Crofton, R. W., Diesselhoff-den Dulk, M. M. C., and van Furth, R., 1978, The origin, kinetics and characteristics of the Kupffer cells in the normal steady state, *J. Exp. Med.* **148**:1.

Currie, G. A., 1978, Activated macrophages kill tumour cells by releasing arginase, *Nature* **273**:758.

Czop, J. K., Fearon, D. T., and Austen, K. F., 1978, Opsonin-independent phagocytosis of activators of the alternative complement pathway by human monocytes, *J. Immunol.* **120**:1132.

Daems, W. T., Wisse, E., Brederoo, P., and Emeis, J. J., 1975, Peroxidatic activity in monocytes and macrophages, in: *Mononuclear Phagocytes in Immunity, Infection and Pathology* (R. van Furth, ed.), pp. 57–77, Blackwell, Oxford.

David, J. R., 1975, Macrophage activation by lymphocyte mediators, *Fed. Proc.* **34**:1730.

Defendi, V., 1976, Macrophage cell lines and their uses in immunobiology, in: *Immunobiology of the Macrophage* (D. Nelson, ed.), p. 275, Academic Press, New York.

Dexter, T. M., Allen, T. D., and Lajtha, L. G., 1977, Conditions controlling the proliferation of haematopoietic stem cells *in vitro, J. Cell. Physiol.* **91**:335.

Diamond, B., Bloom, B. R., and Scharff, M., 1978, The Fc-receptors of primary and cultured phagocytic cells studied with homogeneous antibodies, *J. Immunol.* **121**:1329.

Dicke, K. A., Platenburg, M. G. C., and van Bekkum, D. W., 1971, Colony formation in agar: *In vitro* assay for haematopoietic stem cells, *Cell Tissue Kinet.* **4**:463.

Diesselhoff-den Dulk, M. M. C., Crofton, R. W., and van Furth, R., 1979, Origin and kinetics of Kupffer cells during an acute inflammatory response, *Immunology* **37**:7.

Edelson, P. J., and Cohn, Z. A., 1976a, 5′-nucleotidase activity of mouse peritoneal macrophages. I. Synthesis and degradation in resident and inflammatory populations, *J. Exp. Med.* **144**:1581.

Edelson, P. J., and Cohn, Z. A., 1976b, 5′-nucleotidase activity of mouse peritoneal macrophages. II. Cellular distribution and effects of endocytosis, *J. Exp. Med.* **144**:1596.

Edelson, P. J., and Erbs, C., 1978. Plasma membrane localization and metabolism of alkaline phosphodiesterase I in mouse peritoneal macrophages, *J. Exp. Med.* **147**:77.

Edelson, P. J., Zwiebel, R., and Cohn, Z. A., 1975, The pinocytic rate of activated macrophages, *J. Exp. Med.* **142**:1150.

Einstein, L. P., Schneeberger, E. E., and Colten, H. R., 1976, Synthesis of the second component of complement by long term primary cultures of human monocytes, *J. Exp. Med.* **143**:114.

Emeis, J. J., and Planqué, B., 1976, Heterogeneity of cells isolated from rat liver by pronase digestion: Ultrastructure, cytochemistry, and cell culture, *J. Reticuloendothel. Soc.* **20**:11.

Erb, P., and Feldmann, M., 1975a, Role of macrophages in *in vitro* induction of T helper cells, *Nature (London)* **254**:352.

Erb, P., and Feldmann, M., 1975b, The role of macrophages in the generation of T helper cells. III. Influence of macrophage derived factors in helper cell induction, *Eur. J. Immunol.* **5**:759.

Erb, P., Feldmann, M., and Hogg, N., 1976, The role of macrophages in the generation of T helper cells. IV. Nature of genetically related factor derived from macrophages incubated with soluble antigens, *Eur. J. Immunol.* **6**:365.

Erb, P., Feldmann, M., Gisler, R., Meier, B., Mueller, A., and Vogt, P., 1980, Role of macrophages in the *in vitro* induction and regulation of antibody responses, in: *Mononuclear Phagocytes—Functional Aspects* (R. van Furth, ed.), pp. 1857–1886, Martinus Nijhoff, The Hague.

Evans, R., 1975, Macrophage cytotoxicity, in: *Mononuclear Phagocytes in Immunity, Infection and Pathology* (R. van Furth, ed.), pp. 827–841, Blackwell, Oxford.
Ferluga, J., Schorlemmer, H. U., Baptista, L. C., and Allison, A. C., 1978, Production of the complement cleavage product, C3a, by activated macrophages and its tumorolytic effects, *Clin. Exp. Immunol.* **31**:512.
Fowles, R. E., Fajardo, I. M., Leibowitch, J. L., and David, J. R., 1973, The enhancement of macrophage bacteriostasis by products of activated lymphocytes, *J. Exp. Med.* **138**:952.
Fox, R. A., Gregory, D. G., and Feldman, J. D., 1974, Macrophage receptors for migration inhibitory factor (MIF), migration stimulating factor (MSF) and agglutinating factor, *J. Immunol.* **112**:1867.
Friedland, J., Setton, C., and Silverstein, E., 1977, Angiotensin converting enzyme: Induction by steroids in rabbit alveolar macrophages in culture, *Science* **197**:64.
Friedland, J., Setton, C., and Silverstein, E., 1978, Induction of angiotensin converting enzyme in human monocytes in culture, *Biochem. Biophys. Res. Comm.* **83**:843.
Fujuwara, J., Sakai, T., Oda, R., and Igaraski, S., 1973, The presence of collagenase in Kupffer cells of the rat liver, *Biochem. Biophys. Res. Comm.* **54**:531.
Garvey, J. S., 1961, Separation and *in vitro* culture of cells from liver tissue, *Nature (London)* **191**:972.
Gery, I., and Waksman, B. H., 1972, Potentiation of T lymphocyte response to mitogens. II. The cellular source of potentiating mediators, *J. Exp. Med.* **136**:143.
Glatt, M., Kälin, H., Wagner, K., and Brune, K., 1977, Prostaglandin release from macrophages. An assay system for antiinflammatory drugs *in vitro*, *Agents Actions* **7**:321.
Glynn, A. A., and Parkman, R., 1964, Studies with an antibody to rat lysozyme, *Immunology* **7**:724.
Godal, T., Rees, R. J. W., and Lamvik, J. O., 1971, Lymphocyte-mediated modification of blood-derived macrophage function *in vitro:* Inhibition of growth of intracellular mycobacteria with lymphokines, *Clin. Exp. Immunol.* **8**:625.
Godleski, J., and Brain, J., 1972, The origin of alveolar macrophages in mouse radiation chimeras, *J. Exp. Med.* **136**:630.
Golde, D. W., and Cline, M. J., 1973, Growth of human bone marrow in liquid culture, *Blood* **41**:45.
Golde, D. W., Finley, T. N., and Cline, M. J., 1972, Production of colony-stimulating factor by human macrophages, *Lancet* **2**:1397.
Gordon, S., and Cohn, Z. A., 1978, Bacillus Calmette-Guérin infection in the mouse. Regulation of macrophage plasminogen activator by T lymphocytes and specific antigen, *J. Exp. Med.* **147**:1175.
Gordon, S., Todd, J., and Cohn, Z., 1974a, *In vitro* synthesis and secretion of lysozyme by mononuclear phagocytes, *J. Exp. Med.* **139**:1228.
Gordon, S., Unkeless, J., and Cohn, Z. A., 1974b, Induction of plasminogen activator by endotoxin stimulation and phagocytosis. Evidence for a two stage process. *J. Exp. Med.* **140**:995.
Gordon, S., Werb, Z., and Cohn, Z. A., 1976, Methods for detection of macrophage secretory enzymes, in: *In Vitro Methods in Cell Mediated and Tumor Immunity* (B. Bloom and J. Davids, eds.), pp. 341–352, Academic Press, New York.
Goren, M. B., 1977, Phagocyte lysosomes: Interactions with infectious agents, phagosomes, and experimental pertubations in function, *Annu. Rev. Microbiol.* **31**:507.
Götze, O., Bianco, C., Sundsmo, J. S., and Cohn, Z. A., 1980. The stimulation of mononuclear phagocytes by components of the classical and the alternative pathways of complement activation, in: *Mononuclear Phagocytes—Functional Aspects* (R. van Furth, ed.), pp. 1421–1442, Martinus Nijhoff, The Hague.
Goud, T. J. L. M., 1975, Identification and characterization of the monoblast, Ph.D. Thesis, University of Leiden, The Netherlands.
Goud, T. J. L. M., and van Furth, R., 1975, Proliferative characteristics of monoblasts grown *in vitro*, *J. Exp. Med.* **142**:1200.
Goud, T. J. L. M., Schotte, C., and van Furth, R., 1975, Identification and characterization of the monoblast in mononuclear phagocyte colonies grown *in vitro*, *J. Exp. Med.* **142**:1180.
Granger, G. A., and Weiser, R. S., 1966, Homograft target cells: Contact destruction *in vitro* by immune macrophages, *Science* **151**:97.
Griffin, F. M., Jr., Griffin, J. A., Leider, J. E., and Silverstein, S. C., 1975, Studies on the mechanism of phagocytosis. I. Requirement for circumferential attachment of particle-bound ligands to specific receptors on the macrophage plasma membrane, *J. Exp. Med.* **142**:1263.

Guilbert, L. J., and Iscove, N. N., 1976, Partial replacement of serum by selenite transferrin, albumin and lecithin in haemopoietic cell cultures, *Nature (London)* **263**:594.

Habeshaw, J. A., 1972, A serum-protein-free medium for the culture of macrophages and related cells, *J. Pathol.* **108**:95.

Haeseler, F., Bodel, P., and Atkins, E., 1977, Characteristics of pyrogen production by isolated rabbit Kupffer cells *in vitro, J. Reticuloendothel. Soc.* **22**:569.

Hamilton, J. A., Ralph, P., and Moore, M. A. S., 1978, A macrophage tumor cell line and plasminogen activator: A potential model system for macrophage regulation of enzyme production, *J. Exp. Med.* **148**:811.

Hauser, P., and Vaes, G., 1978, Degradation of cartilage proteoglycans by a neutral proteinase secreted by rabbit bone marrow macrophages in culture, *Biochem. J.* **172**:275.

Heise, E. R., and Myrvik, Q. N., 1967, Secretion of lysozyme by rabbit alveolar macrophages *in vitro, J. Reticuloendothel. Soc.* **4**:510.

Hirsch, J. G., Jones, T. C., and Len, L., 1974, Interactions *in vitro* between *Toxoplasma gondii* and mouse cells. Parasites in the immunized host: Mechanisms of survival, *Ciba Found. Symp.* **25**:205.

Hoff, R., 1975, Killing *in vitro* of *Trypanosoma cruzi* by macrophages from mice immunized with *T. cruzi* or BCG, and absence of cross immunity on challenge *in vivo. J. Exp. Med.* **142**:299.

Hoffmann, M., and Dutton, R. W., 1971, Immune response restoration with macrophage culture supernatants, *Science* **172**:1047.

Holdsworth, S. R., Thomson, N. M., Glasgow, E. E., Dowling, J. P., and Atkins, R. C., 1978, Tissue culture of isolated glomeruli in experimental crescentic glomerulonephritis, *J. Exp. Med.* **147**:98.

Huber, H., and Fudenberg, H. H., 1968, Receptor sites of human monocytes for IgG, *Int. Arch. Allergy* **34**:18.

Humes, J. L., Bonney, R. G., Pelus, L., Daulgren, M. E., Kuehl, F. A., and Davies, P., 1977, Macrophages synthesize and release prostaglandins in response to inflammatory stimuli, *Nature (London)* **269**:149.

Johnson, W. D., Mei, B., and Cohn, Z. A., 1977, The separation, long term cultivation, and maturation of the human monocyte, *J. Exp. Med.* **146**:1613.

Johnston, R. B., Jr., 1978, Oxygen metabolism and the microbicidal activity of macrophages, *Fed. Proc.* **37**:2759.

Johnston, R. B., Jr., Godzik, C. A., and Cohn, Z. A., 1978, Increased superoxide anion production by immunologically activated and chemically elicited macrophages, *J. Exp. Med.* **148**:115.

Karnovsky, M. L., Lazdius, J., Drath, D., and Harper, A., 1975, Biochemical characteristics of activated macrophages, *Ann. N.Y. Acad. Sci.* **257**:266.

Keller, H. U., Wilkinson, P. C., Abercrombie, M., Becher, E. L., Hirsch, J. G., Miller, M. E., Scott Ramsey, W., and Zigmond, S. H., 1977, A proposal for the definition of terms related to locomotion of leucocytes and other cells, *Clin. Exp. Immunol.* **27**:377.

Klimetzek, V., and Sorg, C., 1977, Lymphokine induced secretion of plasminogen activator by murine macrophages, *Eur. J. Immunol.* **7**:185.

Knook, D. L., and Sleyster, E. C., 1976, Separation of Kupffer and endothelial cells of the rat liver by centrifugal elutriation, *Exp. Cell Res.* **99:444**.

Knowles, D. M., Hoffman, H. T., Ferrarini, M., and Kunkel, H. G., 1978, The demonstration of acid alpha-naphthyl acetate esterase activity in human lymphocytes: Usefulness as a T cell marker, *Cell. Immunol.* **35**:112.

Koren, H. S., and Hodes, R. J., 1977, Effect of tumor cells on the generation of cytotoxic T lymphocytes *in vitro.* I. accessory cell functions of mouse tumor cells in the generation of cytotoxic T lymphocytes *in vitro, Eur. J. Immunol.* **7**:394.

Koren, H. S., Handwerger, B. S., and Wunderlich, J. R., 1975, Identification of macrophage like characteristics in a cultured murine tumor line, *J. Immunol.* **114**:894.

Kung, J. T., Brooks, S. B., Jackway, J. P., Leonard, L. L., and Talmage, D. W., 1977, Suppression of *in vitro* cytotoxic response by macrophages due to induced arginase, *J. Exp. Med.* **146**:665.

Kurland, J. I., and Bockman, R., 1978, Prostaglandin E production by human blood monocytes and mouse macrophages, *J. Exp. Med.* **147**:952.

Kurland, J. I., and Moore, M. A. S., 1977, Regulatory role of the macrophage in normal and neoplastic hemopoiesis, in: *Experimental Hematology Today* (S. Baum and G. Ledney, eds.), pp. 51–62, Springer Verlag, New York.

Kurland, J. I., Kincade, P. W., and Moore, M. A. S., 1977, Regulation of B-lymphocyte clonal proliferation by stimulatory and inhibitory macrophage-derived factors, *J. Exp. Med.* **146**:1420.

Lachman, L. B., Hacker, M. P., Blyden, G. T., and Handschumacher, R. E., 1977a, Preparation of lymphocyte activating factor from continuous murine macrophage cell lines, *Cell. Immunol.* **34**:416.

Lachman, L. B., Hacker, M. P., and Handschumacher, R. E., 1977b, Partial purification of human lymphocyte activating factor (LAF) by ultrafiltration and electrophoretic techniques, *J. Immunol.* **119**:2019.

Lai A Fat, R. F. M., and van Furth, R., 1975, *In vitro* synthesis of some complement components (C1q, C3 and C4) by lymphoid tissues and circulating leucocytes in man, *Immunology* **28**:359.

Lay, W. H., and Nussenzweig, V., 1968, Receptors for complement on leucocytes, *J. Exp. Med.* **128**:991.

Lazarus, G. S., Daniels, J. R., Lian, J., and Burleigh, M. C., 1972, Role of granulocyte collagenase in collagen degradation, *Am. J. Pathol.* **68**:565.

Lazdins, J. K., Külmer, A. L., David, J. R., and Karnovsky, M. L., 1978, Alteration of some functional and metabolic characteristics of resident mouse peritoneal macrophages by lymphocyte mediators, *J. Exp. Med.* **148**:746.

Leibovitch, S. J., and Ross, R., 1976, A macrophage dependent factor that stimulates the proliferation of fibroblasts *in vitro*, *Am J. Pathol.* **84**:501.

Lerner, R. G., Goldstein, R., and Cummings, C., 1971, Stimulation of human leucocyte thromboplastin activity by endotoxin, *Proc. Soc. Exp. Biol. Med.* **138:**145.

Lewis, W. H., and Gey, G. O., 1923, Clasmatocytes and tumor cells in cultures of mouse sarcoma, *Johns Hopkins Hosp. Bull.* **34:**269.

Lin, H. S., 1974, Peritoneal exudate cells. II. Kinetics of appearance of colony forming cells, *J. Cell. Physiol.* **84**:159.

Lin, H. S., 1977, Colony formation *in vitro* by mouse blood monocytes, *Blood* **49**:593.

Lin, H. S., and Stewart, C. C., 1974, Peritoneal exudate cells. I. Growth requirement of cells capable of forming colonies in soft agar, *J. Cell. Physiol.* **83**:369.

Lin, H. S., Kuhn, C., and Kuo, T. T., 1975, Clonal growth of hamster free alveolar cells in soft agar, *J. Exp. Med.* **142**:877.

Lin, H. S., Kuhn, C., and Stewart, C. C., 1978, Peritoneal exudate cells. V. Influence of age, sex, strain and species on the induction and the growth of macrophage colony forming cells, *J. Cell. Physiol.* **96**:133.

Littleton, C., Kessler, D., and Burkholder, P. M., 1970, Cellular basis for synthesis of the fourth component of guinea pig complement as determined by a haemolytic plaque technique, *Immunology* **18**:693.

Lohman-Matthes, M-L., and Fischer, H., 1975, Macrophage mediated cytotoxic induction by a specific T cell factor, in: *Mononuclear phagocytes in Immunity, Infection and Pathology* (R. van Furth, ed.), pp. 845–854, Blackwell, Oxford.

Loos, H., Blok-Schut, B., van Doorn, R., Hoksbergen, R., Brutel de la Riviere, A., and Meerhof, L., 1976, A method for the recognition and separation of human blood monocytes on density gradients, *Blood* **48**:731.

Lorbacher, P., Yam, L. T., and Mitus, W. J., 1967, Cytochemical demonstration of β glucuronidase activity in blood and bone marrow cells, *J. Histochem. Cytochem.* **15**:680.

Mackaness, G. B., 1960, Phagocytosis and inactivation of staphylococci by macrophages of normal rabbits, *J. Exp. Med.* **112**:35.

Mackaness, G. B., 1964, The immunological basis of acquired cellular resistance, *J. Exp. Med.* **120**:105.

MacVittie, T. J., and McCarthy, K. F., 1977, The detection of *in vitro* monocyte macrophage colony-forming cells in mouse thymus and lymph nodes, *J. Cell. Physiol.* **92**:203.

Mangornkanok, M., Markowitz, A. S., and Esterly, J. R., 1975, Establishment of replicating long term lines of rabbit macrophages and lymphocytes, *J. Immunol. Methods* **7**:327.

Marbrook, J., 1967, Primary immune response in cultures of spleen cells, *Lancet* **2**:1279.

Mariano, M., and Spector, W. G., 1974, The formation and properties of macrophage polykaryons (inflammatory giant cells), *J. Pathol.* **113**:1.

Mauel, J., and Behin, R., 1974, Cell mediated and humoral immunity to protozoan infections (with special reference to leishmaniasis), *Transplant Rev.* **19**:121.

McClelland, D. B. L., and van Furth, R., 1975, *In vitro* synthesis of lysozyme by human and mouse tissues, *Immunology* **28**:1099.

Meltzer, M. S., and Oppenheim, J. J., 1977, Bidirectional amplification of macrophage-lymphocyte interactions: Enhanced lymphocyte activation factor production by activated adherent mouse peritoneal cells, *J. Immunol.* **118**:77.

Michell, R. H., Pancake, S. J., Noseworthy, J., and Karnovsky, M. L., 1969, Measurement of rates of phagocytosis: The use of cellular monolayers, *J. Cell Biol.* **40**:216.

Michl, J., Ohlbaum, D. J., and Silverstein, S. C., 1976, 2-deoxyglucose selectively inhibits Fc and complement receptor mediated phagocytosis in mouse peritoneal macrophages, *J. Exp. Med.* **144**:1484.

Mills, D. M., and Zucker-Franklin, D., 1969, Electron microscopic study of isolated Kupffer cells, *Am. J. Pathol.* **54**:147.

Mueller, J., Brundel, R. G., Buerki, H., Keller, H-U, Hess, M. W., and Cottier, H., 1975, Non-specific acid esterase activity: A criterion for differentiation of T and B lymphocytes in mouse lymph nodes, *Eur. J. Immunol.* **5**:270.

Munder, P. G., Modolell, M., and Wallach, D. F. H., 1971, Cell propagation on films of polymeric fluorocarbon as a means to regulate pericellular pH and pO_2 in cultured monolayers, *FEBS Let.* **15**:191.

Munthe-Kaas, A. C. T., Berg, T., Seglen, P. O., and Seljelid, R., 1975, Mass isolation and culture of rat Kupffer cells, *J. Exp. Med.* **141**:1.

Muschel, R. J., Rosen, N., Rosen, O. M., and Bloom, B. R., 1977, Modulation of Fc-mediated phagocytosis by cyclic AMP and insulin in a macrophage-like cell line, *J. Immunol.* **119**:1813.

Nathan, C. F., 1980, The release of hydrogen peroxide from mononuclear phagocytes and its role in extracellular cytolysis, in: *Mononuclear Phagocytes—Functional Aspects* (R. van Furth, ed.), pp. 1165–1186, Martinus Nijhoff, The Hague.

Nathan, C. F., and Root, R. K., 1977, Hydrogen peroxide release from mouse peritoneal macrophages: Dependence on sequential activation and triggering, *J. Exp. Med.* **146**:1648.

Nathan, C. F., Karnovsky, M. L., and David, J. R., 1971, Alterations of macrophage functions by mediators from lymphocytes, *J. Exp. Med.* **133**:1356.

Nelson, D. S., 1973, Production by stimulated macrophages of factors depressing lymphocyte transformation, *Nature* **246**:306.

Nogueira, N., and Cohn, Z. A., 1978, *Trypanosoma cruzi: In vitro* induction of macrophage microbicidal activity, *J. Exp. Med.* **148**:288.

Nogueira, N., Gordon, S., and Cohn, Z. A., 1977, *Trypanosoma cruzi:* The immunological induction of macrophage plasminogen activator requires thymus-derived lymphocytes, *J. Exp. Med.* **146**:172.

North, R. J., 1969a, Cellular kinetics associated with the development of acquired cellular resistance, *J. Exp. Med.* **130**:299.

North, R. J., 1969b, The mitotic potential of fixed phagocytes in the liver as revealed during the development of cellular immunity, *J. Exp. Med.* **130**:315.

Ohlsson, K., and Olsson, I., 1973, The neutral proteinases of human granulocytes: Isolation and partial characterization of two granulocyte collagenases, *Eur. J. Biochem.* **36**:474.

Opitz, H. G., Niethammer, D., Jackson, R. C., Lemke, H., Huget, R., and Flad, H. D., 1975, Biochemical characterization of a factor released by macrophages, *Cell. Immunol.* **18**:70.

Ornstein, L., Ansley, H., and Saunders, A., 1976, Improving manual differential white cell counts with cytochemistry, *Blood Cells* **2**:557.

Papadimitriou, J. M., Sporsina, D., and Papelias, L., 1973, Kinetics of multinucleate giant cell formation and their modification by various agents in foreign body reactions, *Am. J. Pathol.* **73**:349.

Parker, R. C., 1932, The races that constitute the group of common fibroblasts. I. The effect of blood plasma, *J. Exp. Med.* **55**:713.

Paul, J., 1972, *Cell and Tissue Culture,* Churchill Livingstone, Edinburgh.

Pierce, C. W., 1969, Immune response *in vitro.* I. Cellular requirements for the immune response by non-primed and primed spleen cells *in vitro, J. Exp. Med.* **130**:345.

Piessens, W. F., Churchill, W. H., and David, J. R., 1974, Non-specific tumor killing by macrophages activated *in vitro* with lymphocyte mediators, *Fed. Proc.* **33**:781.

Polverini, P. J., Cotran, R. S., Gimbrone, M. A., and Unanue, E. R., 1977, Activated macrophages induce vascular proliferation, *Nature (Lond)* **269**:804.

Pluznik, D. H., and Sachs, L., 1965, The cloning of normal mast cells in tissue culture, *J. Cell. Comp. Physiol.* **66**:319.

Rabinovitch, M., 1975, Macrophage spreading *in vitro*, in: *Mononuclear Phagocytes in Infection, Immunity and Pathology* (R. van Furth, ed.), pp. 369–383, Blackwell, Oxford.

Rabinovitch, M., and DeStefano, M. J., 1973a, Macrophage spreading *in vitro*. I. Inducers of spreading, *Exp. Cell Res.* **77**:323.

Rabinovitch, M., and DeStefano, M. J., 1973b, Macrophage spreading *in vitro*. II. Manganese and other metals as inducers or as cofactors for induced spreading, *Exp. Cell Res.* **79**:423.

Rabinovitch, M., and DeStefano, M. J., 1973c, Particle recognition by cultivated macrophages, *J. Immunol.* **110**:695.

Rabinovitch, M., and DeStefano, M. J., 1975, Use of the local anesthetic lidocaine for cell harvesting and subcultivation, *In Vitro* **11**:379.

Rabinovitch, M., Manejias, R. E., Russo, M., and Abbey, E. E., 1977, Increased spreading of macrophages from mice treated with interferon inducers, *Cell. Immunol.* **29**:86.

Rachmilewitz, M., and Schlesinger, M., 1977, Production and release of transcobalamin, II. A vitamin B12 transport protein by mouse peritoneal macrophages, *Exp. Hemat.* **5** (Suppl. 2):108.

Ralph, P., 1980, Functions of macrophage cell lines, in: *Mononuclear Phagocytes—Functional Aspects* (R. van Furth, ed.), pp. 439–456, Martinus Nijhoff, The Hague.

Ralph, P., and Nakoinz, I., 1977a, Direct toxic effects of immune potentiators on monocytic, myelomonocytic and histiocytic tumor cells in culture, *Cancer Res.* **37**:546.

Ralph, P., and Nakoinz, I., 1977b, Antibody-dependent killing of erythrocyte and tumor targets by macrophage related cell lines: Enhancement by PPD and LPS, *J. Immunol.* **119**:950.

Ralph, P., Moore, M. A. S., and Nilsson, K., 1976, Lysozyme synthesis by established human and murine histiocytic lymphoma cell lines, *J. Exp. Med.* **143**:1528.

Ralph, P., Broxmeyer, H. E., and Nakoinz, I., 1977, Immunostimulators induce granulocyte/ macrophage colony-stimulating activity and block proliferation in a monocyte tumor cell line, *J. Exp. Med.* **146**:611.

Ralph, P., Ito, M., Broxmeyer, H. E., and Nakoinz, I., 1978, Corticosteroids block newly induced but not constitutive functions of macrophage cell lines: Myeloid colony stimulating activity production, latex phagocytosis, and antibody-dependent lysis of RBC and tumor targets, *J. Immunol.* **121**:300.

Remington, J. S., Krahenbuhl, J. L., and Hibbs, J. B., Jr., 1975, A role for the macrophage in resistance to tumor development and tumor destruction, in: *Mononuclear phagocytes in Immunity, Infection and Pathology* (R. van Furth, ed.), pp. 869–891, Blackwell, Oxford.

Remold, H. G., 1973, Requirements for L-fucose on the macrophage membrane receptor for MIF, *J. Exp. Med.* **138**:1065.

Remold, H. G., David, R. A., and David, J. R., 1972, Characterization of migration inhibition factor (MIF) from guinea pig lymphocytes stimulated with concanavalin A, *J. Immunol.* **109**:578.

Reynolds, H. Y., Atkinson, J. P., Newballm, H. H., and Frank, M. M., 1975, Receptors for immunoglobulin and complement on human alveolar macrophages, *J. Immunol.* **114**:1813.

Rickles, F. R., and Rick, P. D., 1976, Structural features of *S. typhimurium* lipopolysaccharide required for the activation of monocyte tissue factor, *Fed. Proc.* **35**:804.

Rivers, R. P. A., Hathaway, W. E., and Weston, W. L., 1975, The endotoxin-induced coagulant activity of human monocytes, *Br. J. Haemat.* **30**:311.

Robertson, P. B., Shyn, K. W., Vail, M. S., Taylor, R. E., and Fullmer, H. M., 1973, Collagenase: Demonstration in rabbit macrophages, *J. Dent. Res.* **52**:189.

Rosenthal, A. S., and Shevach, E. M., 1973, The functions of macrophages in antigen recognition by guinea pig T lymphocytes. I. Requirement for histocompatible macrophages and lymphocytes, *J. Exp. Med.* **138**:1194.

Rothburger, H., Zimmerman, T. S., Spiegelberg, H. L., and Vaughan, J. H., 1976, Enhancement of human mononuclear leucocyte procoagulant activity by IgG and antigen-antibody complexes, *Clin. Res.* **24**:318A.

Salthouse, T. N., and Malaga, B. F., 1972, Collagenase associated with macrophage and giant cell activity, *Experientia* **28**:326.

Sanderson, R. J., Shepperdson, F. T., Vatter, A. E., and Talmage, D. W., 1977, Isolation and enumeration of peripheral blood monocytes, *J. Immunol.* **118**:1409.

Schnyder, J., and Baggiolini, M., 1980, Secretion of lysosomal enzymes by macrophages, in: *Mononuclear Phagocytes—Functional Aspects* (R. van Furth, ed.), pp. 1369–1384, Martinus Nijhoff Medical Division, The Hague.

Schwartz, R. H., Bianco, A. R., Handwerger, B. S., and Kalm, C. R., 1975, Demonstration that monocytes rather than lymphocytes are insulin binding cells in preparations of human peripheral blood mononuclear leucocytes, *Proc. Natl. Acad. Sci.* **72**:474.

Sheagren, J. N., Simon, H. B., Tuazon, C. U., and Mehrota, P. P., 1975, Cell mediated immunity *in vitro,* in: *Mononuclear Phagocytes in Immunity, Infection and Pathology* (R. van Furth, ed.), pp. 653–659, Blackwell, Oxford.

Shevach, E. M., and Rosenthal, A. S., 1973, The function of macrophages in antigen recognition of guinea pig T lymphocytes. II. Role of the macrophage in the regulation of genetic control of the immune response, *J. Exp. Med.* **138**:1213.

Silverstein, S. C., Steinman, R. M., and Cohn, Z. A., 1977, Endocytosis, *Ann. Rev. Biochem.* **46**:669.

Smith, T. J., and Wagner, R. R., 1967, Rabbit macrophage interferons. I. Conditions for biosynthesis by virus infected and uninfected cells, *J. Exp. Med.* **125**:559.

Snyderman, R. M., Pike, M. C., Fischer, D. G., and Koren, H. S., 1977. Biological and biochemical activities of continuous macrophage cell lines P388D1 and J774.1, *J. Immunol.* **119**:2060.

Stadecker, M. J., and Unanue, E. R., 1979, The regulation of thymidine secretion by macrophages, *J. Immunol.* **123**:568.

Stadecker, M. J., Calderon, J., Karnovsky, M. J., and Unanue, E. R., 1977, Synthesis and release of thymidine by macrophages, *J. Immunol.* **119**:1738.

Stanley, E. R., and Heard, P. M., 1977, Factors regulating macrophage production and growth. Purification and some properties of the colony stimulating factor from medium conditioned by mouse L cells, *J. Biol. Chem.* **252**:4305.

Stanley, E. R., Cifone, M., Heard, P. M., and Defendi, V., 1976. Factors regulating macrophage production and growth: Identity of colony stimulating factor and macrophage growth factor, *J. Exp. Med.* **143**:631.

Stecher, V. J., 1970, Synthesis of proteins by mononuclear phagocytes, in: *Mononuclear Phagocytes* (R. van Furth, ed.), pp. 133–147, Blackwell, Oxford.

Stecher, V. J., and Thorbecke, G. J., 1967, Sites of synthesis of serum proteins. II. Medium requirements for serum protein production by rat macrophages, *J. Immunol.* **99**:653.

Steigbigel, R. T., Lambert, L. H., and Remington, J. S., 1974, Phagocytic and bactericidal properties of normal human monocytes, *J. Clin. Invest.* **53**:131.

Steinman, R. M., and Cohn, Z. A., 1972, The interaction of soluble horseradish peroxidase with mouse peritoneal macrophages *in vitro, J. Cell Biol.* **55**:186.

Steinman, R. M., and Cohn, Z. A., 1973, Identification of a novel cell type in peripheral lymphoid organs of mice. I. Morphology, quantitation, tissue distribution, *J. Exp. Med.* **137**:1142.

Stewart, C. C., 1980, Formation of colonies by mononuclear phagocytes outside the bone marrow, in: *Mononuclear Phagocytes—Functional Aspects* (R. van Furth, ed.), pp. 377–413, Martinus Nijhoff, The Hague.

Stewart, C. C., Lin, H. S., Adles, C., 1975, Proliferation and colony forming ability of peritoneal exudate cells in liquid culture, *J. Exp. Med.* **141**:1114.

Stuart, A. E., Habeshaw, J. A., and Davidson, A. E., 1977, Phagocytes *in vitro,* in: *Handbook of Experimental Immunology* (D. Weir, ed.), pp. 31.1–31.30, Blackwell, Oxford.

Stulting, R. D., and Berke, G., 1973, Nature of lymphocyte tumor interaction, *J. Exp. Med.* **137**:932.

Sutton, J. S., and Weiss, L., 1966, Transformation of monocytes in tissue culture into macrophages epithelioid cells and multinucleated giant cells, *J. Cell. Biol.* **28**:303.

Syren, E., and Raeste, A. M., 1971, Identification of blood monocytes by demonstration of lysozyme and peroxidase activity, *Acta Haematol. (Basel)* **45**:29.

Testa, N. G., and Lord, B. I., 1970, A technique for the morphological examination of hemopoietic cells grown in agar, *Blood* **36**:586.

Till, J. E., and McCulloch, E. A., 1961, A direct measurement of the radiation sensitivity of normal mouse bone marrow cells, *Radiat. Res.* **14**:213.

Turk, J. L., and Poulter, P. W., 1975, Effect of lymphocyte mediators on macrophages *in vitro:* A correlation of morphological and cytochemical changes, in: *Mononuclear Phagocytes in Immunity, Infection and Pathology* (R. van Furth, ed.), pp. 711–718, Blackwell, Oxford.

Unkeless, J. C., 1977, The presence of two Fc receptors on mouse macrophages: Evidence from a variant cell line and differential trypsin sensitivity, *J. Exp. Med.* **145**:931.

Unkeless, J. C., Gordon, S., and Reich, E., 1974, Secretion of plasminogen activator by stimulated macrophages, *J. Exp. Med.* **139**:834.

van der Meer, J. W. M., Bulterman, D., and van Furth, R., 1977, Long term primary macrophage cultures, *Exp. Hematol.* **5** (Suppl. 106).

van der Meer, J. W. M., Bulterman, D., van Zwet, T. L., Elzenga-Claasen, I., and van Furth, R., 1978, Culture of mononuclear phagocytes on a Teflon surface to prevent adherence, *J. Exp. Med.* **147**:271.

van der Meer, J. W. M., Beelen, R. H. J., Fluitsma, D. M., and van Furth, R., 1979a, Ultrastructure of mononuclear phagocytes developing in liquid bone marrow cultures: A study on peroxidatic activity, *J. Exp. Med.* **149**:17.

van der Meer, J. W. M., van de Gevel, J. S., Elzenga-Claasen, I., and van Furth, R., 1979b, Suspension cultures of mononuclear phagocytes in the Teflon culture bag, *Cell. Immunol.* **42**:208.

van der Meer, J. W. M., van de Gevel, J. S., Diesselhoff-den Dulk, M. M. C., and van Furth, R., 1980, Long term cultures of bone marrow mononuclear phagocytes, in: *Mononuclear Phagocytes—Functional Aspects* (R. van Furth, ed.), pp. 343–362, Martinus Nijhoff, The Hague.

van der Rhee, H. J., Hillebrands, W., and Daems, W. T., 1978, Are Langhans giant cells precursors of foreign body giant cells? *Arch. Dermat. Res.* **263**:13.

van der Zeyst, B. A. M., Stewart, C. C., and Schlesinger, S., 1978, Proliferative capacity of mouse peritoneal macrophages *in vitro, J. Exp. Med.* **147**:1253.

van Furth, R., 1976, An approach to the characterization of mononuclear phagocytes involved in pathological processes, *Agents Actions* **6**:91.

van Furth, R., 1978, Mononuclear phagocytes in inflammation, in: *Handbook of Experimental Pharmacology,* Vol. 50/I, *Inflammation* (J. R. Vane and S. H. Ferreira, eds.), pp. 68–108, Springer Verlag, Berlin.

van Furth, R., and Cohn, Z. A., 1968, The origin and kinetics of mononuclear phagocytes, *J. Exp. Med.* **128**:415.

van Furth, R., and Diesselhoff-den Dulk, M. M. C., 1970, The kinetics of promonocytes and monocytes in the bone marrow, *J. Exp. Med.* **132**:813.

van Furth, R., Hirsch, J. G., and Fedorko, M. E., 1970, Morphology and peroxidase cytochemistry of mouse promonocytes, monocytes, and macrophages, *J. Exp. Med.* **132**:794.

van Furth, R., Cohn, Z. A., Hirsch, J. G., Humphrey, J. H., Spector, W. G., and Langevoort, H. L., 1972, The mononuclear phagocyte system: A new classification of macrophages, monocytes and their precursor cells, *Bull. W. H. O.* **46**:845.

van Furth, R., van Zwet, T. L., and Leijh, P. C. J., 1977, *In vitro* determination of phagocytosis and intracellular killing by polymorphonuclear and mononuclear phagocytes, in: *Handbook of Experimental Immunology* (D. Weir, ed.), pp. 32.1–32.19, Blackwell, Oxford.

van Furth, R., van Zwet, T. L., and Raeburn, J. A., 1979, Characteristics of human mononuclear phagocytes, *Blood* **54**:485.

van Ginkel, C. J. W., van Aken, W. G., Oh, J. I. H., and Vreeken, J., 1977, Stimulation of monocyte procoagulant activity by adherence to different surfaces, *Br. J. Haematol.* **37**:35.

van Ginkel, C. J. W., Thörig, L., Thompson, J., Oh, J. I. H., and van Aken, W. G., 1979, Enhancement of generation of monocyte tissue thromboplastin by bacterial phagocytosis: A possible pathway for fibrin formation on infected vegetations in bacterial endocarditis, *Infect. Immun.* **25**:388.

van Waarde, D., Hulsing-Hesselink, E., Sandkuyl, L. A., and van Furth, R., 1977a, Humoral regulation of monocytopoiesis during an inflammatory reaction caused by particulate substances, *Blood* **5**:141.

van Waarde, D., Hulsing-Hesselink, E., and van Furth, R., 1977b, Properties of a factor increasing monocytopoiesis (FIM) occurring in the serum during the early phase of an inflammatory reaction, *Blood* **5**:727.

Vassalli, J. D., and Reich, E., 1977, Macrophage plasminogen activator induction by products of activated lymphoid cells, *J. Exp. Med.* **145**:429.

Virolainen, M., and Defendi, V., 1967, Dependence of macrophage growth *in vitro* upon interaction with other cell types: Growth regulating substances for animal cells in culture, *Wistar Inst. Symp. Monogr.* **7**:67.

Volkman, A., 1976, Disparity in origin of mononuclear phagocyte populations, *J. Reticuloendothel. Soc.* **19**:249.

Wachsmuth, E. D., 1975, Aminopeptidase as a marker for macrophage differentiation, *Exp. Cell Res.* **96**:409.

Wachsmuth, E. D., and Staber, F. G., 1977, Changes in membrane bound aminopeptidase on bone marrow derived macrophages during their maturation *in vitro*, *Exp. Cell Res.* **109**:269.

Wachsmuth, E. D., and Torhorst, A., 1974, Possible precursors of aminopeptidase and alkaline phosphatase in the proximal tubules of kidney and the crypts of small intestine of mice, *Histochemistry* **38**:43.

Wahl, L. M., Wahl, S. M., Mergenhagen, S. E., and Martin, G. R., 1974, Collagenase production by endotoxin-activated macrophages, *Proc. Natl. Acad. Sci. USA* **71**:3598.

Wahl, L. M., Wahl, S. M., Mergenhagen, S. E., and Martin, G. R., 1975a, Collagenase production by lymphokine activated macrophages, *Science* **187**:261.

Wahl, S. M., Wilton, J. M., Rosenstreich, D. L., and Oppenheim, J. J., 1975b, The role of macrophages in the production of lymphokines by T and B lymphocytes, *J. Immunol.* **114**:1296.

Waldman, S. R., and Gottlieb, A. A., 1973, Macrophage regulation of DNA synthesis in lymphoid cells: Effects of a soluble factor from macrophages, *Cell. Immunol.* **9**:142.

Weaver, J. M., 1958, Destruction of mouse ascites tumor cells *in vivo* and *in vitro* by homologous macrophages, lymphocytes and cell-free antibodies, *Proc. Am. Soc. Cancer Res.* **2**:354.

Weiden, P., Storb, R., and Tsoi, M. S., 1975, Marrow origin of canine alveolar macrophages, *J. Reticuloendothel. Soc.* **17**:342.

Weir, D. M., and Ögmundsdottir, H. M., 1980, Cellular recognition by phagocytes: Role of lectin-like receptor(s), in: *Mononuclear Phagocytes—Functional Aspects* (R. van Furth, ed.), pp. 865–881, Martinus Nijhoff, The Hague.

Werb, Z., and Burleigh, M. C., 1974, A specific collagenase from rabbit fibroblasts in monolayer culture, *Biochem. J.* **137**:373.

Werb, Z., and Cohn, Z. A., 1971, Cholesterol metabolism in the macrophage. II. Alteration of subcellular exchangeable cholesterol compartments and exchange in other cell types, *J. Exp. Med.* **134**:1570.

Werb, Z., and Gordon, S., 1975a, Secretion of a specific collagenase by stimulated macrophages, *J. Exp. Med.* **142**:346.

Werb, Z., and Gordon, S., 1975b, Elastase secretion by stimulated macrophages, *J. Exp. Med.* **142**:361.

Werb, Z., Foley, R., and Munck, A., 1978a, Glucocorticoid receptors and glucocorticoid sensitive secretion of neutral proteinases in a macrophage line, *J. Immunol.* **121**:115.

Werb, Z., Foley, R., and Munck, A., 1978b, Interactions of glucocorticoids with macrophages. Identification of glucocorticoid receptors in monocytes and macrophages, *J. Exp. Med.* **147**:1684.

White, R., Lin, H. S., and Kuhn, C., 1977, Elastase secretion by peritoneal exudative and alveolar macrophages, *J. Exp. Med.* **146**:802.

Wiener, E., and Levanon, D., 1968, Macrophage cultures: An extracellular esterase, *Science* **159**:217.

Wilkinson, P. C., 1974, *Chemotaxis and Inflammation*, Churchill Livingstone, Edinburgh.

Willcox, M. B., Golde, D. W., and Cline, M. J., 1976, Cytochemical reactions of human hematopoietic cells in liquid culture, *J. Histochem. Cytochem.* **24**:979.

Williams, N., Jackson, H., and Rabellino, E. M., 1977, Proliferation and differentiation of normal granulopoietic cells in continuous bone marrow cultures, *J. Cell. Physiol.* **93**:435.

Worton, R. G., McCulloch, E. A., and Till, J. E., 1969, Physical separation of hematopoietic stem cells from cells forming colonies in culture, *J. Cell. Physiol.* **74**:171.

Wyatt, H. V., Colten, H. R., and Borsos, T., 1972, Production of the second (C2) and fourth (C4)

components of guinea pig complement by single peritoneal cells: Evidence that one cell may produce both components, *J. Immunol.* **108**:1609.
Zigmond, S. H., and Hirsch, J. G., 1973, Leucocyte locomotion and chemotaxis. New methods for evaluation and demonstration of cell-derived chemotactic factor, *J. Exp. Med.* **137**:387.
Zucker-Franklin, D., 1974, The percentage of monocytes among "mononuclear" cell fractions obtained from normal blood, *J. Immunol.* **112**:234.

Index